HYDROCARBON CHEMISTRY

Dedicated to Katherine Bogdanovich Loker,
friend and generous supporter of hydrocarbon research.

HYDROCARBON CHEMISTRY

SECOND EDITION

George A. Olah

Loker Hydrocarbon Research Institute and Department of Chemistry
University of Southern California
Los Angeles, California

Árpád Molnár

Department of Organic Chemistry
University of Szeged
Szeged, Hungary

WILEY-INTERSCIENCE

A JOHN WILEY & SONS, INC., PUBLICATION

Library of Congress Cataloging-in-Publication Data:

Olah, George A. (George Andrew), 1927–
    Hydrocarbon chemistry / George A. Olah, Arpad Molnar. – 2nd ed.
        p.   cm.
    Includes index.
        ISBN 0-471-41782-3 (Cloth)
    1. Hydrocarbons.   I. Molnâar. âArpâad, 1942–   II. Title.

    QD305.H5 043 2003
    547′.01–dc21
                                                      2002154116

Printed in the United States of America

10 9 8 7 6 5 4 3 2 1

ac

CONTENTS

12 Metathesis   696

PREFACE TO THE SECOND EDITION

Seven years passed since the publication of the first edition of our book. It is rewarding that the favorable reception of and interest in hydrocarbon chemistry called for a second edition. All chapters were updated (generally considering literature through 2001) by adding sections on recent developments to review new advances and results. Two new chapters were also added on acylation as well as emerging areas and trends (including green chemistry, combinatorial chemistry, fluorous biphase catalysis, solvent-free chemistry, and synthesis via CO_2 recycling from the atmosphere). Because of its importance a more detailed treatment of chemical reduction of CO_2 as a source for hydrocarbons is also included in Chapter 3. The new edition should keep our book current and of continuing use for interested readers.

We hope that *Hydrocarbon Chemistry* will continue to serve its purpose and the goals that we originally intended.

Los Angeles, California
Szeged, Hungary
March 2002

GEORGE A. OLAH
ÁRPÁD MOLNÁR

PREFACE TO THE
FIRST EDITION

The idea of a comprehensive monograph treating the hydrocarbon chemistry as an entity emphasizing basic chemistry, while also relating to the practical aspects of the broad field, originally developed in the late 1970s by G. A. Olah and the late Louis Schmerling, a pioneer of hydrocarbon chemistry. The project was pursued albeit intermittently through the following years, producing a number of draft chapters. It became, however, clear that the task was more formidable than initially anticipated. Progress was consequently slow, and much of the initial writings became outdated in view of rapid progress. The project as originally envisaged became clearly no longer viable. A new start was needed and made in 1992 with Á. Molnár coming to the Loker Hydrocarbon Research Institute for 2 years as Moulton Distinguished Visiting Fellow. We hope that our efforts on *Hydrocarbon Chemistry* will be of use to those interested in this broad and fascinating field, which also has great practical significance.

Los Angeles, California   GEORGE A. OLAH
Szeged, Hungary   ÁRPÁD MOLNÁR
March 1995

The Olahs' grandchildren (Peter, Kaitlyn, and Justin) enlighten an otherwise blank page.

INTRODUCTION

Hydrocarbons and their transformations play a major role in chemistry. Industrial applications, basic to our everyday life, face new challenges from diminishing petroleum supplies, regulatory problems, and environmental concerns. Chemists must find answers to these challenges. Understanding the involved chemistry and finding new approaches is a field of vigorous development.

Hydrocarbon chemistry (i.e., that of carbon- and hydrogen-containing compounds) covers a broad area of organic chemistry that at the same time is also of great practical importance. It includes the chemistry of saturated hydrocarbons (alkanes, cycloalkanes), as well as that of unsaturated alkenes and dienes, acetylenes, and aromatics. Whereas numerous texts and monographs discuss selected areas of the field, a comprehensive up-to-date treatment as an entity encompassing both basic chemistry and practical applications is lacking. The aim of our book is to bring together all major aspects of hydrocarbon chemistry, including fundamental and applied (industrial) aspects in a single volume. In order to achieve this, it was necessary to be selective, and we needed to limit our discussion.

The book is arranged in 14 chapters. After discussing general aspects, separation of hydrocarbons from natural sources and synthesis from C_1 precursors with the most recent developments for possible future applications, each chapter deals with a specific type of transformation of hydrocarbons. Involved fundamental chemistry, including reactivity and selectivity, as well as stereochemical considerations and mechanistic aspects are discussed, as are practical applications. In view of the immense literature, the coverage cannot be comprehensive and is therefore selective, reflecting the authors' own experience in the field. It was attempted nevertheless to cover all major aspects with references generally until the early 1994.

The chemistry of the major processes of the petrochemical industry, including cracking, reforming, isomerization, and alkylation, is covered in Chapters 2, 4, and 5, respectively. The increasingly important C_1 chemistry—that of one-carbon compounds (CO_2, CO, methane, and its derivatives)—is discussed in Chapter 3 (Synthesis from C_1 sources).

Chapter 6 (Addition), Chapter 7 (Carbonylation), Chapter 8 (Acylation), and Chapter 10 (Heterosubstitution) deal with derivatization reactions to form carbon–heteroatom bonds. The important broad field of hydrocarbon oxidations is covered in Chapter 9 (Oxidation–oxygenation). Both the chemistry brought about by conventional oxidizing agents and the most recent developments introducing selectively oxygen functionality into hydrocarbons are discussed. The hydrogenation (catalytic and chemical) and reduction techniques (homogeneous catalytic, ionic, and electrochemical) are similarly discussed in Chapter 11 (Reduction–hydrogenation).

Chapter 12 deals with metathesis; Chapter 13, with oligomerization and polymerization of hydrocarbons. Each of these fields is of substantial practical significance and treated emphasizing basic chemistry and significant practical applications. Challenges in the new century and possible solutions relevant to hydrocarbon chemistry are discussed in Chapter 14 (Emerging areas and trends).

Hydrocarbon Chemistry addresses a wide range of readers. We hope that research and industrial chemists, college and university teachers, and advanced undergraduate and graduate students alike will find it useful. Since it gives a general overview of the field, it should also be useful for chemical engineers and in the chemical and petrochemical industry in general. Finally, we believe that it may serve well as supplementary textbook in courses dealing with aspects of the diverse and significant field.

1

GENERAL ASPECTS

1.1. HYDROCARBONS AND THEIR CLASSES

Hydrocarbons, as their name indicates, are compounds of carbon and hydrogen. As such, they represent one of the most significant classes of *organic* compounds (i.e., of carbon compounds).[1] In methane (CH_4) the simplest saturated alkane, a single-carbon atom, is bonded to four hydrogen atoms. In the higher homologs of methane (of the general formula C_nH_{2n+2}) all atoms are bound to each other by single [(sigma (σ), two-electron two-center] bonds with carbon displaying its tendency to form C—C bonds. Whereas in CH_4 the H : C ratio is 4, in C_2H_6 (ethane) it is decreased to 3; in C_3H_8 (propane), to 2.67; and so on. Alkanes can be straight-chain (with each carbon attached to not more than two other carbon atoms) or branched (in which at least one of the carbons is attached to either three or four other carbon atoms). Carbon atoms can be aligned in open chains (acyclic hydrocarbons) or can form rings (cyclic hydrocarbons).

Cycloalkanes are cyclic saturated hydrocarbons containing a single ring. Bridged cycloalkanes contain one (or more) pair(s) of carbon atoms common to two (or more) rings. In bicycloalkanes there are two carbon atoms common to both rings. In tricycloalkanes there are four carbon atoms common to three rings such as in adamantane (tricyclo[3.3.1]^{1,5,3,7}decane), giving a caged hydrocarbon structure.

Carbon can also form multiple bonds with other carbon atoms. This results in unsaturated hydrocarbons such as olefins (alkenes, C_nH_{2n}), specifically, hydrocarbons containing a carbon–carbon double bond or acetylenes (alkynes, C_nH_{n-2}) containing a carbon–carbon triple bond. Dienes and polyenes contain two or more unsaturated bonds.

Aromatic hydrocarbons (arenes), a class of hydrocarbons of which benzene is parent, consist of cyclic arrangement of formally unsaturated carbons, which,

1

however, give a stabilized (in contrast to their hypothetical cyclopolyenes) delocalized π system.

The H : C ratio in hydrocarbons is indicative of the hydrogen deficiency of the system. As mentioned, the highest theoretical H : C ratio possible for hydrocarbons is 4 (in CH_4), although in carbocationic compounds (the positive ions of carbon compound) such as CH_5^+ and even CH_6^{2+} the ratio is further increased (to 5 and 6, respectively). On the other end of the scale in extreme cases, such as the dihydro or methylene derivatives of recently (at the time of writing) discovered C_{60} and C_{70} fullerenes, the H : C ratio can be as low as ~ 0.03!

An index of unsaturation (hydrogen deficiency) i can be used in hydrocarbons whose value indicates the number of ring and/or double bonds (a triple bond is counted as two double bonds) present (C and H = the number of carbon and hydrogen atoms), $i = 0$ for methane, for ethene $i = 1$ (one double bond), for acetylene (ethyne) $i = 2$, and so on:

$$i = \frac{(2C + 2) - H}{2}$$

The International Union of Pure and Applied Chemistry (IUPAC) established rules to name hydrocarbons. Frequently, however, trivial names are also used and will continue to be used. It is not considered necessary to elaborate here on the question of nomenclature. Systematic naming is mostly followed. Trivial (common) namings are, however, also well extended. *Olefins* or *aromatics* clearly are very much part of our everyday usage, although their IUPAC names are *alkenes* and *arenes*, respectively. Straight-chain saturated hydrocarbons are frequently referred to as *n*-alkanes (normal) in contrast to their branched analogs (isoalkanes, *i*-alkanes). Similarly straight-chain alkenes are frequently called *n*-alkenes as contrasted with branch isoalkenes (or olefins). What needs to be pointed out, however, is that one should not mix the systematic IUPAC and the still prevalent trivial (or common) namings. For example, $(CH_3)_2C=CH_2$ can be called isobutylene or 2-methylpropene. It, however, should not be called *isobutene* as only the common name *butylene* should be affixed by *iso*. On the other hand, isobutane is the proper common name for 2-methylpropane [$(CH_3)_3CH$]. Consequently we discuss isobutane–isobutylene alkylation for production of isooctane: high-octane gasoline (but it should not be called *isobutane–isobutene alkylation*).

1.2. ENERGY–HYDROCARBON RELATIONSHIP

Every facet of human life is affected by our need for energy. The sun is the central energy source of our solar system. The difficulty lies in converting solar energy into other energy sources and also to store them for future use. Photovoltaic devices and other means to utilize solar energy are intensively studied and developed, but at the

Table 1.1. U.S. energy sources (%)

Power Source	1960	1970	1990
Oil	48	46	41
Natural gas	26	26	24
Coal	19	19	23
Nuclear energy	3	5	8
Hydrothermal, geothermal, solar, etc. energy	4	4	4

level of our energy demands, Earth-based major installations by present-day technology are not feasible. The size of collecting devices would necessitate utilization of large areas of the Earth. Atmospheric conditions in most of the industrialized world are unsuitable to provide a constant solar energy supply. Perhaps a space-based collecting system beaming energy back to Earth can be established at some time in the future, but except for small-scale installation, solar energy is of limited significance for the foreseeable future. Unfortunately, the same must be said about wind, ocean waves, and other unconventional energy sources.

Our major energy sources are fossil fuels (i.e., oil, gas, and coal), as well as atomic energy. Fossil energy sources are, however, nonrenewable (at least on our timescale), and their burning causes serious environmental problems. Increased carbon dioxide levels are considered to contribute to the "greenhouse" effect. The major limitation, however, is the limited nature of our fossil fuel resources (see Section 1.5). The most realistic estimates[2] put our overall worldwide fossil resources as lasting for not more than 200 or 300 years, of which oil and gas would last less than a century. In human history this is a short period, and we will need to find new solutions. The United States relies overwhelmingly on fossil energy sources, with only 8% coming from atomic energy and 4% from hydro energy (Table 1.1).

Other industrialized countries utilize to a much higher degree of nuclear and hydroenergy[2] (Table 1.2). Since 1980, concerns about safety and fission byproduct

Table 1.2. Power generated in industrial countries by nonfossil fuels (1990)

Country	Non-Fossil-Fuel Power (%)		
	Hydroenergy	Nuclear Energy	Total
France	12	75	87
Canada	58	16	74
Former West Germany	4	34	38
Japan	11	26	37
United Kingdom	1	23	24
Italy	16	0	16
United States	4	8	12

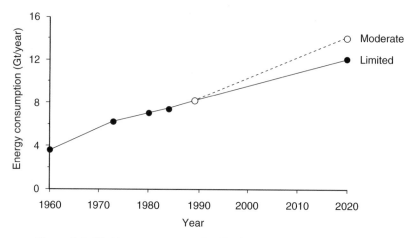

Figure 1.1. World energy consumption (in gigatons per year) projections.

disposal difficulties, however, dramatically limited the growth of the otherwise clean atomic energy industry.

A way to extend the lifetime of our fossil fuel energy reserves is to raise the efficiency of thermal power generation. Progress has been made in this respect, but the heat efficiency even in the most modern power plants is limited. Heat efficiency increased substantially from 19% in 1951 to 38% in 1970, but for many years since then 39% appeared to be the limit. Combined-cycle thermal power generation—a combination of gas turbines—was allowed in Japan to further increase heat efficiency from 35–39% to as high as 43%. Conservation efforts can also greatly contribute to moderate worldwide growth of energy consumption, but the rapidly growing population of our planet (5.4 billion today, but should reach 7–8 billion by 2010) will put enormous pressure on our future needs.

Estimates of the world energy consumption until 2020 are shown graphically in Figure 1.1 in relationship to data dating back to 1960.[2] A rise in global energy consumption of 50–75% for the year 2020 is expected compared with that for 1988. Even in a very limited growth economic scenario the global energy demand is estimated to reach 12 billion tons of oil equivalent (t/oe) by the year 2020.

Our long-range energy future clearly must be safe nuclear energy, which should increasingly free still remaining fossil fuels as sources for convenient transportation fuels and as raw materials for synthesis of plastics, chemicals, and other substances. Eventually, however, in the not too distant future we will need to make synthetic hydrocarbons on a large scale.

1.3. HYDROCARBON SOURCES AND SEPARATION

All fossil fuels (coal, oil, gas) are basically hydrocarbons, deviating, however, significantly in their H : C ratio (Table 1.3).

Table 1.3. H : C ratio of natural hydrocarbon sources

Methane	4.0
Natural gas	3.8
Petroleum crude	1.8
Tar sand bitumen	1.5
Shale oil (raw)	1.5
Bituminous coal	0.8

1.3.1. Natural Gas

Natural gas, depending on its source, contains—besides methane as the main hydrocarbon compound (present usually in concentrations >80–90%)—some of the higher homologous alkanes (ethane, propane, butane). In "wet" gases the amount of C_2–C_6 alkanes is more significant (gas liquids). Typical composition of natural gas of various origin[2,3] is shown in Table 1.4.

Natural-gas liquids are generally of thermal value only but can be used for dehydrogenation to alkenes. Their direct upgrading to gasoline-range hydrocarbons is also pursued. Natural gas as we know it is of biological origin (not unlike petroleum oil). Large gas reservoirs were discovered and utilized in the twentieth century. Increasingly deeper wells are drilled and deposits under the seas are explored and tapped. An interesting but as yet unproved theory by Gold[4] holds that hydrocarbons may also be formed by slow outgassing of methane from vast deep deposits dating back to the origin of our planet. Besides biologically derived oil and gas, "deep" carbon compounds trapped in the Earth's crust are subject to intense heat, causing them to release hydrocarbons that migrate toward the Earth's surface, where they are trapped in different stratas. Seepage observed at the bottom of the oceans and finds of oil during drilling into formations (such as granite) where no "biogenic" oil was expected are cited as proof for "abiogenic" hydrocarbons. If abiogenic methane and other hydrocarbons exist (although most geologists presently disagree), vast new reserves would become available when improved drilling technology is developed to reach deeper into the Earth's crust.[4]

Other vast yet untapped reserves of natural gas (methane) are locked up as hydrates under the permafrost in Siberia. Methane gas hydrates are inclusion

Table 1.4. Composition of natural gas [in weight percent (wt%)]

Location	CH_4	C_2H_6	C_3H_8	C_4H_{10}
United States	89.5–92.5	5.1–2	2.1–0.7	1.6–0.5
Algeria	86.9	9.0	2.6	1.2
Iran	74.9	13.0	7.2	3.1
North Sea	90.8	6.1	0.7	0.1

compounds of $CH_4 \cdot nH_2O$ composition. Their amount is estimated to equal or exceed known conventional natural-gas reserves. Their economical utilization, however, remains a challenge. Significant amounts [$\leq$500 million tons per year (Mt/y)] of natural methane is also released into the atmosphere from varied sources ranging from marsh lands to landfills, to farm animals. Methane in the atmosphere represents only a small component, although its increase can cause a significant greenhouse effect.

1.3.2. Petroleum or Crude Oil

Petroleum or crude oil is a complex mixture of many hydrocarbons.[1b] It is characterized by the virtual absence of unsaturated hydrocarbons consisting mainly of saturated, predominantly straight-chain alkanes, small amounts of slightly branched alkanes, cycloalkanes, and aromatics. Petroleum is generally believed to be derived from organic matter deposited in the sediments and sedimentary rocks on the floor of marine basins. The identification of biological markers such as petroporphyrins provides convincing evidence for the biological origin of oil (see, however, the abovementioned possibility for abiogenic "deep" hydrocarbons). The effect of time, temperature, and pressure in the geological transformation of the organics to petroleum is not yet clear. However, considering the low level of oxidized hydrocarbons and the presence of porphyrins, it can be surmised that the organics were acted on by anaerobic microorganisms and that temperatures were moderate, <200°C. By comparing the elemental composition of typical crude oils with typical bituminous coals, it is clear why crude oil is a much more suitable fuel source in terms of its higher H : C atomic ratio, generally lower sulfur and nitrogen contents, very low ash content, (probably mostly attributable to some suspended mineral matter and vanadium and nickel associated with porphyrins), and essentially no water content.

Finally, it is interesting to mention that the most recent evidence shows that even extraterrestrially formed hydrocarbons can reach the Earth. The Earth continues to receive some 40,000 tons of interplanetary dust every year. Mass-spectrometric analysis revealed the presence of hydrocarbons attached to these dust particles, including polycyclic aromatics such as phenanthrene, chrysene, pyrene, benzopyrene, and pentacene of extraterrestrial origin (indicated by anomalous isotopic ratios[5]).

Petroleum—a natural mineral oil—was referred to as early as in the Old Testament. The word *petroleum* means "rock oil" [from the Greek *petros* (rock) and *elaion* (oil)]. It had been found over the centuries seeping out of the ground, for example, in the Los Angeles basin (practically next door to where this review is written) and what are now the La Brea Tar Pits. Vast deposits were found in varied places ranging from Europe, to Asia, to the Americas, and to Africa. In the United States the first commercial petroleum deposit was discovered in 1859 near Titusville in western Pennsylvania when Edwin Drake and Billy Smith struck oil in their first shallow (~20-m-deep) well.[6] The well yielded 400 gallons (gal) of oil a day (about 10 barrels). The area was known before to contain petroleum that residents skimmed from a local creek's surface, which was thus called "oil creek." The oil-producing first well opened up a whole new industry. The discovery was not unexpected, but provided evidence for oil deposits in the ground that could be reached by drilling into them. Oil was used for many purposes, such as in lamp illumination and even for medical remedies. The newly discovered Pennsylvania petroleum was soon also marketed to degrease wool, prepare paints, fuel steam engines, and power light railroad cars and for many other uses. It was recognized that the well oil was highly impure and had to be refined to separate different fractions for varied uses (see Section 1.4). The first petroleum refinery, a small stilling operation, was established in Titusville in 1860. Petroleum refining was much cheaper than producing coal oil (kerosene), and soon petroleum became the predominant source for kerosene as an illuminant. In the 1910s the popularity of automobiles spurred the production of gasoline as the major petroleum product. California, Texas, Oklahoma, and more recently Alaska provided large petroleum deposits in the United States, whereas areas of the Middle East, Asia, Russia, Africa, South America, and more recently of the North Sea became major world oil production centers.

The daily consumption of crude oil in the United States is about 16–17 million barrels. Most of this is used for the generation of electricity and space heating and as transportation fuel. About 4% of the petroleum and natural gas is used as feedstocks for the manufacturing of chemicals, pharmaceuticals, plastics, elastomers, paints, and a host of other products. Petrochemicals from hydrocarbons provide a great many of the necessities of modern life, to which we have become so accustomed that we do not even notice our increasing dependence on them, and yet the consumption of petrochemicals is still growing at an annual rate of 10%. Advances in the petroleum–hydrocarbon industry, more than anything else, may be credited to the high standard of living we enjoy in the early twenty-first century.

1.3.3. Heavy Oils, Shale, and Tar Sand

Whereas light crudes are preferred in present-day refining operations, increasingly heavier petroleum sources also need to be processed to satisfy our ever-increasing needs. These range from commercially usable *heavy oils* (California, Venezuela, etc.) to the huge petroleum reserves locked up in *shale* or *tar sand* formations.[1b,7a] These more unconventional hydrocarbon accumulations exceed the quantity of oil present in all the rest of the oil deposits in the world taken together. The largest is located in Alberta, Canada, in the form of enormous tar sand and carbonate rock deposits containing some 2.5–6 trillion barrels of extremely heavy oil, called *bitumen*. It is followed by the heavy oil accumulations in the Orinoco Valley, Venezuela and of Siberia. Another vast commercially significant reservoir of oil is the oil shale deposits of the northwestern United States located in Wyoming, Utah, and Colorado. The practical use of these potentially vast reserves will depend on finding economical ways to extract the oil (by thermal retorting or other processes) for further processing.

The quality of petroleum varies and, according to specific gravity and viscosity, we talk about light, medium, heavy, and extraheavy crude oils. Light oils of low

Table 1.5. Compositions (%) of typical light and heavy oils

Fraction	Light Oil	Heavy Oil
Saturates	78	17–21
Aromatics	18	36–38
Resins	4	26–28
Asphaltene	Trace–2	17

specific gravity and viscosity are more valuable than heavy oils with higher specific gravity and viscosity. In general, light oils are richer in saturated hydrocarbons, especially straight-chain alkanes, than are heavy oils and contain ≤75% straight-chain alkanes and ≤95% total hydrocarbons. Extraheavy oils, the bitumens, have a high viscosity, and thus may be semisolids with high levels of heteroatoms (nitrogen, oxygen, and sulfur) and a correspondingly reduced hydrocarbon content, of the order of 30–40%.

Heavy oils, especially bitumens, contain high concentrations of resins (30–40%) and asphaltenes (≤20%). Most heavy oils and bitumens are thought to be derivatives of lighter, conventional crude oils that have lost part or all of their *n*-alkane contents along with some of their low-molecular-weight cyclic hydrocarbons through processes taking place in the oil reservoirs. Heavy oils are also abundant in heteroatom (N,O,S)-containing molecules, organometallics, and colloidally dispersed clays and clay organics. The prominent metals associated with petroleum are nickel, vanadium (mainly in the form of vanadyl, VO^{2+} ions), and iron. The former two are (in part) bound to porphyrins to form metalloporphyrins. Table 1.5 compares the difference in composition of typical light and heavy oils.

Processing heavy oils and bitumens represent challenges for presently used refinery processes, as heavy oils and bitumens can poison the metal catalysts used in the current refineries. The use of superacid catalysts, which are less sensitive to these feeds, is one possible solution[7b] to this problem.

1.3.4. Coal and Its Liquefaction

Coals (the plural is deliberately used as coal has no defined, uniform nature or structure) are fossil sources with low hydrogen content.[1b,8a] The "structure" of coals means only structural models depicting major bonding types and components, relating changes with coal rank. Coal is classified—or ranked—as lignite, subbituminous, bituminous, and anthracite. This is also the order of increased aromaticity and decreased volatile matter. The H:C ratio of bituminous coal is about 0.8, whereas anthracite has H:C ratios as low as 0.2.

From a chemical, as contrasted to a geologic viewpoint, the coal formation (coalification) process can be grossly viewed as a continuum of chemical changes, some microbiological changes, and some thermal changes involving a progression in which woody or cellulosic plant materials (the products of nature's photosynthetic recycling of CO_2) in peat swamps are converted during many millions of years and

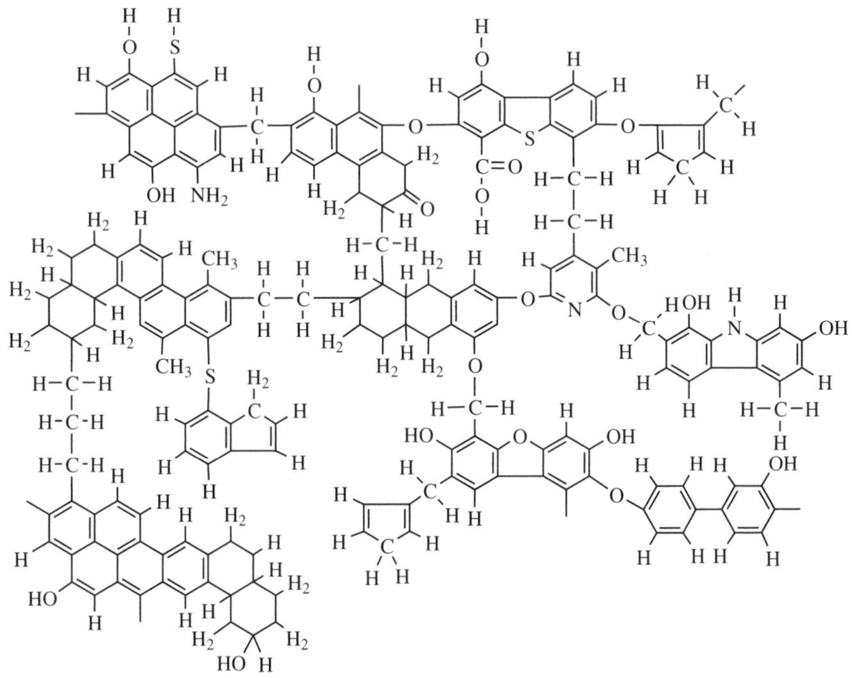

Figure 1.2. Schematic representation of structural groups and connecting bridges in bituminous coal according to Wiser.[8b]

increasingly severe geologic conditions to coals. Coalification is a grossly deoxy-genation–aromatization process. As the "rank" or age of the coal increases, the organic oxygen content decreases and the aromaticity (defined as the ratio of aromatic carbon to total carbon) increases. Lignites are young or "brown" coals containing more organic oxygen functional groups than subbituminous coals, which, in turn have a higher carbon content but fewer oxygen functionalities.

The organic chemical structural types believed to be characteristic of coals can be schematically represented as shown in Figure 1.2 showing probable structural groups and connecting bridges that are present in a typical bituminous coal.[8b]

The principal type of bridging linkages between clusters are short aliphatic groups $(CH_2)_n$ (where $n = 1 - 4$), different types of ether linkages, and sulfide and biphenyl bonds. All except the latter may be considered scissible bonds in that they can readily undergo thermal and chemical cleavage reactions.

The main approaches employed in converting coals to liquid hydrocarbons revolve around breaking down the large, complex "structures" generally by hydro-genative cleavage reactions and increasing the solubility of the organic portion. Alkylation, hydrogenation, and depolymerization—as well combinations of these reactions followed by extraction of the reacted coals—are major routes taken. This can provide clean liquid fuels, such as gasoline and heating oil.

Three types of direct coal liquefaction processes have emerged to convert coals to liquid hydrocarbon fuels:[8]

1. The first is a high-temperature solvent extraction process in which no catalyst is added. The liquids produced are those that are dissolved in the solvent, or solvent mixture. The solvent usually is a hydroaromatic hydrogen donor, while molecular hydrogen is added as a secondary source of hydrogen.

2. The second, catalytic liquefaction process is similar to the first except that there is a catalyst in direct contact with the coal. $ZnCl_2$ and other Friedel–Crafts catalysts, including $AlCl_3$, as well as BF_3–phenol and other complexes catalyze the depolymerization–hydrogenation of coals, but usually forceful conditions (375–425°C, 100–200 atm) are needed. Superacidic HF–BF_3-induced liquefaction of coals[8] involves depolymerization–ionic hydrogenation at relatively modest 150–170°C.

3. The third coal liquefaction approach is direct catalytic hydrogenation (pioneered by Bergius) in which a hydrogenation catalyst is intimately mixed with the coal. Little or no solvent is employed, and the primary source of hydrogen is molecular hydrogen in the latter case. The ultimate "depolymerization" of coal occurs in Fischer–Tropsch chemistry wherein the coal is reacted with oxygen and steam at about 1100°C to break up, or gasify, the coal into carbon monoxide, hydrogen, and carbon dioxide. A water–gas shift reaction is then carried out to adjust the hydrogen : carbon monoxide ratio, after which the carbon monoxide is catalytically hydrogenated to form methanol, or to build up liquid hydrocarbons (see Chapter 3 for discussion of Fischer–Tropsch chemistry).

1.4. PETROLEUM REFINING AND UPGRADING

1.4.1. Distillation of Crude Petroleum

Crude oil (petroleum), a dark, viscous liquid, is a mixture of virtually hundreds of different hydrocarbons. Distillation of the crude oil yields several fractions (Table 1.6),[9] which then are used for different properties.

Table 1.6. Fractions of typical distillation of crude petroleum

Boiling-Point Range (°C)	Carbon Products (Atoms)	
<30	C_1–C_4	Natural gas, methane, ethane, propane, butane, LPG[a]
30–200	C_4–C_{12}	Petroleum ether (C_5, C_6), ligroin (C_7), straight-run gasoline
200–300	C_{12}–C_{15}	Kerosene, heating oil
300–400	C_{15}–C_{25}	Gas oil, diesel fuel, lubricating oil, waxes
>400	>C_{25}	Residual oil, asphalt, tar

[a] Liquefied petroleum gas.

1.4.2. Hydrocarbon Refining and Conversion Processes

The relative amounts of usable fractions obtainable from a crude oil do not coincide with the commercial needs. Also, the qualities of the fractions obtained directly by distillation of the crude oil seldom meet the required specifications for various applications; for example, the octane rating of the naphtha fractions must be substantially upgraded to meet the requirements of internal-combustion engines in today's automobiles. These same naphtha liquids must also be treated to reduce sulfur and nitrogen components to acceptable levels (desulfurization and denitrogenation) in order to minimize automotive emissions and pollution of the environment. Therefore, each fraction must be upgraded in the petroleum refinery to meet the requirements for its end-use application. Hydrocarbon feeds of the refining operations are further converted or upgraded to needed products, such as high-octane alkylates, oxygenates, and polymer. Major hydrocarbon refining and conversion processes include cracking, dehydrogenation (re-forming), alkylation, acylation, isomerization, addition, substitution, oxidation–oxygenation, reduction–hydrogenation, metathesis, oligomerization, and polymerization. These processes and their applications can be schematically characterized as follows:

Cracking[10–16]—to form lower-molecular-weight products and to supply alkenes for alkylation:

$$\text{(1.1)}$$

Dehydrogenation (Reforming)[17–20]—to increase octane number of gasoline, to produce alkenes from alkanes [Eq. (1.2)], as well as aromatics, such as benzene, toluene, and xylenes [Eq. (1.3)]:

$$CH_3CH_3 \xrightarrow{\text{catalyst}} CH_2{=}CH_2 \qquad (1.2)$$

$$\text{(1.3)}$$

Dehydrocyclization[17,19]—to produce aromatics such as toluene:

$$\text{(1.4)}$$

Isomerization [(*of alkanes*, Eq. (1.5); *alkylaromatics*, Eq. (1.6)]:[10,19–22]—to increase the octane number of gasoline, to produce xylenes and so on:

$$CH_3(CH_2)_4CH_3 \xrightleftharpoons[\text{catalyst}]{\text{acid}} \quad \text{(structures)} \quad \text{etc.} \tag{1.5}$$

$$\tag{1.6}$$

Alkylation [*alkenes with alkanes*, Eq. (1.7); *aromatics*, Eq. (1.8)].[19,21,23–26]—to produce high-octane gasoline and jet-fuel components, detergent alkylates, plastics, intermediates, and other products:

$$CH_3-\underset{\underset{CH_3}{|}}{C}=CH_2 \;+\; CH_3-\underset{\underset{CH_3}{|}}{\overset{\overset{CH_3}{|}}{C}}-H \xrightarrow[\text{or HF}]{H_2SO_4} \quad \text{(structures)} \quad \text{etc.} \tag{1.7}$$

$$\bigcirc \;+\; RCH=CH_2 \xrightarrow{\text{AlCl}_3,\text{ etc.}} \text{(structure with } RCHCH_3) \tag{1.8}$$

Metathesis:[26–30]

$$\underset{RCH}{\overset{RCH}{\|}} \;+\; \underset{CHR'}{\overset{CHR'}{\|}} \longrightarrow 2\,RCH=CHR' \tag{1.9}$$

Oligomerization[31,32] and *polymerization*:[33–40]

$$2n\;XCH=CH_2 \Longrightarrow \left[CHX^{-CH_2-}CHX^{-CH_2-} \right]_n \tag{1.10}$$

Further transformation (functionalization) reactions include varied *additions,*[41] *carbonylative conversions,*[42] *acylations,*[43–45] *substitutions,*[43,46–50] *oxidations,*[51–60] and *reductions.*[61–69] Major petroleum refining operations are discussed in Chapter 2, whereas Chapters 4–13 discuss the chemistry of prototype hydrocarbon transformation reactions.

1.5. FINITE, NONRENEWABLE HYDROCARBON RESOURCES

Our oil and natural-gas reserves are finite and not renewable (except on a geologic timescale). The dire predication of the early 1970s following the first Arab oil crises that we will exhaust our oil reserves by the end of the twentieth century, turned out to be overly pessimistic. Our known oil reserves have, in fact, since doubled, and our gas reserves more than tripled[2] (Table 1.7).

If we take into account less accessible petroleum and gas reserves or those locked up in the form of tar sands, oil shale, or methane hydrates, our overall hydrocarbon reserves could be even 3–5 times higher. This would give us a century's supply, whereas our coal reserves may stretch to two or three centuries. Despite this more favorable outlook, it is necessary to point out the existing close relationship between our ever-increasing overall energy needs and our hydrocarbon sources essential in a technological society.

Hydrocarbons are required in our modern-day life not only as energy sources, including convenient transportation fuels for our cars, tracks, airplanes (see Section 1.8.2) but also to produce commonly used products ranging from polymers to textiles to pharmaceuticals. At the beginning of the twenty-first century we can look back with substantial satisfaction at our technological and scientific achievements. We should, however, also realize that we continue to deplete the nonrenewable resources of our planet, particularly fossil fuels and hydrocarbons and at the same time create ecological and environmental problems. As mentioned earlier, dire predictions of the early exhaustion of our natural hydrocarbon sources by the

Table 1.7. Recognized oil and natural-gas reserves (in billion tons) from 1960 to 1990

Year	Oil	Natural Gas
1960	43.0	15.3
1965	50.0	22.4
1970	77.7	33.3
1975	87.4	55.0
1980	90.6	69.8
1986	95.2	86.9
1987	121.2	91.4
1988	123.8	95.2
1989	136.8	96.2
1990	136.5	107.5

end of the twentieth century proved to be grossly exaggerated. Together with the discovery of new oil and gas finds, we are assured of supplies, although with the inevitably higher cost to open up less accessible sources, through the twenty-first century. We must realize, however, that a century of remaining natural hydrocarbon supplies is not changing the basic predicament; we will need to find feasible and economical ways to synthetically produce hydrocarbons on a very large scale to satisfy our future needs. The challenge truly is a global one and will need to unite human efforts to find practical solutions.

1.6. HYDROCARBON SYNTHESIS

Using Cornforth's definition of chemical synthesis[70] as an intentional construction of molecules by means of chemical reactions, hydrocarbon synthesis is a systematic construction of hydrocarbons including alkanes, alkenes, aromatics, and the like. The simplest hydrocarbon, methane itself, can be practically produced from carbon monoxide (or carbon dioxide) and hydrogen (methanation). Generally, however, hydrocarbon synthesis relates to obtaining higher hydrocarbons from one-carbon (C_1) precursors, including syngas (i.e., synthesis gas: $CO + H_2$), methyl alcohol, methyl halides, or methane itself (via oxidative condensation).

The alternative possibilities existing today in C_1 chemistry can be depicted as in Scheme 1.1.

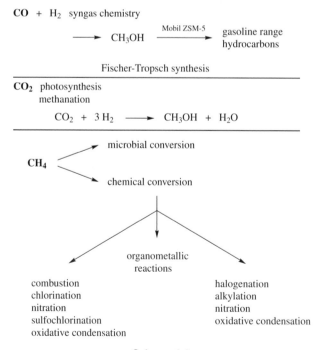

Scheme 1.1

1.6.1. Syngas (CO–H$_2$)-Based Fischer–Tropsch Synthesis

Hydrocarbon synthesis is not a new challenge. Germany, realizing its very limited resources in the 1920s and 1930s, developed a technology for the conversion of coal into liquid hydrocarbons. The work of Bergius[71] and that of Fischer and Tropsch[72] culminated in the development of catalytic coal liquefaction, and more significantly in an industrial process utilizing coal-derived mixtures of carbon monoxide and hydrogen gas (called *synthesis gas* or *syngas*), to catalytically produce hydrocarbons.[73–76] Syngas was obtained from the reaction of coal with steam (but was more recently obtained also by the partial burning of natural gas). The Fischer–Tropsch syngas-based synthesis was used during World War II in Germany on an industrial scale, with a peak production of around 60,000 barrels [1 bl = 42 gal (gallons)] per day. For comparison the present U.S. domestic consumption is about 16–17 million barrels per day. South Africa starting in the 1960s developed an updated Fischer–Tropsch synthetic fuel plant (Sasol) using improved engineering and technology. The capacity of this project is coincidentally estimated to be about the same as the peak World War II Germany production.

In contrast, if the United States would rely overnight solely on synthetic oil for its overall needs, we (i.e., U.S. citizens) would need some 300 Sasol-size plants. Whereas our coal reserves may last for three centuries, mining coal on the scale needed for conversion to hydrocarbons would be a gigantic task. If the United States would today convert to coal-based synthetic fuels for the 300 synthetic fuel plants mentioned above, some 10 million metric tons of coal would be needed daily. Even if our coal reserves could sustain such a demand over a long period of time, not only would an enormous investment be needed (in the trillions of the dollars); we would also need to recruit millions of young people to become coal miners, create an entirely new transportation system, and in general adjust our standard of living and lifestyle to pay for the enormous cost of coal-based synthetic oil—a hardly feasible scenario. It could be said that coal mining in places can be by surface strip mining and that underground gasification of coal to methane (and subsequently pumping it out through pipelines) will become feasible in advantageous locations. Nevertheless, many of the coal reserves would remain difficult to access.

Syngas itself cannot be piped over long distances. The Fischer–Tropsch process is also energetically wasteful overall as it burns half of the coal (or natural gas) to generate syngas in the first step, followed by an equally energetic second step in converting it into a hydrocarbon mixture of such complexity that no existing refinery could handle it. The strongest argument for Fischer–Tropsch chemistry is, of course, that it works and can produce synthetic fuels on an industrial scale. Then you have to disregard economic, labor-related, social, and ecological problems (for which wartime Germany and South Africa in the 1960s and 1970s were hardly acceptable models). With the relative stability of the world oil supplies for the foreseeable near future, it seems to be the right time to look for more feasible alternatives and to explore new chemistry. [During the short-lived alternate fuel research boom of the 1970s and early 1980s (following the two oil crises) extensive research on updating Fischer–Tropsch chemistry was carried out in the United States, western Europe, and Japan.] Because of the proven commercial feasibility of the Fischer–Tropsch synthesis, it is still customary to refer to C_1 chemistry as the conversion of syngas into higher hydrocarbons, although its scope has developed much further.[77]

1.6.2. Methyl Alcohol Conversion

One shortcoming of the Fischer–Tropsch synthesis is its lack of selectivity in giving complex product mixtures. In an attempt to improve the selectivity of syngas-based hydrocarbon synthesis, Mobil researchers developed a process consisting of converting methyl alcohol (itself, however, produced from syngas) to gasoline (or other hydrocarbons) over a shape-selective intermediate-pore-size zeolite catalyst (H-ZSM-5):[22,78]

$$n\ CH_3OH \xrightarrow[\text{--H}_2\text{O}]{\text{H-ZSM-5, 350°C}} \text{hydrocarbons} \tag{1.11}$$

The process, referred to as *methyl alcohol-to-gasoline conversion*, involves initial dehydration of methyl alcohol to dimethyl ether, with a subsequent carbenoid-type reaction. It goes probably through an oxonium ylide-type intermediate that is readily methylated by excess methanol (dimethyl ether) producing the crucial $C_1 \rightarrow C_2$ conversion and then cleaves to ethylene. Once a C_1 precursor is converted into a C_2 derivative (i.e., ethylene), the further conversion to higher hydrocarbons of the gasoline range (or to aromatics) follows well-known chemistry.

Further research has shown that bifunctional acid–base catalysts such as WO_3 on Al_2O_3 or tungstophosphoric acid, lacking shape-selective nature can also bring about the methyl alcohol–hydrocarbon conversions.[79] Shape selectivity of the catalyst thus is important in controlling product distributions and also to limit coking over the catalysts.

The Mobil process as discussed still uses syngas in its initial step. Olah has shown,[79] however, that it is possible to convert methane directly by selective electrophilic halogenation to methyl halides and through their hydrolysis to methyl alcohol [Eq. (1.12)]. When bromine is used, the HBr byproduct of the reaction is readily reoxidized to bromine, allowing a catalytic process in which bromine acts only as a redox catalyst [Eq. (1.13)]:

$$CH_4 \xrightarrow[\text{catalyst}]{Br_2} HBr + CH_3Br \xrightarrow[\text{catalyst}]{H_2O} CH_3OH + HBr \tag{1.12}$$

$$2\ HBr + 0.5\ O_2 \longrightarrow Br_2 + H_2O \tag{1.13}$$

It is even not necessary to hydrolyze methyl halides to methyl alcohol as methyl halides themselves readily condense over bifunctional acid–base catalysts such as zeolite or WO_3 on Al_2O_3 to ethylene (and propylene) and subsequently to higher hydrocarbons (gasoline, aromatics):

$$2\ CH_3Hlg \xrightarrow[-2\ HHlg]{\substack{\text{H-ZSM-5 or} \\ \text{WO}_3/\text{Al}_2\text{O}_3}} CH_2{=}CH_2 \implies \text{hydrocarbons} \qquad (1.14)$$

1.6.3. Carbon Dioxide Conversion

C_1 chemistry can no longer be equated only with syngas chemistry. Nature's own CO_2 *photosynthesis* and *bacterial methane conversion* are also C_1 conversion processes. We are far from approaching these processes for practical synthetic use efficiently. Production of methane from carbon dioxide (similarly to carbon monoxide) and hydrogen is a feasible process (methanation).[80] Similarly, reduction of carbon dioxide with hydrogen to methyl alcohol[81] can be readily carried out, and the method has been industrially developed:

$$CO_2\ +\ 3\,H_2 \longrightarrow CH_3OH\ +\ H_2O \qquad (1.15)$$

If technology one day achieves unlimited cheap energy (as, e.g., from fusion), our long-range hydrocarbon needs will also be assured. We will be able to take CO_2 from the atmosphere (thus also recycling the excess carbon dioxide generated during burning fossil fuels contributing to the greenhouse effect) and hydrogen from electrolysis of seawater and combine them into methane or methyl alcohol and subsequently to hydrocarbons.

1.6.4. Direct Methane Conversion

The direct conversion of *methane* as a building block for higher hydrocarbons and derived products is one of the most promising routes of future hydrocarbon synthesis.[77,79] Methane is still abundant on Earth; it is the major component of natural gas. If we would stop burning natural gas for its energy content, our reserves as a hydrocarbon source could last for a long time. The possibility of large abiogenic methane sources was also mentioned previously. Even after our natural-gas reserves would be exhausted, biomass (sustained as long as there is life on our planet) can be easily and efficiently converted by microorganisms into methane (on a very short timescale, as compared to what is needed to form "natural gas"). Exploratory work has shown the feasibility of growing algae or kelp in the vast expanses of oceans.

These substances can be harvested and biologically converted into methane, and the gas can be pumped ashore. In some of the large U.S. cities landfills refuse is already biologically converted into methane gas and used for energy generation. It is also known that aluminum carbide, which can be produced similarly to calcium carbide in electric arcs, gives primarily methane on hydrolysis. (The similar process from calcium carbide giving acetylene was the foundation of the organic chemical industry in Germany early in the twentieth century.) Thus, atomic power plants in off-peak periods could be producing aluminum carbide and through it methane. Regardless of its source, methane will be available, and thus it is highly practical to consider its conversion to higher hydrocarbons and their functionalized products. Biological conversion of methane to higher hydrocarbons or to methyl alcohol is clearly of great significance for the future but is not discussed further in the scope of the present discussion of C_1 chemistry.

The reactivity of methane (in many organic textbooks still discussed only in a chapter devoted to "paraffins," indicating lack of reactivity) was until recently (at the time of writing) considered mostly only in terms of its *free-radical* reactions. After all, we burn large amounts of natural gas (i.e., methane) to produce energy. Combustion processes are free-radical chain reactions. So is the high-temperature conversion of methane to acetylene. Similarly other radical reactions such as chlorination, nitration, and sulfochlorination of methane are practiced industrially to obtain substituted derivatives. These reactions generally show limited selectivity characteristics of radical reactions (e.g., the chlorination of methane under radical conditions gives all four possible chloromethanes).

More recently, direct catalytic *oxidative condensation* of methane to ethane (with metal oxides),[57,82–84] as well as to ethylene and acetylene (via high-temperature chlorinative conversion) was explored.[76] In all these processes, however, a significant portion of methane is lost by further oxidation and soot formation. The selectivity in obtaining ethane and ethylene (or acetylene), respectively, the first C_2 products, is low. There has, however, been much progress in metal-oxide-catalyzed oxidative condensation to ethane.

Another chemical approach to the chemical conversion of methane involves *organometallic reactions*.[85–89] Interesting work with iridium complexes and other transition metal insertion reactions (rhodium, osmium, rhenium, etc.) were carried out. Even iron organometallics were studied. These reactions take place in the coordination spheres of the metal complexes, but so far the reactions are stoichiometric and noncatalytic.[77] In terms of synthetic hydrocarbon chemistry, these conversions are thus not yet practical, but eventually it is expected that catalytic reactions will be achieved.

The third alternative approach of methane conversion is via electrophilic reactions.[77] The *electrophilic conversion* of methane is based on the feasibility of electrophilic reactions of single bonds and thus saturated hydrocarbons.[90] Both C—H and C—C bonds can act as electron donors against strongly electrophilic reagents or superacids. Olah's studies showed that even methane is readily protonated or alkylated under these conditions. Methane with SbF_5-containing superacids was found to undergo condensation to C_2–C_6 hydrocarbons at 50–60°C.

Combining two methane molecules to ethane and hydrogen is endothermic by some 16 kcal/mol:

$$2\,CH_4 \longrightarrow C_2H_6 + H_2 \tag{1.16}$$

Any condensation of methane to ethane and subsequently to higher hydrocarbons must overcome the unfavorable thermodynamics. This can be achieved in condensation processes of oxidative nature, where hydrogen is removed by the oxidant. SbF_5- or FSO_3H-containing superacid systems also act as oxidants. The oxidative condensation of methane was subsequently found to take place with more economical cooxidants such as halogens, oxygen, sulfur, or selenium:[91]

$$CH_4 \xrightarrow[\text{superacid catalyst}]{\text{halogens, } O_2, S_x, Se} \text{hydrocarbons} \tag{1.17}$$

Significant practical problems, however, remain to carry out the condensation effectively. Conversion was so far achieved only in low yields. Because of the easy cleavage of longer chain alkanes by the same superacids, C_3–C_6 products predominate.

Natural gas instead of pure methane can also be used in condensation reactions.[91] When natural gas is dehydrogenated, the C_2–C_4 alkanes it contain are converted into olefins. The resulting methane–olefin mixture can then without separation be passed through a superacid catalyst, resulting in exothermic alkylative condensation:

$$CH_4 + RCH{=}CH_2 \xrightarrow{H^+} \underset{\overset{|}{CH_3CHCH_3}}{\overset{R}{}} \text{ etc.} \tag{1.18}$$

Chapter 3 discusses in more detail C_1-based synthetic possibilities of hydrocarbons.

Fuel cells use an electrochemical process to convert the energy of a chemical reaction directly into electricity. The chemical reaction is catalytically activated. In previous fuel cells hydrogen (produced in a separate converter, e.g., from natural gas) and oxygen (from air) were reacted to produce electricity, heat, and water. Hydrocarbons (or oxygenated hydrocarbons) were thus used as fuels only *indirectly*, giving through reforming by hydrogen as the fuel. Direct-oxidation fuel cells use liquid oxygenated hydrocarbons (methanol[92]) or hydrocarbon fuels. They are much better suited for transportation applications in view of their lower weight and volume as compared to indirect cells. The weight and volume advantages of direct-oxidation fuel cells are due to the fact that they do not require any preprocessing (converter) equipment. It is expected that they will play a significant role in the future in providing cleaner and more efficient propulsion systems for transportation vehicles.

1.7. CHEMICAL NATURE OF HYDROCARBON CONVERSION REACTIONS

1.7.1. Homolytic (Free-Radical) Reactions

The chemistry of the high-temperature conversion and transformation processes of hydrocarbons is based on homolytic (*free-radical*) processes.[93a]

The combustion of hydrocarbons for energy generation or in propulsion systems is itself a radical-chain process. Thermal cracking, oxygenation, hydrogenation–dehydrogenation, cyclization, and so on proceed through free radicals. In the discussion of the transformation of hydrocarbons the reaction chemistry as well as relevant mechanistic aspects will be treated throughout the book.

Free-radical reactions play an important role not only in high-temperature refining and processing operations (cracking, reforming, hydrocracking, dehydrogenation, etc.) but also in oxidation chemistry (Chapter 9) and in many addition and substitution reactions (Chapters 6 and 10) as well as in polymerization (Chapter 13). The high-pressure polymerization of ethylene, for example, played a key role in the development of the plastics industry.

1.7.2. Heterolytic (Ionic) Reactions

In *acid-catalyzed* reactions of unsaturated hydrocarbons (alkenes, alkynes, arenes) positive hydrocarbon ions—*carbocations*—are formed, which are then responsible for the electrophilic transformations:[93b]

$$RCH=CH_2 \xrightleftharpoons{H^+} R\overset{+}{C}H-CH_3$$

$$RCH=CH_2 \longrightarrow R\overset{+}{C}HCH_2\overset{R}{\underset{|}{C}}HCH_3 \qquad (1.19)$$

$$\text{benzene} \longrightarrow \underset{-H^+}{\longrightarrow} \qquad (1.20)$$

$$CH_3CH_2CH=CH_2 \xrightleftharpoons{H^+} CH_3CH_2\overset{+}{C}HCH_3 \rightleftharpoons \overset{+}{C}H_2\overset{CH_3}{\underset{|}{C}}HCH_3 \rightleftharpoons CH_3\overset{CH_3}{\underset{|}{\underset{+}{C}}}CH_3 \xrightleftharpoons{H^-} CH_3\overset{CH_3}{\underset{|}{C}}HCH_3$$

$$(1.21)$$

The carbocations involved in these reactions are trivalent carbenium ions, of which CH_3^+ is parent. It was Whitmore in the 1930s who first generalized their importance in hydrocarbon transformations based on fundamental studies by Meerwein, Ingold, Pines, Schmerling, Nenitzescu, Bartlett, and others.

More recently it was realized that hydrocarbon ions (carbocations) also encompass five (or higher) coordinate carbonium ions for which CH_5^+ is parent.[94] Alkanes having only saturated C—H and C—C bonds were found to be protonated by very strong acids, specifically, superacids, which are billions of times stronger than concentrated sulfuric acid.[21]

Protolytic reactions of saturated hydrocarbons in superacid media[21] were interpreted by Olah as proceeding through the protonation (protolysis) of the covalent C—H and C—C single bonds. The reactivity is due to the electron donor ability of the σ bonds via two-electron, three-center bond formation. Protolysis of C—H bonds leads via five-coordinate carbocations with subsequent cleavage of H_2 to trivalent ions, which then themselves can further react in a similar fashion:

$$R_3CH + H^+ \rightleftharpoons \left[R_3C - \overset{H}{\underset{H}{\diagup}} \right]^+ \rightleftharpoons R_3C^+ + \overset{H}{\underset{H}{|}} \qquad (1.22)$$

The reverse reaction of carbenium ions with molecular hydrogen, can be considered as alkylation of H_2 through the same pentacoordinate carbonium ions that are involved in C—H bond protolysis. Indeed, this reaction is responsible for the long used (but not explained) role of H_2 in suppressing hydrocracking in acid-catalyzed reactions.

Protolysis can involve not only C—H but also C—C bonds [Eq. (1.23)]; this explains why alkanes can be directly cleaved protolytically by superacids, which is of significance, for example, in hydrocleaving heavy oils, shale oil, tar-sand bitumens, and even coals:

$$\overset{|}{\underset{|}{{>}}}C - C{<} + H^+ \longrightarrow \left[\overset{H}{\underset{}{{>}C \cdots \diagup \cdots C{<}}} \right]^+ \longrightarrow {>}\overset{+}{C}{<} + {>}C - H \qquad (1.23)$$

In contrast, cracking of longer-chain alkanes with conventional acid catalysts is considered to proceed via β scission involving initial formation of trivalent carbocations:

$$-\overset{R}{\underset{|}{\overset{|}{C}}}-\overset{+}{C}{<} \longrightarrow \left[\overset{R}{\underset{}{{>}C \cdots \diagup \cdots C{<}}} \right]^+ \longrightarrow R^+ + {>}C=C{<} \qquad (1.24)$$

Acid-catalyzed alkylation and isomerization processes all proceed through carbocations. Typical is the isobutylene–isobutane alkylation giving high-octane isooctane.

In this and other conventional acid-catalyzed reactions the key is the reactivity of alkenes, giving on protonation alkyl cations that then readily react with excess alkene, giving the alkylate cations. These carbocations then abstract hydrogen from the isoalkane, yielding the product alkylate and forming a new alkyl cation to reenter the reaction cycle. Chapter 5 discusses acid-catalyzed alkylations and their mechanism.

In the acid-catalyzed isomerization of straight-chain alkanes to higher-octane branched ones, after initial protolytic ionization, alkyl and hydrogen shifts in the formed carbocations lead to the most branched and therefore thermodynamically preferred, generally tertiary, carbocations. Intermolecular hydrogen transfer from excess alkane then produces the isomeric isoalkane with the formed new carbocation reentering the reaction cycle.

In alkane–alkene alkylation systems it is always the π-donor alkene that is alkylated by carbocations formed in the system. In the absence of excess alkenes (i.e., under superacidic conditions), however, the σ-donor alkanes themselves are alkylated. Even methane or ethane, when used in excess, are alkylated by ethylene to give propane and *n*-butane, respectively:

$$CH_2{=}CH_2 \xrightarrow{\ H^+\ } CH_3\overset{+}{C}H_2$$

$$\left[CH_3CH_2 - \underset{CH_3}{\overset{H}{\diagup}} \right]^+ \xrightarrow{\ -H^+\ } CH_3CH_2CH_3 \qquad (1.25)$$

$$\left[CH_3CH_2 - \underset{CH_2CH_3}{\overset{H}{\diagup}} \right]^+ \xrightarrow{\ -H^+\ } CH_3(CH_2)_2CH_3 \qquad (1.26)$$

(with CH_4 and CH_3CH_3 as reactants branching off)

Base-catalyzed hydrocarbon conversions, although generally less common than acid-catalyzed reactions, also play a significant role in hydrocarbon chemistry. They proceed through proton abstraction giving intermediate *carbanions*:[92a,94]

$$R{-}H \xrightarrow{\ -H^+\ } R^- \qquad (1.27)$$

Base-catalyzed carbanionic alkylation, isomerization, polymerization reactions are of major significance. Base-catalyzed alkylation of alkylarenes, in contrast to acid-catalyzed ring alkylation, leads to alkylation of the side chain in the benzylic position [Eq. (1.28); see also Chapter 5]; of particular interest is the alkylation of toluene to ethylbenzene (for styrene production).

$$\begin{array}{c} CH_3 \\ \text{C}_6\text{H}_5 \end{array} + CH_3CH{=}CH_2 \xrightarrow[180{-}200°C]{Na\ or\ K} \begin{array}{c} CH_2CH(CH_3)_2 \\ \text{C}_6\text{H}_5 \end{array} + \begin{array}{c} CH_2CH_2CH_2CH_3 \\ \text{C}_6\text{H}_5 \end{array} \qquad (1.28)$$

major   minor

Selective double-bond migration of linear and cyclic alkenes without skeletal rearrangement is achieved by treatment with basic reagents ([Eq. (1.29)]; see also Chapter 4):

$$\text{\raisebox{0pt}{}}\quad\xrightarrow{\text{Na on Al}_2\text{O}_3}\quad\text{\raisebox{0pt}{}}\qquad\qquad(1.29)$$

Base-catalyzed anionic oligomerization and polymerization are also of great significance (see Chapter 13). Styrene derivatives and conjugated dienes are the most suited for anionic polymerization. Generally good control of molecular weight distribution can be achieved, which is of great practical use. An interesting and significant characteristic of anionic polymerization is the ability to obtain "living" polymers that contain active centers for further polymerization (copolymerization) to give unique products.

Reactions of hydrocarbons, particularly those with relatively low ionization potentials (aromatics and π systems but not necessarily alkanes) frequently proceed via *single-electron transfer* (SET) processes involving *radical ions*.[96] These reactions are particularly significant in oxidation reactions (Chapter 9).

1.8. USE OF HYDROCARBONS

1.8.1. Refined Petroleum Products

As discussed above, hydrocarbons (oil and gas) are used primarily as fuels to generate energy and for space heating. Refined petroleum products provide gasoline, diesel fuel, heating oil, lubricating oil, waxes, and asphalt. A relatively small (4%) portion of oil is used as raw material to produce chemical products essential to our everyday life ranging from plastics to textiles to pharmaceuticals, and so on.

1.8.2. Transportation Fuels

A specific role is assigned to fuels to propel our cars, trucks, airplanes, and other vehicles. They are generally referred to as *transportation fuels*. The impact of the internal-combustion engine on our lives is fundamental. It allowed mankind to achieve mobility via easy means of transportation of persons and goods alike. In less than a century this revolution has transformed and eased our lives but at the same time raised new challenges.

Gasoline and *diesel* fuels are used worldwide in enormous amounts and are produced by the petroleum industry by oil refining (see Chapter 2). Liquid transportation fuels are conveniently transported, distributed, and dispensed directly into our vehicles and aircraft. Increased environmental concerns did much to begin to assure cleaner burning of our fuels. In the United States, law requires significant reduction of emissions and resulted in lead-free and more recently reformulated gasoline

(RFG) containing oxygenates for cleaner burning. At the same time they did not fundamentally change the existing technology. To find alternate fuels to supplement our oil reserves and to ensure cleaner fuels, much further effort is needed.

As long as the internal-combustion engine will remain as the mainstay of our transportation technology, liquid fuels by their convenience and ease of distribution through the existing system will be preferred. Liquefied natural gas (LNG) or methane certainly are clean-burning, high-energy-content fuels, but their wide-spread use in transportation systems is limited by the difficulty of handling highly volatile and inflammable gases, necessitating the use of heavy pressure vessels to be used as fuel tanks (with associated safety hazards).

Commercial gasoline must provide sufficiently high octane numbers for present-day engines. Organometallic additives, such as tetraethyl lead used to achieve higher octane numbers, interfere with catalytic converters (which assure cleaner exhausts) and were (or are) phased out. To maintain needed octane numbers, oxygenated additives, such as alcohols or ethers [e.g., *tert*-butyl methyl ether or *tert*-butyl ethyl ether (MTBE and ETBE)] are added to gasoline. These oxygenates are produced on large scale by the petrochemical industry. In the United States about 85% of RFG contained MTBE and 8% contained ethanol in 1999.[97] More recently, however, there has been growing concern about the use of MTBE following its detection in water supplies in California, Maine, and other states.[98] MTBE is soluble in water, moves rapidly in groundwater, and has an unpleasant taste and odor. In addition, it is difficult to remove from water and persistent to microbial degradation. Although its health effects are not exactly known, legislatures are under pressure to scale back or eliminate MTBE use. The main sources of MTBE contamination are leaks from underground tanks and pipeline spills and leaks. Naturally, MTBE is only part of the contamination problem; banning MTBE would remove only MTBE without addressing the problem of the hydrocarbon contamination. The ban of MTBE and, at the same time, maintaining air quality require the increase of the use of other oxygenates. Ethanol could be the best candidate as the main blending component.

Isomerization of straight-chain to branched alkanes also increases the octane number, as do alkylates produced by alkene–isoalkane *alkylation* (such as that of isobutane and propylene, isobutylene, etc.). These large-scale processes are by now an integral part of the petroleum industry. Refining and processing of transportation fuels became probably the largest-scale industrial operation.

A much pursued alternative to the internal-combustion engine in transportation vehicles is electric propulsion. *Electric vehicles* thus far have been used for only specific and limited applications, although they are clean, noiseless, and safe. The main difficulty is that existing batteries are relatively inefficient to provide suitable range at reasonable weights. Pending a major breakthrough in battery technology, this will greatly hinder the wider use of electric vehicles. It also must be kept in mind when considering clean electricity-powered vehicles that we are only transferring the potential environmental pollution from individual vehicles to the electric power plants where electricity is generated by combustion of fossil fuels (if not nuclear). In large power plants, however, pollution can be controlled more efficiently.

1.8.3. Chemicals and Plastics

About 4% of petroleum is used as raw material for the chemical and plastics industries. It was coal tar-based aromatics and calcium carbide-based acetylene that led to the establishment of the chemical and plastic industries. World War II, however, brought about rapid development of the petrochemical industry with ethylene becoming the key starting material. A large variety of the chemical and plastics products are now produced. Starting with petrochemicals throughout the book, the reader will find discussion and references to the practical applications, as well as the underlying chemistry of the diverse and significant use of hydrocarbons.

REFERENCES AND NOTES

1. (a) All textbooks of organic chemistry contain discussion of hydrocarbons and their classes; (b) *Kirk-Othmer Encyclopedia of Chemical Technology*, 3rd ed., M. Grayson and D. Eckroth, eds., Wiley-Interscience, New York, 1978–1984; (c) K. Weissermel and H. J. Arpe, *Industrial Organic Chemistry*, 2nd ed., VCH, Weinheim, 1992.

2. For discussion of energy sources and consumption, see *Science and Sustainability*, Int. Inst. for Appl. Anal. Syst. (IIASA 20th Anniversary), Vienna, 1992, and references cited therein.

3. F. Asinger, *Paraffins, Chemistry and Technology*, Pergamon Press, Oxford, 1968.

4. (a) T. Gold, *Gas Res. Inst. Digest* **4**(3) (1981); (b) *Chem. Week* **131**(24), 19 (1982); *Chem. Eng. News* **69**(43), 48 (1991).

5. (a) S. J. Chemett, C. R. Maechling, R. N. Zare, P. P. Swan, and R. M. Walker, *Science* **262**, 721 (1993); (b) S. G. Loue and D. E. Brownlee, *Science* **262**, 550 (1993); (c) E. J. Zeman, *Physics Today* **47**(3), 17 (1994).

6. P. Lucier, *Invention and Technology*, 1991, p. 56.

7. (a) O. P. Strausz is thanked for communicating a review of his on tar-sand chemistry; (b) O. P. Strausz, T. W. Mojelsky, J. D. Payzant, G. A. Olah, and G. K. S. Prakash, U.S. Patent 5,290,428 (1994): *Superacid Catalyzed Hydrocracking of Heavy Oils and Bitumens*.

8. (a) M. Siskin (Exxon) is thanked for communicating in the mid 1980s a review of his on coal chemistry; (b) W. H. Wiser of the University of Utah is credited to have developed this scheme.

9. P. C. Vollhardt and N. E. Shore, *Organic Chemistry*, 2nd ed., Freeman, New York, 1994.

10. A. P. Bolton, in *Zeolite Chemistry and Catalysis*, ACS (American Chemical Society) Monograph 171, J. A. Rabo, ed., American Chemical Society, Washington, DC, 1976, Chapter 13.

11. C. N. Satterfield, *Heterogeneous Catalysis in Industrial Practice*, McGraw-Hill, New York, 1991, Chapter 9.

12. *Fluid Catalytic Cracking*, ACS Symp. Series 375, M. L. Occelli, ed., American Chemical Society, Washington, DC, 1988.

13. *Fluid Catalytic Cracking II*. ACS Symp. Series 452, M. L. Occelli, ed., American Chemical Society, Washington, DC, 1991.

14. *Fluid Catalytic Cracking: Science and Technology*, Studies in Surface Science and Catalysis, Vol. 76, J. S. Magee and M. M. Mitchell, Jr., eds., Elsevier, Amsterdam, 1993.

15. *Fluid Catalytic Cracking III, Methods and Processes*, ACS Symp. Ser. Vol. 571, M. L. Occelli and P. O'Connor, eds., American Chemical Society, Washington, DC, 1994.

16. *Fluid Cracking Catalysts*, M. L. Occelli and P. O'Connor, eds., Marcell Dekker, New York, 1998.

17. D. M. Little, *Catalytic Reforming*, Penn Well, Tulsa, OK, 1985.

18. *Catalytic Naphtha Reforming*, G. J. Antos, A. M. Aitani, and J. M. Parera, eds., Marcel Dekker, New York, 1995.

19. *Toluene, the Xylenes and Their Industrial Derivatives*, E. G. Hancock, ed., Elsevier, Amsterdam, 1982.

20. H. Pines and N. E. Hoffman, in *Friedel-Crafts and Related Reactions*, Vol. 2, G. A. Olah, ed., Wiley-Interscience, New York, 1964, Chapter 28.

21. G. A. Olah, G. K. S. Prakash, and J. Sommer, *Superacids*, Wiley, New York, 1985.

22. N. Y. Chen, W. E. Garwood, and F. G. Dwyer, *Shape Selective Catalysis in Industrial Applications*, Marcel Dekker, New York, 1996.

23. *Friedel-Crafts and Related Reactions*, Vol. 2, G. A. Olah, ed. Wiley-Interscience, New York, 1964.

24. *Industrial and Laboratory Alkylations*, ACS Symp. Series 55, L. F. Albright and A. R. Goldsby, eds., American Chemical Society, Washington, DC, 1977.

25. R. M. Roberts and A. A. Khalaf, *Friedel-Crafts Alkylation Chemistry*, Marcel Dekker, New York, 1984.

26. R. L. Banks, *Top. Curr. Chem.* **25**, 39 (1972).

27. J. K. Ivin, *Olefin Metathesis*, Academic Press, London, 1983.

28. V. Dragutan, A.T. Balaban, and M. Dimonie, *Olefin Metathesis and Ring-Opening Polymerization of Cyclo-Olefins*, Editura Academiei, Bucuresti, and Wiley, Chichester, 1985.

29. K. J. Ivin and J. C. Mol, *Olefin Metathesis and Metathesis Polymerization*, Academic Press, London, 1997.

30. V. Dragutan and R. Streck, *Catalytic Polymerization of Cyclolefins: Ionic, Ziegler–Natta and Ring-Opening Metathesis Polymerization*, Elsevier, Amsterdam, 2000.

31. D. C. Pepper, in *Friedel-Crafts and Related Reactions*, Vol. 2, G.A. Olah, ed., Wiley-Interscience, New York, 1964, Chapter 30.

32. J. Skupinska, *Chem. Rev.* **91**, 613 (1991).

33. J. Boor, Jr., *Ziegler-Natta Catalysis and Polymerization*, Academic Press, New York, 1979.

34. J. P. Kennedy and E. Maréchal, *Carbocationic Polymerization*, Wiley-Interscience, New York, 1982.

35. M. Morton, *Anionic Polymerization: Principles and Practice*, Academic Press, New York, 1983.

36. *Handbook of Polymer Science and Technology*, N. P. Cheremisinoff, ed., Marcel Dekker, New York, 1989.

37. *Comprehensive Polymer Science*, G. Allen and J. C. Bevington, eds., Pergamon Press, Oxford, 1989.

38. *Handbook of Polymer Synthesis*, H. R. Kricheldorf, ed., Marcel Dekker, New York, 1992.

39. M. Szwarc and M. Van Beylen, *Ionic Polymerization and Living Polymers*, Chapman & Hall, New York, 1993.

40. *Industrial Polymers Handbook: Products, Processes, Applications*, E. S. Wilks, ed., Wiley-VCH, Weinheim, 2001.

41. *Comprehensive Organic Chemistry*, B. M. Trost and I. Fleming, eds., Vol. 4: *Additions and Substitutions at C–C π-Bonds*, M. F. Semmelhack, ed., Pergamon Press, Oxford, 1991.

42. *New Syntheses with Carbon Monoxide*, J. Falbe, ed., Springer, Berlin, 1980.

43. *Friedel-Crafts and Related Reactions*, Vol. 3, G. A. Olah, ed., Wiley-Interscience, New York, 1964.

44. H. Heaney, in *Comprehensive Organic Synthesis*, B. M. Trost and I. Fleming, eds., Pergamon Press, 1991, Vol. 2: *Additions to C–X π-Bonds*, Part 2, C. H. Heatchock, ed., Chapter 3.2, p. 733.

45. S. C. Eyley, in *Comprehensive Organic Synthesis*, B. M. Trost and I. Fleming, eds., Pergamon Press, 1991, Vol. 2: *Additions to C–X π-Bonds*, Part 2, C. H. Heathcock, ed., Chapter 3.1, p. 707.

46. R. O. C. Norman and R. Taylor, *Electrophilic Substitution in Benzenoid Compounds*. Elsevier, Amsterdam, 1965.

47. H. Cerfontain, *Mechanistic Aspects in Aromatic Sulfonation and Desulfonation*, Interscience, New York, 1968.

48. E. S. Huyser, *Free-Radical Chain Reactions*, Wiley-Interscience, New York, 1970, Chapter 5.

49. M. L. Poutsma, in *Free Radicals*, Vol. 2, J. K. Kochi, ed., Wiley-Interscience, New York, 1973, Chapter 15.

50. G. A. Olah, R. Malhotra, and S. C. Narang, *Nitration: Methods and Mechanisms*, VCH, New York, 1989.

51. D. J. Hucknall, *Selective Oxidation of Hydrocarbons*, Academic Press, London, 1974.

52. R. A. Sheldon and J. K. Kochi, *Metal-Catalyzed Oxidations of Organic Compounds*, Academic Press, New York, 1981.

53. *Organic Syntheses by Oxidation with Metal Compounds*, W. J. Mijs and C. R. H. I. de Jonge, eds., Plenum Press, New York, 1986.

54. P. S. Bailey, *Ozonation in Organic Chemistry*, Academic Press, New York, 1978, 1982.

55. *Singlet Oxygen*, H. H. Wasserman and R. W. Murray, eds., Academic Press, New York, 1979.

56. *Singlet O₂*, A. A. Frimer, ed., CRC Press, Boca Raton, FL, 1985.

57. A. Bielanski and J. Haber, *Oxygen in Catalysis*. Marcel Dekker, New York, 1991.

58. *New Developments in Selective Oxidation II*, Studies in Surface Science and Catalysis, Vol. 82, V. Cortés Corberán and S. Vic Bellón, eds., Elsevier, Amsterdam, 1994.

59. *Heterogeneous Hydrocarbon Oxidation*, ACS Symp. Ser. Vol. 638, B. K. Warren and S. T. Oyama, eds., American Chemical Society, Washington, DC, 1996.

60. G. Centi, F. Cavani, and F. Trifirò, *Selective Oxidation by Heterogeneous Catalysis*. Kluwer–Plenum, New York, 2001.

61. P. N. Rylander, *Catalytic Hydrogenation over Platinum Metals*, Academic Press, New York, 1967; *Catalytic Hydrogenation in Organic Syntheses*, Academic Press, New York, 1979.

62. A. P. G. Kieboom and F. van Rantwijk, *Hydrogenation and Hydrogenolysis in Synthetic Organic Chemistry*, Delft University Press, Delft, 1977.

63. M. Bartók, J. Czombos, K. Felföldi, L. Gera, Gy. Göndös, Á. Molnár, F. Notheisz, I. Pálinkó, G. Wittmann, and Á. G. Zsigmond, *Stereochemistry of Heterogeneous Metal Catalysis*, Wiley, Chichester, 1985.

64. B. R. James, *Homogeneous Hydrogenation*, Wiley, New York, 1973.

65. F. J. McQuillin, *Homogeneous Hydrogenation in Organic Chemistry*, D. Reidel, Dordrecht, 1976.

66. D. N. Kursanov, Z. N. Parnes, M. I. Kalinkin, and N. M. Loim, *Ionic Hydrogenation and Related Reactions*. Harwood, Chur, 1985.

67. *Comprehensive Organic Synthesis*, B. M. Trost and I. Fleming, eds., Vol. 8: *Reduction*, I. Fleming, ed., Pergamon Press, Oxford, 1991.

68. R. L. Augustine, *Heterogeneous Catalysis for the Synthetic Chemists*, Marcel Dekker, New York, 1996, Chapter 15.

69. G. V. Smith and F. Notheisz, *Heterogeneous Catalysis in Organic Chemistry*, Academic Press, San Diego, 1999, Chapter 2.

70. J. W. Cornforth, *Austr. J . Chem.* **46**, 157 (1993).

71. F. Bergius, Nobel Lecture, May 21, 1932, in *Nobel Lectures in Chemistry 1922–1941*. Elsevier, Amsterdam, 1966, p. 244.

72. F. Fischer and H. Tropsch, Ger. Patent 484,337 (1929); *Chem. Zbl. I*, 434 (1930), and discussion in Chapter 3.

73. C. D. Frohning, in *New Syntheses with Carbon Monoxide*, J. Falbe, ed., Springer, Berlin, 1980, Chapter 4.

74. M. E. Dry, in *Applied Industrial Catalysis*, Vol. 2, B. E. Leach, ed., Academic Press, 1983, Chapter 5.

75. R. B. Anderson, *The Fischer–Tropsch Synthesis*, Academic Press, Orlando, FL, 1984.

76. C. N. Satterfield, *Heterogeneous Catalysis in Industrial Practice*, McGraw-Hill, New York, 1991, Chapter 10.

77. C_1 chemistry is discussed in Chapter 3 of this volume, where relevant references are also to be found.

78. C. D. Chang, *Hydrocarbons from Methanol*, Marcel Dekker, New York, 1983.

79. G. A. Olah, *Acc. Chem. Res.* **20**, 422 (1987).

80. *Methanation of Synthesis Gas*, Advances in Chemistry Series 146, L. Seglin, ed., American Chemical Society, Washington, DC, 1975.

81. *Chem. Eng. News* **72**(13), 29 (1994); *Chem. Week* **154**(11), 14 (1994).

82. *Methane Conversion*, Studies in Surface Science and Catalysis, Vol. 36, D. M. Bibby, C. D. Chang, R. F. Howe, and S. Yurchak, eds., Elsevier, Amsterdam, 1988.

83. *Natural Gas Conversion*, Studies in Surface Science and Catalysis, Vol. 61, A. Holmen, K.-J. Jens, and S. Kolboe, eds., Elsevier, Amsterdam, 1991.

84. *Methane Conversion by Oxidative Processes: Fundamental and Engineering Aspects*, E. E. Wolf, ed., Van Nostrand Reinhold, New York, 1992.

85. A. E. Shilov, *Activation of Saturated Hydrocarbons by Transition Metal Complexes*, D. Reidel, Dordrecht, 1984, Chapter 5.

86. *Activation and Functionalization of Alkanes*, C. L. Hill, ed., Wiley-Interscience, New York, 1989.

87. *Selective Hydrocarbon Activation*, J. A. Davies, P. L. Watson, A. Greenberg, and J. F. Liebman, eds., VCH, New York, 1990.

88. A. E. Shilov, *Metal Complexes in Biomimetic Chemical Reactions*, CRC Press, Boca Raton, FL, 1997.

89. A. E. Shilov and G. B. Shul'pin, *Activation and Catalytic Reactions of Saturated Hydrocarbons in the Presence of Metal Complexes*, Kluwer, Dordrecht, 2000.

90. (a) G. A. Olah and J. A. Olah, *J. Am. Chem. Soc.* **93**, 1256 (1971); (b) D. T. Roberts, Jr. and E. L. Calihan, *J. Macromol. Sci., Chem.* **7**, 1629, 1641 (1973); (c) M. Siskin, *J. Am. Chem. Soc.* **98**, 5413 (1976); (d) M. Siskin, R. H. Schlosberg, and W. P. Kocsis, *ACS Symp. Ser.* **55**, 186 (1977); (e) J. Sommer, M. Muller, and K. Laali, *New. J. Chem.* **6**, 3 (1982); (f) G. A. Olah, J. D. Felberg, and K. Lammertsma, *J. Am. Chem. Soc.* **105**, 6529 (1983).

91. G. A. Olah, U.S. Patents 4,443 192 (1984); 4,513,164 (1984); 4,465,893 (1984); 4,467,130 (1984); 4,513,164 (1985).

92. G. A. Olah et al., U.S. Patent 5,559,638 (1997).

93. (a) J. March, *Advanced Organic Chemistry*, 4th ed. Wiley, New York, 1992, and references cited therein; (b) *Friedel-Crafts and Related Reactions*, Vols. 1–4, G. A. Olah, ed., Wiley-Interscience, New York, 1963–1965; (c) G. A. Olah, *Friedel–Crafts Chemistry*, Wiley-Interscience, New York, 1973.

94. G. A. Olah, G. K. S. Prakash, R. E. Williams, L. D. Field, and K. Wade, *Hypercarbon Chemistry*, Wiley, New York, 1987, and references cited therein.

95. H. Pines and W. M. Stalick, *Base-Catalyzed Reactions of Hydrocarbons and Related Compunds*, Academic Press, New York, 1977.

96. L. Eberson, *Electron Transfer Reactions in Organic Chemistry*, Springer, New York, 1987.

97. B. Allen, *Green Chem.* **1**, G142 (1999).

98. J. H. Vautrain, *Oil Gas J.* **97**(3), 18 (1999).

2

HYDROCARBONS FROM PETROLEUM AND NATURAL GAS

2.1. CRACKING

Low-molecular-weight olefins, such as ethylene and propylene, are very rare or absent in hydrocarbon sources. Demand for these olefins required their preparation from readily available petroleum sources. Cracking is the process in which higher-molecular-weight hydrocarbons, olefinic or aliphatic, are converted to more useful low-molecular-weight materials through carbon–carbon bond fission. The cracking of petroleum products is of primary importance in the production of gasoline. Indeed, the widespread use of automobiles was made possible by the ready availability of gasoline from petroleum. Cracking provides also gasoline with sufficiently high octane numbers, although frequently further enhancement is needed. Gasoline contains only about 20% of natural crude oil distillate and the remaining comes from petroleum refining processes.

2.1.1. Cracking Processes

Cracking is effected by one of three general methods: thermal cracking, catalytic cracking, or hydrocracking. Each process has its own characteristics concerning operating conditions and product compositions.

Thermal cracking or pyrolysis, the oldest of the these processes,[1] was first carried out by Burton[2] in his treatment of the residue remaining from the distillation of volatile components (so-called straight-run gasoline) of oil. The residue was treated in a horizontal drum by heating to 450–550°C under 5–6 atm. The volatile components were distilled off continuously until only coke remained in the still. The *Burton process* was supplanted by the continuous *Dubbs process*.[3] Operating at 450–725°C and above 70 atm, it gave a higher yield of light products. The accumulation of coke, which required cessation of operation, was eliminated by what was described as "the clean circulation principle." When the Dubbs process was first demonstrated, refiners were astounded by the fact that the plant operated continuously for 10 days, establishing a world record for length of operation, quantity of oil processed, and yield of gasoline produced. Excessive coke formation nevertheless is a serious shortcoming of thermal cracking processes. Since high olefin yields are characteristic of thermal cracking, the process still plays an important role in olefin manufacture. The temperature range 425–525°C and pressures of ∼25–70 atm are used.

The thermal cracking process, however, could not meet the high demand for quality gasoline. During World War I a new industrial process, catalytic cracking, was introduced for the production of gasoline from petroleum based on the Lewis acid aluminum chloride.[4] A nonaromatic oil was slowly distilled from 3–5% $AlCl_3$ at atmospheric pressure until the liquid temperature reached 225°C. The distilled product had a high octane number and a low sulfur content when compared to thermally cracked products. This process was later replaced by treatment of residual oils with acidic clays.[5–8] Known as the *Houdry process*, it used acid-treated montmorillonite at 450°C yielding gasoline of even higher octane number.[8–10] The clay was rapidly deactivated by the formation of coke but could be regenerated by burning off the carbon with hot air.

Natural clay catalysts were replaced by amorphous synthetic silica–alumina catalysts[5,11] prepared by coprecipitation of orthosilicic acid and aluminum hydroxide. After calcining, the final active catalyst contained 10–15% alumina and 85–90% silica. Alumina content was later increased to 25%. Active catalysts are obtained only from the partially dehydrated mixtures of the hydroxides. Silica–magnesia was applied in industry, too.

Since the early 1960s, the synthetic amorphous silica–alumina catalysts have been to a large extent replaced by catalysts containing up to 15% zeolites and crystalline aluminosilicates in an amorphous matrix.[12–19] The first catalysts were synthetic faujasites (Linde X) modified with rare-earth metal ions by ion exchange. These active catalysts give up to 10–50% higher yields of gasoline than the completely amorphous catalysts, with high selectivity and lowered coking. A rare-earth form of zeolite Y (REHY) with improved chemical and hydrothermal stability was introduced later.[14–16] Ultrastable hydrogen-exchanged zeolite (USY) and the use of medium-pore ZSM-5 are significant developments.[14,15,17] USY is employed in a matrix material that can be catalytically inert or active. ZSM-5 is used in small amounts (up to 3%) as an additive to a Y cracking catalyst or as a composite catalyst containing both ZSM-5 and Y zeolite. Zeolite–matrix interactions play a crucial role in overall cracking catalyst performance, ensuring improved gasoline yields, lower SO_x emission, improved catalyst attrition resistance, zeolite stability, and metal tolerance.[20,21] The latter may be improved by using different catalyst additives (passivators) and Bi- and Sb-containing liquid additives.[22–24] Because of deteriorating crude-oil quality a continuing trend is to develop new catalysts suitable for processing heavier feedstocks.[25–29]

Catalytic cracking with first-generation zeolite catalysts was operated at about 500°C, slightly above atmospheric pressure, and the catalysts were regenerated at about 600°C; 50–60% conversion to low-boiling materials was achieved. Low gas and coke yields, less sensitivity to poisoning, lower operating temperature (above 400°C for ZSM-5), and low contact time are the main features of the new generation of cracking catalysts. They also withstand higher temperature, allowing the use of higher regeneration temperatures (above 700°C). This results in less residual carbon after regeneration and thus higher catalytic activity. Conversions up to 80–90% can be achieved. One of the main achievements of catalytic cracking is the highly increased naphtha production. What is more important, the gasoline it produces has much better quality because of the different chemistry of the process. The new-generation zeolite catalysts (REHY, USY, ZSM-5), therefore, are also called *octane-enhancing catalysts*.

Further developments in catalytic cracking are determined by more stringent future requirements for transportation fuel.[30,31] New legislation limits the concentration in gasoline of olefins, benzene (<1%) and other aromatics (<25%), as well as sulfur content. Gasoline volatility must also be lowered necessitating butane removal. To meet these requirements for reformulated gasoline, the development of new catalysts and fine tuning of FCC technologies[32] to produce high-octane-number gasoline together with high yields of C_3–C_5 alkanes and alkenes are needed.[33] The latter products then can be transformed to alkylates and oxygenates for use as octane-enhancing gasoline-blending components in reformulated gasoline. New molecular sieve cracking materials such as offretite, zeolite L, Beta, and Omega have been designed and studied. The most promising, however, are silicoaluminophosphate (SAPO) molecular sieves.[34–37] These are crystalline materials discovered by researchers of Union Carbide.[34–36] They are prepared by isomorphous substitution of silicon into the framework of microporous aluminophosphate.[37,38] SAPO-37 was shown to exhibit promising characteristics.[39]

Hydrocracking,[40–44] or cracking in the presence of added hydrogen under high pressure, may be effected by special catalysts. The essential feature of such catalysts is their bifunctional nature, where the catalyst contains an acidic component as well as a hydrogenation–dehydrogenation component. Large-pore zeolites containing a small amount of noble metals (Pt, Pd, Ni), and Co–Mo- or Ni–Mo-on-alumina catalysts are mainly used. The process operates at 270–450°C and 80–200 atm. It produces no olefinic components whatsoever as the added hydrogen traps the intermediate species to yield saturated products. Such processes are generally expensive because of the considerable amounts of hydrogen used and the high operating pressure. However, feeds that are not suitable for catalytic cracking may be processed by hydrocracking (see Section 2.1.3). Extensive isomerization, skeletal rearrangement, and cracking occur primarily on the acidic component, but cracking (hydrogenolysis) on the hydrogenation component is seldom significant.[45] Its main function is to saturate olefinic intermediates formed in cracking and prevent catalyst deactivation by hydrogenating coke precursors.

Several technological variations for cracking operations were developed over the years. The fixed-bed and moving-bed technologies have been largely replaced by fluidized-bed operations.[5,18,19,46,47] Fluidized-bed reactors were introduced in 1941, ensuring mass production of high-grade military aviation gasoline. In fluid catalytic cracking (FCC) the finely divided catalyst is fed into the reactor concurrently with the oil. It speeds regeneration of the spent catalyst by continuous separation from the product followed by regeneration with hot air. FCC is the dominating industrial process at present, performed in numerous technological variations.[18,19,47,48]

2.1.2. Mechanism of Cracking

Thermal cracking processes generally proceed via radical pathways. The principal reactions of pyrolysis involve hydrogenation or dehydrogenation concurrently with fragmentation of the carbon skeleton. A suggested mechanism (known as the *Rice–Kossiakov concept*[49–51]) is depicted subsequently with hexadecane (cetane) to explain the observed results.[52] The process is initiated by carbon–carbon bond homolysis with formation of two free radicals [Eq. (2.1)]. The initially formed radicals may then abstract a secondary hydrogen [Eqs. (2.2) and (2.3)], as secondary hydrogens have lower bond strengths than do the primary hydrogens. All secondary hydrogens are equivalent in their ease of homolysis in comparison with homolysis of primary hydrogen from a methyl group. The intermediate radicals may undergo C–C bond homolysis (β scission) to form olefins (ethylene and propylene) and a new alkyl radical [Eqs. (2.4) and (2.5)]. Disproportionation leads to a new olefin and an alkane with the same number of carbon atoms [Eq. (2.6)]:

$$CH_3(CH_2)_{14}CH_3 \longrightarrow CH_3(CH_2)_8CH_2CH_2\overset{\cdot}{C}H_2 + \cdot CH_2CH_2CH_2CH_3 \qquad (2.1)$$

$$CH_3(CH_2)_8CH_2CH_2\overset{\cdot}{C}H_2 + CH_3(CH_2)_{14}CH_3 \longrightarrow$$
$$CH_3(CH_2)_8CH_2CH_2CH_3 + CH_3(CH_2)_{11}CH_2CH_2\overset{\cdot}{C}HCH_3 \qquad (2.2)$$

$$\cdot CH_2CH_2CH_2CH_3 + CH_3(CH_2)_{14}CH_3 \longrightarrow$$
$$CH_3CH_2CH_2CH_3 + CH_3(CH_2)_{11}CH_2\overset{\cdot}{C}HCH_2CH_3 \qquad (2.3)$$

$$CH_3(CH_2)_8CH_2CH_2\overset{\cdot}{C}H_2 \longrightarrow CH_3(CH_2)_8\overset{\cdot}{C}H_2 + CH_2{=}CH_2 \qquad (2.4)$$

$$CH_3(CH_2)_{11}CH_2CH_2\overset{\cdot}{C}HCH_3 \longrightarrow CH_3(CH_2)_{11}\overset{\cdot}{C}H_2 + CH_2{=}CHCH_3 \qquad (2.5)$$

$$2\cdot CH_2CH_2CH_2CH_3 \longrightarrow CH_2{=}CHCH_2CH_3 + CH_3CH_2CH_2CH_3 \qquad (2.6)$$

In the temperature range 500–800°C methyl and ethyl radicals are the only stable radicals formed.[49] Apparently all higher radicals decompose into alkenes and either methyl and ethyl radicals or atomic hydrogen. These then react via hydrogen

$$(CH_3)_2CHCH_2CH_3 \longrightarrow \begin{array}{l} CH_4 \quad + \quad (CH_3)_2C{=}CH_2 \\ CH_4 \quad + \quad CH_3CH{=}CHCH_3 \\ C_2H_6 \quad + \quad CH_3CH{=}CH_2 \end{array}$$

Scheme 2.1

abstraction to give methane, ethane, and dihydrogen. The overall result of thermal cracking is changes in hydrogen content and carbon skeleton. The reactions of methylalkanes (Scheme 2.1) are also illustrative of the chemistry occurring.

Cycloalkanes may be pyrolized in a manner similar to that for alicyclic alkanes. Cyclopentane, for instance, yields methane, ethane, propane, ethylene, propylene, cyclopentadiene, and hydrogen at 575°C. Analogous to cracking of alicyclic alkanes, the reaction proceeds by abstraction of a hydrogen atom followed by β scission. The cyclopentyl radical may undergo successive hydrogen abstractions to form cyclopentadiene.

Under certain conditions dehydrogenation is becoming an important reaction during pyrolysis (see further details in Section 2.3). However, only ethane is transformed to ethylene (the corresponding alkene) in high yield without cracking.[53] Propane also gives propylene, but cracking to form methane and ethylene is significant.[53,54] In general, lower-molecular-weight (hydrogen-rich) alkanes undergo dehydrogenation with the hydrogen formed participating in hydrogen transfer reactions. In contrast, C—C bond rupture is the characteristic transformation of higher-molecular-weight (more hydrogen-deficient) hydrocarbons. Paraffins most easily undergo thermal cracking.[52] Olefins, cycloalkanes, and aromatics all exhibit decreasing order of ease of cracking. In each type higher-molecular-weight hydrocarbons are more reactive than hydrocarbons with lower molecular weight.

Very little skeletal rearrangement occurs via pyrolysis, a fact inherent in the failure of free radicals to readily isomerize by hydrogen atom or alkyl group migration. As a result, little branched alkanes are produced. Aromatization through the dehydrogenation of cyclohexanes and condensation to form polynuclear aromatics can take place. Additionally, olefin polymerization also can occur as a secondary process.

In contrast with these results, catalytic cracking yields a much higher percentage of branched hydrocarbons. For example, the catalytic cracking of cetane yields 50–60 mol of isobutane and isobutylene per 100 mol of paraffin cracked. Alkenes crack more easily in catalytic cracking than do saturated hydrocarbons. Saturated hydrocarbons tend to crack near the center of the chain. Rapid carbon–carbon double-bond migration, hydrogen transfer to trisubstituted olefinic bonds, and extensive isomerization are characteristic.[52] These features are in accord with a carbocationic mechanism initiated by hydride abstraction.[43,55–62] Hydride is abstracted by the acidic centers of the silica–alumina catalysts or by already formed carbocations:

$$C_{11}H_{23}(CH_2)_2CH_2CH_2CH_3 \ + \ R^+ \ \longrightarrow \ C_{11}H_{23}(CH_2)_2CH_2\overset{+}{C}HCH_3 \ + \ RH \qquad (2.7)$$

The acidic centers, both the Brønsted and Lewis types, are generated by the isomorphous substitution of trivalent aluminum for a tetravalent silicon in the silica lattice.[61–63] Additionally, carbocations may be formed by protonation of alkenes,[43,52,56,61] which explains their higher reactivity in catalytic cracking.

Formed carbocations can undergo β scission [Eq. (2.8)] to yield propylene (or the corresponding alkene, but not ethylene) and a new primary cation. The primary ion rapidly rearranges to a secondary ion involving a hydride shift [Eq. (2.9)], which, in turn, can continue the process. Isomeric cations may also be formed through intermediate alkenes [Eq. (2.10)]:

$$C_{11}H_{23}(CH_2)_2CH_2\overset{+}{C}HCH_3 \longrightarrow C_{11}H_{23}CH_2\overset{+}{C}H_2 + CH_2{=}CHCH_3 \qquad (2.8)$$

$$C_{11}H_{23}CH_2CH_2\overset{+}{C}H_2 \longrightarrow C_{11}H_{23}\overset{+}{C}HCH_3 \qquad (2.9)$$

$$C_{11}H_{23}\overset{+}{C}HCH_3 \underset{-H^+}{\longrightarrow} C_{10}H_{21}CH{=}CHCH_3 \overset{H^+}{\longrightarrow} C_{10}H_{21}\overset{+}{C}HCH_2CH_3 \quad (2.10)$$

Alternatively, the secondary ion may undergo alkyl group migration followed by hydride shift to form a tertiary cation [Eq. (2.11)]. This cation may also undergo β scission [Eq. (2.12)], proton loss [Eq. (2.13)], or hydride abstraction [Eq. (2.14)] to form the appropriate new products:

$$C_{10}H_{21}\overset{+}{C}HCH_2CH_3 \longrightarrow C_{10}H_{21}\overset{\overset{Me}{|}}{C}H\overset{+}{C}H_2 \longrightarrow C_{10}H_{21}\overset{+}{C}\overset{Me}{\underset{Me}{<}} \qquad (2.11)$$

$$\longrightarrow C_8H_{17}\overset{+}{C}H_2 + CH_2{=}C\overset{Me}{\underset{Me}{<}} \qquad (2.12)$$

$$C_9H_{19}CH_2\overset{+}{C}\overset{Me}{\underset{Me}{<}}$$

$$\underset{-H^+}{\longrightarrow} C_9H_{19}CH{=}C\overset{Me}{\underset{Me}{<}} \qquad (2.13)$$

$$\underset{-R^+}{\overset{R-H}{\longrightarrow}} C_{10}H_{21}CH\overset{Me}{\underset{Me}{<}} \qquad (2.14)$$

Any of the carbocationic intermediates may give products or generate new carbocations via hydride abstraction reactions. The susceptibility of the primary cations toward rearrangement to secondary or tertiary carbocations results in the formation of more branched products. Hydrogen transfer during cracking on zeolites seems to be more important than β scission. This is explained by the ordered micropore structure of zeolites, allowing the reacting molecules to remain in proximity, resulting in a high probability of intermolecular reactions.

In contrast with these observations, it has been found that manganese and iron oxide supported on magnesium oxide catalyze cracking reactions via a free-radical mechanism and ensure enhanced ethylene and ethane production.[64]

Because of the bifunctional nature of hydrocracking catalysts, both cracking via carbocationic intermediates and hydrogenolysis occur during hydrocracking.[65] Additionally, hydrogenation and isomerization reactions take place either simultaneously or sequentially. An important transformation in hydrocracking is the saturation of aromatic hydrocarbons to naphthenes. Cracking in the middle and at the end of the chain of hydrocarbons are equally important reactions.

When a nonacidic catalyst such as nickel on kieselguhr is used, demethylation of alkanes occurs almost exclusively.[66] It is significant that the demethylation is very selective—a methyl group attached to secondary carbon is more readily cleaved off than is one attached to a tertiary, which in turn is more readily eliminated than one at a quaternary carbon. Demethylation of 2,2,3-trimethylpentane yielded product consisting of 90% 2,2,3-trimethylbutane, and only 7% 2,3- and 3% 2,2-dimethylpentane.

When an acidic catalyst is used, isomerization to branched alkanes is significant,[67] and fragments larger than methane are formed in hydrocracking. Hydrocracking was found to take place through olefin intermediates.[68,69] The initial step is the dehydrogenation of an alkane to the corresponding alkene. The alkene is transformed to a carbocation on an acidic site, which then participates in all possible transformations discussed previously. Significant evidence to support this mechanism is the great similarity between the carbon numbers in the products of both catalytic cracking and hydrocracking.[45,70] The metal present in hydrocracking catalysts greatly improves their stability, presumably by hydrogenation, thus eliminating various coke precursors.

2.1.3. Comparison of Cracking Operations

Because of the different chemistry of cracking processes their products have different compositions. The major product of thermal cracking is ethylene, with large amounts of C_1 and C_3 hydrocarbons, and C_4–C_{15} terminal alkenes. Thermal cracking, consequently, is used mainly in olefin manufacture.

Under pyrolytic conditions hydrocarbons are unstable relative to their constituents hydrogen and carbon. Effective cracking therefore requires the rapid input of energy for the breaking of bonds followed by equally rapid quenching to terminate the reaction. Control of the cracking process is by changing operating conditions (temperature, contact time, partial pressure). Dilution of the hydrocarbon mixture with steam results in low hydrocarbon partial pressure promoting olefin production while further decreasing the partial pressure of aromatic reaction products. Steam also tends to diminish the catalytic effect of metals on the walls of the reactor and heat exchanger. This process, called *steam cracking*, is carried out above 800°C and slightly above atmospheric pressure to produce olefin-rich feedstock for further petrochemical utilization. Ethane, butane, and naphtha are the most useful feeds in steam cracking.

Two additional important thermal cracking processes should be mentioned. *Coking* is used to obtain more useful, lower-boiling cracked products from low-value heavy-vacuum resides (residual oils) not suitable for catalytic operations. It is carried out either as a batch operation (*delayed coking*[71–73]) or as a continuous process (*fluid coking*[73]) with the complete conversion of the feed. Coking yields mainly olefinic gas, naphtha and gas oil, and coke. Viscosity breaking or *visbreaking* decreases the viscosity of heavy fuel oil.[74] In contrast with coking, it is a mild thermal cracking producing only 10–30% gas and naphtha.

Because of the increased yield and the high quality of gasoline produced, *catalytic cracking* plays an important role in gasoline manufacture.[14,15] Gasoline produced by catalytic cracking contains less olefinic compounds than that produced by thermal cracking. The principal products in cracking catalyzed by silica–alumina catalysts are mainly branched C_3–C_6 alkanes and alkenes. In contrast, gasolines produced by zeolites always contain more paraffins and aromatics attributed to the higher activity of zeolites in catalytic hydrogen transfer reactions.

Octane ratings of gasoline can be improved by changing cracking operating conditions. Higher reactor temperature and lower contact time lead to increasing importance of cracking reactions relative to hydrogen transfer reactions. As a result, products of higher olefin content, that is, of higher octane number, are formed. In parallel, however, there is an increase in both the gas yield (hydrogen, methane, ethane, ethylene) and coke formation. ZSM-5, in contrast, operated at higher feed rates and higher conversions improves octane numbers without these drawbacks. Gasoline produced by ZSM-5 contains lighter products with higher olefin content.[75,76] Octane enhancement results from the shape-selective action of ZSM-5.[77] The narrow pore opening in ZSM-5 (5.8 Å) compared to that of Y zeolites (13 Å) allows only linear and monomethyl-branched alkanes and alkenes to enter into the zeolite channel. As a result, linear (low-octane) hydrocarbons are removed and selectively transformed through cracking and isomerization to yield a product with high iso-/normal alkane and alkene ratios. A most important change is the decrease in the content of low-octane C_7 olefins and alkanes and a corresponding increase in the amount of high-octane C_5 components.[77] Octane enhancement, therefore, arises from the decrease in concentration of low-octane compounds.[78,79] In addition, there is a relative increase in concentration of high-octane components due to the overall loss in gasoline yield. The use of ZSM-5 is the cheapest way of octane enhancement[75,76] with the additional advantage of the formation of C_3–C_5 alkenes for alkylate feed and for oxygenate (*tert*-butyl methyl ether and *tert*-amyl methyl ether) formation.

The effect of rare-earth HY zeolites is somewhat similar, except that it produces less olefin as a result of enhanced hydrogen transfer reactions. Decreased hydrogen transfer is the main feature of USY zeolites, yielding a product significantly richer in olefins. It is slightly richer in aromatics because of the retardation of their secondary transformations (condensation, coke formation).

Hydrocracking[40–44] is a highly severe hydrotreating operation and a highly flexible and versatile process. It allows the manufacture of products with wide composition range by selecting appropriate feed, catalyst, and operating conditions. Since almost all the initial reactions of catalytic cracking take place (some of the secondary reactions are inhibited), the products of the two operations are similar. The essential difference, however, is that only saturated compounds are formed in hydrocarcking in the presence of hydrogen and the hydrogenating component of the catalyst. Since no alkene is present, the gasoline thus produced needs octane improvement, namely, reforming. The light fraction is usually rich in isobutane that can be used in alkylation. Unlike the other two cracking processes, hydrocracking decreases the amount and molecular weight of aromatics present in the feed and can be specifically used in processing feeds with high aromatic content.

The trend toward the development of hydrocracking is a direct result of their ability to utilize heavier crudes, resids, tar sand, and shale oil.[80–83] Because of their high metal, sulfur, nitrogen, and oxygen contents, these feedstocks are not suitable for direct catalytic cracking. A mild hydrotreating is often carried out before the actual hydrocracking process with the aim of removing metals and sulfur-, nitrogen-, and oxygen-containing compounds. Hydrodesulfurization (HDS), hydrodenitrogenation (HNS), hydrodeoxygenation (HDO), and hydrodemetalation (HDM) are the processes taking place.[44,84,85] Co—Mo-, Ni—Mo-, and Co—Ni—Mo-oxide-on-alumina catalysts are mostly used.[84,85] Hydrotreatment catalysis faces new challenges due to the future new requirements for automobil fuel and the diminishing quality of crude-oil supply. The realization of new active elements and supports, the development of new catalyst combinations and modifiers, and a better understanding of the complicated chemistry taking place are required to meet future demands.[86]

The latest development in the combination of vacuum resid hydrotreating/hydrodesulfurization (VRDS) and resid fluid catalytic cracking (RFCC) gained wide acceptance due to the direct production of gasoline with only a small amount of low-value byproducts.[87] Although not yet developed commercially, cracking of heavy feeds, resid, shale oil, and tar sand with superacidic catalysts not sensitive to ready poisoning is also gaining significance.

Dewaxing is an important specific hydrocracking process used to improve diesel and heating oils by pour-point reduction.[88] This is achieved by shape selectivity of certain zeolites allowing selective hydrocracking of long-chain paraffinic waxes to C_3–C_5 alkanes in the presence of other paraffins.[89] Platinum H-mordenite is used in an industrial process.[90]

2.2. REFORMING

The antiknock value or octane number of a gasoline depends on the hydrocarbons of which it is composed. Straight-chain alkanes have low octane numbers; *n*-heptane by definition is 0. Branched alkanes have higher octane numbers. 2,2,4-Trimethylpentane (isooctane) is 100 by definition. Alkenes generally have higher octane numbers than do the corresponding alkanes: the values are higher for alkenes having an internal double bond compared to the terminal isomers. The octane numbers of cycloalkanes are quite high, whereas aromatic hydrocarbons have very high octane numbers, usually over 100. Alkenes and arenes, however, are also potential carcinogens, and their use as octane enhancers therefore is expected to diminish.

Typical naphthas from the distillation of crude oil in the boiling range 65–200°C contain predominantly straight-chain C_5–C_{11} saturated hydrocarbons and have an octane number of about 50. It is obvious that the fuel quality of a straight-run gasoline can be improved only if hydrocarbons of low octane number are converted to compounds of higher octane number. In other words, it is desirable to rearrange—that is, "re-form"—the structures of at least some of the constituents of the gasoline. Substantial octane enhancement is achieved by producing branched alkanes, alkenes, and aromatics. The processes in which this is done are designated as reforming operations. Reforming is characterized by upgrading the octane number of gasoline without substantial loss of material, that is, without changing the carbon number of the constituents. Catalytic reforming also serves as an important source of aromatics.

Several basic transformations, which are also discussed elsewhere in this chapter, give advantageous results when they occur during the reforming of gasoline. They include isomerization of saturated hydrocarbons (both alkanes and naphthenes); dehydrogenation of alkanes and cycloalkanes to olefinic hydrocarbons and aromatics, respectively; and dehydrocyclization of alkanes to aromatic hydrocarbons. Hydrocracking may also take place, but it is usually undesirable. However, it may play an important role in specific reforming processes (see, for example, discussion of selectoforming in Section 2.2.2). Early reforming operations were dependent on one or two of the abovementioned conversions for increasing the octane number of the gasoline. Processes were later developed in which all transformations occurred.

Catalytic reforming, the dominant process today, is carried out over bifunctional catalysts in the presence of hydrogen. Since dehydrogenation of saturated hydrocarbons, mainly that of cycloalkanes to aromatics, is an important reaction during reforming, the overall process is hydrogen-producing. Reforming usually yields high-octane gasoline (octane number 95–103) that is blended with other refinery gasolines to produce the finished product.

2.2.1. Thermal Reforming

Thermal reforming introduced in 1930 was the first industrial refining process developed for improving the antiknock value of motor fuel. It consists of the thermal cracking of straight-run naphthas at 500–550°C under 35–70 atm. The octane number improvement is due to the formation of alkenes by dehydrogenation of alkanes and by cracking to lower-molecular-weight alkanes plus alkenes. Additionally, there is an increase in the concentration of aromatics in the naphtha because alkanes originally present are removed by conversion to gaseous alkanes and alkenes. Some aromatic hydrocarbons are also formed by dehydrogenation of cycloalkanes. Because thermal reforming involves free-radical reactions, very little or no isomerization of alkanes or cycloalkanes occurs.

Although thermal reforming was a relatively inexpensive process, it had the disadvantage that it produced a low yield of product because of gas formation. Additionally, it did not give a sufficiently high octane-number improvement compared to catalytic processes developed later. For example, thermal reformate was obtained in only 66% yield compared to 88% yield obtained catalytically (platforming process) when a midcontinent straight-run gasoline was reformed to a product having an octane number of 80 in each case.[91] When propene and butenes formed during the thermal cracking process were oligomerized and then added to the thermal reformate, the yield of gasoline increased to 77%.

2.2.2. Catalytic Reforming

Catalytic reforming[92–94] of naphthas occurs by way of carbocationic processes that permit skeletal rearrangement of alkanes and cycloalkanes, a conversion not possible in thermal reforming, which takes place via free radicals. Furthermore, dehydrocyclization of alkanes to aromatic hydrocarbons, the most important transformation in catalytic reforming, also involves carbocations and does not occur thermally. In addition to octane enhancement, catalytic reforming is an important source of aromatics (see BTX processing in Section 2.5.2) and hydrogen. It can also yield isobutane to be used in alkylation.

Hydroforming. The first commercial catalytic reforming plant was built around 1940. The process used a molybdena-on-alumina catalyst at about 475–550°C under 10–20 atm pressure, a large part of which was due to hydrogen.[95] The process was termed *hydroforming*, which refers to reforming in the presence of hydrogen.

The major reactions occurring during hydroforming are dehydrogenation and isomerization of cycloalkanes, and dehydrocyclization.[96,97] In the latter process carbon atoms separated by four carbons in a chain are combined to form a six-membered ring. Thus all C_7 saturated hydrocarbons are transformed to toluene (Scheme 2.2). Besides these reactions, the transformations that occur during hydroforming include hydrocracking (mainly demethylation) and isomerization of straight-chain alkanes to the branched isomers.

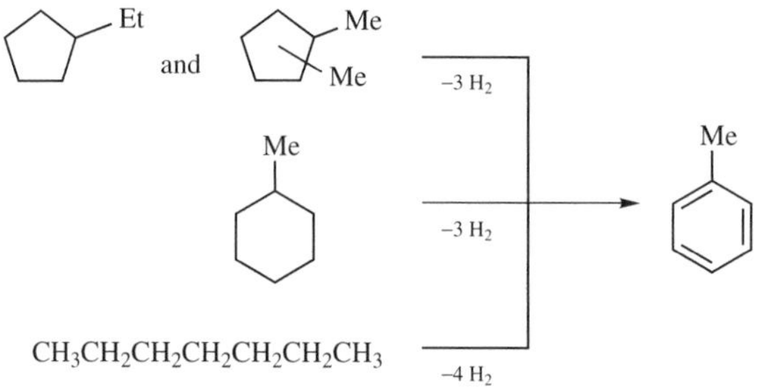

Scheme 2.2

Both chromia-on-alumina and molybdena-on-alumina catalysts are active in hydroforming. Whereas the former gave better results at ambient hydrogen pressure, molybdena-on-alumina catalysts proved to exhibit much better performance at higher hydrogen pressures and were utilized in industrial practice. It is of industrial importance to note that hydroforming in fluid operation gave somewhat higher yields than did hydroforming in fixed-bed operation.[97] This occurred because the fluid operation gave less thermal cracking and operated at lower temperature and pressure.

During World War II, hydroforming found ample use for the production of toluene and of components for aviation gasoline. After the war, hydroforming was used for the formation of high-octane gasoline for automobiles. The realization of the carcinogenic nature of toluene, however, greatly limits its role in gasoline and increasingly emphasizes the importance of isomerization and oxygenate additions in enhancing octane rating of gasoline. The most widely used blending component is MTBE (see, however, a discussion in Section 1.8.2 about the controversy of its use).

Metal-Catalyzed Reforming. In 1949, the first industrial unit employing the now very widely used platinum catalyst was installed. The catalyst developed by Haensel[98,99] at UOP consisted of a commercially feasible small amount of platinum (less than 1%; 0.1% forms a very effective catalyst) impregnated on alumina that also incorporated chlorine or fluorine to the extent of about 0.8%. Halides greatly increase the acidity of the catalyst. The process was called *platforming*, combining the terms *plat*inum and r*eforming*.

The platforming catalyst was the first example of a reforming catalyst having two functions.[43,44,93,100–103] The functions of this bifunctional catalyst consist of platinum-catalyzed reactions (dehydrogenation of cycloalkanes to aromatics, hydrogenation of olefins, and dehydrocyclization) and acid-catalyzed reactions (isomerization of alkanes and cycloalkanes). Hyrocracking is usually an undesirable reaction since it produces gaseous products. However, it may contribute to octane enhancement. *n*-Decane, for example, can hydrocrack to C_3 and C_7 hydrocarbons; the latter is further transformed to aromatics.

Platforming is carried out at temperatures in the range of 450–525°C and under pressure in the range (due in part of hydrogen) of 15–70 atm. The use of hydrogen and the presence of platinum result in the formation of a product containing no alkenes or cycloalkenes. Hydrogen also prevents rapid buildup of carbonaceous material on the catalyst and results in a long-lived catalyst. Studies of the role of hydrogen in metal-catalyzed reforming revealed that the rate of dehydrocyclization (with aromatics production), the most important of the chemical reactions taking place during reforming, goes through a maximum as a function of hydrogen partial pressure.[104] At low hydrogen pressure, hydrogen limits the concentration of highly hydrogen-deficient species (coke precursors). At high hydrogen partial pressure the rate of aromatics formation decreases again. This is due to enhanced hydrogen adsorption, resulting in a decrease in the number of surface sites available for hydrocarbon adsorption. Substantial hydrogen pressures and high hydrogen : hydrocarbon ratios in the feed cause slow deactivation of the catalyst. Runs varying from months to a year or more can be made before it becomes necessary to reactivate the catalyst to its original activity by burning off the carbonaceous material with air.

Subsequent reforming operations were required to produce products with high aromatic contents. The conditions conductive to aromatic production are relatively low hydrogen pressure and high temperature, which also favor coke formation. A new bimetallic catalyst, Pt–Re-on-alumina,[105,106] which exhibited a greatly enhanced coke tolerance, was introduced in 1967. This allowed for the reduction of hydrogen pressure and the hydrogen : hydrocarbon ratio, and an increase in reaction temperature. Under such conditions dehydrogenation is more favored and hydrocracking becomes less important, resulting in higher overall selectivities. The operating interval between regenerations could also be increased.

The advantageous effects of rhenium (and other metals such as germanium, iridium, and tin) are not yet well understood. Rhenium may form an alloy with platinum, thus preventing the increase of the size of platinum crystallites; that is, it keeps platinum in a well-dispersed state. However, it may exist in the ionic state. High coke tolerance may result from the catalytic effect of rhenium to remove coke precursors. Since the second metal was shown to shift the maximum of the dehydrocyclization rate versus hydrogen partial-pressure curves toward lower hydrogen pressure, the adsorption properties of platinum are also changed.[104] The interest in basic research of Pt–Re and other bimetallic, multifunctional catalysts is still high.[106,107]

The octane number of the reformate (platformate) obtained by reforming with the platforming catalyst can vary over a wide range depending on the conditions. When the process was first introduced, the average octane number was 80–85. Later this was increased to about 95; at present some commercial plants yield products having octane numbers higher than 100.

After the platforming process was introduced, many industries developed reforming processes using platinum catalysts.[92–94,108] These processes differ from platforming in operating conditions and procedures, and in the nature of the catalyst. They include *magnaforming*,[109–111] *rheniforming*,[112,113] *powerforming*,[114] and *Houdriforming*.[115,116] *Selectoforming*[65,117,118] is a unique reforming process in that it employs a nonnoble metal with erionite. This process was the first commercial application of shape-selective catalysis.[119] Because of size exclusion this catalyst cracks C_5–C_9 *n*-paraffins to produce predominantly propane and thereby yields an improved high-octane gasoline by selectively removing straight-chain alkanes. The metal (usually sulfided nickel) with weak hydrogenation ability hydrogenates alkenes, and prevents coke formation, but it is not able to hydrogenate aromatics.

The first suggestions concerning the mechanism of catalytic reforming were based on studies with hydrocarbon mixtures that permitted only observation of composition changes.[91,98,120] It was observed, for example, that about 30% of the C_1–C_4 product consisted of methane and ethane. These, however, are not common products in catalytic cracking processes. In fact, when *n*-heptane was hydrocracked, less methane and ethane were formed than expected, according to the stoichiometry of Eqs. (2.15) and (2.16). Therefore, C_5 and C_6 hydrocarbons were not considered to be the products of direct cracking of *n*-heptane. Rather, they are formed through cracking of higher hydrocarbons [Eqs. (2.17) and (2.18)] produced in the interaction of carbocations and alkenes. Methane and ethane, in turn, were supposedly formed through demethylation and deethylation, respectively, of aromatics under the conditions of catalytic reforming, which are different from those of hydrocracking:

$$CH_3(CH_2)_4CH_2CH_3 \ + \ H_2 \ \begin{cases} \longrightarrow CH_3CH_2CH_2CH_2CH_2CH_3 \ + \ CH_4 \quad (2.15) \\ \longrightarrow CH_3CH_2CH_2CH_2CH_3 \ + \ CH_3CH_3 \quad (2.16) \end{cases}$$

$$C_7 \ + \ C_3 \ \longrightarrow \ C_{10} \ \longrightarrow \ 2\,C_5 \ \text{or} \ C_4 \ + \ C_6 \qquad (2.17)$$

$$C_7 \ + \ C_4 \ \longrightarrow \ C_{11} \ \longrightarrow \ C_5 \ + \ C_6 \qquad (2.18)$$

A more detailed interpretation of the chemistry of catalytic cracking was based on studies with pure hydrocarbons.[121-123] A simplified summary put forward by Heinemann and coworkers[123] (Fig. 2.1) shows how C_6 open-chain and cyclic alkanes are transformed to benzene by the action of both the hydrogenating (metal) and acidic (halogenated alumina) functions of the catalyst.

Since the kinetics of the different processes is different, the many reactions taking place in a complex hydrocarbon mixture during catalytic reforming do not

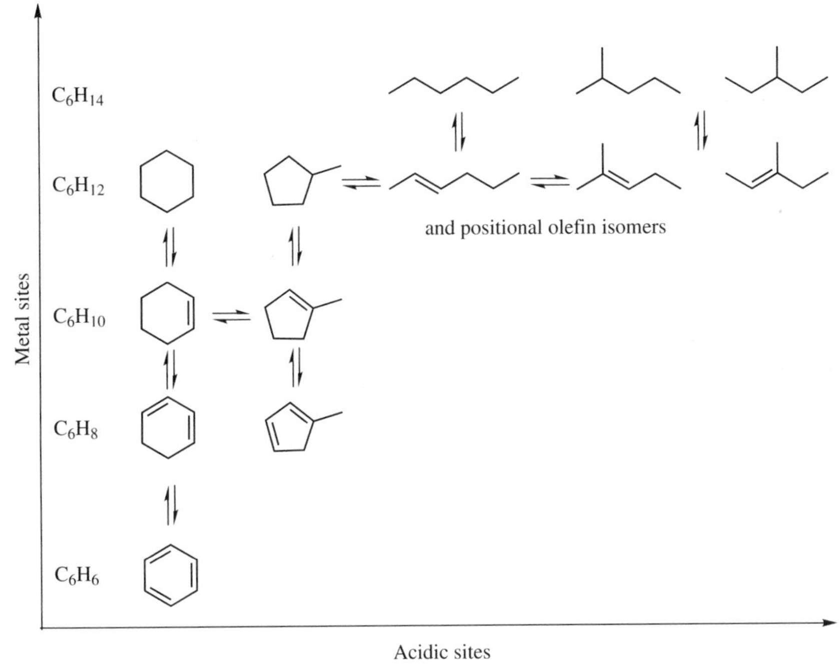

Figure 2.1

contribute equally to the overall transformation.[101,124,125] Whereas *n*-heptane, for instance, is readily transformed to toluene, *n*-hexane undergoes hydrocracking and isomerization much faster and yields methylcyclopentane as the cyclization product. Aromatics are produced primarily from alkylcyclohexanes; alkylcyclopentanes, in turn, require more severe conditions to proceed to aromatics.[120] High temperature favors aromatics production, yielding an equilibrium product composition.[126] Operating conditions and the catalyst used allow quite an effective control of product distribution.

Conventional reforming catalysts are unable to transform light hydrocarbons to valuable products. Specific processes, therefore, have been developed to achieve octane enhancement by selective reforming of light naphthas.[127] H-ZSM-5 may be used to transform C_2–C_{10} olefins. At relatively low temperature and high pressure olefin transmutation–disproportionation[128,129] leads to isoalkenes (the Mobil distillate process).[89,129] H-ZSM-5 at high temperature, a platinum-exchanged L zeolite, and gallium-doped ZSM-5 are able to catalyze complex transformations of C_3 and C_4 hydrocarbons (oligomerization, hydrogen transfer, cyclization, and dehydrogenation, also known as *dehydrocyclodimerization*[130]) to produce gasoline with high aromatic content. These processes are discussed further in Section 2.5.2.

Although aromatics (as well as olefins) increase to octane number of gasoline-range hydrocarbons, because of the potential carcinogenicity, future approaches will increasingly emphasize isomerization and alkylation-based branching of saturated hydrocarbons,[131,132] as well as oxygenate additives[133,134] to achieve the goal. A valuable approach is UOP's "I-forming process."[135] A noble metal on acidic support (IC-2 catalyst) is used in the presence of hydrogen to selectively transform heavy gasoline streams. The IC-2 catalyst exhibits a unique selectivity toward cracking of heavy paraffins and also catalyzes the hydrogenation of aromatics. Selective cracking results in the formation of isobutane and isopentane well in excess of the equilibrium values with only limited amounts of light paraffins. Isobutane is recovered and used in alkylation. The debutanized product has an increased octane number (due mainly to isopentane and cycloparaffins), and its aromatic content is significantly reduced. Reforming using erionite catalysts is another promising possibility.[136] Erionite favors shape-selective hydrocracking of normal paraffins and the formation of cyclopentane derivatives, resulting in the production of high-octane-number gasoline with low amounts of aromatics. The increase in octane number is due to the formation of isopentane and substituted cyclopentanes and the loss of high-molecular-weight straight-chain alkanes. The decrease in aromatic content of motor fuel may also be achieved by undercutting gasoline, that is, distilling out the heavier 10% fraction of FCC gasoline.[30]

2.3. DEHYDROGENATION WITH OLEFIN PRODUCTION

Unsaturated hydrocarbons are important industrial raw materials in the manufacture of chemicals and polymers. Moreover, they are valuable constituents in high-octane gasoline produced mainly through cracking of larger alkanes, although, as mentioned before, their potential carcinogenic nature will greatly reduce such use in the future. Dehydrogenation of saturated compounds also takes place during refining processes and is practiced in industrial synthesis of olefins and dienes.

2.3.1. Thermal Dehydrogenation

Of the saturated hydrocarbons only ethane and propane undergo practical dehydrogenation to yield the corresponding alkenes. All other higher homologs and cycloalkanes undergo cracking when treated at high temperatures.

Under pyrolytic conditions at temperatures above 300°C, generally within 500–800°C, the pyrolysis reaction forms alkenes by carbon–hydrogen bond scissions. An early experiment, where propane was heated to 575°C for 4 min in a silica flask, yielded propylene by dehydrogenation [Eqs. (2.19)–(2.21)] at a somewhat slower rate than it yielded methane and ethylene by cracking:[54]

$$CH_3CH_2CH_3 \longrightarrow CH_3\dot{C}HCH_3 + H\cdot \qquad (2.19)$$

$$CH_3\dot{C}HCH_3 \longrightarrow CH_2{=}CHCH_3 + H\cdot \qquad (2.20)$$

$$CH_3CH_2CH_3 + H\cdot \longrightarrow CH_3\dot{C}HCH_3 + H_2 \qquad (2.21)$$

Similar observations were made in a study of pyrolysis[53] in the temperature range 700–850°C. Of the propane that decomposed, 60% did so by cracking and 35–39% by dehydrogenation.

Dehydrogenation is the chief reaction when ethane is pyrolyzed.[53] The reaction begins at about 485°C and is quite rapid at 700°C. At this temperature about 90% of the reacting ethane is converted to ethylene and hydrogen. Minor byproducts include 1,3-butadiene and traces of liquid products.

The major industrial source of ethylene and propylene is the pyrolysis (thermal cracking) of hydrocarbons.[137–139] Since there is an increase in the number of moles during cracking, low partial pressure favors alkene formation. Pyrolysis, therefore, is carried out in the presence of steam (steam cracking), which also reduces coke formation. Cracking temperature and residence time are used to control product distribution.

Because of their ready availability, natural-gas liquids (mixtures of mainly ethane, propane, and butanes) are used in the manufacture of ethylene and propylene in the United States.[140,141] In Europe and Japan the main feedstock is naphtha. Changing economic conditions have led to the development of processes using heavy oil for olefin production.[138,142]

The ideal feed for ethylene production is ethane, which can give the highest ultimate yield via dehydrogenation. Ethylene selectivities of 85–90% can be achieved at temperatures above 800°C and a steam : ethane ratio of about 0.3.

Other hydrocarbons are steam-cracked at temperatures of 750–870°C with a residence time of about 1 s. Severe cracking may be carried out at 900°C with a shorter residence time (0.5 s). The higher the average molecular weight of the feed, the higher the quantity of the byproducts formed, and they may outweigh the quantity of ethylene and propylene produced. Higher steam : hydrocarbon ratios (up to 1) are used for heavier feedstocks. The reactor effluent is rapidly quenched to prevent secondary reactions, and a series of separation steps follows. Ethane and propane are recovered and recycled, and acetylene, propyne, and propadiene are removed by hydrogenation. 1,3-Butadiene and butenes are separated from the C_4 fraction.

These processes are specifically designed for ethylene production but they also yield C_4 hydrocarbons as coproducts. The amount of C_4 compounds produced depends on the feedstock, the cracking method, and cracking severity. Steam cracking of naphtha provides better yields than does catalytic cracking of gas oil. With more severe steam cracking both butenes and overall C_4 productions decrease, whereas the relative amount of 1,3-butadiene increases. Overall C_4 yields of 4–6% may be achieved.

Because of their very similar boiling points and azeotrope formation, the components of the C_4 fraction cannot be separated by distillation. Instead, other physical and chemical methods must be used. 1,3-Butadiene is recovered by complex formation or by extractive distillation.[143–146] Since the reactivity of isobutylene is higher than that of n-butenes, it is separated next by chemical transformations. It is converted with water or methyl alcohol to form, respectively, *tert*-butyl alcohol and *tert*-butyl methyl ether, or by oligomerization and polymerization. The remaining n-butenes may be isomerized to yield additional isobutylene. Alternatively, 1-butene in the butadiene-free C_4 fraction is isomerized to 2-butenes. The difference between the boiling points of 2-butenes and isobutylene is sufficient to separate them by distillation. n-Butenes and butane may also be separated by extractive distillation.[147]

The cracking of isobutane to isobutylene is of special interest because of the high demand for isobutylene as a feed for the production of oxygenates, mainly MTBE, as octane-enhancing gasoline additives. Isobutane separated from the steam cracking C_4 cut or produced by butane isomerization is cracked in the presence of steam to yield isobutylene and propylene.[148,149]

2.3.2. Catalytic Dehydrogenation to Alkenes

Excellent results (close to equilibrium) are obtained when chromia is used as catalyst for the dehydrogenation of ethane, propane, n-butane, and isobutane.[150] The catalytic dehydrogenation required a markedly lower temperature than did thermal dehydrogenation. Use of too high contact time or temperatures cause cracking as a side reaction, but under proper reaction conditions selective dehydrogenation occurs. Isobutane tends to undergo carbon–carbon bond scission yielding methane and propylene rather than dehydrogenation.

Longer life and better activity were obtained with catalysts composed of chromia and alumina.[151] While pure alumina has little dehydrogenation activity, the incorporation of as little as 3% or as much as 60% chromia provides effective catalysts; the most widely used commercial catalyst usually contains 20% chromia. Chromia–alumina is used in the dehydrogenation of C_2–C_4 hydrocarbons to the corresponding alkenes.[152–154] 1,3-Butadiene may also be manufactured under appropriate conditions (see Section 2.3.3).

When alkanes containing a chain of at least six carbon atoms are reacted in the presence of the chromia–alumina catalyst, the major reaction is not formation of alkenes. Instead, aromatic hydrocarbons are produced by dehydrocyclization (see Section 2.5.). On the other hand, it was discovered that dehydrogenation of hexane and higher straight-chain alkanes to the corresponding alkenes occurs with high selectivity in the presence of a modified reforming catalysts consisting of platinum supported on nonacidic or alkaline alumina.[155,156] The reaction is carried out in the presence of hydrogen at relatively low pressure and moderately high temperature. The selectivity to linear monoalkenes with the same carbon number as the alkane reaches or exceeds 95% at conversion levels somewhat less than equilibrium. The byproducts consist of dienes, aromatics, traces of isoalkanes, and branched alkenes. The linear monoalkenes show random distribution of the double bond along the carbon chain and are believed to be equilibrium mixtures. 1-Alkenes accordingly are less abundant than the internal isomers. This reaction forms the basis of the UOP Pacol process (see Section 2.3.3).

2.3.3. Practical Applications

C_2–C_3 Alkenes. Ethylene and propylene are produced in industry by steam cracking (see Section 2.3.1). In 1991, about 40% of the total world production of ethylene was based on gas cracking (cracking of ethane, propane, or ethane–propane mixtures).[157] Propane cracking, however, is not economical because of the low yield of propylene.[158,159] Cracking of an ethane–propane mixture gives only a 60–66% total olefin yield with a composition of 44–47% ethylene and 14–22% propylene.[159] In comparison, cracking of pure ethane produces ethylene in yields over 80%. An integrated ethane cracking–propane dehydrogenation complex gives both olefins in about 80% yield.

Processes for the catalytic dehydrogenation of ethane and propane have been developed.[138] UOP modified the catalytic system of the Pacol process for use in dehydrogenation of lighter alkanes (the Oleflex process.)[160–162] A moving-bed technology with continuous catalyst regeneration is applied. The catalyst shows good stability and selectivity in the dehydrogenation of ethane and propane. In propane dehydrogenation conducted at 600–700°C propylene yield is 77% (compared to 33% in thermal cracking).[160] Phillips STAR (steam active reforming) process[163,164] uses a supported noble metal catalyst with multiple metal promoters. The use of steam ensures better equilibrium conditions to achieve propylene selectivities as high as 95%. Chromium oxide on alumina is the catalyst in the Catofin process (Air Products).[158]

A technology called the *Ethoxene process* was developed by Union Carbide for the oxidative dehydrogenation of ethane.[165] Under the relatively mild conditions applied (300–400°C, 1–40 atm) cracking is minimal and the only byproduct is acetic acid. The combined efficiency to ethylene and acetic acid is about 90% with an ethylene : acetic acid ratio of 10.

C_4 Alkenes. Several industrial processes have been developed for olefin production through catalytic dehydrogenation[138,166,167] of C_4 alkenes. All four butenes are valuable industrial intermediates used mostly for octane enhancement. Isobutylene, the most important isomer, and its dimer are used to alkylate isobutane to produce polymer and alkylate gasoline (see Section 5.5.1). Other important utilizations include oxidation to manufacture maleic anhydride (see Section 9.5.4) and hydroformylation (see Section 7.1.3).

The high demand for oxygenates, especially for MTBE, in new gasoline formulations called for a substantial increase in the capacity of isobutylene production in the early 1990s. As a result, new processes emerged to manufacture isobutylene through the dehydrogenation of isobutane. Four technologies employing noble-metal-based catalysts or chromia on alumina were developed.[164]

The Oleflex process[160–162] applies platinum on basic alumina and operates at higher severity than the Pacol process from which it is derived. Modification of the original catalyst systems was required to decrease skeletal isomerization and coke formation.[161] Isobutylene is produced with a selectivity of 91–93%. The Phillips STAR dehydrogenation process[163,164] also employs a noble metal catalyst. Users of the FBD-4 (Snamprogetti) fluidized-bed process with a promoted chromium oxide catalyst[168] are located primarily in the former Soviet Union. The Catofin process[169–171] is based on the Houdry Catadiene technology, which was developed in the 1940s to produce 1,3-butadiene from butane over a chromium oxide–alumina catalyst. Dehydrogenation of isobutane to isobutylene is carried out over chromia–alumina at 500–675°C in vacuo to achieve high conversion.

1,3-Butadiene and Isoprene. Butane may be transformed directly to 1,3-butadiene on chromia–alumina (Houdry Catadiene process).[144–146,172] The most significant condition is operation under subatmospheric pressure (0.1–0.4 atm), which provides an improved yield of 1,3-butadiene. Operating at about 600°C, the process produces a mixture of butenes and 1,3-butadiene. After the removal of the latter, the remaining butane–butenes mixture is mixed with fresh butane and recycled. Extensive coke formation requires regeneration of the catalyst after a few minutes of operation. 1,3-Butadiene yields up to 63% are obtained at a conversion level of 30–40%.

Much higher butadiene yields may be obtained in a two-step process developed by Phillips in which butane is first converted to butenes with the chromia–alumina catalyst, and the butenes are then further dehydrogenated to 1,3-butadiene.[144,173] The butene selectivity in the first step is about 80–85% (600°C, atmospheric pressure). The butenes recovered from the reaction mixture undergo further dehydrogenation in the presence of excess steam (10–20 mol) over a mixed iron–magnesium–potassium–chromium oxide catalyst.[173] The Dow process uses a calcium–nickel phosphate catalyst promoted by chromia[174,175] and gives butadiene selectivities of ≤90%. Other catalysts in commercial application operating under slightly different conditions are iron oxide and chromium oxide with K_2O (Shell)[173] and iron oxide with bauxite (Phillips).

Similar to the processes used in the manufacture of 1,3-butadiene, isoprene can be prepared from isopentane, isoamylenes, or a mixed isoC$_5$ feed.[172,176,177] The Shell process[177] dehydrogenates isoamylenes to isoprene in the presence of steam with 85% selectivity at 35% conversion, over a Fe_2O_3–K_2CO_3–Cr_2O_3 catalyst at 600°C.

Direct catalytic dehydrogenation of hydrocarbons is characterized by low conversion per pass and low yields. The process is endothermic and the thermodynamic equilibrium is unfavorable [Eq. (2.22)]. Temperatures above 500°C are required for commercially feasible conversions. At these temperatures, undesirable side reactions limit the yields and coke formation requires frequent catalyst regeneration. In contrast, oxidative dehydrogenation of n-butane and butenes in the presence of air and steam over suitable catalysts to yield 1,3-butadiene is more efficient. By removing the hydrogen in the form of water [Eq. (2.23)], oxidative dehydrogenation proceeds to a greater extent than catalytic dehydrogenation:

$$C_4H_8 \; \rightleftharpoons \; C_4H_6 \; + \; H_2 \qquad\qquad \Delta H = 28 \text{ kcal/mol} \qquad (2.22)$$

$$C_4H_8 \; + \; 0.5\,O_2 \; \rightleftharpoons \; C_4H_6 \; + \; H_2 \qquad \Delta H = -29.7 \text{ kcal/mol} \qquad (2.23)$$

Continuous in situ catalyst regeneration occurs since oxygen efficiently removes carbon deposits. Steam controls the reaction temperature serving as heat sink and enhances selectivity. As a result, oxidative dehydrogenation offers better yields and higher conversion by altering equilibrium conditions and eliminating side reactions due to lower operating temperature.

The processes commercialized are the O-X-D process (Phillips),[178,179] the Oxo-D process (Petro-Tex),[180] and a technology developed by BP applying a tin–antimony oxide catalyst.[181,182] The Phillips technology produces 1,3-butadiene with 88–92% selectivity at 75–80% conversion.[179] A new catalyst used in the Nippon Zeon process offers improved process characteristics.[183]

Higher Olefins. The Pacol–Olex (UOP) process[184–187] is commonly used to manufacture detergent-range (C$_{10}$–C$_{14}$) olefins via dehydrogenation of the corresponding paraffins. In general, it can be applied in the production of olefins in the C$_6$–C$_{20}$ range. It is a fixed-bed process and operates under milder conditions than the Oleflex process, which is an extension of the former. Nonacidic noble metal catalysts, such as platinum on basic alumina, are used at 400–600°C and 3 atm in the presence of hydrogen. At low (10–15%) conversion per pass it yields monoolefins up to 93% selectivity with 96% internal linear olefin content. The double bond is statistically distributed. The olefins formed are recovered by extraction using a selective adsorbent (Olex unit) and converted to detergent alcohols (see Section 7.1.3) or to linear alkylbenzenes (see Section 5.5.6). Alternatively, the olefin–paraffin mixture may directly be used in alkylation of benzene to manufacture detergent alkylate.[188]

Styrene. All commercial processes use the catalytic dehydrogenation of ethylbenzene for the manufacture of styrene.[189] A mixture of steam and ethylbenzene is reacted on a catalyst at about 600°C and usually below atmospheric pressure. These operating conditions are chosen to prevent cracking processes. Side reactions are further suppressed by running the reaction at relatively low conversion levels (50–70%) to obtain styrene yields about 90%. The preferred catalyst is iron oxide and chromia promoted with K_2O, the so-called Shell 015 catalyst.[190]

Several technologies differing in the method of heat supply have been developed.[191–193] When superheated steam is used, it provides the necessary energy for the endothermic dehydrogenation, acts as a diluent to achieve favorable equilibrium conditions, and prevents coke formation by transforming carbon to carbon oxides. A new oxidative dehydrogenation process (Styro-Plus) is under development.[194]

2.4. UPGRADING OF NATURAL-GAS LIQUIDS

Natural gas, depending on its source, is accompanied by varying amounts of liquid hydrocarbons. Natural-gas liquids contain mainly saturated aliphatic hydrocarbons, practically free of olefins, and also some aromatics. After separation these liquids are utilized either as a low-grade fuel or via processing customary in the petroleum industry (refining, hydrocracking, etc.) as hydrocarbon raw materials. Their use as motor gasoline is not feasible because of their low octane ratings, which generally can be improved only by adding high-octane components.

Upgrading of natural-gas liquids to motor fuels with customary technology used in the petroleum industry (isomerization, hydroisomerization, etc.) represents difficulties. One reason is that the volatility of motor fuels cannot be increased, thus limiting treatments such as hydrocracking.

Alternatively, upgrading can also be effected using effective superacid catalysts such as trifluoromethanesulfonic acid or hydrogen fluoride–boron trifluoride systems.[195] These upgradings are carried out at relatively low temperatures (10–50°C), thus maximizing branched alkanes. They can be operated in a batchwise or continuous fashion with relatively short contact times to give motor fuel gasolines of suitable octane rating. Frequently an added olefin activator such as 1–3% butene is added, although the upgrading process is also effective without such activity. There is no appreciable increase of volatility of upgraded fuel, although some isobutane is formed, which is topped off. Aromatics, present in natural-gas liquids (1–3%), are nearly completely removed (via ionic hydrogenation, ring-opening isomerization) in the superacid-catalyzed upgrading producing environmentally safer fuels. The overall upgrading involves both isomerization and transalkylation processes increasing branched C_6–C_8 content, but not volatile C_1–C_3.

2.5. AROMATICS PRODUCTION

Aromatic compounds are the most widely used and one of the most important classes of petrochemicals. They are still an important constituent of high-octane gasoline, although because of their carcinogenic nature their application will decrease. FCC gasoline contains about 29% aromatics, whereas the aromatic content of reformates is about 63%. They also are excellent solvents and constitute an important component of synthetic rubbers and fibers.

Historically aromatic compounds were produced from hard coal by coking. The polyaromatics present in coal are released under the pyrolytic conditions and are absorbed in oil or on activated charcoal to separate them from the other coal gases. The components are freed by codistillation with steam or by simple distillation. The contaminant nitrogen- and sulfur-containing compounds are removed by washing with sulfuric acid or by hydrogenation.

Crude oil, however, has almost completely replaced coal as a source of aromatics. Crude oil contains several percents of benzene, toluene, and xylenes and their cycloalkane precursors. The conversion efficiency for preparing toluene or xylenes from their precursors is nearly 100%. For benzene this efficiency is slightly lower. Moreover, alkanes are also transformed to aromatics during refining processes, allowing efficient production of simple aromatic compounds.

Catalytic reforming has become the most important process for the preparation of aromatics. The two major transformations that lead to aromatics are dehydrogenation of cyclohexanes and dehydrocyclization of alkanes. Additionally, isomerization of other cycloalkanes followed by dehydrogenation (dehydroisomerization) also contributes to aromatic formation. The catalysts that are able to perform these reactions are metal oxides (molybdena, chromia, alumina), noble metals, and zeolites.

2.5.1. Catalytic Dehydrogenation and Dehydrocyclization

Many metals have long been known to catalyze the dehydrogenation of cyclohexane to benzene. These include nickel,[196] palladium,[197] and platinum.[197] These metals are catalysts for both the hydrogenation of aromatic hydrocarbons and the dehydrogenation of cyclohexanes to aromatics. The dehydrogenation requires a higher temperature. For example, benzene and hydrogen passed over palladium black at 100–110°C yielded much cyclohexane; passing cyclohexane over the palladium catalyst at 300°C produced benzene and hydrogen in practically quantitative yield.[197] In other words, higher temperatures favor reversal of the hydrogenation reactions and result in aromatic production.

The dehydrogenation of cyclohexane occurs in a stepwise manner via σ-bonded (**1,2**) or π-adsorbed (**3**) species to form cyclohexene[198] (Scheme 2.3).

Stepwise removal of hydrogen and extension of the π system eventually leads to benzene. Since the removal of either of the first two hydrogen atoms is rate-determining the intermediate cyclohexenic species react further without desorption and are barely detectable in the gas phase. Adsorbed cyclohexene and a C_6H_9 intermediate, however, have been identified on platinum single crystal at low

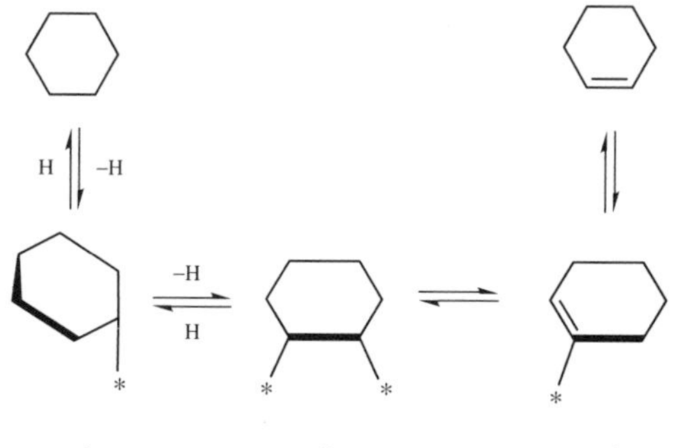

Scheme 2.3

temperature.[199,200] Detailed spectroscopic studies resulted in a better understanding of the nature of active surface sites and a more detailed picture of surface reactions.[201]

Cyclohexanes with *gem*-dialkyl substituents undergo aromatization at higher temperatures.[202] The transformation is accompanied by methyl migration to yield a xylene mixture on platinum on alumina. Seven- and higher-membered rings are similarly less reactive. Ring contraction required to form the six-membered ring was shown to occur through bicycloalkane intermediates, including a cyclopropane ring.[202]

Alkylcyclopentanes do not undergo dehydrogenation under the conditions required for other cycloalkanes. Apparently, strongly adsorbed cyclopentadiene blocks the catalyst surface, bringing about catalyst deactivation. They yield, however, aromatic compounds via skeletal rearrangement and dehydrogenation (dehydroisomerization) when bifunctional catalysts (metals on acidic support) are applied. The acidic alumina used in platforming catalyzes the carbocationic isomerization of alkylcylopentanes, which is a very important reaction in the formation of aromatics. The transformation of methylcyclopentane, for instance, occurs according to Scheme 2.4 and leads in equilibrium to the cyclohexyl carbocation. This is then transformed to cyclohexene and, eventually, to benzene, and may initiate the

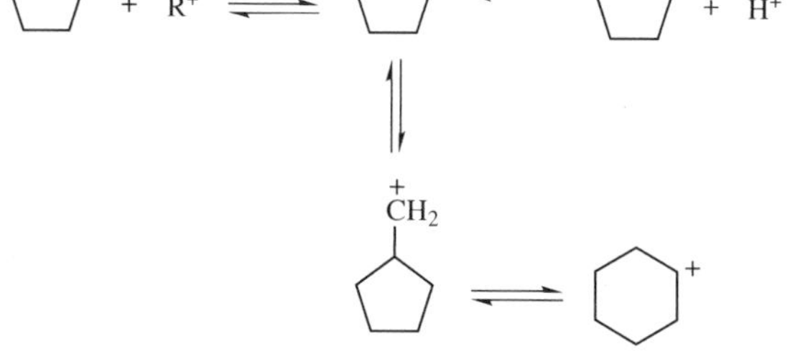

Scheme 2.4

process by hydride removal from methylcyclopentane. Cyclohexane thus formed was indeed detected in the reaction of methylcyclopentane.[98]

Of the metal oxides, chromia–alumina is the preferred and most widely studied catalyst.[202] The temperature required for the aromatization of cyclohexanes is much higher (400–500°C) than that over noble metals.

Chromia and molybdena were found to effect dehydrocyclization of alkanes under reaction conditions similar to those of aromatization of cyclohexanes.[96] Because of its great practical significance in refining (hydrorefining), chromia– alumina was extensively studied in the dehydrocyclization of alkanes.

One of the first mechanisms suggested for the dehydrocyclization on an oxide catalyst was proposed by Herington and Rideal.[203] They suggested that all possible monoolefins are formed in the first step and give α,β-diadsorbed surface intermediates. Cyclization involves one of the surface-bound carbon atoms and another carbon atom in the chain to form a six-membered ring (1,6 carbon–carbon closure). Rapid dehydrogenation follows, resulting in the formation of benzene derivatives. In accordance with this suggestion, 1-heptene (**4**) and 2-heptene (**5**) are transformed to toluene in a fast reaction since they are able to participate in ring closure. 3-Heptene (**6**), in turn, reacts slowly since it must undergo double-bond migration before it can form a six-membered ring.

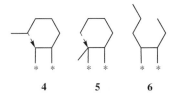

process by hydride removal from methylcyclopentane. Cyclohexane thus formed

In a series of experiments carried out by Pines and coworkers,[202] the intrinsic acidity of alumina was neutralized before use to avoid ionic-type skeletal isomerization. Radiotracer studies showed that toluene formed from [1-^{14}C]-heptane over chromia–alumina contained only 18–32% labeling in the methyl group, that is, less than 50% required by 1,6 carbon–carbon closure.[204] This indicates that direct 1,6 carbon–carbon closure is not the only path of cyclization of n-heptane.

Aromatization of n-octane yielded ethylbenzene and o-xylene, the products of direct 1,6 closure, but m- and p-xylenes were also formed.[205] Isomeric dimethylhexanes, expected to give only one aromatic compound each according to 1,6 carbon–carbon closure, produced all isomeric alkylbenzenes.[206] Alkanes that cannot undergo dehydrogenation did not yield aromatic compounds, which pointed to the importance of alkene intermediates in aromatization.

All these results were interpreted by a free-radical mechanism with the involvement of alkenes, and smaller (C_3,C_4) and larger (C_7,C_8) ring intermediates in aromatization. Skeletal isomerization was found to occur through vinyl shift and C_3,C_4 cyclic intermediates.[202] Transition metals with the exception of Fe and Os, as well as Re, Co, and Cu, are active in aromatization of alkanes.[207–209] Platinum, long recognized to be an active metal,[210] is the most thoroughly studied catalyst yielding benzene derivatives and alkylcyclopentanes. The selectivity of cyclization depends on the reaction conditions. For example, methylcyclopentane and isomeric methylpentanes are formed from *n*-hexane mainly at low temperature, whereas elevated temperatures bring about a shift to dominant benzene formation.[211]

The pathways suggested include direct cyclization similar to the Herington–Rideal mechanism accounting for both C_5 and C_6 cyclization.[207] There are observations, however, that benzene formation does not involve the five-membered ring, that is, that the two ring structures are formed in independent reactions.[208] A stepwise dehydrocyclization with gradual loss of hydrogen to form a conjugated triene followed by ring closure and further dehydrogenation may account for aromatics formation:[212,213]

$$\text{(2.24)}$$

Evidence also suggests, however, that alkanes with only five carbon atoms in a linear chain may undergo aromatization via 1,5 ring closure followed by ring enlargement. Numerous mechanisms were put forward to rationalize these transformations.[209]

Dual-function catalysts possessing both metallic and acidic sites bring about more complex transformations. Carbocationic cyclization and isomerization as well as reactions characteristic of metals occurring in parallel or in subsequent steps offer new reaction pathways. Alternative reactions may result in the formation of the same products in various multistep pathways. Mechanical mixtures of acidic supports (silica–alumina) and platinum gave results similar to those of platinum supported on acidic alumina.[214,215] This indicates that proximity of the active sites is not a requirement for bifunctional catalysis, that is, that the two different functions seem to operate independently.

In the dehydrocyclization of alkanes it is clear that ring closure can take place both in a metal-catalyzed reaction and as a carbocationic process. The interpretation of the reforming process proposed by Heinemann and coworkers,[123] therefore, is not a complete picture of the chemistry taking place. The scheme they presented (Fig. 2.1) attributes cyclization activity solely to acidic sites. The ample evidence available since requires that metal-catalyzed C_5 and C_6 ring-closure possibilities be included in a comprehensive interpretation. Additionally, the metal component plays and important role in carbocationic reactions in that it generates carbocations through the formation of alkenes.

Alkylaromatic compounds possessing a sufficiently long side chain may also undergo dehydrocyclization.[216] In fact, the aromatic ring enhances this reaction; alkylaromatics undergo dehydrocyclization faster than do alkanes. Depending on the side chain, condensed or isolated ring systems may be formed.

Platinum, the most studied metal, catalyzes both C_5 and C_6 cyclizations in parallel reactions. *n*-Butylbenzene, for example, gives methylindan and methylindenes, and naphthalene. Both direct cyclization and stepwise transformation involving olefinic intermediates are operative. Over dual-function catalysts, product distributions are different from those over platinum since the acidic component brings about a new reaction, the acid-catalyzed self-alkylation of the aromatic ring. Phenylalkenes formed in platinum-catalyzed dehydrogenation are protonated to yield carbocations, which participate in intramolecular electrophilic aromatic cyclialkylation. The ratio of C_5/C_6 cyclization is determined by the stability of the intermediate carbocations. n-Butylbenzene, for example, yields mainly methylindan due to the difference in stability of the primary and secondary carbocations:[217]

$$(2.25)$$

$$(2.26)$$

In contrast, methylnaphthalene is the main product in the dehydrocyclization of n-pentylbenzene.[218]

Formation of aromatics also takes place over molecular-sieve catalysts. This is an essential reaction over the newer-generation zeolite-cracking catalysts. They differ from earlier catalysts (amorphous silica–alumina) in that the product gasoline they produce contains less olefins and more aromatics. This is the result of more dominant hydrogen-transfer reactions between cyclic hydrocarbons and alkenes producing aromatics and paraffins. The recent high demand for aromatic compounds, however, has resulted in the development of additional new processes based on zeolites. In contrast with conventional reforming catalysts, zeolites are capable of transforming light hydrocarbons (C_3–C_5 alkanes and alkenes) to aromatics.

H-ZSM-5 was found to be an excellent catalyst with prolonged catalytic activity.[219,220] The first reaction in the multistep process leading ultimately to aromatics is the formation of small alkenes by cracking and dehydrogenation of small alkanes. Alkenes then undergo oligomerization and disproportionation to form a near-equilibrium mixture of higher olefins. Finally, these higher alkenes cyclize to cycloalkenes, which, in turn, participate in dehydrogenation and hydrogen transfer with olefins to give aromatics. The rapid olefin equilibration compared to further processes accounts for the observation that H-ZSM-5 transforms varied feeds (smaller and higher hydrocarbons, methanol, Fischer–Tropsch products, triglycerides) to products of very similar product distribution consisting mainly of C_3–C_4 aliphatics and C_6–C_{10} aromatics.[220,221] As a result of shape selectivity, ZSM-5 does not generate appreciable amounts of aromatics higher than C_{10}. Since they are considered to be coke precursors, the lack of higher aromatic compounds ensures sustained catalytic activity.

Aromatization over H-ZSM-5, however, has the stoichiometric restriction to produce 3 mol of alkanes per one mole of aromatic compound. It was later observed that doping ZSM-5 with gallium or zinc brings about an increase in catalytic activity and selectivity of aromatics production.[222] Extensive studies have been conducted with impregnated[223] and ion-exchanged[224] Ga–H-ZSM-5 catalysts, gallosilicates,[225,226] and mechanical mixtures of ZSM-5 and gallium oxide.[227,228] Increased selectivity over these metal-doped catalysts results in part, from their ability to form a certain amount of molecular hydrogen. The exact role of gallium is still debated, but it appears that it promotes dehydrogenation of alkanes and alkenes, thus accelerating the formation of aromatics.[223,226,229,230]

2.5.2. Practical Applications

There are several processes of industrial significance in the production of aromatics that were valuable components in high-octane gasoline and are used as solvents and chemicals.[231] BTX (benzene, toluene, xylenes) processing, the major source of these important chemicals, is connected to catalytic cracking and metal-catalyzed reforming.[232–234] In the United States steam cracking is the major source of aromatics. Although it is used primarily for ethylene manufacture, it produces a substantial quantity of benzene as well. In Europe and Japan, catalytic reforming is the major process for aromatics production.

In modern refineries separate reforming and BTX units are usually operated. The former process produces gasoline with high aromatics content that is blended with other gasoline streams. The BTX unit uses different feedstocks to maximize aromatics production. Since new regulations severely limit aromatics content of gasoline (less than 1% of benzene, less than 25% of total aromatics), refiners will face either to avoid formation of aromatics or to separate them from gasoline streams. In the latter case undercutting (distilling out the heaviest 10% fraction of FCC gasoline containing over 80% aromatics) will produce an additional pool of aromatics in the future.

Separation of the aromatics from each other and from other hydrocarbons by distillation is not economical because of the limited boiling-point differences and the formation of azeotropic mixtures. Instead, extractive or azeotropic distillation and liquid–liquid extraction are applied.[234,235] The latter process is by far the most often used technique. The three processes are applied according to the aromatic content of the gasoline source. *p*-Xylene, the most valuable of the isomeric xylenes, is isolated by freezing (crystallization) or solid adsorption.

A new effective aromatization called the *Cyclar process*[172,236–238] (UOP-BP) employs a gallium-containing H-ZSM-5 zeolite catalyst that is able to transform light hydrocarbons to aromatics. H-ZSM-5 is the catalyst in M2 forming an aromatization developed by Mobil.[219] Light alkene feedstocks can be transformed at relatively low temperature (370°C), whereas light alkanes require higher temperature (538°C).[219] Both processes may use liquefied petroleum gas (LPG), which consists mainly of propane and butanes. Alternatively the process may be applied to upgrade light paraffinic naphthas. Conventional reforming of these feedstocks usually gives low yields of reformates, whereas increased product yields and very high octane numbers are obtained by M2 forming. The Aromax process,[239] the latest development by Chevron, applies a nonacidic barium-exchanged L zeolite containing highly dispersed platinum.[240] The catalyst is very reactive and highly selective in the aromatization of light paraffins, particularly in the transformation of hexanes and heptanes.

A cracking process, the dealkylation of alkylbenzenes, became an established industrial synthesis for aromatics production. Alkylbenzenes (toluene, xylenes, trimethylbenzenes) and alkylnaphthalenes are converted to benzene and naphthalene, respectively, in this way. The hydrodealkylation of toluene to benzene is the most important reaction,[241,242] but it is the most expensive of all benzene manufacturing processes. This is due to the use of expensive hydrogen rendering hydrodealkylation too highly dependent on economic conditions.

Homogeneous thermal processes and catalytic hydrodealkylations are practiced. Operating temperatures ranging between 500 and 650°C and a pressure of about 50 atm are typical for catalytic hydrodealkylation with alumina-supported cobalt, molybdenum, or nickel catalysts. A mixed cobalt oxide–molybdena–alumina is used most widely. Thermal processes operate at temperatures of ≤800°C. Benzene yields are usually better than 95%, with side reactions giving condensation products (biphenyls) and coke. Coke formation may be minimized by the use of a high excess of hydrogen. Methane is produced as a coproduct. Several technologies for both catalytic hydrodealkylation (Detol,[243–245] Pyrotol,[245] Hydeal[242,246]) and thermal dealkylation[241,242,247–251] are operated industrially.

Dealkylation and isomerization (disproportionation) processes used in the manufacture of benzene and xylenes are discussed in Sections 4.5.2 and 5.5.4.

2.6. RECENT DEVELOPMENTS

2.6.1. Cracking

The history of ZSM-5 fluid catalytic cracking additive development at Mobil until its first commercial trial in 1983 has been disclosed.[252] Despite the long history of catalytic cracking and the technological maturity of cracking processes, some of the basic questions still appear to be unresolved. Significant future improvements in catalytic cracking technologies, however, will require major advances in the fundamental understanding of all aspects of cracking transformations. A detailed treatment of all the open questions, including the analysis of the cracking mechanism, quantification of activity, selectivity and catalyst decay, and a quantitative formulation of all aspects of catalytic cracking, can be found in a 1998 review.[253] An up-to-date review about the cracking mechanism is also available.[254] Consequently, only a few important aspects are briefly discussed.

Some of the unanswered questions are as follows. According to the classical β-cracking mechanism the alkane to alkene ratio must be 1 or less. In fact, it is usually greater than 1. The excess hydrogen resulting in the formation of excess paraffins is suggested to originate from the coke accumulating on the catalyst during cracking operations. Detailed mass balance analysis revealed, however, that coke is not an adequate source of hydrogen. It also follows that fragment ratios—specifically, the ratio of the molar yield of each alkane produced to the molar yield of all alkenes derived from the same alkane—must be 1, which is rarely so. Another problem is coke formation, which results in catalyst decay. It is usually thought to be a side reaction not associated with the main, ionic mechanism of cracking. However, it is reasonable to assume that coke formation is associated with the ionic intermediates involved in the cracking reaction steps. A mechanism integrating the coke formation process with the cracking process providing a mechanistic basis for the kinetics and catalyst decay was formulated.[255]

Our incomplete understanding of the mechanism of catalytic cracking originates from the complexity of the feed with a wide array of reactant molecules participating in cracking reactions. This is further compounded by rapid catalyst deactivation. There have been, however, significant developments to solve many of the questions. One is the time-on-stream concept, which provides a formulation of the kinetics of catalyst decay, allowing the correct evaluation of cracking selectivity. Furthermore, optimum performance envelopes permit to determine the selectivities of a cracking reaction in the pristine state before any catalyst decay using conventional experimental methods. The better understanding of all the important factors in catalytic cracking provides the opportunity to make predictions as to how various experimental variables affect the performance of a new catalyst or a new feedstock on an old catalyst in catalytic cracking. This gives a rational basis for design of new catalysts and feedstocks and selection of suitable reaction conditions. In a large-scale technology such as catalytic cracking even marginal improvements result in significant gaines in the economics of commercial cracking operations.

As mentioned above, catalyst decay or deactivation, which includes coking, poisoning, and sintering, is a serious problem in catalytic cracking and other refinery operations such as hydrotreatment and hydroprocessing. Catalyst decay has become even more important because of the increasing severity of the processes, which leads to increasing catalyst deactivation, and the trend toward the treatment of heavier residues, which typically have higher amounts of impurities. Catalyst deactivation has been widely studied, and there are reviews addressing the problem of decreasing catalyst activity brought about by coking and its fundamental and practical aspects as well as catalyst regeneration, including modeling, the consequence of catalyst deactivation for process design and operation, and process and catalyst development.[256–262] The proceedings of regular symposia on catalyst deactivation offer additional readings.[263–265] Further important issues such as development of technologies of fluid catalytic cracking (FCC) processes aiming at reducing air-pollutant emissions and for changing the composition of fuel products with emphasis on the control of sulfur in gasoline are treated in a review.[266]

Hydrocracking, the other major cracking operation, is predicted to grow approximately 3–5% annually, mainly because it operates at relatively high hydrogen pressure (typically >100 atm), which results in high removal rates of S and N heteroatoms from the feedstock and deep saturation of aromatic compounds. Consequently, it produces middle distillates with excellent product qualities, namely, jet and diesel fractions with very low sulfur content and very good combustion properties.

As a result of this growing significance, new technologies were developed to carry out hydrodesulfurization (HDS), hydrodenitrogenation (HDN), hydrodeoxygenation (HDO), and hydrodemetallation (HDM). These allow to convert various distillate residues into valuable distillates using high-reaction-temperature, high-pressure, and low-contact-time hydroprocessing units. A new technology, the ISAL process, developed jointly by INTEVEP and UOP, applies a selective hydroprocessing method to produce reformulated gasoline with improved naphtha quality by reducing S, N, and olefins without octane loss.[267]

Hydrodesulfurization (HDS) is a well-established upgrading process designed to remove and accumulate metals and to desulfurize the feed. By increasing the reaction temperature and applying suitable catalysts, the conversion of a substantial fraction of the residue into distillate can also be obtained. The residue upgrading, therefore, is switching from HDS to hydrocracking. The major developments in residue hydrocracking processes and catalysts were reviewed.[268–270]

Because of the changing market conditions and requirements, specifically, a steady, further growth of the market and environmental pressure on product quality, hydroprocessing and hydrotreatment operations have growing significance. As a result, new methods and catalysts based on a thorough knowledge of existing commercial operations were developed.[271] Hydroprocessing was originally developed mainly because low-cost hydrogen was available from catalytic reforming. In addition, the oil price hikes in the 1970s and environmental regulations also contributed to its growing importance. In the future, refiners are expected to continue to rely on hydroprocessing to meet new product quality specifications. With the necessity to convert residue and to remove sulfur, refiners usually turn to hydroprocessing. Two approaches have been very popular, and IFP has excellent technologies, which serve both of these.[271] The fixed-bed Hyvahl process is uniquely suited for upgrading atmospheric or low-metal-content vacuum residues for the production of low-sulfur fuel oil or as a feedstock to the residue FCC unit. For processing moderate-to-high metal vacuum residues the H-Oil process based on the ebullated-bed system, is the preferred choice.

New developments in FCC, resid FCC, hydrotreatment, and hydroprocessing operations are treated in review papers[272–275] and symposia proceedings.[276,277] Books addressing various problems of FCC[278,279] and a detailed analysis of both technological aspects and fundamental knowledge of hydrotreating catalysis[280] are also available.

The catalysts applied in hydroprocessing operations are typically sulfided $CoO–MoO_3–Al_2O_3$ or $NiO–MoO_3–Al_2O_3$. The results of relevant studies[281] and the application in refinery processes of these and other transition-metal sulfide catalysts were reviewed.[282] Selection of catalysts and reactors for particular feeds and products is also an important issue.[257,280,283,284]

Extensive studies describe the development of new materials, such as transition-metal nitrides[285,286] and oxynitrides,[287,288] as new potential catalysts for hydroprocessing operations. Studies were performed to acquire information on the effect of high levels of ZSM-5 additive on light olefins and gasoline composition in FCC.[289] It was found that ZSM-5 can substantially increase propylene and butenes, and ethylene yield linearly increases with the amount of ZSM-5 in the system.[290] Increasing naphthene in the feed, however, decreases light alkene production.[291] This is due to the high propensity of naphthenes to donate hydrogen to hydrogen transfer reactions leading to saturation of alkenes or alkene precursors.

2.6.2. Reforming

The petroleum industry faced many challenges during the 1990s. Legislation requiring the reduction in emission of volatile and toxic components in gasoline affects, among other compounds, aromatics. Since reformate is a major source of aromatics in gasoline, reduced reformer severity may be a possible way to reduce aromatics content. However, it adversely affects the reformate octane number. Moreover, hydrogen generated at increased reformer severity could be utilized to increase hydrotreating capacities of refineries. Because of new legislation on diesel and fuel oil sulfur content, the demand for hydrogen continues to increase further. Refiner operating conditions may be changed to maintain hydrogen production and, at the same time, decrease aromatics content by lowering reformer pressure. The downside, however, is increasing catalyst deactivation and shorter cycle lengths. Another option is the aromatic reduction of distillates.[292] Extensive studies with respect to the effect of reaction conditions on yields and quality of reformate,[293,294] coke formation on bimetallic Pt on Al_2O_3 catalysts,[295,296] the chemical state and effect of the second metal,[295,297–300] and kinetic modeling[301,302] are available. The effect of regenerative oxidation–reduction and oxychlorination–reduction cycles on dispersion and catalytic performance is an especially important issue.[303–305] A broad treatment from the point of view of alloy catalysis was published.[306] These and other significant issues of catalytic reforming, including the role of sulfur, the influence of chloriding, the problem of catalyst regeneration, and the effects of hydrogen are discussed in a review.[307]

The interpretation of catalytic reforming presented in Fig. 2.1 is based on the bifunctional nature of the Pt-on-Al_2O_3 catalysts applied in industry. Specifically, all dehydrogenation and hydrogenation reactions are attributed to Pt metal sites and all skeletal reaction steps (ring closure, ring contraction, and enlargement) to carbocation reactions occurring at the acidic sites of alumina. The latter transformations are slow rate-determining processes. The reactions on both metal sites and acidic sites were thought to be necessary to convert methylcyclopentane to benzene, and n-hexane to branched C_6 hydrocarbons and benzene.

Aromatization, however, may also be envisaged as taking place via stepwise dehydrogenation of an unbranched hydrocarbon molecule followed by ring closure of the polyunsaturated intermediates. In fact, the formation of dienes was proved during the aromatization of C_6 and C_7 alkanes to the corresponding aromatics over monofunctional metal oxides and metal black, and bifunctional catalysts.[307,308] Radiotracer studies even allowed the detection in very low concentration of hexatriene during the aromatization of n-hexane over Pt black.[309] It was also proposed that aromatics are formed from the *cis* isomers, whereas *trans* isomers may be coke precursors.[213] Direct experimental evidence has recently been

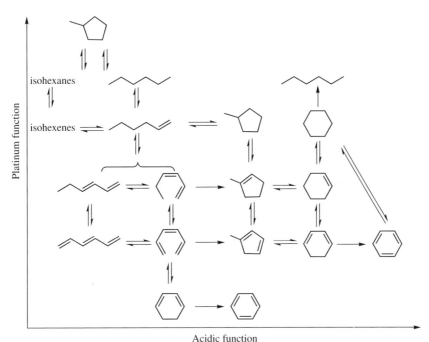

Figure 2.2

provided by spectroscopic studies to show the existence of a flat-lying hexa-σ-
bonded triene intermediate over a $Cu_3Pt(111)$ single-crystal alloy surface.[310] On
the basis of this information, a new, more comprehensive reaction scheme was constructed that now includes also the additional metal-catalyzed reactions (Fig. 2.2).
Cracking reactions leading to low-molecular-weight products and coke formation
pathways are omitted for clarity).[307,311]

Additional evidence to this scheme was reported applying temporal analysis of
products. This technique allows the direct determination of the reaction mechanism
over each catalyst. Aromatization of *n*-hexane was studied on Pt, Pt–Re, and Pd
catalysts on various nonacidic supports, and a monofunctional aromatization pathway was established.[312] Specifically, linear hydrocarbons undergo rapid dehydrogenation to unsaturated species, that is, alkenes and dienes, which is then
followed by a slow 1,6-cyclization step. Cyclohexane was excluded as possible
intermediate in the dehydrocyclization network.

As seen in Figure 2.2 and from the corresponding discussion, dehydrocyclization
is a key reaction in forming aromatic compounds.[307] A study comparing dehydrocyclization over mono- and bifunctional catalysts at atmospheric pressure and high
pressure representative of naphtha reforming conditions concludes that primary
aromatic products at all pressures are formed by direct six-carbon ring formation.[313] Over bifunctional catalysts the acid-catalyzed cyclization is more rapid

than the monofunctional metal cyclization. Pt in the form of single crystals has been extensively studied as catalyst for the dehydrocyclization reaction. Results related to the aromatization reaction, however, were complicated because of the existence of other competing transformations such as isomerization and hydrogenolysis. Alloying Pt with a suitable metal can modify reactivity and selectivity. An especially well suited catalyst is $Cu_3Pt(111)$ allowing significant observations to be acquired. Aromatization reactions were studied in detail with several C_6 hydrocarbons, including 1-hexene,[310,314,315] cyclohexene,[315] polyunsaturated linear and cyclic compounds,[310,315] and 4-phenyl-1-butene.[316] In the case of C_6 hydrocarbons, cyclization is the rate-determining step and mono- and biunsaturated cyclic compounds are not suggested to be intermediates for the dehydrocyclization of 1-hexene to benzene.[315]

2.6.3. Dehydrogenation with Olefin Production

Because of the great practical importance of and increasing demand for the alkenes formed in the dehydrogenation of alkanes, significant attention is still devoted to the fundamental aspects of the catalytic and oxidative dehydrogenation of alkanes to alkenes. Numerous reports were published, new catalysts were reported, and basic questions such as C—H bond activation, active sites and mechanistic problems were addressed. The possibility of using CO_2 as oxidizing agent in the dehydrogenation of ethylbenzene to styrene is gaining importance. General reviews of the subject with respect to thermal and catalytic dehydrogenations[317] as well as oxidative dehydrogenation[318,319] are available. Only selected aspects of the most recent significant developments in the dehydrogenative formation of alkenes of practical importance, namely, ethylene, propylene, butenes, 1,3-butadiene, and styrene, are discussed here.

Catalytic Dehydrogenation. Reviews are available for chromium-based dehydrogenation catalysts, particularly the industrial chromia–alumina used for the dehydrogenation of propane and isobutane,[320] and more recent advances in research and development of commercial catalysts for ethylbenzene dehydrogenation.[321] Studies analyze catalyst deactivation phenomena in styrene production over iron oxide catalysts,[322] and report the development of a new styrene production catalyst.[321] Comparative analyses of various processes used in propylene production, namely, the nonoxidative dehydrogenation process of UOP and autothermal and isothermal oxidative dehydrogenations,[323] and in the production of styrene were reported.[324] A new technology, called the SMART process, a combination of the Lummus process with UOP concept for oxidative reheating by selective catalytic oxidation of hydrogen, was developed for styrene production. This concept was further explored by using a combination of mixed oxides for selective hydrogen combustion and a 0.7% Pt–Sn-on-ZSM-5 catalyst for dehydrogenation.[325]

New results of styrene formation over iron oxide single-crystal model catalysts were reported.[326] In ultra-high-vacuum experiments with $Fe_3O_4(111)$ and α-$Fe_2O_3(0001)$ films combined with batch reaction studies only Fe_2O_3 showed catalytic activity. The activity increased with increasing surface defect concentration revealing that atomic surface defects are the active sites for ethylbenzene dehydrogenation over unpromoted Fe_2O_3. Carbon deposition and partial reduction of Fe_2O_3 lead to the deactivation of the sample. In addition to the above-mentioned single-crystal samples, K-promoted Fe_2O_3 [$KFe_xO_y(111)$] was also investigated in a similar study.[327] On $Fe_3O_4(111)$ almost all chemisorption sites were found to be blocked by the product styrene, on α-$Fe_2O_3(0001)$ 43% of these sites were occupied, and on $KFe_xO_y(111)$ only 17% were occupied. This suggests an increasing trend of catalytic activity from $Fe_3O_4(111)$ over α-$Fe_2O_3(0001)$ to $KFe_xO_y(111)$, which indeed was observed in the previous reactivity studies.[326]

Particular attention has been devoted to study dehydrogenations over supported Pt–Sn bimetallic catalysts. Tin is a widely used modifier in hydrocarbon transformations since it strongly inhibits cracking by decreasing platinum ensemble size (geometric effect) and decreases the deactivation rate. The addition of Sn to Pt on γ-Al_2O_3 notably enhanced catalyst performance in *n*-hexane dehydrogenation[328] and in isobutylene formation by limiting cracking reactions.[329] Selectivities close to 100% at a high level of activity were obtained when Li was added. Silica- and alumina-supported Pt–Sn catalysts prepared by surface organometallic chemistry in water exhibited remarkable catalytic properties.[330,331] Tin increased drastically both activity and selectivity of the dehydrogenation of isobutane to isobutylene. It almost fully inhibited hydrogenolysis leading to lower alkanes and eventually to carbon deposits by isolating the surface Pt atoms. This results in a catalyst that is extremely selective for dehydrogenation.

A commercial Pt–Sn on γ-Al_2O_3 catalyst showed 2–3 times higher activity in the catalytic dehydrogenation of a mixture of C_{10}–C_{12} alkanes to linear monoalkenes when applied in a supercritical phase.[332] The strong shift of the equilibrium under supercritical conditions is believed to be due to the high solubility of the product in supercritical fluids or the rapid desorption of alkenes from the catalyst surface.

As seen from the discussions, there are numerous heterogeneous catalysts suitable for the dehydrogenation of alkanes. In sharp contrast, homogeneous catalysts are rare. Exceptions are the Ir pincer complexes (**7a** and **7b**), which are extraordinarily active and robust catalysts.[333]

$$R = isoPr \quad \textbf{7a}$$
$$R = tert\text{-}Bu \quad \textbf{7b}$$

Complexes **7a** and **7b** are able to induce the transformation of alkanes to α-olefins with very high regioselectivity although selectivity decreases with time due to isomerization:[334]

$$C_6H_{13}-CH_2-CH_3 \xrightarrow[150°C]{\textbf{7a}, \, tert\text{-}Bu\text{-ethylene}} C_6H_{13}-CH=CH_2 \; + \; C_5H_{11}-CH=CH-CH_3$$

15 min	> 95	< 5
120 min	68	32

(2.27)

The complexes also catalyze the dehydrogenation of cycloalkanes to cycloalkenes without the use of sacrificial hydrogen acceptors[335,336] and show high activity in the dehydrogenation of ethylbenzene to styrene.[337] Theoretical calculations for ethane dehydrogenation showed that the transfer dehydrogenation process is initiated by the hydrogen acceptor, which dehydrogenates the Ir dihydrido complex.[338] The 14-electron species thus formed reacts readily with the alkane to regenerate the dihydrido complex and yield ethylene.

Oxidative Dehydrogenation. Specific reviews deal with oxidative dehydrogenation of lower alkanes,[339–343] particularly over vanadium oxide–based catalysts,[339–341] and styrene production.[324] Various metal oxides were intensively studied as catalysts in the dehydrogenation of alkanes to alkenes. The Li^+–MgO catalyst system used and studied in detail in the oxidative coupling of methane (see Section 3.3) exhibited excellent performance in the oxidative dehydrogenation of ethane as well when promoted with chloride ions (58% ethylene yield at 75% conversion at 620°C).[344] The promoted catalyst has a higher surface area and Cl^- appears to prevent poisoning by CO_2. The active centers are believed to be associated with an atomic layer of Li_2O. It is capable of activating ethane, but its basic strength is modified and, therefore, it does not form carbonate at the reaction temperature.

Vanadium pentoxide usually as mixed oxide was mainly investigated in the dehydrogenation of propane to yield propylene.[339,345–351] The support, in most cases, is zirconia.[345–349] For the dehydrogenation of butane V_2O_5–MgO mixed oxides were used.[352,353]

Analysis of structure–activity relationships shows that various species characterized by different reactivities exist on the surface of vanadium oxide–based catalysts.[339] The redox cycle between V^{5+} and V^{4+} is generally accepted to play a key role in the reaction mechanism, although opposite relationships between activity and selectivity, and reducibility were established. More recent studies with zirconia-supported vanadium oxide catalysts showed that vanadium is present in the form of isolated vanadyl species or oligomeric vanadates depending on the loading.[345,346] The maximum catalytic activity was observed for catalysts with vanadia content of 3–5 mol% for which highly dispersed polyvanadate species are dominant.

Various Mo-containing catalysts were also tested in the dehydrogenation of propane[354–358] and butane.[359,360] Alkali promotion was shown to have a beneficial effect, specifically to increase selectivity due to lowering the acidity of the samples.[355,360] With silica-supported binary and ternary molybdates, the best results were achieved with $NiMoO_4$ and $Co_{0.5}Ni_{0.5}MoO_4$ (16% yield at 27% conversion).[354]

Another well-studied catalyst system is Cr_2O_3 loaded onto various supports used in almost exclusively in the oxidative dehydrogenation of isobutane to isobutylene.[361–364] In the case of Cr_2O_3 on $La_2(CO_3)_3$, the surface active center was found to be a chromate species bound to the surface La carbonate.[361] With optimum loadings between 10–15% selectivities exceed 95% below 250°C. Correlations between

Cr_2O_3 coverage and selectivity were also found for other catalyst combinations.[362–364] High activities and selectivities are due to well-dispersed $Cr^{6+}–O_x$ species,[363] however, Cr^{3+} dominates after reaction.[364]

Metal phosphates such as $Ni_2P_2O_7$ (82% selectivity at 550°C),[365] as well as $CePO_4$ and $LaPO_4$ (higher than 80% selectivity),[366] were found to be active in the dehydrogenation of isobutane to isobutylene. Strong acidic sites were suggested to play a key role in the process. Magnesium–cobalt phosphate with a composition of $Co_2Mg(PO_4)_2$ is highly selective (77% at 500°C) in the oxidative dehydrogenation of propane when calcined at 550°C.[367]

High activities in the dehydrogenation of propane to propylene were reported for supporting submonolayer or monolayer $VOPO_4$ on TiO_2 or Al_2O_3 which ensures the formation of highly dispersed $VOPO_4$ giving the best catalytic performance.[368,369]

Very high olefin yields were reported in the oxidative dehydrogenation of ethane,[370] propane and butanes,[371] and isobutane,[372,373] as well as C_5 and C_6 alkanes[374] over Pt-coated α-Al_2O_3 foam monoliths operating adiabatically at very short contact time (a few milliseconds). In the case of ethane and propane selectivities exceeding 60% were obtained at almost complete conversion. According to the latest report, ethylene can be produced with about 85% selectivity at about 70% conversion using a Pt–Sn bimetallic catalyst and by adding large amounts of hydrogen to the reaction mixture at 950°C.[375] New findings show a cooperation between catalytic (deep oxidation) and homolytic (alkene formation) processes.[376] It was demonstrated that the same functions can be equally well accomplished with a Pt-free $BaMnAl_{11}O_{19}$ catalyst that gives ethylene yields higher than 50%.[376]

Early studies with respect to the dehydrogenation of hydrocarbons to alkenes on oxide catalysts indicated that carbonaceous deposits formed in the early stages of the process on the surface of acidic catalysts act as the real active centers for the oxidative dehydrogenation. The hypothesis was later confirmed[377,378] and verified by using carbon molecular sieves. With this catalyst 90% styrene selectivity could be obtained at 80% ethylbenzene conversion.[379] Various coals for the synthesis of isobutylene[380] and activated carbon in the synthesis of styrene[381] were used in further studies.

In a comparative study carbon nanofilaments showed higher activity than soot and graphite in the formation of styrene in oxidative dehydrogenation.[382] In addition, activity increased with time, indicating stability toward burning. Strongly basic, possibly quinoidic, surface sites were identified after the catalytic reaction.

In spite of significant fundamental studies and its significant economic potential as an alternate route to alkenes, the oxidative dehydrogenation of alkanes to alkenes is not currently practiced.[383] The main reason is that the secondary oxidation of the primary alkene products limits severely alkene yields, which becomes more significant with increasing conversion. This is due mainly to the higher energies of the C—H bonds in the reactant alkanes compared to those of the product alkenes. This leads to the rapid combustion of alkenes, that is, the formation of carbon oxides, at the temperatures required for C—H bond activation in alkanes.

Detailed studies were carried out to determine the effect of CO_2 as oxidizing agent in alkane dehydrogenation. Iron oxide when supported on suitable solid carriers was found to exhibit enhanced activity in the oxidative dehydrogenation of ethylbenzene to styrene. For instance, 10% Fe_2O_3 on Al_2O_3,[384] Fe_2O_3 on zeolite,[385] and particularly, Fe_2O_3 on activated carbon promoted by alkali metals[386–388] show good catalytic performance. Among other systems, Cr_2O_3 or CeO_2 on activated carbon[389] or highly tetragonal ZrO_2[390] are also effective catalysts.

Similar improvements of activity and selectivity were reported in the oxidative dehydrogenation by CO_2 of ethane over Ga_2O_3 (18.6% ethylene yield and 94.5% selectivity)[391] and that of propane over rare earth vanadates.[392] Cr_2O_3 shows medium activity in the oxidative dehydrogenation of ethane, but support on SiO_2 enhances the catalytic performance (55.5% ethylene yield at 61% conversion at 650°C).[393]

Li-promoted Fe_2O_3 on activated C is also highly active in the oxidative dehydrogenation of isobutane to isobutylene.[394] Here magnetite is considered to be the active phase. Observations indicated that isobutylene was produced via both simple dehydrogenation and oxidative dehydrogenation through the redox cycle of Fe_3O_4 and Fe.

Two mechanisms were suggested to interpret the results found on Fe catalysts. According to one suggestion, hydrogen formed in dehydrogenation is removed by the reverse water–gas shift reaction [Eq. (3.4), reverse reaction]. The other mechanism suggests a redox cycle: surface dehydrogenation induced by lattice oxygens of ferrite to form water and styrene, then the oxygen defects are subsequently reoxidized by CO_2 to form ferrite.[386]

2.6.4. Aromatics Production

As discussed in Section 2.5.1, extensive research in the field of aromatization have resulted in the development of commercial processes, the Mobil M2 forming,[219] the Cyclar (UOP–BP),[172] the Aroforming (IFP–Salutee),[395] and the Alpha (Sanyo Petrochemical)[396] processes. New reviews cover the fundamental and practical aspects of both monofunctional (acid-catalyzed) and bifunctional routes of aromatization.[397–400] Further readings with respect to industrial applications may be find is symposia proceedings.[401]

Gallium-promoted zeolites are usually preferred to the Zn-promoted catalysts because the reduction of ZnO into metal under reaction conditions results in Zn volatilization. During the Mobil M2 forming process dehydrogenation takes place by hydrogen-transfer reactions, which limits the production of aromatics and significantly reduces hydrogen. Since the production of hydrogen makes the process economical, this is a severe limitation. In the Aroforming process the product distribution from LPG (liquefied petroleum gas, mainly propane and butanes) was reported to be close to that in the Cyclar process. Two major problems emerge in the aromatization of light paraffins: deactivation by coke formation and excess production of methane and ethane by cracking or hydrogenolysis. An interesting observation related to coke formation is that Ga-silicate modified with Pt and CeO_2 exhibited longer catalyst life and the coke deposited was much less than on an unmodified catalyst.[402] CeO_2 decreased the acid sites on the external surface of Ga-silicate, resulting in a decrease in the extensive condensation of aromatic rings. In addition, Pt enhanced catalytic activity and selectivity toward aromatization.[403]

Further research has been performed and is continued to be reported, mostly with zeolites unloaded or loaded with Pt, and Ga- and Zn-promoted H-ZSM-5 or H-[Al]ZSM-5 catalysts to clarify the details of the complex transformations taking place and make further improvements. In addition, new catalysts were studied and reported. Reference should also be made to work addressing the problems of the modification of catalyst features of ZSM-5[404] and the development of a new light naphtha aromatization process using a conventional fixed-bed unit.[405,406]

A comparison of the catalytic performance of H-ZSM-5 and Zn-promoted H-ZSM-5 in the aromatization of propane showed that Zn cations increase propane conversion, hydrogen formation rates, and selectivity to aromatics.[407] Temperature-programmed reduction studies showed that Zn cations in Zn–H-ZSM-5 prepared by ion exchange do not reduce to zerovalent species during propane aromatization at 500°C. Condensation reactions of $(ZnOH)^+$ species with acidic OH groups appear to lead the formation of the active Zn species having Zn^{2+} interacting with two Al sites $(O^--Zn^{2+}-O^-)$.[407,408] In the sample prepared with impregnation instead of ion-exchange extrazeolitic ZnO crystals are present, which reduce to Zn metal and elute from the catalyst under propane aromatization conditions.

Additional evidence acquired by ^{13}C labeling studies has been reported[409] for the propane aromatization pathways over H-ZSM-5 suggested by previous researchers.[398,399,410] The complex reaction sequence starts with propane dehydrogenation to propylene or propane cracking to ethylene and methane. Chain growth involves rapid alkene oligomerization and cracking cycles, and addition of 2-propyl carbocation (Scheme 2.5). These reactions proceed within the channels of the zeolite. Additional transformations taking place are methyl shifts and hydride transfer reactions, and β scission becomes significant with increasing chain length. Ultimately a pool of alkene intermediates and carbocations is established with a broad size distribution. C_{6+} alkenes then undergo cyclization and further dehydrogenation to kinetically stable and thermodynamically favored aromatics.

These conclusions are supported by the observation that products formed in the reaction of propane-2-^{13}C contains binomial ^{13}C isotopomer distributions of

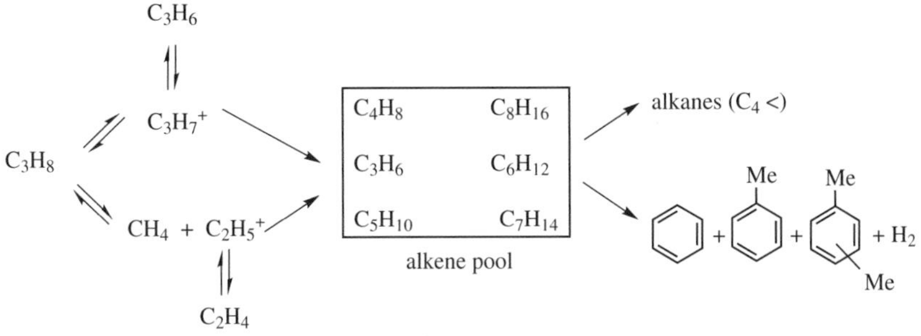

Scheme 2.5

labeling, and similar average ^{13}C content and isotopomer distributions in all aromatic compounds. Benzene, for example, contains one to five ^{13}C atoms instead of the two expected from direct propylene cyclodimerization. Aromatic products, consequently, form from a common pool of gas-phase and adsorbed alkenes. During the aromatization only a small amount of hydrogen is formed. Since hydrogen-transfer reactions are characteristic over H-ZSM-5, the majority of hydrogen formed is used up for the removal of carbonaceous intermediates.

The aromatization processes discussed above transform C_3–C_4 light paraffins and C_5–C_7 liquid paraffins into aromatics. Much less attention has been paid to the aromatization of ethane, although natural gas contains appreciable amounts of ethane (up to 10%). In addition, ethane is also produced in the Cyclar process as an undesirable byproduct. Because of its low reactivity and high thermodynamic barrier to aromatization, the transformation of ethane to aromatics requires high temperature (>600°C). It was shown, however, that conversion and yields of aromatization of ethane over Ga–H-ZSM-5 catalyst at relatively low temperature (400–500°C) can be significantly increased by adding olefins or higher alkanes to the feed.[411] This was attributed to a change in the mechanism; namely, hydrogen transfer between ethane and ethylene or higher olefins becomes operative. More specifically, without additives ethane dehydrogenation was suggested to occur through the interaction with Lewis acid sites of the catalyst (extraframework Ga, denoted G) to form the ethyl cation, which is transformed to ethylene [Eq. (2.28)]. This process, however, is questionable. When higher olefins are present, which are either added to the feed or formed from the more reactive higher alkanes, these are protolytically activated by the zeolite to form carbocations, which then undergo hydride transfer to form the ethyl cation [Eq. (2.29)]:

$$C_2H_6 \; + \; G \; \rightleftharpoons \; C_2H_5^+ \; + \; GH^- \; \rightleftharpoons \; C_2H_4 \; + \; H^+ \tag{2.28}$$

$$C_nH_{2n} \; + \; H^+ \; \rightleftharpoons \; (C_nH_{2n+1})^+ \; \overset{C_2H_6}{\rightleftharpoons} \; C_2H_5^+ \; + \; C_nH_{2n+2} \tag{2.29}$$

This process is much faster than the one depicted in Eq. (2.28). Since aromatization of ethylene is exothermic and ethane dehydrogenation and overall aromatization are endothermic, the coupling of the two processes would be energy-efficient.

In a study the direct aromatization of natural gas over H-gallosilicate, H-gallo-aluminosilicate (Ga–H-[Al]ZSM-5) and Ga–H-ZSM-5 zeolites were compared.[412] Natural gas containing 27.3 wt% C_{2+} hydrocarbons can be converted over Ga–H-[Al]ZSM-5 to aromatics with very high selectivity (~90%) at a high conversion (70%) at 600°C. This catalyst was found to be the best choice for aromatization of propane[413] and n-hexane.[414] This results from the uniform distribution of extra-framework Ga-oxide species in the zeolite channel.[415]

Ga–ZSM-5 was studied in the aromatization of dilute ethylene (3% ethylene in methane) reactant streams, since this allows the removal of ethylene and further utilization of methane, for example, in oxidative coupling.[416] A Ga loading as low as 0.5% was sufficient to obtain maximum yields of aromatic products. At 520°C, an ethylene conversion of 93%, with an aromatics selectivity of 81% was obtained over a 5% Ga–ZSM-5 catalyst. The study also showed that Ga exists as both Ga^{3+} at zeolitic exchange sites and as Ga_2O_3 within the channels and on the external surface of the calcined catalyst. It is also important to point out, that aromatization over all Ga-loaded zeolite catalysts strongly influenced by operating conditions. Specifically, product selectivity in the aromatization of propane depends on conversion,[417–419] that is, space velocity and temperature.[419]

New high-surface-area zirconia–carbon composites were prepared by a sol–gel method followed by a high-temperature treatment in inert gas.[420] The samples proved to be very active and selective in the aromatization of C_{6+} alkanes (n-hexane and n-octane). From n-octane, as main products, only ethylbenzene and o-xylene were formed. The catalysts have low acidity due to significant dehydration of the zirconia surface, and high surface area since the carbon matrix prevents sintering of the ZrO_2 particles.

A pore-size-regulated 5% ZnO–H-ZSM-5 catalyst was studied in the aromatization of C_4–C_6 hydrocarbon stream at 500–540°C.[421] Modification of the pore size was performed by chemical vapour deposition of $(EtO)_4Si$ to form SiO_2. A significant enhancement of p-xylene selectivity as high as 99% was achieved with this catalyst with the production of benzene and toluene remaining unaffected.

A new method has been described to obtain acid or bifunctional catalysts for aromatization by steaming the as-synthesized Ga–H[Al]-ZSM-5 at low temperature (530°C).[422,423] During the steaming process a profound and progressive demetalization (removal of Ga and Al) takes place. The propane aromatization properties of these catalysts are similar to those prepared by means of direct calcination under dry air at higher temperature ($>700°C$) or by conventional impregnation methods. This method provides an alternative synthesis for aromatization catalysts without long and expensive conventional preparation.

Acidic solids loaded with bimetallic Pt alloys were also studied. A synergistic effect was observed using Pt–Ga supported on H-ZSM-5 zeolite in the aromatization of n-butane.[424] Enhanced aromatization activity, significant increase in aromatization selectivities, and suppressed side reactions (hydrogenolysis and hydrogenation) are the main improvements observed. The Pt–Ge intermetallic compound was shown to form on the external surface of H-ZSM-5 during chemical vapor deposition of $Ge(acac)_2Cl_2$.[425] The catalyst exhibited lower dehydrogenation ability, a key step in aromatization, than that of Pt on H-ZSM-5. This, however, is coupled with lower coke formation and, consequently, results in a longer catalyst life. The activity of Pt–Ge on H-ZSM-5 is also lower for hydrogenolysis. These catalytic properties make this sample a favorable catalyst for selective aromatization. A systematic study of the aromatization of propane and butane resulted in the observation that ZSM-5 loaded with Pt and Re when presulfided gives high BTX selectivity, low methane and ethane selectivity, and low coke formation. All of these are attractive features for a commercial process.[426]

In connection with the aromatization of methane induced by Mo–H-ZSM-5 catalysts (see Section 3.5.2), Mo_2C on H-ZSM-5 prepared by carburation of MoO_3 was studied in the aromatization of ethane,[427] ethylene,[428] and propane.[429] The high dehydrogenation activity of Mo_2C and the ability of H-ZSM-5 to induce aromatization of alkenes makes this an active and selective catalyst for aromatics production.

REFERENCES

1. W. P. Ballard, G. I. Cottington, and T. A. Cooper, in *Encyclopedia of Chemical Processing and Design*, Vol. 13, J. J. McKetta and W. A. Cunningham, eds., Marcel Dekker, New York, 1981, p. 144.

2. W. M. Burton, U.S. Patent 1,049,667 (1912).

3. J. A. Dubbs, U.S. Patent 1,123,502 (1915).

4. A. M. McAfee, *J. Ind. Eng. Chem.* **7**, 737 (1915).

5. R. V. Shankland, *Adv. Catal.* **6**, 271 (1954).

6. E. J. Houdry, *Adv. Catal.* **9**, 499 (1957).

7. L. B. Ryland, M. W. Tamele, and J. N. Wilson, in *Catalysis*, Vol. 7, P. H. Emmett, ed., Reinhold, New York, 1960, Chapter 1, p. 1.

8. E. C. Luckenbach, A. C. Worley, A. D. Reichle, and E. M. Gladrow, in *Encyclopedia of Chemical Processing and Design*, Vol. 13, J. J. McKetta and W. A. Cunningham, eds., Marcel Dekker, New York, 1981, p. 1.

9. C. L. Thomas, *ACS Symp. Ser.* **222**, 241 (1983).

10. E. J. Houdry, U.S. Patents 2,078,945, 2,078,951 (1937); *Bull. assoc. franç. techniciens pétrole* **49**, 5 (1939).

11. A. G. Oblad, *ACS Symp. Ser.* **222**, 61 (1983).

12. C. J. Plank, E.J. Rosinski, and W.P. Hawthorne, *Ind. Eng. Chem., Prod. Res. Dev.* **3**, 165 (1964).

13. C. J. Plank, *ACS Symp. Ser.* **222**, 253 (1983).

14. J. Scherzer, *Catal. Rev.-Sci. Eng.* **31**, 215 (1989).

15. J. Scherzer, *Octane-Enhancing Zeolitic FCC Catalysts*. Marcel Dekker, New York, 1990.

16. A. Corma, in *Zeolites: Facts, Figures, Future*, Studies in Surface Science and Catalysis, Vol. 49, P. A. Jacobs and R. A. van Santen, eds., Elsevier, Amsterdam, 1989, p. 49.

17. P. H. Schipper, F. G. Dwyer, P. T. Sparrell, S. Mizrahi, and J. A. Herbst, *ACS Symp. Ser.* **375**, 64 (1988).

18. A. A. Avidan, M. Edwards, and H. Owen, *Oil Gas J.* **88**(2), 33 (1990).

19. A. A. Avidan, in *Fluid Catalytic Cracking: Science and Technology*, Studies in Surface Science and Catalysis, Vol. 76, J. S. Magee and M. M. Mitchell, Jr., eds., Elsevier, Amsterdam, 1993, Chapter 1, p. 1.

20. C.-M. T. Hayward and W. S. Winkler, *Hydrocarbon Process., Int Ed.* **69**(2), 55 (1990).

21. K. Rajagopalan and E. T. Habib, Jr., *Hydrocarbon Process., Int Ed.* **71**(9), 43 (1992).

22. M. L. Occelli, *ACS Symp. Ser.* **452**, 343 (1991).

23. A. S. Krishna, C. R. Hsieh, A. R. English, T. A. Pecoraro, and C. W. Kuehler, *Hydrocarbon Process., Int Ed.* **70**(11), 59 (1991).

24. R. H. Nielsen and P. K. Doolin, in *Fluid Catalytic Cracking: Science and Technology*, Studies in Surface Science and Catalysis, Vol. 76, J. S. Magee and M. M. Mitchell, Jr., eds., Elsevier, Amsterdam, 1993, Chapter 10, p. 339.

25. R. E. Ritter, L. Rheaume, W. A. Welsh, and J. S. Magee, *Oil Gas J.* **79**(27), 103 (1981).

26. J. M. Maselli and A. W. Peters, *Catal. Rev.-Sci. Eng.* **26**, 525 (1984).

27. J.-E. Otterstedt, B. Gevert, and J. Sterte, *ACS Symp. Ser.* **375**, 266 (1988).

28. P. O'Connor, L. A. Gerritsen, J. R. Pearce, P. H. Desai, S. Yanik, and A. Humphries, *Hydrocarbon Process., Int Ed.* **70**(11), 76 (1991).

29. M. M. Mitchell, Jr., J. F. Hoffman, and H. F. Moore, in *Fluid Catalytic Cracking: Science and Technology*, Studies in Surface Science and Catalysis, Vol. 76, J. S. Magee and M. M. Mitchell, Jr., eds., Elsevier, Amsterdam, 1993, Chapter 9, p. 293.

30. G. Yepsen and T. Witoshkin, *Oil Gas J.* **89**(14), 68 (1991).

31. G. Parkinson, *Chem. Eng.* (NY) **99**(4), 35 (1992).

32. M. Schlossman, W. A. Parker, and L. C. Chen, *Chemtech* **24**(2), 41 (1994).

33. A. Corma, *Catal. Lett.* **22**, 33 (1993).

34. B. M. Lok, C. A. Messina, R. L. Patton, R. T. Gajek, T. R. Cannan, and E. M. Flanigen, *J. Am. Chem. Soc.* **106**, 6092 (1984).

35. E. M. Flanigen, B. M. Lok, R. L. Patton, and S. T. Wilson, in *New Developments in Zeolite Science and Technology*, Studies in Surface Science and Catalysis, Vol. 28, Y. Murakami, A. Iijima, and J. W. Ward, eds., Kodansha, Tokyo, Elsevier, Amsterdam, 1986, p. 103.

36. E. M. Flanigen, R. L. Patton, and S. T. Wilson, in *Innovation in Zeolite Materials and Science*, Studies in Surface Science and Catalysis, Vol. 37, P. J. Grobet, W. J. Mortier, E. F. Vansant, and G. Schulz-Ekloff, eds., Elsevier, Amsterdam, 1988, p. 13.

37. R. Szostak, *Molecular Sieves*, Van Nostrand Reinhold, New York, 1989, Chapter 4, p. 205.

38. S. T. Wilson, B. M. Lok, C. A. Messina, T. R. Cannan, and E. M. Flanigen, *J. Am. Chem. Soc.* **104**, 1146 (1982).

39. A. Corma, V. Fornés, M. J. Franco, F. A. Mocholí, and J. Pérez-Pariente, *ACS Symp. Ser.* **452**, 79 (1991).

40. N. Choudhary and D. N. Saraf, *Ind. Eng. Chem. , Prod. Res. Dev.* **14**, 74 (1975).

41. D. W. Breck and R. A. Anderson, in *Kirk-Othmer Encyclopedia of Chemical Technology*, 3rd ed., Vol. 15, M. Grayson and D. Eckroth, eds., Wiley-Interscience, New York, 1981, p. 665.

42. R. F. Sullivan and J. W. Scott, *ACS Symp. Ser.* **222**, 293 (1983).

43. K. Tanabe, M. Misono, Y. Ono, and H. Hattori, *New Solid Acids and Bases*, Studies in Surface Science and Catalysis, Vol. 51, Kodansha, Tokyo, Elsevier, Amsterdam, 1989, Chapter 4, p. 215.

44. C. N. Satterfield, *Heterogeneous Catalysis in Industrial Practice*, McGraw-Hill, New York, 1991, Chapter 9, p. 390.

45. G. E. Langlois and R. F. Sullivan, *Adv. Chem. Ser.* **97**, 38 (1970).

46. C. E. Jahnig, H. Z. Martin, and D. L. Campbell, *ACS Symp. Ser.* **222**, 273 (1983).

47. L. L. Upson, C. L. Hemler, and D. A. Lomas, in *Fluid Catalytic Cracking: Science and Technology*, Studies in Surface Science and Catalysis, Vol. 76, J. S. Magee and M. M. Mitchell, Jr., eds., Elsevier, Amsterdam, 1993, Chapter 11, p. 385.

48. *Hydrocarbon Process., Int. Ed.* **71**(11), 169, 170, 175 (1992).

49. F. O. Rice, *J. Am. Chem. Soc.* **55**, 3035 (1933).

50. F. O. Rice and E. Teller, *J. Chem. Phys.* **6**, 489 (1938); **7**, 199 (1939).

51. A. Kossiakoff and F. O. Rice, *J. Am. Chem. Soc.* **65**, 590 (1943).

52. B. S. Greensfelder, H. H. Voge, and G. M. Good, *Ind. Eng. Chem.* **41**, 2573 (1949).

53. E. N. Hague and R. V. Wheeler, *J. Chem. Soc.* **135**, 378 (1929).

54. F. E. Frey an D. F. Smith, *Ind. Eng. Chem.* **20**, 948 (1928).

55. R. C. Hansford, *Ind. Eng. Chem.* **39**, 849 (1947); *Adv. Catal.* **4**, 1 (1952).

56. C. L. Thomas, *Ind. Eng. Chem.* **41**, 2564 (1949).

57. V. Haensel, *Adv. Catal.* **3**, 179 (1951).

58. A. G. Oblad, T. H. Milliken, Jr., and G. A. Mills, *Adv. Catal.* **3**, 199 (1951).

59. H. H. Voge, in *Catalysis*, Vol. 6, P. H. Emmett, ed., Reinhold, New York, 1958, Chapter 5, p. 407.

60. H. H. Voge, *ACS Symp. Ser.* **222**, 235 (1983).

61. M. L. Poutsma, in *Zeolite Chemistry and Catalysis*, ACS Monograph 171, J. A. Rabo, ed., American Chemical Society, Washington, DC, 1976, Chapter 8, p. 437.

62. A. Corma and B. W. Wojciechowski, *Catal. Rev.-Sci. Eng.* **27**, 29 (1985).

63. R. C. Hansford, *ACS Symp. Ser.* **222**, 247 (1983).

64. J. H. Kolts and G. A. Delzer, *Science* **232**, 744 (1986).

65. A. P. Bolton, in *Zeolite Chemistry and Catalysis*, ACS Monograph 171, J. A. Rabo, ed., American Chemical Society, Washington, DC, 1976, Chapter 13, p. 714.

66. V. Haensel and V. N. Ipatieff, *Ind. Eng. Chem.* **39**, 853 (1947).

67. F. G. Ciapetta and J. B. Hunter, *Ind. Eng. Chem.* **45**, 147, 155 (1953).

68. H. L. Coonradt and W. E. Garwood, *Ind. Eng. Chem., Proc. Design Dev.* **3**, 38 (1964).

69. H. F. Schulz and J. H. Weitkamp, *Ind. Eng. Chem., Prod. Res. Dev.* **11**, 46 (1972).

70. G. E. Langlois, R. F. Sullivan, and C. J. Egan, *J. Phys. Chem.* **70**, 3666 (1966).

71. R. DeBiase and J. D. Elliott, *Hydrocarbon Process., Int. Ed.* **61**(5), 99 (1982).

72. J. D. Elliott, *Hydrocarbon Process., Int. Ed.* **71**(1), 75 (1992).

73. *Hydrocarbon Process., Int. Ed.* **69**(11), 106 (1990).

74. *Hydrocarbon Process., Int. Ed.* **61**(9), 160 (1982).

75. S. J. Yanik, E. J. Demmel, A. P. Humphries, and R. J. Campagna, *Oil Gas J.* **83**(19), 108 (1985).

76. D. A. Pappal and P. H. Schipper, *ACS Symp. Ser.* **452**, 45 (1991).

77. F. G. Dwyer and T. F. Degnan, in *Fluid Catalytic Cracking: Science and Technology*, Studies in Surface Science and Catalysis, Vol. 76, J. S. Magee and M. M. Mitchell, Jr., eds., Elsevier, Amsterdam, 1993, Chapter 13, p. 499.

78. D. J. Rawlence and J. Dwyer, *ACS Symp. Ser.* **452**, 56 (1991).

79. S. J. Miller and C. R. Hsieh, *ACS Symp. Ser.* **452**, 96 (1991).

80. A. Billon, J. P. Franck, J. P. Peries, E. Fehr, E. Gallei, and E. Lorenz, *Hydrocarbon Process.* **57**(5), 122 (1978).

81. S. D. Light, R. V. Bertram, and J. W. Ward, *Hydrocarbon Process., Int Ed.* **60**(5), 93 (1981).

82. B. Schuetze and H. Hofmann, *Hydrocarbon Process., Int. Ed.* **63**(2), 75 (1984).

83. P. H. Desai, D. A. Keyworth, M. Y. Asim, T. Reid, and A. H. Pichel, *Hydrocarbon Process., Int Ed.* **71**(11), 51 (1992).

84. D. M. Little, *Catalytic Reforming*, Penn Well, Tulsa, 1985, Chapter 8, p. 175.

85. D. C. McCulloch, in *Applied Industrial Catalysis*, Vol. 1, B. E. Leach, ed., Academic Press, New York, 1983, Chapter 4, p. 69.

86. B. Delmon, *Catal. Lett.* **22**, 1 (1993).

87. B. E. Reynolds, E. C. Brown, and M. A. Silverman, *Hydrocarbon Process., Int Ed.* **71**(4), 43 (1992).

88. B. W. Burbidge, I. M. Keen, and M. K. Eyles, *Adv. Chem. Ser.* **102**, 400 (1971).

89. N. Y. Chen, W. E. Garwood, and F. G. Dwyer, *Shape Selective Catalysis in Industrial Applications*, Marcel Dekker, New York, 1989, Chapter 5, p. 155.

90. J. D. Hargrove, G. J. Elkes, and A. H. Richardson, *Oil Gas J.* **77**(3), 103 (1979).

91. V. Haensel and M. J. Sterba, *Adv. Chem. Ser.* **5**, 60 (1951).

92. H. Heinemann, *Catal. Rev.-Sci. Eng.* **15**, 53 (1977).

93. J. H. Sinfelt, in *Catalysis, Science and Technology*, Vol. 1, J. R. Anderson and M. Boudart, eds., Springer, Berlin, 1981, Chapter 5, p. 257.

94. D. M. Little, *Catalytic Reforming*, Penn Well, Tulsa, 1985.

95. L. R. Hill, G. A. Vincent, and E. F. Everett, *Trans. Am. Inst. Chem. Eng.* **42**, 611 (1946).

96. A. V. Grosse, J. C. Morrell, and W. J. Mattox, *Ind. Eng. Chem.* **32**, 528 (1940).

97. H. G. McGrath and L. R. Hill, *Adv. Chem. Ser.* **5**, 39 (1951).

98. M. J. Sterba and V. Haensel, *Ind. Eng. Chem., Prod. Res. Dev.* **15**, 2 (1976).

99. V. Haensel, *ACS Symp. Ser.* **222**, 141 (1983).

100. F. G. Ciapetta, R. M. Dobres, and R. W. Baker, in *Catalysis*, Vol. 6, P. H. Emmett, ed., Reinhold, New York, 1958, Chapter 6, p. 495.

101. F. G. Ciapetta and D. N. Wallace, *Catal. Rev.-Sci. Eng.* **5**, 67 (1971).

102. M. Dean Edgar, in *Applied Industrial Catalysis*, Vol. 1, B. E. Leach, ed., Academic Press, New York, 1983, Chapter 5, p. 123.

103. D. M. Little, *Catalytic Reforming*, Penn Well, Tulsa, 1985, Chapter 4, p. 40.

104. J.-P. Bournonville and J.-P. Franck, in *Hydrogen Effects in Catalysis*, Z. Paál and P. G. Menon, eds., Marcel Dekker, New York, 1988, Chapter 25, p. 653.

105. G. C. A. Schuit and B. C. Gates, *Chemtech* **13**, 556 (1983).

106. R. L. Moss, in *Catalysis, a Specialist Periodical Report*, Vol. 1, C. Kemball, senior reporter, The Chemical Society, Burlington House, London, 1977, Chapter 2, p. 73.

107. H. Charcosset, *Int. Chem. Eng.* **23**, 187, 411 (1983).

108. *Hydrocarbon Process., Int. Ed.* **59**(9), 159 (1980).

109. F. W. Kopf, W. H. Decker, W. C. Pfefferle, M. H. Dalson, and J. A. Nevison, *Hydrocarbon Process.* **48**(5), 111 (1969).

110. J. A. Nevison, C. J. Obaditch, and M. H. Dalson, *Hydrocarbon Process.* **53**(6), 111 (1974).

111. *Hydrocarbon Process., Int. Ed.* **71**(11), 144 (1992).

112. T. R. Hughes, R. L. Jacobson, K. R. Gibson, L. G. Schornack, and J. R. McCabe, *Hydrocarbon Process.* **55**(5), 75 (1976).

113. *Hydrocarbon Process., Int. Ed.* **69**(11), 118 (1990).

114. *Hydrocarbon Process., Int. Ed.* **59**(9), 164 (1980).

115. C. G. Kirkbride, *Petrol. Refiner* **30**(6), 95 (1951).

116. *Hydrocarbon Process., Int. Ed.* **59**(9), 162 (1980).

117. N. Y. Chen, J. Maziuk, A. B. Schwartz, and P. B. Weisz, *Oil Gas J.* **66**(47), 154 (1968).

118. S. D. Burd, Jr. and J. Maziuk, *Hydrocarbon Process.* **51**(5), 97 (1972).

119. N. Y. Chen and W. E. Garwood, *Adv. Chem. Ser.* **121**, 575 (1973).

120. P. B. Weisz, *Adv. Catal.* **13**, 137 (1962).

121. V. Haensel and G. R. Donaldson, *Ind. Eng. Chem.* **43**, 2102 (1951).

122. H. Heinemann, G. A. Mills, J. B. Hattman, and F. W. Kirsch, *Ind. Eng. Chem.* **45**, 130 (1953).

123. G. A. Mills, H. Heinemann, T. H. Milliken, and A. G. Oblad, *Ind. Eng. Chem.* **45**, 134 (1953).

124. A. M. Kugelman, *Hydrocarbon Process.* **55**(1), 95 (1976).

125. J. H. Jenkins and T. W. Stephens, *Hydrocarbon Process., Int. Ed.* **59**(11), 163 (1980).

126. G. R. Donaldson, L. F. Pasik, and V. Haensel, *Ind. Eng. Chem.* **47**, 731 (1955).

127. T. R. Hughes, R. L. Jacobson, and P. W. Tamm, in *Catalysis 1987*, Studies in Surface Science and Catalysis, Vol. 38, J. W. Ward, ed., Elsevier, Amsterdam, 1988, p. 317.

128. N. Y. Chen, W. E. Garwood, and F. G. Dwyer, *Shape Selective Catalysis in Industrial Applications*, Marcel Dekker, New York, 1989, Chapter 4, p. 65.

129. N. Y. Chen and W. E. Garwood, *Catal. Rev.-Sci. Eng.* **28**, 185 (1986).

130. O. V. Bragin, *Russ. Chem. Rev.* (Engl. transl.) **50**, 1045 (1981).

131. V. E. Pierce and A. K. Logwinuk, *Hydrocarbon Process., Int. Ed.* **64**(9), 75 (1985).

132. L. F. Albright, *Oil Gas J.* **88**(46), 79 (1990).

133. M. Prezelj, *Hydrocarbon Process., Int. Ed.* **66**(9), 68 (1987).

134. W. J. Piel and R. X. Thomas, *Hydrocarbon Process., Int. Ed.* **69**(7), 68 (1990).

135. R. J. Schmidt, P. L. Bogdan, and N. L. Gilsdorf, *Chemtech* **23**(2), 41 (1993).

136. J. Liers, J. Meusinger, A. Mösch, and W. Reschetilowski, *Hydrocarbon Process., Int. Ed.* **72**(8), 165 (1993).

137. L. Kniel and O. Winter, in *Encyclopedia of Chemical Processing and Design*, Vol. 20, J. J. McKetta and W. A. Cunningham, eds., Marcel Dekker, New York, 1984, p. 88.

138. Y. C. Hu, in *Encyclopedia of Chemical Processing and Design*, Vol. 32, J. J. McKetta and W. A. Cunningham, eds., Marcel Dekker, New York, 1990, p. 194.

139. K. Weissermel and H.-J. Arpe, *Industrial Organic Chemistry*, 2nd ed., VCH, Weinheim, 1993, Chapter 3, p. 59.

140. R. S. Zack and R. O. Skamser, *Hydrocarbon Process., Int. Ed.* **62**(10), 99 (1983).

141. D. L. Waller, *Hydrocarbon Process., Int. Ed.* **62**(11), 187 (1983).

142. Y. C. Hu, *Hydrocarbon Process., Int. Ed.* **61**(11), 109 (1982).

143. *Hydrocarbon Process., Int. Ed.* **72**(3), 170 (1993).

144. T. Rielly, in *Encyclopedia of Chemical Processing and Design*, Vol. 5, J. J. McKetta and W. A. Cunningham, eds., Marcel Dekker, New York, 1977, p. 110.

145. D. P. Tate and T. W. Bethea, in *Encyclopedia of Polymer Science and Engineering*, 2nd ed., Vol. 2, J. I. Kroschwitz, ed., Wiley-Interscience, New York, 1985, p. 544.

146. K. Weissermel and H.-J. Arpe, *Industrial Organic Chemistry*, 2nd ed., VCH, Weinheim, 1993, Chapter 5, p. 103.

147. G. Emmrich and K. Lackner, *Hydrocarbon Process., Int. Ed.* **68**(1), 71 (1989).

148. M. Soudek and J. J. Lacatena, *Hydrocarbon Process., Int. Ed.* **69**(5), 73 (1990).

149. J. L. Monfils, S. Barendregt, S. K. Kapur, and H. M. Woerde, *Hydrocarbon Process., Int. Ed.* **71**(2), 47 (1992).

150. F. E. Frey and W. F. Huppke, *Ind. Eng. Chem.* **25**, 54 (1933).

151. F. E. Frey and W. F. Huppke, U. S. Patent 2,098,959 (1937).

152. I. Ya. Tyuryaev, *Russ. Chem. Rev.* (Engl. transl.) **35**, 59 (1966).

153. O. D. Sterligov, T. G. Olfer'eva, and N. F. Kononov, *Russ. Chem. Rev.* (Engl. transl.) **36**, 498 (1967).

154. S. Carrà and L. Forni, *Catal. Rev.-Sci. Eng.* **5**, 159 (1971).

155. H. S. Bloch, *Detergent Age* **3**(8), 33, 105 (1967).

156. J. F. Roth, J. B. Abell, L. W. Fannin, and A. R. Schaefer, *Adv. Chem. Ser.* **97**, 193 (1970).

157. R. Di Cintio, M. Picciotti, V. Kaiser, and C. A. Pocini, *Hydrocarbon Process., Int. Ed.* **70**(7), 83 (1991).

158. T. Wett, *Oil Gas J.* **83**(35), 46 (1985).

159. A. Wood, *Chem. Week* **154**(1), 28 (1994).

160. R. C. Berg, B. V. Vora, and J. R. Mowry, *Oil Gas J.* **78**(45), 191 (1980).

161. B. V. Vora and T. Imai, *Hydrocarbon Process., Int. Ed.* **61**(4), 171 (1982).

162. P. R. Pujado and B. V. Vora, *Energy Prog.* **4**, 186 (1984); *Hydrocarbon Process., Int. Ed.* **69**(3), 65 (1990).

163. F. M. Brinkmeyer, D. F. Rohr, M. E. Olbrich, and L. E. Drehman, *Oil Gas J.* **81**(13), 75 (1983).

164. P. R. Sarathy and G. S. Suffridge, *Hydrocarbon Process., Int. Ed.* **72**(1), 89 (1993); **72**(2), 43 (1993).

165. K. Brooks, *Chem. Week* **137**(4), 36 (1985).

166. F. Asinger, *Mono-Olefins, Chemistry and Technology*, Pergamon, Press, Oxford, 1968, Chapter 2, p. 70.

167. H. H. Voge, in *Encyclopedia of Chemical Processing and Design*, Vol. 14, J. J. McKetta and W. A. Cunningham, eds., Marcel Dekker, New York, 1982, p. 276.

168. D. Sanfilippo, *Chemtech* **23**(8), 35 (1993).

169. *Hydrocarbon Process., Int. Ed.* **59**(9), 210 (1980).

170. S. Gussow, D. C. Spence, and E. A. White, *Oil Gas J.* **78**(49), 96 (1980).

171. R. G. Craig and E. A. White, *Hydrocarbon Process., Int. Ed.* **59**(12), 111 (1980).

172. *Hydrocarbon Process., Int. Ed.* **72**(3), 168 (1993).

173. I. Kirshenbaum, in *Kirk-Othmer Encyclopedia of Chemical Technology*, 3rd ed., Vol. 4, M. Grayson and D. Eckroth, eds., Wiley-Interscience, New York, 1978, p. 313.

174. E. C. Britton, A. J. Dietzler, and C. R. Noddings, *Ind. Eng. Chem.* **43**, 2871 (1951).

175. C. R. Noddings, S. B. Heath, and J. W. Corey, *Ind. Eng. Chem.* **47**, 1373 (1955).

176. A. A. DiGiacomo, J. B. Maerker, and J. W. Schall, *Chem. Eng. Prog.* **57**(5), 35 (1961).

177. M. L. Senyek, in *Encyclopedia of Polymer Science and Engineering*, 2nd ed., Vol. 8, J. I. Kroschwitz, ed., Wiley-Interscience, New York, 1987, p. 496.

178. P. C. Husen, K. R. Deel, and W. D. Peters, *Oil Gas J.* **69**(31), 60 (1971).

179. T. Hutson, Jr., R. D. Skinner, and R. S. Logan, *Hydrocarbon Process.* **53**(6), 133 (1974).

180. L. Marshall Welch, L. J. Croce, and H. F. Christmann, *Hydrocarbon Process.* **57**(11), 131 (1978).

181. F. C. Newman, *Ind. Eng. Chem.* **62**(5), 42 (1970).

182. *Hydrocarbon Process.* **50**(11), 137 (1971).

183. A. Yoshioka, H. Hokai, R. Sato, H. Yamamoto, and K. Okumura, *Hydrocarbon Proc., Int. Ed.* **63**(11), 97 (1984).

184. H. S. Bloch, *Oil Gas J.* **65**(3), 79 (1967).

185. D. B. Broughton and R. C. Berg, *Hydrocarbon Process.* **48**(6), 115 (1969).

186. *Hydrocarbon Process.* **58**(11), 185 (1979).

187. B. V. Vora, P. R. Pujado, J. B. Spinner, and T. Imai, *Hydrocarbon Process., Int. Ed.* **63**(11), 86 (1984).

188. *Hydrocarbon Process.* **58**(11), 184 (1979).

189. D. H. James, J. B. Gardner, and E. C. Mueller, in *Encyclopedia of Polymer Science and Engineering*, 2nd ed., Vol. 16, J. I. Kroschwitz, ed., Wiley-Interscience, New York, 1989, p. 5.

190. E. H. Lee, *Catal. Rev.-Sci. Eng.* **8**, 285 (1973).

191. C. R. Hagopian, P. J. Lewis, and J. J. McDonald, *Hydrocarbon Process., Int. Ed.* **62**(2), 45 (1983).

192. *Hydrocarbon Process., Int. Ed.* **64**(11), 168, 169 (1985).

193. H. H. Al-Morished and M. Y. Lamb, *Hydrocarbon Process., Int. Ed.* **70**(8), 125 (1991).

194. D. J. Ward, S. M. Black, T. Imai, Y. Sato, N. Nakayama, H. Tokano, and K. Egawa, *Hydrocarbon Process., Int. Ed.* **66**(3), 47 (1987).

195. G. A. Olah, U.S. Patents 4,472,268 (1984), 4,508,618 (1985).

196. P. Sabatier and J. B. Senderens, *Ann. Chim. Phys.* [8], **4**, 339, 457 (1905).

197. N. Zelinsky, *Chem. Ber.* **44**, 3121 (1911).

198. J. E. Germain, *Catalytic Conversion of Hydrocarbons*, Academic Press, London, 1969, Chapter 3, p. 91.

199. D. P. Land, C. L. Pettiette-Hall, R. T. McIver, Jr., and J. C. Hemminger, *J. Am. Chem. Soc.* **111**, 5970 (1989).

200. C. L. Pettiette-Hall, D. P. Land, R. T. McIver, Jr., and J. C. Hemminger, *J. Am. Chem. Soc.* **113**, 2755 (1991).

201. G. A. Somorjai, *Adv. Catal.* **26**, 1 (1977).

202. H. Pines, *The Chemistry of Catalytic Hydrocarbon Conversions*, Academic Press, New York, 1981, Chapter 4, p. 185.

203. E. F. G. Herington and E. K. Rideal, *Proc. Roy. Soc. London, Ser. A* **184**, 434, 447 (1945).

204. H. Pines and C.-T. Chen, *J. Org. Chem.* **26**, 1057 (1961).

205. H. Pines and C. T. Goetschel, *J. Org. Chem.* **30**, 3530 (1965).

206. H. Pines, C. T. Goetschel, and J. W. Dembinski, *J. Org. Chem.* **30**, 3540 (1965).

207. J. R. Anderson, *Adv. Catal.* **23**, 1 (1973).

208. Z. Paál, *Adv. Catal.* **29**, 273 (1980).

209. F. G. Gault, *Adv. Catal.* **30**, 1 (1981).

210. B. A. Kasansky and F. Plate, *Chem. Ber.* **69**, 1862 (1936).

211. V. S. Fadeev, I. V. Gostunskaya, and B. A. Kazanskii, *Dokl. Chem.* (Engl. transl.) **189**, 959 (1969).

212. F. M. Dautzenberg and J. L. Platteeuw, *J. Catal.* **19**, 41 (1970).

213. Z. Paál and P. Tétényi, *J. Catal.* **30**, 350 (1973).

214. P. B. Weisz and, E. W. Swegler, *Science* **126**, 31 (1957).

215. S. G. Hindin, S. W. Weller, and G. A. Mills, *J. Phys. Chem.* **62**, 244 (1958).

216. S. M. Csicsery, *Adv. Catal.* **28**, 293 (1979).

217. S. M. Csicsery, *J. Catal.* **9**, 336 (1967).

218. S. M. Csicsery, *J. Catal.* **15**, 111 (1969).

219. N. Y. Chen and T. Y. Yan, *Ind. Eng. Chem., Proc. Des. Dev.* **25**, 151 (1986).

220. P. B. Weisz, *Ind. Eng. Chem. Fund.* **25**, 53 (1986).

221. P. B. Weisz, W. O. Haag, and P. G. Rodewald, *Science* **206**, 57 (1979).

222. Y. Ono, *Catal. Rev.-Sci. Eng.* **34**, 179 (1992).

223. N. S. Gnep, J. Y. Doyemet, A. M. Seco, F. Ramoa Ribeiro, and M. Guisnet, *Appl. Catal.* **43**, 155 (1988).

224. H. Kitagawa, Y. Sendoda, and Y. Ono, *J. Catal.* **101**, 12 (1986).

225. L. M. Thomas and X.-S. Liu, *J. Phys. Chem.* **90**, 4843 (1986).

226. C. R. Bayense, A. J. H. P. van der Pol, and J. H. C. van Hooff, *Appl. Catal.* **72**, 81 (1991).

227. N. S. Gnep, J. Y. Doyemet, and M. Guisnet, *J. Mol. Catal.* **45**, 281 (1988).

228. V. Kanazirev, G. L. Price, and K. M. Dooley, *J. Chem. Soc., Chem. Commun.* 712 (1990).

229. A. Corma, C. Goberna, J. M. Lopez Nieto, N. Paredes, and M. Perez, in *Zeolite Chemistry and Catalysis*, Studies in Surface Science and Catalysis, Vol. 69, P. A. Jacobs, N. I. Jaeger, L. Kubelková, and B. Wichterlová, eds., Elsevier, Amsterdam, 1991, p. 409.

230. Y. Ono, K. Kanae, K. Osako, and K. Nakashiro, *Mat. Res. Soc. Symp. Proc.* **233**, 3 (1991).

231. S. Field, *Hydrocarbon Process.* **49**(5), 113 (1970).

232. D. L. Ransley, in *Kirk-Othmer Encyclopedia of Chemical Technology*, 3rd ed., Vol. 4, M. Grayson and D. Eckroth, eds., Wiley-Interscience, New York, 1978, p. 264.

233. D. M. Little, *Catalytic Reforming.* Penn Well, Tulsa, 1985, Chapter 6, p. 142.

234. K. Weissermel and H.-J. Arpe, *Industrial Organic Chemistry*, 2nd ed., VCH, Weinheim, 1993, Chapter 12, p. 316.

235. K. J. Day and T. M. Snow, in *Toluene, the Xylenes and their Industrial Derivatives*, E. G. Hancock, ed., Elsevier, Amsterdam, 1982, Chapter 3, p. 50.

236. J. R. Mowry, R. F. Anderson, and J. A. Johnson, *Oil Gas J.* **83**(48), 128 (1985).

237. P. C. Doolan and P. R. Pujado, *Hydrocarbon Process., Int. Ed.* **68**(9), 72 (1989).

238. C. D. Gosling, F. P. Wilcher, L. Sullivan, and R. A. Mountford, *Hydrocarbon Process., Int. Ed.* **70**(12), 69 (1991).

239. P. W. Tamm, D. H. Mohr, and C. R. Wilson, in *Catalysis 1987*, Studies in Surface Science and Catalysis, Vol. 38, J. W. Ward, ed., Elsevier, Amsterdam, 1988, p. 335.

240. T. R. Hughes, W. C. Buss, P. W. Tamm, and R. L. Jacobson, in *New Developments in Zeolite Science and Technology*, Studies in Surface Science and Catalysis, Vol. 28, Y. Murakami, A. Iijima, and J. W. Ward, eds., Kodansha, Tokyo, Elsevier, Amsterdam, 1986, p. 725.

241. *Hydrocarbon Process.* **48**(11), 153 (1969).

242. A. P. Dosset, in *Toluene, the Xylenes and their Industrial Derivatives*, E. G. Hancock, ed., Elsevier, Amsterdam, 1982, Chapter 5, p. 157.

243. *Petrol. Refiner* **40**(6), 228 (1961).

244. *Hydrocarbon Process.* **58**(11), 139 (1979).

245. *Oil Gas J.* **68**(12), 67 (1970).

246. G. F. Asselin and R. A. Erickson, *Chem. Eng. Prog.* **58**(4), 47 (1962).

247. *Hydrocarbon Process Petrol. Refiner* **44**(11), 277 (1965).

248. S. Feigelman, L. M. Lehman, E. Aristoff, and P. M. Pitts, *Hydrocarbon Process Petrol. Refiner* **44**(12), 147 (1965).

249. S. Feigelman and C. B. O'Connor, *Hydrocarbon Process Petrol. Refiner* **45**(5), 140 (1966).

250. S. Masamune, J. Fukuda, and S. Katada, *Hydrocarbon Process.* **46**(2), 155 (1967).

251. M. J. Fowle and P. M. Pitts, *Chem. Eng. Prog.* **58**(4), 37 (1962).

252. T. F. Degnan, G. K. Chitnis, and P. H. Schipper, *Micropor. Mesopor. Mat.* **35**, 245 (2000).

253. B. W. Wojciechowski, *Catal. Rev.-Sci. Eng.* **40**, 209 (1998).

254. Y. V. Kissin, *Catal. Rev.-Sci. Eng.* **43**, 85 (2001).

255. B. W. Wojciechowski and N. M. Rice, *ACS Symp. Ser.* **634**, 134 (1996).

256. P. O'Connor, E. Brevoord, A. C. Pouwels, and H. N. J. Wijngaards, *ACS Symp. Ser.* **634**, 147 (1996).

257. J. P. Janssens, A. D. van Langavald, S. T. Sie, and J. A. Moulijn, *ACS Symp. Ser.* **634**, 238 (1996).

258. J. W. Gosselink and W. H. J. Stork, *Ind. Eng. Chem. Res.* **36**, 3354 (1997).

259. E. Furimsky and F. E. Massoth, *Catal. Today* **52**, 381 (1999).

260. J. W. Gosselink and J. A. R. van Veen, in *Catalyst Decativation 1999*, Studies in Surface Science and Catalysis, Vol. 126, B. Delmon and G. F. Froment, eds., Elsevier, Amsterdam, 1999, p. 3.

261. G. F. Froment, *Appl. Catal., A* **212**, 117 (2001).

262. S. T. Sie, *Appl. Catal., A* **212**, 129 (2001).

263. *Catalyst Decativation 1994*, Studies in Surface Science and Catalysis, Vol. 88, B. Delmon and G. F. Froment, eds., Elsevier, Amsterdam, 1994.

264. *Catalyst Deactivation 1997*, Studies in Surface Science and Catalysis, Vol. 111, C. H. Bartholomew and G. A. Fuentes, eds., Elsevier, Amsterdam, 1997.

265. *Catalyst Deactivation 1999*, Studies in Surface Science and Catalysis, Vol. 126, B. Delmon and G. F. Froment, eds., Elsevier, Amsterdam, 1999.

266. W.-C. Cheng, G. Kim, A. W. Peters, X. Zhao, and K. Rajagopalan, *Catal. Rev.-Sci. Eng.* **40**, 39 (1998).

267. G. J. Antos, B. Solari, and R. Monque, in *Hydrotreatment and Hydrocracking of Oil Fractions*, Studies in Surface Science and Catalysis, Vol. 106, G. F. Froment, B. Delmon, and P. Grange, eds., Elsevier, Amsterdam, 1997, p. 27.

268. J. Scherzer and A. J. Guia, *Hydrocracking Science and Technology*, Marcel Dekker, New York, 1996.

269. F. Morel, S. Kressmann, V. Harlé, and S. Kasztelan, in *Hydrotreatment and Hydrocracking of Oil Fractions*, Studies in Surface Science and Catalysis, Vol. 106, G. F. Froment, B. Delmon, and P. Grange, eds., Elsevier, Amsterdam, 1997, p. 1.

270. J. K. Minderhound, J. A. R. van Veen, and A. P. Hagan, in *Hydrotreatment and Hydrocracking of Oil Fractions*, Studies in Surface Science and Catalysis, Vol. 127, B. Delmon, G. F. Froment, and P. Grange, eds., Elsevier, Amsterdam, 1999, p. 3.

271. L. Wisdom, E. Peer, and P. Bonnifay, *Oil Gas J.* **96**(6), 58 (1998).

272. S. Kressmann, F. Morel, V. Harlé, S. Kasztelan, *Catal. Today* **43**, 203 (1998).

273. P. O'Connor, J. P. J. Verlaan, and S. J. Yanik, *Catal. Today* **43**, 305 (1998).

274. W. P. Hettinger, *Catal. Today* **53**, 367 (1999).

275. E. Furimsky, *Appl. Catal., A* **199**, 147 (2000).

276. *Hydrotreatment and Hydrocracking of Oil Fractions*, Studies in Surface Science and Catalysis, Vol. 106, G. F. Froment, B. Delmon, and P. Grange, eds., Elsevier, Amsterdam, 1997.

277. *Hydrotreatment and Hydrocracking of Oil Fractions*, Studies in Surface Science and Catalysis, Vol. 127, B. Delmon, G. F. Froment, and P. Grange, eds., Elsevier, Amsterdam, 1999.

278. *Fluid Catalytic Cracking III, Methods and Processes*, ACS Symp. Ser. Vol. 571, M. L. Occelli and P. O'Connor, eds., American Chemical Society, Washington, DC, 1994.

279. *Fluid Cracking Catalysts*, M. L. Occelli and P. O'Connor, eds., Marcell Dekker, New York, 1998.

280. H. Topsøe, B. S. Clausen, and F. E. Massoth, *Catalysis, Science and Technology*, Vol. 11, J. R. Anderson and M. Boudart, eds., Springer, Berlin, 1996.

281. R. R. Chianelli, M. Daage, and M. J. Ledoux, *Adv. Catal.* **40**, 177 (1994).

282. J. W. Gosselink, in *Transition Metal Sulphides, Chemistry and Catalysis.* T. Weber, R. Prins, and R. A. Van Santen, eds., Kluwer, Dordrecht, 1998, p. 311.

283. E. Furimsky, *Appl. Catal., A* **171**, 177 (1998).

284. W. I. Beaton and R. J. Bertolacini, *Catal. Rev.-Sci. Eng.* **33**, 281 (1991).

285. S. Ramanathan and S. T. Oyama, *J. Phys. Chem.* **99**, 16365 (1995).

286. D. J. Sajkowski and S. T. Oyama, *Appl. Catal., A* **134**, 339 (1996).

287. C. C. Yu, S. Ramanathan, and S. T. Oyama, *J. Catal.* **173**, 1 (1998).

288. S. Ramanathan, C. C. Yu, and S. T., Oyama, *J. Catal.* **173**, 10 (1998).

289. J. S. Buchanan and Y. G. Adewuyi, *Appl. Catal., A* **134**, 247 (1996).

290. X. Zhao and T. G. Roberie, *Ind. Eng. Chem. Res.* **38**, 3847 (1999).

291. X. Zhao and R. H. Harding, *Ind. Eng. Chem. Res.* **38**, 3854 (1999).

292. B. H. Cooper and B. B. L. Donnis, *Appl. Catal., A* **137**, 203 (1996).

293. K. Moljord, H. G. Hellenes, A. Hoff, I. Tanem, K. Grande, and A. Holmen, *Ind. Eng. Chem. Res.* **35**, 99 (1996).

294. K. Sertic Bionda, *Oil Gas-Eur. Mag.* **23**(3), 35 (1997).

295. N. Macleod, J. R. Fryer, D. Stirling, and G. Webb, *Catal. Today* **46**, 37 (1998).

296. C. L. Li, O. Novaro, X. Bokhimi, E. Muñoz, J. L. Boldú, J. A. Wang, T. López, R. Gómez, and N. Batina, *Catal. Lett.* **65**, 209 (2000).

297. T. F. Garetto, A. Borgna, and C. R. Apesteguía, in *11th International Congress on Catalysis—40th Anniversary*, Studies in Surface Science and Catalysis, Vol. 101, J. W. Hightower, W. N. Delgass, E. Iglesia, and A. T. Bell, eds., Elsevier, Amsterdam, 1996, p. 1155.

298. R. Prestvik, K. Moljord, K. Grande, and A. Holmen, *J. Catal.* **174**, 119 (1998).

299. R. Prestvik, B. Tøtdal, C. E. Lyman, and A. Holmen, *J. Catal.* **176**, 246 (1998).

300. M. C. Rangel, L. S. Carvalho, P. Reyes, J. M. Parera, and N. S. Fígoli, *Catal. Lett.* **64**, 171 (2000).

301. G. Padmavathi and K. K. Chaudhuri, *Can. J. Chem. Eng.* **75**, 930 (1997).

302. J. Ancheyta-Juárez and E. Villafuerte-Macías, *Energy Fuels* **14**, 1032 (2000).

303. G. J. Arteaga, J. A. Anderson, S. M. Becker, and C. H. Rochester, *J. Mol. Catal. A: Chem.* **145**, 183 (1999).

304. G. J. Arteaga, J. A. Anderson, and C. H. Rochester, *J. Catal.* **184**, 268 (1999); **187**, 219 (1999); **189**, 195 (2000).

305. F. K. Chong, J. A. Anderson, and C. H. Rochester, *J. Catal.* **190**, 327 (2000); *Phys. Chem. Chem. Phys.* **2**, 5730 (2000).

306. V. Ponec and G. C. Bond, *Catalysis by Metals and Alloys*, Studies in Surface Science and Catalysis, Vol. 95, Elsevier, Amsterdam, 1995, Chapter 13, p. 583.

307. P. G. Menon and Z. Paál, *Ind. Eng. Chem. Res.* **36**, 3282 (1997).

308. Z. Paál, in *Catalytic Naphtha Reforming*, G. J. Antos, A. M. Aitani, and J. M. Parera, eds., Marcel Dekker, New York, 1995, Chapter 2, p. 19.

309. Z. Paál and P. Tétényi, Acta Chim. *Acad. Sci. Hung.* **58**, 105 (1968).

310. A. V. Teplyakov, A. B. Gurevich, E. R. Garland, B. E. Bent, and J. G. Chen, *Langmuir* **14**, 1337 (1998).

311. Z. Paál, *J. Catal.* **105**, 540 (1987).

312. D. S. Lafyatis, G. F. Froment, A. Pasau-Claerbout, and E. G. Derouane, *J. Catal.* **147**, 552 (1994).

313. B. H. Davis, *Catal. Today* **53**, 443 (1999).

314. A. V. Teplyakov and B. E. Bent, *Catal. Lett.* **42**, 1 (1996).

315. A. V. Teplyakov and B. E. Bent, *J. Phys. Chem. B* **101**, 9052 (1997).

316. A. T. Mathauser and A. V. Teplyakov, *Catal. Lett.* **73**, 207 (2001).

317. F. Buonomo, D. Sanfilippo, and F. Trifirò, in *Handbook of Heterogeneous Catalysis*, G. Ertl, H. Knözinger, and J. Weitkamp, eds., Wiley-VCH, Weinheim, 1997, Chapter 4. 3. 1, p. 2140.

318. H. H. Kung, *Adv. Catal.* **40**, 1 (1994).

319. K. Kochloefl, in *Handbook of Heterogeneous Catalysis*, G. Ertl, H. Knözinger, and J. Weitkamp, eds., Wiley-VCH, Weinheim, 1997, Chapter 4.3.2, p. 2151.

320. B. M. Weckhuysen and R. A. Schoonheydt, *Catal. Today* **51**, 223 (1999).

321. Q. L. Chen, X. Chen, L. S. Mao, and W. C. Cheng, *Catal. Today* **51**, 141 (1999).

322. G. R. Meima and P. G. Menon, *Appl. Catal., A* **212**, 239 (2001).

323. D. Wolf, N. Dropka, Q. Smejkal, and O. Buyevskaya, *Chem. Eng. Sci.* **56**, 713 (2001).

324. F. Cavani and F. Trifirò, *Appl. Catal., A* **133**, 219 (1995).

325. R. K. Grasselli, D. L. Stern, and J. G. Tsikoyiannis, *Appl. Catal., A* **189**, 1 (1999); **189**, 9 (1999).

326. W. Weiss, D. Zscherpel, and R. Schlögl, *Catal. Lett.* **52**, 215 (1998).

327. Sh. K. Shaikhutdinov, Y. Joseph, C. Kuhrs, W. Ranke, and W. Weiss, *Faraday Discuss.* **114**, 363 (1999).

328. J. Llorca, N. Homs, J.-L. G. Fierro, J. Sales, and P. R. de la Piscina *J. Catal.* **166**, 44 (1997).

329. G. J. Siri, M. L. Casella, G. F. Santori, and O. A. Ferretti, *Ind. Eng. Chem. Res.* **36**, 4821 (1997).

330. F. Humblot, J. P. Candy, F. Le Peltier, B. Didillon, and J. M. Basset, *J. Catal.* **179**, 459 (1998).

331. F. Z. Bentahar, J. P. Candy, J. M. Basset, F. Le Peltier, and B. Didillon, *Catal. Today* **66**, 303 (2001).

332. W. Wei, Y. Sun, and B. Zhong, *Chem. Commun.* 2499 (1999).

333. C. M. Jensen, *Chem. Commun.* 2443 (1999).

334. F. Liu, E. B. Pak, B. Singh, C. M. Jensen, and A. S. Goldman, *J. Am. Chem. Soc.* **121**, 4086 (1999).

335. F. Liu and A. S. Goldman, *Chem. Commun.* 655 (1999).

336. W.-W. Xu, G. P. Rosini, M. Gupta, C. M. Jensen, W. C. Kaska, K. Krogh-Jespersen, and A. S. Goldman, *Chem. Commun.* 2273 (1997).

337. M. Gupta, W. C. Kaska, and C. M. Jensen, *Chem. Commun.* 461 (1997).

338. S. Li and M. B. Hall, *Organometallics* **20**, 2153 (2001).

339. E. A. Mamedov and V. Cortés Corberán, *Appl. Catal., A* **127**, 1 (1995).

340. S. Albonetti, F. Cavani, and F. Trifiró, *Catal. Rev.-Sci. Eng.* **38**, 413 (1996).

341. T. Blasco and J. M. L. Nieto, *Appl. Catal., A* **157**, 117 (1997).

342. M. Baerns and O. Buyevskaya, *Catal. Today* **45**, 13 (1998).

343. M. A. Bañares, *Catal. Today* **51**, 319 (1999).

344. D. Wang, M. P. Rosynek, and J. H. Lunsford, *J. Catal.* **151**, 155 (1995).

345. A. Khodakov, J. Yang, S. Su, E. Iglesia, and A. T. Bell, *J. Catal.* **177**, 343 (1998).

346. A. Adamski, Z. Sojka, K. Dyrek, M. Che, G. Wendt, and S. Albrecht, *Langmuir* **15**, 5733 (1999).

347. A. Khodakov, B. Olthof, A. T. Bell, and E. Iglesia, *J. Catal.* **181**, 205 (1999).

348. F. Arena, F. Frusteri, and A. Parmaliana, *Catal. Lett.* **60**, 59 (1999).

349. K. Chen, A. T. Bell, and E. Iglesia, *J. Phys. Chem. B* **104**, 1292 (2000).

350. P. Viparelli, P. Ciambelli, L. Lisi, G. Ruoppolo, G. Russo, and J. C. Volta, *Appl. Catal., A* **184**, 291 (1999).

351. R. Monaci, E. Rombi, V. Solinas, A. Sorrentino, E. Santacesaria, and G. Colon, *Appl. Catal., A* **214**, 203 (2001).

352. D. L. Stern, J. N. Michaels, L. DeCaul, and R. K. Grasselli, *Appl. Catal., A* **153**, 21 (1997).

353. A. Dejoz, J. M. López Nieto, F. Márquez, and M. I. Vázquez, *Appl. Catal., A* **180**, 83 (1999).

354. D. L. Stern and R. K. Grasselli, *J. Catal.* **167**, 550 (1997).

355. M. C. Abello, M. F. Gomez, and L. E. Cadus, *Catal. Lett.* **53**, 185 (1998).

356. R. B. Watson and U. S. Ozkan, *J. Catal.* **191**, 12 (2000).

357. L. Jalowiecki-Duhamel, A. Ponchel, C. Lamonier, A. D'Huysser, and Y. Barbaux, *Langmuir* **17**, 1511 (2001).

358. M. C. Abello, M. F. Gomez, and O. Ferretti, *Appl. Catal., A* **207**, 421 (2001).

359. M. F. Portela, R. M. Aranda, M. Madeira, M. Oliveira, F. Freire, R. Anouchinsky, A. Kaddouri, and C. Mazzocchia, *Chem. Commun.* 501 (1996).

360. L. M. Madeira, R. M. Martín-Aranda, F. J. Maldonado-Hódar, J. L. G. Fierro, and M. F. Portela, *J. Catal.* **169**, 469 (1997).

361. M. Hoang, A. E. Hughes, J. F. Mathews, and K. C. Pratt, *J. Catal.* **171**, 313 (1997).

362. J. Słoczyński, B. Grzybowska, R. Grabowski, A. Kozowska, and K. Wcisło, *Phys. Chem. Chem. Phys.* **1**, 333 (1999).

363. P. Moriceau, B. Grzybowska, L. Gengembre, and Y. Barbaux, *Appl. Catal., A* **199**, 73 (2000).

364. S. M. Al-Zahrani, N. O. Elbashir, A. E. Abasaeed, and M. Abdulwahed, *Ind. Eng. Chem. Res.* **40**, 781 (2001).

365. Y. Takita, K. Sano, K. Kurosaki, N. Kawata, H. Nishiguchi, M. Ito, and T. Ishihara, *Appl. Catal., A* **167**, 49 (1998).

366. Y. Takita, K. Sano, T. Muraya, H. Nishiguchi, N. Kawata, M. Ito, T. Akbay, and T. Ishihara, *Appl. Catal., A* **170**, 23 (1998).

367. A. Aaddane, M. Kacimi, and M. Ziyad, *Catal. Lett.* **73**, 47 (2001).

368. P. Ciambelli, P. Galli, L. Lisi, M. A. Massucci, P. Patrono, R. Pirone, G. Ruoppolo, and G. Russo, *Appl. Catal., A* **203**, 133 (2000).

369. L. Lisi, P. Patrono, and G. Ruoppolo, *Catal. Lett.* **72**, 207 (2001).

370. D. W. Flick and M. C. Huff, *J. Catal.* **178**, 315 (1998).

371. M. Huff and L. D. Schmidt, *J. Catal.* **149**, 127 (1994).

372. M. Huff and L. D. Schmidt, *J. Catal.* **155**, 82 (1995).

373. L. S. Liebmann and L. D. Schmidt, *Appl. Catal., A* **179**, 93 (1999).

374. A. G. Dietz, III, A. F. Carlsson, and L. D. Schmidt, *J. Catal.* **176**, 459 (1996).

375. A. S. Bodke, D. A. Olschki, L. D. Schmidt, and E. Ranzi, *Science* **285**, 712 (1999).

376. A. Beretta and P. Forzatti, *J. Catal.* **200**, 45 (2001).

377. G. E. Vrieland and P. G. Menon, *Appl. Catal.* **77**, 1 (1991).

378. R. S. Drago and K. Jurczyk, *Appl. Catal., A* **112**, 117 (1994).

379. G. C. Grunewald and R. S. Drago, *J. Mol. Catal.* **58**, 227 (1990).

380. F. J. Maldonado-Hódar, L. M. Madeira, and M. F. Portela, *Appl. Catal., A* **178**, 49 (1999).

381. M. F. R. Pereira, J. J. M. Órfão, and J. L. Figueiredo, *Appl. Catal., A* **184**, 153 (1999).

382. G. Mestl, N. I. Maksimova, N. Keller, V. V. Roddatis, and R. Schlögl, *Angew. Chem., Int. Ed. Engl.* **40**, 2066 (2001).

383. C. H. Li and D. S. Hobbell, in *Encyclopedia of Chemical Processing and Design*, Vol. 55, J. J. McKetta and G. E. Weismantel, eds., Marcel Dekker, New York, 1996, p. 197.

384. N. Mimura and M. Saito, *Catal. Lett.* **58**, 59 (1999); *Appl. Organomet. Chem.* **14**, 773 (2000).

385. J.-S. Chang, S.-E. Park, and M. S. Park, *Chem. Lett.* 1123 (1997).

386. M. Sugino, H. Shimada, T. Turuda, H. Miura, N. Ikenaga, and T. Suzuki, *Appl. Catal., A* **121**, 125 (1995).

387. T. Badstube, H. Papp, P. Kustrowski, and R. Dziembaj, *Catal. Lett.* **55**, 169 (1998).

388. T. Badstube, H. Papp, R. Dziembaj, and P. Kustrowski, *Appl. Catal., A* **204**, 153 (2000).

389. N. Ikenaga, T. Tsuruda, K. Senma, T. Yamaguchi, Y. Sakurai, and T. Suzuki, *Ind. Eng. Chem. Res.* **39**, 1228 (2000).

390. J.-N. Park, J. Noh, J.-S. Chang, and S.-E. Park, *Catal. Lett.* **65**, 75 (2000).

391. K. Nakagawa, M. Okamura, N. Ikenaga, T. Suzuki, and T. Kobayashi, *Chem. Commun.* 1025 (1998).

392. B. Zhaorigetu, R. Kieffer, and J.-P. Hindermann, in *11th International Congress on Catalysis—40th Anniversary*, Studies in Surface Science and Catalysis, Vol. 101, J. W. Hightower, W. N. Delgass, E. Iglesia, and A. T. Bell, eds., Elsevier, Amsterdam, 1996, p. 1049.

393. S. B. Wang, K. Murata, T. Hayakawa, S. Hamakawa, and K. Suzuki, *Appl. Catal., A* **196**, 1 (2000).

394. H. Shimada, T. Akazawa, N. Ikenaga, and T. Suzuki, *Appl. Catal., A* **168**, 243 (1998).

395. L. Mank, A. Minkkinen, and J. Shaddick, *Hydrocarbon Tech. Int.* 69 (1992).

396. Y. Nagamori and M. Kawase, *Micropor. Mesopor. Mat.* **21**, 439 (1998).

397. D. Seddon, *Catal. Today* **6**, 350 (1990).

398. M. Guisnet, N. S. Gnep, and F. Alario, *Appl. Catal., A* **89**, 1 (1992).

399. G. Giannetto, R. Monque, and R. Galiasso, *Catal. Rev.-Sci. Eng.* **36**, 271 (1994).

400. P. Mériaudeau and C. Naccache, *Catal. Rev.-Sci. Eng.* **39**, 5 (1997).

401. *Catalysts in Petroleum Refining and Petrochemical Industries 1995*, Studies in Surface Science and Catalysis, Vol. 100, M. Absi-Halabi, J. Beshara, H. Qabazard, and A. Stanislaus, eds., Elsevier, Amsterdam, 1996, pp. 447–488.

402. T. Inui, T. Yamada, A. Matsuako, and S.-B. Pu, *Ind. Eng. Chem. Res.* **36**, 4827 (1997).

403. A. Matsuoka, J.-B. Kim, and T. Inui, *Micropor. Mesopor. Mat.* **35**, 89 (2000).

404. T. S. R. Prasad Rao, in *Recent Advances in Basic and Applied Aspects of Industrial Catalysis*, Studies in Surface Science and Catalysis, Vol. 113, T. S. R. Prasad Rao and G. Murali Dhar, eds., Elsevier, Amsterdam, 1998, p. 3.

405. S. Fukase, N. Igarashi, K. Kato, T. Nomura, and Y. Ishibashi, in *Catalysts in Petroleum Refining and Petrochemical Industries 1995*, Studies in Surface Science and Catalysis, Vol. 100, M. Absi-Halabi, J. Beshara, H. Qabazard, and A. Stanislaus, eds., Elsevier, Amsterdam, 1996, p. 455.

406. S. Fukase, N. Igarashi, K. Aimoto, H. Inoue, and H. Ono, *Progress in Zeolite and Microporous Materials*, Studies in Surface Science and Catalysis, Vol. 105, H. C. Chon, S.-K. Ihm, and Y. S. Uh, eds., Elsevier, Amsterdam, 1997, p. 885.

407. J. A. Biscardi, G. D. Meitzner, and E. Iglesia, *J. Catal.* **179**, 192 (1998).

408. J. A. Biscardi and E. Iglesia, *Phys. Chem. Chem. Phys.* **1**, 5753 (1999).

409. J. A. Biscardi and E. Iglesia, *J. Phys. Chem. B* **102**, 9284 (1998).

410. N. Viswanadham, A. R. Pradhan, N. Ray, S. C. Vishnoi, U. Shanker, and T. S. R. Prasad Rao, *Appl. Catal., A* **137**, 225 (1996).

411. V. R. Choudhary, A. K. Kinage, and T. V. Choudhary, *Angew. Chem., Int. Ed. Engl.* **36**, 1305 (1997).

412. V. R. Choudhary, A. K. Kinage, and T. V. Choudhary, *Appl. Catal., A* **162**, 239 (1997).

413. V. R. Choudhary, A. K. Kinage, C. Sivadinarayana, and M. Guisnet, *J. Catal.* **158**, 23 (1996).

414. J. Kanai and N. Kawata, *Appl. Catal.* **55**, 115 (1989).

415. A. K. Kinage, T. V. Choudhary, and V. R. Choudhary, in *Recent Advances in Basic and Applied Aspects of Industrial Catalysis*, Studies in Surface Science and Catalysis, Vol. 113, T. S. R. Prasad Rao and G. Murali Dhar, eds., Elsevier, Amsterdam, 1998, p. 707.

416. P. Qui, J. H. Lunsford, and M. P. Rosynek, *Catal. Lett.* **52**, 37 (1998).

417. G. Giannetto, A. Montes, N. S. Gnep, A. Florentino, P. Cartraud, and M. Guisnet, *J. Catal.* **145**, 86 (1993).

418. V. R. Choudhary, A. K. Kinage, C. Sivadinarayana, P. Devadas, S. D. Sansare, and M. Guisnet, *J. Catal.* **158**, 34 (1996).

419. V. R. Choudhary and P. Devadas, *J. Catal.* **172**, 475 (1997).

420. D. L. Hoang, H. Preiss, B. Parlitz, F. Krumeich, and H. Lieske, *Appl. Catal., A* **182**, 385 (1999).

421. J. Das, Y. S. Bhat, and A. B. Halgeri, in *Recent Advances in Basic and Applied Aspects of Industrial Catalysis*, Studies in Surface Science and Catalysis, Vol. 113, T. S. R. Prasad Rao and G. Murali Dhar, eds., Elsevier, Amsterdam, 1998, p. 447.

422. A. Montes, Z. Gabélica, A. Rodríguez, and G. Giannetto, *Appl. Catal., A* **169**, 87 (1998).

423. A. Montes and G. Giannetto, *Appl. Catal., A*, **197**, 31 (2000).

424. E. S. Shpiro, D. P. Shevchenko, R. V. Dmitriev, O. P. Tkachenko, and Kh. M. Minachev, *Appl. Catal., A* **107**, 165 (1994).

425. T. Komatsu, M. Mesuda, and T. Yashima, *Appl. Catal., A* **194**, 333 (2000).

426. W. J. H. Dehertog and G. F. Fromen, *Appl. Catal., A* **189**, 63 (1999).

427. F. Solymosi and A. Szőke, *Appl. Catal., A* **166**, 225 (1998).

428. F. Solymosi and A. Szőke, in *Natural Gas Conversion V,* Studies in Surface Science and Catalysis, Vol. 119, A. Parmaliana, D. Sanfilippo, F. Frusteri, A. Vaccari, and F. Arena, eds., Elsevier, Amsterdam, 1998, p. 355.

429. F. Solymosi, R. Németh, L. Óvári, and L. Egri, *J. Catal.* **195**, 316 (2000).

3

SYNTHESIS FROM
C_1 SOURCES

Methane and carbon monoxide are presently the two primary raw materials of practical importance in C_1 hydrocarbon chemistry. According to the present industrial practice, natural gas (methane) or coal can be converted to a mixture of carbon monoxide and hydrogen called *synthesis gas*:

$$CH_4 + H_2O \rightleftharpoons CO + 3H_2 \qquad \Delta H_{298} = -49.2 \text{ kcal/mol} \qquad (3.1)$$

The two approaches, however, are interconvertible, since methane itself is obtained from synthesis gas in the methanation reaction. Both methane and carbon monoxide, in turn, can be converted to hydrocarbons.

Hydrocarbons can be produced from synthesis gas through the Fischer–Tropsch synthesis. Because of poor economics the process was, however, practiced on large scale only under extreme conditions (wartime Germany, South Africa under trade boycott).

Methane (natural gas) reserves, which are potentially larger than those of crude oil, make methane the most promising potential source for synthetic replacement for petroleum-derived hydrocarbons. It has a number of advantages as a raw material. It is cheap and can be easily purified, and its conversion is source-independent. Further, as it has the highest possible H : C ratio of any neutral hydrocarbon (i.e., 4), its conversion to higher hydrocarbons does not necessitate further hydrogen. In fact, these processes are in principle all oxidative condensations eliminating excess hydrogen. The main strategies to convert methane to higher hydrocarbons can be divided into two types: (1) methods involving formation of hydrocarbons formed in a single step (direct conversion or oxidative coupling of methane) and (2) methods in which methane is first converted into a derivative that is then transformed into hydrocarbons. Process 2, however, can be rendered catalytic, representing a selective oxidative way.

Although it is a desirable goal, methods of the direct coupling of methane to C_2 and subsequently to higher hydrocarbons have not yet achieved practical application. Exceptions are the acetylene-based technologies.

According to another important and promising technology, hydrocarbons are produced from methanol, which, in turn, is synthesized from synthesis gas. Called the *methanol-to-gasoline process*, it was practiced on a commercial scale and its practical feasibility was demonstrated. Alternative routes to eliminate the costly step of synthesis gas production may use direct methane conversion through intermediate monosubstituted methane derivatives. An economic evaluation of different methane transformation processes can be found in a 1993 review.[1]

3.1. NATURE'S C_1 CHEMISTRY

Carbon dioxide, the end product of combustion of any organic compound, is the most abundant carbon compound in the atmosphere and the essential building block of reforming organic compounds. In nature's photosynthetic cycle green plants, photosynthetic bacteria and cyanobacteria produce carbohydrates (cellulose) and oxygen from carbon dioxide and water using solar energy. Cellulose is known to condense with strong acids to hydrocarbons. In addition to maintaining a balance of the oxygen : carbon dioxide ratio in the atmosphere, photosynthesis thus is the source of all other carbon compounds. Enzymatic processes, such as hydration of carbon dioxide by carbonic anhydrase and different carboxylations, are also important in carbon dioxide fixation.

Considering the immense and increasing amount of fossil-fuel-derived carbon dioxide and its global effects; it would be highly desirable to find ways to convert carbon dioxide into useful chemicals, thus supplementing nature's photosynthetic recycling. Diminishing petroleum and natural-gas sources in the short run, and decreasing fossil fuel supplies, in general, make chemical utilization of CO_2 eventually a necessity. At present major chemical industrial use of carbon dioxide is limited to its recovery in ammonia plants formed in the water–gas shift reaction and its conversion to urea.

A possibility to replace the demand for natural hydrocarbon deposits could be the use of plant hydrocarbons.[2] Carbohydrates produced by photosynthesis undergo complex transformations in certain plants, forming hydrocarbon-based compounds of very low oxygen content. The best known example is *Hevea brasiliensis*. Other plants (euphorbias) also produce a latex containing about 30% triterpenoids. An even more viable source can be a tropical tree (family Leguminosae, genus *Copaifera*) that can be tapped to collect a diesel-like material in substantial quantities (25 L in 24 h per tree). Similarly, the fruits of another tree (family Pittosporacea, genus *Pittosporum*) are rich in terpenes. They and others, after necessary development, could serve as alternative energy sources in the future.

3.2. THE CHEMICAL REDUCTION AND RECYCLING OF CO_2

As power plant and other industries emit large amounts of carbon dioxide, they contribute to the so-called greenhouse warming effect of our planet which causes an environmental concern. This was first implicated in a paper by Arrhenius in 1898. The warming trend of our Earth can be evaluated only longer time periods, but there is a relationship between increasing CO_2 content of the atmosphere and the temperature. What is not clear is how much of the climate change is due to nature own cycles, on which there is superimposed the human activity. Recycling of excess CO_2 into useful fuels would in any case help not only alleviate the question of our diminishing fuel resources but also mitigate the global warming problem.

It was found that methyl alcohol or related oxygenates, in addition to conventional routes (catalytic hydrogenation, etc.), can be made from carbon dioxide via aqueous electrocatalytic reduction without prior electrolysis of water to produce hydrogen in what is termed a "reversed fuel cell." A fuel cell can convert chemical energy of a fuel directly into electrical energy via catalytic chemical oxidation. Such a direct-oxidation liquid feed methyl alcohol–based fuel cell was developed in a cooperative effort between USC Loker Hydrocarbon Institute and Caltech–Jet Propulsion Laboratory. It reacts methyl alcohol with oxygen or air over a suitable metal catalyst producing electricity while forming CO_2 and H_2O.[3] The reverse fuel cell can convert CO_2 and H_2O electrocatalytically into oxygenated fuels, that is, formic acid or its derivatives, even methyl alcohol depending on the cell potential used.[4]

The reversed fuel cell accomplishes the electrocatalytic reduction of CO_2 outside the potential range of the conventional electrolysis of water. In its reversed mode the fuel cell is charged with electricity and can produce, from carbon dioxide aqueous solution, oxygenated methane derivatives such as formates, formaldehyde, and methanol, which then can be converted to dimethyl ether, dimethoxymethane, trimethoxymethane, trioxymethylene, dimethyl carbonate, and the like. The fuel cell thus acts as a reversible storage device for electric power, much more efficiently than any known battery. Further recycling of CO_2 provides not only the regeneration of fuels but also helps diminish the atmospheric buildup of carbon dioxide, one of the most harmful greenhouse gases. Recycling of CO_2 into methanol or dimethyl ether can subsequently also be used to make, via catalytic conversion, ethylene as well as propylene. These allow ready preparation of gasoline range or aromatic hydrocarbons, as well as the whole wide variety of other hydrocarbons and their derivatives.[5]

Since methane and methyl alcohol are highly versatile feedstocks for hydrocarbons, an efficient process to convert carbon dioxide directly to methyl alcohol or methane would be highly desirable. The present technology uses carbon monoxide in syngas, which is readily methanated or converted to methyl alcohol (see Section 3.5.1). The most successful electrochemical CO_2 reduction, however, is still very inefficient, requiring high energy. Effective catalytic or chemical reduction of CO_2, in contrast, can be readily achieved, including superacid catalyzed ionic reduction, which can be carried out even at room temperature, eventually leading to methane (see Section 3.2.2).

Another problem associated with the reduction of atmospheric CO_2 is its low concentration (0.035%). Even large power plants the main carbon dioxide emitters produce only dilute form of CO_2. Efficient and costly recovery technologies (membranes), therefore, are required. In the long run, however, the reduction of carbon dioxide to methyl alcohol or methane will be the ultimate solution of the carbon equilibrium problem, assuming that unlimited, cheap energy (atomic, including fusion?) would become available to produce hydrogen from the oceans.

Catalytic reductions of CO_2 has attracted great attention and numerous reviews,[6–8] monographs,[9–12] and symposia proceedings[13–16] are available.

3.2.1. Catalytic Reduction

The vast material of the chemical reduction of CO_2 that has accumulated since the 1970s or so will be divided here, according to the reduction methods, into two parts: heterogeneous and homogeneous hydrogenations. Further organization is based on the degree of reduction; specifically, processes to afford formic acid and its derivatives, formaldehyde, methanol, and methane and other hydrocarbons will be treated separately.

Heterogeneous Hydrogenation. The heterogeneous metal-catalyzed hydrogenation of CO_2 goes directly to methanol, or methane which is the result of the drastic reaction conditions applied in hydrogenations in the presence of heterogeneous catalysts. Formate, formaldehyde, and formyl species, however, have been shown by spectroscopic techniques to exist as surface intermediates during the hydrogenation of carbon oxides. Two exceptional examples, however, can be mentioned. When CO_2 was hydrogenated on a Pt–Cu on SiO_2 catalyst with an optimum Pt : Cu ratio of 0.03, formaldehyde rather than methanol was mainly produced.[17] In the other case, CO_2 hydrogenation was performed in the presence of ethanol over $Cu–Cr_2O_3$ or $Pd–Cu–Cr_2O_3$ catalysts.[18] The original goal of this study was the synthesis of methanol via ethyl formate in a three-step, one-pot reaction; specifically reduction of CO_2 to formic acid, esterification of the latter with ethanol, and hydrogenolysis of ethyl formate to yield methanol. At low temperature (150°C) and low conversion level (15%), ethyl formate was formed with high (approximately 90%) selectivity.

The hydrogenation of CO_2 to methanol [Eq. (3.2)] is interrelated to the hydrogenation of CO [Eq. (3.3)] through the water–gas shift reaction [Eq. (3.4)]:

$$CO_2 \; + \; 3\,H_2 \; \rightleftharpoons \; CH_3OH \; + \; H_2O \qquad \Delta H_{298} = -25.5 \text{ kcal/mol} \qquad (3.2)$$

$$CO \; + \; 2\,H_2 \; \rightleftharpoons \; CH_3OH \qquad \Delta H_{298} = -21.7 \text{ kcal/mol} \qquad (3.3)$$

$$CO \; + \; H_2O \; \rightleftharpoons \; CO_2 \; + \; H_2 \qquad\qquad\qquad\qquad\qquad (3.4)$$

A significant disadvantage of the CO_2 hydrogenation process as compared to the hydrogenation of CO is that more hydrogen is consumed because of the formation of water.

It is known through isotope labeling studies (see Section 3.4.1) that in the hydrogenation of CO the primary source of methanol is CO_2. It is also known, however, that traditional three-component Cu catalysts (Cu–ZnO–Al$_2$O$_3$ and Cu–ZnO–Cr$_2$O$_3$) exhibiting high activity and selectivity in the hydrogenation of CO to methanol show low activity and selectivity in the hydrogenation of pure CO$_2$. Because of the water–gas shift reaction, however, the strict differentiation between CO hydrogenation and CO$_2$ hydrogenation is rather arbitrary. With respect to the active sites involved in the reduction of CO$_2$ to methanol over Cu–ZnO-based catalysts, various models have been suggested.

1. The *morphology model* proposed by Campbell,[19] Waugh,[20] and Topsøe[21,22] emphasizes the role of the wetting–nonwetting of the copper particles on ZnO significantly affecting the methanol synthesis activity. Here, the Cu particle size should also be considered because studies on the morphology effect were carried out with catalysts with low copper loading (1–5 wt%), that is, with high copper dispersion. Industrial methanol synthesis catalysts, in turn, have usually high copper loading (50–70 wt%).
2. Fujitani and Nakamura proposed the *Cu–Zn active-site model*.[23–26] According to this model, the role of the Cu–Zn site is to activate the hydrogenation of the formate species. It is known that formate can form directly from CO$_2$ and H$_2$.[27] Originally the active site was suggested to be the Cu–O–Zn site but it was changed because of the results of surface science studies.[23,24,28]
3. According to Poels, the *interfacial active sites* between Cu and ZnO play the key role in the synthesis of methanol.[29]

Because of the pure performance of traditional Cu catalysts in the hydrogenation of CO$_2$, efforts have been made to find new, more effective catalysts for direct CO$_2$ hydrogenation. The problem is to improve selectivity, specifically, to find catalysts that display high selectivity toward methanol formation and, at the same time, show low selectivity in the reverse water–gas shift reaction, that is, in the formation of CO. It appears that copper is the metal of choice for methanol synthesis from CO$_2$ provided suitable promoters may be added. Special synthesis methods have also been described for the preparation of traditional three-component Cu catalysts (Cu–ZnO–Al$_2$O$_3$ and Cu–ZnO–Cr$_2$O$_3$) to improve catalytic performance for CO$_2$ reduction.

It has been widely reported that Cu–ZnO-based catalysts with additional oxides are among the most useful systems for the catalytic hydrogenation of CO$_2$ to methanol. The best example is a high-performance multicomponent Cu–ZnO–ZrO$_2$–Al$_2$O$_3$–Ga$_2$O$_3$ catalyst prepared by coprecipitation.[30] Addition of a small amount (0.6 wt%) of silica greatly improved the long-term stability of the catalyst.[31–33] It was also found that calcination at 600°C and the removal of Na from the catalyst by washing the precipitate are important requirements to get satisfactory catalyst performance.[32] It was shown that ZrO$_2$ and Al$_2$O$_3$ increase the dispersion of copper particles and the total surface area whereas Ga$_2$O$_3$ increases the activity per unit surface area by regulating the Cu$^+$: Cu^0 ratio. The small amount of silica added to the catalysts was able to suppress the crystallization ZnO that is likely brought about by water produced during the reaction. It appears that silica is able to prevent reduction of the surface area and the Cu surface area.

As these findings indicate, ZnO is not an ideal component for CO_2 hydrogenation catalysts. As an alternative, a $Cu–Ga_2O_3$ catalyst was prepared using ZnO as support, and a silica-supported $Cu–ZnO–Ga_2O_3$ sample was also investigated.[34] The latter catalyst exhibited high selectivity to methanol, the conversion to CO was very low, and only negligible amounts of hydrocarbons were formed. Very small particles of Ga_2O_3 on the catalyst surface controlling the oxidation state of copper are suggested to account for enhanced catalyst performance.

Other studies have demonstrated that zirconia present either as a support or an additive strongly enhances the activity of Cu for the synthesis of methanol from CO_2. Baiker's group tested a large number of various catalysts in the hydrogenation of carbon oxides.[8,35] Cu on ZrO_2 and Ag on ZrO_2 proved to be excellent catalysts for both CO and CO_2 hydrogenation. However, the catalytic behavior is greatly influenced by the preparation method used. Specifically, the methods leading to an intimate mixing of the metal with zirconia (coprecipitation, sol-gel synthesis, and controlled oxidation of bicomponent amorphous metal–zirconium alloys) result in better catalysts, which points to the important role of the interfacial contacts.[36] Cu on ZrO_2 aerogel prepared by supercritical drying exhibited significantly higher CO_2 conversion than did the corresponding Ag catalyst, but the methanol selectivity was very similar.[37] In addition, due to favorable textural properties and high copper surface area, the Cu on ZrO_2 aerogel showed higher activity than other similar catalysts prepared by coprecipitation. It was also found that the addition of Ag to Cu on ZrO_2 improved methanol selectivity.[38]

Titania and the combination of titania and zirconia also bring about improvements when added to Cu on Al_2O_3 or Cu on SiO_2 catalysts.[39–41] For example, a $Cu–Al_2O_3–TiO_2$ catalyst has much smaller Cu crystallites, exhibits an amorphouslike structure, and may be reduced at lower temperature.[41] The maximum methanol selectivity for CO_2 hydrogenation was observed when a 1 : 1 mixture of ZrO_2 and TiO_2 was added to Cu on SiO_2.[40] This was attributed to an appropriate concentration of basic adsorption sites on the surface of the mixed oxides desirable for the adsorption of CO_2.

Spectroscopic techniques, namely, in situ IR investigations and vibrational spectroscopies, allowed investigators to acquire information of the adsorbed species involved in the hydrogenation of carbon oxides.[8,35] Although different interpretations exist with respect to the role of surface formates, there is agreement that a bifunctional mechanism is operative; namely, CO_2 adsorbs mainly on ZrO_2 and hydrogen adsorbs and dissociates on Cu.

A mechanism for CO_2 hydrogenation over Cu on ZrO_2 was presented by Bell and coworkers.[39] CO_2 is suggested to adsorb on zirconia as bicarbonate, which undergoes stepwise hydrogenation to produce formate (**1**), dioxomethylene, and, finally, methoxy species:

Hydrogen is suggested to come from the spillover from copper.[42] The last step is the rapid hydrolysis of the surface methoxy to form the product methanol and restore the surface OH species required for the adsorption of the next CO$_2$. The water–gas shift reaction occurs predominantly on Cu and it is not affected by the presence of ZrO$_2$. The suppressed rate of methanol formation over Cu on ZrO$_2$ from CO as compared to CO$_2$ is ascribed to the absence of water formation, which prevents the more facile release of methoxy by hydrolysis. Hydrogenation of formates detected on Cu on TiO$_2$ was also observed to occur.[43]

According to results coming from Baiker's group, the water–gas shift reaction plays a crucial role in CO$_2$ hydrogenation.[44–46] In the presence of surface hydroxyl groups, that is, over a prereduced catalyst, CO$_2$ is converted to surface carbonate, which is first reduced to yield adsorbed CO and water. Methanol is then generated via surface-bound formaldehyde and methoxy species, which is released from the surface as a result of hydrogenolysis or protolysis. In contrast to the suggestion of Bell, surface formate is believed to be an intermediate for methane formation not involved in the reduction of CO$_2$.

The activity of traditional ZnO-based catalysts may be improved by using new preparation methods. Cu–ZnO–Al$_2$O$_3$ catalysts prepared by a new oxalate gel precipitation exhibited higher activity in CO$_2$ reduction than did those made by coprecipitation, which was attributed to the isomorphous substitution of Cu and Zn.[47,48]

New catalysts such as a Raney Cu–based system containing Zr[49] and ultrafine CuB with Cr, Zr, and Th[50] exhibit good characteristics. The improved catalyst performance observed for a Cu–ZnO catalyst with added Pd is explained by the relative ease of hydrogen dissociation by the incorporated Pd particles and then spillover to the Cu–ZnO.[51]

Catalysts with metals other than Cu have also been prepared and studied. Ru on zeolite NaY prepared by ion-exchange and promoted with Co showed high activity for methanol production.[52] Co showed a similar promoting effect when it was added to Rh on SiO$_2$.[53] The beneficial effect of Co was attributed to the stabilization of the CO species derived from the dissociation of CO$_2$.[54] An iron-based Fischer–Tropsch catalyst was mixed with a Cu-based catalyst and further doped with Pd and Ga to give a novel composite catalyst.[55] By selecting appropriate combinations of the components and reaction conditions, significant deviation from the Schulz–Flory distribution could be achieved, which allowed the highly selective synthesis of ethanol.

New catalysts have been developed to afford coproduction of methanol and dimethyl ether, which requires the use of an active methanol synthesis catalyst and a solid acid component. The transformation of methanol to dimethyl ether shifts the methanol formation equilibrium and thereby allows higher per-pass yields. Hybrid catalysts composed of Cu–ZnO–Al$_2$O$_3$ and silica–alumina[56] or Cu–ZnO–Al$_2$O$_3$–Cr$_2$O$_3$ and H-ZSM-5 zeolite[57] are active and stable catalysts. Cu–Mo on H-ZSM-5 prepared by impregnation is also very effective in the production of dimethyl ether.[58] Yields are usually low but high dimethyl ether selectivities are achieved.

Methanation, that is, the selective hydrogenation of CO_2 to methane [Eq. (3.6)], is a potentially important transformation, although at present there is no practical reason for producing methane from carbon dioxide and hydrogen:

$$CO_2 \; + \; 4\,H_2 \; \rightleftharpoons \; CH_4 \; + \; 3\,H_2O \qquad \Delta H_{298} = -55.3 \text{ kcal/mol} \qquad (3.6)$$

Studies, nevertheless, have been carried out to develop suitable catalysts and find effective reaction conditions.

Among all catalysts studied, Ru, Rh, and Ni supported on metal oxides exhibit the highest activity, and Ru has been reported to be the most selective.[59–61] The supports exert a marked influence on both the adsorption and specific activity of the metals. In most cases, TiO_2 is the best support material to achieve good catalyst performance,[62,63] which is attributed to a strong electronic interaction between the metal and the support.

Carbon dioxide usually adsorbs dissociatively on noble metals, and hydrogen facilitates the adsorption.[64,65] On supported Ru, Rh, and Pd catalysts, chemisorption of CO_2 in the presence of hydrogen leads to the formation of surface carbonyl hydride. IR spectroscopic measurements revealed that carbonyl hydride and formate as well as surface carbon are present during the catalytic reaction.[60,62,66,67] On Ru on TiO_2, CO and formate intermediates were identified.[66] Dissociation of CO_2 promoted by hydrogen, the dissociation of surface carbonyl hydride into reactive surface carbon, and the hydrogenation of the latter are proposed as important steps in the hydrogenation process. Surface formate located on the support was shown to be inactive undergoing decomposition, whereas formate located at the interface is the precursor for CO, which eventually leads to methane.[66] Dissociation of intermediate formaldehyde was shown to be the rate-determining step over Rh on SiO_2.[60]

A metal–support interaction between Ru and the support that modifies the electronic state of the metal was found to be important when Ru was supported on H-ZSM-5.[68] Doping Rh on TiO_2 with W^{6+}, again, modifies the electronic state of Ru via electronic interactions. The resulting changes in the adsorption strength and the surface concentration of hydrogen and various intermediates lead to an enhanced catalytic activity.[69] Ru on CeO_2 is also highly active, and promotion of Ru on SiO_2 with CeO_2 results in substantial increase in catalytic activity.[70] On increasing reduction temperature progressive reduction of CeO_2 takes place. Activation of CO_2 occurs with its interaction with Ce^{3+} producing CO and surface carbonaceous species.[71] The role of support was shown in a study using Rh on SiO_2 with different metal loadings.[72] Over the 1% Rh on SiO_2 sample CO formed as a result of dissociative adsorption reacted with surface hydroxyl groups of silica to form Rh carbonyl clusters. Because of the considerably higher metal coverage, the concentration of hydroxyl groups on the 10% Rh on SiO_2 catalyst is much lower. CO, consequently, is not able to form carbonyl clusters, but reacts with hydride species, resulting in methane formation.

It is interesting to note that the methanation of CO_2, which requires the additional step of CO_2 dissociation to adsorbed CO, exhibits higher activity toward methane and has lower activation energy as compared to CO methanation. A partial explanation is that the coverage of the catalyst surface exposed to CO$_2$ and H$_2$ by surface intermediates is significantly lower and, at the same time, the hydrogen coverage is significantly higher than during CO hydrogenation. Higher hydrogen coverage induces enhanced removal of surface carbonaceous species and, consequently, results in better catalyst performance.

Methanation of CO$_2$ over Pd on zirconia and Ni on zirconia catalysts derived form amorphous Pd–Zr-, Ni–Zr-, and Ni-containing multicomponent alloys prepared by controlled oxidation–reduction treatment or generated under reaction conditions have been studied in detail.

In a comprehensive study amorphous Ni-containing alloy precursors underwent oxidation and subsequent reduction pretreatment to form active catalysts that showed exclusive methane formation.[73] Furthermore, the highest methanation rates were observed for catalysts containing 40–50 at% Zr. The methanation rates of all samples were higher than that of 3% Ni on ZrO$_2$ catalyst prepared by impregnation. Two types of zirconia were identified on the working catalyst: metastable tetragonal and stable monoclinic. When 5–10% of rare-earth elements, particularly Sm, were added to the original alloys, remarkable increases in conversion were detected.[74–76] Here, tetragonal zirconia was formed predominantly during the oxidation–reduction pretreatment. The improved catalytic activity was associated with the stabilization of tetragonal nanograined zirconia by Sm, uniform dispersion of Ni, and increased surface area.[77] The development of these highly active and selective Ni catalysts is a part of a project aimed at global carbon recycling.[78]

When used in the hydrogenation of carbon oxides, amorphous Ni–Zr and Pd–Zr alloys are slowly transformed into microporous solids containing metal particles embedded into a ZrO$_2$ matrix.[35,79,80] The derived catalysts exhibit high activity and selectivity in CO$_2$ methanation. Surface formate is believed to be the pivotal intermediate leading to methane formation, although it was detected only on Ni. Interestingly, there is no interconversion between CO$_2$ and CO via the water–gas shift reaction over Ni on ZrO$_2$, in contrast to what was observed for Pd on ZrO$_2$ and Cu on ZrO$_2$.

In a rather similar fashion a LaNi$_5$ intermetallic compound also undergoes marked structural changes under conditions of CO$_2$ hydrogenation to form metallic Ni, La(OH)$_3$, and LaCO$_3$OH.[81] New Ni species, possibly Ni–O–La, are thought to be responsible for the high activity of the resulting catalyst as compared with a Ni on La$_2$O$_3$ sample prepared by coprecipitation. Selective formation of methane was observed over a mixed Raney Fe/Ru on C catalyst used in the methanation of CO$_2$ with water as the hydrogen source. A formate intermediate was suggested to be involved in this unique reaction.[82]

Fischer–Tropsch synthesis catalysts when applied in the hydrogenation of CO$_2$ perform rather poorly, yielding only small amounts of C$_{2+}$ hydrocarbons. The development of new catalysts, therefore, is necessary for the production of higher hydrocarbons through the direct hydrogenation of CO$_2$. Effective hybrid or composite catalyst systems may be used for this purpose by combining CO$_2$ hydrogenation catalysts with an acidic component. Two basic combinations have been studied.

Fe-based Fischer–Tropsch catalysts and metal oxides (TiO_2, ZrO_2, MgO, Al_2O_3)[83–85] or zeolites (HY, H-ZSM-5, H-ZSM-11)[86,87] exhibit promising characteristics. Cu catalysts used in the hydrogenation of carbon oxides combined with zeolites (HY, H-ZSM-5, SAPO), that is, the coupling of methanol synthesis and the methanol-to-gasoline reaction, are also effective.[88–91] The most interesting observation made with the Fe-containing catalysts is that they allow the synthesis of branched hydrocarbons with high selectivity. A multicomponent $Cu–ZnO–Cr_2O_3$ catalyst promoted with K and Fe^{3+}-exchanged ZSM-11 produced mainly $C_5–C_{11}$ hydrocarbons with iso : normal ratios of approximately 4.[91] Fe–Zn–Zr oxide/HY composite catalysts proved to be highly selective in the production of isobutane, whereas the hybrid catalyst with H-ZSM-5 gave mainly iso-C_{5+} hydrocarbons (iso : normal ratio $= 9$).[86] $Cu–ZnO–ZrO_2$ with zeolites showed significant changes in product composition depending on the acidity of the zeolite used; the composite with H-ZSM-5 gave ethane as the main product, while that with SAPO was more selective in the formation of $C_3–C_4$ hydrocarbons.[90] $Cu–ZnO–Cr_2O_3$ with HY, when doped with alkaline metals, especially Cs, showed increasing selectivities toward the formation of alkenes.[89] The addition of K to a $Cu–Fe–Al_2O_3$ catalyst, applied in a slurry reactor, also increases activity and hydrocarbon production, and shifts product distribution to alkenes.[92] An exceptional example is the hydrogenation of CO_2 in the supercritical state over Fe_3O_4 to form multicarbon molecules.[93] The transformation yield of CO_2 to phenol, for example, is 7.6%.

Homogeneous Hydrogenation. The mild reaction conditions used in the homogeneous hydrogenation of CO_2 catalyzed by transition metal complexes allows the partial hydrogenation of CO_2 to yield formic acid and derivatives:

$$CO_2 \ + \ H_2 \ \rightleftharpoons \ HCOOH \qquad \begin{array}{l} \Delta H_{298} = -7.6 \text{ kcal/mol} \\ \Delta G_{298} = -7.9 \text{ kcal/mol} \end{array} \qquad (3.7)$$

Further reduction of formic acid under homogeneous conditions is a more difficult task, and only very few examples are known reporting the formation of methanol and methane catalyzed by Pd[94] and Ru[95] complexes in very low yields.

The reduction of CO_2 is an exothermic but strongly endoergic equilibrium process. The equilibrium lies far to the left, which is due to the involvement of two gaseous reactants and a liquid product with strong intermolecular hydrogen bonds. The solvent used in the reaction is highly important. Solvation not only lowers the entropy but also breaks up the hydrogen bonds. The equilibrium may also be shifted to product formation by trapping formic acid in the form of derivatives. Addition of a base greatly improves the enthalpy of the reaction, while dissolution of the gases improves the entropy.

The first homogeneously catalyzed synthetic CO_2 reduction was reported in 1976 by Inoue[96] to show that [RhCl(PPh$_3$)$_3$] (Wilkinson's complex) in benzene solution in the presence of Et$_3$N produced formic acid. Small amounts of water increased the activity. The accelerating effect of added water in organic solvents has also been observed in other systems.[97,98] According to one study, the reaction barrier to the insertion of CO_2 into the metal–H bond to form metal formate is significantly reduced in the presence of water. This is due to the enhanced electrophilicity of the carbon of CO_2 as a result of the formation of hydrogen bonds between its oxygen atoms and water.[99]

Since the work of Inoue a wide variety of metal complexes have been shown to be highly active in CO_2 reduction. These are usually hydrides or halides with phosphines as neutral ligands, and Rh and Ru proved to be the most active metals.[9,100–104] Variation of the ligand has a considerable effect on catalytic activity.

A significant breakthrough in the early 1990s was the observation by Noyori and coworkers that hydrogenation of CO_2 was markedly enhanced in supercritical CO_2.[105] They demonstrated that dissolving Et_3N and $[RuH_2(PMe_3)_4]$ in supercritical CO_2 leads to rapid formation of formic acid as the 2 : 1 adduct with the amine at a turnover frequency (TOF) of 680 h^{-1} at 50°C.[106] Adding dimethyl sulfoxide to the reaction mixture improves the TOF value by up to 4000 h^{-1}.[98] The synthesis of dimethyl formamide[98,107] and methyl formate[98,108] was also possible when reductions were carried out at 100°C in the presence of Me_2NH or methanol, respectively. Extremely high turnover numbers up to 420,000 were measured for the former process. Because of its better stability at higher temperatures, the complex $[RuCl_2(PMe_3)_4]$ was a better catalyst in these reductions.

Subsequent studies by Baiker and coworkers demonstrated that silica-supported hybrid materials derived by the sol-gel method are very effective for the synthesis of dimethyl formamide.[35,109] They could increase the efficiency by applying bidentate ligands; for instance, a TOF value of 360,000 h^{-1} was measured with the complex $[RuCl_2(dppe)_2]$ [dppe $= Ph_2P(CH_2)_2PPh_2$].[110] Even further improvements were reported when the ligand was dppp [$Ph_2P(CH_2)_3PPh_2$] and the textural properties ware optimized.[111] Further details of the preparation and structural characterization of these materials and their application in the synthesis of dimethyl formamide and methyl formate have been reported.[112,113]

Other effective catalysts to synthesize formic acid are Rh complexes with formate[114] or hexafluoroacetylacetonate[115] ligands. New results of theoretical and mechanistic studies for these and other systems have been disclosed.[98,103,116–121]

Water-soluble transition metal complexes, which are effective catalysts in other hydrogenation processes, were found to be effective catalysts in CO_2 hydrogenation. The first report disclosed the application of Rh complexes with water-soluble phosphine ligands in water–amine mixtures to afford formic acid.[122] Water-soluble Ru–phosphine[121,123,124] and Rh–phosphine[123,124] complexes were used in aqueous solution to hydrogenate CO_2 or HCO_3^- to formic acid or $HCOO^-$, respectively.

Another study describes the synthesis of N,N-dipropyl formamide using the room-temperature ionic liquid [bmim][PF_6] and supercritical CO_2 under biphasic conditions catalyzed by $[RuCl_2(dppe)_2]$ at 80°C.[125]

3.2.2. Other Reductions

Ionic Reduction. The remarkable hydrogenating ability of sodium borohydride–triflic acid was reported in the hydrogenation of CO_2.[126] Methane as the sole product could be synthesized at ambient temperature and modest pressure (50 atm). Mechanistically, the superacid protonates the oxygen of CO_2, forming a carbocationic intermediate that is then quenched by the hydrogen donor source.

Electrochemical and Electrocatalytic Reduction. The one-electron reduction of CO_2 yields the radical anion $CO_2^{\cdot-}$, which reacts with an H$^\cdot$ source to give the formate ion. The reaction, however, is not selective because various other reactions may take place. An alternative and more promising approach is the two-electron reduction of CO_2 in the presence of a proton source to afford formic acid. The latter process requires a considerably lower potential (-0.61 V) than does the one-electron reduction (-1.9 V); consequently, the electrolysis in the presence of catalysts may be performed at lower voltages. The control of selectivity, however, is still a problem, since other two-electron reductions, most importantly reduction to form CO and H$_2$, may also occur.[101,127] The reduction of CO_2 to CO, in fact, is the subject of numerous studies. Electrochemical and electrocatalytic reductions of CO_2 in aqueous solutions have been studied and reviewed.[11,128–130]

A large number of metal complexes have been proved active in electrochemical reduction of CO_2. Among these, certain Re, Ni, and Ru complexes, mostly the type $[Ru(bpy)_2(CO)L]^{n+}$ (bpy = bipyridyne, L = CO or H, $n = 1, 2$), have attracted much attention because of their characteristic reactivity and high efficiency.[129,131] The catalytic cycles have been elucidated.[132] These complexes, however, cannot be used in aqueous solution because of the competing hydrogen evolution. Re complexes, in turn, when incorporated into Nafion membrane, proved to be efficient in formic acid production.[133]

Other metal complexes such as 2,2′-bipyridine complexes of Rh and Ir are efficient electrocatalysts for the reduction of CO_2 in acetonitrile.[134] In the production of formate the current efficiency is up to 80%. Electrochemical reduction catalyzed by mono- and dinuclear Rh complexes affords formic acid in aqueous acetonitrile, or oxalate in the absence of water.[135] The latter reaction, that is, the reduction of CO_2 directed toward C–C bond formation, has attracted great interest.[131] An exceptional example[136] is the use of metal–sulfide clusters of Ir and Co to catalyze selectively the electrochemical reduction of CO_2 to oxalate without the accompanying disproportionation to CO and CO_3^{2-}.

There are certain combinations of electrode materials, solvents and operating conditions, which allow the reduction of CO_2 to afford hydrocarbons. It was concluded in a comparative study using many different metal electrodes in aqueous KHCO$_3$ solution that either CO (Ag, Au) or formic acid (In, Sn, Hg, Pb) are produced as a result of the reduction of CO_2, and Cu has the highest electrocatalytic activity for the production of hydrocarbons, alcohols, and other valuable products.[137] Most of the studies, since the mid-1990s, consequently, have focused on the further elucidation of electrocatalytic properties of copper.

Comparing the effects of alkali cations of various sizes applied in reduction of CO_2 in HCO$_3^-$ solution with a Cu cathode, Na$^+$, K$^+$, and Cs$^+$ were shown to favor the formation of hydrocarbons.[138] The selectivity of ethylene formation surpasses that of methane with increasing cation size. Deactivation of the Cu cathode observed in aqueous solution could be decreased by applying anodic pulses or by addition of Cu^+ ions into the electrolyte.[139] The cation of the supporting electrolyte was also found to play an important role in reduction using a Cu electrode in methanol; in the presence of tetrabutylammonium salts the main product was CO, whereas methyl formate was formed predominantly when Li salts were used.[140] The methyl group of methyl formate derived from methanol and the formyl group originated from CO_2.[141] Formic acid was shown to be the main product when reductions were carried out in KOH–methanol electrolyte with Pb[142] and Ti or hydrogen-storing Ti + H electrode.[143]

Improved yields of oxalate, glyoxalate, and glycolate were detected in the reduction of CO_2 in methanol containing hydroxylamine and Me_4NCl.[144] The formation of oxygen-containing C_2 compounds was also reported in other systems. A unique system applying a dual-film electrode consisting of Prussian blue and polyaniline on Pt doped with an Fe complex using a solar cell as the energy source afforded C_1–C_3 compounds including lactic acid, ethanol, and acetone.[145,146] Hydrogenation on a cathodized WO_3/polyaniline/polyvinylsulfate-modified electrode in aqueous solution occurs through the reaction with the adsorbed hydrogen atoms generated on the cathodized electrode.[147] The products are lactic acid, formic acid, ethanol, and methanol.

Many investigators have actively studied the electrochemical reduction of CO_2 using various metal electrodes in organic solvents because these solvents dissolve much more CO_2 than water. With the exception of methanol, however, no hydrocarbons were obtained. The solubility of CO_2 in methanol is approximately 5 times that in water at ambient temperature, and 8–15 times that in water at temperatures below 0°C. Thus, studies of electrochemical reduction of CO_2 in methanol at -30°C have been conducted.[148–150] In methanol-based electrolytes using Cs^+ salts the main products were methane, ethane, ethylene, formic acid, and CO.[151] This system is effective for the formation of C_2 compounds, mainly ethylene. In the LiOH–methanol system, the efficiency of hydrogen formation, a competing reaction of CO_2 reduction, was depressed to below 2% at relatively negative potentials.[152] The maximum current efficiency for hydrocarbon (methane and ethylene) formation was of 78%.

The reduction of CO_2 was also accomplished on a Ni–zeolite Y catalyst at 600°C.[153] CO_2 conversion and methane yield reached 100% and 80%, respectively. Reduction to formic acid without a protic solvent was also demonstrated using atomic hydrogen penetrating through palladized Pd sheet electrodes.[154] Interestingly, the selectivity of CO_2 reduction can be significantly altered by changing the CO_2 pressure. The reduction of CO_2 under high (<50 atm) pressure on a Pt gas diffusion electrode produced methane at high faradaic efficiency, whereas formic acid was the main product at 1 atm.[155]

Photoreduction. Only very strong reducing agents are able to perform the one-electron reduction of CO_2 to $CO_2^{\cdot-}$, which is difficult to produce by photochemical means. Two-electron reduction, that is, the formation of formic acid (formate), in turn, is energetically highly favorable.

Photocatalytic reduction of CO_2 can be accomplished by suspending photosensitive semiconductor powders in aqueous solutions under irradiation, usually using UV light.[129,156] Photoreduction of CO_2, however, is in competition with H_2 formation due to water decomposition, and leads to mixtures of reduced carbon products. Selectivity, therefore, is one of major problems of these processes.

In an early study TiO_2, ZnO, CdS, GaP, and SiC were applied in the reduction of CO_2 to form organic compounds such as formic acid, formaldehyde, methanol, and methane.[157] Titanium dioxide appears to be the best and, therefore, most studied material for photoreduction. It was applied in supercritical CO_2 for the synthesis of formic acid.[158] Acidic solutions instead of pure water was found to be preferable for formic acid formation. It was shown that mixing metal powders into TiO_2 suspensions promotes the photocatalytic reaction. This was explained to be due to the rapid transfer of excited electrons to the metal particles; consequently, the separation between holes and electrons is enhanced. Further examples of this concept are the use of TiO_2 loaded with Ru/RuO_x,[159] or copper[160–162]. In these cases, hydrocarbons are isolated as the main products. When a Pd–TiO_2 catalyst was supported on acidic oxides, improved selectivity of the formation of C_1 products was observed, whereas basic oxides favored the formation of C_1–C_3 compounds.[163] The best C_1 selectivity was found for Pd–TiO_2–Al_2O_3. Rh on TiO_2 doped with W^{6+} produced mainly formic acid. In sharp contrast, exclusive formation of methanol was observed when it was reduced prior to application.[164] In an early study methane formation was reported through the photocatalytic reaction of gas-phase water and CO_2 adsorbed on $SrTiO_3$ surface in contact with Pt foil.[165]

Since the concentration of CO_2 in water is rather low, increasing pressure may enhance the performance of photoreduction. Studies using TiO_2 in water showed that hydrocarbons such as methane and ethylene, which were not produced at ambient pressure, were obtained under high pressure.[161,166] Methane was formed as the main reduction product when the reduction was performed in isopropyl alcohol, a positive hole scavenger.[167]

As mentioned, the photocatalytic activation of TiO_2 requires UV irradiation, and hence the semiconductor performance in the solar spectrum is inefficient. A solution to switch the photocatalytic activity to the visible spectral region was described by covalent attachment of an eosin dye monolayer to the semiconductor oxide particles of a Pd–TiO_2 catalyst.[168] The improved photocatalytic activities, specifically, the efficient formation of formate, are attributed to the effective injection of electrons from the excited dye into the semiconductor conduction band.

TiO_2 coated with potassium ferrocyanide proved to be an effective catalyst for the reduction of CO_2 to formic acid and formaldehyde.[169] A very stable and reproducible catalytic system was prepared by immobilizing Ni^{2+} and Ru^{2+} complexes into Nafion membrane, which was used for the selective reduction of CO_2 to formic acid.[170] Formic acid was again formed when Zn and Co phthalocyanines were adsorbed onto a Nafion membrane on irradiation with visible light in acidic aqueous solution containing triethanolamine as a hole scavenger.[171] Cobalt corrins (B_{12}) acting as homogeneous catalysts in acetonitrile–methanol solutions induced the formation of formic acid and CO.[172]

Formic acid, methyl formate, and CO were detected when photoreduction was performed in Ti silicate molecular sieve using methanol as electron donor.[173] Mechanistic studies with labeled compounds indicated, however, that CO originates from secondary photolysis of formic acid, whereas methyl formate emerges mainly from the Tishchenko reaction of formaldehyde, the initial oxidation product of methanol.

Aqueous solutions of CO$_2$ containing Me$_4$NCl were photolyzed in the presence of colloidal ZnS to yield tartaric acid, oxalic acid, formic acid, and formaldehyde.[174] Irradiating with visible light of ZnS loaded onto large-surface-area SiO$_2$ (13% coverage) affords formate using 2,5-dihydrofuran as reducing agent.[175] Analogous observations were made with 40% CdS on SiO$_2$. The photocatalytic reduction process may be combined with electrochemical reduction to perform photoelectrochemical reduction of CO$_2$.[129] Among the first examples, CO$_2$ was shown to undergo reduction to formic acid at p-type GaP photocathode in the electrochemical photocell with a fairly high conversion efficiency,[176] whereas formaldehyde and methanol as major products were detected at p-GaP in 0.5 M H$_2$SO$_4$.[157] Pretreatment of p-GaP by dipping in a solution of CuSO$_4$ caused an enhancement in the production of formic acid.[177]

Despite extensive studies, the photovoltage or the solar-to-chemical energy conversion efficiency still remains relatively low. The main reason is that it is very difficult to meet all requirements for high efficiency. For example, high catalytic activity and sufficient passivation at the electrode surface are incompatible. It was found, however, that a semiconductor electrode modified with small metal particles can meet all the requirements and thus becomes an ideal type semiconductor electrode. Cu, Ag, and Au were chosen because they were reported to work as efficient electrocatalysts for the CO$_2$ reduction. p-Si electrodes modified with these metals in CO$_2$-staurated aqueous electrolyte under illumination produce mainly methane and ethylene.[178] This is similar to the metal electrodes but the metal-particle-coated electrodes work at approximately 0.5 V more positive potentials, contrary to continuous metal-coated p-Si electrodes.

Enzymatic Reduction. Enzymatically coupled sequential reduction of carbon dioxide to methanol has been reported by using a series of reactions catalyzed by three different dehydrogenase enzymes. Overall, the process involves an initial reduction of CO$_2$ to formate catalyzed by formate dehydrogenase, followed by the reduction of formate to formaldehyde by formaldehyde dehydrogenase and, finally, formaldehyde is reduced to methanol by alcohol dehydrogenase.[179] Reduced nicotinamide adenine dehydrogenase acts as terminal electron donor for each dehydrogenase-catalyzed reduction. When the results found in solution experiments were compared to those determined by using the four-component enzyme system immobilized in a silica solgel matrix, enhanced production of methanol was found; the yield is low in the solution phase (10–20%), whereas the production of methanol for the immobilized system ranges from 40 to 90%, indicating that the overall equilibrium is shifted more toward the products. The enhancement of methanol production may be attributed to the confinement and matrix effects.

Enzymes may also be utilized in photo- or electrochemical CO_2 reductions. ZnS microcrystallites in the presence of methanol dehydrogenase produced methanol with high (5.9%) quantum efficiency.[180] Formate dehydrogenase, when combined with a p-InP semiconductor, effected the two-electron reduction of CO_2 to formic acid.[181] Electrocatalysis of CO_2-saturated phosphate buffer solutions containing formate dehydrogenase yielded formate with current efficiencies as high as 90%.[182] At relatively high concentrations of the enzyme methanol production also began to occur.

3.3. FISCHER–TROPSCH CHEMISTRY

The reaction that takes its name from the work of Fischer and Tropsch was first described[183,184] in the 1920s. However their work was preceded by two decades by the report of Sabatier and Senderens, who in 1902 reported the formation of methane when a stream of hydrogen and carbon monoxide was passed over a nickel surface.[185] Fischer and Tropsch, in contrast, observed that carbon monoxide and hydrogen reacted over an iron surface to produce a variety of organic compounds. In their first published work they used alkalized iron turnings as catalyst at high temperature and pressure (about 400°C, 100–150 atm) and observed the formation of oxygenated compounds.[183] This became known as the *Synthol* process. By lowering the pressure to 7 atm, they formed increasing amounts of hydrocarbons.[184] Certain iron and cobalt catalysts were later found to be active at atmospheric pressure.

The Fischer–Tropsch synthesis is a rather nonselective process. The products obtained include alkanes and alkenes with a very broad composition, and oxygen-containing compounds, mostly alcohols, carbonyl compounds, acids, and esters. 1-Alkenes, the main isomeric olefins obtained, as well as alcohols, are considered to be the primary products. Most compounds are linear with only a small amount of branched hydrocarbons. Only methyl branching occurs, with the methyl group distributed randomly along the chains. The fraction of dimethyl-substituted compounds is very small, and compounds with quaternary carbon atoms are not formed. Under industrial conditions primarily linear, saturated hydrocarbons are produced with only little oxygenated compounds. Product compositions, however, may be substantially varied by catalysts and reaction conditions. Presently existing refineries could not handle synthetic Fischer–Tropsch feeds for processing adding to the enormous capital cost of the process if it were utilized.[186]

The basic transformations of Fischer–Tropsch synthesis may be generally summarized as in Eqs. (3.8) and (3.9):

$$2n\,CO \;+\; n\,H_2 \;\longrightarrow\; \text{—}[CH_2]_n\text{—} \;+\; n\,CO_2 \qquad \Delta H = -39.8\,\text{kcal/mol} \qquad (3.8)$$

$$n\,CO \;+\; 2n\,H_2 \;\longrightarrow\; \text{—}[CH_2]_n\text{—} \;+\; n\,H_2O \qquad \Delta H = -48.9\,\text{kcal/mol} \qquad (3.9)$$

These indicate two different ways of carbon monoxide conversion; both processes are highly exothermic. Iron catalyzes the transformation according to Eq. (3.8), whereas Eq. (3.9) represents the chemistry characteristic of cobalt. In fact, Eq. (3.9) is the primary synthesis reaction on both metals. Over iron catalyst, however, water reacts further with a second molecule of CO to form CO_2 called the *water–gas shift reaction* [Eq. (3.4)]. Equation (3.8), thus is the sum of Eqs. (3.9) and (3.4).

Other secondary reactions taking place under operating conditions are the *Boudouard reaction* [Eq. (3.10)], coke deposition [Eq. (3.11)], and carbide formation [Eq. (3.12)]:

$$2\,CO \longrightarrow C + CO_2 \qquad (3.10)$$

$$CO + H_2 \rightleftharpoons C + H_2O \qquad (3.11)$$

$$x\,CO + y\,M \longrightarrow M_yC_x \qquad (3.12)$$

Methane formation [Eq. (3.1), reverse process] is an undesirable side-reaction of the Fischer–Tropsch synthesis. Most of the original research on Fischer–Tropsch chemistry was carried out before World War II in Germany, where the lack of crude oil demanded alternative sources for liquid fuels.[187,188] The first commercial catalyst developed in the 1930s was cobalt and thoria supported on kieselguhr operated in a fixed-bed reactor at about 200°C and atmospheric pressure with an H_2 : CO ratio of 2. The catalyst composition was later changed to include magnesia, which resulted in decreased solid paraffin production. At medium pressure (5–20 atm) higher hydrocarbon yields with different product compositions could be achieved. Synthesis gas used in Fischer–Tropsch plants originated from coal gasification processes, as coal was available as raw material. The German brown coal, however, is high in sulfur, and consequently it was necessary to develop catalysts that tolerated it.[189,190]

The first plant came into operation in 1936, and eight others were built until the end of the war in Germany. Plants were also constructed in France and Japan. All plants used the so-called *Ruhrchemie process*. After World War II the commercial Fischer–Tropsch synthesis was terminated in Germany. Research, however, still continued mainly in the United States at the U.S. Bureau of Mines. This included mechanistic studies[191–193] and catalyst[191–198] and process developments.[196–199] A commercial plant with a fluidized-bed iron catalyst was built in 1951 and operated until 1957 (Carthage Hydrocol). At this time cheap Middle East crude oil rendered Fischer–Tropsch oil synthesis uneconomical. South Africa, however, facing a worldwide embargo in the 1950s, established a Fischer–Tropsch industry. The oil embargo of the 1970s brought about the revival of research and development (R&D) interest in Fischer–Tropsch synthesis. Catalyst characterization and chemisorption studies with spectroscopic techniques,[200–204] kinetic and mechanistic studies[202–208] including the use of labeled compounds, and new ideas originating from coordination and organometallic chemistry[209–212] resulted in new, deeper insights into the chemistry of the process. Significant, multi-billion-dollar demonstration plants were built and operated. More recent research efforts are aimed at increasing the selectivity of Fischer–Tropsch synthesis to produce specific products used as chemical feedstock.

At present South Africa is the only country where Fischer–Tropsch synthesis is commercially operated.[213–216] The first plant, known as Sasol One, started producing hydrocarbons in 1955. It uses two different reactors.[215–218] A fixed-bed reactor with a precipitated iron on silica catalyst yields mainly high-molecular-weight saturated hydrocarbons (diesel oil and waxes) at 220–250°C and 27 atm. The other, a transported (circulating) fluidized-bed reactor (also known as *Synthol entrained-bed reactor*), produces primarily gaseous hydrocarbons and gasoline rich in olefin utilizing a fused iron catalyst (320–350°C, 22 atm). Two other, much larger plants (Sasol Two, 1980, Sasol Three, 1982) use the fluidized-bed technology. The Sasol Fischer–Tropsch process provides South Africa with about 50% of its oil need estimated at 60,000 barrels (bl)/day, about the same capacity as World War II production was at its peak in wartime Germany. The investments involved exceed $U.S. 20 billion. In comparison, the U.S. oil consumption is about 16–17 million bl/day. Thus, if Fischer–Tropsch synthetic oil would be considered eventually to replace oil, some 300 Sasol-size plants would be needed, a highly unlikely scenario.

3.3.1. Catalysts

Most Group VIII metals are active in the reduction of carbon monoxide, but they form different products. Methanation, hydrocarbon formation, and methanol formation are the characteristic major transformations. The two classical metals, iron and cobalt, used in commercial Fischer–Tropsch operations are capable of producing hydrocarbons at atmospheric pressure. Nickel is also active in Fischer–Tropsch chemistry, yielding, however, primarily methane. Ruthenium is very active in methane formation at temperatures as low as 100°C and moderate pressures, whereas high-molecular-weight hydrocarbons are formed at higher pressures. In the range of 1000–2000 atm a product termed *polymethylene*, equivalent to high-density polyethylene, is produced.[219,220] Although discovered in the 1930s, the process was never commercialized probably because of the low activity of the catalyst used and the very severe reaction conditions.[221] Iron nitrides are an interesting class of Fischer–Tropsch catalysts.[194] They exhibit high mechanical stability and yield increased amounts of oxygenated compounds, mainly alcohols.

Rhodium is a unique metal since it can catalyze several transformations.[222,223] It is an active methanation catalyst and yields saturated hydrocarbons on an inert support. Methanol is the main product in the presence of rhodium on $Mg(OH)_2$. Transition-metal oxides as supports or promoters shift the selectivity toward the formation of C_2 and higher oxygenates.

Iron catalysts used in Fischer–Tropsch synthesis are very sensitive to conditions of their preparation and pretreatment. Metallic iron exhibits very low activity. Under Fischer–Tropsch reaction conditions, however, it is slowly transformed into an active catalyst. For example, iron used in medium-pressure synthesis required an activation process of several weeks at atmospheric pressure to obtain optimum activity and stability.[188] During this activation period, called *carburization*, phase transformation takes place, and surface carbon, carbides, and oxides are formed. The catalyst undergoes additional slow structural changes during its lifetime with parallel changes in activity and selectivity.

Catalyst preparation is crucial in successful Fischer–Tropsch synthesis. Appropriate catalyst composition and delicate pretreatment and operating conditions are all necessary preconditions to achieve the desired results. Catalyst disintegration brought about by oxidation and carbide formation is a serious problem that can be prevented only by using catalysts with adequate chemical and mechanical stability under appropriate operating conditions.

Both cobalt and iron require chemical promotion to exhibit steady activity and selectivity. These are achieved by adding an optimum amount of potassium to iron. Electron donation from potassium to iron is assumed to weaken the carbon–oxygen and iron–hydrogen bonds, and strengthen the iron–carbon bond.[224] These changes result in increased CO adsorption, increased probability of chain growth, and decreased hydrogenation ability. The overall effect on product distribution is decreased methane and increased oxygenate production. Paraffin formation is also decreased, resulting in enhanced alkene selectivity. The product has a higher average molecular weight. Other possible factors include the enhanced dissociation of CO due to its direct interaction with adsorbed potassium[225] and enhanced migratory insertion in propagation[212] brought about by surface K^+.

Catalyst composition also depends on the type of reactor used. Fixed-bed iron catalysts are prepared by precipitation and have a high surface area. A silica support is commonly used with added alumina to prevent sintering. Catalysts for fluidized-bed application must be more attrition-resistant. Iron catalysts produced by fusion best satisfy this requirement. The resulting catalyst has a low specific surface area, requiring higher operating temperature. Copper, another additive used in the preparation of precipitated iron catalysts, does not affect product selectivity, but enhances the reducibility of iron. Lower reduction temperature is beneficial in that it causes less sintering.

3.3.2. Mechanism

The Fischer–Tropsch synthesis can be considered as the polymerization of carbon monoxide under reductive conditions. Product distributions may be predicted by assuming that chain growth occurs via the addition to the growing chain of one carbon atom at a time. The statistics of chain length distribution usually give a linear correlation between the carbon number n and the mole fraction of molecules containing n carbon atoms when an appropriate mathematical formula is used. Known as the *Schulz–Flory* or *Schulz–Flory–Anderson distribution*, it was developed for free-radical polymerizations. A similar treatment of the Fischer–Tropsch synthesis first applied by Anderson[226] usually leads to satisfactory correlations with the further assumption that unbranched 1-alkenes are the primary products. Exceptions are a maximum for methane and a minimum for C_2 compounds. Monotonous decrease is usually observed for higher hydrocarbons, although deviations from the predicted values are observed for compounds with carbon numbers higher than 10.

Considering the complexity of Fischer–Tropsch reaction, it is remarkable that the Schulz–Flory distribution is observed at all. The Fischer–Tropsch synthesis is not a simple polymerization, since the "monomer" itself is formed in a multistep pathway, discussed later. Further complications may arise from the heterogeneous nature of the process. The deviation of the higher-molecular-weight products, for instance, was explained by invoking the participation of two types of active site with different chain growth probability factors.[227] Transport limitations may also contribute to this phenomenon.[228] Retarded diffusion, for example, enhances the readsorption of olefin products. As a result, they may initiate new chains, eventually leading to the formation of heavier products and a higher fraction of paraffins.

The question of the mechanism of Fischer–Tropsch reaction is of considerable controversy. Three principal routes for product formation have been proposed: the carbide mechanism, the hydroxymethylene mechanism, and the CO insertion mechanism. Numerous modifications were also introduced in attempts to account for some details in the complex chemistry of the process.[205,207,208,211,229–233]

The carbide mechanism was first suggested by Fischer and Tropsch themselves in their first publications.[184] They proposed that a carbide-rich intermediate was the source of the products. This was later elaborated by Craxford and Rideal[234] in a model assuming that surface carbide is hydrogenated to methylene groups that polymerize to form a (CH$_2$)$_n$ macromolecule on the catalyst surface. Cracking would lead to gaseous products and other hydrocarbons. Inconsistent with this model is the small fraction of monomethyl-substituted products and the significant difference between product distributions of Fischer–Tropsch synthesis and hydrocracking. This mechanism does not explain the formation of oxygenated products, either.

An important subsequent observation seemed to indicate that carbides are not reactive under Fischer–Tropsch conditions.[235] When carbon was deposited on a surface by the decomposition of ^{14}CO, labeled carbon was not incorporated into the products. This and other evidence accumulated against the carbide mechanism by the 1950s led to the formulation of other mechanisms. The hydroxymethylene or enolic mechanism[191] assumes the formation via the hydrogenation of carbon monoxide [Eq. (3.13)] of a surface-bound hydroxymethylene species (**2**):

$$
\begin{array}{c}
\underset{M}{\overset{O}{\underset{|}{\overset{||}{C}}}} \ \xrightarrow{\ 2\,H\ } \ \underset{\underset{\mathbf{2}}{M}}{\overset{H}{\diagdown}\underset{||}{\overset{C}{\diagup}}\overset{OH}{}}
\end{array}
\tag{3.13}
$$

$$
\underset{M}{\overset{OH}{\underset{||}{\overset{|}{CH}}}} + \underset{M}{\overset{OH}{\underset{||}{\overset{|}{CH}}}} \ \xrightarrow[-H_2O]{} \ \underset{M\ \ M}{\overset{}{HC-C}}{\overset{OH}{\diagup}} \ \xrightarrow[-M]{\ 2\,H\ } \ \underset{\underset{\mathbf{3}}{M}}{\overset{Me}{\diagdown}\underset{||}{\overset{C}{\diagup}}\overset{OH}{}}
\tag{3.14}
$$

Two hydroxymethylene species react further to form a condensation product via water removal [Eq. (3.14)] and then to yield a C$_2$ unit (**3**) after partial hydrogenolysis. Propagation takes place through the further reaction of **3** with **2**.

Surface intermediates may be hydrogenated to form half-hydrogenated intermediates (**4**), and they may desorb to form aldehydes [Eq. (3.15)] and alcohols
[Eq. (3.16)]:

$$
\begin{array}{c}
R-C\!\!\diagup_{\displaystyle H}^{\displaystyle \lesssim O} \qquad (3.15) \\[2em]
\underset{\displaystyle M}{\overset{\displaystyle \|}{\underset{\displaystyle C}{R\diagdown\diagup OH}}}
\qquad
\underset{\underset{\displaystyle \mathbf{4}}{M}}{\overset{H}{\underset{\displaystyle |}{\underset{\displaystyle CH}{R\diagdown\diagup OH}}}}
\xrightarrow[-M]{H} RCH_2OH \qquad (3.16)
\end{array}
$$

Alkene and alkane formation was suggested to take place through β cleavage and
subsequent hydrogenation [Eq. (3.17)]; chain branching involves the reaction of **1**
with a half-hydrogenated intermediate [Eq. (3.18)]:

$$
\underset{M}{\overset{\|}{\underset{C}{RCH_2CH_2CH_2\diagdown\diagup OH}}}
\longrightarrow
\underset{M}{\overset{\|}{\underset{C}{Me\diagdown\diagup OH}}} + RCH{=}CH_2
\xrightarrow{2\,H} RCH_2CH_3 \qquad (3.17)
$$

$$
\mathbf{2} + \underset{M}{\overset{}{\underset{|}{\underset{CH}{R\diagdown\diagup OH}}}}
\xrightarrow{-H_2O}
\underset{M\ \ M}{\overset{OH}{\underset{|\ \ \|}{R-C-C-H}}}
\xrightarrow[-M]{2\,H}
\underset{M}{\overset{Me}{\underset{|}{R-C-OH}}} \qquad (3.18)
$$

A strong argument in favor of this mechanism is that alcohols can initiate chain
growth.[236] They do not participate in propagation, however, although they should
do so through dehydrocondensation. Additionally, no convincing evidence has been
found to support the existence of the suggested enolic surface species.

By invoking certain analogies with coordination complexes and organometallic
chemistry, a carbon monoxide insertion mechanism was later proposed.[237,238]
Initiation takes place via CO insertion into a metal–hydrogen bond either through
a metal hydrocarbonyl intermediate [Eq. (3.19)] or as adsorbed carbon monoxide
[Eq. (3.20)]. Propagation occurs through CO insertion into a metal–carbon bond
(alkyl migration) to form an acyl intermediate [Eq. (3.21)], which then undergoes
further product-forming transformations:

$$
\underset{M(CO)_n}{\overset{H}{\underset{|}{}}}
\longrightarrow
\underset{M(CO)_{n-1}}{\overset{H\diagdown\diagup O}{C\lesssim}} \qquad (3.19)
$$

$$
\underset{M}{\overset{H}{\underset{|}{}}} + CO_{ads}
\longrightarrow
\underset{M}{\overset{H\diagdown_{\displaystyle C}\diagup O}{}}
\xrightarrow{2\,H}
\underset{M}{\overset{OH}{\underset{|}{CH_2}}}
\xrightarrow[-H_2O]{2\,H}
\underset{M}{\overset{CH_3}{\underset{|}{}}} \qquad (3.20)
$$

$$
\underset{M}{\overset{R}{\underset{|}{}}} + CO_{ads}
\longrightarrow
\underset{M}{\overset{R\diagdown\diagup O}{\underset{|}{C\lesssim}}} \qquad (3.21)
$$

The revival of interest in Fischer–Tropsch chemistry in the 1970s resulted in new observations that eventually led to the formulation of a modified carbide mechanism, the most widely accepted mechanism at present.[202–204,206,214] Most experimental evidence indicates that carbon–carbon bonds are formed through the interaction of oxygen-free, hydrogen-deficient carbon species.[206] Ample evidence shows that carbon monoxide undergoes dissociative adsorption on certain metals to form carbon and adsorbed oxygen:

$$CO_g \longrightarrow CO_{ads} \longrightarrow C + O_{ads} \tag{3.22}$$

The bond strengths of the metal–carbon and metal–oxygen bonds play an important role in the reduction of CO.[239,240] If these bonds are too strong, stable oxides and carbides are formed. Metals that do not dissociate CO easily (Pt, Pd, Ir, Cu) are inactive in the Fischer–Tropsch synthesis and yield methanol instead of hydrocarbons (see Section 3.5.1). Metals that are the most active in CO dissociation (Fe, Co, Ni, Ru) are the most active. Supports and promoters may strongly affect activities and selectivities. Adsorbed oxygen is removed by reacting with CO at low pressure [Eq. (3.23)], and by adsorbed hydrogen at high pressure [Eq. (3.24)]:

$$O_{ads} \begin{cases} \xrightarrow{CO_{ads}} CO_2 & (3.23) \\ \xrightarrow{H_{ads}} HO_{ads} + H_{ads} \longrightarrow H_2O & (3.24) \end{cases}$$

Carbon may exist in different forms, of which carbidic surface carbon (C_s) is the most reactive species. It is readily transformed to carbenic species [Eq. (3.25], which participates in chain growth. This is supported by reevaluation of earlier labeling experiments. New studies showed[241–243] that certain surface carbon species may be very reactive and do incorporate into products mainly into methane. Hydrogenation of adsorbed CO without previous dissociation may also lead to surface carbenic species[244] [Eq. (3.26)]:

$$C_s + x H_{ads} \longrightarrow CH_{x,ads} \tag{3.25}$$

$$CO_{ads} + x{+}2 H_{ads} \xrightarrow{-H_2O} CH_{x,ads} \tag{3.26}$$

Oxygen-containing compounds are suggested to be formed with the participation of nondissociated CO in a parallel pathway.[206]

Efforts have been made over the years to advance a unified concept of Fischer–Tropsch chemistry. The basic problem, however, is that most information comes from studies of different metals. Considering the specificity of metals, it is highly probable that different mechanisms may be operative on different metals. The numerous mechanistic proposals, therefore, may represent specific cases on specific surfaces and may be considered as extremes of a highly complex, widely varied reaction network. A unified concept should include multiple active sites and several surface species participating in parallel product-forming reactions.[208]

3.3.3. Related Processes

The Fischer–Tropsch synthesis to produce hydrocarbons for use as motor fuels is not economical at present (except as mentioned use in South Africa). The synthesis of specific products (light olefins, oxygenates) used as chemical feedstocks, however, may be more attractive.[186] A significant drawback of the Fischer–Tropsch synthesis is its low selectivity due to the Schulz–Flory polymerization kinetics. This sets a serious limitation to produce specialty chemicals since a wide distribution of products is always formed. For example, the maximum obtainable quantity (in weight percents) of C_2–C_4 hydrocarbons, gasoline (C_5–C_{11}), and diesel fuel (C_{12}–C_{17}) is 56%, 47%, and 40%, respectively.[245] In the 1970s and 1980s improved selectivities of specialty chemical production was achieved through catalyst manipulation[228,245–251] and process modifications.[228,249–252] The proper choice of metal loading, dispersion, promoters and supports, alloying, careful control of reaction conditions, and the use of gaseous additives can significantly alter product compositions resulting in higher selectivities.

The direct synthesis of low-molecular-weight olefins and the synthesis of LPG to produce ethylene and propylene by subsequent cracking are the most promising processes.[247,252,253] Promoted iron–manganese catalysts are highly active and selective in alkene production.[221,251,254–257] Iron and iron–manganese supported on silicalite promoted with potassium yield C_2–C_4 olefins with high selectivities.[258,259] One paper has reported the selective synthesis of 1-alkenes over highly reduced zeolite-entrapped cobalt clusters.[260] Mixed metal–zeolite catalysts, in general, produce higher amounts of aromatics and gasoline-range hydrocarbons through the interception of intermediates.[258,261,262] Liquefied petroleum gas, principally ethane and propane, can be produced with a molybdenum-based alkali-promoted catalyst.[263]

The unique selectivity of rhodium to catalyze the formation of oxygen-containing compounds is a potentially promising possibility for practical utilization. When promoted by transition-metal oxides, rhodium is highly selective in the formation of C_2 oxygenates (ethanol, acetaldehyde).[222,264] An even more attractive transformation is the direct synthesis of ethylene glycol from synthesis gas. The reaction was demonstrated using rhodium, cobalt, and ruthenium carbonyls.[261,265] Although highly economical, it is afar from practical application.[266]

Of the technological modifications, Fischer–Tropsch synthesis in the liquid phase (slurry process) may be used to produce either gasoline or light alkenes under appropriate conditions[249,251] in a very efficient and economical way.[267] The slurry reactor conditions appear to establish appropriate redox (reduction–oxidation) conditions throughout the catalyst sample. The favorable surface composition of the catalyst (oxide and carbide phases) suppresses secondary transformations (alkene hydrogenation, isomerization), thus ensuring selective α-olefin formation.[268]

A modification of the Fischer–Tropsch synthesis is the Kölbel–Engelhardt reaction, which converts carbon monoxide and water to hydrocarbons [Eq. (3.27)] by combining two processes:[269]

$$3n\,CO \ + \ n\,H_2O \ \longrightarrow \ -[CH_2]_n- \ + \ 2n\,CO_2 \qquad (3.27)$$

The water–gas shift reaction [Eq. (3.4)] produces hydrogen, which then reacts with carbon monoxide in the normal Fischer–Tropsch reaction [Eq. (3.9)] to yield hydrocarbons. All metals that are active in Fischer–Tropsch synthesis may be employed in the Kölbel–Engelhardt reaction as well. Iron, cobalt, and nickel were found to be the best catalysts. Product composition in the Kölbel–Engelhardt reaction differs significantly from that of Fischer–Tropsch synthesis. Since the hydrogen partial pressure is much lower in the former process, all hydrogen-consuming transformations are retarded. As a result, olefins and alcohols, the primary products in the Fischer–Tropsch synthesis, are not reduced and formed in much larger amounts in the Kölbel–Engelhardt reaction. Even methanation on nickel is suppressed to only about 25% of hydrocarbons produced. Reaction in the liquid phase in a slurry reactor provides the best operating conditions.[251]

Methanation, that is, the transformation of CO to methane[222,270–272] [Eq. (3.1), reverse process], was developed in the 1950s as a purification method in ammonia synthesis. To prevent poisoning of the catalyst, even low levels of residual CO must be removed from hydrogen. This is done by methanation combined with the water–gas shift reaction.[214,273,274] In the 1970s the oil crises spurred research efforts to develop methods for substitute natural-gas production from petroleum or coal via the methanation of synthesis gas.[214,274,275]

The same catalyst compositions used in the more important methane steam reforming [Eq. (3.1), forward reaction], may be used in methanation, too.[222] All Group VIII metals, and molybdenum and silver exhibit methanation activity. Ruthenium is the most active but not very selective since it is a good Fischer–Tropsch catalyst as well. The most widely used metal is nickel usually supported on alumina or in the form of alloys[272,276,277] operating in the temperature range of 300–400°C. A bimetallic ruthenium–nickel catalyst shows excellent characteristics.[278]

The methanation reaction is a highly exothermic process ($\Delta H = -49.2$ kcal/mol). The high reaction heat does not cause problems in the purification of hydrogen for ammonia synthesis since only low amounts of residual CO is involved. In methanation of synthesis gas, however, specially designed reactors, cooling systems and highly diluted reactants must be applied. In adiabatic operation less than 3% of CO is allowed in the feed.[214] Temperature control is also important to prevent carbon deposition and catalyst sintering. The mechanism of methanation is believed to follow the same pathway as that of Fischer–Tropsch synthesis.

Isosynthesis is a potentially important variation of the Fischer–Tropsch synthesis since it yields mainly branched C_4–C_8 paraffins.[187,188,279,280] Most of the pioneering work was carried out by Pichler and coworkers.[281,282] They observed that certain nonreducible tetravalent oxides (thoria, zirconia, ceria) exhibited activity in the reaction producing primarily isobutane. The optimum operating conditions for isosynthesis on thoria are 375–475°C and 300–600 atm. Alcohols are the main products at lower temperature, whereas methane and dimethyl ether predominate at higher temperature. Alumina produces only a minor amount of branched hydrocarbons but enhances the activity of thoria. A 20% alumina–thoria catalyst with 3% K_2CO_3 gave the best results.

Rare-earth oxides (La_2O_3, Dy_2O_3) have been shown to be active in isosynthesis.[283,284] It was also observed that zirconia may catalyze the highly selective formation of isobutylene under appropriate conditions.[285] Two chain growth processes, CO insertion into aldehydic Zr–C bonds and condensation between methoxide and enolate surface species, were invoked to interpret the mechanism on zirconia.[286,287] Acidity of the catalyst was found to correlate with C_4 selectivity.[287] This results from the enhancement of condensation through the stabilization of the enolate species on Lewis acid sites.

Isoalkanes can also be synthesized by using two-component catalyst systems composed of a Fischer–Tropsch catalyst and an acidic catalyst. Ruthenium-exchanged alkali zeolites[288,289] and a hybrid catalyst[290] (a mixture of RuNaY zeolite and sulfated zirconia) allow enhanced isoalkane production. On the latter catalyst 91% isobutane in the C_4 fraction and 83% isopentane in the C_5 fraction were produced. The shift of selectivity toward the formation of isoalkanes is attributed to the secondary, acid-catalyzed transformations on the acidic catalyst component of primary olefinic (Fischer–Tropsch) products.

3.4. DIRECT COUPLING OF METHANE

3.4.1. Catalytic Oxidative Condensation

The oxidative coupling of methane to higher hydrocarbons has received considerable attention. Several review articles,[291–300] symposia proceedings,[301–303] and a collection of papers[304] have been devoted to the subject.

A great variety of oxides are active in the oxidative coupling of methane. In some cases redox (reduction–oxidation) metal oxides are used in stoichiometric reaction with methane. Since cofeeding of methane and oxygen initially was considered to cause nonselective oxidation (i.e., extensive oxidation of methane to carbon oxides), the reaction was operated cyclically.[305–307] With such a reaction strategy, the oxide is first converted to a fully oxidized state by transformation with oxygen. This oxide, when contacted with methane in the absence of oxygen, results in the transformation of methane and partial reduction of the metal oxide. Of the active oxides producing ethane and ethylene as major products above 600°C, supported manganese oxide[307] and lead oxide[308–310] gave the best results. In addition, oxides of Sn, Sb, and Bi also exhibited good characteristics.

Later studies demonstrated that cyclic operation was not necessary for the attainment of high product selectivities. High selectivities could be obtained on suitable catalysts with contemporaneously fed methane and oxygen in the continuous, catalyst-mediated oxidative coupling of methane. A large-scale catalyst screening and evaluation examining the majority of simple oxides demonstrated that a wide range of basic oxides can be effective. The most active catalysts are the nonreducible Group IIA metal oxides.[311,312] Other good catalysts are supported alkali oxides[313,314] and carbonates.[291,315] It was also observed that Group IIA oxides, when promoted with Group IA cations, gave increased activity and selectivity.[316,317] Chloride salts and chlorine compounds have similar beneficial effects.[318–322] Certain lanthanide sesquioxides (La$_2$O$_3$, Sm$_2$O$_3$) also show considerable activity in converting methane to C$_2$ hydrocarbons.[323–327] These catalysts function optimally in the temperature range 650–800°C. The products observed in the reaction are C$_2$H$_6$, C$_2$H$_4$, CO, CO$_2$, H$_2$, H$_2$O, and some higher hydrocarbons. CH$_3$OH and CH$_2$O are also found in trace amounts. Many of these oxides give remarkably similar conversion–selectivity results, suggesting a common active site and reaction mechanism. It was also found that the C$_2$ selectivity decreases almost linearly with increasing methane conversion, usually achieved by increasing the oxygen concentration in the feed. This is formulated as the "100% rule,"[295] indicating that methane conversion and C$_2$ selectivity roughly sum to 100%.

At low conversions, using reaction mixtures of high CH$_4$: O$_2$ ratio, selectivities of ≤70% may be obtained. In fixed-bed flow reactors a C$_2$ selectivity of 50% may be achieved at a conversion level of 40% at atmospheric pressure.[312] C$_2$ yields in the range 20–25% are usually achieved. A C$_2$ yield slightly better than 50% (80% selectivity at 65% methane conversion, above 700°C, Sm$_2$O$_3$ catalyst), however, has been reported.[328] A special technique, a separative chemical reactor simulating a countercurrent chromatographic moving bed has been applied. Another technical modification has resulted in even higher yields.[329] In this case methane coupling is carried out in a gas recycle electrocatalytic or catalytic reactor over silver catalysts (800°C, Ag on Y$_2$O$_3$–ZrO$_2$ or Ag on Sm$_2$O$_3$ doped with CaO) and the recycle gas passes through a Linde 5A molecular sieve at 30°C. The adsorbed C$_2$ hydrocarbons, which are trapped and therefore protected from further oxidation, are subsequently desorbed at 400°C. The method ensures high yields and selectivities (total yields up to 88%, ethylene yields ≤85%, ethylene selectivities ≤88% for methane conversion ≤97%).

At present, there is a general agreement that the primary function of the catalysts in oxidative coupling of methane is to form methyl radicals. Considerable evidence attained by electron paramagnetic resonance,[313,330] mass spectroscopy,[331] and isotopic labeling studies[332–334] indicates that methyl radicals produced in a heterogeneous process are the primary intermediate in the production of ethane. Methyl radicals are formed in a direct route with the involvement of surface oxygen (O$_s{}^-$) [Eq. (3.28)]. Alternatively, an acid–base process with the interaction of oxygen ions of low coordination number (O$_s^{2-}$) with methane may form a methyl anion that then reacts further with molecular oxygen to give methyl radical[335–337] [Eq. (3.29)]:

$$CH_4 \ + \ O_s^- \ \longrightarrow \ \cdot CH_3 \ + \ \cdot OH \tag{3.28}$$

$$CH_4 \ + \ O_s^{2-} \ \longrightarrow \ CH_3^- \ + \ HO^- \ \xrightarrow{\ O_2\ } \ \cdot CH_3 \ + \ \cdot O_2 \tag{3.29}$$

It was also concluded, however, that a homogeneous (noncatalytic) route is also available, which results in product distributions very similar to those obtained with most catalysts. Presumably, this reaction path also involves methyl radicals:

$$CH_4 + O_2 \longrightarrow \cdot CH_3 + HOO\cdot \qquad (3.30)$$

Methyl radicals formed in surface reactions enter the gas phase and form ethane, the primary coupling product:

$$2\cdot CH_3 \longrightarrow C_2H_6 \qquad (3.31)$$

A heterogeneous route, however, cannot be ruled out, either. Similarly, ethane can dehydrogenate to ethylene either in a homogeneous or in a surface reaction.

Most mechanistic studies have focused on elucidation of the role of alkali promoters. The addition of Li^+ to MgO has been shown to decrease the surface area and to increase both methane conversion and selective C_2 production.[338,339] As was mentioned, however, besides this surface-catalyzed process, a homogeneous route also exists to the formation of methyl radicals.[340–342] The surface active species on lithium-doped catalysts is assumed to be the lithium cation stabilized by an anion vacancy. The methyl radicals are considered to be produced by the interaction of methane with O^- of the $[Li^+O^-]$ center[330,343] [Eq. (3.32)]. This is supported by the direct correlations between the concentration of $[Li^+O^-]$ and the concentration of $\cdot CH_3$ and the methane conversion, respectively. The active sites then are regenerated by dehydration [Eq. (3.33)] and subsequent oxidation with molecular oxygen [Eq. (3.34)]:

$$CH_4 + [Li^+O^-] \longrightarrow \cdot CH_3 + [Li^+OH^-] \qquad (3.32)$$

$$2\,[Li^+OH^-] \longrightarrow [Li_2^+O^-] + H_2O \qquad (3.33)$$

$$[Li_2^+OH^-] + 0.5\,O_2 \longrightarrow 2\,[Li^+O^-] \qquad (3.34)$$

Quenching experiments revealed the presence of the superoxide O_2^- on the La_2O_3 surface formed by chemisorption of oxygen,[344] which can also induce hydrogen abstraction according to Eq. (3.35):

$$CH_4 + O_2^- \longrightarrow \cdot CH_3 + HOO^- \qquad (3.35)$$

In this very complicated reaction system, selectivities are determined by many reaction variables affecting different competing homogeneous and surface-catalyzed transformations. For example, the total oxidation of the $\cdot CH_3$ radical to carbon oxides is known to proceed through peroxy radicals formed in the gas phase. The formation of methoxy species in a surface reaction may also play a role in this process.

The overall selectivity to C_2 compounds is thus determined by the relative rates of the combination of $^\bullet CH_3$ radicals [Eq. (3.35)] and their reaction with oxygen molecules [Eq. (3.36)] or with surface O^- [Eq. (3.37)] to form methyl peroxy radical and surface methoxy species, respectively:

$$^\bullet CH_3 + O_2 \longrightarrow CH_3OO^\bullet + CO_x \qquad (3.36)$$

$$^\bullet CH_3 + [Li^+O^-] \longrightarrow [Li^+OCH_3^-] + CO_x \qquad (3.37)$$

Most experimental observations demonstrate that ethylene is formed by further oxidation of ethane:

$$C_2H_6 + O^- \longrightarrow {^\bullet C_2H_5} + {^\bullet OH} \qquad (3.38)$$

$$^\bullet C_2H_5 + O^{2-} \longrightarrow C_2H_5O^- + e^- \qquad (3.39)$$

$$C_2H_5O^- \longrightarrow CH_2{=}CH_2 + OH^- \qquad (3.40)$$

In general, the ethylene to ethane selectivity is governed by the relative rates of the processes shown in Scheme 3.1. Here, combination of methyl radicals (step 2) is a homogeneous process, but all other steps can be both homogeneous and heterogeneous.[294] It was also found that further oxidation of the C_2 products (steps 4 and 6) becomes important only with increasing C_2 concentration in the system.[327]

During the high-temperature operation of methane coupling catalysts, a number of components in the multiphase systems (lead, alkalis, chlorine) are susceptible to evaporation. Efforts in the early 1990s have concentrated on the preparation of stable catalyst formulations with increased activity and selectivity.[322,345,346]

At present, it seems that the doping cation serves to create an active center for hydrogen removal from methane, and to inhibit secondary reactions with the host oxide. Further clarification of many details of oxidative methane coupling is still necessary.[347] As for industrial applications, economic evaluations indicate[348,349] that direct oxidative coupling of methane is feasible at conversions of about 35% and C_2 selectivities in the range of 85% corresponding to yields of ~30%. Since these values have not been achieved yet, commercialization requires further developments. Changing economic conditions, however, may open the possibility of industrial application even with the current process characteristics.

The radical coupling reaction can also be effected by chlorination. Benson found[350,351] that with short contact times (0.03–0.3 s) above 930°C methane gave

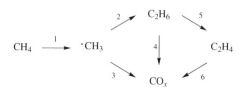

Scheme 3.1

ethylene (as well as aromatics and chlorinated C_2 products). As in the chlorinative coupling, HCl is the byproduct of the reaction, and its oxidative recycling is a major and not satisfactorily resolved problem. Modifications of the original Deacon process are explored to this effect.

Oxidative carbocationic condensation of methane under superacidic conditions was first achieved in the late 1960s. Olah and coworkers[352–355] observed that when methane was introduced into SbF_5-containing superacids, C_4 and higher alkyl cations were obtained. As the initial reaction [Eq. (3.41)] is endothermic, it became apparent that the superacid oxidatively helped to remove H_2 (as HF):

$$2\,CH_4 \longrightarrow C_2H_6 + H_2 \qquad \Delta H_{298} = -15.5\ \text{kcal/mol} \qquad (3.41)$$

The oxidative condensation to yield C_2–C_6 hydrocarbons can be achieved by reacting methane and oxygen over bifunctional oxide catalysts (WO_3 on Al_2O_3),[356,357] or methane is reacted in the presence of ethylene, acetylene, oxygen, or halogens catalyzed by superacids (SbF_5, TaF_5, TaF_5 on Nafion-H).[356,357]

3.4.2. High-Temperature Self-Coupling

Thermal dehydrogenation of methane to yield C_2 hydrocarbons, predominantly acetylene, is favorable only at temperatures above 1200°C. The process takes advantage of the fact that, in contrast to the behavior of other hydrocarbons, the free energy of formation of acetylene decreases at higher temperatures. Although acetylene is more stable than other hydrocarbons under these conditions, this is only a very temporary advantage. Even at this high temperature, acetylene is less stable than its component elements. Hence, the contact time must be very short. The stabilization of acetylene requires rapid cooling to prevent its decomposition to carbon, hydrogen, and condensation products. The decomposition of methane to its elements begins at relatively low temperature, at about 600°C. Hence, this decomposition proceeds in competition with the formation of acetylene.

Several basic processes are used in practical application.[358–360] The electric-arc technology supplies the high cracking temperatures (1500°C) rapidly, relies on short contact time (several milliseconds), and the product mixture is rapidly quenched with water to about 250°C. The conversion of methane to C_2 hydrocarbons is relatively low (10–20 vol%), but it generates a higher quantity of acetylene compared with other processes. The thermal cracking, also called *regenerative furnace pyrolysis* or the *Wulff process*, is more flexible considering operating conditions. It can be made to deliver mainly either acetylene or ethylene. The residual gas after separation of the C_2 hydrocarbons is burned to reheat the pyrolysis furnace.

The most important processes nowadays to produce acetylene are the partial-combustion technologies.[358–360] They are flame processes and involve the combustion of the feedstock or the residual gas to yield the necessary energy to attain the required high temperature. Combined yields of acetylene and ethylene can reach 50% with an ethylene : acetylene ratio of 0.1 : 3.

3.5. HYDROCARBONS THROUGH METHANE DERIVATIVES

3.5.1. Hydrocarbons through Methanol

Methanol Synthesis. The transformation of synthesis gas to methanol [Eq. (3.3)] is a process of major industrial importance. From the point of view of hydrocarbon chemistry, the significance of the process is the subsequent conversion of methanol to hydrocarbons (thus allowing Fischer–Tropsch chemistry to become more selective).

Since other possible transformations, such as, formation of dimethyl ether, higher alcohols, and hydrocarbons, are accompanied with higher negative free-energy change, methanol is thermodynamically a less probable product. Therefore, solely on a thermodynamic basis, these compounds as well as methane should be formed in preference to methanol. To avoid the formation of the former compounds, the synthesis of methanol requires selective catalysts and suitable reaction conditions. Under such conditions, methanol is the predominant product. This indicates that the transformations leading to the formation of the other compounds are kinetically controlled. In the methanol-to-hydrocarbon conversion, dimethyl ether generally is converted similarly to methanol.

An industrial process to produce methanol from carbon monoxide and hydrogen was developed by BASF in 1923 using a zinc oxide–chromia catalyst.[361,362] Since this catalyst exhibited relatively low specific activity, high temperature was required. The low equilibrium methanol concentration at this high temperature was compensated by using high pressures. This so-called high-pressure process was operated typically at 200 atm and 350°C. The development of the process and early results on methanol synthesis were reviewed by Natta.[363]

The high-pressure synthesis was superseded in the 1960s by the ICI low-pressure method after introducing Cu–ZnO–based catalysts.[361,362,364,365] Conditions applied in this technology are 50–100 atm and 220–250°C. A similar process was developed by Lurgi employing a CuO–ZnO catalyst.[361,366,367] Since copper-based catalysts are sensitive to sulfur and chlorine in the feed, a precondition for their industrial application is the use of high-purity synthesis gas prepared from methane. At present the low-pressure technology dominates the worldwide production of methanol. The introduction of copper-based catalysts spurred an intense research activity resulting in several review articles on the topic.[200,368–371]

The low-pressure methanol synthesis process utilizes ternary catalysts based on copper, zinc oxide, and another oxide, such as alumina or chromia, prepared by coprecipitation. Cu–ZnO–Al$_2$O$_3$ and Cu–ZnO–Cr$_2$O$_3$ are usually the most important industrial catalysts. A significant advance was made when a two-stage precipitation was suggested in which ZnAl$_2$O$_4$, a crystalline zinc aluminate spinel, was prepared prior to the main precipitation of copper–zinc species.[372] This alteration resulted in an increase in catalyst stability for long-term performance with respect to deactivation. Catalyst lifetimes industrially are typically about 2 years.

The Cu–ZnO binary catalyst has been known to exhibit high activity in methanol formation at low temperature, but fast deactivation prevented commercialization.[373] The addition of a third component, however, results in stable catalyst activity. The primary functions of the noncopper components are to enhance dispersion of copper, and to maintain the active phase in a stable dispersion and with suitable physical strength.

The low-pressure copper catalysts are very selective for the synthesis of methanol. Under industrial conditions on the $Cu-ZnO-Al_2O_3$ catalyst, the selectivity is typically greater than 99%. The impurities formed include hydrocarbons, higher alcohols, ethers, ketones, and esters. These, as well as any water formed, can easily be removed by distillation to give very pure methanol.

Although numerous investigations have been performed on methanol synthesis catalysts, the structure of the active catalysts, the nature of the active sites, and the reaction mechanism are still subjects of considerable controversy.

Originally it was proposed that the active centers in methanol formation are Cu^+ species dissolved in and strongly interacting with the ZnO (Cr_2O_3) phase with Al_2O_3 playing a role as a dispersant of this active Cu^+-ZnO phase.[374–376] Spectroscopic evidence lends additional support to this suggestion.[377] Photoelectron spectroscopy indicates the development of positive charge on CO adsorbed on Cu^+. This activates the CO for nucleophilic attack by hydride ion present on ZnO surfaces as a results of heterolysis of H_2. The rate-determining step is suggested to be the formation of adsorbed formaldehyde complex. The role of Cu^+ is to lower the activation barrier not only electrostatically but also by further stabilization of subsequent transition states.

In contrast, certain results indicate that the active centers are located exclusively on metallic copper particles, and that the activity is directly proportional to the copper surface of the metal particles.[378–383] This linear relationship is practically independent of the type of support used. However, there are conflicting reports suggesting no general correlation between the activity and the copper surface area.[384–386]

An increasing number of observations also points to the existence of support effects indicating that the support may influence the activity of the catalysts in ways other than by simply increasing the copper surface area.[370,381,385–387] Zinc oxide, for example, has been proposed to play the role of a reservoir for atomic hydrogen. Hydrogen spillover from copper to ZnO and the facile migration of hydrogen atoms from ZnO to the copper surface would enhance the rate of methanol synthesis.[383,388,389] It was also suggested that an electronic interaction between small metal particles and encapsulating oxide creates surface oxygen ion vacancies proposed as sites for methanol formation.[390]

Two basic mechanisms have been proposed to interpret methanol formation in the $CO + H_2$ reaction. When carbon monoxide adsorbed on the copper surface is hydrogenated by the stepwise addition of hydrogen atoms [Eq. (3.42)], the principal intermediate is a surface formyl species (**5**):

$$\underset{\substack{\| \\ \text{C} \\ \| \\ \text{M}}}{\overset{\text{O}}{}} \xrightarrow{\text{H}} \underset{\substack{\text{C} \\ | \\ \text{M}}}{\overset{\text{H}\diagdown\diagup\text{O}}{}} \xrightarrow{\text{H}} \underset{\substack{\| \\ \text{C} \\ \| \\ \text{M}}}{\overset{\text{H}\diagdown\diagup\text{OH}}{}} \xrightarrow{\text{H}} \underset{\substack{\text{CH}_2 \\ | \\ \text{M}}}{\overset{\text{OH}}{}} \xrightarrow[-\text{M}]{\text{H}} \text{CH}_3\text{OH} \qquad (3.42)$$

$$ \mathbf{5}$$

Another possibility is the insertion of CO into a surface OH group [Eq. (3.43)] to form a surface formate (**6**). Subsequent reactions lead to surface methoxy (**7**) and, eventually, to methanol and reformation of the surface OH group:

$$
\underset{M}{\overset{O}{\underset{\|}{\overset{\|}{C}}}} + \underset{M}{OH} \longrightarrow \underset{\underset{6}{M}}{\overset{\overset{H}{\diagdown}\overset{O}{\diagup}}{\underset{|}{\overset{C}{\underset{|}{O}}}}} \xrightarrow{2H} \underset{M}{\overset{CH_2OH}{\underset{|}{O}}} \xrightarrow[-2H_2O]{2H} \underset{\underset{7}{M}}{\overset{CH_3}{\underset{|}{O}}} \xrightarrow{2H} CH_3OH + \underset{M}{OH} \quad (3.43)
$$

To exhibit such an active and selective catalytic effect, the catalyst must be a fairly good hydrogenation catalyst that is able to activate molecular hydrogen. It must also activate carbon monoxide without dissociating it. A nondissociative chemisorption permits the hydrogenation of carbon monoxide to occur on both oxygen and carbon. Considering the formation of surface methoxide in the second mechanism [Eq. (3.43)], a further requirement is that the catalyst not form a too stable metal methoxide.

Most Group VIII metals adsorb carbon monoxide dissociatively, and, consequently, they are good Fischer–Tropsch catalysts.[240] In contrast, Pd, Pt, Ir, and Cu do not dissociate carbon monoxide. Of these metals, copper[368–371] and more recently palladium were found to be excellent methanol-forming catalysts.[391,392] On the basis of a comparison of the characteristics of these two metals, Ponec strongly favors the involvement of Cu⁺ and Pd⁺ centers in methanol synthesis.[393]

Although synthesis gas always contains a certain amount of carbon dioxide, its role in the overall transformation was not taken into account until the 1970s. Considering the fact that the water–gas shift reaction [Eq. (3.4)] is a principal transformation that occurs over copper-based catalysts, three different routes can exist for the formation of methanol.

The implicit assumption was that hydrogenation of carbon monoxide is the main reaction for the formation of methanol. It was observed, however, that an optimum proportion of carbon dioxide was needed to achieve maximum methanol yield and to prevent catalyst deactivation.[369] Klier attributed these effects to the ability of CO_2 to maintain an adequate concentration of Cu⁺ species. Later studies indicated that the optimal $CO_2 : CO$ ratio corresponds to 6% CO_2 in the feed under industrial conditions.

Isotope labeling studies furnished a convincing body of evidence to indicate that even in the presence of carbon monoxide, only carbon dioxide is converted to methanol [Eq. (3.2)].[394–398] In a review article[371] Chinchen considers this as the dominant route, with CO merely serving as a source of CO_2 via the water–gas shift reaction.

Other investigators propose that both direct CO_2 hydrogenation and direct CO hydrogenation may take place.[399–402] The controversy over the significance of the different routes has not yet been properly resolved. It seems clear, however, that certain catalyst compositions are known to be more active toward a carbon dioxide–hydrogen mixture,[401] whereas others exhibit higher activity in the $CO + H_2$ reaction.[386,398,401]

More recent research efforts have focused on the development of other possible catalysts such as promoted Raney copper,[371,403] catalysts prepared from intermetallic precursors,[362,371,386,404–406] and catalysts that tolerate high CO_2 content.[407] Catalyst modifications allowed to shift the selectivity to the formation of higher alcohols.[208,408–410] For example, in a process developed by IFP, a multicomponent oxide catalyst is applied with copper and chromium as the main components.[410] By this method, 70–75% total alcohol selectivities and 30–50% of C_2 and higher alcohol selectivities can be achieved at 12–18% conversion levels (260–320°C, 60–100 atm).

The current two-step industrial route for the synthesis of methanol, from coal or methane to synthesis gas and then from synthesis gas to methanol, has certain drawbacks. The economic viability of the whole process depends on the first step, which is highly endothermic. Thus a substantial amount of the carbon source is burned to provide the heat for the reaction. It would be highly desirable, therefore, to replace this technology with a technically simpler, single-step process. This could be the direct partial oxidation of methane to methanol, allowing an excellent way to utilize the vast natural-gas resources. Although various catalysts, some with reasonable selectivity, have been found to catalyze this reaction (see Sections 9.1.1 and 9.6.1), the very low methane conversion does not make this process economically feasible at present.

Methanol through intermediate halogenation to methyl halide with subsequent gas-phase hydrolysis[411–413] and oxidative recycling of HHlg formed in the reaction[412] was studied by Olah and coworkers [Eq. (3.44), route a]. Bromine was found to be superior to chlorine in the process because of the ease of reoxidation of HBr in a catalytic cycle. A one-step process with the direct formation of oxygenates is also feasible[413] [Eq. (3.44), route b].

$$CH_4 \xrightarrow[\text{acid catalyst}]{a.\ Hlg_2} HHlg + CH_3Hlg \xrightarrow[\text{catalyst}]{H_2O} CH_3OH \text{ and } CH_3OCH_3 \qquad (3.44)$$

O$_2$ (above); *b. Hlg$_2$, H$_2$O* / *acid catalyst* (below)

Methanol Conversion to Hydrocarbons. The conversion of methanol to hydrocarbons requires the elimination of oxygen, which can occur in the form of H_2O, CO, or CO_2. The reaction is an exothermic process; the degree of exothermicity is dependent on product distribution. The stoichiometry for a general case can be written as in Eq. (3.45):

$$n\ CH_3OH \longrightarrow -[CH_2]_n- + n\ H_2O$$
$$n = 1, 2, \ldots \qquad (3.45)$$

The products may be alkenes, cycloalkanes, or mixtures of alkanes and aromatics.

When alkanes are the sole products, Eqs. (3.46)–(3.49) represent the principal reactions with the formation of water, hydrogen, coke, and carbon oxides as byproducts; eq. (3.49) describes the formation of aromatics:

$$(n + 1)\, CH_3OH \longrightarrow C_nH_{2n+2} + C + (n + 1)\, H_2O \tag{3.46}$$

$$(2n + 1)\, CH_3OH \longrightarrow 2\, C_nH_{2n+2} + CO_2 + 2n\, H_2O \tag{3.47}$$

$$(3n + 1)\, CH_3OH \longrightarrow 3\, C_nH_{2n+2} + CO_2 + (3n - 1)\, H_2O \tag{3.48}$$

$$n\, CH_3OH \longrightarrow C_nH_{2n-6} + n\, H_2O + 3\, H_2 \tag{3.49}$$

$$n = 6, 7, \ldots$$

Methanol can be converted to hydrocarbons over acidic catalysts. However, with the exception of some zeolites, most catalysts deactivate rapidly. The first observation of hydrocarbon formation from methanol in molten $ZnCl_2$ was reported in 1880, when decomposition of methanol was described to yield hexamethylbenzene and methane.[414] Significant amounts of light hydrocarbons, mostly isobutane, were formed when methanol or dimethyl ether reacted over $ZnCl_2$ under superatmospheric pressure.[415] More recently, bulk zinc bromide and zinc iodide were found to convert methanol to gasoline range (C_4–C_{13}) fraction (mainly 2,2,3-trimethylbutane) at 200°C with excellent yield (>99%).[416]

P_2O_5 was also reported to decompose methanol to mixtures of hydrocarbons with widely varying compositions.[417,418] Polyphosphoric acid is unique in effecting the transformation of methanol under comparatively mild conditions (190–200°C).[418] Superacidic TaF_5 and related halide catalysts[419,420] condense methanol to saturated hydrocarbons of the gasoline range at 300°C.

Metal molybdates[421] and cobalt-thoria-kieselguhr[422] also catalyze the formation of hydrocarbons. It is believed, however, that methanol is simply a source of synthesis gas via dissociation and the actual reaction leading to hydrocarbon formation is a Fischer–Tropsch reaction. Alumina is a selective dehydration catalyst, yielding dimethyl ether at 300–350°C, but small quantities of methane and C_2 hydrocarbons[423,424] are formed above 350°C. Heteropoly acids and salts exhibit high activity in the conversion of methanol and dimethyl ether.[425–428] Acidity was found to determine activity,[427–430] while hydrocarbon product distribution was affected by several experimental variables.[428–432]

A remarkable discovery made in the 1970s by Mobil researchers was their finding[433] that methanol could be converted to gasoline over the medium-size shape-selective zeolite ZSM-5. The MTG (methanol to gasoline) process, developed by Mobil went on production in New Zealand in 1985. Zeolites, which are crystalline aluminosilicates have a well-defined geometry and reproducible morphology of pores, channels, and cages. They are unique catalysts in their ability to discriminate between reactant molecules. As a result, zeolites are able to control product selectivities depending on molecular size and shape known as *shape-selective catalysis.*[434–438]

In general, medium-pore zeolites (ZSM-5 and ZSM-11) are the best catalysts to produce hydrocarbons from methanol. Below 300°C, the main reaction is dehydration of methanol to dimethyl ether. With increasing temperature, however, the formation of hydrocarbons with a characteristic product distribution may be achieved. The hydrocarbons formed are mainly C_3–C_5 alkanes and C_6–C_{10} aromatics with very little alkenes. The aromatics are mostly methyl-substituted benzenes. The much less than equilibrium concentration of tri- and tetramethyl-benzenes, and the sharp cutoff in product molecular weight, are attributed to zeolite shape selectivity. The decrease in reactant partial pressure results in an increasing selectivity to alkenes. Similarly, small-pore zeolites, which absorb linear hydrocarbons and exclude branched isomers, yield light hydrocarbons, mainly ethylene and propylene. Increasing pressure, in contrast, leads to increasing aromatic substitution, giving durene (1,2,4,5-tetramethylbenzene) with especially high selectivity. Important additional features of the medium-pore zeolite catalyzed process are the high octane number of the product formed and the excellent yields. Medium-pore zeolites exhibit an exceptional resistance to coke deposition, ensuring long catalyst life.

More recently phosphorus-containing zeolites developed by Union Carbide (aluminophosphates, silicoaluminophosphates) were shown to be equally effective in methanol condensation.[439–444] ZSM-5 was also shown to exhibit high activity and selectivity in the transformation of Fischer–Tropsch oxygenates to ethylene and propylene in high yields.[445] Silicalite impregnated with transition-metal oxides, in turn, is selective in the production of C_4 hydrocarbons (15–50% isobutane and 8–15% isobutylene).[446]

Bifunctional acid-based catalysts with no shape selectivity such as WO_3 on Al_2O_3 have been found by Olah and coworkers to readily condense methanol or dimethyl ether (as well as monosubstituted methanes) to hydrocarbons.[420,447] Primary products are ethylene and propylene, as well as light hydrocarbons, and methane and hydrogen. The high yield of methane and hydrogen may be a consequence of some competing thermal reactions. Hutchings and coworkers made further contributions concerning the mechanism.[448,449]

The mechanism of the conversion of methanol to hydrocarbons has been the subject of substantial studies.[450–455] Despite the intensive research, however, many details of this very complex transformation remain unsolved. It now appears generally accepted that methanol undergoes a preliminary Brønsted acid–catalyzed dehydration step and that dimethyl ether, or an equilibrium mixture of dimethyl ether and methanol, acts as the precursor to hydrocarbon formation:[433]

$$
\begin{array}{c}
2\,CH_3OH \\
\Big\updownarrow {\scriptstyle H_2O} \; {\scriptstyle -H_2O} \\
CH_3OCH_3
\end{array}
\;\longrightarrow\;
\underset{\text{alkenes}}{\text{lower}}
\;\longrightarrow\;
\begin{array}{l}
\text{acyclic and cyclic} \\
\text{saturated hydrocarbons,} \\
\text{higher alkenes, aromatics}
\end{array}
\qquad (3.50)
$$

It is also agreed on that the primary products are ethylene and propylene. Once the first alkenes are formed, they can undergo further transformations to higher

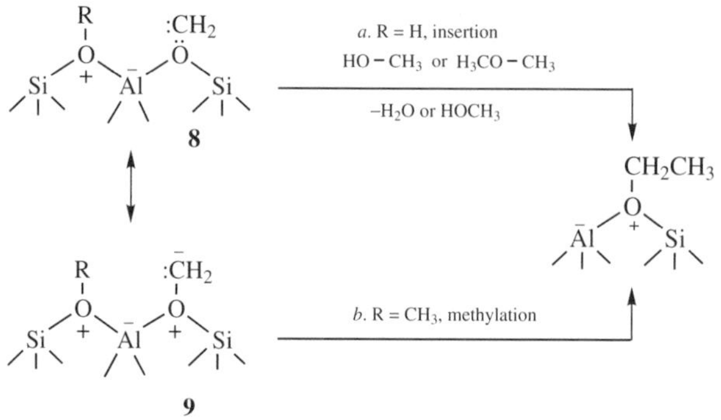

Scheme 3.2

hydrocarbons in one of two ways. Methylation of alkenes will increase the carbon chain length one carbon at a time. Oligomerization–transmutation, on the other hand, has a much more drastic effect on chain lengthening, leading to redistribution of alkenes. Cyclization converts alkenes to cycloalkanes, whereas cyclization and hydride ion transfer reactions yield alkanes and aromatics. The ongoing controversy on the mechanism concerns the nature of the first C_1-to-C_2 conversion, what C_1 species is involved in the formation of ethylene, and how the first carbon–carbon covalent bond is formed.

A carbenoid-type mechanism with free or surface-bound species formed by α elimination from methanol promoted by the strong electrostatic field of zeolites was proposed first.[433,456,457] Hydrocarbons then can be formed by the polymerization of methyl carbene, or by the insertion of a surface carbene (**8**) into a C–O bond[453–455,458,459] (Scheme 3.2, route *a*). If surface methoxyl or methyloxonium species are also present, they may participate in methylation of carbene[454,455,460,461] depicted here as a surface ylide (**9**) (Scheme 3.2, route *b*). A concerted mechanism with simultaneous α elimination and sp^3 insertion into methanol or dimethyl ether was also suggested.[433,454,457]

The difficulty in this mechanism is the highly unfavorable endothermic nature of the α elimination of methanol to carbene [Eq. (3.51)] in contrast with the thermodynamically favored bimolecular dehydration [Eq. (3.52)]:

$$CH_3OH \longrightarrow :CH_2 + H_2O \qquad \Delta H = 349.3 \text{ kcal/mol} \qquad (3.51)$$

$$2\,CH_3OH \longrightarrow CH_3OCH_3 + H_2O \qquad \Delta H = -24.5 \text{ kcal/mol} \qquad (3.52)$$

Other mechanisms with the involvement of an incipient methyl cation were also proposed.[418,460,462] An attack of methyl cation released from extensively polarized methoxy groups on the carbon–hydrogen bond forms pentacoordinated carbocation intermediates that, in turn, yields ethyl methyl ether after the loss of a proton:

$$CH_3OR \xrightarrow{H^+} \overset{+}{C}H_3\overset{|}{\underset{H}{O}}R \xrightarrow[-ROH]{CH_3OCH_3} \left[\begin{matrix} H_3C \\ \diagdown \\ H \end{matrix} \!\!> - CH_2OCH_3 \right]^+ \xrightarrow{-H^+} CH_3OCH_2CH_3 \quad (3.53)$$

However, the methyl cation (in itself a very energetic, unprobable species in the condensed state) is not expected to attack a carbon–hydrogen bond in dimethyl ether (or methanol) in preference to the oxygen atom. The more probable attack on oxygen would lead to the trimethyloxonium ion, which was observed experimentally[447,463] On the action of a basic site the trimethyloxonium ion can then be deprotonated to form dimethyloxonium methylide:

$$
\overset{+}{\underset{H}{CH_3\!O\!R}} + CH_3OCH_3 \xrightarrow[-ROH]{} \underset{CH_3}{\overset{H_3C}{\underset{O}{\diagdown}}\overset{+}{\!O\!}\overset{CH_3}{\diagup}} \xrightarrow[-H^+]{} \underset{CH_3}{\overset{H_3C}{\underset{O}{\diagdown}}\overset{+}{\!O\!}\overset{CH_2^-}{\diagup}} \tag{3.54}
$$

Van den Berg et al. suggested an intramolecular Stevens rearrangement of the oxonium ylide to ethyl methyl ether to interpret carbon–carbon bond formation[463] [Eq. (3.55)]. Olah and coworkers, however, provided evidence (based on isotopic labeling studies) that the oxonium ylide undergoes intermolecular methylation to ethyl dimethyloxonium ion [Eq. (3.56)] instead of Stevens rearrangement:[447]

$$
\underset{CH_3}{\overset{H_3C}{\underset{O}{\diagdown}}\overset{+}{\!O\!}\overset{CH_2^-}{\diagup}}
\begin{cases}
\xrightarrow[\text{rearrangement}]{\text{Stevens}} CH_3OCH_2CH_3 & (3.55) \\[2ex]
\xrightarrow[-CH_3O^-]{CH_3OCH_3} \underset{CH_3}{\overset{H_3C}{\underset{O}{\diagdown}}\overset{+}{\!O\!}\overset{CH_2CH_3}{\diagup}} \xrightarrow[-H^+]{} \underset{CH_3OCH_3}{\overset{CH_2=CH_2}{+}} & (3.56)
\end{cases}
$$

The oxonium ylide mechanism requires a bifunctional acid–base catalyst. The validity of the oxonium ylide mechanism on zeolites was questioned[459,461,464] because zeolites do not necessarily possess sufficiently strong basic sites to abstract a proton from the trimethyloxonium ion to form an ylide. It should, however, be pointed out, as emphasized by Olah,[447,465] that over solid acid–base catalysts (including zeolites) the initial coordination of an electron-deficient (i.e., Lewis acidic) site of the catalysts allows formation of a catalyst-coordinated dimethyl ether complex. It then can act as an oxonium ion forming the catalyst-coordinated oxonium ylide complex (**10**) with the participation of surface bound CH_3O^- ions:

$$
\overset{-}{cat}\!\rightarrow\!\underset{CH_3}{\overset{+}{O}\diagup^{CH_3}} \xrightarrow[-CH_3OH]{cat-\overset{..}{O}CH_3} cat\!\rightarrow\!\underset{CH_2^-}{\overset{+}{O}\diagup^{CH_3}} \xrightarrow[-CH_3O^-]{CH_3OCH_3} cat\!\rightarrow\!\underset{CH_2CH_3}{\overset{+}{O}\diagup^{CH_3}} \xrightarrow{} \underset{CH_3OH}{\overset{CH_2=CH_2}{+}} \tag{3.57}
$$
$$\underset{\mathbf{10}}{}$$

Finally, **10** can undergo methylation followed by elimination of ethylene.

Free radicals were also proposed to play a role in the conversion of methanol to hydrocarbons.[466,467] Although more recent experimental observations did indeed show the presence of some free radicals during methanol transformation, their involvement in the reaction mechanism is doubtful.[468–470] Carbon monoxide[471] and ketene[472] were also suggested to be intermediates in methanol conversion. It has been shown, however, that neither of these plays any significant role in primary product formation.[473]

A demonstration MTG fixed-bed plant has operated in New Zealand since 1985 producing gasoline of an octane number of 93, with a typical composition of 32% aromatics, 60% saturated hydrocarbons, and 8% alkenes. Since the net process is highly exothermic, the reaction was carried out in two stages. First methanol was converted over a dehydration catalyst to an equilibrium mixture of methanol, dimethyl ether, and water. This mixture was then transformed in a second reactor over ZSM-5 to hydrocarbons. During the production the yield of C_{5+} hydrocarbons slowly increases and the composition slowly changes from an aromatic-rich gasoline to an alkene-rich gasoline as a result of catalyst deactivation. A fluidized-bed process, in turn, works under continuous regeneration and produces a constant-quality product with lower C_{5+} liquid yield and higher olefin yield. Word economy changes, and still available cheap oil makes the immediate future of the Mobil or other synthetic hydrocarbon processes uncertain. In the long run, however, it is inevitable that synthetic hydrocarbons will play a major role.

The methanol transformations discussed precedingly can be modified to produce high amounts of light alkenes.[437,454,474,475] The key to achieve this change is to prevent $C_2–C_4$ olefinic intermediates to participate in further transformations. Such decoupling of alkene formation and aromatization can be done by the use of small-pore zeolites or zeolites with reduced acidity. Reduced contact time and increased operating temperature, and dilution of methanol with water to decrease methanol partial pressure, are also necessary to achieve high alkene selectivities. This approach has led to the development of the MTO (methanol-to-olefin) process, which yields $C_2–C_5$ alkenes with about 80% selectivity.

An interesting possibility is the further coupling of the MTO process with the olefin-to-gasoline and distillate (MOGD) technology to provide another synthetic fuel process.[437,475] In this case the light olefins produced in the MTO process are separated and then fed to the MOGD unit. High-molecular-weight isoolefins (distillate process) or gasoline-range hydrocarbons (gasoline process) can be produced through a series of oligomerization, cracking, and polymerization steps over H-ZSM-5 catalyst under appropriate conditions (see Section 13.1.1).

Results have been reported to carry out the two-stage process: methanol synthesis from synthesis gas and its subsequent transformation to hydrocarbons in a single step.[454,475] Dual-function catalysts composed of a metal oxide and an acid component can combine the two necessary functions for a single-pass conversion of synthesis gas to hydrocarbons. By producing methanol selectively in the first stage of the syngas conversion, the selectivity of the overall Fischer–Tropsch synthesis can be significantly improved. However, methanol is still produced from synthesis gas in a Fischer–Tropsch-like process necessitating large capital investment and substantial energy losses in the highly energy demanding conversion of syngas. Thus, there is substantial interest to achieve methane conversion to hydrocarbons without the necessity to first generate syngas.

3.5.2. Hydrocarbons through Methyl Halides

It has been demonstrated that methyl chloride or bromide can be dehydrohalogenatively condensed into hydrocarbons over acidic catalysts:

$$n \, CH_3Cl(Br) \xrightarrow{\text{catalyst}} -[CH_2]_n- \; + \; n \, HCl(Br) \tag{3.58}$$

One important prerequisite to the application of this reaction in hydrocarbon synthesis is the selective monochlorination of methane. Usual radical chlorination of methane is not selective, and high $CH_4 : Cl_2$ ratios are needed to minimize formation of higher chlorinated methanes (see Section 10.2.5). In contrast with radical halogenation, electrophilic halogenation of methane was shown to be a highly selective process.[412]

A range of superacid catalysts such as $NbOF_3-Al_2O_3$, $ZrOF_2-Al_2O_3$, SbF_5- graphite, and TaF_5-Nafion-H effect the chlorination and bromination of methane.[411,413] $ZrOF_2-Al_2O_3$ gives the best results in chlorination (96% selectivity at 34% conversion), and $SbOF_3-Al_2O_3$ is the best for bromination (99% selectivity at 20% conversion). Supported platinum and palladium catalysts are also very effective with selectivities also exceeding 90% (99% selectivity at 36% conversion over 0.5% platinum on alumina at 250°C, with a feed composition of $CH_4 : Cl_2 = 3 : 1$).[411] Solid acids and zeolites have also been found to be very selective in monochlorination.[476,477]

Hydrogen chloride formed during the chlorination of methane must be oxidized to chlorine, either in the oxychlorination of methane[478] or in a separate Deacon-type operation to allow the process to become economical. This is still difficult to achieve. Existing data also show that formation of higher chlorinated methane derivatives is often also a difficulty of the reaction.[479-481] These difficulties can be minimized by using bromine in the catalytic oxidative conversion process.

Hydrocarbon formation from methyl chloride can be catalyzed by ZSM-5[482,483] or bifunctional acid-base catalysts such as WO_3 on alumina.[420,447] The reaction on ZSM-5 gives a product distribution (43.1% aliphatics and 57.1% aromatics at 369°C) that is very similar to that in the transformation of methanol, suggesting a similar reaction pathway in both reactions.[483] WO_3 on Al_2O_3 gives 42.8% C_2-C_5 hydrocarbons at 327°C at 36% conversion.[447] When using methyl bromide as the feed, conversions are comparable. However, in this case, HBr can be very readily air-oxidized to Br_2 allowing a catalytic cycle to be operated. Since bromine is the oxidant, the reaction is economical. The "one step" oxidative condensation of methane to higher hydrocarbons was also achieved in the presence of chlorine or bromine over superacidic catalysts.[357]

3.5.3. Hydrocarbons through Sulfurated Methanes

The ready availability of carbon disulfide from methane and sulfur in oxide-catalyzed reactions[484] [Eq. (3.59)] and its further transformation over zeolites[485] [Eq. (3.60)] or other catalysts offer an alternative way to the production of hydrocarbons from methane:

$$CH_4 + 2S_2 \xrightarrow{\text{oxide catalyst}} CS_2 + 2H_2S \qquad (3.59)$$

$$nCS_2 + 3nH_2 \xrightarrow{\text{zeolite}} -[CH_2]_n- + 2nH_2S \qquad (3.60)$$

The transformation of carbon disulfide over ZSM-5 at 482°C leads to the formation of a mixture of aliphatic (14.8%) and aromatic compounds (20.4% C_6–C_{10} and 12.4% C_{11+}); the balance is methane.[485] It is highly probable that partially reduced species (CH_3SH, CH_3SCH_3) participate in the reaction. It is known that CS_2 is transformed into such species on acidic catalysts.[486] It was also shown that CH_3SH and CH_3SCH_3 are readily converted to hydrocarbons over ZSM-5[482] and bifunctional metal oxides.[420,447] For example, CH_3SCH_3 is transformed over WO_3 on Al_2O_3 to yield 36.2% C_2–C_4 hydrocarbons at 380°C at 32% conversion.[447] The intermediacy of reduced species is strongly supported by the observation that a cobalt-promoted ZSM-5 gives improved yields of C_{2+} hydrocarbons.[485] Cobalt presumably enhances the rate of formation of the reduced intermediates.

It is also significant that the direct oxidative condensation of methane to higher hydrocarbons takes place in the presence of S_8 over superacidic catalysts, such as TaF_5 and the like.[357]

3.6. RECENT DEVELOPMENTS

3.6.1. Fischer–Tropsch Chemistry

Fischer–Tropsch Synthesis. Fischer–Tropsch synthesis still receives great interest. Special sections are devoted to the topic at regular symposia on natural-gas conversion,[487–489] a journal special issues have been published,[490] and there are papers on Fischer–Tropsch technologies,[490–495] and an analysis of present trends.[496] Two reviews about mechanistic studies[497] and the kinetics and the selectivities of Fischer–Tropsch synthesis[498] are available. A few significant new findings are collected here.

A new mechanism to interpret alkene formation in Fischer–Tropsch synthesis has been presented.[499–501] There is a general agreement that hydrocarbon formation proceeds according to the modified carbene mechanism. Specifically, CO decomposes to form surface carbide and then undergoes hydrogenation to form surface methine (=CH), methylene (=CH$_2$), methyl and, finally, methane. Linear hydrocarbons are formed in a stepwise polymerization of methylene species. When chain growth is terminated by β-hydride elimination [Eq. (3.61)], 1-alkenes may be formed,[502] which is also called the *alkyl mechanism*:

$$\underset{M}{\overset{H}{|}} + \underset{M}{\overset{CH_2}{||}} \longrightarrow \underset{M}{\overset{CH_3}{|}} \xrightarrow{2\,M=CH_2} \underset{M}{\overset{CH_2CH_2CH_3}{|}} \xrightarrow{-H-M} H_2C=CH-CH_3 \quad (3.61)$$

However, there are certain anomalies with respect to the alkyl mechanism, namely, the relatively small amounts of C_2 hydrocarbons in the Anderson–

Schulz–Flory distribution and the difficulty in accounting for the formation of branched hydrocarbons. Additionally, since by β-hydride elimination is a reversible process, it is surprising to observe the presence of alkenes under Fischer–Tropsch conditions under hydrogen pressure and on a surface covered with hydride. Using organometallic models and combining ^{12}C and ^{13}C labeling studies with various C_1 and C_2 probe molecules,[503–505] the research group at Sheffield suggested a new mechanism.[499–501]

These new data acquired with double-labeled vinyl probes ($^{13}CH_2=^{13}CHBr$ and $^{13}CH_2=^{13}CH_2$) determined first on Rh, but found to be similar for more common Fischer–Tropsch catalysts (Ru, Fe, Co) showed that these are readily incorporated into the alkene and the alkane products. In addition, an increase in the rate of hydrocarbon formation was observed during vinylic but not ethyl addition. These data indicate that the participation of vinyl intermediates is an integral part of the surface polymerization mechanism, specifically, vinyl (alkenyl) intermediates couple with surface methylene in hydrocarbon formation:

$$
\begin{array}{c}
\underset{M}{\overset{CH}{\underset{|||}{}}} + \underset{M}{\overset{CH_2}{\underset{||}{}}} \longrightarrow \underset{M}{\overset{HC=CH_2}{\underset{|}{}}} \xrightarrow{M=CH_2} \underset{M}{\overset{H_2C-CH=CH_2}{\underset{|}{}}} \rightleftharpoons \underset{M}{\overset{HC=CH-CH_3}{\underset{|}{}}} \longrightarrow \\
\xrightarrow{M=CH_2} \underset{M}{\overset{H_2C-CH=CHCH_3}{\underset{|}{}}} \longrightarrow \underset{M}{\overset{HC=CH-R}{\underset{|}{}}} \xrightarrow{H} H_2C=CH-R \qquad (3.62)
\end{array}
$$

The termination step for 1-alkene formation is now the reaction of the surface alkenyl with surface H instead of the β-elimination step. Chain branching can proceed by the involvement of allylic intermediates. Since this new mechanism involves different types of reactions to form C_2 and $C_{2<}$ hydrocarbons, it is not expected that the amounts of C_2 products will lie on the normal curve of the Anderson–Schulz–Flory distribution.

Further evidence to support this suggestion are data for organometallic complexes showing that metal-bound alkenyl (sp^2 carbon) can couple with metal-bound alkyl (sp^3 carbon) with a lower activation barrier than can two metal-bound alkyl sp^3 carbons. The role of added vinylic probes in the Fischer–Tropsch chain growth mechanism has generated some controversy about the interpretation of data mainly concerning the decreasing level of incorporation of $^{13}C_2$ into the hydrocarbon products with increasing carbon number and the existence of surface-bound vinyl intermediate.[506–508]

A new approach to develop a molecular mechanism for Fischer–Tropsch catalysis based on the use of $[Fe_2Co(CN)_6]$ and $[Fe(HCN)_2]_3$ precursor complexes has been disclosed.[509] The former produced mainly liquid aliphatic hydrocarbons, whereas the latter gave waxy aliphatic products. Results acquired by various techniques were interpreted to imply that chain growth proceeds via the insertion of CO into an established metal–carbon bond, that is, a C_1 catalytic insertion mechanism is operative. It follows that C_2 insertion is an unlikely possibility.

Isotope exchange studies with perdeuterated C_8–C_{15} alkanes led to the surprising observation that H/D exchange did not occur in alkanes under Fischer–Tropsch synthesis conditions (Fe catalyst, 270°C, 11.9 atm).[510] In contrast, under the same experimental conditions, there was H/D exchange when 1-octene-d_{16} was used. 1-Octene was hydrogenated, was incorporated into higher hydrocarbons, and underwent isomerization to isomeric 2-octenes with no detectable d_{16} isotopomers. Isomerization in the Fe-catalyzed Fischer–Tropsch reaction was proposed to likely occur via the metal–hydrogen atom addition–elimination mechanism.

There have been efforts to interpret the deviation form the Anderson–Schulz–Flory ideal distribution for various metals. Product distribution can often be well represented by superposition of two Anderson–Schulz–Flory distributions. According to a new study,[511] the superimposed distributions for Co and alkali-promoted Fe catalysts are the results of two different chain mechanisms: one with a low and the other with a higher growth probability. For cobalt, the product distribution is modified by both secondary chain growth of readsorbed 1-alkenes and by hydrogenolysis of hydrocarbons. For iron, the chain growth of readsorbed 1-alkenes is negligible and the superposition is very strict. It was also found that the products associated with low growth are formed by the well-accepted methylene insertion mechanism.

A new product distribution model for slurry-phase Fischer–Tropsch synthesis on precipitated Fe catalysts is based on the assumption that two types of active sites exist on the catalyst surface: one where growth of hydrocarbon intermediates occurs and another where reversible readsorption of 1-alkenes takes place.[512] The model provides information on hydrocarbon product distribution, 2-alkene content, and total olefin content as a function of carbon number. The curved line for distribution can be fitted well with the proposed selectivity model for the entire range (C_1–C_{50}) of carbon numbers. Product distributions over an industrial Fe–Mn catalyst showed that alkene selectivities were significantly higher than those for alkanes in a broad chain length range from C_2 to about C_{20}.[513] Alkenes with carbon number higher than 27, however, could not be detected in the wax product. Consequently, this implies that primary alkenes with chain longer than C_{27} are dominantly hydrogenated to alkanes because of the long residence time in the catalyst pore.

X-ray adsorption spectroscopy (XRAS), a powerful tool for in situ investigation of solid materials, allowed Iglesia and coworkers to get important new findings about the structure and site evolution of an iron oxide catalyst precursor during the Fischer–Tropsch synthesis.[514] The activation of Fe_2O_3 precursor promoted by Cu and/or K occurs via reduction to Fe_3O_4 followed by carburization to form FeC_x. The reaction rate increases markedly during the initial stages of carburization, suggesting that the conversion of near-surface layers of Fe_3O_4 to FeC_x is sufficient for the evolution of active sites. The promoters increase the initial carburization rates and induce the formation of smaller FeC_x crystallites; that is, they increase the surface area of the active FeC_x phase. Consequently, they act predominantly as structural promoters. The role of alkali promoters on a Ru on SiO_2 catalyst was shown to block low coordination sites needed for dissociative chemisorption of hydrogen, significantly slowing down the kinetics of adsorption-desorption.[515,516] The restricted hydrogen mobility is in agreement with the observed effects of alkali promoters, namely, decreased methane formation rates, increased chain growth probabilities, and increased alkene : alkane ratios.

Related Processes. The propensity of Rh to catalyze the formation of valuable oxygen-containing compounds, alcohols, aldehydes, and acetic acid is of great industrial interest. C_{2+} oxygenates can serve as fuel alternatives or additive to gasoline, as precursors in organic synthesis, and as solvents. Certain studies have shown again that Rh and also Ru exhibit exceptionally high activity and selectivity for oxygenate formation.[517] The total selectivity of C_{2+} oxygenates can sometimes amount to 70%.[518] Dissociative adsorption of CO followed by the formation of CH_x species and insertion of undissociated CO has been proved crucial for the formation of C_{2+} oxygenates. The unique activity of Rh appears to be related to its ability to promote both processes. Many studies have been devoted to find a suitable supports promoter to direct the transformation toward one particular group of products.

A $RhVO_4$ on SiO_2 catalyst was shown to exhibit increased activity and selectivity toward the formation of oxygenates by the addition of MnO.[519] Increasing Rh content of Rh on SiO_2 catalysts prepared using microemulsion brings about an increase of oxygenate production from 12% to more than 40%.[520]

Rh on SiO_2 promoted by vanadium showed a good selectivity for the formation of ethanol.[521] In contrast, promotion with Mn and Li favors the formation of acetaldehyde and acetic acid.[518] The activity for oxygenate formation of these catalysts was correlated with Rh^0, but the high selectivity to acetaldehyde and acetic acid induced by Mn and Li promotion was believed to originate from the presence of a high concentration of Rh ions indicated by a strong IR band of twin adsorbed CO.[518] The difference in selectivities, that is, the low selectivity of ethanol formation, may be related to the lower capacity of the Mn-doped catalyst for hydrogen storage and therefore, a lower possibility for hydrogenation.

A Rh on SiO_2 catalyst promoted with V and Sm showed similar improvements in catalyst performance.[522] Sm^{3+} ions improve the dispersion of Rh and increase CO and H_2 uptakes, enhancing the formation of acetaldehyde and acetic acid. Lower-valence V ions have a good H_2 storage capacity, thereby increasing the hydrogenation ability of the catalyst and promoting the formation of ethanol.

According to Sachtler and coworkers, the role of MnO applied for a Rh on zeolite NaY zeolite catalyst is to promote the formation and stabilization of $C_xH_yO_z$ intermediates, including surface acetate groups.[523] These undergo hydrogenation to form the final oxygenate products if they are located in immediate proximity to a Rh particle, that is, at the Rh–MnO interface.

Results of an unusual selectivity of the direct formation of acetic acid over Rh on NaY zeolite have been disclosed.[524,525] Syngas conversion of high yields of acetic acid is achieved in the absence of any catalyst promoter at relatively low temperature. The selectivity to acetic acid was 45–56% with a syngas of equal moles of CO and H_2, but 67% with a CO-rich syngas (CO : H_2 = 3). Exposure of the catalyst to high pressure was found to be crucial to induce this selectivity, but subsequent lowering of the pressure does not affect activity and selectivity. A startup period with strongly varying activity and selectivity is observed indicating thorough catalyst reconstruction. The original Rh^0 clusters are converted to multi-nuclear $Rh_6(CO)_{16}$ and $CH_3Rh_y(CO)_x$ and mononuclear $CH_3Rh(CO)_x$ carbonyl complexes. This results in a working catalyst that is vastly superior to a fresh catalyst. Stable acetate groups, but not the surface-bound acetyl groups, are formed and detected by FTIR.

On the basis of results acquired by various experimental techniques different mechanisms for the formation of ethanol and acetaldehyde have been suggested.[526] Temperature-programmed desorption (TPD), temperature-programmed surface reaction (TPSR), and XPS showed that the structure of the active sites may be formulated as $(Rh^0_x Rh^+_y)$–O–M^{n+} (M = Mn or Zr). The tilt adsorbed CO species is the main precursor for CO dissociation, which is directly hydrogenated to CH_2–O species followed by CH_2 insertion and hydrogenation to ethanol. Acetaldehyde, in turn, is formed through insertion into surface CH_3–Rh species and subsequent hydrogenation. The role of the promoters is to stabilize the intermediate surface acetyl species.

Studies have been devoted to further elucidate the effects of structural changes of zirconia on activity and selectivity in *isosynthesis* and to acquire information on the mechanism of the formation of isomeric hydrocarbons.

Zirconia catalysts prepared by precipitation exhibit higher activity, while those prepared by a hydrothermal process show higher selectivity toward iso-C_4 products.[527] ZrO_2 with 7% CeO_2 is the most promising catalyst to afford fairly high conversion (25–40%) and reasonable selectivities (around 20%). Higher selectivity, again, is the main characteristic of catalysts made by the sol-gel method, which is due to increased basicity.[528] Basic sites were also found to affect the selectivity of the formation of isobutylene, whereas acidic sites increased the conversion of CO.[529] Large differences in selectivities were observed over catalysts prepared by precipitation from various zirconium salts attributed to the presence of precursor anions in the hydroxide gel.[530] The product distribution of isosynthesis changed with increasing calcination temperature. The highest selectivity to isobutane and isobutylene was observed on catalysts calcined at 500–700°C.

Tracer studies with ^{13}C showed that the carbon of methoxy species formed in CO hydrogenation appeared as the central C atom of isobutylene.[531] Methoxy was shown to be reduced to surface methyl species that was transformed to surface acetyl group via insertion of CO into the methyl–metal bond. Aldol condensation of C_2 oxygenates with formaldehyde affords iso-C_4 products.[532] Earlier results of a different approach for isosynthesis using metal–zeolite catalyst systems are summarized in a 1998 review.[533]

Methanation, that is, the hydrogenation of carbon oxides, is a potentially important transformation. Hydrogenation of CO_2 to methane, which has attracted great interest, is treated in Section 3.2.1. Interestingly, only a few new (at the time of writing) studies have addressed the problem of the reduction of CO to methane. Ni on Al_2O_3 catalyst doped with 1.5% CeO_2 showed enhanced activity in methanation.[534] Doping resulted in improved reducibility of Ni and lowering the reaction temperature. The effects were attributed to an electronic interaction between the dopant and the active sites. The addition of an appropriate amount of copper to nickel was found to prevent the loss of active surface by sintering of Ni without inhibiting the rate of CO hydrogenation.[535]

3.6.2. Direct Coupling of Methane

Catalytic Oxidative Condensation. Extensive research efforts have been devoted since the early 1990s to the oxidative coupling of methane to produce ethylene but have not resulted in a commercial process. Serious engineering problems are also limiting factors hampering practical application, especially the large exothermic nature of the reaction. This results in a substantial increase (150–300°C) in the temperature of the catalyst zone.[536,537] Furthermore, metals used in reactor construction catalyze the total combustion of methane, limiting productivity. Reviews[538–540] and symposia proceedings[489,541,542] give adequate coverage of the most recent developments.

Studies conducted since the mid-1990s have lent additional support to the reaction network shown in Scheme 3.1 and emphasized the interesting class of heterogeneous–homogeneous radical nature of the system.[539] Isotope labeling experiments have demonstrated that at small conversions levels most of the CO_2 is derived from methane, whereas at high conversion levels required for practical applications, ethylene is the dominant source of CO_2.[543]

Efforts have also been made to find new catalysts with improved catalyst performance. The main goal is to inhibit the secondary reactions of ethylene, whereas the activation of methane still occurs at a high level. $LiSbTe_2$, when subjected to oxidative pretreatment prior to use, proved to be highly selective for C_2 product formation (13% yield with a selectivity of 88% at 800°C).[544] Under similar conditions 22% yield (56–59% selectivity) was achieved with a $CaCl_2$-promoted Ca chlorophosphate catalyst.[545] A multicomponent catalyst (Na–S–W–P–Zr–Mn) developed by an artificial neural network-aided design gave a 20.77% C_2 yield.[546] The best results, a C_2 selectivity of 80% at a methane conversion level of 20% with ethane and ethylene formed in approximately equal amounts using $Mn-Na_2WO_4-SiO_2$ and Na_2WO_4-MgO, are still not satisfactory for commercial applications.[543] On the basis of XAFS and XPS characterization, the outstanding properties of these catalysts were attributed to the combination of two active centers with different oxidation states capable of transferring electrons and oxygens simultaneously: W is responsible for activation of methane, while Mn is active in the transformation and transportation of oxygens.[547] Mn-based binary oxides are active with CO_2 as oxidant.[548]

A new catalyst formulation containing alkali metals and W on a silica support gives more promising results.[549] Alkali metals are able to lower the phase transition temperature from amorphous silica to α-crystoballite, shown to be critically important for an effective catalyst, while incorporation of W enhances catalytic activity to ensure high methane conversion and excellent ethylene selectivity. An alkali-stabilized tungsten oxo species is thought to be the active site.

New results on the improvement of membrane reactors developed and applied to prevent nonselective oxidation by oxygen are still reported.[550,551] For example, yttria-doped Bi_2O_3 gives C_2 selectivity about 30% higher at the same C_2 yield than that found in the co-feed fixed-bed reactor.[550] The best values are 17% yield with 70% selectivity.

Characterization of the active catalyst systems to identify the catalytically active surface sites and active form of surface oxygen has also attracted interest. Observations made by IR spectroscopy showed the presence of O^{2-} species on fluoride-containing rare-earth-based systems.[552] A study applying a Ni catalyst supported on La$_2$O$_3$ promoted by Sr^{2+} demonstrated that Ni shifts the main reaction to syngas formation instead of oxidative coupling.[553] In contrast, Ni supported on CeO$_2$ or Sm$_2$O$_3$ doped with Li showed high selectivity in hydrocarbon formation.[554] Anionic vacancies and coordinatively unsaturated surface oxygen atoms were suggested to be involved in hydrocarbon formation. Structural analysis by XRD, XPS, XANES, and SEM techniques of Li-doped MgO, the most frequently studied catalyst, led to the conclusion that the defect species containing Li$^+$–O$^-$ in both surface and bulk can act as the active species for producing C$_2$ compounds with high selectivity.[555] Contrary to earlier suggestions, new EPR and UV–vis spectroscopic studies demonstrated that Li^0 and F centers are not required for hydrocarbon formation.[556] Further support has been obtained for the existence of a new precursor, most probably O$_3^{2-}$, formed from reversible redox coupling of an O$_2$ adspecies at an anionic vacancy with a neighboring O^{2-} in the surface lattice.[557,558]

It was demonstrated in several studies that high product yields may be achieved during the oxidative coupling of methane by applying a recycle reactor with continuous removal of ethylene similar to those reported in two earlier studies.[328,329] In a continuous adsorption–desorption system the ethylene yield was 60%.[559] Complexation of ethylene with aqueous Ag$^+$ ions through a membrane contactor and thermal regeneration allows higher than 70% product yields.[560,561] Similar product yields were reported when the membrane contactor was replaced by a Ga/H-ZSM-5 catalyst operating at 520°C that converted ethylene to aromatics, mainly benzene and toluene.[560,562] Steady-state conversion of ethylene to C$_{4+}$ aliphatics was also demonstrated.[563] In this case, oxidative coupling of methane was performed using a Mn–Na$_2$WO$_4$–SiO$_2$ catalyst at 800°C. This reaction is coupled with the subsequent conversion of ethylene over an H-ZSM-5 zeolite catalyst. The total yields of C$_{4+}$ hydrocarbons were in the range of 60–80%, and the yields of nonaromatics were about 50–60%.

Oxidative coupling of methane to form C$_2$ hydrocarbons may also be accomplished by using plasma technology. This may be performed in the presence of catalysts, such as zeolites,[564,565] metal oxides,[564] or La$_2$O$_3$ promoted by Sr.[566] The process may also be carried out without a catalyst.[567]

Other Processes.   Direct nonoxidative conversion of methane to higher hydrocarbons has been studied to improve the yield and selectivity of the desired products.

Selective synthesis of acetylene (>90%) from methane was accomplished by microwave plasma reactions.[568] Conversion of methane to acetylene by using direct current pulse discharge was performed under conditions of ambient temperature and atmospheric pressure.[569] The selectivity of acetylene was >95% at methane conversion levels ranging from 16 to 52%. In this case oxygen was used to effectively remove deposited carbon and stabilize the state of discharge. Similar high selectivities can be achieved by high-power pulsed radiofrequency and microwave catalytic processes over carbon as catalyst.[570]

Microwave heating was also used to induce catalytic oligomerization of methane to afford C_2–C_4 hydrocarbons.[571] Changing the catalysts and the applied power and the use of diluent gas (He) allowed significant alteration of product selectivities. Selectivity to benzene over nickel powder or activated carbon was 24 and 33%, respectively.

Photoinduced coupling of methane was reported over TiO_2 at 100–200°C in the presence of oxygen, albeit in a yield of coupling products of about 0.5%.[572] Better yields (3.02–5.9%) were achieved in nonoxidative photoinduced coupling at room temperature on silica–alumina or alumina evacuated at 800°C.[573]

Two-step homologation of methane, also called *low-temperature coupling*, is based on the ability of transition metals to catalyze the dissociative adsorption of methane to produce CH_x surface species and hydrogen. Subsequently the carbonaceous surface intermediates thus formed are hydrogenated to form higher hydrocarbons.[574] The original studies were performed using Pt[575] and Co, Ru, and Rh.[576] The carbonaceous surface residues were formed between 175 and 527°C, followed by a hydrogenation to form hydrocarbons up to C_5 at a temperature range of 27–127°C. A total C_{2+} yield of 13% on 5% Ru on SiO_2 at 100% methane conversion was achieved.[576] Later other precious metals supported on silica were applied to have higher yields of ethane but still low efficiency.[577] Catalysts also used in further studies include Co on SiO_2,[578] Ru on SiO_2,[579] Rh on CeO_2+SiO_2,[580] and EUROPT-1 (6.3% Pt on SiO_2).[581,582]

Co–Pt on NaY zeolite[583] and Ru–Co on NaY bimetallic catalysts[584] gave exceptionally high yields and high selectivities in the formation of hydrocarbons at 250°C. This synergistic effect can be interpreted by insertion of Co into the other metal inside the zeolite cages, causing a preferential coupling of CH_x species versus its hydrogenation to methane. IR studies confirmed the formation of CH_x species on a catalyst containing Co clusters encapsulated into NaY.[585] Methane can be mostly converted to cyclohexane and its methyl-substituted derivatives using Ni–Cu on SiO_2.[586] Increasing the hydrogen pressure strongly enhances the formation of heavier products. The reasons for this unique selectivity are the especially good fit of the C_6 ring to the Ni (111) planes and the decrease in hydrogenolysis reactions due to Cu.

The two-step homologation of methane can be performed using a dual-temperature sequence to favor thermodynamics of each step, specifically, a fast methane adsorption at about 400°C, then hydrogenation at lower temperature (approximately 100°C).[576] Alternatively, experiments may be conducted under isothermal conditions; that is, both steps can be performed at the same temperature in the range of 200–320°C.[575,580,582] Both methane conversion and the yield of higher hydrocarbons could be significantly enhanced at a lower temperature in a Ru-catalyzed process using a Pd–Ag membrane reactor.[587]

A heterogeneous hydrogen-accumulating system containing porous Ti with 0.4 wt% Ni combined with high-purity Ti chips was tested for methane activation.[588] Methane conversion to C_1–C_4 hydrocarbons reached a value of 20% with 50–55% of ethylene at 450°C and 10 atm. The hydrogen formed was accumulated as TiH_2.

In 1993 methane was reported to undergo transformation to aromatics under nonoxidative conditions at 700°C and atmospheric pressure by using molecular sieves modified by transition metals.[589] Since then this catalyst system and reaction, called *dehydroaromatization*, have been thoroughly investigated.[539,590] It was found that the Mo–H-ZSM-5 catalyst with 2–6% Mo loading is the best of all catalysts tested, which can give a methane conversion of about 10% and benzene selectivities of 50–60%.[591–594] Other acidic solids such as a pentasil-type high-silica zeolite modified with P and rare-earth oxides[595] and HMCM-22[596] were also shown to be effective. Fe–H-ZSM-5 also exhibits good characteristics[597] and modification of Mo–H-ZSM-5 with Fe[598] or Co[598,599] improves activity and stability. Dealumination by steaming results in high coke resistance and higher selectivity toward aromatics.[600] The best results have been disclosed in a recent study: 13.4% conversion of methane could be achieved with a Mo–H-ZSM-5 catalyst promoted with 0.1% Ga, which was shown to induce the reduction of molybdenum.[601]

Important features of this catalyst are the presence of Brønsted acid sites, the state of Mo species (Mo_2C or MoO_xC_y), and the channel structure of the zeolite component, which play a key role in methane activation and aromatization.[591,602–605] The catalysts usually exhibit a rapid initial decline of activity followed by a much slower deactivation. A constant selectivity of benzene is achieved, but naphthalene selectivity goes through a maximum, which is attributed to shape selectivity. It becomes more pronounced when the zeolite channels become partially coked, but small amounts of CO or CO_2 in the feed inhibits deactivation.[598] The use of a new concept, the continuous removal of hydrogen from the system by H-transport membrane reactor of a dense $SrZr_{0.95}Y_{0.05}O_3$ allowed Iglesia and coworkers to achieve near-equilibrium conversion (10–12%) with higher than 90% selectivity.[606]

A part of Mo species migrates into the channels of the zeolite during the process of calcination.[607] An induction period is usually recognized in which Mo^{6+} species possibly located on the external surface of the support are partially reduced possibly to $(Mo_2O_5)^{2+}$ dimers.[607] One study revealed, however, a slight catalytic activity of Mo^{6+} species for the methane–benzene conversion.[608] Interaction between Mo and the support is necessary for the high dispersion of Mo species required to achieve optimum catalytic performance. Highly dispersed MoC_x sites were detected on the working catalyst.[609] It is accepted that the Mo–H-ZSM-5 catalyst acts like a bifunctional catalyst.[539,590] Experimental evidence supports the view that methane is dissociated on the Mo carbide clusters supported on the zeolite to form CH_x ($x > 1$) and C_2 species as the primary intermediates. These subsequently convert to aromatics via oligomerization, cracking, and cyclization reactions at the interface of Mo and H-ZSM-5 having the optimum Brønsted acidity. The highest activity was observed at intermediate Mo content (Mo to Al ~ 0.4)[607] and on zeolite samples with SiO_2 to Al_2O_3 ratios around 40.[603]

3.6.3. Hydrocarbons through Methane Derivatives

Methanol Synthesis. Cu–ZnO-based catalyst are well known to be highly active for the methanol synthesis, that is, for the hydrogenation of CO.[610] The active site, the role of ZnO, and the reaction mechanism are still subjects of considerable controversy, even though efforts have been made to obtain consensus by considering the experimental conditions.[611] The problems originate from the fact that various research groups use rather different experimental conditions. The significant features of conditions used by the ICI group are high total pressure (typically 50 bar), high $H_2 : CO_2$ ratio, and high space velocity, which gives low conversions, well below equilibrium.[611] Under these conditions, good correlations of rate of methanol production with Cu metal area were obtained. Most other groups carried out experiments at low–medium total pressure (typically 1–10 bar), low $H_2 : CO_2$ ratio, and high conversion, sometimes approaching equilibrium. Under these conditions, strong dependence of catalytic activity on support compositions is seen, with catalysts containing ZnO giving the highest rate values. More recent studies also support the view that CO_2 rather than CO is the direct precursor to methanol on copper-containing catalysts. This is based on IR spectra of adsorbed intermediates,[39] and on transient kinetic studies.[612,613] Correlation between methanol synthesis activity and CO_2 content also supports this conclusion.[614]

The problem of the oxidation state of copper, although difficult to establish, still does not seem to be resolved. X-ray photoelectron spectroscopy (XPS) studies have detected significant amounts of Cu^+ and Cu^{2+} concentrations in materials prepared under well-controlled conditions. Naturally, this approach cannot detect Cu species present during catalysis because they are likely to be modified by the quenching protocol, by oxidation with air during transfer into the vacuum system, or by auto-oxidation by protons formed during the initial reduction of the copper oxide catalyst precursor. It is also reported that both Cu^+ and Cu^0 species are essential for methanol formation and the ratio of Cu^+ to Cu^0 determines the specific activity.[615] A new study lends further support to this suggestion by disclosing a correlation between catalytic activity and the concentration of a new copper oxide phase formed in the calcination of Cu–Zn–ZrO_2 catalysts by the dissolution of Zr ions in Cu oxide.[616] When the catalyst is reduced, the Cu^{2+} species present in the ZrO_2 lattice are transformed to Cu^+ species. The addition of ZrO_2 to Cu–ZnO, therefore, gives rise to the formation of Cu^+ species that are related to the methanol synthesis activity in addition to Cu metal particles. Cu–O–Zn sites have also been suggested to be active in methanol synthesis.[25] These are created when Zn dissolves into Cu particles to form Cu–Zn alloy. Zn over the dissolution limit becomes oxidized to form the Cu–O–Zn sites, which are located at the interface with the Cu particles.

All in situ X-ray absorption spectroscopy (XAS) studies, however, indicate that Cu metal is the principal copper species present. Results disclosed in a new study[27] are consistent with earlier work and findings[617] that Cu^0 is the dominant copper species under synthesis conditions. There is also evidence indicating a bifunctional mechanism involving support hydroxyl groups. An induction period with fivefold increase in methanol production rate was found on a Cu on SiO_2 catalyst without detectable chemical or morphological changes in Cu. This may indicate a slow accumulation of a species on the surface of support. Surface hydroxyl groups were suggested to be involved in the formation of formate from CO–H_2 on the support. Furthermore, a transient increase in methanol production was observed when CO_2 was introduced to the feed. This supports earlier observations that formate is formed directly from CO_2 not requiring the participation of the support. An important new information, however, is that formate was detected on unsupported polycrystalline Cu during methanol synthesis using sum frequency generation.[618]

In a new study of a series of binary Cu–ZnO catalysts a correlation was found between methanol synthesis activity and strain in the Cu metal phase.[619] Structural defects of Cu resulting from ZnO dissolved in Cu, incomplete reduction, or epitaxial orientation to ZnO are believed to bring about strain, which modifies the Cu surface and, consequently, affects the catalytic activity. The higher amount of water formed in methanol synthesis from a CO_2-rich feed compared to a CO-rich feed brings about significant catalyst deactivation by inducing crystallization of both Cu and ZnO.[620]

With respect to other catalyst systems, zirconia has attracted much attention because it shows a good activity for both CO–H_2 and CO_2–H_2 reactions.[32,39,45,621,622] High specific surface area, uniform particle diameter, and the presence of ultrafine particles were correlated with high methanol formation activity.[623] Studies of the mechanism of methanol synthesis have revealed that zirconia is an active component of the catalyst, based on the information that all intermediates were detected to be adsorbed on ZrO_2. It was found that CO hydrogenation involves the spillover of H atoms formed on Cu to the surface of ZrO_2.[42] The atomic H then participates in the hydrogenation of carbon-containing species to form formate species (**1**), which then undergoes sequential hydrogenation to form methoxy species [Eq. (3.5)]. The enhanced rate of methanol synthesis by the addition of ZrO_2 to Cu on SiO_2 is attributed to a change in the reaction pathway from a high-energy formyl mechanism to the lower-energy formate route.[39] A comparison of the dynamics of H–D exchange and the formation of surface methoxy species revealed that spillover from Cu is more than an order of magnitude faster than the rate of methanol formation. Spillover of hydrogen, consequently, is not a rate-limiting step in the synthesis of methanol over Cu on ZrO_2.

Although other experimental evidence also exists for a formate to methoxy mechanism, Baiker argues that formate intermediates are not involved in the reduction of carbon oxides to methanol on ZrO_2-based catalysts.[8,35,624] Rather, formates are intermediates in methanation (see Section 3.2.1).

Ca and La oxides, when added to Cu on ZrO_2, exhibit a strong promoting effect on methanol synthesis.[614] Addition of ZrO_2 to Cu on SiO_2 increases the rate of methanol synthesis, which is attributed to the presence of acidic bridging OH groups on the surface of ZrO_2, which can adsorb Lewis basic CO to form formate groups, a key intermediate along the pathway from CO to methanol.[40]

Pd on ceria catalysts prepared by various methods were shown to exhibit varying characteristics depending on the preparation methods.[625] Specifically, Pd on ceria samples prepared by the deposition–precipitation method are highly active in comparison with the catalyst prepared by the conventional impregnation method. Cationic palladium species are present in the former samples after reduction with hydrogen at 300°C, suggesting that the active species are produced by strong interaction between palladium particles and the support. The results of a comparative study using Pd supported on various oxides with Pd on ZrO_2 exhibiting the highest activity also point to the participation of cationic Pd species formed through metal–support interaction.[626] As an exception, methanol synthesis from syngas has been accomplished using the $Ni(CO)_4$ and KOMe catalyst system in the homogeneous phase.[627]

The technological improvements of methanol synthesis by comparing 25 technologies are treated in a review paper.[628]

Since the early 1990s there has been intensive research centered on the development of active and selective catalysts for the synthesis of higher alcohols.[629,630] Earlier work with ZnO–Cr_2O_3 catalysts promoted by alkali metals showed increased production of higher alcohols. The process, however, was rather nonselective, although alcohol production could be increased by heavy alkali ions.[631,632] Later the low-temperature Cu–ZnO-based catalysts were found to be more active and selective toward the synthesis of higher alcohols when promoted by heavy alkali ions. For example, the addition of 4 wt% Cs^+ to Cu–ZnO increases the ethanol formation by an order of magnitude [from 3.0 to 24.6 g $(kg_{cat} h)^{-1}$], whereas the methanol formation rates increased only slightly [from 200 to 271 g $(kg_{cat} h)^{-1}$].[633] In addition to methanol, principally C_2–C_4 alcohols, mainly isobutyl alcohol, are formed. A unique carbon chain growth mechanism, referred to as *oxygen retention reversal aldol condensation*, centered on β carbon (carbon adjacent to the carbon bonded to oxygen) addition, was suggested to be operative in the formation of C_3–C_4 alcohols.[629] The slow, rate-determining step in forming the higher alcohols from synthesis gas was proved to be the C_1–C_2 step, that is, the formation of the first C–C bond.[634] Subsequent formation of C_3 and C_4 alcohols proceeds at a higher rate, and this leads to a depletion of ethanol. Branched alcohols are terminal products, and this leads to an enrichment in the product mixture of branched alcohols, mainly isobutyl alcohol.[630]

Another class of catalysts for higher alcohol production is MgO-based Cu catalysts promoted by K.[634–636] The mechanism, that is, the initial C–C bond formation pathway, however, is different from those of other catalysts. Studies with labeled ^{13}CO and $^{12}CH_3OH$ mixtures show that direct reactions of CO without significant involvement of labeled methanol in the feed is the predominant pathway for ethanol formation. Pathways involve coupling reactions of surface formate and methyl formate species formed directly from ^{13}CO account for formation of the initial C–C bond in ethanol.[636,637] Formation of 1-propanol and isobutyl alcohol proceeds also via aldol condensation of ethanol-derived species with methanol-derived C_1 species following the rules of base-catalyzed aldol condensation. The role of alkali metals in these catalysts is to block acid sites that form dimethyl ether and hydrocarbons.[636] In contrast, the C–C bond in ethanol is formed predominantly via a faster route on Cs-promoted Cu–ZnO–Al_2O_3. This involves two methanol-derived C_1 species.

Attempts were also made to increase the productivity of isobutyl alcohol by introducing ZrO_2 to the catalyst that is known to be an isosynthesis component (see Sections 3.2 and 3.5.1) and by adding redox oxides and strong bases.[630]

High productivities and selectivities were observed with these catalysts such as Li–Pd–Zr–Zn–Mn. A disadvantage of these catalysts is the high operation temperature required to achieve high alcohol productivities.

Advances in higher alcohol synthesis that need to be made include improving productivity and selectivity. Higher spacetime yields are also required which may be achieved by dual catalyst bed reactors, the use of slurry phase synthesis process to increase CO conversion levels, and injection of lower alcohols into the reactant stream to increase the rate of higher alcohol formation.[630]

Methanol Conversion to Hydrocarbons. Methanol-to-gasoline (MTG) chemistry on zeolite H-ZSM-5 and the closely related process of methanol-to-olefin (MTO) chemistry also catalyzed by H-ZSM-5 attracted attention because it is one of the most successful routes for the conversion of methanol to synthetic fuel.[638] It has also generated great interest in fundamental studies. Two excellent reviews on catalytic materials[639] and process[640] technology are available. MTG/MTO chemistry has been intensely studied, and new basic information is still acquired by using a wide variety of methods, both theoretical studies[641–644] and spectroscopic techniques.[645–647] It has been established that methanol is first dehydrated to dimethyl ether and an equilibrium mixture of these two compounds and water is formed. A study using an in situ flow variable-temperature MAS NMR revealed, however, that there is no equilibrium between methanol and dimethyl ether at high temperature and that surface methoxy groups do not exist at high temperature.[648] Methanol and dimethyl ether then converted to olefins, aliphatics, and aromatics up to C_{10}. It is also known that there is an induction period before the onset of extensive hydrocarbon synthesis. The mechanism of these transformations, especially the formation of the first C–C bond and the nature of the intermediates involved, is still a matter of significant debate. Further fundamental questions are how methanol or dimethyl ether is converted to alkenes, why propylene is formed with higher initial selectivity than other olefins, and how aromatics are produced.

Theoretical investigations indicated[641,642] that two main pathways are possible for the first, dehydration step. One proceeds via a methoxy surface intermediate (**10**) formed as a result of the adsorption of methanol on acidic site of the zeolite (HO–Z) to give dimethyl ether [Eqs (3.63) and (3.64)]. The other, direct route is a reaction between two methanol molecules adsorbed simultaneously on the same active site [Eqs (3.65) and (3.66)]:

$$CH_3-OH \ + \ HO-Z \ \longrightarrow \ CH_3-\overset{+}{O}H_2 \ + \ {}^-O-Z \ \longrightarrow \ CH_3-OZ \ + \ H_2O \quad (3.63)$$
$$\textbf{10}$$

$$CH_3-OH \ + \ CH_3-OZ \ \longrightarrow \ CH_3-O-CH_3 \ + \ HO-Z \quad\quad\quad (3.64)$$

$$CH_3-\overset{+}{O}H_2 \ + \ CH_3-OH \ + \ {}^-O-Z \ \longrightarrow \ CH_3-\overset{+}{O}H-CH_3 \ + \ {}^-O-Z \ + \ H_2O \quad (3.65)$$

$$CH_3-\overset{+}{O}H-CH_3 \ + \ {}^-O-Z \ \longrightarrow \ CH_3-O-CH_3 \ + \ HO-Z \quad\quad (3.66)$$

A complex and comprehensive simulation of the formation of dimethyl ether, the first intermediate in the MTG process, has been reported.[649]

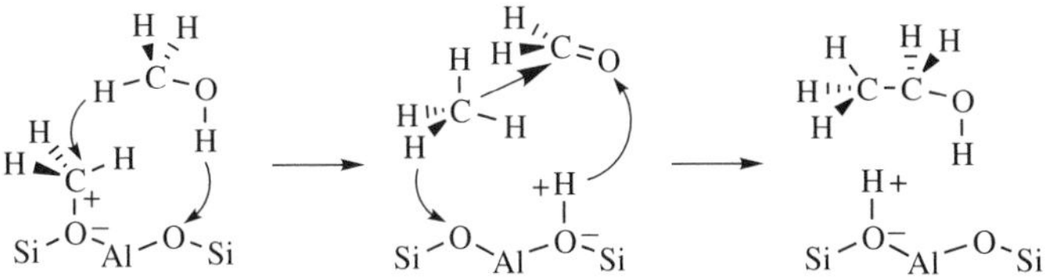

Scheme 3.3

A new mechanism, called the *methane–formaldehyde mechanism*, has been put forward for the transformation of the equilibrium mixture of methanol and dimethyl ether, that is, for the formation of the first C–C bond.[643] This, actually, is a modification of the carbocation mechanism that suggested the formation of ethanol by methanol attaching to the incipient carbocation CH_3^+ from surface methoxy.[460,462] This mechanism (Scheme 3.3) is consistent with experimental observations and indicates that methane is not a byproduct and ethanol is the initial product in the first C–C bond formation. Trimethyloxonium ion, proposed to be an intermediate in the formation of ethyl methyl ether,[447] was proposed to be excluded as an intermediate for the C–C bond formation.[641] The suggested role of impurities in methanol as the reason for ethylene formation is highly speculative and unsubstantiated.

Experimental work using a pulse-quench catalytic reactor[650] to probe transition between induction reactions and hydrocarbon synthesis on a working H-ZSM-5 catalyst has resulted in the suggestion that stable cyclopentenyl cations are formed during the induction period from small amounts of olefins formed in an induction reaction.[647] One study reports a surprising observation, namely, enhanced aromatic formation over the physical mixture of Ga_2O_3 and H-ZSM-5 (1 : 1) (18.2% of benzene and methylbenzenes).[651]

REFERENCES

1. J. M. Fox, III, *Catal. Rev.-Sci. Eng.* **35**, 169 (1993).

2. M. Calvin, *Cell Biophys.* **9**, 189 (1986); *J. Chem. Educ.* **64**, 335 (1987).

3. G. A. Olah et al., U.S. Patent 5,559,638 (1997).

4. G. A. Olah and G. K. S. Prakash, U.S. Patent 5,928,806 (1999).

5. G. A. Olah, *Acc. Chem. Res.* **20**, 422 (1987).

6. B. Denise and R. P. A. Sneeden, *Chemtech* **12**, 108 (1982).

7. Y. Borodko and G. A. Somorjai, *Appl. Catal., A* **186**, 355 (1999).

8. A. Baiker, *Appl. Organomet. Chem.* **14**, 751 (2000).

9. A. Behr, *Carbon Dioxide Activation by Metal Complexes*, VCH, Weinheim, 1988.

10. *Catalytic Activation of Carbon Dioxide*, ACS Symp. Ser. Vol. 363, W. M. Ayers, ed., American Chemical Society, Washington, DC, 1988.

11. M. M. Halmann, *Chemical Fixation of Carbon Dioxide: Methods for Recycling CO$_2$ into Useful Products*, CRC Press, Boca Raton, FL, 1993.

12. M. M. Halmann, *Greenhouse Gas Carbon Dioxide Mitigation: Science and Technology*, Lewis Publishers, Boca Raton, FL, 1999.

13. *Carbon Dioxide Chemistry: Environmental Issues*, J. Paul and C.-M. Pradier, eds., Royal Society of Chemistry, Cambridge, UK, 1994.

14. *Advances in Chemical Conversions for Mitigating Carbon Dioxide*, Studies in Surface Science and Catalysis, Vol. 114, T. Inui, M. Anpo, K. Izui, S. Yanagida, and T. Yamaguchi, eds., Elsevier, Amsterdam, 1998.

15. *Greenhouse Gas Control Technologies*, B. Eliasson, P. Riemer, and A. Wokaun, eds., Elsevier, Amsterdam, 1999.

16. *5th Int. Conf. Carbon Dioxide Utilization*, Karlsruhe, 1999, E. Dinjus and O. Walter, eds., *Appl. Organomet. Chem.* **14**, 749–873 (2000); **15**, 87–150 (2001).

17. D.-K. Lee, D.-S. Kim, and S.-W. Kim, *Appl. Organomet. Chem.* **15**, 148 (2001).

18. L. Fan, Y. Sakaiya, and K. Fujimoto, *Appl. Catal., A* **180**, L11 (1999).

19. J. Yoshihara and C. T. Campbell, *J. Catal.* **161**, 776 (1996).

20. R. A. Hadden, B. Sakakini, J. Tabatabaei, and K. C. Waugh, *Catal. Lett.* **44**, 145 (1997).

21. C. V. Ovesen, B. S. Clausen, P. Schiøtz, P. Stoltze, H. Topsøe, and J. K. Nørskov, *J. Catal.* **168**, 133 (1997).

22. N.-Y. Topsøe and H. Topsøe, *J. Mol. Catal. A: Chem.* **141**, 95 (1999).

23. T. Fujitani, T. Matsuda, Y. Kushida, S. Ogihara, T. Uchijima, and J. Nakamura, *Catal. Lett.* **49**, 175 (1997).

24. T. Fujitani and J. Nakamura, *Appl. Catal., A* **191**, 111 (2000).

25. Y. Choi, K. Futagami, T. Fujitani, and J. Nakamura, *Catal. Lett.* **73**, 27 (2001).

26. Y. Choi, K. Futagami, T. Fujitani, and J. Nakamura, *Appl. Catal., A* **208**, 163 (2001).

27. G. Meitzner and E. Iglesia, *Catal. Today* **53**, 433 (1999).

28. T. Fujitani and J. Nakamura, *Catal. Lett.* **63**, 245 (1999).

29. E. K. Poels and D. S. Brands, *Appl. Catal., A* **191**, 83 (2000).

30. M. Saito, T. Fujitani, M. Takeuchi, and T. Watanabe, *Appl. Catal., A* **138**, 311 (1996).

31. S. Luo, J. Wu, J. Toyir, M. Saito, M. Takeuchi, and T. Watanabe, in *Advances in Chemical Conversions for Mitigating Carbon Dioxide*, Studies in Surface Science and Catalysis, Vol. 114, T. Inui, M. Anpo, K. Izui, S. Yanagida, and T. Yamaguchi, eds., Elsevier, Amsterdam, 1998, p. 549.

32. J. Wu, S. Luo, J. Toyir, M. Saito, M. Takeuchi, and T. Watanabe, *Catal. Today* **45**, 215 (1998).

33. M. Saito, M. Takeuchi, T. Fujitani, J. Toyir, S. Luo, J. Wu, H. Mabuse, K. Ushikoshi, K. Mori, and T. Watanabe, *Appl. Organomet. Chem.* **14**, 763 (2000).

34. J. Toyir, P. R. de la Piscina, J. L. G. Fierro, and N. Homs, *Appl. Catal., B* **29**, 207 (2001).

35. J. Wambach, A. Baiker, and A. Wokaun, *Phys. Chem. Chem. Phys.* **1**, 5071 (1999).

36. R. A. Köppel, A. Baiker, and A. Wokaun, *Appl. Catal., A* **84**, 77 (1992).

37. R. A. Köppel, C. Stöcker, and A. Baiker, *J. Catal.* **179**, 515 (1998).

38. C. Fröhlich, R. A. Köppel, A. Baiker, M. Kilo, and A. Wokaun, *Appl. Catal.* **106**, 275 (1993).

39. I. A. Fisher and A. T. Bell, *J. Catal.* **172**, 222 (1997); **178**, 153 (1998).

40. T. C. Schilke, I. A. Fisher, and A. T. Bell, *Catal. Lett.* **54**, 105 (1998).

41. G.-X, Qi, X.-M. Zheng, J.-H. Fei, and Z.-Y. Hou, *Catal. Lett.* **72**, 191 (2001).

42. K.-D. Jung and A. T. Bell, *J. Catal.* **193**, 207 (2000).

43. K. K. Bando, K. Sayama, H. Kusama, K. Okabe, and H. Arakawa, *Appl. Catal., A* **165**, 391 (1997).

44. J. Weigel, C. Fröhlich, A. Baiker, and A. Wokaun, *Appl. Catal., A* **140**, 29 (1996).

45. J. Weigel, R. A. Koeppel, A. Baiker, and A. Wokaun, *Langmuir* **12**, 5319 (1996).

46. E. E. Ortelli, J. M. Weigel, and A. Wokaun, *Catal. Lett.* **54**, 41 (1998).

47. Y. Ma, Q. Sun, D. Wu, W.-H. Fan, Y.-L. Zhang, and J.-F. Deng, *Appl. Catal., A* **171**, 45 (1998).

48. W. Ning, H. Shen, and H. Liu, *Appl. Catal., A* **211**, 153 (2001).

49. J. Toyir, M. Saito, I. Yamauchi, S. C. Luo, J. G. Wu, I. Takahara, and M. Takeuchi, *Catal. Today* **45**, 245 (1998).

50. B. J. Liaw and Y. Z. Chen, *Appl. Catal., A* **206**, 245 (2001).

51. I. Melian-Cabrera, M. L. Granados, P. Terreros and J. L. G. Fierro, *Catal. Today* **45**, 251 (1998).

52. K. K. Bando, H. Arakawa, and N. Ichikuni, *Catal. Lett.* **60**, 125 (1999).

53. H. Kusama, K. Okabe, K. Sayama, and H. Arakawa, *Appl. Organomet. Chem.* **14**, 836 (2000).

54. H. Kusama, K. Okabe, and H. Arakawa, *Appl. Catal., A* **207**, 85 (2001).

55. T. Inui, T. Yamamoto, M. Inoue, H. Hara, T. Takeguchi, and J.-B. Kim, *Appl. Catal., A* **186**, 395 (1999).

56. K.-W. Jun, W.-J. Shen, and K.-W. Lee, in *Greenhouse Gas Control Technologies*, B. Eliasson, P. Riemer, and A. Wokaun, eds., Elsevier, Amsterdam, 1999, p. 421.

57. J.-L. Tao, K.-W. Jun, and K.-W. Lee, *Appl. Organomet. Chem.* **15**, 105 (2001).

58. G.-X. Qi, J.-H. Fei, X.-M. Zheng, and Z.-Y. Hou, *Catal. Lett.* **72**, 121 (2001).

59. F. Solymosi and A. Erdöhelyi, *J. Mol. Catal.* **8**, 471 (1980).

60. I. A. Fisher and A. T. Bell, *J. Catal.* **162**, 54 (1996).

61. M. A. Henderson and S. D. Worley, *J. Phys. Chem.* **89**, 1417 (1985).

62. F. Solymosi, A. Erdöhelyi, and T. Bánsági, *J. Catal.* **68**, 371 (1981).

63. S. Ichikawa, *J. Mol. Catal.* **53**, 53 (1989).

64. F. Solymosi, *J. Mol. Catal.* **65**, 337 (1991).

65. H.-J. Freund and M. W. Roberts, *Surf. Sci. Reports* **25**, 225 (1996).

66. M. Marwood, R. Doepper, and A. Renken, *Appl. Catal., A* **151**, 223 (1997).

67. F. Solymosi, A. Erdöhelyi, and M. Kocsis, *J. Chem. Soc., Faraday Trans. 1* **77**, 1003 (1981).

68. S. Scirè, C. Crisafulli, R. Maggiore, S. Minicò, and S. Galvagno, *Catal. Lett.* **51**, 41 (1998).

69. Z. Zhang, A. Kladi, and X. E. Verykios, *J. Catal.* **156**, 37 (1995).

70. A. Trovarelli, C. de Leitenburg, G. Dolcetti, and J. Llorca, *J. Catal.* **151**, 111 (1995).

71. C. de Leitenburg, A. Trovarelli, and J. Kašpar, *J. Catal.* **166**, 98 (1997).

72. H. Kusama, K. K. Bando, K. Okabe, and H. Arakawa, *Appl. Catal., A* **197**, 255 (2000).

73. M. Yamasaki, H. Habazaki, T. Yoshida, E. Akiyama, A. Kawashima, K. Asami, K. Hashimoto, M. Komori, and K. Shimamura, *Appl. Catal., A* **163**, 187 (1997).

74. H. Habazaki, T. Yoshida, M. Yamasaki, M. Komori, K. Shimamura, E. Akiyama, A. Kawashima, and K. Hashimoto, in *Advances in Chemical Conversions for Mitigating*

Carbon Dioxide, Studies in Surface Science and Catalysis, Vol. 114, T. Inui, M. Anpo, K. Izui, S. Yanagida, and T. Yamaguchi, eds., Elsevier, Amsterdam, 1998, p. 261.

75. M. Yamasaki, M. Komori, E. Akiyama, H. Habazaki, A. Kawashima, K. Asami, and K. Hashimoto, *Mater. Sci. Eng. A* **267**, 220 (1999).

76. H. Habazaki, M. Yamasaki, A. Kawashima, and K. Hashimoto, *Appl. Organomet. Chem.* **14**, 803 (2000).

77. M. Yamasaki, H. Habazaki, T. Yoshida, M. Komori, K. Shimamura, E. Akiyama, A. Kawashima, K. Asami, and K. Hashimoto, in *Advances in Chemical Conversions for Mitigating Carbon Dioxide*, Studies in Surface Science and Catalysis, Vol. 114, T. Inui, M. Anpo, K. Izui, S. Yanagida, and T. Yamaguchi, eds., Elsevier, Amsterdam, 1998, p. 451.

78. K. Hashimoto, H. Habazaki, M. Yamasaki, S. Meguro, T. Sasaki, H. Katagiri, T. Matsui, K. Fujimura, K. Izumiya, N. Kumagai, and E. Akiyama, *Mater. Sci. Eng. A* **304–306**, 88 (2001).

79. C. Schild, A. Wokaun, and A. Baiker, *J. Mol. Catal.* **69**, 347 (1991).

80. C. Schild, A. Wokaun, R. A. Koeppel, and A. Baiker, *J. Phys. Chem.* **95**, 6341 (1991).

81. H. Ando, M. Fujiwara, Y. Matsumura, M. Tanaka, and Y. Souma, *J. Mol. Catal. A: Chem.* **144**, 117 (1999).

82. K. Kudo and K. Komatsu, *J. Mol. Catal. A: Chem.* **145**, 257 (1999).

83. Z.-H. Suo, Y. Kou, J.-Z. Niu, W.-Z. Zhang, and H.-L. Wang, *Appl. Catal., A* **148**, 301 (1997).

84. S.-S. Nam, G. Kishan, M.-J. Choi, and K.-W. Lee, in *Greenhouse Gas Control Technologies*, B. Eliasson, P. Riemer, and A. Wokaun, eds., Elsevier, Amsterdam, 1999, p. 415.

85. S.-S. Nam, G. Kishan, M.-W. Lee, M.-J. Choi, and K.-W. Lee, *Appl. Organomet. Chem.* **14**, 794 (2000).

86. Y. Tan, M. Fujiwara, H. Ando, Q. Xu, and Y. Souma, *Ind. Eng. Chem. Res.* **38**, 3225 (1999).

87. S.-S. Nam, H. Kim, G. Kishan, M.-J. Choi, and K.-W. Lee, *Appl. Catal., A* **179**, 155 (1999).

88. J.-K. Jeon, K.-E. Jong, Y.-K. Park, and S.-K. Ihm, *Appl. Catal., A* **124**, 91 (1995).

89. M. Fujiwara, H. Ando, M. Tanaka, and Y. Souma, *Appl. Catal., A* **130**, 105 (1995).

90. Y.-K. Park, K.-C. Park, and S.-K. Ihm, *Catal. Today* **44**, 165 (1998).

91. N. K. Lunev, Yu. I. Shmyrko, N. V. Pavlenko, and B. Norton, *Appl. Organomet. Chem.* **15**, 99 (2001).

92. S.-R. Yan, K.-W. Jun, J.-S. Hong, M.-J. Choi, and K.-W. Lee, *Appl. Catal., A* **194–195**, 63 (2000).

93. Q. Chen and Y. Qian, *Chem. Commun.* 1402 (2001).

94. B. Denise and R. P. A. Sneeden, *J. Organomet. Chem.* **221**, 111 (1981).

95. K. Tominaga, Y. Sasaki, M. Kawai, T. Watanabe, and M. Saito, *J. Chem. Soc., Chem. Commun.* 629 (1993).

96. Y. Inoue, H. Izumida, Y. Sasaki, and H. Hashimoto, *Chem. Lett.* 863 (1976).

97. J.-C. Tsai and K. M. Nicholas, *J. Am. Chem. Soc.* **114**, 5117 (1992).

98. P. G. Jessop, Y. Hsiao, T. Ikariya, and R. Noyori, *J. Am. Chem. Soc.* **118**, 344 (1996).

99. C. Yin, Z. Xu, S.-Y. Yang, S. M. Ng, K. Y. Wong, Z. Lin, and C. P. Lau, *Organometallics* **20**, 1216 (2001).

100. D. J. Darensbourg and C. Ovalles, *Chemtech* **15**, 636 (1985).

101. I. Willner and B. Willner, *Top. Curr. Chem.* **159**, 153 (1991).

102. P. G. Jessop, T. Ikariya, and R. Noyori, *Chem. Rev.* **95**, 259 (1995).

103. W. Leitner, *Angew. Chem., Int. Ed. Engl.* **34**, 2207 (1995).

104. E. Dinjus, in *Advances in Chemical Conversions for Mitigating Carbon Dioxide*, Studies in Surface Science and Catalysis, Vol. 114, T. Inui, M. Anpo, K. Izui, S. Yanagida, and T. Yamaguchi, eds., Elsevier, Amsterdam, 1998, p. 127.

105. P. G. Jessop, T. Ikariya, and R. Noyori, *Chem. Rev.* **99**, 475 (1999).

106. P. G. Jessop, T. Ikariya, and R. Noyori, *Nature* **368**, 231 (1994).

107. P. G. Jessop, Y. Hsiao, T. Ikariya, and R. Noyori, *J. Am. Chem. Soc.* **116**, 8851 (1994).

108. P. G. Jessop, Y. Hsiao, T. Ikariya, and R. Noyori, *J. Chem. Soc., Chem. Commun.* 707 (1995).

109. O. Kröcher, R. A. Köppel, and A. Baiker, *Chem. Commun.* 1497 (1996).

110. O. Kröcher, R. A. Köppel, and A. Baiker, *Chem. Commun.* 453 (1997).

111. L. Schmid, M. Rohr, and A. Baiker, *Chem. Commun.* 2303 (1999).

112. O. Kröcher, R. A. Köppel, M. Fröba, and A. Baiker, *J. Catal.* **178**, 284 (1998).

113. O. Kröcher, R. A. Köppel, and A. Baiker, *J. Mol. Catal. A: Chem.* **140**, 185 (1999).

114. T. Burgemeister, F. Kastner, and W. Leitner, *Angew. Chem., Int. Ed. Engl.* **32**, 739 (1993).

115. R. Fornika, H. Görls, B. Seemann, and W. Leitner, *J. Chem. Soc., Chem. Commun.* 1479 (1995).

116. F. Gassner, E. Dinjus, H. Görls, and W. Leitner, *Organometallics* **15**, 2078 (1996).

117. F. Hutschka, A. Dedieu, M. Eichberger, R. Fornika, and W. Leitner, *J. Am. Chem. Soc.* **119**, 4432 (1997).

118. C. S. Pomelli, J. Tomasi, M. Sola, *Organometallics* **17**, 3164 (1998).

119. Y. Musashi and S. Sakaki, *J. Am. Chem. Soc.* **122**, 3867 (2000).

120. Y. Gao, J. K. Kuncheria, H. A. Jenkins, R. J. Puddephatt, and G. P. A. Yap, *J. Chem. Soc., Dalton Trans.* 3212 (2000).

121. G. Laurenczy, F. Joó, and L. Nádasdi, *Inorg. Chem.* **39**, 5083 (2000).

122. F. Gassner and W. Leitner, *J. Chem. Soc., Chem. Commun.* 1465 (1993).

123. F. Joó, G. Laurenczy, L. Nádasdi, and J. Elek, *Chem. Commun.* 971 (1999).

124. F. Joó, G. Laurenczy, P. Kárády, J. Elek, L. Nádasdi, and R. Roulet, *Appl. Organomet. Chem.* **14**, 857 (2000).

125. F. Liu, M. B. Abrams, R. T. Baker, and W. Tumas, *Chem. Commun.* 433 (2001).

126. G. A. Olah and A. Wu, *Synlett* 599 (1990).

127. F. R. Keene and B. P. Sullivan, in *Electrochemical and Electrocatalytic Reactions of Carbon Dioxide*, B. P. Sullivan, K. Krist, and H. E. Guard, eds., Elsevier, New York, 1993, p. 118.

128. *Electrochemical and Electrocatalytic Reactions of Carbon Dioxide*, B. P. Sullivan, K. Krist, and H. E. Guard, eds., Elsevier, New York, 1993.

129. I. Taniguchi, in *Modern Aspects of Electrochemistry*, Vol. 20, J. O'M Bockris, R. E. White, and R. E. Conway, eds., Plenum Press, New York, 1989, Chapter 5, p. 327.

130. M. Jitaru, D. A. Lowy, M. Toma, B. C. Toma, and L. Oniciu, *J. Appl. Electrochem.* **27**, 875 (1997).

131. K. Tanaka, *Bull. Chem. Soc. Jpn.* **71**, 17 (1998).

132. J. R. Pugh, M. R. M. Bruce, B. P. Sullivan, and T. J. Meyer, *Inorg. Chem.* **30**, 86 (1991).

133. T. Yoshida, K. Tsutsumida, S. Teratani, K. Yasufuku, and M. Kaneko, *J. Chem. Soc., Chem. Commun.* 631 (1993).

134. C. M. Bolinger, N. Story, B. P. Sullivan, and T. J. Meyer, *Inorg. Chem.* **27**, 4582 (1988).

135. Md. M. Ali, H. Sato, T. Mizukawa, K. Tsuge, M. Haga, and K. Tanaka, *Chem. Commun.* 249 (1998).

136. Y. Kushi, H. Nagao, T. Nishioka, K. Isobe, and K. Tanaka, *J. Chem. Soc., Chem. Commun.* 1223 (1995).

137. H. Noda, S. Ikeda, Y. Oda, K. Imai, M. Maeda, and K. Ito, *Bull. Chem. Soc. Jpn.* **63**, 2459 (1990).

138. A. Murata and Y. Hori, *Bull. Chem. Soc. Jpn.* **64**, 123 (1991).

139. P. Friebe, P. Bogdanoff, N. Alonso-Vante, and H. Tributsch, *J. Catal.* **168**, 374 (1997).

140. T. Saeki, K. Hashimoto, N. Kimura, K. Omata, and A. Fujishima, *J. Electroanal. Chem.* **390**, 77 (1995).

141. T. Saeki, K. Hashimoto, A. Fujishima, N. Kimura, and K. Omata, *J. Phys. Chem.* **99**, 8440 (1995).

142. S. Kaneco, R. Iwao, K. Iiba, K. Ohta, and T. Mizuno, *Energy* **23**, 1107 (1998).

143. T. Mizuno, M. Kawamoto, S. Kaneco, and K. Ohta, *Electrochim. Acta* **43**, 899 (1998).

144. B. R. Eggins, C. Ennis, R. McConnell, and M. Spencer, *J. Appl. Electrochem.* **27**, 706 (1997).

145. K. Ogura, M. Yamada, M. Nakayama, and N. Endo, in *Advances in Chemical Conversions for Mitigating Carbon Dioxide*, Studies in Surface Science and Catalysis, Vol. 114, T. Inui, M. Anpo, K. Izui, S. Yanagida, and T. Yamaguchi, eds., Elsevier, Amsterdam, 1998, p. 207.

146. K. Ogura, N. Endo, and M. Nakayama, *J. Electrochem. Soc.* **145**, 3801 (1998).

147. N. Endo, Y. Miho, and K. Ogura, *J. Mol. Catal. A: Chem.* **127**, 49 (1997).

148. T. Mizuno, A. Naitoh, and K. Ohta, *J. Electroanal. Chem.* **391**, 199 (1995).

149. T. Mizuno, K. Ohta, M. Kawamoto, and A. Saji, *Energy Sources* **19**, 249 (1997).

150. S. Kaneco, K. Iiba, K. Ohta, and T. Mizuno, *Energy Sources* **21**, 643 (1999).

151. S. Kaneco, K. Iiba, K. Ohta, and T. Mizuno, *Energy Sources* **22**, 127 (2000).

152. S. Kaneco, K. Iiba, S. Suzuki, K. Ohta, and T. Mizuno, *J. Phys. Chem. B* **103**, 7456 (1999).

153. S. Furukawa, M. Okada, and Y. Suzuki, *Energy Fuels* **13**, 1074 (1999).

154. C. Iwakura, S. Takezawa, H. Inoue, *J. Electroanal. Chem.* **459**, 167 (1998).

155. K. Hara and T. Sakata, *J. Electrochem. Soc.* **144**, 539 (1997).

156. H. Yoneyama, *Catal. Today* **39**, 169 (1997).

157. T. Inoue, A. Fujishima, S. Konishi, and K. Honda, *Nature* **277**, 637 (1979).

158. S. Kaneco, H. Kurimoto, Y. Shimizu, K. Ohta, and T. Mizuno, *Energy* **24**, 21 (1999).

159. K. R. Thampi, J. Kiwi, and M. Grätzel, *Nature* **327**, 506 (1987).

160. K. Hirano, K. Inoue, and T. Yatsu, *J. Photochem. Photobiol. A: Chem.* **64**, 255 (1992).

161. K. Adachi, K. Ohta, and T. Mizuno, *Solar Energy* **53**, 187 (1994).

162. M. Anpo, H. Yamashita, Y. Ichihashi, and S. Ehara, *J. Electroanal. Chem.* **396**, 21 (1995).

163. M. Subrahmanyam, S. Kaneco, and N. Alonso-Vante, *Appl. Catal., B* **23**, 169 (1999).

164. F. Solymosi and I. Tombácz, *Catal. Lett.* **27**, 61 (1994).

165. J. C. Hemminger, R. Carr, and G. A. Somorjai, *Chem. Phys. Lett.* **57**, 100 (1978).

166. T. Mizuno, K. Adachi, K. Ohta, and A. Saji, *J. Photochem. Photobiol. A: Chem.* **98**, 87 (1996).

167. S. Kaneco, Y. Shimizu, K. Ohta, and T. Mizuno, *J. Photochem. Photobiol. A: Chem.* **115**, 223 (1998).

168. V. Heleg and I. Willner, *Chem. Commun.* 2113 (1994).

169. B. K. Sharma, R. Ameta, J. Kaur, and S. C. Ameta, *Int. J. Energy Res.* **21**, 923 (1997).

170. R. Ramaraj and J. R. Premkumar, *Curr. Sci. Ind.* **79**, 884 (2000).

171. J. R. Premkumar and R. Ramaraj, *Chem. Commun.* 343 (1997).

172. J. Grodkowski and P. Neta, *J. Phys. Chem. A* **104**, 1848 (2000).

173. N. Ulagappan and H. Frei, *J. Phys. Chem. A* **104**, 7834 (2000).

174. B. R. Eggins, P. K. J. Robertson, J. H. Stewart, and E. Woods, *J. Chem. Soc., Chem. Commun.* 349 (1993).

175. P. Johne and H. Kisch, *J. Photochem. Photobiol. A: Chem.* **111**, 223 (1997).

176. M. Halmann, *Nature* **275**, 155 (1978).

177. H. Flaisher, R. Tenne, and M. Halmann, *J. Electroanal. Chem.* **402**, 97 (1996).

178. R. Hinogami, Y. Nakamura, S. Yae, and Y. Nakato, *J. Phys. Chem. B* **102**, 974 (1998).

179. R. Obert and B. C. Dave, *J. Am. Chem. Soc.* **121**, 12192 (1999).

180. S. Kuwabata, K. Nishida, R. Tsuda, H. Inoue, and H. Yoneyama, *J. Electrochem. Soc.* **141**, 1498 (1994).

181. B. A. Parkinson and P. F. Weaver, *Nature* **309**, 148 (1984).

182. S. Kuwabata, R. Tsuda, and H. Yoneyama, *J. Am. Chem. Soc.* **116**, 5437 (1994).

183. F. Fischer and H. Tropsch, *Brennstoff-Chem.* **4**, 276 (1923).

184. F. Fischer and H. Tropsch, *Brennstoff-Chem.* **7**, 97 (1926); *Chem. Ber.* **59**, 830, 832, 923 (1926).

185. P. Sabatier and J. B. Senderens, *Compt. Rend.* **134**, 514 (1902).

186. J. H. Gregor, *Catal. Lett.* **7**, 317 (1990/1991).

187. H. H. Storch, *Adv. Catal.* **1**, 115 (1948).

188. H. Pichler, *Adv. Catal.* **4**, 271 (1952).

189. J. Varga, *Magy. Chem. Foly.* **34**, 65 (1928).

190. E. E. Donath, *Adv. Catal.* **8**, 239 (1956).

191. H. H. Storch, N. Golumbic, and R. B. Anderson, *The Fischer–Tropsch and Related Syntheses*, Wiley, New York, 1951.

192. R. B. Anderson, in *Catalysis*, Vol. 4, P. H. Emmett, ed., Reinhold, New York, 1956, Chapter 3, p. 257.

193. R. B. Anderson, *The Fischer–Tropsch Synthesis*, Academic Press, Orlando, FL, 1984.

194. R. B. Anderson, *Adv. Catal.* **5**, 355 (1953); *ACS Symp. Ser.* **222**, 389 (1983).

195. R. B. Anderson, in *Catalysis*, Vol. 4, P. H. Emmett, ed., Reinhold, New York, 1956, Chapter 2, p. 29.

196. F. Asinger, *Paraffins, Chemistry and Technology*, Pergamon Press, Oxford, 1968, Chapter 2, p. 89.

197. R. D. Srivastava, V. U. S. Rao, G. Cinquegrane, and G. J. Stiegel, *Hydrocarbon Process., Int. Ed.* **69**(2), 59 (1990).

198. M. J. Baird, R. R. Schehl, W. P. Haynes, and J. T. Cobb, Jr., *Ind. Eng. Chem., Prod. Res. Dev.* **19**, 175 (1980).

199. J. H. Crowell, H. E. Benson, J. H. Field, and H. H. Storch, *Ind. Eng. Chem.* **42**, 2376 (1950).

200. P. J. Denny and D. A. Whan, in *Catalysis, a Specialist Periodical Report*, Vol. 2, C. Kemball and D. A. Dowden, senior reporters, The Chemical Society, Burlington House, London, 1978, Chapter 3, p. 46.

201. M. A. Vannice, *Catal. Rev.-Sci. Eng.* **14**, 153 (1976).

202. M. A. Vannice, in *Catalysis, Science and Technology*, Vol. 3, J. R. Anderson and M. Boudart, eds., Springer, Berlin, 1982, Chapter 3, p. 139.

203. V. Ponec, in *Catalysis, a Specialist Periodical Report*, Vol. 5, G. C. Bond, G. Webb, senior reporters, The Royal Society of Chemistry, Burlington House, London, 1982, Chapter 2, p. 48.

204. A. T. Bell, *Catal. Rev.-Sci. Eng.* **23**, 203 (1981).

205. V. Ponec, *Catal. Rev.-Sci. Eng.* **18**, 151 (1978).

206. P. Biloen and W. M. H. Sachtler, *Adv. Catal.* **30**, 165 (1981).

207. C. K. Rofer-DePoorter, *Chem. Rev.* **81**, 447 (1981).

208. J. P. Hindermann, G. J. Hutchings, and A. Kiennemann, *Catal. Rev.-Sci. Eng.* **35**, 1 (1993).

209. C. Masters, *Adv. Organomet. Chem.* **17**, 61 (1979).

210. E. L. Muetterties and J. Stein, *Chem., Rev.* **79**, 479 (1979).

211. W. A. Herrmann, *Angew. Chem., Int. Ed. Engl.* **21**, 117 (1982).

212. G. Henrici-Olivé and S. Olivé, in *The Chemistry of the Metal-Carbon Bond*, Vol. 3, F. R. Hartley and S. Patai, eds., Wiley, Chichester, 1985, Chapter 9, p. 391.

213. M. E. Dry, in *Catalysis, Science and Technology*, Vol. 1, J. R. Anderson and M. Boudart, eds., Springer, Berlin, 1981, Chapter 4, p. 159.

214. C. N. Satterfield, *Heterogeneous Catalysis in Industrial Practice*, McGraw-Hill, New York, 1991, Chapter 10, p. 419.

215. M. E. Dry and J. C. Hoogendoorn, *Catal. Rev.-Sci. Eng.* **23**, 265 (1981).

216. M. E. Dry, in *Applied Industrial Catalysis*, Vol. 2, B. E. Leach, ed., Academic Press, 1983, Chapter 5, p. 167.

217. M. E. Dry, *Chemtech* **12**, 744 (1982).

218. B. Büssemeier, C.-D. Frohning, and B. Cornils, in *Encyclopedia of Chemical Processing and Design*, Vol. 22, J. J. McKetta and W. A. Cunningham, eds., Marcel Dekker, New York, 1985, p. 81.

219. H. Pichler, *Brennstoff-Chem.* **19**, 226 (1938).

220. H. Pichler, B. Firnhaber, D. Kioussis, and A. Dawallu, *Makromol. Chem.* **70**, 12 (1964).

221. C. D. Frohning, in *New Syntheses with Carbon Monoxide*, J. Falbe, ed., Springer, Berlin, 1980, Chapter 4, p. 309.

222. J. R. H. Ross, in *Catalysis, a Specialist Periodical Report*, Vol. 7, G. C. Bond and G. Webb, senior reporters, The Royal Society of Chemistry, Burlington House, London, 1985, Chapter 1, p. 1.

223. G. van der Lee, A. G. T. M. Bastein, J. van den Boogert, B. Schuller, H.-Y. Luo, and V. Ponec, *J. Chem. Soc., Faraday Trans. 1* **83**, 2103 (1987).

224. M. E. Dry, T. Shingles, L. J. Boshoff, and G. J. Oosthuizen, *J. Catal.* **15**, 190 (1969).

225. J. Benziger and R. J. Madix, *Surf. Sci.* **94**, 119 (1980).

226. R. A. Friedel and R. B. Anderson, *J. Am. Chem. Soc.* **72**, 1212, 2307 (1950).

227. G. A. Huff, Jr. and C. N. Satterfield, *J. Catal.* **85**, 370 (1984).

228. E. Iglesia, S. C. Reyes, R. J. Madon, and S. L. Soled, *Adv. Catal.* **39**, 221 (1993).

229. Ya. T. Eidus, *Russ. Chem. Rev.* (Engl. Transl.) **36**, 338 (1967).

230. G. Blyholder and L. D. Neff, *J. Phys. Chem.* **70**, 893 (1966).

231. G. Blyholder, D. Shihabi, W. V. Wyatt, and R. Bartlett, *J. Catal.* **43**, 122 (1976).

232. A. Deluzarche, J. P. Hindermann, R. Kieffer, A. Muth, M. Papadopoulos, and C. Tanielian, *Tetrahedron Lett.* 797 (1977).

233. R. S. Sapienza, M. J. Sansone, L. D. Spaulding, and J. F. Lynch, *Fundam. Res. Homogeneous Catal.* **3**, 197 (1979).

234. S. R. Craxford and E. K. Rideal, *J. Chem. Soc.* 1604 (1939).

235. J. T. Kummer, T. W. deWitt, and P. H. Emmett, *J. Am. Chem. Soc.* **70**, 3632 (1948).

236. J. T. Kummer and P. H. Emmett, *J. Am. Chem. Soc.* **75**, 5177 (1953).

237. H. Pichler and H. Schulz, *Brennstoff-Chem.* **48**, 78 (1967); *Chem.-Ing.-Techn.* **42**, 1162 (1970).

238. G. Henrici-Olivé and S. Olivé, *Angew. Chem., Int. Ed., Engl.* **15**, 136 (1976); *J. Mol. Catal.* **18**, 367 (1983).

239. V. Ponec and W. A. van Barneveld, *Ind. Eng. Chem., Prod. Res. Dev.* **18**, 268 (1979).

240. V. Ponec, in *New Trends in CO Activation*, Studies in Surface Science and Catalysis, Vol. 64, L. Guczi, ed., Elsevier, Amsterdam, 1991, Chapter 4, p. 117.

241. P. R. Wentrcek, B. J. Wood, and H. Wise, *J. Catal.* **43**, 363 (1976).

242. M. Araki and V. Ponec, *J. Catal.* **44**, 439 (1976).

243. J. A. Rabo, A. P. Risch, and M. L. Poutsma, *J. Catal.* **53**, 295 (1978).

244. P. Biloen, *Recl. Trav. Chim. Pays-Bas* **99**, 33 (1980).

245. C. H. Bartholomew, *Catal. Lett.* **7**, 303 (1990/1991).

246. R. Snel, *Catal. Rev.-Sci. Eng.* **29**, 361 (1987).

247. R. A. Sheldon, *Chemicals from Synthesis Gas*, Reidel, Dordrecht, 1983, Chapter 3, p. 64.

248. D. L. King, J. A. Cusumano, and R. L. Garten, *Catal. Rev.-Sci. Eng.* **23**, 233 (1981).

249. J. Haggin, *Chem. Eng. News* **59**(43), 22 (1981).

250. C. H. Bartholomew, in *New Trends in CO Activation*, Studies in Surface Science and Catalysis, Vol. 64, L. Guczi, ed., Elsevier, Amsterdam, 1991, Chapter 5, p. 158.

251. H. Kölbel and M. Ralek, *Catal. Rev.-Sci. Eng.* **21**, 225 (1980).

252. C. D. Frohning and B. Cornils, *Hydrocarbon Process.* **53**(11), 143 (1974).

253. B. Büssemeier, C. D. Frohning, and B. Cornils, *Hydrocarbon Process.* **55**(11), 105 (1976).

254. K. B. Jensen and F. E. Massoth, *J. Catal.* **92**, 109 (1985).

255. T. Grzybek, H. Papp, and M. Baerns, *Appl. Catal.* **29**, 335 (1987).

256. J. M. Stencel, J. R. Diehl, S. R. Miller, R. A. Anderson, M. F. Zarochak, and H. W. Pennline, *Appl. Catal.* **33**, 129 (1987).

257. J. J. Venter and M. A. Vannice, *Catal. Lett.* **7**, 219 (1990/1991).

258. V. U. S. Rao and R. J. Gormley, *Hydrocarbon Process., Int. Ed.* **59**(11), 139 (1980).

259. D. Das, G. Ravichandran, D. K. Chakrabarty, S. N. Piramanayagam, and S. N. Shringi, *Appl. Catal., A* **107**, 73 (1993).

260. D.-J. Koh, J. S. Chung, and Y. G. Kim, *J. Chem. Soc., Chem. Commun.* 849 (1991).

261. *Chem. Eng. News* **58**(36), 44 (1980).

262. P. K. Coughlin and J. A. Rabo, U.S. Patents 4,556,645 (1986); 4,652,538 (1987).

263. C. B. Murchison and D. A. Murdick, *Hydrocarbon Process., Int. Ed.* **60**(1), 159 (1981).

264. H. Y. Luo, A. G. T. M. Bastein, A. A. J. P. Mulder, and V. Ponec, *Appl. Catal.* **38**, 241 (1988).

265. M Röper, in *New Trends in CO Activation*, Studies in Surface Science and Catalysis, Vol. 64, L. Guczi, ed., Elsevier, Amsterdam, 1991, Chapter 9, p. 381.

266. A. Aquiló, J. S. Alder, D. N. Freeman, and R. J. H. Voorhoeve, *Hydrocarbon Process., Int. Ed.* **62**(3), 57 (1983).

267. J. M. Fox, III, *Catal. Lett.* **7**, 281 (1990/1991).

268. S. Soled, E. Iglesia, and R. A. Fiato, *Catal. Lett.* **7**, 271 (1990/1991).

269. H. Kölbel and M. Ralek, in *The Fischer–Tropsch Synthesis*, R. B. Anderson ed., Academic Press, Orlando, 1984, Chapter 7, p. 265.

270. V. M. Vlasenko and G. E. Yuzefovich, *Russ. Chem. Rev.* (Engl. transl.) **38**, 728 (1969).

271. G. A. Mills and F. W. Steffgen, *Catal. Rev.-Sci. Eng.* **8**, 159 (1973).

272. B. B. Pearce, M. V. Twigg, and C. Woodward, in *Catalyst Handbook*, M. V. Twigg, ed., Wolfe, London, 1989, Chapter 7, p. 340.

273. R. F. Probstein and R. E. Hicks, *Synthetic Fuels*, pH Press, Cambridge, UK, 1990, Chapter 5, p. 210.

274. J. A. Cusumano, R. A. Dalla Betta, and R. B. Levy, *Catalysis in Coal Conversion*, Academic Press, New York, 1978, Chapter 17, p. 221.

275. L. Seglin, R. Geosits, B. R. Franko, and G. Gruber, *Adv. Chem. Ser.* **146**, 1 (1975).

276. F. W. Moeller, H. Roberts, and B. Britz, *Hydrocarbon Process.* **53**(4), 69 (1974).

277. C. Woodward, *Hydrocarbon Process.* **56**(1), 136 (1977).

278. E. R. Tucci and R. C. Streeter, *Hydrocarbon Process., Int. Ed.* **59**(4), 107 (1980).

279. E. M. Cohn, in *Catalysis*, Vol. 4, P. H. Emmett, ed., Reinhold, New York, 1956, Chapter 5, p. 443.

280. Y. T. Shah and A. J. Perrotta, *Ind. Eng. Chem., Prod. Res. Dev.* **15**, 123 (1976).

281. H. Pichler and K.-H. Ziesecke, *Brennstoff-Chem.* **30**, 13, 60, 81 (1949).

282. H. Pichler, K.-H. Ziesecke, and E. Titzenthaler, *Brennstoff-Chem.* **31**, 333 (1949); **31**, 361 (1950).

283. R. Kieffer, J. Valera, and A. Deluzarche, *J. Chem. Soc., Chem. Commun.* 763 (1983).

284. R. Kieffer, G. Cherry, and R. El Bacha, *C₁ Mol. Chem.* **2**, 11 (1987).

285. K. Maruya, T. Maehashi, T. Haraoka, S. Narui, Y. Asakawa, K. Domen, and T. Onishi, *Bull. Chem. Soc. Jpn.* **61**, 667 (1988).

286. S. C. Tseng, N. B. Jackson, and J. G. Ekerdt, *J. Catal.* **109**, 284 (1988).

287. N. B. Jackson and J. G. Ekerdt, *J. Catal.* **126**, 46 (1990).

288. R. Oukaci, A. Sayari, and J. G. Goodwin, Jr., *J. Catal.* **102**, 126 (1986).

289. R. Oukaci, J. C. S. Wu, and J. G. Goodwin, Jr., *J. Catal.* **107**, 471 (1987).

290. X. Song and A. Sayari, *Appl. Catal., A* **110**, 121 (1994).

291. J. S. Lee and S. T. Oyama, *Catal. Rev.- Sci. Eng.* **30**, 249 (1988).

292. Kh. M. Minachev, N. Ya. Usachev, V. N. Udut, and Yu. S. Khodakov, *Russ. Chem. Rev.* (Engl. transl.) **57**, 221 (1988).

293. M. Yu. Sinev, V. N. Korshak, and O. V. Krylov, *Russ. Chem. Rev.* (Engl. transl.) **58**, 22 (1989).

294. J. R. Anderson, *Appl. Catal.* **47**, 177 (1989).

295. G. J. Hutchings, M. S. Scurrell, and J. R. Woodhouse, *Chem. Soc. Rev.* **18**, 251 (1989).

296. Y. Amenomiya, V. I. Birss, M. Goledzinowski, J. Galuszka, and A. R. Sanger, *Catal. Rev.-Sci. Eng.* **32**, 163 (1990).

297. A. Bielanski and J. Haber, *Oxygen in Catalysis*. Marcel Dekker, New York, 1991, Chapter 9, p. 423.

298. A. M. Maitra, *Appl. Catal., A* **104**, 11 (1993).

299. R. D. Srivastava, P. Zhou, G. J. Stiegel, V. U. S. Rao, and G. Cinquegrane, in *Catalysis, Specialist Periodical Reports*, Vol. 9, J. Spivey, senior reporter, The Royal Society of Chemistry, Thomas Graham House, Cambridge, UK, 1992, Chapter 4, p. 183.

300. Z. Kalenik and E. E. Wolf, in *Catalysis, Specialist Periodical Reports*, Vol. 10, J. Spivey and S. K. Agarwal, senior reporters, The Royal Society of Chemistry, Thomas Graham House, Cambridge, UK, 1993, Chapter 5, p. 154.

301. *Proc. 9th Int. Congr. Catalysis,* Calgary, 1988, M. J. Phillips and M. Ternan, eds., The Chemical Institute of Canada, Ottawa, 1988, pp. 883–997.

302. *1st Workshop on the Catalytic Methane Conversion*, Bochum, 1988. *Catal. Today* **4**, 271–500 (1989).

303. *Natural Gas Conversion*, Studies in Surface Science and Catalysis, Vol. 61, A. Holmen, K.-J. Jens, and S. Kolboe, eds., Elsevier, Amsterdam, 1991, pp. 1–221.

304. *Methane Conversion by Oxidative Processes: Fundamental and Engineering Aspects*, E. E. Wolf, ed., Van Nostrand Reinhold, New York, 1992.

305. C. A. Jones, J. J. Leonard, and J. A. Sofranko, *Energy Fuels* **1**, 12 (1987).

306. G. E. Keller and M. M. Bashin, *J. Catal.* **73**, 9 (1982).

307. J. A. Sofranko, J. J. Leonard, and C. A. Jones, *J. Catal.* **103**, 302 (1987).

308. M. Baerns, *Catal. Today* **1**, 357 (1987).

309. J. A. S. P. Carreiro, G. Follmer, L. Lehmann, and M. Baerns, in *Proc. 9th Int. Congr. Catalysis*, Calgary, 1988, M. J. Phillips and M. Ternan, eds., The Chemical Institute of Canada, Ottawa, 1988, p. 891.

310. K. Asami, S. Hashimoto, K. Fujimoto, and H. Tominaga, in *Methane Conversion*, Studies in Surface Science and Catalysis, Vol. 36, D. M. Bibby, C. D. Chang, R. F. Howe, and S. Yurchak, eds., Elsevier, Amsterdam, 1988, p. 403.

311. T. Ito and J. H. Lunsford, *Nature* (London) **314**, 721 (1985).

312. J. H. Lunsford, *Catal. Today* **6**, 235 (1990).

313. J.-X. Wang and J. H. Lunsford, *J. Phys. Chem.* **90**, 5883 (1986).

314. C.-H. Lin, T. Ito, J.-X. Wang, and J. H. Lunsford, *J. Am. Chem. Soc.* **109**, 4808 (1987).

315. C.-H. Lin, J.-X. Wang, and J. H. Lunsford, *J. Catal.* **111**, 302 (1988).

316. T. Doi, Y. Utsumi, and I. Matsuura, in *Proc. 9th Int. Congr. Catalysis*, Calgary, 1988, M. J. Phillips and M. Ternan, eds., The Chemical Institute of Canada, Ottawa, 1988, p. 937.

317. G. S. Lane and E. E. Wolf, in *Proc. 9th Int. Congr. Catalysis*, Calgary, 1988, M. J. Phillips and M. Ternan, eds., The Chemical Institute of Canada, Ottawa, 1988, p. 944.

318. R. Burch and S. C. Tsang, *Appl. Catal.* **65**, 259 (1990).

319. K. Otsuka, M. Hatano, and T. Kotmatsu, *Catal. Today* **4**, 409 (1989).

320. S. Ahmed and J. B. Moffat, *J. Catal.* **125**, 54 (1990).

321. D. I. Bradshaw, P. T. Coolen, R. W. Judd, and C. Komodromos, *Catal. Today* **6**, 427 (1990).

322. P. G. Hinson, A. Clearfield, and J. H. Lunsford, *J. Chem. Soc., Chem. Commun.* 1450 (1991).

323. C.-H. Lin, K. O. Campbell, J.-X. Wang, and J. H. Lunsford, *J. Phys. Chem.* **90**, 534 (1986).

324. Y. Tong, M. P. Rosynek, and J. H. Lunsford, *J. Phys. Chem.* **93**, 2896 (1989).

325. K. Otsuka, Q. Liu, and A. Morikawa, *J. Chem. Soc., Chem. Commun.* 586 (1986).

326. K. Otsuka, K. Jinno, and A. Morikawa, *J. Catal.* **100**, 353 (1986).

327. A. Ekstrom, in *Methane Conversion by Oxidative Processes: Fundamental and Engineering Aspects*, E. E. Wolf, ed., Van Nostrand Reinhold, New York, 1992, Chapter 4, p. 99.

328. A. L. Tonkovich, R. W. Carr, and R. Aris, *Science* **262**, 221 (1993).

329. Y. Jiang, I. V. Yentekakis, and C. G. Vayenas, *Science* **264**, 1563 (1994).

330. D. J. Driscoll, W. Martir, J.-X. Wang, and J. H. Lunsford, *J. Am. Chem. Soc.* **107**, 58 (1985).

331. V. T. Amorebieta and A. J. Colussi, *J. Phys. Chem.* **93**, 5155 (1989).

332. A. Ekstrom and J. A. Lapszewicz, *J. Am. Chem. Soc.* **110**, 5226 (1988).

333. C. A. Mimms, R. B. Hall, K. D. Rose, and G. R. Myers, *Catal. Lett.* **2**, 361 (1990).

334. P. F. Nelson and N. W. Cant, *J. Phys. Chem.* **94**, 3756 (1990).

335. E. Garrone, A. Zecchina, and F. S. Stone, *J. Catal.* **62**, 396 (1980).

336. T. Ito, T. Watanabe, T. Tashiro, and K. Toi, *J. Chem. Soc., Faraday Trans. 1* **85**, 2381 (1989).

337. T. Ito, T. Tashiro, M. Kawasaki, T. Watanabe, K. Toi, and H. Kobayashi, *J. Phys. Chem.* **95**, 4476 (1991).

338. T. Ito, J.-X. Wang, C.-H. Lin, and J. H. Lunsford, *J. Am. Chem. Soc.* **107**, 5062 (1985).

339. G. J. Hutchings, M. S. Scurrell, and J. R. Woodhouse, *J. Chem. Soc., Chem. Commun.* 1862 (1987).

340. K. Asami, K. Omata, K. Fujimoto, and H. Tominaga, *Energy Fuels* **2**, 574 (1988).

341. D. J. C. Yates and N. E. Zlotin, *J. Catal.* **111**, 317 (1988).

342. G. S. Lane and E. E. Wolf, *J. Catal.* **113**, 144 (1988).

343. D. J. Driscoll, W. Martir, J.-X. Wang, and J. H. Lunsford, in *Adsorption and Catalysis on Oxide Surfaces*, Studies in Surface Science and Catalysis, Vol. 21, M. Che and G. C. Bond, eds., Elsevier, Amsterdam, 1985, p. 403.

344. J.-X. Wang and J. H. Lunsford, *J. Phys. Chem.* **90**, 3890 (1986).

345. G. A. Martin and C. Mirodatos, in *Methane Conversion by Oxidative Processes: Fundamental and Engineering Aspects*, E. E. Wolf, ed., Van Nostrand Reinhold, New York, 1992, Chapter 10, p. 351.

346. V. R. Choudhary, S. T. Chaudhari, and M. Y. Pandit, *J. Chem. Soc., Chem. Commun.* 1158 (1991).

347. J.-L. Dubois and C. J. Cameron, *Appl. Catal.* **67**, 49 (1990).

348. E. E. Wolf, in *Methane Conversion by Oxidative Processes: Fundamental and Engineering Aspects*, E. E. Wolf, ed., Van Nostrand Reinhold, New York, 1992, Chapter 16, p. 527.

349. J. L. G. Fierro, *Catal. Lett.* **22**, 67 (1993).

350. S. W. Benson, U.S. Patent 4,199,533 (1980).

351. M. Weissman and S. W. Benson, *Int. J. Chem. Kinet.* **16**, 307 (1984).

352. G. A. Olah and R. H. Schlosberg, *J. Am. Chem. Soc.* **90**, 2726 (1968).

353. G. A. Olah, G. Klopman, and R. H. Schlosberg, *J. Am. Chem. Soc.* **91**, 3261 (1969).

354. G. A. Olah, Y. Halpern, J. Shen, and Y. K. Mo, *J. Am. Chem. Soc.* **93**, 1251 (1971).

355. G. A. Olah, *J. Am. Chem. Soc.* **94**, 808 (1972).

356. G. A. Olah, U.S. Patent 4,467,130 (1984).

357. G. A. Olah, U.S. Patents 4,433,192 (1984); 4,465,893 (1984).

358. D. A. Duncan, in *Kirk-Othmer Encyclopedia of Chemical Technology*, Vol. 1, M. Grayson and D. Eckroth, eds., Wiley-Interscience, New York, 1978, p. 211.

359. L. R. Roberts, in *Encyclopedia of Chemical Processing and Design*, Vol. 1, J. J. McKetta and W. A. Cunningham, eds., Marcel Dekker, New York, 1976, p. 363.

360. R. J. Tedeschi, *Acetylene-Based Chemicals from Coal and Other Natural Resources*, Marcel Dekker, New York, 1982, Chapter 1, p. 10.

361. F. Marschner and F. W. Moeller, in *Applied Industrial Catalysis*, Vol. 2, B. E. Leach, ed., Academic Press, 1983, Chapter 6, p. 215.

362. G. W. Bridger and M. S. Spencer, in *Catalyst Handbook*, M. V. Twigg, ed., Wolfe, London, 1989, Chapter 9, p. 441.

363. G. Natta, in *Catalysis*, Vol. 3, R. H. Emmett, ed., Reinhold, New York, 1955, Chapter 8, p. 349.

364. P. Davis and F. F. Snowdon, U.S. Patent 3,326,956 (1967).

365. D. H. Bolton, *Chem.-Ing.-Techn.* **41**, 129 (1969).

366. H. Hiller and F. Marschner, *Hydrocarbon Process.* **49**(9), 281 (1970).

367. E. Supp, *Chemtech* **3**, 430 (1973).

368. H. H. Kung, *Catal. Rev.- Sci. Eng.* **22**, 235 (1980).

369. K. Klier, *Adv. Catal.* **31**, 243 (1982).

370. J. C. J. Bart and R. P. A. Sneeden, *Catal. Today* **2**, 1 (1987).

371. G. C. Chinchen, P. J. Denny, J. R. Jennings, M. S. Spencer, and K. C. Waugh, *Appl. Catal.* **36**, 1 (1988).

372. D. Cornthwaite, B. Patent 1,296,212 (1972).

373. C. Lormand, *Ind. Eng. Chem.* **17**, 430 (1925).

374. R. G. Herman, K. Klier, G. W. Simmons, B. F. Finn, J. B. Bulko, and T. P. Kobylinski, *J. Catal.* **56**, 407 (1979).

375. K. Klier, *Appl. Surf. Sci.* **19**, 267 (1984).

376. G. R. Apai, J. R. Monnier, and M. J. Hanrahan, *J. Chem. Soc., Chem. Commun.* 212 (1984).

377. E. I. Solomon, P. M. Jones, and J. A. May, *Chem. Rev.* **93**, 2623 (1993).

378. J. B. Friedrich, M. S. Wainwright, and D. J. Young, *J. Catal.* **80**, 1 (1983).

379. R. H. Höppener, E. B. M. Doesburg, and J. J. F. Scholten, *Appl. Catal.* **25**, 109 (1986).

380. G. C. Chinchen, K. C. Waugh, and D. A. Whan, *Appl. Catal.* **25**, 101 (1986).

381. B. Denise, R. P. A. Sneeden, B. Beguin, and O. Cherifi, *Appl. Catal.* **30**, 353 (1987).

382. G. C. Chinchen, M. S. Spencer, K. C. Waugh, and D. A. Whan, *J. Chem. Soc., Faraday Trans. 1* **83**, 2193 (1987).

383. R. Burch and S. E. Golunski, M. S. Spencer, *Catal. Lett.* **5**, 55 (1990).

384. G. E. Parris and K. Klier, *J. Catal.* **97**, 374 (1986).

385. G. J. J. Bartley and R. Burch, *Appl. Catal.* **43**, 141 (1988).

386. G. Owen, C. M. Hawkes, D. Lloyd, J. R. Jennings, R. M. Lambert, and R. M. Nix, *Appl. Catal.* **33**, 405 (1987).

387. W. R. A. M. Robinson and J. C. Mol, *Appl. Catal.* **76**, 117 (1991).

388. R. Burch, R. J. Chappell, and S. E. Golunski, *Catal. Lett.* **1**, 439 (1988); *J. Chem. Soc., Faraday Trans. 1* **85**, 3569 (1989).

389. R. Burch, S. E. Golunski, and M. S. Spencer, *Catal. Lett.* **6**, 152 (1990).

390. J. C. Frost, *Nature* (London) **334**, 577 (1988).

391. M. L. Poutsma, L. F. Elek, P. A. Ibarbia, A. P. Risch, and J. A. Rabo, *J. Catal.* **52**, 157 (1978).

392. J. M. Driessen, E. K. Poels, J. P. Hindermann, and V. Ponec, *J. Catal.* **82**, 26 (1983).

393. V. Ponec, *Surf. Sci.* **272**, 111 (1992).

394. A. Ya. Rozovskii, Yu. B. Kagan, G. I. Lin, E. V. Slivinskii, S. M. Loktev, L. G. Liberov, and A. N. Bashkirov, *Kinet. Catal.* (Engl. transl.) **16**, 706 (1975).

395. V. D. Kuznetsov, F. S. Shub, and M. I. Temkin, *Kinet. Catal.* (Engl. transl.) **23**, 788 (1982).

396. G. C. Chinchen, P. J. Denny, D. G. Parker, M. S. Spencer, and D. A. Whan, *Appl. Catal.* **30**, 333 (1987).

397. M. Bowker, R. A. Hadden, H. Houghton, J. N. K. Hyland, and K. C. Waugh, *J. Catal.* **109**, 263 (1988).

398. B. Denise, O. Cherifi, M. M. Bettahar, and R. P. A. Sneeden, *Appl. Catal.* **48**, 365 (1989).

399. R. Kieffer, E. Ramaroson, A. Deluzarche, and Y. Trambouze, *React. Kinet. Catal. Lett.* **16**, 207 (1981).

400. G. Liu, D. Willcox, M. Garland, and H. H. Kung, *J. Catal.* **96**, 251 (1985).

401. B. Denise and R. P. A. Sneeden, *Appl. Catal.* **28**, 235 (1986).

402. M. Takagawa and M. Ohsugi, *J. Catal.* **107**, 161 (1987).

403. H. E. Curry-Hyde, M. S. Wainwright, and D. J. Young, *Appl. Catal.* **77**, 75 (1991).

404. E. G. Baglin, G. A. Atkinson, and L. J. Nicks, *Ind. Eng. Chem., Prod. Res. Dev.* **20**, 87 (1981).

405. F. P. Daly, *J. Catal.* **89**, 131 (1984).

406. R. M. Nix, T. Rayment, R. M. Lambert, J. R. Jennings, and G. Owen, *J. Catal.* **106**, 216 (1987).

407. J. Ladebeck, *Hydrocarbon Process., Int. Ed.* **72**(3), 89 (1993).

408. G. Natta, U. Colombo, and I. Pasquon, in *Catalysis*, Vol. 5, R. H. Emmett, ed., Reinhold, New York, 1957, Chapter 3, p. 131.

409. R. G. Herman, in *New Trends in CO Activation*, Studies in Surface Science and Catalysis, Vol. 64, L. Guczi, ed., Elsevier, Amsterdam, 1991, Chapter 7, p. 265.

410. Ph. Courty, J. P. Arlie, A. Convers, P. Mikitenko, and A. Sugier, *Hydrocarbon Process., Int. Ed.* **63**(11), 105 (1984).

411. G. A. Olah, B. Gupta, M. Farina, J. D. Felberg, W. M. Ip, A. Husain, R. Karpeles, K. Lammertsma, A. K. Melhotra, and N. J. Trivedi, *J. Am. Chem. Soc.* **107**, 7097 (1985).

412. G. A. Olah, *Acc. Chem. Res.* **20**, 422 (1987).

413. G. A. Olah, U.S. Patent 4,523,040 (1985).

414. J. A. LeBel and W. H. Greene, *Am. Chem. J.* **2**, 20 (1880).

415. A. V. Grosse and J. C. Snyder, U.S. Patent 2,492,984 (1950).

416. L. Kim, M. M. Walsd, and S. G. Brandenberger, *J. Org. Chem.* **43**, 3432 (1978).

417. E. Sernagiotto, *Gazz. Chim. Ital.* **44**, 587 (1914); *Chem. Abstr.* **9**, 69 (1915).

418. D. E. Pearson, *J. Chem. Soc., Chem. Commun.* 397 (1974).

419. G. Salem, Ph.D. thesis, Univ. Southern California, 1980.

420. G. A. Olah, U.S. Patent 4,373,109 (1983).

421. F. S. Fawcett and B. W. Howk, U.S. Patent 2,744,151 (1956).

422. Ya. T. Eidus, *Izv. Akad. Nauk SSSR, Otd. Khim. Nauk* 65 (1943); *Chem. Abstr.* **38**, 1342[1] (1944).

423. H. Adkins and P. D. Perkins, *J. Phys. Chem.* **32**, 221 (1928).

424. P. Ciambelli, G. Bagnasco, and P. Corbo, in *Successful Design of Catalysts*, Studies in Surface Science and Catalysis, Vol. 44, T. Inui, ed., Elsevier, Amsterdam, 1989, p. 239.

425. Y. Ono, T. Baba, J. Sakai, and T. Keii, *J. Chem. Soc., Chem. Commun.* 400 (1981).

426. T. Okuhara, T. Hibi, K. Takahashi, S. Tatematsu, and M. Misono, *J. Chem. Soc., Chem. Commun.* 697 (1984).

427. T. Baba and Y. Ono, *Appl. Catal.* **8**, 315 (1983).

428. T. Hibi, K. Takahashi, T. Okuhara, M. Misono, and Y. Yoneda, *Appl. Catal.* **24**, 69 (1986).

429. H. Ehwald, W. Fiebig, H.-G. Jerschkewitz, G. Lischke, B. Parlitz, P. Reich, and G. Öhlmann, *Appl. Catal.* **34**, 23 (1987).

430. H. Hayashi and J. B. Moffat, *J. Catal.* **81**, 61 (1983).

431. H. Hayashi and J. B. Moffat, *J. Catal.* **83**, 192 (1983).

432. H. Ehwald, W. Fiebig, H.-G. Jerschkewitz, G. Lischke, B. Parlitz, E. Schreier, and G. Öhlmann, *Appl. Catal.* **34**, 13 (1987).

433. C. D. Chang and A. J. Silvestri, *J. Catal.* **47**, 249 (1977).

434. P. B. Weisz and V. J. Frilette, *J. Phys. Chem.* **64**, 382 (1960).

435. S. M. Csicsery, in *Zeolite Chemistry and Catalysis*, J. A. Rabo, ed., ACS Monograph 171, American Chemical Society, Washington, DC, 1976, Chapter 12, p. 680.

436. P. B. Weisz, *Pure Appl. Chem.* **52**, 2091 (1980).

437. S. M. Csicsery, *Pure Appl. Chem.* **58**, 841 (1986).

438. N. Y. Chen, W. E. Garwood, and F. G. Dwyer, *Shape Selective Catalysis in Industrial Applications*, Marcel Dekker, New York, 1989.

439. S. W. Kaiser, U.S. Patent 4,524,234 (1985).

440. T. Inui, H. Matsuda, H. Okaniwa, and A. Miyamoto, *Appl. Catal.* **58**, 155 (1990).

441. J. Liang, H. Li, S. Zhao, W. Guo, R. Wang, and M. Ying, *Appl. Catal.* **64**, 31 (1990).

442. Y. Xu, C. P. Grey, J. M. Thomas, and A. K. Cheetham, *Catal. Lett.* **4**, 251 (1990).

443. O. V. Kikhtyanin, E. A. Paukshtis, K. G. Ione, and V. M. Mastikhin, *J. Catal.* **126**, 1 (1990).

444. S. Hocevar, J. Batista, and V. Kaucic, *J. Catal.* **139**, 351 (1993).

445. R. A. Stowe and C. B. Murchison, *Hydrocarbon Process., Int. Ed.* **61**(1), 147 (1982).

446. R. G. Anthony and P. E. Thomas, *Hydrocarbon Process., Int. Ed.* **63**(11), 95 (1984).

447. G. A. Olah, H. Doggweiler, J. D. Felberg, S. Frohlich, M. J. Grdina, R. Karpeles, T. Keumi, S. Inaba, W. M. Ip, K. Lammertsma, G. Salem, and D. C. Tabor, *J. Am. Chem. Soc.* **106**, 2143 (1984).

448. L. J. van Rensburg, R. Hunter, and G. J. Hutchings, *Appl. Catal.* **42**, 29 (1988).

449. G. J. Hutchings, R. Hunter, W. Pickl, and L. J. van Rensburg, in *Methane Conversion*, Studies in Surface Science and Catalysis, Vol. 36, D. M. Bibby, C. D. Chang, R. F. Howe, and S. Yurchak, eds., Elsevier, Amsterdam, 1988, p. 183.

450. N. Y. Chen and W. E. Garwood, *Catal. Rev.-Sci. Eng.* **28**, 185 (1986).

451. *Methane Conversion*, Studies in Surface Science and Catalysis, Vol. 36, D. M. Bibby, C. D. Chang, R. F. Howe, and S. Yurchak, eds., Elsevier, Amsterdam, 1988.

452. N. Y. Chen, W. E. Garwood, and F. G. Dwyer, *Shape Selective Catalysis in Industrial Applications*, Marcel Dekker, New York, 1989, Chapter 4, p. 117.

453. G. J. Hutchings and R. Hunter, *Catal. Today* **6**, 279 (1990).

454. C. D. Chang, *Catal. Rev.-Sci. Eng.* **25**, 1 (1983); *Hydrocarbons from Methanol*, Marcel Dekker, New York, 1983.

455. C. D. Chang, in *Natural Gas Conversion*, Studies in Surface Science and Catalysis, Vol. 61, A. Holmen, K.-J. Jens, and S. Kolboe, eds., Elsevier, Amsterdam, 1991, p. 393.

456. P. B. Venuto and P. S. Landis, *Adv. Catal.* **18**, 259 (1968).

457. C. D. Chang, *J. Catal.* **69**, 244 (1981).

458. F. X. Cormerais, G. Perot, F. Chevalier, and M. Guisnet, *J. Chem. Res. (S)* 362 (1980).

459. C. D. Chang, in *Methane Conversion*, Studies in Surface Science and Catalysis, Vol. 36, D. M. Bibby, C. D. Chang, R. F. Howe, and S. Yurchak, eds., Elsevier, Amsterdam, 1988, p. 127.

460. Y. Ono and T. Mori, *J. Chem. Soc., Faraday Trans. 1* **77**, 2209 (1981).

461. R. Hunter and G. J. Hutchings, *J. Chem. Soc., Chem. Commun.* 1643 (1985).

462. D. Kagi, *J. Catal.* **69**, 242 (1981).

463. J. P. van den Berg, J. P. Wolthuizen, and J. H. C. van Hooff, in *Proc. 5th Int. Conf. Zeolites*, Naples, 1980, L. V. C. Rees, ed., Heyden, London, 1980, p. 649.

464. R. Hunter and G. J. Hutchings, *J. Chem. Soc., Chem. Commun.* 886 (1985).

465. G. A. Olah, G. K. S. Prakash, R. W. Ellis, and J. A. Olah, *J. Chem. Soc., Chem. Commun.* 9 (1986).

466. W. Zatorski and S. Krzyzanowski, *Acta Phys. Chem.* **24**, 347 (1978).

467. J. K. A. Clarke, R. Darcy, B. F. Hegarty, E. O'Donoghue, V. Amir-Ebrahimi, and J. J. Rooney, *J. Chem. Soc., Chem. Commun.* 425 (1986).

468. S. Kolboe, in *Natural Gas Conversion*, Studies in Surface Science and Catalysis, Vol. 61, A. Holmen, K.-J. Jens, and S. Kolboe, eds., Elsevier, Amsterdam, 1991, p. 413.

469. C. D. Chang, S. D. Hellring, and J. A. Pearson, *J. Catal.* **115**, 282 (1989).

470. G. J. Hutchings, L. J. van Rensburg, R. G. Copperthwaite, R. Hunter, J. Dwyer, and J. Dewing, *Catal. Lett.* **4**, 7 (1990).

471. M. W. Anderson and J. Klinowski, *J. Am. Chem. Soc.* **112**, 10 (1990).

472. J. E. Jackson and F. M. Bertsch, *J. Am. Chem. Soc.* **112**, 9085 (1990).

473. G. J. Hutchings, R. Hunter, P. Johnston, and L. J. van Rensburg, *J. Catal.* **142**, 602 (1993).

474. C. D. Chang, *Catal. Rev.-Sci. Eng.* **26**, 323 (1984).

475. N. Y. Chen, W. E. Garwood, and F. G. Dwyer, *Shape Selective Catalysis in Industrial Applications*, Marcel Dekker, New York, 1989, Chapter 7, p. 219.

476. I. Bucsi and G. A. Olah, *Catal. Lett.* **16**, 27 (1992).

477. P. Batamack, I. Bucsi, G. A. Olah, and Á. Molnár, *Catal. Lett.* **25**, 11 (1994).

478. Yu. A. Treger and V. N. Rozanov, *Russ. Chem. Rev.* (Engl. transl.) **58**, 84 (1989).

479. A. G. Aglulin, Yu. M. Bakshi, and A. I. Gel'bshtein, *Kinet. Catal.* (Engl. transl.) **17**, 581 (1976).

480. Wm. C. Conner, Jr., W. J. M. Pieters, W. Gates, and J. E. Wilkalis, *Appl. Catal.* **11**, 49 (1984).

481. C. E. Taylor and R. P. Noceti, in *Proc. 9th Int. Congr. Catalysis*, Calgary, 1988, M. J. Phillips and M. Ternan, eds., The Chemical Institute of Canada, Ottawa, 1988, p. 990.

482. S. A. Butter, A. T. Jurewicz, and W. W. Kaedig, U.S. Patent 3,894,107 (1975).

483. V. N. Romannikov and K. G. Ione, *Kinet. Catal.* (Engl. transl.) **25**, 75 (1984).

484. H. O. Folkins, E. Miller, and H. Hennig, *Ind. Eng. Chem.* **42**, 2202 (1950).

485. C. D. Chang, W. H. Lang, and A. J. Silvestri, U.S. Patent 4,480,143 (1984).

486. R. T. Bell, U.S. Patent 2,565,195 (1959).

487. *Natural Gas Conversion II*, Studies in Surface Science and Catalysis, Vol. 81, H. E. Curry-Hyde and R. F. Howe, eds., Elsevier, Amsterdam, 1994, pp. 413–489.

488. *Natural Gas Conversion IV*, Studies in Surface Science and Catalysis, Vol. 107, M. dePontes, R. L. Espinoza, C. P. Nicolaides, J. H. Scholtz, and M. S. Scurrell, eds., Elsevier, Amsterdam, 1997, pp. 151–254.

489. *Natural Gas Conversion V*, Studies in Surface Science and Catalysis, Vol. 119, A. Parmaliana, D. Sanfilippo, F. Frusteri, A. Vaccari, and F. Arena, eds., Elsevier, Amsterdam, 1998, pp. 227–402.

490. (a) H. Schulty, ed., *Appl. Catal., A* **186**, 1–431 (1999); (b) E. van Santen, ed., *Catal. Today* **71**, 224–445 (2002).

491. B. Jager and R. Espinoza, *Catal. Today* **23**, 17 (1995).

492. B. Jager, in *Natural Gas Conversion V*, Studies in Surface Science and Catalysis, Vol. 119, A. Parmaliana, D. Sanfilippo, F. Frusteri, A. Vaccari, and F. Arena, eds., Elsevier, Amsterdam, 1998, p. 25.

493. R. L. Espinoza, A. P. Steynberg, B. Jager, and A. C. Vosloo, *Appl. Catal., A* **186**, 13 (1999).

494. J. J. C. Geerlings, J. H. Wilson, G. J. Kramer, H. P. C. E. Kuipers, A. Hoek, and H. M. Huisman, *Appl. Catal., A* **186**, 27 (1999).

495. A. P. Steynberg, R. L. Espinoza, B. Jager, and A. C. Vosloo, *Appl. Catal., A* **186**, 41 (1999).

496. H. Schulz, *Appl. Catal., A* **186**, 3 (1999).

497. A. Raje and B. H. Davis, in *Catalysis, a Specialist Periodical Report*, Vol. 12, J. J. Spivey, senior reporter, The Royal Society of Chemistry, Cambridge, UK, 1996, Chapter 3, p. 52.

498. G. P. van der Laan and A. A. C. M. Beenackers, *Catal. Rev.-Sci. Eng.* **41**, 255 (1999).

499. M. L. Turner, H. C. Long, A. Shenton, P. K. Byers, and P. M. Maitlis, *Chem. Eur. J.* **1**, 549 (1995).

500. P. M. Maitlis, H. C. Long, R. Quyoum, M. L. Turner, and Z.-Q. Wang, *Chem. Commun.* 1 (1996).

501. P. M. Maitlis, R. Quyoum, H. C. Long, and M. L. Turner, *Appl. Catal., A* **186**, 363 (1999).

502. R. C. Brady, III and R. Pettit, *J. Am. Chem. Soc.* **102**, 6181 (1980); **103**, 1287 (1981).

503. H. C. Long, M. L. Turner, P. Fornasiero, J. Kašpar, M. Graziani, and P. M. Maitlis, *J. Catal.* **167**, 172 (1997).

504. R. Quyoum, V. Berdini, M. L. Turner, H. C. Long, and P. M. Maitlis, *J. Catal.* **173**, 355 (1998).

505. B. E. Mann, M. L. Turner, R. Quyoum, N. Marsih, and P. M. Maitlis, *J. Am. Chem. Soc.* **121**, 6497 (1999).

506. B. Shi and B. H. Davis, *Catal. Today* **58**, 255 (2000).

507. P. M. Maitlis and M. L. Turner, *Catal. Today* **65**, 91 (2001).

508. B. Shi and B. H. Davis, *Catal. Today* **65**, 95 (2001).

509. M. K. Carter, *J. Mol. Catal. A: Chem.* **172**, 193 (2001).

510. B. Shi, R. J. O'Brien, S. Bao, and B. H. Davis, *J. Catal.* **199**, 202 (2001).

511. J. Patzlaff, Y. Liu, C. Graffmann, and J. Gaube, *Appl. Catal., A* **186**, 109 (1999).

512. L. Nowicki, S. Ledakowicz, and D. B. Bukur, *Chem. Eng. Sci.* **56**, 1175 (2001).

513. Y.-Y. Ji, H.-W. Xiang, J.-L. Yang, Y.-Y. Xu, Y.-W. Li, and B. Zhong, *Appl. Catal., A* **214**, 77 (2001).

514. S. Li, G. D. Meitzner, and E. Iglesia, *J. Phys. Chem. B* **105**, 5743 (2001).

515. D. O. Uner, N. Savargoankar, M. Pruski, and T. S. King, in *Dynamics of Surfaces and Reaction Kinetics in Heterogeneous Catalysis*, Studies in Surface Science and Catalysis, Vol. 109, G. F. Froment and K. C. Waugh, eds., Elsevier, Amsterdam, 1997, p. 315.

516. D. O. Uner, *Ind. Eng. Chem. Res.* **37**, 2239 (1998).

517. S. A. Hedrick, S. S. C. Chuang, A. Pant, and A. G. Dastidar, *Catal. Today* **55**, 247 (2000).

518. H.-Y. Luo, P.-Z. Lin, S.-B. Xie, H.-W. Zhou, C.-H. Xu, S.-Y. Huang, L.-W. Lin, D.-B. Liang, P.-L. Yin, and Q. Xin, *J. Mol. Catal. A: Chem.* **122**, 115 (1997).

519. S. Ito, S. Ishiguro, K. Nagashima, and K. Kunimori, *Catal. Lett.* **55**, 197 (1998).

520. T. Tago, T. Hanaoka, P. Dhupatemiya, H. Hayashi, M. Kishida, and K. Wakabayashi, *Catal. Lett.* **64**, 27 (2000).

521. H. Luo, H. Zhou, L. Lin, D. Liang, C. Li, D. Fu, and Q. Xin, *J. Catal.* **145**, 232 (1994).

522. H. Y. Luo, W. Zhang, H. W. Zhou, S. Y. Huang, P. Z. Lin, Y. J. Ding, and L. W. Lin, *Appl. Catal., A* **214**, 161 (2001).

523. H. Treviño, T. Heyon, and W. M. H. Sachtler, *J. Catal.* **170**, 236 (1997).

524. B.-Q. Xu and W. M. H. Sachtler, *J. Catal.* **180**, 194 (1998).

525. B.-Q. Xu, K.-Q. Sun, Q.-M. Zhu, and W. M. H. Sachtler, *Catal. Today* **63**, 453 (2000).

526. Y. Wang, H. Luo, D. Liang, and X. Bao, *J. Catal.* **196**, 46 (2000).

527. W. S. Postula, Z. Feng, C. V. Philip, A. Akgerman, and R. G. Anthony, *J. Catal.* **145**, 126 (1994).

528. Z. Feng, W. S. Postula, C. Erkey, C. V. Philip, A. Akgerman, and R. G. Anthony, *J. Catal.* **148**, 84 (1994).

529. C. Su, J. Li, D. He, Z. Cheng, and Q. Zhu, *Appl. Catal., A* **202**, 81 (2000).

530. S. Caili, H. Dehua, L. Junrong, C. Zhenxing, and Z. Quiming, *J. Mol. Catal. A: Chem.* **153**, 139 (2000).

531. K. Maruya, A. Takasawa, T. Haraoka, K. Domen, and T. Onishi, *J. Mol. Catal. A: Chem.* **112**, 143 (1996).

532. K. Maruya, M. Kawamura, M. Aikawa, M. Hara, and T. Arai, *J. Organomet. Chem.* **551**, 101 (1998).

533. A. L. Lapidus and A.Yu. Krylova, *Russ. Chem. Rev.* **67**, 941 (1988).

534. K. O. Xavier, R. Sreekala, K. K. A. Rashid, K. K. M. Yusuff, and B. Sen, *Catal. Today* **49**, 17 (1999).

535. M. Agnelli and C. Mirodatos, *J. Catal.* **192**, 204 (2000).

536. D. Schweer, L. Meeczko, and M. Baerns, *Catal. Today* **21**, 357 (1994).

537. S. Pak and J. H. Lunsford, *Appl. Catal., A* **168**, 131 (1998).

538. E. N. Voskresenskaya, V. G. Roguleva, and A. G. Anshits, *Catal. Rev.-Sci. Eng.* **37**, 101 (1995).

539. J. H. Lunsford, *Angew. Chem., Int. Ed. Engl.* **34**, 970 (1995); *Catal. Today* **63**, 165 (2000).

540. P. J. Gellings and H. J. M. Bouwmeester, *Catal. Today* **58**, 1 (2000).

541. *Natural Gas Conversion II*, Studies in Surface Science and Catalysis, Vol. 81, H. E. Curry-Hyde and R. F. Howe, eds., Elsevier, Amsterdam, 1994, pp. 125–283.

542. *Natural Gas Conversion IV*, Studies in Surface Science and Catalysis, Vol. 107, M. dePontes, R. L. Espinoza, C. P. Nicolaides, J. H. Scholtz, and M. S. Scurrell, eds., Elsevier, Amsterdam, 1997, pp. 291–394.

543. S. Pak, P. Qui, and J. H. Lunsford, *J. Catal.* **179**, 222 (1998).

544. J. I. Kong, J. S. Jung, J. G. Choi, and S. H. Lee, *Appl. Catal., A* **204**, 241 (2000).

545. J. H. Hong and K. J. Yoon, *Appl. Catal., A* **205**, 253 (2001).

546. K. Huang, F.-Q. Chen, and D.-W. Lü, *Appl. Catal., A* **219**, 61 (2001).

547. Y. Kou, B. Zhang, J. Niu, S. Li, H. Wang, T. Tanaka, and S. Yoshida, *J. Catal.* **173**, 309 (1998).

548. Y. Wang and Y. Ohtsuka, *Appl. Catal., A* **219**, 183 (2001).

549. A. Palermo, J. P. H. Vazquez, and R. M. Lambert, *Catal. Lett.* **68**, 191 (2000).

550. Y. Zeng and Y. S. Lin, *J. Catal.* **193**, 58 (2000); *AIChE J.* **47**, 436 (2001).

551. Y. P. Lu, A. G. Dixon, W. R. Moser, Y. H. Ma, and U. Balachandran, *Catal. Today* **56**, 297 (2000).

552. H. L. Wan, X. P. Zhou, W. Z. Weng, R. Q. Long, Z. S. Chao, W. D. Zhang, M. S. Chen, J. Z. Luo, and S. Q. Zhou, *Catal. Today* **51**, 161 (1999).

553. S. L. González-Cortés, J. Orozco, and B. Fontal, *Appl. Catal., A* **213**, 259 (2001).

554. A. Djaidja, A. Barama, and M. M. Bettahar, *Catal. Today* **61**, 303 (2000).

555. H. Aritani, H. Yamada, T. Nishio, T. Shiono, S. Imamura, M. Kudo, S. Hasegawa, T. Tanaka, and S. Yoshida, *J. Phys. Chem. B* **104**, 10133 (2000).

556. J. L. Boldú, E. Muñoz, X. Bokhimi, O. Novaro, T. López, and R. Gómez, *Langmuir* **15**, 32 (1999).

557. K. R. Tsai, D. A. Chen, H. L. Wan, H. B. Zhang, G. D. Lin, and P. X. Zhang, *Catal. Today* **51**, 3 (1999).

558. H. B. Zhang, G. D. Lin, H. L. Wan, Y. D. Liu, W. Z. Weng, J. X. Cai, Y. F. Shen, and K. R. Tsai, *Catal. Lett.* **73**, 141 (2001).

559. A. Machocki, *Appl. Catal., A* **146**, 391 (1996).

560. J. H. Lunsford, E. M. Cordi, P. Qiu, and M. P. Rosynek, in *Natural Gas Conversion V*, Studies in Surface Science and Catalysis, Vol. 119, A. Parmaliana, D. Sanfilippo, F. Frusteri, A. Vaccari, and F. Arena, eds., Elsevier, Amsterdam, 1998, p. 227.

561. E. M. Cordi, S. Pak, M. P. Rosynek, and J. H. Lunsford, *Appl. Catal., A* **155**, L1 (1997).

562. P. Qui, J. H. Lunsford, and M. P. Rosynek, *Catal. Lett.* **48**, 11 (1997).

563. S. Pak, T. Rades, M. P. Rosynek, and J. H. Lunsford, *Catal. Lett.* **66**, 1 (2000).

564. C. J. Liu, A. Marafee, R. Mallinson, and L. Lobban, *Appl. Catal., A* **164**, 21 (1997).

565. C. J. Liu, R. Mallinson, and L. Lobban, *J. Catal.* **179**, 326 (1998).

566. A. Marafee, C. Liu, G. Xu, R. Mallinson, and L. Lobban, *Ind. Eng. Chem. Res.* **36**, 632 (1997).

567. C. Liu, A. Marafee, B. Hill, G. Xu, R. Mallinson, and L. Lobban, *Ind. Eng. Chem. Res.* **35**, 3295 (1996).

568. K. Onoe, A. Fujie, T. Yamaguchi, and Y. Hatano, *Fuel* **76**, 281 (1997).

569. S. Kado, Y. Sekine, and K. Fujimoto, *Chem. Commun.* 2485 (1999).

570. M. S. Ioffe, S. D. Pollington, and J. K. S. Wan, *J. Catal.* **151**, 349 (1995).

571. C. Marún, L. D. Conde, and S. L. Suib, *J. Phys. Chem. A* **103**, 4332 (1999).

572. K. Okabe, K. Sayama, H. Kusama, and H. Arakawa, *Chem. Lett.* 457 (1997).

573. Y. Kato, H. Yoshida, and T. Hattori, *Chem. Commun.* 2389 (1998).

574. L. Guczi, R. A. van Santen, and K. V. Sarma, *Catal. Rev.-Sci. Eng.* **38**, 249 (1996).

575. M. Belgued, A. Amariglio, P. Pareja, and H. Amariglio, *Nature* **352**, 789 (1991).

576. T. Koerts, M. J. A. G. Deelen, and R. A. van Santen, *J. Catal.* **138**, 101 (1992).

577. F. Solymosi, A. Erdóhelyi, and J. Cserényi, *Catal. Lett.* **16**, 399 (1992).

578. J. S. Mohammad, J. S. M. Zadeh, and K. J. Smith, *J. Catal.* **183**, 232 (1999).

579. M. Belgued, A. Amariglio, L. Lefort, P. Paréja, and H. Amariglio, *J. Catal.* **161**, 282 (1996).

580. P. Pareja, S. Molina, A. Amariglio, and H. Amariglio, *Appl. Catal., A* **168**, 289 (1998).

581. E. Marceau, J.-M. Tatibouët, M. Che, and J. Saint-Just, *J. Catal.* **183**, 384 (1999).

582. M. Belgued, A. Amariglio, P. Paréja, and H. Amariglio, *J. Catal.* **159**, 449 (1996).

583. L. Guczi, K. V. Sarma, and L. Borkó, *Catal. Lett.* **39**, 43 (1996).

584. L. Guczi, K. V. Sarma, and L. Borkó, *J. Catal.* **167**, 495 (1997).

585. G.-C. Shen and M. Ichikawa, *J. Chem. Soc., Faraday Trans.* **93**, 1185 (1997).

586. F. Simon, S. Molina, A. Amariglio, P. Paréja, H. Amariglio, and G. Szabo, *Catal. Today* **46**, 217 (1998).

587. O. Garnier, J. Shu, and B. P. A. Grandjean, *Ind. Eng. Chem. Res.* **36**, 553 (1997).

588. M. V. Tsodikov, Ye. V. Slivinskii, V. P. Mordovin, O. V. Bukhtenko, G. Colón, M. C. Hidalgo, and J. A. Navío, *Chem. Commun.* 943 (1999).

589. L. Wang, L. Tao, M. Xie, G. Xu, J. Huang, and Y. Xu, *Catal. Lett.* **21**, 35 (1993).

590. Y. Xu and L. Lin, *Appl. Catal., A* **188**, 53 (1999).

591. D. Wang, J. H. Lunsford, and M. P. Rosynek, *J. Catal.* **169**, 347 (1997).

592. F. Solymosi, A. Szöke, and J. Cserényi, *Catal. Lett.* **39**, 157 (1996).

593. S. Liu, Q. Dong, R. Ohnishi, M. Ichikawa, *Chem. Commun.* 1455 (1997).

594. D. Ma, Y. Shu, X. Bao, and Y. Xu, *J. Catal.* **189**, 314 (2000).

595. Y. Shu, D. Ma, X. Bao, and Y. Xu, *Catal. Lett.* **66**, 161 (2000).

596. D. Ma, Y. Shu, X. Han, X. Liu, Y. Xu, and X. Bao, *J. Phys. Chem. B* **105**, 1786 (2001).

597. B. M. Weckhuysen, D. Wang, M. P. Rosynek, and J. H. Lunsford, *Angew. Chem., Int. Ed. Engl.* **36**, 2374 (1997).

598. R. Ohnishi, S. Liu, Q. Dong, L. Wang, and M. Ichikawa, *J. Catal.* **182**, 92 (1999).

599. S. Tang, H. Chen, J. Lin, and K. L. Tan, *Catal. Commun.* 31 (2001).

600. Y. Lu, D. Ma, Z. Xu, Z. Tian, X. Bao, and L. Lin, *Chem. Commun.* 2048 (2001).

601. B. Liu, Y. Yang, and A. Sayari, *Appl. Catal., A* **214**, 95 (2001).

602. F. Solymosi, J. Cserényi, A. Szöke, T. Bánsági, and A. Oszkó, *J. Catal.* **165**, 150 (1997).

603. S. Liu, L. Wang, R. Ohnishi, and M. Ichikawa, *J. Catal.* **181**, 175 (1999).

604. F. Solymosi, L. Bugyi, A. Oszkó, and I. Horváth, *J. Catal.* **185**, 160 (1999).

605. D. Ma, Y. Shu, W. Zhang, X. Han, Y. Xu, and X. Bao, *Angew. Chem., Int. Ed. Engl.* **39**, 2928 (2000).

606. R. W. Borry, III, E. C. Lu, Y.-H. Kim, and E. Iglesia, in *Natural Gas Conversion V*, Studies in Surface Science and Catalysis, Vol. 119, A. Parmaliana, D. Sanfilippo, F. Frusteri, A. Vaccari, and F. Arena, eds., Elsevier, Amsterdam, 1998, p. 403.

607. R. W. Borry, III, Y. H. Kim, A. Huffsmith, J. A. Reimer, and E. Iglesia, *J. Phys. Chem.* **103**, 5787 (1999).

608. L. Men, L. Zhang, Y. Xie, Z. Liu, J. Bai, G. Sha, and J. Xie, *Chem. Commun.* 1750 (2001).

609. W. Ding, S. Li, G. D. Meitzner, and E. Iglesia, *J. Phys. Chem. B* **105**, 506 (2001).

610. T. Matsuhisa, in *Catalysis, a Specialist Periodical Report*, Vol. 12, J. J. Spivey, senior reporter, The Royal Society of Chemistry, Cambridge, UK, 1996, Chapter 1, p. 1.

611. M. S. Spencer, *Catal. Lett.* **50**, 37 (1998); **60**, 45 (1999).

612. J. L. Robbins, E. Iglesia, C. P. Kelkar, and B. DeRites, *Catal. Lett.* **10**, 1 (1991).

613. M. Muhler, E. Törnquist, L. P. Nielsen, B. S. Clausen, and H. Topsøe, *Catal. Lett.* **25**, 1 (1994).

614. A. Gotti and R. Prins, *J. Catal.* **178**, 511 (1998).

615. Y. Kanai, T. Watanbe, T. Fujitani, T. Uchijima, and J. Nakamura, *Catal. Lett.* **38**, 157 (1996).

616. Y.-W. Suh, S.-H. Moon, and H. K. Rhee, *Catal. Today* **63**, 447 (2000).

617. O.-S. Joo, K.-D. Jung, S.-H. Han, S.-J. Uhm, D.-K. Lee, and S.-K. Ihm, *Appl. Catal., A* **135**, 273 (1996).

618. S. Lin, A. Oldfield, and D. Klenerman, *Surf. Sci.* **464**, 1 (2000).

619. M. M. Günter, T. Ressler, B. Bems, C. Büscher, T. Genger, O. Hinrichsen, M. Muhler, and R. Schlögl, *Catal. Lett.* **71**, 37 (2001).

620. J. G. Wu, M. Saito, M. Takeuchi, and T. Watanabe, *Appl. Catal., A* **218**, 235 (2001).

621. I. A. Fisher, H. C. Woo, and A. T. Bell, *Catal. Lett.* **44**, 11 (1997).

622. Z. Liu, W. Ji, L. Dong, and Y. Chen, *J. Catal.* **172**, 243 (1997).

623. J. Liu, J. Shi, D. He, Q. Zhang, X. wu, Y. Liang, and Q. Zhu, *Appl. Catal., A* **218**, 113 (2001).

624. J. Weigel, K. Horbaschek, A. Baiker and A. Wokaun, *Ber. Bunsenges. Phys. Chem.* **101**, 1097 (1997).

625. W.-J. Shen, Y. Ichihashi, M. Okumura, and Y. Matsumura, *Catal. Lett.* **64**, 23 (2000).

626. W.-J. Shen, M. Okumara, Y. Matsumura, and M. Haruta, *Appl. Catal., A* **213**, 225 (2001).

627. K. Li and D. Jiang, *J. Mol. Catal. A: Chem.* **147**, 125 (1999).

628. J.-P. Lange, *Catal. Today* **64**, 3 (2001).

629. K. Klier, A. Beretta, Q. Sun, O. C. Feeley, and R. G. Herman, *Catal. Today* **36**, 3 (1997).

630. R. G. Herman, *Catal. Today* **55**, 233 (2000).

631. W. S. Epling, G. B. Hoflund, and D. M. Minahan, *J. Catal.* **175**, 175 (1998).

632. M. M. Burcham, R. G. Herman, and K. Klier, *Ind. Eng. Chem. Res.* **37**, 4657 (1998).

633. J. M. Campos-Martín, J. L. G. Fierro, A. Guerrero-Ruiz, R. G. Herman, and K. Klier, *J. Catal.* **163**, 418 (1996).

634. A.-M. Hilmen, M. Xu, M. J. L. Gines, and E. Iglesia, *Appl. Catal., A* **169**, 355 (1998).

635. C. R. Apesteguía, B. DeRites, S. Miseo, and S. Soled, *Catal. Lett.* **44**, 1 (1997).

636. M. Xu and E. Iglesia, *Catal. Lett.* **51**, 47 (1998).

637. M. Xu and E. Iglesia, *J. Catal.* **188**, 125 (1999).

638. *Catalysis: Methanol to Hydrocarbons*, M. Stöcker and J. Weitkamp, eds., *Micropor. Mesopor. Mat.* **29**, 1–218 (1999).

639. M. Stöcker, *Micropor. Mesopor. Mat.* **29**, 3 (1999).

640. F. J. Keil, *Micropor. Mesopor. Mat.* **29**, 49 (1999).

641. S. R. Blaszkowski and R. A. van Santen, *J. Am. Chem. Soc.* **119**, 5020 (1997).

642. I. Štich, J. D. Gale, K. Terakura, and M. C. Payne, *J. Am. Chem. Soc.* **121**, 3292 (1999).

643. N. Tajima, T. Tsuneda, F. Toyama, and K. Hirao, *J. Am. Chem. Soc.* **120**, 8222 (1998).

644. G. J. Hutchings, G. W. Watson, and D. J. Willock, *Micropor. Mesopor. Mat.* **29**, 67 (1999).

645. M. Hunger and T. Horvath, *J. Am. Chem. Soc.* **118**, 12302 (1996).

646. S. M. Campbell, X.-Z. Jiang, and R. F. Howe, *Micropor. Mesopor. Mat.* **29**, 91 (1999).

647. J. F. Haw, J. B. Nicholas, W. Song, F. Deng, Z. Wang, T. Xu, and C. S. Heneghan, *J. Am. Chem. Soc.* **122**, 4763 (2000).

648. L. K. Carlson, P. K. Isbester, and E. J. Munson, *Solid State Nucl. Mag. Resonance* **16**, 93 (2000).

649. M. Hytja, I. Štich, J. D. Gale, K. Terakura, and M. C. Payne, *Chem. Eur. J.* **7**, 2521 (2001).

650. J. F. Haw, *Top. Catal.* **8**, 81 (1999).

651. D. Freeman, R. P. K. Wells, and G. J. Hutchings, *Chem. Commun.* 1754 (2001).

4

ISOMERIZATION

Isomerizations are reversible conversions, leading ultimately to a thermodynamic equilibrium mixture of isomers, namely, of compounds of the same molecular formula (same number and kind of atoms) but with different arrangements. There are three major classes of isomers: structural, stereochemical, and conformational. Isomerization of saturated hydrocarbons includes skeletal rearrangements and *cis–trans* isomerization of substituted cycloalkanes. Skeletal isomerizations may be classified as chain branching, that is, transformation of straight-chain compounds into branched isomers; shift of substituents, usually methyl groups along a chain or in the ring; ring contraction and enlargement; and ring opening of cycloalkanes to form alkenes. Isomerization of straight-chain alkanes to branched ones for octane-number enhancement of gasoline and that of dialkylbenzenes, such as xylenes, are examples of practical importance. A special case of isomerization of saturated hydrocarbons is the racemization of asymmetric compounds. Besides skeletal rearrangements, alkenes may undergo double-bond migration and *cis–trans* isomerization. Dienes and polyenes may be converted to cyclic unsaturated hydrocarbons. Thermal and photochemical pericyclic rearrangements are specific, noncatalytic isomerization processes.

Different catalysts bring about different types of isomerization of hydrocarbons. Acids are the best known and most important catalysts bringing about isomerization through a carbocationic process. Brønsted and Lewis acids, acidic solids, and superacids are used in different applications. Base-catalyzed isomerizations of hydrocarbons are less frequent, with mainly alkenes undergoing such transformations. Acetylenes and allenes are also interconverted in base-catalyzed reactions. Metals with dehydrogenating–hydrogenating activity usually supported on oxides are also used to bring about isomerizations. Zeolites with shape-selective characteristics exhibit high activity in the isomerization of hydrocarbons and play an important role in refining processes (see Sections 2.1 and 2.2).

4.1. ACID-CATALYZED ISOMERIZATION

The first published report of the isomerization of a saturated hydrocarbon appeared in 1902, when Aschan reported[1] that cyclohexane undergoes isomerization to methylcyclopentane with aluminum chloride. Later researchers were, however, unable to repeat this work, and it was not until 1933 that their failures were, as discussed subsequently, explained by Nenitzescu and coworkers.[2,3]

Ipatieff and Grosse[4] showed in 1936 that n-butane is isomerized to isobutane by properly promoted aluminum chloride. The importance of isobutane in the alkylation of alkenes and the possibility of converting n-alkanes of low octane values into branched high-octane alkanes for gasoline quickly resulted in much research that furnished information of both theoretical and industrial importance.[5–11]

4.1.1. Alkanes

Alkane isomerization equilibria are temperature-dependent, with the formation of branched isomers tending to occur at lower temperatures (Table 4.1). The use of superacids exhibiting high activity allows to achieve isomerization at lower temperature (as discussed below). As a result, high branching and consequently higher octane numbers are attained. Also, thermodynamic equilibria of neutral hydrocarbons and those of derived carbocations are substantially different. Under appropriate conditions (usual acid catalysts, longer contact time) the thermodynamic

Table 4.1. Temperature-dependent isomerization equilibria of C_4–C_6 alkanes[12]

Alkane	Temperatures (°C)			
	21	100	149	204
Butanes				
n-Butane	15	25	35	43
Isobutane	85	75	65	57
Pentanes				
n-Pentane	5	15	22	29
2-Methylbutane	95	85	78	71
2,2-Dimethylpropane	0	0	0	0
Hexanes				
n-Hexane	4	11	14	17
2-Methylpentane	20	28	34	36
3-Methylpentane	8	13	15	17
2,2-Dimethylbutane	57	38	28	21
2,3-Dimethylbutane	11	10	9	9

equilibrium mixture of hydrocarbons can be reached. In contrast, quenching a reaction mixture in contact with excess of strong (super)acid can give a product distribution approaching the thermodynamic equilibrium of the corresponding carbocations. The two equilibria can be very different. There is a large energy difference in the stability of primary < secondary < tertiary carbocations. Thus, in excess of superacid solution, generally the most stable tertiary cations predominate—allowing, for example, isomerization of n-butane to isobutane to proceed past the equilibrium concentrations of the neutral hydrocarbons, as the *tert*-butyl cation is by far the most stable butyl cation.

When aluminum chloride is used as a catalyst in the presence of a promoter or initiator (called *promoted aluminum chloride*) for the isomerization of n-butane, the product is isobutane accompanied by little or no byproducts. On the other hand, isomerization of higher-molecular-weight alkanes is accompanied by much disproportionation or cracking. n-Pentane yields not only isopentane but also butanes and hexanes; similarly, n-hexane yields not only isomeric methylpentanes but also butanes, pentanes, and heptanes. With higher alkanes, little isomerization product is obtained; the principal reaction is cracking. In the case of n-pentane and higher alkanes, isomerization accompanied by little side reaction can be achieved by adding certain cracking inhibitors, such as benzene, isobutane, methylcyclopentane, or hydrogen under pressure.

With active halide catalysts and cocatalysts, cycloalkanes containing five or six carbon atoms in the ring undergo isomerization quite analogous to those discussed for the various types of alkanes. Shift of alkyl groups attached to the ring occurs. In addition, there is interconversion of five- and six-membered ring cycloalkanes. At 100°C either cyclohexane or methylcyclopentane is isomerized to an equilibrium mixture containing 33.5% methylcyclopentane.[13]

Cyclopropanes and cyclobutanes are not produced by isomerization of cyclopentane. Cycloheptane is irreversibly isomerized to methylcyclohexane by promoted aluminum chloride[14] or bromide.[15]

Methylcyclohexane undergoes little isomerization when treated with aluminum halides because the equilibrium between the dimethylcyclopentanes and methylcyclohexane greatly favors the latter.

Isomerization of alkylcyclopentanes or alkylcyclohexanes in which the alkyl group consists of at least two carbon atoms yields methyl-substituted cyclohexanes. For example, at 50°C, *sec*-butylcyclopentane produced trimethylcyclohexanes but no propylcyclohexane.[16] In another study, n-, *sec*-, or *tert*-butylcyclohexanes at 150°C were almost quantitatively converted to tetramethylcyclohexanes.[17]

Concentrated sulfuric acid does not catalyze the isomerization of n-alkanes; neither does silica–alumina. On the other hand, each does isomerize methylpentanes and higher alkanes that possess tertiary carbon atoms, the isomerization occurring by formation of alkanes containing tertiary carbon atoms in new positions:

$$
\underset{\underset{\displaystyle CH_3CHCH_2CH_2CH_3}{|}}{CH_3} \xrightarrow[]{\;H_2SO_4\;} \underset{\underset{\displaystyle CH_3CH_2CHCH_2CH_3}{|}}{CH_3} \qquad (4.1)
$$

The isomerization activity of sulfuric acid strongly depends on its concentration,[18] with the highest activity at 99.8%.

Neither isomerization initiators nor cracking inhibitors need to be present when isomerization occurs in the presence of these acidic catalysts. Disproportionation does not accompany the isomerization.

A tertiary carbon atom is also a prerequisite in the isomerization of cycloalkanes in the presence of sulfuric acid. Cyclopentane and cyclohexane, consequently, do not react. The ring seldom contracts or enlarges. Ethylcyclopentane, however, undergoes ring enlargement to form methylcyclohexane.[19] The reaction is limited chiefly to shifting of alkyl (usually methyl) groups around the ring. The *cis* and *trans* isomers of 1,2-, 1,3-, and 1,4-dimethylcyclohexane undergo rapid epimerization and slow methyl shift at 25°C when treated with 99.8% sulfuric acid.[18]

Two studies in 1985 and 1992 focused on the use of superacids in isomerization.[20,21] An important advantage of these catalysts is that they may be used at lower temperature because of their greatly increased ability to bring about carbocationic process. For example, aluminum chloride, still one of the important isomerization catalysts in industry, can be used at about 80–100°C. In comparison, superacids are active in alkane isomerization at room temperature or below. In addition to avoid side reactions under such conditions, lower reaction temperatures favor thermodynamic equilibria with higher proportion of branched isomers, which is fundamental in increasing the octane number of gasoline.

Isomerization of alkanes not possessing tertiary hydrogens is slow when relatively weak superacids such as CF_3SO_3H are used. In such cases electrochemical oxidation or the addition of alkenes can initiate isomerization through promotion of the initial formation of carbocations.[22] Solid superacids have the further advantages of easy handling and convenient separation. Over a SbF_5-intercalated graphite catalyst below 0°C, isomeric hexanes undergo highly selective isomerization without significant cracking to give an equilibrium mixture of isomers.[23,24] The interconversion of cyclohexane and methylcyclopentane takes place similarly at room temperature.[25] Oxides treated with Lewis acids can also be very active and selective.[20,21] $SbF_5–TiO_2$ and $SbF_5–TiO_2–SiO_2$ exhibited the best activity and selectivity in the isomerization of butane.[26] Isomerization of straight-chain alkanes (as well as cycloalkanes and alkylaromatics) can be effected by catalysts consisting of a Lewis acid (SbF_5, TaF_5, NbF_5) and a Brønsted acid (CF_3COOH, CF_3SO_3H).[27] n-Hexane reacts at 25°C and 1 atm to yield isomeric heptanes (21%) and hexanes (50%), with the latter containing 30% of dimethylbutanes.

Fluorosulfuric acid and related superacids were found by Olah to be efficient in the important n-butane–isobutane isomerization.[28,29] The isomerization of n-butane in HSO_3F containing 2–5% HF gives a 70% conversion to isobutane under very mild conditions (flow system, room temperature, 10–30 s contact time).[29] Initiators such as butyl fluorosulfates or butyl fluorides (also formed in situ when small amounts of butenes are added as promoters) accelerate initial isomerization. A unique characteristic of this isomerization is the limitation of C_4 carbocations to undergo β cleavage resulting in cracking. In contrast with these observations, the even stronger $HF–SbF_5$ superacid is an inefficient isomerization catalyst for butane; it causes extensive hydrocracking through direct carbon–carbon bond protolysis.[30]

A related superacidic method was also developed to upgrade natural-gas liquids (accompanying methane in many natural gases in significant amounts). Isomerization of the straight-chain alkanes they contain to highly branched isomers is achieved by HF–BF$_3$ catalyst.[31] At the same time there also take place alkylative processes enhancing branched C$_6$–C$_8$ components without significant hydrocracking (which would increase unwanted volatility). Small amounts of alkenes (preferably butenes) accelerate the upgrading process. This is explained by facilitating the initial formation of essential concentration of carbocations. Further improvement was achieved when applying some hydrogen gas pressure, which substantially suppresses side reactions (cracking, disproportionation). Since hydrogen controls the amount of hydrocarbons entering into the catalyst phase, secondary transformations of product hydrocarbons are also minimized. Finally, the volatility of the HF–BF$_3$ catalyst system allows easy recovery and recycling.

All C$_{10}$ polycylic hydrocarbons under strongly acidic conditions undergo a unique isomerization to give adamantane.[32–34] The adamantane rearrangement was discovered by Schleyer in connection with the catalytic interconversion of *endo*- and *exo*-trimethylenenorbornane:[35–37]

$$(4.2)$$

Although the yield was initially low, the reaction offered a very convenient way for obtaining adamantane. Diamantane[38] and triamantane[39] were later obtained by a similar approach from C$_{14}$ and C$_{18}$ precursors, respectively.

Improved yields of adamantane up to 65% were achieved by using superacidic systems [HF–SbF$_5$, CF$_3$SO$_2$OH–B(OSO$_2$CF$_3$)$_3$, various solid superacids][40–42] and chlorinated Pt on alumina.[43] More recently near-quantitative isomerization with conjugate superacids promoted by 1-haloadamantanes as the source of 1-adamantyl cation and sonication was reported.[44]

The interconversion of possible C$_{10}$H$_{16}$ isomers to adamantane, the thermodynamically most stable isomer, is a very complex system. It involves practically thousands of carbocationic intermediates via 1,2-hydride, alkyde shifts. The interconversion possibilities were summarized in a complex map of rearrangement graph ("adamantaneland").[45,46] The adamantane rearrangement is now recognized to be general for polycyclic hydrocarbons with the formula C$_{4n+6}$H$_{4n+12}$ ($n = 1, 2, 3$).

Significant efforts were carried out to obtain dodecahedrane via the similar carbocationic isomerization of appropriate C$_{20}$H$_{20}$ polycyclic precursors.[47,48] Despite the favorable overall thermodynamics, however, most such efforts failed and only trace amounts of dodecahedrane were isolated. Cyclization of the dehydration product **1** was, however, successful to yield 1,16-dimethyldodecahedrane under strongly acidic conditions:[49,50]

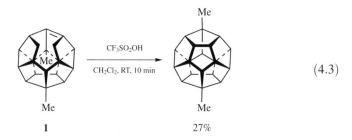

$$(4.3)$$

The unsubstituted parent compound could also be synthesized[51] through the metal-catalyzed isomerization in the presence of H_2 of the saturated polycyclic system [1.1.1.1]pagodane in 8% yield (0.1% Pt on Al_2O_3, 315°C). This reaction, however, is considered to follow a radical pathway[52] through a series of hydrogenation–dehydrogenation steps.[48,53]

Mechanism. The proven acidity of the effective alkane isomerization catalysts suggests that carbocations are involved in acid-catalyzed alkane isomerization. Such a mechanism was first proposed by Schmerling and coworkers[54] on the basis of the pioneering ideas of Whitmore[55] for the skeletal isomerization of alkanes and cycloalkanes in the presence of aluminum chloride and a trace of olefin or other promoter. Subsequently these concepts were used to explain the mechanism of the acid-catalyzed isomerizations in general.

The mechanism can be considered as a chain reaction including an initiation step to form the first carbocation [Eq. (4.4)], the isomerization of the intermediate carbocation [Eq. (4.5)], and propagation (chain transfer) through hydride ion transfer [Eq. (4.6)]:

$$R{-}H \; \rightleftharpoons \; R^+ \qquad (4.4)$$

$$R^+ \; \rightleftharpoons \; \text{iso-}R^+ \qquad (4.5)$$

$$\text{iso-}R^+ \; + \; R{-}H \; \rightleftharpoons \; \text{iso-}R{-}H \; + \; R^+ \qquad (4.6)$$

The involvement of carbocations accounts for the side reactions that accompany isomerization. Carbocations are known to undergo β scission to yield low-molecular-weight cracking products. They can also undergo proton elimination to form alkenes that, in turn, participate in condensation (oligomerization), cyclization, and disproportionation reactions.

Formation of the initial carbocation depends on the catalyst used. It was shown by Nenitzescu[2] that no isomerization occurred when pure cyclohexane was contacted with pure aluminum chloride, even at refluxing temperature. However, when a trace of water was added to the reaction mixture, isomerization occurred. Careful studies also demonstrated that butane, methylcyclopentane, and cyclohexane underwent isomerization with $HCl-AlCl_3$ or $HBr-AlBr_3$ as the catalysts only when a small amount (less than 0.1%) of olefin promoter[56,57] or alkyl halide[57,58] was added. The added alkene serves as a source of carbocations [Eq. (4.7)] to initiate the chain reaction according to Eq. (4.8):

$$\text{⬡ || } + \ AlHlg_3 \ + \ HHlg \ \rightleftharpoons \ \text{⬡}^+ \ AlHlg_4^- \tag{4.7}$$

$$\text{⬡}^+ \ + \ CH_3CH_2CH_2CH_3 \ \rightleftharpoons \ CH_3CH_2\overset{+}{C}HCH_3 \ + \ \text{⬡} \tag{4.8}$$

Water participates in carbocation formation[59,60] by generating the strong acid $H[Al(OH)Hlg_3]$,[6,59] which forms a carbocation through an alkyl halide intermediate[60] or via direct hydride ion abstraction.[6] Experiments with heavy water by Pines further proved its cocatalytic effect indicating that HCl (HBr) formed is not the real activator.[60,61]

Sulfuric acid, a relatively weak protic acid catalyst, generates carbocations through the oxidation of tertiary hydrogen atoms:

$$CH_3-\overset{\overset{\displaystyle CH_3}{|}}{C}H-CH_3 \ + \ 2\,H_2SO_4 \ \xrightarrow[-2\,H_2O]{} \ CH_3-\overset{\overset{\displaystyle CH_3}{|}}{\underset{+}{C}}-CH_3 \ + \ SO_2 \ + \ HSO_3O^- \tag{4.9}$$

The basic information comes from isotope exchange studies of isobutane.[62] Exchange of only the primary methyl hydrogens was shown to occur when isobutane reacts with D_2SO_4 yielding eventually $(CD_3)_3CH$. Deuterium exchange was suggested to take place via the redeuteration of isobutylene formed in equilibrium with the *tert*-butyl cation [Eq. (4.10)], whereas the tertiary hydrogen derives only from intermolecular hydride transfer with isobutane [Eq. (4.11)]:

$$CH_3-\overset{\overset{\displaystyle CH_3}{|}}{\underset{+}{C}}-CH_3 \ \xrightleftharpoons{-H^+} \ (CH_3)_2C=CH_2 \ \xrightleftharpoons{D_2SO_4} \ (CH_3)_2\overset{+}{C}CH_2D \ \Longrightarrow \ (CD_3)_3C^+ \tag{4.10}$$

$$(CD_3)_3C^+ \ \xrightleftharpoons{(CH_3)_3CH} \ (CD_3)_3CH \ + \ (CH_3)_3C^+ \tag{4.11}$$

In superacids protonation of isobutane leads to pentacoordinate carbocations, the interconversion of which scrambles all hydrogens inclusive of the methine hydrogen with initial preference for tertiary C–H exchange:[63]

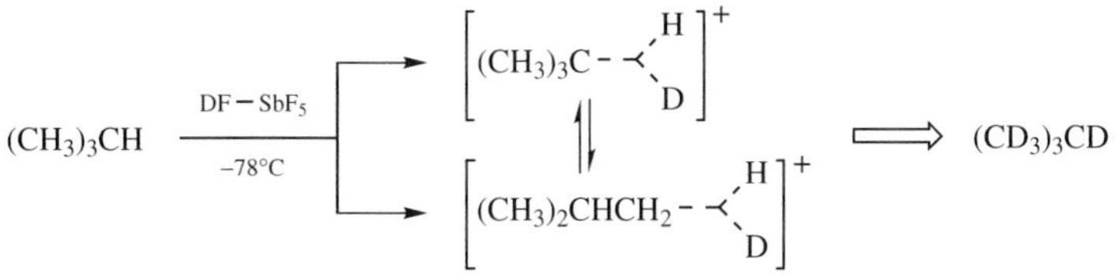

$$(CH_3)_3CH \xrightarrow[-78°C]{DF-SbF_5} \left\{ \begin{array}{l} \left[(CH_3)_3C - \overset{H}{\underset{D}{\overset{\diagup}{\diagdown}}} \right]^+ \\ \\ \left[(CH_3)_2CHCH_2 - \overset{H}{\underset{D}{\overset{\diagup}{\diagdown}}} \right]^+ \end{array} \right. \Longleftrightarrow (CD_3)_3CD \qquad (4.12)$$

The *n*-butane–isobutane isomerization similarly involves initial protonation of *n*-butane to five-coordinate carbocation (**2**, Scheme 4.1) followed by loss of H_2, giving the first trivalent carbocation (**3**). Compound **3** gives isobutane via 1,2-alkyde shift involving protonated cyclopropane and 1,2-hydride shift[21] (Scheme 4.1). The acid-catalyzed isomerization of *n*-butane thus proceeds through the relatively high-energy primary carbocation **4**, but this ion de facto is substantially delocalized. Additionally, little hydrocracking occurs due to limitation of β cleavage of the intermediate butyl cations.

Cyclohexane–methylcyclopentane isomerization[21] can be depicted as in Scheme 4.2. The isomerization of substituted cyclopentanes and cyclohexanes to polymethylcyclohexanes similarly occurs by way of a series of consecutive steps

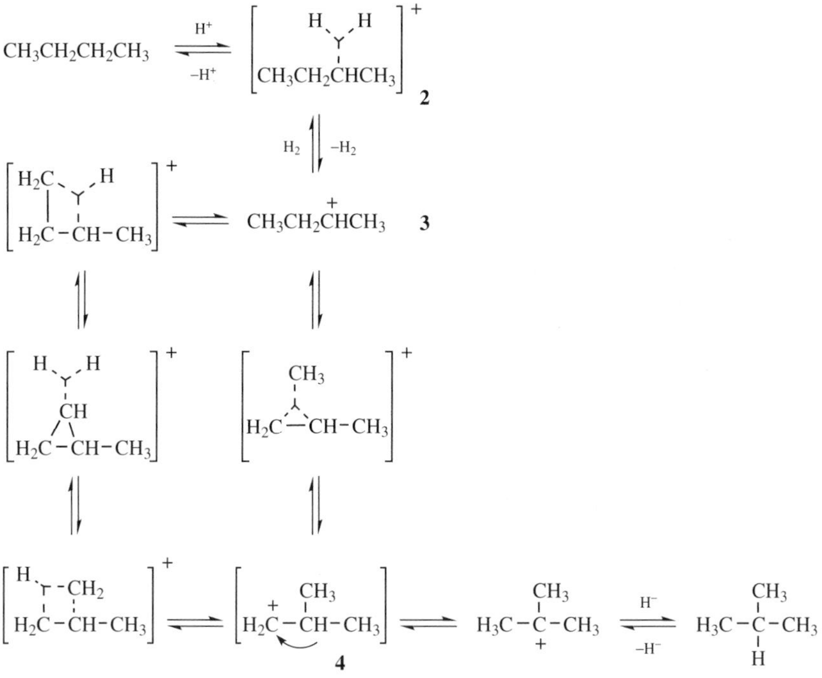

Scheme 4.1

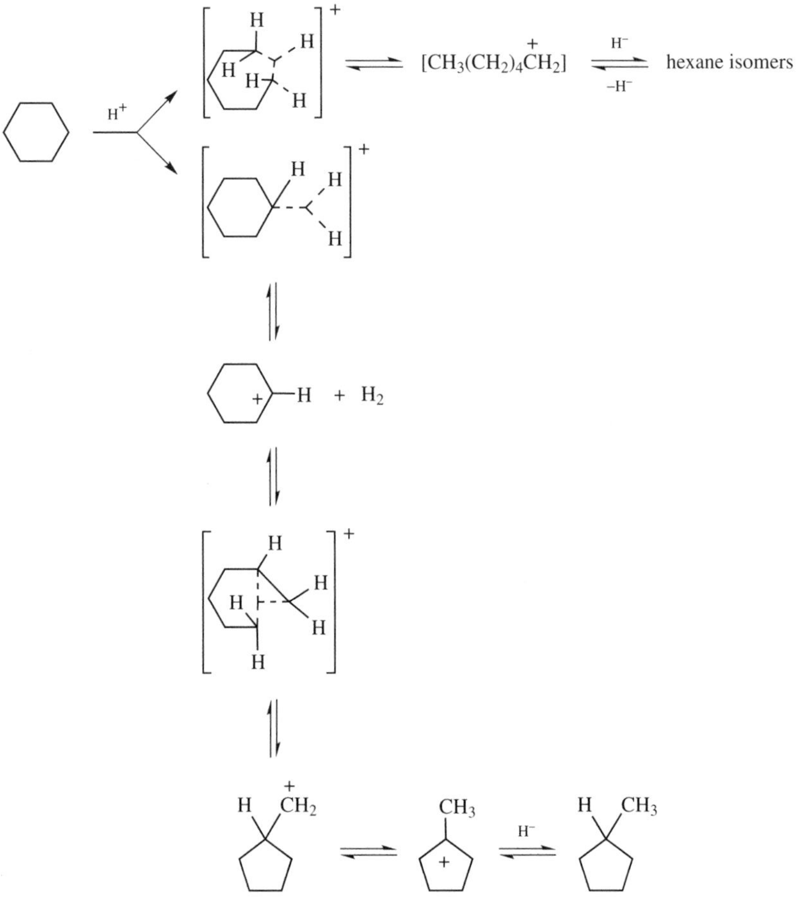

Scheme 4.2

including ring expansion and contraction. For example, the isomerization of *sec*-butylcyclopentane in the presence of aluminum chloride at 50°C yields trimethyl-cyclohexane[16] according to the following:

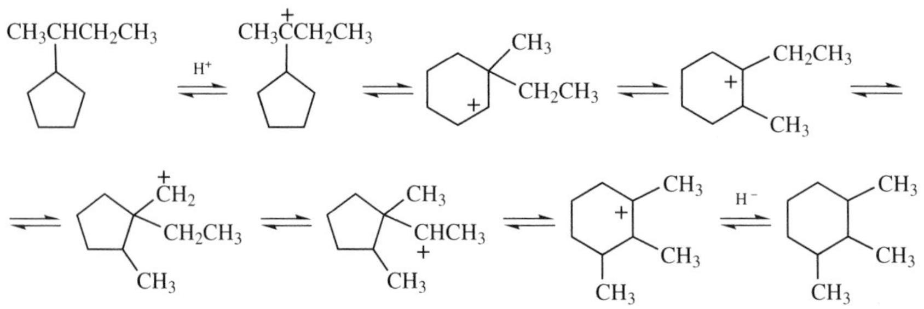

$$(4.13)$$

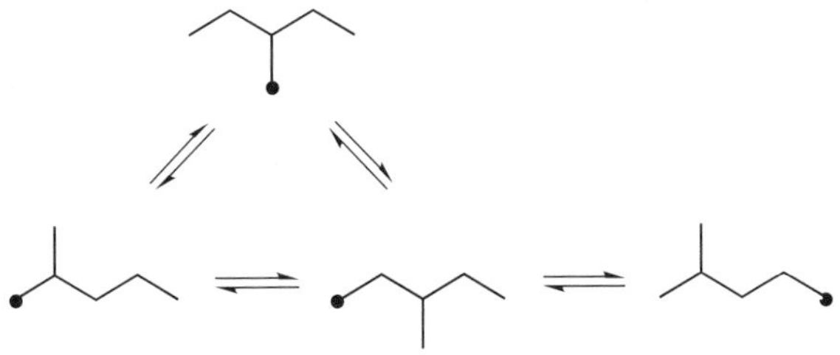

Scheme 4.3

The intramolecular nature of most carbocationic isomerization was proved by means of labeling experiments. $[1-^{13}C]$-Propane was isomerized in the presence of aluminum bromide promoted by hydrogen bromide to form $[2-^{13}C]$-propane. None of the propane product contained more than one ^{13}C atom per molecule.[64] Similarly, very little label scrambling was observed in the isomerization of labeled hexanes over SbF_5-intercalated graphite.[65] Thus simple consecutive 1,2-methyl shifts can account for the isomerization of ^{13}C-labeled methylpentanes (Scheme 4.3).

Novel aprotic superacid systems have been found by Vol'pin and coworkers to be highly active in alkane isomerization.[66] $RCOHlg-2AlHlg_3$ (R = alkyl, aryl, Hlg = Cl, Br) complexes, called *aprotic organic superacids*, were shown to catalyze isomerization of n-butane and n-pentane at room temperature with high selectivity. The selective isobutane formation in the presence of $AcBr-2AlBr_3$, for example, exceeds 88% even when it is close to equilibrium in the reaction mixture. Catalysts prepared by the combination of $AlBr_3$ with tri- and tetrahalomethanes ($CHlg_4-nAlBr_3$ and $CHHlg_3-nAlBr_3$, Hlg = Cl, Br, n = 1,2) effectively isomerize n-alkanes into isoalkanes.[67]

A special case of isomerization is the racemization of optically active compounds catalyzed, for example, by sulfuric acid or promoted $AlCl_3$. Thus, treatment of (+)-(S)-3-methylhexane at 60°C with 96% sulfuric acid yields a mixture of racemic 2- and 3-methylhexane[68] (Scheme 4.4). At lower temperature (0 or 30°C), racemization occurs, but shift of the methyl group does not take place. It can be concluded that at 60°C methyl migration is faster than hydride abstraction to yield isomeric alkanes. At 0 or 30°C, hydride transfer occurs before methyl

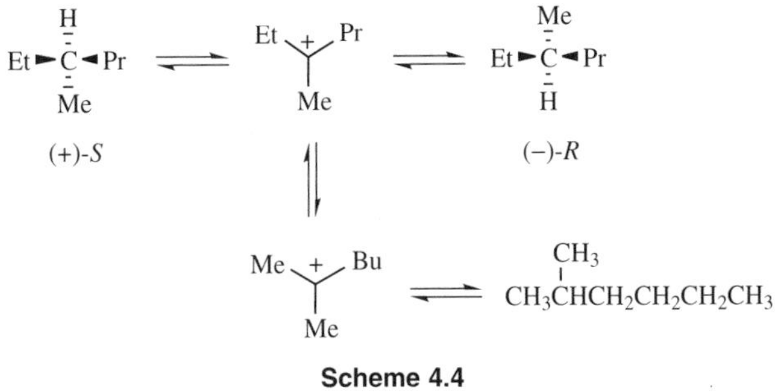

Scheme 4.4

migration and only racemic 3-methylhexane is produced. Halosulfuric acids cause selective racemization at even lower temperature.[69]

4.1.2. Arylalkanes

Side-Chain Isomerization. Arylalkanes undergo acid-catalyzed isomerization in the side chain in a way similar to the skeletal rearrangement of alkanes.[70–72] There are, however, notable differences. Propylbenzene, for instance, yields only a small amount (a few percentages) of isopropylbenzene.[73] Similarly, *sec*-butyl- and iso-butylbenzene are interconverted at 100°C with wet AlCl$_3$, but only a negligible amount of *tert*-butylbenzene is formed.[74] In the transformation of labeled propylbenzene the recovered starting material was shown to have equal amounts of labeling in the α and β positions of the side chain, but none in the γ position.[73]

These experimental observations were explained by invoking the formation of the most stable benzylic carbocation that then undergoes rearrangement via an intermediate phenonium ion[72] (Scheme 4.5). This mechanism accounts for α–β isotope scrambling, and the low amount of isopropylbenzene is in harmony with the involvement of the primary cation. Other observations, in turn, suggest the involvement of the π complex of the arene and the side-chain carbocation.[73–75]

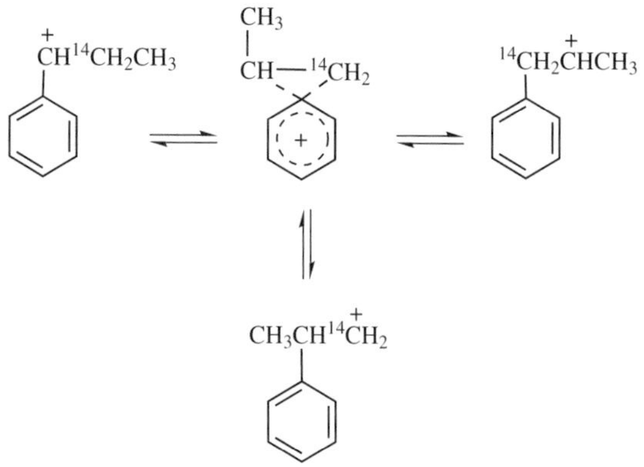

Scheme 4.5

Positional Isomerization. A different type of isomerization, substituent migration, takes place when di- and polyalkylbenzenes (naphthalenes, etc.) are treated with acidic catalysts. Similar to the isomerization of alkanes, thermodynamic equilibria of neutral arylalkanes and the corresponding carbocations are different. This difference permits the synthesis of isomers in amounts exceeding thermodynamic equilibrium when appropriate reaction conditions (excess acid, fast hydride transfer) are applied. Most of these studies were carried out in connection with the alkylation of aromatic hydrocarbons, and further details are found in Section 5.1.4.

The AlCl$_3$-catalyzed isomerization of dialkylbenzenes was studied in detail by Allen and coworkers[76–79] and Olah and coworkers.[80–82] The equilibrium isomer

Table 4.2. Equilibrium composition of dialkylbenzenes

Compound (Reaction Conditions)	% ortho	% meta	% para	Ref.
Xylenes				
(HCl–AlCl$_3$, toluene, 50°C)	18	61	21	75
(6 M HF, 0.06–0.10 M BF$_3$, 80°C)	19	60	21	83
Calculated	18	58	24	84
Ethyltoluenes (HCl–AlCl$_3$, toluene, RT)	7	66	27	76
Isopropyltoluenes (HCl–AlCl$_3$, toluene, 0°C)	1.5	69	29.5	77
tert-Butyltoluenes (H$_2$O–AlCl$_3$, benzene, RT)	0	64	36	79
Diethylbenzenes (H$_2$O–AlCl$_3$, RT)	3	69	28	80
Diisopropylbenzenes (H$_2$O–AlCl$_3$, RT)	0	68	32	81
Di-tert-butylbenzenes (H$_2$O–AlCl$_3$, CS2, RT)	0	52	48	82

distributions of representative compounds are given in Table 4.2. A striking feature is the lack of *ortho* isomer formation in the isomerization of dialkylbenzenes possessing bulky substituents.

Both intramolecular isomerization (1,2 shift) and an intermolecular process (disproportionation–transalkylation) were observed. Allen found that in the isomerization of alkyltoluenes with HCl-promoted AlCl$_3$ transalkylation becomes the dominant process with increasing branching of the substituent.[79] Exclusive or dominant 1,2 shift is characteristic of xylenes and ethyltoluenes. Isomerization of *tert*-butyltoluene, in contrast, involves both disproportionation and transalkylation: *tert*-butyltoluene transalkylates toluene, which is formed in disproportionation. Isomerization of dialkylbenzenes with H$_2$O-promoted AlCl$_3$ (applied neat or in benzene, nitromethane, or CS$_2$) was found to occur by a predominant 1,2 shift accompanied by substantial disproportionation. *o-tert*-Butylalkylbenzenes isomerize at particularly high rates because of steric factors. In addition to a less important 1,2 shift [Eq. (4.14)], they undergo a fast, kinetically controlled *ortho–para* conversion [Eq. (4.15)]:

$$(4.14)$$

$$(4.15)$$

This process probably takes place through a π-complex-type intermediate without the complete detachment of the migrating *tert*-butyl group from the aromatic ring.[80,82]

Interconversion of isomeric xylenes is an important industrial process achieved by $HF-BF_3$ or zeolite catalysts (see Section 4.5.2). Studies of xylenes and tri- and tetramethylbenzenes showed that the amount of catalyst used has a pronounced effect on the composition of isomeric mixtures.[83] When treated with small amounts of $HF-BF_3$, isomeric xylenes yield equilibrium mixtures (Table 4.2). Using a large excess of the superacid, however, *o*- and *p*-xylenes can be isomerized to *m*-xylene, which eventually becomes the only isomer. Methylbenzenes are well known to form stable σ complexes (arenium ions) in superacids, such as $HF-BF_3$. Since the most stable arenium ion formed in superacids is the 2,4-dimethylbenzenium ion (proto- nated *m*-xylene, **5**), all other isomers rapidly rearrange into this ion. The equili- brium concentration of protonated *m*-xylene in the acidic phase, consequently, approaches 100%.

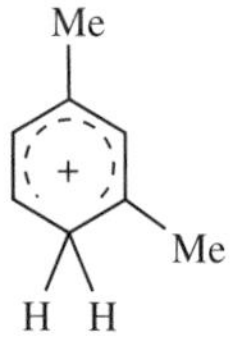

5

As discussed previously for the isomerization of butanes, differentiation must be made between thermodynamic equilibria of hydrocarbons and their derived carbo- cations. The acidic isomerization of xylenes further emphasizes this point.

When discussing isomerization of alkylbenzenes, it is useful to recollect that the alkylbenzenium ion (σ complex or arenium ion) intermediates involved are identi- cal to those of the alkylation reaction of aromatics:

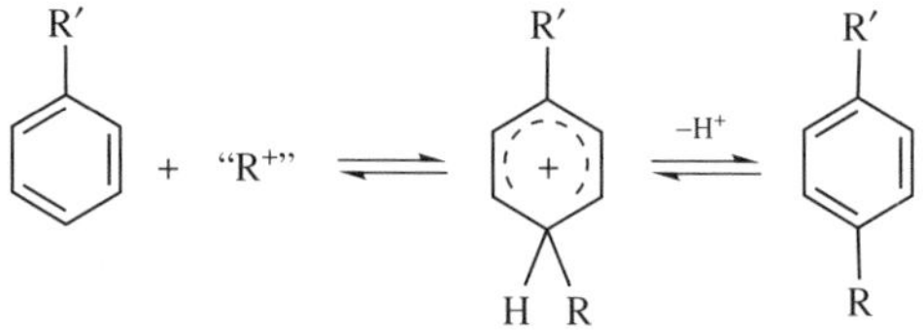

(4.16)

As pointed out (see Section 5.1.4) alkyl group migration (i.e., isomerization) in these carbocationic intermediates can take place with ease, even under conditions where the product dialkylbenzenes themselves do not undergo further isomeriza- tion. Kinetically controlled alkylation free of thermodynamically controlled alkyl group migration therefore cannot be judged on absence of product alkylbenzene isomerization.

A different product distribution is obtained by shape-selective zeolites.[85,86] Since the diffusion coefficient of *p*-xylene with its preferred shape into the zeolite

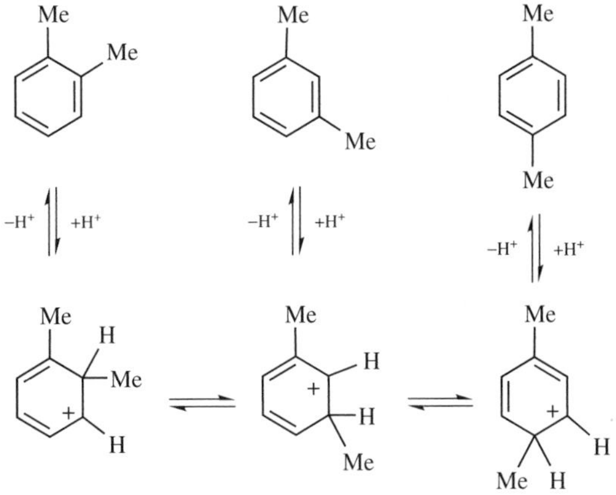

Scheme 4.6

channels is much higher than those of the other two isomers, isomeric mixtures rich in *p*-xylene can be obtained. Intramolecular 1,2-methyl shifts account for methyl migration under such conditions (Scheme 4.6). Silica–alumina catalysts can be used to establish a near-equilibrium composition of xylenes and ethylbenzene in a heterogeneous gas-phase reaction.[87] C_6–C_5 ring interconversions and 1,2 hydride shifts are involved in this rearrangement[70,72] (Scheme 4.7).

Studies of Lewis acid-catalyzed isomerization of alkylnaphthalenes were carried out by Olah and Olah in connection with alkylation of naphthalene with alkyl halides.[88] Methyl-, ethyl-, isopropyl- and *tert*-butylnaphthalene readily undergo isomerization with $AlCl_3$ in CS_2 solution. The rate of isomerization and the equilibrium concentration of the β isomer were found to increase with increasing branching of the alkyl substituent (Table 4.3).

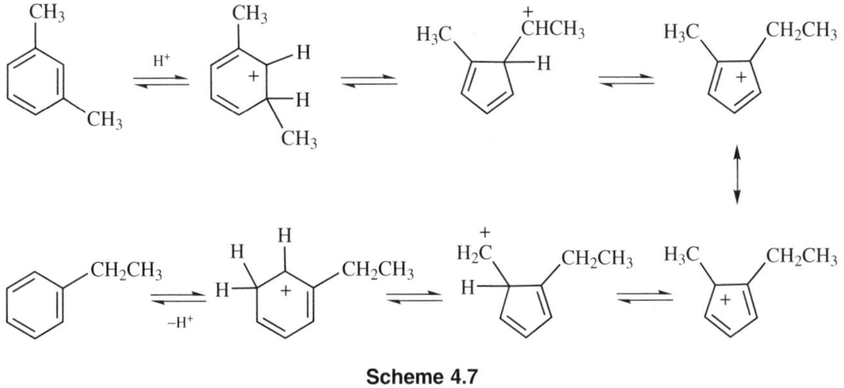

Scheme 4.7

Table 4.3. Equilibration of α-alkylnaphthalenes (AlCl$_3$, CS$_2$)[88]

Compound	Equilibrium Composition (%)		Reaction Conditions	
	α isomer	β isomer	Temperature (°C)	Time (h)
α-Methylnaphthalene	24.5	75.5	Reflux	10
α-Ethylnaphthalene	9.5	90.5	Reflux	30
α-Isopropylnaphthalene	1.5	98.5	0	3
α-tert-Butylnaphthalene	0	100	0	0.25

HF—BF$_3$ is also active in the isomerization of dimethylnaphthalenes. Facile intramolecular methyl migration is accounted for by the involvement of intermediate dimethylnaphthalenium ions. Intermolecular methyl migration, however, does not take place[72] since it would require transfer of unfavorable CH$_3^+$.

4.1.3. Alkenes and Dienes

Double-bond migration, *cis–trans* isomerization, and skeletal isomerization are the characteristic isomerization transformations of alkenes in the presence of acidic catalysts.[89,90] Mineral acids, strong organic acids, metal halides, acidic oxides, and zeolites all exert catalytic activity. Double-bond migration and *cis–trans* isomerization are facile transformations that can be induced under mild conditions (relatively mild acids, low temperature). Since the thermodynamic stability of alkenes increases with increasing substitution on the sp^2 carbon atoms, terminal alkenes exhibit the highest reactivity in double-bond migration. More forcing conditions, in turn, are required to bring about skeletal rearrangement, which may also lead to side reactions (polymerization, cracking).

Migration of the double bond of terminal alkenes to internal position is favored by the equilibria. Thus 1-butene in the presence of activated clay, silica gel, alumina, or phosphoric acid on pumice may yield equilibrium product mixtures comprised of about 20% 1-butene and 80% 2-butenes.[91] The main transformation of branched 1-alkenes under mild conditions is also double-bond migration. For example, 2,4,4-trimethyl-1-pentene is isomerized to the equilibrium mixture[92] with 20% 2,4,4-trimethyl-2-pentene when treated with silica gel at 25°C.

The isomerization is believed to occur by a carbocation mechanism[90] initiated by a proton furnished by the catalyst or by an ion (such as an alkyl cation formed by cracking) present in the reaction mixture:

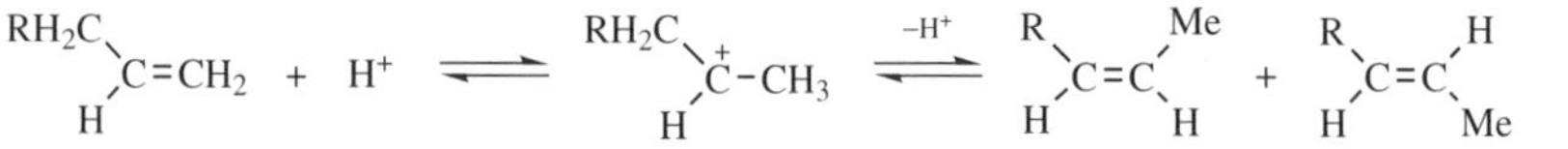

6

$$\tag{4.17}$$

The proton formed in step 2 may restart the cycle. This mechanism accounts for *cis–trans* isomerization as well, since carbocation **6** is the common intermediate of all three isomeric compounds. Earlier mechanisms included a proton addition–elimination pathway[93] and the so-called hydrogen switch mechanism.[94] It described only double-bond migration through a concerted hydrogen donation–hydrogen abstraction by a single acid molecule. In contrast to expectations, the isomerization of 1-butene over a silica–alumina catalyst leads to a reaction mixture rich in the thermodynamically unfavored *cis*-2-butene, particularly at low conversions.[95] This was explained by invoking a bridged carbocation that cannot adopt the conformation necessary to the formation of the *trans* isomer and gives exclusively the *cis* compound. Protonation of the latter after readsorption yields the *trans* compound. Higher-than-unity *cis–trans* ratios were observed over certain zeolites as well.[96]

A comparative study of the isomerization of labeled 1-alkenes over oxide catalysts (CaO, La$_2$O$_3$, ZrO$_2$, MoO$_3$, Al$_2$O$_3$) indicated that surface hydroxyl groups do not participate in isomerization.[97] Instead, double-bond migration was found to result from an intramolecular hydrogen shift occurring in the alkene adsorbed on a metal ion–metal oxide ion pair.

Skeletal isomerization requires higher temperature and stronger acid catalysts than do double-bond migration and *cis–trans* isomerization. Butylenes, for example, are transformed to isobutylene over supported phosphoric acid catalysts.[98] The equilibrium mixture at 300°C contains approximately equal amounts of straight-chain and branched butenes. Similar studies were carried out with pentene isomers.[99] Side reactions, however, may become dominant under more severe conditions.[100]

In contrast, 3,3-dimethylbutene readily undergoes skeletal rearrangement (over specially prepared alumina catalyst free of alkali ions possessing intrinsic acidic sites, at 350°C).[101] The extent of isomerization strongly depends on reaction conditions. At low contact time, isomeric 2,3-dimethylbutenes are the main products (Scheme 4.8, *a*) in accordance with the involvement of tertiary carbocation **7**.

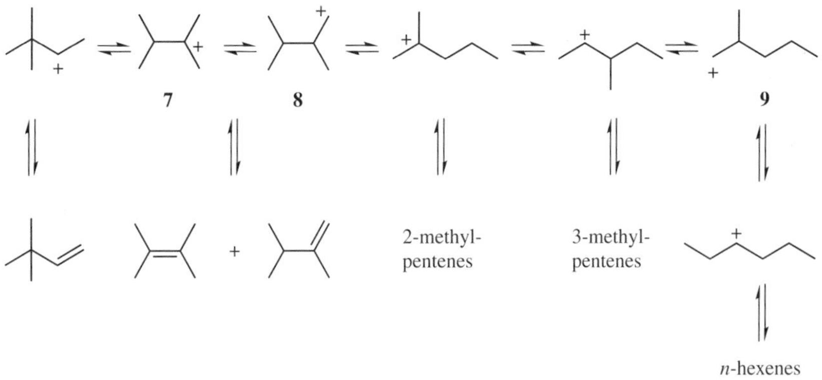

a. 21	73	6		
b. 1.9	23	42.6	31	1.5

Scheme 4.8

The formation of 2-methyl and 3-methylpentenes requires long contact times due to the involvement of the primary cation **8**. Under such conditions the concentration of the latter two compounds is near the equilibrium and only a very small amount of *n*-hexenes is detected (Scheme 4.8, *b*). This, again, indicates that further isomerization of methylpentenes to *n*-hexenes occurs through a primary cation (**9**). These distinct changes in product distribution strongly indicate that only 1,2 shifts are involved in skeletal rearrangement under these conditions.

There has been considerable interest in the selective isomerization of butylenes to isobutylene, to satisfy the demand for reformulated gasoline that requires, among others, oxygenate addition. Since MTBE is one of the best oxygenates to boost octane number and it is produced from isobutylene, refineries face the problem of converting large surplus butylenes into isobutylene. One study showed that Group VIA (Cr, Mo, W) oxides and alumina are the best catalysts,[102] effecting isomerization probably through the formation of π-allylic carbanions on Lewis acid sites.

The acid-catalyzed isomerization of cycloalkenes usually involves skeletal rearrangement if strong acids are used. The conditions and the catalysts are very similar to those for the isomerization of acyclic alkenes. Many alkylcyclohexenes undergo reversible isomerization to alkylcyclopentenes. In some cases the isomerization consists of shift of the double bond without ring contraction. Side reactions, in this case, involve hydrogen transfer (disproportionation) to yield cycloalkanes and aromatics. In the presence of activated alumina cyclohexene is converted to a mixture of 1-methyl- and 3-methyl-1-cyclopentene:[103]

$$\text{(4.18)}$$

Similar isomerization occurs in the presence of silica–alumina–thoria.[104] As it might be expected, this reaction is similar to the isomerization of cyclohexane to methylcyclopentane. Both processes involve the same intermediate cyclohexyl carbocation, which is formed, however, in different reactions. It may be formed from cyclohexane by hydride ion transfer, or by protonation of cyclohexene. Bicyclic alkenes undergo complex interconversions via carbocations over acidic catalysts.[105]

Dienes undergo isomerization due to shifts of the double bonds. The reversible isomerization of allenes to acetylenes is catalyzed characteristically by basic reagents (see Section 4.2.2). Nonconjugated alkadienes tend to isomerize to conjugated alkadienes; the conversion is usually accompanied by polymerization. Among other catalysts, activated alumina and chromia–alumina may be used to catalyze the formation of conjugated dienes.[89,106–108]

Polymerization and disproportionation are the characteristic transformations of cyclohexadienes. On the other hand, alkylidenecyclohexadienes undergo isomerization to alkylbenzenes when treated under mild conditions:[109]

$$(4.19)$$

4.2. BASE-CATALYZED ISOMERIZATION

4.2.1. Alkenes

Selective double-bond migration of linear and cyclic alkenes without skeletal rearrangement can be brought about by basic reagents.[110–113] The catalysts required to promote double-bond migration depend on the acidity of the hydrocarbon. Both homogeneous and heterogeneous catalysts can be applied in a very wide temperature range.

Of the heterogeneous catalysts, organoalkali metal compounds, metal amides, and hydrides are used most frequently. Metal hydroxides are effective at rather high temperature. 1-Butene, for instance, is transformed on LiOH (440°C), NaOH (400°C) and KOH (320°C) to yield 2-butenes with high *cis* stereoselectivity (initial selectivities are better than 73%, 78%, and 94%, respectively)[114]. Organoalkali metal compounds require lower reaction temperature. For example, sodium–anthracene[115] gives *cis*-2-butene as the main product at 145°C. The preferential formation of *cis* isomers, indicating that the initial product distributions are kinetically controlled, is a characteristic feature of base-catalyzed isomerizations.

The most active heterogeneous catalysts that are effective at room temperature are alkali metals supported on activated alumina. Simple C_4–C_8 alkenes, for example, were shown to yield equilibrium mixtures in short contact time over sodium on alumina.[112,116] Partial conversion of 1-butene at low temperature and in very short contact time ($-60°C$, 14 s) led to the stereoselective formation of *cis*-2-butene.[112] The changes in isomeric composition in the transformation of 4-methyl-1-pentene[112] are as follows:

$$(4.20)$$

	10 min	60.3	29.1	10.6
	2 h	75	1.6	3.4

Of the various homogeneous systems, metals and amides in amine solvents and alkoxides in aprotic solvents are the most active catalysts. Solvents have a profound effect on the transformations by affecting aggregation of ions in solution. The most widely used systems are lithium organoamides in hexamethylphosphoramide and potassium *tert*-butoxide in dimethyl sulfoxide.

Cycloalkenes may undergo facile double-bond migration[117] catalyzed by both homogeneous and heterogeneous catalysts [Eq. (4.21)] establishing equilibrium compositions:

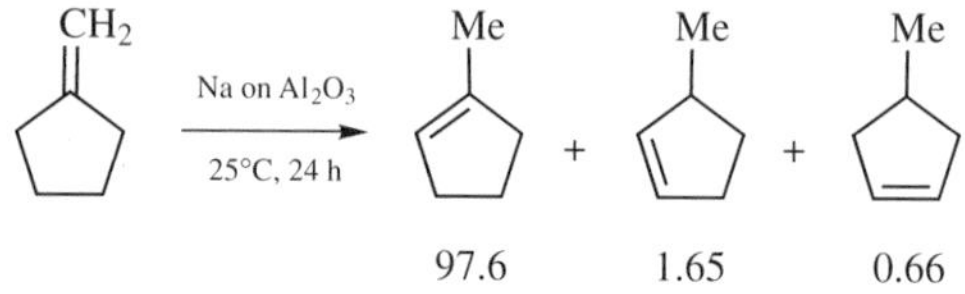

$$\begin{array}{ccc} & Me & Me & Me \end{array}$$

$$\xrightarrow[25°C, 24\,h]{Na\ on\ Al_2O_3}$$

(4.21)

97.6   1.65   0.66

The rates of *exo* to *endo* double-bond migration of methylenecycloalkanes in *tert*-BuOK–DMSO are found to decrease with ring size, with methylenecyclohexane exhibiting the lowest reactivity (the relative reactivity values for methylenecyclobutane, methylenecyclopentane, and methylenecyclohexane are 1070, 454, and 1, respectively).[118,119] Higher rings display a slight rate increase. These observations are consistent with a rate-determining proton removal and the formation of a planar carbanionic transition state. Different steric and torsional strains arising in rings of different size affect p–π overlap and, therefore, the relative rates of double-bond migration may differ considerably.

Since alkenylarenes are more acidic than simple alkenes, they are more reactive to yield isomers with the double bond in conjugation with the aromatic ring:[120]

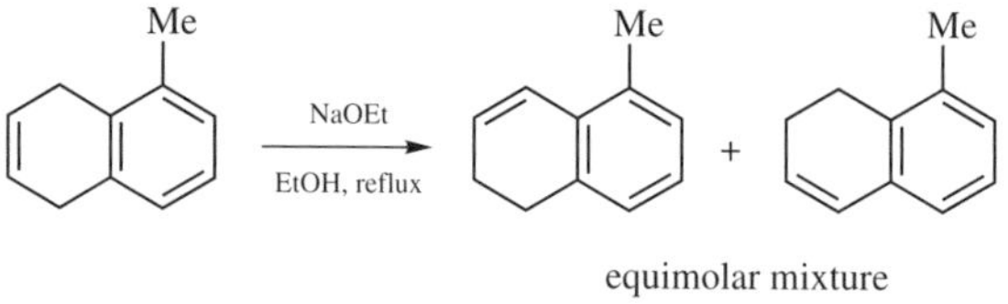

(4.22)

equimolar mixture

Most studies concerning the mechanism of base-catalyzed alkene isomerization were conducted with alkenylarenes.[121]

Nonconjugated dienes and polyenes undergo double-bond migration to form the most stable conjugated system in the presence of basic catalysts. Dienes in which the two double bonds are separated by only one carbon atom exhibit the highest reactivity. Double-bond shifts in dienes with more separated double bonds take place in consecutive steps under catalytic conditions; that is, only the monoanion is formed.[122]

Conjugated linear trienes were reported to undergo cyclization to form seven-membered ring systems.[112] Cyclic polyenes, in turn, isomerize to yield bicyclic compounds. In the transformation of nonatriene (**10**), the role of the base was demonstrated to catalyze the formation of the conjugated triene [Eq. (4.23)], which then undergoes a thermal electrocyclic reaction to give the bicyclic **11** in a disrotatory manner:[123,124]

$$(4.23)$$

10   **11**

The experimental observations presented can be explained by a carbanionic mechanism with prototropic rearrangement. On heterogeneous catalysts this involves the formation of an allylic carbanion by proton abstraction[112,125] [Eq. (4.24)]. The carbanion then participates in transmetalation to yield a new anion and the isomerized product [Eq. (4.25)]:

$$(4.24)$$

$$(4.25)$$

In homogeneous systems the tight ion pair **12** is formed, the collapse of which depends on the electronic effects and the steric interactions of the substituents. The stability of this ion pair determines the intramolecularity of the rearrangement.[121] For example, when (+)-3-phenyl-1-butene was isomerized in *tert*-BuOK–*tert*-BuOD, the recovered starting material was deuterium-free and exhibited the same rotation as the starting material, and the product *cis*-2-phenyl-2-butene contained 0.46 deuterium at C(4). This indicates that the isomerization is at least 54% intramolecular.[126] Proton migration in perdeutero-1-pentene in *tert*-BuOK–DMSO was demonstrated to proceed almost exclusively in an intramolecular manner.[127]

Much work has been done to interpret the characteristic *cis* selectivity of base-catalyzed isomerizations. In homogeneous systems no correlation was found between the rates, the nature of cations, the base, and the solvent. The higher thermodynamic stability of the *cis* allylic anion, therefore, was suggested to account for the stereoselectivity observed.[128] Additionally, a stabilizing effect by the migrating proton via electrostatic interaction as in **13** may contribute to the high *cis* selectivity. A somewhat similar resonance stabilized structure (**14**) was suggested to account for the preferential formation of the cis allylic carbanion over heterogeneous catalysts, namely, on sodium on alumina.[112,116]

12   **13**   **14**

Detailed studies were conducted with ZnO, which is a hydrogenation catalyst and effects isomerization without hydrogen.[129,130] Spectroscopic observations[130–132] led to the postulation of surface allylic species and, consequently, a 1,3 hydrogen shift was suggested to account for isomerization. The π-allylic species is formed and stabilized on adjacent acid–base (Zn^{2+}–O^{2-}) site pairs. The characteristic *cis* selectivity also supports a mechanism similar to that suggested to be effective in homogeneous systems.

4.2.2. The Reversible Acetylene–Allene Transformation

The first rearrangement involving an acetylenic compound was reported by Favorskii[133] as early as 1887. He found that when 1-butyne and higher-molecular-weight straight-chain 1-alkynes are heated with an alcoholic solution of potassium hydroxide, they are isomerized to 2-alkynes:

$$RCH_2-C\equiv CH \xrightarrow[\text{EtOH, 170°C}]{\text{KOH}} R-C\equiv C-CH_3 \qquad (4.26)$$

$$R = Me, Et, \textit{n}\text{-Pr}$$

The process may be reversed by treating 2-alkynes with sodium to get 1-alkynes after acidification of the product. The presence of an alkyl substituent at C(3) of a 1-alkyne prevents its isomerization to the corresponding 2-alkyne. Instead, an allene is formed [Eq. (4.27)]. This observation led to the suggestion of the involvement of allene intermediates to interpret the shift of the triple bond in the interconversion of acetylenes [Eq. (4.28)]:

$$\begin{matrix} Me \\ \diagdown \\ Me \diagup \end{matrix} CH-C\equiv CH \xrightarrow[\text{EtOH, 170°C}]{\text{KOH}} \begin{matrix} Me \\ \diagdown \\ Me \diagup \end{matrix} C=C=CH_2 \qquad (4.27)$$

$$RCH_2-C\equiv CH \rightleftharpoons RCH=C=CH_2 \rightleftharpoons R-C\equiv C-CH_3 \qquad (4.28)$$

The transformation termed *propargylic rearrangement* attracted much attention, and detailed discussions are available in review papers.[134–141] Suitable reagents to bring about this rearrangement are metal alkoxides and metal amides in alcohols and dipolar aprotic solvents (DMSO, HMPT), and metal amides in ammonia. Reactivities are strongly dependent on the base employed, the solvent, and the reaction conditions.

Considering the thermodynamic stability of isomeric acetylenes and dienes 1,3-dienes, by far the most stable compounds involved in propargylic rearrangements should be formed in the highest amount. They are, however, rarely observed since they are formed in a slow reaction. The product distribution, consequently, is kinetically controlled.

The order of stability of other isomers is 2-alkyne > 2,3-diene > 1,2-diene > 1-alkyne.[142] The equilibrium product distribution in the C_5 system is 1.3%

1-pentyne, 3.5% 1,2-pentadiene, and 95.2% 2-pentyne.[143] Substitution at any of the *sp* and *sp*^{2} hybride carbons increases the stability. Kinetic data seem to indicate that 1,2-dienes exhibit the highest rate of isomerization. 2-Alkynes, in contrast, isomerize only very slowly to 2,3-dienes, and the isomerization of 1-alkynes, therefore, stops at the 2-alkyne stage.[141]

The molecular structure has a profound effect on isomer distribution. In cyclic systems the equilibrium constant strongly depends on the ring size. Cyclononyne was found to be far less stable than 1,2-cyclononadiene:[142]

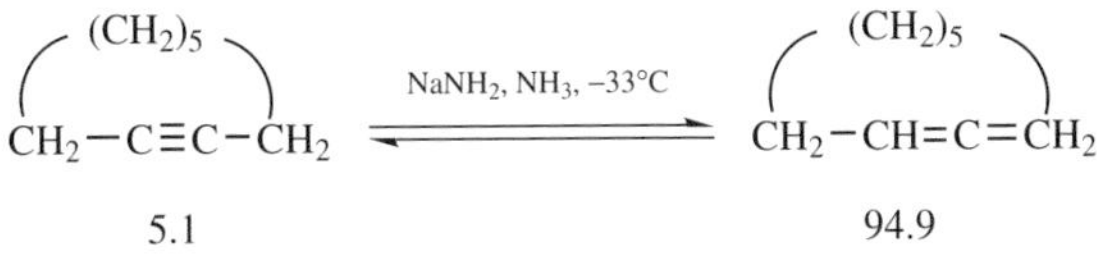

$$
\underset{5.1}{\overset{(CH_2)_5}{CH_2-C\equiv C-CH_2}} \quad \underset{NaNH_2,\ NH_3,\ -33°C}{\rightleftharpoons} \quad \underset{94.9}{\overset{(CH_2)_5}{CH_2-CH=C=CH_2}} \qquad (4.29)
$$

In the equilibration of the C_{11} and C_{12} ring systems, in contrast, the corresponding alkynes are the main products.[142,144]

The driving force for conjugation may be a significant factor to stabilize the allene structure in the isomerization of aryl-substituted alkynes[145] [Eq. (4.30)[146]]:

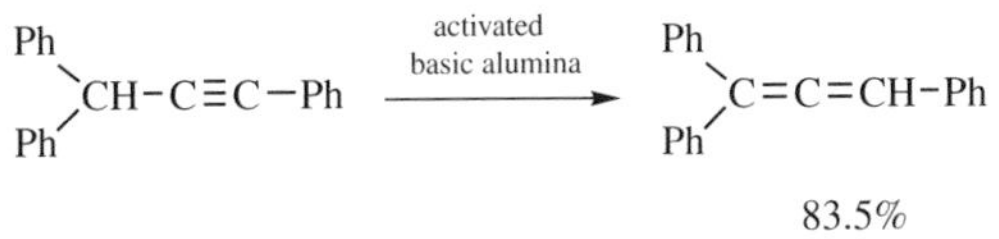

$$
\overset{Ph}{\underset{Ph}{>}}CH-C\equiv C-Ph \quad \xrightarrow{\substack{activated\\basic\ alumina}} \quad \overset{Ph}{\underset{Ph}{>}}C=C=CH-Ph \qquad (4.30)
$$
$$
83.5\%
$$

The mechanism Favorskii envisioned involved the initial attack of the ethoxy anion on the triple bond to form a vinyl ether. The now accepted carbanionic mechanism assumes the formation of resonance-stabilized anions, allowing the stepwise interconversion of 1- and 2-alkynes, and allenes[143,147] (Scheme 4.9).

Similar to the base-catalyzed double-bond shift of alkenes, the question of intramolecularity arises in the propargylic rearrangement as well. Intramolecularity was

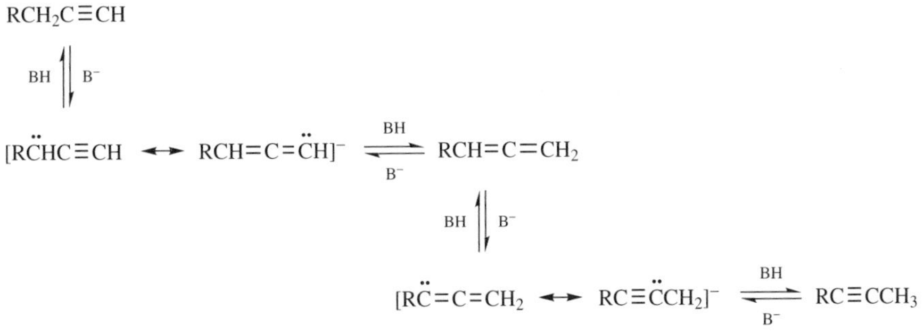

$$RCH_2C\equiv CH$$

$$BH \updownarrow B^-$$

$$[R\ddot{C}HC\equiv CH \longleftrightarrow RCH=C=\ddot{C}H]^- \underset{B^-}{\overset{BH}{\rightleftharpoons}} RCH=C=CH_2$$

$$BH \updownarrow B^-$$

$$[R\ddot{C}=C=CH_2 \longleftrightarrow RC\equiv\ddot{C}CH_2]^- \underset{B^-}{\overset{BH}{\rightleftharpoons}} RC\equiv CCH_3$$

Scheme 4.9

found to change widely depending on the solvent and the base used.[121] The highest degree of intramolecular hydrogen migration (about 90%) was obtained in the rearrangement of phenyl-substituted alkynes in DMSO catalyzed by triethylenediamine[148] or the methylsulfinyl anion.[149]

4.3. METAL-CATALYZED ISOMERIZATION

4.3.1. Alkanes

Platinum on charcoal was the first catalyst known to bring about skeletal isomerization of saturated hydrocarbons. The process observed is the transformation of butyl- and pentylcyclopentenes to the corresponding alkylarenes[150] that must involve of C_5–C_6 ring enlargement. Besides platinum, which is by far the most active catalyst operating above 250°C, other metals (Pd, Ir, Ru, Au) also exhibit some, substantially lower, activity. The main transformation over these metals is hydrocracking. The possible isomerization reactions are chain branching, substituent migration, and C_5–C_6 ring interconversion. Since these transformations usually take place in the presence of hydrogen, they are often called *hydroisomerizations*.

Basic information concerning the mechanism of skeletal rearrangement was provided by labeling experiments and kinetic studies. The use of specifically prepared catalysts, such as metal films and alloys, and structure sensitivity studies supplied additional data. The information resulted in establishing two basic processes: the bond shift and the cyclic mechanisms.[151–154]

Low-molecular-weight hydrocarbons (C_4 and C_5 alkanes) usually undergo isomerization through a simple bond shift. The transformation of $[1-^{13}C]$-butane, for instance, yields isobutane via skeletal isomerization and the isotopomer $[2-^{13}C]$-butane:[155]

$$\text{Pt film, 281–315°C} \qquad (4.31)$$

No scrambling occurs; thus, none of the products contain two labeled carbons indicating purely intramolecular rearrangements. The ratio of the two compounds (1 : 4) is independent of the temperature pointing to the existence of a common surface intermediate.

Several mechanisms were proposed to interpret bond shift isomerization, each associated with some unique feature of the reacting alkane or the metal. Palladium, for example, is unreactive in the isomerization of neopentane, whereas neopentane readily undergoes isomerization on platinum and iridium. Kinetic studies also revealed that the activation energy for chain branching and the reverse process is higher than that of methyl shift and isomerization of neopentane.

The Anderson–Avery mechanism[151,155] assumes the involvement of the **15** α,α,γ triadsorbed surface species:

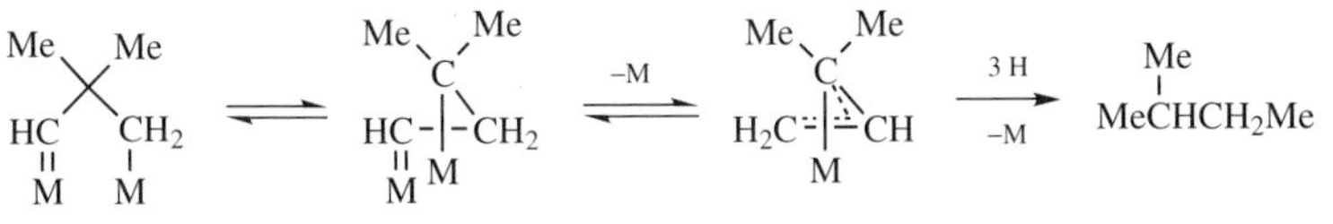

$$(4.32)$$

15

This mechanism does not allow the formation of products with a quaternary carbon atom. Nevertheless, small but significant amounts of such compounds were often observed as products.

A dissociatively adsorbed cyclopropane intermediate (**16**) stabilized by methyl substituents was proposed by Muller and Gault to give a more general interpretation:[156]

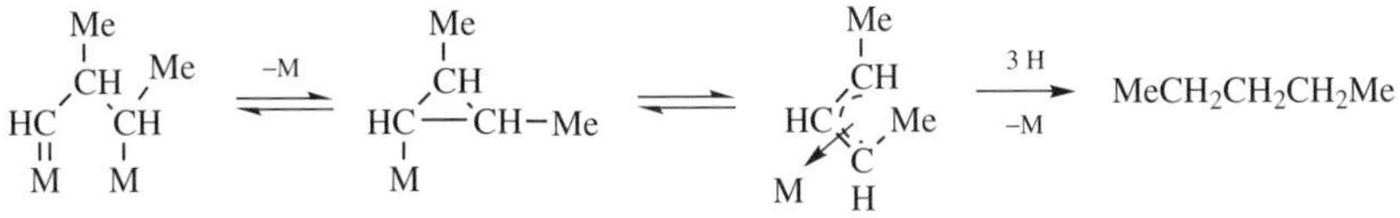

16

$$(4.33)$$

The facile isomerization of strained bridged molecules was inconsistent with the involvement of triadsorbed species **15**. Instead, a σ-alkyl adsorbed species was proposed. Its further transformation leads to a π-complex-like species (**17**) with a three-center, two-electron bonding with its simultaneous attachment to the surface[152,157] (McKervey–Rooney–Samman mechanism):

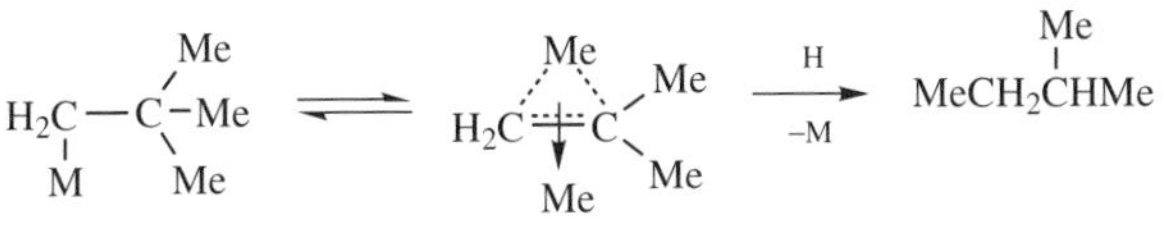

$$(4.34)$$

17

In order to interpret the remarkably high activity of platinum to promote isomerization of neopentane to isopentane, the direct formation and involvement of metallacyclobutane intermediate **18** was suggested.[152,158] This α,γ diadsorbed species bonded to a single platinum atom is in accordance with the existence of platinum complexes and the ability of platinum to catalyze α–γ exchange.[156,159]

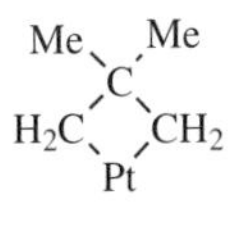

18

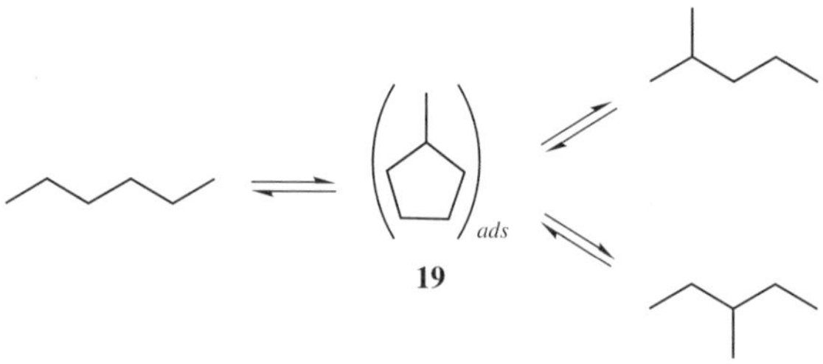

Scheme 4.10

Larger molecules are suggested to undergo isomerization by a different mechanism. The almost identical initial product distributions observed in the isomerization of *n*-hexane, 2- and 3-methylpentane, and the hydrogenolysis of methylcyclopentane strongly indicates a common intermediate.[160] This led to the formulation of the C_5 cyclic mechanism.[151–154] Accordingly, alkanes first undergo 1,5 dehydrocyclization to form the adsorbed C_5 intermediate **19** (Scheme 4.10). Interconversion of isomeric C_5 species attached to the surface at different carbon atoms proved by multiple hydrogen–deuterium exchange may take place before ring cleavage results in the formation of isomerized products.

Since dehydrocyclization and ring cleavage are reversible processes, the same mechanisms may be operative in both transformations. Two basic cleavage patterns—a selective and a nonselective ring opening—are known. In the selective process only secondary–secondary carbon–carbon bonds are broken. Conversely, only the formation of carbon–carbon bonds between methyl groups is allowed in ring formation. In contrast, all bonds are cleaved equally in the nonselective reaction. Bonds adjacent to quaternary carbon atoms are exceptions, since they are not cleaved at all.

Different approaches to the C_5 cyclic mechanism have been put forward. Comparative studies of the isomerization of 2,2,3-trimethylpentane, 2,2,4-trimethylpentane, and 2,2,4,4-tetramethylpentane indicated[161] that the latter did not display any appreciable isomerization on palladium below 360°C. Since this compound cannot undergo dehydrogenation without rearrangement to form an alkene, the cyclic mechanism on palladium is suggested to occur via the 1,2,5 triadsorbed species (**20**) bonded to a single surface atom.

In contrast, comparable rates were determined over platinum of low dispersion suggesting that isomerization occurs without alkene formation.[161] The carbene–alkyl species (**21**) formed with the involvement of terminal carbon atoms is a probable surface intermediate in this selective mechanism. Highly dispersed platinum catalysts are active in nonselective isomerization in which the precursor species is the **22** dicarbene allowing ring closure between methyl and methylene groups. On iridium a pure selective mechanism is operative,[162] which requires a dicarbyne surface species (**23**).

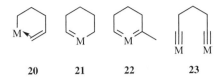

In most cases, the two types of mechanisms, the bond shift and cyclic mechanisms, are not exclusive but parallel pathways. With increasing molecular weight, the contribution of the cyclic mechanism increases and may become dominant. The pure selective mechanism on iridium is a unique exception. Hydrogenolysis, however, is the characteristic transformation on this metal. The nature of possible surface intermediates in metal-catalyzed alkane reactions, the role of electronic and geometric effects in their formation, and the relation of isomerization and hydrogenolysis have been reviewed.[163]

4.3.2. Alkenes

The hydrogenation of alkenes is often accompanied by double-bond migration and *cis–trans* isomerization.[89,164–169] Isomerizations, however—either double-bond migration or *cis–trans* isomerization—may not be observable, unless the isomer is less reactive, or the isomerization results in other structural changes in the molecule, such as racemization. Deuterium exchange taking place during saturation of alkenes is also a strong indication of isomerization processes. Process variables, however, may allow for selective isomerizations. Under conditions of low hydrogen availability hydrogenation is retarded and isomerization that does not consume hydrogen may become the main reaction. Some characteristic examples involving isomerization during hydrogenation of stereoisomeric cycloalkenes are given in Section 11.1.2.

The majority of observations concerning the isomerization of alkenes are in harmony with the Horiuti–Polanyi associative mechanism,[170,171] which involves the reversible formation of the **24** half-hydrogenated state (Scheme 4.11) (see in more detail in Section 11.1.2).

The half-hydrogenated state, which is the common intermediate in both isomerization and hydrogen addition, may revert to the starting alkene or yield isomeric olefins with double-bond migration. This mechanism explains *cis–trans* isomerization as well. Since the formation of the half-hydrogenated state requires hydrogen, isomerization can be observed only in the presence of molecular hydrogen. Numerous investigations have proved the role of hydrogen in these so-called hydro-isomerizations. When deuterium instead of hydrogen is used, reversal of the half-hydrogenated intermediate accounts for deuterium exchange in alkenes and the formation of saturated hydrocarbons with more than two deuterium atoms in the molecule.

The other possibility to interpret isomerization originally suggested by Rooney and Webb[172] invokes π-allyl species (**25a, 25b**, Scheme 4.12) formed as a result of

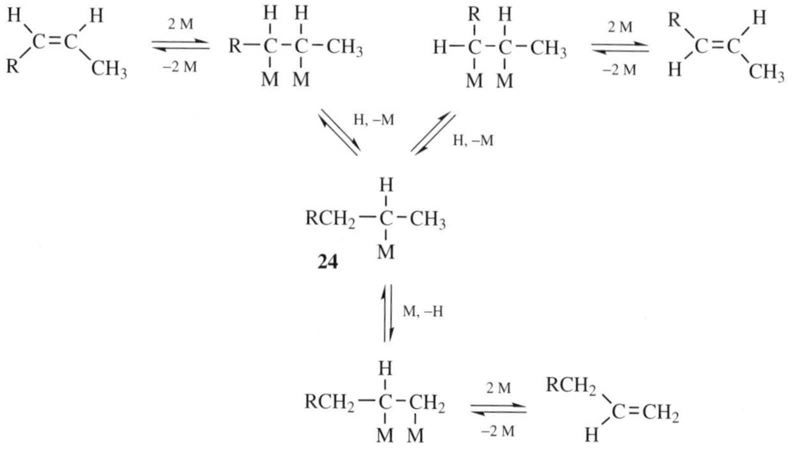

Scheme 4.11

dissociative adsorption.[173] Since direct interconversion between **25a** and **25b** is not possible, the π-allyl intermediate alone cannot account for *cis–trans* isomerization, but only in combination with double-bond migration.[172,174] In contrast with the Horiuti–Polanyi mechanism, isomerization and hydrogen addition are separate processes in the dissociative mechanism.

Metals differ in their ability to catalyze isomerizations. Both the relative rates of isomerization of individual alkenes and the initial isomer distribution vary with the metal. The rates of isomerization of the three *n*-butenes on ruthenium and osmium, for example, are *cis*-2- > *trans*-2- > 1-butene,[175] whereas on platinum and iridium they are *cis*-2- > 1- > *trans*-2-butene.[176] These observations are in accordance with the fact that the rates of formation of the 1- and 2-butyl intermediates are different on the different metals. The order of decreasing activity of platinum metals in catalyzing the isomerization of dimethylcyclohexenes was found to be Pd ≫ Rh,

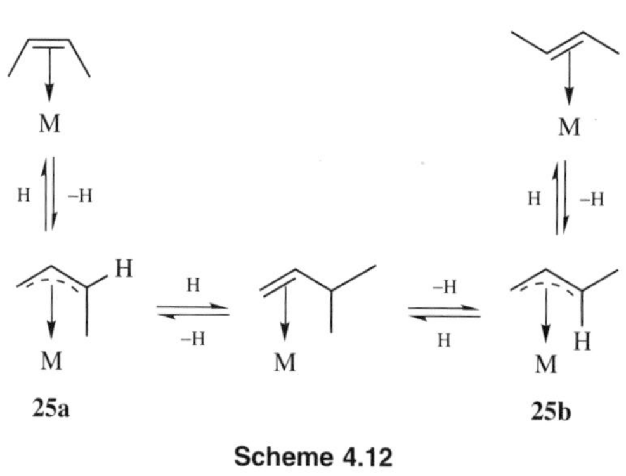

Scheme 4.12

Ru, Pt > Os > Ir.[177] A slightly different order of activity Pd > Ni > Rh $\geq$ Ru > Os, Ir, Pt is based on the isomerization of hexenes.[178] Platinum is generally the preferable catalyst if isomerization is to be avoided, whereas palladium is the metal of choice in isomerization. The most active Raney nickel preparations rival palladium in their isomerizing activity.

The relative contribution of the two mechanisms to the actual isomerization process depends on the metals and the experimental conditions. Comprehensive studies of the isomerization of *n*-butenes on Group VIII metals demonstrated[179–181] that the Horiuti–Polanyi mechanism, the dissociative mechanism with the involvement of π-allyl intermediates, and direct intramolecular hydrogen shift may all contribute to double-bond migration. The Horiuti–Polanyi mechanism and a direct 1,3 sigmatropic shift without deuterium incorporation may be operative in *cis–trans* isomerization.

Substantial experimental evidence has accumulated to indicate the involvement of 1,3 hydrogen shift in isomerization of alkenes over metal catalysts. This was first suggested by Smith and Swoap to explain deuterium exchange of cyclohexene on palladium and platinum catalysts.[182] The preferred formation of *cis*-2-butene from 1-butene over palladium without hydrogen was also accounted for by an intramolecular shift.[183] Microwave spectroscopy employed in the study of deuterium exchange during isomerization of simple alkenes over metal catalysts provided direct evidence of the occurrence of 1,3 intramolecular hydrogen shift.[179–181,184,185] Suprafacial 1,3 sigmatropic shifts as it was originally envisioned, however, are symmetry-forbidden both in the gas phase and on a metal surface. According to the present view supported by theoretical calculations, the 1,3 hydrogen shift occurs via a coadsorbed allyl species and hydrogen atom.[186]

Two major mechanisms have been recognized for isomerization of alkenes in the presence of homogeneous metal complexes.[187–191] The so-called metal hydride addition–elimination mechanism was originally advanced to interpret alkene isomerization in the presence of [HCo(CO)$_4$] during hydroformylation.[187,192] This mechanism is similar to the Horiuti–Polanyi mechanism in that it involves the reversible formation of alkylmetal (σ-alkyl) intermediates (Scheme 4.13, **26**) analogous to the half-hydrogenated state. β-Hydrogen elimination can give either the original alkene or isomers; thus, this mechanism may account for both double-bond migration and *cis–trans* (symbols *c*, *t* in Scheme 4.13) isomerization.

The reversibility of the first hydrogen transfer strongly depends on the metal complex used. It is not important with the Wilkinson catalyst. This step, which is rate-determining, is considerably slower than the final reductive elimination, yielding the saturated alkane end product during hydrogenation. Double-bond migration and isotope scrambling, accordingly, are not characteristic of the Wilkinson catalyst or, in general, of dihydride complexes.

The addition–elimination mechanism, however, is strongly preferred for monohydride systems such as [HCo(CO)$_4$][187] and the Vaska complex[193,194] promoting extensive isomerization. Hydroformylation of 2-pentenes in the presence of [HCo(CO)$_4$], for instance, yields mainly the nonbranched aldehyde resulting from double-bond migration.[195] Nickel hydride complexes are one of the most active

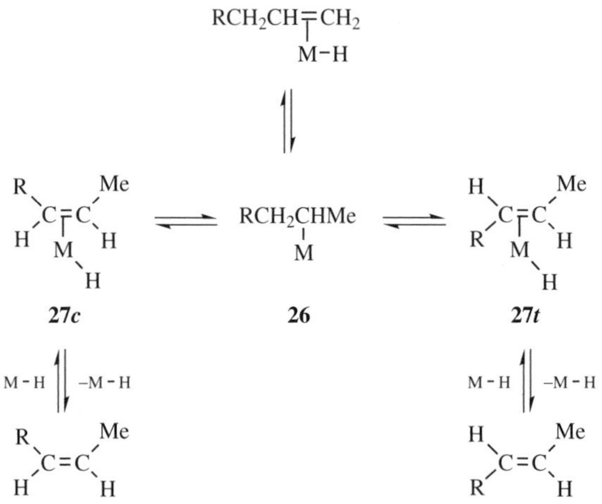

Scheme 4.13

isomerization catalysts. For example, they bring about isomerization of 1-butene to 2-butenes in a few minutes at room temperature.[196] Isomerization in the presence of such complexes is faster than hydrogenation and is most apparent in hydrogenations of terminal olefins when internal isomers accumulate. In addition, complexes that selectively catalyze the hydrogenation of terminal alkenes can bring about the migration of the double bond into internal position without hydrogen:[197]

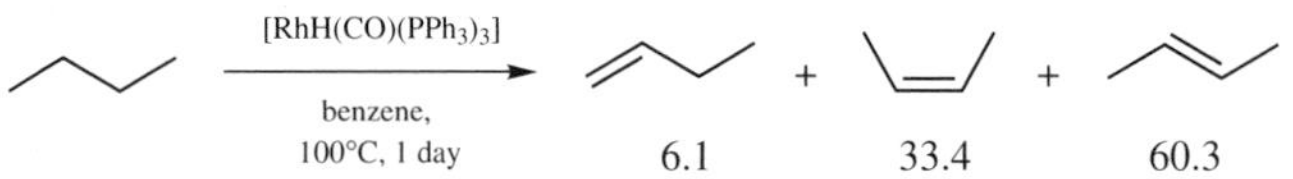

$$\text{(4.35)}$$

Alkene isomerization can also occur via π-allyl hydride intermediates[187,189,191,198,199] [Scheme 4.14, **28s** (*syn*), **28a** (*anti*)] demonstrated mainly on palladium and platinum complexes.

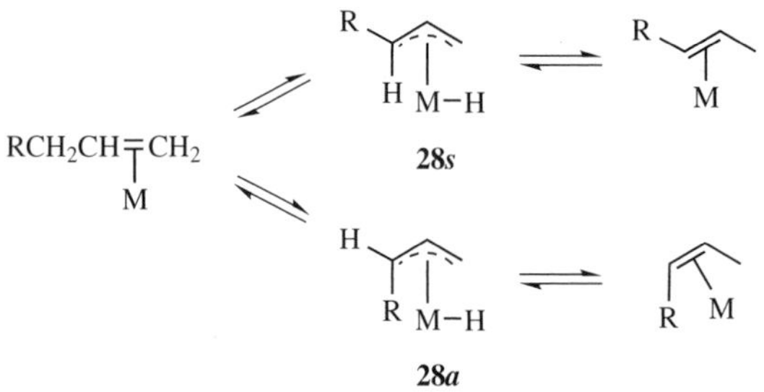

Scheme 4.14

Isomer distributions may be substantially different depending on the mechanism. In many cases higher-than-equilibrium *cis* : *trans* ratios are detected especially at low conversions.[191,200–202] This is accounted for by the greater stability of the **27c** *cis* alkene complex relative to the **27t** *trans* alkene complex (Scheme 4.13) participating in the addition–elimination mechanism. In contrast, high *trans* : *cis* ratios[203–205] are usually compatible with the π-allyl mechanism and the higher stability of the **28s** *syn*-allyl complex relative to the **28a** *anti*-allyl complex[205–207] (Scheme 4.14). Results and interpretations, however, are often contradictory.[189]

The two mechanisms may result in substantial and characteristic differences in deuterium distribution. The metal hydride addition–elimination mechanism usually leads to a complex mixture of labeled isomers.[195,198,208–210] Hydride exchange between the catalyst and the solvent may further complicate deuterium distribution. Simple repeated intramolecular 1,3 shifts, in contrast, result in deuterium scrambling in allylic positions when the π-allyl mechanism is operative.[198,205,211,212] No exchange with the solvent is expected.

The driving force for conjugation may ensure selective isomerizations. Thus, allylbenzene is transformed with high selectivity to *trans*-β-methylstyrene in the presence of metal complexes.[204,213,214] Metal complexes catalyze the isomerization of 1,5-cyclooctadiene to 1,3-cyclooctadiene with the intermediacy of the 1,4 isomer.[215–217] Iron pentacarbonyl, however, gives selectively 1,3-cyclooctadiene in excellent yield:[218]

$$(4.36)$$

The reverse process, in turn, is catalyzed by $RhCl_3$, which preferentially forms a stable complex with 1,5-cyclooctadiene.[197]

4.4. PERICYCLIC REARRANGEMENTS

Pericyclic reactions are unimolecular, concerted, uncatalyzed transformations. They take place in a highly stereoselective manner governed by symmetry properties of interacting orbitals.[219–223] Characteristic of all these rearrangements is that they are reversible and may be effected thermally or photochemically. The compounds in equilibrium are usually interconverted through a cyclic transition state,[224] although biradical mechanisms may also be operative. A few characteristic examples of pericyclic rearrangements relevant to hydrocarbon isomerizations are presented here.

Thermal sigmatropic 1,5 hydrogen shifts are quite common in certain allene and diene systems. *cis*-1,3-Dienes have the proper geometric arrangement to undergo a thermally allowed suprafacial hydrogen shift. *cis*-1,3-Hexadiene, for instance, gives predominantly *cis,trans*-2,4-hexadiene in the gas phase[225] via cyclic transition state **29**:[224,226]

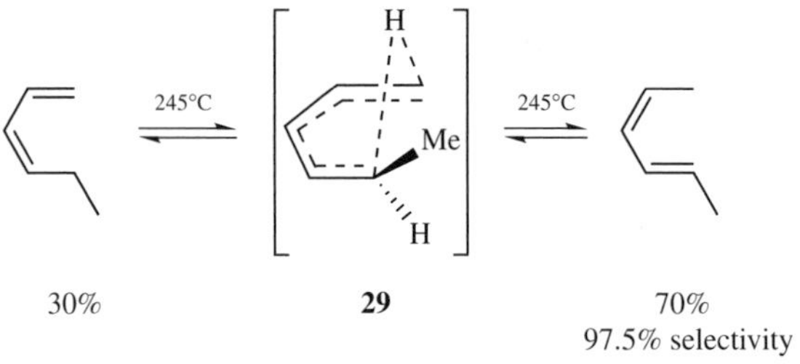

$$(4.37)$$

30%   **29**   70%
                                        97.5% selectivity

Thermal 1,5 hydrogen shift of cyclopentadiene and 1,3,5-cycloheptatriene, and methyl shifts in the corresponding methyl-substituted derivatives and in methyl-1,3-cyclohexadienes are well-documented sigmatropic rearrangements.[226,227]

cis-1-Alkyl-2-vinylcyclopropanes may also rearrange through 1,5 shifts since a hydrogen from the alkyl group can migrate to the *cis* vinyl group. Vinylcyclopropanes, however, undergo more characteristically the vinylcyclopropane–cyclopentene rearrangement[228–230] [Eq. (4.38)[231]]:

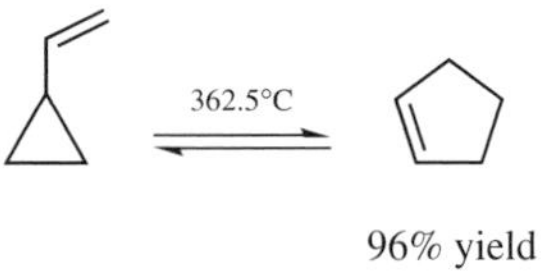

$$(4.38)$$

96% yield

The Cope and Claisen rearrangements are important sigmatropic reactions. Both have high significance in the synthesis of oxygen-containing compounds.[232,233] The basic reaction of the Cope rearrangement is the thermal interconversion of isomeric 1,5-dienes:[234]

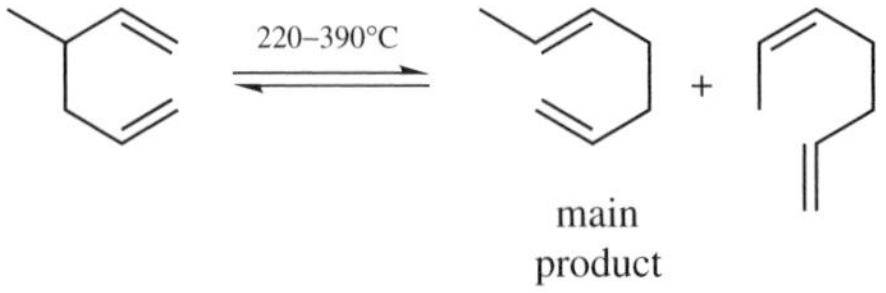

$$(4.39)$$

main
product

Examples of other hydrocarbons demonstrate that the equilibrium is determined by the relative stabilities of the olefinic bonds; in other words, the isomer with more highly substituted double bonds is formed preferentially. Conjugation with an aromatic ring may be a similar driving force. Ring strain may also be an important factor in the Cope rearrangement. Divinylcyclopropanes, for instance, are thermally transformed to cycloheptadienes.[228,235,236] *cis*-1,2-Divinylcyclopropane is a highly unstable compound (its half-life at 15.5°C is 11 min)[237] and can adopt a conformation, allowing the formation of boat-like transition state **30**,[235,238] which eventually leads to 1,4-cycloheptadiene:[237,239]

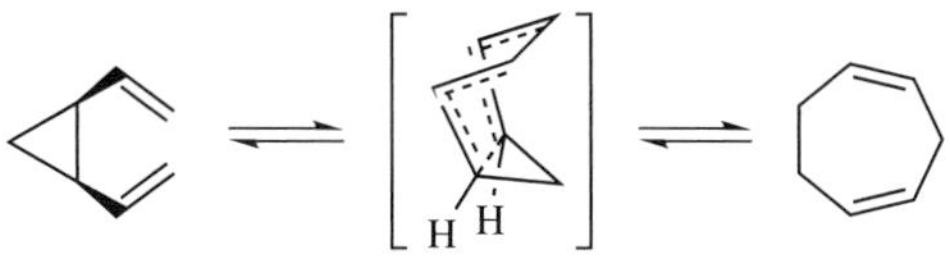

$$
\text{(structure)} \rightleftharpoons \left[\begin{array}{c} \text{(structure)} \\ \text{H H} \end{array} \right] \rightleftharpoons \text{(structure)}
$$

(4.40)

30

The *trans* compound, in contrast, reacts at much higher reaction temperature (190°C) and undergoes first a radical isomerization to give the *cis* isomer, which then rearranges to yield 1,4-cycloheptadiene.[240] Substituents at the terminal vinylic carbons considerably decrease the rate of rearrangement.[235,241]

Rearrangement of the stereoisomers in the cyclobutane series leads to different products. *cis*-1,2-Divinylcyclobutane rearranges through a transition state similar to **30** to give *cis,cis*-1,5-cyclooctadiene[242] [Eq. (4.41)].[243] The radical transformation of the *trans* isomer, in contrast, leads to 4-vinylcyclohexene[242] [Eq. (4.42)]:

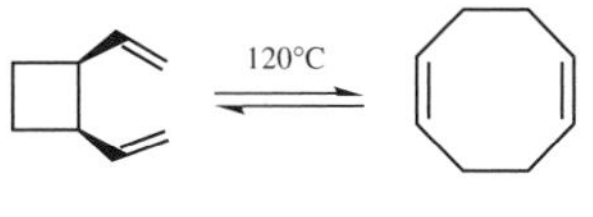

$$
\text{(structure)} \xrightleftharpoons{120°C} \text{(structure)}
$$

(4.41)

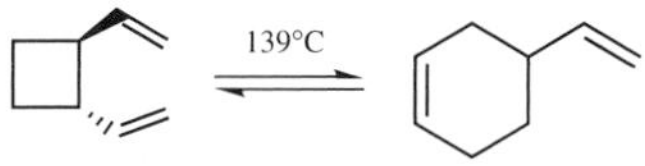

91% yield

$$
\text{(structure)} \xrightleftharpoons{139°C} \text{(structure)}
$$

(4.42)

71.3% selectivity

Dienes may be involved in electrocyclization reactions as well. Two well-documented examples are the cyclobutene ring opening[244] and the 1,3-cyclohexadiene formation[245] reactions. Predictions regarding the stereochemical outcome of these rearrangements can be made applying the orbital symmetry rules. The thermally allowed clockwise conrotatory ring opening of *cis*-3,4-dimethylcyclobutene,[219,223,246] for example, yields *cis,trans*-2,4-hexadiene[247] [Eq. (4.43)],[248] whereas the *trans* isomer gives the *trans,trans* diene[247] [Eq. (4.44)]:

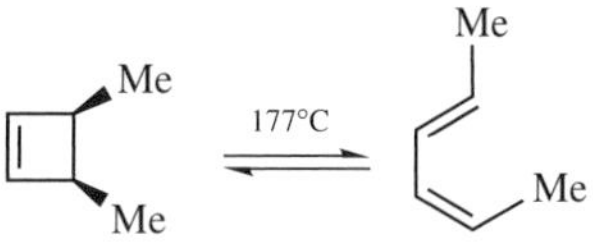

$$
\begin{array}{c} \text{Me} \\ \text{(structure)} \\ \text{Me} \end{array} \xrightleftharpoons{177°C} \begin{array}{c} \text{Me} \\ \text{(structure)} \\ \text{Me} \end{array}
$$

(4.43)

14%   86%

$$
\begin{array}{c} \text{Me} \\ \text{(structure)} \\ \text{Me} \end{array} \xrightleftharpoons{175°C} \begin{array}{c} \text{Me} \\ \text{(structure)} \\ \text{Me} \end{array}
$$

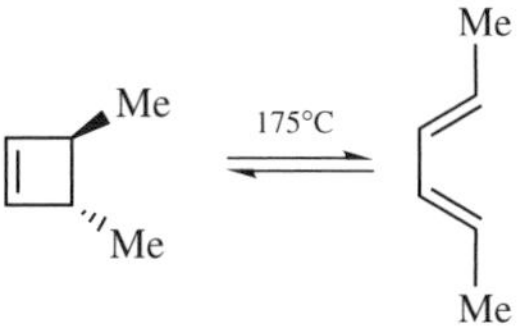

(4.44)

The 1,3,5-triene–1,3-cyclohexadiene interconversion is a six-electron electrocyclization that requires a *cis* central double bond to occur.[245] An important application of this rearrangement is the photocyclization of *cis*-stilbene to dihydrophenanthrene [Eq. (4.45)], which is usually further oxidized to phenanthrene:[249]

$$\tag{4.45}$$

4.5. PRACTICAL APPLICATIONS

Many of the petroleum refinery operations involve isomerization of hydrocarbons as part of the complex chemistry taking place (see Chapter 2).

4.5.1. Isomerization of C_4–C_6 Hydrocarbons

Alkanes. Processes for the isomerization of C_4–C_6 hydrocarbons were developed to increase the octane number of motor fuels. In the World War II era several technologies were developed[250,251] for the isomerization of paraffins by HCl–$AlCl_3$. Isobutane, one product of these processes, is alkylated with C_3–C_4 refinery olefins or dehydrogenated to isobutylene. The latter is used in alkylation of isobutane itself to produce isooctane and may be transformed to *tert*-butyl alcohol or *tert*-butyl methyl ether, which all are blending components for high-octane gasoline. Isopentane and dimethylbutanes, in contrast, are directly blended to gasoline to improve performance.

Corrosion problems and process difficulties associated with aluminum chloride led to the development of hydroisomerization processes applying noble metals on high-surface-area acidic supports (bifunctional catalysts) in the presence of hydrogen. In a new generation of such catalysts zeolites and other supports with enhanced acidity (chloride-treated alumina) are used that permit low operation temperature (100–200°C).[251] These catalysts are employed in the Penex (UOP),[252] Hysomer (Shell),[253] and the C_5/C_6 Isom (BP)[254,255] isomerization technologies. All three were developed to process C_5 or C_6 hydrocarbons or their mixtures, but the BP process was adapted to the isomerization of butane as well (Butamer technology).[256–258] Butane–isobutane isomerization is also performed over a chloride-treated alumina (Lummus process) at low temperature.[259] Because of extensive cracking on the highly reactive acidic catalysts, isomerization of C_7 and higher alkanes is seldom practiced.

A noble metal on zeolite is applied in the Hysomer process to isomerize C_5/C_6 alkanes in the presence of hydrogen.[260,261] The once-through isomerization increases the octane numbers by 10–12 numbers. When unreacted straight-chain alkanes, however, are removed (most economically by selective absorption with molecular sieve zeolites) and recycled, then complete isomerization can result in an increase in 20 octane numbers.[262,263]

Environmental considerations, including removal of lead additives as well as carcinogenic alkenes and aromatics from gasoline, increase the significance of isomerization to provide gasoline of suitable octane numbers.

Alkenes. At present alkene isomerization is an important step in the production of detergent alkylates (Shell higher olefin process; see Sections 12.3 and 13.1.3).[264,265] Ethylene oligomerization in the presence of a nickel(0) catalyst yields terminal olefins with a broad distribution range. C_4–C_6 and C_{20+} alkenes, which are not suitable for direct alkylate production, are isomerized and subsequently undergo metathesis. Isomerization is presumably carried out over a MgO catalyst.

Double-bond isomerization was once used in the multistep synthesis of isoprene developed by Goodyear.[266–268] 2-Methyl-1-pentene produced by the dimerization of propylene was isomerized to 2-methyl-2-pentene over a silica–alumina catalyst at 100°C. The product was cracked to isoprene and methane. Because of the lower cost of isoprene isolated from naphtha or gas oil-cracking streams, synthetic isoprene processes presently are not practiced commercially.

Although not a separate process, isomerization plays an important role in pretreatment of the alkene feed in isoalkane–alkene alkylation to improve performance and alkylate quality.[269–273] The FCC C_4 alkene cut (used in alkylation with isobutane) is usually hydrogenated to transform 1,3-butadiene to butylenes since it causes increased acid consumption. An additional benefit is brought about by concurrent 1-butene to 2-butene hydroisomerization. Since 2-butenes are the ideal feedstock in HF alkylation, an optimum isomerization conversion of 70–80% is recommended.[273]

4.5.2. Isomerization of Xylenes

Isomeric xylenes are important raw materials of the petrochemical industry.[251,274,275] The diacids prepared by the oxidation of *m*- and *p*-xylene are used in the synthesis of polyesters. *o*-Xylene, in turn, is oxidized to phthalic anhydride, which, in turn, is transformed to plasticizers. *p*-Xylene the most important of the three isomers may be separated from the aromatics formed in BTX processing (see Section 2.5.2). Specific processes, however, were designed to increase *p*-xylene production by further treatments of the aromatics formed in cracking or reforming (the equilibrium concentration of xylenes at 200°C is about 20% *ortho*, 55% *meta*, 25% *para*). These are toluene disproportionation and transalkylation (see Section 5.5.4), and xylene isomerization. Numerous processes have been developed for the isomerization of nonequilibrium xylene mixtures[251,274–276] over noble metals or in the presence of silica–alumina catalysts. Noble metals,[277–281] such as platinum on acidic supports, operate in the presence of hydrogen. They have the advantage of isomerizing ethylbenzene to xylenes, too, thus eliminating its costly separation before xylene isomerization.

Silica–alumina, either amorphous catalysts or zeolites, are used in several processes.[261,282–285] Mobil developed several technologies employing a medium-pore ZSM-5 zeolite.[85,86,285] They use a xylene mixture from which ethylbenzene is removed by distillation, operate without hydrogen, and yield *p*-xylene in amounts exceeding equilibrium. This favorable selectivity is due to shape selectivity of the catalyst[286] bringing about differences in diffusion rates in the zeolite structure of the three xylenes.[287] Because of its molecular geometry, *p*-xylene diffuses faster and leaves the catalyst. The other two isomers, on the other hand, undergo isomerization to reestablish equilibrium. The high selectivity of *p*-xylene formation is also attributed to the inability of ZSM-5 to catalyze transalkylation (disproportionation) of xylenes under the conditions applied. As a result, side reactions are suppressed; polymethylbenzenes, for example, are not formed in significant amounts.

Isomerization of xylenes is always coupled with separation processes. In most cases, *p*-xylene is removed from the reaction mixture by crystallization or selective adsorption. The recovered *o-/m*-xylene mixture, in turn, is usually recycled for reequilibration. Because of cost of separation, the highly selective carbonylation of toluene to *p*-tolualdehyde gained significance. Subsequent reduction gives *p*-xylene.

A unique liquid-phase isomerization to produce *m*-xylene in the presence of HF–BF$_3$ was developed by Mitsubishi[288] on the basis of earlier work by Lien and McCaulay,[83] who found that *m*-xylene forms the most stable σ complex in HF–BF$_3$ at low temperature. This complex selectively dissolves in the acid phase, allowing easy separation from the less stable other two isomers. The complex then is decomposed to produce pure *m*-xylene.

4.6. RECENT DEVELOPMENTS

4.6.1. Acid-Catalyzed and Bifunctional Isomerization

Reviews on solid acids[289,290] and heteropoly compounds, applied in both the heterogeneous[289–292] and homogeneous phases,[293] briefly discuss the use of these materials as catalysts in various isomerization processes.

Alkanes. The assumed superacidity of sulfated metal oxides, particularly that of sulfated zirconia, reported in early studies attracted considerable attention to these materials. This is due to the practical significance of their possible applications in industrial processes such as cracking, isomerization, and alkylation. Unfortunately, sulfated zirconia samples deactivate rapidly, exhibit decreasing selectivity in the isomerization of higher alkanes, and tend to lose sulfur during reaction, and there is no general regeneration method available, although an effective reversible regeneration by oxidative treatment has been reported.[294] Reviews about sulfated zirconia are available.[295–298]

According to an early report, sulfated zirconia promoted with 1.5% Fe and 0.5% Mn increased the rate of isomerization of *n*-butane to isobutane by several orders of magnitude at modest temperature (28°C).[299] This reactivity is surprising, since the isomerization of *n*-butane in strong liquid acids takes place at a rate much lower than that of higher alkanes, which is due to the involvement of the primary carbocationic intermediate. In addition, other solid acids, such as zeolites, did not show activity under such mild conditions. Evidence by isotope labeling studies with double-labeled *n*-butane unequivocally shows, however, that the isomerization of

n-butane proceeds via an intermolecular mechanism with 2-butene involved inter-mediately.[300–303] The role of the transition metal promoters such as Fe and Mn was shown to increase the surface concentration of the intermediate butene.[304] The formation of butene is speculated to occur through an oxidative dehydrogenation on the metal site[305] or by one-electron oxidation.[306]

Isobutane formation is accounted for by alkylation to form C_8^+ surface inter-mediate species followed by β scission.[300,305] One possible pathway is depicted as follows:

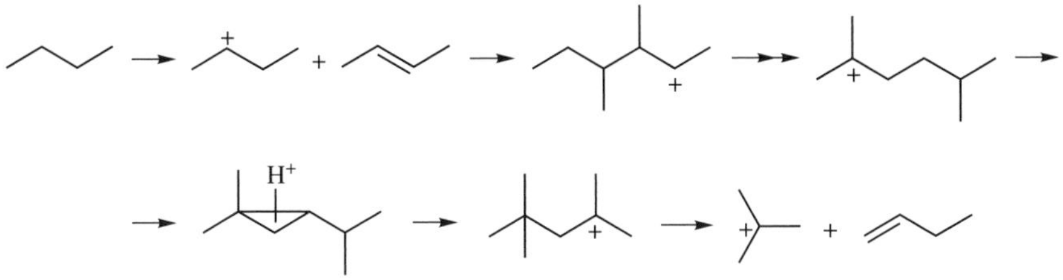

31

$$(4.46)$$

Kinetic analysis[307,308] and the identification of the $C_3 + C_5$ coproducts gave addi-tional support to this mechanism.

Studies with sulfated zirconia promoted with Pt[309] and industrial chlorinated Pt on Al_2O_3 isomerization catalysts[310] led to the same conclusion, namely, the inter-molecular mechanism operative for n-butane isomerization. A significant differ-ence, however, is that on the industrial catalysts extensive hydride and methyl shifts taking place in the intermediates prior to β scission do not lead to a random distribution of the labels. Instead, a binomial distribution with one and three ^{13}C atoms is observed.[310] This is indicative of the involvement of the **31** carbocationic intermediate.

A new study with n-hexane, however, suggests a monomolecular mechanism with the participation of Lewis acid–Lewis base site pairs generating the carboca-tion. The role of Pt is to activate hydrogen which spills over to form hydride species for the desorption of the isocarbocation.[311]

In conclusion, extensive research has revealed that the Lewis and Brønsted acid sites on the promoted sulfated zirconia catalysts are not necessarily stronger acids than the corresponding sites in zeolites, but sulfated zirconia circumvents the ener-getically unfavorable monomolecular reaction path by following a bimolecular mechanism. The question of superacidity of sulfated zirconia, however, is still debated.[312]

Studies with unpromoted and Pt-promoted solid acids, specifically, H morde-nite,[308] Pt Beta-zeolite,[313] and acidic polyoxometalate cesium salts[314,315] applied in the presence of hydrogen, showed that the isomerization of n-butane follows a monomolecular pathway. This consists of dehydrogenation on the metal site and isomerization on acidic sites. A new study using isotope labeling provided direct evidence about a carbocationic mechanism over Pt–$Cs_{2.5}H_{0.5}[PW_{12}O_{40}]$.[316] It was also reported that in the isomerization of C_6–C_9 alkanes over Pt–HY and Pt–USY zeolites carbocations exist as true intermediates rather than transition states or not completely separated from the acid site.[317]

Another interesting and potentially useful class of solid acids is materials based on WO_3. When dispersed on oxides, such as zirconia, which is called *tungstated zirconia*, and adding a metal component, usually less than 1% Pt, the resulting materials are stable and active catalysts with high selectivity to isomers in *n*-alkane isomerization in the presence of hydrogen. Selectivities in the isomerization of *n*-heptane are much higher over 0.3% Pt–WO_3–ZrO_2 than over sulfated oxides or zeolites.[318,319] Saturation coverage of WO_x species was shown to form during high-temperature calcination of samples that stabilize tetragonal ZrO_2 crystallites and inhibit sintering. High selectivity in the isomerization of *n*-heptane is attributed to a decrease in residence time of surface intermediates because of efficient hydrogen transfer steps. As a result, desorption of isomers occurs before β scission can take place. Mechanistic studies with ^{13}C-labeled *n*-pentane showed[320] that at high conversions both mono- and bimolecular pathways are operative as indicated in earlier studies.[321]

Addition of $FeSO_4$ to Pt-promoted tungstated zirconia catalysts results in increased activity and selectivity (>98% selectivity at 50–70% conversion at 250°C).[322]

Alkenes. Isobutylene is an important raw material for the production of lubricating oils, butyl rubber, polybutylenes, methacrolein, and methacrylic acid. The skeletal isomerization of butenes to isobutylene is a promising route to increase isobutylene supply; therefore, this transformation has been the focus of many studies.[323–325] In general, medium-pore zeolites and aluminophosphates exhibit good catalyst performance.[324] Particularly high selectivities and yields of isobutylene close to the thermodynamic equilibrium under typical conditions can be obtained over H-ferrierite (ZSM-35), a zeolite with a bidimensional pore system (10-ring/8-ring system of channels). Clipnotilolite, a natural zeolite with similar pore structure, has similar characteristics.[326,327] Studies in the 1990s concluded that there are three mechanisms to account for product formation.[323–325]

It was observed that selectivity of isobutylene formation on a fresh ferrierite catalyst is low, yielding propene and pentenes as main products. In fact, when compared to other 10-ring zeolites and silicoaluminophosphate microporous materials (SAPO-11), exhibiting the highest pore constraint has the highest selectivity and ferrierite is the least selective.[328] Selectivity, however, improves significantly with time on stream, which is due to a change in the mechanism.[329,330] On the fresh catalyst, a bimolecular mechanism, that is, dimerization of butenes to octenes followed by isomerization and cracking to isobutylene is operative.[329–331] As the catalyst is aging, the formation of all byproducts rapidly decreases. Deactivation is attributed to the formation and buildup of substantial amounts of carbonaceous materials. Under these conditions isobutylene is formed via direct isomerization of butenes with the involvement of the primary carbocation and the protonated cyclopropane intermediate.[330–332]

The role of coke inducing these substantial and beneficial changes has been the subject of much debate. It is suggested to modify the pore structure and poison the strong, nonselective acid sites on the external surface of the catalyst.[332–335] It was also suggested that the active sites are modified from the acidic structural zeolitic proton to benzylic carbocations formed from carbonaceous deposits trapped in the pore near the outer surface of the crystallites. These active sites induce the so-called pseudomonomolecular mechanism.[336] A new role of coke deposited in the pores has been proposed.[337]

New instrumental studies[338–341] and molecular modeling[342] lent further support to earlier observations. In conclusion, pore topology, acid strength, acid site density, and the location of acid sites are the main factors controlling catalyst performance.[324,325,341,343]

A further point of interest is that, in addition to ferrierite, high selectivity for isobutylene is also obtained on alumina catalysts modified with halides[323,325,344,345] and over mesoporous materials with low Al content.[325,346] Selective formation of isobutylene on these samples is attributed to the low concentration of active sites. This results in distant location of activated 1-butene molecules, thereby suppressing the bimolecular mechanism.[339,346,347]

Xylenes. Because of the practical significance of xylenes, isomerization of xylenes over zeolites is frequently studied.[348] The aim is to modify zeolite properties to enhance shape selectivity, that is, to increase the selectivity of the formation of the _para_ isomer, which is the starting material to produce terephthalic acid. In addition, _m_-xylene isomerization is used as a probe reaction to characterize acidic zeolites.[349,350]

As discussed in Section 4.1.2, the isomerization of xylenes over shape-selective zeolites follows an intramolecular methyl shift mechanism and gives a high _para_ : _ortho_ ratio due to the high diffusion coefficient of _p_-xylene. In contrast, over silica–alumina and non-shape-selective zeolites, such as zeolite Y, L and unidimensional, high-silica zeolites (UTD-1, SSZ-24, ZSM-12, ZSM-48), the _para_ : _ortho_ ratio is approximately 1 or smaller.[351–353] This is indicative of a bimolecular isomerization mechanism. In this case xylenes undergo disproportionation to give toluene and trimethylbenzenes followed by transalkylation of toluene to yield xylenes. The external surface plays a major role in selectivity because the nonselective bimolecular mechanism takes place on the surface. In agreement with this conclusion, the _para_ selectivity in ethylbenzene isomerization was correlated with increasing crystallite size of H-ZSM-5, that is, with decreasing surface area.[354]

Various methods have been developed to maximize _p_-xylene formation. A mild hydrothermal treatment of H-ZSM-5 and the incorporation of Pt greatly reduce disproportionation activities and, consequently, increase _para_ selectivity.[355] Suppressing the activity of the external surface can also lead to enhanced _para_ selectivity. By judicious choice of reaction conditions and organometallic reagents

[Ge(*n*-Bu)₄, ZrNp₄], the external surface, but not the internal surface, of mordenite was modified. The treatment greatly reduced disproportionation and significantly enhanced *para* selectivity.[356] A change in selectivity of disproportionation versus isomerization, namely, decreasing *para* selectivity, was observed on HY zeolite, which was induced by coke formation.[351] This unfavorable change in selectivity can be related to a greater selectivity of the strong acid sites inducing monomolecular isomerization. Since coke is preferentially formed on the strong acid sites, a decrease in their population inevitably leads to an increase in the relative rate of disproportionation.

A particular selectivity was observed in the isomerization of *m*-xylene over MCM-41 mesoporous aluminosilicate.[357,358] When compared to silica–alumina, MCM-41 exhibits very similar acidity characteristics and activities, and the disproportionation to isomerization ratios are not very different. However, there exists a large difference in the relative rates of formation of the *para* and the *ortho* isomers; namely, the *ortho* isomer is preferentially formed (*ortho* : *para* ≥ 2.5). This was attributed to the presence of the regular, noninterconnected channels of MCM-41, in which xylene molecules undergo, before desorption, successive reactions of disproportionation and transalkylation.

The Acetylene–Allene Rearrangement. In Section 4.2.2 we discussed the base-catalyzed acetylene–allene transformation. In a study by Kitagawa et al. a useful acid-catalyzed example was reported.[359]

7-Ethynylcycloheptatriene (**32**) and its derivatives cleanly isomerize to phenylallenes in the presence of acids according to the carbocationic mechanism depicted in Scheme 4.15. The first step of the transformation is the protonation of the norcaradiene tautomer, which is in equilibrium with 7-ethynylcycloheptatriene.

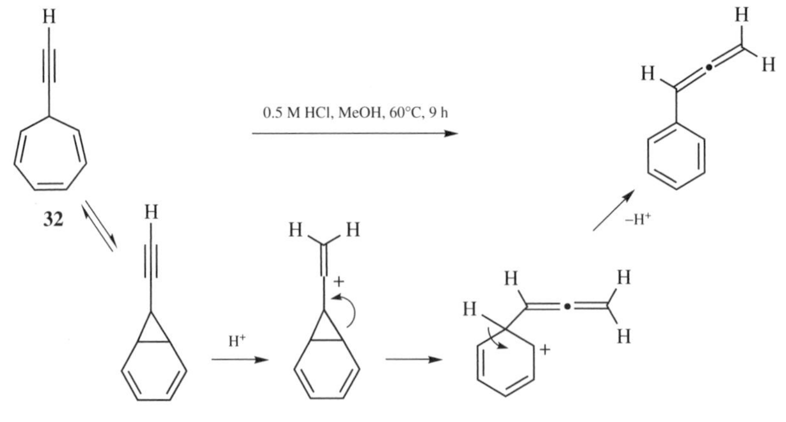

Scheme 4.15

4.6.2. Metal-Catalyzed Isomerization

A unique mechanism was suggested to interpret the difference observed in the isomerization and hydrogenation of 1-butene and *cis*-2-butene over a stepped Pt(775) surface.[360] It was observed that the hydrogenation rates were insensitive to surface structure for both 1-butene and *cis*-2-butene. The isomerization rates of *cis*-2-butene to give only *trans*-2-butene on the stepped Pt(775) surface, however, was double that of 1-butene to yield both *cis*- and *trans*-2-butenes. The Horiuti–Polanyi associative mechanism, that is, the involvement of the 2-butyl intermediate (see Section 4.3.2), cannot explain this difference. However, a facile dehydrogenation of *cis*-2-butene to 2-butyne followed by a rehydrogenation is consistent with the experimental observations:

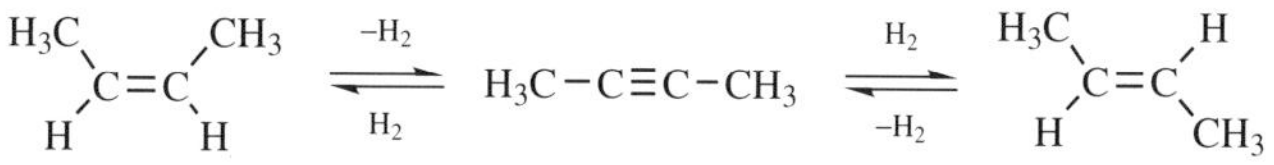

$$
\begin{array}{c}
\underset{H}{\overset{H_3C}{\diagdown}}C=C\underset{H}{\overset{CH_3}{\diagup}} \quad \underset{H_2}{\overset{-H_2}{\rightleftharpoons}} \quad H_3C-C\equiv C-CH_3 \quad \underset{-H_2}{\overset{H_2}{\rightleftharpoons}} \quad \underset{H}{\overset{H_3C}{\diagdown}}C=C\underset{CH_3}{\overset{H}{\diagup}}
\end{array}
\tag{4.47}
$$

4.6.3. Pericyclic Rearrangements

There appears to be much interest in the mechanism of various pericyclic transformations,[361–365] particularly of the Cope rearrangement.[366–371] A pair of interacting allyl radicals, an aromatic species, or a 1,4-cyclohexanediyl diradical are the possible intermediates and transition states for the rearrangement represented here as resonance hybrids in the transformation of 1,5-hexadiene (Scheme 4.16). Two high-order theoretical studies indicate that the Cope rearrangement is concerted and proceeds via an aromatic chair transition state (**33**).[362,364]

According to new data for the vinylcyclopropane–cyclopentadiene rearrangement,[372] particularly concerning the stereochemistry of the process, the [1,3] sigmatropic carbon shift proceeds through all four stereochemical reaction paths,

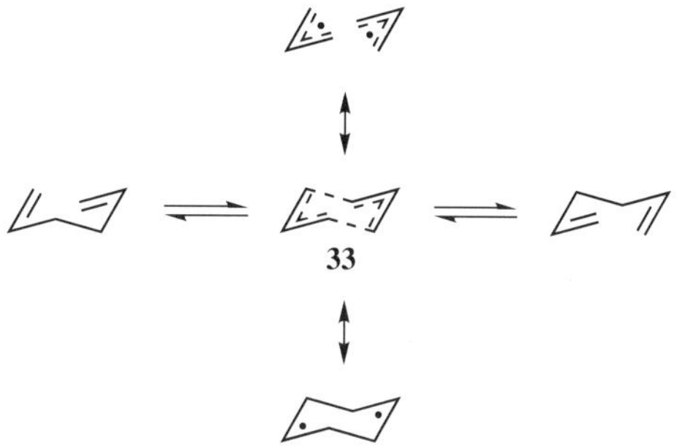

Scheme 4.16

including the allowed suprafacial, inversion (*si*) or antarafacial, retention (*ar*) and forbidden *sr* or *ai* possibilities:[373–377]

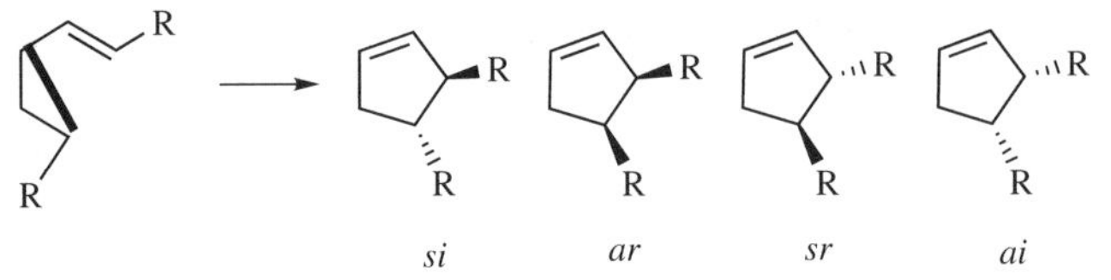

$$(4.48)$$

The thermal rearrangement of the parent trideutero compound, for example, gave an isomer mixture that indicated only a mild preference for the Woodward–Hoffmann-allowed reaction ($si + ar = 53\%$, $sr + ai = 47\%$).[378]

A unique example is a report about new experimental evidence of a "no reaction" reaction.[379] The term was coined by Berson and Willcott for the attempted isomerization of 1-vinylnortricyclene to the corresponding cyclopentene derivative, which could not be performed even after prolonged heating.[380] The term may be used for any transformation when the product has a much higher energy reverting at once back to the starting material and, therefore, is undetectable. Evidence was found when the labeled **34** compound was isomerized (Scheme 4.17).[379] The rate constants measured at two different temperatures are inconsistent with a direct *E–Z* equilibration or an *E*-to-*Z* isomerization through a diradical process. A surface-catalyzed reaction (protonation–deprotonation) could also be ruled out since almost no deuterium was lost during the reaction. The approach used here, namely, a stereochemical redisposition of label, was capable of detecting an "undetectable" reaction.

A related family of reactions is automerization, that is, rearrangements that are degenerate in the absence of a label. The term was introduced by Balaban, who first described the isotope scrambling of the ^{14}C label from position 1 to positions 9 and 10 of naphthalene under the influence of $AlCl_3$ (60°C, 2 h).[381] Catalytic auto-merization of phenanthrene[382] and biphenyl[383] were also reported. Thermal automerization of aromatics, including benzene[384,385] and naphthalene,[386] is also

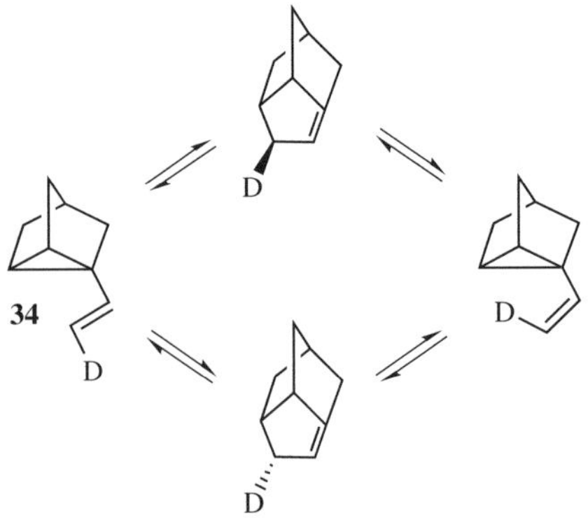

Scheme 4.17

possible,[387] although it requires high temperature (750–1100°C). More recent theoretical calculations suggest that a 1,2 hydrogen shift with the intermediacy of fulvene (**35**) is the most probable mechanism:[388]

$$(4.49)$$

The most fascinating classical example is the automerization of bulvallene,[389] a degenerative Cope rearrangement, that lets every C atom in the molecule eventually occupy every possible position in the skeleton, even in the solid state.[390]

Cubane was found to undergo a rearrangement via the hitherto unknown **36** intermediate:[391]

$$(4.50)$$

The process is induced photochemically and involves the single-electron transfer oxidation of cubane then completed with a backward electron transfer to the transient radical cations. A Li^+ salt with a weakly coordinating anion is able to induce pericyclic transformations, including the rearrangement of cubane to cuneane, quadricyclane to norbornadiene, and basketene to Nenitzescu's hydrocarbon:[392]

$$(4.51)$$

Moisture strongly promotes the reactions.

REFERENCES

1. A. Aschan, *Justus Liebigs Ann. Chem.* **324**, 10 (1902).
2. C. D. Nenitzescu and I. P. Cantuniari, *Chem. Ber.* **66**, 1097 (1933).

3. C. D. Nenitzescu and A. Dragan, *Chem. Ber.* **66**, 1892 (1933).

4. V. N. Ipatieff and A. V. Grosse, *Ind. Eng. Chem.* **28**, 461 (1936).

5. H. Pines and N. E. Hoffman, in *Friedel-Crafts and Related Reactions*, Vol. 2, G. A. Olah, ed., Wiley-Interscience, New York, 1964, Chapter 28, p. 1211.

6. C. D. Nenitzescu, in *Carbonium Ions*, Vol. 2, G. A. Olah and P.v.R. Schleyer, eds., Wiley-Interscience, New York, 1970, Chapter 13, p. 490.

7. F. Asinger, *Paraffins, Chemistry and Technology*. Pergamon Press, New York, 1968, Chapter 8, p. 693.

8. H. Pines, *Adv. Catal.* **1**, 201 (1948).

9. H. Pines, *The Chemistry of Catalytic Hydrocarbon Conversions*, Academic Press, New York, 1981, Chapter 1, p. 12.

10. F. E. Condon, in *Catalysis*, Vol. 6, P. H. Emmett, ed., Reinhold, New York, 1958, Chapter 2, p. 43.

11. N. N. Lebedev, *Usp. Khim.* **21**, 1399 (1952).

12. B. L. Evering, N. Fragen, and G. S. Weems, *Chem. Eng. News* **22**, 1898 (1944).

13. D. P. Stevenson and J. H. Morgan, *J. Am. Chem. Soc.* **70**, 2773 (1948).

14. M. B. Turova-Pollak and F. P. Sidel'kovskaya, *Zhur. Obshch. Khim.* **11**, 817 (1941); *Chem. Abstr.* **36**, 4099[3] (1942).

15. H. Pines, F. J. Pavlik, and V. N. Ipatieff, *J. Am. Chem. Soc.* **74**, 5544 (1952).

16. H. Pines and V. N. Ipatieff, *J. Am. Chem. Soc.* **61**, 1076 (1939).

17. V. Grignard and R. Stratford, *Compt. Rend.* **178**, 2149 (1924).

18. A. K. Roebuck and B. L. Evering, *J. Am. Chem. Soc.* **75**, 1631 (1953).

19. H. Pines, F. J. Pavlik, and V. N. Ipatieff, *J. Am. Chem. Soc.* **73**, 5738 (1951).

20. G. A. Olah, G. K. S. Prakash, and J. Sommer, *Superacids*, Wiley, New York, 1985, Chapter 5, p. 254.

21. G. A. Olah and G. K. S. Prakash, in *The Chemistry of Alkanes and Cycloalkanes*, S. Patai and Z. Rappoport, eds., Wiley, Chichester, 1992, Chapter 13, p. 615.

22. H. Choukroun, A. Germain, D. Brunel, and A. Commeyras, *New J. Chem.* **7**, 83 (1983).

23. F. Le Normand, F. Fajula, and J. Sommer, *New J. Chem.* **6**, 291 (1982).

24. F. Le Normand, F. Fajula, F. Gault, and J. Sommer, *New J. Chem.* **6**, 411 (1982).

25. K. Laali, M. Muller, and J. Sommer, *J. Chem. Soc., Chem. Commun.* 1088 (1980).

26. K. Tanabe and H. Hattori, *Chem Lett.* 625 (1976).

27. G. A. Olah, U.S. Patent 3,766,286 (1973).

28. G. A. Olah, U.S. Patent 4,613,723 (1986).

29. G. A. Olah, O. Farooq, A. Husain, N. Ding, N. J. Trivedi, and J. A. Olah, *Catal. Lett.* **10**, 239 (1991).

30. R. M. Brouwer, *Recl. Trav. Chim. Pays-Bas* **87**, 1435 (1968).

31. G. A. Olah, U.S. Patents 4,472,268 (1984); 4,508,618 (1985).

32. M. A. McKervey, *Chem. Soc. Rev.* **3**, 479 (1974); *Tetrahedron* **36**, 971 (1980).

33. R. C. Fort, Jr., *Adamantane, the Chemistry of Diamond Molecules*, Marcel Dekker, New York, 1976, Chapter 2, p. 35.

34. G. A. Olah, in *Cage Hydrocarbons*, G. A. Olah, ed., Wiley-Interscience, New York, 1990, Chapter 4, p. 103.

35. P. v. R. Schleyer, *J. Am. Chem. Soc.* **79**, 3292 (1957).

36. P. v. R. Schleyer and M. M. Donaldson, *J. Am. Chem. Soc.* **82**, 4645 (1960).

37. P. v. R. Schleyer, in *Cage Hydrocarbons*, G. A. Olah, ed., Wiley-Interscience, New York, 1990, Chapter 1, p. 1.

38. T. M. Gund, W. Thielecke, and P. v. R. Schleyer, *Org. Synth., Coll. Vol.* **6**, 378 (1988).

39. F. S. Hollowood, M. A. McKervey, and R. Hamilton, J. J. Rooney, *J. Org. Chem.* **45**, 4954 (1980).

40. J. A. Olah and G. A. Olah, *Synthesis* 488 (1973).

41. G. A. Olah, K. Laali, and O. Farooq, *J. Org. Chem.* **49**, 4591 (1984).

42. G. A. Olah and O. Farooq, *J. Org. Chem.* **51**, 5410 (1986).

43. D. A. Johnston, M. A. McKervey, and J. J. Rooney, *J. Am. Chem. Soc.* **93**, 2798 (1971).

44. O. Farooq, S. M. F. Farnia, M. Stephenson, and G. A. Olah, *J. Org. Chem.* **53**, 2840 (1988).

45. H. W. Whitlock, Jr. and M. W. Siefken, *J. Am. Chem. Soc.* **90**, 4929 (1968).

46. E. M. Engler, M. Farcasiu, A. Sevin, J. M. Cense, and P. v. R. Schleyer, *J. Am. Chem. Soc.* **95**, 5769 (1973).

47. L. A. Paquette, in *Cage Hydrocarbons*, G. A. Olah, ed., Wiley-Interscience, New York, 1990, Chapter 9, p. 313.

48. W.-D. Fessner and H. Prinzbach, in *Cage Hydrocarbons*, G. A. Olah, ed., Wiley-Interscience, New York, 1990, Chapter 10, p. 353.

49. L. A. Paquette, D. W. Balogh, R. Usha, D, Kountz, and G. G. Cristoph, *Science* **211**, 575 (1981).

50. L. A. Paquette and D. W. Balogh, *J. Am. Chem. Soc.* **104**, 774 (1982).

51. W.-D. Fessner, B. A. R. C. Murty, J. Wörth, D. Hunkler, H. Fritz, H. Prinzbach, W. D. Roth, P. v. R. Schleyer, A. B. McEwen, and W. F. Meier, *Angew. Chem., Int. Ed. Engl.* **26**, 452 (1987).

52. V. Amir-Ebrahimi and J. J. Rooney, *J. Chem. Soc., Chem. Commun.* 260 (1988).

53. M. A. McKervey, J. J. Rooney, in *Cage Hydrocarbons*, G. A. Olah, ed., Wiley-Interscience, New York, 1990, Chapter 2, p. 39.

54. H. S. Bloch, H. Pines, and L. Schmerling *J. Am. Chem. Soc.* **68**, 153 (1946).

55. F. C. Whitmore, *Chem. Eng. News* **26**, 668 (1948).

56. H. Pines and R. C. Wackher, *J. Am. Chem. Soc.* **68**, 595 (1946).

57. H. Pines, B. M. Abraham, and V. N. Ipatieff, *J. Am. Chem. Soc.* **70**, 1742 (1948).

58. H. Pines, E. Aristoff, and V. N. Ipatieff, *J. Am. Chem. Soc.* **71**, 749 (1949).

59. C. D. Nenitzescu, I. Necsoiu, A. Glatz, and M. Zalman, *Chem. Ber.* **92**, 10 (1959).

60. R. C. Wackher and H. Pines, *J. Am. Chem. Soc.* **68**, 1642 (1946).

61. H. Pines and R. C. Wackher, *J. Am. Chem. Soc.* **68**, 2518 (1946).

62. J. W. Otvos, D. P. Stevenson, C. D. Wagner, and O. Beeck, *J. Am. Chem. Soc.* **73**, 5741 (1951).

63. G. A. Olah, Y. Halpern, J. Shen, and Y. K. Mo, *J. Am. Chem. Soc.* **93**, 1251 (1971).

64. O. Beeck, J. W. Otvos, D. P. Stevenson, and C. D. Wagner, *J. Chem. Phys.* **16**, 255 (1948).

65. F. Le Normand, F. Fajula, F. Gault, and J. Sommer, *New J. Chem.* **6**, 417 (1982).

66. M. Vol'pin, I. Akhrem, and A. Orlinkov, *New J. Chem.* **13**, 771 (1989).

67. I. Akhrem, A. Orlinkov, and M. Vol'pin, *J. Chem. Soc., Chem. Commun.* 671 (1993).

68. R. L. Burwell, Jr., R. B. Scott, L. G. Maury, and A. S. Hussey, *J. Am. Chem. Soc.* **76**, 5822 (1954).

69. R. L. Burwell, Jr., L. G. Maury, and R. B. Scott, *J. Am. Chem. Soc.* **76**, 5828 (1954).

70. D. A. McCaulay, in *Friedel-Crafts and Related Reactions*, Vol. 2, G. A. Olah, ed., Wiley-Interscience, New York, 1964, Chapter 24, p. 1049.

71. C. D. Nenitzescu, in *Carbonium Ions*, Vol. 2, G. A. Olah and P. v. R. Schleyer, eds., Wiley-Interscience, New York, 1970, Chapter 13, p. 501.

72. H. Pines, *The Chemistry of Catalytic Hydrocarbon Conversions*, Academic Press, New York, 1981, Chapter 1, p. 24.

73. R. M. Roberts and S. G. Brandenberger, *J. Am. Chem. Soc.* **79**, 5484 (1957).

74. R. M. Roberts, Y. W. Han, C. H. Schmid, and D. A. Davis, *J. Am. Chem. Soc.* **81**, 640 (1959).

75. R. W. Holman and M.L. Gross, *J. Am. Chem. Soc.* **111**, 3560 (1989).

76. R. H. Allen and L. D. Yats, *J. Am. Chem. Soc.* **81**, 5289 (1959).

77. R. H. Allen, L. D. Yats, and D. S. Erley, *J. Am. Chem. Soc.* **82**, 4853 (1960).

78. R. H. Allen, T. Alfrey, Jr., and L. D. Yats, *J. Am. Chem. Soc.* **81**, 42 (1959).

79. R. H. Allen, *J. Am. Chem. Soc.* **82**, 4856 (1960).

80. G. A. Olah, M. W. Meyer, and N. A. Overchuk, *J. Org. Chem.* **29**, 2310 (1964).

81. G. A. Olah, M. W. Meyer, and N. A. Overchuk, *J. Org. Chem.* **29**, 2313, 2315 (1964).

82. G. A. Olah, C. G. Carlson, and J. C. Lapierre, *J. Org. Chem.* **29**, 2687 (1964).

83. D. A. McCaulay and A. P. Lien, *J. Am. Chem. Soc.* **74**, 6246 (1952).

84. W. J. Taylor, D. D. Wagman, M. G. Williams, K. S. Pitzer, and F. D. Rossini, *J. Research Natl. Bur. Stand.* **37**, 95 (1946).

85. N. Y. Chen and W. E. Garwood, *Catal. Rev.-Sci. Eng.* **28**, 251 (1986).

86. N. Y. Chen, W. E. Garwood, and F. G. Dwyer, *Shape Selective Catalysis in Industrial Applications*, Marcel Dekker, New York, 1989, Chapter 4, p. 90; Chapter 6, p. 207.

87. E. R. Boedeker and W. E. Erner, *J. Am. Chem. Soc.* **76**, 3591 (1954).

88. G. A. Olah and J. A. Olah, *J. Am. Chem. Soc.* **98**, 1839 (1976).

89. A. J. Hubert and H. Reimlinger, *Synthesis* 405 (1970).

90. H. Pines, *The Chemistry of Catalytic Hydrocarbon Conversions*, Academic Press, New York, 1981, Chapter 1, p. 6.

91. V. R. Zharkova and B. L. Moldavskii, *Zhur. Obshch. Khim.* **17**, 1268 (1947); *Chem. Abstr.* **42**, 1869i (1948).

92. W. S. Gallaway and M. J. Murray, *J. Am. Chem. Soc.* **70**, 2584 (1948).

93. F. C. Whitmore, *J. Am. Chem. Soc.* **54**, 3274 (1932).

94. J. Turkevich and R. K. Smith, *J. Chem. Phys.* **16**, 466 (1948).

95. P. J. Lucchesi, D. L. Baeder, and J. P. Longwell, *J. Am. Chem. Soc.* **81**, 3235 (1959).

96. J. Dewing, M. S. Spencer, and T. V. Whittam, *Catal. Rev.-Sci. Eng.* **27**, 461 (1985).

97. G. Gáti and G. Resofszki, *J. Mol. Catal.* **51**, 295 (1989).

98. E. K. Serebryakova and A. V. Frost, *Zhur. Obshch. Khim.* **7**, 122 (1937); *Chem. Abstr.* **31**, 4569^9 (1937).

99. V. R. Zharkova and B. L. Moldavskii, *Zhur. Obshch. Khim.* **18**, 1674 (1948); *Chem. Abstr.* **43**, 3343g (1949).

100. G. Egloff, J. C. Morrell, C. L. Thomas, and H. S. Bloch, *J. Am. Chem. Soc.* **61**, 3571 (1939).

101. W. O. Haag and H. Pines, *J. Am. Chem. Soc.* **82**, 2488 (1960).

102. Z. X. Cheng and V. Ponec, *Catal. Lett.* **25**, 337 (1994).

103. H. Adkins and A.K. Roebuck, *J. Am. Chem. Soc.* **70**, 4041 (1948).

104. H. S. Bloch and C. L. Thomas, *J. Am. Chem. Soc.* **66**, 1589 (1944).

105. J. E. Germain and M. Blanchard, *Adv. Catal.* **20**, 267 (1969).

106. S. V. Lebedev and Ya. M. Slobodin, *Zhur. Obshch. Khim.* **4**, 23 (1934); *Chem. Abstr.* **28**, 5399[1] (1934).

107. A. L. Henne and A. Turk, *J. Am. Chem. Soc.* **64**, 826 (1942).

108. A. L. Henne and H. H. Chanan, *J. Am. Chem. Soc.* **66**, 395 (1944).

109. K. Auwers and K. Ziegler, *Justus Liebigs Ann. Chem.* **425**, 217 (1921).

110. H. Pines and L. A. Schaap, *Adv. Catal.* **12**, 117 (1960).

111. A. J. Hubert and H. Reimlinger, *Synthesis* 97 (1969).

112. H. Pines and W. M. Stalick, *Base-Catalyzed Reactions of Hydrocarbons and Related Compunds*, Academic Press, New York, 1977, Chapter 2, p. 25.

113. H. Pines, *The Chemistry of Catalytic Hydrocarbon Conversions*, Academic Press, New York, 1981, Chapter 2, p. 125.

114. N. F. Foster and R. J. Cvetanovic, *J. Am. Chem. Soc.* **82**, 4274 (1960).

115. H. Pines and W. O. Haag, *J. Org. Chem.* **23**, 328 (1958).

116. W. O. Haag and H. Pines, *J. Am. Chem. Soc.* **82**, 387 (1960).

117. J. Herling, J. Shabtai, and E. Gil-Av, *J. Am. Chem. Soc.* **87**, 4107 (1965).

118. A. Schriesheim, R. J. Muller, and C. A. Rowe, Jr., *J. Am. Chem. Soc.* **84**, 3164 (1962).

119. A. Schriesheim, C. A. Rowe, Jr., and L. Naslund, *J. Am. Chem. Soc.* **85**, 2111 (1963).

120. W. Hückel and C.-M. Jennewein, *Chem. Ber.* **95**, 350 (1962).

121. D. J. Cram, *Fundamentals of Carbanion Chemistry*, Academic Press, New York, 1965, Chapter 5, p. 175.

122. W. Smadja, C. Prevost, and C. Georgoulis, *Bull. Soc. Chim. Fr.* 925 (1974).

123. J. W. H. Watthey and S. Winstein, *J. Am. Chem. Soc.* **85**, 3715 (1963).

124. D. S. Glass, J. W. H. Watthey, and S. Winstein, *Tetrahedron Lett.* 377 (1965).

125. H. Pines, J. A. Vesely, and V. N. Ipatieff, *J. Am. Chem. Soc.* **77**, 347 (1955).

126. D. J. Cram and R. T. Uyeda, *J. Am. Chem. Soc.* **84**, 4358 (1962).

127. S. Bank, C. A. Rowe, Jr., and A. Schriesheim, *J. Am. Chem. Soc.* **85**, 2115 (1963).

128. S. Bank, A. Schriesheim, and C. A. Rowe, Jr., *J. Am. Chem. Soc.* **87**, 3244 (1965).

129. C. S. John, in *Catalysis, a Specialist Periodical Report*, Vol. 3, C. Kemball and D. A. Dowden, senior reporters, The Chemical Society, Burlington House, London, 1980, Chapter 7, p. 173.

130. R. J. Kokes and A. L. Dent, *Adv. Catal.* **22**, 29, 41 (1972).

131. A. L. Dent and R. J. Kokes, *J. Am. Chem. Soc.* **92**, 6709 (1970).

132. T. T. Nguyen and N. Sheppard, *J. Chem. Soc., Chem. Commun.* 868 (1978).

133. A. Favorskii, *Chem. Zentr.* **18**, 1539 (1887); *J. Prakt. Chem.* **37**, 384, 417 (1888).

134. D. R. Taylor, *Chem. Rev.* **67**, 323 (1967).

135. J. H. Wotiz, in *Chemistry of Acetylenes*, H. G. Viehe, ed., Marcel Dekker, New York, 1969, Chapter 7, p. 371.

136. I. Iwai, in *Mechanisms of Molecular Migrations*, Vol. 2, B. S. Thyagarajan, ed., Interscience, New York, 1969, p. 78.

137. R. J. Bushby, *Quart. Rev., Chem. Soc.* **24**, 585 (1970).

138. H. Pines and W. M. Stalick, *Base-Catalyzed Reactions of Hydrocarbons and Related Compunds*, Academic Press, New York, 1977, Chapter 3, p. 124.

139. F. Théron, M. Verny, and R. Vessière, in *The Chemistry of the Carbon-Carbon Triple Bond*, S. Patai, ed., Wiley, Chichester, 1978, Chapter 10, p. 382.

140. W. D. Huntsman, in *The Chemistry of Ketenes, Allenes and Related Compounds*, S. Patai, ed., Wiley, Chichester, 1980, Chapter 15, p. 521.

141. P. D. Landor, in *The Chemistry of the Allenes*, Vol. 1, S. R. Landor, ed., Academic Press, London, 1982, Chapter 2, p. 44.

142. W. R. Moore and H. R. Ward, *J. Am. Chem. Soc.* **85**, 86 (1963).

143. T. J. Jacobs and R. Akawie, R. G. Cooper, *J. Am. Chem. Soc.* **73**, 1273 (1951).

144. W. Ziegenbein and W. M. Schneider, *Chem. Ber.* **98**, 824 (1965).

145. T. L. Jacobs and D. Dankner, *J. Org. Chem.* **22**, 1424 (1957).

146. T. J. Jacobs, D. Dankner, and S. Singer, *Tetrahedron* **20**, 2177 (1964).

147. M. D. Carr, L. H. Gan, and I. Reid, *J. Chem. Soc., Perkin Trans.* 2 668, 672 (1973).

148. D. J. Cram, F. Willey, H. P. Fischer, H. M. Relles, and D. A. Scott, *J. Am. Chem. Soc.* **88**, 2759 (1966).

149. J. Klein and S. Brenner, *J. Chem. Soc., Chem. Commun.* 1020 (1969).

150. B. A. Kasansky and A. F. Plate, *Chem. Ber.* **69**, 1862 (1936).

151. J. R. Anderson, *Adv. Catal.* **23**, 1 (1973).

152. J. K. A. Clarke and J. J. Rooney, *Adv. Catal.* **25**, 125 (1976).

153. V. Amir-Ebrahimi, F. Garin, F. Weisang, and F. G. Gault, *New J. Chem.* **3**, 529 (1979).

154. F. G. Gault, *Adv. Catal.* **30**, 1 (1981).

155. J. R. Anderson and N. R. Avery, *J. Catal.* **5**, 446 (1966); **7**, 315 (1967).

156. J. M. Muller and F. G. Gault, *Proc. 4th Int. Congr. Catal.*, Moscow, 1968; Symposium on Kinetics and Mechanism of Complex Catalytic Reactions, Paper 15.

157. M. A. McKervey, J. J. Rooney, and N. G. Samman, *J. Catal.* **30**, 330 (1973).

158. F. Garin and F. G. Gault, *J. Am. Chem. Soc.* **97**, 4466 (1975).

159. F. G. Gault, J. J. Rooney, and C. Kemball, *J. Catal.* **1**, 255 (1962).

160. Y. Barron, G. Maire, J. M. Muller, and F. G. Gault, *J. Catal.* **5**, 428 (1966).

161. J. M. Muller and F. G. Gault, *J. Catal.* **24**, 361 (1972).

162. F. Weisang and F. G. Gault, *J. Chem. Soc., Chem. Commun.* 519 (1979).

163. F. Garin and G. Maire, *Acc. Chem. Res.* **22**, 100 (1989).

164. T. I. Taylor, in *Catalysis*, Vol. 5, P.H. Emmett, ed., Reinhold, New York, 1957, Chapter 5, p. 257.

165. G. C. Bond, *Catalysis by Metals*, Academic Press, London, 1962.

166. G. C. Bond and P. B. Wells, *Adv. Catal.* **15**, 105 (1964).

167. P. N. Rylander, *Catalytic Hydrogenation in Organic Syntheses*, Academic Press, New York, 1979, Chapter 3, p. 31.

168. M. Bartók and Á. Molnár, in *Stereochemistry of Heterogeneous Metal Catalysis*, Wiley, Chichester, 1985, Chapter 3, p. 61.

169. S. Siegel, in *Comprehensive Organic Synthesis*, B. M. Trost and I. Fleming, eds., Vol. 8: *Reduction*, I. Fleming, ed., Pergamon Press, 1991, Chapter 3.1, p. 422.

170. J. Horiuti, G. Ogden, and M. Polanyi, *Trans. Faraday Soc.* **30**, 663 (1934).

171. J. Horiuti and M. Polanyi, *Trans. Faraday Soc.* **30**, 1164 (1934).

172. J. J. Rooney and G. Webb, *J. Catal.* **3**, 488 (1964).

173. J. J. Rooney, F. G. Gault, and C. Kemball, *Proc. Chem. Soc.* 407 (1960).

174. P. B. Wells and G. R. Wilson, *Disc. Faraday Soc.* **41**, 237 (1966).

175. G. C. Bond, G. Webb, and P. B. Wells, *Trans. Faraday Soc.* **64**, 3077 (1968).

176. G. C. Bond, J. J. Phillipson, P. B. Wells, and J. M. Winterbottom, *Trans. Faraday Soc.* **60**, 1847 (1964).

177. S. Nishimura, H. Sakamoto, and T. Ozawa, *Chem. Lett.* 855 (1973).

178. I. V. Gostunskaya, V. S. Petrova, A. I. Leonova, V. A. Mironova, M. Abubaker, and B. A. Kazanskii, *Neftekhimiya* **7**, 3 (1967); *Chem. Abstr.* **67**, 21276t (1967).

179. M. J. Ledoux, F. G. Gault, A. Bouchy, and G. Roussy, *J. Chem. Soc., Faraday Trans. 1* **74**, 2652 (1978); **76**, 1547 (1980).

180. M. J. Ledoux and F. G. Gault, *J. Catal.* **60**, 15 (1979).

181. M. J. Ledoux, *J. Catal.* **70**, 375 (1981).

182. G. V. Smith and J. R. Swoap, *J. Org. Chem.* **31**, 3904 (1966).

183. S. Carrá and V. Ragaini, *J. Catal.* **10**, 230 (1968).

184. R. Touroude, L. Hilaire, and F. G. Gault, *J. Catal.* **32**, 279 (1974).

185. S. Naito and M. Tanimoto, *J. Catal.* **102**, 377 (1986).

186. A. B. Anderson, D. B. Kang, and Y. Kim, *J. Am. Chem. Soc.* **106**, 6597 (1984).

187. M. Orchin, *Adv. Catal.* **16**, 1 (1966).

188. E. W. Stern, *Catal. Rev.-Sci. Eng.* **1**, 108 (1968).

189. P. A. Chaloner, *Handbook of Coordination Catalysis in Organic Chemistry*, Butterworths, London, 1986, Chapter 5, p. 405.

190. G. W. Parshall and S. D. Ittel, *Homogeneous Catalysis*, Wiley-Interscience, New York, 1992, Chapter 2, p. 12.

191. F. Conti, L. Raimondi, G. F. Pregaglia, and R. Ugo, *J. Organomet. Chem.* **70**, 107 (1974).

192. R. F. Heck and D. S. Breslow, *J. Am. Chem. Soc.* **83**, 4023 (1961).

193. H. van Gaal, H. G. A. M. Cuppers, and A. van der Ent, *J. Chem. Soc., Chem. Commun.* 1694 (1970).

194. A. J. Birch and D. H. Williamson, *Org. React.* (NY) **24**, 29 (1976).

195. F. Piacenti, P. Pino, R. Lazzaroni, and M. Bianchi, *J. Chem. Soc. C*, 488 (1966).

196. C. A. Tolman, *J. Am. Chem. Soc.* **94**, 2994 (1972).

197. R. A. Schunn, *Inorg. Chem.* **9**, 2567 (1970).

198. R. E. Rinehart and J. S. Lasky, *J. Am. Chem. Soc.* **86**, 2516 (1964).

199. J. F. Harrod and A. J. Chalk, *J. Am. Chem. Soc.* **88**, 3491 (1966).

200. H. Kanai, *J. Chem. Soc., Chem. Commun.* 203 (1972).

201. M. A. Cairns and J. F. Nixon, *J. Organomet. Chem.* **87**, 109 (1975).

202. W. Strohmeier and W. Rehder-Stirnweiss, *J. Organomet. Chem.* **22**, C27 (1970).

203. F. D'Amico, J. v. Jouanne, and H. Kelm, *J. Mol. Catal.* **6**, 327 (1979).

204. B. Corain, *Gazz. Chim. Ital.* **102**, 687 (1972).

205. D. Bingham, B. Hudson, D. E. Webster, and P. B. Wells, *J. Chem. Soc., Dalton Trans.* 1521 (1974).

206. J. W. Faller, M. T. Tully, and K. J. Laffey, *J. Organomet. Chem.* **37**, 193 (1972).

207. J. Lukas, J. E. Ramakers-Blom, T. G. Hewitt, and J. J. De Boer, *J. Organomet. Chem.* **46**, 167 (1972).

208. R. Cramer, *J. Am. Chem. Soc.* **88**, 2272 (1966).

209. W. T. Hendrix and J. L. von Rosenberg, *J. Am. Chem. Soc.* **98**, 4850 (1976).

210. M. J. D'Aniello, Jr. and E. K. Barefield, *J. Am. Chem. Soc.* **100**, 1474 (1978).

211. C. P. Casey and C. R. Cyr, *J. Am. Chem. Soc.* **95**, 2248 (1973).

212. C. A. Tolman and L. H. Scharpen, *J. Chem. Soc., Dalton Trans.* 584 (1973).

213. B. I. Cruikshank and N. R. Davies, *Aust. J. Chem.* **19**, 815 (1966); **26**, 2635 (1973).

214. E. O. Sherman, Jr. and M. Olson, *J. Organomet. Chem.* **172**, C13 (1979).

215. J. K. Nicholson and B.L. Shaw, *Tetrahedron Lett.* 3533 (1965).

216. M. Gargano, P. Giannoccaro, and M. Rossi, *J. Organomet. Chem.* **84**, 389 (1975).

217. G. P. Pez and S. C. Kwan, *J. Am. Chem. Soc.* **98**, 8079 (1976).

218. J. E. Arnet and R. Pettit, *J. Am. Chem. Soc.* **83**, 2954 (1961).

219. R. Hoffmann and R. B. Woodward, *Acc. Chem. Res.* **1**, 17 (1968).

220. R. B. Woodward and R. Hoffmann, *Angew. Chem., Int. Ed. Engl.* **8**, 781 (1969).

221. K. Fukui and H. Fujimoto, in *Mechanisms of Molecular Migrations*, Vol. 2, B. S. Thyagarajan, ed., Interscience, New York, 1969, p. 117.

222. K. Fukui, *Acc. Chem. Res.* **4**, 57 (1971).

223. G. B. Gill, *Quart. Rev.* (London) **22**, 338 (1968).

224. K. N. Houk, Y. Li, and J. D. Evanseck, *Angew. Chem., Int. Ed. Engl.* **31**, 682 (1992).

225. H. M. Frey, and B. M. Pope, *J. Chem. Soc. A* 1701 (1966).

226. H. M. Frey, and R. Walsh, *Chem. Rev.* **69**, 114 (1969).

227. C. W. Spangler, *Chem. Rev.* **76**, 187 (1976).

228. E. M. Mil'vitskaya, A. V. Tarakanova, and A. F. Plate, *Russ. Chem. Rev.* (Engl. transl.) **45**, 469 (1976).

229. T. Hudlicky, T. M. Kutchan, and S. M. Naqvi, *Org. React.* (NY) **33**, 247 (1985).

230. T. Hudlicky and J. W. Reed, in *Comprehensive Organic Synthesis*, B. M. Trost and I. Fleming, eds., Vol. 5: *Combining C—C π-Bonds*, L. A. Paquette, ed., Pergamon Press, Oxford, 1991, Chapter 8.1, p. 899.

231. C. A. Wellington, *J. Phys. Chem.* **66**, 1671 (1962).

232. S. J. Rhoads and N. R. Raulins, *Org. React.* (NY) **22**, 1 (1975).

233. R. K. Hill, in *Comprehensive Organic Synthesis*, B. M. Trost and I. Fleming, eds., Vol. 5: *Combining C–C π-Bonds*, L. A. Paquette, ed., Pergamon Press, Oxford, 1991, Chapter 7.1, p. 785.

234. H. M. Frey and R. K. Solly, *Trans. Farad. Soc.* **64**, 1858 (1968).

235. E. Piers, in *Comprehensive Organic Synthesis*, B. M. Trost and I. Fleming, eds., Vol. 5: *Combining C–C π-Bonds*, L. A. Paquette, ed., Pergamon Press, Oxford, 1991, Chapter 8.2, p. 971.

236. T. Hudlicky, R. Fan, J.W. Reed, and K.G. Gadamasetti, *Org. React.* (NY), **41**, 1 (1992).

237. J. M. Brown, B. T. Golding, and J. J. Stofko, Jr., *J. Chem. Soc., Perkin Trans.* 2 436 (1978).

238. M. Simonetta, G. Favini, C. Mariani, and P. Gramaccioni, *J. Am. Chem. Soc.* **90**, 1280 (1968).

239. M. Schneider, *Angew. Chem., Int. Ed. Engl.* **14**, 707 (1975).

240. E. Vogel, *Angew. Chem., Int. Ed. Engl.* **2**, 1 (1963).

241. M. P. Schneider and A. Rau, *J. Am. Chem. Soc.* **101**, 4426 (1979).

242. G. S. Hammond and C. D. DeBoer, *J. Am. Chem. Soc.* **86**, 899 (1964).

243. E. Vogel, *Justus Liebigs Ann. Chem.* **615**, 1 (1958).

244. T. Durst and L. Breau, in *Comprehensive Organic Synthesis*, B. M. Trost and I. Fleming, eds., Vol. 5: *Combining C–C π-Bonds*, L. A. Paquette, ed., Pergamon Press, Oxford, 1991, Chapter 6.1, p. 675.

245. W. H. Okamura and A. R. de Lera, in *Comprehensive Organic Synthesis*, B. M. Trost and I. Fleming, eds., Vol. 5: *Combining C–C π-Bonds*, L. A. Paquette, ed., Pergamon Press, Oxford, 1991, Chapter 6.2, p. 699.

246. H. C. Longuet-Higgins and E. W. Abrahamson, *J. Am. Chem. Soc.* **87**, 2045 (1965).

247. R. E. K. Winter, *Tetrahedron Lett.* 1207 (1965).

248. R. Srinivasan, *J. Am. Chem. Soc.* **91**, 7557 (1969).

249. F. B. Mallory and C. W. Mallory, *Org. React.* (NY) **30**, 1 (1984).

250. B. L. Evering, *Adv. Catal.* **6**, 197 (1954).

251. R. E. Conser, T. Wheeler, and F. G. McWilliams, in *Encyclopedia of Chemical Processing and Design*, Vol. 27, J. J. McKetta and W. A. Cunningham, eds., Marcel Dekker, New York, 1988, p. 435.

252. *Hydrocarbon Process.* **57**(9), 172 (1978).

253. *Hydrocarbon Process.* **57**(9), 171 (1978).

254. T. C. O'May and D. L. Knights, *Hydrocarbon Process Petrol. Refiner* **41**(11), 229 (1962).

255. *Hydrocarbon Process.* **57**(9), 170 (1978).

256. H. W. Grote, *Oil Gas J.* **56**(13), 73 (1958).

257. K. J. Ware and A. H. Richardson, *Hydrocarbon Process.* **51**(11), 161 (1972).

258. *Hydrocarbon Process.* **57**(9), 168 (1973).

259. *Hydrocarbon Process., Int. Ed.* **72**(3), 188 (1993).

260. H. W. Kouwenhoven and W. C. van Zijll Langhout, *Chem. Eng. Prog.* **67**(4), 65 (1971).

261. A. P. Bolton, in *Zeolite Chemistry and Catalysis*, ACS Monograph 171, J. A. Rabo, ed., American Chemical Society, Washington DC, 1976, Chapter 13, p. 714.

262. M. F. Symoniak, *Hydrocarbon Process., Int. Ed.* **59**(5), 110 (1980).

263. M. F. Symoniak and T. C. Holcombe, *Hydrocarbon Process., Int. Ed.* **62**(5), 62 (1983).

264. G. R. Lappin, in *Alpha Olefins Applications Handbook*, G. R. Lappin and J. D. Sauer, eds., Marcel Dekker, New York, 1989, Chapter 3, p. 35.

265. E. R. Freitas and C. R. Gum, *Chem. Eng. Prog.* **75**(1), 73 (1979).

266. V. J. Anhorn, *Petrol. Refiner* **39**(11), 227 (1960).

267. J. Habeshaw, in *Propylene and its Industrial Derivatives*, E. G. Hancock, ed., Wiley, New York, 1973, Chapter 4, p. 128.

268. M. L. Senyek, in *Encyclopedia of Polymer Science and Engineering*, 2nd ed., Vol. 8, J. I. Kroschwitz, ed., Wiley-Interscience, New York, 1987, p. 497.

269. C. L. Rogers, *Oil Gas J.* **69**(45), 60 (1971).

270. A. E. Eleazar, R. M. Heck, and M. P. Witt, *Hydrocarbon Process* **58**(5), 112 (1979).

271. R. M. Heck, G. R. Patel, W. S. Breyer, and D. D. Merrill, *Oil Gas J.* **81**(3), 103 (1983).

272. S. Novalany and R. G. McClung, *Hydrocarbon Process., Int. Ed.* **68**(9), 66 (1989).

273. G. Chaput, J. Laurent, J. P. Boitiaux, J. Cosyns, and P. Sarrazin, *Hydrocarbon Process., Int. Ed.* **71**(9), 51 (1992).

274. K. J. Day and T. M. Snow, in *Toluene, the Xylenes and their Industrial Derivatives*, E. G. Hancock, ed., Elsevier, Amsterdam, 1982, Chapter 3, p. 50.

275. D. L. Ransley, in *Kirk-Othmer Encyclopedia of Chemical Technology*, 3rd ed., Vol. 24, M. Grayson and D. Eckroth, eds., Wiley, New York, 1984, p. 709.

276. R. S. Atkins, *Hydrocarbon Process.* **49**(11), 127 (1970).

277. P. M. Pitts, Jr., J. E. Connor, Jr., and L. N. Leum, *Ind. Eng. Chem.* **47**, 770 (1955).

278. *Hydrocarbon Process Petrol. Refiner* **44**(11), 247 (1965).

279. H. F. Uhlig and W. C. Pfefferle, *Adv. Chem. Ser.* **97**, 204 (1970).

280. C. V. Berger, *Hydrocarbon Process.* **52**(9), 173 (1973).

281. *Hydrocarbon Process., Int. Ed.* **62**(11), 160 (1983).

282. L. R. Aalund, *Oil Gas J.* **66**(29), 102 (1968).

283. *Hydrocarbon Process.* **48**(8), 109 (1969); **48**(11), 253 (1969).

284. *Hydrocarbon Process., Int. Ed.* **60**(11), 240 (1981).

285. P. Grandio, F. H. Schneider, A. B. Schwartz, and J. J. Wise, *Hydrocarbon Process.* **51**(8), 85 (1972).

286. N. Y. Chen, W. E. Garwood, and F. G. Dwyer, *Shape Selective Catalysis in Industrial Applications*, Marcel Dekker, New York, 1989, Chapter 3, p. 39.

287. D. H. Olson and W. O. Haag, *ACS Symp. Ser.* **248**, 275 (1984).

288. *Hydrocarbon Process., Int. Ed.* **60**(11), 241 (1981).

289. A. Corma, *Chem. Rev.* **95**, 559 (1995).

290. A. Corma and H. García, *Catal. Today* **38**, 257 (1997).

291. N. Mizuno and M. Misono, *Chem. Rev.* **98**, 199 (1998).

292. M. Misono, *Chem. Commun.* 1141 (2001).

293. I. V. Kozhevnikov, *Chem. Rev.* **98**, 171 (1998).

294. T. Buchholz, U. Wild, M. Muhler, G. Resofszki, and Z. Paál, *Appl. Catal., A* **189**, 225 (1999).

295. B. H. Davis, R. A. Keogh, and R. Srinivasan, *Catal. Today* **20**, 219 (1994).

296. X. Song and A. Sayari, *Catal. Rev.-Sci. Eng.* **38**, 329 (1996).

297. K. Tanabe and H. Hattori, in *Handbook of Heterogeneous Catalysis*, G. Ertl, H. Knözinger, and J. Weitkamp, eds., Wiley-VCH, Weinheim, 1997, Chapter 2.4, p. 404.

298. G. D. Yadav and J. J. Nair, *Catal. Today* **33**, 1 (1999).

299. C.-Y. Hsu, C. R. Heimbuch, C. T. Armes, and B. C. Gates, *J. Chem. Soc., Chem. Commun.* 1645 (1992).

300. W. M. H. Sachtler, in *Recent Advances in Basic and Applied Aspects of Industrial Catalysis*, Studies in Surface Science and Catalysis, Vol. 113, T. S. R. Prasad Rao and G. Murali Dhar, eds., Elsevier, Amsterdam, 1998, p. 41.

301. V. Adeeva, G. D. Lei, and W. M. H. Sachtler, *Appl. Catal., A* **118**, L11 (1994).

302. F. Garin, L. Seyfried, P. Girard, G. Maire, A. Abdulsamad, and J. Sommer, *J. Catal.* **151**, 26 (1995).

303. J. Sommer, R. Jost, and M. Hachoumy, *Catal. Today* **38**, 309 (1997).

304. J.E. Tábora and R. J. Davis, *J. Catal.* **162**, 125 (1996).

305. K. T. Wan, C. B. Khouw, and M. E. Davis, *J. Catal.* **158**, 311 (1996).

306. D. Fărcaşiu, A. Ghenciu, and J. Q. Li, *J. Catal.* **158**, 116 (1996).

307. T. K. Cheung, J. L. D'Itri, and B. C. Gates, *J. Catal.* **151**, 464 (1995).

308. H. Liu, G. D. Lei, and W. M. H. Sachtler, *Appl. Catal., A* **137**, 167 (1996).

309. H. Liu, V. Adeeva, G. D. Lei, and W. M. H. Sachtler, *J. Mol. Catal. A: Chem.* **100**, 35 (1995).

310. V. Adeeva and W. M. H. Sachtler, *Appl. Catal., A* **163**, 237 (1997).

311. J. C. Duchet, D. Guillaume, A. Monnier, C. Dujardin, J. P. Gilson, J. van Gestel, G. Szabo, and P. Nascimento, *J. Catal.* **198**, 328 (2000).

312. N. Katada, J. Endo, K. Notsu, N. Yasunobu, N. Naito, and M. Niwa, *J. Phys. Chem. B* **104**, 10321 (2000).

313. S. Siffert, J.-L. Schmitt, J. Sommer, and F. Garin, *J. Catal.* **184**, 19 (1999).

314. K. Na, T. Okuhara, and M. Misono, *J. Catal.* **170**, 96 (1997).

315. C. Travers, N. Essayem, M. Delage, and S. Quelen, *Catal. Today* **65**, 355 (2001).

316. T. Suzuki and T. Okuhara, *Catal. Lett.* **72**, 111 (2001).

317. J. F. Denayer, G. V. Baron, G. Vanbutsele, P. A. Jacobs, and J. A. Martens, *J. Catal.* **190**, 469 (2000).

318. E. Iglesia, D. G. Barton, S. L. Soled, S. Miseo, J. E. Baumgartner, W. E. Gates, G. A. Fuentes, and G. D. Meitzner, in *11th Int. Congr. Catalysis—40th Anniversary*, Studies in Surface Science and Catalysis, Vol. 101, J. W. Hightower, W. N. Delgass, E. Iglesia, and A. T. Bell, eds., Elsevier, Amsterdam, 1996, p. 533.

319. D. G. Barton, S. L. Soled, G. D. Meitzner, G. A. Fuentes, and E. Iglesia, *J. Catal.* **181**, 57 (1999).

320. S. V. Filimonova, A. V. Nosov, M. Scheithauer, and H. Knözinger, *J. Catal.* **198**, 89 (2001).

321. J. C. Yori, C. L. Pieck, and J. M. Parera, *Appl. Catal., A* **181**, 5 (1999).

322. S. Kuba, B. C. Gates, R. K. Grasselli, and H. Knözinger, *Chem. Commun.* 321 (2001).

323. A. C. Butler and C. P. Nicolaides, *Catal. Today* **18**, 443 (1993).

324. P. Mériaudeau and C. Naccache, *Adv. Catal.* **44**, 505 (2000).

325. S. van Donk, J. H. Bitter, and K. P. de Jong, *Appl. Catal., A* **212**, 97 (2001).

326. G. Seo, M.-W. Kim, J.-H. Kim, B. J. Ahn, S. B. Hong, and Y. S. Uh, *Catal. Lett.* **55**, 105 (1998).

327. H. C. Lee, H. C. Woo, R. Ryoo, K. H. Lee, and J. S. Lee, *Appl. Catal., A* **196**, 135 (2000).

328. P. Mériaudeau, A. V. Tuan, N. L. Hung, and G. Szabo, *Catal. Lett.* **47**, 71 (1997).

329. M. Guisnet, P. Andy, N. S. Gnep, C. Travers, and E. Benazzi, *J. Chem. Soc., Chem. Commun.* 1685 (1995).

330. P. Meriaudeau, R. Bacaud, L. N. Hung, and A. T. Vu, *J. Mol. Catal. A: Chem.* **110**, L177 (1996).

331. M. Guisnet, P. Andy, N. S. Gnep, E. Benazzi, and C. Travers, *J. Catal.* **158**, 551 (1996).

332. J. Houžvička and V. Ponec, *Ind. Eng. Chem. Res.* **36**, 1424 (1997).

333. W.-Q. Xu, Y.-G. Yin, S. L. Suib, J. C. Edwards, and C.-L. O'Young, *J. Catal.* **163**, 232 (1996).

334. P. Mériaudeau, C. Naccache, H. N. Le, T. A. Vu, and G. Szabo, *J. Mol. Catal. A: Chem.* **123**, L1 (1997).

335. G. Seo, H. S. Jeong, D.-L. Jang, D. L. Cho, and S. B. Hong, *Catal. Lett.* **41**, 189 (1996).

336. M. Guisnet, P. Andy, Y. Boucheffa, N. S. Gnep, C. Travers, and E. Benazzi, *Catal. Lett.* **50**, 159 (1998).

337. L. Domokos, L. Lefferts, K. Seshan, and J. A. Lercher, *J. Catal.* **197**, 68 (2001).

338. C. Pazè, B. Sazak, A. Zecchina, and J. Dwyer, *J. Phys. Chem. B* **103**, 9978 (1999).

339. G. Seo, M.-Y. Kim, and J.-H. Kim, *Catal. Lett.* **67**, 207 (2000).

340. P. Ivanov and H. Papp, *Langmuir* **16**, 7769 (2000).

341. D. Rutenbeck, H. Papp, H. Ernst, and W. Schwieger, *Appl. Catal., A* **208**, 153 (2001).

342. P. Andy, D. Martin, M. Guisnet, R. G. Bell, and C. R. A. Catlow, *J. Phys. Chem. B* **104**, 4827 (2000).

343. L. Domokos, L. Lefferts, K. Seshan, and J. A. Lercher, *J. Mol. Catal. A: Chem.* **162**, 147 (2000).

344. Z. X. Cheng and V. Ponec, *J. Catal.* **148**, 607 (1994).

345. G. Seo, N.-H. Kim, Y.-H. Lee, and J.-H. Kim, *Catal. Lett.* **51**, 101 (1998).

346. G. Seo, N.-H. Kim, Y.-H. Lee, and J.-H. Kim, *Catal. Lett.* **57**, 209 (1999).

347. P. Mériaudeau, V. A. Tuan, L. N. Hung, C. Naccache, and G. Szabo, *J. Catal.* **171**, 329 (1997).

348. J. S. Beck and W. O. Haag, in *Handbook of Heterogeneous Catalysis*, G. Ertl, H. Knözinger, and J. Weitkamp, eds., Wiley-VCH, Weinheim, 1997, Chapter 4.2, p. 2136.

349. A. Corma, C. Corell, F. Llopis, A. Martínez, J. Pérez-Pariente, *Appl. Catal., A* **115**, 121 (1994).

350. M. Guisnet, N. S. Gnep, and S. Morin, *Microp. Mesop. Mat.* **35–36**, 47 (2000).

351. S. Morin, N. S. Gnep, and M. Guisnet, *J. Catal.* **159**, 296 (1996); *Appl. Catal., A* **168**, 63 (1998).

352. R. F. Lobo, M. Tsapatsis, C. C. Freyhardt, S. Khodabandeh, P. Wagner, C.-Y. Chen, K. J. Balkus, Jr., S. I. Zones, and M. E. Davis, *J. Am. Chem. Soc.* **119**, 8474 (1997).

353. C. W. Jones, S. I. Zones, and M. E. Davis, *Appl. Catal., A* **181**, 289 (1999).

354. N. Arsenova-Härtel, H. Bludau, R. Schumacher, W. O. Haag, H. G. Karge, E. Brunner, and U. Wild, *J. Catal.* **191**, 326 (2000).

355. A. K. Aboul-Gheit, S. M. Abdel-Hamid, and D. S. El-Desouki, *Appl. Catal., A* **209**, 179 (2001).

356. F. Lefebvre, A. de Mallmann, and J.-M. Basset, *Eur. J. Inorg. Chem.* 361 (1999).

357. S. Morin, P. Ayrault, S. El Mouahid, N. S. Gnep, and M. Guisnet, *Appl. Catal., A* **159**, 317 (1997).

358. M. Guisnet, N. S. Gnep, S. Morin, J. Patarin, F. Loggia, and V. Solinas, in *Mesoporous Molecular Sieves 1998*, Studies in Surface Science and Catalysis, Vol. 117, L. Bonneviot, F. Béland, C. Danumach, S. Giasson, and S. Kaliaguine, eds., Elsevier, Amsterdam, 1998, p. 591.

359. T. Kitagawa, J. Kamada, S. Minegishi, and K. Takeuchi, *Org. Lett.* **2**, 3011 (2000).

360. C. Yoon, M. X. Yang, and G. A. Somorjai, *J. Catal.* **176**, 35 (1998).

361. O. Wiest and K. N. Houk, *Top. Curr. Chem.* **183**, 1 (1996).

362. H. Jiao and P. v. R. Schleyer, *J. Phys. Org. Chem.* **11**, 655 (1998).

363. T. Miyashi, H. Ikeda, and Y. Takahashi, *Acc. Chem. Res.* **32**, 815 (1999).

364. R. V. Williams, *Chem. Rev.* **101**, 1185 (2001).

365. N. J. Saettel, J. Oxgaard, and O. Wiest, *Eur. J. Org. Chem.* 1429 (2001).

366. K. A. Black, S. Wilsey, K. N. Houk, *J. Am. Chem. Soc.* **120**, 5622 (1998).

367. W. R. Roth, R. Gleiter, V. Paschmann, U. E. Hackler, G. Fritzsche, and H. Lange, *Eur. J. Org. Chem.* 961 (1998).

368. D. A. Hrovat, J. A. Duncan, and W. T. Borden, *J. Am. Chem. Soc.* **121**, 169 (1999).

369. V. N. Staroverov and E. R. Davidson, *J. Am. Chem. Soc.* **122**, 186 (2000).

370. S. Sakai, *Int. J. Quantum Chem.* **80**, 1099 (2000).

371. J. A. Duncan and M. C. Spong, *J. Org. Chem.* **65**, 5720 (2000).

372. J. E. Baldwin, *J. Comput. Chem.* **19**, 222 (1998).

373. J. E. Baldwin, S. J. Bonacorsi, Jr., and R. C. Burrell, *J. Org. Chem.* **63**, 4721 (1998).

374. J. E. Baldwin and R. C. Burrell, *J. Org. Chem.* **64**, 3567 (1999).

375. J. E. Baldwin and R. Shukla, *J. Am. Chem. Soc.* **121**, 11018 (1999).

376. M. Nendel, D. Sperling, O. Wiest, and K. N. Houk, *J. Org. Chem.* **65**, 3259 (2000).

377. C. Doubleday, *J. Phys. Chem. A* **105**, 6333 (2001).

378. J. E. Baldwin, K. A. Villarica, D. I. Freedberg, F. A. L. Anet, *J. Am. Chem. Soc.* **116**, 10845 (1994).

379. J. E. Baldwin and D. A. Dunmire, *J. Org. Chem.* **65**, 6791 (2000).

380. J. A. Berson and M. R. Willcott, *J. Org. Chem.* **30**, 3569 (1965).

381. A. T. Balaban and D. Fǎrcaşiu, *J. Am. Chem. Soc.* **89**, 1958 (1967).

382. A. T. Balaban, M. D. Gheorghiu, A. Schiketanz, and A. Necula, *J. Am. Chem. Soc.* **111**, 734 (1989).

383. A. Necula, A. Racoveanu-Schiketanz, M. D. Gheorghiu, and L. T. Scott, *J. Org. Chem.* **60**, 3448 (1995).

384. L. T. Scott, N. H. Roelofs, and T.-H. Tsang, *J. Am. Chem. Soc.* **109**, 5456 (1987).

385. G. Zimmermann, M. Nüchter, H. Hopf, K. Ibrom, and L. Ernst, *Liebigs Ann.* 1407 (1996).

386. L. T. Scott, M. M. Hashemi, T. H. Schultz, and M. B. Wallace, *J. Am. Chem. Soc.* **113**, 9692 (1991).

387. L. T. Scott, *Acc. Chem. Res.* **15**, 52 (1982).

388. K. M. Merz, Jr. and L. T. Scott, *J. Chem. Soc., Chem. Commun.* 412 (1993).

389. W. E. Doering and W. R. Roth, *Angew. Chem., Int. Ed. Engl.* **2**, 115 (1963); *Tetrahedron*, **19**, 715 (1963).

390. B. H. Meier and W. L. Earl, *J. Am. Chem. Soc.* **107**, 5553 (1985).

391. P. R. Schreiner, A. Wittkopp, P. A. Gunchenko, A. I. Yaroshinsky, S. A. Peleshanko, and A. A. Fokin, *Chem. Eur. J.* **7**, 2739 (2001).

392. S. Moss, B. T. King, A. de Meijere, S. I. Kozhushkov, P. E. Eaton, and J. Michl, *Org. Lett.* **3**, 2375 (2001).

5

ALKYLATION

Reactions of alkanes and arenes with different alkylating agents (alkenes, alkynes, alkyl halides, alcohols, ethers, esters) are usually performed with acid catalysts. Brønsted and Lewis acids and, more recently, zeolites are applied. Superacids allow much improved alkylation and give important information regarding the carbocationic mechanism of alkylation. Acid-catalyzed alkylations, particularly alkane–alkene reactions, are of great practical importance in upgrading motor fuels. Acid-catalyzed alkylation produces alkylaromatics for manufacture of plastics (styrenes), detergents, and chemicals (phenol and acetone from cumene). Alkylation of alkanes was also shown to take place thermally. Thermal alkylation, however, usually gives lower yields, and is of less practical significance than catalytic alkylation. Basic catalysts may be used to carry out carbanionic alkylation of alkylaromatics in reactive benzylic positions and in alkane-alkene systems.

5.1. ACID-CATALYZED ALKYLATION

5.1.1. Alkylation of Alkanes with Alkenes

The belief that paraffins are compounds of *parum affinis* (limited affinity) was further disproved in 1932 when Ipatieff and his coworkers found[1] that alkanes undergo reaction with alkenes in the presence of aluminum chloride and other acid catalysts. Since the alkylation of gaseous isobutane with gaseous olefins yields saturated alkylate boiling in the gasoline range and possessing high octane numbers, the reaction was studied intensively. It served as a major source of aviation gasoline in World War II and has been the basis of producing high-octane gasoline since.

The major catalysts applied in acid-catalyzed alkylations fall into two classes:[2–6] metal halides of the Friedel–Crafts type, specifically, Lewis acids (promoted by hydrogen halide); and protic acids. Zeolites[7] and superacids[8–10] have gained increasing attention as Friedel–Crafts catalysts. Some of the most active metal halides are aluminum chloride, aluminum bromide, boron trifluoride, and zirconium chloride. Frequently used protic acid catalysts are hydrogen fluoride and sulfuric acid, which are preferred over the metal halides because they are handled more conveniently and can be reused. However, unlike Friedel–Crafts catalysts, these acids do not catalyze alkylations with ethylene, although with stronger superacids alkylation takes place readily. The reaction of ethylene with isobutane does give excellent yield in the presence of aluminum chloride, serving for the commercial production of 2,3-dimethylbutane. The catalysts loose activity with use; aluminum chloride forms a viscous red-brown liquid (the "lower" layer), and hydrogen fluoride and particularly sulfuric acid form sludges (acid-soluble oils). The lower layer and sludges are complexes of the catalysts with cyclic polyalkapolyenes formed during the alkylation.

Depending on the reactants and the catalysts, the temperature of acid-catalyzed alkylations may be as low as $-30°C$ or as high as $100°C$. To avoid oxidation, H_2SO_4 is used under $30°C$. The pressure is often the vapor pressure of the reaction mixture and ranges from 1 to 15 atm. Thermal alkylation, in contrast, requires high temperature ($500°C$) and pressure (150–300 atm).[5] There are substantial differences in the reactivity of alkanes and alkenes in thermal and acid-catalyzed alkylations. Propane and higher alkanes, both straight-chain and branched, undergo reaction under thermal conditions, most readily with ethylene. Alkanes with a tertiary carbon–hydrogen bond and more substituted alkenes are the most reactive in the presence of acidic catalysts. Reactivity, however, depends on the catalyst, too. BF_3 (and a protic coacid) catalyzes only the alkylation of branched alkanes,[11] whereas promoted $AlCl_3$ is effective in the alkylation of straight-chain alkanes as well.[12]

The main products formed by the catalytic alkylation of isobutane with ethylene ($HCl–AlCl_3$, 25–35°C) are 2,3-dimethylbutane and 2-methylpentane with smaller amounts of ethane and trimethylpentanes.[13] Alkylation of isobutane with propylene ($HCl–AlCl_3$, $-30°C$) yields 2,3- and 2,4-dimethylpentane as the main products and propane and trimethylpentanes as byproducts.[14] This is in sharp contrast with product distributions of thermal alkylation that gives mainly 2,2-dimethylbutane (alkylation with ethylene)[15] and 2,2-dimethylpentane (alkylation with propylene).[16]

Markedly different product distributions are obtained in alkylation of isobutane with butenes depending on the catalyst used (Table 5.1). A 1 : 1 crystalline complex of aluminum chloride with methyl alcohol promoted with hydrogen chloride is a more selective catalyst by causing little side reaction.[17] Alkylation with both 1- and 2-butenes gives high yields of octanes but with striking differences in composition. Dimethylhexanes are the main products with 1-butene, and trimethylpentanes are formed with high selectivity with 2-butenes. Hydrogen fluoride is the best catalyst to produce 2,2,4-trimethylpentane (isooctane) with minor differences in the product distributions of butenes.[3,18] The results with sulfuric acid[19] are very similar to those of hydrogen fluoride.

Table 5.1. Product composition (wt%) in the alkylation of isobutane with C$_4$ alkenes

	1-Butene			2-Butene				Isobutylene	
	AlCl$_3$ + HCl[a]	AlCl$_3$–MeOH + HCl[a]	HF[b]	AlCl$_3$ + HCl[a]	AlCl$_3$–MeOH + HCl[a]	HF[b]	H$_2$SO$_4$[c]	HF[d]	H$_2$SO$_4$[c]
Temperature (°C)	30	55	10	30	28	10	10	10	10
				Alkylate Composition					
C$_5$	17.1	—	3.2	13.1	—	5.2	—	—	7–9
C$_6$	14.2	3.2	1.8	13.9	3.8	2.6	4–6	1.5	8–10
C$_7$	11.2	5.7	3.3	10.2	5.0	3.9	—	5.0	—
C$_8$	21.4	69.4	60.1	23.9	68.9	68.1	94–96	75.3	54–62
>C$_8$	36.1	18.6	31.8	38.9	18.5	20.2	—	14.9	—
				C$_8$ Composition					
Methylheptanes	16.4	—	—	11.1	—	—	—	—	—
2,2-Dimethylhexane	1.7	—	—	1.4	—	—	—	—	—
2,3-Dimethylhexane	6.5	25.4	5.8	2.2	0.7	1.9	—	3.2	—
2,4-; 2,5-Dimethylhexane	45.2	49.8	12.3	24.9	4.7	7.5	—	8.2	—
3,4-Dimethylhexane	—	11.1	—	4.0	—	—	—	—	—
2,2,3-Trimethylpentane	3.4	—	—	4.0	0.8	—	—	2.0	—
2,2,4-Trimethylpentane	13.4	9.7	45.8	39.5	41.0	58.3	36–41	65.0	39–45
2,3,3-Trimethylpentane	6.5	—	13.0	8.0	20.3	13.8	55–59	9.0	48–55
2,3,4-Trimethylpentane	6.9	4.0	20.3	8.9	32.6	18.5	—	12.5	—

[a] Refs. 5, 17.
[b] Ref. 18.
[c] Ref. 19.
[d] Refs. 3, 6.

Cycloalkanes possessing a tertiary carbon atom may be alkylated under conditions similar to those applied for the alkylation of isoalkanes. Methylcyclopentane and methylcyclohexane were studied most.[5] Methylcyclopentane reacts with propylene and isobutylene in the presence of HF (23–25°C), and methylcyclohexane can also be reacted with isobutylene and 2-butene under the same conditions.[20] Methylcyclopentane is alkylated with propylene in the presence of HBr–AlBr₃ (−42°C) to produce 1-ethyl-2-methylcyclohexane.[21] $C_{12}H_{22}$ bicyclic compounds are also formed under alkylation conditions.[21,22] Cyclohexane, in contrast, requires elevated temperature, and only strong catalysts are effective. HCl–AlCl₃ catalyzes the cyclohexane–ethylene reaction at 50–60°C to yield mainly dimethyl- and tetramethylcyclohexanes (rather than mono- and diethylcyclohexanes). The relatively weak boron trifluoride, in turn, is not active in the alkylation of cyclohexane.[23]

These observations demonstrate that multiplicity of products is a characteristic feature of acid-catalyzed alkylation with alkenes indicating the involvement of isomerizations. Some of the compounds isolated cannot be formed by the reaction of one molecule of the alkane with one or more of the alkenes. This points to secondary reactions accompanying alkylation.

The results discussed were accounted for by the carbocationic mechanism of Schmerling[4,5] incorporating the Barlett[24]–Nenitzescu[25] intermolecular hydride transfer concept. The reaction is initiated by the carbocation formed by protonation of the alkene [by the protic acid or the hydrogen halide (water)–promoted metal halide catalysts] [Eqs. (5.1) and (5.2)]. Intermolecular hydride transfer between the isoalkane and the cation then generates a new carbocation [Eq. (5.3)]. Compound **1** adds to the alkene to form the cation **2** [Eq. (5.4)], which then, through intermolecular hydride transfer, forms the product [Eq. (5.5)]. In this final product forming step **1** is regenerated from the isoalkane and then starts a new cycle:

$$AlCl_3 \; + \; HCl \; \rightleftharpoons \; [H^+ AlCl_4^-] \tag{5.1}$$

$$RCH{=}CH_2 \; + \; H^+ \; \rightleftharpoons \; R\overset{+}{C}HCH_3 \tag{5.2}$$

$$\underset{\underset{CH_3}{|}}{\overset{\overset{CH_3}{|}}{CH_3CH}} \; + \; R\overset{+}{C}HCH_3 \; \rightleftharpoons \; \underset{\underset{CH_3}{|}}{\overset{\overset{CH_3}{|}}{CH_3\overset{+}{C}}} \; + \; RCH_2CH_3 \tag{5.3}$$

1

$$\underset{\underset{CH_3}{|}}{\overset{\overset{CH_3}{|}}{CH_3\overset{+}{C}}} \; + \; RCH{=}CH_2 \; \longrightarrow \; \underset{\underset{CH_3}{|}}{\overset{\overset{CH_3}{|}}{CH_3C}}{-}CH_2\overset{+}{C}HR \tag{5.4}$$

2

$$\mathbf{2} \; + \; \underset{\underset{CH_3}{|}}{\overset{\overset{CH_3}{|}}{CH_3CH}} \; \rightleftharpoons \; \underset{\underset{CH_3}{|}}{\overset{\overset{CH_3}{|}}{CH_3C}}{-}CH_2CH_2R \; + \; \mathbf{1} \tag{5.5}$$

As is apparent, the reaction is properly designated as one in which the alkene is alkylated by a carbocation, the isoalkane serving only as the reservoir for the

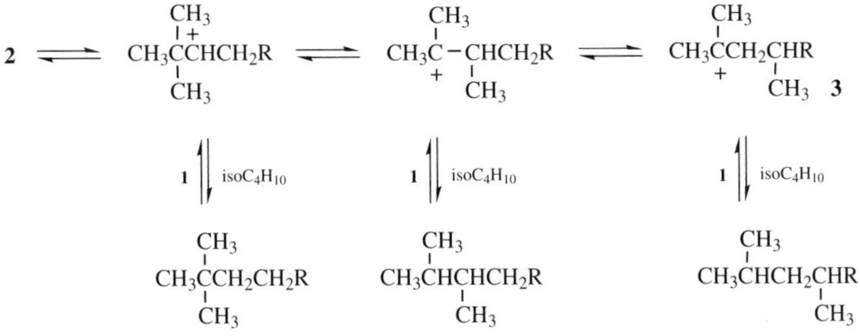

$$
2 \; \rightleftharpoons \; \underset{\overset{|}{CH_3}}{\overset{\overset{CH_3}{|\,+}}{CH_3\overset{}{C}CHCH_2R}} \; \rightleftharpoons \; \underset{\overset{+}{\underset{CH_3}{|}}}{\overset{\overset{CH_3}{|}}{CH_3\overset{}{C}-CHCH_2R}} \; \rightleftharpoons \; \underset{\overset{+}{\underset{CH_3}{}}}{\overset{\overset{CH_3}{|}}{CH_3\overset{}{C}CH_2CHR}} \quad 3
$$

$$1 \big\Updownarrow\, isoC_4H_{10} \qquad\qquad 1 \big\Updownarrow\, isoC_4H_{10} \qquad\qquad 1 \big\Updownarrow\, isoC_4H_{10}$$

$$
\underset{\overset{|}{CH_3}}{\overset{\overset{CH_3}{|}}{CH_3\overset{}{C}CH_2CH_2R}} \qquad\qquad \underset{\overset{|}{CH_3}}{\overset{\overset{CH_3}{|}}{CH_3\overset{}{C}HCHCH_2R}} \qquad\qquad \underset{\overset{|}{CH_3}}{\overset{\overset{CH_3}{|}}{CH_3\overset{}{C}HCH_2CHR}}
$$

Scheme 5.1

alkylating carbocation. However, frequently the reaction is still referred to as "alkylation of alkanes by alkenes." In fact, however, alkanes are alkylated directly only under superacidic conditions (see Section 5.1.2) where no free alkenes are present and therefore the σ-donor alkanes do not compete for the carbocationic alkylating agent with the much higher nucleophilicity π-donor alkenes. This is, as stated by Olah, the fundamental difference between conventional acid-catalyzed alkylation chemistry in which alkenes play the key role and superacidic chemistry, which proceeds on saturated alkanes in absence of alkenes.

Since carbocations do not readily abstract hydride ion from secondary (or primary) carbon atoms, *n*-alkanes seldom enter the alkylation reaction unless they first isomerize to isoalkanes containing tertiary carbon atoms.

The formation of isomeric alkylated products is explained by the ability of carbocations to undergo rearrangements. Ion **2** may participate in a series of hydride and alkyde shifts, and each carbocation thus formed may react via hydride ion transfer to form isomeric alkanes (Scheme 5.1).

The formation of 1-ethyl-2-methylcyclohexane in the alkylation of methylcyclopentane with propylene is also explained by the carbocationic mechanism.[21] Carbocation **4** formed according to the general pattern given in Eq. (5.4) undergoes rearrangement [Eq. (5.6)] with the resulting **5** cyclohexyl cation further reacting by intermolecular hydride transfer [Eq. (5.7)]. The low reactivity of cyclohexane in alkylation with alkenes is accounted for by the necessary isomerization to methylcyclopentane requiring more forcing reaction conditions:

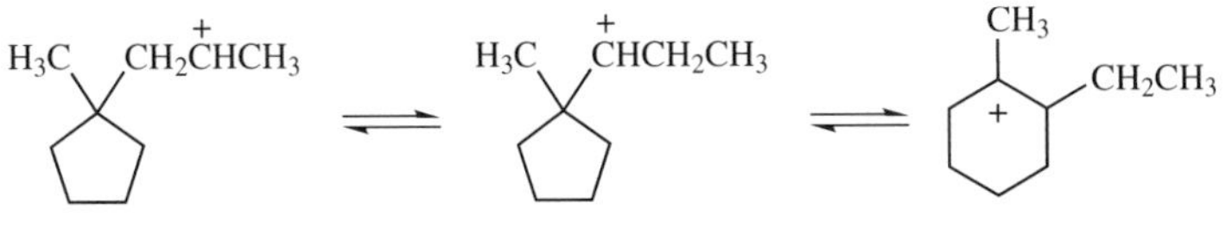

$$\tag{5.6}$$

4   **5**

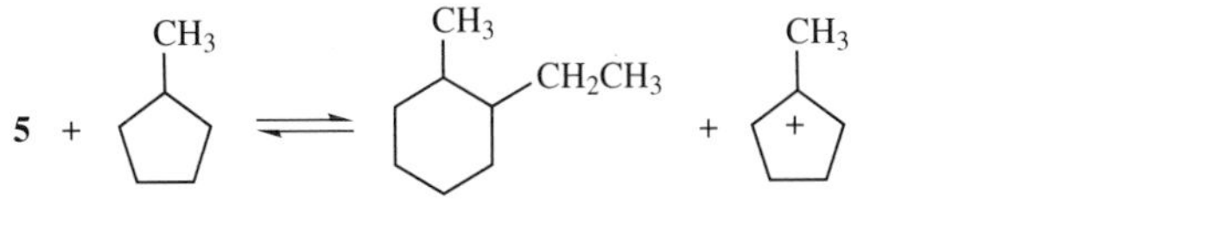

$$\tag{5.7}$$

Besides the rearrangement of carbocations resulting in the formation of isomeric alkylated products, alkylation is accompanied by numerous other side reactions. Often the alkene itself undergoes isomerization prior to participating in alkylation and hence, yields unexpected isomeric alkanes. The similarity of product distributions in the alkylation of isobutane with n-butenes in the presence of either sulfuric acid or hydrogen fluoride is explained by a fast preequilibration of n-butenes. Alkyl esters (or fluorides) may be formed under conditions unfavorable for the hydride transfer between the protonated alkene and the isoalkane.

Isopentane and trimethylpentanes (C_8 alkanes) are inevitably formed whenever isobutane is alkylated with any alkene. They are often called "abnormal" products. Formation of isopentane involves the reaction of a primary alkylation product (or its carbocation precursor, e.g., **3**) with isobutane. The cation thus formed undergoes alkyl and methyl migration and, eventually, β scission:

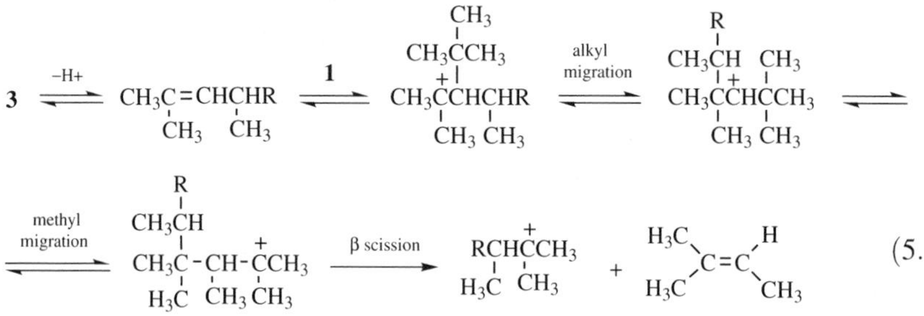

$$\textbf{3} \; \underset{-H+}{\rightleftharpoons} \; \underset{\substack{| \quad | \\ CH_3 \; CH_3}}{CH_3C=CHCHR} \; \overset{\textbf{1}}{\rightleftharpoons} \; \underset{\substack{+| \\ | \quad | \\ CH_3 \; CH_3}}{\overset{\substack{CH_3 \\ | \\ CH_3CCH_3}}{CH_3CCHCHR}} \; \overset{\substack{alkyl \\ migration}}{\rightleftharpoons} \; \underset{\substack{| + \quad | \\ CH_3 \; CH_3}}{\overset{\substack{R \\ | \\ CH_3CH \; CH_3}}{CH_3CCHCCH_3}} \; \rightleftharpoons$$

$$\overset{\substack{methyl \\ migration}}{\rightleftharpoons} \; \underset{\substack{| \quad | \quad | \\ H_3C \; CH_3 CH_3}}{\overset{\substack{R \\ | \\ CH_3CH \\ | \quad + }}{CH_3C-CH-CCH_3}} \; \overset{\beta \; scission}{\longrightarrow} \; \underset{\substack{| \quad | \\ H_3C \; CH_3}}{\overset{+}{RCHCCH_3}} \; + \; \underset{H_3C}{\overset{H_3C}{\diagdown}}C=C\overset{H}{\underset{CH_3}{\diagup}} \quad (5.8)$$

Formation of C_8 alkanes in the alkylation of isobutane even when it reacts with propene or pentenes is explained by the ready formation of isobutylene in the systems (by olefin oligomerization–cleavage reaction) (Scheme 5.2). Hydrogen transfer converting an alkane to an alkene is also a side reaction of acid-catalyzed alkylations. Isobutylene thus formed may participate in alkylation; C_8 alkanes, therefore, are formed via the isobutylene–isobutane alkylation.

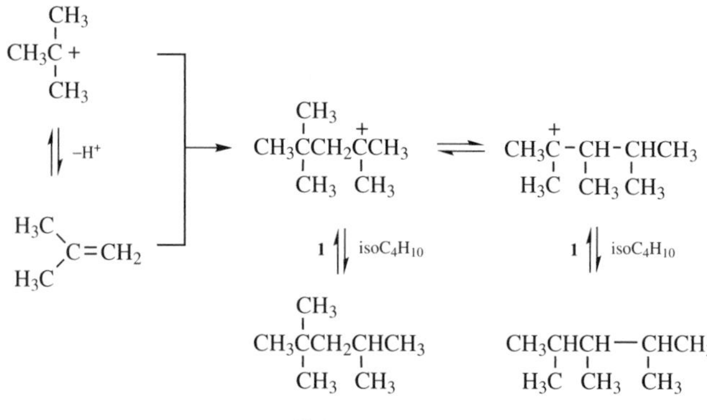

Scheme 5.2

Fast side reactions under the conditions of acid-catalyzed alkylation are oligo-merization (polymerization) and conjunct polymerization. The latter involves polymerization, isomerization, cyclization, and hydrogen transfer to yield cyclic polyalkapolyenes. To suppress these side reactions, relatively high (10 : 1) alkane : alkene ratios are usually applied in commercial alkylations.

Sulfuric acid and anhydrous hydrogen fluoride are most widely used in industrial alkene–isoalkane alkylations. The former results in significant amounts of acid-soluble oil formation, which together with the more difficult acid regeneration led to wider use of the hydrogen fluoride alkylation. Environmental considerations of the effect of the toxicity of volatile HF if accidentally released into the atmo-sphere necessitated development of modified HF alkylation. For example, relatively stable (and therefore less volatile) environmentally safe onium polyhydrogen fluoride complexes, such as pyridine–$(HF)_x$ complexes, are used for alkylation (see Section 5.5.1). Future generation of catalysts probably will be solids, but their use will necessitate higher temperatures with associated difficulties, and large capital investment is needed to start new plants.

Zeolites tend to lose their catalytic activity during alkylation and give rather complex product distribution.[7] The latter originates, in part, from the strong adsorp-tion of hydrocarbons resulting in alkane self-alkylation and extensive cracking. Furthermore, preferential adsorption of alkenes can bring about oligomerization. Despite these difficulties, the alkylate produced on zeolites (and other solid catalysts) is very similar to that formed in liquid-phase alkylation. In both cases trimethylpentanes are the main products and isopentane is always present. This suggests that the carbocationic mechanism discussed above is also operative with zeolites and other solid acid catalysts. New types of solid acid catalysts (H_3PO_4–BF_3–H_2SO_4 supported on SiO_2 or ZrO_2) were reported to exhibit only small deactivation in the isobutane–butene reaction.[26] H_2SO_4 was shown to bring about a remarkable increase in the number of acid sites, whereas BF_3 enhances acid strength.

5.1.2. Alkylation of Alkanes under Superacidic Conditions

Except in some special cases involving concurrent isomerization, n-alkanes cannot be alkylated by conventional acid catalysts. It was, however, found that even methane and ethane react with alkenes in the presence of superacids (e.g., SbF_5 or TaF_5 dissolved in HF, HSO_3F, or CF_3SO_3H).[8,9,27,28] Passage of a mixture of excess methane and ethylene gas through a reactor containing 10 : 1 HF–TaF_5 at 40°C resulted in formation of propane with 58% selectivity.[29,30] Under similar conditions, n-butane was produced with 78% selectivity as the only C_4 product by the reaction of ethane with ethylene. Alkylation of n-butane with ethylene[29,31] and that of methane with propylene[30] were also reported. Similar alkylations of C_1–C_3 alkanes with ethylene were carried out in a static reactor under pressure.[32] Heterogeneous gas-phase ethylation of methane (including $^{13}CH_4$) with ethylene was performed over solid superacids (SbF_5 intercalated into graphite, TaF_5 on AlF_3).[33] The experimental conditions used for alkylation of alkanes require an excess of the superacid and a high alkane : alkene ratio to prevent alkene oligomerization.

In the ethane–ethylene reaction in a flow system with short contact time, exclusive formation of n-butane takes place (longer exposure to the acid could result in isomerization). This indicates that a mechanism involving a trivalent butyl cation depicted in Eqs. (5.1)–(5.5) for conventional acid-catalyzed alkylations cannot be operative here. If a trivalent butyl cation were involved, the product would have included, if not exclusively, isobutane, since the 1- and 2-butyl cations would preferentially isomerize to the *tert*-butyl cation and thus yield isobutane [Eq. (5.9)]. It also follows that the mechanism cannot involve addition of ethyl cation to ethylene. Ethylene gives the ethyl cation on protonation, but because it is depleted in the excess superacid, no excess ethylene is available and the ethyl cation will consequently attack ethane via a pentacoordinated (three-center, two-electron) carbocation [Eq. (5.10)]:

$$CH_3\overset{+}{CH_2} + CH_2{=}CH_2 \longrightarrow CH_3CH_2CH_2\overset{+}{CH_2} \rightleftharpoons CH_3\overset{CH_3}{\underset{CH_3}{\overset{|}{\underset{|}{C}}}}{}^+ \underset{-CH_3CH_2{}^+}{\overset{CH_3CH_3}{\longrightarrow}} CH_3\overset{CH_3}{\underset{CH_3}{\overset{|}{\underset{|}{CH}}}}$$

$$(5.9)$$

$$CH_2{=}CH_2 \underset{-H^+}{\rightleftharpoons} CH_3CH_2{}^+ \overset{CH_3CH_3}{\rightleftharpoons} \left[\begin{array}{c} CH_3CH_2 \\ \diagdown \\ H \end{array} {-}CH_2CH_3 \right]^+ \underset{-H^+}{\longrightarrow} CH_3CH_2CH_2CH_3$$

$$(5.10)$$

Mechanistic studies with ^{13}C-labeled methane demonstrated the formation of monolabeled propane. This is again in accordance with a process initiated by protonation of ethylene to form a reactive ethyl cation:[33]

$$CH_2{=}CH_2 \underset{-H^+}{\rightleftharpoons} CH_3CH_2{}^+ \overset{^{13}CH_4}{\rightleftharpoons} \left[CH_3CH_2{-}\overset{^{13}CH_3}{\underset{H}{\diagup}} \right]^+ \underset{-H^+}{\longrightarrow} CH_3CH_2{}^{13}CH_3 \quad (5.11)$$

The superacid-catalyzed reaction is thus alkylation of the methane with the ethyl cation.

There is clear differentiation of the alkylation of alkenes (π-donor nucleophiles) and alkanes (σ donors). The former follows Markovnikov addition, giving a trivalent carbocation and derived branched products. The latter proceeds through a five-coordinate carbocation without involvement of trivalent carbenium ions and thus without necessary branching.

The reaction of adamantane with lower alkenes (ethylene, propylene, and butylenes) in the presence of superacids [CF$_3$SO$_3$H or CF$_3$SO$_3$H–B(OSO$_2$CF$_3$)$_3$] shows the involvement of both alkylations.[34] In the predominant reaction the 1-adamantyl cation formed through protolytic C–H bond ionization of adamantane adamantylates the alkenes (π alkylation). In a parallel but lesser reaction, the alkyl cations formed by olefin protonation insert themselves into the bridgehead carbon–hydrogen bond generating a pentacoordinated cation, which, via deprotonation, yields the alkylated products:

$$(5.12)$$

This process is called *direct σ alkylation*. Even the formation of *tert*-butyladamantane was observed in the superacid-catalyzed alkylation of adamantane with isobutylene necessitating a highly crowded, unfavorable tertiary–tertiary interaction. This provides further evidence for the σ-alkylation reaction, as adamantylation of isobutylene with any subsequent isomerization does not lead to this isomer.

Lewis superacid–catalyzed direct alkylation of alkanes is also possible with alkyl cations prepared from alkyl halides and SbF_5 in sulfuryl chloride fluoride solution.[35,36] Typical alkylation reactions are those of propane and butanes by *sec*-butyl and *tert*-butyl cations. The $CH_3F–SbF_5$ and $C_2H_5F–SbF_5$ complexes acting as incipient methyl and ethyl cations besides alkylation preferentially cause hydride transfer.[37] Since intermolecular hydride transfer between different carbocations and alkanes are faster than alkylation, a complex mixture of alkylated products is usually formed. A significant amount of 2,3-dimethylbutane was, however, detected when propane was propylated with the 2-propyl cation at low temperature:[36]

$$(5.13)$$

26% of C_6 hydrocarbons

Byproducts include isomeric C_4 and C_5 hydrocarbons formed through C–C bond alkylation (alkylolysis). No 2,2-dimethylbutane, the main product of conventional acid-catalyzed alkylation, was detected, which is a clear indication of predominantly nonisomerizing reaction conditions.

The presence of other hexane isomers and a typical hexane isomer distribution of 26% 2,3-dimethylbutane, 28% 2-methylpentane, 14% 3-methylpentane, 32% *n*-hexane, far from equilibrium, indicate that the 1-propyl cation (although significantly delocalized with protonated cyclopropane nature) is also involved in alkylation. It yields *n*-hexane and 2-methylpentane through primary or secondary C–H bond insertion, respectively (Scheme 5.3).

Particularly revealing is the alkylation of isobutane with *tert*-butyl cation.[35] The product showed, inter alia, the presence of 2,2,3,3-tetramethylbutane (although in low 2% yield). Thus, despite unfavorable steric hindrance in the tertiary–tertiary

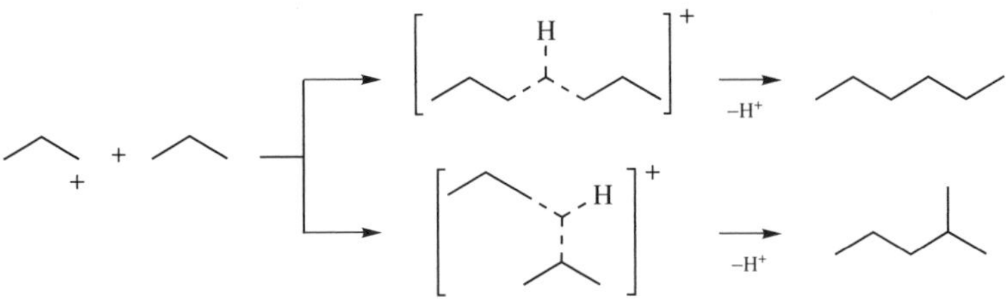

Scheme 5.3

system, the five-coordinate carbocationic transition state is not entirely collinear, and despite predominant intermolecular hydride transfer, direct alkylation also takes place:

$$(CH_3)_3CH \ + \ \overset{+}{C}(CH_3)_3 \ \rightleftharpoons \ \left[(CH_3)_3C \cdots \overset{\overset{\displaystyle H}{|}}{\cdots} \cdots C(CH_3)_3\right]^+ \xrightarrow[-H^+]{} (CH_3)_3C-C(CH_3)_3$$

$$(5.14)$$

Superacidic systems are powerful enough to catalyze self-alkylative condensation of alkanes, an endothermic reaction made feasible by oxidative removal of hydrogen either by the oxidizing superacid ($FSO_3H–SbF_5$) or by added oxidants. Oxidative condensation of methane in "Magic Acid" forms $C_4–C_6$ alkanes through a carbocationic pathway demonstrating the involvement of the **6** and **7** pentacoordinated cations[38–40] [Eqs. (5.15) and (5.16)]:

$$CH_4 \xrightarrow{H^+} \left[CH_3 - \overset{\overset{\displaystyle H}{|}}{\underset{\displaystyle H}{\diagdown}}\right]^+ \longrightarrow \overset{+}{[CH_3]} + H_2$$

6

$$(5.15)$$

$$[CH_3] \ + \ CH_4 \longrightarrow \left[\overset{\overset{\displaystyle H}{|}}{H_3C} \overset{}{\diagup} CH_3\right]^+ \xrightarrow{-H_2} \overset{+}{CH_3CH_2} \xrightarrow{H^-} CH_3CH_3 \Rrightarrow C_4-C_6$$

7

$$(5.16)$$

Condensation of $C_1–C_4$ alkanes to produce highly branched oligomeric and polymeric hydrocarbons[41] was also achieved by their condensation in $FSO_3H–SbF_5$. Block methylene units in the polymeric chain were observed even when methane was brought into reaction with alkenes under similar conditions.[42]

Novel highly active aprotic superacid systems have been more recently found by Vol'pin and coworkers to initiate rapid transformation of alkanes as well as $C_5–C_6$ cycloalkanes.[43] The reaction, for example, of excess cycloalkanes without solvent in the presence of $RCOHlg-2AlBr_3$ ($R = Me$, n-Pr, $Hlg = Cl$, Br) complexes (called *aprotic organic superacids*) results in a facile oxidative coupling of cycloalkanes. The product is a mixture of isomeric dimethyldecalins formed with high selectivities:

$$\text{(5.17)}$$

near-quantitative
yields

Alkylolysis (Alkylative Cleavage). In superacidic alkylation of alkanes not only C—H but also C—C bonds are attacked resulting in alkylolysis (alkylative cleavage). This became apparent in the reaction of alkanes with long-lived alkylcarbenium fluoroantimonates.[36] For steric reasons, however, formation of five-coordinate carbocationic transition states (intermediates) (**8**) are much less preferred than the analogs (**9**) involved in C—H bond alkylations. Methyl and ethyl fluoride–SbF_5 complexes, which are powerful methylating and ethylating agents, respectively, also were shown to bring about limited (>5%) alkylolysis in reactions with alkanes.[37]

$$\begin{array}{cc} \mathbf{8} & \mathbf{9} \end{array}$$

5.1.3. Alkylation of Alkenes and Alkynes

Alkylation of Alkenes with Organic Halides. Schmerling has shown that simple alkyl halides can alkylate alkenes in the presence of Lewis acids to form new alkyl halides.[44,45] With active catalysts, such as aluminum halides, low reaction temperatures (below 0°C) are usually satisfactory. With less active catalysts, such as bismuth chloride and zinc chloride, higher temperatures (20–100°C) are necessary. The best yields are obtained with tertiary halides, whereas secondary halides usually give low yields. Allyl halides[46,47] and α-haloalkylarenes[48] are reactive enough to alkylate alkenes in the presence of weak Lewis acids such as $ZnCl_2$ or $SnCl_4$. Dihaloalkanes also react, provided there is at least one halogen atom attached to a tertiary carbon atom.[44] Methyl and ethyl halides do not undergo reaction, and other primary halides isomerize to secondary halides before addition. Other transformations complicating these alkylations are polymerization, telomerization, and further reaction of the new alkyl halides in alkyaltion.

tert-Butyl chloride reacts with propene to yield three isomeric chlorohep-tanes:[49–51]

$$tert\text{-BuCl} \; + \; CH_2\!=\!CHMe \; \xrightarrow{\text{Lewis acid}}$$

$$\xrightarrow{} \quad \underset{\substack{| \\ Me}}{\overset{\substack{Me \\ |}}{MeCCH_2CHMe}} \;+\; \underset{\substack{| \\ Cl}}{\overset{\substack{MeMe \\ | \; |}}{MeC\!-\!CHCH_2Me}} \;+\; \underset{\substack{| \\ Cl}}{\overset{\substack{Me \; Me \\ | \quad |}}{MeCH\!-\!CCH_2Me}} \qquad (5.18)$$

$$\mathbf{10} \qquad\qquad\qquad \mathbf{11} \qquad\qquad\qquad \mathbf{12}$$

The isomer distribution strongly depends on the catalyst used. $ZnCl_2$ affords the best selectivity, giving exclusively 4-chloro-2,2-dimethylpentane (**10**).

A mechanism with the involvement of the *tert*-butyl cation attacking the less substituted carbon of the double bond accounts for the formation of the main product[44] [Eqs (5.19)–(5.21)]. The other two products (**11**, **12**) are formed by the isomerization of the **13** carbocation through methyl and hydride migration to give the more stable tertiary carbocations:

$$\underset{\substack{| \\ Me}}{\overset{\substack{Me \\ |}}{Me\!-\!C\!-\!Cl}} \;+\; AlCl_3 \;\longrightarrow\; \underset{\substack{| \\ Me}}{\overset{\substack{Me \\ |}}{Me\!-\!C+}} \;+\; AlCl_4^- \qquad (5.19)$$

$$\underset{\substack{| \\ Me}}{\overset{\substack{Me \\ |}}{Me\!-\!C+}} \;+\; CH_2\!=\!CHMe \;\longrightarrow\; \underset{\substack{| \\ Me}}{\overset{\substack{Me \\ |}}{MeCCH_2\overset{+}{C}HMe}} \qquad (5.20)$$

$$\mathbf{13}$$

$$\mathbf{13} \;+\; \underset{\substack{| \\ Me}}{\overset{\substack{Me \\ |}}{Me\!-\!C\!-\!Cl}} \;\longrightarrow\; \underset{\substack{| \quad | \\ Me \; Cl}}{\overset{\substack{Me \\ |}}{MeCCH_2CHMe}} \;+\; \mathbf{1} \qquad (5.21)$$

The reaction of the carbocations through their reaction with the starting alkyl halide [Eq. (5.21)] to form the addition products competes with quenching by the metal halide anion to form halogenated products. In mentioned example, the chlorine atom in compounds **10**, **11**, and **12** originates from $AlCl_4^-$. This was shown in the reaction of ethylene[52] with *tert*-BuBr in the presence of $AlCl_3$, or with *tert*-BuCl in the presence of $AlBr_3$. In both reactions, both 4-bromo- and 4-chloro-2,2-dimethyl-butanes were formed. The main product-forming reaction, however, is that accord-ing to Eq. (5.21) with the generation of a new carbocation.

The complex nature of alkylation of alkenes is illustrated by the reaction of 1- and 2-chloropropane with ethylene,[53] which does not yield the expected isopen-tyl chloride. Instead, 1-chloro-3,3-dimethylpentane is formed by isomerization of the primary product to give *tert*-amyl chloride, which adds to ethylene more readily than do starting propyl halides (Scheme 5.4). The same product is also obtained[54] when ethylene reacts with neopentyl chloride in the presence of $AlCl_3$.

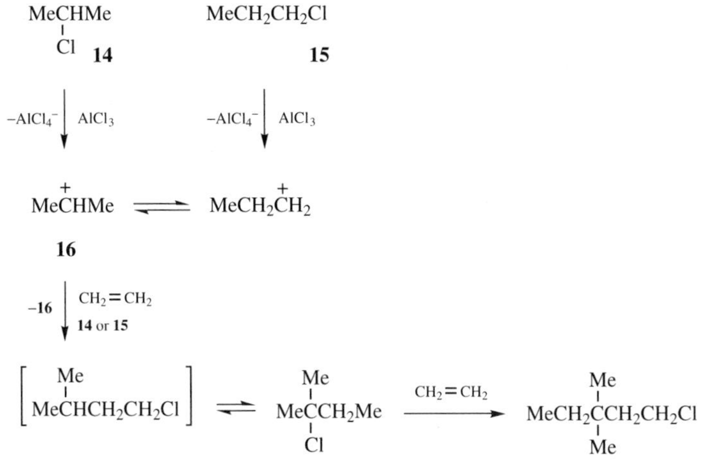

MeCHMe   MeCH$_2$CH$_2$Cl
|
Cl **14**   **15**

−AlCl$_4^-$ | AlCl$_3$ −AlCl$_4^-$ | AlCl$_3$

+   +
MeCHMe $\rightleftharpoons$ MeCH$_2$CH$_2$

16

−**16** | CH$_2$=CH$_2$
       | **14** or **15**

$\left[\begin{array}{c} Me \\ | \\ MeCHCH_2CH_2Cl \end{array}\right]$ $\rightleftharpoons$ $\begin{array}{c} Me \\ | \\ MeCCH_2Me \\ | \\ Cl \end{array}$ $\xrightarrow{CH_2=CH_2}$ $\begin{array}{c} Me \\ | \\ MeCH_2CCH_2CH_2Cl \\ | \\ Me \end{array}$

Scheme 5.4

The scope and limitations of the Lewis acid–catalyzed additions of alkyl chlorides to carbon–carbon double bonds were studied.[51] Since Lewis acid systems are well-known initiators in carbocationic polymerizations of alkenes, the question arises as to what factors govern the two transformations. The prediction was that alkylation products are expected if the starting halides dissociate more rapidly than the addition products.[55] In other words, addition is expected if the initial carbocation is better stabilized than the one formed from the dissociation of the addition product. This has been verified for the alkylation of a range of alkyl- and aryl-substituted alkenes and dienes with alkyl and aralkyl halides. Steric effects, however, must also be taken into account in certain cases, such as in the reactions of trityl chloride.[51]

Alkylation of Alkynes. Organic halides can alkylate acetylenes in the presence of Lewis acids. In most cases, however, the products are more reactive than the starting acetylenes. This and the ready polymerization of acetylenes under the reaction conditions result in the formation of substantial amounts of byproducts. Allyl, benzyl, and *tert*-alkyl halides giving stable carbocations under mild conditions are the best reagents to add to acetylenes.[44,56]

The alkylation of phenylacetylene with *tert*-butyl chloride, benzyl chloride, and diphenylmethyl chloride follows an Ad$_E$2 mechanism[57–59] with the involvement of the open vinyl cationic intermediate **17**:

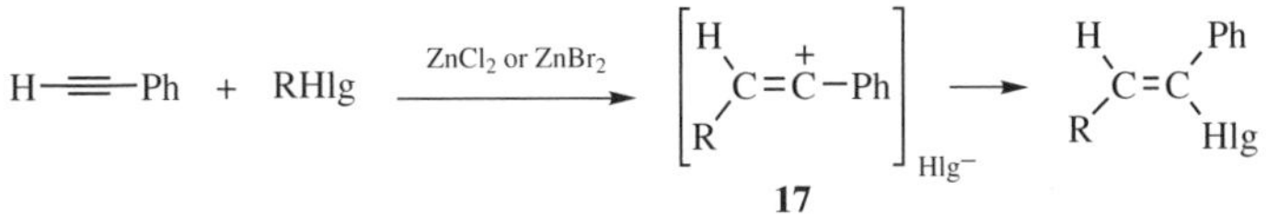

$$H\text{---}Ph + RHlg \xrightarrow{ZnCl_2 \text{ or } ZnBr_2} \left[\begin{array}{c} H \\ \diagdown \\ C=\overset{+}{C}-Ph \\ \diagup \\ R \end{array}\right]Hlg^- \longrightarrow \begin{array}{cc} H & Ph \\ \diagdown & \diagup \\ & C=C \\ \diagup & \diagdown \\ R & Hlg \end{array} \quad (5.22)$$

17

The *anti* stereoselectivity originates from the preferential attack of the chloride ion on **17** from the less hindered side. Stereoselectivity, therefore, is determined by the relative effect of the two β substituents. Alkyl-substituted alkynes show decreased stereoselectivity compared with that of aryl-substituted acetylenes.[59,60]

Alkylation with Carbonyl Compounds: The Prins Reaction. Carbonyl compounds react with alkenes in the presence of Brønsted acids to form a complex mixture of products known as the *Prins reaction*. The use of appropriate reaction conditions, solvents, and catalysts allows one to perform selective syntheses. Characteristically formaldehyde is the principal aldehyde used. Mineral acids (sulfuric acid, phosphoric acid), *p*-toluenesulfonic acid, and ion exchange resins are the most frequent catalysts. Certain Lewis acids (BF_3, $ZnCl_2$, $SnCl_4$) are, however, also effective.

The Prins reaction was investigated extensively.[61–65] It proceeds by the electrophilic attack of protonated formaldehyde to the double bond. Markovnikov addition is followed, and alkenes exhibit increasing reactivity with increasing alkyl substitution. Although different intermediates were suggested, the 3-hydroxy carbocation intermediate **18** seems to be consistent with the experimental data[61,65,66] (Scheme 5.5). Intermediate **18** can deprotonate to give allylic alcohol (**19**) or homoallylic alcohol, react further with nucleophiles to yield β-substituted alcohols (**20**), or by reacting with a second molecule of aldehyde, form 1,3-dioxacyclohexane (**21**).

Simple mono- and disubstituted alkenes react to yield 1,3-diols, when the Prins reaction is carried out at elevated temperature. Diols originate from the attack of water on carbocation **18**, or through the acidolysis of dioxanes under the reaction conditions. When the reaction is conducted in acetic acid, monoacetates are formed by acetate attack on **18**. Dienes resulting from the dehydration of intermediate diols are the products of the transformation of more substituted alkenes. Monoacetates and diols may react further to yield 1,3-diol diacetates. When the Prins reaction

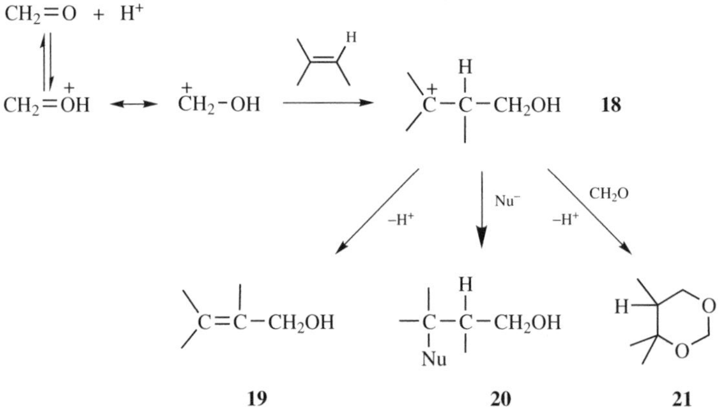

Scheme 5.5

is performed with hydrochloric acid as catalyst, chloride ion serves as the nucleophile affording 1,3-chlorohydrins.

1,3-Dioxacyclohexanes can be produced in excellent yields from aliphatic or aryl-substituted alkenes.[64] Dilute sulfuric acid at or above room temperature with paraformaldehyde appears to give the best results. Dioxane–water or acetic acid as solvent was found to afford increased yields in the Prins reaction of arylalkenes.

The stereochemistry of the Prins reaction is complex. In the transformation of cyclohexene and 2-butenes *anti* stereoselective addition was observed,[67–69] whereas *syn* addition of two formaldehyde units takes place in the formation of 1,3-dioxanes from substituted styrenes.[70] Most of the transformations are, however, nonstereoselective,[71,72] accounted for by carbocation **18**.

Few examples of the Prins reaction of homologous aldehydes and ketones are known.[73,74] Acetaldehyde was shown to react smoothly with isobutylene under mild conditions, whereas other, less substituted alkenes were found to be very unreactive.[73]

5.1.4. Alkylation of Aromatics

Friedel and Crafts discovered[75] the aluminum chloride-catalyzed alkylation of benzene with alkyl halides in 1877. During the more than a century that has passed, the reaction has been exhaustively investigated. The results showed that the reaction is far more complicated than it first appeared. In their early work, Friedel and Crafts observed that passage of methyl chloride through moderately heated benzene containing aluminum chloride resulted in the evolution of hydrogen chloride and the formation of toluene together with smaller amounts of xylenes, mesitylene, and higher-boiling polymethylbenzenes:

$$\tag{5.23}$$

The product consisted of two liquid layers. The upper layer comprised benzene and methylated benzenes. The lower layer was a red-brown oily complex of the catalyst and the same methylbenzenes as those found in the upper layer, and which could be liberated by hydrolysis.

Similarly, a product believed to be *n*-amylbenzene was formed by the reaction of amyl chloride with benzene in the presence of aluminum chloride. It is quite probable, however, that the product reported at boiling at 185–190°C contained much *sec*-amylbenzenes and possibly even a small amount of *tert*-amylbenzene. In fact, isomerization often accompanies alkylation of aromatic hydrocarbons.

Shortly after the work of Friedel and Crafts, Balsohn discovered that ethylene also reacted with benzene in the presence of $AlCl_3$ to give ethylbenzene.[76] Other alkenes act as similar alkylating agents. The alkylations are accompanied by dialkylation (polyalkylation) and isomerization processes. Two types of isomerization—rearrangement of the reactant alkylating agent and positional isomerization (reorientation of substituents into new ring position)—may occur during alkylation. Friedel–Crafts alkylations are often also complicated by other side reactions. These include formation of saturated hydrocarbons, dealkylation, and disproportionation (transalkylation). The nature of the catalyst and reaction conditions exert a strong effect on the extent of these transformations.

Any aromatic hydrocarbon that contains a replaceable hydrogen can be alkylated unless steric effects prevent introduction of the alkyl group. For example, little or no *ortho* substitution of an alkylbenzene takes place when the entering group is tertiary or when the aromatic hydrocarbon contains a tertiary alkyl group. Alkylable aromatic hydrocarbons include benzene and its homologs, fused polycyclic compounds (naphthalene, anthracene, phenanthrene, indene), and polycyclic ring assemblies (e.g., biphenyl, diphenylmethane, bibenzyl, tetraphenyls). Polycyclic aromatic hydrocarbons readily undergo alkylation but usually give lower yields of alkylated products because of their own high reactivity under the Friedel–Crafts conditions.

Since alkyl groups attached to the benzene ring would be expected to direct entering substituents into the *ortho* and *para* positions, the occurrence of significant *meta* substitution when alkylbenzenes are monoalkylated (or benzene is dialkylated) was initially unexpected. It was subsequently established that such *meta* substitution is seldom caused by a low-selectivity kinetic alkylation, but by subsequent (or concurrent) thermodynamically controlled isomerization. The relative amounts of the three isomeric dialkylbenzenes that are produced depend on the aromatic hydrocarbon, the alkylating agent, the catalyst, and the reaction conditions. When considering mechanistic aspects, it is essential to differentiate kinetically controlled direct alkylation and thermodynamically affected isomerizative reactions.

Besides aluminum chloride, the most often used and studied Friedel–Crafts catalyst, many other acid catalysts are effective in alkylation. Although Friedel–Crafts alkylation was discovered and explored mainly with alkyl halides, from a practical point of view, alkenes are the most important alkylating agents. Others include alcohols, ethers, and esters.

Detailed coverage of Friedel–Crafts-type alkylations is to be found in relevant reviews and monographs,[77–86] and the reader is advised to consult these for detailed information.

Catalysts. It was shown by Friedel and Crafts that while aluminum chloride is a very active alkylation catalyst, other metal chlorides (ferric chloride, zinc chloride, etc.) are also effective. They also showed that other aluminum halides (aluminum bromide and aluminum iodide) are excellent catalysts. Indeed, Kline et al., 1959[87] claimed that for the isopropylation of benzene with isopropyl chloride, aluminum iodide is the most active catalyst, although this claim may be overstated and AlI_3 also can cause side reactions.

Boron trifluoride is another important, reactive Friedel–Crafts catalyst that has been widely used. Because it is a volatile gas but forms many complexes for some applications, it is preferred and can be readily recovered for reuse.

Zinc chloride is much less reactive than aluminum chloride and usually requires higher reaction temperatures. However, it has the advantage that, unlike aluminum chloride, it is less sensitive to moisture and can sometimes even be used in aqueous media.[88] Concentrated aqueous solutions of ferric chloride, bismuth chloride, zinc bromide, stannous chloride, stannic chloride, and antimony chloride are also alkylation catalysts, particularly in the presence of hydrochloric acid.[89]

Aluminum chloride loses its activity when dissolved in excess methanol but yields a catalyst when converted to its monomethanolate ($AlCl_3$–CH_3OH). On the other hand, the dimethanolate has no catalytic activity. Nitroalkanes (such as nitromethane) form 1 : 1 addition complexes with aluminum chloride that modify its reactivity and are used for more selective alkylations.[90]

Attempts to establish a reliable reactivity order of Friedel–Crafts catalysts have been unsuccessful because the activity of Lewis acid metal halides in Friedel–Crafts reactions also depends on the reagents and conditions of the particular reaction. A purely empirical order of some reactive metal halide alkylation catalysts was given as $AlHlg_3 > GaHlg_3 > ZrCl_4 > NbCl_5 > TaCl_5 > MoCl_5$ established for the benzylation of benzene with methylbenzyl chlorides.[91] These catalysts give high yields and cause extensive isomerization. Moderate activity (sufficient yields without significant isomerization) was exhibited by $InHlg_3$, $SbCl_5$, $NbCl_5$, $FeCl_3$, and the $AlCl_3$–RNO_2 complex. BCl_3, $SnCl_4$, and $TiCl_4$, in turn, displayed weak activity, giving products in low yields, whereas BF_3 and $ZnCl_2$ proved to be very weak Friedel–Crafts catalysts. BF_3 as subsequently found catalyzes readily the reaction of alkyl (benzyl) fluorides but not of chlorides (or bromides). An activity order based on the alkylation of benzene with *sec*-butyl chloride is somewhat different[92] (very active catalysts—$AlBr_3$, $AlCl_3$, $MoCl_5$, $SbCl_5$, and $TaCl_5$; moderately active catalysts—$NbCl_5$, $FeCl_3$, $AlCl_3$–CH_3NO_2; weak catalysts—$ZrCl_4$, $TiCl_4$, $AlCl_3$–nitrobenzene, WCl_6).

Whereas *n*-donor haloalkanes such as alkyl halides are activated by coordination (ionization) with Lewis acid halides for aromatic alkylations, there is much evidence showing that these Friedel–Crafts catalysts in alkylations with π-donor alkenes are active only in the presence of a cocatalyst (promoter) such as hydrogen chloride, hydrogen bromide, or water, allowing them to act as strong conjugate protic acids. While many successful alkylations have been reported without added promoters, it seems probable that moisture was always unintentionally present (either in incompletely dried reactants or in the apparatus) or that the reactions were promoted by hydrogen halide formed as byproduct during the reaction; that is, the reactions were autocatalytic.

Acids capable of efficiently protonating olefins catalyze the alkylation of aromatics. The most active practical protic acid catalysts are hydrogen fluoride and concentrated sulfuric acid, which usually catalyze alkylation at 0–50°C. Phosphoric acid, particularly at higher temperatures, is also an active catalyst and has substantial practical importance (see Section 5.5.3). Other protic acids that are active in alkylation include trifluoromethanesulfonic acid, polyphosphoric acid, fluorophosphoric acid, fluoro- and chlorosulfuric acids, and *p*-toluenesulfonic acid (or other alkane- and arenesulfonic acids).

Several new effective Friedel–Crafts catalysts have been developed. These include triflate[93] (trifluoromethanesulfonate) derivatives of boron, aluminum, and gallium [$M(OSO_2CF_3)_3$]. Trichlorolanthanides have also been proved to be active reusable catalysts in benzylation.[94] Superacids as catalysts are also very efficient in many Friedel–Crafts alkylations.[95]

Protic acids are usually used as catalysts for alkylation with olefins rather than with alkyl halides. Anhydrous hydrogen chloride, in the absence of metal halides, is a catalyst for the alkylation of benzene with *tert*-butyl chloride only at elevated temperatures[96] where an equilibrium with isobutylene may exist. Other alkyl halides and cyclohexene were also reacted with benzene and toluene using this catalyst.

Acidic mixed oxides, including alumina and silica, as well as natural clays, and natural or synthetic aluminosilicates, are sufficiently (although mildly) hydrated to be effective as solid protic acids for the alkylation of aromatic hydrocarbons with olefins. The most studied of these catalysts are zeolites that are used in industrial processes (see Section 5.5).

Frequently substantially more than catalytic amounts of a Lewis acid metal halide are required to effect Friedel–Crafts alkylation. This is due partly to complex formation between the metal halide and the reagents or products, especially if they contain oxygen or other donor atoms. Another reason is the formation of "red oils." Red oils consist of protonated (alkylated) aromatics (i.e., arenium ions) containing metal halides in the counterions or complexed with olefin oligomers. This considerable drawback, however, can be eliminated when using solid acids such as clays,[97,98] zeolites (H-ZSM-5),[99,100] acidic cation-exchange resins, and perfluoroalkanesulfonic acid resins (Nafion-H).[101–104]

More details about Friedel–Crafts catalysts can be found in the review literature.[79,82,86,105–107]

Alkylation with Alkyl Halides. Friedel–Crafts alkylation with alkyl halides is catalyzed by metal halides since they are able to form complexes with the *n*-donor alkyl halides and ionize them to alkyl cations.[86,106–109] The most effective and most frequently used catalysts are $AlCl_3$, $GaCl_3$, $FeCl_3$, and BF_3. Hydrogen fluoride and sulfuric acid, in turn, are less effective. When the reaction is carried out without solvent, excess of the aromatics is used as the medium. Primary halides are the least reactive alkylating agents, requiring the use of strong Friedel–Crafts catalysts ($AlCl_3$, $GaCl_3$). Moderately active or weak catalysts ($FeCl_3$, BF_3, $AlCl_3$–RNO_2) can be used in alkylation with *tert*-alkyl, allyl, and benzyl halides. The reactivity of alkyl halides[110–112] in alkylation generally shows the decreasing order $F > Cl > Br > I$.

Friedel and Crafts apparently assumed that the alkyl group in the alkylated product had the same structure as that of the alkyl halide. However, it is seldom possible to predict the configuration of the alkyl group or the orientation in alkylation. Indeed, shortly after the discovery by Friedel and Crafts of the alkylation of aromatics, Gustavson[113] and then Silva[114] reported that the aluminum chloride–induced reaction of either propyl or isopropyl halides with benzene yields the same isopropylbenzene. These and other early observations were, however, sometimes conflicting. Ipatieff and coworkers, for example, reported[115] that the product composition in alkylation of benzene with n-propyl chloride in the presence of $AlCl_3$ depends on the reaction temperature. Carrying out the reaction at $-6°C$ gave a $60:40$ mixture of propylbenzene and isopropylbenzene, whereas a composition of $40:60$ was identified in the reaction at $35°C$.

More recent work applying modern analytic techniques for reliable identification of products revealed that the product distribution was almost independent of the temperature (the normal:isopropyl ratio was $34:66$ at $-18°C$, and $30:70$ at $80°C$).[116] In contrast, alkylation of p-xylene exhibited strong temperature-dependent regioselectivity[117] [Eq. (5.24)] due to obvious isomerization equilibria:

$-17°C$	73	27	19% yield
$50°C$	53	31	16 68% yield

Di- and polyalkylation can occur during alkylation with alkyl halides since the product alkylbenzenes are more reactive, although the reactivity difference with reactive alkylation systems is small. Toluene, for example, reacts only about 2–5 times faster in some benzylations than benzene.[118,119] As alkylbenzenes, however, dissolve preferentially in the catalyst containing layer, heterogeneous systems can cause enhanced polysubstitution. The use of appropriate solvents and reaction conditions as well as of an excess of aromatics allow the preparation of monoalkylated products in high yields.

The mechanism of Friedel–Crafts alkylation with alkyl halides involves initial formation of the active alkylating agent, which then reacts with the aromatic ring. Depending on the catalyst, the solvent, the reaction conditions, and the alkyl halide, the formation of a polarized donor–acceptor complex or real carbocations (as either an ion pair or a free entity) may take place:

$$R-Hlg + MHlg_n \; \rightleftharpoons \; \overset{\delta+}{R} \cdots Hlg \cdots \overset{\delta-}{MHlg_n} \; \rightleftharpoons \; R^+[MHlg_{n+1}]^- \qquad (5.25)$$

$$22$$

It is usually considered that the carbocation formed in the interaction between the alkyl halide and the metal halide catalyst attacks the aromatic ring and forms the product through a σ complex or arenium ion (Wheland intermediate) (**23**) in a typical aromatic electrophilic substitution:

$$
\text{(benzene)} + R^+ \rightleftharpoons \text{(23)} \xrightarrow{-H^+} \text{(R-benzene)} \tag{5.26}
$$

23

Dewar,[120] as well as Brown and Olah, respectively, raised the importance of initial π complexing of aromatics in alkylations.[109,121] Relevant information was derived from both substrate selectivities (usually determined in competitive alkylations of benzene and toluene, or other alkylbenzenes), and from positional selectivities in the alkylation of substituted benzenes. Olah realized that with reactive alkylating agents substrate and positional selectivities are determined in two separate steps.

Substantial evidence indicates that in alkylation with primary alkyl halides (and some secondary alkyl halides), it is the highly polarized donor–acceptor complex **22** that undergoes rate-determining nucleophilic displacement by the aromatic (acting as π-nucleophile).[118,122] This assumption is supported, for example, by the overall third-order kinetics of the reaction (first-order each in the aromatic, the halide, and the catalyst).[118,123] Further evidence is the finding that methylation of toluene by methyl bromide and methyl iodide occurred at different rates and gave different xylene distributions.[111] The involvement of a common intermediate (CH_3^+) should result in identical rate and product distribution. Further, the formation of primary alkyl cations, particularly that of CH_3^+, is very unfavorable. Strong evidence against the participation of primary carbocations is also the lack of scrambling in aluminum chloride–catalyzed alkylation of benzene with [2-^{14}C]-ethyl chloride.[124] The transition state of the rate-determining displacement step suggested by Brown[122] can be depicted as σ complex **24** involving a partially formed carbon–carbon bond and a partially broken carbon–chlorine bond. As mentioned, Olah and coworkers have found that alkylation by strongly electrophilic reagents involves a separate initial rate-determining step suggested to be an oriented σ complex (**25**), preceding the formation of the π complex.[119,125,126]

$$
\begin{array}{cc}
\text{24} & \text{25}
\end{array}
$$

24   **25**

The nature of the transition state of highest energy depends on the reagents.[91,126,127] A transition state lying early on the reaction coordinate will resemble starting aromatics (thus of π-complex nature). This may happen when strongly electrophilic reagents or strongly basic aromatics are involved in alkylation. Since the subsequently formed σ complexes (separate for each position) also lie relatively early on the reaction coordinate, *ortho* over *para* substitution is favored. Weak electrophiles or weakly basic aromatics, in turn, tend to proceed through transition states lying late on the reaction coordinate. Weaker alkylating agents result in increasing selectivity of the formation of the *para* isomer (assuming that kinetic conditions are maintained).

When good correlations are found between rates and σ-complex stabilities of arenes, Brown's one-step mechanism with rate-determining σ-complex formation determines both substrate and positional selectivities and Brown's selectivity relationship is valid. Care must be exercised, however, to ascertain that kinetic conditions are always maintained. If such correlation does not hold, Olah's two-step mechanism with separate π- and σ-complex formation may be operative. The nature of the initial "first complex" is still argued. In some instances, encounter complexes or charge transfer complexes are suggested to be involved but there is increasing agreement about the necessity of two separate steps to explain low substrate but high positional selectivities. Of course, it must be emphasized again, that if isomerizing conditions prevail no kinetic treatment of selectivities is possible.

Studies by Nakane et al. also support the two-step mechanism when alkylation is carried out with alkyl halides under substantially nonionizing conditions.[128,129] It was further shown that in nonpolar organic solvents carbocations rather than the polarized complexes participate directly in the formation of the first π complex. BF_3–H_2O catalyzes ethylation,[130] isopropylation,[131] and benzylation[132] through the corresponding carbocations. Accordingly, ethylbenzene equally labeled in both α and β positions was obtained when [2-^{14}C]-ethyl halides were reacted in hexane solution in the presence of boron trifluoride,[130] BF_3–H_2O,[133] or aluminum bromide.[134] The benzyl cation was, however, shown to be involved in $TiCl_4$- and $SbCl_5$-catalyzed benzylation in polar solvents.[135]

The stereochemical outcome of Friedel–Crafts alkylation may be either inversion (nucleophilic displacement) or racemization (involvement of a trivalent flat carbocation). Most transformations were shown to occur with complete racemization. In a few instances, inversion or retention was observed. For example, the formation of (−)-1,2-diphenylpropane [(−)-**28**] from (−)-**26** was interpreted to take place through the **27** nonsymmetrically π-bridged carbocation, ensuring 50–100% retention of configuration:[136]

$$PhCH_2-C\overset{H}{\underset{Cl}{\diagup}}Me \quad \xrightarrow[-20\ to\ 40°C]{AlCl_3\ or\ FeCl_3} \quad H_2C-C\overset{+\diagdown H}{\underset{Me}{}} \quad \xrightarrow{benzene} \quad PhCH_2-C\overset{H}{\underset{Ph}{\diagup}}Me \quad (5.27)$$

(−)-**26**   **27**   (−)-**28**

As was already mentioned, isomerizations (rearrangement of the reactant alkyl halide, as well as positional isomerization in the aromatic ring, or disproportionative transalkylation) may occur during alkylation. The ease of rearrangement of alkylating agents follows the order primary < secondary < tertiary. As a rule, straight-chain secondary halides yield mixtures in which the aromatic ring is attached to all possible secondary carbon atoms in the chain. Branched alkyl halides usually form tertiary alkylates under nonisomerizing conditions[137] [Eq. (5.28), $AlCl_3$–$MeNO_2$). Under isomerizing conditions ($AlCl_3$), however, different product mixtures are isolated:

$$(5.28)$$

$AlCl_3$–$MeNO_2$, 25°C, 0.25 h	100	
$AlCl_3$, 0–5°C, 0.25 h	43	57
$AlCl_3$, 0–5°C, 4 h	18	82

Selective isopropylation and *tert*-butylation may be accomplished by alkylation with isopropyl and *tert*-butyl halides, respectively.

Isomerization can take place in the arenium ion (σ-complex) intermediates of the alkylation reactions prior to their deprotonation, or the product alkylbenzenes may undergo isomerization to yield rearranged and ring isomerized products. Alkylation is further frequently a reversible reaction. Strong catalysts, more elevated temperatures, and prolonged reaction times favor isomerizations. When substituted benzenes, such as toluene, is *tert*-butylated, widely varying isomer distributions can be obtained.[125,138,139] Under nonisomerizing (kinetically controlled) conditions selective formation of *p-tert*-butyltoluene occurs, giving 95% of this isomer:[125]

$$(5.29)$$

$AgClO_4$, 5 min	94.4	5.6
$AgBF_4$, 15 min	38.2	62.8

In contrast, the *meta* isomer is formed under isomerizing conditions ($AlCl_3$, $FeCl_3$) in yields up to 67%, which reflects the thermodynamically controlled isomer distribution.[138] As a result of severe steric interaction, little or no *ortho* isomer is formed.

As mentioned, positional isomerization can also take place in the σ complexes through intramolecular 1,2 shifts prior to deprotonation:[126]

$$(5.30)$$

It is noteworthy that the thermodynamic equilibria of the neutral dialkylbenzenes and their protonated arenium ions (σ complexes) are significantly different (see Section 4.1.2). In the latter case the 1,3-dialkyl-substituted arenium ion is the most stable.

Solid Lewis acid halide catalysts seem to have advantages in some Friedel–Crafts reactions. $ZnCl_2$ and $NiCl_2$, supported on K10 montmorillonite exhibit high activity and selectivity in benzylation:[140]

$$
\text{C}_6\text{H}_6 + \text{PhCH}_2\text{Cl} \xrightarrow[\text{RT, 15 min}]{\text{MCl}_2 \text{ on K10}} \text{C}_6\text{H}_5\text{CH}_2\text{Ph} \tag{5.31}
$$

$MCl_2 = ZnCl_2$ 100% conversion 80%
$MCl_2 = NiCl_2$ 96% conversion 67%

Alkylation with di- (poly)-functional alkylating agents, such as di- and polyhaloalkanes, is of synthetic significance. Despite the increased probability of additional intra- and intermolecular side reactions, selective alkylations may be possible. Carbon tetrachloride, for example, can be reacted with benzene to form triphenylmethyl chloride in high yield (84–86%).[141] The difference in reactivity of different halogens allows selective haloalkylation. Boron trihalides generally catalyze only reaction of C—F but not C—Cl or C—Br bonds[112] [Eq. (5.32)]. Observed isomerization in the presence of BF_3 is attributed to the formation of hydrogen fluoride, which forms the strong conjugate acid HF—BF_3 capable of effecting extensive isomerization[112] [Eq. (5.33)]:

$$
\text{C}_6\text{H}_6 + \text{FCH}_2\text{CH}_2\text{Cl} \xrightarrow[-10^\circ\text{C, 30 min}]{\text{BHlg}_3} \text{C}_6\text{H}_5\text{CH}_2\text{CH}_2\text{Cl} \tag{5.32}
$$

50–55%

$$
\text{C}_6\text{H}_6 + \text{FCH}_2\text{CH}_2\text{CH}_2\text{Cl} \xrightarrow{0-10^\circ\text{C}} \text{C}_6\text{H}_5\text{CH}(\text{CH}_3)\text{CH}_2\text{Cl} + \text{C}_6\text{H}_5\text{CH}_2\text{CH}_2\text{CH}_2\text{Cl} \tag{5.33}
$$

	CH_3CHCH_2Cl	$CH_2CH_2CH_2Cl$	
BF_3, 60 min	100		90% yield
BCl_3 or BBr_3, 30 min	10	90	63% yield

A systematic study of the alkylation of benzene with optically active 1,2-, 1,3-, 1,4-, and 1,5-dihaloalkanes showed that the stereochemical outcome depends on the halogens and the chain length.[142] Retention of configuration in alkylation with (S)-1,2-dichloropropane was explained by invoking double inversion and the involvement of a bridged cation such as **27**. (S)-1-Bromo-2-chloropropane, in contrast, reacts with inversion. The involvement of a cyclic halonium ion intermediate ensures retention of configuration in alkylation with (S)-1,4-dichloropentane.

An important application of alkylation with di- and polyhalides is cyclialkylative ring formation. Of different α,ω-dihaloalkanes studied, 1,4-dichlorobutane was shown to react at exceptional rate and with least rearrangement.[143] This was ratio- nalized in terms of participation by the cyclic chloronium ion **29** formed with anchimeric assistance by the second chlorine atom in the molecule:

$$\text{(5.34)}$$

Cyclialkylative ring closure is more frequently performed with haloalkylarenes. Such cyclialkylation reactions are, however, outside the scope of this discussion. Details of these and other haloalkylations are found in the literature.[106,144–147]

Alkylation with Alkenes. Besides alkyl halides, alkenes,[86,148–151] alkynes,[152] alcohols and ethers,[86,151,153] and esters[86,151,154] are most frequently used for alkyla- tion of arenes. Of all alkylating agents, alkenes are practically the most important.

When an alkene is used in the alkylation of arenes, metal halides with hydrogen halide or water as cocatalyst, protic acids, and acidic oxides can be used as cata- lysts. Both linear and cyclic alkenes are used in alkylations. Alkylation with alkenes is usually preferred in industry because the processes are simpler and olefins are readily and cheaply available in pure form from petroleum refining processes.

The alkylation of benzene with ethylene is a very important reaction for the man- ufacture of ethylbenzene (see Section 5.5.2), which is then dehydrogenated to styrene. The alkylation is not readily catalyzed by hydrogen fluoride or sulfuric acid because the symmetric π bond of ethylene is not as easily protonated as those of more polar alkenes (propylene, etc.). Further, the ethyl cation can also be readily quenched to ethyl fluoride and ethyl hydrogen sulfate (or diethyl sulfate), respec- tively. Sulfuric acid also sulfonates benzene and alkylbenzenes. Superacidic trifluoromethanesulfonic acid (triflic acid) does not react with arenes, but is a much more expensive catalyst. Promoted aluminum chloride,[155,156] phosphoric acid,[157–159] silica–alumina,[160] and zeolites[99,161] are applied in industrial processes. The solid acid-catalyzed manufacture of ethylbenzene and p-ethyltoluene by alkyl- ation of benzene and toluene with ethylene, respectively, is also of industrial sig- nificance (see Sections 5.5.2 and 5.5.5). A characteristic feature of intermediate- pore-size H-ZSM-5 zeolite catalyst in the alkylation of alkylaromatics is that it brings about the formation of the *para* isomers with high selectivity.[162] Hydrogen mordenite, in contrast, yields a near-equilibrium mixture due to isomerization of the initial product.

The related manufacture of cumene (isopropylbenzene) through the alkylation of benzene with propylene is a further industrially important process, since cumene is used in the synthesis of phenol and acetone. Alkylation with propylene occurs more readily (at lower temperature) with catalysts (but also with hydrogen fluoride and acidic resins) similar to those used with ethylene, as well as with weaker acids, such as supported phosphoric acid (see further discussion in Section 5.5.3).

In alkylation of benzene with both ethylene and propylene di- and polyalkylates are also formed. In alkylation with propylene 1,2,4,5-tetraisopropylbenzene is the most highly substituted product; steric requirements prevent formation of penta- and hexaisopropylbenzene. On the other hand, alkylation of benzene with ethylene readily even yields hexaethylbenzene. Alkylation with higher alkenes occurs more readily than with ethylene or propylene, particularly when the alkenes are branched. Both promoted metal chlorides and protic acids catalyze the reactions.

Carbocations formed through protonation of alkenes by proton acids are usually assumed as intermediates in alkylation with alkenes. Metal halides, when free of protic impurities, do not catalyze alkylation with alkenes except when a cocatalyst is present. It was shown that no neat conjugate Friedel–Crafts acids such as $HAlCl_4$ or HBF_4 are formed from 1 : 1 molar compositions in the absence of excess HCl or HF, or another proton acceptor.[163–166] In the presence of a proton acceptor (alkene), however, the Lewis acid halides—hydrogen halide systems are readily able to generate carbocations:

$$RCH=CH_2 + HHlg + MHlg_n \rightleftharpoons R\overset{+}{C}HCH_3 \ [MHlg_{n+1}]^- \qquad (5.35)$$

It was observed that all three straight-chain butenes give similar product distribution in alkylation of naphthalene:

1-butene	74	26	76% yield
cis-2-butene	72	28	81% yield
trans-2-butene	70	30	83% yield

$$(5.36)$$

This suggests the intermediacy of a common carbocationic intermediate, the 2-butyl cation.[167]

Alkylation with isobutylene under similar conditions[167] and alkylation with propylene catalyzed by perfluorosulfonic acids,[168] in turn, lead to 2-substituted naphthalenes. As was shown by Olah and Olah,[169] the kinetic alkylation of naphthalene, even in case of *tert*-butylation, gives preferentially the 1-isomers, which, however, isomerize very fast to the thermodynamically preferred 2-alkylnaphthalenes. Great care is therefore needed to evaluate kinetic regioselectivities.

Selective mono- or dicyclohexylation of naphthalene with cyclohexene (and cyclohexyl bromide) was achieved over HY zeolites.[170] Monocyclohexylnaphthalenes are formed with 99% selectivity at 80°C. The amount of dicyclohexylnaphthalenes increases dramatically with increasing temperature. Due to zeolite shape selectivity, mainly the β,β-disubstituted derivatives (2,6- and 2,7-dicyclohexylnaphthalene) are formed at 200°C.

Reaction conditions in alkylation with alkenes exert a strong effect on product distribution[151] [Eq. (5.37)][171] similar to that in alkylation with alkyl halides:

$$
\text{(5.37)}
$$

		CH3CCH2CH3	CH3CHCHCH3
−40°C, 50 min		100		36% yield
21°C, 34 min		55	45	42% yield

Linear alkenes yield all possible *sec*-alkyl-substituted arenes when strong acid catalysts (AlCl₃, AlBr₃, HF−BF₃) are used, allowing isomerization to also occur.[172,173] When branched alkenes are used as alkylating agents, the corresponding *tert*-alkyl-substituted products can be obtained at low temperature using hydrogen fluoride or sulfuric acid as catalyst under limited or nonisomerizing conditions. Isomerizing conditions at more elevated temperatures cause rearrangement to yield alkylated product with the aromatic ring attached to a secondary carbon atom.

Positional isomerization occurs similarly as during alkylation with alkyl halides. HF and H_2SO_4, which are weaker catalysts than the conjugate Friedel–Crafts acids, however, do not bring about ready positional isomerization of the alkylated products. Rearrangement in the side chain always takes place before the attachment of the substituent to the aromatic ring when these catalysts are used.

An acidic cesium salt of 12-tungstophosphoric acid ($Cs_{2.5}H_{0.5}PW_{12}O_{40}$) was found to be remarkably more active then the parent acid, zeolites, Nafion-H, sulfated zirconia, H_2SO_4, and HCl−AlCl₃ in the liquid-phase alkylation of *m*-xylene and 1,3,5-trimethylbenzene with cyclohexene at 100°C.[174] Although the acid strength of the salt determined by different methods (Hammett indicators, temperature-programmed desorption of ammonia) was lower than that of sulfated zirconia, it is presumed that the high activity is brought about by the large number of surface acid sites.

Phenylalkenes also undergo intramolecular alkylation under Friedel–Crafts conditions to yield five- or six-membered cyclic products (indan or tetralin derivatives, respectively). This cyclialkylation[145,146,175] may actually also be considered as an isomerization reaction.

Formation of indan derivatives is usually expected only when benzylic or tertiary carbocations are involved in cyclialkylation.[146] Tetralin formation through secondary and tertiary carbocations is always favored over indan formation, and it may even take place through an assisted primary carbocation:[176]

$$(5.38)$$

Steric factors play a determining role in directing the course of ring closure. They can be invoked to rationalize cyclialkylation of the **30** arylalkene[177] [Eq. (5.39)]. The less severe 1,4 interaction between the benzyl and methyl groups during cyclization contrasted with the 1,3 interaction between the ethyl and phenyl groups ensures the predominant formation of **31**:

$$(5.39)$$

Alkylation with Alkanes. Alkylation of aromatic hydrocarbons with alkanes, although possible, is more difficult than with other alkylating agents (alkyl halides, alkenes, alcohols, etc.).[178] This is due to the unfavorable thermodynamics of the reaction in which hydrogen must be oxidatively removed.

Ipatieff and coworkers observed first that $AlCl_3$ catalyzes the "destructive" alkylation of aromatics with branched alkanes.[179] For example, *tert*-butylbenzene (35%), *p*-di-*tert*-butylbenzene (25%), and considerable isobutane are the main products when benzene is reacted at 20–50°C with 2,2,4-trimethylpentane. Toluene and biphenyl are alkylated at 100°C in a similar way.[180] Straight-chain alkanes required more severe reaction conditions. *n*-Pentane reacted at 175°C to yield 8% propylbenzene, 25% ethylbenzene, and 20% toluene.[181] Phosphoric acid afforded similar products at higher temperature (450°C).[182] Pentasil zeolites and dealuminated pentasils have been found to promote alkylation of benzene with C_2–C_4 alkanes to form toluene and xylenes.[183,184]

Alkylation without cracking was shown by Schmerling et al. to take place when saturated hydrocarbons react with benzene in the presence of $AlCl_3$ and a "sacrificial" alkyl halide to form monoalkylbenzenes.[185–187] For example, a mixture of 1,1-dibromoethane, 2,3-dimethylbutane, benzene, and $AlCl_3$ provides ethylbenzene (34%) and isomeric hexylbenzenes (21%), mainly 3-phenyl-2,2-dimethylbutane.[187]

Detailed studies with isoalkane–alkyl chloride mixtures showed that the relative ability of the isoalkane and the halide in participating in alkylation of benzene depends on the reactivity of the alkyl chloride.[188] Alkylation with isopentane in the presence of *tert*-butyl chloride gave mainly isomeric pentylbenzenes. Alkylation with methylcyclopentane was shown to be the principal reaction when a mixture of benzyl chloride, methylcyclopentane, and benzene was reacted in the presence of $AlCl_3$. In these cases, alkylation with the alkane took place [Eq. (5.43)]. In contrast, mainly isopropylation occurred when isopropyl chloride was applied [Eq. (5.41)]. These observations indicate that hydrogen transfer from the isoalkane to the carbocation [Eq. (5.42)] formed through the interaction of the alkyl chloride with $AlCl_3$ [Eq. (5.40)] takes place with the derived carbocation alkylating the aromatics; consequently, alkylation with the alkane is the dominant reaction when stable (tertiary, benzyl) carbocations are involved:

$$R{-}Cl \;+\; AlCl_3 \;\rightleftharpoons\; R^+ \;+\; AlCl_4^- \tag{5.40}$$

$$(5.41)$$

$$isoR'^+ \;+\; R{-}H \tag{5.42}$$

$$(5.43)$$

Alkylation of aromatics occurs when a mixture of aromatic and saturated hydrocarbons are treated at room temperature with a mixture of a Lewis acid ($AlCl_3$, $FeCl_3$, $ZnCl_2$) and a higher-valent metal halide as oxidant ($CuCl_2$, $FeCl_3$).[189] The reaction of isopentane with benzene, for example, resulted in the formation of 40% of isomeric pentylbenzenes (neopentylbenzene, 2-methyl-2-phenylbutane, 2-methyl-3-phenylbutane) together with smaller amounts of isopropylbenzene, ethylbenzene, and 1,1-diphenylethane. *p*-Polyphenyl was also detected (12%). Similar results were obtained in alkylation of benzene with ethylbenzene (40–45% yield of 1,1-diphenylethane), whereas only a trace of cyclohexylbenzene was found in the reaction with cyclohexane. When, however, the latter reaction was carried out in the presence of isopentane, cyclohexylbenzene was isolated in 20% yield. Isopentane promoted alkylation of benzene with other saturated hydrocarbons as well. Increased yields were achieved when $CuCl_2$ was regenerated in situ by carrying out the alkylation reaction in the presence of oxygen (air).

Oxidative removal of hydrogen by cupric chloride was suggested by Schmerling to explain the observations:[189]

$$CH_3CHCH_2CH_3 \ (CH_3) \ + \ CuCl_2 \cdot AlCl_3 \ \longrightarrow \ CH_3\overset{+}{C}CH_2CH_3 \ (CH_3) \ + \ HCl \ + \ CuCl \ + \ AlCl_3 \tag{5.44}$$

$$isoC_5H_{11}$$

$$C_6H_6 \ + \ CH_3\overset{+}{C}CH_2CH_3 \ (CH_3) \ \xrightarrow{-H^+} \ C_6H_5\text{–}isoC_5H_{11} \tag{5.45}$$

$$H^+ \ + \ CuCl_2 \ \longrightarrow \ HCl \ + \ CuCl \tag{5.46}$$

A related explanation suggested by Olah[190] involves the protonolysis of isopentane (with protic impurities acting as coacids for $AlCl_3$) forming the corresponding *tert*-amyl cation. $CuCl_2$ oxidizes the hydrogen formed in the protonolysis, thus making the reaction thermodynamically more feasible. The HCl formed in the reaction could increase the HCl–$AlCl_3$ conjugate Friedel–Crafts acid concentration, thus facilitating subsequent protolytic activation.

Superacids were shown to have the ability to effect the protolytic ionization of σ bonds to form carbocations even in the presence of benzene.[190] The formed carbocations then alkylate benzene to form alkylbenzenes. The alkylation reaction of benzene with C_1–C_5 alkanes (methane, ethane, propane, butane, isobutane, isopentane) are accompanied by the usual acid-catalyzed side reactions (isomerization, disproportionation). Oxidative removal of hydrogen by SbF_5 is the driving force of the reaction:

$$R_3CH \ \underset{HF-SbF_5}{\rightleftharpoons} \ \left[R_3C\text{-}\text{-}\overset{+}{\underset{H}{\overset{H}{\diagdown}}} \right]^+ \ \underset{-H_2}{\rightleftharpoons} \ R_3C^+ \ \underset{C_6H_6}{\rightleftharpoons} \ \mathbf{32} \ \underset{-H^+}{\rightleftharpoons} \ \mathbf{33} \tag{5.47}$$

A comparison of C—C and C—H bond reactivities in protolytic reactions indicated that under the conditions of alkylation C—H cleavage is much more pronounced (in the reaction of ethane, the relative ratio of C—C : C—H cleavages is 9 : 1 in the absence of benzene and 1 : 10 in the presence of benzene). This indicates that overall alkylation can also involve the attack of protonated benzene (benzenium ion) on the C—H bond of the alkane [Eq. (5.48)] or interaction of benzene with the five-coordinate alkonium ion prior to its cleavage to the trivalent cation:

$$C_6H_6 \ \underset{HF-SbF_5}{\rightleftharpoons} \ [\text{benzenium ion}] \ \xrightarrow{R_3CH} \ \underset{-H_2}{\longrightarrow} \ \mathbf{32} \ \underset{-H^+}{\rightleftharpoons} \ \mathbf{33} \tag{5.48}$$

When cycloalkanes (cyclopentane, cyclohexane) alkylate benzene, cycloalkylbenzenes, as well as bicyclic compounds (indan and tetralin derivatives) and products of destructive alkylation, are formed.[191,192] Cyclohexane reacts with the highest selectivity in the presence of HF–SbF$_5$ to yield 79% cyclohexylbenzene and 20–21% isomeric methylcyclopentylbenzenes.[191]

FeCl$_3$-doped K10 montmorillonite was used to promote adamantylation of benzene to form 1-phenyl and 1,3-diphenyladamantanes in ratios depending on the amount of catalyst:[193]

$$(5.49)$$

In adamantylation of toluene a close to 2 : 1 *meta*- and *para*-substituted isomer distribution was observed.

Alkylation with Alcohols. Alkylation of aromatic hydrocarbons with alcohols[86,151,153] in the presence of protic catalysts yields the products expected from typical electrophilic alkylation:

$$(5.50)$$

With Friedel–Crafts halides (usually AlCl$_3$ and BF$_3$) it is necessary to use equimolar or excess catalyst when alcohols are the alkylating agents. With primary alcohols usually 2 mol of catalyst per mole of alcohol must be used. Complexing the alcohol, as well as binding the water formed in the reaction explains the experimental findings. With secondary and tertiary alcohols, lesser amounts of catalyst are needed. The reactivity of different alcohols follows the order methyl < primary < secondary < tertiary, allyl, benzyl. The ease of carbocation formation according to Eq. (5.51) is most probably responsible for this reactivity order:

$$ROH \; + \; AlCl_3 \; \xrightarrow[-HCl]{} \; ROAlCl_2 \; \longrightarrow \; R^+ \; + \; {}^-OAlCl_2 \qquad (5.51)$$

Protonated (or complexed) alcohols themselves, however, can also be directly displaced by the aromatics, which is the case with primary alcohols and methanol.

The effect of catalysts on product distribution is shown in case of alkylation of benzene with neopentyl alcohol:[194,195]

$$(5.52)$$

Further studies revealed, however, that unrearranged **34** is not a primary product[196,197] but formed in a two-step rearrangement according to the following equation:

$$(5.53)$$

The first step involves hydride abstraction and a phenonium ion intermediate.[198] The low yields are accounted for by the ready dealkylation of *sec-* and *tert*-pentyl-benzenes under the reaction conditions (see discussion below).

An interesting example of alkylation with alcohols is the selective transformation of isomeric phenylpropanols under appropriate reaction conditions:[199]

$$(5.54)$$

Acidic clay catalysts can also be used in alkylation with alcohols.[98] The main advantages of these catalysts are the reduced amount necessary to carry out alkylation compared with conventional Friedel–Crafts halides, possible regeneration, and good yields. Natural montmorillonite (K10 clay) doped with transition metal cations was shown to be an effective catalyst:[200]

$$(5.55)$$

Alkylation of toluene with methanol, a process of practical importance, was investigated over several acidic clay catalysts.[98] Cation-exchanged synthetic saponites[201] and fluor–terasilicic mica modified by La^{3+} ions[202] were found to exhibit increased *para* selectivity compared with H-ZSM-5.

Substantial progress has been made to carry out alkylation in the gas phase over solid superacid catalysts. Nafion-H, a perfluorinated resin–sulfonic acid, for example, catalyzes the methylation of benzene and methylbenzenes with methyl alcohol under relatively mild conditions. The reaction shows low substrate selectivity.[203]

There is also much interest to apply shape-selective zeolites in the selective alkylation of aromatics.[99] H-ZSM-5, after pretreatment to properly adjust acidity and modify pore and channel characteristics, was found to be an excellent catalyst giving high *para* selectivity. In the alkylation of toluene with methanol the production of p-xylene was achieved with over 90% selectivity when H-ZSM-5 was modified by impregnation with phosphorus or boron compounds.[204] *p*-Ethyltoluene can also be obtained through the alkylation of toluene with ethanol (see Section 5.5.5). Pentasil zeolites modified with oxides, coking, or steaming were shown to exhibit even higher *para* selectivities in the ethylation of ethylbenzene with ethanol than were the parent zeolites.[205] Shape-selective zeolites were also found to provide products with high selectivity in the alkylation of polycyclic aromatics. Monomethylnaphthalenes were obtained with 90% selectivity in 20% yield by alkylation of naphthalene with methanol at 460°C over H-ZSM-5.[100]

Transalkylation and Dealkylation. In addition to isomerizations (side-chain rearrangement and positional isomerization), transalkylation (disproportionation) [Eq. (5.56)] and dealkylation [Eq. (5.57)] are side reactions during Friedel–Crafts alkylation; however, they can be brought about as significant selective hydrocarbon transformations under appropriate conditions. Transalkylation (disproportionation) is of great practical importance in the manufacture of benzene and xylenes (see Section 5.5.4):

$$(5.56)$$

$$(5.57)$$

Transalkylation[86,206,207] can be effected by Lewis and Brønsted acids and zeolites. Methylbenzenes exhibit exceptional behavior since they undergo isomerization without appreciable transalkylation in the presence of conventional Friedel–Crafts catalysts. Penta- and hexamethylbenzene do readily transmethylate benzene and toluene over Nafion-H.[208] More characteristically, however, zeolites may be used for transalkylation. Since benzene and xylenes are generally more valuable than toluene, there is substantial significance in the production of benzene and xylenes from toluene or mixtures of toluene and trimethylbenzenes.[86,207,209–211]

Disproportionation (transalkylation) and positional isomerization usually take place simultaneously when either linear or branched alkylbenzenes are treated with conventional Friedel–Crafts catalysts[212–215] or with Nafion-H.[216] The reactivity of alkyl groups to participate in transalkylation increases in the order ethyl, propyl < isopropyl < *tert*-butyl.[117,207,217]

n-Alkylbenzenes undergo transalkylation without concomitant side-chain rearrangement.[216,218,219] This led to the suggestion that transalkylation occurs through 1,1-diarylalkane intermediates.[219,220] Proof was provided by Streitwieser and Reif,[221] who observed that in the transalkylation of [α-^{2}H-1-^{14}C]-ethylbenzene catalyzed by HBr–GaBr$_3$, the rates of deuterium scrambling, and racemization and the loss of radioactivity were almost the same. This requires a mechanism in which all three processes have the same rate-determining step, that is, the formation of the 1-phenylethyl cation. In the transalkylation path, carbocation **35** reacts with a second molecule of alkylbenzene to form the 1,1-diarylalkane intermediate:

$$(5.58)$$

This then readily undergoes cleavage to produce benzene and a dialkylbenzene. The initial R$^+$ cation initiating the reaction might arise from some impurity present in the reaction mixture. Consistent with this mechanism is the observed very low reactivity of methylbenzenes due to the necessary involvement of the primary benzyl cations. At the same time 1,1-diarylalkanes undergo cleavage with great ease.

The transalkylation of *tert*-alkylbenzenes follows a different route, since they have no abstractable benzylic hydrogen. They were shown to transalkylate by a dealkylation–transalkylation mechanism with the involvement of free *tert*-alkyl cations.[213,222] The exceptional ability of *tert*-alkyl groups to undergo transalkylation led to the extensive utilization of these groups, especially the *tert*-butyl group, as positional protective groups in organic synthesis pioneered by Tashiro.[223]

Nafion-H was also found to be a very convenient, highly active catalyst for the trans-*tert*-butylation of *tert*-butylalkylbenzenes in excellent yields:[224]

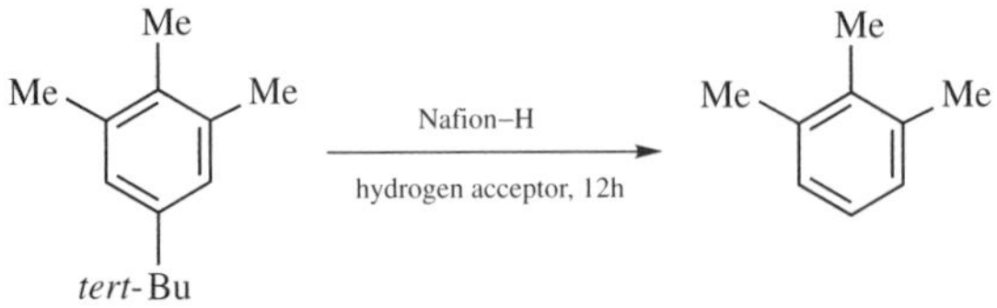

(5.59)

toluene, reflux   100%
biphenyl, 130–135°C   80%

Dealkylation of alkylbenzenes[225] [Eq. (5.57)] was first studied systematically by Ipatieff and coworkers.[226] Alkylbenzenes heated with $AlCl_3$ in the presence of a hydrogen donor (cyclohexane, decalin) yielded benzene and the corresponding alkanes with the observed reactivity *tert*-butyl > *sec*-butyl > isopropyl. Primary alkylbenzenes did not undergo dealkylation. Later studies confirmed these observations[227] and led to the finding that dealkylation is a facile reaction even without added hydrogen donors, suggesting that the side chains themselves may be a sacrificial hydride source. Reaction conditions to induce dealkylation are sufficiently severe to bring about other transformations usually occurring under Friedel–Crafts alkylation conditions. Dealkylation of *sec*-butylbenzene, for example, produced significant amounts of isobutane early in the reaction, which indicates that fast isomerization of the side chain precedes dealkylation.[227]

The reactivity of aromatic side chains to undergo dealkylation is in line with the stability of the corresponding carbocations. This indicates the possible involvement of carbocations in dealkylation, which was proved to be the case. The intermediacy of the *tert*-butyl cation in superacid solution was shown by direct spectroscopic observation.[228,229] Additional proof was provided by trapping the *tert*-butyl cation with carbon monoxide during dealkylation:[230]

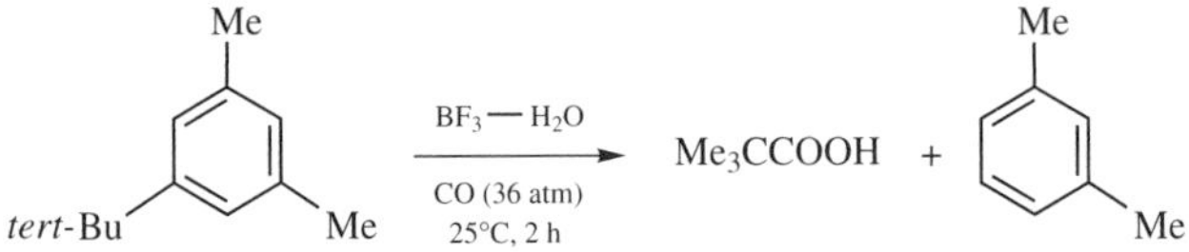

(5.60)

89.5%   98.4%
93% conversion

5.2. BASE-CATALYZED ALKYLATION

In contrast to acid-catalyzed alkylation of arenes, where alkylation of the aromatic nucleus occurs, the base-catalyzed reaction of alkylarenes leads to alkylation of the side chain.[231–234] Alkenes (straight-chain 1-alkenes, isobutylene), conjugated dienes (1,3-butadiene, isoprene), and styrene and its derivatives can react with alkyl-arenes possessing a benzylic hydrogen. In contrast, arylalkanes that do not have reactive benzylic hydrogens yield only ring-alkylated products in very low yields.

Sodium and potassium are the two most frequently used alkali metals in side-chain alkylation. Sodium usually requires a promoter (*o*-chlorotoluene, anthracene) to form an organosodium intermediate that is the true catalyst of the reaction. A temperature range of 150–250°C is usually required for alkylation with monoolefins, whereas dienes and styrenes are reactive at lower temperatures.

The reaction of alkylbenzenes with ethylene in the presence of sodium occurs with successive replacement of one or two benzylic hydrogens to produce mono- and diethylated compounds.[235] Propylene reacts to form isomeric products with the preponderance of the branched isomer[236] [Eq. (5.61)]. When potassium is used as catalyst, indan derivatives can also be formed[237] [Eq. (5.62)]:

$$ (5.61) $$

$$ (5.62) $$

Alkylnaphthalenes are more reactive than alkylbenzenes and react at temperatures as low as 90°C. Sodium is a selective catalyst for side-chain monoalkylation. More elevated temperatures and prolonged reaction times allow the replacement in ethylation of all α-hydrogen atoms of alkylnaphthalenes with high selectivity.[238]

1,3-Dienes react quite readily with alkylbenzenes to form monoalkenylbenzenes under controlled experimental conditions.[239–241] Sodium and potassium deposited on calcium oxide were found to be very suitable catalysts for these alkenylation reactions.[240] Naphthalene–sodium in tetrahydrofuran is a very effective catalyst even at room temperature.[242]

Styrene reacts with arylalkanes to yield mono- and diadducts with the concomitant formation of polymeric byproducts.[243] α-Methylstyrene and β-methylstyrene, in turn, give monoadducts in high yields.[244,245]

The nature of the catalyst and certain experimental observations, such as product distribution [Eq. (5.61)], indicate a carbanionic reaction mechanism. A benzylic carbanion formed through proton abstraction by an organoalkali metal compound

[Eq. (5.63)] initiates the transformation.[235] The next step, the attack of anion **36** on the alkene [Eq. (5.64)], is the rate-determining step since it results in the transformation of a resonance-stabilized anion (**36**) into anions (**37** and **38**) that are not stabilized. Transmetalation of **37** and **38** forms the end products and restores the benzylic carbanion [Eqs. (5.65) and (5.66)]:

$$\text{PhCH}_2\text{R} + \text{B}^- \text{M}^+ \; \rightleftharpoons \; \underset{\textbf{36}}{\text{PhC}\overset{..}{\text{H}}\text{R}^- \text{M}^+} + \text{BH} \tag{5.63}$$

$$\textbf{36} + \text{R}'\text{CH}{=}\text{CH}_2 \; \rightleftharpoons \; \underset{\textbf{37}}{\overset{\text{R} \quad \text{R}'}{\text{PhCH}{-}\text{CHCH}_2}} \text{M}^+ \; + \; \underset{\textbf{38}}{\overset{\text{R}}{\text{PhCH}{-}\text{CH}_2\overset{..}{\text{C}}\text{HR}'}} \text{M}^+ \tag{5.64}$$

$$\underset{\textbf{37}}{} \xrightarrow{\text{PhCH}_2\text{R}} \underset{}{\overset{\text{R} \quad \text{R}'}{\text{PhCHCHCH}_3}} + \textbf{36} \tag{5.65}$$

$$\underset{\textbf{38}}{} \xrightarrow{\text{PhCH}_2\text{R}} \overset{\text{R}}{\text{PhCHCH}_2\text{CH}_2\text{R}'} + \textbf{36} \tag{5.66}$$

The higher stability of primary anion **37** as compared to secondary anion **38** explains the predominant formation of branched isomers. The high reactivity of conjugated dienes and styrenes compared with that of monoolefins is accounted for by the formation of new resonance-stabilized anions (**39** and **40**). Base-catalyzed alkylation with conjugated dienes may be accompanied by telomerization. The reason for this is that the addition of a second molecule of diene to the **39** monoadduct anion competes with transmetallation, especially at lower temperatures.[233]

$$\underset{\textbf{39}}{\text{PhCH}_2{-}\text{CH}_2{-}\text{CH}\overset{..}{=}\overset{..}{\text{C}}\text{H}\overset{..}{=}\text{CH}_2} \qquad \underset{\textbf{40}}{\text{PhCH}_2{-}\text{CH}_2{-}\overset{..}{\text{C}}\text{H}{-}\text{Ph}}$$

5.3. ALKYLATION THROUGH ORGANOMETALLICS

Hydrocarbons with suitably acidic (allylic, benzylic, propargylic, acetylenic) hydrogens can be transformed to organometallic derivatives, which then can react with alkylating agents to yield alkylated products.

Alkali metal alkyls, particularly *n*-butyl lithium, are the most frequently used reagents to form metallated intermediates.[246,247] In certain cases (di- and triphenylmethane, acetylene and 1-alkynes, cyclopentadiene) alkali metals can be directly applied. Grignard reagents are used to form magnesium acetylides and cyclopentadienyl complexes.[248] Organolithium compounds with a bulky alkoxide, most notably *n*-BuLi–*tert*-BuOK in THF/hexane mixture, known as the *Lochmann–Schlosser reagent* or LICKOR superbase, are more active and versatile reagents.[249–252]

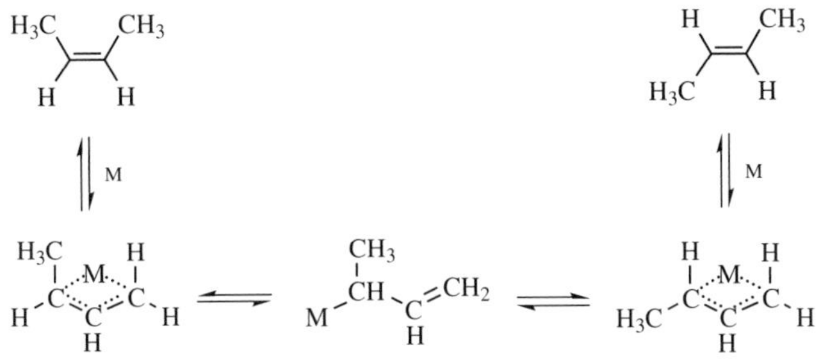

41   **42**

Scheme 5.6

Trimethylsilylmethyl potassium (Me_3SiCH_2K) is likewise a very good reagent in allylic and benzylic metalations.[253] Strongly coordinating solvents, such as THF and *N,N,N′N′*-tetramethylethylenediamine (TMEDA) make these reagents even more active. Alkyl halides are the most frequently used alkylating agents.

The reactivity of different hydrogens to participate in metalation decreases in the order primary > secondary > tertiary. The subsequent alkylation step, however, can be controlled to give isomeric compounds with high selectivity. When Grignard reagents are formed they usually undergo alkylation with the shift of the double bond [Eq. (5.67)], whereas alkali metal compounds in THF give predominantly the unrearranged product[251,254] [Eq. (5.68)]:

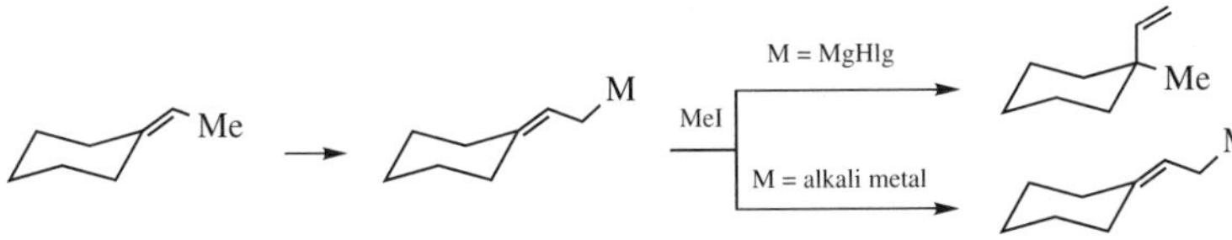

$$ (5.67) $$

$$ (5.68) $$

Stereoselective alkylations can be carried out by using appropriate metals and reaction conditions. Isomeric 2-alkenes react to give stereoisomeric metalated products, which can retain their original configuration. The two intermediates, however, are in equilibrium. In the case of the isomeric 2-butenes, lithium and magnesium reagents give a roughly 1 : 1 mixture of the corresponding *endo* (**41**) and *exo* (**42**) derivatives[251,254] (Scheme 5.6). Sodium, potassium, and cesium derivatives, in turn, are increasingly more stable in the sterically hindered *endo* form attributed to the formation of intramolecular hydrogen bond.[255]

Taking advantage of this favorable equilibrium, 2-methyl-2-heptene was transformed to (Z)-3-methyl-3-octene with good selectivity:[256]

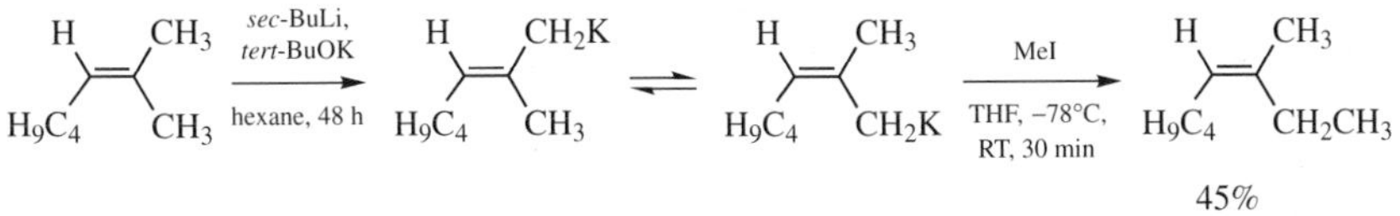

45%

$$ (5.69) $$

Alkylation of isoprene at the methyl group was achieved through the reaction of the corresponding metal derivative (*tert*-BuOK–lithium tetramethylpiperidine, THF) with 1-bromoheptane (60% yield).[257] Regioselective monoalkylation of isopropenylacetylene can be performed in high yields through the dilithium derivative:[258]

$$HC\equiv CC=CH_2 \xrightarrow[\text{ii. LiBr, THF, 20°C}]{\substack{\text{i. } n\text{-BuLi, } tert\text{-BuOK,} \\ \text{THF, hexane, } -80°C}} LiC\equiv CC=CH_2 \xrightarrow[\text{ii. H}_2\text{O}]{\text{i. RBr}} HC\equiv CC=CH_2 \quad (5.70)$$

(with CH$_3$, CH$_2$Li, CH$_2$R substituents respectively)

R = n-C$_5$H$_{11}$, n-C$_6$H$_{13}$, cyclohexyl   70–83% yield

Alkylation of aromatics at benzylic positions can also be performed. Me$_3$SiCH$_2$K is a highly selective reagent for this process:[253]

$$\xrightarrow[\text{ii. MeI}]{\substack{\text{i. Me}_3\text{SiCH}_2\text{K,} \\ \text{THF, } -48°C}} \quad (5.71)$$

46% yield
99% selectivity

Dialkyl-substituted aromatics can be alkylated at both alkyl groups under appropriate conditions. *m*-Xylene and 1,3-dimethylnaphthalene were transformed into the corresponding α,α′-dianions with *n*-pentylsodium in the presence of catalytic amount of *N*,*N*,*N*′,*N*′-tetramethylethylenediamine in quantitative yield.[259] The dianions were subsequently alkylated with methyl iodide to yield *m*-diethylbenzene and 1,3-diethylnaphthalene, respectively. *p*-Xylene and 1,4-dimethylnaphthalene, however, could not be alkylated. This was explained to result from the very low reactivity of the corresponding monoanions to participate in a second metallation due to the accumulation of the negative charge at the *para* position of the ring.[260] Even *p*-xylene could be dialkylated, however, when metallation was carried out with the *n*-BuLi–*tert*-BuOK reagent followed by reaction with dimethyl sulfate.[259] Unsymmetric dialkylation was achieved by stepwise alkylation.[261] *p*-Xylene was first monomethylated to yield *p*-ethyltoluene, which, in turn, underwent ethylation to give *p*-propylethylbenzene in 74% combined yield. The reaction of dianions with α,ω-dihalides afforded the synthesis of cyclophanes in low yields.[261]

In addition to the abovementioned alkylations at activated positions, powerful metalating agents permit the removal of protons from olefinic carbons or aromatic ring carbons. As a result, alkylation at vinylic position or at the aromatic ring is also possible. *n*-BuK[262] or the LICKOR superbase when used in pentane[253,263] are nonselective reagents in the alkylation of aromatics, producing a mixture of products methylated at both the benzylic position and at the ring. The combinations of different alkyllithiums and potassium alcoholates proved to be somewhat more selective.[263] However, when the LICKOR reagent is applied in THF, subsequent methylation gives ring-methylated products with high selectivity.[253,263] Even better (95%) selectivities in ring methylation were achieved with *n*-pentylsodium.[253]

Selective ring monoalkylation of phenylacetylene[264,265] was achieved through dimetal compound **43** [Eq. (5.72)], which does not lead to alkylation of the acetylenic carbon even when excess alkylating agents are used, provided the reaction is carried out below 0°C:

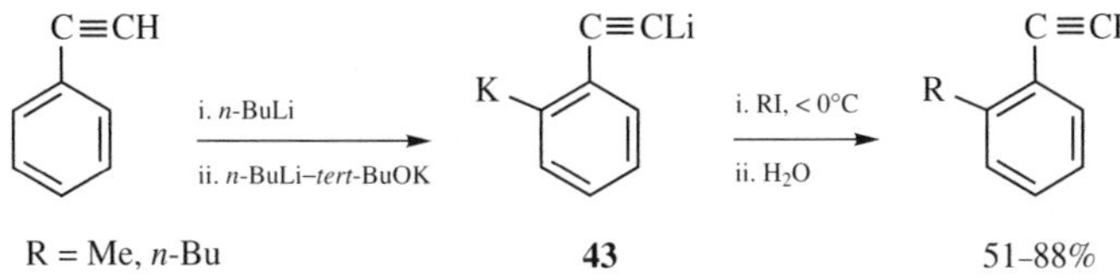

$$R = Me, n\text{-}Bu \qquad \textbf{43} \qquad 51\text{--}88\%$$

(5.72)

o-Allylation was carried out via the corresponding Grignard compound.[265]

The easy metallation of the acetylenic hydrogen and the subsequent reaction of the formed alkali metal and alkylmagnesium acetylides allow the alkylation of alkynes at the triple bond.[266] Such alkylations are valuable methods for the synthesis of alkynes, although there are limitations. Only primary alkyl halides unbranched in the β position give satisfactory yields when alkali metal acetylides are used since other halides undergo elimination under the basic conditions used. Magnesium acetylides, in turn, react only with activated (allylic, benzylic) halides. When an excess of a strong base is used, 1-alkynes can be transformed to the corresponding 1,3-dianion. This can be either monoalkylated at the propargylic position[267,268] or dialkylated at both the propargylic and acetylenic carbons[268] (Scheme 5.7).

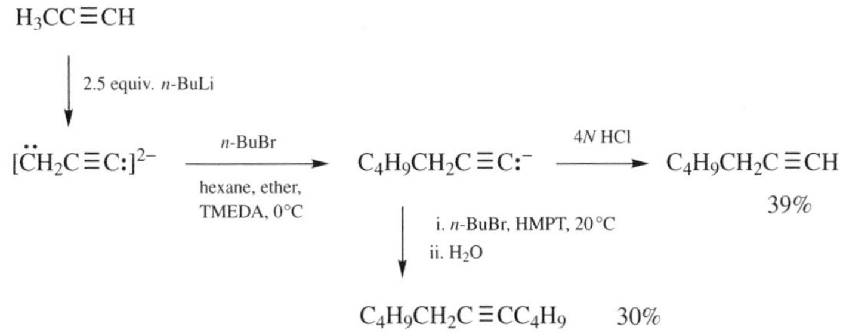

Scheme 5.7

5.4. MISCELLANEOUS ALKYLATIONS

A unique example of alkane–alkene reaction is the homologation of olefins with methane in a stepwise manner over transition-metal catalysts.[269] First methane is adsorbed dissociatively on rhodium or cobalt at 327–527°C; then an alkene (ethylene or propylene) is coadsorbed at lower temperature (27–127°C). On hydrogenation, carbon–carbon bond formation occurs. The mechanism appears to be related to that occurring in the Fischer–Tropsch reaction.

There is increasing interest in studying the alkylation of methylbenzenes in the side chain with methyl alcohol over zeolite catalysts.[270] These reactions may lead to new nontraditional technologies in the synthesis of styrene from toluene and that of *p*-methylstyrene from *p*-xylene. A one-step route from toluene or *p*-xylene to the corresponding styrenes would be of great practical importance compared with the presently practiced two-step syntheses (alkylation followed by dehydrogenation).

Side-chain alkylation with methyl alcohol can be catalyzed by X and Y zeolites exchanged with alkali cations. Observations by Itoh et al. indicate that both acidic and basic sites are necessary to achieve alkylation in the side chain.[271] A mechanism[272] involving formaldehyde as the alkylating agent formed by the dehydrogenation of methyl alcohol appears to be operative. Among other observations, it is supported by the finding that both methyl alcohol and formaldehyde as alkylating agents give very similar results. The hydrogen generated may affect selectivities by hydrogenating the styrene formed.

Rb^+- and Cs^+-impregnated X zeolites were found to exhibit the highest activity and selectivity in these transformations.[272–275] A CsX zeolite treated with boric acid, for example, gave better than 50% overall selectivity in the formation of styrene and ethylbenzene (410°C, 60% conversion).[275] Treatment of these catalysts with copper or silver nitrate resulted in further improvements in catalyst performance.[276] The promoting role of these metals was suggested to be their involvement in dehydrogenation of methyl alcohol.

The favored side-chain alkylation over ring alkylation is interpreted by the geometric constraints exerted by the large rubidium and cesium cations within the zeolite supercage.[274,275] These result in steric restriction of the formation of transition states that would lead to ring alkylation products.

5.5. PRACTICAL APPLICATIONS

5.5.1. Isoalkane–Alkene Alkylation

The development of alkylation processes during World War II was spurred by the need for high-octane aviation gasoline.[277–280] Part of the success of the Royal Air Force in the Battle of Britain is credited to the performance of high-octane fuels sent to England in the summer of 1940 by the United States, which allowed the British fighter planes to outperform their German opponents. The first commercial plants came into production in 1938 (H_2SO_4 catalyst) and 1942 (HF catalyst). Commercial alkylation technologies are still based on using sulfuric acid or hydrogen fluoride catalysts.[7,281–284] The main advantages of protic acids over other Friedel–Crafts catalysts are more easy handling, fewer side reactions, and longer catalyst lifetime. Over the years numerous technologies applying different reactors have been developed.[7,277,284–289] Because of their rapidly declining activity, zeolites have not reached commercial application in alkane–alkene alkylation.[7]

Because of regulations concerning transportation fuel quality (lowering olefin and aromatics content, lowering volatility), alkylation (in addition to isomerization and oxygenate production) is the most important process in the manufacture of reformulated gasoline.[288]

Alkylation processes usually combine isobutane with an alkene or with mixed alkene streams (C_3–C_5 olefins from FCC units). The best octane ratings are attained when isobutane is alkylated with butylenes. Alkylation of higher-molecular-weight hydrocarbons (>C_5) is less economic because of increased probability of side reactions. Phillips developed a technology that combines its *triolefin process* (metathesis of propylene to produce ethylene and 2-butenes) with alkylation since 2-butenes yield better alkylate than propylene.[290] Since ethylene cannot be readily used in protic acid-catalyzed alkylations, a process employing $AlCl_3$ promoted by water was also developed.[291]

In all alkylation processes high isobutane : olefin ratios are applied. It is necessary to achieve sufficiently high isobutane concentration in the acid phase to suppress side reactions and to minimize polymerization of alkenes. The two processes employing H_2SO_4 or HF catalysts operate at different temperatures under the necessary pressure to keep the reactants in the liquid phase. Both acids are diluted during alkylation due to the formation of polyunsaturated hydrocarbons. As a result, catalyst activity decreases and polymerization increases markedly. Catalyst activity must be maintained by adding fresh acid to the system. Optimal catalyst concentrations are 88–95% for sulfuric acid and 80–95% for hydrogen fluoride. Of the spent catalysts, only hydrogen fluoride is regenerated by distillation.

There are also other notable differences between the two alkylation processes. The quality of alkylates in terms of octane number formed by using sulfuric acid from the isomeric butenes follows the order 2-butene (96–98) > 1-butene (95.7–97.7) > isobutylene (91–92).[292] In the process catalyzed by HF the order is 2-butene (96–98) > isobutylene (93–95) > 1-butene (87–89). Isobutylene gives a lower-quality alkylate in alkylation with H_2SO_4 due to substantial oligomerization.[293] C_{12}–C_{20} carbocations thus formed undergo fragmentation to produce C_5–C_9 carbocations and alkenes. Subsequent hydride transfer results in the formation of C_5–C_9 isoalkanes, mostly dimethylhexanes. Such processes have less significance with *n*-butenes that oligomerize less readily.

The best product quality in the process catalyzed by sulfuric acid can be achieved by running the alkylation at the temperature range 0–10°C. Since solubility of isobutane in the catalyst phase is very low, effective mixing of the two-phase system is necessary. Isobutane : olefin ratios of 5–8 : 1 and contact times of 20–30 min are usually applied. Since the rate of double-bond migration under such conditions is comparable to that of alkylation, high-quality alkylate may be manufactured from 1-butene. Substantial oligomerization of isobutylene giving acid-soluble oils and sludges and the high cost of refrigeration are the main drawbacks of the sulfuric acid–catalyzed process. High acid consumption is usually experienced too, when alkenes other than butenes are used.

Hydrocarbon solubilities in HF are more favorable than in H_2SO_4, which allows the use of shorter contact times and higher reaction temperatures (10–40°C), but require higher isobutane : alkene ratios (10–15 : 1). Alkylation in the presence of hydrogen fluoride is fast compared with double-bond migration, resulting in the formation of higher amounts of undesirable dimethylhexanes from 1-butene. Alkylation of isobutane with 2-butenes, therefore, gives the best results in this case. HF technologies do not tolerate much water since it deactivates the catalyst. Hence, hydrocarbon streams must be dried before use.

It is advantageous to pretreat butene feeds before alkylation.[294–298] 1,3-Butadiene is usually hydrogenated (to butenes or butane) since it causes increased acid consumption. The additional benefit of this process is that under hydrogenation conditions alkene isomerization (hydroisomerization) takes place, too. Isomerization, or the transformation of 1-butene to 2-butenes, is really attractive for HF alkylation since 2-butenes give better alkylate (higher octane number) in HF-catalyzed alkylation. Excessive 1,3-butadiene conversion, therefore, ensuring 70–80% isomerization, is carried out for HF alkylation. In contrast, approximately 20% isomerization is required at lower butadiene conversion for alkylation with H_2SO_4.

Further improvements in alkylation can be achieved when an MTBE unit (acid-catalyzed addition of methanol to isobutylene to form *tert*-butyl methyl ether) is added before the alkylation unit.[299] The MTBE unit removes the lowest-octane-producing isomer, isobutylene, from the C_4 stream (and produces a very-high-octane number component, MTBE). The H_2SO_4 alkylation unit then can be operated under better conditions to produce an alkylate with a somewhat increased octane number (0.75–1). Higher acid consumption, however, may be experienced as a result of methanol, MTBE and dimethyl ether impurities in the olefin feed. The combination of MTBE with HF alkylation offers no apparent advantages.

Anhydrous HF is volatile [boiling point (bp) 19.6°C], and when accidentally released into the atmosphere, forms toxic aerosol clouds. A number of industrial accidents showed that the use of anhydrous HF represents a significant environmental hazard; hence the need to either replace HF or find a remedy for the problem.[292,300] Olah found[301] that *n*-donor bases (pyridine, alkylamines, etc.) form with HF remarkably stable liquid onium poly(hydrogen fluoride) complexes capable of effecting the isoalkane–alkene alkylation similarly to HF itself. At the same time these complexes are substantially less volatile than hydrogen fluoride itself and when accidentally released, can be readily diluted with water and neutralized with caustic treatment without causing the problems associated with anhydrous HF. Additionally, because of their lower vapor pressure, the operating pressure in the alkylation process is decreased. The environmentally safe modified HF alkylation process is commercialized (Texaco–UOP).[302]

Over the years H_2SO_4 alkylation capacity has declined relative to HF alkylation.[303,304] Reasons for this trend are a significantly lower acid consumption, overall lower operation costs, and less expensive regeneration plant cost and operating expenses of the HF alkylation process. Because of the mentioned environmental hazard associated with anhydrous HF and the need for further improvement of alkylate quality, improved technologies for sulfuric acid alkylation emerged, although it is improbable that there will be a major turnback to sulfuric acid alkylation.

A two-step (two-reactor) process[292,293,305,306] may be operated at lower operating costs (lower ratios of isobutane to *n*-butenes, lower levels of agitation and acid consumption). More importantly, it affords alkylates of higher quality (99–101 octane number). In the first step *sec*-butyl sulfate is produced using a limited amount of acid. This then is used to alkylate isobutane with additional acid added. The two-step alkylation can be carried out in the temperature range of −20 to 0°C with the second reactor usually operated at lower temperature.

In the long run solid catalysts are expected to be used, which would reduce the safety problems of liquid-phase alkylations. However, much further work is needed to develop such processes,[7] and their introduction will be costly. The startup of a pilot plant to demonstrate a solid acid catalyst alkylation technology jointly developed by Catalytica, Conoco, and Neste Oy has been announced.[307]

5.5.2. Ethylbenzene

Alkylation of benzene with ethylene gives ethylbenzene,[283,284,308,309] which is the major source of styrene produced by catalytic dehydrogenation. High benzene : ethylene ratios are applied in all industrial processes to minimize polyethylation. Polyethylbenzenes formed are recycled and transalkylated with benzene. Yields better than 98% are usually attained. Reactants free of sulfur impurities and water must be used.

The oldest of the industrial processes (Dow, and a similar process developed by Union Carbide) employs aluminum chloride catalyst.[155,156] In the original process a small amount of HCl (or ethyl chloride) is added to the feed (on reacting with benzene, ethyl chloride supplies the promoter hydrogen chloride). $AlCl_3$ concentration in the reactors can reach 25%. Most of the $AlCl_3$ serves as the counterion ($AlCl_4^-$ or $Al_2Cl_7^-$) for the ethylbenzene–$AlCl_3$–HCl σ complex or higher ethylated complexes (including the heptaethylbenzenium ion), which with hydrocarbons form a separate lower "red oil." Makeup fresh $AlCl_3$ is regularly added to maintain catalytic activity. The process is run at about 100°C and at slightly above atmospheric pressure. A modification of this technology developed by Monsanto[310,311] is a more economic liquid-phase homogeneous process. It employs a small amount of $AlCl_3$ dissolved in benzene with ethyl chloride or anhydrous hydrogen chloride promoters. Under proper operating conditions (160–180°C, 1 atm) the reaction is virtually instantaneous and complete. Transalkylation of polyethylbenzenes is carried out in a separate unit.

A gas-phase alkylation over an alumina-supported BF_3 catalyst developed by UOP (Alkar process)[312,313] reduces the corrosion problems associated with liquid-phase alkylation processes. An advantage of this technology is that it utilizes very dilute (8–10%) ethylene streams, such as refinery fuel gases (cracked gas streams). At the same time, to properly support BF_3 is difficult.

A more recent development in ethylbenzene technology is the Mobil–Badger process,[161,314–316] which employs a solid acid catalyst in the heterogeneous vapor-phase reaction (400–450°C, 15–30 atm). A modified H-ZSM-5 catalyst that is regenerable greatly eliminates the common problems associated with

Friedel–Crafts catalysts (corrosion, continuous catalyst makeup). Although it gives a broader spectrum of alkylated products, recycling and transalkylation ensure high ethylbenzene yields. Steamed ZSM-5 and chrysozeolite ZSM-5 were shown by Union Carbide to afford ethylbenzene with high selectivity in the alkylation of benzene with ethanol.[317]

Monsanto disclosed the manufacture of ethylbenzene through a different approach by the methylation of toluene in the side chain.[318] A cesium-exchanged faujasite promoted by boron or phosphorus is used as the catalyst. Toluene and methanol (5 : 1) reacting at 400–475°C produce an ethylbenzene–styrene mixture at very high toluene conversion. About 50% of the methanol is converted to carbon monoxide and hydrogen, which is a disadvantage since such a plant should operate in conjunction with a methanol synthesis plant.

5.5.3. Cumene

Cumene is manufactured by the alkylation of benzene with propylene[283,284,308,309] in either the liquid or gas phase. Most of the cumene produced is used in the production of phenol and acetone via its hydroperoxide. High benzene : propylene ratios are used to minimize polyalkylation. The main byproducts (*n*-propylbenzene, di- and triisopropylbenzenes, and butylbenzenes if the propylene stream contains butenes) are separated by distillation.

Lewis and protic acids, usually $AlCl_3$ and H_2SO_4, are used in the liquid phase at temperatures of 40–70°C and at pressures of 5–15 atm. Phosphoric acid on kieselguhr promoted with BF_3 (UOP process)[309,319] is used in gas-phase alkylation (175–225°C, 30–40 atm). In addition to the large excess of benzene, propane as diluent is also used to ensure high (better than 94%) propylene conversion. This solid phosphoric acid technology accounts for 80–90% of the world's cumene production.

Three companies (Lummus,[320] Mobil–Badger,[321] Dow Chemical[322,323]) have announced new zeolite-based cumene technologies. Improved selectivity with lower benzene : propylene ratio, higher yields, and decreased production costs are the main characteristics of these processes. A dealuminated three-dimensional mordenite (3-DDM) is the catalyst in the Dow process. The dealumination of mordenite is carried out in such a way as to transform two-dimensional tubular pores into a controlled three-dimensional structure.[323] The catalyst exhibits low coking and high shape selectivity, allowing highly selective formation of cumene and *p*-diisopropylbenzene. The latter and minor amounts of *m*-diisopropylbenzene are transalkylated with benzene to form more cumene.

5.5.4. Xylenes

Toluene disproportionation and transalkylation are important industrial processes in the manufacture of *p*-xylene. Toluene disproportionation [Eq. (5.73)] transforms toluene into benzene and an equilibrium mixture of isomeric xylenes. The theoretical conversion of toluene is 55%. Commercial operations are usually run to attain 42–48% conversions. In conventional processes[308,309,324,325] alumina-supported noble metal or rare-earth catalysts are used in the presence of hydrogen (350–

530°C, 10–50 atm). Xylene : benzene ratios of 1–10 may be obtained. Metal catalysts were later replaced by zeolites.[210,211,326–328] The most recent development is the Mobil selective toluene disproportionation process,[329] which takes advantage of the high *para* shape selectivity of a zeolite catalyst.[210] The catalyst activated by a novel procedure ensures a *p*-xylene content of up to 95%. After the successful commercialization at an Enichem refinery in Italy,[329] the process is now licensed.[330] The catalysts and technologies applied in toluene disproportionation may be also used for transalkylation[324,325,331] [Eq. (5.74)]:

$$(5.73)$$

$$(5.74)$$

In this case the feed is a mixture of toluene and higher aromatics, preferably trimethylbenzenes. Disproportionation and transalkylation occur simultaneously, permitting the manufacture of benzene and xylenes.

5.5.5. *p*-Ethyltoluene

p-Ethyltoluene is made for subsequent dehydrogenation to *p*-methylstyrene (*p*-vinyltoluene) monomer. Poly(*p*-methylstyrene) has improved characteristics over polystyrene. Commercial methylstyrene polymer is produced by the polymerization of vinyltoluene. Vinyltoluene used in the latter polymerization is a 65% : 35% mixture of *m*- and *p*-vinyltoluene. It is produced by dehydrogenation of a mixture of ethyltoluenes, which, in turn, are manufactured by a method similar to that of ethylbenzene[332] (alkylation of toluene with ethylene in the presence of HCl–AlCl₃). The search for a selective synthesis of *p*-methylstyrene has led to the development of a process to manufacture *p*-ethyltoluene through the shape-selective alkylation of toluene with ethylene by Mobil.[99,333]

Mobil ZSM-5 zeolite catalysts can be modified to reduce the effective pore and channel dimensions. These modified zeolites allow discrimination between molecules of slightly different dimensions. Because of this shape-selective action, *p*-ethyltoluene is able to diffuse out of the catalyst pores at a rate about three orders of magnitude greater than the two regioisomers. As a result, *p*-ethyltoluene is formed with very high (97%) selectivity.[333]

Ethanol instead of ethylene can also be used in alkylation of toluene[334] with *para* selectivities up to 90%. Anhydrous ethanol was shown to undergo dehydration to ethylene, which, in turn, alkylated the aromatic hydrocarbon. The alkylation step was suggested to take place inside the zeolite channels to yield the *para* isomer while isomerization of the latter toward the thermodynamic equilibrium mixture proceeds on the external surface of the catalyst. Steam-treated ZSM-5 was found to be selective catalyst when an aqueous (10%) solution of ethanol produced by fermentation was used in alkylation.[317] Thus, costly separation, concentration, or purification of fermentation broth is not necessary.

5.5.6. Detergent Alkylates

Detergent alkylates are long-chain monoalkylbenzenes.[308,309,335] The first commercial process used propylene tetramer to alkylate benzene. The formed product dodecylbenzenes underwent sulfonation followed by neutralization with NaOH to give the alkylbenzene sulfonate detergent. Propylene tetramer is a mixture of highly branched isomeric C_{12} alkenes. These detergents, however, were found to be nonbiodegradable because of side-chain branching, which contributed to serious environmental pollution.

Linear side chains, in turn, are highly biodegradable, and technologies to produce linear alkylbenzenes were developed accordingly. Linear alkyl halides, linear alkenes[336] produced by dehydrogenation, wax cracking, or dehydrochlorination of alkyl halides, and 1-alkenes (ethylene oligomers) can be used as alkylating agents. Since alkenes undergo isomerization (double-bond migration) during alkylation, the linear alkylbenzenes manufactured from different alkene sources (internal alkenes or 1-alkenes) have similar characteristics. HF, HF–BF_3, or $AlCl_3$ catalysts are used in the liquid-phase (40–70°C) alkylation.

A technology developed by UOP[335,337,338] combines UOP's Molex and Pacol processes to produce linear alkenes. The former process separates long-chain (C_{10}–C_{15}) *n*-alkanes from other homologs, isoparaffins, and cycloalkanes by selective absorption employing molecular sieves. The alkane mixture thus produced undergoes dehydrogenation (Pacol process; see Section 2.3.3) to yield monoalkenes with an internal double bond randomly distributed along the chain. This alkene mixture is used as the alkylating agent in alkylation of benzene catalyzed by hydrogen fluoride.[337–339] A mixture of alkylbenzenes with almost no 2-phenyl isomer is formed, which gives a high-quality detergent after sulfonation.

In other technologies[340,341] a mixture of *n*-alkyl chlorides formed by the chlorination of *n*-alkanes is used directly in alkylation with aluminum chloride.

5.6. RECENT DEVELOPMENTS

In addition to a detailed review,[342] alkylations are briefly addressed in reviews about homogeneous and heterogeneous electrophilic catalysts.[343–349]

5.6.1. Alkane–Alkene Alkylation

The acidity dependence of the isobutane–isobutylene alkylation was studied using triflic acid modified with trifluoroacetic acid (TFA) and water in the range of acidity between $H_0 = -10.1$ and -14.1.[350] For both systems the best alkylation conditions were those at an acid strength of about $H_0 = -10.7$, giving calculated research octane numbers of 89.1 (triflic acid–TFA) and 91.4 (triflic acid–water). Isobutane–isobutylene alkylation was studied in liquid CO_2 with various acids (anhydrous HF, PPHF, conc. H_2SO_4, triflic acid) in the temperature range of 0–50°C.[351] For HF and triflic acid, higher optimum research octane numbers were obtained in CO_2 than in the neat acids (95.6 for HF and 88.0 for triflic acid). Moreover, these reactions required less acid catalyst and afforded increased selectivities to trimethylpentanes. A novel liquid acid catalyst composed of a heteropoly acid (phophotungstic acid or silicotungstic acid) and acetic acid was used in the alkylation isobutane with butenes.[352] A viscous, oily phase separating during the early stages of the reaction showed catalytic performance comparable to that of sulfuric acid.

Concentrated sulfuric acid and hydrogen fluoride are still mainly used in commercial isoalkane–alkene alkylation processes.[353] Because of the difficulties associated with these liquid acid catalysts (see Section 5.1.1), considerable research efforts are still devoted to find suitable solid acid catalysts for replacement.[354–356] Various large-pore zeolites, mainly X and Y, and more recently zeolite Beta were studied in this reaction. Considering the reaction scheme [(Eqs (5.3)–(5.5) and Scheme 5.1)] it is obvious that the large-pore zeolitic structure is a prerequisite, since many of the reaction steps involve bimolecular bulky intermediates. In addition, the fast and easy desorption of highly branched bulky products, such as trimethylpentanes, also requires sufficient and adequate pore size. Experiments showed that even with large-pore zeolite Y, alkylation is severely diffusion limited under liquid-phase conditions.[357]

Brønsted acid sites serve as the active sites for the alkylation reaction. Higher acid site density, obviously, increases the reaction rate, providing higher concentration of carbocations and, hence, hydride transfer necessary for the desorption of the alkylated carbocationic intermediates becomes faster. Since alkane formation is a more demanding reaction requiring two adjacent acid sites, the alkane : alkene ratio was found to increase with increasing acid site density.[355] The ratio of trimethylpentanes to dimethylpentanes also increases with increasing acid site density. This is due to the fact that the formation of the desired highly branched trimethylpentanes includes two hydride transfer steps whereas that of dimethylpentanes has only one.

The major problem of the application of zeolites in alkane–alkene alkylation is their rapid deactivation by carbonaceous deposits. These either strongly adsorb on acidic sites or block the pores preventing the access of the reactants to the active sites. A further problem is that in addition to activity loss, the selectivity of the zeolite-catalyzed alkylation also decreases severely. Specifically, alkene formation through oligomerization becomes the dominant reaction. This is explained by decreasing ability of the aging catalyst to promote intermolecular hydride transfer. These are the main reasons why the developments of several commercial processes reached only the pilot plant stage.[356] New observations with Y zeolites reconfirm the problems found in earlier studies.[358,359]

Studies with sulfated zirconia also show similar fast catalyst deactivation in the alkylation of isobutane with butenes. It was found, however, that original activities were easily restored by thermal treatment under air without the loss of selectivity to trimethylpentanes.[360–362] Promoting metals such as Fe, Mn, and Pt did not have a marked effect on the reaction.[362,363] Heteropoly acids supported on various oxides have the same characteristics as sulfated zirconia.[364] Wells–Dawson heteropoly acids supported on silica show high selectivity for the formation of trimethylpentanes and can be regenerated with O_3 at low temperature (125°C).[365]

5.6.2. The Prins Reaction

Various solid acids were studied in the Prins reaction mainly because of their potential use in the production of isoprene. Ion-exchanged silicates of MFI structure[366] and a wide range of other molecular sieves[367] were tested in the synthesis of isoprene form isobutylene and formaldehyde. The best results (conversions up to 40% and selectivities above 99%) were obtained with H-boralites. Because of shape selectivity, appropriately selected zeolites can suppress the formation of 4,4-dimethyl-1,3-dioxane, thereby increasing isoprene selectivity. The results show unambiguously that high selectivities to isoprene are obtained only in the presence of weak and medium Brønsted acid sites ensuring only protonation of formaldehyde, whereas isobutylene is mainly π-adsorbed. Similar observations were made by using Fe-MCM-22 zeolites.[368]

Other solid catalysts such as cation-exchanged montmorillonites,[369] zeolites,[370] and heteropoly acids[371] were used in the Prins reaction of styrene to produce 1,3-dioxanes. Ce^{3+}- and Fe^{3+}-exchanged montmorillonites proved to be highly effective in the transformation of a wide range of substituted styrene derivatives with formaldehyde to yield the corresponding 4-phenyl-1,3-dioxanes.[369] The use of other aldehydes resulted in decreasing yields with increasing chain length. The used catalysts could be regenerated with a slight loss of activity. The Prins reaction catalyzed by various heteropoly acids and acidic cesium salts of phosphotungstic and phosphomolybdic acid was found to be a valuable method for the synthesis of substituted 4-phenyl-1,3-dioxanes under both thermal activation and microwave irradiation.[371]

The microenvironmental effect of various sulfonated polystyrene beads was studied using the Prins reaction as a probe.[372] Reaction rates were found to be lower when carbomethoxy or carbobutoxy neighboring groups were present compared to phenyl. A less ionic microenvironment appears to allow for a higher concentration of styrene within the polymer and leads to an immediate reaction with protonated formaldehyde.

5.6.3. Alkylation of Aromatics

Solid Acid Catalysts. There have been commercial alkylation processes in operation that apply solid acids (viz., zeolites) in the manufacture of ethylbenzene (see Section 5.5.2), cumene (see Section 5.5.3), and xylenes (see Section 5.5.4). Research still continues to explore new, recyclable solid materials with better catalyst performance.[373] In fact, new commercial processes using large-pore zeolites, such as USY, zeolite Beta, and MCM-22, were developed and commercialized for the manufacture of ethylbenzene. The MCM-22 catalyst applied in the Mobil EBMax process is substantially more selective than the other two catalysts (89% selectivity of ethylbenzene at greater than 95% conversion of ethylene).[374] This excellent selectivity is due to the 12-ring pore system and the facile desorption of ethylbenzene from the surface pockets of MCM-22. A new study shows that alkylation takes place on the external surface of the catalyst.[375]

A new process for the manufacture of p-diethylbenzene developed by Indian Petrochemicals Corporation was commercialized.[376] p-Diethylbenzene is produced by alkylating ethylbenzene with ethanol over a highly shape-selective, pore-size-regulated, high-silica zeolite. The catalyst exhibits a steady activity of 6–8% conversion with 97–98% selectivity.

New studies with zeolites were reported; the main concern was selectivity improvements.

Regioselective dialkylation of biphenyl and naphthalene are of interest because the corresponding dicarboxylic acids prepared from 4,4-disubstituted biphenyls and 2,6-disubstituted naphthalenes are used to manufacture commercially valuable polyesters. High selectivities were achieved in the shape-selective isopropylation of biphenyl with propylene to form 4,4-diisopropylbiphenyl over H-mordenites having high Si : Al ratios.[377–379] On these samples alkylation proceeds in the pores and transition state control ensures high selectivity. When the Si : Al ratio is low, the concentration of surface sites with strong acidity is high, resulting in enhanced coke formation. Coke deposition blocks the pore entrance and, therefore, nonselective alkylation occurs on the external surface of the catalyst.

Highly regioselective dialkylation of naphthalene was performed with 2-propanol[380] and tert-butyl alcohol[381] over dealuminated H-mordenite, and with tert-butyl alcohol over HY and Beta zeolites.[382] Selectivities are within 60–84%. Computational modeling showed that the kinetic diameter of the 2,6-disubstituted isomer is smaller than that of the 2,7 isomer, which may explain the selective formation of the former within the appropriate zeolitic framework.[382]

Clays are efficient electrophilic catalysts and were applied in Friedel–Crafts alkylations. In most studies K10 montmorillonite, an acid-treated bentonite, was used. The catalytic activity of clays can be further enhanced by various modifications. Ion exchange with F^{3+} or Zn^{2+} brings about outstanding catalytic properties. Pillaring is another possibility to modify the properties of clay materials. Pillared clays (PILCs) exhibit a microporous structure and acidic properties that make them suitable for application in Friedel–Crafts alkylations. Clays pillared with F^{3+} gave alkylated products in the reaction of alkylbenzenes with benzyl chloride with near-quantitative conversion.[383] With toluene, monomethylation without the formation of the *meta* isomer was observed. Materials pillared with Al_{13} and $GaAl_{12}$ polyoxocations showed high catalytic activity in the alkylation of toluene with methanol yielding xylenes with 60% and 74% selectivities, respectively.[384]

When montmorillonite is treated with a metal salt in organic solvents, impregnation occurs instead of ion exchange. The first highly successful catalytic material prepared by supporting $ZnCl_2$ from methanol solution to K10 montmorillonite was called "clayzic."[140] Clayzic exhibits activities which are several orders of magnitude higher than those of the ion-exchanged materials and can induce Friedel–Crafts alkylations with alcohols, alkyl halides and aldehydes.[385]

Since the development of clayzic, other montmorillonite-supported catalysts were also prepared and studied. New examples are montmorillonites modified with $FeCl_3$ or $SbCl_3$ from acetonitrile solution that show good characteristics in the alkylation of arenes with benzyl chloride.[386,387] The performance of the catalyst modified with $SbCl_3$ is especially surprising considering the low activity of this weak Lewis acid under homogeneous conditions. $ZnCl_2$ supported on hydroxyapatite promotes selective monoalkylation of benzene, toluene, and p-xylene by benzyl chloride,[388] whereas $AlCl_3$ and $ZnCl_2$ supported on MCM-41 are promising catalysts in the formation of diphenylmethane.[389]

An immobilized form of $AlCl_3$ using porous materials, including SiO_2, Al_2O_3, and K10, which displays high activity in the liquid phase, was reported.[390] Alkenes and chloroalkanes react smoothly under mild conditions. Activities are comparable to that of $AlCl_3$, and selectivities are often better. The catalysts can be recycled without loss of activity and selectivity. $AlCl_3$ could also be immobilized on MCM-41.[391] When used in the alkylation of benzene with C_6–C_{16} linear 1-alkenes, this novel shape-selective material with optimal pore size always exhibited better selectivities toward monosubstituted linear alkylbenzenes than $AlCl_3$ in the homogeneous phase. A significant increase in the selectivity toward 2,6-diisopropylnaphthalene could be achieved with a similar catalyst prepared with external surface-silylated MCM-41.[392] MCM-41 was found to be an ideal support to prepare highly active and reusable supported Ga_2O_3, In_2O_3, $GaCl_3$, and $InCl_3$ catalysts for the benzylation of methylbenzenes.[393]

$ZnCl_2$ supported in sol-gel-derived silica[394] or aluminosilicates[395] is effective in Friedel–Crafts alkylations. Significantly higher activities than over commercially available clayzic were observed, which were correlated with high total pore volume.

Sulfated zirconia catalysts showed selectivities comparable to those of $AlCl_3$ in the alkylation of benzene with 1-alkenes to linear alkylbenzenes.[396] The mesoporous sulfated zirconia sample could be regenerated by solvent extraction or thermal treatment. Phosphotungstic acid supported on sulfated zirconia doped with Fe proved to be very active and highly selective in the alkylation of benzene with propylene at 100–150°C to produce cumene; both monoalkylation and cumene formation have better than 90% selectivity.[397] It can be regenerated at moderate temperature (350°C).

The solid acid catalyzed adamantylation of substituted benzene derivatives was studied with the aim to achieve high *para* regioselectivities. Of various acidic resins, zeolite HY, sulfated zirconia, perfluorolkanesulfonic acids, and phosphotungstic acid, Amberlyst XN-1010 was found to be the catalyst of choice to afford *para*-substituted alkylbenzenes with selectivities exceeding 70%.[398] In further systematic studies with a Nafion-H–silica nanocomposite catalyst,[399] and various dodecaheteropoly acids of Keggin type, either neat[400] or immobilized in a silica matrix,[401] *para* isomers were shown to form under kinetic control. In the reaction of toluene, Mo-containing acids exhibited high selectivities in the formation of *para*-adamantyltoluene. The best result (99.2% selectivity) was obtained with $H_4[SiMo_{12}O_{40}]$.[401] Regioselectivity, however, showed strong acid dependence. Increasing acid strength prefers the formation of the *meta* isomer through subsequent isomerization of the primary *para* product.

Specific Examples. With the availability of extensive literature, only a few selected examples of interest with respect to the alkylating agent and the catalysts used are discussed.

Silica–alumina particles coated with a permselective silicalite membrane is almost completely selective in the formation of *p*-xylene in the disproportionation of toluene.[402] Friedel–Crafts alkylations were performed in ionic liquids. The strong polarity and high electrostatic fields of these materials usually bring about enhanced activity.[403,404] Easy recycling is an additional benefit. Good characteristics in the alkylation of benzene with dodecene were reported for catalysts immobilized on silica or MCM-41.[405]

Mesitylene was alkylated with propylene or 2-propanol in supercritical CO_2 using Deloxane, a polysiloxane-supported solid acid catalyst in a continuous flow reactor.[406] Monoisopropylation with 100% selectivity occurred with 2-propanol.

A binuclear Ir^{3+} complex used in the reaction of benzene with alkenes induces the formation of straight-chain alkylbenzenes with higher selectivity than in the branched isomers.[407] The selectivity, unusual in conventional acid-catalyzed Friedel–Crafts alkylations, is suggested to result from the complex activating the C—H bond of benzene instead of the alkene.

There are reports disclosing the alkylation of benzene with small alkanes to produce alkylbezenes. Methylation of aromatics was achieved over molecular sieves using methane as the methylating agent.[408,409] It was shown by isotope labeling that over zeolite Beta a part of the methyl groups in the product methyl-substituted aromatics did not originate from methane but from benzene fragmentation. In turn, it was demonstrated in a more recent study that methane can be the real methylating agent when the process is carried out in the presence of oxygen.[410] In this case a two-step mechanism involving the partial oxidation of methane to methanol followed by methylation with methanol was postulated.[411] Trace amounts of ethylene impurities in the methane feed bring about enhanced benzene conversion over ZSM-5 and zeolite Beta.[412] Alkylation of benzene with ethane was performed over zeolites with Pt-loaded H-ZSM-5 giving the best results.[413] When the alkylating agent is propane, the main products are methylbenzene and ethylbenzene and cumene is not observed. Mechanistic studies by means of in situ MAS NMR spectroscopy over Ga—H-ZSM-5 allowed Derouane et al. to conclude that bifunctional propane activation yields protonated pseudocyclopropane intermediate.[414] This decomposes preferentially to yield incipient methyl and ethyl cations, which are the apparent alkylating species. The formation of cycloproponium cation that would lead to *n*-proylbenzene and cumene is less favorable.

Friedel–Crafts alkenylation of arenes with various phenylacetylenes to yield 1,1-diarylalkenes was reported. Alkyl-substituted benzenes and naphthalene react with phenylacetylene in the presence of zeolite HSZ-360 to afford the products in good to excellent yields and selectivities:[415]

$$R = H, Me, m\text{–}diMe, p\text{–}diMe \qquad 62\text{–}92\%$$
$$1,3,5\text{-triMe}$$

$$(5.75)$$

1,1-Diarylalkenes are formed in high to excellent yields when both terminal and internal alkynes react with benzene and methyl-substituted benzenes in the presence of catalytic amounts of metal (Sc, Zr, In) triflates.[416]

Various triflate salts, namely, scandium triflate,[417–419] lanthanum triflates,[417,419] copper triflate, and hafnium triflate, are effective, water-tolerant, recyclable catalysts in Friedel–Crafts alkylations. The alkylating agents are allyl and benzyl alcohols,[420] arencarbaldehydes and arencarbaldehyde acetals,[420] secondary alcohol methanesulfonates,[421,422] benzyl chloride,[423,424] and alkyl chlorides.[424] The activity of hafnium triflate could be greatly increased by the addition of $LiClO_4$.[424] A more recent study gave detailed characterization of p-toluenesulfonic acid monohydrate as a convenient, recoverable, safe, and selective catalyst for Friedel–Crafts alkylation.[425] In comparison to conventional Friedel–Crafts catalysts such as $AlCl_3$, BF_3, HF, and concentrated H_2SO_4, the extent of the formation of undesired products from side reactions such as transalkylation and oligomerization was minimal with this catalyst. Triflic acid was an effective catalyst in the tandem alkylation–cyclialkylation of benzene, toluene, and xylenes with various diols to afford tetralin derivatives.[426]

Benzaldehyde reacts readily with benzene under superacid conditions to give triphenylmethane in high yields.[427,428] The reaction is slow in triflic acid, giving a yield of 90% in 24 h, whereas with the strong superacid $CF_3SO_3H_2^+$ $B(OSO_2CF_3)_4^-$ the product is formed instantaneously in 83% yield.[427] Experimental evidence supports the involvement of diprotonated benzaldehyde in the reaction [Eq. (5.76)]. Theoretical calculations show that the **44** O,C(aromatic)-diprotonated cation is the reactive intermediate and not the **45** O,O-diprotonated cation:

$$(5.76)$$

45

In a similar fashion, reductive alkylation of benzene, toluene, and xylenes with benzaldehyde, aromatic methyl ketones, and cycloalkanones can be induced by $InCl_3$ in the presence of Me_2SiClH.[429]

Side-Chain Alkylation. There is continued interest in the alkylation of toluene with methanol because of the potential of the process in practical application to produce styrene.[430] Basic catalysts, specifically, alkali cation-exchanged zeolites, were tested in the transformation. The alkali cation acts as weak Lewis acid site, and the basic sites are the framework oxygen atoms. The base strength and catalytic activity of these materials can be significantly increased by incorporating alkali metal or alkali metal oxide clusters in the zeolite supercages. Results up to 1995 are summarized in a review.[430]

A comparative study with various types of zeolite showed that Cs-exchanged X and Y zeolites were active for toluene alkylation but primarily catalyzed the decomposition of methanol to CO.[431] L and Beta zeolites, in turn, were less active and required higher reaction temperature but were much more selective, providing only very little CO. Adding boron to Cs-exchanged zeolites promotes the alkylation reaction.[432] It appears that boron reduces the decomposition of methanol to CO without inhibiting active sites for side-chain alkylation.

Apparently, zeolites with low basic site densities and appropriate base strength are the selective systems catalyzing toluene side-chain alkylation.[432,433] In agreement with these observations, KNaX was found to show enhanced catalytic selectivity on proper ball milling.[434] Ball milling was suggested to moderately decrease the concentration of both Lewis base and Lewis acid sites. At the same time, the density of the strong Brønsted acid sites that are responsible for ring alkylation decreases significantly.

Mechanistic studies using solid-state NMR technique with alkali-exchanged X and Y zeolites showed strong interaction between Cs^+ ions and toluene[435] and the formation of formaldehyde, the apparent alkylating agent.[435,436] There is a new report on CsNaX impregnated with CsOH studied under flow conditions by in situ NMR spectroscopy.[437] It shows that formate species, earlier thought to be responsible for catalyst poisoning, are reactive and contribute as alkylating agent to the process. IR spectroscopic investigations indicated that large cations have higher preference to adsorb toluene over methanol.[438] The methyl group of toluene interacts with the lattice oxygen and is polarized, which is essential for participation in the alkylation process. Rb and Cs are the two cations best satisfying these requirements.

Some observations indicate that the zeolite framework is a prerequisite for good catalyst performance via physical constraints and proximity of acid–base sites within the molecular sieves environment.[431] Other observations show that rather microporosity plays an important role in the alkylation.[432]

Alkenes may also be used as reagents in side-chain alkylation. Zeolites with occluded alkali metal and alkali metal oxide clusters were tested in alkylation of toluene with ethylene.[439,440] Alkali metal clusters created by decomposing Na or

Cs azide were active catalysts in the alkylation reaction, whereas alkali metal oxide clusters showed no activity at all.

A further possibility for side-chain alkylation of toluene is oxidative methylation with methane. Catalysts with occluded alkali metal oxides, prepared by impregnating zeolites with alkali metal hydroxides followed by calcination, usually exhibit better performance.[441] Further enhancement was achieved by impregnating ion-exchanged zeolites.[442] Significant improvements in stability and the yields of C_8 hydrocarbons were also observed when NaX was impregnated with 13% MgO which was found to increase the amounts of active sites.[443]

According to another study, Cs entrapped in nanoporous carbon is an active catalyst in the side-chain alkylation with propylene either in the liquid or in the gas phase to form isbutylbenzene as the main product.[444]

The LICKOR superbase was shown to be an effective and chemoselective reagent to obtain sterically demanding alkylbenzenes from methylbenzenes.[445] When a methyl group is flanked by two neighboring methyl groups, it adds only one molecule of ethylene:

$$(5.77)$$

75%

In other cases (toluene, the xylenes, mesitylene, durene), two ethylene molecules react with each methyl group to form the corresponding 3-pentyl-substituted derivatives in good yields (60–79%).

Transalkylation. Advances in research and developments with respect to disproportionation and transalkylation of alkylbenzenes, particularly the transformation of toluene to produce xylenes, are adequately treated in reviews.[446–449] A review is available for de-*tert*-butylation.[450]

5.6.4. Miscellaneous Alkylations

Allylindiums were found to be highly active reagents to react with terminal alkynes to yield allylated products.[451] In a similar fashion efficient allylation can be mediated by indium metal under mild conditions:[452]

$$(5.78)$$

R = n-Bu, C_6H_{13}, Ph   85–90%

Ethylaluminum sesquichloride induces the alkylation of alkenes with isopropyl chloroformate.[453] Terminal alkenes [Eq. (5.79)] and cycloalkenes react in the presence of Et_3SiH:

$$R^1 = H, CH_3$$
$$R^2 = C_6H_{13}, C_9H_{19}$$

47–55%

$$(5.79)$$

REFERENCES

1. V. N. Ipatieff, *Catalytic Reactions at High Pressures and Temperatures*, Macmillan, New York, 1936, p. 673.

2. J. H. Simons, *Adv. Catal.* **2**, 197 (1950).

3. R. M. Kennedy, in *Catalysis*, Vol. 6, P. H. Emmett, ed., Reinhold, New York, 1958, Chapter 1, p. 1.

4. V. N. Ipatieff and L. Schmerling, *Adv. Catal.* **1**, 27 (1948).

5. L. Schmerling, in *Friedel–Crafts and Related Reactions*, Vol. 2, G. A. Olah, ed., Interscience, New York, 1964, Chapter 25, p. 1075.

6. H. Pines, *The Chemistry of Catalytic Hydrocarbon Conversions*, Academic Press, New York, 1981, Chapter 1, p. 50.

7. A. Corma and A. Martínez, *Catal. Rev.-Sci. Eng.* **35**, 483 (1993).

8. G. A. Olah, G. K. S. Prakash, and J. Sommer, *Superacids*, Wiley-Interscience, New York, 1985, Chapter 5, p. 270.

9. G. A. Olah and G. K. S. Prakash, in *The Chemistry of Alkanes and Cycloalkanes*, S. Patai and Z. Rappoport, eds., Wiley, Chichester, 1992, Chapter 13, p. 624.

10. J. Sommer and J. Bukala, *Acc. Chem. Res.* **26**, 370 (1993).

11. V. N. Ipatieff and A. V. Grosse, *J. Am. Chem. Soc.* **57**, 1616 (1935).

12. V. N. Ipatieff, A. V. Grosse, H. Pines, and V. I. Komarewsky, *J. Am. Chem. Soc.* **58**, 913 (1936).

13. A. V. Grosse and V. N. Ipatieff, *J. Org. Chem.* **8**, 438 (1943).

14. H. Pines, A. V. Grosse, and V. N. Ipatieff, *J. Am. Chem. Soc.* **64**, 33 (1942).

15. F. E. Frey and H. J. Hepp, *Ind. Eng. Chem.* **28**, 1439 (1936).

16. J. E. Knap, E. W. Comings, and H. G. Drickamer, *Ind. Eng. Chem.* **46**, 708 (1954).

17. L. Schmerling, *J. Am. Chem. Soc.* **68**, 275 (1946).

18. C. B. Linn and H. G. Nebeck, in *The Science of Petroleum*, Vol 5, Part II, B. T. Brooks and M. E. Dansten, eds., Oxford Univ. Press, London, 1953, p. 302.

19. S. H. McAllister, J. Anderson, S. A. Ballard, and W. E. Ross, *J. Org. Chem.* **6**, 647 (1941).

20. A. Schneider, *J. Am. Chem. Soc.* **76**, 4938 (1954).

21. H. Pines and V. N. Ipatieff, *J. Am. Chem. Soc.* **70**, 531 (1948).

22. W. K. Conn and A. Schneider, *J. Am. Chem. Soc.* **76**, 4578 (1954).

23. V. N. Ipatieff, V. I. Komarewsky, and A. V. Grosse, *J. Am. Chem. Soc.* **57**, 1722 (1935).

24. P. D. Bartlett, F. E. Condon, and A. Schneider, *J. Am. Chem. Soc.* **66**, 1531 (1944).

25. C. D. Nenitzescu, in *Carbonium Ions*, Vol. 2, G. A. Olah and P. v. R. Schleyer, eds., Wiley-Interscience, New York, 1970, Chapter 13, p. 463.

26. C. Guo, S. Liao, Z. Quian, and K. Tanabe, *Appl. Catal., A* **107**, 239 (1994).

27. G. A. Olah, *Carbocations and Electrophilic Reactions*, Verlag-Wiley, 1974, p. 108.

28. G. A. Olah, R. Krishnamurti, and G. K. S. Prakash, in *Comprehensive Organic Synthesis*, B. M. Trost and I. Fleming, eds., Vol. 3: *Carbon–Carbon σ-Bond Formation*, G. Pattenden, ed., Pergamon Press, Oxford, 1991, Chapter 1.8, p. 332.

29. M. Siskin, *J. Am. Chem. Soc.* **98**, 5413 (1976).

30. M. Siskin, R. H. Schlosberg, and W. P. Kocsi, *ACS Symp. Ser.* **55**, 186 (1977).

31. G. A. Olah, U.S. Patent 3,708,553 (1973).

32. J. Sommer, M. Muller, and K. Laali, *New J. Chem.* **6**, 3 (1982).

33. G. A. Olah, J. D. Felberg, and K. Lammertsma, *J. Am. Chem. Soc.* **105**, 6529 (1983).

34. G. A. Olah, O. Farooq, V. V. Krishnamurthy, G. K. S. Prakash, and K. Laali, *J. Am. Chem. Soc.* **107**, 7541 (1985).

35. G. A. Olah and J. A. Olah, *J. Am. Chem. Soc.* **93**, 1256 (1971).

36. G. A. Olah, Y. K. Mo, and J. A. Olah, *J. Am. Chem. Soc.* **95**, 4939 (1973).

37. G. A. Olah, J. R. DeMember, and J. Shen, *J. Am. Chem. Soc.* **95**, 4952 (1973).

38. G. A. Olah and R. H. Schlosberg, *J. Am. Chem. Soc.* **90**, 2726 (1968).

39. G. A. Olah, G. Klopman, and R. H. Schlosberg, *J. Am. Chem. Soc.* **91**, 3261 (1969).

40. G. A. Olah, Y. Halpern, J. Shen, and Y. K. Mo, *J. Am. Chem. Soc.* **95**, 4960 (1973).

41. D. T. Roberts, Jr. and L. E. Calihan, *J. Macromol. Sci., Chem.* **7**, 1629 (1973).

42. D. T. Roberts, Jr. and L. E. Calihan, *J. Macromol. Sci., Chem.* **7**, 1641 (1973).

43. M. Vol'pin, I. Akhrem, and A. Orlinkov, *New J. Chem.* **13**, 771 (1989).

44. L. Schmerling, in *Friedel–Crafts and Related Reactions*, Vol. 2, G. A. Olah, ed., Wiley, New York, 1964, Chapter 26, p. 1133.

45. G. A. Olah, *Friedel–Crafts Chemistry*, Wiley-Interscience, New York, 1973, Chapter 2, p. 80.

46. H. Klein, A. Erbe, and H. Mayr, *Angew. Chem., Int. Ed. Engl.* **21**, 82 (1982).

47. H. Mayr, H. Klein, and G. Kolberg, *Chem. Ber.* **117**, 2555 (1984).

48. G. A. Olah, S. J. Kuhn, and D. G. Barnes, *J. Org. Chem.* **29**, 2685 (1964).

49. V. A. Miller, *J. Am. Chem. Soc.* **69**, 1764 (1947).

50. L. Schmerling and E. E. Meisinger, *J. Am. Chem. Soc.* **75**, 6217 (1953).

51. H. Mayr and W. Stripe, *J. Org. Chem.* **48**, 1159 (1983).

52. L. Schmerling and J. P. West, *J. Am. Chem. Soc.* **74**, 3592 (1952).

53. L. Schmerling, *J. Am. Chem. Soc.* **67**, 1152 (1945).

54. L. Schmerling and E. E. Meisinger, *J. Am. Chem. Soc.* **71**, 753 (1949).

55. H. Mayr, *Angew. Chem., Int. Ed. Engl.* **20**, 184 (1981).

56. G. A. Olah, *Friedel–Crafts Chemistry*, Wiley-Interscience, New York, 1973, Chapter 2, p. 84.

57. R. Maroni, G. Melloni, and G. Modena, *J. Chem. Soc., Perkin Trans. 1* 2491 (1973); 353 (1974).

58. F. Marcuzzi, and G. Melloni, *J. Am. Chem. Soc.* **98**, 3295 (1976).

59. G. Melloni, G. Modena, and U. Tonellato, *Acc. Chem. Res.* **8**, 227 (1981).

60. F. Marcuzzi and G. Melloni, *J. Chem. Soc., Perkin Trans. 2* 1517 (1976).

61. E. Arundale and L. A. Mikeska, *Chem. Rev.* **51**, 505 (1952).

62. C. W. Roberts, in *Friedel–Crafts and Related Reactions*, Vol. 2, G. A. Olah, ed., Wiley, New York, 1964, Chapter 27, p. 1175.

63. V. I. Isagulyants, T. G. Khaimova, V. R. Melikyan, and S. V. Pokrovskaya, *Russ. Chem. Rev.* (Engl. transl.) **37**, 17 (1968).

64. D. R. Adams and S. P. Bhatnagar, *Synthesis* 661 (1977).

65. B. B. Snider, in *Comprehensive Organic Synthesis*, B. M. Trost and I. Fleming, eds., Pergamon Press, 1991, Vol. 2: *Additions to C–X π-Bonds*, Part 2, C. H. Heathcock, ed., Chapter 2.1, p. 527.

66. J. W. Baker, *J. Chem. Soc.* 296 (1944).

67. A. T. Blomquist and J. Wolinsky, *J. Am. Chem. Soc.* **79**, 6025 (1957).

68. L. J. Dolby, *J. Org. Chem.* **27**, 2971 (1962).

69. B. Fremaux, M. Davidson, M. Hellin, and F. Coussemant, *Bull. Soc. Chim. Fr.* 4250 (1967).

70. E. E. Smissman, R. A. Schnettler, and P. S. Portoghese, *J. Og. Chem.* **30**, 797 (1965).

71. N. A. LeBel, R. N. Liesemer, and E. Mehmedbasich, *J. Og. Chem.* **28**, 615 (1963).

72. L. J. Dolby, C. Wilkins, and T. G. Frey, *J. Og. Chem.* **31**, 1110 (1966).

73. M. I. Farberov, Ya. I. Rotshtein, A. M. Kut'in, and N. K. Shemyakina, *J. Gen. Chem. USSR* (Engl. transl.) **27**, 2841 (1957).

74. G. Fodor and I. Tömösközi, *Suomen Kemistilehti* **B31**, 22 (1958).

75. C. Friedel and J. M. Crafts, *Compt. Rend.* **84**, 1450 (1877); **85**, 74 (1877).

76. M. Balsohn, *Bull. Soc. Chim.* [2]**31**, 539 (1879).

77. C. C. Price, *Org. React.* (NY) **3**, 1 (1946).

78. A. W. Francis, *Chem. Rev.* **43**, 257 (1948).

79. *Friedel–Crafts and Related Reactions*, G. A. Olah, ed., Wiley-Interscience, New York, 1964, Vol. 2, Part 1.

80. R. O. C. Norman and R. Taylor, *Electrophilic Substitution in Benzenoid Compounds*, Elsevier, Amsterdam, 1965, Chapter 6, p. 156.

81. G. A. Olah, *Friedel–Crafts Chemistry*, Wiley-Interscience, New York, 1973.

82. G. G. Yakobson and G. G. Furin, *Synthesis* 345 (1980).

83. H. Pines, *The Chemistry of Catalytic Hydrocarbon Conversions*, Academic Press, New York, 1981, Chapter 1, p. 59.

84. G. A. Olah and D. Meidar, in *Kirk-Othmer Encyclopedia of Chemical Technology*, 3rd ed., Vol. 11, M. Grayson and D. Eckroth, eds., Wiley-Interscience, New York, 1980, p. 269.

85. R. M. Roberts and A. A. Khalaf, *Friedel–Crafts Alkylation Chemistry*, Marcel Dekker, New York, 1984.

86. G. A. Olah, R. Krishnamurti, and G. K. S. Prakash, in *Comprehensive Organic Synthesis*, B. M. Trost and I. Fleming, eds., Vol. 3: *Carbon–Carbon σ-Bond Formation*, G. Pattenden, ed., Pergamon Press, Oxford, 1991, Chapter 1.8, p. 293.

87. E. R. Kline, B. N. Campbell, Jr., and E. C. Spaeth, *J. Org. Chem.* **24**, 1781 (1959).

88. R. Jenny, *Compt. Rend.* **246**, 3477 (1958).

89. R. Jenny, *Compt. Rend.* **250**, 1659 (1960).

90. L. Schmerling, *Ind. Eng. Chem.* **40**, 2072 (1948).

91. G. A. Olah, S. Kobayashi, and M. Tashiro, *J. Am. Chem. Soc.* **94**, 7448 (1972).

92. M. Segi, T. Nakajima, and S. Suga, *Bull. Chem. Soc. Jpn.* **53**, 1465 (1980).

93. G. A. Olah, O. Farooq, S. M. F. Farnia, and J. A. Olah, *J. Am. Chem. Soc.* **110**, 2560 (1988).

94. N. Mine, Y. Fujiwara, and H. Taniguchi, *Chem. Lett.* 357 (1986).

95. G. A. Olah, G. K. S. Prakash, and J. Sommer, *Superacids*. Wiley-Interscience, New York, 1985, Chapter 5, p. 278.

96. J. H. Simons and H. Hart, *J. Am. Chem. Soc.* **66**, 1309 (1944).

97. P. Laszlo, *Acc. Chem. Res.* **19**, 121 (1986).

98. M. Balogh and P. Laszlo, *Organic Chemistry Using Clays*. Springer, Berlin, 1993, Chapter 1, p. 5.

99. W. W. Kaeding, G. C. Barile, and M. M. Wu, *Catal. Rev.-Sci. Eng.* **26**, 597 (1984).

100. W. Hölderich, M. Hesse, and F. Näumann, *Angew. Chem., Int. Ed. Engl.* **27**, 226 (1988).

101. G. A. Olah, P. S. Iyer, and G. K. S. Prakash, *Synthesis* 513 (1986).

102. F. J. Waller, *ACS Symp. Ser.* **308**, 42 (1986).

103. S. J. Sondheimer, N. J. Bunce, and C. A. Fyfe, *J. Macromol. Sci., Rev. Macromol. Chem. Phys.* **C26**, 353 (1986).

104. G. K. S. Prakash and G. A. Olah, in *Acid-Base Catalysis*, K. Tanabe, H. Hattori, T. Yamaguchi, and T. Tanaka, eds., Kodansha, Tokyo, 1989, p. 59.

105. G. A. Olah, in *Friedel–Crafts and Related Reactions*, Vol. 1, G. A. Olah, ed., Wiley-Interscience, New York, 1963, Chapter 4, p. 201.

106. F. A. Drahowzal, in *Friedel–Crafts and Related Reactions*, Vol. 2, G. A. Olah, ed., Wiley-Interscience, New York, 1964, Chapter 17, p. 417.

107. G. A. Olah, *Friedel–Crafts Chemistry*, Wiley-Interscience, New York, 1973, Chapter 4, p. 215.

108. G. A. Olah and M. W. Meyer, in *Friedel–Crafts and Related Reactions*, Vol. 1, G. A. Olah, ed., Wiley-Interscience, New York, 1963, Chapter 8, p. 623.

109. R. M. Roberts and A. A. Khalaf, *Friedel–Crafts Alkylation Chemistry*, Marcel Dekker, New York, 1984, Chapter 3, p. 122.

110. N. O. Calloway, *J. Am. Chem. Soc.* **59**, 1474 (1937).

111. H. C. Brown and H. Jungk, *J. Am. Chem. Soc.* **77**, 5584 (1955).

112. G. A. Olah and S. J. Kuhn, *J. Org. Chem.* **29**, 2317 (1964).

113. G. Gustavson, *Chem. Ber.* **11**, 1251 (1878).

114. R. D. Silva, *J. Chem. Soc.* **48**, 1054 (1885).

115. V. N. Ipatieff, H. Pines, and L. Schmerling, *J. Org. Chem.* **5**, 253 (1940).

116. R. M. Roberts and D. Shiengthong, *J. Am. Chem. Soc.* **82**, 732 (1960).

117. R. M. Roberts and D. Shiengthong, *J. Am. Chem. Soc.* **86**, 2851 (1964).

118. H. C. Brown and M. Grayson, *J. Am. Chem. Soc.* **75**, 6285 (1953).

119. G. A. Olah, S. J. Kuhn, and S. H. Flood, *J. Am. Chem. Soc.* **84**, 1688 (1962).

120. M. J. S. Dewar, *J. Chem. Soc.* 406, 777 (1946).

121. R. Taylor, *Compr. Chem. Kinet.* **13**, 139 (1972).

122. H. C. Brown, H. W. Pearsall, L. P. Eddy, W. J. Wallace, M. Grayson, and K. L. Nelson, *Ind. Eng. Chem.* **45**, 1462 (1953).

123. S. U. Choi and H. C. Brown, *J. Am. Chem. Soc.* **85**, 2596 (1963).

124. R. M. Roberts, G. A. Ropp, and O. K. Neville, *J. Am. Chem. Soc.* **77**, 1764 (1955).

125. G. A. Olah, S. H. Flood, and M. E. Moffatt, *J. Am. Chem. Soc.* **86**, 1060 (1964).

126. G. A. Olah, S. H. Flood, S. J. Kuhn, M. E. Moffatt, and N. A. Overchuck, *J. Am. Chem. Soc.* **86**, 1046 (1964).

127. G. A. Olah, *Acc. Chem. Res.* **4**, 240 (1971).

128. R. Nakane, A. Natsubori, and O. Kurihara, *J. Am. Chem. Soc.* **87**, 3597 (1965).

129. R. Nakane and A. Natsubori, *J. Am. Chem. Soc.* **88**, 3011 (1966).

130. A. Natsubori and R. Nakane, *J. Org. Chem.* **35**, 3372 (1970).

131. R. Nakane, O. Kurihara, and A. Takematsu, *J. Org. Chem.* **36**, 2753 (1971).

132. A. Takematsu, K. Sugita, and R. Nakane, *Bull. Chem. Soc. Jpn.* **51**, 2082 (1978).

133. T. Oyama, T. Hamano, K. Nagumo, and R. Nakane, *Bull. Chem. Soc. Jpn.* **51**, 1441 (1978).

134. R. Nakane, O. Kurihara, and A. Natsubori, *J. Am. Chem. Soc.* **91**, 4528 (1969).

135. F. P. DeHaan, G. L. Delker, W. D. Covey, J. Ahn, M. S. Anisman, E. C. Brehm, J. Chang, R. M. Chicz, R. L. Cowan, D. M. Ferrara, C. H. Fong, J. D. Harper, C. D. Irani, J. Y. Kim, R. W. Meinhold, K. D. Miller, M. P. Roberts, E. M. Stoler, Y. J. Suh, M. Tang, and E. L. Williams, *J. Am. Chem. Soc.* **106**, 7038 (1984).

136. S. Masuda and T. Nakajima, S. Suga, *Bull. Chem. Soc. Jpn.* **56**, 1089 (1983).

137. R. M. Roberts and S. E. McGuire, *J. Org. Chem.* **35**, 102 (1970).

138. M. J. Schlatter and R. D. Clark, *J. Am. Chem. Soc.* **75**, 361 (1953).

139. R. H. Allen and L. D. Yats, *J. Am. Chem. Soc.* **83**, 2799 (1961).

140. J. H. Clark, A. P. Kybett, D. J. Macquarrie, S. J. Barlow, and P. Landon, *J. Chem. Soc., Chem. Commun.* 1353 (1989).

141. C. R. Hauser and B. E. Hudson, Jr., *Org. Synth., Coll. Vol.* **3**, 842 (1955).

142. S. Masuda, M. Segi, T. Nakajima, and S. Suga, *J. Chem. Soc., Chem. Commun.* 86 (1980).

143. D. L. Ransley, *J. Org. Chem.* **33**, 1517 (1968).

144. G. A. Olah and W. S. Tolgyesi, in *Friedel–Crafts and Related Reactions*, Vol. 2, G. A. Olah, ed., Wiley-Interscience, New York, 1964, Chapter 21, p. 659.

145. L. R. C. Barclay, in *Friedel–Crafts and Related Reactions*, Vol. 2, G. A. Olah, ed., Wiley-Interscience, New York, 1964, Chapter 22, p. 785.

146. R. M. Roberts and A. A. Khalaf, *Friedel–Crafts Alkylation Chemistry*, Marcel Dekker, New York, 1984, Chapter 6, p. 521.

147. R. M. Roberts and A. A. Khalaf, *Friedel–Crafts Alkylation Chemistry*, Marcel Dekker, New York, 1984, Chapter 5, p. 404.

148. S. H. Patinkin and B. S. Friedman, in *Friedel–Crafts and Related Reactions*, Vol. 2, G. A. Olah, ed., Wiley-Interscience, New York, 1964, Chapter 14, p. 1.

149. R. Koncos and B. S. Friedman, in *Friedel–Crafts and Related Reactions*, Vol. 2, G. A. Olah, ed., Wiley-Interscience, New York, 1964, Chapter 15, p. 289.

150. A. V. Topchiev, S. V. Zavgorodnii, and V. G. Kryuchkova, *Alkylation with Olefins*, Elsevier, Amsterdam, 1964, Chapter 2, p. 58.

151. R. M. Roberts and A. A. Khalaf, *Friedel–Crafts Alkylation Chemistry*, Marcel Dekker, New York, 1984, Chapter 4, p. 228.

152. V. Franzen, in *Friedel–Crafts and Related Reactions*, Vol. 2, G. A. Olah, ed., Wiley-Interscience, New York, 1964, Chapter 16, p. 413.

153. A. Schriesheim, in *Friedel–Crafts and Related Reactions*, Vol. 2, G. A. Olah, ed., Wiley-Interscience, New York, 1964, Chapter 18, p. 477.

154. F. A. Drahowzal, in *Friedel–Crafts and Related Reactions*, Vol. 2, G. A. Olah, ed., Wiley-Interscience, New York, 1964, Chapter 20, p. 641.

155. J. E. Mitchell, Jr., *Trans. Am. Inst. Chem. Eng.* **42**, 293 (1946).

156. A. W. Francis and E. E. Reid, *Ind. Eng. Chem.* **38**, 1194 (1946).

157. V. N. Ipatieff, H. Pines, and V. I. Komarewsky, *Ind. Eng. Chem.* **28**, 222 (1936).

158. Wm. J. Mattox, *Trans. Am. Inst. Chem. Eng.* **41**, 463 (1945).

159. V. N. Ipatieff and L. Schmerling, *Ind. Eng. Chem.* **38**, 400 (1946).

160. A. A. O'Kelly, J. Kellett, and J. Plucker, *Ind. Eng. Chem.* **39**, 154 (1947).

161. F. G. Dwyer, P. J. Lewis, and F. H. Schneider, *Chem. Eng. (NY)* **83**(1), 90 (1976).

162. I. Akervold, J. M. Bakke, and E. Stensvik, *Acta Chim. Scand.* **B39**, 437 (1985).

163. H. C. Brown and H. Pearsall, *J. Am. Chem. Soc.* **73**, 4681 (1951).

164. D. A. McCaulay, B. H. Shoemaker, and A. P. Lien, *Ind. Eng. Chem.* **42**, 2103 (1950).

165. G. A. Olah and S. J. Kuhn, *J. Am. Chem. Soc.* **80**, 6535 (1958).

166. G. A. Olah and A. E. Pavlath, *J. Am. Chem. Soc.* **80**, 6540 (1958).

167. H. M. Friedman and A. L. Nelson, *J. Org. Chem.* **34**, 3211 (1969).

168. G. A. Olah, Fr. Patent 2,498,177 (1982); Can. Patent 1,152,111 (1983).

169. G. A. Olah and J. A. Olah, *J. Am. Chem. Soc.* **98**, 1839 (1976).

170. P. Moreau, A. Finiels, P. Geneste, F. Moreau, and J. Solofo, *J. Org. Chem.* **57**, 5040 (1992).

171. B. S. Friedman and F. L. Morritz, *J. Am. Chem. Soc.* **78**, 2000 (1956).

172. A. C. Olson, *Ind. Eng. Chem.* **52**, 833 (1960).

173. H. R. Alul, *J. Org. Chem.* **33**, 1522 (1968).

174. T. Okuhara, T. Nishimura, H. Watanabe, and M. Misono, *J. Mol. Catal.* **74**, 247 (1992).

175. E. Bergman, *Chem. Rev.* **29**, 529 (1941).

176. E. Giovannini and K. Brandenberger, *Helv. Chim. Acta*, **56**, 1775 (1973).

177. A. A. Khalaf and R. M. Roberts, *J. Org. Chem.* **37**, 4227 (1972).

178. S. H. Patinkin and B. S. Friedman, in *Friedel–Crafts and Related Reactions*, Vol. 2, G. A. Olah, ed., Wiley-Interscience, New York, 1964, Chapter 14, p. 110.

179. A. V. Grosse and I. N. Ipatieff, *J. Am. Chem. Soc.* **57**, 2415 (1935).

180. A. V. Grosse, J. M. Mavity, and I. N. Ipatieff, *J. Org. Chem.* **3**, 448 (1938).

181. A. V. Grosse, J. M. Mavity, and I. N. Ipatieff, *J. Org. Chem.* **3**, 137 (1938).

182. I. N. Ipatieff, V. I. Komarewsky, and H. Pines, *J. Am. Chem. Soc.* **58**, 918 (1936).

183. O. V. Bragin, T. V. Vasina, S. A. Isaev, and Kh. M. Minachev, *Bull. Acad. Sci. USSR, Div. Chem. Sci.* (Engl. transl.) **36**, 1562 (1987).

184. S. A. Isaev, T. V. Vasina, and O. V. Bragin, *Izv. Akad. Nauk, Ser. Khim.* 2228 (1991); 2708 (1992).

185. L. Schmerling, J. P. Luvisi, and R. W. Welch, *J. Am. Chem. Soc.* **77**, 1774 (1955).

186. L. Schmerling, R. W. Welch, and J. P. West, *J. Am. Chem. Soc.* **78**, 5406 (1956).

187. L. Schmerling, R. W. Welch, and J. P. Luvisi, *J. Am. Chem. Soc.* **79**, 2636 (1957).

188. L. Schmerling, *J. Am. Chem. Soc.* **97**, 6134 (1975).

189. L. Schmerling and J. A. Vesely, *J. Org. Chem.* **38**, 312 (1973).

190. G. A. Olah, P. Schilling, J. S. Staral, Y. Halpern, and J. A. Olah, *J. Am. Chem. Soc.* **97**, 6807 (1975).

191. R. Miethchen, S. Steege, and C.-F. Kröger, *J. Prakt. Chem.* **325**, 823 (1983).

192. R. Miethchen, C. Roehse, and C.-F. Kröger, *Z. Chem.* **24**, 145 (1986).

193. S. Chalais, A. Cornélis, A. Gertsmans, W. Kolodziejski, P. Laszlo, and A. Mathy, *Helv. Chim. Acta* **68**, 1196 (1985).

194. H. Pines, L. Schmerling, and V. N. Ipatieff, *J. Am. Chem. Soc.* **62**, 2901 (1940).

195. A. Streitwieser, Jr., D. P. Stevenson, and W. D. Schaeffer, *J. Am. Chem. Soc.* **81**, 1110 (1959).

196. C. D. Nenitzescu, I. Necsoiu, and M. Zalman, *Chem. Ber.* **92**, 10 (1959).

197. R. M. Roberts and Y. W. Han, *J. Am. Chem. Soc.* **85**, 1168 (1963).

198. A. A. Khalaf and R. M. Roberts, *J. Org. Chem.* **35**, 3717 (1970).

199. K. Ichikawa, K. Chano, M. Inoue, and T. Sugita, *Bull. Chem. Soc. Jpn.* **55**, 3039 (1982).

200. P. Laszlo and A. Mathy, *Helv. Chim. Acta* **70**, 577 (1987).

201. K. Urabe, H. Sakurai, and Y. Izumi, *J. Chem. Soc., Chem. Commun.* 1074 (1986); 1520 (1988).

202. H. Sakurai, K. Urabe, and Y. Izumi, *J. Chem. Soc., Chem. Commun.* 1519 (1988).

203. J. Kaspi, D. D. Montgomery, and G. A. Olah, *J. Org. Chem.* **43**, 3147 (1978).

204. W. W. Kaeding, C. Chu, L. B. Young, B. Weinstein, and S. A. Butter, *J. Catal.* **67**, 159 (1981).

205. J.-H. Kim, S. Namba, and T. Yashima, in *Zeolites as Catalysts, Sorbents and Detergent Builders*, Studies in Surface Science and Catalysis, Vol. 46, H. G. Karge and J. Weitkamp, eds., Elsevier, Amsterdam, 1989, p. 71.

206. D. A. McCaulay, in *Friedel–Crafts and Related Reactions*, Vol. 2, G. A. Olah, ed., Wiley-Interscience, New York, 1964, Chapter 24, p. 1066.

207. R. M. Roberts and A. A. Khalaf, *Friedel–Crafts Alkylation Chemistry*, Marcel Dekker, New York, 1984, Chapter 7, p. 673.

208. G. A. Olah and J. Kaspi, *New J. Chem.* **2**, 581 (1978).

209. K. J. Day and T. M. Snow, in *Toluene, the Xylenes and their Industrial Derivatives*, E. G. Hancock, ed., Elsevier, Amsterdam, 1982, Chapter 3, p. 50.

210. N. Y. Chen and W. E. Garwood, *Catal. Rev.-Sci. Eng.* **28**, 253 (1986).

211. N. Y. Chen, W. E. Garwood, and F. G. Dwyer, *Shape Selective Catalysis in Industrial Applications*, Marcel Dekker, New York, 1989, Chapter 4, p. 93.

212. R. L. Burwell, Jr. and A. D. Shields, *J. Am. Chem. Soc.* **77**, 2766 (1955).

213. R. H. Allen, *J. Am. Chem. Soc.* **82**, 4856 (1960).

214. G. A. Olah, C. G. Carlson, and J. C. Lapierre, *J. Org. Chem.* **29**, 2687 (1964).

215. G. G. Moore and A. P. Wolf, *J. Org. Chem.* **31**, 1106 (1966).

216. G. A. Olah and J. Kaspi, *New J. Chem.* **2**, 585 (1978).

217. V. G. Lipovich, A. V. Vysotskii, and V. V. Chenets, *Neftekhimiya* **11**, 656 (1971).

218. R. E. Kinney and L. A. Hamilton, *J. Am. Chem. Soc.* **76**, 786 (1954).

219. D. A. McCaulay and A. P. Lien, *J. Am. Chem. Soc.* **75**, 2411 (1953).

220. H. Pines and J. T. Arrigo, *J. Am. Chem. Soc.* **80**, 4369 (1958).

221. A. Streitwieser, Jr. and L. Reif, *J. Am. Chem. Soc.* **86**, 1988 (1964).

222. P. C. Myhre, T. Rieger, and J. T. Stone, *J. Org. Chem.* **31**, 3425 (1966).

223. M. Tashiro, *Synthesis* 921 (1979).

224. G. A. Olah, G. K. S. Prakash, P. S. Iyer, M. Tashiro, and T. Yamato, *J. Org. Chem.* **52**, 1881 (1987).

225. R. M. Roberts and A. A. Khalaf, *Friedel–Crafts Alkylation Chemistry*, Marcel Dekker, New York, 1984, Chapter 8, p. 726.

226. V. N. Ipatieff and H. Pines, *J. Am. Chem. Soc.* **59**, 56 (1937).

227. R. M. Roberts, E. K. Baylis, and G. J. Fonken, *J. Am. Chem. Soc.* **85**, 3454 (1963).

228. D. M. Brouwer, *Recl. Trav. Chim. Pays-Bas* **87**, 210 (1968).

229. G. A. Olah, R. H. Schlosberg, R. D. Porter, Y. K. Mo, D. P. Kelly, and G. D. Mateescu, *J. Am. Chem. Soc.* **94**, 2034 (1972).

230. H. M. Knight, J. T. Kelly, and J. R. King, *J. Org. Chem.* **28**, 1218 (1963).

231. H. Pines and L. A. Schaap, *Adv. Catal.* **12**, 126 (1960).

232. H. Pines, *Synthesis* 309 (1974); *Acc. Chem. Res.* **7**, 155 (1974).

233. H. Pines and W. M. Stalick, *Base-Catalyzed Reactions of Hydrocarbons and Related Compounds*, Academic Press, New York, 1977, Chapter 5, p. 240.

234. H. Pines, *The Chemistry of Catalytic Hydrocarbon Conversions*, Academic Press, New York, 1981, Chapter 1, p. 139.

235. H. Pines, J. A. Vesely, and V. N. Ipatieff, *J. Am. Chem. Soc.* **77**, 554 (1955).

236. R. M. Schramm and G. E. Langlois, *J. Am. Chem. Soc.* **82**, 4912 (1960).

237. L. Schaap and H. Pines, *J. Am. Chem. Soc.* **79**, 4967 (1957).

238. B. Stipanovic and H. Pines, *J. Org. Chem.* **34**, 2106 (1969).

239. H. Pines and N. C. Sih, *J. Org. Chem.* **30**, 280 (1965).

240. G. G. Eberhardt and H. J. Peterson, *J. Org. Chem.* **30**, 82 (1965).

241. G. G. Eberhardt, *Organomet. Chem. Rev.* **1**, 491 (1966).

242. S. Watanabe, K. Suga, and T. Fujita, *Synthesis* 375 (1971).

243. H. Pines and D. Wunderlich, *J. Am. Chem. Soc.* **80**, 6001 (1958).

244. J. Shabtai and H. Pines, *J. Org. Chem.* **26**, 4225 (1961).

245. J. Shabtai, E. M. Lewicki, and H. Pines, *J. Org. Chem.* **27**, 2618 (1962).

246. J. L. Wardell, in *Comprehensive Organometallic Chemistry*, Vol. 1, G. Wilkinson, F. G. A. Stone, and E. W. Abel, eds., Pergamon Press, Oxford, 1982, Chapter 2, p. 54.

247. J. L. Wardell, in *The Chemistry of the Metal–Carbon Bond*, Vol. 4, F. R. Hartley, ed., Wiley, Chichester, 1987, Chapter 1, p. 27.

248. W. E. Lindsell, in *Comprehensive Organometallic Chemistry*, Vol. 1, G. Wilkinson, F. G. A. Stone, and E. W. Abel, eds., Pergamon Press, Oxford, 1982, Chapter 4, p. 163.

249. L. Lochmann, J. Pospísil, and D. Lím, *Tetrahedron Lett.* 257 (1966).

250. M. Schlosser, *J. Organomet. Chem.* **8**, 9 (1967).

251. M. Schlosser, *Pure Appl. Chem.* **60**, 1627 (1988).

252. A. Mordini, *Adv. Carbanion Chem.* **1**, 1 (1992).

253. J. Hartmann and M. Schlosser, *Helv. Chim. Acta* **59**, 453 (1976).

254. M. Schlosser, *Angew. Chem., Int. Ed. Engl.* **13**, 701 (1974).

255. M. Schlosser and J. Hartmann, *J. Am. Chem. Soc.* **98**, 4674 (1976).

256. M. Schlosser, J. Hartmann, and V. David, *Helv. Chim. Acta* **57**, 1567 (1974).

257. P. A. A. Klusener, H. H. Hommes, H. D. Verkruijsse, and L. Brandsma, *J. Chem. Soc., Chem. Commun.* 1677 (1985).

258. P. A. A. Klusener, W. Kulik, and L. Brandsma, *J. Org. Chem.* **52**, 5261 (1987).

259. G. B. Trimitsis, A. Tuncay, R. D. Beyer, and K. J. Ketterman, *J. Org. Chem.* **38**, 1491 (1973).

260. V. R. Sandel and H. H. Freedman, *J. Am. Chem. Soc.* **85**, 2328 (1963).

261. R. B. Bates and C. A. Ogle, *J. Org. Chem.* **47**, 3949 (1982).

262. C. D. Broaddus, *J. Org. Chem.* **29**, 2689 (1964).

263. M. Schlosser and S. Strunk, *Tetrahedron Lett.* **25**, 741 (1984).

264. H. Hommes, H. D. Verkruijsse, and L. Brandsma, *J. Chem. Soc., Chem. Commun.* 366 (1981).

265. L. Brandsma, H. Hommes, H. D. Verkruijsse, and R. L. P. de Jong, *Recl. Trav. Chim. Pays-Bas* **104**, 226 (1985).

266. D. A. Ben-Efraim, in *The Chemistry of the Carbon–Carbon Triple Bond*, S. Patai, ed., Wiley, Chichester, 1978, Chapter 18, p. 790.

267. A. J. Quillinan and F. Scheinmann, *Org. Synth., Coll. Vol.* **6**, 595 (1988).

268. S. Bhanu and F. Scheinmann, *J. Chem. Soc., Perkin Trans. 1* 1218 (1979).

269. T. Koerts, P. A. Leclercq, and R. A. van Santen, *J. Am. Chem. Soc.* **114**, 7272 (1992).

270. A. M. Browstein, in *Catalysis of Organic Recations*, W. R. Moser, ed., Marcel Dekker, New York, 1981, p. 3.

271. H. Itoh, T. Hattori, K. Suzuki, and Y. Murakami, *J. Catal.* **79**, 21 (1983).

272. Yu. N. Sidorenko, P. N. Galich, V. S. Gutyrya, V. G. Il'in, and I. E. Neimark, *Dokl. Akad. Nauk, SSSR* **173**, 132 (1967).

273. T. Yashima, K. Sato, T. Hayasaka, and N. Hara, *J. Catal.* **26**, 303 (1972).

274. M. D. Sefcik, *J. Am. Chem. Soc.* **101**, 2164 (1979).

275. M. L. Unland and G. E. Barker, in *Catalysis of Organic Recations*, W. R. Moser, ed., Marcel Dekker, New York, 1981, p. 51.

276. C. Yang and Z. Meng, *Appl. Catal.* **71**, 45 (1991).

277. R. E. Payne, *Petrol. Refiner* **37**(9), 316 (1958).

278. E. K. Jones, *Adv. Catal.* **10**, 165 (1968).

279. W. L. Lafferty, Jr. and R. W. Stokeld, *Adv. Chem. Ser.* **103**, 130 (1971).

280. *Catalysis Looks to the Future*, National Research Council Reports. National Academy Press, Washington, DC, 1992, p. 11.

281. F. Asinger, *Mono-Olefins, Chemistry and Technology*, Pergamon Press, Oxford, 1968, Chapter 5, p. 464.

282. A. Schriesheim and I. Kirshenbaum, *Chemtech* **8**, 310 (1978).

283. R. H. Rosenwald, in *Kirk-Othmer Encyclopedia of Chemical Technology*, 3rd ed., Vol. 2, M. Grayson and D. Eckroth, eds., Wiley-Interscience, New York, 1978, p. 50.

284. G. Stefanidakis and J. E. Gwyn, in *Encyclopedia of Chemical Processing and Design*, Vol. 2, J. J. McKetta and W. A. Cunningham, eds., Marcel Dekker, New York, 1977, p. 357.

285. D. H. Putney and O. Webb, Jr., *Petrol. Refiner* **38**(9), 166 (1959).

286. *Hydrocarbon Process.* **47**(9), 164 (1968).

287. *Hydrocarbon Process., Int. Ed.* **59**(9), 174–176 (1980).

288. L. F. Albright, *Oil Gas J.* **88**(46), 79 (1990).

289. H. Lerner and V. A. Citarella, *Hydrocarbon Process., Int. Ed.* **70**(11), 89 (1991).

290. R. S. Logan and R. L. Banks, *Hydrocarbon Process.* **47**(6), 135 (1968); *Oil Gas J.* **66**(21), 131 (1968).

291. *Hydrocarbon Process.* **47**(9), 165 (1968).

292. L. F. Albright, *Oil Gas J.* **88**(48), 70 (1990).

293. L. F. Albright, M. A. Spalding, J. Faunce, and R. E. Eckert, *Ind. Eng. Chem. Res.* **27**, 391 (1988).

294. C. L. Rogers, *Oil Gas J.* **69**(45), 60 (1971).

295. A. E. Eleazar, R. M. Heck, and M. P. Witt, *Hydrocarbon Process.* **58**(5), 112 (1979).

296. R. M. Heck, G. R. Patel, W. S. Breyer, and D. D. Merrill, *Oil Gas J.* **81**(3), 103 (1983).

297. S. Novalany and R. G. McClung, *Hydrocarbon Process., Int. Ed.* **68**(9), 66 (1989).

298. G. Chaput, J. Laurent, J. P. Boitiaux, J. Cosyns, and P. Sarrazin, *Hydrocarbon Process., Int. Ed.* **71**(9), 51 (1992).

299. K. R. Masters and E. A. Prohaska, *Hydrocarbon Process., Int. Ed.* **67**(8), 48 (1988).

300. R. L. Van Zele, and R. Diener, *Hydrocarbon Process., Int. Ed.* **69**(6), 92 (1990); **69**(7), 77 (1990).

301. G. A. Olah, U.S. Patent 5,073,674 (1991).

302. K. R. Comey, III, L. K. Gilmer, G. P. Partridge, and D. W. Johnson, *AIChE Summer Natl. Meeting*, 1993.

303. H. U. Hammershaimb and B. R. Shah, *Hydrocarbon Process., Int. Ed.* **64**(6), 73 (1985).

304. L. E. Chapin, G. C. Liolios, and T. M. Robertson, *Hydrocarbon Process., Int. Ed.* **64**(9), 67 (1985).

305. L. F. Albright, B. M. Doshi, M. A. Ferman, and A. Ewo, *ACS Symp. Ser.* **55**, 96, 109 (1977).

306. L. F. Albright, M. A. Spalding, J. A. Nowinski, R. M. Ybarra, and R. E. Eckert, *Ind. Eng. Chem. Res.* **27**, 381 (1988).

307. *Appl. Catal., A* **103**, N15 (1993).

308. F. Asinger, *Mono-Olefins, Chemistry and Technology*, Pergamon Press, Oxford, 1968, Chapter 11, p. 948.

309. K. Weissermel and H.-J. Arpe, *Industrial Organic Chemistry*, 2nd ed., VCH, Weinheim, 1993, Chapter 13, p. 333.

310. A. C. MacFarlane, *ACS Symp. Ser.* **55**, 341 (1977).

311. *Hydrocarbon Process., Int. Ed.* **66**(11), 74 (1987).

312. H. W. Grote, *Oil Gas J.* **56**(13), 73 (1958).

313. *Hydrocarbon Process.* **58**(11), 159 (1979).

314. P. J. Lewis and F. G. Dwyer, *Oil Gas J.* **75**(40), 55 (1977).

315. F. G. Dwyer, in *Catalysis of Organic Reactions*, W. R. Moser, ed., Marcel Dekker, New York, 1981, p. 39.

316. *Hydrocarbon Process., Int. Ed.* **70**(3), 154 (1991).

317. P. Lévesque, *Chemtech* **20**, 756 (1990).

318. M. C. Hoff, in *Kirk-Othmer Encyclopedia of Chemical Technology*, 3rd ed., Vol. 23, M. Grayson and D. Eckroth, eds., Wiley-Interscience, New York, 1983, p. 269.

319. *Hydrocarbon Process., Int. Ed.* **70**(3), 152 (1991).

320. D. Rotman, *Chem. Week* **152**(1), 48 (1993).

321. D. Rotman, *Chem. Week* **152**(7), 9 (1993).

322. A. Wood and D. Rotman, *Chem. Week* **153**(20), 18 (1993).

323. P. G. Menon, *Appl. Catal., A* **110**, N22 (1994).

324. S. Otani, *Chem. Eng. (NY)* **77**(16), 118 (1970).

325. *Hydrocarbon Process., Int. Ed.* **62**(11), 159 (1983).

326. K. P. Menard, *Oil Gas J.* **85**(11), 46 (1987).

327. P. Grandio, F. H. Schneider, A. B. Schwartz, and J. J. Wise, *Hydrocarbon Process.* **51**(8), 85 (1972).

328. S. Han, D. S. Shihabi, R. P. L. Absil, Y. Y. Huang, S. M. Leiby, D. O. Marler, and J. P. McWilliams, *Oil Gas J.* **87**(34), 83 (1989).

329. F. Gorra, L. L. Breckenridge, W. M. Guy, and R. A. Sailor, *Oil Gas J.* **90**(41), 60 (1992).

330. *Hydrocarbon Process., Int. Ed.* **72**(3), 166 (1993).

331. *Hydrocarbon Process., Int. Ed.* **66**(11), 66 (1987).

332. K. E. Coulter, H. Kehde, and B. F. Hiscock, in *Vinyl and Diene Monomers*, Part 2, *High Polymers*, Vol. 24, E. C. Leonard, ed., Wiley-Interscience, New York, 1971, Chapter 1, p. 546.

333. W. W. Kaeding, B. Young, and A. G. Prapas, *Chemtech* **12**, 556 (1982).

334. G. Paparatto, E. Moretti, G. Leofanti, and F. Gatti, *J. Catal.* **105**, 227 (1987).

335. R. C. Berg and B. V. Vora, in *Encyclopedia of Chemical Processing and Design*, Vol. 15, J. J. McKetta and W. A. Cunningham, eds., Marcel Dekker, New York, 1982, p. 266.

336. D. G. Demianiw, in *Kirk-Othmer Encyclopedia of Chemical Technology*, 3rd ed., Vol. 16, M. Grayson and D. Eckroth, eds., Wiley-Interscience, New York, 1981, p. 480.

337. H. S. Bloch, *Oil Gas J.* **65**(3), 79 (1967).

338. B. V. Vora, P. R. Pujado, J. B. Spinner, and T. Imai, *Hydrocarbon Process., Int. Ed.* **63**(11), 86 (1984).

339. *Hydrocarbon Process., Int. Ed.* **66**(11), 63 (1987).

340. *Hydrocarbon Process., Int. Ed.* **64**(11), 127 (1985).

341. *Hydrocarbon Process.* **58**(11), 183 (1979).

342. J. S. Beck and W. O. Haag, in *Handbook of Heterogeneous Catalysis*, G. Ertl, H. Knözinger, and J. Weitkamp, eds., Wiley-VCH, Weinheim, 1997, Chapter 4.1, p. 2123.

343. A. Corma, *Chem. Rev.* **95**, 559 (1995).

344. A. Corma and H. García, *Catal. Today* **38**, 257 (1997); *J. Chem. Soc., Dalton Trans.* 1381 (2000).

345. T. Okuhara, N. Mizuno, and M. Misono, *Adv. Catal.* **41**, 113 (1996).

346. N. Mizuno and M. Misono, *Chem. Rev.* **98**, 199 (1998).

347. M. Misono, *Chem. Commun.* 1141 (2001).

348. I. V. Kozhevnikov, *Chem. Rev.* **98**, 171 (1998).

349. G. D. Yadav and J. J. Nair, *Catal. Today* **33**, 1 (1999).

350. G. A. Olah, P. Batamack, D. Deffieux, B. Török, Q. Wang, Á. Molnár, and G. K. S. Prakash, *Appl. Catal., A* **146**, 107 (1996).

351. G. A. Olah, E. Marinez, B. Török, and G. K. S. Prakash, *Catal. Lett.* **61**, 105 (1999).

352. Z. Zhao, W. Sun, X. Yang, X. Ye, and Y. Wu, *Catal. Lett.* **65**, 115 (2000).

353. J. Weitkamp and Y. Traa, in *Handbook of Heterogeneous Catalysis*, G. Ertl, H. Knözinger, and J. Weitkamp, eds., Wiley-VCH, Weinheim, 1997, Chapter 3.14, p. 2039.

354. P. S. E. Dai, *Catal. Today* **26**, 3 (1995).

355. M. Guisnet and N. S. Gnep, *Appl. Catal., A* **146**, 33 (1996).

356. J. Weitkamp and Y. Traa, *Catal. Today* **49**, 193 (1999).

357. M. F. Simpson, J. Wei and S. Sundaresan, *Ind. Eng. Chem. Res.* **35**, 3861 (1996).

358. Y. Zhuang and F. T. T. Ng, *Appl. Catal., A* **190**, 137 (2000).

359. C. A. Querini, *Catal. Today* **62**, 135 (2000).

360. D. Das and D. K. Chakrabarty, *Energy Fuels* **12**, 109 (1998).

361. R. B. Gore and W. J. Thomson, *Appl. Catal., A* **168**, 23 (1998).

362. X. Xiao, J. W. Tierney, and I. Wender, *Appl. Catal., A* **183**, 209 (1999).

363. K. Satoh, H. Matsuhashi, and K. Arata, *Appl. Catal., A* **189**, 35 (1999).

364. A. de Angelis, P. Ingallina, D. Berti, L. Montanari, and M. G. Clerici, *Catal. Lett.* **61**, 45 (1999).

365. G. Baronetti, H. Thomas, and C. A. Querini, *Appl. Catal., A* **217**, 131 (2001).

366. E. Dumitriu, V. Hulea, I. Fechete, C. Catrinescu, A. Auroux, J.-F. Lacaze, and C. Guimon, *Appl. Catal., A* **181**, 15 (1999).

367. E. Dumitriu, D. T. On, and S. Kaliaguine, *J. Catal.* **170**, 150 (1997).

368. T. Yashima, Y. Katoh, and T. Komatsu, in *Porous Materials in Environmentally Friendly Processes*, Studies in Surface Science and Catalysis, Vol. 125, I. Kiricsi, G. Pál-Borbély, J. B. Nagy, and H. G. Karge, eds., Elsevier, Amsterdam, 1999, p. 507.

369. J. Tateiwa, K. Hashimoto, T. Yamauchi, and S. Uemura, *Bull. Chem. Soc. Jpn.* **69**, 2361 (1996).

370. M. A. Aramendía, V. Borau, C. Jiménez, J. M. Marinas, F. J. Romero, and F. J. Urbano, *Catal. Lett.* **73**, 203 (2001).

371. Á. Molnár, Cs. Keresszegi, T. Beregszászi, B. Török, and M. Bartók, in *Catalysis of Organic Reactions*, F. E. Herkes, ed., Marcel Dekker, New York, 1998, pp. 507.

372. S. D. Alexandratos and D. H. J. Miller, *Macromolecules* **33**, 2011 (2000).

373. Y. Sugi and Y. Kubota, in *Catalysis, a Specialist Periodical Report*, Vol. 13, J. J. Spivey, senior reporter, The Royal Society of Chemistry, Cambridge, 1997, Chapter 3, p. 55.

374. J. C. Cheng, T. F. Degnan, J. S. Beck, Y. Y. Huang, M. Kalyanaraman, J. A. Kowalski, C. A. Loehr, and D. N. Mazzone, in *Science and Technology in Catalysis 1998*, Studies in Surface Science and Catalysis, Vol. 121, H. Hattori and K. Otsuka, eds., Kodansha, Tokyo and Elsevier, Amsterdam, 1999, p. 53.

375. A. Corma, V. Martínez-Soria, and E. Schnoeveld, *J. Catal.* **192**, 163 (2000).

376. A. B. Halgeri, *Appl. Catal., A* **115**, N4 (1994).

377. D. Vergani, R. Prins, and H. W. Kouwenhoven, *Appl. Catal., A* **163**, 71 (1997).

378. T. Hanaoka, K. Nakajima, Y. Sugi, T. Matsuzaki, Y. Kubota, S. Tawada, K. Kunimori, and A. Igarashi, *Catal. Lett.* **50**, 149 (1998).

379. Y. Sugi, S. Tawada, T. Sugimura, Y. Kubota, T. Hanaoka, T. Matsuzaki, K. Nakajima, and K. Kunimori, *Appl. Catal., A* **189**, 251 (1999).

380. A. D. Schmitz and C. Song, *Catal. Lett.* **40**, 59 (1996).

381. K. Smith and S. D. Roberts, *Catal. Today* **60**, 227 (2000).

382. P. Moreau, Z. Liu, F. Fajula, and J. Joffre, *Catal. Today* **60**, 235 (2000).

383. B. M. Choudary, M. L. Kantam, M. Sateesh, K. K. Rao, and P. L. Santhi, *Appl. Catal., A* **149**, 257 (1997).

384. I. Benito, A. del Riego, M. Martínez, C. Blanco, C. Pesquera, and F. González, *Appl. Catal., A* **180**, 175 (1999).

385. J. H. Clark and D. J. Macquarrie, *Org. Process Res. Dev.* **1**, 149 (1997).

386. S. G. Pai, A. R. Bajpai, A. B. Deshpande, and S. D. Samant, *J. Mol. Catal. A: Chem.* **156**, 233 (2000).

387. A. B. Deshpande, A. R. Bajpai, and S. D. Samant, *Appl. Catal., A* **209**, 229 (2001).

388. S. Sebti, R. Tahir, R. Nazih, and S. Boulaajaj, *Appl. Catal., A* **218**, 25 (2001).

389. X. Hu, G. K. Chuah, and S. Jaenicke, *Appl. Catal., A* **217**, 1 (2001).

390. J. H. Clark, K. Martin, A. J. Teasdale, and S. J. Barlow, *J. Chem. Soc., Chem. Commun.* 2037 (1995).

391. X. Hu, M. L. Foo, G. K. Chuah, and S. Jaenicke, *J. Catal.* **195**, 412 (2000).

392. X. S. Zhao, G. Q. Lu, and C. Song, *Chem. Commun.* 2306 (2001).

393. V. R. Choudhary, S. K. Jana, and B. P. Kiran *J. Catal.* **192**, 257 (2000); *Catal. Lett.* **64**, 223 (2000).

394. J. M. Miller, D. Wails, J. S. Hartman, K. Schebesh, and J. L. Belelie, *Can. J. Chem.* **76**, 382 (1998).

395. J. M. Miller, M. Goodchild, J. L. Lakshmi, D. Wails, and J. S. Hartman, *Catal. Lett.* **63**, 199 (1999).

396. J. H. Clark, G. L. Monks, D. J. Nightingale, P. M. Price, and J. F. White, *J. Catal.* **193**, 348 (2000).

397. A. de Angelis, S. Amarilli, D. Berti, L. Montanari, and C. Perego, *J. Mol. Catal. A: Chem.* **146**, 37 (1999).

398. G. A. Olah, B. Török, T. Shamma, M. Török, and G. K. S. Prakash, *Catal. Lett.* **42**, 5 (1996).

399. B. Török, I. Kiricsi, Á. Molnár, and G. A. Olah, *J. Catal.* **193**, 132 (2000).

400. T. Beregszászi, B. Török, Á. Molnár, G. A. Olah, and G. K. S. Prakash, *Catal. Lett.* **48**, 83 (1997).

401. Á. Molnár, Cs. Keresszegi, and B. Török, *Appl. Catal., A* **189**, 217 (1999).

402. N. Nishiyama, M. Miyamoto, Y. Egashira, and K. Ueyama, *Chem. Commun.* 1746 (2001).

403. K. Qiao and Y. Deng, *J. Mol. Catal. A: Chem.* **171**, 81 (2001).

404. C. E. Song, W. H. Shim, E. J. Roh, and J. H. Choi, *Chem. Commun.* 1695 (2000).

405. M. H. Valkenberg, C. deCastro, and W. F. Hölderich, *Top. Catal.* **14**, 139 (2001).

406. M. G. Hitzler, F. R. Smail, S. K. Ross, and M. Poliakoff, *Chem. Commun.* 359 (1998).

407. T. Matsumoto. D. J. Taube, R. A. Periana, H. Taube, and H. Yoshida, *J. Am. Chem. Soc.* **122**, 7414 (2000).

408. E. M. Kennedy, F. Lonyi, T. H. Ballinger, M. P. Rosynek, and J. H. Lunsford, *Energy Fuels* **8**, 846 (1994).

409. S. J. X. He, M. A. Long, M. A. Wilson, M. L. Gorbaty, and P. S. Maa, *Energy Fuels* **9**, 616 (1995).

410. M. Adebajo, M. A. Long, and R. F. Howe, *Res. Chem. Intermed.* **26**, 185 (2000).

411. M. O. Adebajo, R. F. Howe, and M. A. Long, *Catal. Today* **63**, 471 (2000).

412. M. O. Adebajo, R. F. Howe, and M. A. Long, *Catal. Lett.* **72**, 221 (2001).

413. S. Kato, K. Nakagawa, N. Ikenaga, and T. Suzuki, *Catal. Lett.* **73**, 175 (2001).

414. I. I. Ivanova, N. Blom, and E. G. Derouane, *J. Mol. Catal. A: Chem.* **109**, 157 (1996).

415. G. Sartori, F. Bigi, A. Pastorio, C. Porta, A. Arienti, R. Maggi, N. Moretti, and G. Gnappi, *Tetrahedron Lett.* **36**, 9177 (1995).

416. T. Tsuchimoto, T. Maeda, E. Shirakawa, and Y. Kawakami, *Chem. Commun.* 1573 (2000).

417. S. Kobayashi, *Synlett* 689 (1994).

418. S. Kobayashi, *Eur. J. Org. Chem.* 15 (1999).

419. R. W. Marshman, *Aldrichimica Acta* **28**, 77 (1995).

420. T. Tsuchimoto, K. Tobita, T. Hiyama, and S. Fukuzawa, *J. Org. Chem.* **62**, 6997 (1997).

421. H. Kotsuki, T. Ohishi, M. Inoue, and T. Kojima, *Synthesis* 603 (1999).

422. R. P. Singh, R. M. Kamble, K. L. Chandra, P. Saravanan, and V. K. Singh, *Tetrahedron* **57**, 241 (2001).

423. A. Kawada, S. Mitamura, J. Matsuo, T. Tsuchiya, and S. Kobayashi, *Bull. Chem. Soc. Jpn.* **73**, 2325 (2000).

424. I. Hachiya, M. Moriwaki, and S. Kobayashi, *Bull. Chem. Soc. Jpn.* **68**, 2053 (1995).

425. M. P. D. Mahindaratne and K. Wimalasena, *J. Org. Chem.* **63**, 2858 (1998).

426. Á. Molnár, B. Török, I. Bucsi, and A. Földvári, *Top. Catal.* **6**, 9 (1998).

427. G. A. Olah, G. Rasul, C. York, and G. K. S. Prakash, *J. Am. Chem. Soc.* **117**, 11211 (1995).

428. S. Saito, T. Ohwada, and K. Shudo, *J. Am. Chem. Soc.* **117**, 11081 (1995); *J. Org. Chem.* **61**, 8089 (1996).

429. T. Miyai, Y. Onishi, and A. Baba, *Tetrahedron* **55**, 1017 (1999).

430. D. Barthomeuf, *Catal. Rev.-Sci. Eng.* **38**, 521 (1996).

431. W. S. Wieland, R. J. Davies, and J. M. Garces, *J. Catal.* **173**, 490 (1998).

432. W. S. Wieland, R. J. Davies, and J. M. Garces, *Catal. Today* **28**, 443 (1996).

433. R. J. Davis, *Res. Chem. Intermed.* **26**, 21 (2000).

434. J. Xie and S. Kaliaguine, *Appl. Catal., A* **148**, 415 (1997).

435. M. Hunger, U. Schenk, and J. Weitkamp, *J. Mol. Catal. A: Chem.* **134**, 97 (1998).

436. A. Philippou and M. W. Anderson, *J. Am. Chem. Soc.* **116**, 5774 (1994).

437. M. Hunger, U. Schenk, M. Seiler, and J. Weitkamp, *J. Mol. Catal. A: Chem.* **156**, 153 (2000).

438. A. E. Palomares, G. Eder-Mirth, and J. A. Lercher, *J. Catal.* **168**, 442 (1997).

439. S. V. Bordawekar and R. J. Davis, *J. Catal.* **189**, 79 (2000).

440. R. J. Davis, E. J. Doskocil, and S. Bordawekar, *Catal. Today* **62**, 241 (2000).

441. K. Arishtirova, P. Kovacheva, and N. Davidova, *Appl. Catal., A* **167**, 271 (1998).

442. K. Arishtirova, P. Kovacheva, and N. Davidova, in *Porous Materials in Environmentally Friendly Processes*, Studies in Surface Science and Catalysis, Vol. 125, I. Kiricsi, G. Pál-Borbély, J. B. Nagy, and H. G. Karge, eds., Elsevier, Amsterdam, 1999, p. 489.

443. P. Kovacheva, K. Arishtirova, and S. Vassilev, *Appl. Catal., A* **210**, 391 (2001).

444. M. G. Stevens, M. R. Anderson, and H. C. Foley, *Chem. Commun.* 413 (1999).

445. B. R. Steele and C. G. Screttas, *J. Am. Chem. Soc.* **122**, 2391 (2000).

446. N. Y. Chen, W. E. Garwood, and F. G. Dwyer, *Shape Selective Catalysis in Industrial Applications*, Marcel Dekker, New York, 1996, Chapter 6.

447. Q. Chen, X. Chen, L. Mao, and W. Cheng, *Catal. Today* **51**, 141 (1999).

448. T.-C. Tsai, S.-B. Liu, and I. Wang, *Appl. Catal., A* **181**, 355 (1999).

449. M. Guisnet, N. S. Gnep, and S. Morin, *Micropor. Mesopor. Mat.* **35**, 47 (2000).

450. S. A. Saleh and H. I. Tashtoush, *Tetrahedron* **54**, 14157 (1998).

451. N. Fujiwara and Y. Yamamoto, *J. Org. Chem.* **62**, 2318 (1997).

452. B. C. Ranu and A. Majee, *Chem. Commun.* 1225 (1997).

453. U. Biermann and J. O. Metzger, *Angew. Chem., Int. Ed. Engl.* **38**, 3675 (1999).

6

ADDITION

Numerous compounds can add to carbon–carbon multiple bonds by electrophilic addition mechanism. These include halogens, interhalogens, pseudohalogens, H–X compounds (water, inorganic and organic acids, alcohols, thiols, ammonia, and amines), Hlg–X reagents (hypohalous acids and hypohalites, electrophilic sulfur, selenium and tellurium compounds, inorganic and organic acyl halides), metal hydrides, metal halides, and organometallics. Of these reactions only those having great chemical and practical significance are discussed here. A short discussion about cycloaddition is also included. Electrophilic oxygenation of alkenes (epoxidation by peroxyacids), in turn, is treated in Chapter 9.

Other additions, such as addition of alkyl halides and carbonyl compounds, are discussed in Chapter 5, whereas Chapter 7 covers addition reactions involving carbon monoxide (hydroformylation, carboxylations). Hydrogen addition is discussed in Chapter 11. The nucleophilic addition of organometallics to multiple bonds is of great significance in the anionic polymerization of alkenes and dienes and is treated in Chapter 13.

6.1. HYDRATION

Addition of water to unsaturated hydrocarbons (hydration) and the related addition of alcohols are important processes from both practical and chemical points of view. Alcohols and ethers are manufactured in industry by the addition of water and alcohols to alkenes. Hydration of acetylene to produce acetaldehyde, a once important process, has lost its practical importance.

6.1.1. Alkenes and Dienes

Strong inorganic acids (H_2SO_4, H_3PO_4, $HClO_4$) can be used to catalyze the addition of water to alkenes.[1,2] The reaction is a reversible, exothermic process [Eq. (6.1)] favored by lower temperature and elevated pressure. General acid catalysis with rate-determining proton transfer to the alkene, the so-called A–S_E2 mechanism, is well established[2–4] (for a definition of A–S_E reactions, see Ref. 5):

$$\text{C=C} \xrightarrow{H_3O^+} \left[\begin{array}{c} \overset{\delta+}{C}-C \\ H \\ \overset{\delta+}{O}-H \\ H \end{array} \right] \xrightarrow{} \overset{+}{C}-\underset{H}{C} \xrightarrow[-H^+]{H_2O} \underset{HO}{C}-\underset{H}{C} \quad (6.1)$$

Hydration is regioselective and follows the Markovnikov rule. The involvement of carbocationic intermediate can result in rearrangements.

Since reactivity of alkenes increases with increasing alkyl substitution, hydration is best applied in the synthesis of tertiary alcohols. Of the isomeric alkenes, *cis* compounds are usually more reactive than the corresponding *trans* isomers, but strained cyclic isomeric olefins may exhibit opposite behavior. Thus, for example, *trans*-cyclooctene is hydrated 2500 times faster than *cis*-cyclooctene.[6] Similar large reactivity differences were observed in the addition of alcohols to strained *trans* cycloalkenes compared with the *cis* isomers. *trans*-Cycloheptene, an extremely unstable compound, for instance, reacts with methanol 10^9 faster at $-78°C$ than does the *cis* compound.[7]

When alkenes react with concentrated sulfuric acid, the corresponding sulfate esters (alkyl hydrogen sulfate, dialkyl sulfate) are formed. Depending on the alkene structure and the acid concentration, ester formation is accompanied by polymer formation (see Section 13.1.1). Hydrolysis of the sulfate esters allows the synthesis of alcohols. The reaction, called *indirect hydration*, is applied in the manufacture of ethanol and 2-propanol (see Section 6.1.3).

In addition to mineral acids, other acidic reagents may also catalyze hydration and related additions. Heteropoly acids are used mainly in industry.[8] Ion-exchange resins applied primarily in industrial processes were studied in hydration[9] and in the addition of alcohols to alkenes.[10,11]

Studies on the use of proton-exchanged zeolites have also been conducted.[12–14] Since *tert*-butyl alcohol and *tert*-butyl methyl ether (MTBE) are important gasoline octane enhancers, their synthesis by the addition of water and methanol to isobutylene attracted special attention. Among other processes, clay-catalyzed additions have been studied.[15]

Solid superacidic Nafion-H was also found to be effective in the hydration of acyclic alkenes.[16] Isopropyl alcohol was produced with 97% selectivity in hydrating propylene[17] at 150°C, whereas isobutylene yielded *tert*-butyl alcohol[18] with 84% selectivity at 96°C.

Acid-catalyzed photohydration of styrenes[19] and additions to cyclohexenes[20] leading exclusively to the Markovnikov products are also possible. Sensitized photoaddition, in contrast, results in products from anti-Markovnikov addition. The process is a photoinduced electron transfer[21] taking place usually in polar solvents.[22,23] Enantiodifferentiating addition in nonpolar solvents has been reported.[24] The addition of MeOH could be carried out in a stereoselective manner to achieve solvent-dependent product distribution:[25]

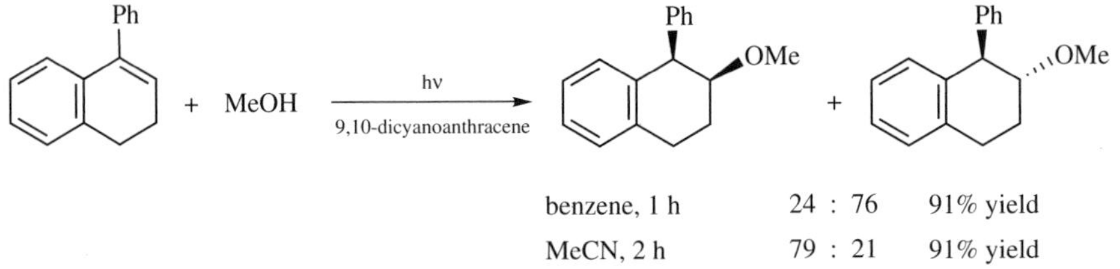

benzene, 1 h	24 : 76	91% yield
MeCN, 2 h	79 : 21	91% yield

(6.2)

Montmorillonites containing interlamellar water associated with cations (Cu^{2+}, Fe^{3+}, Al^{3+}) react under mild conditions with terminal alkenes to give the corresponding di-*sec*-alkyl ethers.[26] The alkene–water reaction is a stoichiometric process, while the reaction of the second molecule of alkene with the intermediate 2-alkanol is catalytic.

Very few data are available concerning the stereochemistry of these additions. Hydration of 1,2-dimethylcyclohexene was found to be nonstereoselective.[27] In contrast, predominant or exclusive *exo–syn* addition was shown to take place in the reaction of 2,3-dideuteronorbornene with water and alcohols:[28]

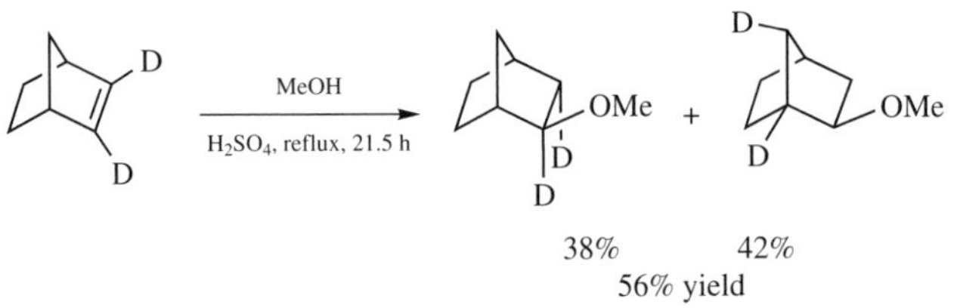

38%   42%
56% yield

(6.3)

Hydration of allene and mono- and disubstituted allenes leads to ketones through the rearrangement of the intermediate enols.[29] Further details of the mechanism are not known, but protonation of the terminal carbon in monosubstituted allenes is probable. Although the formation of isomeric ketones may be expected, only ketones possessing the keto function on the central carbon of the allene bond system were found to form.[30] When alcohols add to allenes, enol ethers of the corresponding ketones are usually the products. They may further react to form acetals. Hg^{2+} salts may be used as catalysts.[31]

The few data available on conjugated dienes indicate that 1,3-cycloalkadienes undergo 1,2 addition,[32] and acyclic dienes yield isomeric allylic alcohols resulting from 1,2 and 1,4 addition.[33]

6.1.2. Alkynes

Both acid and metal catalysis are usually required to accomplish hydration of alkynes to yield carbonyl compounds.[34] The addition is usually regioselective, allowing for conversion of terminal alkynes to ketones. Hydration of acetylene to produce acetaldehyde used to be an industrially significant process but was replaced by the Wacker synthesis.

The generally accepted pathway for the hydration of alkynes are the generation and subsequent tautomerization of an intermediate enol. The use of fairly concentrated acids, usually H_2SO_4, is necessary to achieve suitable reaction rates. Addition of catalytic amounts of metal salts, however, greatly accelerates product formation. In most cases mercury(II) salts are used. Mercury-impregnated Nafion-H [with 25% of the protons exchanged for Hg(II)] is a very convenient reagent for hydration:[35]

$$R\!\!-\!\!\equiv\!\!-\!\!R' \xrightarrow[\text{EtOH or AcOH, RT, 90 min}]{\text{Hg(II)/Nafion-H}} \underset{\underset{O}{\|}}{R\!-\!C\!-\!CH_2\!-\!R'} \tag{6.4}$$

R = H, Me, *n*-Pr, *n*-Bu, Ph   65–94%
   n-C$_5$H$_{11}$, cyclohexyl
R' = H, Ph

Other cations (Cu^{2+}, Pd^{2+}, Ru^{3+}, Ni^{2+}, Rh^{3+}) incorporated into Nafion-H have been found to promote hydration.[36] Other metals that catalyze hydration of alkynes include gold(III),[37] ruthenium(III),[38] and platinum(II) (Zeise's salt[39,40] and halides[40]). *p*-Methoxybenzenetellurinic acid is very effective in the hydration of terminal alkynes.[41]

Similar to the hydration of alkenes, photochemical acid-catalyzed hydration of alkynes is possible:

$$\text{Ph}\!\!-\!\!\equiv\!\!-\!\!\text{H} \xrightarrow{\text{hv, H}^+} \text{Ph}\!-\!\!\underset{+}{\overset{\text{CH}_2}{\diagup}} \xrightarrow[-\text{H}^+]{\text{H}_2\text{O}} \text{Ph}\!-\!\!\underset{\text{OH}}{\overset{\text{CH}_2}{\diagup}} \rightleftharpoons \text{Ph}\!-\!\!\underset{\text{O}}{\overset{\text{CH}_3}{\diagup}} \tag{6.5}$$

Since photoexcitation greatly enhances the reactivity of acetylenes, formation of the enol intermediate becomes faster than its rearrangement to ketone. As a result, the intermediate acetophenone enol in the hydration of phenylacetylene could be directly observed.[42]

Enol ethers and acetals are formed when alcohols add to alkynes. The reaction of alcohols may be catalyzed by mercury(II) salts.[43,44] Hg(OAc)$_2$ with or without TosOH allows the synthesis of enol ethers, acetals, or ketones under appropriate reaction conditions.[44] Au^{3+} was found to be an effective catalyst in the synthesis of acetals:[37]

$$R\!\!-\!\!\equiv\!\!-\!\!R' \xrightarrow[\text{MeOH, reflux, 1–10 h}]{\text{NaAuCl}_4} \underset{\underset{\text{OMe}}{|}}{\overset{\overset{\text{OMe}}{|}}{R\!-\!C\!-\!CH_2\!-\!R'}} \tag{6.6}$$

R = *n*-C$_6$H$_{13}$, *n*-C$_{10}$H$_{21}$, Ph   85–96%
R' = H, *n*-C$_6$H$_{13}$

6.1.3. Practical Applications

Production of Alcohols by Hydration of Alkenes. Several alcohols (ethyl alcohol, isopropyl alcohol, *sec*-butyl alcohol, *tert*-butyl alcohol) are manufactured commercially by the hydration of the corresponding olefins.[2,45,46] Ethanol, an industrial solvent and a component of alcohol–gasoline blends, and isopropyl alcohol—a solvent and antiknock additive—are the most important compounds. Isopropyl alcohol is often considered the first modern synthetic petrochemical since it was produced on a large scale in the United States in the 1920s.

Indirect hydration, the traditional process, is still used in many factories.[47–50] Counterflows of sulfuric acid and alkene react at elevated temperature and under pressure to form sulfate esters. The more reactive propylene requires milder reaction conditions than ethylene (70–85% H_2SO_4, 40–60°C, 20–30 atm, vs. 95–98% H_2SO_4, 65–85°C, 10–35 atm). The rate of absorption of alkenes is increased by agitation of the liquid. In the second step sulfate esters are hydrolyzed to alcohol:

$$R{-}CH{=}CH_2 \;+\; H_2SO_4 \;\longrightarrow\; \underset{OSO_2OH}{RCHCH_3} \;+\; \overset{CH_3}{(RCHO)_2SO_2} \;\xrightarrow[H_2SO_4]{H_2O}\; \underset{OH}{RCHCH_3} \quad (6.7)$$

In the manufacture of ethanol, for instance, this is done in two stages at 70°C and 100°C, respectively, with added water to get a final concentration of H_2SO_4 of about 40–50%. This technology is designed to suppress ether formation the principal side reaction.

The costly reconcentration of sulfuric acid recovered after hydrolysis and its corrosive nature led to the search for more economic alternatives. These are catalytic hydration processes.[2,46] The Veba process[51] is a gas-phase hydration using phosphoric acid supported on silica gel, alumina, or diatomaceous earth. Ethylene and water in a molar ratio up to 1 : 0.8 react at temperatures of 170–190°C and at pressures of 25–45 atm. Because of equilibrium limitations only 4–6% ethylene conversion per pass is achieved. Diethyl ether, the main byproduct, is formed up to 2%. Acid loss from the support requires continuous replenishment of the catalyst by adding small amounts of H_3PO_4 to the stream. Higher temperature and pressure was applied in the process with a catalyst composed of tungsten oxide on silica promoted by zinc oxide and other oxides.[2,47] Cation-exchange resins are used as catalysts in a mixed gas–liquid-phase hydration of propylene.[52] The reaction is most effective at 130–150°C and at pressures of 60–100 atm. The use of excess water allows more favorable equilibrium conditions, and as a result, much higher conversion levels, about 75% per pass, are attained. Several percents of diisopropyl ether byproduct are formed. When Nafion-H, a solid superacid, is used as catalyst at 170°C, diisopropyl ether can become the major product.

Another catalytic process developed by Tokuyama Soda applies heteropolytungstic acid in the liquid phase at high temperature and pressure (240–290°C, 150–300 atm).[53,54] Long catalyst lifetime, high conversion levels (60–70% per pass), and high selectivity (over 99%) are the very attractive features of the process.

C$_4$ alcohols (*sec-* and *tert*-butyl alcohol) can be manufactured by the sulfuric acid process.[45,55–57] According to the Markovnikov rule, all three *n*-butenes give *sec*-butyl alcohol by using 75–80% H$_2$SO$_4$. Isobutylene, which is more reactive, may be hydrated with more dilute sulfuric acid. The difference in reactivity may be applied for the separation of isobutylene from the linear isomers. The mixture of C$_4$ alkenes is first reacted at 0°C by 50–60% H$_2$SO$_4$ to form *tert*-butyl hydrogen sulfate. *n*-Butenes are then separated and transformed with 75–80% H$_2$SO$_4$ at 40–50°C to *sec*-butyl hydrogen sulfate. Finally, both esters are hydrolyzed to the corresponding alcohols. Hydration of isobutylene is often carried out with the aim of separating it from butenes. In this case the product *tert*-butyl alcohol is readily transformed back to isobutylene by dehydration. The direct hydration of C$_4$ alkenes is also practiced industrially. Cation-exchange resins are used in the production of *sec*-butyl alcohol.[58,59]

Production of Octane-Enhancing Oxygenates. Because of legislative actions to improve gasoline quality, alkenes and aromatics content of motor gasoline will be limited, requiring additional sources of octane enhancing components.[60,61] Besides alkylation and isomerization to yield highly branched (high-octane) hydrocarbons, oxygenate additives are the most important way to satisfy these requirements.[62–64] The most important gasoline octane enhancer is *tert*-butyl methyl ether (MTBE)[65–67] (see, however, a discussion in Section 1.8.2 of the controversy about its use). MTBE is manufactured by the acid-catalyzed addition of methanol to isobutylene.[68,69] The main source of isobutylene is FCC C$_4$ cuts, a mixture of isomeric butylenes. Isobutylene can originate from dehydration of *tert*-butyl alcohol a major byproduct of propylene oxide manufacture,[70] and from dehydrogenation of isobutane[71] (see Section 2.3.3). The addition of methanol to isobutylene is catalyzed by an acidic ion-exchange resin. Preheated methanol and isobutylene-containing C$_4$ fraction freed from 1,3-butadiene are fed to a reactor at a pressure of 20 atm. The temperature is kept below 120°C. Since the reaction is exothermic, this may require cooling, depending on the initial isobutylene concentration in the feed. Single- or two-stage processes are employed; the latter allow isobutylene conversions better than 99%. Since linear butene isomers are unreactive under these operating conditions, high selectivities are achieved.

Over the years several technological variations were introduced to make the process more economical (particularly the separation of methyl alcohol from MTBE forming an azeotropic mixture).[72–75] Other ethers such as *tert*-butyl ethyl ether (ETBE) and *tert*-amyl methyl ether (TAME) are manufactured similarly from ethanol and isobutylene, and methanol and isoamylenes, respectively.[74–77] Etherification of the entire light FCC gasoline (C$_4$–C$_7$ branched olefins) was also suggested.[78] Octane numbers of higher ethers decrease more steeply than do those of the corresponding alkenes, and thus the gain in octane numbers is not significant. Such a process, however, improves gasoline quality by decreasing olefin content and vapor pressure. A new catalyst to transform branched alkenes in FCC naphtha fraction to a mixture of methyl ethers has been introduced by Erdölchemie[79] and BP (Etherol process).[80] A palladium-on-cation-exchange resin catalyst promotes double-bond migration (isomerization of terminal alkenes to trisubstituted olefins) and hydrogenates dienes to monoolefins. Finally, it catalyzes addition of methanol to the alkenes.

Besides being used as gasoline additives, the formed ethers may also have other uses. Etherification of refinery streams is carried out to separate butylenes, a difficult problem because of azeotrope formation. Cracking of MTBE in the gas phase over solid acid catalysts yields pure isobutylene[81,82] used in polymer synthesis and to produce methyl methacrylate through direct oxidation (see Section 9.5.2). A similar separation was also developed through the formation and cracking of *tert*-butyl phenyl ether.[83] An MTBE unit may be incorporated ahead of an alkylation unit to improve alkylate characteristics and alkylation reaction conditions (see Section 5.5.1) through the selective removal of isobutylene from butylene mixtures.[84,85] The environmental and health problems found with MTBE seeping into ground water, however, recently resulted in regulations phasing out its use in the U.S.

Acetaldehyde. The industrial production of acetaldehyde by the hydration of acetylene has lost its importance with the introduction of more economical petrochemical processes (dehydrogenation of ethanol, oxidation of ethylene; see Section 9.5.2). At present it is practiced only in a few European countries where relatively cheap acetylene is still available.[86–88]

In the original German process acetylene is injected into an aqueous solution of mercuric sulfate acidified with sulfuric acid at 90–95°C and about 1–2 atm. As a result of side reactions, the catalytically active mercury(II) ions are reduced to mercury. To prevent this process ferric sulfate is continuously added to the reactor. Since ferric ions are reduced to ferrous ions, the catalyst solution requires reactivation, which is accomplished by hot nitric acid and air. Excess acetylene and acetaldehyde formed are removed, cooled, absorbed in water, and then separated by distillation. Excess acetylene is recycled. Conversion per pass is bout 55%. The Montecatini process[89] operates at 85°C and provides 95% overall yield. A modification developed by Chisso[90] allows lower operating temperature (70°C) without excess acetylene. Since side reactions are less important under these conditions, higher yields may be achieved.

6.2. HX ADDITION

6.2.1. Hydrohalogenation

Hydrogen halides add to carbon–carbon double and triple bonds to yield halogenated hydrocarbons. The reactivity of the four hydrogen halides in the addition is in the order HI > HBr > HCl > HF. HI and HBr, the two most reactive compounds, add readily at room temperature to unsaturated hydrocarbons. HCl requires heating or catalyst, while HF often exhibits irregular behavior. Most studies of hydrohalogenations have focused on the use of HCl and HBr.

Alkenes. The reaction of hydrogen halides with alkenes is generally an electrophilic addition. It occurs by a variety of mechanisms and involves a carbocation intermediate in many cases. As a result, rearrangements may be observed, and solvent incorporation may occur. The regioselectivity of the addition to unsymmetric double bounds is in accord with the Markovnikov rule; specifically, the hydrogen adds to the site with the highest electron density. Electrophilic addition of hydrogen halides to alkenes may be *syn-* or *anti*-stereoselective. The reaction conditions, the nature of the solvent and the structure of the reacting alkene affect the stereochemical outcome of the reaction.

In contrast to HCl and HBr addition, there are no such systematic studies concerning the addition of HF and HI to alkenes. Aqueous HF exhibits rather low reactivity toward alkenes, and anhydrous liquid HF is used instead to effect hydrofluorination.[91,92] Because of its low boiling point, high toxicity, and corrosive nature, alternative reagents were developed. The combination of HF with organic bases gives more suitable reagents.[93] They react with alkenes in typical Markovnikov-type addition. Of these, pyridinium poly(hydrogen fluoride)[94–96] (PPHF) and melamine–HF solution[97] are used most frequently. An effective solid fluorinating agent, a polymer-supported HF reagent, is prepared from poly-4-vinylpyridine and anhydrous HF. Easy handling and convenient workup make polyvinylpyridinium poly(hydrogen fluoride)[98,99] (PVPHF) an attractive reagent:

$$(6.8)$$

Concentrated HI in water or acetic acid is usually applied in hydroiodination of alkenes. Markovnikov products are exclusively formed and rearrangements may be observed.

Detailed wide-ranging studies are available on the addition of HCl and HBr to alkenes.[3,92,100] The most useful procedure is to react dry HCl gas and the alkene neat or in an inert organic solvent. Water or acetic acid may also be used. Alkenes yielding tertiary or benzylic alkyl chlorides react most readily. Styrene, however, adds HCl only at −80°C to give α-chloroethylbenzene without polymerization.[101] At more elevated (room) temperature polymerization prevails. HBr adds to alkenes in an exothermic process more rapidly than does HCl. Rearrangements may occur during addition indicating the involvement of a carbocation intermediate:[102]

$$(6.9)$$

A similar conclusion can be drawn from the higher reactivity of 1,1-dialkylalkenes compared with 1,2-dialkylalkenes. Norbornene and related bicyclic olefins were shown to react with HCl via the classical norbornyl cations prior to full equilibration.[103,104] The stereochemistry of addition may be either *syn* or *anti*. Product distributions can be strongly affected by the solvent used:[105]

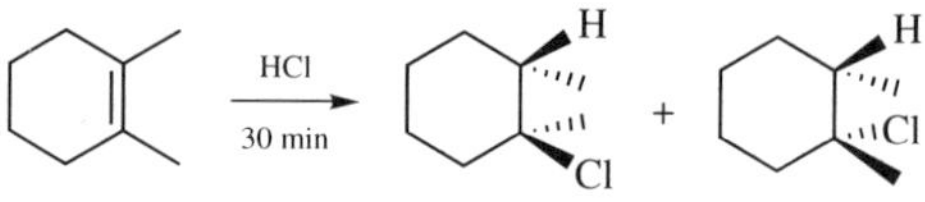

(6.10)

CH_2Cl_2, −98°C	88	12
Et_2O, 0°C	5	95

Acenaphthylene, indene, and *cis*- and *trans*-1-phenylpropene, all possessing the double bond in conjugation with an aromatic system, were found to react with hydrogen halides in nonpolar solvents to give predominantly the products of *syn* addition.[106] This stereochemistry is consistent with an Ad_E2 mechanism in which the undissociated hydrogen halide reacts with the alkene (Scheme 6.1). An ion pair (**1**) is first formed in which the halide ion is retained on the same side of the alkene as the hydrogen. Collapse of **1** gives exclusively the *syn* product. If the ion pair has a sufficiently long lifetime, the halide ion may migrate to the opposite side of the planar carbocation to give ion pair **2**. Collapse of the latter leads to the *anti* product. A similar ion-pair mechanism is operative in the hydrochlorination of 3,3-dimethylbutene and styrene in acetic acid.[107]

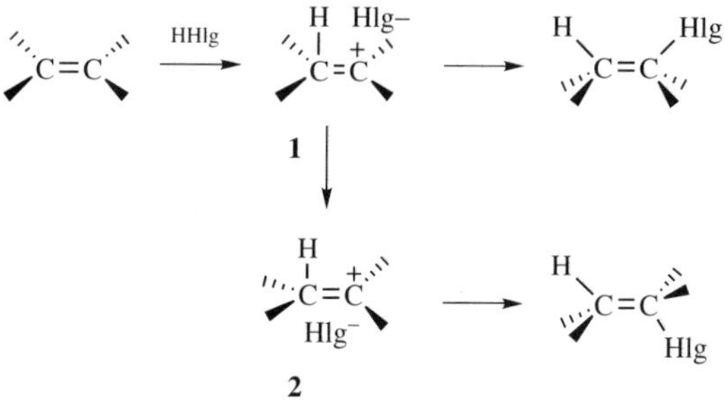

Scheme 6.1

In contrast, kinetic studies indicate that a different mechanism may also be operative. It was found that the rate law for the addition of HCl and HBr to alkenes in nonprotic solvents (hydrocarbons, diethyl ether, nitromethane) is typically overall third-order[3,100,102,108] or even higher.[109,110] It is usually first-order in alkene and second-order in hydrogen halide. A slow irreversible proton transfer from the nondissociated molecule to the alkene was shown to be rate-determining. The second molecule of HHlg assists this proton transfer by bonding to the developing halide ion. This is a typical Ad_E3 mechanism:

$$\text{R–CH=CH}_2 + 2\,\text{HHlg} \xrightarrow{\text{slow}} \overset{+}{\text{R}}\text{CHCH}_3 + \text{HHlg}_2^- \xrightarrow{\text{fast}} \underset{\text{Hlg}}{\text{RCHCH}_3} + \text{HHlg} \qquad (6.11)$$

A more detailed picture could be obtained from the results of stereochemical studies.[3,111] Since 1,2-dimethylcyclohexene, 1,6-dimethylcyclohexene, and 2-methylmethylenecyclohexane give different proportions of *cis*- and *trans*-1,2-dimethylbromocyclohexanes, simultaneous formation of the C–H and the C–Br bonds was suggested.[112] A similar conclusion was arrived at on the basis of the stereoselective, predominantly *anti* addition of HCl and HBr to 1,2-dimethylcyclopentene in pentane.[113]

The addition of DBr in AcOD to cyclopentene yields deuterocyclopentyl bromide and deuterocyclopentyl acetate, both formed with better than 96% *anti* stereoselectivity.[114] A concerted *anti* Ad_E3 transition state (**3**) is consistent with these and other experimental observations. A similar but *syn* transition state (**4**) was evoked to explain the somewhat decreased degree of *anti* addition (84–85%) in the hydrohalogenation of linear isomeric alkenes.[114] The reaction of $[1,3,3-^2H_3]$-cyclohexene[115] and 1,2-dimethylcyclohexene[116] with HCl in AcOH was shown to form the corresponding chlorides resulting from both *syn* and *anti* additions. In these cases an Ad_E2 mechanism competes with a concerted *anti* Ad_E3 process. High HCl concentration and addition of chloride salts such as Me_4NCl favor the Ad_E3 mechanism.

3   **4**

Exceptions to the Markovnikov rule when hydrogen bromide reacts with unsymmetric alkenes have long been known.[117,118] The reaction for this anti-Markovnikov addition was explained as being a chain reaction with the involvement of bromine atoms influenced by the presence of peroxides.[119–121] Both added peroxides and peroxides formed by the action of oxygen (air) on the alkene are effective.

The chain reaction is initiated by the interaction of a free radical [formed according to Eq. (6.12)] with hydrogen bromide to form a bromine atom [Eq. (6.13)], which, after reacting with the alkene, yields a bromoalkyl radical [Eq. (6.14)]. The reaction of this radical with HBr yields the alkyl bromide and regenerates the bromine atom [Eq. (6.15)]:

$$ROOH \longrightarrow RO\cdot + HO\cdot \qquad (6.12)$$

$$RO\cdot + HBr \longrightarrow ROH + Br\cdot \qquad (6.13)$$

$$R-CH=CH_2 + Br\cdot \longrightarrow R-\overset{\cdot}{C}HCH_2-Br \qquad (6.14)$$

$$R-\overset{\cdot}{C}HCH_2-Br + HBr \longrightarrow R-CH_2CH_2-Br + Br\cdot \qquad (6.15)$$

Anti-Markovnikov free-radical-induced addition of HBr to alkenes can be prevented by carrying out the reaction in the presence of small amounts of antioxidants that inhibit the reaction of oxygen with the alkene to form peroxides.

It is significant that no other hydrogen halides add to alkenes, contrary to the Markovnikov rule, even in the presence of free radicals. This appears to be due to the difficulty with which other halogen atoms, particularly fluorine and chlorine, are produced by reaction of the hydrogen halides with radicals:

$$RO\cdot \; + \; H-Hlg \; \rightleftharpoons \; ROH \; + \; Hlg\cdot \qquad (6.16)$$

It is only in the case of bromine that the equilibrium in this reaction is in the direction of the formation of the halogen atom.

Addition of HBr in nonpolar solvents to terminal alkenes was found to give anti-Markovnikov products even under nonradical conditions.[122] Products formed in both the normal and abnormal additions may be obtained in near-quantitative yields by changing the temperature and the reagent : reactant ratio:[122]

$$n\text{-}C_6H_{13}CH{=}CH_2 \; \xrightarrow[\text{hexane, }-78°C]{\text{HBr}} \; n\text{-}C_6H_{13}CH_2CH_2Br \; + \; n\text{-}C_6H_{13}\underset{\underset{Br}{|}}{C}HCH_3 \qquad (6.17)$$

HBr : 1-octene = 0.125	100	
HBr : 1-octene = 15	2	98

On the basis of theoretical calculations and spectroscopic evidence, a molecular mechanism with the formation of HBr–alkene complexes was suggested. A 2 : 1 HBr–alkene complex forms the product of the normal addition. A 1 : 2 complex, in contrast, accounts for the anti-Markovnikov product.

Special reagents introduced more recently allow significant improvements in hydrohalogenation of alkenes. HCl, HBr, and HI in aqueous–organic two-phase systems under phase-transfer conditions readily add to alkenes affording halides in nearly quantitative yields.[123] Appropriately prepared silica and alumina have been found to mediate the addition of HCl, HBr, and HI to alkenes.[124] The method is very convenient since it uses hydrogen halides prepared in situ:[124,125]

$$n\text{-}C_6H_{13}CH{=}CH_2 \; \xrightarrow[\text{CH}_2\text{Cl}_2, \, 25°C]{} \; n\text{-}C_6H_{13}\underset{\underset{HIg}{|}}{C}HCH_3 \qquad (6.18)$$

(COCl)$_2$, Al$_2$O$_3$, 1 h	98%
HBr, SiO$_2$, 0.7 h	96%
Me$_3$Sil, SiO$_2$, 1 h	98%

Competing radical addition in hydrobromination is not observed. It is suggested that surface-bound hydrogen halides with greatly increased acidity through polarization of the hydrogen–halogen bond protonate the double bond. A rapid transfer of the halide ion from the surface follows, affording predominantly the *syn* addition products. Highly acidic solid catalysts (K10 montmorillonite, ZF520 zeolite) were found to yield the Markovnikov addition product with high selectivity when methylcyclohexene was reacted with SOCl$_2$ as the HCl precursor.[126]

Dienes. Reaction of allenes with HCl and HBr was mostly studied.[3,92] Depending on the reagent, reaction conditions, and the structure of the reacting molecule, markedly different product distributions may be obtained.[29,127] Monoalkyl-substituted allenes give predominantly the corresponding 2-halo-2-alkenes. Phenylallene, in contrast, is converted exclusively to cynnamyl chloride:[128]

$$PhCH{=}C{=}CH_2 \xrightarrow[\text{AcOH, RT, 2 days}]{\text{HCl}} [Ph\overset{+}{C}H{-}CH{=}CH_2] \longrightarrow PhCH{=}CH{-}CH_2{-}Cl \quad (6.19)$$

5   95%

Mechanistic studies established the formation of carbocation intermediate **5** stabilized by the conjugation with the phenyl ring.

Mixtures of isomeric allylic chlorides are formed when 1,1-disubstituted allenes react with HCl. 1,3-Disubstituted allenes yield products of both central and terminal carbon attacks. In contrast, selective transformations occur with HBr. The gas-phase, photocatalytic addition of HBr gives selectively vinylic bromides[129] [Eq. (6.20)], while hydrobromination in solution phase yields allylic bromides[130] [Eq. (6.21)]:

$$
\begin{array}{c}
\overset{Me}{\underset{Me}{\diagdown}}C{=}C{=}CH_2
\end{array}
\quad
\begin{cases}
\xrightarrow{\text{HBr}} \overset{Me}{\underset{Me}{\diagup}}C{=}C\overset{Br}{\underset{Me}{\diagdown}} \quad 97\% & (6.20)\\[2em]
\xrightarrow[\text{Et}_2\text{O, RT, 5 days}]{\text{HBr}} \overset{Me}{\underset{Me}{\diagup}}C{=}CH{-}CH_2Br \quad 80\% & (6.21)
\end{cases}
$$

Similar differences may be observed in hydrobromination of tetrasubstituted allenes. Ring size and reaction conditions determine the outcome of hydrobromination of cyclic allenes.[92]

In gas-phase hydrobromination, where a radical mechanism is operative, the bromine atom always adds to the central carbon atom of the allenic system. As a result, vinylic bromides are formed through the stable allylic radical. In the solution phase under ionic addition conditions, either the vinylic or the allylic cation may be the intermediate, resulting in nonselective hydrobromination. Allylic rearrangement or free-radical processes may also affect product distributions.

Conjugated dienes may yield mixtures of products of 1,2 or 1,4 addition. Product of 1,4-addition may be initial product or may actually be formed by 1,2 addition followed by isomerization. In any event, however, formation of the 1,4-product involves isomerization of a carbocation.

Addition of HCl to 1,3-butadiene in the gas phase was shown to be a surface-catalyzed reaction occurring at the walls between a multilayer of adsorbed HCl and gaseous or weakly adsorbed 1,3-butadiene.[131] The initial proton transfer to the terminal carbon and the chloride attack occur almost simultaneously. Chloride attack to C(2) or C(4) has equal probability, thus producing equal amounts of 1,2 and 1,4 addition.

The addition of one mole of HCl or HBr to isoprene yields the corresponding primary halide produced through the rearrangement of the initially formed tertiary halide:[92]

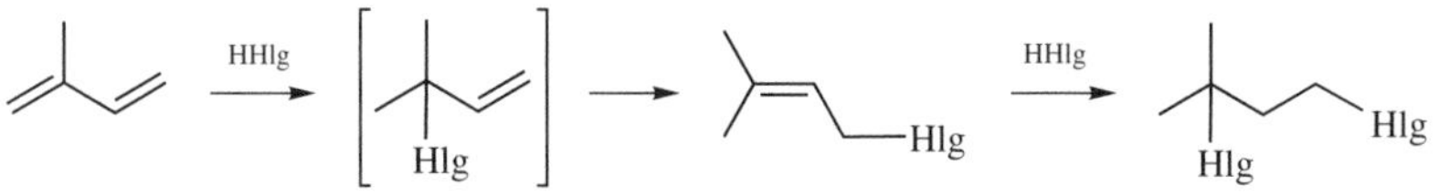

(6.22)

The addition of a second molecule of hydrogen halide gives the 1,3-dihalo compounds in good yields.[132]

Addition of DBr to 1,3-cyclohexadiene occurs by a 1,2-*anti* process and a 1,4-*syn* process,[133] a conclusion also predicted by molecular orbital calculations.[134] Stereospecific allylic rearrangement of the 1,2-*trans* product, however, eventually produces a reaction mixture containing predominantly *cis* products.

A single chloro compound, *trans*-1-phenyl-3-chloro-1-butene (**7**), is formed in hydrochlorination of isomeric 1-phenyl-1,3-butadienes in AcOH[135] (Scheme 6.2). The observation is interpreted by the formation of different isomeric allylic carbocations (**6c** and **6t**). Rapid rotation of **6c** to **6t** before captured by the chloride ion ensures selective formation of **7**.

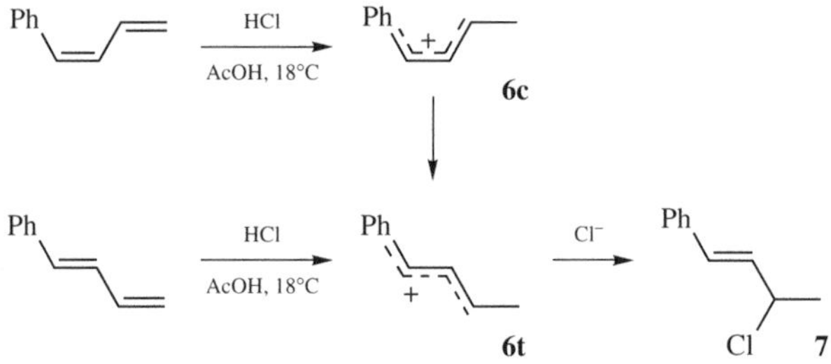

Scheme 6.2

Alkynes. Because of their less nucleophilic character, alkynes react less readily with hydrogen halides than do alkenes and often require the use a metal halide catalyst. Vinyl halides are formed in the reaction with one equivalent of HHlg. They may react further in an excess of the reagent to yield geminal dihalides. High yields of these compounds can be achieved. The addition of HCl to acetylene was studied in detail because of the practical importance of the product vinyl chloride (see Section 6.2.4).

In accordance with the Markovnikov rule, terminal alkynes are converted to 2-halo-1-alkenes. The reagents prepared in the reaction of anhydrous HF with organic bases and used in hydrofluorination of alkenes can add to the carbon–carbon triple bond producing Markovnikov products.[94–96,98,99] However, 1-bromo-1-alkenes and

1,2-dibromoalkanes are formed when hydrobromination is carried out in the presence of peroxides.[136,137] Reactions with reagents or reactants labeled appropriately with deuterium indicate that the products are formed in nonselective additions.[138,139] The mercury(II)-catalyzed addition of HCl and HBr to propyne, however, takes place with *anti* stereoselectivity.[136]

Dialkyl- and alkylarylacetylenes react in a stereoselective manner, yielding *anti* and *syn* products, respectively.[34,138–140] 3-Hexyne, for instance, gives mainly *trans*-3-chloro-3-hexene in hydrochlorination through an *anti* Ad$_E$3 transition state (**8**). In contrast, formation of (*E*)-1-phenyl-1-chloropropene (**9**), the *syn* product in hydrochlorination of 1-phenylpropyne in AcOH [Eq. (6.23)], involves an ion-pair intermediate[138] (**12**). The two mechanisms compete, and reaction conditions can affect selectivities.[138] For example, stereoisomeric compound **10** is formed through the *anti* Ad$_E$3 transition state **13** in the presence of a large amount of Me$_4$NCl. Since the positive charge in this transition state is not stabilized by the phenyl group as in **12**, an increased amount of the regioisomeric **11** is also formed through transition state **14**:

$$Ph \!\!\equiv\!\! Me \xrightarrow[\text{AcOH, 25°C}]{0.75\,M\,\text{HCl}} \begin{array}{c}Ph\\Cl\end{array}\!\!=\!\!\begin{array}{c}Me\\\end{array} + \begin{array}{c}Ph\\Cl\end{array}\!\!=\!\!\begin{array}{c}\\Me\end{array} + \begin{array}{c}Ph\\Me\end{array}\!\!=\!\!\begin{array}{c}Cl\\\end{array} \qquad (6.23)$$

	9	**10**	**11**
without Me$_4$NCl	80%	12%	0.3%
with 1 M Me$_4$NCl	30%	51%	10%

8 **12** **13** **14**

The surface-mediated addition of HCl, HBr, and HI to 1-phenylpropyne in the presence of silica or alumina yields initially the corresponding *E* isomer as a result of *syn* addition.[141] Rapid isomerization in excess reagents allows the formation of the thermodynamically more stable *Z* isomers.

6.2.2. Hypohalous Acids and Hypohalites

Hypohalous acids either preformed or prepared in situ readily add to carbon–carbon multiple bonds.[142–144] The addition is an electrophilic reaction with the positive halogen attaching itself predominantly to the less substituted carbon. The Markovnikov rule, therefore, is followed in most cases. The addition of the elements of hypohalous acids may be carried out by the use of reagents serving as positive halogen sources in the presence of water. Most of the data published concern additions to monoolefins, and very little is known about the transformations of dienes and alkynes. The addition of hypochlorous acid to alkenes was an important industrial reaction to manufacture chlorohydrins (see Section 6.2.4), which were then transformed to oxiranes. The so-called chlorohydrination process once accounted for about 50% of the U.S. production of oxiranes but has largely been replaced by direct epoxidation processes.

Limited data are available on hypofluorous acid, which is a very reactive, explosive reagent.[145] Exceptionally among hypohalous acids, it is polarized in the sense $HO^{\delta+}-F^{\delta-}$. It converts alkenes into fluoroalkanols with regioselectivites opposite to other hypohalous acids [Eq. (6.24)]; the large amount of HF present may affect the actual mechanism of addition:

$$
\begin{array}{c}
\text{(structure: 1-methylcyclopentene)} \xrightarrow[\text{CCl}_4,\ -50°\text{C}]{\text{HOF}} \text{(structure: 1-methyl-2-fluoro-cyclopentanol with OH)} \\
90\%
\end{array}
\tag{6.24}
$$

Hypochlorous acid and hypobromous acid react with acyclic alkenes to give Markovnikov products. In striking contrast, exclusive anti-Markovnikov orientation was observed in the transformation of methylenecycloalkanes with HOBr, and mixtures of chlorohydrins were formed with HOCl:[146]

$$
(CH_2)_n\ C=CH_2 \xrightarrow[\text{H}_2\text{O, below 15°C}]{\text{HOHlg}} (CH_2)_n\ C\overset{CH_2OH}{\underset{Hlg}{\diagup}} + (CH_2)_n\ C\overset{CH_2OHlg}{\underset{OH}{\diagup}}
\tag{6.25}
$$

$n = 3–6$	Hlg = Cl	35–67	33–65	63–91% yield
	Hlg = Br	100		61–87% yield

This was attributed to the fact that the open tertiary carbocation leading to Markovnikov products are unfavorable in these ring systems.

A convenient synthesis of chlorohydrins is based on the use of Chlorine T (TosNClNa) as the positive chlorine source in water–acetone.[147] It adds to a variety of alkenes to form Markovnikov and anti-Markovnikov products in a ratio of 4 : 1.

Since the reaction of $I_2 + H_2O$ with alkenes is an equilibrium process, the use of oxidizing agents allows one to shift the equilibrium by oxidizing the I^- formed.[148,149] Substituted cyclohexenes can be converted to iodohydrins in regioselective and stereoselective manner with I_2 and pyridinium dichromate.[149] High yields of iodohydrins may be achieved by reacting alkenes with iodine in the presence of moist tetramethylene sulfone—CHCl$_3$.[150]

Alkyl and acyl hypohalites, when adding to carbon–carbon double bond, afford halohydrin ethers and esters, respectively.[151] Regioselective and *syn* stereoselective addition of CF_3OF, CF_3CF_2OF, and CF_3COOF to stilbenes was reported.[152–154] The stereochemistry was explained to originate from the formation and immediate

collapse of the tight ion pair **15**, which might exist in equilibrium with the phenonium ion **16** [Eq. (6.26), where R_F stands for perfluoroalkyl groups]:

$$\left[\text{Ph}-\text{CH}-\overset{+}{\text{CH}}-\text{Ph} \atop \underset{\text{F}}{} \quad {}^-\text{OR}_F \right] \quad \rightleftharpoons \quad \left[\begin{array}{c} \text{(+)} \\ \text{F} \qquad \text{Ph} \\ \text{H} \qquad \text{H} \end{array} \right]_{{}^-\text{OR}_F} \tag{6.26}$$

$$\qquad\qquad\quad \textbf{15} \qquad\qquad\qquad\qquad\qquad \textbf{16}$$

Other alkyl hypohalites usually add to carbon–carbon multiple bonds in a free-
radical process.[155–158] Ionic additions may be promoted by oxygen, BF₃, or
B(OMe)₃.[156–160] While the BF₃-catalyzed reaction of alkyl hypochlorites and hypo-
bromites gives mainly halofluorides,[159] haloethers are formed in good yields but
nonstereoselectively under other ionic conditions.[156–158,160] In contrast, *tert*-BuOI
reacts with alkenes in the presence of a catalytic amount of BF₃ to produce 2-
iodoethers.[161] Since the addition is stereoselective, this suggests the participation
of a symmetric iodonium ion intermediate without the involvement of carbocationic
intermediates.

   1,2-Halohydrin derivatives may also be formed when the cationic intermediates
in any electrophilic halogenations are intercepted by appropriate oxygen nucleophiles.[144] In alcohol and carboxylic acid solvents, halohydrin ethers and esters,
respectively, may be formed. Such solvent incorporated products often observed
in halogenations provide valuable mechanistic information. Halogenation with
elemental halides, however, requires proper adjustment of reaction conditions to
achieve satisfactory yields and selectivities. Chemoselectivity—the probability of
the attack of different nucleophiles (Br⁻ vs. MeOH) present—was shown to depend
on charge distribution on the two carbon atoms in the cyclic bromonium ion intermediate.[162] A study of bromination of acyclic olefins in MeOH revealed that
geminal dialkylsubstituted and trisubstituted alkenes give the best selectivities of
the 2-bromomethoxy product.[163] Other positive halogen sources, such as *tert*-
BuOHlg,[144] I(pyridine)₂BuF₄,[164,165] and *N*-bromosuccinimide,[166] may afford better
selectivities.

6.2.3. Hydrogen Cyanide

Because of its low acidity, hydrogen cyanide seldom adds to nonactivated multiple
bonds. Catalytic processes, however, may be applied to achieve such additions.
Metal catalysts, mainly nickel and palladium complexes, and [Co(CO)₄]₂ are
used to catalyze the addition of HCN to alkenes known as *hydrocyanation*.[167–174]
Most studies usually apply nickel triarylphosphites with a Lewis acid promoter. The
mechanism involves the insertion of the alkene into the Ni–H bond of a hydrido
nickel cyanide complex to form a σ-alkylnickel complex[173–176] (Scheme 6.3).
The addition of DCN to deuterium-labeled compound **17** was shown to take place

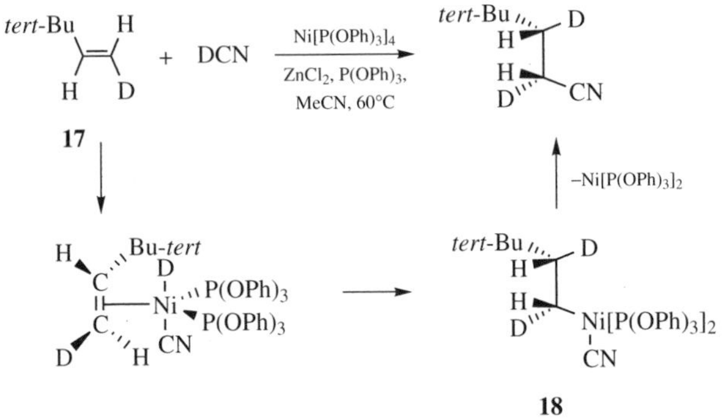

Scheme 6.3

with dominant (>90%) *syn* addition.[175] The σ-nickel complex **18** is known to be formed in a *syn* process. The overall *syn* addition requires reductive elimination to occur with retention of configuration.

Hydrocyanation of alkenes usually gives anti-Markovnikov products. Interestingly, however, addition of HCN to styrene yields mostly the branched (Markovnikov) adduct. This was suggested to result from stabilization of the branched alkylnickel cyanide intermediate by interaction of nickel with the aromatic ring.[176]

Hydrocyanation of dienes, a process of industrial importance (see Section 6.2.4), yields 1,4-addition products when conjugated dienes are reacted. The addition involves η^3-allyl intermediates (**19**):

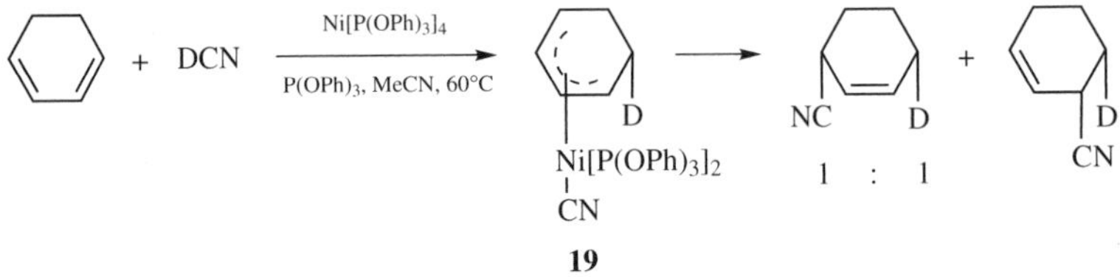

$$(6.27)$$

The stereochemistry of hydrocyanation of 1,3-cyclohexadiene was shown to occur with *syn* stereoselectivity, indicating that *cis* migration of the cyanide anion follows the formation of the π-allylnickel complex.[177]

Nonconjugated dienes (1,4-pentadiene, 1,5-hexadiene) are transformed mainly to products originating from conjugated dienes formed by isomerization.[178] In contrast, 1,7-octadiene in which the double bonds are separated by four methylene groups preventing isomerization to conjugated dienes, yields mainly isomeric mononitriles.

HCN adds more readily to alkynes than to alkenes.[179] The addition of HCN to acetylene catalyzed by Cu^+ ions was once a major industrial process to manufacture acrylonitrile carried out in the presence of copper(I) chloride, NH_4Cl, and HCl[180] (see Section 6.2.4). Zerovalent Ni and Pd complexes are effective catalysts in hydrocyanation of alkynes as well.[168] Diphenylacetylene gives excellent yields in producing the *syn* addition products:[181]

$$Ph\text{———}Ph \quad + \quad HCN \quad \xrightarrow[\text{P(OPh)}_3]{\text{Ni[P(OPh)}_3]_4} \quad \underset{H}{\overset{Ph}{\diagdown}}\diagup\diagdown\underset{CN}{\overset{Ph}{\diagup}} \qquad (6.28)$$

benzene, 120°C   93%

toluene, ZnCl$_2$, 60°C   82%

Much lower yields are achieved when terminal alkynes react with HCN. Terminal nitriles formed due to mainly steric factors are the main products. Regio- and stereoselectivities similar to those in hydrocyanation of alkenes indicate a very similar mechanism.

6.2.4. Practical Applications

Ethyl Chloride. Hydrochlorination of ethylene with HCl is carried out in either the vapor or the liquid phase, in the presence of a catalyst.[182–184] Ethyl chloride or 1,2-dichloroethane containing less than 1% AlCl$_3$ is the reaction medium in the liquid-phase process operating under mild conditions (30–90°C, 3–5 atm). In new plants supported AlCl$_3$ or ZnCl$_2$ is used in the vapor phase. Equimolar amounts of the dry reagents are reacted in a fluidized- or fixed-bed reactor at elevated temperature and pressure (250–400°C, 5–15 atm). Both processes provide ethyl chloride with high (98–99%) selectivity.

A highly economical production of ethyl chloride combines radical ethane chlorination and ethylene hydrochlorination.[185,186] Called the *Shell integrated process*, it uses the hydrogen chloride produced in the first reaction to carry out the second addition step:

$$CH_3CH_3 \ + \ Cl_2 \ \longrightarrow \ CH_3CH_2Cl \ + \ HCl \ \xrightarrow{CH_2=CH_2} \ CH_3CH_2Cl \qquad (6.29)$$

Ethyl chloride was used in the manufacture of the antiknock agent tetraethyllead, which is, however, phased out.

Hydrochlorination of 1,3-Butadiene. A historic process for the synthesis of adiponitrile was the transformation of 1,3-butadiene through 1,4-dichlorobutene.[187] Addition of HCl was carried out in the vapor phase (125–230°C) in an excess of 1,3-butadiene. A mixture of isomeric dichlorobutenes (3,4-dichloro-1-butene, *cis*- and *trans*-1,4-dichloro-2-butene) was formed. They underwent further transformations to form the dicyanides and then, after hydrogenation, to give adiponitrile.

Vinyl Chloride. Vinyl chloride is an important monomer in the manufacture of polyvinyl chloride and vinyl polymers. Two basic transformations are in commercial use.[188–190] The catalytic hydrochlorination of acetylene, once an important industrial route, is now of limited use. Because of the high energy requirements of acetylene production, this process has been largely replaced by the ethylene chlorination–oxychlorination reaction (see Section 6.3.4).

The catalytic hydrochlorination of acetylene[188–190] demands high-quality reactants. Acetylene must be free of catalyst poisons (S, P, As), and HCl must not contain chlorine. Since water reacts with the product vinyl chloride to give acetaldehyde and HCl, the reactants must be dry. In the more widely used vapor-phase synthesis, the premixed reactants with a slight excess of HCl are fed into a multitube fixed-bed reactor kept at about 200°C. The catalyst is almost invariably the superior mercury(II) chloride supported on active carbon. Certain additives (Ce, Th, and Cu chlorides) are used to suppress the loss of the volatile $HgCl_2$. High conversion (near 100%) and high selectivity (98%) are achieved. The addition of a second molecule of HCl to vinyl chloride may produce 1,1-dichloroethane, the only significant byproduct.

Ethylene Chlorohydrin. Two industrial processes were used in the synthesis of ethylene chlorohydrin,[182,191] which, in turn, was transformed to ethylene oxide. Since the direct oxidation of ethylene to ethylene oxide is more economical, these technologies are being abandoned.

Ethylene was reacted with chlorine water, or with a mixture of hydrated lime and chlorine. In the latter case the $Ca(OCl)_2$ formed decomposes to yield HOCl. The aqueous opening of the intermediate chloronium ion leads to the formation of the product. Ethylene chlorohydrin then was cyclized to ethylene oxide by addition of calcium hydroxide.

Propylene Chlorohydrin. Propylene chlorohydrin is synthesized with the aim of producing propylene oxide. Although the latter is manufactured commercially mainly by the direct oxidation of propylene, the chlorohydrination process is still in limited use.

In an older version of the synthesis, propylene and chlorine react in an aqueous solution to form propylene chlorohydrin.[192–194] The slightly exothermic reaction maintains the 30–40°C reaction temperature to yield isomeric propylene chlorohydrins (1-chloro-2-propanol/2-chloro-1-propanol = 9 : 1). The main byproduct is 1,2-dichloropropane formed in amounts up to 10%. The product propylene chlorohydrin then undergoes saponification to propylene oxide with calcium hydroxide or sodium hydroxide.

Because of technological difficulties of this second step (large amount of wastewater, concentration of the NaCl solution before electrolysis to produce chlorine), a modified synthesis was developed by Lummus.[195] Instead of HOCl, *tert*-butyl hypochlorite generated by reacting *tert*-butyl alcohol and chlorine in sodium hydroxide solution is used for the chlorohydrination step. An organic phase containing the reagent reacts with propylene and water to produce propylene chlorohydrin and *tert*-butyl alcohol. The latter is recycled to the first stage to regenerate the reagent.

Adiponitrile. Adiponitrile is an important intermediate in polyamide manufacture. 1,6-Hexamethylenediamine formed by the hydrogenation of adiponitrile is used in the production of nylon-6,6, one of the most important polyamides in commercial production.

Among other nonaddition processes, adiponitrile may be manufactured by the direct hydrocyanation of 1,3-butadiene (DuPont process).[169,172,187,196] A homogeneous Ni(0) complex catalyzes both steps of addition of HCN to the olefinic bonds (Scheme 6.4). The isomeric monocyano butenes (**20** and **21**) are first formed in a ratio of approximately 1:2. All subsequent steps, the isomerization of **20** to the desired 1,4-addition product (**21**), a further isomerization step (double-bond migration), and the addition of the second molecule of HCN, are promoted by Lewis acids (ZnCl$_2$ or SnCl$_2$). Without Lewis acids the last step is much slower then the addition of the first molecule of HCN. Reaction temperatures below 150°C are employed.

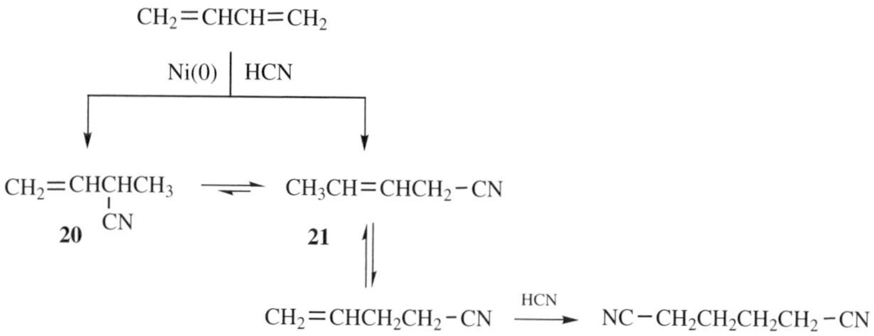

$$CH_2=CHCH=CH_2$$

Ni(0) | HCN

$$CH_2=CHCHCH_3 \quad \rightleftharpoons \quad CH_3CH=CHCH_2-CN$$
$$\underset{\textbf{20}}{\overset{|}{CN}} \qquad\qquad\qquad \textbf{21}$$

$$CH_2=CHCH_2CH_2-CN \quad \xrightarrow{HCN} \quad NC-CH_2CH_2CH_2CH_2-CN$$

Scheme 6.4

Acrylonitrile. Acrylonitrile, an important monomer in numerous polymerization processes, is produced mainly from propylene by ammoxidation (see Section 9.5.3). In the traditional process, the major industrial route in the 1950s and 1960s, an aqueous solution of copper(I) chloride, NH$_4$Cl and HCl was reacted with acetylene and hydrogen cyanide:[180,197]

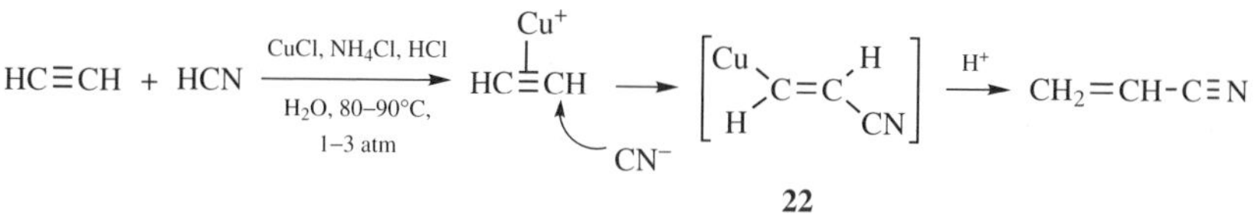

$$HC\equiv CH + HCN \xrightarrow[\substack{H_2O,\ 80-90°C,\\ 1-3\ atm}]{CuCl,\ NH_4Cl,\ HCl} \overset{Cu^+}{\underset{CN^-}{HC\equiv CH}} \longrightarrow \left[\underset{H}{\overset{Cu}{C}}C=C\overset{H}{\underset{CN}{}} \right] \xrightarrow{H^+} CH_2=CH-C\equiv N$$

22

$$(6.30)$$

Coordination of the metal ion activates the carbon–carbon triple bond toward nucleophilic attack to yield a σ-vinyl complex (**22**), which is a characteristic pathway of metal-catalyzed additions to the acetylenic bond. Protolysis of **22** gives the end product.

6.3. HALOGEN ADDITION

6.3.1. Alkenes

Elemental fluorine is rarely used in addition to multiple bonds. This is due mainly to its extreme reactivity, resulting in violent reactions and difficult handling.[198–200] Since it readily undergoes homolytic dissociation, radical processes may interfere with ionic processes, resulting in nonselective transformations. This is probably why early observations indicated that addition of fluorine to alkenes, although it gave predominantly products of *syn* addition, was nonstereoselective.[201,202] However, it has been shown[203] that good results can be achieved by working with a very dilute stream of fluorine (1%) in an inert gas, in the presence of a proton donor in solution, at low temperature. This method permits the addition of fluorine to alkenes with high stereoselectivity in a controlled manner.

In contrast to other halogens, the addition of fluorine is predominantly *syn*. This is compatible with the formation and rapid collapse of a 2-fluorocarbocation–fluoride tight ion pair [Eq. (6.31)]. Extra stabilization for the 2-fluorocarbocation, as in stilbene, decreases stereoselectivity; this mechanism also accounts for the formation of 1,1,2-trifluoro compounds [Eq. (6.32)] and solvent incorporation:

$$RCH{=}CH_2 \; + \; F_2 \; \longrightarrow \; \left[\overset{+}{RCH}{-}\underset{|}{CH_2} \; \overset{F^-}{} \right] \longrightarrow$$

$$\begin{aligned} &\longrightarrow \; \underset{F \quad F}{RCH{-}CH_2} \qquad\qquad (6.31)\\ &\underset{-HF}{\longrightarrow} \; RCH{=}CHF \; \underset{F_2}{\longrightarrow} \; \underset{F \quad F}{RCH{-}CH} \quad (6.32) \end{aligned}$$

A four-center molecular *syn* addition ([2 + 2]-cycloaddition) is also consistent with experimental observations.[202] Molecular-orbital calculations and experimental observations by NMR spectroscopy in superacidic systems rule out the existence of the bridged fluoronium ion.[204–206]

Xenon fluorides may add to alkenes in the presence of HF or CF_3COOH as catalyst.[207–209] The addition, though, takes place via predominant *anti* attack and is not stereoselective.[210] The formation of trifluoro and trifluoromethyl derivatives via radical intermediates, and trifluoroacetate incorporation were also observed.[211,212]

Chlorination, bromination, and iodination of alkenes with the corresponding elemental halogens to give vicinal dihalides take place in any solvent that does not react with halogens. Skeletal rearrangements and solvent incorporation sometimes accompany addition. A stepwise electrophilic addition with the involvement of a halogen–alkene complex and a cationic intermediate are the general aspects of the mechanism.[3,200,213,214] The addition is usually *anti*-stereoselective. This and other evidence indicate a bridged halonium ion intermediate. Under certain conditions a free-radical mechanism may occur. Besides these general features there exist, however, some characteristic differences in the addition of differing halogens.

The rate law for the addition of chlorine across a carbon–carbon double bond is first-order in both alkene and chlorine independently on the polarity of solvent used.[215,216] Strong acceleration of the reaction rate is brought about by increasing alkyl substitution about the double bond.[217] An Ad_E2 mechanism involving a molecular (donor–acceptor) complex,[218] or an ionic intermediate is operative.[217] The predominant *anti* stereoselectivity of addition is compatible with a cyclic chloronium ion intermediate.[111] This is also supported by the observation that the position of alkyl substituents does not affect the rate.[217] Thus, the reactivities of isobutylene and 2-butenes are almost identical. This indicates that the positive charge in the rate-determining transition state is distributed over both carbon atoms in a bridged structure. Quantum-mechanical calculations[213] and direct experimental observations of chloronium ions by NMR spectroscopy[204–206,219] are further proofs of the bridged structure of this intermediate.

Other reagents used to synthesize 1,2-dichloro compounds include NCl_3, which ensures better yields than does SO_2Cl_2, PCl_5, or chlorine.[220,221] In contrast to the usual predominant *anti* addition in chlorination using chlorine, dominant *syn* addition can be accomplished with $SbCl_5$,[222] $MoCl_5$,[223] or tetrabutylammonium octamolybdate.[224] The lack of selectivity in the chlorination with $CuCl_2$ in polar solvents (AcOH, alcohols, nitromethane) is indicative of an open ionic intermediate.[225]

The addition of bromine to alkenes is a rapid, exothermic reaction usually taking place at room temperature. In contrast to chlorination, the rate law in bromination depends on the solvent used. On passing from hydroxylic to nonpolar aprotic solvents, the overall second-order changes to a rate law that is first-order in alkene and second-order in bromine.[226] Alkene–bromine complexes with varying compositions were shown to form under reaction conditions[3,218,227,228] (Scheme 6.5). At low bromine concentration in protic solvents the reaction proceeds via a 1 : 1 complex (**23**). A 1 : 2 alkene–bromine complex (**25**) is involved at high bromine concentration in nonprotic solvents. The ionic intermediates (**24, 26**) were shown to exist as contact ion pairs, solvent-separated ions, or dissociated ions.

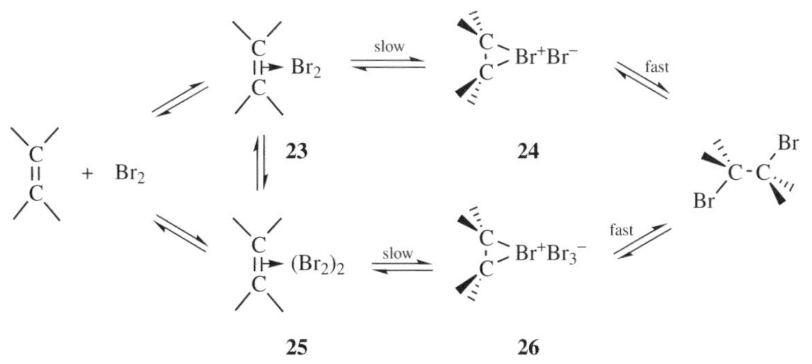

Scheme 6.5

The formation of the ionic intermediates, long considered to be irreversible, has been shown to be reversible.[229–231] The 1 : 1 complex (**23**) dominating in protic solvents may be converted to **24** in a solvent-assisted process.[232,233] The ionic intermediate was first postulated as a cyclic bromonium cation by Roberts and Kimball.[234] This was later supported by stereochemical results indicating near-exclusive *anti* addition stereochemistry,[111] by kinetic studies,[235] and more recently by calculations.[236–238] Cyclic bromonium ions were first prepared by Olah and coworkers under superacidic stable ion conditions and studied by NMR spectroscopy.[219,239–241]

The actual mechanism of bromination and the structure of the ionic intermediate, however, are more complex than Scheme 6.5 indicates.[240–242] It can involve several competing pathways through the 2-bromocarbocation (**27**) and partially bridged intermediates (**28**). The bridged bromonium ion and the 2-bromocarbocation can be considered as extremes only in a multipathway scheme.[242,243] Electronic effects and the symmetry of the starting alkenes may affect the structure of the intermediate.[236,238,241] In nonpolar solvents, as in CCl$_4$, the direct rearrangement of a $1:2$ complex through the **29** nonpolar six-membered ring transition state was suggested.[218]

Since the alkene structure and solvent affect the mechanism, product distribution and stereochemistry may change accordingly. Arylalkenes, for example, can form stable carbocationic intermediates. The predominating carbocationic structure results in nonstereoselective but highly regioselective products. In contrast, alkyl-substituted olefins without extra stabilization of the carbocation are transformed through the bridged bromonium ion, ensuring *anti* stereoselectivity. In comparison with chlorine, bromine has better bridging properties. *Anti* stereoselectivity, therefore, is usually more pronounced in bromination. When competition between the cyclic halonium and open carbocationic intermediates exists, as in the bromination of certain arylalkenes, solvent-dependent stereochemistry may be observed.

Certain convenient brominating agents, such as pyridine hydrobromide perbromide (PyHBr$_3$) and tetramethylammonium tribromide, may be used to transform alkenes to vicinal dibromides. They often give better yields than does liquid bromine, but may react by different mechanisms. Isomeric 1-phenylpropenes, for instance, react nonstereoselectively with bromine, but exhibit near-exclusive *anti* addition with the other reagents:[244]

$$(6.33)$$

	74.5	25.5
Br$_2$		
PyHBr$_3$	100	

Alkenes react with CuBr$_2$ in acetonitrile under mild conditions to produce vicinal dibromoalkanes in an exclusive *anti* addition.[245]

Shape-selective bromination of either linear alkenes or branched and cyclic alkenes in the presence of zeolites has been reported.[246] In the so-called out-of-pore bromination, a solution of bromine is added to the preequilibrated mixture of the alkenes and zeolite, resulting in selective bromination of cyclic (branched) olefins [Eq. (6.34)]. When the solution of alkenes is added to the bromine–zeolite mixture (in-pore bromination), linear olefins are brominated preferentially:

$$\text{(6.34)}$$

without zeolite	39%	61%
out-of-pore	89%	6%
in-pore	6%	79%

Iodine, the least reactive halogen, adds more slowly to alkenes in an equilibrium process. The alkene structure, solvent, and temperature determine the position of the equilibrium. The rate law is very complex[3,111] with overall fourth-order in nonpolar solvents, and third-order in other solvents such as AcOH. Electrophilic attack by complexing iodine molecules or additional iodine molecules participating in iodine–iodine bond breaking in the rate-determining transition state explains the kinetics of iodination. Although less documented than in chlorination and bromination, molecular complexes[218] and bridged iodonium ions[3,111] are the general features of iodination as well.

Halogenations may also occur by a free-radical mechanism.[121,218] Besides taking place in the gas phase, halogenation may follow a free-radical pathway in the liquid phase in nonpolar solvents. Radical halogenation is initiated by the alkene and favored by high alkene concentrations. It is usually retarded by oxygen and yields substitution products, mainly allylic halides.

Cyclohexene was shown to react partly by a radical mechanism when chlorinated in the absence of oxygen even in the dark.[247] The reaction is slightly less stereoselective than the ionic process (96% *trans*-1,2-dichlorocyclohexane vs. 99%). Isomeric butenes under similar conditions give products nonstereoselectively.[248] Branched alkenes are less prone to undergo free-radical halogenation.

Free-radical bromination was observed to take place in the presence of oxygen on illumination.[249] In a competitive reaction under ionic conditions (CCl₄, room temperature), cyclohexene reacted 5 times faster than styrene. In contrast, the yield of 1,2-dibromoethylbenzene was 10 times as high as that of 1,2-dibromocyclohexane when bromination was carried out in the presence of light.

Radical iodination of simple alkenes was found to take place more readily than the ionic process. The addition is *anti*-stereoselective,[250,251] resulting from an attack of iodine atom or molecule on charge-transfer complex between the alkene and an iodine molecule or atom.

Numerous reagents may react with alkenes to yield mixed 1,2-dihalogenated compounds. Interhalogens prepared on mixing two different halides or their stoichiometric equivalents are used in mixed halogenations.

6.3.2. Dienes

Relatively few data and no systematic studies of the halogenation of dienes have been reported. Monoaddition products are usually formed through the attack on the more substituted double bond. Interhalogens add to allenes to give products in which the positive part of the reagent is attached to the central allenic carbon. Rearrangements, however, may affect product distributions. Ionic mechanisms are usually established, but very few details are available.

An interesting observation of the chlorination of allenes is that besides the addition products, monochlorinated compounds are also formed.[252–254] Thus allene gave propargyl chloride, and tetramethylallenes gave 3-chloro-2,4-dimethyl-1,3-pentadiene. Since the reactions were carried out in the presence of oxygen, that is, under ionic conditions, proton loss from the intermediate chloronium ion or 2-chlorocarbocation explains these results.

No similar reaction takes place during bromination. Transformation of the optically active $(-)$-2,3-pentadiene, the most studied compound, gave a mixture of dibromides in inert solvents through a stereoselective *anti* addition:[255]

$$(6.35)$$

Since the products are optically active, this excludes the symmetric, resonance-stabilized allylic ion intermediate. The main (*trans*) product arises from the preferential attack of Br^- on the more stable, less strained, partially bridged **30** intermediate (as compared to **31**). Similar conclusions were drawn from the results of the iodination and haloiodination of 2,3-pentadiene[256] and from bromination of 1,2-cycloalkadienes.[257]

A comparative study of addition of halogens to phenylallene revealed that isomer distributions depend strongly on the halogen and temperature.[258] In bromination, for example, the 1,2-dibromide, formed initially at low temperature, was found to rearrange completely to the 2,3-addition product at higher temperature:

$$(6.36)$$

$-70°C$	25	75
$0°$	<2	>98

Analogous to the addition of hydrogen halides, conjugated dienes may be converted through 1,2 and 1,4 addition in halogenation. Since bridged and partially bridged halonium ions and open carbocations may all be involved, each permitting different stereochemical outcome, very complex product distributions may result. Isomerization of products and the involvement of radical processes may further complicate halogenations. These and the lack of sufficient data do not allow investigators to arrive at unambiguous conclusions in most cases.

1,3-Butadiene reacts with chlorine under radical and ionic conditions to give both 1,2- and 1,4-addition products.[259] The predominant formation of the *trans* isomer in radical 1,4-addition [Eq. (6.37)] was explained to result from the predominant transoid form of the starting compound and the configurational stability of the resonance-stabilized allylic radical intermediate:

$$
\text{(6.37)}
$$

| illumination under N_2 | 22 | 78 |
| in the presence of O_2 | 55 | 45 |

Ionic chlorination was found to be insensitive to solvent polarity.[259] In contrast, significant changes in product distributions were reported in bromination in different solvents:[260]

$$
\text{(6.38)}
$$

| in *n*-pentane | 69 | 31 |
| in nitromethane | 26 | 74 |

Comparative studies[260–263] with other acyclic and cyclic conjugated dienes revealed a predominantly *syn* stereochemistry in 1,4 addition. Chlorination, however, is generally less stereoselective than bromination. In 1,2 addition, in turn, both reagents are usually nonstereoselective. A striking difference between chlorination and bromination of 1,3-cycloalkadienes is that chlorination produces appreciable *syn* 1,2-addition products, whereas bromination does not give detectable *syn* products.[263] The differences in the addition of the two halogens were interpreted as resulting from different degrees of charge delocalization in the intermediate chloronium and bromonium ions.

Since the reactivity of double bonds in electrophilic bromination increases drastically with increasing alkyl substitution, selective monoaddition to the more substituted double bond in nonconjugated dienes can be accomplished with pyridin hydrobromide perbromide.[264] Chlorination[265] and bromination[266] of *cis,cis*-1,5-cyclooctadiene lead to the expected dihalogen and tetrahalogen derivatives.

Iodination, surprisingly, differs profoundly to give isomeric diiodo compounds by transannular cyclization:[266]

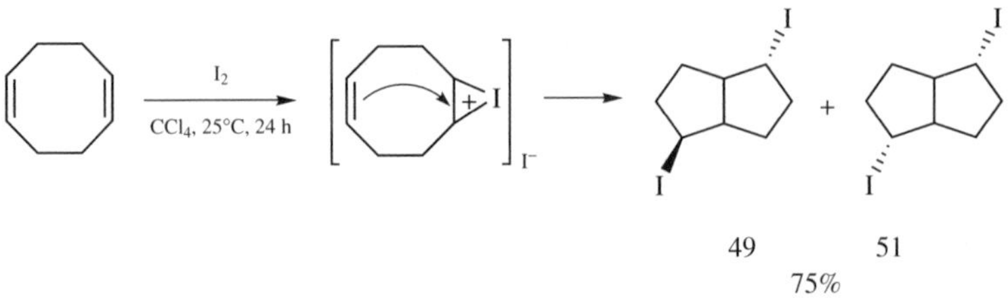

$$(6.39)$$

49   51

75%

6.3.3. Alkynes

Alkynes are usually much less reactive in halogenation than alkenes.[267] Relative reactivities (k_{alkene}/k_{alkyne}) in the range 10^2–10^5 were found in comparative studies of chlorination and bromination in acetic acid.[215] 1,2-Monoaddition products are usually formed except in fluorination, which yields tetrafluoro and solvent incorporated derivatives. In fluorination and chlorination the open β-halovinyl carbocations were calculated to be more stable than the corresponding cyclic halonium ions.[34] The formation of nonstereoselective 1,2-dichloroalkenes and regioselective (Markovnikov) solvent incorporated products are in agreement with these calculations.

The rate law of bromination of acetylenes is complicated and strongly depends on reaction conditions.[268] Both reaction conditions and alkyne structure affect stereochemistry. Bromination of alkylacetylenes is a stereoselective *anti* addition process occurring through the cyclic bromonium ion. Phenylacetylenes, in contrast, react according to an Ad$_E$2 mechanism via a vinyl cation intermediate in a nonstereoselective process. Added bromide ions bring about a new, selective bromide ion-catalyzed process. It proceeds through transition state **32** and results in stereoselective *anti* addition.[268] Much higher steroselectivites can be achieved by bromination with CuBr$_2$.[269]

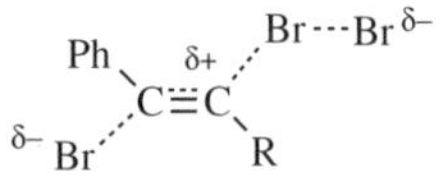

32

In contrast to the reversible reaction of iodine with alkenes, iodination of alkynes yields stable 1,2-diiodoalkenes formed in *anti* addition.[270,271] Very good yields were achieved when iodination was carried out in the presence of γ-alumina.[125]

6.3.4. Practical Applications

Vinyl Chloride. The present-day most important process in the industrial maufacture of vinyl chloride is the chlorination–oxychlorination of ethylene.[188–190, 272–275]

It consists of three basic steps: direct chlorination of ethylene to form 1,2-dichloro-ethane [Eq. (6.40)], cracking of 1,2-dichloroethane to vinyl chloride and HCl [Eq. (6.41)], and oxychlorination of ethylene with HCl [Eq. (6.42)] formed in the second step. The net reaction is the oxychlorination of ethylene to vinyl chloride [Eq. (6.43)]:

$$CH_2{=}CH_2 \ + \ Cl_2 \ \longrightarrow \ \underset{\underset{Cl \ \ Cl}{| \ \ |}}{CH_2CH_2} \qquad \Delta H = -50 \ \text{kcal/mol} \quad (6.40)$$

$$2 \underset{\underset{Cl \ \ Cl}{| \ \ |}}{CH_2CH_2} \ \longrightarrow \ 2 \underset{\underset{Cl}{|}}{CH_2{=}CH} \ + \ 2 \, HCl \quad \Delta H = 17 \ \text{kcal/mol} \quad (6.41)$$

$$CH_2{=}CH_2 \ + \ 2 \, HCl \ + \ 0.5 \, O_2 \ \longrightarrow \ \underset{\underset{Cl \ \ Cl}{| \ \ |}}{CH_2CH_2} \ + \ H_2O \quad \Delta H = -57 \ \text{kcal/mol} \quad (6.42)$$

$$2 \, CH_2{=}CH_2 \ + \ Cl_2 \ + \ 0.5 \, O_2 \ \longrightarrow \ 2 \underset{\underset{Cl}{|}}{CH_2{=}CH} \ + \ H_2O \qquad (6.43)$$

This so-called balanced oxychlorination vinyl chloride production is the most economical commercial process for the manufacture of vinyl chloride.

The first addition step—specifically, the direct catalytic chlorination of ethylene [Eq. (6.40)]—is almost always conducted in the liquid phase.[188–190,272,273] 1,2-Dichloroethane is used as solvent with ferric chloride, an efficient and selective catalyst. Ionic addition predominates at temperatures of $\sim$50–70°C. The conversion is usually 100% with 1,2-dichloroethane selectivity higher than 99%. The primary byproduct is 1,1,2-trichloroethane, believed to be formed by subsequent radical chlorination of 1,2-dichloroethane. A low amount of oxygen (below 1%), therefore, is added to the chlorine feed to suppress radical side reactions.

Thermal dehydrochlorination of 1,2-dichloroethane[188–190,272,273] takes place at temperatures above 450°C and at pressures about 25–30 atm. A gas-phase free-radical chain reaction with chlorine radical as the chain-transfer agent is operative. Careful purification of 1,2-dichloroethane is required to get high-purity vinyl chloride. Numerous byproducts and coke are produced in the process. The amount of these increases with increasing conversion and temperature. Conversion levels, therefore, are kept at about 50–60%. Vinyl chloride selectivities in the range of 93–96% are usually achieved.

The third most crucial stage in the balanced process is the oxychlorination step.[188–190,272,273] In this reaction ethylene and HCl are converted to 1,2-dichloro-ethane in an oxidative, catalytic process. The reaction proceeds at temperatures of 225–325°C and pressures 1–15 atm. Pure oxygen or air is used as oxidant.[276–278] Numerous, somewhat different industrial processes were developed independently.[272–274] However, the reaction is generally carried out in the vapor phase, in fixed-bed or fluidized-bed reactors.

Copper chloride is universally applied as catalyst.[279–281] Known as the *modified Deacon catalyst*, $CuCl_2$ is supported on alumina and contains KCl. Under operating conditions a $CuCl_2$–Cu_2Cl_2–KCl ternary mixture, possibly in the molten state, is the active catalyst. $CuCl_2$ is believed to be the active chlorinating agent, producing 1,2-dichloroethane directly without the involvement of elemental chlorine [Eq. (6.44)]. The coupled oxidoreduction steps are indicated in Eqs. (6.45) and (6.46):

$$CH_2\!=\!CH_2 \;+\; 2\,CuCl_2 \;\longrightarrow\; \underset{\underset{Cl}{|}}{CH_2}\underset{\underset{Cl}{|}}{CH_2} \;+\; Cu_2Cl_2 \qquad (6.44)$$

$$CuCl_2 \;+\; 0.5\,O_2 \;\longrightarrow\; CuO.CuCl_2 \qquad (6.45)$$

$$CuO.CuCl_2 \;+\; 2\,HCl \;\longrightarrow\; 2\,CuCl_2 \;+\; H_2O \qquad (6.46)$$

A modified vinyl chloride process[282] combines Monsanto's direct chlorination, cracking, and purification technologies[283] and Kellog's proprietary technology for oxidation of HCl to Cl_2. In this process the inefficient hazardous oxychlorination step is eliminated. Instead, chlorine is recovered after the oxidation of HCl and used again in the direct chlorination of ethylene. This improved technology is more economical, less polluting and yields high-purity vinyl chloride.

Chlorination of 1,3-Butadiene. Gas-phase chlorination of 1,3-butadiene is the first step of a commercial synthesis of chloroprene.[284–287] At an elevated temperature (290–330°C), excess 1,3-butadiene is reacted with chlorine to yield a near-equilibrium mixture of the three isomeric dichlorobutenes:

$$CH_2\!=\!CHCH\!=\!CH_2 \;+\; Cl_2 \;\longrightarrow\; \underset{\underset{Cl}{|}\;\underset{Cl}{|}}{CH_2CHCH}\!=\!CH_2 \;+\; \underset{\underset{Cl}{|}}{CH_2}CH\!=\!\underset{\underset{Cl}{|}}{CHCH_2} \qquad (6.47)$$

At low conversions (10–25%) high yields can be achieved (85–95%). 1,4-Dichloro-2-butenes formed may be used in the manufacture of other products or isomerized to give the desired 3,4-dichloro-1-butene. This is done in the presence of a catalytic amount of $CuCl_2$ with the equilibrium shifted to the formation of 3,4-dichloro-1-butene by distilling it off the reaction mixture. Finally, 3,4-dichloro-1-butene undergoes alkaline dehydrochlorination to produce chloroprene (5–15% NaOH, 80–110°C).

6.4. AMMONIA AND AMINE ADDITION

The addition of ammonia and amines to unsaturated hydrocarbons, called *hydroamination*, is a desirable but difficult reaction. Only activated multiple bonds react readily in hydroamination to yield amines in an equilibrium reaction. It is necessary, therefore, to use catalysts in the transformation of nonactivated compounds.

6.4.1. Alkenes

One possible way to promote hydroamination is to activate the reacting ammonia or amine.[288–291] Alkali metals were found to be useful catalysts. Even in the catalyzed addition of ammonia to simple alkenes, however, drastic reaction conditions are required and a mixture of amines is formed. Ethylene, for instance, reacts with ammonia at 175–200°C and 800–1000 atm to yield ethylamines.[292] Alkali metals reacting with ammonia form the amide ion, which is believed to promote addition:[291]

$$H_2\overset{\cdot\cdot}{N}{:} \;\; + \;\; CH_2{=}CH_2 \;\;\longrightarrow\;\; H_2N{-}CH_2{-}\overset{\cdot\cdot}{C}H_2 \;\;\xrightarrow{\;NH_3\;}\;\; H_2N{-}CH_2{-}CH_3 \qquad (6.48)$$

The reactivity of alkenes decreases with increasing alkyl substitution, with ethylene exhibiting the highest activity. The addition to terminal alkenes follows the Markovnikov rule. Formation of high-molecular-weight nitrogen-containing compounds complicates the hydroamination of higher alkenes.[292,293] Because of its lower acidity, however, aniline participates only in monoaddition.[293]

The addition is more effective, allowing less severe reaction conditions if the metal amides are preformed,[294] or generated in situ from alkyllithium and the amine in the presence of N,N,N',N'-tetramethylethylenediamine.[295] Styrenes react under very mild conditions:[296]

$$Ph{-}CH{=}CH_2 \;\; + \;\; Et_2NH \;\; \xrightarrow[\text{cyclohexane, 50°C, 23 h}]{sec\text{-}BuLi} \;\; Et_2NCH_2CH_2Ph \qquad (6.49)$$
$$57\%$$

The feasibility of acid-catalyzed direct hydroamination has been demonstrated. Acidic zeolites afford, at low conversions, highly selective formation of ethylamine,[297,298] isopropylamine,[298] and *tert*-butylamine,[298–301] in the reaction of ammonia with ethylene, propylene, and isobutylene, respectively. Amine formation is explained as a reaction of surface carbocation intermediates with adsorbed or gas-phase ammonia.[298,299]

A few examples are known using homogeneous transition-metal-catalyzed additions. Rhodium(III) and iridium(III) salts catalyze the addition of dialkylamines to ethylene.[302] These complexes are believed to activate the alkene, thus promoting hydroamination. A cationic iridium(I) complex, in turn, catalyzes the addition of aniline to norbornene through the activation of the H—N bond.[303] For the sake of comparison it is of interest to note that dimethylamino derivatives of Nb, Ta, and Zr can be used to promote the reaction of dialkylamines with terminal alkenes.[304] In this case, however, *C*-alkylation instead of *N*-alkylation occurs.

There is a rich patent literature about the use of transition metal complexes in photochemical hydroamination, and the application of heterogeneous catalysts and the ammonium ion as catalyst.[290]

6.4.2. Dienes

Base-catalyzed addition of secondary amines to conjugated dienes usually give, under mild conditions, unsaturated amines formed by 1,4 addition.[291] Alkali metals,[305] metal amides,[296] and alkali naphthalenides[306] are used as catalysts.

The reaction of 1,3-butadiene with dialkylamines initiated by n-BuLi gives selectively 1-dialkylamino-cis-2-butenes[307] when the reagents are applied in a ratio of $3:1$:

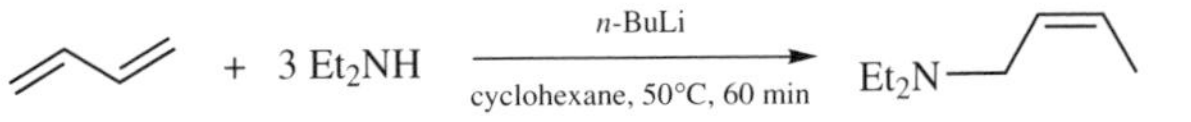

$$\text{[structure]} + 3 \text{ Et}_2\text{NH} \xrightarrow[\text{cyclohexane, 50°C, 60 min}]{n\text{-BuLi}} \text{Et}_2\text{N} \text{[structure]} \qquad (6.50)$$

50% yield
98–99% selectivity

A $2:1$ dialkylamine–lithium dialkylamide complex was found to play an important role in the addition process.

Hydroamination of allenes and 1,3-dienes in the presence of Ni(II), Pd(II), and Rh(III) complexes yields product mixtures composed of simple addition products and products formed by addition and telomerization.[288] Nickel halides[308] and rhodium chloride[309] in ethanol [Eq. (6.51)] and Pd(II) diphosphine complexes[310] are the most selective catalysts in simple hydroamination, while phosphine complexes favor telomerization:[288]

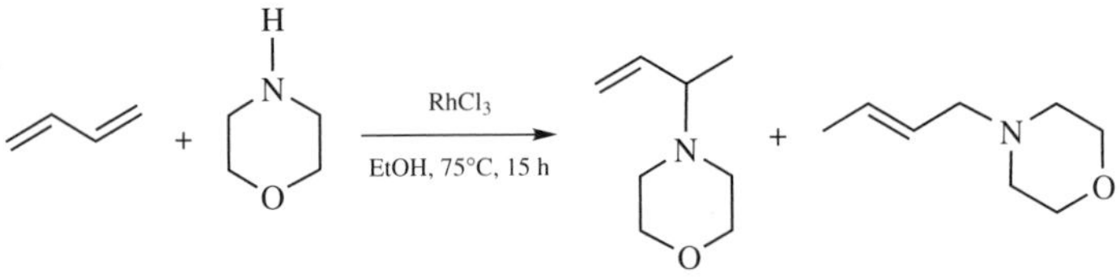

$$\text{[structure]} + \text{[morpholine structure]} \xrightarrow[\text{EtOH, 75°C, 15 h}]{\text{RhCl}_3} \text{[structure]} + \text{[structure]} \qquad (6.51)$$

70   30
85% conversion

Steric effects play a role in determining product distribution since substitution of 1,3-butadiene markedly increases selectivity of hydroamination.[308]

6.4.3. Alkynes

High temperature, pressure, and catalyst are required to achieve addition of ammonia to alkynes. Acetylene and ammonia yield a complex mixture of heterocyclic nitrogen bases.[311,312] Ethylideneimines, thought to form through the intermediate enamines, are the products of the reaction of acetylene with primary alkylamines in the presence of catalysts.[313]

$$\text{R}_2\text{NH} + \text{HC}\equiv\text{CH} \xrightarrow[\substack{120-140°C, \ 13-15 \ \text{atm}, \\ 35-74 \ \text{h}}]{\text{Cd(OAc)}_2 + \text{Zn(OAc)}_2} [\text{RNHCH}=\text{CH}_2] \rightleftharpoons \text{RNH}=\text{CHCH}_3 \qquad (6.52)$$

Since tautomeric rearrangement is not possible when secondary amines react with acetylene, the intermediate enamines can be isolated.[312,314] Vinylation with acetylene of heterocyclic nitrogen bases, such as pyrrole, indole, carbazole, was achieved in this way.[312,315] However, when terminal alkynes react with secondary amines, the intermediate enamine reacts further with a second molecule of alkyne[316] to give compound **33**:

$$R_2NH \ + \ MeC{\equiv}CH \ \xrightarrow[\text{120°C, 5–39 h}]{Cd(OAc)_2 + Zn(OAc)_2} \ \underset{NH_2}{MeC{=}CH_2} \ \xrightarrow{MeC{\equiv}CH} \ \underset{\underset{\mathbf{33}}{NH_2}}{\overset{Me}{MeCC{\equiv}CMe}} \qquad (6.53)$$

R = Me, Et, *n*-Bu

71–88%

Mercury(II) salts[317] and Tl(OAc)$_3$[318] may be used as catalyst to add aliphatic and aromatic amines to alkynes to yield imines or enamines. Selective addition to the carbon–carbon triple bond in conjugated enynes was achieved by this reaction:[319]

$$R^1{-}CH{=}\overset{R^2}{\underset{}{C}}{-}C{\equiv}CH \ + \ \underset{R^3}{\overset{NH_2}{\bigodot}} \ \xrightarrow[\text{THF, H}_2\text{O, 70°C, 3 h}]{HgCl_2, K_2CO_3} \ R^1 \diagup\!\!\!\diagup \overset{R^2 \quad Me}{\underset{N}{\diagdown}}\!\!\!-\!\!\!\underset{R^3}{\bigodot} \qquad (6.54)$$

R^1, R^2 = H, Me

R^1–R^2 = —(CH$_2$)$_3$—, —(CH$_2$)$_4$—

R^3 = 2-Me, 3-Me

45–80%

6.5. HYDROMETALLATION

The addition of metal–hydrogen bonds across carbon–carbon multiple bonds, called *hydrometallations*, are very important, versatile transformations in organic synthesis. First, they allow the synthesis of new organometallic compounds. The products thus formed may be further transformed into other valuable compounds. The two most important reactions, hydroboration and hydrosilylation, will be treated here in detail, whereas other hydrometallation reactions (hydroalanation, hydrozirconation) will be discussed only briefly. Hydrostannation, a very important transformation of substituted unsaturated compounds, has no significance in the chemistry of hydrocarbons possessing nonactivated multiple bonds.

6.5.1. Hydroboration

The addition of borane and organoboranes to multiple bonds to form new organoboranes called *hydroboration*, pioneered and developed mainly by H. C. Brown, is an extremely valuable organic transformation. Mono- and dialkylboranes formed may be themselves utilized as hydroborating agents. Among other reactions organoboranes may be oxidized with H$_2$O$_2$ and NaOH to form alcohols, diols, and polyols, depending on the structure of the starting unsaturated compounds. Hydroboration, therefore, combined with oxidation, provides a convenient indirect way of hydration of the carbon–carbon double bond in an anti-Markovnikov fashion.[320–325] Treatment with carboxylic acids converts organoboranes to hydrocarbons, which is an indirect way of saturating carbon–carbon multiple bonds.[323–326] Further possibilities to transform organoboranes are abundant.[322–325,327–330] These reactions, however, are beyond the scope of the present discussion.

Many different organoboranes have been developed and are in use at present to control selectivities, which makes hydroboration a particularly valuable and flexible process. A further important advantage of this method is that a wide range of functional groups tolerate hydroboration. Catalytic hydroboration, a recent development, is a new way of asymmetric hydroborations. Detailed treatment of hydroboration can be found in monographs[321,323,324,331] and review papers.[320,325,332–335]

Alkenes. Addition of diborane and organoboranes to carbon–carbon double bonds is a rapid, quantitative, and reversible transformation. Reversibility, however, is not a severe limitation under the usual reaction conditions, since thermal dissociation of organoboranes becomes significant only above 100°C. The addition is *syn* and occurs in an anti-Markovnikov manner; that is, boron adds preferentially to the less substituted carbon atom. The attack of the reagent takes place on the less hindered side of the reacting alkene molecule.

Early studies in hydroboration were carried out with in situ–prepared diborane.[336,337] Although gas-phase addition of diborane to alkenes is very slow, it was found that weak Lewis bases such as ethers and sulfides facilitate the reaction. Under such conditions addition is instantaneous and takes place at or below room temperature. Solution of diborane in tetrahydrofuran $(H_3B–THF)$[320] and the borane–dimethyl sulfide complex $(H_3B–SMe_2)$[338,339] are now commercially available reagents. Borane–amine complexes may also be used in hydroboration, although they react only at higher temperatures (above 100°C), where thermal dissociation of organoboranes may be a severe limitation. In nonbonded electron-pair solvents monomeric borane is present; otherwise it is dimeric. To achieve hydroboration, the dimer must dissociate [Eq. (6.55)] or probably only one of the hydrogen bridges must open to produce an electron-deficient boron center [Eq. (6.56)]:

$$B_2H_6 \rightleftharpoons 2\,BH_3 \qquad (6.55)$$

$$(6.56)$$

Borane may react sequentially with 3 mol of alkene to form mono-, di-, and trialkylboranes. Both the alkene structure and reaction conditions affect product distribution. Trialkylboranes are usually formed from terminal olefins [Eq. (6.57)] and unhindered disubstituted alkenes such as cyclopentene irrespective of the reactant ratio.[340] The reaction cannot be stopped at the mono- or dialkylborane stage. In contrast, hindered disubstituted olefins (e.g., cyclohexene) and trisubstituted alkenes are converted mainly to dialkylboranes [Eq. (6.58)]. Careful control of temperature and molar ratio of the reactants may allow selective synthesis of di-alkylboranes.[341] Tetrasubstituted alkenes, in turn, form monoalkylboranes as initial products and react further only very slowly[341] [Eq. (6.59)]:[342]

$$R-CH=CH_2 \ + \ BH_3 \ \xrightarrow{\text{ether solvent}} \ (R-CH_2-CH_2)_3B \qquad (6.57)$$

$$2 \ \diagup\!\!\!=\!\!\!\diagup \ + \ H_3B-THF \ \xrightarrow[\text{THF, 0°C, 6–9 h}]{} \ \left(\!\diagdown\!\!\diagup\!\!\!\diagdown\right)_{\!2}\!BH \qquad (6.58)$$

34

$$\diagup\!\!\!=\!\!\!\diagdown \ + \ H_3B-THF \ \xrightarrow[\text{0°C, 1 h}]{} \ \diagup\!\!\!\diagdown\!\!\!-BH_2 \qquad (6.59)$$

35

Hindered alkenes, however, may undergo hydroboration at high pressure and may even yield trialkylboranes.[343]

It is clear from these representative results that regioselectivity in hydroboration is controlled by steric effects. As a result, nonsymmetric internal olefins usually yield a mixture of regioisomeric alkylboranes when they react with borane. Several hindered mono- and diakylboranes with sterically demanding alkyl groups, how-ever, have been developed for use in selective hydroboration. Disiamylborane [bis(3-methyl-2-butyl)borane, **34**], thexylborane (2,3-dimethyl-2-butylborane, **35**), and 9-BBN (9-borabicyclo[3.3.1]nonane) are the most frequently used reagents.[337,339,342] Improvements of regioselectivities in hydroboration of both terminal and internal alkenes can be achieved with such hindered dialkylbor-anes.[340,344] Disiamylborane shifts the selectivity to the formation of the regioisomer substituted at the less hindered olefinic carbon:[340]

$$\begin{array}{c}\text{isoPr}\diagdown\quad\diagup\text{H}\\ C=C\\ \text{H}\diagup\quad\diagdown\text{Me}\end{array} \xrightarrow[\text{ii. oxidation}]{\text{i. diglyme, 25°C}} \ \underset{\underset{\text{OH}}{|}}{\text{isoPrCHCH}_2\text{Me}} \ + \ \underset{\underset{\text{OH}}{|}}{\text{isoPrCH}_2\text{CHMe}} \qquad (6.60)$$

BH$_3$, 1 h	43	57
34, 12 h	5	95

Thexylborane (**35**) is one of the most useful monoalkylboranes.[337,342] It reacts with 1-alkenes to form dialkylthexylboranes, or yields monoalkylthexylboranes with di- or trisubstituted olefins.[345] Mixed trialkylboranes may also be available with this reagent in stepwise hydroboration, provided the more hindered alkene is hydro-borated in the first step[346] [Eq. (6.61)][347]:

$$\bigcirc\!\!\!\!= \ + \ \diagup\!\!\!\diagdown\!\!\!-BH_2\!\cdot\!NEt_2Ph \ \xrightarrow[\text{0°C, 2 h}]{\text{THF}} \ \overset{\bigcirc}{\underset{H}{B}}\!\!\diagdown\!\!\!\diagup \ \xrightarrow[\text{THF, 0°C, 2 h}]{\text{1-octene}} \ \overset{\bigcirc}{\underset{H_{17}C_8}{B}}\!\!\diagdown\!\!\!\diagup \qquad (6.61)$$

Since thexylborane reacts about 100 times faster with Z alkenes than with the corresponding E isomers, it may also be used to separate isomeric mixture.[348]

The stereochemical outcome of hydroboration, *syn* steroselectivity, was originally based on the stereochemistry of alcohols isolated after hydroboration–oxidation.[349,350] Thus, for example, $(+)$-α-pinene yields $(-)$-isopinocampheol after the oxidation of $(-)$-diisopinocampheylborane (**36**):

$$(6.62)$$

The *syn* steroselectivity was based on the assumption that the oxidation step—the transformation of the carbon–boron bond to carbon–oxygen bond—took place with retention of configuration. More recent NMR studies of alkylboranes formed in hydroborating labeled alkenes indeed confirmed the validity of the earlier conclusion.[351]

Hydroboration of α-pinene [Eq. (6.62)] is an excellent example demonstrating that the addition takes place on the less hindered side of the double bond. Similarly norbornene, when undergoing hydroboration with 9-BBN, gives 2-*exo*-norbornanol with 99.5% selectvity,[344] whereas the *gem*-dimethyl group in 7,7-dimethylnorbornene brings about the reversal of the side of attack to yield the corresponding *endo* alcohol with 97% selectvity.[352] Furthermore, even in the hydroboration of such sensitive compounds as α-pinene and norbornene, no skeletal rearrangement has ever been observed during hydroboration.

The characteristic features of hydroboration of alkenes—namely, regioselectivity, stereoselectivity, *syn* addition, and lack of rearrangement—led to the postulation of a concerted $[2+2]$ cycloaddition of borane[353,354] via four-center transition state **37**. Kinetic studies, solvent effects, and molecular-orbital calculations are consistent with this model. As four-center transition states are unfavorable, however, the initial interaction of borane [or mentioned monobridged dimer, Eq. (6.56)] with the alkene probably involves an initial two-electron, three-center interaction[355,356] (**38, 39**).

On the basis of overall second-order kinetics of hydroboration with disiamylborane, a two-step mechanism was suggested involving the interaction between the nondissociated reagent and alkene in the rate-determining first step[354,357] to produce a mixed diborane–monomeric **34**:

$$(6.63)$$

$$\text{fast}$$

The latter then reacts with the alkene in a fast reaction.

An overall second-order rate law is also operative in hydroboration with 9-BBN of terminal alkenes,[354,358] but a slow rate-determining dissociation of the dimer [Eq. (6.64)] precedes the addition step [Eq. (6.65)]:

$$(9\text{-BBN})_2 \;\underset{}{\overset{\text{slow}}{\rightleftharpoons}}\; 2\ 9\text{-BBN} \qquad\qquad (6.64)$$

$$9\text{-BBN} \;+\; 1\text{-alkene} \;\xrightarrow{\text{fast}}\; \text{product} \qquad\qquad (6.65)$$

Since more substituted alkenes are less reactive, the now much slower addition becomes rate-determining and a much more complicated rate law is operative.

Among the many hydroborating agents developed over the years,[337] catecholborane, a boric ester derivative, exhibits decreased reactivity.[359] It has since been observed, however, that the addition can be greatly accelerated by transition-metal complexes.[360,361] Most of the studies focused on Rh(I) catalysts.[361]

Terminal alkenes react almost instantaneously in rhodium-catalyzed hydroboration, whereas disubstituted alkenes require the use of excess reagent and increased amounts of catalyst (2–5%).[362] Other alkenes are unreactive. High regioselectivities can be achieved.[363,364]

A multistep pathway analogous to the mechanism of alkene hydrogenation has been shown to be operative in the rhodium-catalyzed hydroboration of alkenes.[363] Deuterium labeling studies furnished evidence that the reversibility of the elementary steps is strongly substrate-dependent. The key step is hydride rather than boron migration to the rhodium-bound alkene.

Dienes. The bifunctionality of dienes makes their hydroboration more complex than that of simple alkenes. Competing hydroboration of the two double bonds may lead to mixtures of products arising from mono- or diaddition. Additionally, cyclic or polymeric organoboranes may be formed. Differences in the reactivity of the two double bonds and the use of appropriate hydroborating agents, however, may allow selective hydroboration.[29,330]

Conjugated dienes and α,ω-dienes usually undergo dihydroboration with borane. 1,3-Butadiene, for instance, yields a mixture of 1,3- and 1,4-butanediol when the organoborane intermediates formed through inter- and intramolecular hydroboration are oxidized.[365–367] This product distribution points to the formation of regioisomers in the first hydroboration steps. The data also indicate that the isolated double bond formed after the addition of 1 mol of borane is more reactive than the starting diene.

Conjugated dienes exhibit a marked decrease in reactivity toward 9-BBN relative to alkenes and nonconjugated dienes. Structurally different double bonds, however, may allow clean monohydroboration.[368] Homoallylic boranes are usually formed, but certain dienes yield allylic boranes.

A characteristic transformation of nonconjugated dienes is their hydroboration to form boraheterocycles.[369] For example, the favored hydroborating agent, 9-BBN, is synthesized in such cyclization reaction[370,371] (Scheme 6.6). An approximately 1 : 3 mixture of 1,4- and 1,5-addition products is formed in a second, intramolecular hydroboration step. The 1,4-addition product (**40**), however, can be readily isomerized under mild conditions (65°C, 1 h) through a dehydroboration–hydroboration step to yield pure 9-BBN.

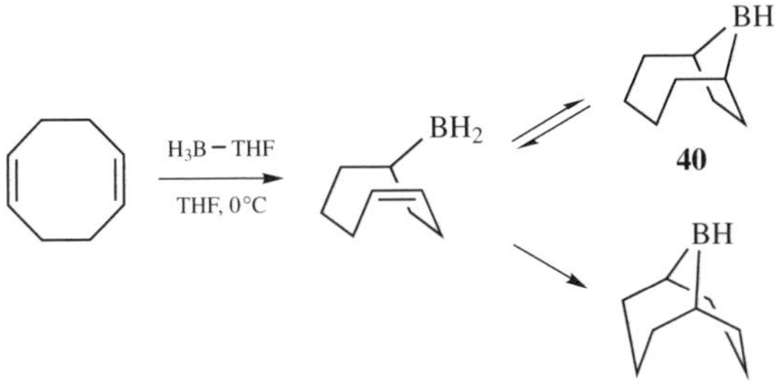

Scheme 6.6

Alkynes. The selectivity of hydroboration of alkynes depends on the hydroborating reagents. Diborane usually forms complex mixtures of organoboranes, including polymers.[372,373] Other reagents, however, may be applied to perform selective transformations.

Terminal alkynes undergo clean monohydroboration with disiamylborane[372] and thexylborane[374] to yield vinylboranes. The addition of a second molecule of dialkylborane to form *gem*-diboryl compounds is usually much slower. The less hindered 9-BBN, in turn, produces monoaddition products only when the reaction is carried out in excess alkyne.[375] Mono- versus dihydroboration, however, can be controlled by temperature with increasing selectivity of monohydroboration brought about by decreasing temperature.[376] Boron is placed almost exclusively at the terminal carbon of the triple bond. More recently the thexylborane–dimethyl sulfide complex was shown to afford clean monoboration with exceptional regioselectivity (>97%).[377] Direct spectroscopic evidence was presented to show selective *syn* addition.[378,379]

The same reagents may be used to conduct monohydroboration of internal acetylenes as well. Unsymmetrically disubstituted compounds usually produce a mixture of regioisomers. The best reagent for the regioselective hydroboration of alkynes is dimesitylborane. It gives, for example, 90% selectivity of boron attack at C-2 in the hydroboration of 2-hexyne.[380] Competitive steric and electronic effects control the selectivities in the hydroboration of phenyl-substituted alkynes.[375,380]

6.5.2. Hydroalanation

Aluminum hydrides add across carbon–carbon multiple bonds to form organo-alanes[381–385] (hydroalanation or hydroalumination), which are usually further reacted with suitable electrophiles. The addition is thermal or may be catalyzed by metals, mainly titanium and zirconium compounds.[386,387]

Hydroalanation was first carried out by Ziegler and coworkers by reacting alkenes with activated aluminum in the presence of hydrogen.[388] Preformed AlH_3 was also shown to add to alkenes.[389,390] Dialkylaluminum hydrides in alkane solvents and $LiAlH_4$ in diethyl ether are more convenient and the most frequently applied reagents. An interesting application is, however, the use of triisobutylalu-minum.[391] Since hydroalumination of alkenes is a reversible reaction, isoBu$_3$Al, when heated above 100°C, dissociates to form isobutylene and isoBu$_2$AlH; the latter then adds to alkenes.

Many of the characteristic features of hydroalanation of alkenes (reactivities, selectivities) are very similar to those of hydrosilylation. Terminal alkenes react readily in hydroalumination, whereas internal alkenes are much less reactive. Aluminum usually adds selectively to the terminal carbon. Hydroalumination of styrene, however, leads to a mixture of regioisomers.[392] When hydroalumination of alkenes is followed by protolysis, saturated hydrocarbons are formed; that is, net hydrogenation of the carbon–carbon double bond may be achieved. The difference in reactivity of different double bonds allows selective hydroalumination of the less hindered bond in dienes:[393]

$$(6.66)$$

97%

Alkynes are more reactive in hydroalumination than alkenes. Hence, unlike internal alkenes, internal acetylenes readily undergo hydroalumination. Side reactions, however, may occur. Proton–aluminum exchange in terminal alkynes, for instance, leads to substituted products, and geminal dialuminum derivatives are formed as a result of double hydroalumination. In addition, nonsymmetric alkynes usually give mixtures of regioisomers. Appropriate reaction conditions, however, allow selective hydroalumination. Thus, the addition of isoBu$_2$AlH to alkynes is a selective *syn* process[394] [Eq. (6.67)].[395] In contrast, LiAlH$_4$ and its derivatives add to alkynes in a selective *anti* fashion[396] [Eq. (6.68)]:[397]

$$Ph\!\!=\!\!\!=\!\!Me \; + \; isoBu_2AlH \xrightarrow[50\,°C,\,26\,h]{} \underset{isoBu_2Al \quad H}{\overset{Ph \quad Me}{>\!\!=\!\!<}} \xrightarrow{H_2O} \underset{H \quad H}{\overset{Ph \quad Me}{>\!\!=\!\!<}} \quad (6.67)$$

$$Ph\!\!=\!\!\!=\!\!Me \; + \; LiAlH_4 \xrightarrow[THF,\,66\,°C,\,13\,h]{} \underset{Ph \quad H}{\overset{LiH_3Al \quad Me}{>\!\!=\!\!<}} \xrightarrow{H_2O} \underset{Ph \quad H}{\overset{H \quad Me}{>\!\!=\!\!<}} \quad (6.68)$$

>99% selectivity

Note that regioselectivities are identical with the two reagents. The formation of aluminum hydrides (deuterides), followed by protolysis (deuterolysis) of the intermediate vinylalanes, allows the synthesis of selectively labeled alkenes.[394,395,397]

6.5.3. Hydrosilylation

Hydrosilylation, the addition of a silicon–hydrogen bond to multiple bonds, is a valuable laboratory and industrial process in the synthesis of organosilicon compounds. The addition to carbon–carbon multiple bonds can be accomplished as a radical process initiated by ultraviolet (UV) light, γ irradiation, or peroxides. Since the discovery in the 1950s that chloroplatinic acid is a good catalyst to promote the addition, metal-catalyzed transformations have become the commonly used hydrosilylation process. Numerous reviews give detailed coverage of the field.[171,381,398–406] Hydrosilylation finds practical application as the curing (crosslinking) process in the manufacture of silicone rubbers.[407]

Radical hydrosilylation takes place according to a usual free-radical mechanism with silyl radicals as chain carriers. Products are formed predominantly through the most stable radical intermediate. Even highly hindered alkenes undergo radical hydrosilylation. This process, however, is not stereoselective, and alkenes that are prone to free-radical polymerization may form polymers.

None of these difficulties arise when hydrosilylation is promoted by metal catalysts. The mechanism of the addition of silicon–hydrogen bond across carbon–carbon multiple bonds proposed by Chalk and Harrod[408,409] includes two basic steps: the oxidative addition of hydrosilane to the metal center and the *cis* insertion of the metal-bound alkene into the metal–hydrogen bond to form an alkylmetal complex (Scheme 6.7). Interaction with another alkene molecule induces the formation of the carbon–silicon bond (route *a*). This rate-determining reductive elimination completes the catalytic cycle. The addition proceeds with retention of configuration.[410] An alternative mechanism, the insertion of alkene into the metal–silicon bond (route *b*), was later suggested to account for some side reactions (alkene reduction, vinyl substitution).[411–414]

Detailed studies concerning the active species in hydrosilylation catalyzed by chloroplatinic acid have pointed to the involvement of several oxidation states. It

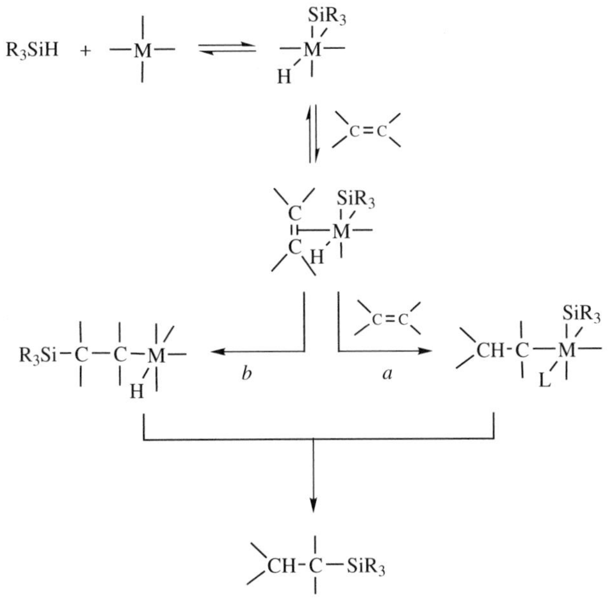

Scheme 6.7

was observed, for example, that Pt(IV) undergoes partial reduction to Pt(II) in the H_2PtCl_6—2-propanol solution, which is used in most cases.[415,416] Oxygen strongly affects the reaction, but its effect remains largely unclear.[400,401,417–419] An induction period is usually observed, and carefully purified reactants and H_2PtCl_6 react only very slowly.[401] Formation of colloidal platinum and rhodium on the action of hydrosilanes was observed, and their intermediacy in hydrosilylation has been suggested.[419–421]

Alkenes. Most Group VIII metals, metal salts, and complexes may be used as catalyst in hydrosilylation of alkenes. Platinum and its derivatives show the highest activity. Rhodium, nickel, and palladium complexes, although less active, may exhibit unique selectivities. The addition is exothermic and it is usually performed without a solvent. Transition-metal complexes with chiral ligands may be employed in asymmetric hydrosilylation.[406,422]

Chloroplatinic acid, used as a solution in 2-propanol and referred to as *Speier's catalyst*,[423] is a unique and very active catalyst for hydrosilylation. It catalyzes the addition of chloro- and alkylhydrosilanes to terminal alkenes to form almost exclusively the corresponding terminally silylated products in high yields (anti-Markovnikov addition). Alkylsilanes are much less reactive hydrosilylating reagents. The addition of di- and trihydroalkylsilanes to form tetraalkylsilanes, however, were found to be highly accelerated by oxygen.[417,418]

Internal alkenes and cycloalkenes react much more slowly. If double-bond migration, however, can lead to an isomeric terminal olefin, slow transformation to the terminally substituted alkylsilane is possible.[401,423–426] All three isomeric ethylcyclohexenes, for example, form 2-cyclohexylethyltrichlorosilane as the sole product:[426]

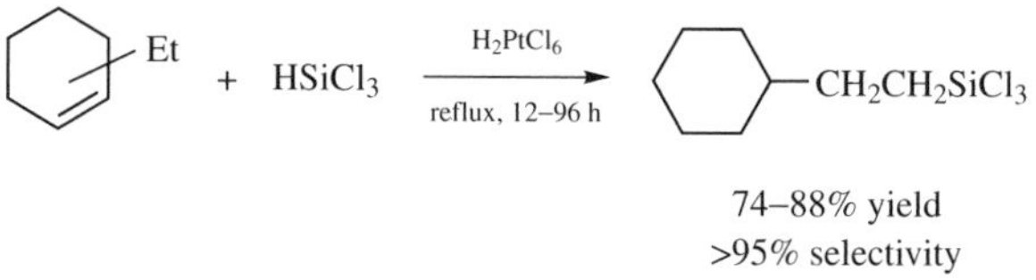

$$\text{(6.69)}$$

74–88% yield
>95% selectivity

Alkene isomerization is common during hydrosilylation. At partial conversion in the reaction of ethylcyclohexenes, the recovered olefin contained all possible isomers except vinylcyclohexane.[426] Even in the transformation of the most reactive terminal alkenes, the recovered unreacted olefin is a mixture of isomers.[425,427] Isomerization thus may prevent complete conversion of 1-alkenes. Similar double-bond migration is not observed in free-radical addition:[428]

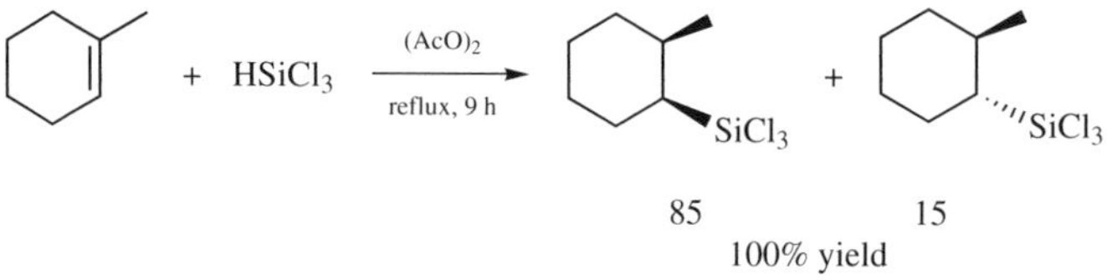

$$\text{(6.70)}$$

85   15
100% yield

The regioselectivity of hydrosilylation of arylalkenes strongly depends on the catalyst. Hydrosilylation of styrene catalyzed by H_2PtCl_6 is not selective and produces a mixture of regioisomers.[429] Cocatalysts, such as PPh_3, however, increase the selectivity of the formation of the anti-Markovnikov product [Eq. (6.71)]. Nickel catalysts, in contrast, bring about the formation of the Markovnikov adduct[430] [Eq. (6.72):[429]]

$$PhCH{=}CH_2 \ + \ HSiCl_2Me$$

$$\xrightarrow[\substack{80°C,\ 2\,h}]{H_2PtCl_6,\ PPh_3} PhCH_2CH_2SiCl_2Me \qquad \text{(6.71)}$$

99% yield

$$\xrightarrow[\substack{reflux,\ 4\,h}]{Ni(CO)_4} \underset{\underset{SiCl_2Me}{|}}{PhCHCH_3} \quad 88\% \text{ yield} \qquad \text{(6.72)}$$

In addition to the formation of regioisomers, vinyl substitution and hydrogenation also take place in the presence of rhodium catalysts.[413]

More substituted alkenes are usually unreactive in metal-catalyzed hydrosilylation. $AlCl_3$, however, promotes the addition of dialkylsilanes to tetrasubstituted alkenes under mild conditions.[431]

Dienes. Hydrosilylation of conjugated dienes often leads to a complex mixture of products composed of regio- and stereoisomers of 1 : 1 addition, 1 : 2 addition products, and oligomers. Dihydrosilanes may form silacycloalkanes as well. In contrast with its lack of activity in catalyzing the addition of hydrosilanes to alkenes, palladium exhibits excellent activity in the hydrosilylation of conjugated dienes with chlorosilanes. The reaction gives *syn* 1,4-addition products in high yields with high selectivity[432–434] [Eq. (6.73)[435]]:

$$
\begin{array}{c}
\text{\Large$\gamma\!\!\!\!\diagdown$} + R_3SiH \xrightarrow[\text{60–80°C, 6 h}]{(PhCN)_2PdCl_2 + PPh_3}
\left[\begin{array}{c} \\ \text{Pd(II)} \\ R_3Si \end{array} \right]
\longrightarrow \begin{array}{c} \\ R_3Si \end{array}
\end{array}
\qquad (6.73)
$$

R$_3$SiH = Cl$_3$SiH, MeCl$_2$SiH,
        Me$_2$ClSiH

quantitative

The addition presumably takes place via a π-allyl intermediate.[435]

Nickel catalysts exhibit similar activity but are usually less selective than palladium.[435–438] In addition to chlorosilanes, however, alkylsilanes also add in the presence of nickel catalysts. Nickel vapor showed an exceptionally high activity and selectivity in hydrosilylation.[435] The photocatalytic hydrosilylation of 1,3-dienes with a chromium(0) complex occurs at room temperature to afford regioisomeric 1,4-adducts in quantitative yield.[439]

Alkynes. Hydrosilanes react with acetylene to form vinylsilanes, and subsequently, 1,2-disilylethane.[406] Product distribution can be affected by catalyst composition and temperature.[440,441]

Hydrosilylation of terminal alkynes may yield three isomeric vinylsilanes:

$$
R\text{---}\!\!\!\equiv\!\!\!\text{---}H \xrightarrow{R'_3SiH} R\diagup\!\!\diagdown SiR'_3 + \begin{array}{c} R \\ \diagdown\diagup \\ SiR'_3 \end{array} + \begin{array}{c} R \\ \diagup\!\!= \\ R'_3Si \end{array} \qquad (6.74)
$$

R = *n*-Bu, HSiCl$_3$, H$_2$PtCl$_6$, 5°C		95	5
R = *n*-Bu, HSiMe$_2$Et, [RhCl(PPh$_3$)$_3$], 40°C	81	19	
R = Ph, HSiCl$_3$, [RhCl(PPh$_3$)$_3$], 100°C	14	55	28

The ratio of the three products depends on the reacting silane and alkyne, the catalyst, and the reaction conditions. Platinum catalysts afford the anti-Markovnikov adduct as the main product formed via *syn* addition.[442–446] Rhodium usually is a nonselective catalyst[404] and generally forms products of *anti* addition.[447–451] Minor amounts of the Markovnikov adduct may be detected. Complete reversal of stereoselectivity has been observed.[452] [Rh(COD)Cl]$_2$-catalyzed hydrosilylation with Et$_3$SiH of 1-hexyne is highly selective for the formation of the Z-vinylsilane in EtOH or DMF (94–97%). In contrast, the E-vinylsilane is formed with similar selectivity in the presence of [Rh(COD)Cl]$_2$–PPh$_3$ in nitrile solvents.

Hydrosilanes add slowly to disubstituted alkynes.[453] Exclusive *syn* addition occurs in the platinum-catalyzed process, usually with low regioselectivity.[445,454] Rhodium, in contrast, affords *anti*-addition products.[451] These were suggested to form through radical intermediates,[447,451] or via isomerization of the initial *syn* adducts.[448,449]

Group III and IV halides enhance the activity of chloroplatinic acid in hydrosilylation of phenylacetylene,[455] whereas aluminum halides alone promote the addition of hydrosilanes to terminal acetylenes.[456]

6.5.4. Hydrozirconation

Of other main-group and transition-metal hydrides, until now only the addition of bis(cyclopentadienyl)zirconium(IV) hydrochloride [Cp$_2$ZrHCl]—that is, hydrozirconation, a relatively new process—is of wider importance.[457,458] This is mainly because both [Cp$_2$ZrHCl] and the organozirconum intermediates formed in hydrozirconation are relatively stable compounds. Other suitable metal hydride reagents are not readily available, and few organometallics are stable enough to apply in organic synthesis. Organozirconium compounds, in turn, are useful intermediates and can be transformed in subsequent reactions to other valuable products (protolysis, deuterolysis, transmetallations, additions, coupling reactions).[458,459]

Hydrozirconation usually takes place readily at or slightly above room temperature, in benzene or toluene. The reactivity pattern of alkenes in hydrozirconation is the same as in hydroboration, except that tetrasubstituted alkenes are unreactive.[457,460] A unique feature of hydrozirconation of alkenes is that, with some exceptions, terminally zirconated addition products are formed exclusively. Since isomeric alkenes cannot be isolated, zirconium moves rapidly along the chain without dissociation. Tertiary zirconium intermediates seem to be so unstable that the isopropylidene group never undergoes hydrozirconation [Eq. (6.75)],[460] and hence, 1-methylcyclohexene fails to react.[460] The reason is apparently steric, since the alkene must fit into the somewhat bent sandwich structure of the reagent (**41**):

$$(6.75)$$

41

Exceptions are cycloalkenes, which do form secondary zirconium derivatives,[460] as well as arylalkenes[461] [Eq. (6.76)] and conjugated dienes:[462,463]

$$(6.76)$$

In the latter cases, steric and electronic factors facilitate the formation of small amounts of secondary substituted products.

Terminal vinylzirconium products are formed in the addition of [Cp$_2$ZrHCl] to terminal alkynes. Since zirconium migration does not occur in these organozirconium derivatives, internal alkynes give stable internal vinylzirconium regioisomers. Product distributions are controlled by steric effects; specifically, zirconium attaches itself to the carbon bearing the less bulky group:[464]

$$R\!\!-\!\!\equiv\!\!-\!\!R' + [Cp_2ZrHCl] \xrightarrow[\text{benzene, RT, 2 h}]{}$$

$$\qquad\qquad\qquad\qquad\qquad\qquad\qquad\qquad\qquad\qquad\qquad\qquad (6.77)$$

	initial ratio	in excess reagent
a R = Me, R′ = Et	55 : 45	89 : 12
b R = Me, R′ = isoPr	84 : 15	>98 : 2
c R = H, R′ = *tert*-Bu	>98 : 2	

The kinetic product distribution can be further shifted in excess reagent to get organozirconium derivatives with high regioselectivity. Even subtle steric differences such as in 2-pentyne ensure high selectivity [Eq. (6.77), *a*]. As Eq. (6.77) demonstrates, hydrozirconation of alkynes is an exclusive *syn* process.

6.6. HALOMETALLATION

Reactive metal halides are known to add to unsaturated compounds to form 1,2 halometal compounds. Addition of boron and aluminum halides is of particular interest.

Boron trihalides, particularly BBr$_3$, add to alkenes in a reaction (haloboration) that is usually more complicated than hydroboration.[465] Several compounds can be isolated as a result of secondary transformation of the primary addition product:[466]

$$\qquad\qquad\qquad\qquad\qquad\qquad\qquad\qquad\qquad\qquad\qquad\qquad (6.78)$$

BCl$_3$ reacts only with more reactive alkenes. Norbornadiene gives a mixture of isomeric (6-chloro-2-nortricyclyl)dichloroboranes, whereas cycloheptatriene and cyclooctatetraene yield the rearranged products benzyldichloroborane and *trans*-β-styryldichloroborane, respectively.[466,467] Addition of haloboranes to alkynes leads to the formation of halovinylboranes.[468–470] The difference in reactivity between BBr$_3$ and BCl$_3$ can also be observed in this addition: BCl$_3$ reacts with phenylacetylene, but does not add to dipehylacetylene.[468] The addition follows the Markovnikov rule, and when BCl$_3$ and BBr$_3$ add to substituted acetylenes,

Z compounds are formed[469] (the halogen and boron substituents are in *cis* position):[470]

$$R\text{---}\!\!\equiv\!\!\text{---}H \;+\; BBr_3 \xrightarrow[\substack{\text{dichloromethane,}\\ -78°C,\, 2\,h}]{} \begin{array}{c} R \qquad H \\ \diagdown \quad \diagup \\ C\!=\!C \\ \diagup \qquad \diagdown \\ Br \qquad BBr_2 \end{array} \qquad (6.79)$$

$$R = n\text{-Bu},\; n\text{-}C_6H_{13},\; n\text{-}C_8H_{17},$$
$$\text{cyclohexyl, Ph}$$

With increasing reactivity of the boron halide, however, trans addition can also be observed: BI_3 yields a $1:1$ mixture of Z and E compounds in the addition to phenylacetylene.[469] However, the *trans* isomer is formed when BBr_3 adds to acetylene.[468]

B-Bromo and *B*-iodo-9-borabicyclo[3.3.1]nonane add similarly in a *cis* fashion to terminal triple bonds.[471] They do not react, however, with alkenes and internal acetylenic bonds. In contrast to the results mentioned above, phenyl-substituted chloroboranes ($PhBCl_2$, Ph_2BCl) do not participate in haloboration. Instead, the C—B bond adds across the multiple bond to form phenylalkyl-(alkenyl)boranes.[466,468]

Tetrahalodiboranes, B_2Hlg_4, react with alkenes to form 1,2-bisdihaloboron derivatives (diboration).[472] The first example was the addition of B_2Cl_4 to ethylene to yield the stable liquid product 1,2-bis(dichloroboryl)ethane:[473]

$$CH_2\!=\!CH_2 \;+\; B_2Cl_4 \xrightarrow[-80°C,\, 16\,h]{} Cl_2BCH_2CH_2BCl_2 \qquad (6.80)$$
$$95\%$$

Further studies with other alkenes and cycloalkenes proved the general nature of the reaction.[474,475] B_2F_4 is less reactive but forms more stable products.[474]

B_2Cl_4 can add to dienes (1,3-butadiene, 1,3- and 1,4-cyclohexadienes) to yield either mono- or diadducts depending on the reagent : reactant ratio.[474–477] 1,3-Cyclohexadiene, for example, forms the unstable 1,2,3,4-tetrakis(dichloroboryl)cyclohexane with high selectivity when it reacts with a threefold excess of B_2Cl_4.[476] Naphthalene yields a similar diadduct.[476,477]

Acetylene reacts rapidly with B_2Cl_4 to form $1:1$ and $1:2$ addition products.[478] The transformation of phenylacetylene is much slower and only the monoaddition product is formed.[479] 2-Butyne, however, undergoes trimerization to give hexamethylbenzene when less than a stoichiometric amount of the reagent is used,[475] but reacts with an equimolar amount of B_2Cl_4 to yield the monoadduct.[480]

The addition of B_2Cl_4 to multiple bonds was shown to be stereoselective. The initial diborane products isolated after the addition of B_2Cl_4 to *cis*- and *trans*-2-butene were oxidized to yield *meso*- and racemic-2,3-butanediols, respectively.[478,480] A similar treatment of the addition product of cyclohexene resulted in *cis*-1,2-cyclohexanediol.[480] These observations are consistent with *syn* addition and the involvement of four-center transition state **42**. Similar studies with addition products of acetylenes[480] as well as spectroscopic evidence[478] also support stereo-selective *syn* addition to the triple bond.

42

Aluminum subhalides, $(AlCl_2)_2$ and $(AlBr_2)_2$, formed in the reaction between $AlCl_3$ ($AlBr_3$) and aluminum add to ethylene in their dimeric form. The reaction was first studied by Hall and Nash, and more recently by Olah and coworkers. Hall and Nash observed the formation of ethylaluminum sesquichloride ($Et_2AlCl \cdot EtAlCl_2$) when $AlCl_3$, Al, ethylene, and hydrogen were heated at 155°C under pressure (60 atm).[481] Olah and coworkers found that, besides ethylaluminum chlorides, addition products [1,1- and 1,2-bis(dihaloaluminio)ethane] were also formed:[482,483]

$$CH_2{=}CH_2 \xrightarrow[\substack{\text{n-hexane, 65°C, 3 atm, or} \\ \text{methylcyclohexane, 95°C}}]{AlHlg_3 + Al \text{ or } K} Hlg_2AlCH_2CH_2AlHlg_2 \longrightarrow (Hlg_2Al)_2CHCH_3 \quad (6.81)$$

$$\phantom{CH_2{=}CH_2 \xrightarrow{AlHlg_3 + Al} } \mathbf{43} \mathbf{44}$$

Compound **43** was shown to quantitatively isomerize to the more stable isomer **44**. Similar results were obtained when the reaction was carried out with preformed $(AlCl_2)_2$. The treatment of cyclohexene with a reagent prepared from $AlBr_3$ and potassium gave a product that on D_2O quench gave a 11% yield of $[1,2\text{-}^2H_2]$-cyclohexane indicative of the formation of bis(dihaloaluminio)cyclohexane.[483]

6.7. SOLVOMETALLATION

Certain metal salts can react with unsaturated compounds with the concomitant addition of a solvent molecule or other nucleophile to yield β-substituted metal derivatives. These reactions, called *solvometallations* or *oxymetallations*, have a number of applications in organic synthesis. A short treatment of certain classic and important reactions is given here.

6.7.1. Solvomercuration

Solvomercuration,[3,111,484–487] or the addition of mercury(II) salts, is a convenient route to organomercurials. On the other hand, replacement of mercury with hydrogen allows Markovnikov functionalization of alkenes.[488] A method called *mercuration–demercuration*, for instance, has been developed for the Markovnikov hydration of alkenes under mild conditions[489] [Eq. (6.82)[490]]:

$$(6.82)$$

70–75% yield

$Hg(OAc)_2$ and $Hg(OOCCF_3)_2$ are the most frequently used reagents. Water and alcohols are the most common nucleophiles, and these processes are termed *oxymercuration*. Other additions are aminomercuration and peroxymercuration.

Alkenes participate readily in oxymercuration with increasing alkyl-substitution resulting in decreasing reactivity.[489,491,492] The oxymercuration of alkenes is usually an *anti* process.[493-495] Exceptions are strained bicyclic olefins, such as norbornene[495,496] [Eq. (6.83)],[497] and *trans*-cyclooctene and *trans*-cyclononene.[494]

$$\text{(structure)} + Hg(OAc)_2 \xrightarrow[\text{ii. NaCl, } H_2O]{\text{i. THF, } H_2O} \text{(structure) } -OH(OAc), -HgCl \qquad (6.83)$$

99.8% selectivity

The *anti* addition and the lack of rearrangements are compatible with a mechanism with the involvement of the cyclic mercurinium ion.[493,495] Mercurinium ions are known to exist, for example, in superacidic media.[498]

Surprisingly, the stereochemistry of addition to 7,7-dimethylnorbornene does not change compared with that of norbornene (i.e., oxymercuration yields the *exo* product from both compounds),[497] although it always happens in additions taking place via cyclic intermediates.[352] This was taken as evidence against the formation of the mercurinium ion in these systems. Rather, the formation of a 2-mercury-substituted carbocation was suggested.[497] It reacts rapidly with the solvent from the *cis* side if the *trans* side of the molecule is hindered. Since much of the positive charge is retained on mercury, this mechanism is also consistent with the lack of carbocationic rearrangements. Other possibilities, such as a concerted process, were also suggested.[499]

Only scattered information is available on oxymercuration of alkynes. Alkylphenylacetylenes were found to undergo stereoselective *anti* addition.[500] Diphenylacetylene, in contrast, gives only *syn* adducts.[501]

6.7.2. Oxythallation

Oxythallation is another important solvometallation process.[502] The intermediate organothallium compounds formed during addition, however, are seldom isolable, and tend to undergo spontaneous rapid decomposition to form oxidation products. Some examples are discussed in Sections 9.3 and 9.4.1.

6.8. CARBOMETALLATION

Organometallics may add to carbon–carbon multiple bonds to form new organometallic compounds. These reactions are called *carbometallations* since the carbon–metal bond adds across the multiple bond. In most cases the newly formed organometallics are further transformed to produce other valuable organic compounds. Carbometallations, when carried out in a controlled manner, therefore, are extremely useful selective transformations in syntheses. Because of the versatility of organometallics in subsequent reactions, carbometallation is a burgeoning field in the synthesis of organic compounds. Detailed treatment of the topic can be found in reviews,[381,384,385,503–508] and only a few of the most widely used characteristic examples are mentioned here.

When ethylene reacts with triethyl- or tripropylaluminum, multiple carbometalation takes place, resulting in the formation of oligomers.[509] Oxidation of the products followed by hydrolysis yields alcohols, whereas displacement reaction produces terminal alkenes that are of commercial importance.[510] Transition-metal compounds promote the addition to form polymers (Ziegler–Natta polymerization; see Section 13.2.4).

The addition of trialkylaluminums to acetylenes is a slow, nonselective process.[384,385,507,508] Zirconium compounds, however, catalyze the addition to produce the corresponding vinylalanes in highly regio- and stereoselective manner.[385–387,511] The addition is *syn*, and aluminum attaches itself to the terminal carbon. Deuterium-labeled alkenes may be formed after deuterolysis of the organoaluminum intermediates[512] (Scheme 6.8). A more important use of the organoalanes is the substitution of aluminum with carbon electrophiles to synthesize trisubstituted alkenes. Trimethylaluminum seems to be the best reagent for carboalumination since higher homologs react less selectively.

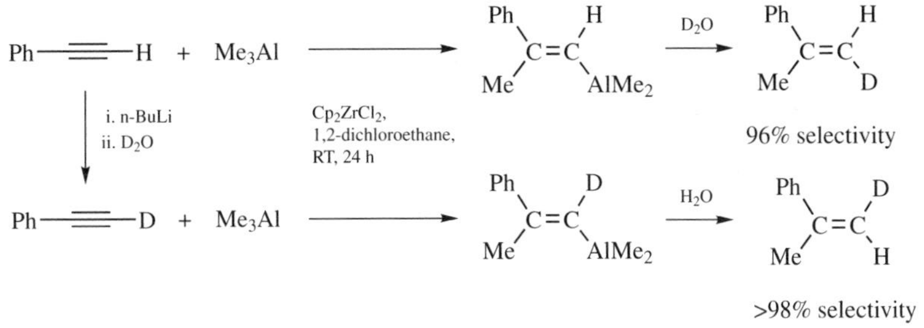

Scheme 6.8

Addition of the Grignard reagent to the carbon–carbon double bond requires severe reaction conditions.[508,513] Zirconium-based compounds, in contrast, catalyze carbomagnesation of 1-alkenes to occur selectively under mild conditions. $[NiCl_2(PPh_3)_2]$ is the best catalyst to promote *syn* addition to disubstituted acetylenes, although usually in low yields:[514]

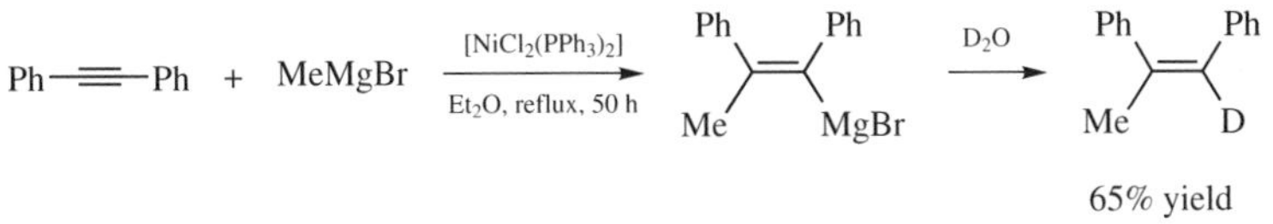

$$(6.84)$$

Carbocupration is one of the most useful carbometalations.[504,508] Organocopper and organocuprate reagents add to the carbon–carbon triple bond to form new copper derivatives with usually high regio- and stereoselectivity (*syn* addition in the Markovnikov fashion)[515,516] [Eq. (6.85)[517]]:

$$n\text{-Bu}\!\!\equiv\!\!\text{H} \ + \ \text{EtCu.MgBr}_2 \quad \xrightarrow[\substack{\text{Et}_2\text{O},\\-10°\text{C, 1 h}}]{} \quad \underset{\underset{\text{Et}}{}\overset{\overset{n\text{-Bu}}{}}{}\!\!=\!\!\underset{\underset{\text{Cu.MgBr}}{}}{\overset{\overset{\text{H}}{}}{}} \quad \xrightarrow{\text{D}_2\text{O}} \quad \underset{\underset{\text{Et}}{}\overset{\overset{n\text{-Bu}}{}}{}\!\!=\!\!\underset{\underset{\text{D}}{}}{\overset{\overset{\text{H}}{}}{}} \quad (6.86)$$

95% D content

The best copper reagents are RCu.MgHlg$_2$ and R$_2$Cu.MgHlg derived from Grignard reagents, and R$_2$CuLi (R = primary alkyl). Acetylene is the most reactive alkyne,[515] whereas internal acetylenes do not react.

Acetylene may undergo two subsequent additions to form conjugated (Z,Z)-dienylcuprates.[515,518] High yields are achieved when acetylene reacts with α-substituted vinylcuprates.[518] The addition is faster in tetrahydrofuran than in diethyl ether, and even secondary and tertiary alkylcopper reagents participate in addition.[519] Regioisomers, though, may be formed in some cases.

When the CuBr–SMe$_2$ complex is used in preparing the organocopper reagent, and Me$_2$S is a cosolvent in the addition, higher yields may be achieved. Under such conditions even the otherwise unreactive methylcopper adds to terminal alkynes to generate 2-methyl-susbtituted alkenylcopper derivatives.[520]

6.9. CYCLOADDITION

The Diels–Alder reaction is one of the most important carbon–carbon bond forming reactions,[521,522] which is particularly useful in the synthesis of natural products. Examples of practical significance of the cycloaddition of hydrocarbons, however, are also known. Discovered in 1928 by Diels and Alder,[523] it is a reaction between a conjugated diene and a dienophile (alkene, alkyne) to form a six-membered carbocyclic ring. The Diels–Alder reaction is a reversible, thermally allowed pericyclic transformation or, according to the Woodward–Hoffmann nomenclature,[524] a [4 + 2]-cycloaddition. The prototype reaction is the transformation between 1,3-butadiene and ethylene to give cyclohexene:

$$(6.86)$$

20%

This and other similar cycloadditions, however, when unactivated hydrocarbons without heteroatom substituents participate in Diels–Alder reaction, are rarely efficient, requiring forcing conditions (high temperature, high pressure, prolonged reaction time) and giving the addition product in low yield. Diels–Alder reactions work well if electron-poor dienophiles (α,β-unsaturated carbonyl compounds, esters, nitriles, nitro compounds, etc.) react with electron-rich dienes. For example, compared to the reaction in Eq. (6.86), 1,3-butadiene reacts with acrolein at 100°C to give formyl-3-cyclohexene in 100% yield.

According to the frontier orbital theory,[525] electron-withdrawing substituents lower the energies of the lowest unoccupied molecular orbital (LUMO) of the dienophile thereby decreasing the highets occupied molecular orbital (HOMO)–LUMO energy difference and the activation energy of the reaction. 1,3-Butadiene itself is sufficiently electron-rich to participate in cycloaddition. Other frequently used dienes are methyl-substituted butadienes, cyclopentadiene, 1,3-cyclohexadiene, and 1,2-dimethylenecyclohexane.

Reaction conditions markedly affect the Diels–Alder reaction.[521,522] First observed by Yates and Eaton,[526] AlCl$_3$ accelerates cycloadditions. Other Lewis acids as well as acidic solids (clays, zeolites) were later shown to exhibit similar effects attributed to the coordination of the catalyst with the dienophile. High pressure[521,522] and water[521,522,527] were also found to show beneficial effects. Reaction conditions also enhance selectivities of the Diels–Alder reaction (see discussion below).

More than one mechanism can account for the experimental observation of the Diels–Alder reaction.[521,522,528] However, most thermal [4 + 2]-cycloadditions are symmetry-allowed, one-step concerted (but not necessarily synchronous) process with a highly ordered six-membered transition state.[529] Two-step mechanisms with the involvement of biradical or zwitterion intermediates can also be operative.[522,528]

In [4 + 2]-cycloadditions the diene reacts in the cisoid conformation,[530] which is essential for p-orbital overlapping in the diene and the dienophile and to form the new double bond in the six-membered ring. Diels–Alder reactions are regioselective. Of the two possible isomers formed in the cycloaddition of nonsymmmetrically substituted reactants [Eq. (6.87)], usually one predominates depending on the structure of the reacting molecules and reaction conditions:

$$(6.87)$$

As a consequence of the concerted mechanism, the Diels–Alder reaction is also stereoselective, implying that the relative configuration of the groups of the reactants is retained. Besides the numerous examples of heterosubstituted compounds,[521,522] this was also proved by 1,3-butadiene and ethylene labeled with deuterium [Eqs. (6.88) and (6.89)]:[531]

$$(6.88)$$

$$(6.89)$$

Moreover, of the two possible parallel approaches of the reacting molecules [*endo* or *exo* approach; Eqs. (6.90) and (6.91)], the *endo* approach usually prevails:

$$(6.90)$$

$$(6.91)$$

Known as the "*endo* rule," it was first formulated by Alder and Stein[530,532] and explained as the maximum accumulation of double bonds. Now *endo* selectivity is considered to be the result of stabilizing secondary orbital interactions.[533] The *endo* rule, however, strictly applies only for cyclic dienophiles.

The abovementioned rate acceleration and selectivity enhancement brought about by catalysts are particularly marked when unactivated dienes and dienophiles are involved. Two molecules of 1,3-butadiene can react in a Diels–Alder reaction, one acting as diene and the other as a dienophile to produce 4-vinylcyclohexene (in 0.1% yield at 250°C in the absence of a catalyst). Cs^+, $Cu,^+$ and trivalent transition-metal exchanged montmorillonites[534] as well as large-pore sodium zeolites (Na ZSM-20, NaY) and carbon molecular sieves,[535] result in 20–35% yields with 95% selectivity. Large rate enhancement was observed when 1,3-cyclohexadiene underwent a similar cycloaddition[536] in the presence of K10 montmorillonite doped with Fe^{3+}:

$$(6.92)$$

200°, 20 h	4	1	30% yield
K10–Fe^{3+} montmorillonite,			
4-*tert*-butylphenol, CH_2Cl_2, 0°C, 1 h	4	1	77% yield

Cycloadditions can also be catalyzed by cation radicals generated from aminium salts.[537–539] In fact, these are cation-radical chain reactions mimicking Diels–Alder reaction. Interestingly, 2,4-dimethyl-1,3-pentadiene undergoes either a Brønsted acid–catalyzed cycloaddition [Eq. (6.93)] or cation-radical-catalyzed cycloaddition [Eq. (6.94)] to yield different addition products:[539]

$$(6.93)$$

$$(6.94)$$

Transition metals can also catalyze Diels–Alder reaction of norbornadiene,[540,541] and acetylenes can also serve as dienophiles.[542]

Much less information is available about [2 + 2]-cycloadditions. These allow the formation of cyclobutane derivatives in the reaction between two alkenes, or that of cyclobutenes from alkenes and alkynes. The reaction can be achieved thermally via biradical intermediates,[543] by photoreaction,[544] and there are also examples for transition-metal-catalyzed transformations. An excellent example is a ruthenium-catalyzed reaction between norbornenes and alkynes to form cyclobutenes with *exo* structure:[545]

$$(6.95)$$

As was mentioned, cycloaddition of unactivated hydrocarbons, namely, that of cyclopentadiene, has practical significance. 5-Vinyl-2-norbornene is produced by the cycloaddition of cyclopentadiene and 1,3-butadiene[546,547] [Eq. (6.96)] under conditions where side reactions (polymerization, formation of tetrahydroindene) are minimal. The product is then isomerized to 5-ethylidene-2-norbornene, which is a widely used comonomer in the manufacture of an EPDM (ethylene–propylene–diene monomer) copolymer (see Section 13.2.6). The reaction of cyclopentadiene (or dicyclopentadiene, its precursor) with ethylene leads to norbornene[548,549] [Eq. (6.97)]:[550]

$$(6.96)$$

$$(6.97)$$

Among others, it can be used in ring-opening metathesis polymerization to produce polynorbornene the oldest polyalkenamer (see Section 12.3). Cycloaddition of cyclopentadiene and acetylene to produce norbornadiene can be carried out similarly.[551]

Finally, the valence isomerization (an intramolecular [2 + 2]-cycloaddition) of norbornadiene to quadricyclane [Eq. (6.98)] has a potential as a solar-energy-storage system:[552]

$$\begin{array}{c} \text{hv} \\ \underset{\underset{\Delta H = -26\ \text{kcal/mol}}{\text{metal catalyst}}}{\rightleftharpoons} \end{array} \tag{6.98}$$

Photocyclization of norbornadiene takes place in high yields to give quadricyclane,[553] which is a sufficiently stable compound in the absence of metals and their ions. The process is accompanied by a large enthalpy change, and when the reverse process is effected by metal catalysis, the stored chemical energy is liberated producing a large amount of heat. A disadvantage of the system is that norbornadiene itself absorbs only in the ultraviolet light range. Research efforts in the late 1970s therefore focused on the development of substituted derivatives that undergo valence isomerization by visible light and the use of sensitizers[554] and catalysts.[555] Of these, copper(I) compounds were found to exhibit promising properties.[556–558]

6.10. RECENT DEVELOPMENTS

6.10.1. Hydration

The use of silica-supported $Zn(BH_4)_2$ is a useful procedure for the hydration of unactivated alkenes and alkynes.[559] The main products are usually formed as a result of anti-Markovnikov addition. In contrast to acid-catalyzed hydration (see Section 6.1.2), this procedure allows the transformation of alkynes to alcohols.

A $RhCl_3$–Dowex 1 anion exchanger ion pair is an efficient and recyclable catalyst to induce hydration of acetylenes.[560] Whereas the soluble ion pair mediates oligomerization, this solid catalyst transforms phenylacetylenes to the corresponding methyl ketones in high yields.

A highly regioselective, efficient, and clean anti-Markovnikov hydration of terminal acetylenes has been realized through the use of catalytic amounts of Ru complexes.[561] Typically, [CpRu(dppm)Cl] catalyzes the reaction at 100°C to give aldehydes in high yields (81–94%). Triflic acid or trifluoromethanesulfonimide effectively catalyzes the hydration of alkynes without a metal catalyst to afford Markovnikov products (ketones).[562]

6.10.2. Hydrohalogenation

A study applying core-electron spectroscopy on the addition of HCl to ethylene, propylene, and isobutylene in the gas phase concludes that a significant portion of the difference between Markovnikov and anti-Markovnikov addition is also due to the charge distribution in the initial state and not to different ability of the molecules to delocalize the added charge in the transition state.[563] The increase in reactivity with increasing substitution of the double bond is also strongly influenced by the initial-state charge distribution.

Mixtures of HCl and 2-butyne were reacted in the gas phase in a Pyrex IR cell between 23 and 63°C to yield only (Z)-2-chloro-2-butene.[564] The process is suggested to be initiated and the rate determined by surface-assisted proton–alkyne interaction. Donation of the proton occurs from within a multilayer found on the wall, that is, from an HCl molecule not directly attached to the wall.

Hydrogen halides may add to acetylenes in a similar way to afford alkenyl halides.[565] The use of silica and alumina, in this case, provides a simple means for facilitating addition of hydrogen halides to alkynes that does not occur readily in solution. Arylalkylacetylenes yield the corresponding *syn*-addition products (**45**) which undergo isomerization on extended treatment:

$$ \text{(6.99)} $$

0.3 h	85%	14%
3 h	3%	96%

Addition of HBr or HI to alkylacetylenes, such as 4-octyne, in turn, affords primarily the *anti*-addition products. The change in mechanism may be due to the difficulty in forming an alkyl cation lacking stabilization of the Ph substituent, thereby requiring simultaneous nucleophilic trapping from the opposite side of the alkyne by a HHlg molecule (Ad_E3-type addition).

The Me_3SiCl and water system is a convenient reagent for the selective hydrochlorination of alkenes. The *anti* selective manner of addition was shown to occur by reacting 9(10)-octalin to afford 4a-chloro-*trans*-decahydronaphthaline exclusively.[566]

6.10.3. Halogen Addition

More than six decades after the postulation of the cyclic bromonium ion intermediate in electrophilic bromination of alkenes important, new findings are still emerging.[567] Updated general treatments of the halogenation of alkenes became available.[568,569]

In another study correlations between ionization potentials and HOMO energies versus relative reactivities of Cl_2, Br_2, and I_2 with acyclic alkenes were reported.[570] Chlorination and bromination data each show a single line of correlation with positive slope for all alkenes regardless of the steric requirements. Furthermore, increasing substitution at the double bond increases the rate indicating an electrophilic reaction. In the case of iodination data calculated for adsorption separate into to groups of similarly substituted alkenes in which increased substitution reduces the rate. Within each group, a good-to-excellent correlation is observed, with a lower ionization potential generally corresponding to a higher relative rate. The results indicate that the relative magnitude of the steric requirements about the double bond is similar to that of the electrophilic effects in iodination. This difference found for iodination can easily be explained. Chlorination and bromination are addition reactions that go to completion. The reaction with iodine, in turn, is unfavorable entropically and is endothermic; it is a reversible complexation reaction in which the equilibrium favors the reactants.

Applying new experimental techniques, the 1 : 1 molecular complex of chlorine and ethylene could be characterized; the Cl_2 molecule lies along the C_2 axis of ethylene, which is perpendicular to the molecular plane and interacts only weakly with the π bond.[571]

In accordance with an earlier suggestion by Olah and Hockswender,[240] the initial 1 : 1 adducts have been shown to be π complexes[572] and a second species, a 2 : 1 bromine–alkene complex was directly observed.[573] In the course of the bromination of crowded alkenes, such as (E)-2,2,3,4,5,5-hexamethylhex-3-ene (**46**), the bromonium ion intermediate was shown to be reversibly formed.[574] However, steric hindrance prevents the formation of the usual dihalogenated product. Instead, it loses a β proton to give an allylic bromo derivative, which is then dehydrobrominated to diene **47**:

$$(6.100)$$

A similar observation was made with 1,2,3-tri-*tert*-butyl-1,3-butadiene.[575]

The first direct detection of a 1 : 1 Br_2–alkyne complex as an intermediate during the addition of bromine to alkynes has been disclosed.[576] Brominations with negative apparent activation parameters, such as the transformation of arylalkylacetylenes, pass through the energetically favored open β-bromovinyl cation leading to mixtures of E and Z vinyl dibromides. Deactivated phenylacetylenes and dialkylacetylenes reacting with positive apparent activation parameters, in turn, form the bromonium ion and give (E)-dibromoalkenes.

Room-temperature ionic liquids may be used as "green" recyclable alternatives to chlorinated solvents for stereoselective halogenation.[577] The bromination of alkenes and alkynes in [bmim][Br] is *anti*-stereospecific, whereas that of 1,3-dienes gives selectively the 1,4-addition products. The reactions of arylacetylenes, however, are not selective when carried out in [bmim][PF$_6$]. Tetraethylammonium trichloride, a stable crystalline solid may be used in the chlorination of alkenes and alkynes to afford the products with exclusive *anti* stereoselectivity.[578] It has been demonstrated, that a new phenyl-functionalized hexagonal mesoporous silica (Ph-HMS) can affect the relative ratio of addition and substitution, that is, retard bromination in benzylic position:[579]

$$(6.101)$$

−	10%	82%
HMS	0%	69%
Ph-HMS	77%	23%

A novel manganese reagent generated from $KMnO_4$ and Me_3SiCl in the presence of a quaternary ammonium salt smoothly dichlorinates alkenes in high yields in CH_2Cl_2 at room temperature (81–95%).[580] According to the reaction protocol, HCl or HBr reacts readily with alkenes in the presence of aqueous H_2O_2 or tert-BuOOH to yield vicinal dihalides.[581] The reaction with acetylenes under similar conditions gives isomeric mixtures with the *trans* isomer dominating. An in situ–generated halonium ion is believed to be involved in the electrophilic addition.

The fluorination of alkenes with XeF_2 has been reviewed.[582] Furthermore, $PhTeF_5$ and $PhSeF_5$ have been shown to be powerful difluorinating reagents.[583]

6.10.4. Hydroamination

Summaries of results of hydroamination mediated with Rh(I) amide complexes[584] and comprehensive reviews giving detailed information of the field are available.[585–587] Therefore, only the more important relatively new findings are presented here. In most of the transformations reported transition metals are applied as catalysts. The feasibility of the use of *tert*-BuOK was demonstrated in the base-catalyzed amination of styrenes with aniline.[588]

Ruthenium complexes mediate the hydroamination of ethylene with pyridine.[589] The reaction, however, is not catalytic, because of strong complexation of the amine to metal sites. Iridium complexes with chiral diphosphine ligands and a small amount of fluoride cocatalyst are effective in inducing asymmetric alkene hydroamination reaction of norbornene with aniline [the best enantiomeric excess (ee) values exceed 90%].[590] Strained methylenecyclopropanes react with ring opening to yield isomeric allylic enamines:[591]

$$(6.102)$$

R = $PhCH_2CH_2CH_2$, R'_2NH = Bu_2NH	91%	−
R = Ph, R'_2NH = phthalimide	−	84%

An efficient hydroamination of vinylarenes with arylamines catalyzed by Pd complexes affords *sec*-phenylethylamine products.[592] The reaction requires the use of protic acids (CF_3COOH, triflic acid), which are suggested to oxidize Pd(0) to Pd(II), thereby generating the active catalytic species.

The reaction of norbornadiene and secondary amines (morpholine, piperidine, pyrrolidine, N- methylbutylamine) results in the formation of interesting aminated products:[593]

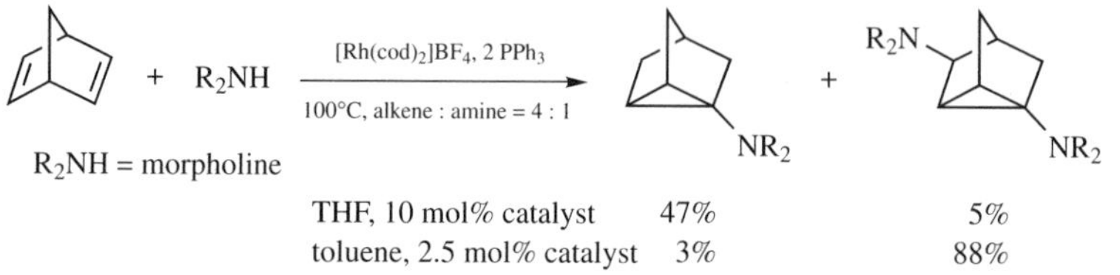

$$(6.103)$$

R_2NH = morpholine

THF, 10 mol% catalyst	47%	5%
toluene, 2.5 mol% catalyst	3%	88%

A combinatorial approach was applied to evaluate various catalysts for the amination of 1,3-dienes.[594] Complexes formed from [(η^3-C_3H_5)PdCl]$_2$ and PPh$_3$ were the most active to induce the reaction of a broad range of primary and secondary arylamines and 1,3-cyclohexadiene, 1,3-cycloheptadiene, or 2,3-dimetyl-1,3-butadiene to give allylamines. The enantioselective version of the transformation is also very effective:

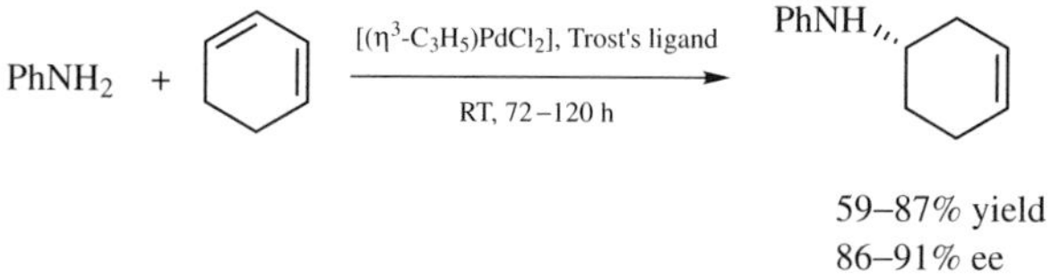

$$(6.104)$$

59–87% yield
86–91% ee

Catalytic amination of phenylacetylene and its derivatives may be performed by using [Ru$_3$(CO)$_{12}$] catalyst to afford the corresponding enamines in high yields:[595]

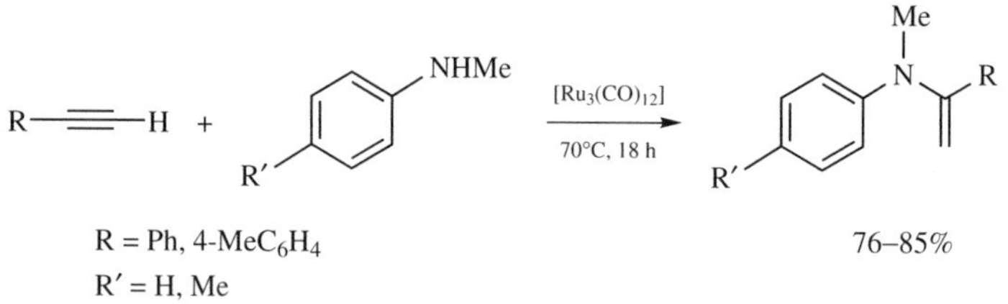

$$(6.105)$$

R = Ph, 4-MeC$_6$H$_4$
R′ = H, Me

76–85%

In sharp contrast, amination of diphenylacetylene with primary amines induced by dimethyltitanocene[596] or the reaction of terminal acetylenes with various primary amines in the presence of [Ru$_3$(CO)$_{12}$][597] results in the formation of the corresponding imines:

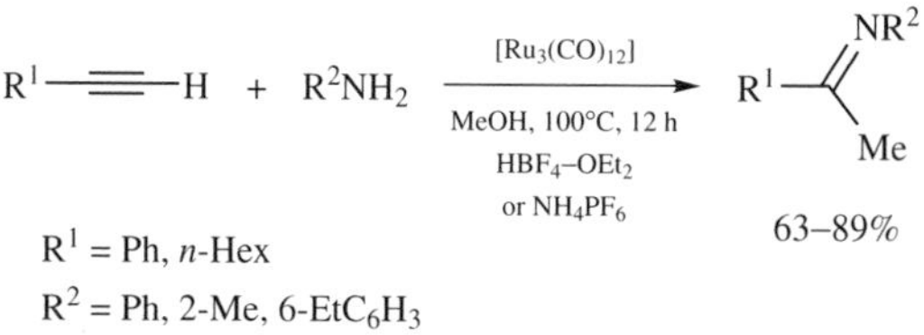

$$(6.106)$$

R^1 = Ph, n-Hex
R^2 = Ph, 2-Me, 6-EtC$_6$H$_3$

63–89%

The latter transformation requires the use of a small amount of an acid or its ammonium salt. By using [Cp$_2$TiMe$_2$] as the catalyst, primary anilines as well as sterically hindered *tert*-alkyl- and *sec*-alkylamines can be reacted.[596] Hydroamination with sterically less hindered amines are very slow. This was explained by a mechanism in which equilibrium between the catalytically active [L^{1}L^{2}Ti=NR] imido complex and ist dimer for sterically hindered amines favors a fast reaction. Lanthanade metallocenes catalyze the regiospecific addition of primary amines to alkenes, dienes, and alkynes.[598] The rates, however, are several orders of magnitude lower than those of the corresponding intramolecular additions.

6.10.5. Hydrometallation

Because of the versatility in organic synthesis of intermediate organometallic compounds, the wide variety of hydrometalation transformations is still attracting high interest.

Hydroboration. Although hydroboration seldom requires a catalyst, hydroboration with electron-deficient boron compounds, such as boric esters, may be greatly accelerated by using transition-metal catalysts. In addition, the chemo-, regio- and stereoslectivity of hydroboration could all be affected. Furthemore, catalyzed hydroboration may offer the possibility to carry out chiral hydroboration by the use of catalysts with chiral ligands. Since the hydroboration of alkynes is more facile than that of alkenes the main advantage of the catalytic process for alkynes may be to achieve better selectivities. Hydroboration catalyzed by transition-metal complexes has become the most intensively studied area of the field.[599]

According to more recent studies,[599] the catalytic cycle shown here for the Wilkinson complex and a boronic ester includes the most probable **48** Rh(III) intermediate formed by the addition of the B—H bond (Scheme 6.9).

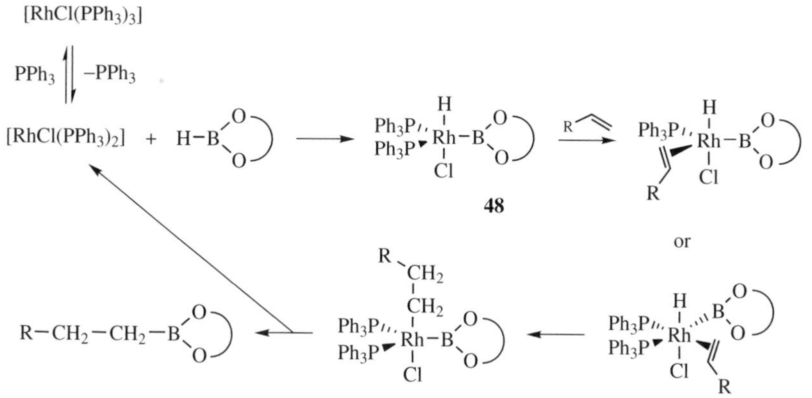

Scheme 6.9

Asymmetric hydroboration with chiral ligands attached to the metal has yielded promising results. Hydroboration with ligands such as DIOP [2,3-*O*-isopropylidene-2,3-dihydroxy-1,4-bis(diphenylphosphino)butane] derivatives, BINAP [2,2′-bis(diphenylphosphino)-1,1′-binaphthyl], CHIRAPHOS [2,3-bis(diphenylphosphino) butane], and phosphine–amine ferrocyl complexes, can be performed with high ee values in most cases.[600] With respect to noncatalytic asymmetric hydroboration, versatile α-pinene-based borane reagents, such as mono- and diisopinocampheyl boranes, are available to afford boranes and alcohols with good to excellent asymmetric induction.[601] Immobilized, preformed chiral Rh catalysts show high activity and regio- and enantioselectivity in the hydroboration of styrene and can be recycled.[602]

New mechanistic studies with [Cp$_2$Ti(CO)$_2$] led to the observation that the titanocene bis(borane) complex [Cp$_2$Ti(HBcat)$_2$] (Hbcat = catecholborane) generated in situ is the active catalyst.[603] It is highly active in the hydroboration of vinylarenes to afford anti-Markovnikov products exclusively, which is in contrast to that of most Rh(I)–catalyzed vinylarene hydroboration. Catecholborane and pinacolborane hydroborate various terminal alkynes in the presence of Rh(I) or Ir(I) complexes in situ generated from [Rh(COD)Cl$_2$] or [Ir(COD)Cl$_2$] and trialkylphosphines.[604] The reaction yields (*Z*)-1-alkenylboron compounds [Eq. (6.107)]; that is, *anti* addition of the B—H bond occurs, which is opposite to results found in catalyzed or uncatalyzed hydroboration of alkynes:

$$
R\!\!=\!\!=\!\!-H \; + \; HBcat \quad \xrightarrow[\text{Et}_3\text{N, hexane, RT, 1--2 h}]{\text{[Rh(COD)Cl}_2\text{], P(iPr)}_3} \quad R\diagup \diagdown Bcat \qquad (6.107)
$$

R = C$_8$H$_{17}$, *tert*-Bu, Ph

62–81% yield
89–99% selectivity

HBcat = HB$\diagup$$\diagup$ (O, O)

This was explained by the involvement of a vinylidene complex that is also in agreement with the migration of the acetylenic hydrogen to C–2 observed by deuterium labeling. The stereoselective reaction requires the use of Et$_3$N and a slight excess of the alkyne.

The synthesis of complex [RhCl{P[CH$_2$CH$_2$(CF$_2$)$_5$CF$_3$]$_3$}$_3$], the fluorous analog of the Wilkinson catalyst, allowed investigators to perform heterogeneous biphasic hydroboration of a variety of alkenes and terminal alkynes with catecholborane in the fluorous solvent CF$_3$C$_6$F$_{11}$.[605] Alcohols were obtained in high yields (76–95%) with satisfactory reaction rates (TON values are approximately 300–2000). An unusual rapid hydroboration of alkenes using diborane in chlorohydrocarbon solvents has been described.[606] When dissolved in dichloromethane, 1,2-dichloroethane, or 1,1,2,2-tetrachloroethane, diborane is in equilibrium with solvent–BH$_3$ adducts (8–15%), and these solutions hydroborate alkenes almost instantaneously, even at −16°C.

Hydrosilylation. In addition to the two most prominent catalysts, Speier's catalyst (H$_2$PtCl$_6$ in 2-propanol) and the catalyst developed by Karstedt ([Pt(CH$_2$=CHSi Me$_2$)$_3$], and Group VIII metals, new families of catalysts have been developed and applied in hydrosilylation since the early 1990s. New aspects of the mechanism have also been disccussed. Reviews give comprehensive treatment of the field.[607–612]

Detailed mechanistic studies with respect to the application of Speier's catalyst on the hydrosilylation of ethylene showed that the process proceeds according to the Chaik–Harrod mechanism and the rate-determining step is the isomerization of Pt(silyl)(alkyl) complex formed by the ethylene insertion into the Pt–H bond.[613] In contrast to the platinum-catalyzed hydrosilylation, the complexes of the iron and cobalt triads (iron, ruthenium, osmium and cobalt, rhodium, iridium, respectively) catalyze dehydrogenative silylation competitively with hydrosilylation. Dehydrogenative silylation occurs via the formation of a complex with σ-alkyl and σ-silylalkyl ligands:

$$R'_3Si-M-H + 2\,RHC{=}CH_2 \longrightarrow$$

hydrosilylation: $RCH_2CH_2SiR'_3 + RCH{=}CH_2$

dehydrogenative hydrosilylation: $RCH{=}CHSiR'_3 + RCH_2CH_3$

$$(6.108)$$

Two new mechanisms have been suggested to account for the observation. Dehydrogenative silylation has become a useful method for the synthesis of unsaturated organosilicon compounds.[611,614]

Contrary to previous reports suggesting colloidal metal as the active species in Pt-catalyzed hydrosilylations, the catalyst was found to be a monomeric platinum compound with silicon and carbon in the first coordination sphere.[615] The platinum end product at excess olefin concentration contains only platinum–carbon bonds, whereas at high hydrosilane concentration, it is multinuclear and also contains platinum–silicon bonds. An explanation of the oxygen effect in hydrosilylation was also given to show that oxygen serves to disrupt multinuclear platinum species that are formed when poorly stabilizing olefins are employed.

Organolanthanide and Group III metallocene complexes have been found to be effective catalysts in hydrosilylation.[616] One of their attractive features is the ability to vary their reactivity by changing the metal or by altering the ligands on the metal.[617,618] Larger metals and sterically less crowded ligand substitution surrounding the metal increase the open space about the catalytic center, thereby increasing access to more sterically demanding alkenes. Certain catalyst modifications provide increased yield of the secondary silane, although synthetically useful selectivities are not achieved. Asymmetric hydrosilylation was also effected with medium enantioselectivites.[619] Organoactinide complexes of the type $[Cp_2^*AnMe_2]$ (An = Th, U) are also useful catalysts in hydrosilylation.[620]

Lewis acid–catalyzed hydrosilylation of alkenes with trialkylsilanes allows the synthesis of tetrasubstituted silanes formed with high regio- and stereoselectivities.[621] $AlBr_3$ exhibits the highest activity, and triorganochlorosilanes bring about a drastic rate increase in $AlCl_3$-catalyzed reactions. Lewis acids such as $AlCl_3$ or $EtAlCl_2$ also catalyze the hydrosilylation of alkynes and allenes.[622,623] Hydrosilylation of acetylenes proceeds in a regio- and anti-stereoselective manner to produce the corresponding *cis*-vinylsilanes. According to the proposed mechanism, π-basic alkynes coordinate to the Lewis acids, making an electron-deficient unsaturated carbon center, and nucleophiles attack the carbon from the phase opposite to the Lewis acid coordination site.[623] Hydrosilylation mediated by [RuCl$_2$(*p*-cymene)]$_2$ complex is highly stereoselective and efficient: β-(Z)-vinylsilanes are obtained in excellent yields (79–98%, $Z : E = 95 : 5$).[624] Layer-segregated Pt–Ru cluster complexes exhibit high activity in the hydrosilylation of diphenylacetylene with triethylsilane.[625] Certain Pt(0) complexes generated from the Karstedt catalyst by using electron-deficient alkene ligands, especially [Pt(η^2-NQ)]{(η^4-CH$_2$=CHSiOMe$_2$)$_2$O}] (NQ = naphthoquinone derivatives), were found to catalyze very efficiently the hydrosilylation of alkenes.[626] The addition of excess ligand to the catalyst solution greatly extended the lifetime and productivity of the catalysts, which were more efficient in both rate and overall product yield than the original Karstedt catalyst solution. This was suggested to result from the electron-deficient ligand bound to Pt throughout the catalytic cycle.

Studies on the immobilization of Pt-based hydrosilylation catalysts have resulted in the development of polymer-supported Pt catalysts that exhibit high hydrosilylation and low isomerization activity, high selectivity, and stability in solventless alkene hydrosilylation at room temperature.[627] Results with Rh(I) and Pt(II) complexes supported on polyamides[628] and Mn-based carbonyl complexes immobilized on aminated poly(siloxane) have also been published.[629] A supported Pt–Pd bimetallic colloid containing Pd as the core metal with Pt on the surface showed a remarkable shift in activity in the hydrosilylation of 1-octene.[630]

Hydrozirconation. Detailed information about hydrozirconation can be found in a review paper by Wipf and Jahn.[631]

A new concept, transfer hydrozirconation, has been developed by Negishi[632] that eliminates the need of the cumbersome preparation of [Cp$_2$ZrClH] used in direct hydrozirconation. Terminal acetylenes react readily with [Cp$_2$Zr(isoBu)Cl] to yield vinylzirconium products, which is a satisfactory and convenient alternative to hydrozirconation with [Cp$_2$ZrClH].[632] Monosubstituted alkenes, in contrast, may react very sluggishly with [Cp$_2$Zr(isoBu)Cl] to yield the **49** zirconium derivative, the product of a formal hydrozirconation. The reaction, however, is greatly accelerated by Lewis acids, most notably AlCl$_3$, Pd complexes, or Me$_3$SiI:[633]

$$\text{C}_8\text{H}_{17}\text{CH=CH}_2 \xrightarrow[\text{catalyst, solvent, 50°C, 3 h}]{\text{[Cp}_2\text{Zr(isoBu)Cl]}} \underset{\textbf{49}}{\text{C}_{10}\text{H}_{21}\text{ZrCp}_2\text{Cl}} \longrightarrow \begin{cases} \xrightarrow{\text{HCl}} \text{C}_{10}\text{H}_{22} \quad \textbf{50} \\ \xrightarrow{\text{I}_2} \text{C}_{10}\text{H}_{21}\text{I} \quad \textbf{51} \end{cases} \qquad (6.109)$$

	49	**50**	**51**
		% yield	
AlCl$_3$, benzene	87	91	84
Me$_3$SiI, benzene–ether	82	85	82
[Cl$_2$Pd(PPh$_3$)$_2$], benzene–ether	76	74	72

The organozirconium intermediate may be further transformed to the corresponding alkane by quenching with HCl or 1-iodoalkane by reacting with I_2.

Other Hydrometallations. *Hydrostannation* is a method to synthesize organos-tannanes, which are of great utility as building blocks in organic chemistry.[623,634] Radical hydrostannation suffers from the drawback of poor stereoselectivity. Hydrostannation catalyzed by transition-metal complexes, in turn, is usually a selective process. Palladium is the most popular catalyst, and tri(*n*-butyl) tin hydride (Bu_3SnH) is the most commonly used reagent. A method has been described to generate Bu_3SnH from various Bu_3SnHlg (Hlg = F, Cl) reagents by reduction with poly(methylhydrosiloxane).[635] This allows the use of catalytic amounts of Bu_3SnHlg, which increases the importance of metal-catalyzed hydro-stannation reactions.

The hydrostannation of alkynes is the most widely used transformation of acces-sing vinyl stannanes. The reaction occurs via a stereospecific *syn* addition with the regiochemistry, in general, controlled by substrate steric considerations. Exception is the α-regioselectivity obtained with substituted phenylacetylenes. Hydrostanna-tion of unactivated alkenes can also be accomplished by using the heterogeneous $Pd(OH)_2$ on C catalyst.[636] With the exception of 1,3-pentadiene, conjugated dienes react regio- and stereoselectively, yielding 1,4-addition products.[634] This is believed to be due to an *s-cis*-coordination of the diene to the Pd(0) center.

Since the 1980s or so, organoaluminium compounds have gained popularity in organic and organometallic synthesis. *Hydroalumination* of mono-, di- and trisub-stituted alkenes with organoaluminium compounds containing an Al—H bond is still one of the most popular methods to prepare higher organoaluminium com-pounds. Hydroalumination is thoroughly treated in two reviews.[637,638]

6.10.6. Halometallation

A most useful haloboration agents, B-bromo-9-borabicyclo[3.3.1]nonane and BBr_3, react readily with terminal alkynes via Markovnikov *syn* addition of the Hlg–B moiety to the carbon–carbon triple bond in an uncatalyzed process. The halobora-tion reaction occurs regio- and chemoselectively at terminal triple bonds, but for other types of unsaturated compounds it is nonselective.[639]

Phenylselenium trifluoride ($PhSeF_3$) is an effective reagent used to perform sele-nofluorination of alkenes (cyclohexene, norbornene, styrene).[583] Vicynal (*E*)-fluor-oalkenylselenides may be prepared by reacting acetylenes with XeF_2–Ph_2Se_2 or XeF_2–$PhSeSIMe_2R$.[640] The best results were attained with dialkylacetylenes:

$$R\text{---}\!\!\equiv\!\!\text{---}R \xrightarrow[\text{ii. addition of alkyne, } -20°C, \text{ 30 min, RT, 1h}]{\text{i. } Ph_2Se_2, XeF_2, CH_2Cl_2, -20°C, \text{ 15 min}} \begin{matrix} F & R \\ \diagdown\!\!=\!\!\diagup \\ R & SePh \end{matrix} \quad (6.110)$$

R = Me, Et, Pr, *n*-Bu

72–87%

The ratio of regioisomers formed in the reaction of nonsymmetric alkynes can be shifted by bulky substituents. A benzeneselenenyl fluoride equivalent generated in situ is the apparent reagent.[640,641]

6.10.7. Solvometallation

Chiral solvomercuration was accomplished by carrying out the reaction of alkenes with Hg(OAc)$_2$ in the presence of chiral quaternary ammonium salts synthesized from natural ephedrine.[642] Chiral secondary alcohols may be isolated with ee values up to 96%. Chiral nitrogen-containing diselenides are transformed by peroxodisulfate to selenium electrophiles, which may add to alkenes to form oxyselenylation products. These are, however, not isolated but oxidized to induce oxidative β-hydride elimination to afford chiral allyl methyl ethers with ee values up to 75%.[643]

6.10.8. Carbometallation

Carbometallation of alkenes and alkynes is attracting considerable interest as a potential new method for C—C bond formation. A general treatment is available by Negishi a leading expert of the field.[632]

Group IV metals catalyze *carboalumination* of alkenes and alkynes. Two reviews give historical background and present the most recent results.[637,638] An interesting example about the role of water in asymmetric carboalumination is described here. The first asymmetric alkylalumination of alkenes applying Erker's chiral zirconium catalyst [bis(1-neomentylindenyl)zirconium dichloride] was reported by Negishi and coworkers to afford chiral primary alcohols in good yields and with good enantioselectivities.[644] It was found that reaction rates may be increased dramatically by the addition of water.[645] Styrene, for example, underwent methylation very slowly (30% yield after 22 days at room temperature) but could be transformed easily to the desired product in the presence of water:

$$\text{Me}_3\text{Al (3 equiv.), Erker's catalyst}$$
$$\text{H}_2\text{O (1 equiv.), CH}_2\text{Cl}_2, -5°\text{C, 12 h}$$

89% ee
73% yield

(6.111)

Methylalumoxane (MAO) also accelerates the reaction, albeit, at a lower rate.

New catalysts have been described,[646] and ab initio MO calculations have shown that the transformation takes place through a four-center transition state.[647] In addition, the anomalous relative reactivities of substrates, specifically, the higher reactivity of alkynes compared to those of alkenes, can be explained by considering the reaction to essentially be a nucleophilic attack by an alkyl anion, rather than an electrophilic one.

Since the early 1990s catalytic *carbomagnesation* of alkenes has become an established method for the synthesis of special Grignard reagents useful for preparing functionalized organic molecules.[637] Detailed studies with respect to reactivates, and regio- and stereoselectivities allowed the identification of a mechanism[648] that has been further supported by the isolation and identification of various complexes involved in carbomagnesation reactions catalyzed by $[Cp_2ZrCl_2]$.[649]

The first Ni-catalyzed *carbozinconation* reaction was described by adding dialkylzincs or diphenylzinc to phenylalkylacetylenes in the presence of $Ni(acac)_2$ to form trisubstituted alkenes through *syn* addition.[650,651] When the *syn*-carbozincated products formed from alkenes in the Zr-catalyzed process are reacted with electrophiles, tetrasubstituted alkenes are obtained:[652]

$$C_8H_{17}\diagup\diagup \xrightarrow[\text{THF, RT}]{[Cp_2ZrCl_2],\ EtMgBr} \left(C_8H_{17}\diagdown\diagup\diagdown_{Et}^{}Zn\right)_2 \xrightarrow[\text{DME, RT, 12 h}]{RX,\ [Cl_2Pd(dppp)]} C_8H_{17}\diagdown\diagup\diagdown_{Et}^{}R$$

$$\begin{array}{ll} RX = PhI & 80\%\ \text{selectivity} \\ CH_2=CHBr & 76\%\ \text{selectivity} \end{array}$$

$$(6.112)$$

Organozirconium compounds generated by hydrozirconation add in selective *syn* fashion to terminal and internal alkynes in the presence of $[(C_6H_5)_3C]^+$ $[B(C_6F_5)_4]^-$.[653] Quenching the alkylzirconium intermediate results in the formation of di- or trisubstituted alkenes in high yields. Zirconoesterification occurs when diphenylacetylene reacts with $[Cp_2ZrEt_2]$ and methyl or ethyl chloroformates.[654] The product is reacted with various electrophiles to form highly substituted propenoates.

6.10.9. Cycloaddition

As discussed in Section 6.9 1, 3-dienes and dienophiles in which multiple bonds are not activated by electron-withdrawing or electron-releasing substituents fail to undergo cycloaddition except under the most severe conditions. Particular difficulty is encountered in the cycloaddition of two unactivated species since homodimerization can be a competitive and dominant reaction pathway. The use of transition-metal catalysts, however, has proved to be a valuable solution. Complexation of unactivated substrates to such catalysts promotes both inter- and intramolecular cycloadditions. Consequently, the cycloaddition of such unactivated compounds, that is, simple unsubstituted dienes and alkenes, catalyzed by transition metals is a major, important area of study.[655] In addition, theoretical problems of the transformation have frequently been addressed in the more recent literature.

In a study Rh-catalyzed cycloaddition of unactivated substrates, specifically, vinylallenes and acetylenes, is described.[656] Appropriate catalysts were selected on the basis of a study of the bonding interactions of the reactants and transition metals. For this reaction a complex prepared in situ from $[Rh(COD)_2]OTf$ and $P[OCH(CF_3)_2]_3$, one of the most strongly electron-accepting ligands available, was used. Under the conditions applied, even ethylene, one of the most sluggish dienophiles, undergoes cycloaddition to give a substituted cyclohexene, albeit in low yield:

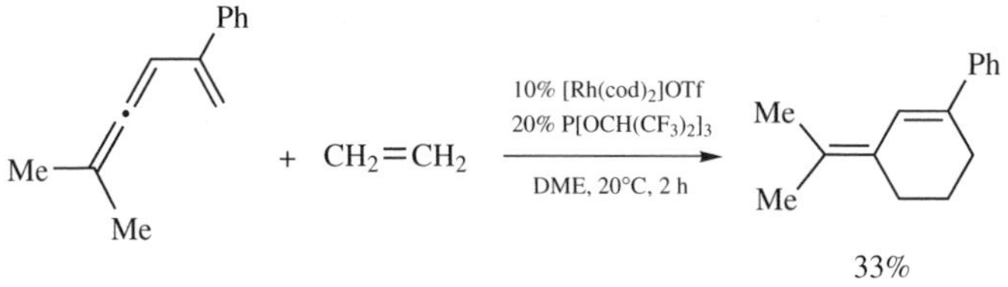

$$(6.113)$$

33%

A major question still often debated in the literature is the problem of whether cycloadditions proceed according to a concerted or stepwise mechanism. Interest in this problem was reawakened by a paper by Zewail and coworkers.[657] The very short lifetime of an intermediate in the retro-Diels–Alder reaction of norbornene observed by the femtosecond time-resolved method was thought to allow a stereospecific stepwise mechanism. Nevertheless, calculation predicts a synchronous concerted transition state for this reaction. At the same time, both theory and experiment provide evidence that a stepwise diradical mechanism lies only a few kilocalories per mole (kcal/mol) higher in energy than the concerted pathway.[658] According to a new experimental and computational study using isotope effects to distinguish between synchronous, asynchronous and stepwise reactions, a concerted transition state is supported for the retro-Diels–Alder reaction of norbornene.[659]

Studies of strained allenes, such as 1,2-cyclohexadiene, show that they undergo facile Diels–Alder reactions with otherwise unreactive dienes. A comparison of the calculated transition structures and intermediates along the reaction paths of 1,2-cyclohexadiene with 1,3-butadiene as well as propadiene and 1,3-butadiene show

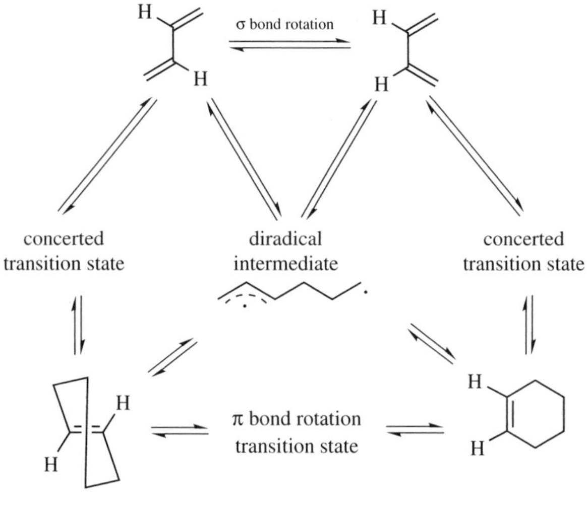

Scheme 6.10

that the diradical stepwise pathways are preferred over the concerted paths. At the same time, the concerted transition structures are extremely asynchronous.[660]

Every discussion of the Diels–Alder reaction for 1,3-butadiene includes the observation that cycloaddition should occur only from the *s-cis* conformer to produce *cis*-cyclohexene. This conformational selectivity, however, further implies that cycloaddition of *s-trans*-1,3-butadiene should lead to *trans*-cyclohexene, but the *s-trans* region of this potential surface has remained unexplored. A study of this problem shows that the concerted and stepwise reaction paths exist for both diene conformers, connecting them to the respective cyclohexene isomers.[661] It is also demonstrated that the usual paradigm for the Diels–Alder reactions is incomplete; a thorough understanding of this archetypal reaction requires consideration of the full range of processes shown in Scheme 6.10, not just those involving the *s-cis* conformer.

REFERENCES

1. M. Liler, *Reaction Mechanisms in Sulphuric Acid and Other Strong Acid Solutions*, Academic Press, London, 1971, Chapter 5, p. 167.

2. F. Asinger, *Mono-Olefins, Chemistry and Technology*, Pergamon Press, Oxford, 1968, Chapter 7, p. 628.

3. G. H. Schmid and D. G. Garratt, in T*he Chemistry of Functional Groups, Supplement A: The Chemistry of Double-Bonded Functional Groups*, S. Patai, ed., Wiley, London, 1977, Chapter 9, p. 725.

4. V. J. Nowlan and T. T. Tidwell, *Acc. Chem. Res.* **10**, 252 (1977).

5. F. A. Long and M. A. Paul, *Chem. Rev.* **57**, 942 (1957).

6. Y. Chiang and A. J. Kresge, *J. Am. Chem. Soc.* **107**, 6363 (1985).

7. Y. Inoue, T. Ueoka, T. Kuroda, and T. Hakushi, *J. Chem. Soc., Perkin Trans. 2*, 983 (1983).

8. M. Misono and N. Nojiri, *Appl. Catal.* **64**, 1 (1990).

9. J. Meuldijk, G. E. H. Joosten, and E. J. Stamhuis, *J. Mol. Catal.* **37**, 75 (1986).

10. J. Tejero, F. Cunill, and M. Iborra, *J. Mol. Catal.* **42**, 257 (1987).

11. S. Randriamahefa, R. Gallo, G. Raoult, and P. Mulard, *J. Mol. Catal.* **49**, 85 (1988).

12. F. Fajula, R. Ibarra, F. Figueras, and C. Gueguen, *J. Catal.* **89**, 60 (1984).

13. M. Iwamoto, M. Tajima, and S. Kagawa, *J. Chem. Soc., Chem. Commun.* 228 (1985).

14. K. Eguchi, T. Tokiai, and H. Arai, *Appl. Catal.* **34**, 275 (1987).

15. M. Balogh and P. Laszlo, *Organic Chemistry Using Clays*, Springer, Berlin, 1993, Chapter 2, p. 37.

16. F. J. Waller, *ACS Symp. Ser.* **308**, 42 (1986).

17. G. A. Olah, Belg. Patent 889,943 (1982).

18. W. R. Cares, U. S. Patent 4,065,512 (1977).

19. J. McEwen and K. Yates, *J. Am. Chem. Soc.* **109**, 5800 (1987).

20. J. A. Marshall, *Acc. Chem. Res.* **2**, 33 (1969).

21. G. J. Kavarnos and N. J. Turro, *Chem. Rev.* **86**, 401 (1986).

22. R. A. Neunteufel and D. R. Arnold, *J. Am. Chem. Soc.* **95**, 4080 (1973).

23. Y. Shigemitsu and D. R. Arnold, *J. Chem. Soc., Chem. Commun.* 407 (1975).

24. Y. Inoue, T. Okano, N. Yamasaki, and A. Tai, *J. Chem. Soc., Chem. Commun.* 718 (1993).

25. K. Mizuno, I. Nakanishi, N. Ichinose, and Y. Otsuji, *Chem. Lett.* 1095 (1989).

26. J. M. Adams, J. A. Ballantine, S. H. Graham, R. J. Laub, J. H. Purnell, P. I. Reid, W. Y. M. Shaman, and J. M. Thomas, *J. Catal.* **58**, 238 (1979).

27. C. H. Collins and G. S. Hammond, *J. Org. Chem.* **25**, 911 (1960).

28. J. K. Stille and R. D. Hughes, *J. Org. Chem.* **36**, 340 (1971).

29. T. L. Jacobs, in *The Chemistry of the Allenes*, Vol. 2: *Reactions*, S. R. Landor, ed., Academic Press, London, 1982, Chapter 5.4, p. 417.

30. A. V. Fedorova and A. A. Petrov, *J. Gen. Chem. USSR* (Engl. transl.) **32**, 1740 (1962).

31. E. V. Brown and A. C. Plasz, *J. Org. Chem.* **32**, 241 (1967).

32. J. L. Jensen and V. Uaprasert, *J. Org. Chem.* **41**, 649 (1976).

33. W. K. Chwang, P. Knittel, K. M. Koshy, and T. T. Tidwell, *J. Am. Chem. Soc.* **99**, 3395 (1977).

34. G. H. Schmid, in *The Chemistry of the Carbon-Carbon Triple Bond*, S. Patai, ed., Wiley, Chichester, 1978, Chapter 8, p. 275.

35. G. A. Olah and D. Meidar, *Synthesis* 671 (1978).

36. I. K. Meier and J. A. Marsella, *J. Mol. Catal.* **78**, 31 (1993).

37. Y. Fukuda and K. Utimoto, *J. Org. Chem.* **56**, 3729 (1991).

38. M. M. Taqui Khan, S. B. Halligudi, and S. Shukla, *J. Mol. Catal.* **58**, 299 (1990).

39. W. Hiscox and P. W. Jennings, *Organometallics* **9**, 1997 (1990).

40. J. W. Hartman, W. C. Hiscox, and P. W. Jennings, *J. Org. Chem.* **58**, 7613 (1993).

41. N. X. Hu, Y. Aso, T. Otsubo, and F. Ogura, *Tetrahedron Lett.* **27**, 6099 (1986).

42. Y. Chiang, A. J. Kresge, M. Capponi, and J. Wirz, *Helv. Chim. Acta* **69**, 1331 (1986).

43. J. Barluenga, F. Aznar, and M. Bayod, *Synthesis* 144 (1988).

44. M. Bassetti, and B. Floris, *J. Chem. Soc., Perkin Trans. 2* 227 (1988).

45. W. J. Tapp, *Ind. Eng. Chem.* **40**, 1619 (1948); **42**, 1698 (1950); **44**, 2020 (1952).

46. K. Weissermel and H.-J. Arpe, *Industrial Organic Chemistry*, 2nd ed., VCH, Weinheim, 1993, Chapter 8, p. 189.

47. J. C. Fielding, in *Propylene and Its Industrial Derivatives*, E. G. Hancock, ed., Wiley, New York, 1973, Chapter 6, p. 214.

48. P. D. Sherman, Jr. and P. R. Kavasmaneck, in *Kirk-Othmer Encyclopedia of Chemical Technology*, 3rd ed., Vol. 9, M. Grayson and D. Eckroth, eds., Wiley-Interscience, New York, 1980, p. 338.

49. A. E. Sommer and R. Bücker, in *Encyclopedia of Chemical Processing and Design*, Vol. 19, J. J. McKetta and W. A. Cunningham, eds., Marcel Dekker, New York, 1983, p. 445.

50. M. R. Schoenberg, J. W. Blieszner, and C. G. Papadopulos, in *Kirk-Othmer Encyclopedia of Chemical Technology*, 3rd ed., Vol. 19, M. Grayson and D. Eckroth, eds., Wiley-Interscience, New York, 1982, p. 241.

51. *Hydrocarbon Process.* **56**(11), 176 (1977).

52. W. Neier and J. Woellner, *Hydrocarbon Process.* **51**(11), 113 (1972); *Chemtech* **3**, 95 (1973).

53. *Hydrocarbon Process.* **50**(11), 172 (1971).

54. Y. Onoue, Y. Mizutani, S. Akiyama, and Y. Izumi, *Chemtech* **8**, 432 (1978).

55. *Petrol. Refiner* **38**(11), 272 (1959).

56. *Hydrocarbon Process Petrol. Refiner* **42**(11), 194 (1963).

57. C. E. Loeffler, L. Stautzenberger, and J. D. Unruh, in *Encyclopedia of Chemical Processing and Design*, Vol. 5, J. J. McKetta and W. A. Cunningham, eds., Marcel Dekker, New York, 1977, p. 358.

58. M. Prezelj, W. Koog, and M. Dettmer, *Hydrocarbon Proc., Int. Ed.* **67**(11), 75 (1988).

59. *Hydrocarbon Process.* **52**(11), 141 (1973).

60. G. Yepsen and T. Witoshkin, *Oil Gas J.* **89**(14), 68 (1991).

61. G. Parkinson, *Chem. Eng.* (NY) **99**(4), 35 (1992).

62. V. E. Pierce and A. K. Logwinuk, *Hydrocarbon Process., Int. Ed.* **64**(9), 75 (1985).

63. M. Prezelj, *Hydrocarbon Process., Int. Ed.* **66**(9), 68 (1987).

64. W. J. Piel and R. X. Thomas, *Hydrocarbon Process., Int. Ed.* **69**(7), 68 (1990).

65. R. W. Reynolds, J. S. Smith, and I. Steinmetz, *Oil Gas J.* **73**(24), 50 (1975).

66. R. Csikos, I Pallay, J. Laky, E. D. Radcsenko, B. A. Englin, and J. A. Robert, *Hydrocarbon Process.* **55**(7), 121 (1976).

67. G. Pecci and T. Floris, *Hydrocarbon Process.* **56**(12), 98 (1977).

68. L. S. Bitar, E. A. Hazbun, and W. J. Piel, in *Encyclopedia of Chemical Processing and Design*, Vol. 30, J. J. McKetta and W. A. Cunningham, eds., Marcel Dekker, New York, 1989, p. 82.

69. K. Weissermel and H.-J. Arpe, *Industrial Organic Chemistry*, 2nd ed., VCH, Weinheim, 1993, Chapter 3, p. 71.

70. O. C. Abraham and G. F. Prescott, *Hydrocarbon Process., Int. Ed.* **71**(2), 51 (1992).

71. G. R. Muddarris and M. J. Pettman, *Hydrocarbon Process., Int. Ed.* **59**(10), 91 (1980).

72. L. A. Smith and M. N. Huddleston, *Hydrocarbon Process., Int. Ed.* **61**(3), 121 (1982).

73. L. S. Bitar, E. A. Hazbun, and W. J. Piel, *Hydrocarbon Process., Int. Ed.* **63**(10), 63 (1984).

74. J. D. Chase and B. B. Galvez, *Hydrocarbon Process., Int. Ed.* **60**(3), 89 (1981).

75. J. Herwig, B. Schleppinghoff, and S. Schulwitz, *Hydrocarbon Process., Int. Ed.* **63**(6), 86 (1984).

76. H. L. Brockwell, P. R. Sarathy, and R. Trotta, *Hydrocarbon Process., Int. Ed.* **70**(9), 133 (1991).

77. K. Rock, *Hydrocarbon Process., Int. Ed.* **71**(5), 86 (1992).

78. E. Pescarollo, R. Trotta, and P. R. Sarathy, *Hydrocarbon Process., Int. Ed.* **72**(2), 53 (1993).

79. P. M. Lange, F. Martinola, and S. Oeckl, *Hydrocarbon Process., Int. Ed.* **64**(12), 51 (1985).

80. H. A. Wittcoff, *Chemtech*, **20**, 48 (1990).

81. V. Fattore, M. M. Mauri, G. Oriani, and G. Paret, *Hydrocarbon Process., Int. Ed.* **60**(8), 101 (1981).

82. A. Convers, B. Juguin, and B. Trock, *Hydrocarbon Process., Int. Ed.* **60**(3), 95 (1981).

83. M. Miranda, *Hydrocarbon Process., Int. Ed.* **66**(8), 51 (1987).

84. A. Clementi, G. Oriani, F. Ancillotti, and G. Pecci, *Hydrocarbon Process.* **58**(12), 109 (1979).

85. K. R. Masters and E. A. Prohaska, *Hydrocarbon Process., Int. Ed.* **67**(8), 48 (1988).

86. A. Aguiló and J. D. Penrod, in *Encyclopedia of Chemical Processing and Design*, Vol. 1, J. J. McKetta and W. A. Cunningham, eds., Marcel Dekker, New York, 1976, p. 114.

87. H. J. Hagemeyer, in *Kirk-Othmer Encyclopedia of Chemical Technology*, 4th ed., Vol. 1, J. I. Kroschwitz and M. Howe-Grant, eds., Wiley-Interscience, New York, 1991, p. 94.

88. K. Weissermel and H.-J. Arpe, *Industrial Organic Chemistry*, 2nd ed., VCH, Weinheim, 1993, Chapter 7, p. 161.

89. *Petrol. Refiner* **40**(11), 207 (1961).

90. D. F. Othmer, K. Kon, and T. Igarashi, *Ind. Eng. Chem.* **48**, 1258 (1956).

91. C. M. Sharts and W. A. Sheppard, *Org. React.* (NY) **21**, 125 (1974).

92. R. C. Larock and W. W. Leong, in *Comprehensive Organic Synthesis*, B. M. Trost and I. Fleming, eds., Vol. 4: *Additions to and Substitutions at C–C π-Bonds*, M. F. Semmelhack, ed., Pergamon Press, Oxford, 1991, Chapter 1.7, p. 269.

93. N. Yoneda, *Tetrahedron* **47**, 5329 (1991).

94. G. A. Olah, J. T. Welch, Y. D. Vankar, M. Nojima, I. Kerekes, and J. A. Olah, *J. Org. Chem.* **44**, 3872 (1979).

95. G. A. Olah, M. Nojima, and I. Kerekes, *Synthesis* 779 (1973).

96. G. A. Olah, J. G. Shih, and G. K. S. Prakash, in *Fluorine: The First Hundred Years*, R. E. Banks, D. W. A. Sharp, and J. C. Tatlow, eds., Elsevier, Lausanne, 1986, Chapter 14, p. 377.

97. N. Yoneda, T. Abe, T. Fukuhara, and A. Suzuki, *Chem. Lett.* 1135 (1983).

98. G. A. Olah and X.-Y. Li, *Synlett* 267 (1990).

99. G. A. Olah, X.-Y. Li, Q. Wang, and G. K. S. Prakash, *Synthesis* 693 (1993).

100. G. B. Sergeev, V. V. Smirnov, and T. N. Rostovshchikova, *Russ. Chem. Rev.* (Engl. transl.) **52**, 259 (1983).

101. M. S. Kharasch and M. Kleiman, *J. Am. Chem. Soc.* **65**, 11 (1943).

102. Y. Pocker and K. D. Stevens, *J. Am. Chem. Soc.* **91**, 4205 (1969).

103. J. K. Stille, F. M. Sonnenberg, and T. H. Kinstle, *J. Am. Chem. Soc.* **88**, 4922 (1966).

104. H. C. Brown and K.-T. Liu, *J. Am. Chem. Soc.* **97**, 600 (1975).

105. K. B. Becker and C. A. Grob, *Synthesis* 789 (1973).

106. J. S. Dewar and R. C. Fahey, *Angew. Chem., Int. Ed. Engl.* **3**, 245 (1964).

107. R. C. Fahey and C. A. McPherson, *J. Am. Chem. Soc.* **91**, 3865 (1969).

108. Y. Pocker and R. F. Buchholz, *J. Am. Chem. Soc.* **92**, 4033 (1970).

109. F. R. Mayo and J. J. Katz, *J. Am. Chem. Soc.* **69**, 1339 (1947).

110. F. R. Mayo and M. G. Savoy, *J. Am. Chem. Soc.* **69**, 1348 (1947).

111. R. C. Fahey, *Top. Stereochem.* **3**, 237 (1968).

112. G. S. Hammond and T. D. Nevitt, *J. Am. Chem. Soc.* **76**, 4121 (1954).

113. G. S. Hammond and C. H. Collins, *J. Am. Chem. Soc.* **82**, 4323 (1960).

114. D. J. Pasto, G. R. Meyer, and B. Lepeska, *J. Am. Chem. Soc.* **96**, 1858 (1974).

115. R. C. Fahey and M. W. Monahan, *J. Am. Chem. Soc.* **92**, 2816 (1970).

116. R. C. Fahey and C. A. McPherson, *J. Am. Chem. Soc.* **93**, 2445 (1971).

117. F. R. Mayo and C. Walling, *Chem. Rev.* **27**, 351 (1940).

118. F. W. Stacey and J. F. Harris, Jr., *Org. React.* (NY) **13**, 150 (1963).

119. M. S. Kharasch, H. Engelmann, and F. R. Mayo, *J. Org. Chem.* **2**, 288 (1937).

120. D. H. Hey and W. A. Waters, *Chem. Rev.* **21**, 169 (1937).

121. P. I. Abell, in *Free Radicals*, Vol. 2, J. K. Kochi, ed., Wiley-Interscience, New York, 1973, Chapter 13, p. 63.

122. G. B. Sergeev, N. F. Stepanov, I. A. Leenson, V. V. Smirnov, V. I. Pupyshev, L. A. Tyurina, and M. N. Mashyanov, *Tetrahedron* **38**, 2585 (1982).

123. D. Landini and F. Rolla, *J. Org. Chem.* **45**, 3527 (1980).

124. P. J. Kropp, K. A. Daus, M. W. Tubergen, K. D. Kepler, V. P. Wilson, S. L. Craig, M. M. Baillargeon, and G. W. Breton, *J. Am. Chem. Soc.* **115**, 3071 (1993).

125. R. M. Pagni, G. W. Kabalka, R. Boothe, K. Gaetano, L. J. Stewart, R. Conaway, C. Dial, D. Gray, S. Larson, and T. Luidhardt, *J. Org. Chem.* **53**, 4477 (1988).

126. L. Delaude and P. Laszlo, *Tetrahedron Lett.* **32**, 3705 (1991).

127. W. Smadja, *Chem. Rev.* **83**, 263 (1983).

128. T. Okuyama, K. Izawa, and T. Fueno, *J. Am. Chem. Soc.* **95**, 6749 (1973).

129. R. Y. Tien and P. I. Abell, *J. Org. Chem.* **35**, 956 (1970).

130. A. V. Fedorova and A. A. Petrov, *J. Gen. Chem. USSR* (Engl. transl.) **31**, 3273 (1961).

131. L. M. Mascavage, H. Chi, S. La, and D. R. Dalton, *J. Org. Chem.* **56**, 595 (1991).

132. L. Schmerling and J. P. West, *J. Am. Chem. Soc.* **74**, 2885 (1952).

133. G. S. Hammond and J. Warkentin, *J. Am. Chem. Soc.* **83**, 2554 (1961).

134. K. Fukui, *Tetrahedron Lett.* 2427 (1965).

135. K. Izawa, T. Okuyama, T. Sakagami, and T. Fueno, *J. Am. Chem. Soc.* **95**, 6752 (1973).

136. H. Hunziker, R. Meyer, and Hs. H. Günthard, *Helv. Chim. Acta* **49**, 497 (1966).

137. M. S. Kharasch, J. G. McNab, and M. C. McNab, *J. Am. Chem. Soc.* **57**, 2463 (1935).

138. R. C. Fahey, M. T. Payne, and D.-J. Lee, *J. Org. Chem.* **39**, 1124 (1974).

139. F. Marcuzzi and E. Melloni, *J. Am. Chem. Soc.* **98**, 3295 (1976).

140. R. C. Fahey and D. J. Lee, *J. Am. Chem. Soc.* **88**, 5555 (1966); **90**, 2124 (1968).

141. P. J. Kropp, K. A. Daus, S. D. Crawford, M. W. Tubergen, K. D. Kepler, S. L. Craig, and V. P. Wilson, *J. Am. Chem. Soc.* **112**, 7433 (1990).

142. R. Stroh, in *Methoden der Organischen Chemie (Houben-Weyl)*, Vol. 5/3: *Halogenverbindungen.* Thieme, Stuttgart, 1962, p. 768.

143. A. Roedig, in *Methoden der Organischen Chemie (Houben-Weyl)*, Vol. 5/4: *Halogenverbindungen.* Thieme, Stuttgart, 1960, p. 132, p. 540.

144. L. S. Boguslavskaya, *Russ. Chem. Rev.* (Engl. transl.) **41**, 740 (1972).

145. K. G. Migliorese, E. H. Appelman, and M. N. Tsangaris, *J. Org. Chem.* **44**, 1711 (1979).

146. J. G. Traynham and O. S. Pascual, *Tetrahedron*, **7**, 165 (1959).

147. B. Damin, J. Garapon, and B. Sillion, *Synthesis* 362 (1981).

148. J. W. Cornforth and D. T. Green, *J. Chem. Soc. C* 846 (1970).

149. R. Antonioletti, M. D'Auria, A. De Mico, G. Piancatelli, and A. Scettri, *Tetrahedron* **39**, 1765 (1983).

150. R. C. Cambie, W. I. Noall, G. J. Potter, P. S. Rutledge, and P. D. Woodgate, *J. Chem. Soc., Perkin Trans. 1* 226 (1977).

151. M. Anbar and D. Ginsburg, *Chem. Rev.* **54**, 925 (1954).

152. D. H. R. Barton, R. H. Hesse, G. P. Jackman, L. Ogunkoya, and M. M. Pechet, *J. Chem. Soc., Perkin Trans. 1* 739 (1974).

153. O. Lerman and S. Rozen, *J. Org. Chem.* **45**, 4122 (1980).

154. S. Rozen and O. Lerman, *J. Org. Chem.* **45**, 672 (1980).

155. C. Walling, L. Heaton, and D. D. Tanner, *J. Am. Chem. Soc.* **87**, 1715 (1965).

156. V. L. Heasley, C. L. Frye, G. E. Heasley, K. A. Martin, D. A. Redfield, and P. S. Wilday, *Tetrahedron Lett.* 1573 (1970).

157. G. E. Heasley, V. M. McCully, R. T. Wiegman, V. L. Heasley, and R. A. Skidgel, *J. Org. Chem.* **41**, 644 (1976).

158. G. E. Heasley, W. E. Emery, III, R. Hinton, D. F. Shellhamer, V. L. Heasley, and S. L. Rodgers, *J. Org. Chem.* **43**, 361 (1978).

159. V. L. Heasley, R. K. Gipe, J. L. Martin, H. C. Wiese, M. L. Oakes, D. F. Shellhamer, G. E. Heasley, and B. L. Robinson, *J. Org. Chem.* **48**, 3195 (1983).

160. G. E. Heasley, M. Duke, D. Hoyer, J. Hunnicutt, M. Lawrence, M. J. Smolik, V. L. Heasley, and D. F. Shellhamer, *Tetrahedron Lett.* **23**, 1459 (1982).

161. V. L. Heasley, B. R. Berry, S. L. Holmes, L. S. Holstein, III, K. A. Milhoan, A. M. Sauerbrey, B. R. Teegarden, D. F. Shellhamer, and G. E. Heasley, *J. Org. Chem.* **53**, 198 (1988).

162. J.-E. Dubois and J. R. Chrétien, *J. Am. Chem. Soc.* **100**, 3506 (1978).

163. J. R. Chrétien, J.-D. Coudert, and M.-F. Ruasse, *J. Org. Chem.* **58**, 1917 (1993).

164. J. Barluenga, J. M. González, P. J. Campos, and G. Asensio, *Angew. Chem., Int. Ed. Engl.* **24**, 319 (1985).

165. J. Barluenga, J. M. González, P. J. Campos, and G. Asensio, *Tetrahedron Lett.* **27**, 1715 (1986).

166. R. W. Nagorski and R. S. Brown, *J. Am. Chem. Soc.* **114**, 7773 (1992).

167. D. T. Mowry, *Chem. Rev.* **42**, 189 (1948).

168. B. R. James, in *Comprehensive Organometallic Chemistry*, Vol. 8, G. Wilkinson, F. G. A. Stone, and E. W. Abel, eds., Pergamon Press, Oxford, 1982, Chapter 51, p. 285.

169. C. A. Tolman, R. J. McKinney, W. C. Seidel, J. D. Druliner, and W. R. Stevens, *Adv. Catal.* **33**, 1 (1985).

170. A. Spencer, in *Comprehensive Coordination Chemistry*, Vol. 6: *Applications*, G. Wilkinson, R. D. Gillard, and J. A. McCleverty, eds., Pergamon Press, Oxford, 1987, Chapter 61.2, p. 229.

171. J. P. Collman, L. S. Hegedus, J. R. Norton, and R. G. Finke, *Principles and Applications of Organotransition Metal Chemistry*, University Science Books, Mill Valley, CA, 1987, Chapter 10, p. 523.

172. R. J. McKinney, in *Homogeneous Catalysis*, G. W. Parshall and S. D. Ittel, eds., Wiley-Interscience, New York, 1993, Chapter 3, p. 42.

173. B. W. Taylor and H. E. Swift, *J. Catal.* **26**, 254 (1972).

174. E. S. Brown, in *Organic Syntheses via Metal Carbonyls*, Vol. 2, I. Wender and P. Pino, eds., Wiley-Interscience, New York, 1977, p. 655.

175. J.-E. Bäckvall and O. S. Andell, *J. Chem. Soc., Chem. Commun.* 1098 (1981).

176. C. A. Tolman, W. C. Seidel, J. D. Druliner, and P. J. Domaille, *Organometallics* **3**, 33 (1984).

177. J.-E. Bäckvall and O. S. Andell, *J. Chem. Soc., Chem. Commun.* 260 (1984).

178. W. Keim, A. Behr, H.-O. Lühr, and J. Weisser, *J. Catal.* **78**, 209 (1982).

179. K. Friedrich and K. Wallenfels, in *The Chemistry of the Cyano Group*, Z. Rappoport, ed., Interscience, London, 1970, Chapter 2, p. 67.

180. J. F. Brazdil, in *Kirk-Othmer Encyclopedia of Chemical Technology*, 4th ed., Vol. 1, J. I. Kroschwitz and M. Howe-Grant, eds., Wiley, New York, 1991, p. 352.

181. W. R. Jackson and C. G. Lovel, *J. Chem. Soc., Chem. Commun.* 1231 (1982).

182. F. Asinger, *Mono-Olefins, Chemistry and Technology*, Pergamon Press, Oxford, 1968, Chapter 6, p. 506.

183. R. G. Striling, in *Encyclopedia of Chemical Processing and Design*, Vol. 20, J. J. McKetta and W. A. Cunningham, eds., Marcel Dekker, New York, 1984, p. 68.

184. K. Weissermel and H.-J. Arpe, *Industrial Organic Chemistry*, 2nd ed., VCH, Weinheim, 1993, Chapter 8, p. 193.

185. A. W. Fleer, A. J. Johnson, and C. R. Nelson, *Ind. Eng. Chem.* **47**, 982 (1955).

186. *Petrol. Refiner* **38**(11), 245 (1959).

187. V. D. Luedeke, in *Encyclopedia of Chemical Processing and Design*, Vol. 2, J. J. McKetta and W. A. Cunningham, eds., Marcel Dekker, New York, 1977, p. 146.

188. J. S. Naworski and E. S. Velez, in *Applied Industrial Catalysis*, Vol. 1, B. E. Leach, ed., Academic Press, New York, 1983, Chapter 9, p. 239.

189. M. W. Newman, in *Encycylopedia of Polymer Science and Engineering*, 2nd ed., supplement volume, J. I. Kroschwitz, ed., Wiley, New York, 1989, p. 822.

190. K. Weissermel and H.-J. Arpe, *Industrial Organic Chemistry*, 2nd ed., VCH, Weinheim, 1993, Chapter 9, p. 213.

191. C. E. Rowe, in *Encyclopedia of Chemical Processing and Design*, Vol. 8, J. J. McKetta and W. A. Cunningham, eds., Marcel Dekker, New York, 1979, p. 160.

192. A. C. Fyvie, *Chem. Ind.* (London) 384 (1964).

193. A. J. Gait, in *Propylene and Its Industrial Derivatives*, E. G. Hancock, ed., Wiley, New York, 1973, Chapter 7, p. 273.

194. K. Weissermel and H.-J. Arpe, *Industrial Organic Chemistry*, 2nd ed., VCH, Weinheim, 1993, Chapter 11, p. 264.

195. *Hydrocarbon Process.* **58**(11), 239 (1979).

196. G. W. Parshall, *J. Mol. Catal.* **4**, 243 (1978).

197. D. J. Hadley, in *Propylene and Its Industrial Derivatives*, E. G. Hancock, ed., Wiley, New York, 1973, Chapter 11, p. 418.

198. S. T. Purrington, B. S. Kagen, and T. B. Patrick, *Chem. Rev.* **86**, 997 (1986).

199. S. Rozen, *Acc. Chem. Res.* **21**, 307 (1988).

200. E. Block and A. L. Schwan, in *Comprehensive Organic Synthesis*, B. M. Trost and I. Fleming, eds., Vol. 4: *Additions and Substitutions at C–C π-Bonds*, M. F. Semmelhack, ed., Pergamon Press, Oxford, 1991, Chapter 1.8, p. 329.

201. R. F. Merritt and F. A. Johnson, *J. Org. Chem.* **31**, 1859 (1966).

202. F. R. Merritt, *J. Am. Chem. Soc.* **89**, 609 (1967).

203. S. Rozen and M. Brand, *J. Org. Chem.* **51**, 3607 (1986).

204. G. A. Olah and J. M. Bollinger, *J. Am. Chem. Soc.* **89**, 4744 (1967); **90**, 947 (1968).

205. G. A. Olah, Y. K. Mo, and Y. Halpern, *J. Org. Chem.* **37**, 1169 (1972).

206. G. A. Olah, D. A. Beal, and P. W. Westerman, *J. Am. Chem. Soc.* **95**, 3387 (1973).

207. R. Filler, *Isr. J. Chem.* **17**, 71 (1978).

208. M. Zupan, in *The Chemistry of Functional Groups,* Supplement D: *The Chemistry of Halides, Pseudo-Halides and Azides*, S. Patai and Z. Rappoport, eds., Wiley, Chichester, 1983, Chapter 15, p. 657.

209. J. A. Wilkinson, *Chem. Rev.* **92**, 505 (1992).

210. A. Gregorcic and M. Zupan, *J. Org. Chem.* **44**, 4120 (1979).

211. T. C. Shieh, E. D. Feit, C. L. Chernick, and N. C. Yang, *J. Org. Chem.* **35**, 4020 (1970).

212. M. Zupan and A. Pollak, *J. Org. Chem.* **42**, 1559 (1977).

213. K. A. V'yunov and A. I. Ginak, *Russ. Chem. Rev.* (Engl. transl.) **50**, 151 (1981).

214. G. H. Schmid, in *The Chemistry of Functional Groups,* Supplement A2: *The Chemistry of Double-Bonded Functional Groups,* S. Patai, ed., Wiley, Chichester, 1989, Chapter 11, p. 679.

215. K. Yates, G. H. Schmid, T. W. Regulski, D. G. Garratt, H.-W. Leung, and R. McDonald, *J. Am. Chem. Soc.* **95**, 160 (1973).

216. Yu. A. Serguchev and V. P. Konyushenko, *J. Org. Chem. USSR* (Engl. transl.) **11**, 1339 (1975).

217. M. L. Poutsma, *J. Am. Chem. Soc.* **87**, 4285 (1965).

218. G. B. Sergeev, Yu.A. Serguchev, and V. V. Smirnov, *Russ. Chem. Rev.* (Engl. transl.) **42**, 697 (1973).

219. G. A. Olah, *Halonium Ions*. Wiley-Interscience, New York, 1975, Chapter 7, p. 98.

220. K. W. Field, P. Kovacic, *Synthesis* 135 (1969).

221. J. W. Strand and P. Kovacic, *Synth. Commun.* **2**, 129 (1972).

222. S. Uemura, A. Onoe, and M. Okano, *Bull. Chem. Soc. Jpn.* **47**, 692 (1974).

223. S. Uemura, A. Onoe, and M. Okano, *Bull. Chem. Soc. Jpn.* **47**, 3121 (1974).

224. W. A. Nugent, *Tetrahedron Lett.* 3427 (1978).

225. T. Koyano, *Bull. Chem. Soc. Jpn.* **43**, 3501 (1970).

226. S. Fukuzumi and J. K. Kochi, *Int. J. Chem. Kinet.* **15**, 249 (1983).

227. J. E. Dubois and F. Garnier, *Spectrochim. Acta, Part A* **23**, 2279 (1967).

228. G. Bellucci, R. Bianchini, and R. Ambrosetti, *J. Am. Chem. Soc.* **107**, 2464 (1985).

229. R. S. Brown, R. Gedye, H. Slebocka-Tilk, J. M. Buschek, and K. R. Kopecky, *J. Am. Chem. Soc.* **106**, 4515 (1984).

230. G. Bellucci, C. Chiappe, and F. Marioni, *J. Am. Chem. Soc.* **109**, 515 (1987).

231. R. S. Brown, H. Slebocka-Tilk, A. J. Bennet, G. Bellucci, R. Bianchini, and R. Ambrosetti, *J. Am. Chem. Soc.* **112**, 6310 (1990).

232. F. Garnier, R. H. Donnay, and J. E. Dubois, *J. Chem. Soc., Chem. Commun.* 829 (1971).

233. A. Modro, G. H. Schmid, and K. Yates, *J. Org. Chem.* **44**, 4221 (1979).

234. I. Roberts and G. E. Kimball, *J. Am. Chem. Soc.* **59**, 947 (1937).

235. E. Bienvenue-Goëtz and J.-E. Dubois, *Tetrahedron* **34**, 2021 (1978).

236. B. Galland, E. M. Evleth, and M.-F. Ruasse, *J. Chem. Soc., Chem. Commun.* 898 (1990).

237. T. P. Hamilton and H. F. Schaefer, III, *J. Am. Chem. Soc.* **112**, 8260 (1990).

238. S. Yamabe and T. Minato, *Bull. Chem. Soc. Jpn.* **66**, 3339 (1993).

239. G. A. Olah and A. M. White, *J. Am. Chem. Soc.* **91**, 5801 (1969).

240. G. A. Olah and T. R. Hockswender, Jr., *J. Am. Chem. Soc.* **96**, 3574 (1974).

241. G. A. Olah, P. W. Westerman, E. G. Melby, and Y. K. Mo, *J. Am. Chem. Soc.* **96**, 3565 (1974).

242. M.-F. Ruasse, *Acc. Chem. Res.* **23**, 87 (1990).

243. F. Freeman, *Chem. Rev.* **75**, 439 (1975).

244. G. E. Heasley, J. M. Bundy, V. L. Heasley, S. Arnold, A. Gipe, D. McKee, R. Orr, S. L. Rodgers, and D. F. Shellhamer, *J. Org. Chem.* **43**, 2793 (1978).

245. W. C. Baird, Jr., J. H. Surridge, and M. Buza, *J. Org. Chem.* **36**, 3324 (1971).

246. K. Smith and K. B. Fry, *J. Chem. Soc., Chem. Commun.* 187 (1992).

247. M. L. Poutsma, *J. Am. Chem. Soc.* **85**, 3511 (1963).

248. M. L. Poutsma, *J. Am. Chem. Soc.* **87**, 2172 (1965).

249. R. M. Dessau, *J. Am. Chem. Soc.* **101**, 1344 (1979).

250. P. S. Skell and R. R. Pavlis, *J. Am. Chem. Soc.* **86**, 2956 (1964).

251. R. L. Ayres, C. J. Michejda, and E. P. Rack, *J. Am. Chem. Soc.* **93**, 1389 (1971).

252. H. G. Peer, *Recl. Trav. Chim. Pays-Bas* **81**, 113 (1962).

253. W. H. Mueller, P. E. Butler, and K. Griesbaum, *J. Org. Chem.* **32**, 2651 (1967).

254. M. L. Poutsma, *J. Org. Chem.* **33**, 4080 (1968).

255. W. L. Waters, W. S. Linn, and M. C. Caserio, *J. Am. Chem. Soc.* **90**, 6741 (1968).

256. M. C. Findlay, W. L. Waters, and M. C. Caserio, *J. Org. Chem.* **36**, 275 (1971).

257. L. R. Byrd and M. C. Caserio, *J. Am. Chem. Soc.* **93**, 5758 (1971).

258. T. Okuyama, K. Ohashi, K. Izawa, and T. Fueno, *J. Org. Chem.* **39**, 2255 (1974).

259. M. L. Poutsma, *J. Org. Chem.* **31**, 4167 (1966).

260. V. L. Heasley, G. E. Heasley, R. A. Loghry, and M. R. McConnell, *J. Org. Chem.* **37**, 2228 (1972).

261. V. L. Heasley, G. E. Heasley, S. K. Taylor, and C. L. Frye, *J. Org. Chem.* **35**, 2967 (1970).

262. G. E. Heasley, V. L. Heasley, S. L. Manatt, H. A. Day, R. V. Hodges, P. A. Kroon, D. A. Redfield, T. L. Rold, and D. E. Williamson, *J. Org. Chem.* **38**, 4109 (1973).

263. G. E. Heasley, D. C. Hayse, G. R. McClung, D. K. Strickland, V. L. Heasley, P. D. Davis, D. M. Ingle, K. D. Rold, and T. S. Ungermann, *J. Org. Chem.* **41**, 334 (1976).

264. U. Husstedt and H. J. Schäfer, *Synthesis*, 966 (1979).

265. H. J. Franz, W. Höbold, R. Höhn, G. Müller-Hagen, R. Müller, W. Pritzkow, and H. Schmidt, *J. Prakt. Chem.* **312**, 622 (1970).

266. S. Uemura, S. Fukuzawa, A. Toshimitsu, M. Okano, H. Tezuka, and S. Sawada, *J. Org. Chem.* **48**, 270 (1983).

267. G. Melloni, G. Modena, and U. Tonellato, *Acc. Chem. Res.* **14**, 227 (1981).

268. J. A. Pincock and K. Yates, *Can. J. Chem.* **48**, 3332 (1970).

269. S. Uemura, A. Onoe, and M. Okano, *J. Chem. Soc., Perkin Trans. 1*, 1278 (1978).

270. R. A. Hollins and M. P. A. Campos, *J. Org. Chem.* **44**, 3931 (1979).

271. V. L. Heasley, D. F. Shellhamer, L. E. Heasley, D. B. Yaeger, and G. E. Heasley, *J. Org. Chem.* **45**, 4649 (1980).

272. R. W. McPherson, C. M. Starks, and G. J. Fryar, *Hydrocarbon Process.* **58**(3), 75 (1979).

273. J. A. Cowfer and A. J. Magistro, in *Kirk-Othmer Encyclopedia of Chemical Technology*, 3rd ed., Vol. 23, M. Grayson and D. Eckroth, eds., Wiley-Interscience, New York, 1983, p. 865.

274. D. P. Keane, R. B. Stobaugh, and P. L. Townsend, *Hydrocarbon Process.* **52**(2), 99 (1973).

275. L. F. Albright, *Chem. Eng.* (NY), **74**(7), 123 (1967); **74**(8), 219 (1967).

276. W. E. Wimer and R. E. Feathers, *Hydrocarbon Process.* **55**(3), 81 (1976).

277. P. Reich, *Hydrocarbon Process.* **55**(3), 85 (1976).

278. R. G. Markeloff, *Hydrocarbon Process., Int. Ed.* **63**(11), 91 (1984).

279. C. N. Kenney, *Catal. Rev.-Sci. Eng.* **11**, 197 (1975).

280. J. Villadsen and H. Livbjerg, *Catal. Rev.-Sci. Eng.* **17**, 203 (1978).

281. E. Cavaterra, *Hydrocarbon Process., Int. Ed.* **67**(12), 63 (1988).

282. E. W. Wong, C. P. Ambler, W. J. Baker, and J. C. Parks, Jr., *Hydrocarbon Process., Int. Ed.* **71**(8), 129 (1992).

283. *Hydrocarbon Process.* **54**(11), 217 (1975).

284. C. E. Hollis, *Chem. Ind.* (London) 1030 (1969).

285. P. R. Johnson, in *Kirk-Othmer Encyclopedia of Chemical Technology*, 3rd ed., Vol. 5, M. Grayson and D. Eckroth, eds., Wiley-Interscience, New York, 1979, p. 773.

286. C. A. Stewart, Jr., T. Takeshita, and M. L. Coleman, in *Encyclopedia of Polymer Science and Engineering*, 2nd ed., Vol. 3, J. I. Kroschwitz, ed., Wiley-Interscience, New York, 1985, p. 441.

287. K. Weissermel and H.-J. Arpe, *Industrial Organic Chemistry*, 2nd ed., VCH, Weinheim, 1993, Chapter 5, p. 118.

288. M. B. Gasc, A. Lattes, and J. J. Perie, *Tetrahedron* **39**, 703 (1983).

289. J.-J. Brunet, D. Neibecker, and F. Niedercorn, *J. Mol. Catal.* **49**, 235 (1989).

290. D. M. Roundhill, *Chem. Rev.* **92**, 1 (1992).

291. H. Pines and W. M. Stalick, *Base-Catalyzed Reactions of Hydrocarbons and Related Compounds*, Academic Press, New York, 1977, Chapter 10, p. 423.

292. B. W. Howk, E. L. Little, S. L. Scott, and G. M. Whitman, *J. Am. Chem. Soc.* **76**, 1899 (1954).

293. R. D. Closson, J. P. Napolitano, G. G. Ecke, and A. J. Kolka, *J. Org. Chem.* **22**, 646 (1957).

294. G. P. Pez and J. E. Galle, *Pure Appl. Chem.* **57**, 1917 (1985).

295. H. Lehmkuhl and D. Reinehr, *J. Organomet. Chem.* **55**, 215 (1973).

296. R. J. Schlott, J. C. Falk, and K. W. Narducy, *J. Org. Chem.* **37**, 4243 (1972).

297. M. Deeba, M. E. Ford, and T. A. Johnson, *J. Chem. Soc., Chem. Commun.* 562 (1987).

298. M. Deeba, M. E. Ford, and T. A. Johnson, in *Catalysis, 1987*, Studies in Surface Science and Catalysis, Vol. 38, J. W. Ward, ed., Elsevier, Amsterdam, 1988, p. 221.

299. M. Deeba and M. E. Ford, *J. Org. Chem.* **53**, 4594 (1988).

300. M. Tabata, N. Mizuno, and M. Iwamoto, *Chem. Lett.* 1027 (1991).

301. N. Mizuno, M. Tabata, T. Uematsu, and M. Iwamoto, *J. Catal.* **146**, 249 (1994).

302. D. R. Coulson, *Tetrahedron Lett.* 429 (1971).

303. A. L. Casalnuovo, J. C. Calabrese, and D. Milstein, *J. Am. Chem. Soc.* **110**, 6738 (1988).

304. M. G. Clerici and F. Maspero, *Synthesis* 305 (1980).

305. D. I. Hoke, D. L. Surbey, and W. R. Oviatt, *J. Polym. Sci., Polym. Chem. Ed.* **10**, 595 (1972).

306. T. Fujita, K. Suga, and S. Watanabe, *Aust. J. Chem.* **27**, 531 (1974).

307. T. Narita, N. Imai, and T. Tsuruta, *Bull. Chem. Soc. Jpn.* **46**, 1242 (1973).

308. R. Baker, A. H. Cook, D. E. Halliday, and T. N. Smith, *J. Chem. Soc., Perkin Trans. 2* 1511 (1974).

309. R. Baker and D. E. Halliday, *Tetrahedron Lett.* 2773 (1972).

310. K. Takahashi, A. Miyake, and G. Hata, *Bull. Chem. Soc. Jpn.* **45**, 1183 (1972).

311. I. A. Chekulaeva and L. V. Kondrat'eva, *Russ. Chem. Rev.* (Engl. transl.) **34**, 669 (1965).

312. W. Reppe, *Justus Liebigs Ann. Chem.* **601**, 81 (1956).

313. C. W. Kruse and R. F. Kleinschmidt, *J. Am. Chem. Soc.* **83**, 213 (1961).

314. P. W. Hickmott, *Tetrahedron* **38**, 1975 (1982).

315. M. F. Shostakovskii, G. G. Skvortsova, and E. S. Domnina, *Russ. Chem. Rev.* (Engl. transl.) **38**, 407 (1969).

316. C. W. Kruse and R. F. Kleinschmidt, *J. Am. Chem. Soc.* **83**, 216 (1961).

317. J. Barluenga, F. Aznar, R. Liz, and R. Rodes, *J. Chem. Soc., Perkin Trans. 1* 2732 (1980); 1087 (1983).

318. J. Barluenga and F. Aznar, *Synthesis* 195 (1977).

319. J. Barluenga, F. Aznar, R. Liz, and M.-P. Cabal, *J. Chem. Soc., Chem. Commun.* 1375 (1985).

320. G. Zweifel and H. C. Brown, *Org. React.* (NY) **13**, 1 (1963).

321. G. M. L. Cragg, *Organoboranes in Organic Synthesis*, Vol. 1. Marcel Dekker, New York, 1973.

322. M. Zaidlewich, in *Comprehensive Organometallic Chemistry*, Vol. 7, G. Wilkinson, F. G. A. Stone, and E. W. Abel, eds., Pergamon Press, Oxford, 1982, Chapter 45.4, p. 199.

323. H. C. Brown, G. W. Kramer, A. B. Levy, and M. M. Midland, *Organic Syntheses via Boranes*. Wiley-Interscience, New York, 1975.

324. B. M. Mikhailov and Yu. N. Bubnov, *Organoboron Compounds in Organic Synthesis*, Harwood, Chur, UK, 1984.

325. D. S. Matteson, in *The Chemistry of the Metal-Carbon Bond*, Vol. 4, F. R. Hartley, ed., Wiley, Chichester, 1987, Chapter 3, p. 307.

326. K. Avasthi, D. Devaprabhakara, and A. Suzuki, in *Organometallic Chemistry Reviews, Journal of Organometallic Chemistry Library 7*, D. Seyferth, A. G. Davies, E. O. Fischer, J. F. Normant, and O. A. Reutov, eds., Elsevier, Amsterdam, 1979, p. 1.

327. J. Weill-Raynal, *Synthesis* 633 (1976).

328. A. Pelter, *Chem. Soc. Rev.* **11**, 191 (1982).

329. H. C. Brown, in *Comprehensive Organometallic Chemistry*, Vol. 7, G. Wilkinson, F. G. A. Stone, and E. W. Abel, eds., Pergamon Press, Oxford, 1982, Chapter 45.1, p. 111.

330. A. Suzuki and R. S. Dhillon, *Top. Curr. Chem.* **130**, 23 (1986).

331. H. C. Brown, *Hydroboration*, Benjamin, New York, 1962.

332. A. Pelter and K. Smith, in *Comprehensive Organic Chemistry*, Vol. 3, D. H. R. Barton and W. D. Ollis, eds., Pergamon, Oxford, 1979, Chapter 14.2, p. 695.

333. K. Smith and A. Pelter, in *Comprehensive Organic Synthesis*, B. M. Trost and I. Fleming, eds., Vol. 8: *Reduction*, I. Fleming, ed., Pergamon Press, Oxford, 1991, Chapter 3.10, p. 703.

334. E. Negishi, *Organometallics in Organic Synthesis*, Vol. 1, Wiley-Interscience, New York, 1980, Chapter 5, p. 287.

335. M. Zaidlewich, in *Comprehensive Organometallic Chemistry*, Vol. 7, G. Wilkinson, F. G. A. Stone, and E. W. Abel, eds., Pergamon Press, Oxford, 1982, Chapter 45.2, p. 143.

336. H. C. Brown and B. C. Subba Rao, *J. Am. Chem. Soc.* **78**, 5694 (1956); *J. Org. Chem.* **22**, 1136 (1957).

337. M. Zaidlewich, in *Comprehensive Organometallic Chemistry*, Vol. 7, G. Wilkinson, F. G. A. Stone, and E. W. Abel, eds., Pergamon Press, Oxford, 1982, Chapter 45.3, p. 161.

338. L. M. Braun, R. A. Braun, H. R. Crissman, M. Opperman, and R. M. Adams, *J. Org. Chem.* **36**, 2388 (1971).

339. H. C. Brown, A. K. Mandal, and S. U. Kulkarni, *J. Org. Chem.* **42**, 1392 (1977).

340. H. C. Brown and G. Zweifel, *J. Am. Chem. Soc.* **83**, 1241 (1961).

341. H. C. Brown and A. W. Moerikofer, *J. Am. Chem. Soc.* **84**, 1478 (1962).

342. E. Negishi and H. C. Brown, *Synthesis* 77 (1974).

343. J. E. Rice and Y. Okamoto, *J. Org. Chem.* **47**, 4189 (1982).

344. H. C. Brown, E. F. Knights, and C. G. Scouten, *J. Am. Chem. Soc.* **96**, 7765 (1974).

345. H. C. Brown, E. Negishi, and J.-J. Katz, *J. Am. Chem. Soc.* **97**, 2791 (1975).

346. H. C. Brown, J.-J. Katz, C. F. Lane, and E. Negishi, *J. Am. Chem. Soc.* **97**, 2799 (1975).

347. A. Pelter, D. J. Ryder, and J. H. Sheppard, *Tetrahedron Lett.* 4715 (1978).

348. J. A. Sikorski and H. C. Brown, *J. Org. Chem.* **47**, 872 (1982).

349. H. C. Brown and G. Zweifel, *J. Am. Chem. Soc.* **83**, 2544 (1961).

350. G. Zweifel and H. C. Brown, *J. Am. Chem. Soc.* **86**, 393 (1964).

351. G. W. Kabalka, R. J. Newton, Jr., and J. Jacobus, *J. Org. Chem.* **43**, 1567 (1978).

352. H. C. Brown, J. H. Kawakami, and K.-T. Liu, *J. Am. Chem. Soc.* **95**, 2209 (1973).

353. H. C. Brown and G. Zweifel, *J. Am. Chem. Soc.* **82**, 4708 (1960).

354. H. C. Brown, J. Chandrasekharan, and K. K. Wang, *Pure Appl. Chem.* **55**, 1387 (1983).

355. R. E. Williams, *Inorg. Chem.* **1** 971 (1962).

356. G. A. Olah, *Angew. Chem., Int. Ed. Engl.* **12**, 173 (1973).

357. H. C. Brown and A. W. Moerikofer, *J. Am. Chem. Soc.* **83**, 3417 (1961).

358. H. C. Brown, C. G. Scouten, and K. K. Wang, *J. Org. Chem.* **44**, 2589 (1979).

359. C. F. Lane and G. W. Kabalka, *Tetrahedron* **32**, 981 (1976).

360. D. Männig and H. Nöth, *Angew. Chem., Intl. Ed. Engl.* **24**, 878 (1985).

361. K. Burgess and M. J. Ohlmeyer, *Chem. Rev.* **91**, 1179 (1991).

362. D. A. Evans, G. C. Fu, and A. H. Hoveyda, *J. Am. Chem. Soc.* **114**, 6671 (1992).

363. D. A. Evans, G. C. Fu, and B. A. Anderson, *J. Am. Chem. Soc.* **114**, 6679 (1992).

364. T. Hayashi, Y. Matsumoto, and Y. Ito, *Tetrahedron: Asymmetry* **2**, 601 (1991).

365. G. Zweifel, K. Nagase, and H. C. Brown, *J. Am. Chem. Soc.* **84**, 183 (1962).

366. H. C. Brown, E. Negishi, and S. K. Gupta, *J. Am. Chem. Soc.* **92**, 2460 (1970).

367. H. C. Brown, E. Negishi, and P. L. Burke, *J. Am. Chem. Soc.* **93**, 3400 (1971).

368. H. C. Brown, R. Liotta, and G. W. Kramer, *J. Org. Chem.* **43**, 1058 (1978).

369. H. C. Brown and E. Negishi, *Tetrahedron* **33**, 2331 (1977).

370. E. F. Knights and H. C. Brown, *J. Am. Chem. Soc.* **90**, 5280 (1968).

371. J. A. Soderquist and H. C. Brown, *J. Org. Chem.* **46**, 4599 (1981).

372. H. C. Brown and G. Zweifel, *J. Am. Chem. Soc.* **83**, 3834 (1961).

373. G. Zweifel and H. Arzoumanian, *J. Am. Chem. Soc.* **89**, 291 (1967).

374. G. Zweifel, G. M. Clark, and N. L. Polston, *J. Am. Chem. Soc.* **93**, 3395 (1971).

375. H. C. Brown, C. G. Scouten, and R. Liotta, *J. Am. Chem. Soc.* **101**, 96 (1979).

376. H. C. Brown, D. Basavaiah, and S. U. Kulkarni, *J. Organomet. Chem.* **225**, 63 (1982).

377. J. S. Cha, S. J. Min, J. M. Kim, and O. O. Kwon *Tetrahedron Lett.* **34**, 5113 (1993).

378. W. G. Woods and P. L. Strong, *J. Am. Chem. Soc.* **88**, 4667 (1966).

379. H. C. Brown and S. K. Gupta, *J. Am. Chem. Soc.* **97**, 5249 (1975).

380. A. Pelter, S. Singaram, and H. Brown, *Tetrahedron Lett.* **24**, 1433 (1983).

381. P. A. Chaloner, *Handbook of Coordination Catalysis in Organic Chemistry*, Butterworths, London, 1986, Chapter 4, p. 307.

382. H. Reinheckel, K. Haage, and D. Jahnke, *Organometal. Chem. Rev., Sect. A* **4**, 47 (1969).

383. G. Zweifel and J. A. Miller, *Org. React.* (NY) **32**, 375 (1984).

384. P. A. Chaloner, in *The Chemistry of the Metal-Carbon Bond*, Vol. 4, F. R. Hartley, ed., Chichester, Wiley, 1987, Chapter 4, p. 411.

385. K. Maruoka and H. Yamamoto, *Tetrahedron* **44**, 5001 (1988).

386. U. M. Dzhemilev, O. S. Vostrikova, and G. A. Tolstikov, *J. Organomet. Chem.* **304**, 17 (1986).

387. J. J. Eisch, in *Comprehensive Organic Synthesis*, B. M. Trost and I. Fleming, eds., Vol. 8: *Reduction*, I. Fleming, ed., Pergamon Press, Oxford, 1991, Chapter 3.11, p. 733.

388. K. Ziegler, H.-G. Gellert, H. Lehmkuhl, W. Pfohl, and K. Zosel, *Justus Liebigs Ann. Chem.* **629**, 1 (1960).

389. K. Ziegler, H.-G. Gellert, H. Martin, and K. Nagel, *Justus Liebigs Ann. Chem.* **589**, 91 (1954).

390. K. Ziegler, *Angew. Chem.* **64**, 323 (1952).

391. K. Ziegler, H. Martin, and F. Krupp, *Justus Liebigs Ann. Chem.* **629**, 14 (1960).

392. G. Natta, P. Pino, G. Mazzanti, P. Longi, and F. Bernardini, *J. Am. Chem. Soc.* **81**, 2561 (1959).

393. F. Sato, S. Sato, and M. Sato, *J. Organomet. Chem.* **131**, C26 (1977).

394. G. Wilke and H. Müller, *Justus Liebigs Ann. Chem.* **618**, 267 (1958); **629**, 222 (1960).

395. J. J. Eisch and W. C. Kaska, *J. Am. Chem. Soc.* **88**, 2213 (1966).

396. G. Zweifel and R. B. Steele, *J. Am. Chem. Soc.* **89**, 5085 (1967).

397. E. F. Magoon and L. H. Slaugh, *Tetrahedron* **23**, 4509 (1967).

398. R. N. Meals, *Pure Appl. Chem.* **13**, 141 (1966).

399. E. Lukevics, Z. V. Belyakova, M. G. Pomerantseva, and M. G. Voronkov, in *Organometallic Chemistry Reviews, Journal of Organometallic Chemistry Library 5*, D. Seyferth, A. G. Davies, E. O. Fischer, J. F. Normant, O. A. Reutov, eds., Elsevier, Amsterdam, 1977, p. 1.

400. J. F. Harrod and A. J. Chalk, in *Organic Syntheses via Metal Carbonyls*, Vol. 2, I. Wender and P. Pino, eds., Wiley-Interscience, New York, 1977, p. 673.

401. J. L. Speier, *Adv. Organomet. Chem.* **17**, 407 (1979).

402. K. A. Andrianov, J. Soucek, and L. M. Khananashvili, *Russ. Chem. Rev.* (Engl. transl.) **48**, 657 (1979).

403. E. Negishi, *Organometallics in Organic Synthesis*, Vol. 1. Wiley-Interscience, New York, 1980, Chapter 6, p. 394.

404. F. H. Jardine, in *The Chemistry of the Metal-Carbon Bond*, Vol. 4, F. R. Hartley, ed., Wiley, Chichester, 1987, Chapter 8, p. 784.

405. I. Ojima, in *The Chemistry of Organic Silicon Compounds*, S. Patai and Z. Rappoport, eds., Wiley, Chichester, 1989, Chapter 25, p. 1479.

406. T. Hiyama and T. Kusumoto, in *Comprehensive Organic Synthesis*, B. M. Trost and I. Fleming, eds., Vol. 8: *Reduction*, I. Fleming, ed., Pergamon Press, Oxford, 1991, Chapter 3.12, p. 763.

407. B. Hardman and A. Torkelson, in *Encyclopedia of Polymer Science and Engineering*, 2nd ed., Vol. 15, J. I. Kroschwitz, ed., Wiley-Interscience, New York, 1989, p. 204.

408. A. J. Chalk and J. F. Harrod, *J. Am. Chem. Soc.* **87**, 16 (1965).

409. J. F. Harrod and A. J. Chalk, *J. Am. Chem. Soc.* **87**, 1133 (1965); **88**, 3491 (1966).

410. L. H. Sommer, J. E. Lyons, and H. Fujimoto, *J. Am. Chem. Soc.* **91**, 7051 (1969).

411. M. A. Schroeder and M. S. Wrighton, *J. Organomet. Chem.* **128**, 345 (1977).

412. A. Millan, M.-J. Fernandez, P. Bentz, and P. M. Maitlis, *J. Mol. Catal.* **26**, 89 (1984).

413. A. Onopchenko, E. T. Sabourin, and D. L. Beach, *J. Org. Chem.* **48**, 5101 (1983).

414. M. Bookhart and B. E. Grant, *J. Am. Chem. Soc.* **115**, 2151 (1993).

415. M. G. Voronkov, V. B. Pukhnarevich, S. P. Sushchinskaya, L. I. Kopylova, and B. A. Trofimov, *J. Gen. Chem. USSR* (Engl. transl.) **41**, 2120 (1971).

416. R. A. Benkeser and J. Kang, *J. Organomet. Chem.* **185**, C9 (1980).

417. A. Onopchenko and E. T. Sabourin, *J. Org. Chem.* **52**, 4118 (1987).

418. L. N. Lewis and R. J. Uriarte, *Organometallics* **9**, 621 (1990).

419. L. N. Lewis, *J. Am. Chem. Soc.* **112**, 5998 (1990).

420. L. N. Lewis and N. Lewis, *J. Am. Chem. Soc.* **108**, 7228 (1986).

421. L. N. Lewis, R. J. Uriarte, and N. Lewis, *J. Mol. Catal.* **66**, 105 (1991).

422. I. Ojima, K. Yamamoto, and M. Kumada, in *Aspects of Homogeneous Catalysis*, Vol. 3, R. Ugo, ed., Reidel, Dordrecht, 1977, Chapter 3, p. 185.

423. J. L. Speier, J. A. Webster, and G. H. Barnes, *J. Am. Chem. Soc.* **79**, 974 (1957).

424. T. G. Selin and R. West, *J. Am. Chem. Soc.* **84**, 1863 (1962).

425. H. M. Bank, J. C. Saam, and J. L. Speier, *J. Org. Chem.* **29**, 792 (1964).

426. R. A. Benkeser, S. Dunny, G. S. Li, P. G. Nerlekar, and S. D. Work, *J. Am. Chem. Soc.* **90**, 1871 (1968).

427. R. N. Haszeldine, R. V. Parish, and R. J. Taylor, *J. Chem. Soc., Dalton Trans.*, 2311 (1974).

428. T. G. Selin and R. West, *J. Am. Chem. Soc.* **84**, 1860 (1962).

429. M. Capka, P. Svoboda, and J. Hetflejs, *Coll. Czech. Chem. Commun.* **38**, 3830 (1973).

430. E. W. Bennett and P. J. Orenski, *J. Organomet. Chem.* **28**, 137 (1971).

431. K. Oertle and H. Wetter, *Tetrahedron Lett.* **26**, 5511 (1985).

432. J. Tsuji, M. Hara, and K. Ohno, *Tetrahedron* **30**, 2143 (1974).

433. I. Ojima and M. Kumagai, *J. Organomet. Chem.* **157**, 359 (1978).

434. A. J. Cornish, M. F. Lappert, J. J. MacQuitty, and R. K. Maskell, *J. Organomet. Chem.* **177**, 153 (1979).

435. I. Ojima, *J. Organomet. Chem.* **134**, C1 (1977).

436. M. Capka and J. Hetflejs, *Coll. Czech. Chem. Commun.* **40**, 3020 (1975).

437. V. P. Yur'ev, I. M. Salimgareeva, O. Zh. Zhebarov, and G. A. Tolstikov, *J. Gen. Chem. USSR* (Engl. transl.) **46**, 368 (1976).

438. V. P. Yur'ev, I. M. Salimgareeva, O. Zh. Zhebarov, and L. M. Khalilov, *J. Gen. Chem. USSR* (Engl. transl.) **47**, 1414 (1977).

439. M. S. Wrighton and M. A. Schroeder, *J. Am. Chem. Soc.* **96**, 6235 (1974).

440. L.-Z. Wang and Y.-Y. Jiang, *J. Organomet. Chem.* **251**, 39 (1983).

441. C.-Y. Hu, X.-M. Han, and Y.-Y. Jiang, *J. Mol. Catal.* **35**, 329 (1986).

442. R. A. Benkeser, R. F. Cunico, S. Dunny, P. R. Jones, and P. G. Nerlekar, *J. Org. Chem.* **32**, 2634 (1967).

443. M. Green, J. L. Spencer, F. G. A. Stone, and C. A. Tsipis, *J. Chem. Soc., Dalton Trans.* 1525 (1977).

444. J. Yoshida, K. Tamao, M. Takahashi, and M. Kumada, *Tetrahedron Lett.* 2161 (1978).

445. C. A. Tsipis, *J. Organomet. Chem.* **187**, 427 (1980).

446. L. N. Lewis, K. G. Sy, G. L. Bryant, Jr., and P. E. Donahue, *Organometallics* **10**, 3750 (1991).

447. I. Ojima, M. Kumagai, and Y. Nagai, *J. Organomet. Chem.* **66**, C14 (1974).

448. H. Watanabe, T. Kitahara, T. Motegi, and Y. Nagai, *J. Organomet. Chem.* **139**, 215 (1977).

449. H. M. Dickers, R. N. Haszeldine, A. P. Mather, and R. V. Parish, *J. Organomet. Chem.* **161**, 91 (1978).

450. V. B. Pukharevich, L. I. Kopylova, M. Capka, J. Hetflejs, É. N. Satsuk, M. V. Sigalov, V. Chvalovsky, and M. G. Voronkov, *J. Gen. Chem. USSR* (Egnl. transl.) **50**, 1259 (1980).

451. M. F. Lappert and R. K. Maskell, *J. Organomet. Chem.* **264**, 217 (1984).

452. R. Takeuchi and N. Tanouchi, *J. Chem. Soc., Chem. Commun.* 1319 (1993).

453. C. A. Tsipis, *J. Organomet. Chem.* **188**, 53 (1980).

454. J. W. Ryan and J. L. Speier, *J. Org. Chem.* **31**, 2698 (1966).

455. M. G. Voronkov and S. P. Sushchinskaya, *J. Gen. Chem. USSR* (Egnl. transl.) **56**, 555 (1986).

456. M. G. Voronkov, S. N. Adamovich, L. V. Sherstyannikova, and V. B. Pukharevich, *J. Gen. Chem. USSR* (Egnl. transl.) **53**, 706 (1983).

457. J. Schwartz and J. A. Labinger, *Angew. Chem., Int. Ed. Engl.* **15**, 333 (1976).

458. J. A. Labinger, in *Comprehensive Organic Synthesis*, B. M. Trost and I. Fleming, eds., Vol. 8: *Reduction*, I. Fleming, ed., Pergamon Press, Oxford, 1991, Chapter 3.9, p. 667.

459. E. Negishi and T. Takahashi, *Synthesis* 1 (1988).

460. D. W. Hart and J. Schwartz, *J. Am. Chem. Soc.* **96**, 8115 (1974).

461. J. E. Nelson, J. E. Bercaw, and J. A. Labinger, *Organometallics* **8**, 2484 (1989).

462. C. A. Bertelo and J. Schwartz, *J. Am. Chem. Soc.* **98**, 262 (1976).

463. S. L. Buchwald and S. J. LaMaire, *Tetrahedron Lett.* **28**, 295 (1987).

464. D. W. Hart, T. F. Blackburn, and J. Schwartz, *J. Am. Chem. Soc.* **97**, 679 (1975).

465. J. D. Odom, in *Comprehensive Organometallic Chemistry*, Vol. 1, G. Wilkinson, F. G. A. Stone, and E. W. Abel, eds., Pergamon Press, Oxford, 1982, Chapter 5.1, p. 276.

466. F. Joy, M. F. Lappert, and B. Prokai, *J. Organomet. Chem.* **5**, 506 (1966).

467. M. F. Lappert, *Angew. Chem.* **72**, 36 (1960).

468. M. F. Lappert and B. Prokai, *J. Organomet. Chem.* **1**, 384 (1964).

469. J. R. Blackborow, *J. Chem. Soc., Perkin Trans.* 2 1989 (1973); *J. Organomet. Chem.* **128**, 161 (1977).

470. S. Hara, T. Kato, H. Shimizu, and A. Suzuki, *Tetrahedron Lett.* **26**, 1065 (1985).

471. S. Hara, H. Dojo, S. Takinami, and A. Suzuki, *Tetrahedron Lett.* **24**, 731 (1983).

472. T. D. Coyle and J. J. Ritter, *Adv. Organomet. Chem.* **10**, 237 (1972).

473. G. Urry, J. Kerrigan, T. D. Parsons, and H. I. Schlesinger, *J. Am. Chem. Soc.* **76**, 5299 (1954).

474. P. Ceron, A. Finch, J. Frey, J. Kerrigan, T. Parsons, G. Urry, and H. I. Schlesinger, *J. Am. Chem. Soc.* **81**, 6368 (1959).

475. H. K. Saha, L. J. Glicenstein, and G. Urry, *J. Organomet. Chem.* **8**, 37 (1967).

476. M. Zeldin and T. Wartik, *J. Am. Chem. Soc.* **88**, 1336 (1966).

477. W. B. Fox and T. Wartik, *J. Am. Chem. Soc.* **83**, 498 (1961).

478. R. W. Rudolph, *J. Am. Chem. Soc.* **89**, 4216 (1967).

479. C. N. Welch and S. G. Shore, *Inorg. Chem.* **8**, 2810 (1969).

480. M. Zeldin, A. R. Gatti, and T. Wartik, *J. Am. Chem. Soc.* **89**, 4217 (1967).

481. F. C. Hall and A. W. Nash, *J. Inst. Petroleum Tech.* **23**, 679 (1937); **24**, 471 (1938).

482. G. A. Olah, M. Bruce, F. Clouet, S. M. F. Farnia, O. Farooq, G. K. S. Prakash, J. Welch, and J. L. Koenig, *Macromolecules* **20**, 2972 (1987).

483. G. A. Olah, O. Farooq, S. M. F. Farnia, M. R. Bruce, F. L. Clouet, P. R. Morton, G. K. S. Prakash, R. C. Stevens, R. Bau, K. Lammertsma, S. Suzer, and L. Andrews, *J. Am. Chem. Soc.* **110**, 3231 (1988).

484. J. Chatt, *Chem. Rev.* **48**, 7 (1951).

485. N. S. Zefirov, *Russ. Chem. Rev.* (Engl. transl.) **34**, 527 (1965).

486. W. Kitching, *Organometal. Chem. Rev., Sect. A.* **3**, 61 (1968).

487. J. L. Wardell, in *Comprehensive Organometallic Chemistry*, Vol. 2, G. Wilkinson, F. G. A. Stone, and E. W. Abel, eds., Pergamon Press, Oxford, 1982, Chapter 17, p. 863.

488. R. C. Larock, *Angew. Chem., Int. Ed. Engl.* **17**, 27 (1978).

489. H. C. Brown and P. J. Geoghegan, Jr., *J. Org. Chem.* **35**, 1844 (1970).

490. J. M. Jerkunica and T. G. Traylor, *Org. Synth., Coll. Vol.* **6**, 766 (1988).

491. H. C. Brown and P. J. Geoghegan, Jr., *J. Org. Chem.* **37**, 1937 (1972).

492. H. J. Bergmann, G. Collin, G. Just, G. Müller-Hagen, and W. Pritzkow, *J. Prakt. Chem.* **314**, 285 (1972).

493. D. J. Pasto and J. A. Gontarz, *J. Am. Chem. Soc.* **93**, 6902 (1971).

494. W. L. Waters, T. G. Traylor, and A. Factor, *J. Org. Chem.* **38**, 2306 (1973).

495. T. G. Traylor and A. W. Baker, *J. Am. Chem. Soc.* **85**, 2746 (1963).

496. F. R. Jensen, J. J. Miller, S. J. Cristol, and R. S. Beckley, *J. Org. Chem.* **37**, 4341 (1972).

497. H. C. Brown and J. H. Kawakami, *J. Am. Chem. Soc.* **95**, 8665 (1973).

498. G. A. Olah and P. R. Clifford, *J. Am. Chem. Soc.* **95**, 6067 (1973).

499. T. G. Traylor, *Acc. Chem. Res.* **2**, 152 (1969).

500. S. Uemura, H. Miyoshi, and M. Okano, *J. Chem. Soc., Perkin Trans. 1* 1098 (1980).

501. G. Drefahl, G. Heublein, and A. Wintzer, *Angew. Chem.* **70**, 166 (1958).

502. A. McKillop and E. C. Taylor, in *Comprehensive Organometallic Chemistry*, Vol. 7, G. Wilkinson, F. G. A. Stone, and E. W. Abel, eds., Pergamon Press, Oxford, 1982, Chapter 47, p. 465.

503. E. Negishi, *Organometallics in Organic Synthesis*, Vol. 1., Wiley-Interscience, New York, 1980, Chapter 5, p. 350.

504. J. F. Normant and A. Alexakis, *Synthesis* 841 (1981).

505. L. Miginiac, in *The Chemistry of the Metal-Carbon Bond*, Vol. 3, F. R. Hartley and S. Patai, eds., Wiley, Chichester, 1985, Chapter 2, p. 99.

506. W. Carruthers, in *Comprehensive Organometallic Chemistry*, Vol. 7, G. Wilkinson, F. G. A. Stone, and E. W. Abel, eds., Pergamon Press, Oxford, 1982, Chapter 49, p. 661.

507. J. R. Zietz, Jr., G. C. Robinson, and K. L. Lindsay, in *Comprehensive Organometallic Chemistry*, Vol. 7, G. Wilkinson, F. G. A. Stone, and E. W. Abel, eds., Pergamon Press, Oxford, 1982, Chapter 46, p. 365.

508. P. Knochel, in *Comprehensive Organic Chemistry*, B. M. Trost and I. Fleming, eds., Vol. 4: *Additions and Substitutions at C–C π-Bonds*, M. F. Semmelhack, ed., Pergamon Press, Oxford, 1991, Chapter 4.4, p. 865.

509. H. Lehmkuhl and K. Ziegler, in *Methoden der Organischen Chemie (Houben-Weyl)*, Vol. 13/4: *Metallorganische Verbindungen*, Thieme, Stuttgart, 1970, p. 184.

510. G. R. Lappin, in *Alpha Olefins Application Handbook*, G. R. Lappin and J. D. Sauer, eds., Marcel Dekker, New York, 1989, Chapter 3, p. 35.

511. E. Negishi, *Pure Appl. Chem.* **53**, 2333 (1981).

512. E. Negishi, D. E. Van Horn, and T. Yoshida, *J. Am. Chem. Soc.* **107**, 6639 (1985).

513. H. Lehmkuhl, *Bull. Soc. Chim. Fr.* II-87 (1981).

514. J. G. Duboudin and B. Jousseaume, *J. Organomet. Chem.* **162**, 209 (1978).

515. J. F. Normant, G. Cahiez, M. Bourgain, C. Chuit, and J. Villieras, *Bull. Soc. Chim. Fr.* 1656 (1974).

516. J. F. Normant, G. Cahiez, C. Chuit, and J. Villieras, *J. Organomet. Chem.* **77**, 269 (1974).

517. J. F. Normant and M. Bourgain, *Tetrahedron Lett.* 2583 (1971).

518. A. Alexakis and J. F. Normant, *Tetrahedron Lett.* **23**, 5151 (1982).

519. H. Westmijze, J. Meijer, H. J. T. Bos, and P. Vermeer, *Recl. Trav. Chim. Pays-Bas* **95**, 299, 304 (1976).

520. A. Marfat, P. R. McGuirk, and P. Helquist, *J. Org. Chem.* **44**, 3888 (1979).

521. F. Fringuelli and A. Taticchi, *Dienes in the Diels-Alder Reaction*, Wiley-Interscience, New York, 1990.

522. W. Oppolzer, in *Comprehensive Organic Synthesis*, B. M. Trost and I. Fleming, eds., Vol. 5: *Combining C–C π-Bonds*, L. A. Paquette, ed., Pergamon Press, Oxford, 1991, Chapter 4.1, p. 315.

523. O. Diels and K. Alder, *Justus Liebigs Ann. Chem.* **460**, 98 (1928).

524. R. B. Woodward and R. Hoffmann, *Angew. Chem., Int. Ed. Engl.* **8**, 781 (1969).

525. K. Fukui, *Acc. Chem. Res.* **4**, 57 (1971).

526. P. Yates and P. Eaton, *J. Am. Chem. Soc.* **82**, 4436 (1960).

527. D. C. Rideout and R. Breslow, *J. Am. Chem. Soc.* **102**, 7816 (1980).

528. J. Sauer and R. Sustmann, *Angew. Chem., Int. Ed. Engl.* **19**, 779 (1980).

529. K. N. Houk, Y. Li, and D. Evanseck, *Angew. Chem., Int. Ed. Engl.* **31**, 682 (1992).

530. J. G. Martin and R. K. Hill, *Chem. Rev.* **61**, 537 (1961).

531. K. N. Houk, Y.-T. Lin, and F. K. Brown, *J. Am. Chem. Soc.* **108**, 554 (1986).

532. K. Alder and G. Stein, *Angew. Chem.* **50**, 510 (1937).

533. D. Ginsburg, *Tetrahedron*, **39**, 2095 (1983).

534. M. Balogh and P. Laszlo, *Organic Chemistry Using Clays*. Springer, Berlin, 1993, Chapter 8, p. 95.

535. R. M. Dessau, *J. Chem. Soc., Chem. Commun.* 1167 (1986).

536. P. Laszlo and J. Lucchetti, *Tetrahedron Lett.* **25**, 1567 (1984).

537. D. J. Bellville, D. D. Wirth, and N. L. Bauld, *J. Am. Chem. Soc.* **103**, 718 (1981).

538. P. G. Gassman and D. A. Singleton, *J. Am. Chem. Soc.* **106**, 7993 (1984).

539. D. W. Reynolds, K. T. Lorenz, H.-S. Chiou, D. J. Bellville, R. A. Pabon, and N. L. Bauld, *J. Am. Chem. Soc.* **109**, 4960 (1987).

540. G. N. Schrauzer, *Adv. Catal.* **18**, 377 (1968).

541. J. E. Lyons, H. K. Myers, and A. Schneider, *Ann. NY Acad. Sci.* **333**, 273 (1980).

542. H. tom Dieck and R. Diercks, *Angew. Chem., Int. Ed. Engl.* **22**, 778 (1983).

543. J. E. Baldwin, in *Comprehensive Organic Synthesis*, B. M. Trost and I. Fleming, eds., Vol. 5: *Combining C–C π-Bonds*, L. A. Paquette, ed., Pergamon Press, Oxford, 1991, Chapter 2.1, p. 63.

544. M. T. Crimmins, in *Comprehensive Organic Synthesis*, B. M. Trost and I. Fleming, eds., Vol. 5: *Combining C–C π-Bonds*, L. A. Paquette, ed., Pergamon Press, Oxford, 1991, Chapter 2.3, p. 123.

545. T. Mitsudo, H. Naruse, T. Kondo, Y. Ozaki, and Y. Watanabe, *Angew. Chem., Int. Ed. Engl.* **33**, 580 (1994).

546. A. F. Plate and N. A. Belikova, *Bull. Acad. Sci. USSR, Div. Chem. Sci.* (Engl. transl.) 1237 (1958).

547. W. Donike, H. Ulrich, H. Thiemer, and J. Pieper, Ger. Offen. 3,307,486 (1983).

548. L. M. Joshel and L. W. Butz, *J. Am. Chem. Soc.* **63**, 3350 (1941).

549. C. L. Thomas, *Ind. Eng. Chem.* **36**, 310 (1944).

550. J. Meinwald and N. J. Hudak, *Org. Synth., Coll. Vol.* **4**, 738 (1963).

551. J. Hyman, E. Freireich, and R. E. Lidov, U. S. Patent 2,875,256 (1959).

552. H.-D. Scharf, J. Fleischhauer, H. Leismann, I. Ressler, W. Schleker, and R. Weitz, *Angew. Chem., Int. Ed. Engl.* **18**, 652 (1979).

553. W. G. Dauben and R. L. Cargill, *Tetrahedron* **15**, 197 (1961).

554. R. R. Hautala, J. Little, and E. Sweet, *Solar Energy* **19**, 503 (1977).

555. C. Kutal, *Adv. Chem. Ser.* **168** 158 (1978).

556. C. Kutal and P. A. Grutsch, *Adv. Chem. Ser.* **173**, 325 (1979).

557. P. A. Grutsch and C. Kutal, *J. Am. Chem. Soc.* **101**, 4228 (1979).

558. K. Maruyama, K. Terada, and Y. Yamamoto, *J. Org. Chem.* **46**, 5294 (1981).

559. B. C. Ranu, A. Sarkar, M. Saha, and R. Chakraborty, *Tetrahedron* **50**, 6579 (1994).

560. M. Setty-Fichman, Y. Sasson, and J. Blum, *J. Mol. Catal. A: Chem.* **126**, 27 (1997).

561. T. Suzuki, M. Tokunaga, and Y. Wakatsuki, *Org. Lett.* **3**, 735 (2001).

562. T. Tsuchimoto, T. Joya, E. Shirakawa, and Y. Kawakami, *Synlett* 1777 (2000).

563. L. J. Sæthre, T. D. Thomas, and S. Svensson, *J. Chem. Soc., Perkin Trans. 2*, 749 (1997).

564. L. M. Mascavage, F. Zhang, and D. R. Dalton, *J. Org. Chem.* **59**, 5048 (1994).

565. P. J. Kropp and S. D. Crawford, *J. Org. Chem.* **59**, 3102 (1994).

566. P. Boudjouk, B.-K. Kim, and B.-H. Han, *Synth. Commun.* **26**, 3479 (1996).

567. R. Herges, *Angew. Chem., Int. Ed. Engl.* **34**, 51 (1995).

568. M.-F. Ruasse, *Adv. Phys. Org. Chem.* **28**, 207 (1993).

569. L. Forlani, in *The Chemistry of Functional Groups*, Supplement A3: *The Chemistry of Double Bonded Functional Groups*, S. Patai, ed., Wiley, London, 1997, Chapter 8, p. 367.

570. D. J. Nelson, R. Li, and C. Brammer, *J. Org. Chem.* **66**, 2422 (2001).

571. H. I. Bloemink, K. Hinds, A. C. Legon, and J. C. Thorn, *Chem. Commun.* 1321 (1994).

572. G. Bellucci, C. Chiappe, R. Bianchini, D. Lenoir, and R. Herges, *J. Am. Chem. Soc.* **117**, 12001 (1995).

573. R. Bianchini, C. Chiappe, D. Lenoir, P. Lemmen, R. Herges, and J. Grunenburg, *Angew. Chem., Int. Ed. Engl.* **36**, 1284 (1997).

574. G. Bellucci, R. Bianchini, C. Chiappe, D. Lenoir, and A. Attar, *J. Am. Chem. Soc.* **117**, 6243 (1995).

575. H. Hopf, R. Hänel, P. G. Jones, and P. Bubenitschek, *Angew. Chem., Int. Ed. Engl.* **33**, 1369 (1994).

576. R. Bianchini, C. Chiappe, G. Lo Moro, D. Lenoir, P. Lemmen, and N. Goldberg, *Chem. Eur. J.* **5**, 1570 (1999).

577. C. Chiappe, D. Capraro, V. Conte, and D. Pieraccini, *Org. Lett.* **3**, 1061 (2001).

578. T. Schlama, K. Gabriel, V. Gouverneur, and C. Mioskowski, *Angew. Chem., Int. Ed. Engl.* **36**, 2342 (1997).

579. A. Itoh and Y. Masaki, *Synlett* 1450 (1997).

580. I. E. Markó, P. R. Richardson, M. Bailey, A. R. Maguire, and N. Coughlan, *Tetrahedron Lett.* **38**, 2339 (1997).

581. N. B. Barhate, A. S. Gajare, R. D. Wakharkar, and A. V. Bedekar, *Tetrahedron* **55**, 11127 (1999).

582. M. A. Tius, *Tetrahedron* **51**, 6605 (1995).

583. S. A. Lermontov, S. I. Zavorin, I. V. Bakhtin, A. N. Pushin, N. S. Zefirov, and P. J. Stang, *J. Fluorine Chem.* **87**, 75 (1998).

584. J.-J. Brunet, *Gazz. Chim. Ital.* **127**, 111 (1997).

585. R. Taube, in *Applied Homogeneous Catalysis with Organometallic Complexes*, B. Cornils and W. A. Herrmann, eds., VCH, Weinheim, 1996, Chapter 2.7, p. 507.

586. D. M. Roundhill, *Catal. Today* **37**, 155 (1997).

587. T. E. Müller and M. Beller, *Chem. Rev.* **98**, 675 (1998).

588. M. Beller, C. Breindl, T. H. Riermeier, H. Eichberger, and H. Trauthwein, *Angew. Chem., Int. Ed. Engl.* **37**, 3389 (1998).

589. H. Schaffrath and W. Keim, *J. Mol. Catal. A: Chem.* **168**, 9 (2001).

590. R. Dorta, P. Egli, F. Zürcher, and A. Togni, *J. Am. Chem. Soc.* **119**, 10857 (1997).

591. I. Nakamura, H. Itagaki, and Y. Yamamoto, *J. Org. Soc.* **63**, 6458 (1998).

592. M. Kawatsura and J. F. Hartwig, *J. Am. Chem. Soc.* **122**, 9546 (2000).

593. H. Trauthwein, A. Tillack, and M. Beller, *Chem. Commun.* 2029 (1999).

594. O. Löber, M. Kawatsura, and J. F. Hartwig, *J. Am. Chem. Soc.* **123**, 4366 (2001).

595. T. Uchimaru, *Chem. Commun.* 1133 (1999).

596. E. Haak, I. Bytschkov, and S. Doye, *Angew. Chem., Int. Ed. Engl.* **38**, 3389 (1999).

597. M. Tokunaga, M. Eckert, and Y. Wakatsuki, *Angew. Chem., Int. Ed. Engl.* **38**, 3223 (1999).

598. Y. Li and T. J. Marks, *Organometallics* **15**, 3770 (1996).

599. I. Beletskaya and A. Pelter, *Tetrahedron* **53**, 4957 (1997).

600. K. Burgess and W. A. van der Donk, *Inorg. Chim. Acta* **220**, 93 (1994).

601. H. C. Brown and P. V. Ramachandran, *J. Organomet. Chem.* **500**, 1 (1995).

602. A. M. Segarra, R. Guerrero, C. Claver, and E. Fernandez, *Chem. Commun.* 1808 (2001).

603. J. F. Hartwig and C. N. Muhoro, *Organometallics* **19**, 30 (2000).

604. T. Ohmura, Y. Yamamoto, and N. Miyaura, *J. Am. Chem. Soc.* **122**, 4990 (2000).

605. J. J. J. Juliette, I. T. Horváth, and J. A. Gladysz, *Angew. Chem., Int. Ed. Engl.* **36**, 1610 (1997).

606. J. V. B. Kanth and H. C. Brown, *Tetrahedron Lett.* **41**, 9361 (2000).

607. B. Marciniec, *Comprehensive Handbook on Hydrosilylation*, Pergamon, Oxford, 1992, Chapter 2.

608. M. A. Brook, *Silicon in Organic, Organometallic, and Polymer Chemistry*, Wiley, New York, 2000, p. 401.

609. B. Marciniec, in *Applied Homogeneous Catalysis with Organometallic Complexes*, B. Cornils and W. A. Herrmann, eds., VCH, Weinheim, 1996, Chapter 2.6, p. 487.

610. I. Ojima, Z. Li, and J. Zhu, in *The Chemistry of Organic Silicon Compounds*, S. Rappoport and Y. Apeloig, eds., Wiley, New York, 1998, Chapter 29.

611. J. A. Reichl and D. H. Berry, *Adv. Organomet. Chem.* **43**, 197 (1999).

612. B. Marciniec, *Appl. Organomet. Chem.* **14**, 527 (2000).

613. S. Sakaki, N. Mizoe, and M. Sugimoto, *Organometallics* **17**, 2510 (1998).

614. B. Marciniec, *New J. Chem.* **21**, 815 (1997).

615. J. Stein, L. N. Lewis, Y. Gao, and R. A. Scott, *J. Am. Chem. Soc.* **121**, 3693 (1999).

616. R. Anwander, in *Applied Homogeneous Catalysis with Organometallic Complexes*, B. Cornils and W. A. Herrmann, eds., VCH, Weinheim, 1996, Chapter 3.2.5, p. 866.

617. G. A. Molander, E. D. Dowdy, and B. C. Noll, *Organometallics* **17**, 3754 (1998).

618. G. A. Molander and E. E. Knight, *J. Org. Chem.* **63**, 7009 (1998).

619. P.-F. Fu, L. Brard, Y. Li, and T. J. Marks, *J. Am. Chem. Soc.* **117**, 7157 (1995).

620. A. K. Dash, J. Q. Wang, and M. S. Eisen, *Organometallics* **18**, 4724 (1999).

621. Y.-S. Song, B. R. Yoo, G.-H. Lee, and I. N. Jung, *Organometallics* **18**, 3109 (1999).

622. T. Sudo, N. Asao, V. Gevorgyan, and Y. Yamamoto, *J. Org. Chem.* **64**, 2494 (1999).

623. N. Asao and Y. Yamamoto, *Bull. Chem. Soc. Jpn.* **73**, 1071 (2000).

624. Y. Na and S. Chang, *Org. Lett.* **2**, 1887 (2000).

625. R. D. Adams and T. S. Barnard, *Organometallics* **17**, 2567 (1998).

626. P. Steffanut, J. A. Osborn, A. DeCian, and J. Fisher, *Chem. Eur. J.* **4**, 2008 (1998).

627. R. Drake, R. Dunn, D. C. Sherrington, and S. J. Thomson, *Chem. Commun.* 1931 (2000).

628. Z. M. Michalska, K. Strzelec, and J. W. Sobczak, *J. Mol. Catal. A: Chem.* **156**, 91 (2000).

629. H. S. Hilal, M. A. Suleiman, W. J. Jondi, S. Khalaf, and M. M. Masoud, *J. Mol. Catal. A: Chem.* **144**, 47 (1999).

630. G. Schmid, H. West, H. Mehles, and A. Lehnert, *Inorg. Chem.* **36**, 891 (1997).

631. P. Wipf and H. Jahn, *Tetrahedron* **52**, 12853 (1996).

632. E. Negishi, *Chem. Eur. J.* **5**, 411 (1999).

633. H. Makabe and E. Negishi, *Eur. J. Org. Chem.* 969 (1999).

634. N. D. Smith, J. Mancuso, and M. Lautens, *Chem. Rev.* **100**, 3257 (2000).

635. R. E. Maleczka, Jr., L. R. Terrel, D. H. Clark, S. L. Whitehead, W. P. Gallagher, and I. Tersteige, *J. Org. Chem.* **64**, 5958 (1999).

636. M. Lautens, S. Kumanovic, and C. Meyer, *Angew. Chem., Int. Ed. Engl.* **35**, 1329 (1996).

637. E. Negishi and D. Y. Kondakov, *Chem. Soc. Rev.* **25**, 417 (1996).

638. U. M. Dzhemilev and A. G. Ibragimov, *Russ. Chem. Rev.* **69**, 121 (2000).

639. A. Suzuki, *Rev. Heteroatom Chem.* **17**, 271 (1997).

640. H. Poleschner, M. Heydenreich, K. Spindler, and G. Haufe, *Synthesis*, 1043 (1994).

641. Y. Uski, M. Iwaoka, and S. Tomoda, *Chem. Lett.* 1507 (1992).

642. Y. Zhang, W. Bao, and H. Dong, *Synth. Commun.* **23**, 3029 (1993).

643. T. Wirth, S. Häuptli, and M. Leuenberger, *Tetrahedron: Asymmetry* **9**, 547 (1998).

644. D. Y. Kondakov and E. Negishi, *J. Am. Chem. Soc.* **117**, 10771 (1995); **118**, 1577 (1996).

645. P. Wipf and S. Ribe, *Org. Lett.* **2**, 1713 (2000).

646. D. B. Millward, A. P. Cole, and R. M. Waymouth, *Organometallics* **19**, 1870 (2000).

647. J. W. Bundens, J. Yudenfreund, and M. M. Francl, *Organometallics* **18**, 3913 (1999).

648. A. F. Houri, M. T. Didiuk, Z. Xu, N. R. Horan, A. H. Hoveyda, *J. Am. Chem. Soc.* **115**, 6614 (1993).

649. R. Fischer, D. Walther, P. Gebhardt, and H. Görls, *Organometallics* **19**, 2532 (2000).

650. T. Stüdemann and P. Knochel, *Angew. Chem., Int. Ed. Engl.* **36**, 93 (1997).

651. T. Stüdemann, M. Ibrahim-Ouali, and P. Knochel, *Tetrahedron* **54**, 1299 (1998).

652. S. Gagneur, J.-L. Montchamp, and E. Negishi, *Organometallics* **19**, 2417 (2000).

653. S. Yamanoi, H. Ohrui, K. Seki, T. Matsumoto, and K. Suzuki, *Tetrahedron Lett.* **40**, 8407 (1999).

654. T. Takahashi, C. Xi, Y. Ura, and K. Nakajima, *J. Am. Chem. Soc.* **122**, 3228 (2000).

655. M. Lautens, W. Klute, amd W. Tam, *Chem. Rev.* **96**, 49 (1996).

656. M. Murakami, M. Ubukata, K. Itami, and Y. Ito, *Angew. Chem., Int. Ed. Engl.* **37**, 2248 (1998).

657. B. A. Horn, J. L. Herek, and A. H. Zewail, *J. Am. Chem. Soc.* **118**, 8755 (1996).

658. B. R. Beno, S. Wilsey, and K. N. Houk, *J. Am. Chem. Soc.* **121**, 4816 (1999).

659. D. A. Singleton, B. E. Schulmeier, C. Hang, A. A. Thomas, S.-W. Leung, and S. R. Merrigan, *Tetrahedron* **57**, 5149 (2001).

660. M. Nendel, L. M. Tolbert, L. E. Herring, Md. N. Islam, and K. N. Houk, *J. Org. Chem.* **64**, 976 (1999).

661. A. Z. Bradley, M. G. Kociolek, and R. P. Johnson, *J. Org. Chem.* **65**, 7134 (2000).

CARBONYLATION

7.1. HYDROFORMYLATION

During a study of the origin of oxygenates in Fischer–Tropsch synthesis in the presence of a cobalt catalyst, Roelen observed the formation of propanal and 3-pentanone when ethylene was added to the feed.[1] The process now termed *hydroformylation* or *oxo reaction* is the metal-catalyzed transformation of alkenes with carbon monoxide and hydrogen to form aldehydes:

$$R\diagdown\diagdown + CO + H_2 \xrightarrow[\substack{\text{temperature,}\\\text{pressure}}]{\text{catalyst}} R\diagup\diagdown CHO + R\underset{CHO}{\overset{Me}{\diagup\diagdown}} \qquad (7.1)$$

Ketone formation, as in the original Roelen experiment, is significant only with ethylene. Hydroformylation gained substantial industrial importance in the manufacture of butyraldehyde and certain alcohols (see Section 7.1.3). Because of its practical significance, commercial applications were followed by substantial research efforts[2–16] resulting in a good understanding of the chemistry of the reaction.

All Group VIII metals, as well as Mn, Cr, and Cu, exhibit some activity in hydroformylation.[6,11] Cobalt, the catalyst in the original discovery, is still used mainly in industry; rhodium, introduced later, is one of the most active and studied catalysts. The metal catalysts may be applied as homogeneous soluble complexes, heterogenized metal complexes, or supported metals.

$[CoH(CO)_4]$, formed in situ when cobalt metal or cobalt(II) salts are reacted with synthesis gas at elevated temperature under pressure, is the active species in the cobalt-catalyzed reaction.[14,17,18] It is also capable of stoichiometric hydroformylation.[6,14,17] Since [CoH(CO)$_4$] is quite unstable and readily decomposes by CO dissociation, the cobalt-catalyzed hydroformylation requires high CO partial pressure. It was observed later that phosphine ligands improve both the stability of the catalyst the and selectivity of the transformation.[11,14,19,20]

The activity of rhodium complexes with phosphine or phosphite ligands is about three to four orders of magnitude higher than that of cobalt catalysts.[21-23] [RhH(CO)(PPh$_3$)$_3$] preformed or prepared in situ has proved to be the active species when triphenylphosphine is used as ligand. Despite the high cost of rhodium, the mild reaction conditions and high selectivities make rhodium complexes the catalyst of choice in hydroformylation.

Of other Group VIII metal complexes, ruthenium was studied quite extensively.[11,12,23,24] Platinum salts and complexes show negligible activity in hydroformylation. Highly active and selective catalysts are formed, however, by adding Group IVB halides to platinum compounds.[25-27] The preferred catalyst composition, [PtCl$_2$(PPh$_3$)$_2$] and SnCl$_2$, for example, forms [PtH(SnCl$_3$)(CO)(PPh$_3$)$_2$] as the active species.[26]

The use of cobalt carbonyls, modified cobalt catalysts, and ligand-modified rhodium complexes, the three most important types of catalysts in hydroformylation, is discussed here.

7.1.1. Alkenes

In 1961 Heck and Breslow presented a multistep reaction pathway to interpret basic observations in the cobalt-catalyzed hydroformylation.[28] Later modifications and refinements aimed at including alternative routes and interpreting side reactions.[6] Although not all the fine details of hydroformylation are equally well understood, the Heck–Breslow mechanism is still the generally accepted basic mechanism of hydroformylation.[6,17,19,29] Whereas differences in mechanisms using different metal catalysts do exist,[30] all basic steps are essentially the same in the phosphine-modified cobalt- and rhodium-catalyzed transformations as well.

Studies of stoichiometric hydroformylation, spectroscopic identification, isolation, and transformation of intermediates provided valuable information of the understanding of the catalytic reaction. Despite the complexity of the process, important conclusions were also drawn from kinetic studies.

Evidence was presented that cobalt precursors under the reaction conditions are transformed into cobalt carbonyls.[31] Additives such as Lewis bases accelerate the formation of the catalyst.[11] [CoH(CO)$_4$] the key catalytic species was shown by infrared (IR) spectroscopy to be formed under hydroformylation conditions[32] and was isolated in the reaction of [Co(CO)$_4$]$_2$ and hydrogen.[33] [CoH(CO)$_4$] dissociates carbon monoxide to create [CoH(CO)$_3$] [Eq. (7.2)], which is capable of olefin complexation because of a ligand vacancy:

$$[Co_2(CO)_4] + H_2 \rightleftharpoons 2\,[CoH(CO)_4] \rightleftharpoons 2\,[CoH(CO)_3] + 2\,CO \quad (7.2)$$

The first basic step in a simplified version of the hydroformylation mechanism is alkene coordination to the cobalt center:

$$[CoH(CO)_3] + RCH=CH_2 \rightleftharpoons \underset{\mathbf{1}}{\overset{H}{\underset{CO}{OC-Co\overset{,,CO}{\underset{CH}{\overset{CH_2}{\diagdown}}}}}} \rightleftharpoons \underset{\mathbf{2}}{[RCH_2CH_2Co(CO)_3]} \overset{CO}{\rightleftharpoons}$$

$$\rightleftharpoons [RCH_2CH_2Co(CO)_4] \rightleftharpoons \underset{\mathbf{3}}{[RCH_2CH_2\underset{O}{\overset{\parallel}{C}}Co(CO)_3]} \overset{CO}{\rightleftharpoons} \underset{\mathbf{4}}{[RCH_2CH_2\underset{O}{\overset{\parallel}{C}}Co(CO)_4]} \qquad (7.3)$$

The π complex **1** has not been isolated or observed directly, but its involvement is strongly supported by indirect evidence. In the second step the alkene inserts into the cobalt–hydrogen bond to yield an alkylcobalt complex (**2**), which is transformed via the migratory insertion of CO into a coordinatively unsaturated acylcobalt complex (**3**).

The last, product-forming, step is the only reaction that is not an equilibrium reaction. It involves oxidative addition followed by reductive elimination, and may take place with the participation of [CoH(CO)_4] [Eq. (7.4)] or hydrogen [Eq. (7.5)]:

$$\mathbf{3} \xrightarrow{[CoH(CO)_4]} RCH_2CH_2CHO + [Co_2(CO)_7] \qquad (7.4)$$

$$-CO \Big\Updownarrow CO$$

$$\mathbf{4} \xrightarrow{H_2} RCH_2CH_2CHO + [CoH(CO)_4] \qquad (7.5)$$

Product formation with the involvement of [CoH(CO)_4] was shown to be the predominant route in the catalytic reaction.[34] Modification of cobalt carbonyls by substituting phosphine ligand for carbon monoxide increases the stability of the complex. Tributyl phosphine proved to be the best ligand. The increased stability is attributed to the increased electron density of cobalt giving rise to a stronger bonding between the metal and CO. IR studies proved the existence of complex multi-step equilibria in these systems,[32,35] and [CoH(CO)_3(PBu_3)] was shown to be the major species in hydroformylation of cyclohexene.[32] Hydroformylation with modified cobalt catalysts can be performed under milder conditions (lower CO partial pressure), although these complexes exhibit lower catalytic activity.[20]

The basic steps in the hydroformylation mechanism do not change in the presence of ligand-modified cobalt, but the kinetics of the reaction is affected.[30] First-order dependence of the rate on hydrogen partial pressure and an inverse first-order dependence on CO partial pressure were observed in the unmodified system.[36] At high CO pressures the rate is first-order in both alkene and cobalt concentrations. The last, product-forming, step—the cleavage of the acylcobalt complex—was found to be rate-determining in most observations.[13] In the modified system the rate-determining step depends on the ligand and the alkene. Since the increased electron density of cobalt brings about an increase in the rate of the last step, and the presence of a bulky phosphine on cobalt hinders olefin complexation, this latter step [Eq. (7.3), formation of **1**] becomes rate-determining in most cases.

Two important observations concerning cobalt-catalyzed hydroformylations not reconcilable with the Heck–Breslow mechanism are worth mentioning. On the basis of deuterium labeling studies, Pino concluded[37] that (at least at low temperature and pressure) [CoH(CO)$_4$] is an unlikely intermediate and suggested the involvement of more reactive Co complexes such as [Co$_3$H(CO)$_9$]. More recent results of NMR studies carried out under the conditions of usual industrial processes (100–180°C, 370 atm) have indicated the presence of thermally generated ·CoH(CO)$_4$ radicals.[38] This intermediate not considered in the conventional Heck–Breslow mechanism is also suggested to play a role in hydroformylation.

Despite some differences, the mechanism of rhodium-catalyzed hydroformylation is very similar to that of the cobalt-catalyzed process.[39–42] Scheme 7.1 depicts the so-called associative route which is operative when the ligand is in excess. Rhodium metal and many Rh(I) compounds serve as precursor to form[21,22]—in the presence of triphenylphosphine, CO and H$_2$—the active species [RhH(CO)(PPh$_3$)$_3$] (**5**). At high CO partial pressure and low catalyst concentration without added PPh$_3$, the [RhH(CO)$_2$(PPh$_3$)] monotriphenylphosphine complex instead of **6** coordinates the alkene and participates in the so-called dissociative route.[21,39]

The kinetics of hydroformylation by phosphine- or phosphite-modified complexes is even more complex than that of the cobalt-catalyzed reaction. Depending on the reaction conditions, either alkene complexation (Scheme 7.1, **6** to **7**) or oxidative addition of hydrogen (Scheme 7.1, **9** to **10**) may be rate-determining.

Straight-chain terminal olefins react most readily in hydroformylation. Internal alkenes exhibit decreased reactivity, whereas branched olefins are the least reactive.[6,11,43] Rhodium modified with phosphite ligands, however, was shown to

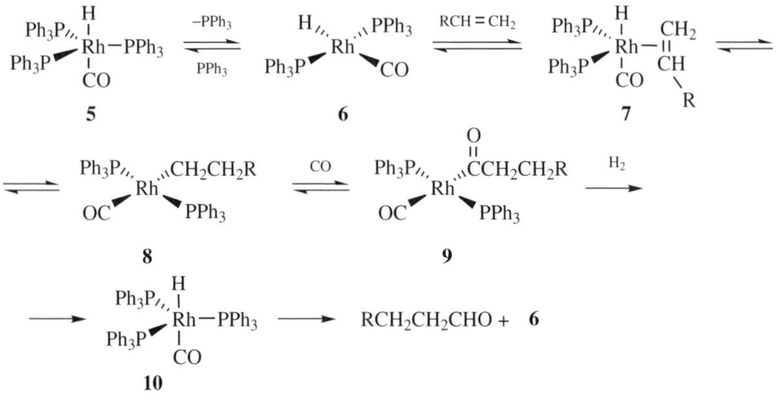

Scheme 7.1

catalyze hydroformylation of relatively less reactive olefins (2-methyl-1-hexene, limonene, cyclohexene) under mild conditions (90°C, 10 atm).[44] Styrene possessing electron-deficient carbons is the most reactive alkene.[43]

Of the isomeric aldehydes indicated in Eq. (7.1), the linear aldehyde corresponding to anti-Markovnikov addition is always the main product. The isomeric branched aldehyde may arise from an alternative alkene insertion step to produce the $[RCH(Me)Co(CO)_3]$ or $[RCH(Me)Rh(CO)(PPh_3)_2]$ complexes, which are isomeric to **2** and **8**, respectively. Alternatively, hydroformylation of isomerized internal alkenes also give branched aldehydes. The ratio of the linear and branched aldehydes, called *linearity*, may be affected by reaction conditions, and it strongly depends on the catalyst used. Unmodified cobalt and rhodium carbonyls yield about 3–5 : 1 mixtures of the normal and iso products.

The selectivity of the commercially more valuable normal aldehydes can be increased by ligand modification. The linearity increases up to 7 in the presence of alkylphosphine-modified cobalt complexes. This increase is due mainly to the steric effect of the bulky phosphine ligand in the metal coordination sphere, which results in the preferred formation of the less congested *n*-alkylcobalt complex. Electronic effects of the ligands also contribute to regioselectivity. The partial charge of the metal affected by the ligands stabilizes the charge developing on the olefinic carbon during the insertion step thus affecting regioselectivity.

Even higher linearities are characteristic of modified rhodium catalysts. The effect is most pronounced when hydroformylation is carried out in molten triphenylphosphine,[41] resulting in linearities as high as 9. The active catalytic species with two phosphine ligands (**6**) ensures the increased selectivity through steric effects. Platinum complexes with bidentate ligands exhibit the highest linearity to form linear aldehydes with 99% selectivity.[27]

Steric effects in the alkene structure also affect linearity. As a result, quaternary carbon atoms are rarely formed in hydroformylation.[45] In contrast, electronic effects in hydroformylation of arylalkenes often result in the predominant formation of the branched aldehyde.[6,40,43,46–48] Styrene has a marked tendency to form 2-phenylpropanal when hydroformylated in the presence of rhodium catalysts. Rhodium complexes modified by biphosphine[49] or mixed amino phosphine oxide ligands[50] were shown to give the branched aldehyde with high reactivity and selectivity (iso : normal ratios ≤61.5).

A number of side-reactions may be observed during hydroformylation. Double-bond migration[51] resulting in the formation of more than two aldehydes may be significant:

$$EtCH=CHMe \xrightarrow[\substack{80\ atm\ H_2,\ 93\ atm\ CO \\ benzene,\ 100°C}]{[Co(CO)_4]_2} C_5H_{11}CHO \ + \ \underset{\underset{Et}{|}}{PrCHCHO} \ + \ \underset{\underset{Et}{|}}{EtCHCHO} \qquad (7.6)$$

$$76 \qquad\qquad 19 \qquad\qquad 5$$

Hydroformylation of 2-pentene, for instance, yields predominantly hexanal.[52] In some cases, isomeric alkenes form identical product mixtures.

In the hydroformylation of optically active alkenes, all aldehydes were shown to be formed with predominant retention of configuration.[53,54] This was interpreted to prove, in accordance with the transformation of deuterium labeled alkenes,[55,56] that stereoselective hydrogen shifts of the coordinated olefins occur without dissociation of the isomeric alkenes.[30]

Hydrogenation of the starting alkene and the product aldehydes may take place during hydroformylation. Phosphine-modified cobalt complexes are the most active in these hydrogenations. This activity is attributed to the increased hydridic character of hydrogen in $[CoH(CO)_3(PBu_3)]$ compared to that in $[CoH(CO)_4]$.[14] The increase in hydridic character enhances migratory insertion via hydride transfer in the intermediate complexes.[13,14] The lower CO partial pressure employed in the case of modified cobalt complexes also leads to increased alcohol formation, since the increased $H_2 : CO$ ratio prefers hydrogenation. Other secondary reactions involve aldolization, acetal formation, and the Tishchenko reaction.

The few data available on the stereochemistry indicate predominant *syn* attachment of hydrogen and the formyl group:[57]

$$\text{(7.7)}$$

85%   1.5%

This stereochemistry results from *syn* metal hydride addition and CO insertion with retention of configuration; thus, the net *syn* addition is not the result of double *anti* addition.[57]

Asymmetric hydroformylation, a process with high potential in the synthesis of biologically active molecules, allows the transformation of prochiral alkenes to chiral aldehydes.[58] Generally, however, the severe reaction conditions, the multistep nature of transformation, and the racemization of the product chiral aldehyde through enolization, do not enable such high optical yields as in other asymmetric syntheses. Since branched aldehyde formation from styrene is prevalent, many studies were carried out with this substrate.[16,46,47,58] Chelating chiral diphosphine ligands give the best results,[59,60] but even with these, enantiomeric excesses could not exceed 50% with rhodium catalysts.[58] Exceptionally high (70–80%) enantiomeric excesses were, however, achieved with platinum complexes with diphosphine-substituted chiral pyrrolidine ligands[61,62] and benzophosphindole ligands[63] in the presence of $SnCl_2$. Even these reactions suffer from the disadvantage of limited scope, rather slow rates, low iso : normal ratio, competitive hydrogenation, and racemization. A more recent report, however, described highly enantioselective hydroformylation of styrene [94% ee (enantiomeric excess)], *p*-methylstyrene (95% ee) and 1-hexene (75% ee) in the presence of phosphinephosphite–Rh(I) complexes derived from 1,1'-binaphthalene-2,2'-diols.[64] The same catalyst promoted asymmetric hydroformylation of internal alkenes[65] with enantiomeric excesses up to 97%.

Asymmetric hydroformylation of the three isomeric straight-chain butenes provided evidence that asymmetric induction in both rhodium- and platinum-catalyzed reactions occurs during and before the formation of the alkylmetal intermediate.[66,67]

7.1.2. Dienes and Alkynes

As a result of extensive isomerization, nonconjugated cyclic dienes give complex product mixtures[11] when subjected to hydroformylation in the presence of [Co(CO)$_4$]$_2$. In contrast, hydroformylation at the less-hindered double bond usually takes place in the presence of rhodium catalysts to give unsaturated aldehydes[68,69] [Eq. (7.8)[44]]; unsubstituted α,ω-dienes, in turn, form dialdehydes:[70]

$$\text{} + \text{CO} + \text{H}_2 \quad \xrightarrow[\substack{\text{benzene, 80°C, 10 atm} \\ \text{H}_2 : \text{CO} = 2 : 1}]{[\text{Rh(COD)OAc}], \text{P(OC}_6\text{H}_4o\text{-Ph)}} \quad \text{CHO} \qquad (7.8)$$

Since conjugated dienes form stable π-allyl complexes with [Co(CO)$_4$]$_2$, they undergo hydroformylation very slowly to give saturated monoaldehydes in low yields.[8] Mixtures of mono- and dialdehydes are usually formed in rhodium-catalyzed hydroformylations.[71,72] Saturated monoaldehydes were isolated, however, when 1,3-butadiene and 1,3-pentadiene were hydroformylated in the presence of rhodium dioxide.[70]

Few studies have been conducted on the hydroformylation of alkynes. Terminal acetylenes give saturated isomeric aldehydes supposedly through the initially formed alkenes.[73] Conclusive evidence was obtained that internal alkynes, in contrast, first undergo hydroformylation to yield α,β-unsaturated aldehydes, which are subsequently hydrogenated to give the saturated aldehyde end products.[74]

7.1.3. Synthesis of Aldehydes and Alcohols by the Oxo Reaction

Aldehydes in the range C$_3$–C$_{18}$ are produced by commercial oxo processes.[4,11,75–82] Large quantities of linear "oxo" alcohols are produced and used in manufacture of plasticizers, detergents, and lubricants.[77,78] These oxo alcohols are produced by the hydrogenation of aldehydes formed in hydroformylation. They also include 1,3-diols and alcohols prepared by the base-catalyzed aldol condensation of aldehydes followed by hydrogenation or dehydration and hydrogenation, respectively. The most important of these compounds is butyraldehyde. It is a starting material in the manufacture of 2-ethylhexanol through aldol condensation followed by hydrogenation. 2-Ethylhexanol, in turn, is used primarily in the production of plasticizers. Butyraldehyde may also be transformed to 1-butanol, which is used as solvent and esterification agent.

In the industrial synthesis of butyraldehyde, propylene and synthesis gas react at high temperature and pressure in the presence of a homogeneous catalyst, in the liquid phase.[11,79] Three traditional processes use nonmodified cobalt catalyst. In the original Rurhchemie process[75,83] equimolar amounts of CO and H_2 with $[Co(CO)_4]_2$ as catalyst are used, whereas the Kuhlmann[84] and the BASF process[85,86] employ preformed $[CoH(CO)_4]$. Operating at high temperature and pressure (130–190°C, 200–300 atm) these catalysts exhibit low linearity, producing n-aldehydes with selectivities of about 70–75%.

Shell developed a process utilizing the more stable but less active ligand-modified $[CoH(CO)_3PBu_3]$ complex.[11,75,79,82] The strong hydrogenation ability associated with the lower operating pressures (50–100 atm) and increased linearity result in relatively high selectivity for linear alcohols. Since this process is operated to produce alcohols, excess hydrogen is used (the $H_2 : CO$ ratio is $2 : 1$). Hydrogenation of the starting alkenes, however, may not be significant. Because of the sensitivity of the catalyst to oxygen and sulfur poisoning, careful purification of the reactants is necessary. The Shell hydroformylation process is an integral part of the SHOP (Shell higher olefin process) unit to manufacture higher linear oxo alcohols.[87] Linear terminal alkenes produced in ethylene oligomerization undergo isomerization and disproportionation to yield detergent range internal alkenes used as feedstock in hydroformylation. Since the catalyst system rapidly isomerizes internal alkenes into terminal alkenes and exhibits strong hydrogenation ability, linear detergent alcohols are produced.[88]

2-Ethylhexanol is usually produced by subsequent aldolization of butyraldehyde produced in the oxo reaction followed by hydrogenation of the intermediate unsaturated aldehyde.[89] In Esso's Aldox process, however, in situ aldol condensation is effected by suitable promoters.[11] Magnesium ethoxide and soluble zinc compounds are recommended to promote controlled aldolization during the oxo reaction. The Shell variant uses potassium hydroxide. Serious disadvantages (mixed aldolization with the branched aldehyde, problems associated with recycling of the additives), however, prevented wider use of the Aldox process.

Union Carbide invented the industrial use of highly active ligand-modified rhodium complexes.[90–93] $[RhH(CO)(PPh_3)_3]$, the most widely used catalyst, operates under mild reaction conditions (90–120°C, 10–50 atm). This process, therefore, is also called *low-pressure oxo process*. Important features of the rhodium-catalyzed hydroformylation are the high selectivity to n-aldehydes (about 92%) and the formation of very low amounts of alcohols and alkanes. Purification of the reactants, however, is necessary because of low catalyst concentrations.

The latest development in industrial alkene hydroformylation is the introduction by Rurhchemie of water-soluble sulfonated triphenylphosphine ligands.[94] Hydroformylation is carried out in an aqueous biphasic system in the presence of Rh(I) and the trisodium salt of tris(m-sulfophenyl)phosphine (TPPTN). High butyraldehyde selectivity (95%) and simple product separation make this process more economical than previous technologies.

Mitsubishi Kasei introduced a process to manufacture isononyl alcohol, an important PVC (polyvinyl chloride) plasticizer, via the hydroformylation of octenes (a mixture of isomers produced by dimerization of the C_4 cut of naphtha cracker or FCC processes).[95] First a nonmodified rhodium complex exhibiting high activity and selectivity in the formation of the branched aldehyde is used. After the oxo reaction, before separation of the catalyst, triphenylphosphine is added to the reaction mixture and the recovered rhodium–triphenylphosphine is oxidized under controlled conditions. The resulting rhodium–triphenylphosphine oxide with an activity and selectivity similar to those of the original complex, is recycled and used again to produce isononanal.

A key economic problem in all industrial oxo processes is the recovery of the homogeneous catalysts. It is important in the case of both the original, relatively unstable cobalt and the very expensive rhodium complexes. A number of special procedures were developed for catalyst recovery and recycling.[75,79]

7.2. CARBOXYLATION

Unsaturated hydrocarbons (alkenes, dienes) react with carbon monoxide and a proton source (H_2O, alcohols, amines, acids) under strong acidic conditions to form carboxylic acids or carboxylic acid derivatives. Since a carbocationic mechanism is operative, not only alkenes but also other compounds that can serve as the carbocation source (alcohols, saturated hydrocarbons) can be carboxylated. Metal catalysts can also effect the carboxylation of alkenes, dienes, alkynes, and alcohols.

7.2.1. Koch Reaction

Strong mineral acids under forcing reaction conditions catalyze the addition of carbon monoxide and water to alkenes to form carboxylic acids:[96,97]

$$R\diagup\!\!\!\!\diagdown\diagup + CO + H_2O \xrightarrow[\substack{100-350°C, \\ 500-1000\ atm}]{H^+} R\diagup\!\!\!\!\overset{\overset{\displaystyle Me}{|}}{\diagdown}COOH \qquad (7.9)$$

The reaction is also called *hydrocarboxylation*. According to a later modification, the alkene first reacts with carbon monoxide in the presence of the acid to form an acyl cation, which then is hydrolyzed with water to give the carboxylic acid.[97] The advantage of this two-step synthesis is that it requires only medium pressure (100 atm). Aqueous HF (85–95%) gave good results in the carboxylation of alkenes and cycloalkenes.[98] Phosphoric acid is also effective in the carboxylation of terminal alkenes and isobutylene, but it causes substantial oligomerization as well.[99,100] Neocarboxylic acids are manufactured industrially with this process (see Section 7.2.4). The addition may also be performed with formic acid as the source of CO (Koch–Haaf reaction).[101,102] The mechanism involves carbocation formation via protonation of the alkene[97,103] [Eq. (7.10)]. It then reacts with carbon monoxide to form an acyl cation (**11**) [Eq. (7.11)], which is subsequently quenched with water [Eq. (7.12)]:

$$R-CH=CH_2 \xrightarrow{H^+} R-\overset{+}{C}H-CH_3 \tag{7.10}$$

$$R-\overset{+}{C}H-CH_3 + CO \rightleftharpoons \underset{\textbf{11}}{R-\overset{\overset{\displaystyle CH_3}{|}}{C}H-CO^+} \tag{7.11}$$

$$\textbf{11} + H_2O \rightleftharpoons R-\overset{\overset{\displaystyle CH_3}{|}}{C}H-COOH_2{}^+ \overset{-H^+}{\rightleftharpoons} R-\overset{\overset{\displaystyle CH_3}{|}}{C}H-COOH \tag{7.12}$$

A further development includes the use of Cu(I) or Ag(I) catalysis in concentrated sulfuric acid,[104,105] H_2O-BF_3,[106,107] or BF_3 mixed with H_2SO_4, H_3PO_4, or organic acids,[108] allowing synthesis at atmospheric pressure. The advantageous effect of these ions is due to the formation of $Cu(CO)_3{}^+$ and $Ag(CO)_3{}^+$ ions ensuring an increased concentration of CO in the liquid phase. Since carbocations are involved extensive rearrangements occur resulting in the formation of tertiary (neo) carboxylic acids:[104]

$$\text{(cyclohexene)} + CO + H_2O \xrightarrow[\text{20–50°C, 1 atm, 1–2 h}]{\text{Cu}_2\text{O, conc. H}_2\text{SO}_4} \underset{63\%}{\text{(cyclopentane with Me and COOH)}} \tag{7.13}$$

Superacid systems, such as $HF-BF_3$ and CF_3SO_3H, allow further improvement of the Koch reaction.[109] Triflic acid proved to be far superior to 95% H_2SO_4 at atmospheric pressure.[110] This was attributed to the higher acidity of triflic acid and the higher solubility of CO in this acid. FSO_3H-SbF_5 combined with Cu(I) is also effective.[111]

Alcohols instead of olefins can also be used in these reactions. Although secondary and tertiary alcohols react preferentially in the presence of H_2SO_4,[112] $SbCl_5-SO_2$,[113] and acids with Cu(I) or Ag(I) ions,[105,107,108,114] even methanol was shown under superacidic conditions to give acetic acid.[115]

The preparation of acetic acid represents a special case. Olah and coworkers as well as Hogeveen and coworkers have demonstrated that CO can react with methane under superacidic conditions, giving the acetyl cation and by subsequent quenching acetic acid or its derivatives (see Section 7.2.3). Monosubstituted methanes, such as methyl alcohol (or dimethyl ether), can be carbonylated to acetic acid.[115] Similarly, methyl halides undergo acid-catalyzed carbonylation.[115,116] Whereas the acid-catalyzed reactions can be considered as analogs of the Koch reaction, an efficient Rh-catalyzed carbonylation of methyl alcohol in the presence of iodine (thus in situ forming methyl iodide) was developed by Monsanto and became the dominant industrial process (see Section 7.2.4).

7.2.2. Carboxylation Catalyzed by Transition Metals

The transformation of alkenes in the presence of Group VIII metal catalysts affords carboxylic acids or carboxylic acid derivatives depending on the protic reagent used:[10,14–16,23,117–123]

$$R\diagdown\diagup + CO + HX \xrightarrow{\text{metal catalyst}} R\diagup\diagdown\diagup COX + R\diagup\overset{Me}{\underset{}{\diagdown}}COX \qquad (7.14)$$

HX = H$_2$O, R'OH, R'$_2$NH,
R'COOH, HCl

A similar addition to alkynes results in the formation of the corresponding unsaturated acids and derivatives.[14,23,121–124] Cobalt, nickel, and iron carbonyls, as well as palladium complexes, are the most often used catalysts.[14]

The syntheses of carboxylic acids and esters are widely studied processes. Since the first examples of carboxylation in the presence of metal carbonyls were reported by Reppe, these reactions are sometimes referred to as the *Reppe reactions*. In his pioneering work[125–127] stoichiometric or catalytic amounts of [Ni(CO)$_4$] and ethylene or acetylene were reacted in the presence of water or alcohols to form saturated and unsaturated acids and esters. Commercial processes are practiced in the manufacture of propionic acid, acrylic acid and acrylates (see Section 7.2.4).

Alkenes and Dienes. Alkenes exhibit lower reactivity in metal-catalyzed carboxylation than in hydroformylation. The general reactivity pattern of different alkenes, however, is the same; terminal linear alkenes are the most reactive substrates. Cycloalkenes are the least reactive, but strained compounds may react under very mild conditions:[128]

$$(7.15)$$

72%

This transformation also indicates that carboxylation is a net *syn* process. The same was proved for the palladium-catalyzed reaction as well.[129]

[Ni(CO)$_4$] may be used in stoichiometric carboxylations under mild conditions. Hydrogen halides[118] or UV irradiation[130] promotes the addition that usually leads to branched carboxylic acids. Catalytic reactions, however, require more severe reaction conditions (250°C, 200 atm).[122] Cobalt carbonyls, in turn, are more effective catalysts, and usually give more linear product. Linear acids are also formed from 2-alkenes via double-bond migration.[122] The activity and selectivity may be substantially increased by adding pyridine or other nitrogen bases to the reaction mixture.[131]

Palladium complexes exhibit even higher catalytic activity and produce branched acids preferentially.[132,133] The selectivity, however, can be shifted to the formation of linear acids by increasing the phosphine concentration.[134] Temperature, catalyst concentration, and solvent may also affect the isomer ratio.[135] Marked increase in selectivity was achieved by the addition of Group IVB metal halides to palladium[136] and platinum complexes.[137] Linear acids may be prepared with selectivities up to 99% in this way. The formic acid–Pd(OAc)$_2$–1,4-bis(diphenylphosphino)butane system has been found to exhibit similar regioselectivities.[138] Significant enhancements of catalytic activity of palladium complexes in carbomethoxylation by use of perfluoroalkanesulfonic acid resin cocatalysts was reported.[139,140]

The formation of branched acids from 1- and 2-alkenes in a palladium-catalyzed carboxylation was found to be completely regioselective, when the reaction was carried out in the presence of CuCl$_2$, oxygen, and limited amounts of water and hydrochloric acid, under remarkably mild conditions:[141]

$$R-CH{=}CH-R' + CO + H_2O \xrightarrow[\text{O}_2,\ \text{RT, 1 atm, 4-18 h}]{\text{PdCl}_2,\ \text{CuCl}_2,\ \text{HCl}} \overset{\displaystyle Me}{\underset{\displaystyle \underset{30\text{--}100\%\ \text{yield}}{}}{R''{-}\overset{|}{C}H{-}COOH}} \tag{7.16}$$

$$R = isoPr,\ n\text{-}C_7H_{15},\ n\text{-}C_8H_{17},\ p\text{-}MeC_6H_4$$
$$R' = H,\ Me$$

Styrene characteristically yields the branched acid in the presence of palladium and monodentate phosphine ligands,[132,142] and in the [Fe(CO)$_5$]-promoted process.[143] Palladium with certain bidentate phosphines, in turn, produces more linear acid.[142] Asymmetric hydrocarboxylations with palladium complexes and chiral ligands with enantiomeric excesses up to 84% have been reported.[144,145]

Carboxylation of dienes and trienes, which takes place in a stepwise fashion, affords mono- or dicarboxylated products.[146] Cobalt carbonyl,[147] palladium chloride,[148,149] and palladium complexes[150] were used. 1,4 Addition to 1,3-butadiene gives the corresponding unsaturated *trans* ester (methyl *trans*-3-pentenoate) in the presence of [Co(CO)$_4$]$_2$ and a pyridine base.[147] The second carboxylation step requires higher temperature than the first one to yield dimethyl adipate. In a direct synthesis (110°C, 500 atm, then 200°C, 530 atm) 51% selectivity was achieved.[147]

Concerning the mechanism of metal-catalyzed carboxylations, two basic pathways—the metal hydride and the metal carboxylate mechanisms—gained general acceptance.[14,16,118,119,122,151] The actual mechanism depends on the catalyst used and the reaction conditions particularly in the presence of additives (acids, bases). It seems reasonable to assume that both mechanisms may be operative under appropriate conditions.

The metal hydride mechanism was first described for the cobalt-carbonyl-catalyzed ester formation by analogy with hydroformylation.[152] It was later adapted to carboxylation processes catalyzed by palladium[136,153,154] and platinum complexes.[137] As in the hydroformylation mechanism, the olefin inserts itself into the metal–hydrogen bond, and only the final step of the catalytic cycle is different in that the acylmetal complex is cleaved by reacting with the protic reagent:

$$[RCH_2CH_2\underset{\underset{O}{\|}}{C}Co(CO)_4] \; + \; R'OH \;\longrightarrow\; RCH_2CH_2COOR' \; + \; [CoH(CO)_4] \qquad (7.17)$$

4

The beneficial effect of hydrogen halides in nickel-catalyzed carboxylations was also explained as forming a hydridonickel complex as the true catalyst.[155]

Despite less direct evidence, the metal carboxylate mechanism has also been widely proposed:[134,156–159]

$$M-C\overset{\displaystyle /\!/ O}{\underset{\displaystyle OR'}{\diagdown}} \; + \; RCH=CH_2 \;\longrightarrow\; M-\overset{R}{\overset{|}{C}}HCH_2COOR' \qquad (7.18)$$

According to this mechanism olefin insertion takes place into the initially formed carboxylate complex.

Alkynes. Alkynes tend to undergo a number of transformations when reacted under carboxylation conditions. Varied ring closure products (hydroquinones, ketones, unsaturated lactones) may be formed.[124,160] Under appropriate reaction conditions, however, clear carboxylations to give unsaturated acids or esters can be performed.

Acetylene forms acrylic acid or esters in excellent yield in the presence of $[Ni(CO)_4]$.[126] The reaction can be catalytic or may require the use of a stoichiometric amount of $[Ni(CO)_4]$ under mild reaction conditions.

Substituted acetylenes react more slowly than does acetylene. The Markovnikov rule is usually followed, that is, terminal acetylenes form 2-substituted acrylic acids[122,124,161] [Eq. (7.19)[162]]:

$$C_5H_{11}-\!\!\equiv\!\!-H \xrightarrow[\substack{isobutyl\ methyl\\ketone}]{CO,\ MeOH} \underset{H_2C}{\overset{C_5H_{11}}{\diagup}}\!\!\diagdown COOMe \; + \; \underset{H_{11}C_5}{\diagup}\!\!\diagdown^{COOMe} \qquad (7.19)$$

$[PdCl_2(PPh_3)_3]$, 80°C, 240 atm CO	81	19
$\{PdCl_2[P(C_6H_4\ p\text{-MeO})_3]_2\}$–$SnCl_2$	3	97
22°C, 1 atm, CO		

In contrast, when palladium is combined with $SnCl_2$, linear products are formed with high selectivity.[162] The addition of hydrogen and the carboxylic group usually takes place in a *syn* manner.[124,162]

Acetylene and substituted acetylenes may form dicarboxylated products in the presence of $[Co(CO)_4]_2$. The product distribution is highly dependent on the solvent, with the highest succinic acid selectivities observed in tetrahydrofuran.[124]

Alcohols. Similarly to acids metals can also catalyze the transformation of alcohols to produce carboxylic acids. Cobalt, rhodium, and iridium were studied most[119,122,163–165] because of their practical importance in the transformation of methyl alcohol to acetic acid (see Section 7.2.4). Iodine compounds were shown to play an important role in this reaction by transforming methyl alcohol to the more reactive methyl iodide. $[Rh(CO)_2I_2]^-$ proved to be the active species in the rhodium-catalyzed reaction.[163–165]

When methanol is reacted with CO in the presence of oxygen and copper salts, oxidative carbonylation takes place, resulting in the formation of dimethyl carbonate.[166] Since dimethyl carbonate is a good methylating and carbonylating agent, it can substitute the highly toxic dimethyl sulfate and phosgene in industrial applications. The process was commercialized by EniChim (see Section 7.2.4).

7.2.3. Carboxylation of Saturated Hydrocarbons

It follows from the carbocationic mechanism discussed before that saturated hydrocarbons can also be carboxylated provided appropriate reaction conditions are applied to generate carbocations. In early studies C_3–C_6 alkanes were shown to react with carbon monoxide in the presence of HCl and stoichiometric amounts of $AlCl_3$ at 20–150°C in the pressure range of 100–150 atm.[167–169] HF–SbF_5 was recently found to be very effective at 0–30°C under atmospheric CO pressure.[170] Branched alkanes with six or fewer carbon atoms react readily to preferentially form secondary (α-branched) carboxylic acids. Exclusive β scission results in the formation of acids of lower molecular weight when C_7 and higher alkanes are treated under identical reaction conditions. Methylcycloalkanes, in turn, yield tertiary carboxylic acids as well as ketones.[171] The carboxylation can be facilitated by carrying out the reaction in the presence of alkenes or alcohols. Since these additives form carbocations more readily, subsequent hydride transfer from the alkane yields the corresponding cations. The usual acidic reaction conditions [H_2SO_4,[172,173] anhydrous HF,[174] acids in the presence of Cu(I) or Ag(I) ions[105,107,175]] can be applied.

Even methane, the least reactive alkane, was shown to undergo carboxylation under superacidic conditions.[115,176] The formation of carboxylated products (acetic acid and methyl acetate) from methane was first observed by Hogeveen and coworkers by trapping methyl cation formed in SbF_5 (60°C, 50 atm CO pressure) followed by quenching with H_2O or MeOH. The intermediate methylcarboxonium ion (CH_3CO^+) and $CH_3CH_2CO^+$ formed in a similar reaction of ethane were identified by NMR spectroscopy.[176,177]

A different approach, the direct conversion of methane to acetic acid, has been revealed.[178,179] The process is an oxidative carboxylation in the presence of $RhCl_3$ in an aqueous medium under mild conditions (100°C).

7.2.4. Practical Applications

Neocarboxylic Acids. The acid-catalyzed carboxylation of alkenes with carbon monoxide in the presence of water is used commercially to produce neocarboxylic acids.[82] Pivalic acid from isobutylene is the most important product. A number of technologies were developed[82,180,181] applying mixed Brønsted and Lewis acid catalysts (H_2SO_4, H_3PO_4, HF and SbF_5, BF_3) under varying experimental conditions (20–80°C, 20–100 atm) to achieve high selectivities (80–100%).[82] Because of the highly corrosive nature of these catalyst systems and the environmental hazards associated with the use of HF, as well as the complicated workup of the reaction mixture (hydrolysis and costly separation), solid catalysts have been developed by BASF.[182] Pentasil zeolites doped with transition metals (copper and cobalt) were tested and shown to be active catalysts in the production of pivalic acid, although under high pressure (300–450 atm).

Hydrocarboxymethylation of Long-Chain Alkenes. An industrial process to carry out hydrocarboxymethylation of olefins to produce methyl esters particularly in the C_{12}–C_{14} range for use as a surfactant feedstock was developed by Huels.[183] A promoted cobalt catalyst in the form of fatty acid salts (preferably those formed in the reaction) is used. With high promoter : catalyst ratio (5 : 1–15 : 1) at 180–190°C and pressure of 150–200 atm, the rate of alkene isomerization (double-bond migration) exceeds the rate of hydrocarboxymethylation. As a result, even internal olefins give linear products (the yield of normal products is about 75% at 50–80 % conversion). Secondary transformations of aldehydes (product of olefin hydroformylation) lead to byproducts (ethers and esters) in small amounts.

Propionic Acid. The Reppe reaction is used to transform ethylene, carbon monoxide, and water to propionic acid.[82] [$Ni(CO)_4$] is formed in situ from nickel salts under reaction conditions (270–320°C, 200–240 atm). The addition of halogens and phosphine ligands allows milder reaction conditions (Halcon process, 170–225°C, 10–35 atm). Propionic acid yields are around 95%.

Acrylic Acid and Acrylates. Acrylic acid and acrylates may be produced commercially by the Reppe reaction of acetylene.[76,184–187] However, the industrial significance of these processes has diminished since acetylene is no longer a viable source and was replaced by ethylene. Acrylic acid and acrylates are now produced by propylene oxidation (see Section 9.5.2).

The industrial catalytic Reppe process is usually applied in the production of acrylic acid. The catalyst is $NiBr_2$ promoted by copper halides used under forcing conditions. The BASF process, for example, is operated at 225°C and 100 atm in tetrahydrofuran solvent.[188] Careful control of reaction conditions is required to avoid the formation of propionic acid, the main byproduct, which is difficult to separate. Small amounts of acetaldehyde are also formed. Acrylates can be produced by the stoichiometric process [Eq. (7.20)], which is run under milder conditions (30–50°C, 1–7 atm). The byproduct $NiCl_2$ is recycled:

$$4\,HC{\equiv}CH \;+\; 4\,ROH \;+\; [Ni(CO)_4] \;+\; 2\,HCl \;\longrightarrow\; 4\,CH_2{=}CHCOOR \;+\; H_2 \;+\; NiCl_2$$

$$(7.20)$$

Acetic Acid. Carbonylation of methanol is the most important reaction in the production of acetic acid.[189–192] BASF developed a process applying CoI_2 in the liquid phase under extreme reaction conditions (250°C, 650 atm).[122,193] The Monsanto low-pressure process, in contrast, uses a more active catalyst combining a rhodium compound, a phosphine, and an iodine compound (in the form of HI, MeI, or I_2).[122,194–196] Methanol diluted with water to suppress the formation of methyl acetate is reacted under mild conditions (150–200°C, 33–65 atm) to produce acetic acid with 99% selectivity at 100% conversion.

Dimethyl Carbonate. An industrial process to manufacture dimethyl carbonate through the oxidative carbonylation of methanol catalyzed by cuprous chloride has been developed and commercialized by EniChem.[197,198] The reaction occurs in two steps. Cupric methoxy chloride is formed in the first oxidation step [Eq. (7.21)], which is then reduced to yield dimethyl carbonate and regenerate cuprous chloride [Eq. (7.22)]:

$$2\,MeOH \;+\; 0.5\,O_2 \;+\; 2\,CuCl \;\longrightarrow\; 2\,Cu(OMe)Cl \;+\; H_2O \tag{7.21}$$

$$2\,Cu(OMe)Cl \;+\; CO \;\longrightarrow\; (MeO)_2CO \;+\; 2\,CuCl \tag{7.22}$$

A continuous slurry process operated at moderate temperature and pressure (below 100°C, 10–20 atm) gives dimethyl carbonate with better than 98% selectivity.

7.3. AMINOMETHYLATION

Formation of amines from alkenes, carbon monoxide, water, and a nitrogen base was discovered by Reppe who used [Fe(CO)$_5$] to catalyze the process to form amines in low yields:[125,199]

$$R\diagup\!\!\!\diagdown \;+\; CO \;+\; H_2O \;+\; H-N\!\!<\; \xrightarrow{[Fe(CO)_5]}\; R\diagup\!\!\diagdown\!\!\diagup\!\!\diagdown N\!\!< \tag{7.23}$$

Rhodium oxide,[200] cobalt carbonyl,[201] rhodium and ruthenium carbonyls,[202] and rhodium compounds[203,204] were later found to be effective catalysts. A three-step mechanism with hydroformylation of the alkene to yield an aldehyde in the first step can be written [Eq. (7.24)]. Condensation to form an imine [Eq. (7.25)] (or enamine) and hydrogenation of this intermediate leads to the product amine:

$$R\diagup\!\!\!\diagdown \;+\; 2\,CO \;+\; H_2O \;\xrightarrow[-CO_2]{catalyst}\; RCH_2CH_2CHO \tag{7.24}$$

$$RCH_2CH_2CHO \;+\; H-N\!\!<^{H} \;\xrightarrow[-H_2O]{}\; RCH_2CH_2CH{=}N- \;\xrightarrow{H_2}\; RCH_2CH_2CH_2N\!\!<^{H}$$

$$\tag{7.25}$$

The hydrogen necessary in steps 1 and 3 is produced by the water–gas shift reaction. Hydrogen instead of water may also be used with similar results.[205,206]

When ammonia or primary amines are used, the product amines may participate in further aminomethylation, resulting in the formation of a mixture of amines. Other byproducts (aldehydes, alcohols, carboxamides) may also be formed. The reaction to produce tertiary amines from secondary amines, however, is fairly selective. Aminomethylation of ethylene with piperidine was reported to form N-propylpiperidine with 75% yield when the reaction was carried out in the presence of [Fe(CO)$_5$] and water without an external source of CO (170°C, 14 h).[207]

7.4. RECENT DEVELOPMENTS

7.4.1. Hydroformylation

Because of its high industrial importance, hydroformylation has remained a widely studied area with numerous new review papers covering various aspects of the topic,[208–217] including the annual surveys of the application of transition metals.[218] A chapter written in 1996 covers hydroformylation catalyzed by organometallic complexes in detail,[219] whereas a review written 5 years later gives a summary of the advances on hydroformylation with respect to synthetic applications.[220] A selection of papers in a special journal issue has been devoted to carbonylation reactions.[221] A major area of the research has been the development of fluorous biphasic catalysis and the design of new catalysts for aqueous/organic biphasic catalysis to achieve high activity and regioselectivity of linear or branched aldehyde formation.

The crucial problem associated with the use of homogeneous rhodium catalysts in industrial hydroformylation is catalyst recovery. Because of the high cost of rhodium, it is necessary to recover rhodium at the ppm level to ensure economical operation. A highly successful solution to this problem was the development and application of the aqueous biphasic catalysis concept.

The prototype industrial process based on this concept is the Ruhrchemie–Rhône Poulenc process for the hydroformylation of propylene to butanal[94,219,220] (see Section 7.3.1). Because of the use of appropriately modified water-soluble ligands, the catalyst resides and operates in the aqueous phase. The particular features of this process are the positive energy balance and easy catalyst recovery, namely, the simply circulation of the aqueous catalyst solution. New types of water-soluble Ir and Rh complexes with tris(hydroxymethyl)phosphine[222] were described, and the biphasic hydroformylation of 1-hexene was accomplished in ionic liquids.[223] A cationic sugar-substituted Rh complex displays high regioselectivity to branched aldehydes.[224]

The catalyst recovery problem has been approached in various other ways. The application of heterogeneous catalysts offers a possible solution.[225] Heterogeneous catalysts, such as highly dispersed Rh on SiO$_2$[226] and Co on SiO$_2$,[227] show high activity in the gas-phase hydroformylation of ethylene.

Heterogenized metal complexes such as [$Rh(CO)_2Cl_2$] anchored to polyamidoamine dendritic phosphine ligands on beads[228] or on silica[229] have been found to exhibit promising characteristics. The first catalyst prepared by the sol-gel technique having a diphosphane ligand and anchored [$Rh(CO)_2(acac)$] (acac = acetylacetone) is remarkably stable under catalytic conditions and shows high selectivity in the hydroformylation of 1-octene (95.5% linear aldehyde).[230] [$Rh(\mu\text{-}Cl)(COD)$]$_2$ supported on functionalized and activated carbon is stable and active after recycling.[231]

Organic–inorganic hybrid catalysts may offer the possibility to avoid the major drawbacks of the anchored complexes, namely, leaching of the metal and the limited long-term stability of organic matrices. New catalysts prepared by the cocondensation of the Rh(I) complex $HRh(CO)[Ph_2P(CH_2)_xSi(OMe)_3]_3$ and alkoxysilanes show high selectivities, high turnover numbers, and the ability to hydroformylate higher alkenes than does 1-hexene.[232]

An interesting new concept of catalyst immobilization is the use of supported aqueous phase catalysts. Here, the catalyst is immobilized in a thin water layer adhered within the pores of a high-surface-area porous support.[233–235] A new Rh catalyst of this class with ligand **11** is stable, recyclable, and highly selective in the hydroformylation of higher alkenes to linear aldehydes.[236]

11 R = SO_3Na, R^1, R^2 = Ph

12 R = *tert*-Bu,
   R^1R^2P = dibenzophospholyl
       phenoxaphosphanyl

Another major feature of progress in hydroformylation reaction is the design of new modifying ligands to ensure high activity, stability, and selectivity. The most innovative example is the use of fluorous phosphorous ligands in a fluorous multiphase system. The advantages of this new concept were demonstrated in the hydroformylation of 1-decene carried out in perfluoromethylcyclohexane and toluene in the presence of a catalyst prepared in situ form [$Rh(CO)_2(acac)$] and the ligand $P[OCH_2CH_2(CF_2)_5CF_3]$.[237] At the reaction temperature (100°C) the system is homogeneous, but on cooling it separates into two phases with the catalyst in the fluorous phase.[237,238] The process is advantageous for the hydroformylation of higher alkenes and offers clean separation and recycling of the catalyst. Catalysts prepared by using fluorous polymer ligands showed similar characteristics.[239]

Because of their low solubilities in the aqueous phase, the hydroformylation of higher alkenes ($>C_7$) is still a challenging problem. In addition to fluorous biphasic catalysis, possible solutions, which have been addressed, include the addition of surfactants[240,241] or the use of amphiphilic ligands[242–244] to enhance mutual solubility or mobility of the components across the phase boundary and thereby increase the rate of reaction. The use of polar solvents such as alcohols,[245] β-cyclodextrin,[246] cyclodextrin ligands,[247,248] thermoregulated phase-transfer ligands,[249,250] and new water-soluble polymeric Co^{251} and Rh catalysts[252,253] may provide similar improvements. Promising results were observed in supercritical $CO_2^{254-256}$ or in ionic liquids.[257,258] Hydroformylation of 2-vinylnaphthalene in subcritical CO_2^{259} as well as the use of CO_2 as reagent instead of CO in Ru-catalyzed hydroformylation[260] have also been described.

A particularly interesting problem is to develop catalysts that exhibit satisfactory activity in the hydroformylation of internal alkenes and produce linear aldehydes preferentially. The transformation of a technical mixture of octenes containing all isomers into octanal is of practical importance:

$$(7.26)$$

The introduction of additional oxy groups into monodentate phosphorous ligands ensures high activities in both isomerization and the consecutive hydroformylation step.[261] Rh complexes with diphosphine ligands of large bite angles such as **12** show unprecedented activities and selectivities.[262] In the hydroformylation of *trans*-4-octene selectivity to octanal exceeds 80%, although the process requires three consecutive double-bond migrations. A dual catalytic system composed of Ru for isomerization and Rh for consecutive hydroformylation has been developed.[263]

Attention has focused on trying to find consistence structure–activity relationships in hydroformylation, especially for the new Rh diphosphine complexes.[264] It was earlier observed that the bite angle of bidentate diphosphines has a dramatic influence on regioselectivity; for the bis-equatorially coordinated BISBI ligand (**13**), a linear : branched aldehyde ratio as high as 66 : 1 was reported.[265] An increase in selectivity for linear aldehyde formation and activity with increasing bite angle was observed by three systematic studies.[266-268] The bite angle concept has also been extended to aqueous biphasic catalysis.[269]

Further novel observations are the hydroformylation of ethylene over graphite nanofiber-supported Rh catalysts,[270] the transformation of a mixture of isomeric octenes to C_9-aldehydes,[271] and the preparation of linear long-chain dialdehydes by the hydroformylation of linear α,ω-dienes.[272]

Alkynes. New catalysts have been reported to effectively catalyze hydroformylation of alkynes. [Rh(CO)$_2$(acac)] and a special biphosphate ligand enables the hydroformylation of internal akynes to occur under mild conditions and with good selectivity.[273] Selective hydroformylation was reported to take place with [PdCl$_2$(PCy$_3$)$_3$], which also showed a remarkable synergistic effect with Co.[274] Regioselective hydroformylation of enynes can be carried out by the zwitterionic rhodium complex [(η^6-C$_6$H$_5$BPh$_3$)$^-$Rh$^+$(1,5-COD)] and triphenyl phosphite. This catalyst system affords formyl dienes in high stereoselectivity under mild conditions.[275]

Asymmetric Hydroformylation. Asymmetric hydroformylation was not considered the method of choice in the synthesis of chiral aldehydes. Platinum–tin complexes with chiral diphosphines were shown to be more enantioselective than the classical Rh systems, but they also induce hydrogenation and racemization, and the linear : branched ratio is usually low.[209] Classical Rh complexes, in turn, display much better regioselectivities but lower eneantioselectivities.[209] However, significant improvements have been achieved since 1997 by the development of new types of chiral bidentate ligands. Rh catalysts based on the use of asymmetric diphosphite[276] or phosphine–phosphite ligands[277] ensure higher selectivities. (*R*,*S*)-BINAPHOS (**14**) may be used for a wide range of substrates allowing outstanding levels of enantiocontrol but unsatisfactory regioselectivities. Similar characteristics were observed with the polymer-supported catalyst in both liquid[278] and gas phases.[279]

14

A modified **14** with perfluoralkylphenyl groups increased the selectivity up to 96%.[280] Both high enantioselectivity (91%) and high regioselectivity (98.8%) were achieved under mild conditions with a new chiral diphosphite ligand derived from D-(+)-glucose.[281]

7.4.2. Formylation of Alkanes

Formylation of isobutane with carbon monoxide in the presence of an excess of AlCl$_3$ was first reported by Nenitzescu to yield, among others, methyl isopropyl ketone (31%).[168] A new highly efficient superelectrophilic formylation–rearrangement of isoalkanes by Olah and coworkers has been described.[282] Selective formation of branched ketone in high yield with no detectable branched acids, that is, the Koch products, was achieved. A particularly suitable acid is HF–BF$_3$, which transforms, for example, isobutane to methyl isopropyl ketone in 91% yield. The transformation is suggested to occur with the involvement of the protosolvated formyl cation **15**:

$$CO + H^+ \rightleftharpoons HC\equiv\overset{+}{O} \rightleftharpoons H-\overset{+}{C}=\overset{+}{O}-H \qquad (7.27)$$
$$\textbf{15}$$

This is sufficiently reactive to attack the C—H bond forming the five-coordinate carbocation **16** to give protonated pivalaldehyde (Scheme 7.2). The latter undergoes the aldehyde–ketone rearrangement to yield the corresponding branched ketone.

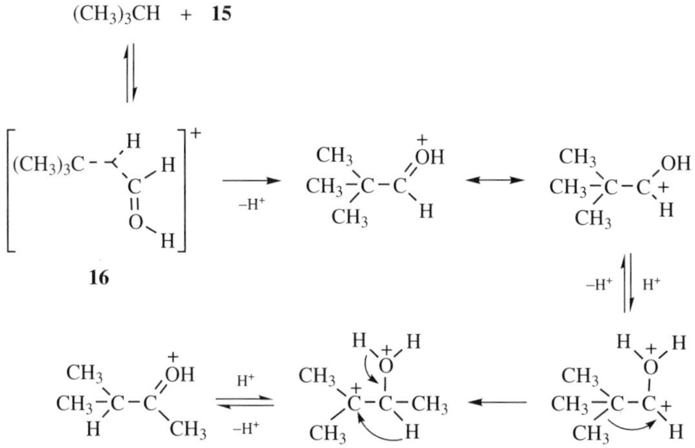

Scheme 7.2

Benzaldehyde as the product of carbonylation was detected by ^{13}C MAS NMR when benzene and CO were adsorbed on sulfated zirconia[283] or HY zeolite promoted with AlCl$_3$.[284] Formylation of hexane, cyclohexane, and cyclooctane could also be achieved in a radical process applying benzophenone and polyoxotungstate under the photocatalytic conditions.[285] Photochemical carbonylation of alkanes catalyzed by [Rh(PMe$_3$)$_2$(CO)Cl] gives linear aldehydes with high photoefficiency and selectivity (98%).[286]

7.4.3. Carboxylation

Alkenes and Alkynes. A series of metal carbonyl cations, such as [Au(CO)$_n$]$^+$,[287,288] [c-Pd(μ-CO)$_2$]$^{2+}$,[289] [Rh(CO)$_4$]$^+$,[290] and [(Pt(CO)$_3$)$_2$]$^{2+}$,[291] was found to induce the formation of carboxylic acids from alkenes and CO in the presence of H$_2$SO$_4$ under mild conditions. A novel water-soluble Pd catalyst[292] and Pd complexes of calixarene-based phosphine ligands[293] showed high activity in the regioselective carboxylation of vinyl arenes to yield 2-arylpropionic acids or esters in the presence of an acid. New observations made by multinuclear NMR spectroscopy and ^{13}C labeling provided strong evidence to support the metal hydride pathway in the Pd-catalyzed methoxycarbonylation of ethylene.[294] According to a report applying a molybdenum catalyst promoted by a halide in the carboxylation of ethylene, either propionic acid (in the presence of water) or propionic anhydride (in the presence of propionic acid) can be produced under mild conditions (30–70 atm, 150–200 °C).[295] A simple method using formic acid as a CO source and IrI_4 as the catalyst is very effective in the carboxylation of 1-hexene and cycloalkenes.[296a] Multigram-scale photochemical methoxycarbonylation of mono- and disubstituted alkenes using $Co(acac)_2$ as catalyst precursors could be performed to give the corresponding methyl esters in good yield (69–85%).[296b] A tandem isomerization–carboxylation induced by highly active adamantyl cage diphosphine Pd(II) complexes allows the transformation of internal alkenes to linear esters.[297]

Formation of carboxylic acids from ethylene, isobutylene, and 1-octene was observed by in situ ^{13}C solid-state MAS NMR over H-ZSM-5 zeolite at 23–100°C.[298] A systematic study with various Pd complexes revealed that styrene is transformed into 2-phenylpropionic acid as the major product when monophosphine ligands were applied, whereas 3-phenylpropionic acid was obtained in the presence of diphosphines.[299]

The aqueous–organic two-phase system was successfully applied to perform hydrocarboxylation.[300] Palladium complexes of trisulfonated triphenylphosphine ligands were shown to exhibit high activity.[301–303] The application of cosolvents and modified cyclodextrins allow to eliminate solubility problems associated with the transformation of higher alkenes.[304]

Highly selective transformation of terminal acetylenes to either linear or branched carboxylic acids or esters may be achieved by appropriately selected catalyst systems. Branched esters are formed with high selectivity when the acetylenes are reacted with 1-butanol by the catalyst system Pd(dba)$_2$/PPh$_3$/TsOH (dba = dibenzylideneacetone)[305] or palladium complexes containing PPh$_3$.[306] Pd(acac)$_2$ in combination with various N- and O-containing phosphines and methanesulfonic acid is also an efficient catalyst for the alkoxycarbonylation of 1-alkynes to yield the branched product with almost complete selectivity.[307,308]

Catalytic systems to afford linear esters selectively are scant.[306,309] A report in 1995 disclosed that palladium complexes based on 1,1'-bis(diphenylphosphine)ferrocene showed excellent regioselectivity for the formation of linear α,β-unsaturated esters.[309] The results with phenylacetylene are remarkable because this compound is known to exhibit a distinct preference for the formation of the branched products on palladium-catalyzed carboxylations. Mechanistic studies indicate that the alkoxycarbonylation of alkynes proceeds via the protonation of a Pd(0)–alkyne species to give a Pd–vinyl complex, followed by CO insertion and alcoholysis.[310]

Alkanes. A review in 1997 summarizes the electrophilic functionalization of alkanes in liquid superacids including carboxylation.[311] When alkanes are treated with superacids in the presence of CO, the highly reactive carbocations are trapped,

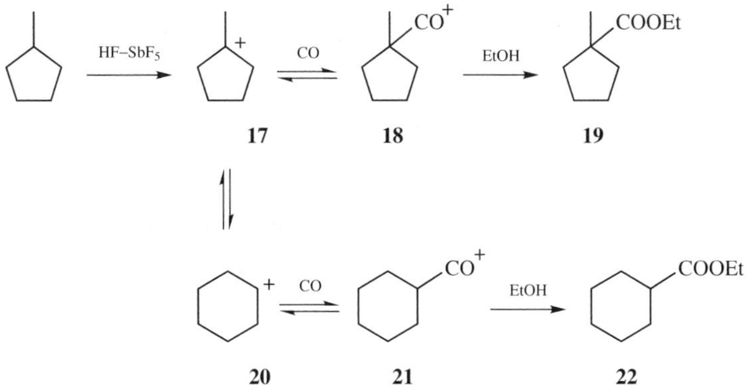

HF–SbF$_5$ CO EtOH

CO$^+$ COOEt

17 **18** **19**

+ CO CO$^+$ EtOH COOEt

20 **21** **22**

Scheme 7.3

preventing side reactions. This trapping technique permits the formation of carboxylic acids and allows one to draw important mechanistic conclusions. Temperature, alkane : CO ratios, and additives (Br$^-$, CHC$_3$, or CCl$_4$) were found to exert strong influence on yields and selectivities. One particularly interesting example is the carboxylation of methylcyclopentane. When methylcyclopentane is treated in HF–SbF$_5$ at 0°C, CO is passed through the solution at −40°C and ethyl methylcyclopentanecarboxylate (**19**) is isolated in 80% yield after quenching with NaHCO$_3$/EtOH. However, when CO is passed through from the start of the reaction at 0°C, ethyl cyclohexanecarboxylate (**22**) is isolated in 60% yield. At −40°C tertiary carbocation **17**, which is in equilibrium with cation **20**, is carbonylated irreversibly to yield the **18** acyl cation (Scheme 7.3). The latter by quenching gives ester **19**. At higher temperature, however, carbonylation is reversible. Since carbocation **20** has a much higher affinity for CO under such equilibrium conditions, the solution becomes enriched in acyl ion **21**. On quenching the latter gives ester **22**. Trapping of the intermediate carbocations and the detection of the corresponding carboxylic acids were also successful in zeolite H-ZSM-5[312] and sulfated zirconia.[313]

Alkylcyclohexanes, cycloheptane, and cyclooctane were also transformed into the esters of the corresponding tertiary cyclohexanecarboxylic acids in the presence of the superelectrophile CBr$_4$–2AlBr$_3$ in good yields.[314,315] Such organic superacids generated from polyhalomethanes with aluminum halides are capable of promoting the selective and effective carboxylation of linear alkanes (propane, butane) and adamantane as well.[316]

Further possibilities of carrying out carboxylation of alkanes have been also pursued. The most important aim of these studies is to achieve the direct oxidative carbonylation of methane to acetic acid. In the method developed by Sen,[176,317] the reagent combination RhCl$_3$, CO, I$^-$, and O$_2$ in perfluorobutyric acid[318] is used under high pressure (>20 bar) in the liquid phase. A similar system is RhCl$_3$–NaCl–KI and O$_2$ used in trifluoroacetic acid.[319] K$_2$S$_2$O$_8$ in trifluoroacetic acid in combination with $Pd(OAc)_2$ and/or $Cu(OAc)_2$ is also effective in this process.[320] This system is capable of the oxidative carbonylation of higher alkanes and aromatics,[321] whereas $Mg/K_2S_2O_8$ is most effective in the carboxylation of cyclohexane.[322] Direct carboxylation of methane can be performed with $Yb(OAc)_3/Mn(OAc)_2/NaClO$ in water.[323] Subsequently it was claimed that a new, highly efficient catalytic system composed of $CaCl_2$ and $K_2S_2O_8$ in trifluoroacetic acid transforms methane and CO into acetic acid almost quantitatively[324] and produces isobutyric acid from pentane.[325] A radical mechanism was suggested to account for the observations just as in the carboxylation of adamantane with CO and O_2 catalyzed by N-hydroxyphthalimide.[326]

Acetic acid is formed when methane reacts with CO or CO_2 in aqueous solution in the presence of O_2 or H_2O_2 catalyzed by vanadium complexes.[327] A Rh-based $FePO_4$ catalyst applied in a fixed-bed reactor operating at atmospheric pressure at 300–400°C was effective in producing methyl acetate in the presence of nitrous oxide.[328] The high dispersion of Rh at sites surrounded by iron sites was suggested to be a key factor for the carbonylation reaction.

7.4.4. Aminomethylation

A review is available for aminomethylation, also called *hydroaminomethylation* or *aminocarbonylation*, including more recent developments.[329]

The application of various ruthenium compounds in acetonitrile[330] or rhodium and ruthenium catalysts or their mixtures[331] was found to show significant improvements (high product selectivities, high linearities) in the aminomethylation of terminal alkenes to produce tertiary amines.

New examples, particularly those by Eilbracht, clearly demonstrate the synthetic potential of this process. Symmetrically and unsymmetrically substituted secondary and tertiary amines are produced selectively from alkenes, CO, and hydrogen in the presence of $[Rh(COD)Cl]_2$ catalyst precursor.[332–334]

There have been developments with respect to the synthesis of aliphatic primary amines from terminal alkenes, synthesis gas, and ammonia.[335,336] The first efficient aminomethylation allows the production of industrially important, low-molecular-weight linear primary amines.[336] The use of a Rh—Ir dual-catalyst system in the aminomethylation of propene, 1-butene, and 1-pentene under hydroformylation conditions in an aqueous two-phase system gives amines in better than 70% yield. Selectivity to the primary amines could be improved to above 90% by lowering the polarity of the organic phase. This was explained by a higher efficiency of a solvent of low polarity to remove the primary amine from the catalyst-containing aqueous medium. As a result, the primary amine no longer competes with ammonia. The improved catalyst performance is also attributed to the high activity of Ir to hydrogenate the C—N double bond of the imine intermediate. Excellent (99 : 1) normal to iso selectivities could be achieved by applying the ligand BINAS [prepared by the direct sulfonation of 1,1'-binaphthaline-2,2-diylbis(methylene)bis(diphenylphosphane)].

Added CO_2 allows the synthesis of substituted maleic anhydrides. According to a novel aminomethylation, ethane or propane is reacted with N,N-dialkylmethyl-amine N-oxides in the presence of trifluoroacetic acid and a catalytic amount of $Cu(OAc)_2$ to afford N,N-dialkylaminomethylated alkanes in good yields:[321,337,338]

$$RCH_2CH_3 \quad + \quad \underset{\underset{O}{\overset{\displaystyle R'}{\underset{\mid}{\overset{\mid}{CH_3-N-R'}}}}}{} \quad \xrightarrow[CF_3COOH]{Cu(OAc)_2} \quad \underset{H_3C}{\overset{R}{}}\!\!CH-CH_2-N\!\!\overset{R'}{\underset{R'}{}} \qquad (7.28)$$

$$R = H, Me$$

REFERENCES

1. O. Roelen, *Angew. Chem.* **60**, 62 (1948).

2. I. Wender, H. W. Sternberg, and M. Orchin, in *Catalysis*, Vol. 5, P. H. Emmett, ed., Reinhold, New York, 1957, Chapter 1, p. 1.

3. C. W. Bird, *Chem. Rev.* **62**, 283 (1962).

4. F. Asinger, *Mono-Olefins, Chemistry and Technology*, Pergamon Press, Oxford, 1968, Chapter 9, p. 785.

5. R. F. Heck, *Organotransition Metal Chemistry*, Academic Press, New York, 1974, Chapter 9, p. 201.

6. P. Pino, F. Piacenti, and M. Bianchi, in *Organic Synthesis via Metal Carbonyls*, Vol. 2, I. Wender and P. Pino, eds., Wiley-Interscience, New York, 1977, p. 43.

7. J. K. Stille and D. E. James, in *The Chemistry of Functional Groups*, Supplement A: *The Chemistry of Double-Bonded Functional Groups*, S. Patai, ed., Wiley, London, 1977, Chapter 12, p. 1099.

8. P. J. Davidson, R. R. Hignett, and D. T. Thompson, in *Catalysis, A Specialist Periodical Report*, Vol. 1, C. Kemball, senior reporter, The Chemical Society, Burlington House, London, 1977, Chapter 10, p. 369.

9. R. L. Pruett, *Adv. Organomet. Chem.* **17**, 1 (1979).

10. J. P. Collman, L. S. Hegedus, J. R. Norton, and R. G. Finke, *Principles and Applications of Organotransition Metal Chemistry*. University Science Books, Mill Valley, CA, 1987, Chapter 12, p. 619.

11. B. Cornils, in *New Syntheses with Carbon Monoxide*, J. Falbe, ed., Springer, Berlin, 1980, Chapter 1, p. 1.

12. C. Masters, *Homogeneous Transition-Metal Catalysis: A Gentle Art*. Chapman & Hall, London, 1981, Chapter 2.4, p. 102.

13. J. A. Davies, in *The Chemistry of the Metal-Carbon Bond*, Vol. 3, F. R. Hartley, and S. Patai, eds., Wiley, Chichester, 1985, Chapter 8, p. 361.

14. I. Tkatchenko, in *Comprehensive Organometallic Chemistry*, Vol. 8, G. Wilkinson, F. G. A. Stone, and E. W. Abel, eds., Pergamon Press, Oxford, 1982, Chapter 50.3, p. 101.

15. P. A. Chaloner, *Handbook of Coordination Catalysis in Organic Chemistry*, Butter-worths, London, 1986, Chapter 3, p. 217.

16. J. K. Stille, in *Comprehensive Organic Synthesis*, B. M. Trost and I. Fleming, eds., Vol. 4: *Additions to and Substitutions at C—C π-Bonds*, M. F. Semmelhack, ed., Pergamon Press, Oxford, 1991, Chapter 4.5, p. 913.

17. A. J. Chalk and J. F. Harrod, *Adv. Organomet. Chem.* **6**, 119 (1968).

18. M. Orchin, *Acc. Chem. Res.* **14**, 259 (1981).

19. F. E. Paulik, *Catal. Rev.-Sci. Eng.* **6**, 49 (1972).

20. L. H. Slaugh and R. D. Mullineaux, *J. Organomet. Chem.* **13**, 469 (1968).

21. F. H. Jardine, *Polyhedron* **1**, 569 (1982).

22. F. H. Jardine, in *The Chemistry of the Metal-Carbon Bond*, Vol. 4, F. R. Hartley, ed., Wiley, Chichester, 1987, Chapter 8, p. 733.

23. A. Spencer, in *Comprehensive Coordination Chemistry*, Vol. 6: *Applications*, G. Wilkinson, R. D. Gillard, and J. A. McCleverty, eds., Pergamon Press, Oxford, 1987, Chapter 61.2, p. 229.

24. P. Kalck, Y. Peres, and J. Jenck, *Adv. Organomet. Chem.* **32**, 121 (1991).

25. C.-Y. Hsu and M. Orchin, *J. Am. Chem. Soc.* **97**, 3553 (1975).

26. I. Schwager and J. F. Knifton, *J. Catal.* **45**, 256 (1976).

27. Y. Kawabata, T. Hayashi, and I. Ogata, *J. Chem. Soc., Chem. Commun.* 462 (1979).

28. R. F. Heck and D. S. Breslow, *J. Am. Chem. Soc.* **83**, 4023 (1961).

29. M. Orchin and W. Rupilius, *Catal. Rev.-Sci. Eng.* **6**, 85 (1972).

30. P. Pino, *J. Organomet. Chem.* **200**, 223 (1980).

31. F. Calderazzo, R. Ercoli, and G. Natta, in *Organic Synthesis via Metal Carbonyls*, Vol. 1, I. Wender and P. Pino, eds., Wiley-Interscience, New York, 1968, p. 1.

32. R. Whyman, *J. Organomet. Chem.* **81**, 97 (1974).

33. M. Orchin, L. Kirch, and I. Goldfarb, *J. Am. Chem. Soc.* **78**, 5450 (1956).

34. N. H. Alemdaroglu, J. L. M. Penninger, and E. Oltay, *Monatsh. Chem.* **107**, 1153 (1976).

35. M. van Boven, N. Alemdaroglu, and J. L. M. Penninger, *J. Organomet. Chem.* **84**, 65 (1975).

36. G. Natta, R. Ercoli, S. Castellano, and F. H. Barbieri, *J. Am. Chem. Soc.* **76**, 4049 (1954).

37. P. Pino, *Ann. NY Acad. Sci.* **415**, 111 (1983).

38. R. J. Klingler and J. W. Rathke, *J. Am. Chem. Soc.* **116**, 4772 (1994).

39. D. Evans, J. A. Osborn, and G. Wilkinson, *J. Chem. Soc. A* 3133 (1968).

40. G. Yagupsky, C. K. Brown, and G. Wilkinson, *J. Chem. Soc. A* 1392 (1970).

41. C. K. Brown and G. Wilkinson, *J. Chem. Soc. A* 2753 (1970).

42. W. R. Moser, C. J. Papile, D. A. Brannon, R. A. Duwell, and S. J. Weiniger, *J. Mol. Catal.* **41**, 271 (1987).

43. B. Heil and L. Markó, *Chem. Ber.* **102**, 2238 (1969).

44. P. W. N. M. van Leeuwen, and C. F. Roobeek, *J. Organomet. Chem.* **258**, 343 (1983).

45. A. I. M. Keulemans, A. Kwantes, and Th. van Bavel, *Recl. Trav. Chim. Pays-Bas* **67**, 298 (1948).

46. M. Tanaka, Y. Watanabe T. Mitsudo, and Y. Takegami, *Bull. Chem. Soc. Jpn.* **47**, 1698 (1974).

47. H. Siegel and W. Himmele, *Angew. Chem., Int. Ed. Engl.* **19**, 178 (1980).

48. D. Neibecker and R. Réau, *Angew. Chem., Int. Ed. Engl.* **28**, 500 (1989).

49. J. M. Brown and S. J. Cook, *Tetrahedron* **42**, 5105 (1986).

50. C. Abu-Gnim and I. Amer, *J. Chem. Soc., Chem. Commun.* 115 (1994).

51. M. Orchin, *Adv. Catal.* **16**, 1 (1966).

52. F. Piacenti, P. Pino, R. Lazzaroni, and M. Bianchi, *J. Chem. Soc.* C 488 (1966).

53. F. Piacenti, S. Pucci, M. Bianchi, R. Lazzaroni, and P. Pino, *J. Am. Chem. Soc.* **90**, 6847 (1968).

54. F. Piacenti, M. Bianchi, and P. Frediani, *Adv. Chem. Ser.* **132**, 283 (1974).

55. C. P. Casey and C. R. Cyr, *J. Am. Chem. Soc.* **93**, 1280 (1971); **95**, 2240 (1973).

56. D. A. von Bézard. G. Consiglio, F. Morandini, and P. Pino, *J. Mol. Catal.* **7**, 431 (1980).

57. A. Stefani, G. Consiglio, C. Botteghi, and P. Pino, *J. Am. Chem. Soc.* **95**, 6504 (1973); **99**, 1058 (1977).

58. G. Consiglio and P. Pino, *Top. Curr. Chem.* **105**, 77 (1982).

59. M. Tanaka, Y. Ikeda, and I. Ogata, *Chem. Lett.* 1115 (1975).

60. T. Hayashi, M. Tanaka, Y. Ikeda, and I. Ogata, *Bull. Chem. Soc. Jpn.* **52**, 2605 (1979).

61. G. Parrinello and J. K. Stille, *J. Am. Chem. Soc.* **109**, 7122 (1987).

62. J. K. Stille, H. Su, P. Brechot, G. Parrinello, and L. S. Hegedus, *Organometallics* **10**, 1183 (1991).

63. G. Consiglio, S. C. A. Nefkens, and A. Borer, *Organometallics* **10**, 2046 (1991).

64. N. Sakai, S. Mano, K. Nozaki, and H. Takaya, *J. Am. Chem. Soc.* **115**, 7033 (1993).

65. N. Sakai, K. Nozaki, and H. Takaya, *J. Chem. Soc., Chem. Commun.* 395 (1994).

66. P. Pino, G. Consiglio, C. Botteghi, and C. Salomon, *Adv. Chem. Ser.* **132**, 295 (1974).

67. P. Haelg, G. Consiglio, and P. Pino, *J. Organomet. Chem.* **296**, 281 (1985).

68. R. Grigg, G. J. Reimer, and A. R. Wade, *J. Chem. Soc., Perkin Trans. 1* 1929 (1983).

69. P. Salvadori, G. Vitulli, A. Raffaelli, and R. Lazzaroni, *J. Organomet. Chem.* **258**, 351 (1983).

70. M. Morikawa, *Bull. Chem. Soc. Jpn.* **37**, 379 (1964).

71. B. Fell and W. Rupilius, *Tetrahedron Lett.* 2721 (1969).

72. A. Spencer, *J. Organomet. Chem.* **124**, 85 (1977).

73. B. Fell and M. Beutler, *Tetrahedron Lett.* 3455 (1972).

74. C. Botteghi and Ch. Salomon, *Tetrahedron Lett.* 4285 (1974).

75. J. Falbe, in *Propylene and Its Industrial Derivatives*, E. G. Hancock, ed., Wiley, New York, 1973, Chapter 9, p. 333.

76. J. Falbe, *J. Organomet. Chem.* **94**, 213 (1975).

77. G. W. Parshall, *J. Mol. Catal.* **4**, 243 (1978).

78. A. Lundeen and R. Poe, in *Encyclopedia of Chemical Processing and Design*, Vol. 2, J. J. McKetta and W. A. Cunningham, eds., Marcel Dekker, New York, 1977, p. 465.

79. C. E. Loeffler, L. Stautzenberger, and J. D. Unruh, in *Encyclopedia of Chemical Processing and Design*, Vol. 5, J. J. McKetta and W. A. Cunningham, eds., Marcel Dekker, New York, 1977, p. 358.

80. I. Kirshenbaum and E. J. Inchalik, in *Kirk-Othmer Encyclopedia of Chemical Technology*, 3rd ed., Vol. 16, M. Grayson and D. Eckroth, eds., Wiley-Interscience, New York, 1981, p. 637.

81. P. W. Allen, R. L. Pruett, and E. J. Wickson, in *Encyclopedia of Chemical Processing and Design*, Vol. 33, J. J. McKetta and W. A. Cunningham, eds., Marcel Dekker, New York, 1990, p. 46.

82. K. Weissermel and H.-J. Arpe, *Industrial Organic Chemistry*, 2nd ed., VCH, Weinheim, 1993, Chapter 6, p. 123.

83. *Hydrocarbon Process.* **52**(11), 134 (1973).

84. H. Lemke, *Hydrocarbon Process Petrol. Refiner* **45**(2), 148 (1966).

85. R. Kummer, H. J. Nienburg, H. Hohenschutz, and M. Strohmeyer, *Adv. Chem. Ser.* **132**, 19 (1974).

86. *Hydrocarbon Process.* **56**(11), 172 (1977).

87. E. R. Freitas and C. R. Gum, *Chem. Eng. Prog.* **75**(1), 73 (1979).

88. R. E. Vincent, *ACS Symp. Ser.* **159**, 159 (1981).

89. B. Cornils and A. Mullen, in *Encyclopedia of Chemical Processing and Design*, Vol. 20, J. J. McKetta and W. A. Cunningham, eds., Marcel Dekker, New York, 1984, p. 366.

90. B. Cornils, R. Payer, and K. C. Traenckner, *Hydrocarbon Process.* **54**(6), 83 (1975).

91. R. Fowler, H. Connor, and R. A. Baehl, *Hydrocarbon Process.* **55**(9), 247 (1976); *Chemtech* **6**, 772 (1976).

92. *Hydrocarbon Process.* **56**(11), 190 (1977).

93. C. E. O'Rourke, P. R. Kavasmaneck, and R. E. Uhl, *ACS Symp. Ser.* **159**, 71 (1981).

94. E. G. Kuntz, *Chemtech* **17**, 570 (1987).

95. T. Onoda, *Chemtech* **23**(9), 34 (1993).

96. G. A. Olah and J. A. Olah, in *Friedel-Crafts and Related Reactions*, Vol. 3, Part 2, G. A. Olah, ed., Wiley, New York, 1964, Chapter 39, p. 1257.

97. H. Bahrmann, in *New Syntheses with Carbon Monoxide*, J. Falbe, ed., Springer, Berlin, 1980, Chapter 5, p. 372.

98. J. R. Norell, *J. Org. Chem.* **37**, 1971 (1972).

99. Ya. T. Éidus, K. V. Puzitskii, and S. D. Pirozhkov, *J. Org. Chem. USSR* (Engl. transl.) **4**, 32 (1968).

100. Ya. T. Éidus, S. D. Pirozhkov, and K. V. Puzitskii, *J. Org. Chem. USSR* (Engl. transl.) **4**, 369 (1968).

101. H. Koch and W. Haaf, *Justus Liebigs Ann. Chem.* **618**, 251 (1958).

102. W. Haaf, *Chem. Ber.* **99**, 1149 (1966).

103. H. Hogeveen, *Adv. Phys. Org. Chem.* **10**, 29 (1973).

104. Y. Souma, H. Sano, and J. Iyoda, *J. Org. Chem.* **38**, 2016 (1973).

105. Y. Souma and H. Sano, *Bull. Chem. Soc. Jpn.* **47**, 1717 (1974).

106. N. Yoneda, T. Fukuhara, Y. Takahashi, and A. Suzuki, *Chem. Lett.* 607 (1974).

107. Y. Souma and H. Sano, *Bull. Chem. Soc. Jpn.* **49**, 3296 (1976).

108. S. D. Pirozhkov, A. S. Stepanyan, T. N. Myshenkova, M. B. Ordyan, and A. L. Lapidus, *Bull. Acad. Sci. USSR, Div. Chem. Sci.* (Engl. transl.) **31**, 1852 (1982).

109. G. A. Olah, G. K. S. Prakash, and J. Sommer, *Superacids*, Wiley-Interscience, New York, 1985, Chapter 5, p. 295.

110. B. L. Booth and T. A. El-Fekky, *J. Chem. Soc., Perkin Trans. 1* 2441 (1979).

111. Y. Souma and H. Sano, *Bull. Chem. Soc. Jpn.* **49**, 3335 (1976).

112. Y. Takahashi, N. Tomita, N. Yoneda, and A. Suzuki, *Chem. Lett.* 997 (1975).

113. M. Yoshimura, M. Nojima, and N. Tokura, *Bull. Chem. Soc. Jpn.* **46**, 2164 (1973).

114. Y. Souma, and H. Sano, *Bull. Chem. Soc. Jpn.* **46**, 3237 (1973).

115. A. Bagno, J. Bukala, and G. A. Olah, *J. Org. Chem.* **55**, 4284 (1990).

116. G. A. Olah and J. Bukala, *J. Org. Chem.* **55**, 4293 (1990).

117. C. W. Bird, *Transition Metal Intermediates in Organic Synthesis*. Logos Press, London, 1967, Chapter 7, p. 149.

118. P. Pino, F. Piacenti, and M. Bianchi, in *Organic Synthesis via Metal Carbonyls*, Vol. 2, I. Wender and P. Pino, eds., Wiley-Interscience, New York, 1977, p. 233.

119. G. K. Anderson and J. A. Davies, in *The Chemistry of the Metal-Carbon Bond*, Vol. 3, F. R. Hartley and S. Patai, eds., Wiley, Chichester, 1985, Chapter 7, p. 335.

120. D. M. Fenton and E. L. Moorehead, in *The Chemistry of the Metal-Carbon Bond*, Vol. 3, F. R. Hartley and S. Patai, eds., Wiley, Chichester, 1985, Chapter 10, p. 435.

121. Ya. T. Eidus, K. V. Puzitskii, A. L. Lapidus, and B. K. Nefedov, *Russ. Chem. Rev.* (Engl. transl.) **40**, 429 (1971).

122. A. Mullen, in *New Syntheses with Carbon Monoxide*, J. Falbe, ed., Springer, Berlin, 1980, Chapter 3, p. 243.

123. P. W. Jolly, in *Comprehensive Organometallic Chemistry*, Vol. 8, G. Wilkinson, F. G. A. Stone and E. W. Abel, eds., Pergamon Press, Oxford, 1982, Chapter 56.6, p. 773.

124. P. Pino and G. Braca, in *Organic Synthesis via Metal Carbonyls*, Vol. 2, I. Wender and P. Pino, eds., Wiley-Interscience, New York, 1977, p. 419.

125. W. Reppe, *Experientia* **5**, 93 (1949).

126. W. Reppe, *Justus Liebigs Ann. Chem.* **582**, 1 (1953).

127. W. Reppe and H. Kröper, *Justus Liebigs Ann. Chem.* **582**, 38 (1953).

128. C. W. Bird, R. C. Cookson, J. Hudec, and R. O. Williams, *J. Chem. Soc.* 410 (1963).

129. G. Consiglio and P. Pino, *Gazz. Chim. Ital.* **105**, 1133 (1975).

130. B. Fell and J. M. J. Tetteroo, *Angew. Chem., Int. Ed. Engl.* **4**, 790 (1965).

131. A. Matsuda and H. Uchida, *Bull. Chem. Soc. Jpn.* **38**, 710 (1965).

132. K. Bittler, N. v. Kutepow, D. Neubauer, and H. Reis, *Angew. Chem., Int. Ed. Engl.* **7**, 329 (1968).

133. J. Tsuji, *Organic Synthesis with Palladium Compounds*, Springer, Berlin, 1980, Chapter 4, p. 81.

134. D. M. Fenton, *J. Org. Chem.* **38**, 3192 (1973).

135. G. Cavinato and L. Toniolo, *J. Mol. Catal.* **6**, 111 (1979); **10**, 161 (1981).

136. J. F. Knifton, *J. Org. Chem.* **41**, 2885 (1976).

137. J. F. Knifton, *J. Org. Chem.* **41**, 793 (1976).

138. B. El Ali and H. Alper, *J. Mol. Catal.* **77**, 7 (1992); *J. Org. Chem.* **58**, 3595 (1993).

139. F. J. Waller, *Br. Polymer J.* **16**, 239 (1984).

140. B.-H. Chang, *Inorg. Chim. Acta* **150**, 245 (1988).

141. H. Alper, J. B. Woell, B. Despeyroux, and D. J. H. Smith, *J. Chem. Soc., Chem. Commun.* 1270 (1983).

142. Y. Sugi and K. Bando, *Chem. Lett.* 727 (1976).

143. J.-J. Brunet, D. Neibecker, and R. S. Srivastava, *Tetrahedron Lett.* **34**, 2759 (1993).

144. G. Consiglio and P. Pino, *Adv. Chem. Ser.* **196**, 371 (1982).

145. H. Alper and N. Hamel, *J. Am. Chem. Soc.* **112**, 2803 (1990).

146. Ya. T. Eidus, A. L. Lapidus, K. V. Puzitskii, and B. K. Nefedov, *Russ. Chem. Rev.* (Engl. transl.) **42**, 199 (1973).

147. A. Matsuda, *Bull. Chem. Soc. Jpn.* **46**, 524 (1973).

148. J. Tsuji, S. Hosaka, J. Kiji, and T. Susuki, *Bull. Chem. Soc. Jpn.* **39**, 141 (1966).

149. J. Tsuji and T. Nogi, *Bull. Chem. Soc. Jpn.* **39**, 146 (1966).

150. J. Tsuji, Y. Mori, and M. Hara, *Tetrahedron* **28**, 3721 (1972).

151. G. W. Parshall and S. D. Ittel, *Homogeneous Catalysis*, Wiley-Interscience, New York, 1993, Chapter 5, p. 100.

152. R. F. Heck and D. S. Breslow, *J. Am. Chem. Soc.* **85**, 2779 (1963).

153. J. Tsuji, *Acc. Chem. Res.* **2**, 144 (1969).

154. R. Bardi, A. Del Pra, A. M. Piazzesi, and L. Toniolo, *Inorg. Chim. Acta* **35**, L345 (1979).

155. R. F. Heck, *J. Am. Chem. Soc.* **85**, 2013 (1963).

156. O. L. Kaliya, O. N. Temkin, N. G. Mekhryakova, and R. M. Flid, *Dokl. Chem.* (Engl. transl.) **199**, 706 (1971).

157. R. F. Heck, *J. Am. Chem. Soc.* **94**, 2712 (1972).

158. D. Milstein and J. L. Huckaby, *J. Am. Chem. Soc.* **104**, 6150 (1982).

159. D. Milstein, *Acc. Chem. Res.* **21**, 428 (1988).

160. A. Mullen, in *New Syntheses with Carbon Monoxide*, J. Falbe, ed., Springer, Berlin, 1980, Chapter 6, p. 414.

161. J. Happel, S. Umemura, Y. Sakakibara, H. Blanck, and S. Kunichika, *Ind. Eng. Chem., Process Des. Dev.* **14**, 44 (1975).

162. J. F. Knifton, *J. Mol. Catal.* **2**, 293 (1977).

163. D. Foster, *Adv. Organomet. Chem.* **17**, 255 (1979).

164. R. S. Dickson, *Homogeneous Catalysis with Compounds of Rhodium and Iridium*, Reidel, Dordrecht, 1985, Chapter 4, p. 129.

165. T. W. Dekleva and D. Forster, *Adv. Catal.* **34**, 81 (1986).

166. U. Romano, R. Tesei, M. M. Mauri, and P. Rebora, *Ind. Eng. Chem., Prod. Res. Dev.* **19**, 396 (1980).

167. H. Hopff, *Chem. Ber.* **64B**, 2739 (1931); **65B**, 482 (1932); *Angew. Chem.* **60**, 245 (1948).

168. H. Hopff, C. D. Nenitzescu, D. A. Isacescu, and I. P. Cantuniari, *Chem. Ber.* **69B**, 2244 (1936).

169. H. Pines and V. N. Ipatieff, *J. Am. Chem. Soc.* **69**, 1337 (1947).

170. N. Yoneda, T. Fukuhara, Y. Takahashi, and A. Suzuki, *Chem. Lett.* 17 (1983).

171. R. Paatz and G. Weisgerber, *Chem. Ber.* **100**, 984 (1967).

172. W. Haaf and H. Koch, *Justus Liebigs Ann. Chem.* **638**, 122 (1960).

173. H. Koch and W. Haaf, *Angew. Chem.* **72**, 628 (1960).

174. B. S. Friedman and S. M. Cotton, *J. Org. Chem.* **27**, 481 (1962).

175. Y. Souma and H. Sano, *J. Org. Chem.* **38**, 3633 (1973).

176. H. Hogeveen, J. Lukas, and C. F. Roobeek, *J. Chem. Soc., Chem. Commun.* 920 (1969).

177. H. Hogeveen and C. F. Roobeek, *Recl. Trav. Chim. Pays-Bas* **91**, 137 (1972).

178. M. Lin and A. Sen, *Nature* (London) **368**, 613 (1994).

179. D. Rotman, *Chem. Week* **154**(15), 8 (1994).

180. W. J. Ellis and C. Roming, Jr., *Hydrocarbon Process Petrol. Refiner* **44**(6), 139 (1965).

181. *Hydrocarbon Process.* **58**(11), 198 (1979).

182. W. F. Hoelderich, *Mater. Res. Soc. Symp. Proc.* **233**, 27 (1991).

183. P. Hofmann and W. H. E. Müller *Hydrocarbon Process., Int. Ed.* **60**(10), 151 (1981).

184. F. T. Maher and W. Bauer, Jr., in *Encyclopedia of Chemical Processing and Design*, Vol. 1, J. J. McKetta and W. A. Cunningham, eds., Marcel Dekker, New York, 1976, p. 401.

185. W. Bauer, Jr., in *Kirk-Othmer Encyclopedia of Chemical Technology*, 4th ed., Vol. 1, J. I. Kroschwitz and M. Howe-Grant, eds., Wiley-Interscience, New York, 1991, p. 287.

186. J. W. Nemec and W. Bauer, Jr., in *Encycylopedia of Polymer Science and Engineering*, Vol. 1, J. I. Kroschwitz, ed., Wiley-Interscience, New York, 1985, p. 211.

187. K. Weissermel and H.-J. Arpe, *Industrial Organic Chemistry*, 2nd ed., VCH, Weinheim, 1993, Chapter 11, p. 286.

188. *Hydrocarbon Process.* **50**(11), 120 (1971).

189. K. S. McMahon, in *Encyclopedia of Chemical Processing and Design*, Vol. 1, J. J. McKetta and W. A. Cunningham, eds., Marcel Dekker, New York, 1976, p. 216.

190. F. S. Wagner, Jr., in *Kirk-Othmer Encyclopedia of Chemical Technology*, 4th ed., Vol. 1, J. I. Kroschwitz and M. Howe-Grant, eds., Wiley-Interscience, New York, 1991, p. 121.

191. K. Weissermel and H.-J. Arpe, *Industrial Organic Chemistry*, 2nd ed., VCH, Weinheim, 1993, Chapter 7, p. 172.

192. R. P. Lowry and A. Aguilo, *Hydrocarbon Process.* **53**(11), 103 (1974).

193. H. Hohenschutz, N. von Kutepow, and W. Himmele, *Hydrocarbon Process Petrol. Refiner* **45**(11), 141 (1966).

194. J. F. Roth, J. H. Craddock, A. Hershman, and F. E. Paulik, *Chemtech* **1**, 600 (1971).

195. H. D. Grove, *Hydrocarbon Process.* **51**(11), 76 (1972).

196. R. T. Eby and T. C. Singleton, in *Applied Industrial Catlysis*, Vol. 1, B. E. Leach, ed., Academic Press, New York, 1983, Chapter 10, p. 275.

197. U. Romano and R. Tesei, Ger. Patent 2,743,690 (1978).

198. J. Haggin, *Chem. Eng. News* **65**(44), 27 (1987).

199. W. Reppe and H. Vetter, *Justus Liebigs Ann. Chem.* **582**, 133 (1953).

200. A. F. M. Iqbal, *Helv. Chim. Acta* **54**, 1440 (1971).

201. K. Murata, A. Matsuda, and T. Masuda, *J. Mol. Catal.* **23**, 121 (1984).

202. R. M. Laine, *J. Org. Chem.* **45**, 3370 (1980).

203. F. Jachimowicz and J. W. Raksis, *J. Org. Chem.* **47**, 445 (1982).

204. F. Jachimowicz and J. W. Raksis, in *Catalysis of Organic Reactions*, J. R. Kosak, ed., Marcel Dekker, New York, 1984, p. 429.

205. T. Baig and P. Kalck, *J. Chem. Soc., Chem. Commun.* 1373 (1992).

206. J. F. Knifton and J. J. Lin, *J. Mol. Catal.* **81**, 27 (1993).

207. J. J. Brunet, D. Neibecker, F. Agbossou, and R. S. Srivastava, *J. Mol. Catal.* **87**, 223 (1994).

208. B. Cornils and E. Wiebus, *Chemtech* **25**(1), 33 (1995).

209. F. Agbossou, J.-F. Carpentier, and A. Mortreux, *Chem. Rev.* **95**, 2485 (1995).

210. M. Beller, B. Cornils, C. D. Frohning, and Ch. W. Kohlpaintner, *J. Mol. Catal. A: Chem.* **104**, 17 (1995).

211. V. A. Likholobov and B. L. Moroz, in *Handbook of Heterogeneous Catalysis*, G. Ertl, H. Knözinger, and J. Weitkamp, eds., Wiley-VCH, Weinheim, 1997, Chapter 4.5, p. 2231.

212. F. Joó and Á. Kathó, *J. Mol. Catal. A: Chem.* **116**, 3 (1997).

213. F. Joó, É. Papp, and Á. Kathó, *Top. Catal.* **5**, 113 (1998).

214. B. Cornils, R. W. Eckl, and W. A. Herrmann, *J. Mol. Catal. A: Chem.* **116**, 27 (1997).

215. B. Cornils, *Org. Process Res. Dev.* **2**, 121 (1998); *Top. Curr. Chem.* **206**, 133 (1999).

216. B. Fell, *Tenside Surf. Det.* **35**, 326 (1998).

217. A. M. Trzeciak and J. J. Ziółkowski, *Coord. Chem. Rev.* **190–192**, 883 (1999).

218. F. Ungváry, *Coord. Chem. Rev.* **147**, 547 (1996); **160**, 129 (1997); **167**, 233 (1997); **170**, 245 (1998); **188**, 263 (1999).

219. C. D. Frohning and Ch. W. Kohlpaintner, in *Applied Homogeneous Catalysis with Organometallic Complexes*, B. Cornils and W. A. Herrmann, eds., VCH, Weinheim, 1996, Chapter 2.1.1, p. 29.

220. B. Breit and W. Seiche, *Synthesis* 1 (2001).

221. *Recent Advances in Carbonylation Reactions*, P. Kalck, ed., *J. Mol. Catal. A: Chem.* **143**, 1–330 (1999).

222. A. Fukuoka, W. Kosugi, F. Morishita, M. Hirano, L. McCaffrey, W. Henderson, and S. Komiya, *Chem. Commun.* 489 (1999).

223. F. Favre, H. Olivier-Bourbigou, D. Commereuc, and L. Saussine, *Chem. Commun.* 1360 (2001).

224. S. U. Son, J. W. Han, and Y. K. Chung, *J. Mol. Catal. A: Chem.* **135**, 35 (1998).

225. M. Lenarda, L. Storaro, and R. Ganzerla, *J. Mol. Catal. A: Chem.* **111**, 203 (1996).

226. L. Huang, Y. Xu, W. Gao, A. Liu, D. Li, and X. Guo, *Catal Lett.* **32**, 61 (1995).

227. T. A. Kainulainen, M. K. Niemelä, and A. O. I. Krause, *Catal Lett.* **53**, 97 (1998).

228. P. Arya, N. V. Rao, J. Singkhonrat, H. Alper, S. C. Bourque, and L. E. Manzer, *J. Org. Chem.* **65**, 1881 (2000).

229. S. C. Bourque, F. Maltais, W.-J. Xiao, O. Tardif, H. Alper, P. Arya, and L. E. Manzer, *J. Am. Chem. Soc.* **121**, 3035 (1999).

230. A. J. Sandee, L. A. van der Veen, J. N. H. Reek, P. C. J. Kamer, M. Lutz, A. L. Spek, and P. W. N. M. van Leeuwen, *Angew. Chem., Int. Ed. Engl.* **38**, 3231 (1999).

231. J. A. Díaz-Auñón, M. C. Román-Martínez, and C. Salinas-Martínez de Lecea, *J. Mol. Catal. A: Chem.* **170**, 81 (2001).

232. E. Lindner, F. Auer, A. Baumann, P. Wegner, H. A. Mayer, H. Bertagnolli, U. Reinöhl, T. S. Ertel, and A. Weber, *J. Mol. Catal. A: Chem.* **157**, 97 (2000).

233. M. E. Davis, *Chemtech* **22**, 498 (1992).

234. I. Tóth, I. Guo, and B. E. Hanson, *J. Mol. Catal. A: Chem.* **116**, 217 (1997).

235. M. S. Anson, M. P. Leese, L. Tonks, and J. M. J. Williams, *J. Chem. Soc., Dalton Trans.* 3529 (1998).

236. A. J. Sandee, V. F. Slagt, J. N. H. Reek, P. C. J. Kamer, and P. W. N. M. van Leeuwen, *Chem. Commun.* 1633 (1999).

237. I. T. Horváth and J. Rábai, *Science* **266**, 72 (1994).

238. I. T. Horváth, G. Kiss, R. A. Cook, J. E. Bond, P. A. Stevens, J. Rábai, and E. J. Mozeleski, *J. Am. Chem. Soc.* **120**, 3133 (1998).

239. W. Chen, L. Xu, and J. Xiao, *Chem. Commun.* 839 (2000).

240. H. Chen, Y. Li, J. Chen, P. Cheng, Y.-E. He, and X. Li, *J. Mol. Catal. A: Chem.* **149**, 1 (1999).

241. F. Van Vyve and A. Renken, *Catal. Today* **48**, 237 (1999).

242. S. Bischoff and M. Kant, *Ind. Eng. Chem. Res.* **39**, 4908 (2000).

243. B. E. Hanson, H. Ding, and C. W. Kohlpaintner, *Catal. Today* **42**, 421 (1998).

244. E. A. Karakhanov, Yu. S. Kardasheva, E. A. Runova and V. A. Semernina, *J. Mol. Catal. A: Chem.* **142**, 339 (1999).

245. P. Purwanto and H. Delmans, *Catal. Today* **24**, 135 (1995).

246. T. Mathivet, C. Méliet, Y. Castanet, A. Mortreux, L. Caron, S. Tilloy, and E. Monflier, *J. Mol. Catal. A: Chem.* **176**, 105 (2001).

247. M. T. Reetz and S. R. Waldvogel, *Angew. Chem., Int. Ed. Engl.* **36**, 865 (1997).

248. E. Monflier, G. Fremy, Y. Castanet, and A. Montreux, *Angew. Chem., Int. Ed. Engl.* **34**, 2269 (1995).

249. Z. Jin, X. Zheng, and B. Fell, *J. Mol. Catal. A: Chem.* **116**, 55 (1997).

250. Y. Wang, J. Jiang, R. Zhang, X. Liu, and Z. Jin, *J. Mol. Catal. A: Chem.* **157**, 111 (2000).

251. U. Ritter, N. Winkhofer, H.-G. Schmidt, and H.W. Roesky, *Angew. Chem., Int. Ed. Engl.* **35**, 524 (1996).

252. A. N. Ajjou and H. Alper, *J. Am. Chem. Soc.* **120**, 1466 (1998).

253. T. Borrmann, H. W. Roesky, and U. Ritter, *J. Mol. Catal. A: Chem.* **153**, 31 (2000).

254. D. R. Palo and C. Erkey, *Ind. Eng. Chem. Res.* **37**, 4203 (1998).

255. N. J. Meehan, A. J. Sandee, J. N. H. Reek, P. C. J. Kamer, P. W. N. M. van Leeuwen, and M. Poliakoff, *Chem. Commun.* 1497 (2000).

256. S. Kainz, D. Koch, W. Baumann, and W. Leitner, *Angew. Chem., Int. Ed. Engl.* **36**, 1628 (1997).

257. N. Karodia, S. Guise, C. Newlands, and J.-A. Andersen, *Chem. Commun.* 2341 (1998).

258. P. Wasserscheid and H. Waffenschmidt, *J. Mol. Catal. A: Chem.* **164**, 61 (2000).

259. P. Jessop, D. C. Wynne, S. DeHaai, and D. Nakawatase, *Chem. Commun.* 693 (2000).

260. K. Tominaga and Y. Sasaki, *Catal. Commun.* **1**, 1 (2000).

261. L. A. van der Veen, P. C. J. Kamer, and P. W. N. M. van Leeuwen, *Angew. Chem., Int. Ed. Engl.* **38**, 336 (1999).

262. D. Selent, K.-D. Wiese, D, Röttger, and A. Börner, *Angew. Chem., Int. Ed. Engl.* **39**, 1639 (2000).

263. M. Beller, B, Zimmermann, and H. Geissler, *Chem. Eur. J.* **5**, 1301 (1999).

264. C. P. Casey, E. L. Paulsen, E. W. Beuttenmueller, B. R. Proft, L. M. Petrovich, B. A. Matter, and D. R. Powell, *J. Am. Chem. Soc.* **119**, 11817 (1997).

265. C. P. Casey, G. T. Whiteker, M. G. Melville, L. M. Petrovich, J. A. Gavney, Jr., and D. R. Powell, *J. Am. Chem. Soc.* **114**, 5535 (1992).

266. M. Kranenburg, Y. E. M. van der Burgt, P. C. J. Kamer, P. W. N. M. van Leeuwen, K. Goubitz, and J. Fraanje, *Organometallics* **14**, 3081 (1995).

267. L. A. van der Veen, P. H. Keeven, G. C. Schoemaker, J. N. H. Reek, P. C. J. Kamer, P. W. N. M. van Leeuwen, M. Lutz, and A. L. Spek, *Organometallics* **19**, 872 (2000).

268. P. C. J. Kamer, P. W. N. M. van Leeuwen, and J. N. H. Reek, *Acc. Chem. Res.* **34**, 895 (2001).

269. M. S. Goedheijt, P. C. J. Kamer, and P. W. N. M. van Leeuwen, *J. Mol. Catal. A: Chem.* **134**, 243 (1998).

270. R. Gao, C. D. Tan, and R. T. K. Baker, *Catal. Today* **65**, 19 (2001).

271. D. He, D. Pang, T. Wang, Y. Chen, Y. Liu, J. Liu, and Q. Zhu, *J. Mol. Catal. A: Chem.* **174**, 21 (2001).

272. C. Botteghi, C. D. Negri, S. Paganelli, and M. Marchetti, *J. Mol. Catal. A: Chem.* **175**, 17 (2001).

273. J. R. Johnson, G. D. Cuny, and S. L. Buchwald , *Angew. Chem., Int. Ed. Engl.* **34**, 1760 (1995).

274. Y. Ishii, K. Miyashita, K. Kamita, and M. Hidai, *J. Am. Chem. Soc.* **119**, 6448 (1997).

275. B. G. Van den Hoven and H. Alper, *J. Org. Chem.* **64**, 3964 (1999).

276. G. J. H. Buisman, L. A. van der Veen, A. Klootvnijk, W. G-J. de Lange, P. C. J. Kamer, P. W. N. M. van Leeuwen, and D. Vogt, *Organometallics* **16**, 2929 (1997).

277. K. Nozaki, N. Sasaki, T. Nanno, T. Higashijima, S. Mano, T. Horiuchi, and H. Takaya, *J. Am. Chem. Soc.* **119**, 4413 (1997).

278. K. Nozaki, Y. Itoi, F. Shibahara, E. Shirakawa, T. Ohta, H. Takaya, and T. Hiyama, *J. Am. Chem. Soc.* **120**, 4051 (1998).

279. K. Nozaki, F. Shibahara, and T. Hiyama, *Chem. Lett.* 694 (2000).

280. G. Franciò and W. Leitner, *Chem. Commun.* 1663 (1999).

281. M. Diéguez, O. Pámies, A. Ruiz, S. Castillón, and C. Claver, *Chem. Commun.* 1607 (2000).

282. G. A. Olah, G. K. S. Prakash, T. Mathew, and E. R. Marinez, *Angew. Chem., Int. Ed. Engl.* **39**, 2547 (2000).

283. T. H. Clingenpeel, T. E. Wessel, and A. I. Biaglow, *J. Am. Chem. Soc.* **119**, 5469 (1997).

284. T. H. Clingenpeel and A. I. Biaglow, *J. Am. Chem. Soc.* **119**, 5077 (1997).

285. B. S. Jaynes and C. L. Hill, *J. Am. Chem. Soc.* **117**, 4704 (1995).

286. G. P. Rosini, K. Zhu, and A. S. Goldman, *J. Organomet. Chem.* **504**, 115 (1995).

287. Q. Xu, Y. Imamura, M. Fujiwara, and Y. Souma, *J. Org. Chem.* **62**, 1594 (1997).

288. Q. Xu and Y. Souma, *Top. Catal.* **6**, 17 (1998).

289. Q. Xu, Y. Souma, J. Umezawa, M. Tanaka, and H. Nakatani, *J. Org. Chem.* **64**, 6306 (1999).

290. Q. Xu, Y. Souma, and H. Nakatani, *J. Org. Chem.* **65**, 1540 (2000).

291. Q. Xu, M. Fujiwara, M. Tanaka, and Y. Souma, *J. Org. Chem.* **65**, 8105 (2000).

292. S. Jayasree, A. Seayad, and R. V. Chaudhari, *Chem. Commun.* 1239 (2000).

293. Z. Csók, G. Szalontai, G. Czira, and L. Kollár, *J. Organomet. Chem.* **570**, 23 (1998).

294. G. R. Eastham, B. T. Heaton, J. A. Iggo, R. P. Tooze, R. Whyman, and S. Zacchini, *Chem. Commun.* 609 (2000).

295. J. R. Zoeller, E. M. Blakely, R. M. Moncier, and T. J. Dickson, *Catal. Today* **36**, 227 (1997).

296. (a) J.-P. Simonato, T. Walter, and P. Métivier, *J. Mol. Catal. A: Chem.* **171**, 91 (2001); (b) V. Dragojlovic, D. B. Gao, and Y. L. Chow, *J. Mol. Catal. A: Chem.*, **171**, 43 (2001).

297. R. I. Pugh, E. Drent, and P. G. Pringle, *Chem. Commun.* 1476 (2001).

298. A. G. Stepanov, M. V. Luzgin, V. N. Romannikov, V. N. Sidelnikov, and K. I. Zamaraev, *J. Catal.* **164**, 411 (1996).

299. I. del Río, N. Ruiz, C. Claver, L. A. van der Veen, and P. W. N. M. van Leeuwen, *J. Mol. Catal. A: Chem.* **161**, 39 (2000).

300. F. Bertoux, E. Monflier, Y. Castanet, and A. Mortreux, *J. Mol, Catal. A: Chem.* **143**, 11 (1999).

301. S. Tilloy, E. Monflier, F. Bertoux, Y. Castanet, and A. Mortreux, *New J. Chem.* **21**, 529 (1997).

302. G. Papadogianakins, G. Verspui, L. Maat, and R. A. Sheldon, *Catal. Lett.* **47**, 43 (1997).

303. G. Verspui, J. Feiken, G. Papadogianakis, and R. A. Sheldon, *J. Mol. Catal. A: Chem.* **146**, 299 (1999).

304. E. Monflier, S. Tilloy, F. Bertoux, Y. Castanet, and A. Mortreux, *New J. Chem.* **21**, 857 (1997).

305. Y. Kushino, K. Itoh, M. Miura, and M. Nomura, *J. Mol. Catal.* **89**, 151 (1994).

306. M. Akao, S. Sugawara, K. Amino, and Y. Inoue, *J. Mol. Catal. A: Chem.* **157**, 117 (2000).

307. E. Drent, P. Arnoldy, and P. H. M. Budzelaar, *J. Organomet. Chem.* **455**, 247 (1993).

308. A. Serivanti, V. Beghetto, M. Zanato, and U. Matteoli, *J. Mol. Catal. A: Chem.* **160**, 331 (2000).

309. B. El Ali and H. Alper, *J. Mol. Catal. A: Chem.* **96**, 197 (1995).

310. A. Scrivanti, V. Beghetto, E. Campagna, M. Zanato, and U. Matteoli, *Organometallics* **17**, 630 (1998).

311. J. Sommer, in *Stable Carbocation Chemistry*, G. K. S. Prakash and P. v. R. Schleyer, eds., Wiley, New York, 1997, Chapter 15, p. 497.

312. M. V. Luzgin, A. G. Stepanov, A. Sassi, and J. Sommer, *Chem. Eur. J.* **6**, 2368 (2000).

313. A. G. Stepanov, M. V. Luzgin, A. V. Krasnoslobodtsev, V. P. Shmachkova, and N. S. Kotsarenko, *Angew. Chem., Int. Ed. Engl.* **39**, 3658 (2000).

314. I. S. Akhrem, A. Orlinkov, and M. E. Vol'pin, *Uspekhi Khimii* **65**, 920 (1996).

315. I. Akhrem, L. Afanas'eva, P. Petrovskii, S. Vitt, and A. Orlinkov, *Tetrahedron Lett.* **41**, 9903 (2000).

316. I. Akhrem, *Top. Catal.* **6**, 27 (1998).

317. A. Sen, *Acc. Chem. Res.* **31**, 550 (1998).

318. M. Lin, T. E. Hogan, and A. Sen, *J. Am. Chem. Soc.* **118**, 4574 (1996).

319. E. G. Chepaikin, A. P. Bezruchenko, A. A. Leshcheva, G. N. Boyko, I. V. Kuzmenkov, E. H. Grigoryan, and A. E. Shilov, *J. Mol. Catal. A: Chem.* **169**, 89 (2001).

320. K. Nakata, Y. Yamaoka, T. Miyata, Y. Taniguchi, K. Takaki, and Y. Fujiwara, *J. Organomet. Chem.* **473**, 329 (1994).

321. Y. Fujiwara, K. Takaki, and Y. Taniguchi, *Synlett* 591 (1996).

322. M. Asadullah, T. Kitamura, and Y. Fujiwara, *Appl. Organomet. Chem.* **13**, 539 (1999).

323. M. Asadullah, Y. Taniguchi, T. Kitamura, and Y. Fujiwara, *Appl. Organomet. Chem.* **12**, 277 (1998).

324. M. Asadullah, T. Kitamura, and Y. Fujiwara, *Angew. Chem., Int. Ed. Engl.* **39**, 2475 (2000).

325. M. Asadullah, T. Kitamura, and Y. Fujiwara, *J. Catal.* **195**, 180 (2000).

326. S. Kato, T. Iwahama, S. Sakaguchi, and Y. Ishii, *J. Org. Chem.* **63**, 222 (1998).

327. G. V. Nizova, G. Süss-Fink, S. Stanislas, and G. B. Shul'pin, *Chem. Commun.* 1885 (1998).

328. Y. Wang, M. Katagiri, and K. Otsuka, *Chem. Commun.* 1187 (1997).

329. D. M. Roundhill, *Chem. Rev.* **92**, 1 (1992).

330. H. Schaffrath and W. Keim, *J. Mol. Catal. A: Chem.* **140**, 107 (1999).

331. M. M. Schulte, J. Herwig, R. W. Fischer, and C. W. Kohlpaintner, *J. Mol. Catal. A: Chem.* **150**, 147 (1999).

332. T. Rische and P. Eilbracht, *Synthesis* 1331 (1997).

333. C. L. Kranemann and P. Eilbracht, *Synthesis* 71 (1998).

334. T. Rische, B. Kitsos-Rzychon, and P. Eilbracht, *Tetrahedron* **54**, 2723 (1998).

335. J. F. Knifton, *Catal. Today* **36**, 305 (1997).

336. B. Zimmermann, J. Herwig, and M. Beller, *Angew. Chem., Int. Ed. Engl.* **38**, 2372 (1999).

337. Y. Taniguchi, S. Horie, K. Takaki, and Y. Fujiwara, *J. Organomet. Chem.* **504**, 137 (1995).

338. Y. Taniguchi, T. Kitamura, Y. Fujiwara, S. Horie, and K. Takaki, *Catal. Today* **36**, 85 (1997).

8

ACYLATION

The acylation of aromatic hydrocarbons was first described by Friedel and Crafts in 1877.[1] Since then the reaction has been widely and thoroughly studied. It is one of the most important reactions in synthetic organic chemistry and a widely applied method to prepare aromatic ketones. It is also of considerable practical significance in the chemical industry[2] since the products are intermediates in the manufacture of fine chemicals and other intermediates. Related topics, including the Hueben–Hoesch reaction and aldehyde synthesis (formylation of aromatics), and the acylation of aliphatic compounds, in contrast, are less important, and consequently, will be treated accordingly. The acylation of aromatic[2-10] and aliphatic compounds[10-13] and the related processes[10,14-17] are covered in reviews, and discussions of acylations can be found in other review papers about homogeneous and heterogeneous electrophilic catalysts.[18-23]

8.1. ACYLATION OF AROMATICS

8.1.1. General Characteristics

Friedel–Crafts acylation is an electrophilic aromatic substitution to afford ketones by replacing one of the hydrogens of an aromatic ring. Carboxylic acid derivatives, characteristically acid halides and anhydrides, serve as acylating agents and Lewis acid metal halides are the characteristic catalysts required to induce the transformation. Esters, in general, are not satisfactory reagents since they give both alkyl- and acyl-substituted products.

In Friedel–Crafts acylation of aromatics with acid chlorides and Lewis acid metal halides the reactive electrophile is considered to be formed in the interaction of the reagent and the catalyst. First the highly polarized donor–acceptor complex **1** is formed, which can further give other complexes and ion pairs.[24] The various

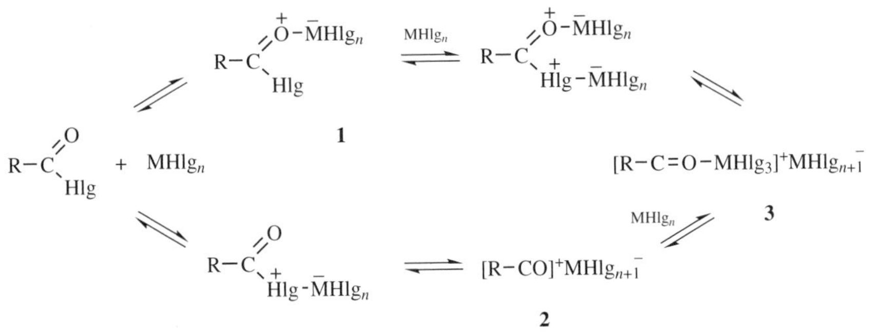

Scheme 8.1

possible intermediates are depicted in Scheme 8.1. Spectroscopic and kinetic data show the presence of these species in the reaction mixture. The scheme includes the **2** acyl cation, which is usually regarded as the reacting species in aromatic Friedel–Crafts acylations and forms with the aromatic the σ complex.[6,25,26]

In fact, stable, well-defined acylium salts, such as acetyl hexafluoro- and hexachloroantimonate, prepared by Olah could be used as effective acylating agents.[27–29] The acyl cation, however, is a weak electrophile because it is stabilized by resonance by the acylium ion (**4**) with the acyloxonium ion (**5**).[30] Excess Lewis acid, however, was suggested to transform the **2** cation into the **3** superelectrophilic acylating agent by decreasing neighboring O participation. Both experimental evidence and theoretical calculations support the existence of such species.[31–33] Under superacidic conditions the **6** protoacylium dication (or the protosolvated acylium ion) is the de facto acylating agent. Kinetic evidence showed that, using aroyl triflates, the acylium ions are the reacting electrophiles.[34]

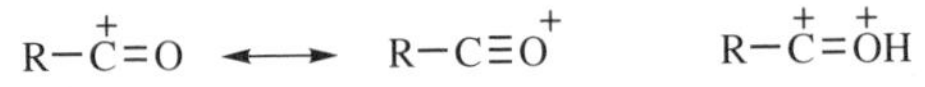

Friedel–Crafts acylations usually exhibit high substrate selectivities. Toluene, for example, is about 100 times more reactive than benzene.[8,25] The predominant formation of the *para* isomer is also characteristic. This is effected by the steric requirement of the acylating agent and the relatively weak acylating electrophile interacting with the aromatic in the σ complex.[25] With increasingly stronger electrophilic acylating agents, substrate selectivity decreases, indicative of more significant initial π-type of interactions.

In a kinetic study of the acylation of toluene, with *p*-xylene and the corresponding perdeutero compounds with aroyl triflates, correlation was found between the primary kinetic isotope effect and the *ortho* : *para* ratio.[35] Different conformations of the bent σ complexes[36] for the two isomers resulted in a much higher rate of deprotonation and rearomatization for the *para* isomer. By appropriately selecting reaction conditions and thereby affecting the ratio of the two conformations, unusually high amounts of *ortho* products may be obtained.[37,38]

8.1.2. Catalysts and Reaction Conditions

$AlCl_3$ is the most frequently used catalyst in aromatic Friedel–Crafts acylations, but other Lewis acid metal halides also show high activity and are applied accordingly. Considering relative efficiencies of metallic halides (most of which were discussed in reference to Friedel–Crafts alkylations), these are also relevant for Friedel–Crafts acylations (see Section 5.1.4). In short, since the activity of Lewis acid metal halides depends on the reagents and reaction conditions, relative reactivity orders may be established for a given reagent only under given reaction conditions. On the basis of their activity in the acetylation with acetyl chloride of toluene, $SbCl_5$, $FeCl_3$, $SnCl_4$, and $TiCl_4$ are also efficient catalysts, whereas $ZnCl_2$ is usually a relatively weak Lewis acid in Friedel–Crafts acylations.[3,4] It is interesting to note that in the solvent-free aroylation of benzene and toluene under microwave irradiation with various Lewis acid chlorides, including $AlCl_3$, $FeCl_3$ exhibited by far the best activity.[39]

Boron trifluoride complexes are also often applied. BF_3 when used with acyl fluorides showed in some cases distinct differences compared to other catalyst–reagent combinations. For example, acylation of 2-methylnaphthalene with isoBuCOF and BF_3 gives high yield (83%) of the 6-substituted isomer in contrast to $AlCl_3$ (30%).[40] A similar example is shown here:

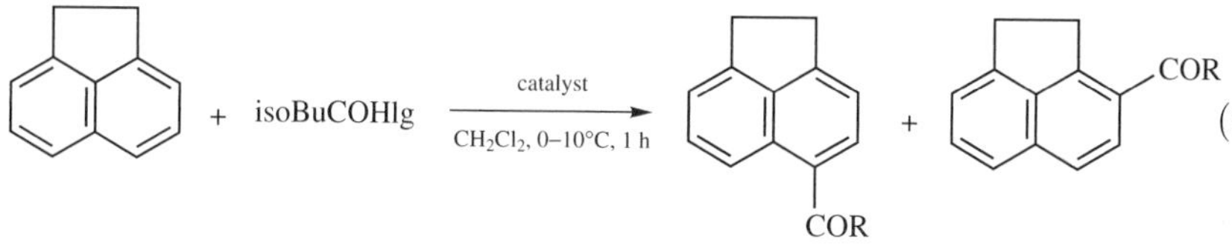

$$\text{(8.1)}$$

$AlCl_3$, Hlg = Cl	80	20
BF_3, Hlg = F	15	85

Strong Brønsted acids are also available to induce acylations.[3,8,9] Perfluoroalkanesulfonic acids were shown to be highly effective. Certain metal powders, such as Zn, Cu, Al, and Fe, were also found to effect acylations with acyl chlorides. The de facto catalysts are the in situ formed corresponding metal halides.[3,8] A number of other catalysts were developed over the years; however, many of these are effective only for the acylation of highly reactive aromatics, such as heterocycles.[9]

A considerable difference between Friedel–Crafts alkylation and acylation is the amount of the Lewis acid necessary to induce the reaction. Friedel–Crafts alkylation requires the use of only catalytic amounts of the catalyst. Lewis acids, however, form complexes with the aromatic ketones, the products in Friedel–Crafts acylations, and the catalyst is thus continuously removed from the system as the reaction proceeds. To achieve complete conversion, therefore, it is necessary to use an equimolar amount of Lewis acid catalyst when the acylating agent is an acyl halide. Optimum yields can be obtained using a 1.1 molar excess of the catalyst. With acid anhydrides as acylating agents, up to 2 mol of the Lewis acid is required to form the acyl halide which is the apparent reagent and the acidic salt:

$$(RCO)_2O \ + \ 2 \, MHlg_n \ \longrightarrow \ RCOHlg\text{–}MHlg_n \ + \ RCOOMHlg_{n-1} \qquad (8.2)$$

$$\underset{MHlg_n}{\underline{\qquad\qquad\qquad}}$$

Applying 3 mol of the Lewis acid, the carboxylic acid itself can also serve as the acylating agent. In fact, this observation led to the successful application of carboxylic acids as acylating agents.

It is known that certain Friedel–Crafts acylations of reactive aromatics may proceed with a small amount of catalyst or even, to some extent, with no catalyst at all.[41] Effective catalysts are Fe, FeCl$_3$, ZnCl$_2$, iodine, and hydroiodic acid. However, satisfactory yields are usually obtained with activated and polynuclear aromatics, and heterocycles. The mixed anhydrides acyl triflates are also powerful acylating agents without catalyst.[34]

Selection of an appropriate solvent for Friedel–Crafts acylation is an important question since solvents are known to affect regioselectivities.[8,9] In many cases acylation is carried out in an excess of the reacting aromatic compound. Aromatics, however, are poor solvents for most Lewis acids and therefore, they merely serve as diluent in biphase systems. Carbon disulfide is a reasonably good solvent just as dichloromethane and dichloroethane. Although AlCl$_3$ is insoluble in chlorinated hydrocarbons, they dissolve many of the complexes formed between acyl halides and AlCl$_3$. Nitrobenzene and nitromethane are also suitable solvents. Moreover, the 1 : 1 addition complexes they form with AlCl$_3$ allow acylations to be performed under mild conditions often without side reactions.

The sequence of addition of the reagents also affect Friedel–Crafts acylations. The most satisfactory procedure, which is actually the same what Friedel and Crafts originally used, is the addition of the catalyst as the last reactant. Another possibility also used very often is the Perrier method. It involves the prior preparation of the complex in a solvent followed by the addition of the aromatic reagent.

8.1.3. Recent Developments

Since its discovery[1] the scope of Friedel–Crafts acylation, considering the nature of the varied acylating agents and the catalysts applied, has substantially widened. Some more important new findings are only discussed.

New Soluble Catalysts. Trifluoromethansulfonic acid (triflic acid, TfOH)[42] and acyl triflates, that is mixed anhydrides of carboxylic acids and triflic acid,[43,44] were first reported to be effective for Friedel–Crafts acylation in 1972. Significantly lower yields (<30%) were obtained with other Brønsted acids. High activities were also observed for perfluorobutanesulfonic acid.[37]

More recent studies reconfirmed the excellent properties of these acylating systems and showed that aroyl triflates are able to acylate aromatics without catalysts in organic solvents.[34] Furthermore, even methyl benzoate, an ester, could be used as efficient benzoylating agent in the presence of triflic acid [Eq. (8.3)];[45] the reaction possibly involves the diprotonated esters or the dicationic **7a** and **7b** intermediates:

$$(8.3)$$

The acylation of alkylbenzenes with benzoyl chloride and a catalytic amount of perfluorobutanesulfonic acid affords the corresponding alkyl-substituted benzophenones with unusually high amounts of *ortho* isomeric products (up to 30%):[38]

R = Me, Et, isoPr, *tert*-Bu   65–86%

$$(8.4)$$

Even *tert*-butylbenzene reacts satisfactorily (86% yield) without de-*tert*-butylation. Acylation of *p*-xylene with benzoic acid gave 71% yield with continuous removal of water. Water removal was also the decisive factor in producing anthraquinones with phthalic anhydride in satisfactory yields (52–89%).

Triflates of boron, aluminum, and gallium were found to be efficient catalysts in Friedel–Crafts acylations.[46] However, these are water-sensitive materials and were required to be used in equimolar quantities to be effective. More recently various water-tolerant and recyclable triflate salts, which were also tested in alkylation, were found to exhibit similar good characteristics in Friedel–Crafts acylations. Although benzene cannot be acylated, Sc triflate,[47,48] lanthanum triflates,[48–51] and Hf triflate[52] usually give high yields of aryl ketones in acylation with acid anhydrides. In many cases, Li perchlorate was found to accelerate the reactions.[48,52]

Bi(OTf)$_3$ (OTf = CF$_3$SO$_2$O) is an even more effective catalyst since it induces the acylation of benzene and even that of deactivated aromatics.[53,54] When applied with benzoic anhydride, the mixed anhydride is generated, which is the active benzoylating species. Ga nonafluorobutanesulfonate has similar good characteristics.[55]

The combined catalyst system of TiCl(OTf)$_3$ and TfOH allows the acylation of benzene and toluene with hexanoic anhydride (66% and 61% yields, respectively).[56]

Further examples include the use of U^{4+} and U^{6+} salts, which are efficient catalysts under dry air,[57] and the homogeneous complex [ReBr(CO)$_5$], which catalyzes the acylation of toluene and *m*-xylene with various acid chorides.[58]

Fluorous biphase catalysis was also applied in Friedel–Crafts acylation with Yb tris(perfluoroalkanesulfonyl)methide catalysts with acid anhydrides.[59] Of the aromatics studied, activated compounds and naphthalene (95% conversion) showed satisfactory reactivity.

Solid Catalysts. Nafion-H is an active catalyst for acylation with aroyl halides and anhydrides.[60,61] The reaction is carried out at the boiling point of the aromatic hydrocarbons. Yields with benzoyl chloride using 10–30% Nafion-H for benzene, toluene, and *p*-xylene are 14%, 85% and 82%, respectively. Attempted acylation with acetyl chloride, however, led to HCl evolution and ketene formation. Nafion resin–silica nanocomposite materials containing a dispersed form of the resin within silica exhibits significantly enhanced activity in Friedel–Crafts acylations.[62,63]

A range of other solid acids was also tested and they usually exhibited good characteristics. These include graphite powders,[64,65] NdCl$_3$ supported on K10 montmorillonite,[66] Ga$_2$O$_3$ and In$_2$O$_3$ supported on MCM-41,[67] and supported ionic liquids.[68]

In acylation with acidic ammonium and cesium salts of phosphotungstic acid, direct correlation was found between activity and the number of accessible Brønsted acid sites.[69,70] Moreover, cesium salts were found to be more active, than HY and H-Beta zeolites.[70]

Since the late 1990s most attention, with respect to the application of solid acid catalysts, has been devoted to sulfated zirconia[71] and zeolites.[2,18] In comparative studies sulfated zirconia and various solid acids, including K10 montmorillonite,[72] ion-exchange resins,[72] H-Beta zeolite,[73] and mixed oxides,[74] were tested with benzoyl chloride or benzoic anhydride as acylating agents. Sulfated zirconia was always found to be the most active catalyst. A significant enhancement of catalytic activity could be obtained by promotion with Al.[75] Comparing activities and physicochemical properties, it was concluded that surface acidity and sufficiently large average pore diameters are preconditions for high catalytic activities.[73]

Numerous acylations mediated by zeolite materials were described since 1986, when the first results were reported.[76] The acylation of toluene and *p*-xylene with C$_2$–C$_{20}$ alkanoic acids over Ce-exchanged NaY at 150°C in a batch reactor was reported. Shape selectivity was observed; maximum activity was found for C$_{12}$ and C$_{14}$ acids. Other studies showed that Brønsted acid centers are involved in acylation[77] and an electrophilic mechanism is operative.[78] A comprehensive review[2] of the subject is available and, therefore, only the more recent results are discussed here.

A conclusion of early studies is that zeolite Y and zeolite Beta are the most promising catalysts for Friedel–Crafts acylation. This is attributed to their large pores, which allow access of the aromatic compound to the catalytically active sites. More recently mainly the medium-pore H-ZSM-5 zeolite[79–81] and the large-pore H-Beta[81–83] were studied. In vapor-phase acylation of benzene with acetic acid, H-ZSM-5 showed the highest activity and HY showed no activity at all.[79] Aceto-phenone was obtained efficiently (95% selectivity at 86% conversion at 250°C) over Ce–H-ZSM-5 using acetic anhydride.[80] Fe-exchanged NaY gives only limited leaching and high yields in benzoylation of alkyl-substituted benzenes.[84]

Acetylation of toluene over Beta zeolite with acetic anhydride at 150°C using a high toluene : anhydride ratio (10–20) ensured reasonable yields of 4-methylaceto-phenone with selectivities close to 100%.[83] However, severe catalyst poisoning occurred. This is caused by the adsorption of the product and coke deposition, resulting in severe diffusion limitations within the catalyst pores. A nanocrystalline sample with high acidity allowed faster product diffusion and showed improved cat-alyst activity (80% conversion) and longer catalyst use time. In another study zeo-lite Beta showed excellent shape-selective behavior in the benzoylation of toluene and naphthalene.[82] Specifically, it exhibited the highest selectivities to form the cor-responding *para*-substituted product from toluene and the 2-substituted isomer from naphthalene among the zeolites studied. Molecular mechanics calculations indicated that zeolite Beta is the best suited in terms of fitting the reactant and pro-duct molecules inside the zeolite cages.

8.2. RELATED ACYLATIONS

8.2.1. Formylation

The only formic acid derivative that allows the direct formylation of aromatics is formyl fluoride[16,17] since others (halides and the anhydride) that could be used in Friedel–Crafts-type acylations are quite unstable. Other related methods, however, are available to transform aromatic hydrocarbons to the corresponding aldehydes. The most frequently used such formylations are the Gattermann–Koch reaction[16,17] and the Gattermann synthesis.[10,16,17]

The Gattermann–Koch Reaction. Gattermann and Koch discovered that CO and HCl in the presence of $AlCl_3$ and Cu_2Cl_2 react with toluene as formyl chloride equivalent at ambient temperature to afford *p*-tolualdehyde in 50% yield.[85] Benzene was unreactive but gave a 90% yield of benzaldehyde with $AlBr_3$. The reaction requires the use of an equimolar amount of the Al halide. Improvements were later achieved by working under elevated CO pressure (>100 atm). In this case, however, the presence of Cu_2Cl_2 is not necessary. Side reactions, such as isomerization and disproportionation, can occur under the Gattermann–Koch reaction conditions. The CO–HF–BF_3 system was found to be effective, allowing much faster reaction, usually at 0°C and 60 atm.[16] Superacidic systems also proved to be highly efficient. These include triflic acid;[86,87] triflic acid combined with HF–BF_3, SbF_5, or TaF_5;[87] and HSO_3F–SbF_5.[88] With triflic acid, a relatively weak superacid, the reaction is carried out under CO pressure and by using excess of the acid. Sulfonylation competes with formylation in $HSO_3F–SbF_5$, but selective formylation can be obtained at high acidity. $HF–SbF_5$ is also an active catalyst, allowing easy, one-pot diformylation of polynuclear aromatics.[89]

The Gattermann–Koch synthesis is considered a typical electrophilic aromatic substitution with high *para* regioselectivity. In fact, among the most frequently used formylation methods, the traditional Gattermann–Koch reaction shows the highest selectivity. It gives both high substrate selectivities ($k_{toluene} : k_{benzene}$ ratios = 155–860) and high degree of *para* substitution (89–96%).[90,91] Selectivities, however, can vary with the used acid systems. The lowest selectivity was observed in the $HF–SbF_5$-catalyzed reaction ($k_{toluene} : k_{benzene} = 1.6$ with 45.2% *ortho*, 2.7% *meta*, and 52.1% *para* isomers at $-95°C$ in SO_2ClF solvent).[90] CO undergoes protonation in the superacidic system to give protosolvated or diprotonated species that are very reactive electrophiles.[31] The reaction is much more selective at $0°C$ in excess toluene as the solvent yielding 89.9% of *para* isomer.

Studies by Tanaka and coworkers showed that the regioselectivity of the Gattermann–Koch reaction in superacidic systems is influenced by the protonation of the aromatic compound.[92–95] For example, in the formylation of 1-methylnaphthalene with $HF–SbF_5$, the regioselectivity drastically changed by changing the SbF_5 to aromatic ratio; specifically, the selective formation of the 4-substituted isomer changed and a significant amount of the 2-substituted isomer was detected when excess SbF_5 was used:[92]

SbF_5 : substrate			yield
1 (0°C)	100	0	24%
1.5 (−40°C)	68	32	13%

In addition, high *ortho* regioselectivity (32%) was observed for bibenzyl, whereas biphenyl, diphenylmethane and 1,3-diphenylpropane exhibited the usual high *para* regioselectivity (100%, 98%, and 95%, respectively).[95]

In these cases the formation of a π complex was suggested to occur from the protonated aromatic (σ complex) and the proelectrophile (P) followed by the formation of the σ complex as suggested by Cacace:[96]

$$(8.6)$$

A similar initial interaction between the arene and the Lewis acid catalyst was suggested recently on the basis of theoretical calculations.[97]

Kinetic and regioselectivity studies provided evidence in favor of a reaction, where the formylation electrophile HCO^+ is generated by CO protonation by the arenium ion.[94] High *para* selectivity stems from such intracomplex reaction, and the observed regioselectivity reflects the ratio between the intracomplex and conventional reaction routes. In the formylation of bibenzyl, semiempirical calculation indicated remarkable preference for *ortho* monoprotonation. High *ortho* regioselectivity results from the intracomplex reaction, whereby CO is protonated by the *ortho* σ complex undergoing rapid proton transfer to generate formyl cation in the vicinity of the *ortho* position:

$$(8.7)$$

The formyl cation could for long not be directly observed by NMR spectroscopy, although its intermediacy was well recognized in aromatic formylations.[86,87,98] In a more recent study, however, using $HF–SbF_5$ (1 : 1) the strongest known superacid together with pressurized NMR and IR cells, which allowed application of high CO pressure (up to 100 atm) during the measurements, evidence was presented to show the existence of HCO^+ associated with equilibrating anions of the type $[Sb_xF_{5x+1}]^-$.[99,100]

The Gattermann Reaction. In the Gattermann aldehyde synthesis anhydrous hydrogen cyanide is reacted in the presence of $HCl–AlCl_3$ to afford aromatic aldehydes.[101,102] The method was originally developed to produce aldehydes of phenols and phenolic ethers that could not be prepared under Gattermann–Koch conditions. Benzene is unreactive at 40°C, the temperature used in Gattermann's original studies, and, therefore, was often used as solvent. Benzene, alkylenzenes, and polycyclic aromatics, however, can also be formylated at more elevated temperatures (>100°C). For example, the yields of *p*-tolualdehyde from toluene are 14% and 100% at 40°C and 100°C, respectively. The modification by Adams eliminates the necessity to use anhydrous HCN.[103] $Zn(CN)_2$ provides HCN and $ZnCl_2$ as the catalyst in situ in the presence of HCl.

The Gattermann synthesis is somewhat less selective than the Gattermann–Koch reaction. In the $HCN–HCl–AlCl_3$ system, the substrate selectivity is 49.1, with 56.4% of *para*-isomer formation, whereas the corresponding values for the $Zn(CN)_2–HCl–AlCl_3$ system in nitromethane solvent are 128 and 63.9%, respectively.[90]

The NH=CH—N=CHCl dimeric species is regarded as the reacting intermediate in the $AlCl_3$-catalyzed reaction, whereas in other systems the formation of imidoformyl chloride was suggested. New studies with superacidic systems (TfOH, TfOH—SbF_5) showed a clear correlation between the reactivity of benzene and acidity.[33,104] These results and the observations that highly acidic media are necessary to obtain sufficiently high yields led to the suggestion that the de facto species that reacts with the aromatics is not the stable monoprotonated ion (**8**) but the highly reactive **9** diprotonated superelectrophile:[31,33]

$$H-C\equiv N \xrightarrow{\;H^+\;} \begin{matrix} H-C\equiv \overset{+}{N}-H \\ \updownarrow \\ H-\overset{+}{C}=N-H \end{matrix} \xrightarrow{\;H^+\;} H-\overset{+}{C}=\overset{+}{N}\overset{H}{\underset{H}{\diagdown}} \qquad (8.8)$$

$$\qquad\qquad\qquad\quad \mathbf{8} \qquad\qquad\qquad \mathbf{9}$$

Protonated HCN (**8**) is resonance-stabilized, shows only limited imidocarbocation character and reacts only with activated benzene derivatives but not with benzene.

Other Formylations. Formyl fluoride, the only known stable formic acid derivative, can be used to perform Friedel–Crafts-type acylation to form aromatic aldehydes. The method was developed by Olah and Kuhn.[105] Although a number of Lewis acids may be used, BF_3 is the best catalyst. It is dissolved in the aromatic compound to be formylated; then formyl fluoride is introduced at low temperature and the reaction mixture is allowed to warm up to room temperature. The aldehydes of benzene, methylbenzenes, and naphthalene were isolated in 56–78% yields. Selectivities are similar to those in the Gattermann synthesis ($k_{toluene} : k_{benzene} =$ 34.6, 53.2% *para* isomer). The reacting electrophile was suggested to be the activated $HCOF \cdot BF_3$ complex and not the free formyl cation. Clearly there is close relationship with the discussed CO—HF—BF_3 system.

There are other methods using reagents, such as dichloromethylalkyl ethers (the Gross formylation)[10,16] and cyanogen bromide (the Koch method).[10,16] A widely used method, the Vilsmeier formylation, gives good results for activated, electron-rich aromatic compounds, but less so for simple aromatics.[10,16] Dimethylformamide and $POCl_3$ as the condensing agent, called the *Vilsmeier–Haack reagent*, is used most frequently.[106] Compound **10** is the apparent reagent. Many improvements were achieved more recently. Polycyclic aromatic compounds were formylated with silica-supported Vilsmeier–Haack reagent in solvent-free conditions under microwave irradiation to yield formylated products in high yields (72–92%).[107] The use of triflic anhydride instead of $POCl_3$ gives the very reactive **11** iminium salt, which allows the formylation of relatively less activated aromatics under mild conditions.[108] Formyl derivatives of naphthalene, acenaphthene, anthracene, and phenanthrene were obtained in 50–98% yields.

$$
\begin{array}{cc}
\underset{H}{\overset{Cl}{\underset{}{\underset{}{\overset{|}{C}}}}}\overset{+}{\underset{N}{\nwarrow}}\overset{Me}{\underset{Me}{}} \ Cl^- & \underset{H}{\overset{CF_3SO_2O}{\underset{}{\overset{|}{C}}}}\overset{+}{\underset{N}{\nwarrow}}\overset{Me}{\underset{Me}{}} \ CF_3SO_2O^- \\
\mathbf{10} & \mathbf{11}
\end{array}
$$

8.2.2. The Houben–Hoesch Synthesis

When a nitrile is reacted with an aromatic compound in the presence of HCl, in the additional presence of $ZnCl_2$ or $AlCl_3$, a ketimine salt is formed, which after subsequent hydrolysis gives an aromatic ketone:

$$
RCN \ + \ ArH \ \xrightarrow{\ HCl, \ ZnCl_2 \ or \ AlCl_3\ } \ Ar-C\overset{\nwarrow NH.HCl}{\underset{R}{}} \ \longrightarrow \ Ar-C\overset{\nwarrow O}{\underset{R}{}} \tag{8.9}
$$

The process is called the Houben–Hoesch reaction.[14]

The reacting species is a rather weak electrophile and, therefore, only particularly reactive aromatics are suitable substrates. Consequently, the reaction is restricted for polyhydric phenols and their ethers, and reactive heterocycles. With strong acidic systems, such as $AlCl_3$ in halogenated benzene solvents, alkylbenzenes can react at more elevated temperatures ($>50°C$). Trichloroacetonitrile works also well with nonactivated aromatics. The **12** chloroimine, the protonated nitrile (**13**), or the nitrile coordinated with the Lewis acid are possible involved electrophiles.

$$
\underset{Cl}{\overset{R}{\diagdown}}C\overset{+}{=}\overset{+}{N}H_2 \qquad R-C\equiv\overset{+}{N}H \ \longleftrightarrow \ R-\overset{+}{C}=NH \qquad R-\overset{+}{\underset{}{C}}=\overset{+}{N}H_2
$$

$$
\quad \mathbf{12} \qquad\qquad\qquad \mathbf{13} \qquad\qquad\qquad\qquad \mathbf{14}
$$

More recent studies with superacidic systems (TfOH, TfOH–SbF_5), used also in the Gattermann reaction, indicated that strong acids significantly increase reactivities of benzene with benzonitrile.[33,104] It is concluded that the superelectrophilic **14** dication formed as a results of protonation of **13** is the reactive species in the Houben–Hoesch reaction.

8.3. ACYLATION OF ALIPHATIC COMPOUNDS

In principle, the acylation of aliphatic compounds is analogous with the Friedel–Crafts acylation of aromatics in the sense that a hydrogen of the reacting alkanes, alkenes, or alkynes is replaced by an acyl group to yield ketones, unsaturated ketones, or conjugated acetylenic ketones, respectively. As discussed subsequently, however, the reactions are more complex. The acylation of aliphatics is an important but less frequently used and studied process.[11-13]

8.3.1. Acylation of Alkenes

Of the acylation of aliphatics, the acylation of alkenes pioneered by Nenitzescu was mainly explored, because the resulting unsaturated ketones are intermediates in synthesis. This acylation of alkenes, however, has its difficulties. First, the product unsaturated ketones, which are reactive compounds themselves, can undergo various further transformations, such as addition, elimination, and isomerization, often resulting in complex product mixtures. The acylation of alkenes, therefore, is less selective and often yields products other than expected by a simple substitution of one of the vinylic hydrogens.

The frequent and sometimes exclusive formation of β,γ-unsaturated ketones was long an intriguing aspect of the Friedel–Crafts acylation of alkenes.[11,13] In fact, it is often found that the more stable α,β-unsaturated ketone is produced by isomerization of the initially formed nonconjugated isomer. This can be understood by the carbocationic mechanism as shown in Eq. (8.10). The reaction is interpreted to proceed through **15** the most stable β-keto carbocationic intermediate. The loss of a proton usually results in the nonconjugated unsaturated ketone that is formed by a [1,5] hydrogen shift followed by an enol–oxo tautomerization. Because of the lack of driving force for rearomatization by a proton loss, as is the case in acylation of aromatics, the intermediate carbocation may stabilize by reacting with the halide of the acid halide reagent, often leading to the formation of β-halo ketones [Eq. (8.11)]:

$$(8.10)$$

$$(8.11)$$

15

Under nonequilibrium conditions the β,γ-unsaturated ketone tends to predominate, which may be transformed to the more stable conjugated isomer. The intermediacy of **15** may allow further rearrangements to occur. However, acylation with preformed acylium ions, such as acetyl hexachloroantimonate, and in the presence of a base to prevent isomerization, affords exclusive formation of β,γ-unsaturated ketones.[109] This led to the suggestion of an ene reaction between the acylium ion and the alkene as a possible mechanistic pathway:

$$(8.12)$$

AlCl$_3$ is the strongest and most frequently used Lewis acid catalyst in aliphatic Friedel–Crafts acylations usually applied with acid chlorides, whereas ZnCl$_2$ is used most often with acid anhydrides as the acylating agent.[11,13,110] An interesting and novel variation is acylation with acid halides by active Zn compounds, which are prepared from a Zn–Cu couple and alkyl iodides.[111] $TiCl_4$ and $SnCl_4$ are also useful catalysts.[11,13] The ease of alkene acylation and the propensity of alkenes to undergo oligomerization under acidic conditions nessesitates that Friedel–Crafts acylation of alkenes usually be carried out at lower temperature.

Alkene acylation may be used in the synthesis of cyclic products by appropriately selecting the alkene and the acid derivative. When α,β-unsaturated acid halides react with cycloalkenes, the intermediate undergoes cyclization to yield octalones and indanones:[112]

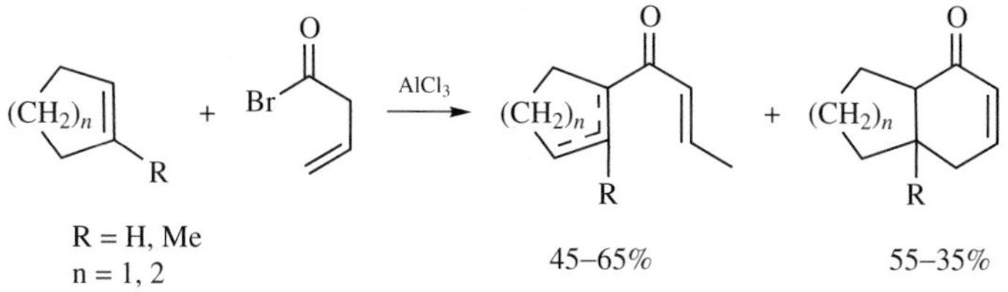

$$R = H, Me$$
$$n = 1, 2$$

45–65%   55–35%   (8.13)

Acylation with cyclopentenyl- or cyclohexenylacetyl chloride gave three-membered ring systems.[113,114] An interesting example is the reaction of ethylene with acid chlorides complexed with excess $AlCl_3$ to give the **17** β,γ-unsaturated ketones:[115]

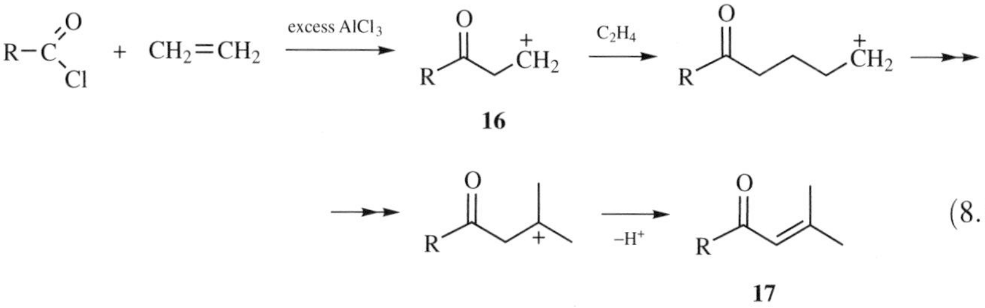

(8.14)

This was explained by the intermediate **16** reacting with a second molecule of ethylene to form a new carbocation that undergoes a series of hydride and methyl shifts to yield the end product.

Since unsaturated ketones themselves are highly reactive compounds, they may react with excess of acid chloride in an overall diacylation process. The unsaturated diketones readily undergo cyclization to form pyrilium salts. In fact, this is often the best way to synthesize pyrilium salts.[116]

The acylation of alkenes with trifluoroacetic anhydride having only weak electrophilicity was reported to give α,β-unsaturated ketones in 19–49% yields.[117,118] The reaction requires the use of the Me_2S–BF_3 complex as catalyst and takes place only with alkenes that readily form stable carbocations. Alkenes were also electrolyzed in CH_2Cl_2 solution of Et_4NCl as the electrolyte at -5 to $-10°C$ in the presence of acetyl chloride or acetic anhydride.[119] The catalyst was electrochemically generated using a sacrificial Al anode. Terminal alkenes and cycloalkenes afforded conjugated ketones, whereas 1-methylcyclopentene and 1-methylcyclohexene gave mixtures of unsaturated ketones.

In the acylation of alkenes with acetyl chloride[120] or acetic anhydride[120,121] over zeolites (HX, HY, H-mordenite, H-ZSM-5, H-Beta) usually complex product mixtures including isomeric unsaturated ketones, chloroketones, and acetates were obtained. Zeolite HY proved to be the most active catalyst under mild experimental conditions (25–60°C).[120,121]

An interesting example of an indirect acylation of alkenes is the reaction of small olefins (ethylene, propylene, isobutylene) over H-ZSM-5 in the presence of CO at 23°C to give unsaturated ketones.[122] In this case the acylium ion is formed in situ from the alkene and CO in the pores of the zeolite at ambient temperature:

$$
\begin{array}{c}
\text{(see scheme)} \\
\textbf{19}
\end{array}
$$

$$
\text{alkene} \underset{-H^+}{\rightleftharpoons} R^{1+} \underset{CO}{\rightleftharpoons} R^1-\overset{+}{C}=O \underset{\text{alkene}}{\rightleftharpoons} \underset{\textbf{18}}{R^2 \overset{+}{\frown}_O R^1} \underset{H^+}{\overset{-H^+}{\rightleftharpoons}} R^2 \frown_O R^1
$$

$$(8.15)$$

This reacts with a second molecule of alkene to form the **18** carbocation which stabilizes by a proton loss to form the β,γ-unsaturated ketone. The **19** cyclic carboxonium ion was observed by CP-MAS NMR spectroscopy.

8.3.2. Acylation of Alkynes

Alkynes are readily acylated with acid chlorides under Friedel–Crafts conditions to form, in most cases, *trans*-β-chlorovinyl ketones through the corresponding vinyl cation intermediate [Eq. (8.16)]. The first study in 1935 reported low yields.[11] Later in acylations with acyl triflates, β-keto vinyl triflates were obtained in satisfactory yields.[123] When aroyl derivatives are used, the intermediate can undergo cyclization to form indenones. Chlorovinyl ketones formed from terminal alkynes may also react further losing hydrogen chloride to yield conjugated acetylenic ketones:[11,13]

$$
R^1\!\!=\!\!R^2 + R^3-C\overset{O}{\underset{X}{\diagdown}} \xrightarrow{\text{Lewis acid}} \underset{X \quad R^2}{\overset{R^1 \diagup\diagdown R^3}{\diagdown C \diagup}} \xrightarrow[-HCl]{\substack{X=Cl \\ R^1=H}} R^1\!\!=\!\!\overset{O}{\overset{\|}{C}}-R^3 \quad (8.16)
$$

$$X = Cl, CF_3SO_2O$$

Clean formation of α,β-acetylenic ketones could be achieved by carrying out the reaction with a catalytic amount of CuI in the presence of Et_3N [Eq. (8.17)][124] or inducing the acylation with a Pd complex and CuI in the presence of Et_2NH:[125]

$$R^1 {-}\!\!\equiv\!\!{-} H \ + \ R^2{-}C\underset{X}{\overset{O}{\diagup}} \quad \xrightarrow[\text{Et}_3\text{N, RT, 30 h}]{\text{CuI}} \quad R^1 {-}\!\!\equiv\!\!{-} \overset{\overset{\textstyle O}{\|}}{C}{-}R^2 \qquad (8.17)$$

$$44\text{--}83\%$$

$$R^1 = n\text{-Bu, Ph, 1-naphthyl}$$
$$R^2 = \text{isoPr, Ph, 4-MeC}_6\text{H}_4$$

An interesting example is the acylation of various terminal acetylenes with pre-formed acylium tetrafluoroborates in the presence of aromatics:[126]

$$R^1 {-}\!\!\equiv\!\!{-} H \ + \ R^2CO^+\,BF_4^- \ + \ Ar{-}F \quad \xrightarrow[-55°C,\ 10\ min]{\substack{\text{ClCH}_2\text{CH}_2\text{Cl,} \\ \text{CH}_2\text{Cl}_2,\ \text{benzene}}} \quad \underset{\underset{\textstyle 40\text{--}78\%}{Ar \qquad H}}{\overset{\overset{\textstyle O}{\|}}{R^1 \diagdown\!\!\diagup R^2}} \ \mathbf{20} \qquad (8.18)$$

$$R^1 = \text{Me, } n\text{-Bu, Ph, 2-, 3-, or 4-MeC}_6\text{H}_4$$
$$R^2 = \text{Et, } tert\text{-Bu}$$
$$Ar = \text{Ph, 2-, 3-, or 4-MeC}_6\text{H}_4, \text{ 2,4,6-triMeC}_6\text{H}_2$$

Under the reaction conditions the direct acylation of the aromatic hydrocarbons with the acylium ion does not take place. Instead, the intermediate vinyl carbocation reacts with the aromatics to form the **20** aryl-substituted vinyl ketones.

8.3.3. Acylation of Alkanes

The Friedel–Crafts acylation of alkanes requires hydride abstraction, which can be induced by the acylium ion itself, to form the corresponding carbocation. This may undergo carbocationic rearrangements prior to a proton loss to form an alkene, which then reacts with the acylating agent. Similar to the acylation of alkenes, the product is an unsaturated ketone. The reaction is limited to alkanes that are prone to undergo hydride transfer.

Cyclohexane can be acylated with acetyl chloride and $AlCl_3$ to yield 1-acetyl-2-methylcyclopentene in 37% yield.[127] Alkanes, such as 2-methylbutane, cyclohexane, methylcyclopentane, and methylcyclohexane, are easily acylated with α,β-unsaturated acyl chlorides in the presence of $AlCl_3$ and a hydride acceptor to afford mono- or bicyclic products after secondary transformations.[128] Isoalkanes (isopentane, 2-methylpentane, 3-methylpentane, 2,3-dimethylbutane) undergo diacylation and eventually form pyrilium salts under Friedel–Crafts acylation conditions with acetyl chloride or acetic anhydride.[129]

The formation of 1-acetyl-2-methylcyclopentene was also observed when either cyclohexane or cyclopentene was reacted by Vol'pin and coworkers with an excess of the aprotic organic superacid $MeCOCl{-}2AlBr_3$:[130,131]

$$(8.19)$$

However, when a 1 : 1 molar ratio was applied, the corresponding saturated ketone was obtained, albeit in low yield. Cyclopentane, in turn, gives acetylcyclopentane in 60–80% yields.

REFERENCES

1. C. Friedel and J. M. Crafts, *Compt. Rend.* **84**, 1392 (1877); **84**, 1450 (1877).

2. H. W. Kouwenhoven and H. van Bekkum, in *Handbook of Heterogeneous Catalysis*, G. Ertl, H. Knözinger, and J. Weitkamp, eds., Wiley-VCH, Weinheim, 1997, Chapter 4.9, p. 2358.

3. P. H. Gore, *Chem. Rev.* **55**, 229 (1955).

4. P. H. Gore, in *Friedel–Crafts and Related Reactions*, Vol. 3, G. A. Olah, ed., Interscience, New York, 1964, Chapter 31, p. 1.

5. A. G. Peto, in *Friedel–Crafts and Related Reactions*, Vol. 3, G. A. Olah, ed., Interscience, New York, 1964, Chapter 34, p. 535.

6. F. R. Jensen and G. Goldman, in *Friedel–Crafts and Related Reactions*, Vol. 3, G. A. Olah, ed., Interscience, New York, 1964, Chapter 36, p. 1003.

7. C.-W. Schellhammer, in *Methoden der Organischen Chemie (Houben–Weyl)*, Vol. 7/2a: *Ketone*, E. Müller, ed., Thieme, Stuttgart, 1973, p. 15–370.

8. G. A. Olah, *Friedel–Crafts Chemistry*, Wiley-Interscience, New York, 1973.

9. H. Heaney, in *Comprehensive Organic Synthesis*, B. M. Trost and I. Fleming, eds., Pergamon Press, 1991, Vol. 2: *Additions to C–X π-Bonds*, Part 2, C. H. Heathcock, ed., Chapter 3.2, p. 733.

10. D. P. N. Satchell and R. S. Satchell, in *The Chemistry of the Carbonyl Group*, S. Patai, ed., Wiley, London, 1966, Chapter 5, p. 233.

11. C. D. Nenitzescu and A. T. Balaban, in *Friedel–Crafts and Related Reactions*, Vol. 3, G. A. Olah, ed., Interscience, New York, 1964, Chapter 37, p. 1033.

12. J. K. Groves, *Chem. Soc. Rev.* **1**, 73 (1972).

13. S. C. Eyley, in *Comprehensive Organic Synthesis*, B. M. Trost and I. Fleming, eds., Pergamon Press, 1991, Vol. 2: *Additions to C–X π-Bonds*, Part 2, C. H. Heathcock, ed., Chapter 3.1, p. 707.

14. O. Bayer, in *Methoden der Organischen Chemie (Houben–Weyl)*, Vol. 7/1: *Aldehyde*, E. Müller, ed., Thieme, Stuttgart, 1954, p. 15–36.

15. W. Ruske, in *Friedel–Crafts and Related Reactions*, Vol. 3, G. A. Olah, ed., Interscience, New York, 1964, Chapter 32, p. 383.

16. G. A. Olah and S. J. Kuhn, in *Friedel–Crafts and Related Reactions*, Vol. 3, G. A. Olah, ed., Interscience, New York, 1964, Chapter 38, p. 1153.

17. G. A. Olah, L. Ohannesian, and M. Arvanaghi, *Chem. Rev.* **87**, 671 (1987).

18. A. Corma and H. García, *Catal. Today* **38**, 257 (1997).

19. M. G. Clerici, *Top. Catal.* **13**, 373 (2000).

20. P. Métivier, in *Fine Chemicals through Heterogeneous Catalysis*, R. A. Sheldon and H. van Bekkum, eds., Wiley–VCH, Weinheim, 2001, Chapter 4.4, p. 161.

21. T. Okuhara, N. Mizuno, and M. Misono, *Adv. Catal.* **41**, 113 (1996).

22. M. Misono, *Chem. Commun.* 1141 (2001).

23. I. V. Kozhevnikov, *Chem. Rev.* **98**, 171 (1998).

24. B. Chevrier and R. Weiss, *Angew. Chem., Int. Ed. Engl.* **13**, 1 (1974).

25. G. A. Olah, *Acc. Chem. Res.* **4**, 240 (1971).

26. C. Galli and S. Fornarini, *J. Chem. Soc., Perkin Trans. 2* 1147 (1993).

27. G. A. Olah, S. J. Kuhn, S. H. Flood, and B. A. Hardie, *J. Am. Chem. Soc.* **86**, 2203 (1964).

28. G. A. Olah, H. C. Lin, and A. Germain, *Synthesis* 895 (1974).

29. G. A. Olah, J. Lukas, and E. Lukas, *J. Am. Chem. Soc.* **91**, 5319 (1969).

30. G. A. Olah, *Angew. Chem., Int. Ed. Engl.* **12**, 173 (1973).

31. G. A. Olah, *Angew. Chem., Int. Ed. Engl.* **32**, 767 (1993).

32. N. Hartz, G. Rasul, and G. A. Olah, *J. Am. Chem. Soc.* **115**, 1277 (1993).

33. Y. Sato, M. Yato, T. Ohwada, S. Saito, and K. Shudo, *J. Am. Chem. Soc.* **117**, 3037 (1995).

34. F. Effenberger, J. K. Eberhard, and A. H. Maier, *J. Am. Chem. Soc.* **118**, 12572 (1996).

35. F. Effenberger and A. H. Maier, *J. Am. Chem. Soc.* **123**, 3429 (2001).

36. F. Effenberger, F. Reisinger, K. H. Schönwälder, P. Bäuerle, J. J. Stezowski, K. H. Jogun, K. Schöllkopf, and W.-D. Stohrer, *J. Am. Chem. Soc.* **109**, 882 (1987).

37. F. Effenberger, E. Sohn, and G. Epple, *Chem. Ber.* **116**, 1195 (1983).

38. F. Effenberger, F. Buckel, A. H. Maier, and J. Schmider, *Synthesis* 1427 (2000).

39. J. Marquié, C. Laporte, A. Laporterie, J. Dubac, J.-R. Desmurs, and N. Roques, *Ind. Eng. Chem. Res.* **39**, 1124 (2000).

40. J. A. Hyatt and P. W. Raynolds, *J. Org. Chem.* **49**, 384 (1984).

41. D. E. Pearson and C. A. Buehler, *Synthesis* 533 (1972).

42. F. Effenberger and G. Epple, *Angew. Chem., Int. Ed. Engl.* **11**, 300 (1972).

43. F. Effenberger and G. Epple, *Angew. Chem., Int. Ed. Engl.* **11**, 299 (1972).

44. R. D. Howells and J. D. Mc Cown, *Chem. Rev.* **77**, 69 (1977).

45. J. P. Hwang, G. K. S. Prakash, and G. A. Olah, *Tetrahedron* **56**, 7199 (2000).

46. G. A. Olah, O. Farooq, S. M. F. Farnia, and J. A. Olah, *J. Am. Chem. Soc.* **110**, 2560 (1988).

47. S. Kobayashi, *Eur. J. Org. Chem.* 15 (1999).

48. A. Kawada, S. Mitamura, and S. Kobayashi, *Chem. Commun.* 183 (1996).

49. S. Kobayashi, *Synlett* 689 (1994).

50. R. W. Marshman, *Aldrichimica Acta* **28**, 77 (1995).

51. A. Kawada, S. Mitamura, J. Matsuo, T. Tsuchiya, and S. Kobayashi, *Bull. Chem. Soc. Jpn.* **73**, 2325 (2000).

52. I. Hachiya, M. Moriwaki, and S. Kobayashi, *Bull. Chem. Soc. Jpn.* **68**, 2053 (1995).

53. J. R. Desmurs, M. Labrouillére, C. Le Roux, H. Gaspard, A. Laporterie, and J. Dubac, *Tetrahedron Lett.* **38**, 8871 (1997).

54. S. Répichet, C. Le Roux, J. Dubac, and J.-R. Desmurs, *Eur. J. Org. Chem.* 2743 (1998).

55. J. Matsuo, K. Odashima, and S. Kobayashi, *Synlett* 403 (2000).

56. J. Izumi and T. Mukaiyama, *Chem. Lett.* 739 (1996).

57. D. Barbier-Baudry, A. Bouazza, J. R. Desmurs, A. Dormond, and S. Richard, *J. Mol. Catal. A: Chem.* **164**, 195 (2000).

58. H. Kusama and K. Narasaka, *Bull. Chem. Soc. Jpn.* **68**, 2379 (1995).

59. A. G. M. Barrett, D. C. Braddock, D. Catterick, D. Chadwick, J. P. Henschke, and R. M. McKinnell, *Synlett* 847 (2000).

60. G. A. Olah, R. Malhotra, S. C. Narang, and J. A. Olah, *Synthesis* 672 (1978).

61. G. A. Olah, G. K. S. Prakash, J. Sommer, *Superacids*. Wiley-Interscience, New York, 1985, Chapter 5.3.

62. M. A. Harmer, W. E. Farneth, and Q. Sun, *J. Am. Chem. Soc.* **118**, 7708 (1996).

63. M. A. Harmer, Q. Sun, A. J. Vega, W. E. Farneth, A. Heidekum, and W. F. Hoelderich, *Green Chem.* **2**, 7 (2000).

64. M. Kodomari, Y. Suzuki, and K. Yoshida, *Chem. Commun.* 1567 (1997).

65. C. Laporte, P. Baules, A. Laporterie, J. R. Desmurs, and J. Dubac, *Compt. Rend. Acad. Sci. Paris* **1**(ser. IIc), 141 (1998).

66. D. Baudry-Barbier, A. Dormond, and F. Duriau-Montagne, *J. Mol. Catal. A: Chem.* **149**, 215 (1999).

67. V. R. Choudary, S. K. Jana, and B. P. Kiran, *J. Catal.* **192**, 257 (2000).

68. M. H. Valkenberg, C. deCastro, and W. F. Hölderich, *Appl. Catal., A* **215**, 185 (2001).

69. Y. Izumi, M. Ogawa, and K. Urabe, *Appl. Catal., A* **132**, 127 (1995).

70. C. De Castro, J. Primo, and A. Corma, *J. Mol. Catal. A: Chem.* **134**, 215 (1998).

71. G. D. Yadav and J. J. Nair, *Micropor. Mesopor. Mat.* **33**, 1 (1999).

72. G. D. Yadav and A. A. Pujari, *Green Chem.* **1**, 69 (1999).

73. J. Deutsch, V. Qaschning, E. Kemnitz, A. Auroux, H. Ehwald, and H. Lieske, *Top. Catal.* **13**, 281 (2000).

74. K. Arata, H. Nakamura, and M. Shouji, *Appl. Catal., A* **197**, 213 (2000).

75. Y. Xia, W. Hua, and Z. Gao, *Catal. Lett.* **55**, 101 (1998).

76. B. Chiche, A. Finiels, C. Gauthier, P. Geneste, J. Graille, and D. Pioch, *J. Org. Chem.* **51**, 2128 (1986).

77. C. Gauthier, B. Chiche, A. Finiels, and P. Geneste, *J. Mol. Catal.* **50**, 219 (1989).

78. B. Chiche, A. Finiels, C. Gauthier, and P. Geneste, *J. Mol. Catal.* **30**, 365 (1987).

79. A. P. Singh and A. K. Pandey, *J. Mol. Catal. A: Chem.* **123**, 141 (1997).

80. P. R. Reddy, M. Subrahmanyam, and V. D. Kumari, *Catal. Lett.* **61**, 207 (1999).

81. A. K. Pandey and A. P. Singh, *Catal. Lett.* **44**, 129 (1997).

82. A. Chatterjee, D. Bhattacharya, T. Iwasaki, and T. Ebina, *J. Catal.* **185**, 23 (1999).

83. P. Botella, A. Corma, J. M. López-Nieto, S. Valencia, and R. Jacquot, *J. Catal.* **195**, 161 (2000).

84. P. Laidlaw, D. Bethell, S. M. Brown, and G. J. Hutchings, *J. Mol. Catal. A: Chem.* **174**, 187 (2001).

85. L. Gattermann and J. A. Koch, *Chem. Ber.* **30**, 1622 (1897).

86. B. L. Booth, T. A. El-Fekky, and G. F. M. Noori, *J. Chem. Soc., Perkin Trans. 1* 181 (1980).

87. G. A. Olah, K. Laali, and O. Farooq, *J. Org. Chem.* **50**, 1483 (1985).

88. M. Tanaka, J. Iyoda, and Y. Souma, *J. Org. Chem.* **57**, 2677 (1992).

89. M. Tanaka, M. Fujiwara, H. Ando, and Y. Souma, *J. Org. Chem.* **58**, 3213 (1993).

90. G. A. Olah, F. Pelizza, S. Kobayashi, and J. A. Olah, *J. Am. Chem. Soc.* **98**, 296 (1976).

91. G. A. Olah, G. K. S. Prakash, and J. Sommer, *Superacids*, Wiley-Interscience, New York, 1985, Chapter 5.5.

92. M. Tanaka, M. Fujiwara, H. Ando, and Y. Souma, *Chem. Commun.* 159 (1996).

93. M. Tanaka, M. Fujiwara, and H. Ando, *J. Org. Chem.* **60**, 2106, 3846 (1995).

94. M. Tanaka, M. Fujiwara, Q. Xu, Y. Souma, H. Ando, and K. K. Laali, *J. Am. Chem. Soc.* **119**, 5100 (1997).

95. M. Tanaka, M. Fujiwara, Q. Xu, H. Ando, and T. J. Raeker, *J. Org. Chem.* **63**, 4408 (1998).

96. M. Aschi, M. Attiná, and F. Cacace, *Angew. Chem., Int. Ed. Engl.* **34**, 1589 (1995); *J. Am. Chem. Soc.* **117**, 12832 (1995).

97. P. Tarakeshwar, J. Y. Lee, and K. S. Kim, *J. Phys. Chem. A* **102**, 2253 (1998).

98. O. Farooq, M. Marcelli, G. K. S. Prakash, and G. A. Olah, *J. Am. Chem. Soc.* **110**, 864 (1988).

99. P. J. F. de Rege, J. A. Gladysz, and I. T. Horvath, *Science* **276**, 776 (1997).

100. T. S. Sorensen, *Angew. Chem., Int. Ed. Engl.* **37**, 603 (1998).

101. L. Gattermann, *Chem. Ber.* **31**, 1149 (1898).

102. L. Gattermann and W. Berchelmann, *Chem. Ber.* **31**, 1765 (1898).

103. R. Adams and I. Levine, *J. Am. Chem. Soc.* **45**, 2373 (1923).

104. M. Yato, T. Ohwada, and K. Shudo, *J. Am. Chem. Soc.* **113**, 691 (1991).

105. G. A. Olah and S. J. Kuhn, *J. Am. Chem. Soc.* **82**, 2380 (1960).

106. A. Vilsmeier and A. Haack, *Chem. Ber.* **60**, 119 (1927).

107. S. Paul, M. Gupta, and R. Gupta, *Synlett* 1115 (2000).

108. A. G. Martínez, R. M. Alvarez, J. O. Barcina, S. de la Moya Cerero, E. T. Vilar, A. G. Fraile, M. Hanack, and L. R. Subramanian, *J. Chem. Soc., Chem. Commun.* 1571 (1990).

109. H. M. R. Hoffmann and T. Tsushima, *J. Am. Chem. Soc.* **99**, 6008 (1977).

110. P. Beak and K. R. Berger, *J. Am. Chem. Soc.* **102**, 3848 (1980).

111. T. Shono, I. Nishiguchi, M. Sasaki, H. Ikeda, and M. Kurita, *J. Org. Chem.* **48**, 2503 (1983).

112. Z. Bounkhala and S. Hacini, *J. Chem. Soc., Chem. Commun.* 263 (1979).

113. A. Tubul and M. Santelli, *J. Chem. Soc., Chem. Commun.* 191 (1988).

114. R. Faure, A. Pommier, J.-M. Pons, M. Rajzmann, and M. Santelli, *Tetrahedron* **48**, 8419 (1992).

115. F. X. Bates, J. A. Donnelly, and J. R. Keegan, *Tetrahedron* **47**, 4991 (1991).

116. A. T. Balaban, A. Dinculescu, G. N. Dorofeenko, G. W. Fischer, A. V. Koblik, V. V. Mezheritskii, and W. Schroth, *Adv. Heterocycl. Chem.* 1982, Suppl. 2.

117. V. G. Nenajdenko, I. F. Leshcheva, and E. S. Balenkova, *Tetrahedron* **50**, 775 (1994).

118. V. G. Nenajdenko, I. D. Gridnev, and E. S. Balenkova, *Tetrahedron* **50**, 11023 (1994).

119. R. D. Vućkievic, L. Joksovic, S. Konstantinovic, Z. Markovic, and M. Lj. Mihailovic, *Bull. Chem. Soc. Jpn.* **71**, 899 (1998).

120. E. Armengol, A. Corma, L. Fernández, H. García, and J. Primo, *Appl. Catal., A* **158**, 323 (1997).

121. K. Smith, Z. Zhenhua, L. Delaude, and P. K. G. Hodgson, in *Heterogeneous Catalysis and Fine Chemicals IV*, Studies in Surface Science and Catalysis, Vol. 108, H. U. Blaser, A. Baiker, and R. Prins, eds., Elsevier, Amsterdam, 1997, p. 99.

122. M. V. Luzgin, V. N. Romannikov, A. G. Stepanov, and K. I. Zamaraev, *J. Am. Chem. Soc.* **118**, 10890 (1996).

123. H. Martens, F. Janssens, and G. Hoornaert, *Tetrahedron* **31**, 177 (1975).

124. C. Chowdhury and N. G. Kundu, *Tetrahedron Lett.* **37**, 7323 (1996).

125. Y. Tohda, K. Sonogashira, and N. Hagihara, *Synthesis* 777 (1977).

126. A. A. Schegolev, W. A. Smit, S. A. Khurshudyan, V. A. Chertkov, and V. F. Kucherov, *Synthesis* 324 (1977).

127. K. E. Harding, K. S. Clement, J. C. Gilbert, and B. Wiechman, *J. Org. Chem.* **49**, 2049 (1984).

128. C. Morel-Fourrier, J.-P. Dulcère, and M. Santelli, *J. Am. Chem. Soc.* **113**, 8062 (1991).

129. M. Arnaud, A. Pedra, C. Roussel, and J. Metzger, *J. Org. Chem.* **44**, 2972 (1979).

130. I. S. Akhrem, A. V. Orlinkov, E. I. Mysov, and M. E. Vol'pin, *Tetrahedron Lett.* **22**, 3891 (1981).

131. M. Vol'pin, I. Akhrem, and A. Orlinkov, *New J. Chem.* **13**, 771 (1989).

9

OXIDATION–OXYGENATION

We are, in general, continuing the practice of calling *oxidation of hydrocarbons* those processes in which hydrocarbons are transformed to oxygen-containing (oxygenated, oxyfunctionalized) products. It should be noted, however, that these processes more properly should be called *oxygenation* or *oxyfunctionalization*. The term *oxidation*, in its proper (literal) use, refers to the increase of the oxidation number of an element to a more positive value. Thus dehydrogenation of alkanes to alkenes is oxidation but not oxyfunctionalization (oxygenation). The differentiation was suggested[1] and is indeed used, but only to a limited degree. We discuss oxidation of hydrocarbons to oxygenated products with this caveat.

9.1. OXIDATION OF ALKANES

9.1.1. Oxidation to Alcohols and Carbonyl Compounds

Autoxidation of Alkanes. The slow oxidation of alkanes with ground-state (triplet) molecular oxygen in the liquid phase under mild conditions, called *autoxidation*, affords organic hydroperoxides as primary products through a free-radical chain reaction.[2–5] The oxidation requires initiation by a radical ($I^{.}$) [Eq. (9.1)]. This may be formed by the thermal decomposition of suitable organic compounds (azoalkanes, organic peroxides) added to the reaction mixture. Propagation, or addition of the alkyl radical to oxygen, is usually a rapid, diffusion-controlled reaction that leads to alkylperoxy radicals [Eq. (9.2)]. A slow, in most cases rate-determining, hydrogen abstraction follows to yield alkyl hydroperoxides [Eq. (9.3)]:

$$R-H \ + \ I \cdot \ \longrightarrow \ R \cdot \ + \ I\,H \tag{9.1}$$

$$R \cdot \ + \ O_2 \ \longrightarrow \ ROO \cdot \tag{9.2}$$

$$ROO \cdot \ + \ R-H \ \longrightarrow \ ROOH \ + \ R \cdot \tag{9.3}$$

Equations (9.4) and (9.5) illustrate termination reactions. Depending on the structure of the alkyl group, disproportionation to alcohols and carbonyl compounds may take place:

$$R \cdot \ + \ ROO \cdot \ \longrightarrow \ ROOR \tag{9.4}$$

$$2\,ROO \cdot \ \longrightarrow \ RO_4R \ \longrightarrow \ \text{molecular products} \ + \ O_2 \tag{9.5}$$

Autoxidation without added initiator may also occur. This process is generally characterized by an induction period, and a much lower reaction rate since formation of alkyl radicals according to Eq. (9.6) is thermodynamically and kinetically unfavorable:

$$R-H \ + \ O_2 \ \longrightarrow \ R \cdot \ + \ HOO \cdot \tag{9.6}$$

The reactivity of different C–H bonds in autoxidation changes in the order tertiary > secondary > primary. As a result, alkanes possessing tertiary hydrogen atoms may be selectively oxidized in the liquid phase to the corresponding *tert*-alkyl hydroperoxides.[6–8] This selectivity difference may also be attributed to the significantly higher rate of termination of primary and secondary alkylperoxy radicals, resulting in a rather slow rate of autoxidation. The best example is the oxidation of isobutane to *tert*-BuOOH, which is a widely studied process, due to the commercial importance of the product.[9–12] *tert*-BuOOH is used as an initiator in radical polymerizations and as an oxidizing agent in metal-catalyzed epoxidations (see Section 9.2.1). Autoxidation of isobutane is usually initiated by the addition of di-*tert*-butyl peroxide or by *tert*-BuOOH itself. The reaction is carried out in the liquid phase, in the temperature range 100–140°C. Selectivities higher than 90% can be achieved at low conversions. HBr was also found to be a powerful agent[13] acting both as a coinitiator [Eq. (9.7)] and a chain-transfer agent [Eq. (9.8)]:

$$HBr \ + \ O_2 \ \longrightarrow \ HOO \cdot \ + \ Br \cdot \ \xrightarrow{Me_3CH} \ Me_3C \cdot \ + \ HBr \tag{9.7}$$

$$Me_3C \cdot \ + \ O_2 \ \longrightarrow \ Me_3COO \cdot \ \xrightarrow{HBr} \ Me_3COOH \ + \ Br \cdot \tag{9.8}$$

At elevated temperature alkyl hydroperoxides undergo thermal decomposition to alcohols [Eqs. (9.9)–(9.11)]. This decomposition serves as a major source of free radicals in autoxidation. Because of side reactions, such as β scission of alkylperoxy radicals, this process is difficult to control. Further transformation of the alkoxy radical with the substrate or with the solvent eventually also leads to an alcohol [Eq. (9.12)] or the radical decomposes with β scission to give a ketone and an alkyl radical:[14]

$$ROOH \longrightarrow RO\cdot + HO\cdot \tag{9.9}$$

$$RO\cdot + ROOH \longrightarrow ROO\cdot + ROH \tag{9.10}$$

$$2\,ROO\cdot \longrightarrow 2RO\cdot + O_2 \tag{9.11}$$

$$RO\cdot + R{-}H \longrightarrow ROH + R\cdot \tag{9.12}$$

Selectivity of alcohol formation can be substantially increased by carrying out the autoxidation in the presence of a stoichiometric quantity of boric acid that reacts with the intermediate hydroperoxide to form alkyl borate. This observation gained practical importance in the commercial oxidation of alkanes (see Section 9.5.1).

Alkanes possessing two tertiary C$-$H bonds in β or γ position may form dihydroperoxides. The product is formed in an intramolecular peroxy radical attack that is highly efficient in the transformation of 2,4-dimethylpentane:[15,16]

$$\tag{9.13}$$

Autoxidation of alkanes may be carried out by metal catalysis.[2,14,17] Although metal ions participate in all oxidation steps, their main role in autoxidation is not in their ability to generate free radicals directly by one-electron oxidation [Eq. (9.14)] but rather their activity to catalyze the homolytic decomposition of the intermediate hydroperoxide according to Eqs. (9.15) and (9.16). As a result of this decomposition, metal ions generate chain-initiating radicals. The overall reaction is given in Eq. (9.17):

$$R{-}H + M^{(n+1)+} \longrightarrow R\cdot + M^{n+} + H^+ \tag{9.14}$$

$$ROOH + M^{n+} \longrightarrow RO\cdot + M^{(n+1)+} + HO\cdot \tag{9.15}$$

$$ROOH + M^{(n+1)+} \longrightarrow ROO\cdot + M^{n+} + H^+ \tag{9.16}$$

$$2\,ROOH \xrightarrow{[M]} ROO\cdot + RO\cdot + H_2O \tag{9.17}$$

At low metal ion concentration alkanes possessing tertiary C$-$H bond may be selectively converted to tertiary alcohols.[9]

Oxidation of Methane. The direct selective transformation of methane to methanol and formaldehyde, two valuable industrial chemicals is a goal that is extremely difficult to accomplish. At present a three-step process is used in industry to manufacture formaldehyde from methane:

$$CH_4 + H_2O \xrightarrow{\text{Ni}} CO + 3H_2 \xrightarrow{\text{Cu}} CH_3OH \xrightarrow{-H_2} H_2C=O \qquad (9.18)$$

It includes the steam reforming of methane over a nickel catalyst to synthesis gas followed by the copper-catalyzed transformation of the latter to methanol (see Section 3.5.1). Finally, formaldehyde is produced by oxidative dehydrogenation of methane.

One disadvantage of this technology is the large energy requirement of the first, highly endothermic, step. The process is also inefficient in the sense that it first transforms methane in an oxidative reaction to carbon monoxide, which, in turn, is reduced to methanol, and the latter is oxidized again to formaldehyde. The direct conversion of methane, therefore, would be a more efficient way in the production of methanol and formaldehyde.

The main problems associated with the direct oxidation of methane are the higher reactivity of the products (methanol and formaldehyde) compared to methane, and the thermodynamically more favorable complete combustion of methane to carbon oxides and water:

$$CH_4 \begin{cases} \xrightarrow{0.5\,O_2} CH_3OH & \Delta H = -30.4\ \text{kcal/mol} \qquad (9.19) \\ \xrightarrow{O_2} H_2C=O + H_2O & \Delta H = -66.0\ \text{kcal/mol} \qquad (9.20) \\ \xrightarrow{1.5\,O_2} CO + 2H_2O & \Delta H = -124.1\ \text{kcal/mol} \qquad (9.21) \\ \xrightarrow{2\,O_2} CO_2 + 2H_2O & \Delta H = -191.9\ \text{kcal/mol} \qquad (9.22) \end{cases}$$

Therefore, finely tuned reaction conditions to achieve high conversion without total oxidation are required. At present, however, there is no commercially viable process satisfying these requirements.

A number of possibilities exist for the direct one-step oxidation of methane. These include homogeneous gas-phase oxidation, heterogeneously initiated homogeneous reaction, heterogeneous catalytic oxidation, and photochemical and electrophilic oxidations. Several review articles give detailed treatment of the homogeneous and catalytic oxidations.[18–27] Some important conclusions are as follows.

Numerous studies of homogeneous gas-phase oxidation at high temperature established an optimum pressure for selective oxidation of methane to methanol.[19,21] It was also observed that methanol yields increased with decreasing oxygen concentration. When natural gas instead of methane was oxidized, much lower temperature was required to achieve the same conversion. Ethane and propane present in natural gas are considered to sensitize the oxidation of methane presumably by providing free radicals through splitoff of their more easily abstractable hydrogen atoms. In general, high pressure (30–200 atm), high temperature (200–400°C), and low oxygen concentration favor high methanol selectivity.

The presently best result (75–80% selectivity in methanol formation at 8–10% conversion) was achieved in a flow system under cool flame conditions (450°C, 65 atm, <5% O_2 content).[28] The authors suspect that the glass-lined reactor suppressing secondary reactions may account for the result.

A large number of catalysts have been studied in the catalytic oxidation of methane, primarily oxides and mixed oxides.[18,21,26] Metals were also tested, but they tend to catalyze complete oxidation. Mixed oxides based on MoO_3 have been studied most, and $FeMoO_4$ was found to be the most active.[18] Gaseous chlorine is a very effective promoter in oxidation over a Cr_2O_3-on-pumice catalyst at low pressure (430°C, 1.5 atm), substantially increasing both methane conversion and methanol selectivity.[18] Significant increase in methane conversion is brought about by organic chlorine compounds (CH_2Cl_2, CCl_4).[29,30] Hydrocarbon impurities in methane are beneficial in catalytic oxidation as well. ZnO doped with copper and iron[31] and boron-containing mixed-metal oxides[32,33] have been studied. In a 1 : 2 : 2 Fe—Nb—B oxide system $FeNbO_4$ was found to supply reactive lattice oxygen atoms for breaking the C—H bonds leading to formaldehyde.[33] Boron, in turn, improved selectivity by suppressing decomposition and further oxidation of formaldehyde.

There has been considerable interest in the study of silica-supported MoO_3 and V_2O_5 catalysts.[34–38] Bare silica itself exhibits a unique activity in oxidation of methane to formaldehyde.[35–37,39,40] This is attributed to surface sites possessing donor properties to activate molecular oxygen. These sites may be Si^{3+} ions on structure defects created at high temperature (1000°C).[39] Molybdena in about one-tenth of a monolayer on silica has a strong promoting effect increasing activity by one order of magnitude.[35] In one study, however, MoO_3 was found to depress the activity of precipitated silica.[36] In contrast, V_2O_5 significantly promoted the activity towards the formation of formaldehyde[36–38] with the highest selectivity over catalysts of the highest dispersion.[38]

According to Lunsford, most of the observations on methane oxidation over oxide catalysts may be interpreted in terms of methyl radical chemistry.[41] Most experimental data support the role of surface O^- ions in the formation of methyl radicals. The latter are transformed by reductive addition to methoxide ions, which decompose to formaldehyde or react with water to form methanol. Methyl radicals may desorb to the gas phase and participate in free-radical reactions to yield nonselective oxidation products.

All these catalytic results, however, were usually achieved at very low (2–3%) conversions. The only exception is a paper reporting up to 80% selectivity at 20% conversion over a $MoCl_5$—R_4Sn-on-silica olefin metathesis catalyst (700°C, 1 atm, CH_4 : air = 1).[42] In general, higher temperature and lower—about ambient— pressure compared to homogeneous oxidation, and high excess of methane are required for the selective formation of formaldehyde in catalytic oxidations.[43] The selectivity, however, decreases dramatically at conversions above 1%, which is attributed to the decomposition and secondary oxidation of formaldehyde.[43,44] It is a common observation that about 30% selectivity can be achieved at about 1% conversion.

In order to increase the conversion level of methane to the oxygenated products, the concentration of oxygen in the reaction mixture has to be raised. This can be done if temperature and pressure are lowered to remain outside the upper explosion limit of the methane–oxygen mixture. In a comparative study[45] the only catalyst that allowed substantial temperature decrease compared to the homogeneous gas-phase process was SnO_2, which, however, readily deactivated. In contrast, considerable reduction of reaction temperature in homogeneous oxidation could be achieved by using different sensitizers. High methanol selectivity under such conditions appears to be associated with the reduced operating temperature.

A promising development in methane oxidation is the use of N_2O as an oxidant in the presence of MoO_3 and V_2O_5 on silica,[43,46–50] Bi_2O_3–SnO_2,[51] or heteropoly acids.[52,53] In the presence of water at 600°C, a 65.8% combined selectivity of methanol and formaldehyde at 16.4% conversion was observed over MoO_3 on silica.[46] An even higher selectivity was reported later,[47] but at a conversion level of only 3%. Kinetic studies indicated that on the same catalyst at temperatures below 540°C, methanol and formaldehyde were probably formed in parallel reactions from the same intermediate. Above this temperature, however, consecutive oxidation of methanol is the major route to the formation of formaldehyde.[48] The O^- ion formed by surface decomposition of N_2O was postulated as the active species initiating the activation process via hydrogen abstraction.[47] The generation of O^{2-} via two-electron oxidation accounted for nonselective oxidation.

An ozone-sensitized oxidative conversion of methane to methanol has been reported.[54] A double-layered Sr on La_2O_3 then MoO_3 on a silica catalyst bed exhibited significantly higher yields of formaldehyde from a methane–air mixture than did MoO_3 on silica alone.[55]

The photocatalytic activation of methane to yield oxygen-containing compounds is possible by carrying out the reaction in the presence of water.[56–58] Addition of oxygen at 100°C enhances the formation of methanol and formaldehyde.[57] Hydroxyl radicals formed by the photolysis of water vapor [Eq. (9.23)] activate methane by hydrogen abstraction[56,57] to form methyl radicals [Eq. (9.24)], which react further with water in the product-forming step [Eq. (9.25)]. Formation of methyl radicals in a different route may be important depending on temperature and methane concentration[58] [Eq. (9.26)]:

$$H_2O \xrightarrow{h\nu} H\cdot \ + \ HO\cdot \tag{9.23}$$

$$HO\cdot \ + \ CH_4 \longrightarrow CH_3\cdot \ + \ H_2O \tag{9.24}$$

$$CH_3\cdot \ + \ H_2O \longrightarrow CH_3OH \ + \ H\cdot \tag{9.25}$$

$$CH_3O\cdot \ + \ CH_4 \longrightarrow CH_3OH \ + \ CH_3\cdot \tag{9.26}$$

A silica-supported MoO_3 was also used[59] in photoinduced oxidation of methane to formaldehyde at 190–220°C.

An important problem inherent in all these studies is the reproducibility of the experimental results. Besides the widely varying reaction conditions, this results from the fact that homogeneous and catalytic oxidations may occur simultaneously. Gas-phase reactions always interfere with catalytic oxidations and nominally inert materials exhibit significant activity or, in turn, may inhibit methane activation by quenching gas-phase radicals. It is also concluded that in the complete absence of a catalyst it is possible to exceed the yields of oxygenated products in catalytic oxidations.[44] Although gas-phase reactions are more difficult to control, the use of catalysts offers only little advantage. It is also realized that reactor design has a profound effect on oxygenate selectivity.[26,28,60]

In contrast to the radical processes described before, high selectivity is characteristic of electrophilic oxidation of methane.[61] It reacts, for instance, with H_2O_2 in superacidic media to give methanol.[1,62] The reaction is best explained by electrophilic insertion of the hydrogen peroxonium ion ($H_3O_2^+$) into the methane C—H bond:

$$CH_4 \ + \ HO\text{-}\overset{+}{\underset{H}{O}}\diagdown^{H} \ \xrightarrow{-H_2O} \ \left[H_3C\text{-}\diagdown{\overset{H}{\underset{OH}{}}} \right]^+ \ \longrightarrow \ CH_3\overset{+}{O}H_2 \qquad (9.27)$$

The formation of protonated methanol (methyloxonium ion) protected from further oxidation accounts for high selectivity of methanol formation.

An electrophilic mechanism was also suggested for the low-temperature (80–90°C), solution-phase oxidation of methane. Both $Pd(OOCCF_3)_2$ in stoichiometric amount[63] and the palladium(II)-catalyzed oxidation by peroxytrifluoroacetic acid[64] yield methyl trifluoroacetate. The catalytic reaction is run in excess $(CF_3CO)_2O$ to remove water generated, thereby preventing the hydrolysis of the product ester. Stoichiometric and catalytic oxidations of methane to methyl trifluoroacetate at 150–180°C with $Co(OOCCF_3)_3$ were also reported.[65]

A mercury-catalyzed, high-yield system oxidizes methane by conc. H_2SO_4 to produce methanol:[66]

$$CH_4 \ + \ 2\,H_2SO_4 \ \xrightarrow{\text{Hg}^{2+}} \ CH_3HgOSO_3H \ \longrightarrow \ CH_3OSO_3H \ + \ 2\,H_2O \ + \ SO_2 \qquad (9.28)$$
$$\textbf{1}$$

The oxidation takes place through the observable intermediate **1** to yield methyl bisulfate, which may be readily hydrolyzed to methanol. At a methane conversion of 50%, an 85% selectivity to methyl bisulfate was achieved. The second molecule of H_2SO_4 reoxidizes Hg^+ to Hg^{2+} completing the catalytic cycle.

When methane is reacted with ozone in superacidic media,[61,67] formaldehyde is directly formed through a pathway that is considered attack by ^+O_3H into a C—H bond, followed by cleavage of H_2O_2 to give very reactive methyloxenium ion (**2**), which instantly rearranges to protonated formaldehyde:

$$CH_4 \ + \ ^+O_3H \ \rightleftharpoons \ \left[H_3C\text{-}\diagdown{\overset{H}{\underset{O\text{-}O\text{-}OH}{}}} \right]^+ \ \xrightarrow{-H_2O_2} \ [CH_3O^+] \ \longrightarrow \ CH_2{=}\overset{+}{O}H \qquad (9.29)$$
$$\textbf{2}$$

This study indicates the possibility of selectively converting methane into formaldehyde without first going to methanol.

It should be mentioned that with superacidic electrophilic oxygenation of methane either to methanol (with protonated hydrogen peroxide) or to formaldehyde (with protonated ozone), the products formed are indeed the protonated products ($CH_3OH_2^+$ and $CH_2{=}OH^+$, respectively), which are protected from further electrophilic oxygenation, which happens only too readily in conventional oxidations.

Oxidation of Other Saturated Hydrocarbons

Oxidation with Stoichiometric Oxidants. Certain peracids reacting with alkanes yield alcohols. Peracetic acid,[68,69] perbenzoic acid,[70] *m*-CPBA,[71,72] and nitroperbenzoic acids[73–75] may be used. Alcohols[76] or an equilibrium mixture of the alcohol and the trifluoroacetate[77] are formed on the action of pertrifluoroacetic acid. A high degree of regioselectivity (better than 97%), specifically, preferential attack at the tertiary C—H bonds, is usually observed:

$$\text{38 : 62}$$
$$\text{38 \% conversion}$$

$$(9.30)$$

Regioselectivities of tertiary vs. secondary attack are in the order 90–500. The oxidation is also stereoselective ensuring a high degree of retention in the transformation of epimeric cycloalkanes[73,75] and optically active acyclic alkanes[74,75] [Eq. (9.30)].

When nonactivated peracids such as peracetic acid or perbenzoic acid are used in oxidation of alkanes, the formation of methane and benzene, respectively, as well as evolution of CO_2 are observed. This indicates a radical mechanism:[68,78]

$$\text{PhCOOOH} \longrightarrow \text{PhCOO·} + \text{HO·} \qquad (9.31)$$

$$\text{PhCOO·} \longrightarrow \text{Ph·} + CO_2 \qquad (9.32)$$

$$\text{R—H} \begin{cases} \xrightarrow{\text{Ph·}} \text{PhH} + \text{R·} & (9.33) \\ \xrightarrow{\text{PhCOO·}} \text{PhCOOH} + \text{R·} & (9.34) \end{cases}$$

$$\text{R·} + \text{PhCOOOH} \longrightarrow \text{ROH} + \text{PhCOO·} \qquad (9.35)$$

Since formation of the alcohol is highly stereospecific under such conditions as well, a rapid reaction between the alkyl radical and the reagent in the product-determining step [Eq. (9.35)] was suggested.[78]

Studies with *p*-nitroperbenzoic acid, a reactive peracid, favor a different mechanism. The high degree of retention observed in oxidation of epimeric cycloalkanes and $(-)$-2,6-dimethyloctane [Eq. (9.30)], as well as substituent and solvent effects, agree with a cyclic transition state (**3**) originally suggested by Bartlett[79] to interpret epoxidation by peracids:

$$\underset{\textbf{3}}{\text{R}\overset{\delta+}{\underset{\text{R}}{\text{C}}} \cdots \overset{\text{R}}{\underset{\text{H}}{\text{C}}} \cdots \overset{\text{H}}{\underset{\text{O}}{\cdots}} \overset{\text{O}}{\underset{\text{O}}{\text{C}}} \text{R}'}$$

Steric effects, namely, the observation that with the exception of adamantane, reactions at bridgehead positions are undetectable or slower, also support this mechanism.

A reagent that gives results very similar to the those in oxidation with peracids is fluorine passed through wet acetonitrile at $-10°C$. A relatively stable complex $HOF \cdot CH_3CN$ is presumably formed that participates in highly regioselective and stereoselective electrophilic hydroxylations under mild conditions.[80] Oxidation of *cis*- and *trans*-decalin, for example, leads to the corresponding 9-hydroxy compounds in yields higher than 80%.

Two new reactive, very powerful organic peroxides, dimethyldioxirane and methyl(trifluoromethyl)dioxirane (**4**), have been introduced.[81–83] The latter is more reactive and can be used more conveniently.[84,85] Acyclic alkanes give a mixture of isomeric ketones on oxidation with methyl(trifluoromethyl)dioxirane,[84,85] while cyclohexanone is the sole product in the oxidation of cyclohexane (99% selectivity at 98% conversion).[85] With the exception of norbornane, which undergoes oxidation at the secondary C–2 position, highly selective tertiary hydroxylations can be carried out with regioselectivities in the same order of magnitude as in oxidations by peracids.[85–87] A similar mild and selective tertiary hydroxylation by perfluorodialkyloxaziridines was also reported.[88] Oxidation with dioxiranes is highly stereoselective:[85]

$$\overset{\text{Me}}{\underset{\text{Me}}{\bigwedge\!\!\bigwedge}} + \underset{\textbf{4}}{\overset{\text{Me}}{\underset{F_3C}{\bigtimes\!\!\overset{O}{\underset{O}{|}}}}} \xrightarrow[\substack{CH_2Cl_2, \\ 4\ min,\ -22°C}]{\text{1,1,1-trifluoroacetone,}} \underset{\substack{98\%\ \text{selectivity} \\ \geq 97\%\ \text{yield}}}{\overset{\text{Me}}{\bigwedge\!\!\bigwedge}\underset{\text{Me}}{\text{OH}}} \tag{9.36}$$

Selective polyhydroxylation of adamantane is possible.[87] A concerted insertion mechanism closely resembling to the mechanism of ozonation has been suggested.[85]

H_2O_2 in superacids at $-78°C$ converts simple straight-chain alkanes into primary alcohol (ethane), or secondary alcohols and ketones (propane, butane).[1,62,89,90] Electrophilic hydroxylation of the secondary C—H bond by the incipient hydroxyl cation formed through the protolytic cleavage of hydroperoxonium ion accommodates these observations:

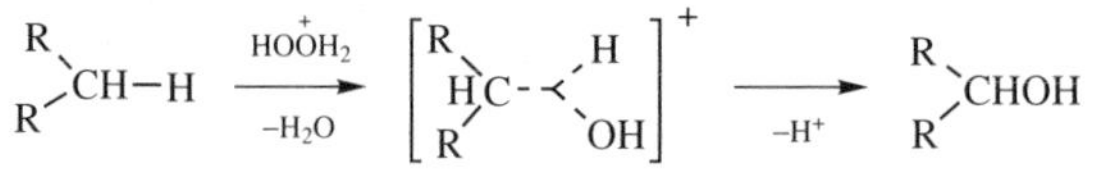

$$\begin{matrix} R \\ \diagdown \\ CH-H \\ \diagup \\ R \end{matrix} \quad \xrightarrow[-H_2O]{HO\overset{+}{O}H_2} \quad \left[\begin{matrix} R & & H \\ \diagdown & & \diagup \\ H\overset{+}{C}-\cdots< \\ \diagup & & \diagdown \\ R & & OH \end{matrix} \right]^{+} \quad \xrightarrow{-H^+} \quad \begin{matrix} R \\ \diagdown \\ CHOH \\ \diagup \\ R \end{matrix} \qquad (9.37)$$

Alcohols may also be formed when these alkanes react with equimolar amount of ozone in superacidic media (i.e., with ^+O_3H) at $-78°C$, followed by raising the temperature of the reaction mixture.[91] In this case H_2O_2 formed in situ in primary ozonolysis participates in hydroxylation according to Eq. (9.37). Even 2,2-dimethylpropane and 2,2,3,3-tetramethylbutane, which undergo characteristic C—C bond cleavage with ozone in superacids, yield a small amount of alcohols in Magic Acid–SO_2ClF solution at $-78°C$ according to this mechanism.[91]

Ozone reacts slowly with saturated hydrocarbons usually to give alcohols.[92,93] The reactivity of alkanes toward ozone is several orders of magnitude less than that of alkenes. Oxidation of saturated hydrocarbons takes place preferentially at the tertiary carbon. In liquid-phase ozonation[94] the order of reactivity of the primary, secondary and tertiary C—H bonds is 1 : 13 : 110. The formation of tertiary alcohols occurs with high degree (60–94%) of stereoselectivity.[94–96]

The mechanism of ozonation of the C—H σ bond is still controversial. The regioselectivity points to electrophilic interaction of ozone with the tertiary C—H bond. The high degree of configurational retention, in turn, is indicative of an insertion reaction. On reacting with ozone, alkanes presumably form an ozone–hydrocarbon complex with partially dissociated C—H bond. Warming, irradiation, or reducing agents bring about the decomposition of the complex to yield the product tertiary alcohol.[69,95] An insertion mechanism with a transition state possessing radical character was suggested.[94] Thermochemical calculations of the energy involved in different mechanistic possibilities led to the suggestion of hydride abstraction,[97] specifically, the possible ionic character of the transition state.[93,96,98] A free-radical mechanism was also postulated.[98,99] On the basis of observations in 1986 and 1988, a concerted insertion mechanism with the transition state **6** seems to have a high probability[100,101] (Scheme 9.1). The transition state may have contribution

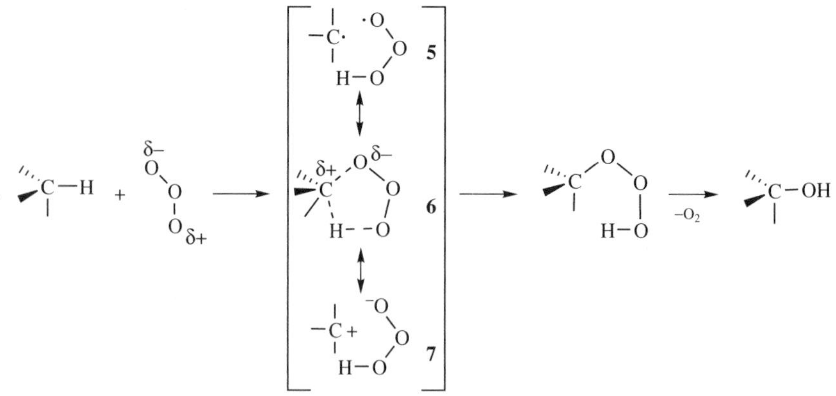

Scheme 9.1

from radical (**5**) or ionic (**7**) resonance forms accounting for certain loss of stereoselectivity.

A severe limitation of the effective use of ozone in organic synthesis is its low solubility in organic solvents. The so-called dry ozonation technique of Mazur solves this problem.[102] Dry silica gel with the preadsorbed organic substrate is saturated with ozone at $-78°C$. After warming up to room temperature, the products are eluted in the usual way. Under these conditions ozone exhibits an enhanced reactivity presumably due to the slightly acidic nature of silica gel:[103]

(9.38)

81–84%

The results of dry ozonation, namely, regioselectivities and stereoselectivities, are very similar to those in superacidic liquid-phase ozonation. Tertiary C—H bonds in strained systems such as norbornane are inert to dry ozonation.[93] Such compounds are oxidized at the secondary carbon to yield a mixture of alcohols and ketones.[93,104] Similarly, substituted cyclopropanes exhibit a general preference for the oxidation of the secondary C—H bond in the α-position to the ring:[104]

(9.39)

R = Me   95%
R = Pr   87%

Branched acyclic alkanes also exhibit a slightly different behavior toward ozone on silica gel.[105] Although tertiary alcohols are usually the main products, C—C bond cleavage to yield ketones always occurs and may become the predominant reaction. Atomic oxygen generated by microwave discharge of a CO_2/He mixture is a more selective reagent in the transformation of these compounds to tertiary alcohols.[106]

Ozone is quite unreactive towards saturated hydrocarbons under conventional ozonation conditions (acting as a dipolar reagent), but its reactivity can be greatly increased by carrying out ozonation in superacids (acting as ^+O_3H, a strong electrophile).[1,67] Ethane gives protonated acetaldehyde as the major reaction product, while simple unbranched alkanes are oxidized at the secondary carbon atom in FSO_3H–SbF_5–SO_2ClF solution at $-78°C$. Propane and butane are transformed mainly to the corresponding protonated 2-alkanones, and pentane gives protonated 3-pentanone. Protonated cyclolkanones are formed from cyclopentane and cyclohexane.[1] The product distribution rules out the formation of carbocations by protolysis prior to their ozone quenching. Instead, electrophilic attack by protonated ozone resulting in ozone insertion into the C—H σ bond, occurs:

$$\begin{array}{c}
R' \\
{\diagdown} \\
{\diagup} \\
R
\end{array}
CH-H \ + \ {}^+OOOH \ \longrightarrow \
\left[
\begin{array}{c}
R' \quad\quad H \\
{\diagdown}{\diagup} \\
CH{\cdot}{\prec} \\
{\diagup}{\diagdown} \\
R \quad\quad OOOH
\end{array}
\right]^{+}
\xrightarrow[-H^+]{}
\begin{array}{c}
R' \\
{\diagdown} \\
CH-OOOH \\
{\diagup} \\
R
\end{array}
\xrightarrow[-H_2O_2]{H^+}
\begin{array}{c}
R' \\
{\diagdown} \\
C{=}\overset{+}{O}-H \\
{\diagup} \\
R
\end{array}$$

$$(9.40)$$

Product distributions in weaker acids ($HF–BF_3–SO_2ClF$, $HF–SO_2ClF$, $FSO_3H–SO_2ClF$, $H_2SO_4–SO_2ClF$) are very similar to those in Magic Acid–SO_2ClF solution. This is interpreted as a further proof of the electrophilic insertion mechanism.

In electrophilic oxygenation with ozone in superacidic media or in dry ozonation over silica gel, protonated ozone, namely, ${}^+O_3H$, was suggested to be the de facto electrophile. These reactions are fundamentally different from conventional ozonations with O_3, which is a highly dipolar molecule. In O_3H^+ the dipole is removed by protonation. Cacace and Sporenza were able to directly identify ${}^+O_3H$ in the gas phase and also measure the proton affinity of ozone.[107]

Metallic oxidants, namely, chromic acid and potassium permanganate, may be used to oxidize alkanes to alcohols or ketones, but these reagents have only limited synthetic value. Alkaline $KMnO_4$ is rather ineffective, mainly because of the insolubility of alkanes in the aqueous solution of the reagent. Oxidations in acidic solutions such as aqueous CF_3COOH,[108] or the use of special reagents such as benzyltriethylammonium permanganate[109] may give better results.

Primary C–H bonds are unreactive toward these oxidants.[108,110] The oxidation of secondary C–H bonds leads to the corresponding ketones. The latter, however, which are more reactive than the parent hydrocarbon, are further oxidized to carboxylic acids by C–C bond breaking.[108] The rate of oxidation of straight-chain alkanes by CrO_3 is proportional to the number of CH_2 groups in the molecule.[111] This means that all CH_2 groups are equally attacked by the reagent resulting in nonselective oxidation.[108]

In contrast, bridgehead (i.e., tertiary) C–H bonds can be selectively oxidized to tertiary alcohols using chromic acid in acetic acid.[112] Since the reactivity of these compounds depends on the ability of the bridgehead carbon to become sp^2-hybridized, norbornane and bicyclooctanes are unreactive at the bridgehead positions. Adamantane, bicyclo[3.3.1]nonane, bicyclo[3.2.2]nonane, and bicyclo-[3.3.2]decane, in turn, afford the corresponding tertiary alcohols in 40–50% yields.

Chromic acid oxidation of saturated hydrocarbons starts with hydrogen abstraction to give a caged radical pair.[113,114] The collapse of the latter leads to a chromium(IV) ester, which hydrolyzes to the product tertiary alcohol. The postulation of the caged pair was necessary to explain the high degree of retention in oxidation of (+)-3-methylheptane:[113]

$$(9.41)$$

70–85% retention

Cobaltic and manganic acetate used as catalysts in autoxidation may be employed as reagents in acetoxylation.[115,116] The reactivity of Co(OAc)$_3$ can be enhanced by strong organic (AcOH, CF$_3$COOH, MeSO$_3$H, PhSO$_3$H) or inorganic acids (H$_2$SO$_4$, H$_3$PO$_4$, HClO$_4$) to carry out oxidations at low temperature (40°C). Acetates are formed with high selectivity under nitrogen, whereas ketones are the main products in the presence of oxygen.[117]

Partial anodic oxidation of *n*-alkanes on a smooth Pt electrode in CF$_3$COOH gives isomeric *sec*-alkyl trifluoroacetates in 50–80% yield.[118]

Oxidation Catalyzed by Metalloporphyrins. Much attention has been devoted to the metal-catalyzed oxidation of unactivated C—H bonds in the homogeneous phase. The aim of these studies is to elucidate the molecular mechanism of enzyme-catalyzed oxygen atom transfer reactions. Additionally, such studies may eventually allow the development of simple catalytic systems useful in functionalization of organic compounds, especially in the oxidation of hydrocarbons. These methods should display high efficiency and specificity under mild conditions characteristic of enzymatic oxidations.

Many of these studies focus on modeling the oxidation by cytochrome P-450. Heme-containing monooxygenases known collectively as *cytochrome P-450* mediate a number of biochemical processes by incorporating one oxygen atom into a substrate. Among others, they catalyze selective hydroxylation of nonactivated hydrocarbons. The oxygen atom incorporated into the substrate may derive from O$_2$ in the presence of a reducing agent, or results from single oxygen atom donors.

Cytochrome P-450, however, is such a powerful oxidant that it can bring about self-degradation of its own porphyrin ligands. Because of this disadvantage, metal complexes of synthetic porphyrin have been designed and used as chemical models of this enzyme.[119–124] Of the different porphyrin complexes, Fe(III) and Mn(III) porphyrins exhibit the highest catalytic activity. The single oxygen atom donors most frequently used are iodosylbenzene (PhIO)[119–123] and potassium hydrogen persulfate.[125] Data for hydrogen peroxide,[126] alkylhydroperoxides,[126] hypohalites,[127] sodium chlorate,[128] tertiary amine *N*-oxides,[129,130] and oxaziridine[131] are also available.

The metalloporphyrin–PhIO system catalyzes the oxidation of alkanes mainly to alcohols under mild conditions. High selectivity for the hydroxylation at the tertiary carbon is observed.[132] Yields up to 40% based on the oxidant consumed are obtained. Acyclic alkanes usually exhibit very poor reactivity. A large isotope effect[132] and retention of configuration in the oxidation of *cis*-decalin[132,133] are additional important characteristics of the process.

A stepwise radical mechanism developed first for iron porphyrins[121,132] and then found to be valid for manganese porphyrins[121,134–136] as well accommodates the above observations (Scheme 9.2). This mechanism is essentially the same as the one proposed for the oxidation of alkanes by cytochrome P-450.[121,137,138] The active oxidizing species is the **8** oxoiron complex formed on the action of the single-oxygen-atom donor. The interaction of **8** with the alkane results in hydrogen

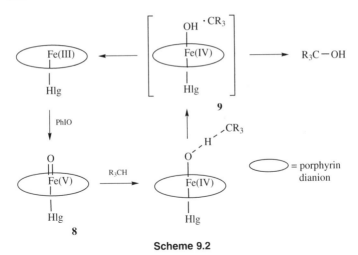

Scheme 9.2

abstraction and the formation of caged radical **9**. The isotope effect, the high regio-selectivity, rearrangement of certain alkanes during oxidation,[134] and the occasional decrease in stereoselectivity testify to the radical nature of the mechanism. The reverse transfer of the OH to the carbon radical before diffusing out from the cage, known as the "oxygen rebound mechanism," yields the product alcohol. The exact nature of this recombination step and alcohol formation, however, is not well understood. Multiple fates, including intramolecular electron transfer to form an ion pair, have been suggested.[136]

It appears that oxidation with alkyl hydroperoxides[139,140] has a different active species that does not include the metal. For example, in oxidation with cumyl hydroperoxide the cumyloxy radical was proposed to participate in hydrogen abstraction.[140] H_2O_2, in turn, can serve as a single oxygen atom donor only in oxidation with manganese porphyrins because of the activity of iron porphyrins to dismutate hydrogen peroxide.[136] Imidazole when used in catalytic amounts was found to be an excellent agent to increase the activity of the manganese porphyrin–H_2O_2 system.[141] Under these conditions, for instance, adamantane is preferentially converted to 1-adamantanol (63% selectivity at 93% conversion).

Simple tetraphenyl metalloporphyrins are not particularly stable under oxidizing conditions. This is due to the oxidative destruction of hemin and the formation of μ-oxodimeric species that are not active in oxidation. It was observed, however, that electron-withdrawing substituents on the phenyl rings substantially increase stability of metalloporphyrins. At the same time improved selectivities of oxidations could also be achieved. Compared with iron(III) tetraphenylporphyrin chloride [Fe(TPP)Cl], perhalogenated metalloporphyrins, such as iron(III) tetra(perfluoro-phenyl)porphyrin chloride [Fe(TPFPP)Cl], are unusually efficient catalysts for high-turnover, high-yield alkane hydroxylations[142–144] [Eq. (9.42)[142]]:

$$
\text{(cyclohexane)} \xrightarrow[\text{CH}_2\text{Cl}_2,\ 22°\text{C},\ 2\text{–}3\ \text{h}]{\text{PhIO}} \text{(cyclohexanol)} \tag{9.42}
$$

Fe(TPP)Cl	5%
Fe(TPFPP)Cl	71%

Iron tetra(o-nitrophenyl)porphyrin is such a powerful catalyst that the simplest alkanes, even methane and ethane, are converted to the corresponding alcohols.[145]

Perhalogenation of the pyrrole rings brings about similar increase in catalytic activity, also affecting regioselectivity.[146] Remarkable shape selectivity was observed by the use of sterically hindered porphyrins.[147–149] The unhindered manganese(III) tetraphenylporphyrin acetate [Mn(TPP)OAc] exhibits the usual regioselectivity to yield mainly the corresponding secondary alcohol from 2,2-dimethylbutane:[147,148]

$$
\xrightarrow[\text{benzene, } 25°\text{C, } 7\ \text{h}]{\text{PhIO}} \quad + \quad + \tag{9.43}
$$

Mn(TPP)OAc	91	9	<1
Mn(TTPPP)OAc	25	6	69

The extremely hindered manganese(III) tetra(2,4,6-triphenylphenyl)porphyrin acetate [Mn(TTPPP)OAc], in contrast, gives the less hindered 3,3-dimethylbutane-1-ol with exceptionally high selectivity. Hindered metalloporphyrins also exhibit an increasing reactivity in the oxidation of the ω–1 CH_2 site in straight-chain alkanes.[148]

Certain manganese porphyrins adsorbed on silica or alumina, or intercalated in different mineral matrices such as montmorillonite, are also very efficient in alkane hydroxylation.[150,151] Alcohol yields and alcohol : ketone ratios are remarkably higher than those obtained with the corresponding soluble manganese porphyrins. Manganese tetra(4-N-methylpyridiniumyl)porphyrin supported on montmorillonite, for instance, is efficient in the hydroxylation of compounds of low reactivity such as n-pentane and n-heptane.[151]

Molecular oxygen may also serve as the oxygen source if applied in the presence of a reducing agent:

$$
O_2 + 2\,e^- + 2\,H^+ \longrightarrow O + H_2O \tag{9.44}
$$

In different models for cytochrome P-450, Zn,[152] Zn amalgam,[153] and sodium ascorbate[154] are used as the reductant. In contrast with all the abovementioned metal complex-catalyzed oxidations, the Mn(TPP)Cl–ascorbate system oxidizes alkanes with predominant formation of ketones.[154] Certain complexes, however, such as iron tetra(pentafluorophenyl)porphyrins,[155] perhaloporphyrin complexes,[156] and Cr(III), Fe(III) and Mn(III) porphyrins possessing azide as axial ligand,[155] hydroxylate propane, and isobutane under mild conditions without added coreductant.

Other Homogeneous Metal-Catalyzed Oxidations. Investigations on the functionalization of nonactivated alkanes using nonporphyrinic catalysts led to the discovery of efficient catalytic systems. Known as the "Gif" and "Gif–Orsay" systems, they operate under oxygen or air under mild conditions.[157,158] The Gif system, discovered by Barton and coworkers at Gif-sur-Yvette, comprises hydrogen sulfide and iron powder in wet pyridine containing a proton source (CH_3COOH).[159] Hydrogen sulfide serving to catalyze the dissolution of iron powder is not necessary in reactions at 40°C. A crystalline complex, the $Fe(II)Fe(III)O(OAc)_6(pyridine)_{3.5}$ cluster (**10**), was isolated and used to develop a catalytic system.[160] In the Gif^{IV} system metallic zinc serves as the reductant.[161] In the electrochemical modification of the Gif–Orsay system a cathode replaces zinc as the electron source.[162]

The Gif and Gif–Orsay systems exhibit some unusual characteristics. Most importantly, they oxidize secondary carbon preferentially and ketones not alcohols are the main products:[160]

$$7.7\% \qquad 4.6\% \qquad 1.6\% \tag{9.45}$$

It is interesting to note that in a somewhat similar catalytic system consisting of iron powder, acetic acid, and heptanal, aerobic oxidation of adamantane gave 1-adamantanol with high selectivity, while cycloalkanes were transformed to ketones as the main products.[163]

It was found that radicals are not involved in ketone formation in Gif oxidation.[164,165] As a minor route, in contrast, tertiary alcohols are formed through carbon radicals. Key intermediates[164] are the Fe(V)=O species **11**, formed from a μ-oxoiron dimer that inserts into the C—H bond, and the alkoxy–iron species **12**:

$$\text{(9.46)}$$

Alcohols are not intermediates in the formation of ketones.

A new series, the so-called GoAgg systems, has been introduced.[158,166,167] The most practical member is GoAgg[II] using iron(III) chloride and H_2O_2. Alkyl hydroperoxides were found to be the intermediates in these oxidations.[168]

A number of other nonporphyrinic metal catalysts have been developed and tested in alkane oxidation. Most of these are iron-based[169-174] or ruthenium-containing catalysts,[163,175-180] and the first Rh-catalyzed reaction has also been reported.[181] H_2O_2 is the preferred single oxygen atom source, but *tert*-BuOOH[169,173,175-177] or oxygen[163,172,178,180] may also be used. RuO_4 prepared in situ with $NaIO_4$ as the reoxidant,[182] as well as a ruthenium heteropolyanion with the usual oxidants ($KHSO_5$, $NaIO_4$, PhIO, *tert*-BuOOH),[183] are also effective. Some of these catalytic systems are very efficient in direct ketonization of methylene groups.[170,173,176] Methane monooxygenase, a nonheme monooxygenase, exhibits catalytic activity for oxygenation of simple alkanes and is believed to contain binuclear iron. This is the reason why these studies include several binuclear iron complexes[169-172] with the aim of mimicking methane monooxygenase.

Miscellaneous Oxidations. Titanium silicalites (TSs) are molecular sieves that incorporate titanium in the framework. They are able to perform oxygenation of various hydrocarbons under mild conditions by hydrogen peroxide.[184,185]

TS-1, which is a ZSM-5 isomorph, exhibits high activity in the oxidation of straight-chain alkanes.[186-189] For example, in the oxidation of *n*-hexane to yield secondary alcohols and ketones[188,189] [Eq. (9.47)], the efficiency of H_2O_2 use is generally higher than 80%. H_2O_2 may be generated in situ on a palladium-containing titanium silicalite in an oxygen–hydrogen atmosphere:[190]

$$\text{TS-1, 33\% } H_2O_2 \atop 55\,^{\circ}\text{C, 75–165 min}$$

		$CH_3CHC_4H_9$	$CH_3CC_4H_9$	$C_2H_5CHC_3H_7$	$C_2H_5CC_3H_7$
		OH	O	OH	O
H_2O		1.5	54.0	8.8	35.6
CH_3CN		16.7	12.3	43.8	27.2

$$(9.47)$$

TS-2 belonging to a different structure group exhibits a similar activity and similar high selectivity in oxidation of hexane.[191] Branched and cyclic alkanes react much more slowly,[186,187,189] indicating that the oxidation occurs preferentially inside the channels of the catalyst structure.[189]

Interesting solvent effects have also been observed[189] [Eq. (9.47)]. Product distribution at secondary sites in water is almost statistical, and ketones are the main products. In acetonitrile, the γ position is oxidized with the preferential formation of alcohols.

The latest experimental evidence suggest the presence of titanyl groups (Ti=O) in the structure[188] (Scheme 9.3). H_2O_2 is activated via chemisorption on these groups with the formation of a surface titanium peroxo complex (**13**). It may exist in the hydrated or open diradical form and initiates hydrogen abstraction. Rapid

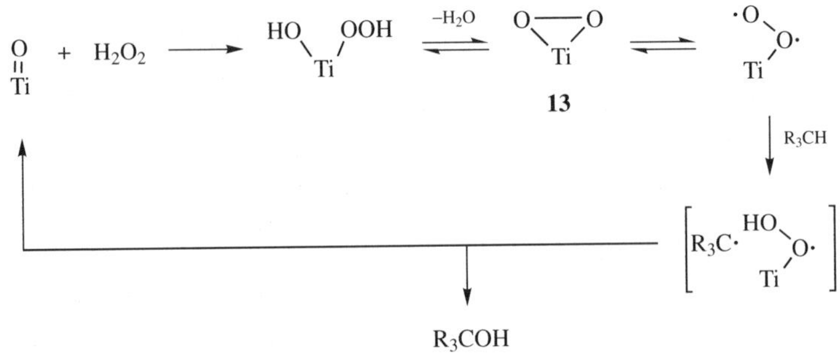

Scheme 9.3

recombination of the radicals allows the formation of the alcohol and regeneration of the titanyl species. Efficiency of H_2O_2 use for oxygenation and selectivity strongly depend on catalyst structure. TiO_2 impurities catalyze the decomposition of H_2O_2, while residual acidity may catalyze secondary reactions thus decreasing selectivity.[188]

Some studies have been conducted on the photochemical oxygenation of alkanes, including the low-temperature conversion of ethane and water to ethanol[192] and the hydroxylation of different acyclic and cyclic alkanes using H_2O_2 or performic acid.[193] In both cases hydroxyl radicals serve as the hydroxylating agent. Photochemical oxidation of cycloalkanes is promoted by ceric ammonium nitrate in acetonitrile.[194] Conversion of ethane to acetaldehyde on UV irradiation over MoO_3 on SiO_2 was observed.[195]

Cycloalkanes can be oxygenated when irradiated in the presence of nitrobenzene.[196] A 50% yield of cyclohexanol and cyclohexanone is achieved from cyclohexane. Since the product ratio is independent of reaction time, the alcohol is not an intermediate in ketone formation. Isomeric 1,2-dimethylcyclohexanes give an identical mixture of the isomeric tertiary alcohols, indicative of conformational equilibration and the presence of a radical intermediate.

Partial oxidation of cyclohexane at the cathode of an O_2/H_2 fuel cell takes place at ambient temperature.[197] Catalytic oxidation with 100% selectivity of the formation of cyclohexanol and cyclohexanone is achieved. Of the cathode materials comprising a mixture of alkaline-earth or rear-earth metal chlorides and graphite, the one which contains $SmCl_3$ exhibits the highest activity.

9.1.2. Oxidations Resulting in Carbon–Carbon Bond Cleavage

Compounds arising from carbon–carbon bond cleavage may be observed in many oxidation reactions as minor byproducts. Cleavage reactions, however, may be the major route when certain saturated hydrocarbons are oxidized by some reagents under specific reaction conditions.

Metal Oxidants. When oxidation of alkanes is carried out in the presence of large amounts of cobalt salts, direct oxidation to carboxylic acids through carbon–carbon bond cleavage can be achieved. Butane, for example, is oxidized at 110°C with Co(OAc)$_2$ promoted with 2-butanone to acetic acid with high selectivity (>83% at 80% conversion).[198] When catalytic amounts of metal salts (cobalt or manganese) are employed, temperatures 170°C and higher are required. Oxidation of cyclohexane[115] and alkyl-substituted cyclohexanes[199] under similar conditions yields directly adipic acid. Significant amounts of glutaric and succinic acid (near 30% at about 90% conversion) are also formed from cyclohexane.

In contrast to autoxidation, tertiary C—H bonds are less reactive under these conditions. A reversible electron-transfer mechanism to form a radical was suggested:[115]

$$RH \; + \; Co^{3+} \; \underset{-Co^{2+}}{\rightleftharpoons} \; \begin{bmatrix} electron \\ deficient \\ species \end{bmatrix}^+ \; \longrightarrow \; R\cdot \; + \; H^+ \tag{9.48}$$

C—H bond breaking was not found to be rate-controlling.

Co(II) acetate in acetic acid with acetaldehyde also gives good selectivity in the formation of adipic acid (73% at 88% conversion at 90°C).[200] The role of acetaldehyde is to promote oxidation of Co(II) to Co(III) and maintain a steady-state concentration of Co(III). Later during the reaction, however, cyclohexanone formed as an intermediate in oxidation serves as promoter. Zirconium ions also exhibit a significant promoter effect.[201]

The liquid-phase oxidation of butane is utilized in the commercial synthesis of acetic acid (see Section 9.5.1). Despite research efforts, however, there is no commercially viable route to the direct one-step oxidation of cyclohexane to adipic acid.

Treatment of alkanes with chromic acid under drastic conditions results in oxidation of methylene and methine groups and ultimately in carbon—carbon bond cleavage. Methyl groups, however, are inert to oxidation, which allows their quantitative determination through the analysis of acetic acid formed. A micro method[202,203] employing a mixture of CrO$_3$ and H$_2$SO$_4$ at 120°C was developed by Kuhn and Roth based on the titration of AcOH distilled off the reaction mixture. The method underwent several modifications to improve accuracy and applicability.[204–207] These included increased reaction temperature and prolonged reaction time to oxidize natural compounds possessing oxidation-resistant geminal dimethyl groups.[206] A submicroscale operation was developed by the addition of V$_2$O$_5$ and MnSO$_4$ to the acid mixture.[207] Spectroscopic methods, however, have replaced this analytic procedure.

Electrophilic Reagents. Electrophilic oxygenation of alkanes can be carried out with hydrogen peroxide or ozone in superacids[89,90] and by using the so-called dry ozonation technique. H$_2$O$_2$ in superacids brings about oxygenation of branched alkanes also involving C—C bond cleavage.[1,62] Isobutane and related isoalkanes are transformed to dialkylalkylcarboxonium ions in Magic Acid—SO$_2$ClF solution at −78°C. Isobutane, for example, gives dimethylmethylcarboxonium ion (**15**). An electrophilic hydroxylation mechanism rationalizes the observations

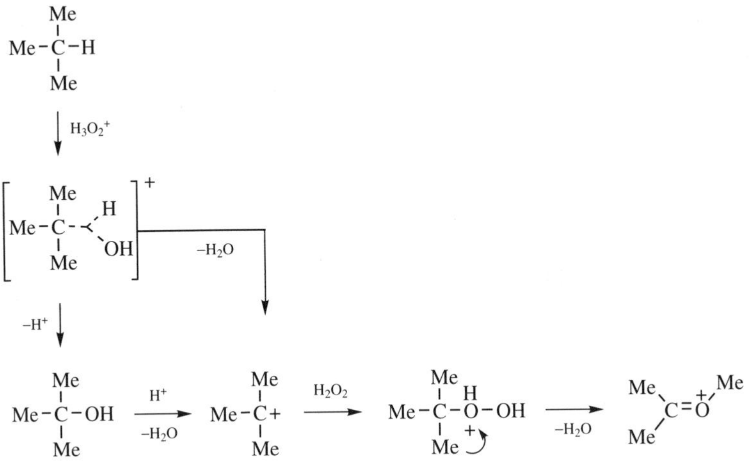

$$
\begin{array}{c}
\text{Me} \\
| \\
\text{Me}-\overset{|}{\underset{|}{\text{C}}}-\text{H} \\
\text{Me}
\end{array}
$$

$\downarrow \, H_3O_2^+$

$$
\left[
\begin{array}{c}
\text{Me} \\
| \quad\quad \text{H} \\
\text{Me}-\overset{|}{\text{C}}-\prec \\
| \quad\quad \text{OH} \\
\text{Me}
\end{array}
\right]^+
$$

$-H_2O$

$-H^+ \downarrow$

$$
\begin{array}{ccc}
\begin{array}{c}
\text{Me} \\
| \\
\text{Me}-\overset{|}{\underset{|}{\text{C}}}-\text{OH} \\
\text{Me}
\end{array}
& \overset{H^+}{\underset{-H_2O}{\longrightarrow}} &
\begin{array}{c}
\text{Me} \\
| \\
\text{Me}-\overset{|}{\underset{|}{\text{C}}}+ \\
\text{Me}
\end{array}
\end{array}
$$

$\overset{H_2O_2}{\longrightarrow}$

$$
\begin{array}{c}
\text{Me} \\
| \quad \text{H} \\
\text{Me}-\overset{|}{\text{C}}-\text{O}-\text{OH} \\
| \\
\text{Me} \; +
\end{array}
\quad \overset{}{\underset{-H_2O}{\longrightarrow}} \quad
\begin{array}{c}
\text{Me} \quad\quad \text{Me} \\
\diagdown \quad + \diagup \\
\text{C}=\text{O} \\
\diagup \\
\text{Me}
\end{array}
$$

14   **15**

Scheme 9.4

(Scheme 9.4). The insertion of hydroxonium ion formed from protonated hydrogen peroxide yields the corresponding tertiary carbocation (**14**) in different pathways; **14** then undergoes further transformations to give **15**.

At higher temperatures hydrolysis of the carboxonium ion and Baeyer–Villiger oxidation lead to the formation of acetone, methanol, and methyl acetate:

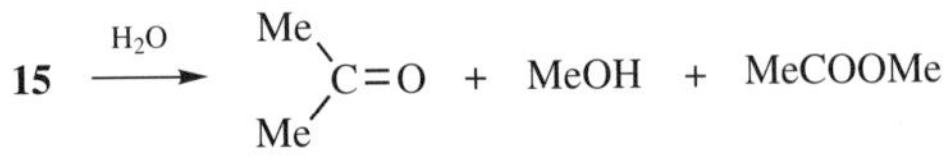

$$
\textbf{15} \quad \overset{H_2O}{\longrightarrow} \quad
\begin{array}{c}
\text{Me} \\
\diagdown \\
\text{C}=\text{O} \\
\diagup \\
\text{Me}
\end{array}
+ \; \text{MeOH} \; + \; \text{MeCOOMe} \tag{9.49}
$$

Because of the enhanced nucleophilicity of water formed under reaction conditions, these hydrolysis products predominate in weaker acids (FSO_3H, HF, H_2SO_4).

Although the carbon–carbon σ bond is quite resistant to ozone, C–C bond cleavage can occur during dry ozonation of branched alkanes and it may become the main reaction path in certain cases. In the ozonation of neopentane acetone is the sole product.[105] Direct insertion of ozone into the C–C bonds through a transition state analogous to that postulated by Olah for alkane ozonation in superacids was proposed. The highly reactive dialkyl trioxide thus formed decomposes to give the cleavage products.

The cyclopropane ring is known to exhibit high stability toward ozone, and oxidation of the secondary C–H bond in the α position to the ring takes place.[104] In contrast, highly strained bicyclo[n.1.0]alkanes undergo C–C bond cleavage in the cyclopropane ring:[208]

$$
\square\!\!\triangleright \quad \overset{O_3 \text{ on silica}}{\underset{-50°C \text{ to } -30°C}{\longrightarrow}} \quad \triangleright\!\!-\text{CH}_2\text{COOH} \; + \; \text{OHC}-\text{CH}_2\text{CH}_2-\text{CHO} \tag{9.50}
$$

28.5   71.5

On the basis of deuterium labeling, 1,3-dipolar cycloaddition to the bridge C—C bond to form a cyclic trioxide was suggested. Rearrangements similar to those in alkene ozonation yield the products.

Carbon—carbon bond cleavage is also characteristic of branched saturated hydrocarbons reacting with ozone in superacid media.[1] Depending on the structure of the reacting alkanes, different mechanisms can be operative.

Isobutane gives dimethylmethylcarboxonium ion (**15**), and related isoalkanes yield the corresponding dialkylalkylcarboxonium ions when treated with ozone in superacid[67] (Magic Acid—SO_2ClF solution at $-78°C$). Cyclic carboxonium ions are formed from methylcyclopentane and methylcyclohexane.[1] There is every evidence suggesting that the reaction proceeds via electrophilic attack by protonated ozone into the alkane tertiary C—H bond through two-electron, three-center bonded pentacoordinated carbenium ion (**16**) [Eq. (9.51)]. Cleavage of H_2O_2 from the **17** protonated trioxide results in product formation:

$$Me-\underset{\underset{Me}{|}}{\overset{\overset{Me}{|}}{C}}-H \ + \ ^+OOOH \ \longrightarrow \ \left[Me-\underset{\underset{Me}{|}}{\overset{\overset{Me}{|}}{C}}\overset{OOOH}{\underset{H}{\diagdown}} \right]^+ \ \underset{-H^+}{\longrightarrow} \ Me-\underset{\underset{Me}{|}}{\overset{\overset{Me}{|}}{C}}-O-O-OH \ \overset{H^+}{\longrightarrow}$$

16

$$\longrightarrow \ Me-\underset{\underset{Me}{|}}{\overset{\overset{Me}{|}}{C}}-O-\overset{H}{\underset{(+)}{O}}-OH \ \longrightarrow \ \overset{Me}{\underset{Me}{\diagup}}\overset{}{\diagdown}C=\overset{+}{O}\overset{Me}{\diagup} \ + \ H_2O_2 \quad (9.51)$$

17   **15**

The alternative cleavage pathway giving M_3C^+ and H_2O_2 is reversible under superacidic conditions.

2,2-Dimethylbutane and 2,3-dimethylbutane in Magic Acid—SO_2ClF solution at $-78°C$ give the same distribution of products, a roughly $3:2$ mixture of dimethylisopropylcarboxonium ion (**19**) and protonated acetone (**20**)[67] (Scheme 9.5). This clearly demonstrates that, in contrast with the transformation of the abovementioned compounds, ozonation in this case proceeds via the intermediate carbocation **18**.

2,2-Dimethylpropane[91] (neopentane) is transformed to a mixture of ethyldimethylcarboxonium ion (**21**) and dimethylmethylcarboxonium ion (**15**). At $-78°C$ **21** is formed exclusively. At temperatures higher than $-20°C$, **15** becomes the predominant product. Formation of **21** can be best explained by a reaction path that involves insertion of protonated ozone into the C—H σ bond, formation of the *tert*-amyl cation through rearrangement, and its quenching by ozone:

$$Me-\underset{\underset{Me}{|}}{\overset{\overset{Me}{|}}{C}}-H_2C-H \ + \ ^+O_3H \ \longrightarrow \ \left[Me-\underset{\underset{Me}{|}}{\overset{\overset{Me}{|}}{C}}-H_2C-\overset{H}{\underset{O_3H}{\diagdown}} \right]^+ \ \underset{-H_2O_3}{\longrightarrow} \ Me-\underset{\underset{Me}{|}}{\overset{\overset{Me}{|}}{C}}-\overset{+}{C}H_2 \ \longrightarrow$$

$$\longrightarrow \ Me-\underset{+}{\overset{\overset{Me}{|}}{C}}-CH_2-Me \ \overset{O_3}{\longrightarrow} \ \left[Me-\underset{\underset{O-O=O}{|}}{\overset{\overset{Me}{|}}{C}}-CH_2-Me \right]^+ \ \underset{-O_2}{\longrightarrow} \ \overset{Me}{\underset{Me}{\diagup}}\overset{}{\diagdown}C=\overset{+}{O}\overset{CH_2Me}{\diagup} \quad (9.52)$$

21

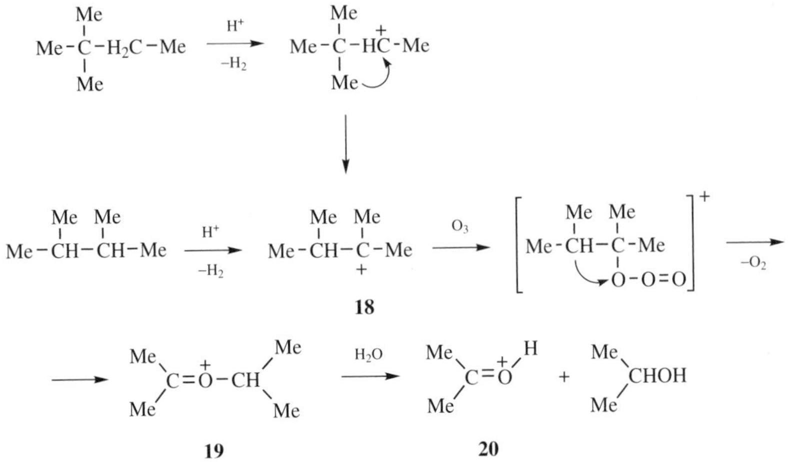

Scheme 9.5

Equation (9.53) depicts the most probable path of the formation of **15** through protolysis of the C—C σ bond:

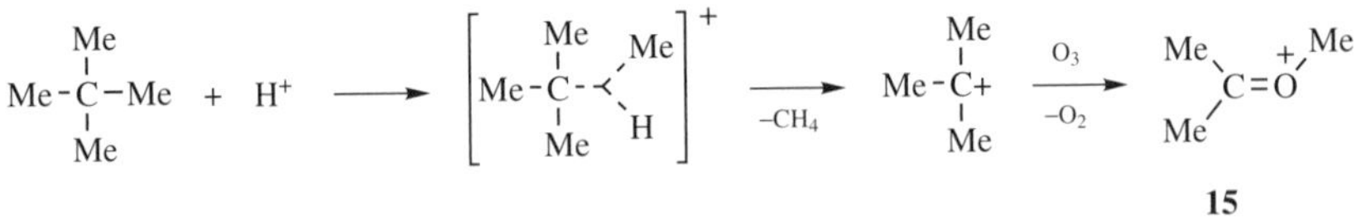

$$(9.53)$$

Similar mechanisms account for the formation of the same products from 2,2,3,3-tetramethylbutane. Besides these predominant pathways, other reactions such as insertion of protonated ozone into the C—C bond and protolysis of the C—H bond may also take place.

Bis(trimethylsilyl)peroxide–triflic acid was also found to be an efficient electrophilic oxygenating agent.[209] Adamantane gives 4-oxahomoadamantane obtained through C—C σ-bond insertion:

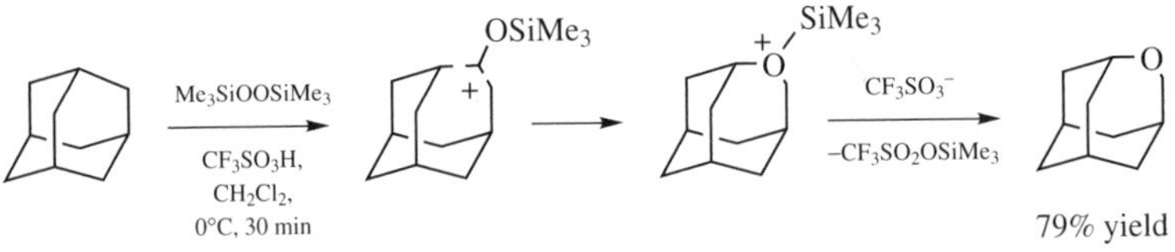

$$(9.54)$$

1-Adamantanol, the C—H insertion product is formed in a very small amount. 4-Oxahomoadamantane was isolated in a similar good yield (84%) when oxygenation was carried out with sodium percarbonate–triflic acid.[210] Two isomeric oxahomodiamantanes (C—C insertion) and bridgehead diamantanols (C—H insertion) are produced from diamantane.[209]

Oxygenolysis. In the course of electrophilic oxygenation of alkanes with hydrogen peroxide or ozone in superacidic media, the strong oxygen electrophiles $O_2H_3^+$ and ^+O_3H, respectively, can insert not only into C—H bonds but also to a lesser extent into C—C bonds.[1] Electrophilic oxygenation of C—C bonds leads to cleavage of these bonds (oxygenolysis), similar to related protolysis, hydrogenolysis, halogenolysis, nitrolysis, and so on. The relative order of reactivity of single bonds in alkanes toward $O_2H_3^+$ from the products obtained in the reaction is $R_3C–H > R_2CH–H > RH_2C–H > C–C$. This order is similar to that obtained for electrophilic insertion of protonated ozone, ^+O_3H, into single bonds of alkanes. Representative is the reaction of neopentane with ^+O_3H [Eqs. 9.55 and 9.56)], indicating 90% C—H and 10% C—C bond attack:[1,91]

$$(9.55)$$

$$(9.56)$$

9.2. OXIDATION OF ALKENES

Depending on the oxidizing agents, reaction conditions, and the structure of the molecule, alkenes may be oxidized to give products formed through the reaction of the π bond (epoxidation, bis-hydroxylation). Strong oxidizing conditions may afford complete cleavage of the double bond to yield two oxygenated molecules (carbonyl compounds and carboxylic acids). Special reaction conditions can lead to the cleavage of vinylic or allylic C—H bonds to produce vinyl- or allyl-substituted products, respectively. The formation of carbonyl compounds and carboxylic acids through such bond breaking is also possible.

9.2.1. Epoxidation

Direct Oxidation with Stoichiometric Oxidants. Discovered by Prilezhaev in 1909,[211] the typical epoxidation reaction of alkenes is their oxidation with organic peracids.[212–221] Of the large number of different peroxycarboxylic acids used in epoxidations, commercially available *m*-chloroperbenzoic acid (*m*-CPBA) is the most favored, but the water-soluble and more stable magnesium monoperoxyphthalate has gained more widespread application.[222] Aliphatic peracids (peroxyformic, peroxyacetic, peroxytrifluoroacetic acid) generated mostly in situ are used preferentially in large-scale syntheses.

A simple generally accepted mechanism known as the "butterfly" mechanism first suggested by Bartlett[79] and Lynch and Pausacker[223] involves the nearly non-polar cyclic transition state **22**:

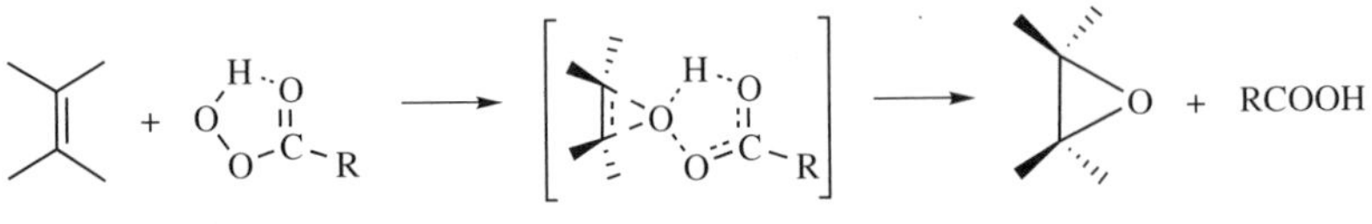

22

$$(9.57)$$

It is formed as a result of the nucleophilic attack on the monomeric, intramolecularly hydrogen-bonded peracid by the π electrons of the alkene double bond. In support of this mechanism are the higher reactivity of peracids possessing electron-withdrawing group, and that of alkenes with more nucleophilic (more substituted) double bond permitting regioselective oxidation of dienes[224,225] [Eq. (9.58)[226]]:

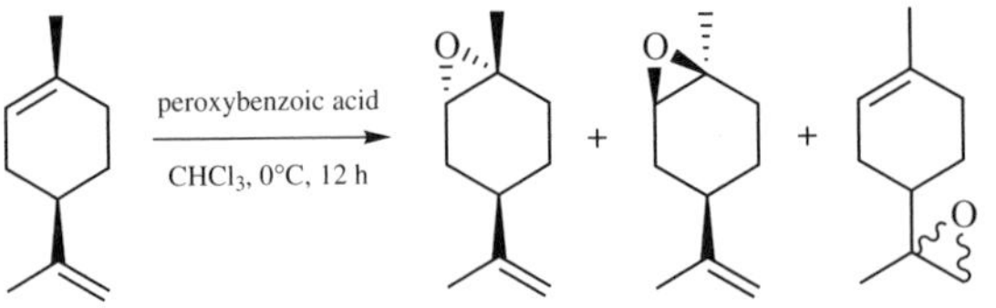

$$(9.58)$$

1:1

92.3   7.7

Also, in accord with the mechanism is the increasing reaction rate with increasing dielectric constant of the solvent and the complete *syn* stereoselectivity.[224]

A closely related 1,3-dipolar cycloaddition mechanism with an 1,2-dioxolane intermediate[227] could not be experimentally proved.[228] Further studies concerning details of the mechanism, including molecular-orbital calculations[229,230] and solvent effects,[231–234] have been carried out, leading, for instance, to the suggestion of the formation of a nonsymmetric transition state[230] and an electron donor–acceptor complex.[234]

The more energetic *cis* double bond in acyclic alkenes is epoxidized faster than the *trans* double bond.[215] In contrast, opposite reactivity is observed for the stereoisomers of cycloalkenes, where the *trans* isomers are more reactive as a result of higher ring strain. This was demonstrated in the selective monoepoxidation of *cis*, *trans*-1,5-cyclodecadiene:[235]

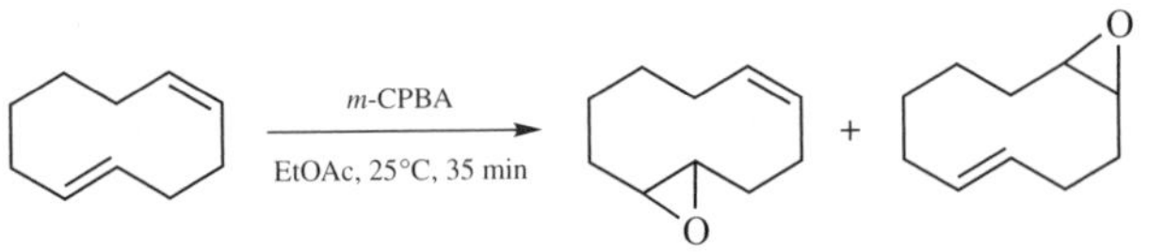

$$(9.59)$$

85–90%   10–15%
91% conversion

The lack of formation of transannular addition products was taken as evidence of the single-step formation of a transition state with little or no ionic character.

Since peroxy acids have a relatively low steric requirement, steric hindrance mainly arises in the epoxidation of bridged cycloalkenes.[224] Marked difference in *exo/endo* selectivity was observed in the reaction of norbornene and 7,7-dimethyl-norbornene with *m*-CPBA[236] [Eqs. (9.60) and (9.61)]:

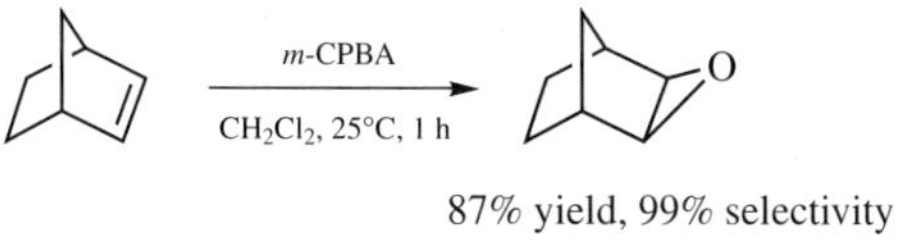

$$
\text{(9.60)}
$$

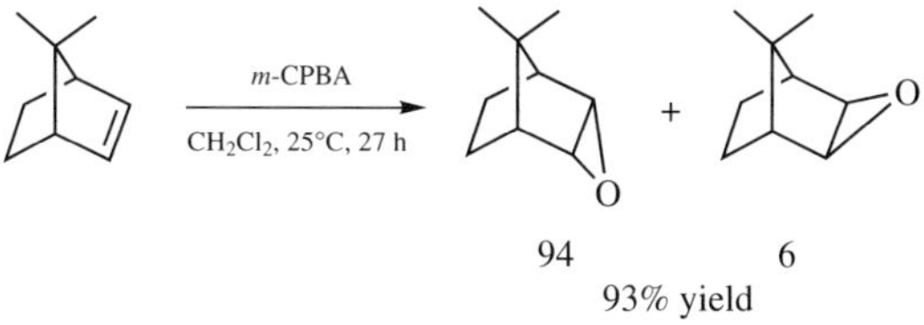

87% yield, 99% selectivity

$$
\text{(9.61)}
$$

94   6

93% yield

In competitive epoxidation norbornene reacts at a rate approximately 100 times that of 7,7-dimethylnorbornene.

Besides peracids, other carboxylic acid derivatives such as peroxycarboximidic acids (**23**), peroxycarbamoic acids (**24**), and peroxycarbonic acids (**25**) also gained importance.[215,237,238] Usually prepared in situ in the reaction of the corresponding acid derivatives with hydrogen peroxide[239] [Eq. (9.62)], they all have the common general formula **26**, and contain the HOO—moiety in conjugation with a double bond:

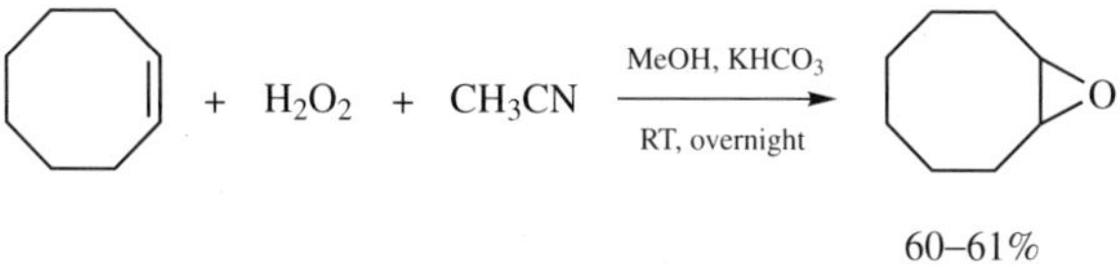

$$
\text{(9.62)}
$$

60–61%

They decompose with ease to transfer oxygen to the double bond. α-Hydroperoxy ethers, amines, ketones, acids, and acid derivatives with the general formula **27** have properties comparable to those of organic peracids.[238,240,241]

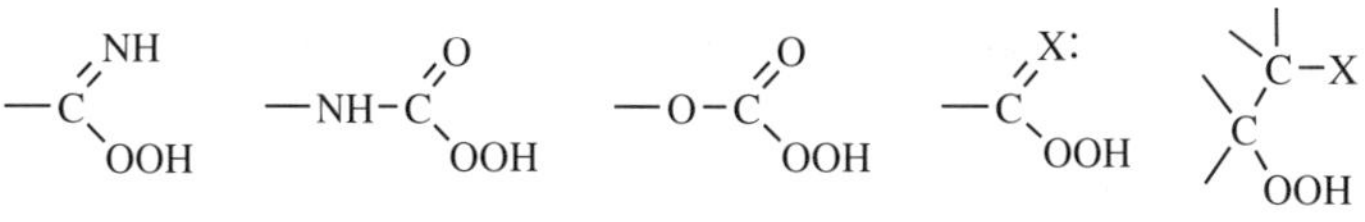

23   24   25   26   27

Further useful stoichiometric oxidizing agents include metal–peroxo complexes of Mo(VI) and W(VI), and high-valence metal–oxo compounds.[242–245] $CrO_2(NO_3)_2$ in aprotic solvents with a basic cosolvent may yield epoxides with high selectivity.[246] The actual oxidizing agent was found to be an oxochromium(V) compound formed by one-electron oxidation of solvent. Molybdenum(VI)–oxoperoxo complexes $MoO(O_2)_2L$ are more selective oxidizing agents.[247–251] Epoxidation with these reagents proceeds via reversible complexation of the alkene to the metal followed by irreversible oxygen transfer to the alkene[247,248] or through direct oxygen transfer to the noncoordinated olefin.[252] An analogous W(VI) complex[253] and polymolybdate-peroxo[251] and polytungstate-peroxo[254] compounds exhibit higher activity.

Direct oxidation of alkenes with molecular oxygen[11,255,256] initiated by free radicals to yield epoxides occurs through addition of peroxy radicals to produce the more stable β-peroxy alkyl radicals (**28**):[257]

$$\text{ROO}\cdot \; + \; \diagup\hspace{-0.3em}=\hspace{-0.3em}\diagdown \quad \longrightarrow \quad \text{ROO}-\overset{|}{\underset{|}{C}}-\overset{|}{\underset{|}{C}}\cdot \quad \longrightarrow \quad \text{RO}\cdot \; + \; \diagdown\hspace{-0.4em}\overset{\displaystyle}{\underset{\displaystyle O}{\diagup}}\hspace{-0.4em}\diagdown \tag{9.63}$$

28

The latter are known to cyclize rather readily to give epoxides and alkoxy radicals.[258]

Maximum yield of epoxides can be expected with molecules, first 1,1-dimethylalkenes, which prefer addition to abstraction (allyl hydrogen removal).[11,259,260] These oxidations are not steroselective; mixtures of the isomeric oxiranes are formed in the reaction of isomeric *n*-octenes.[261] This was interpreted as additional evidence of the formation of radical intermediate **28** allowing the rotation about the carbon–carbon bond.

In cooxidation of alkenes and aldehydes, and in photosensitized epoxidations acyl peroxy radicals are the epoxidizing agents. The mechanism of the former oxidation, on the basis of kinetic measurements and the nonstereospecificity of the reaction,[262–264] involves alkyl peroxy radicals formed through the σ-bonded radical **29** as the intermediate:

$$\text{RCOO}-\overset{|}{\underset{\underset{O}{\|}}{C}}-\overset{|}{\underset{|}{C}}\cdot$$

29

In contrast to these observations, stereoselective epoxidation has been described.[265] Internal olefins reacting with a stoichiometric amount of pivalaldehyde yield quantitatively the corresponding epoxides. The higher reactivity of the internal double bond permits selective epoxidation of dienes:

$$\text{(diene)} \quad \xrightarrow[\text{CCl}_4,\ 40°\text{C},\ 6\ \text{h}]{O_2,\ Me_2CHCHO} \quad \text{(epoxide)} \tag{9.64}$$

78%

The stereoselective nature of the reaction supports the suggestion that epoxidation in this case does not occur by acylperoxy radicals but rather by peracids generated from autoxidation of aldehydes.

Photoepoxidation of alkenes with oxygen in the presence of photosensitizers[266–268] can be used as a synthetic method. The complete absence of any nucleophile in the reaction mixture prevents secondary transformations of the formed oxirane and thus allows the preparation of sensitive epoxides.

The acylperoxy radical was found to epoxidize olefins much faster than peracids also formed under reaction conditions. The result ruled out the role of the latter.[267] The addition of RCO_3· was observed to occur 10^5 faster than that of ROO·. The relative reactivity of alkenes suggests a strongly electrophilic radical forming the polar transition state **30**:

$$RCOO\overset{\delta-}{\cdots}\underset{\underset{O}{\overset{\|}{}}}{C}\overset{\delta+}{=}C$$

30

The rotation about the carbon–carbon bond in intermediate **30** is fast compared to cyclization. As a result isomeric alkenes are transformed to oxiranes of the same, predominantly *trans* stereochemistry. A practical process for the epoxidation of a wide range of alkenes was described in the presence of biacetyl as sensitizer to yield epoxides in about 90% yield.[268]

There are examples in the literature of the formation of epoxides during ozonolysis of olefins:[269]

$$\underset{tert\text{-BuH}_2C}{\overset{tert\text{-BuH}_2C}{}}C=C\underset{R}{\overset{H}{}} \quad \xrightarrow[CH_2Cl_2, -78°C]{O_3} \quad \underset{tert\text{-BuH}_2C}{\overset{tert\text{-BuH}_2C}{}}\underset{O}{C-C}\underset{R}{\overset{H}{}} \qquad (9.65)$$

$$\begin{array}{ll} R = H & 29\% \\ R = tert\text{-Bu} & 91\% \end{array}$$

These epoxides, termed *partial cleavage products*, are formed when the reacting alkenes are sterically hindered.[270]

A detailed study[271] on the reaction of 2,3-dimethyl-2-butene with ozone revealed that epoxide formation strongly depends on alkene concentration and temperature. Under appropiate reaction conditions (neat alkene, 0°C) the corresponding tetramethyloxirane was the main product. Dimethyldioxirane formed from energy-rich acetone oxide (a cleavage product of the alkene) was postulated to be responsible for epoxidation.

Several new powerful oxidizing agents have recently been discovered and used in alkene epoxidation. Dioxiranes [dimethyldioxirane (**31**), methyl(trifluoromethyl)dioxirane 4] react readily with alkenes to give epoxides in high yields.[81–83] **31** was first generated *in situ* from potassium caroate and acetone and was used as such.[272] The rate of oxidation of *cis* dialkylalkenes with **31** was found to be 10 times higher than that of the corresponding *trans* isomers.[273] In contrast, aryl-substituted alkene isomers exhibited similar reactivity. These observations and the exclusive retention of configuration[274] demonstrated for both *cis*-stilbene [Eq. (9.66)] and *trans*-stilbene testify an electrophilic O transfer and the "butterfly" transition state analogous to epoxidation with peracids:

$$(9.66)$$

The mild neutral reaction conditions permit to transform allenes to sensitive spiro-epoxides[275] and norbornadiene to *exo* monoepoxide or *exo,exo* diepoxide:[276]

$$(9.67)$$

less than
stoichiometric **31**   97% selectivity
excess **31**   93% yield

The relatively stable complex HOF•CH$_3$CN generated in the reaction of F$_2$ with water and CH$_3$CN can epoxidize various alkenes quickly and efficiently.[277] Because of the partial positive charge on oxygen generated by the strongly electronegative fluorine, HOF is a strongly electrophilic reagent. This fact as well as the very high reaction rate and the full retention of configuration [Eq. (9.68)] suggest a fast two-step process involving formation of the highly unstable β-oxacarbocation **32**:

$$(9.68)$$

32   >90% yield

Metal-Catalyzed Epoxidation. Hydrogen peroxide is able to convert alkenes to epoxides in the presence of metal catalysts. Several metal oxides (MoO$_3$, WO$_3$, SeO$_2$, V$_2$O$_5$) are known to catalyze such epoxidations.[2,245,278,279] All these catalysts form stable inorganic peracids, and these peracids are supposedly involved in epoxidation in a process similar to organic peracids.

Tungsten(VI) complexes are among the best transition metal catalysts for epoxidation with H$_2$O$_2$. Pertungstenic acid and pertungstate, which are highly stable in aqueous solution, decompose slowly, permitting selective epoxidation under mild conditions.[245] Epoxidations under phase-transfer conditions were found to give the best results.[280–282]

The use of heteropoly acids such as $H_3PM_{12}O_{40}$ [M = Mo(VI), W(VI)] with dilute H_2O_2 is of increasing interest in selective synthesis of epoxides.[254,283,284] Reactions with phase-transfer catalysis again proved to be efficient and synthetically valuable. Even the least reactive simple terminal alkenes are converted under mild conditions, in short reaction time to epoxides in high yields:[283]

$$R\diagup\!\!\!= \xrightarrow[\text{ClCH}_2\text{CH}_2\text{Cl, 70°C, 45–60 min}]{\{[\text{WO(O}_2)_2]_4\text{PO}_4\}^{3-}[(\text{C}_8\text{H}_{17})_3\text{NMe}]^+} \quad R\diagup\!\!\!\triangle_O \tag{9.69}$$

$$\begin{array}{ll} R = n\text{-}C_8H_{17} & 88\% \\ R = n\text{-}C_{10}H_{21} & 94\% \end{array}$$

These reactions, however, have the disadvantage that H_2O_2 is usually employed in aqueous solution. It is known that water seriously retards epoxidation, and it may further transform the product epoxides to the corresponding 1,2-diols. This often results in low selectivity, which is the main reason why these reactions do not have the same broad synthetic utility as do epoxidations with alkyl hydroperoxides.[245,285]

The use of alkyl hydroperoxides in the metal-catalyzed epoxidation of alkenes is their most important synthetic application. High-valence d^0 metals such as Mo(VI), W(VI), V(V), and Ti(IV) are the most effective catalysts.[243–245,278,279,285,286] Used as soluble complexes or as heterogenized supported catalysts, they can give epoxides in near-quantitative yields.

tert-Butyl hydroperoxide the most stable, commercially available alkyl hydroperoxide, is used most frequently.[287] A key question to ensure high selectivity in epoxidation with the ROOH/transition-metal systems is to avoid homolytic cleavage of the hydroperoxide since this may lead to nonstereoselective epoxidation and allylic oxidation.[243,244] Aralkyl hydroperoxides (ethylbenzene hydroperoxide and cumene hydroperoxide) with increased electrophilicity satisfy this requirement and are used accordingly.

Comparative studies concerning the activity of different metal catalysts in epoxidation of cyclohexene[288] led to the findings that metals in the highest oxidation state with strong Lewis acidity and poor oxidizing ability exhibit good catalytic activity and high selectivity in epoxidation.[285,288] Mo(VI) and W(VI) meet best these requirements:

$$\bigcirc \xrightarrow[\text{benzene, 90°C}]{\textit{tert}\text{-BOOH}} \quad \bigcirc\!\!\triangle_O \tag{9.70}$$

	time (h)	% conversion	% yield
Mo(CO)$_6$	2	98	94
W(CO)$_6$	18	95	89
VO(acac)$_2$	2.5	96	13
Ti(On-Bu)$_4$	20	80	66

Because V(V) is a better oxidant, it exhibits low selectivity in the oxidation of simple nonsubstituted alkenes and is used mainly in the epoxidation of unsaturated alcohols, just as Ti(IV).[244]

The best of these catalysts are soluble molybdenum(VI) complexes, of which $Mo(CO)_6$ and $MoO_2(acac)_2$ are the most frequently used and most thoroughly tested and characterized.[245,278] Different molybdenum complexes catalyze the epoxidation of a wide variety of alkenes in nonpolar solvents (benzene, polychlorinated hydrocarbons), at moderate temperature (80–120°C). Of the simplest alkenes, ethylene can be transformed with *tert*-BuOOH + MoO_2(8-hydroxyquinoline)$_2$ to ethylene oxide in high yield,[289] and propylene oxide is currently manufactured on a large scale by Mo-catalyzed epoxidation (see Section 9.5.2). Many different heterogeneous molybdenum catalysts have been used in epoxidations.[245] The ROOH/transition-metal reagents display regularities very similar to those in epoxidation with organic peracids. Increasing nucleophilicity of alkenes brings about increasing reactivity to allow selective monoepoxidation of dienes:[290]

$$ (9.71) $$

$$ 11:1 \qquad 90\% \text{ yield} $$

Similar differences in the reactivity of *cis* and *trans* alkenes (higher reactivity of *cis* compounds), and similar regio- and stereoselectivities are observed in both systems, but slight differences may occur[291] [Eq. (9.72); cf. Eq. (9.58)]:

$$ (9.72) $$

$$ 30:70 \qquad 71\% \text{ yield} $$

One important advantage of the use of ROOH/transition-metal reagents over the traditional organic peracids is that they are particularly useful in the synthesis of acid-sensitive epoxides.

Mechanistic investigations concerning the Mo(VI)-catalyzed epoxidations with alkyl peroxides as oxidants point to the similarities to the V(V)- and Ti(IV)-catalyzed processes. Of the numerous suggestions, two mechanisms are consistent with most experimental observations.[244,245,278]

Mimoun proposed a mechanism that is general for both stoichiometric epoxidations with peroxo complexes and for catalytic systems employing alkyl hydroperoxides.[292–294] It involves an alkylperoxidic species with the alkene complexed through the metal:

$$\text{(9.73)}$$

Insertion to the metal–oxygen bond produces a peroxometallacycle, which decomposes to give the product epoxide and metal alkoxide.

According to the other mechanism,[288,295,296] the intact alkyl hydroperoxide or alkyl hydroperoxidic species activated by coordination to the metal through the oxygen distal from the alkyl group is involved in the epoxidation step:

$$\text{(9.74)}$$

33

The epoxide is produced via the transition state **33** resembling that suggested for epoxidation by peroxomolybdenum reagents[252] and by organic peroxides. The initially formed epoxide is coordinated to the metal. Complexation of the alkene in different orientations to the metal, followed by coordination to the peroxide oxygen, was suggested on the basis of frontier orbitals.[297]

Nonstereospecific aerobic epoxidations of alkenes in the presence of aldehydes catalyzed by nickel(II),[298] iron(III),[299] and cobalt(II)[300] complexes, and clay-supported nickel acetylacetonate[301,302] have been reported. A radical mechanism has been postulated.[300,302] The involvement of active copper species and peracids were suggested in a similar reaction catalyzed by copper salts.[303]

A unique titanium(IV)–silica catalyst prepared by impregnating silica with $TiCl_4$ or organotitanium compounds exhibits excellent properties with selectivities comparable to the best homogeneous molybdenum catalysts.[285] The new zeolite-like catalyst titanium silicalite (TS-1) featuring isomorphous substitution of Si(IV) with Ti(IV) is a very efficient heterogeneous catalyst for selective oxidations with H_2O_2.[184,185] It exhibits remarkable activities and selectivities in epoxidation of simple olefins.[188,304–306] Propylene, for instance, was epoxidized[304] with 97% selectivity at 90% conversion at 40°C. Shape-selective epoxidation of 1- and 2-hexenes was observed with this system that failed to catalyze the transformation of cyclohexene.[306] Surface peroxotitanate **13** is suggested to be the active species.[184,185,188] It forms a complex with the alkene to undergo an intramolecular oxygen transfer to the C–C double bond.[188] A different, heterolytic peracid-like mechanism was put forward in protic solvents with the participation of **34** as the active species:[305]

$$(9.75)$$

Epoxidation Catalyzed by Metalloporphyrins. Metalloporphyrins, which have thoroughly studied as catalysts in alkane oxygenations, have also been tested as epoxidation catalysts.[119,122,244,245,307] Iodosylbenzene (PhIO), sodium hypochlorite, alkyl hydroperoxides, potassium hydrogen persulfate, and molecular oxygen are the oxygen sources used most frequently in these oxidations.[119]

Epoxidation with PhIO in the presence of manganese(III) tetraphenylporphyrine chloride [Mn(TPP)Cl] has been extensively studied.[244,245,307] The main characteristic of these reactions is the loss of stereochemistry at the double bond with product distributions depending on the porphyrin structure:[134]

$$(9.76)$$

Mn(II) tetraphenylporphyrin chloride	1 : 1.6	88%
Mn(III) tetra(*o*-tolyl)porphyrin chloride	2.8 : 1	87%

Epoxidation of norbornene with PhIO in the presence of $H_2{}^{18}O$ resulted in the formation of labeled norbornene oxide, indicating the exchange of a labile oxo ligand in high-valence manganese–oxo species. The reactive complex **35** has been isolated.[134,308] These observations can be explained by a stepwise radical transformation with the involvement of the manganese–oxo species **36** and radical intermediate **37**.[134] It allows rotation around the carbon–carbon double bond before ring closure, resulting in the formation of isomeric oxiranes as follows:

$$(9.77)$$

Epoxidation with NaOCl is of synthetic interest although this system is not stereoselective, either.[244,245] Reactivity, chemo- and stereoselectivity, however, are largely increased by added nitrogen bases (pyridine, imidazole)[309–311] that behave as axial ligand of the metalloporphyrin. Highly regioselective epoxidation of the less hindered double bond in dienes with the sterically hindered manganese(III) tetra(2,4,6-triphenylphenyl)porphyrine acetate [Mn(TTPPP)OAc] catalyst was demonstrated:[311]

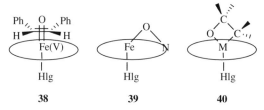

$$\tag{9.78}$$

Dramatic shape selectivities in competitive olefin epoxidation was observed with "picnic basket" metalloporphyrins[312,313] designed to exclude bulky axial ligands on one sterically protected porphyrin face. When oxidized with PhIO in acetonitrile in the presence of the rigid *p*-xylyl-strapped porphyrin, *cis*-2-octene reacted selectively versus *cis*-cyclooctene or 2-methyl-2-pentene, giving >1000 reactivity ratios.[313,314] Some immobilized manganese(III) porphyrins proved to be as efficient as their homogeneous equivalents in epoxidation with PhIO.[151,315]

In contrast with manganese(III) porphyrins, stereoselective epoxidation with retention of configuration can be achieved with iron(III) porphyrins.[244,245] The reactivity of alkenes parallels that observed for other epoxidizing reagents; the rate increases with increasing alkyl substitution at the double bond. The reactivity of *cis* alkenes is higher than that of the corresponding *trans* isomers. A side-on approach of alkene to the reactive iron(V)–oxo intermediate with the plane of the alkene molecule parallel to the porphyrin ring (**38**) was proposed to interpret the experimental observations.[316,317] In these systems, where several epoxidation mechanisms may be operative, intermediate **39** with the oxygen inserted into the iron–nitrogen bond is a viable alternative.[316–319] A metallaoxetane intermediate (**40**) consistent with early results with hindered porphyrins[320–322] was discounted later:[323]

Certain robust manganese porphyrins [e.g., manganese(III) tetra(2',6'-dichlorophenyl)porphyrin chloride] are able to catalyze stereoselective epoxidations, when applied in the presence of imidazole[324–327] or other heterocyclic nitrogen bases.[326] NaIO$_4$[324] or even H$_2$O$_2$[325–327] can be used as oxidant.

Epoxidation with dioxygen is also possible if it is carried out in the presence of added reducing agents to transform the second oxygen atom into H$_2$O.[307] Since the first successful epoxidation with H_2–colloidal Pt–*N*-methylimidazole catalyzed by Mn(TPP)Cl,[328] several other reducing agents have been introduced,[124] but only the Mn(TPP)Cl–O_2–ascorbate[154] and the Mn(TPP)Cl–Zn–imidazole[152] systems yield epoxides exclusively.

Of other related systems, molybdenum(V) porphyrin exhibits very high stereo-selectivity with *tert*-BuOOH as the oxygen source (97% *cis*-2-hexene oxide and 99% *trans*-2-hexene oxide from *cis*- and *trans*-2-hexene, respectively).[329] Nonporphyrin complexes of iron were found to be stereoselective in the epoxidation of stilbene isomers. Iron cyclam, a nonporphyrin iron complex, gives the corresponding *cis* and *trans* epoxides in epoxidation with H_2O_2.[330] Fe(acac)$_3$, in contrast, yields the *trans* epoxide from both stilbene isomers.[331]

Asymmetric Epoxidation. Asymmetric epoxidation of nonfunctionalized alkenes manifests a great synthetic challenge. The most successful method of asymmetric epoxidation, developed by Katsuki and Sharpless,[332] employs a Ti(IV) alkoxide [usually Ti(OisoPr)$_4$], an optically active dialkyl tartrate, and *tert*-BuOOH. This procedure, however, was designed to convert allylic alcohols to epoxy alcohols, and the hydroxyl group plays a decisive role in attaining high degree of enantiofacial selectivity.[333,334] Without such function, the asymmetric epoxidation of simple olefins has been only moderately successful:[335]

$$(9.79)$$

(1*R*,2*S*)
10.2% ee

In most cases only very low asymmetric induction was achieved with chiral epoxidation reagents (chiral peroxy acids, H_2O_2 or *tert*-BuOOH in the presence of chiral catalysts),[336,337] or with metal complexes with chiral ligands.[336–339] However, if the stereogenic center in the oxidizing agent is close to the reactive oxygen, as in the case of chiral sulfanyloxaziridines, enantioface selectivity may be higher (64.7% ee in the epoxidation of *trans*-β-methylstyrene).[340]

Chiral metalloporphyrins[317,341–343] and salen [*N,N'*-bis(salicylideneamino) ethane][344–347] complexes constitute the most enantioselective nonenzymatic olefin epoxidation catalysts, yielding epoxides with 50–80% enantiomeric excess. PhIO,[317,341–344] iodosylmesitylene,[345] NaOCl[346] [Eq. (9.80)], or O_2 + pivalaldehyde[347] are used as oxidants:

$$(9.80)$$

(2*S*,3*R*)
82% ee, 87% yield

Enzymatic epoxidation has proven to be the best method to achieve high enantio-selectivity[348–350] [Eqs. (9.81) and (9.82)[351,352]].

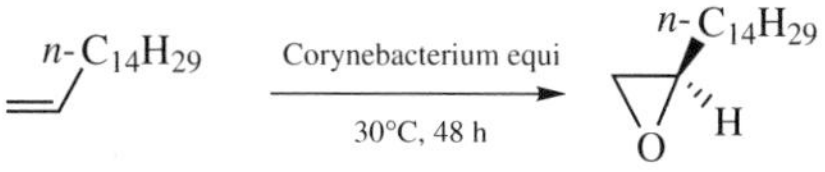

$$
n\text{-}C_{14}H_{29} \quad \xrightarrow[\text{30°C, 48 h}]{\text{Corynebacterium equi}} \quad n\text{-}C_{14}H_{29}\triangle_O H \tag{9.81}
$$

(R)
41% yield, 100% ee

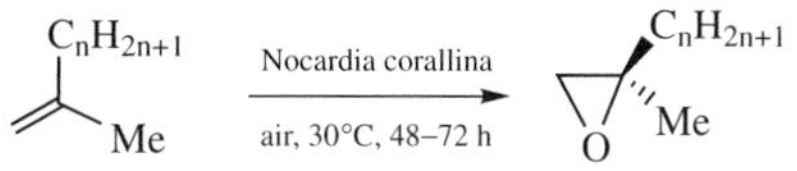

$$
\underset{\text{Me}}{\overset{C_nH_{2n+1}}{\bigtriangleup}} \quad \xrightarrow[\text{air, 30°C, 48–72 h}]{\text{Nocardia corallina}} \quad \underset{O}{\overset{C_nH_{2n+1}}{\triangle}}\text{Me} \tag{9.82}
$$

(R)

n	% yield	% ee
3	32	76
4	55	90
5	56	88

The disadvantage of this method is that only terminal alkenes give epoxides with high enantiomeric excess. For example, in microsomal aerobic epoxidation propene, *trans*-2-butene and 2-methyl-2-butene are transformed to the corresponding epoxides with 40%, 14% and 0% ee, respectively.[350] Chloroperoxidase, however, was shown to be very effective in the epoxidation of *cis* alkenes.[353]

A one-step, metal-catalyzed direct hydroxyepoxidation [ene reaction under irradiation in the presence of triplet oxygen (see Section 9.2.2) and a transition-metal catalyst] to produce epoxy alcohols with high diastereoselectivities is a useful synthetic procedure.[354]

9.2.2. Reactions with Molecular Oxygen

Autoxidation. The slow oxidation of alkenes with ground-state (triplet) molecular oxygen in the liquid phase under mild conditions (autoxidation) affords the formation of organic hydroperoxides as primary products through a free-radical process[3,11,355,356] (Scheme 9.6). The ready abstraction of an allylic hydrogen (initiation) leads to a resonance-stabilized allylic radical. The latter reacts with oxygen to yield the **41** alkenyl peroxy radical. Propagation includes reaction between

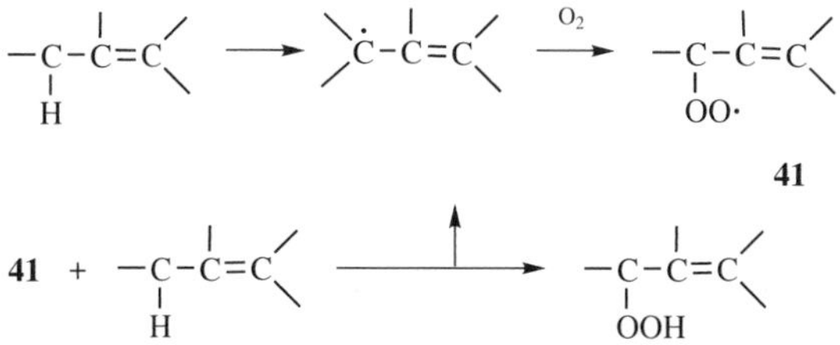

Scheme 9.6

the alkene and **41** resulting in the formation of a new radical and the product alkenyl hydroperoxide. Besides this reaction, the peroxy radical may add to the alkene to form a dialkylperoxy radical. The latter may participate in polymerization or epoxide formation [see Eq. (9.63)]:

$$
\underset{\text{Me}}{\overset{\text{Me}}{\diagdown}}C=C\underset{\text{Me}}{\overset{\text{Me}}{\diagdown}} + {}^1O_2 \xrightarrow{\text{MeOH}} \quad (9.83)
$$

rose bengal   54%
H₂O₂ + NaOCl   64%

Hydroperoxide formation is characteristic of alkenes possessing tertiary allylic hydrogen. Allylic rearrangement resulting in the formation of isomeric products is common. Secondary products (alcohols, carbonyl compounds, carboxylic acids) may arise from the decomposition of alkenyl hydroperoxide at higher temperature.

Reactions with Singlet Oxygen. Singlet oxygen generated by photosensitization of ground-state triplet oxygen,[357] prepared by chemical methods,[357,358] or resulting from electrical discharge[359] reacts with unsaturated hydrocarbons to give oxygenated products. Depending on the structure of the reacting molecules, allylic hydroperoxides, 1,2-dioxetanes, and endoperoxides may be formed. Oxygenations with singlet oxygen are attractive synthetic reactions and have been studied extensively.[360–374] The main reason for this is that the products formed may be readily transformed to functionalized organic compounds.[364,366,374–377]

Monoalkenes possessing allylic hydrogen are transformed to give allylic hydroperoxides[360–367] [Eq. (9.83)[378]]. The latter may undergo secondary transformations to give epoxides and divinyl ethers.[367]

The reaction discovered by Schenck[379] is analogous to the ene reaction and involves oxygen attack at one of the olefinic carbons, abstraction of an allylic hydrogen, and the shift of the double bond. The double-bond shift clearly distinguishes the ene reaction from the radical autoxidation of alkenes. Radical autoxidation can yield hydroperoxides, too, but it usually gives a complex reaction mixture. In contrast, oxidation with singlet oxygen introduces oxygen in a highly selective manner into organic compounds.

The reactivity of alkenes towards singlet oxygen strongly depends on the electron density of the π bond, with increasing alkyl substitution bringing about increasing reactivity[380] [Eq. (9.84)[381]]:

$$
\xrightarrow[\text{MeOH, 10°C, 4.5 h}]{\text{O}_2, \text{hv, methylene blue}} \quad (9.84)
$$

55%

This equation also indicates the findings of early works, that reactivity of allylic hydrogens decreases in the order methyl > methylene ≫ methine. Later, however, anomalous reactivity of the more crowded site was recognized.[382–385]

Steric effects, however, may alter this reactivity pattern. Bulky groups, for instance, may hinder singlet oxygen from approaching the double bond and the allylic hydrogen involved in the ene reaction. In cycloalkenes singlet oxygen prefers to attack from the less hindered face of the alkene double bond. This was demonstrated in the transformation of α-pinene [Eq. (9.85)], where the approach occurs exclusively from the side opposite to the *gem*-dimethyl group:[386,387]

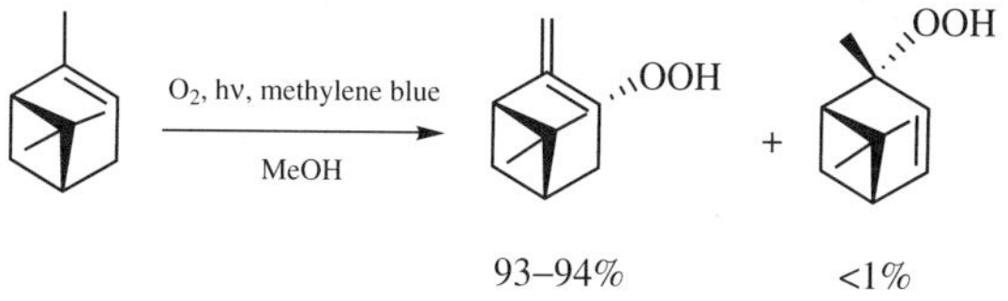

$$93\text{–}94\% \qquad <1\% \tag{9.85}$$

The reaction also features high regioselectivity.

Conformational effects may affect the transformation of both cyclic and acyclic alkenes. The unreactivity of the methine hydrogen in the oxidation of **43** was interpreted[362] as being caused by the unfavorable **44** conformation required for the formation of the tertiary hydroperoxide **45** (Scheme 9.7).

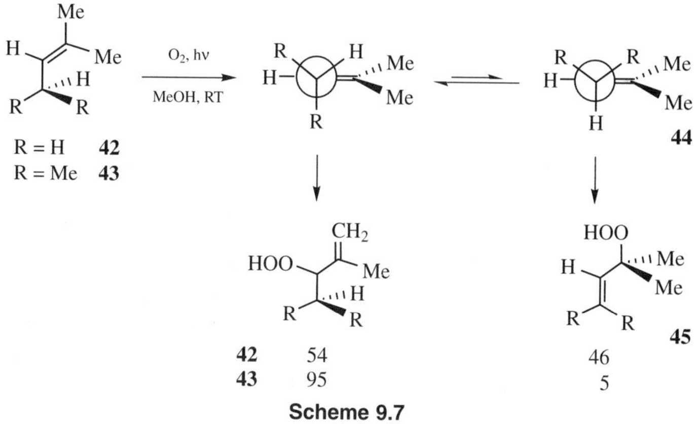

| | 42 | 54 | 46 |
| 43 | 95 | 5 |

Scheme 9.7

On the basis of the electronic, steric, and conformational effects a classical one-step mechanism with transition state **46** was originally proposed:[388]

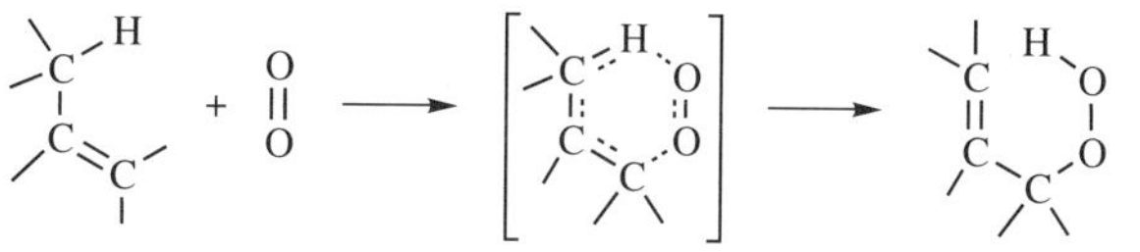

$$\tag{9.86}$$

46

An increasing number of experimental observations have accumulated, however, indicating the involvement of intermediate steps. The conformationally fixed **47** alkene, for example, gives products formed on attack at both sides of the π bond, although only the equatorial side contains correctly oriented allylic hydrogens required for a concerted process:[389]

$$(9.87)$$

Perepoxide (**48**),[383,389–391] 1,2-dioxetane (**49**),[392,393] diradical (**50**),[389,391] and dipolar (**51**)[389,391] intermediates, each accounting for some experimental observations, were suggested, but all were excluded on the basis of later studies:[360,384,394,395]

The involvement of **48** and **51** was, however, concluded by trapping experiments in special cases.[396]

For example, the ene reaction proceeds stereoselectively in a suprafacial manner; specifically, oxygen attack and hydrogen removal take place on the same side of the double bond.[384,397] As a result, neither racemization of optically active compounds[387,398] nor cis–trans isomerization[394] occurs. Diradical or dipolar intermediates are not consistent with these observations. The lack of a Markovnikov-type directing effect also rules out dipolar intermediates.[360,399] Isotope effects observed in the reaction of different deuterium-labeled 2,3-dimethyl-2-butenes indicated that not all hydrogens (deuteriums) are equally competitive in the transformation.[384,397] This led to the suggestion of a symmetric transition state in which oxygen interacts only with the cis grouped C—H(D) bonds.[384,397]

Although 1,2-dioxetanes are excluded as intermediates in the formation of allylic hydroperoxides, they may be formed as primary products in the reaction of singlet oxygen with alkenes not possessing allylic hydrogen:[360,363,366,368–370,374]

$$(9.88)$$

Formation of 1,2-dioxetanes is usually characteristic of alkenes activated by heteroatom in the vinylic position. Most dioxetanes are unstable and decompose with concomitant chemiluminescence. The products are carbonyl compounds resulting from oxidative cleavage of the double bond.

Certain hindered alkenes display exceptional behavior. They readily undergo cycloaddition to yield 1,2-dioxetanes despite the presence of allylic hydrogen. The oxidation of 2,2′-biadamantylidene was the first successful synthesis of the dioxetane of an unactivated alkene:[400,401]

$$(9.89)$$

The product exhibits a remarkable thermal stability decomposing only at its melting point (164–165°C). 7,7′-Binorbornylidene,[402] camphenylideneadamantane,[403] and, surprisingly, an isolable three-ring cyclobutadiene[404] and 2,5-dimethyl-2,4-hexadiene[405] also give relatively stable dioxetanes. Dioxetanes can also be isolated in good yields in the low-temperature oxidation in polar solvent of alkenylarenes,[406,407] such as indenes.

The retention of olefin geometry in the oxidation of *cis* and *trans* alkenes[408] suggests a one-step concerted [2 + 2]-cycloaddition process, suprafacial in the alkene, and antarafacial in singlet oxygen.[363] The strong solvent dependence, however, observed in many cases, points to a stepwise mechanism involving a perepoxide intermediate.[403,409]

The third important reaction of singlet oxygen is 1,4-cycloaddition to conjugated dienes, resulting in the formation of endoperoxides.[360–364,371–373] Acyclic dienes give monocyclic endoperoxides that exist in rapidly equilibrating half-chair conformations.[410,411]

An isolated, less substituted double bond in trienes that is less reactive does not participate in oxidation allowing selective oxygenation:[412]

$$(9.90)$$

This indicates that the dienophilic reactivity of singlet oxygen surpasses the other two modes of reaction (formation of hydroperoxides and 1,2-dioxetanes).

Conjugated cycloalkadienes[413–417] are converted to bicyclic endoperoxides that may undergo thermal or photochemical rearrangement to yield diepoxides.[374–377] 1,3-Cyclopentadiene and 1,3-cyclohexadiene give endoperoxides exclusively. The first synthesis of the naturally occurring ascaridol (**52**) was also carried out in this way:[418,419]

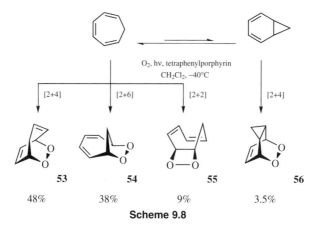

$$(9.91)$$

52

In contrast, 1,3-cycloheptadiene and 1,3-cyclooctadiene whose double bonds are twisted out of coplanarity are considerably less reactive and give endoperoxides in lower yields.[420]

Cycloheptatriene represents a challenging substrate for cycloaddition. It can react either in a [2 + 4]- or [2 + 6]-cycloaddition to yield isomeric endoperoxides **53** and **54**, respectively (Scheme 9.8). Moreover, it may give a 1,2-dioxetane (**55**) in a [2 + 2] process. Norcaradiene, its valence isomer, may also yield a [2 + 4] adduct (**56**). All four products were detected in the transformation of the nonsubstituted molecule.[421,422]

O$_2$, hv, tetraphenylporphyrin
CH$_2$Cl$_2$, –40°C

[2+4]   [2+6]   [2+2]   [2+4]

53   **54**   **55**   **56**

48%   38%   9%   3.5%

Scheme 9.8

Steric and electronic factors operating in valence isomerization equilibria of the 7-substituted compounds greatly affect product distribution. They react to yield only the corresponding compounds **53** and **56**. The **56** : **53** ratio increases in the order Me < Et < isoPr < tert-Bu < Ph.[423] Alkenylarenes, such as stilbenes,[424] indenes,[425] vinylnaphthalenes[426] [Eq. (9.92)], and vinylphenanthrenes,[427] react with singlet oxygen with the participation of the aromatic π electrons in the transformation:[424,425]

O$_2$, hv, tetraphenylporhyrin
CCl$_4$, 2 h

$$(9.92)$$

R^1, R^2 = H, Me, Ph   34–78%

Singlet oxygen addition to the aromatic rings often follows (see Section 9.4.1).

Because of the considerable resemblance of endoperoxide formation to the Diels–Alder reaction, a concerted mechanism with a six-membered ring transition state was suggested.[393] Substituent effects, namely, the higher reactivity of dienes possessing electron-donating substituents toward singlet oxygen, support the assumption stated above. Since a concerted process requires coplanar cisoid conformation of the reacting molecules, highly stereospecific reactions lend additional support to the abovementioned mechanism.[362,426,428,429]

Certain experimental observations, however, are not in accord with a concerted reaction, but are in favor of a two-step process. Solvent effects, such as large differences in rates of gas-phase and solution-phase reactions,[430] indicate the involvement of some polar intermediates, such as a zwitterion[422] or a charge-transfer complex.[430] Compounds that are unable to adopt the required cisoid conformation, such as polysubstituted acyclic 1,3-dienes[405,431] and seven- and eight-membered 1,3-cycloalkadienes,[420] are likely to follow a two-step mechanism. It is also unusual that singlet oxygen, unlike many other conventional dienophiles, exhibits only moderate *endo* selectivity in the transformation of condensed systems.[432] Energetic factors, unusually weak orbital interactions, were found to be responsible for the lack of stereochemical control. Theoretical calculations support a perepoxide intermediate.[433]

The experimental observations available suggest that the mechanism of endoperoxidation is likely to be substrate dependent.[428,433]

9.2.3. Bis-hydroxylation

$KMnO_4$ and OsO_4 have long been used to achieve *syn* hydroxylation of the carbon–carbon double bond. The *syn* stereoselectivity results from the formation of cyclic ester intermediates formed by addition of the reagent from the less hindered side of the double bond.

If used under mild, alkaline conditions to prevent further oxidation of the 1,2-diol formed, $KMnO_4$ can be a selective oxidizing agent.[434,435] However, proper experimental conditions (extensive stirring in dilute solution,[436] phase-transfer catalysis[437–440]), or other permanganate salts[441] [Eq. (9.93)] are required to achieve high yields:

$$
(CH_2)_n \quad \xrightarrow[\substack{\textit{tert}\text{-BuOH} + H_2O \text{ or } CH_2Cl_2, \\ 20°C, 1h}]{C_{16}H_{33}N^+Me_3MnO_4^-} \quad (CH_2)_n \overset{OH}{\underset{OH}{}} \tag{9.93}
$$

$$n = 2, 4, 8 \qquad\qquad 65\text{–}86\%$$

The cyclic manganate(V) diester anion **57** formed in a [3 + 2]-cycloaddition is believed to be the intermediate in the oxidation process[434] (Scheme 9.9). This intermediate may be formed through the rearrangement of oxametallacyclobutane **58** resulting from a [2 + 2]-cycloaddition.[442,443] Experimental observations indicate that the product 1,2-diol is formed as a result of the gradual reduction of the cyclic diester to Mn(IV).[444,445] The novel idea of the formation of soluble, colloidal MnO_2 via rapid decomposition of **57** by abstraction of a hydrogen atom from the solvent[442,443,446] and its role in oxidation seems to be ruled out.[444,445]

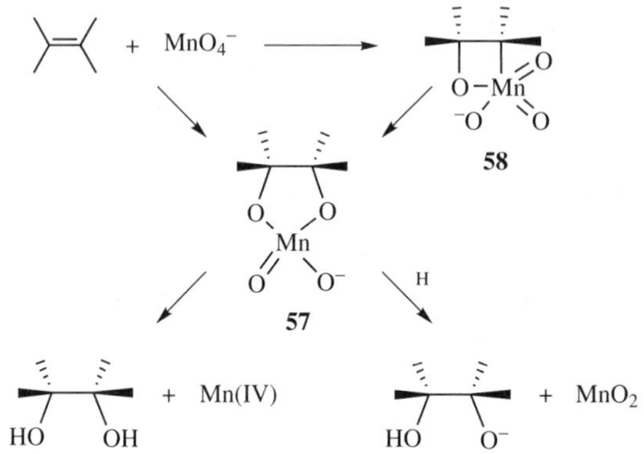

Scheme 9.9

OsO_4, which gives much better yields than $KMnO_4$, is the reagent of choice for *syn* hydroxylation of alkenes.[435,447–449] In the conventional method OsO_4 is used in a stoichiometric amount. Tertiary amines or diamines result in a dramatic rate increase.[447] A catalytic amount of OsO_4 has been successfully employed with suitable cooxidants that continuously regenerate the reagent. H_2O_2 used originally[450,451] has been replaced by *tert*-BuOOH[447,448,452] or amine N-oxides[447,448,452] such as N-methylmorpholine-N-oxide:[453]

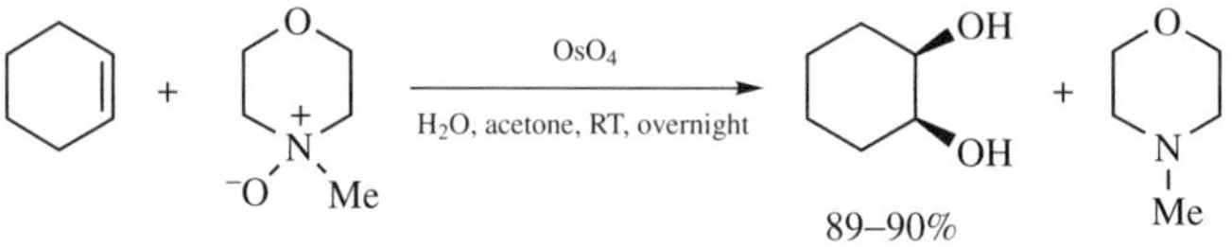

$$(9.94)$$

89–90%

Hexacyanoferrate(III) ion has been found to be very effective.[454] Since overoxidation often takes place in the catalytic process, stoichiometric oxidation generally gives better yields.

An asymmetric osmylation method has been developed by Sharpless and coworkers. OsO_4 modified by a dihydroquinidine auxiliary (cinchona alkaloid derivatives)[449,455–458] or chiral diamines[449,457–460] such as **59** and **60** used in stoichiometric oxidation may yield *cis* diols with excellent optical purity:[460]

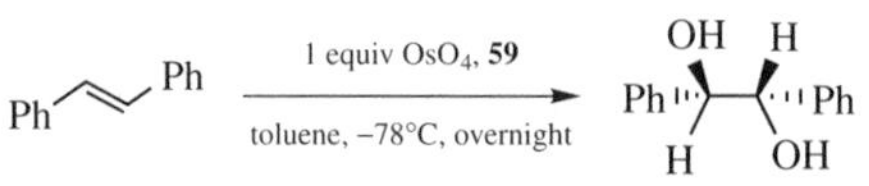

$$(9.95)$$

(S,S)-diol
96% yield, 100% ee

59   **60**

Catalytic asymmetric osmylation can also be carried out in the presence of chiral amine ligands. Since the first examples of this ligand-accelerated catalysis published by Sharpless,[461] steady improvements in selectivities have been achieved. Slow addition of alkene,[462] the successful use of $K_3[Fe(CN)_6]$ as cooxidant,[463] the introduction of 9-O-aromatic dihydroquinidine and dihydroquinine ligands,[464] phthalazine-modified ligands,[465–467] and the fast osmate ester hydrolysis in the presence of organic sulfonamides[466] are the latest important findings. The Sharpless AD-mix (asymmetric dihydroxylation) reagents offer high levels of enantioselectivity in introducing chiral 1,2-diol functionality into olefinic substrates.[466–468]

The mechanism of oxidation with OsO_4 is a much studied process. Tetrahedral osmium(VI) dimeric monoester[448,469] or monomeric diester[447] species are the intermediates that yield the product 1,2-diols through reductive or oxidative hydrolysis. Amines affect the rate by transforming tetrahedral OsO_4 to hexacoordinated octahedral Os(VIII) by complexation.[462,470,471] Similarly as in permanganate oxidation, [2 + 2]-cycloaddition giving oxametallacyclobutane has been suggested,[455,472,473] but most of the experimental evidence support direct [3 + 2]-cycloaddition[470,471,474] and species such as **61**, **62**, and **63** (L = amine ligand) were identified.[460,462,475] The results of new mechanistic studies have been published.[476,477]

61   **62**   **63**

Hydroxylations with H_2O_2 catalyzed by tungsten peroxo complexes[478] or by Re_2O_7[479] provide 1,2-diols with very good selectivity. In a related procedure developed by Woodward,[480] iodine and silver acetate are used in moist acetic acid. The key step determining *syn* stereochemistry is the Ag^+-assisted transformation of the intermediate *trans*-iodoacetate to the 1,3-dioxolan-2-ylium ion (see also Scheme 9.10). Other reagent combinations such as I_2–KIO_3–KOAc[481] or TlOAc–I_2[482] were introduced later.

Alkenes may also react with certain oxidizing agents to result in *anti* hydroxylation. Treatment with peroxycarboxylic acids[435] leads initially to an epoxide. Ring scission of the latter via an S_N2 reaction in an *anti* manner with the corresponding carboxylic acid or water gives the *trans* monoester or *trans* diol, respectively. Complete *anti* stereoselectivity and high yields in the oxidation of cycloalkenes are

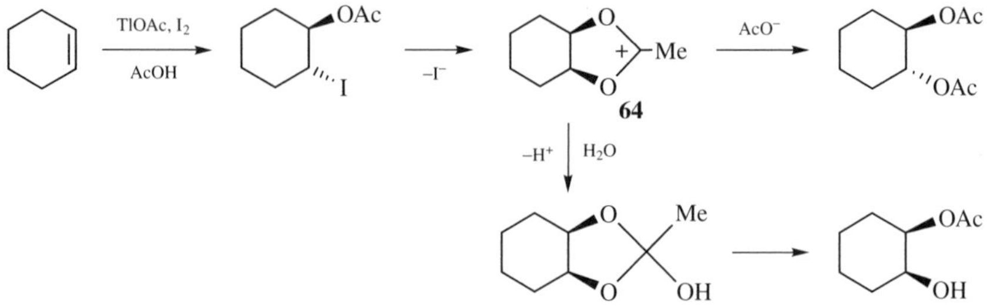

Scheme 9.10

achieved with *m*-chloroperoxybenzoic acid in a one-pot synthesis:[483]

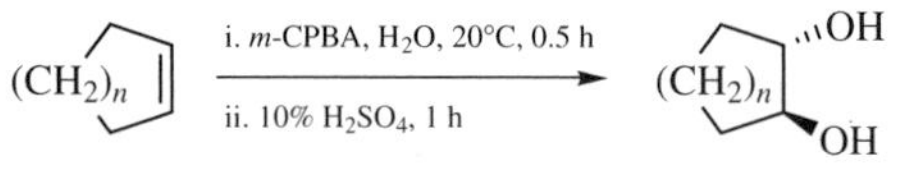

$$n = 1 \ 95\%$$
$$n = 2 \ 80\%$$

The same products are formed with sodium perborate, a cheap industrial chemical when used in $Ac_2O + H_2SO_4$.[484]

The oxidation of alkenes with H_2O_2 in the presence of oxide catalysts (WO_3,[485] H_2WO_4,[486] SeO_2[487,488]) is believed to involve the corresponding inorganic peroxy acids as the oxidizing agent and an epoxide intermediate. WO_3, the most efficient reagent, is used in aqueous medium (for oxidation of water insoluble alkenes), or in acetic acid.[485] Oxidations with SeO_2 are carried out in alcohols.[488] Polymer-bound phenylseleninic acid catalyzes the oxidation of cyclohexene with H_2O_2 to *trans*-1,2-cyclohexanediol in excellent yield.[489]

Oxidation with lead tetraacetate is a far less selective process.[490,491] Studied mainly in the oxidation of cycloalkenes, it gives stereoisomeric 1,2-diol diacetates, but side reactions (allylic acetoxylation, skeletal rearrangement) often occur. A change in reaction conditions in the oxidation of cyclopentadiene allows the synthesis of different isomeric mono- and diesters.[492]

The iodine–silver carboxylate reagent combination modified by Woodward[480] was originally used by Prévost to oxidize alkenes to *trans*-1,2-diesters.[493] Thallium(I) acetate, mentioned above, as a reagent for *syn* dihydroxylation can also be used for the synthesis of *trans*-1,2-diols. If the reaction is carried out in glacial acetic acid, the intermediate 1,3-dioxolan-2-ylium ion **64** formed from the iodoacetate undergoes ring opening by attack of the acetate ion to yield the *trans* diacetate[482] (Scheme 9.10). In contrast, an aqueous workup affords orthoacetate intermediate, which rearranges to give the *cis* monoacetate.

9.2.4. Vinylic Oxidation

Alkenes can be transformed to carbonyl compounds through the oxidation of the vinylic carbon atom. A special case of vinylic oxidation is acetoxylation of alkenes and dienes.

Oxidation to Carbonyl Compounds

Oxidation with Palladium in the Homogeneous Phase. The most thoroughly studied reaction concerning the transformation of alkenes to carbonyl compounds is their oxidation catalyzed by palladium in homogeneous aqueous media.[243,244,494–503] As a rule, ethylene is oxidized to acetaldehyde, and terminal alkenes are converted to methyl ketones.[504,505]

Although the oxidation of ethylene to acetaldehyde was known for a number of years,[506] its utility depended on the catalytic regeneration of Pd(0) in situ with copper(II) chloride discovered by Smidt and coworkers.[507,508] Air oxidation of Cu(I) to Cu(II) makes a complete catalytic cycle. This coupled three-step transformation is known as the *Wacker process* [Eqs. (9.97)–(9.99)]. The overall reaction [Eq. (9.100)] is the indirect oxidation with oxygen of alkenes to carbonyl compounds:

$$CH_2{=}CH_2 \ + \ PdCl_2 \ + \ H_2O \ \longrightarrow \ CH_3CHO \ + \ Pd \ + \ 2\,HCl \qquad (9.97)$$

$$Pd \ + \ 2\,CuCl_2 \ \longrightarrow \ PdCl_2 \ + \ 2\,CuCl \qquad (9.98)$$

$$2\,CuCl \ + \ 2\,HCl \ + \ 0.5\,O_2 \ \longrightarrow \ 2\,CuCl_2 \ + \ H_2O \qquad (9.99)$$

$$CH_2{=}CH_2 \ + \ 0.5\,O_2 \ \longrightarrow \ CH_3CHO \qquad (9.100)$$

Although some details are still debated, there is a general agreement about the basic steps of the mechanistic scheme. The first two steps are consistent with the kinetics of the reaction in that the oxidation is inhibited by Cl^- and step 3 accounts for acid inhibition (Scheme 9.11). Key features of the process are the nucleophilic attack on the π-complexed alkene, resulting in its transformation to a σ-bonded intermediate (**65** to **66**), and β-hydride elimination (**66** to **67**). The disagreement is in the nature of nucleophilic attack called the *hydroxypalladation step.*

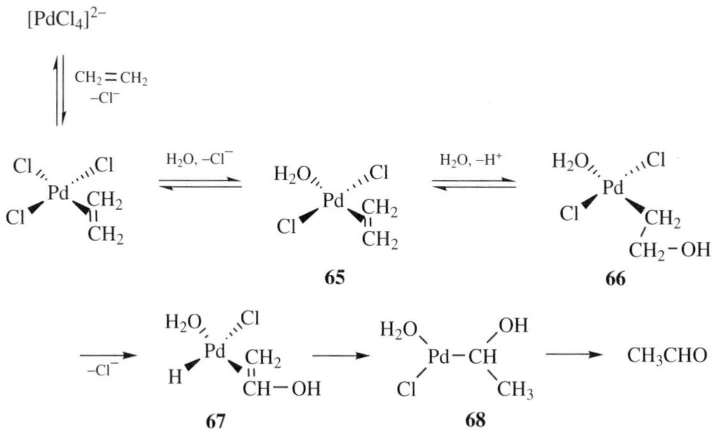

Scheme 9.11

Scheme 9.11 depicts *trans* addition of the external nucleophile water resulting in a transoid β-hydroxyethylpalladium intermediate (**66**).[498] Studies with stereoisomeric DHC=CHD molecules support this view.[509,510] Results with D_2O, however, indicates no deuterium incorporation into the product molecule. This is consistent with an intramolecular *cis* addition of HO^-, namely, ethylene insertion into the Pd–O bond [**70** to **71**, Eq. (9.101)]:

$$
\mathbf{65} \; \overset{-H^+}{\rightleftharpoons} \;
\begin{bmatrix} HO_{/\!/,} & _{\backslash\backslash}Cl \\ & Pd \; CH_2 \\ Cl^{\blacktriangleleft} & \underset{CH_2}{\overset{\|}{}} \end{bmatrix}^{-}
\;\rightleftharpoons\;
\begin{bmatrix} Cl_{/\!/,} & _{\backslash\backslash}OH \\ & Pd \; CH_2 \\ Cl^{\blacktriangleleft} & \underset{CH_2}{\overset{\|}{}} \end{bmatrix}^{-}
\;\overset{H_2O}{\rightleftharpoons}\;
\begin{bmatrix} Cl_{/\!/,} & _{\backslash\backslash}OH_2 \\ & Pd \\ Cl^{\blacktriangleleft} & CH_2 \\ & \underset{CH_2-OH}{} \end{bmatrix}^{-}
$$

$$\qquad\qquad \mathbf{69} \qquad\qquad\qquad \mathbf{70} \qquad\qquad\qquad \mathbf{71}$$

$$(9.101)$$

This step, yielding a cisoid β-hydroxyethylpalladium intermediate (**71**), is preceded by a *trans–cis* isomerization (**69** to **70**).[498] A series of hydride shifts (**66** to **67** and **67** to **68**) eventually leads to the product carbonyl compound.

Of the other olefins, short-chain terminal alkenes are oxidized to methyl ketones. Propylene gives predominantly acetone with a negligible amount of propionaldehyde. 1-Butene and 2-butenes are converted to 2-butanone in about 90% yield. The rate of oxidation decreases in the order ethylene > propylene > butenes, which is consistent with the nucleophilic attack on the coordinated alkene. Higher homologs are oxidized very slowly and afford some chlorinated products and a mixture of isomeric ketones resulting from the palladium-catalyzed isomerization of the alkene facilitated by high temperature. Cyclodextrins, however, used in a two-phase system greatly improve the yields and selectivities of the oxidation of C_8–C_{10} alkenes.[511] The $(RCN)_2PdCl_2$ complex in the presence of CuCl, a chloride salt in *tert*-BuOH was shown to exhibit unusually high selectivity for aldehyde formation (up to 57%).[512]

An important drawback of the original Wacker process is the highly corrosive nature of the aqueous acidic $PdCl_2$–$CuCl_2$ system. Attempts were made to apply electrochemical reoxidation of palladium,[513,514] and to use other oxidants,[495] such as Fe(III) salts, MnO_2, quinones,[514–516] peroxides,[517,518] and more recently, heteropoly acids.[516,519–522]

Successful modifications to improve the oxidation of nonterminal alkenes include the use of organic solvents. Oxidation in anhydrous[523,524] or aqueous alcohols,[525] *N,N*-dimethylformamide,[526] and sulfolane[527] may convert internal alkenes and cycloalkenes into ketones in high yields with high selectivities. Cyclohexene, for example, is transformed to cyclohexanone in 95% yield (ethanol, 90°C, 2 h).[525] It was also found that the addition of moderate or large amounts of inorganic acids ($HClO_4$, HNO_3, HBF_4) to a chloride-free Pd(II)–*p*-benzoquinone system gave rate enhancements to a factor of ≤ 50 in the oxidation of cycloalkenes.[516]

A multistep catalytic system consisting of $Pd(OAc)_2$, hydroquinone, and an oxygen-activating complex [iron phthalocyanine, Fe(Pc)] in aqueous DMF in the

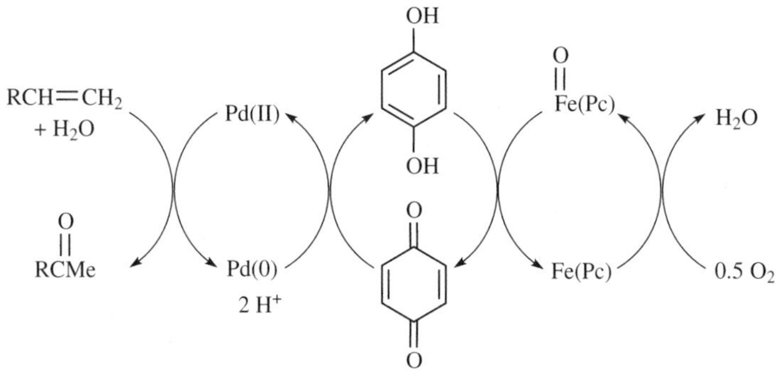

Scheme 9.12

presence of a small amount of strong acid (HClO$_4$) has been described.[528,529] In this chloride-free medium the mild and selective oxidation by molecular oxygen of terminal olefins to methyl ketones can be achieved. The coupled electron-transfer processes are indicated in Scheme 9.12.

Pd(II) *tert*-butyl peroxidic complexes of the general formula [RCOOPdOO*tert*-Bu]$_4$ are very efficient reagents for the selective stoichiometric oxidation of terminal alkenes in anhydrous nonbasic organic solvents at ambient temperature.[518] They also catalyze the ketonization by *tert*-BuOOH. Internal olefins are not reactive at all. A peroxometallacycle intermediate (see discussion below) was suggested.

Oxidation with Other Reagents. Other metals and other oxidizing systems are also active in the transformation of alkenes to carbonyl compounds. Rh(I) compounds that form peroxo complexes are able to transfer oxygen to organic compounds. Terminal alkenes, when treated with rhodium trichloride[530–532] or rhodium complexes[533–535] such as [RhCl(PPh$_3$)$_3$] in the presence of oxygen, are transformed to methyl ketones in good yields. Cu(II) salts may be used, but they do not play a direct role in the oxidation of the alkene.[532] Studies with preformed peroxo complexes[536–538] combined with ^{18}O-labeling experiments led to the finding that water is not involved in the oxidation. A direct oxygen transfer[535–537] to the organic molecule through a peroxometallacycle (**72**) was suggested:[534–536]

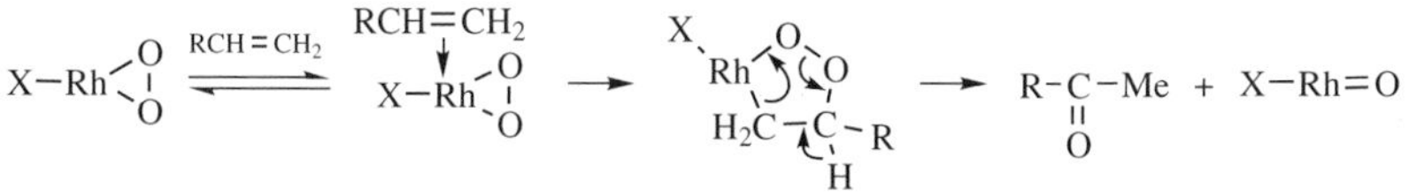

$$(9.102)$$

On the basis of deuterium labeling a β-hydride shift in the decomposition of **72** was ruled out.[539]

It was found, however, that in the RhCl$_3$-catalyzed reaction a complementary Wacker cycle operates as well.[292,531,540] The **73** oxorhodium(III) complex formed in the peroxometalation step [Eq. (9.102)] is transformed to a hydroxy species. This carries out hydroxymetalation similar to that discussed for palladium [Scheme 9.11, Eq. (9.101)] to produce a second molecule of methyl ketone. In phosphine-containing systems methyl ketone formation takes place with the concomitant oxidation of phosphine to phosphine oxide.[533–535]

In one of the few examples known for the oxidation of dienes,[541,542] Rh and Ru complexes under aerobic conditions, and the corresponding peoroxo complexes transform 1,5-cyclooctadiene selectively to 1,4-cyclooctanedione.[541] The monoketone is not an intermediate in the formation of the diketone end product.

A Co(II) Schiff-base complex converts 1- and 2-alkenes into methyl ketones and the corresponding secondary alcohols in the presence of oxygen or H_2O_2 in primary alcohol solvent.[543] A radical oxidation with cobalt hydroperoxide through the formation and subsequent decomposition of alkyl hydroperoxide was suggested.[543] An efficient conversion of alkenylarenes to ketones was achieved by the use of molecular oxygen and Et_3SiH in the presence of a catalytic amount of Co(II) porphyrin in 2-propanol.[544]

Co(III)-nitro,[545] Rh(III)-nitro,[546] and Pd(II)-nitro[547] complexes were shown to transfer oxygen to alkenes in stoichiometric[545,546] or catalytic[545,547] processes. In some cases (Co and certain Rh complexes), when the alkene cannot coordinate to the metal, Pd(II) is required, which serves as an alkene activator through coordination.[545,546]

Relatively few examples are known for heterogeneous catalytic oxidations. Gasphase[548] and liquid-phase[549] oxidations with a Pd on charcoal catalyst and oxygen, as well as with a heterogenized Wacker system [$PdCl_2 + CuCl_2 + Cu(NO_3)_2$ supported on alumina],[550] were studied. A palladium- and copper-exchanged Y-type zeolite catalyzes the oxidation of ethylene to acetaldehyde in exactly the same way as the homogeneous Wacker system.[551] The active center was found to be a partially ammoniated Pd(II) ion.

The single-step production of acetone by the catalytic oxidation of propylene in the gas phase is a desirable goal, which can be achieved mainly by binary oxides.[552] Acetone is obtained with better than 90% selectivity at 100–160°C when propylene is oxidized with H_2O-O_2 on SnO_2-MoO_3.[553] Ketone formation proceeds via hydration of the carbocation intermediate to form an adsorbed alcoholic species followed by oxydehydrogenation:[553,554]

$$CH_3CH=CH_2 \xrightarrow[\text{ii. } H_2O, -H^+]{\text{i. } H^+} \left[\begin{matrix} CH_3CHCH_3 \\ | \\ OH \end{matrix}\right]_{ads} \xrightarrow{O^- \text{ or } O_2} \underset{\underset{O}{\|}}{CH_3CCH_3} + H_2O \quad (9.103)$$

The selectivity of acetone formation on a $V_2O_5-TiO_2$ catalyst active in the transformation of propylene to acetic acid could be enhanced by adding water vapor to the feed.[555]

Varied oxidizing agents may be employed in the laboratory synthesis of carbonyl compounds. Styrenes undergo oxidative rearrangement when treated with $Pb(OAc)_4$ in CF_3COOH, but only aldehydes are produced in good yields:[556]

$$R-\langle\text{benzene ring}\rangle-CH{=}CH_2 \quad\xrightarrow[\text{CH}_2\text{Cl}_2\text{, RT, 30 min}]{\text{Pb(OAc)}_4\text{, CF}_3\text{COOH}}\quad R-\langle\text{benzene ring}\rangle-CH_2-\overset{\displaystyle O}{\underset{\displaystyle H}{C}} \qquad (9.104)$$

$$R = H \quad 98\%$$
$$R = Me \quad 86\%$$

Similar oxidative rearrangements to produce carbonyl compounds have been carried out with thallium(III) nitrate[557] or peroxytrifluoroacetic acid–BF$_3$.[558] The Jones reagent (CrO$_3$–H$_2$SO$_4$) in the presence of a catalytic amount of Hg(II) is also effective.[559] Chromyl chloride oxidation, when combined with reductive workup, is a simple, convenient one-step method to convert 2,2-disubstituted 1-alkenes to aldehydes.[560,561]

Permanganate oxidations may be used for the direct conversion of olefins to α-hydroxy ketones and 1,2-diketones in moderate yields usually under acidic conditions.[562–566]

Vinylic Acetoxylation. When alkenes are treated with Pd(II) compounds in the presence of acetic acid in a nonaqueous medium, acetoxylation takes place.[495,498,499,501,503,567–569] Ethylene is converted to vinyl acetate in high yields and with high selectivity with PdCl$_2$[568,569] in the presence of added bases (NaOAc,[568] Na$_2$HPO$_4$[569]) or with Pd(OAc)$_2$:[570]

$$CH_2{=}CH_2 \;+\; 2\,AcO^- \;+\; Pd(II) \;\longrightarrow\; CH_2{=}CH{-}OAc \;+\; AcOH \;+\; Pd(0)$$

$$(9.105)$$

The oxidation of 1-alkenes usually gives 2-acetoxy-1-alkenes.[571,572] Oxidative acetoxylation of propylene with Pd(OAc)$_2$ may yield allylic or vinylic acetates depending on reaction conditions[573] (see Section 9.2.6).

The mechanism of vinyl acetate formation is closely related to that of the Wacker oxidation (Scheme 9.11); that is, acetoxypalladation–palladium hydride elimination takes place.[498,503] The coordinated alkene is attacked by the external nucleophile acetate ion, or the attack may occur within the coordination sphere. β-Hydride elimination followed by dissociation of the coordinated molecule yields directly the vinyl acetate end product.

When reoxidation of Pd(0) with the Cu(II) + O$_2$ system (used in the Wacker process) is applied, the process is less selective. Besides vinyl acetate, ethylene yields acetaldehyde, ethylidene diacetate, ethylene glycol mono- and diacetate, and chlorine-containing compounds as well.[574] A complex reaction mixture consisting of isomeric vinylic acetates and allylic acetates, and 1,2-diol monoacetates is usually formed from terminal alkenes.[499] Because of the easy formation of palladium π-allyl complexes through oxidation, higher alkene homologs tend to give allylic acetates as the main product. The oxidation of cyclohexene[575] and internal alkenes[571] gives nearly exclusively allylic acetates, although the results are somewhat contradictory.[499]

Diacetoxylation of 1,3-butadiene is a process that drew much attention since the product 1,4-diacetoxy-2-butene may be converted to 1,4-butanediol and tetrahydrofuran by further transformations (see Section 9.5.2). The liquid-phase acetoxylation of 1,3-butadiene in a Wacker-type system yields isomeric 1,2- and *cis*- and *trans*-1,4-diacetoxybutenes:

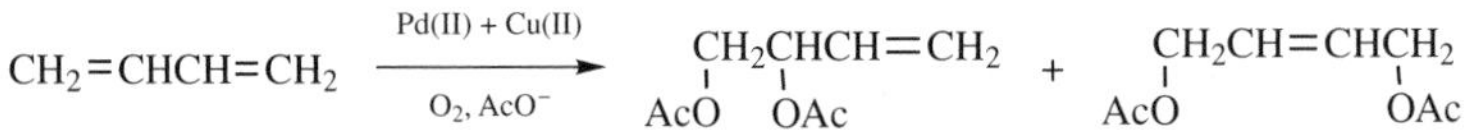
CH2=CHCH=CH2
Pd(II) + Cu(II)
O2, AcO−
CH2CHCH=CH2
AcO OAc
+
CH2CH=CHCH2
AcO OAc
(9.106)

1,3-Cycloalkadienes exhibit a similar behavior. Thorough studies by Bäckvall and coworkers[576] revealed that the proper choice of ligand permits regio- and stereoselective 1,4-difunctionalization. LiCl and LiOAc were found to exhibit a profound effect on the stereoselectivity:[577]

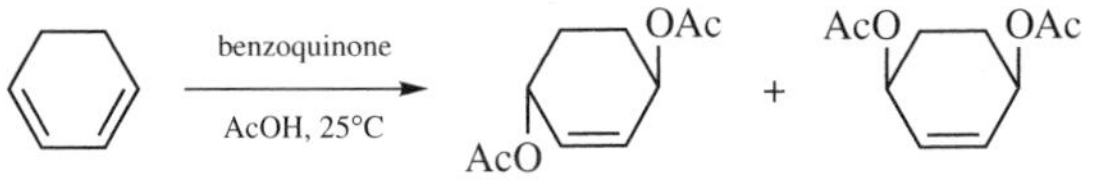
benzoquinone
AcOH, 25°C
OAc
AcO
+
AcO
OAc
(9.107)

$Pd(OAc)_2$	50	50
$Pd(OAc)_2$ + LiOAc	>90%	
Li_2PdCl_4 + LiOAc		>95%

Other conjugated cyclic dienes undergo a similar palladium-catalyzed stereoselective 1,4-diacetoxylation.[578]

The stereoselectivity is explained by a change in the mode of acetate attack on the intermediate π-allylpalladium complex. The coordination of acetate to palladium is effectively blocked by chloride ions, thus hindering *cis*-migration. As a result, path *c* is operative (Scheme 9.13). In the absence of chloride both *cis* and *trans* attacks (paths *a* and *b*) can occur depending on acetate concentration.

Palladium–hydroquinone electrochemical 1,4-oxidation,[579] aerobic oxidation via the Pd(II)–hydroquinone–metal macrocycle triple catalytic system,[529] and MnO_2 as a reoxidant[578] were also applied in the 1,4-diacetoxylation of 1,3-dienes.

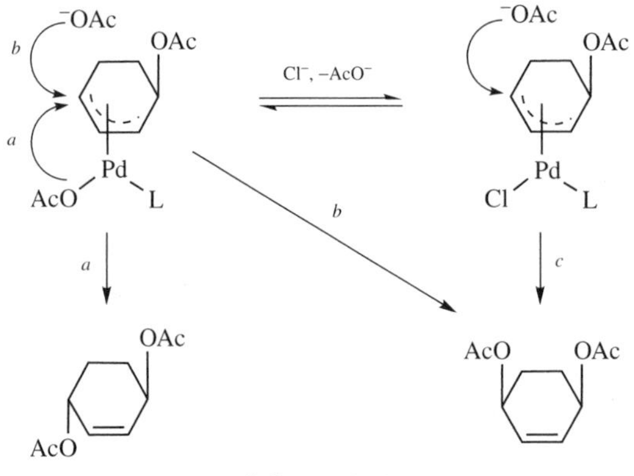
−OAc
OAc
b
a
Pd
AcO
L
Cl−, −AcO−
−OAc
OAc
Pd
Cl
L
b
a
c
OAc
AcO
AcO
OAc

Scheme 9.13

Heterogeneous systems have been developed for the synthesis of vinyl acetate and 1,4-diacetoxy-2-butene. A bimetallic catalyst, Pd—Te (10 : 1) on charcoal, yields isomeric 1,4-diacetoxy-2-butenes with 89% selectivity in the liquid phase.[580] A Pd—Sb—V—KOAc multicomponent system when doped with CsCl proved to have improved lifetime and selectivity in the gas-phase synthesis of 1,4-diacetoxy-2-butene.[581] A tellurium-promoted palladium catalyst provides good yields of 1,4-diacetoxy-2-butene in either liquid-phase or gas-phase acetoxylation.[567] See further details in Section 9.5.2.

9.2.5. Oxidative Cleavage

Ozonation. Alkenes react readily with ozone to give oxidized compounds via cleavage of the π or the σ,π bonds.[582–592] Ozonides, the direct product of the reaction, are rarely isolated but are transformed in situ into synthetically useful products. Depending on the solvent used and the actual workup of the reaction mixture different products may be isolated.[582–584,592] Although its use is inconvenient, ozone is widely employed in selective syntheses. A classical use of ozone is in the structure determination of natural compounds and polymers.

Mechanism. According to the classical Criegee mechanism,[593,594] the reaction of alkenes with ozone takes place in three steps (Scheme 9.14). An unstable 1,2,3-trioxolane (**74**), also referred to as the *initial* or *primary ozonide*, is formed in the first step. **74** readily decomposes to give a zwitterionic carbonyl oxide (**75**) and a carbonyl compound. In step 3 recombination of the two cleavage products yields the **76** 1,2,4-trioxolane (Criegee intermediate, final, or normal ozonide). Steps 1 and 3 are [2 + 3]-cycloadditions, and step 2 can be regarded as a cycloreversion. Criegee did not specify the structure of **75** in his original suggestion. Later, however, 1,2-diols were isolated after the reduction of a crystalline intermediate, indicating only partial cleavage of the carbon—carbon double bond in the early stage of the reaction.[595,596]

The first normal ozonide, the ozonide of 1,2-dimethylcyclopentene, was isolated and characterized as early as 1953 by Criegee.[597] Ozonides were later observed by

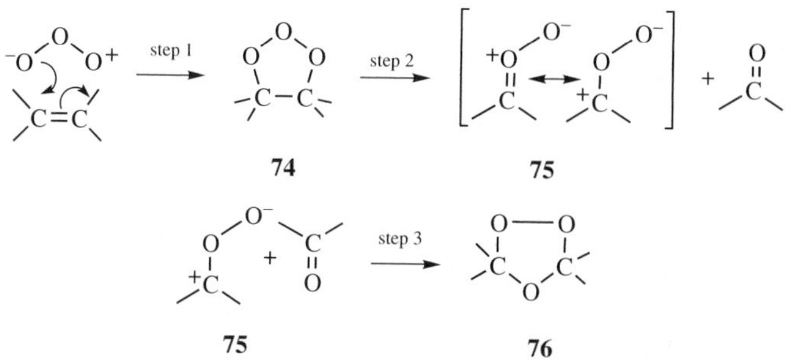

Scheme 9.14

low-temperature IR[583,598,599] and NMR[583,599–601] spectroscopy and were characterized recently by microwave spectroscopy.[589,599,602]

The highly reactive carbonyl oxide could be trapped by added carbonyl compounds to form normal ozonides.[593,603] Isolation of dimeric (**77**) and polymeric
(**78**) peroxides and that of products formed with reactive, participating solvents
(**79**) also proved the existence of **75** (Scheme 9.15). The reactive carbonyl oxide
was also identified spectroscopically.[590]

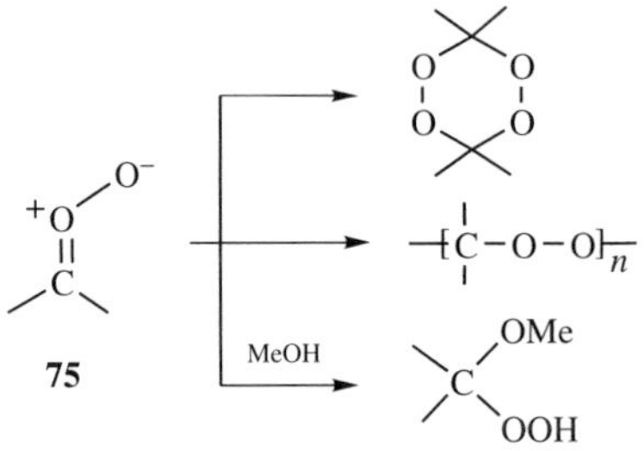

Scheme 9.15

Aldehydes tend to be more reactive toward carbonyl oxide than ketones in the
recombination reaction (Scheme 9.14, step 3). As a result, tetrasubstituted alkenes
generally fail to give normal ozonides[604] but rather, **75** undergoes the side reactions
indicated above. A new method of ozonation carried out with alkenes adsorbed on
polyethylene, however, could yield ozonides of tetrasubstituted alkenes, and even
diozonides of dienes could be isolated.[605]

In agreement with the Criegee mechanism, unsymmetric alkenes,[606,607] such as
2-pentene[606] [Eq. (9.108)], or a pair of symmetrical olefins, such as a mixture of
$CH_2=CH_2$ and $CD_2=CD_2$,[608] yield three different ozonides, including two symmetric cross-ozonation products (**80, 81**):

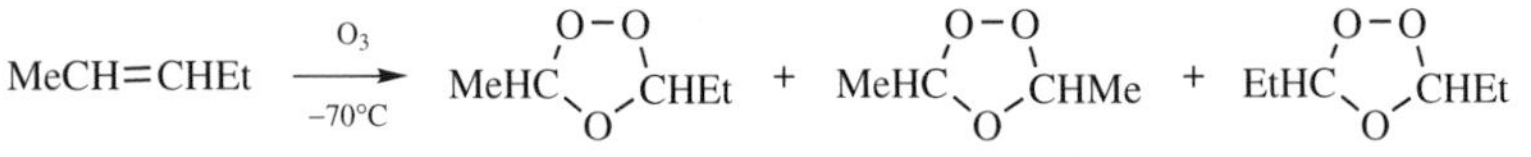

$$\text{MeCH=CHEt} \xrightarrow[-70°C]{O_3} \text{MeHC} \underset{O}{\overset{O-O}{\diagup \diagdown}} \text{CHEt} + \text{MeHC} \underset{O}{\overset{O-O}{\diagup \diagdown}} \text{CHMe} + \text{EtHC} \underset{O}{\overset{O-O}{\diagup \diagdown}} \text{CHEt} \quad (9.108)$$

                                          80   **81**

Since each ozonide exists as a pair of *cis–trans* isomers, six isomeric ozonides can
be detected in such cases.

Studies in the 1960s and 1970s led to new findings resulting in clarification of
the details of the mechanism. Ozonation in the presence of ^{18}O-labeled aldehydes
leads almost exclusively to ether oxygen incorporation of the label, lending additional support to the validity of the Criegee mechanism.[609–611] A number of new
observations were reported, however, which are not consistent with the original
mechanism. In the first study, concerning the stereochemistry of alkene ozonation,
cis-di-*tert*-butylethylene gave a 70 : 30 *cis* : *trans* ozonide ratio, whereas the *trans*
isomer formed 100% *trans* ozonide.[612] Since the stereochemistry is lost in step 2
(Scheme 9.14), the *cis* : *trans* ratio of the symmetric cross-ozonide would also be

expected to remain invariant with the olefin geometry, which is not the case[607,613] [Eq. (9.109)[614]]:

$$(9.109)$$

Also inconsistent with the mechanism is that *cis* alkenes give higher yield of ozonides than do the *trans* isomers. The bulk of alkene susbstituents was also observed to strongly affect the ozonide *cis* : *trans* ratio. These observations made necessary the incorporation of stereochemistry into the Criegee mechanism.

In the refined Criegee mechanism[615,616] that includes orbital symmetry considerations as well,[617] all three steps are stereospecific. Ozonation of *cis*- and *trans*-[1,2-²H₂]ethylene, for instance, gave the *exo* and *endo* isomers according to Eqs. (9.110) and (9.111) (the *cis-exo* and *cis-endo* ozonides from the *cis* compound, and the *trans* ozonide form the *trans* isomer):[602]

$$(9.110)$$

$$(9.111)$$

This indicates no evidence of stereorandomization about the carbon–carbon double bond. It means that **74** cleaves in a concerted fashion to yield *syn* and *anti* zwitterions produced in different amounts depending on the alkene geometry. The stereoisomeric zwitterions remain attached to each other like an ion pair and recombine stereoselectively with aldehydes. The subtle nature of ozonation, however, may be illustrated by the observation that ozonide formation may be nonconcerted under some circumstances.[618,619]

Synthetic Applications. Ozonolysis, that is, the cleavage of the carbon–carbon double bond, is usually carried out in solvents, such as alcohols (methanol or ethanol), chlorinated hydrocarbons (CH_2Cl_2, chloroform), ethyl acetate, and THF, preferably at low temperature (at or below 0°C).[582–584] The nature of the solvent used in ozonolysis has a strong effect on product distribution. In aprotic, nonparti-cipating solvents under anhydrous conditions monomeric ozonide, polymeric per-oxides, and/or polymeric ozonides may be isolated. In participating and usually protic solvents, α-alkoxyalkyl hydroperoxides are primarily formed. Ozonolysis in water usually yields carboxylic acids. The same products may be isolated in the acid-catalyzed decomposition of ozonides. If strong reducing agents, such as metal hydrides,[582,620] diborane,[592] or BH_3–SMe_2[621] are used, primary and/or sec-ondary alcohols can be synthesized depending on the structure of the starting alkene.[592]

Intermediate ozonides may undergo hydrolysis to give carbonyl compounds:

$$\underset{\text{ii. } H_2O}{\overset{\text{i. } O_3}{\longrightarrow}} \quad 2 \; \overset{\displaystyle\diagdown}{\underset{\displaystyle\diagup}{C}}=O \; + \; H_2O_2 \tag{9.112}$$

To avoid further oxidation of the aldehyde by the byproduct H_2O_2 formed, hydro-lysis is usually carried out in the presence of mild reducing agents.[584,592] Catalytic hydrogenation, $Zn + CH_3COOH$, and NaI or KI are the most frequently used con-ventional reductants.[582,584,592] Dimethyl sulfide,[622] triphenyl phosphine,[607] and thiourea,[623] introduced later, are likewise effective and selective reducing agents.

Despite earlier observations,[624,625] ozonolysis of cyclic olefins in MeOH or in mixed solvents (ethers or esters and MeOH) followed by isomerization in the pre-sence of Lindlar Pd and hydrogen does not give directly dicarboxylic acids. Instead, mainly the corresponding dialdehydes are formed.[626] The best method of synthesis of dicarboxylic acids is ozonolysis of cycloalkenes in a mixture of acetic acid and formic acid followed by further oxidation with oxygen:[626]

$$(CH_2)_n \quad \underset{\text{ii. } O_2,\ 110°C,\ 8–10\ h}{\overset{\text{i. } O_3,\ AcOH–HCOOH\ (10:1),\ 10°C}{\xrightarrow{\hspace{3cm}}}} \quad HOOC(CH_2)_{n+2}COOH \tag{9.113}$$

$$n = 1 \quad 52\%$$
$$n = 2 \quad 77\%$$

Oxidative hydrolysis transforms the intermediate ozonide into ketone(s) and/or carboxylic acid(s) in good yields. H_2O_2 in water, in sodium hydroxide solution, or in formic acid is the best proven oxidant.[582,584,592] Peroxy acids and silver oxide are also employed. Rearrangement and overoxidation may be undesirable side-reactions. A simple two-step ozonation in MeOH yields methyl esters without added oxidizing agent.[627]

Interesting variations in product distributions could be observed when ozonoly-sis was carried out with alkenes adsorbed on silica gel.[628] The low-temperature ozonolysis on dry silica gel of cyclopentene led to the formation of the normal, monomeric ozonide in high yield [Eq. (9.114)]. This corresponds to the product formed in aprotic and nonparticipating solvents. In contrast, reaction of cyclopen-tene on wet silica gel resulted in the formation of the corresponding oxo acid [Eq. (9.115)]:

$$\text{(9.114)}$$

$$\text{(9.115)}$$

No cross ozonide was formed from unsymmetrical alkenes. The authors theorized[628] that the carbonyl oxide zwitterionic species formed on wet silica gel immediately adds water followed by rapid decomposition of the intermediate hydroxyalkyl hydroperoxide to carboxylic acid and water. It means that water on silica gel acts as participating solvent. In the absence of adsorbed water, rapid recombination of the adsorbed aldehyde and carbonyl oxide due to a favorable proximity effect gives normal ozonide. The low mobility of adsorbed species on the silica surface accounts for the absence of cross ozonides.

Further variations make ozonolysis a versatile tool in organic synthesis. Equations (9.116)–(9.118) illustrate the possibility of preparing terminally differentiated products from cycloalkenes:[629,630]

$$\text{(9.116)}$$

$$\text{MeOOC(CH}_2)_4\text{CHO} \qquad 65\text{–}72\% \qquad \text{(9.117)}$$

$$\text{(9.118)}$$

Dienes and polyenes usually react at all olefinic sites, but special reaction conditions may allow selective stepwise ozonolysis. When monoozonolysis of 1,3-cycloalkadienes is followed by reduction with Me_2S, unsaturated dialdehydes are the primary products.[631] Selective cleavage of one of the double bonds in the presence of MeOH leads to two isomeric aldehyde acids isolated in the form of the methyl esters.[632] Selectivity is determined by the conformationally controlled regioselective fragmentation of the primary monoozonide. The lack of selectivity in ozonolysis of acyclic dienes[632–634] lends additional support to the conclusion that regioselectivity in ozonolysis of cyclic dienes is controlled by conformational effects. Nonconjugated cyclic dienes can also be oxidized through selective mono-ozonolysis:[635]

$$\text{(9.119)}$$

Because of the electrophilic nature of ozone, selective ozonolysis of the olefinic double bond in enynes occurs.[634]

Other Oxidants. A number of other oxidizing agents are able to cleave the carbon–carbon double bond to form ketones and aldehydes or carboxylic acids. Since aldehydes formed as a result of the cleavage reaction are prone to undergo further oxidation under reaction conditions, only few reagents are capable of producing aldehydes selectively. Among these the permanganate ion, certain Cr(VI) and ruthenium compounds, and $NaIO_4$–OsO_4 are the most useful reagents.

Oxidation with permanganate[588,592,636,637] usually requires the use of a suitable solvent mixture, such as aqueous THF,[638] which dissolves both permanganate and the alkene. Better results may be obtained by phase-transfer agents.[440] Reaction with tetraalkylammonium permanganate may allow selective cleavage of a phenyl-substituted double bond in dienes.[639]

The permanganate ion selectively attacks the double bond according to Scheme 9.9. In a basic medium hydrolysis of **57** yields a 1,2-diol, while further redox reactions take place under neutral or acidic conditions:[640]

$$(9.120)$$

One-electron reduction by hydrogen abstraction from the solvent leads to **82**, which decomposes to diol monoanion **83**. The latter is oxidatively cleaved to carbonyl compounds with the concomitant reduction of manganese.

Oxidation with hexavalent chromium usually leads to complex product mixtures.[641] However, certain Cr(VI) reagents [$CrO_2(OOCCCl_3)_2$,[642] CrO_3 supported on silica,[643] bistriphenylsilyl chromate[644]], and a Cr(V) complex,[645] may be used to form carbonyl compounds.

Alkali periodates (in aqueous solution)[646] and ammonium periodate (in anhydrous aprotic solvent),[647] when applied in the presence of a catalytic amount of OsO_4, are selective oxidizing agents used to form carbonyl compounds. OsO_4 apparently transforms the alkene to an 1,2-diol that is subsequently cleaved by the periodate to the carbonyl compound end products. The periodate also serves to regenerate OsO_4.

According to the patent literature, RuO_2 can be used as a stoichiometric oxidant in aldehyde formation.[648] $RuCl_3$[649] and a ruthenium-substituted heteropoly anion [$SiRu(H_2O)W_{11}O_{39}^{5-}$][650,651] have been found to serve as a selective catalyst in this oxidation. H_2O_2[649] and $NaIO_4$[650,651] as oxidants, or electrochemical oxidation[651] have been used.

Less rigorous reaction conditions are required to cleave the double bond to form ketones and carboxylic acids. Phase-transfer-assisted permanganate oxidations in the presence of quaternary ammonium salts,[439,652–654] crown ethers,[439,655] or polyethers[566] usually ensure high yields. Terminal alkenes are transformed to carboxylic acids with one carbon atom less than the starting compound.[653,654]

Excellent yields of cleavage products (carboxylic acids from acyclic alkenes and dicarboxylic acids from cycloalkenes) were obtained by using $KMnO_4$ supported on silica gel.[656] Carboxylic acids are also formed in good yield when acyclic terminal alkenes are reacted with a heterogeneous $KMnO_4–CuSO_4 \cdot H_2O$ reagent in the presence of catalytic amounts of *tert*-BuOH and H_2O.[563] $KMnO_4$ adsorbed on moist alumina, in turn, is a good oxidant to transform cycloalkenes to α,ω-dialdehydes.[657] Oxidation with the mixture of $NaIO_4$ and a catalytic amount of $KMnO_4$ known as the *Lemieux–von Rudloff* reagent[658] is carried out in mixed solvents.[592] Permanganate ion the actual oxidant is regenerated in situ by $NaIO_4$.

The cleavage reaction with RuO_2 to produce carboxylic acids requires alkaline or acidic conditions.[592,648,659] The oxidation is preferentially carried out with co-oxidants such as $NaOCl$[439,660] and $NaIO_4$:[661]

$$RCH=CHR' \xrightarrow[\text{CH}_3\text{CN, CCl}_4, \text{H}_2\text{O, 25°C}]{\text{RuCl}_3 \cdot (\text{H}_2\text{O})_n, \text{NaIO}_4} R-C{\overset{\displaystyle O}{\underset{\displaystyle OH}{\big\|}}} \qquad (9.121)$$

R, R' = H, *n*-Bu, *n*-C_8H_{17} 75–89%
R-R' = $(CH_2)_6$

Addition of CH_3CN to the conventional solvent mixture was found to permit a rapid, mild reaction.[661] Experimental observations were consistent with a one-electron oxidation to form a cyclic Ru(VI) diester analogous to that formed in OsO_4 oxidation.[662] Because it is less stable, it readily decomposes to cleavage products and RuO_2.

The nitric acid oxidation of cycloalkenes yields, among other products, α,ω-dicarboxylic acids.[663,664] The reaction is catalyzed by vanadium(V) ion.[665,666] Oxidations with H_2O_2,[284,667] PhIO,[650] or $KHSO_5$[650] catalyzed by heteropoly acids[284,650] or tungstic acid[667] have been recently described.

9.2.6. Allylic Oxidation

Oxidation of the allylic carbon of alkenes may lead to allylic alcohols and derivatives or α,β-unsaturated carbonyl compounds. Selenium dioxide is the reagent of choice to carry out the former transformation. In the latter process, which is more difficult to accomplish, Cr(VI) compounds are usually applied. In certain cases, mixture of products of both types of oxidation, as well as isomeric compounds resulting from allylic rearrangement, may be formed. Oxidation of 2-alkenes to the corresponding α,β-unsaturated carboxylic acids, particularly the oxidation of propylene to acrolein and acrylic acid, as well as ammoxidation to acrylonitrile, has commercial importance (see Sections 9.5.2 and 9.5.3).

Allylic Hydroxylation and Acyloxylation. Selenium dioxide is a standard, commonly used stoichiometric oxidant to convert alkenes to allylic alcohols.[668–671] In aqueous solution it reacts as seleniuos acid (**84**), whereas in other hydroxylated solvents (alcohols, carboxylic acids) the alkyl selenite (**85**) and the mixed anhydride (**86**), respectively, are the actual reagents:

| **84** | **85** | **86** |

If carried out in acetic acid, the oxidation leads to the corresponding allylic acetoxy derivatives.

General observations of allylic oxidation with SeO_2 were summarized as early as 1939.[672] According to these rules,[671,672] the order of reactivity of different allylic carbons is $CH_2 > CH_3 > CH$. The oxidation usually takes place at the more substituted side of the double bond, and a strong preference of the formation of the corresponding (*E*)-allylic alcohols is observed[673] [Eq. (9.122)[674]]:

$$(9.122)$$

70% yield
>98% selectivity

Substituted cycloalkenes usually react in the ring and not in the side chain. Internal alkenes with CH_2 groups in both allylic positions yield a mixture of isomers, whereas terminal alkenes give primary alcohols as a result of allylic rearrangement. Later studies revealed, however, that the reactivity depends on both the structure of the alkene and reaction conditions.[674,675] In alcoholic solutions, for example, racemic products are formed. Geminally disubstituted alkenes may exhibit a reactivity sequence $CH > CH_2 > CH_3$.[675,676]

Concerning the mechanism of the reaction, a radical process,[672] and the involvement of allylseleninic acid,[676,677] selenium(II) ester,[678] and oxaselenocyclobutane[679] intermediates were suggested on the basis of early results. The now generally accepted mechanism put forward by Sharpless and coworkers[676,680,681] constitutes an ene reaction to give, after dehydration, the allylseleninic acid intermediate **87**:

| | **87** | **88** |

$$(9.123)$$

Compound **87** isolated in a trapping experiment[681] undergoes a [2,3] sigmatropic rearrangement to yield a selenite ester (**88**). Hydrolysis of the latter gives the product allylic alcohol.

This mechanism explains the formation of racemic products under certain reaction conditions, but does not fully account for (*E*)-allylic alcohol formation [Eq. (9.122)]. In fact, it predicts complete regio- and stereoselective formation of allylic alcohols, which is not commonly observed. Later, deuterium labeling studies[682] led to the conclusion that another pathway also exists. The preference for (*E*)-allylic alcohol formation is due to the steric effect in a six-membered chair transition state in the [2,3] sigmatropic migration in basic solution. In alcoholic solutions, a minor stereorandom path through a stepwise ene reaction involving carbocation formation explains decreased selectivities. A limited stereoselectivity of the ene reaction on the basis of ^{13}C-labeling experiments was also suggested.[683]

The use of a catalytic amount of SeO_2 with reoxidants such as H_2O_2[684] and *tert*-BuOOH[685] is a useful modification of the conventional process producing allylic alcohols in yields comparable to or better than stoichiometric SeO_2 alone. Cycloalkenes, however, give significant amounts of allylic *tert*-butyl ethers and *tert*-butyl peroxides.[686]

Copper-catalyzed[687–689] or photochemical[690] reaction of alkenes with peroxyesters, usually with *tert*-butyl peracetate (or *tert*-BuOOH in acetic acid), may be used to carry out acyloxylation or the synthesis of the corresponding allylic esters in good yields. In contrast to the oxidation with SeO_2, preferential formation without rearrangement of the 3-substituted esters takes place from terminal alkenes:[691]

$$RCH_2CH=CH_2 \xrightarrow[\substack{chlorobenzene, \\ 70°C, 20\ h}]{\textit{tert}\text{-BuOOAc, CuCl}} \underset{\substack{| \\ OAc}}{RCHCH=CH_2} + RCH=CHCH_2OAc \qquad (9.124)$$

$$R = Et,\ \textit{n}\text{-Pr},\ \textit{n}\text{-}C_5H_{11}$$

$$84\text{–}88 \qquad 12\text{–}16$$
$$88\text{–}89\%\ yield$$

A number of metal compounds, primarily acetates, including $Pb(OAc)_4$, $Hg(OAc)_2$, $Tl(OAc)_3$, and $Mn(OAc)_3$, can bring about the formation of allylic acetates.[671,692,693] These reagents, however, with the exception of $Hg(OAc)_2$, are not particularly selective and have only limited synthetic use. Oxymetallation to yield allylmetal species followed by solvolysis is usually believed to occur.[496]

Palladium-catalyzed allylic oxidations, in contrast, are synthetically useful reactions. Palladium compounds are known to give rise to carbonyl compounds or products of vinylic oxidation via nucleophilic attack on a palladium alkene complex followed by β-hydride elimination (Scheme 9.16, path *a*; see also Section 9.2.4). Allylic oxidation, however, can be expected if C–H bond cleavage precedes nucleophilic attack.[694] A poorly coordinating weak base, for instance, may remove a proton, allowing the formation of a palladium π-allyl complex intermediate (**89**, path *b*).[694–696] Under such conditions, oxidative allylic substitution can compete

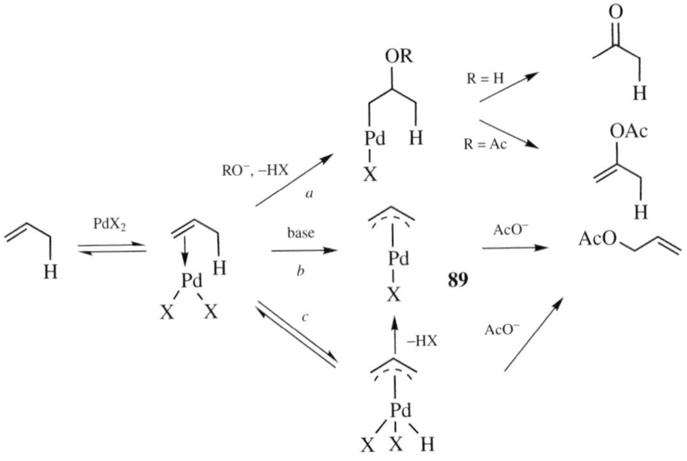

Scheme 9.16

with vinylic oxidation:[573]

$$CH_3CH=CH_2 \xrightarrow[\text{AcOH, 25°C}]{Pd(OAc)_2} \underset{\underset{OAc}{|}}{CH_3C=CH_2} + CH_3CH=CHOAc + CH_2=CHCH_2OAc$$

without NaOAc	96	2	2
with 0.9 M NaOAc	6	<0.5	94

$$(9.125)$$

Pd(II) complexes with strongly electron-withdrawing ligands can insert into the allylic C–H bond (path c) to form directly the π-allyl complex via oxidative addition.[502,694,697] Pd(OOCCF$_3$)$_2$ in acetic acid, for example, ensures high yields of allylic acetoxylated products.[698] The delicate balance between allylic and vinylic acetoxylation was observed to depend on substrate structure, too. For simple terminal alkenes the latter process seems to be the predominant pathway.[571]

Several Pd-catalyzed oxidations with different reoxidants have been developed. In these reactions PdCl$_2$ or Pd(OAc)$_2$ in acetic acid is usually employed, with *tert*-BuOOH and TeO$_2$,[699] or p-benzoquinone and MnO$_2$:[700]

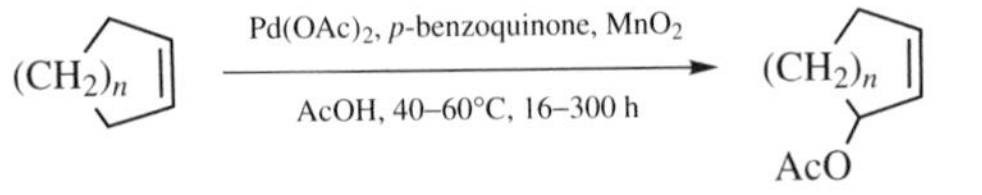

$$(CH_2)_n \text{⬡} \xrightarrow[\text{AcOH, 40–60°C, 16–300 h}]{Pd(OAc)_2, \text{ } p\text{-benzoquinone, MnO}_2} (CH_2)_n \text{⬡}_{OAc} \qquad (9.126)$$

$$n = 1\text{–}3, 6, 8 \quad 60\text{–}82\%$$
$$n = 4 \qquad 35\%$$

Pd-catalyzed aerobic oxidations with hidroquinone–Cu(OAc)$_2$[701] and hydroquinone–iron phthalocyanine[529] were also achieved. On the basis of the transformation of 1,2-dideuterocyclohexene, conclusive evidence for the involvement of the π-allyl intermediate in the quinone-based system has been reported.[702]

Palladium(II) and nitrate ion with oxygen as final oxidant give an excellent yield of cyclohexenyl acetate from cyclohexene (92% at 50°C).[703] The catalyst reoxidation sequence includes a palladium nitro–nitrosyl redox couple.

Heterogeneous palladium catalysts proved to be active in the conversion of simple alkenes to the corresponding allylic acetates, carbonyl compounds, and carboxylic acids.[694,704] Allyl acetate or acrylic acid from propylene was selectively produced on a palladium on charcoal catalyst depending on catalyst pretreatment and reaction conditions.[694] Allylic oxidation with singlet oxygen to yield allylic hydroperoxides is discussed in Section 9.2.2.

Oxidation to α,β-Unsaturated Carbonyl Compounds. Chromium(VI) reagents are the most valuable oxidants in the allylic oxidation of alkenes to obtain α,β-unsaturated carbonyl compounds.[110,671,705] CrO_3 and $Na_2Cr_2O_7$ in AcOH or Ac_2O were found to be useful in organic synthesis, although yields are sometimes low and isomeric compounds are often formed. Detailed studies with the chromium oxide–pyridine complex employed under mild reaction conditions (CH_2Cl_2, room temperature) led to certain generalizations.[706] Allylic methyl groups are not readily oxidized with this reagent. If more than one allylic methylene group is present in a conformationally flexible molecule, mixture of isomeric ketones is formed, whereas selectivity is found in conformationally rigid molecules. Rearrangements occur when an allylic methine group is present, or with sterically hindered methylene systems. The preferential removal of methine hydrogen to form the most stable tertiary radical explains the product distribution of oxidation of 3-methylcyclohexene (Scheme 9.17).

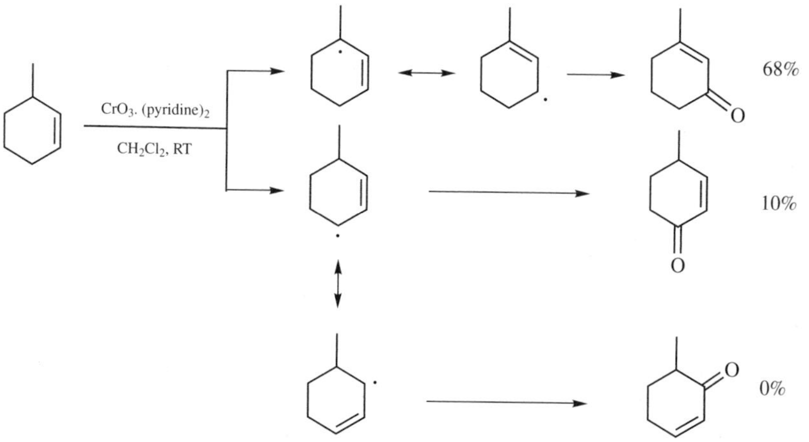

Scheme 9.17

Concerning the mechanism of Cr(VI) oxidations, initial attack of CrO_3 to form a symmetric intermediate was proposed.[677] Hydrogen atom or hydride abstraction from the allylic position leads to resonance-stabilized allylic radical or carbocation, respectively, which is eventually converted to the unsaturated carbonyl compound.

An α-hydroxy ketone intermediate transformed via dehydration was also suggested.[110]

Other chromium(VI) reagents may also be used in similar oxidations.[671,705] A common feature of all of these processes is that a large excess of reagent is usually necessary. Because of the environmental hazard associated with chromium compounds, efforts have been made to develop chromium-catalyzed transformations. Oxidations with tert-BuOOH catalyzed by CrO_3[707] or pyridinium dichromate[708] afford rather low yields, whereas an efficient method was described using catalytic amounts of chromia pillared clay (Cr–PILC).[709] The oxidation of terminal alkenes [Eq. (9.127)] and cycloalkenes is selective and gives very good yields but isomeric enones are formed from internal alkenes:

$$R\diagdown\diagup\diagdown \xrightarrow[\text{CH}_2\text{Cl}_2,\ 34\text{–}38\ \text{h, RT}]{\text{Cr-PILC}} R\diagdown\overset{\displaystyle\diagdown\diagup}{\underset{\displaystyle O}{\diagup}} \qquad (9.127)$$

$$R = \text{Pr, } n\text{-Bu, } n\text{-C}_7\text{H}_{15} \qquad\qquad 88\text{–}91\%$$

In contrast with other reagents, selective oxidation of methylcyclohexenes can also be achieved.

Some reagents, such as SeO_2 (the best oxidant for allylic hydroxylation) and palladium(II) (best known for oxidation of alkenes to allylic alcohols and carbonyl compounds), can be used in the formation of α,β-unsaturated carbonyl compounds under special reaction conditions.[671,692]

9.3. OXIDATION OF ALKYNES

The two major characteristic oxidation processes of alkynes are their transformation to 1,2-dicarbonyl compounds and their cleavage reaction to carboxylic acids.[710] The structure of the starting compounds has a decisive effect on the selectivity of oxidation. Since 1,2-dicarbonyl compounds proved to be intermediates in further oxidations, carefully controlled reaction conditions are often necessary to achieve selective synthesis. Certain oxidizing agents such as peroxyacids and ozone are nonselective oxidants.

9.3.1. Oxidation to Carbonyl Compounds

Several oxidizing agents, such as RuO_4,[711] PhIO with ruthenium catalysts,[712] $KMnO_4$,[713–715] $Tl(NO_3)_3$,[716] NBS–DMS,[717] OsO_4[718] in the presence of tertiary amines, and SeO_2 in the presence of a catalytic amount of H_2SO_4,[719] can be applied to transform alkynes to 1,2-dicarbonyl compounds. Most of these reagents can be used only to convert internal alkynes to the corresponding 1,2-diketones, since terminal alkynes usually undergo cleavage to carboxylic acids. Because of the facile oxidation of 1,2-dicarbonyl compounds, the oxidation with $KMnO_4$ requires a careful control of pH to obtain good yields. The best results are achieved if oxidation is carried out in CH_2Cl_2 in the presence of phase transfer agents:[715]

$$R-C\equiv C-R' \xrightarrow[\text{CH}_2\text{Cl}_2, \text{AcOH, reflux, 4–6 h}]{\text{KMnO}_4, \text{Adogen 464}} R-\underset{\underset{O}{\|}}{C}-\underset{\underset{O}{\|}}{C}-R' \qquad (9.128)$$

$$R, R' = n\text{-Pr}, n\text{-Bu}, n\text{-C}_6\text{H}_{13}, n\text{-C}_7\text{H}_{15}, \text{Ph} \qquad 41–93\%$$

By varying the amount of the reagents, iodine in DMSO[720] or [bis(trifluoroacetoxy) iodo]benzene[721] selectively oxidizes one or both triple bonds in aromatic diynes.

In contrast to the preceding reagents, oxidation with H_2O_2 catalyzed by molybdate and tungstate salts or $Hg(OAc)_2$,[722,723] and the $Hg(OAc)_2$-promoted oxidation with a molybdenum peroxo complex[724] can be applied to transform acetylene and terminal acetylenes to glyoxal and α-ketoaldehydes, respectively, in fair to good yields.

Other, less characteristic transformations include α-monohydroxylation of terminal and α,α'-dihydroxylation of internal alkenes with *tert*-BuOOH + SeO$_2$,[725] and α-oxidation with the CrO$_3$–pyridine complex to yield conjugated ynones.[726,727] A novel oxidation with H_2O_2 catalyzed by peroxotungstophosphates gives α,β-epoxy ketones as well as minor amounts of α,β-unsaturated ketones:[728]

$$R-C\equiv C-CH_2R' \xrightarrow[\text{H}_2\text{O}_2, \text{CHCl}_3, \text{reflux, 24 h}]{\substack{\text{tris(cetylpyridinium)}\\ \text{peroxotungstophosphate}}} \underset{O}{\overset{R}{\diagup}}\!\!\!\diagdown\!\!\!\overset{R'}{\diagdown} + \underset{O}{\overset{R}{\diagup}}\!\!\!\diagdown\!\!\!\overset{R'}{\diagdown} \qquad (9.129)$$

$$R, R' = \text{Me, Et, } n\text{-Pr}, n\text{-Bu} \qquad 40–62\% \qquad 2–15\%$$

α,β-Unsaturated ketones may be the main product in the oxidation of dialkylacetylenes with dioxiranes.[729] In contrast, phenylacetylene and diphenylacetylene are converted to benzaldehyde and benzophenone, respectively. Oxirene **90**, suggested to be the intermediate in all oxidations of acetylenes with peroxide compounds, supposedly participates in a rearrangement and to yield carbonyl compounds with the loss of one carbon atom:[729]

$$Ph-C\equiv C-R \xrightarrow[0°\text{C, 6–7 min}]{} \left[\underset{\underset{\textbf{90}}{}}{\underset{Ph\quad R}{\triangle}} \longrightarrow \underset{R}{\overset{Ph}{>}}C=C=O \right] \xrightarrow[-CO_2]{[O]} \underset{R}{\overset{Ph}{>}}C=O$$

$$R = H \quad > 96\%$$
$$R = Ph \quad 60\%$$
$$(9.130)$$

$Tl(NO_3)_3$ is a versatile oxidant since it converts diarylacetylenes to 1,2-diketones and dialkylacetylenes to α-hydroxy ketones:[716]

$$RC\equiv CR \xrightarrow[\text{H}_3\text{O}^+]{\text{Tl(NO}_3)_3} \underset{\underset{}{\overset{HO\quad Tl-}{}}}{RC=CR} \rightleftharpoons R-\underset{\underset{}{\overset{\|\quad |}{O\quad Tl-}}}{C}-CHR \xrightarrow{\text{H}_3\text{O}^+} R-\underset{\underset{}{\overset{\|\quad |}{O\quad OH}}}{C}-CH-R \quad (9.131)$$

$$R = \text{Et}, n\text{-Pr} \qquad 70–90\%$$

The mixed arylalkylacetylenes, in turn, undergo a smooth oxidative rearrangement in MeOH to yield disubstituted methyl acetates:

$$
\text{PhC} \equiv \text{CR} \xrightarrow[\substack{\text{MeOH,}\\ \text{reflux, 2 h}}]{\text{Tl(NO}_3)_3} \quad \text{Ph-C}\!=\!\text{C-R} \longrightarrow \quad \underset{\text{MeO}}{\overset{\text{MeO}}{}}\text{C}\!=\!\text{C}\underset{R}{\overset{Ph}{}} \xrightarrow[-\text{MeOH}]{\text{H}_3\text{O}^+} \quad \underset{R}{\overset{Ph}{}}\text{CHCOOMe}
$$

$$
\begin{array}{l}
R = \text{Me, Et, } n\text{-Pr, } n\text{-Bu,}\\
\quad\ \text{PhCH}_2\text{, PhCH}_2\text{CH}_2
\end{array}
$$

65–83%

(9.132)

As in all oxidations with Tl(NO$_3$)$_3$ oxythallation is the first step of the reactions.

9.3.2. Oxidative Cleavage

Terminal alkynes are prone to undergo facile oxidative cleavage to yield carboxylic acids with loss of the terminal carbon atom. In fact, most of the oxidizing agents that can be used in the selective oxidation of internal alkenes to 1,2-diketones [RuO$_4$,[711] PhIO with Ru catalysts,[712] KMnO$_4$,[713] Tl(NO$_3$)$_3$,[716] OsO$_4$[718]] convert terminal alkynes to carboxylic acids.

Besides the possibilities described above, KMnO$_4$[715] in aqueous solution, [bis (trifluoroacetoxy)iodo]pentafluorobenzene,[730] and H$_2$O$_2$ in the presence of Mo(VI) or W(VI) polyoxometalates[731] cleave the triple bond in internal alkynes. A Cr(V) complex seems to be the only reagent capable of transforming phenylacetylene and diphenylacetylene to benzaldehyde.[645]

Compared to ozonation of alkenes, much less is known about the ozonation of alkynes,[710] which yields 1,2-dicarbonyl compounds, carboxylic acids, and anhydrides. 1,2,3-Trioxolene (**91**), analogous to **74** in alkene ozonation mechanism (Scheme 9.14), and zwitterionic intermediates (**92**) were formulated on the basis of IR studies[732] and trapping experiments:[733–735]

$$
\text{Ph-C}\!\equiv\!\text{C-Ph} \xrightarrow{\text{O}_3} \quad \underset{\underset{\text{Ph}}{}}{\overset{}{}}\text{C}\!=\!\text{C}\underset{\text{Ph}}{\overset{}{}} \longrightarrow \left[\overset{+}{\text{O}}\text{-C}\!-\!\text{C}\cdots \leftrightarrow \text{C}\!=\!\text{C} \right]
$$

91   **92**

(9.133)

Oxidation with peroxy acids gives a variety of products,[217,710] some of which are indicated in Scheme 9.18. All the compounds formed could be interpreted via further transformation of the **93** intermediate oxirene.[736,737]

Unsaturated ketones, ring-contracted ketones, and bicyclic ketones resulting from transannular insertion are the main products isolated in the oxidation of cycloalkynes.[738] Product distribution was interpreted in terms of an oxirene–α-ketocarbene equilibrium.

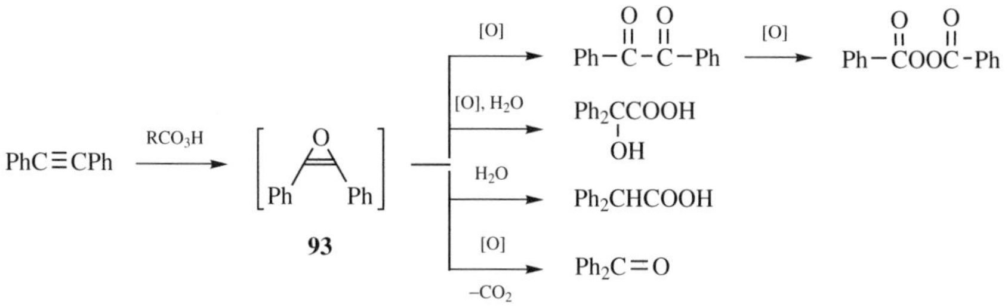

Scheme 9.18

9.4. OXIDATION OF AROMATICS

9.4.1. Ring Oxygenation

Depending on the oxidation conditions, benzene and its substituted derivatives, and polycyclic aromatic hydrocarbons may be converted to phenols and quinones. Alkoxylation and acyloxylation are also possible. Addition reactions may afford dihydrodiols, epoxides, and endoperoxides.

Oxidation to Phenols. Direct hydroxylation of benzene to phenol can be achieved in a free-radical process with H_2O_2 or O_2 as oxidants.[739–744] Metal ions [Fe(II), Cu(II), Ti(III)] may be used to catalyze oxidation with H_2O_2. Of these reactions, the so-called Fenton-type oxidation is the most widely studied process.[742] Oxidation in the presence of iron(II) sulfate was reported in early studies to yield phenol. Since phenol exhibits higher reactivity than benzene, varying amounts of isomeric dihydroxybenzenes were also formed.

Hydroxy radicals produced in an aqueous medium [Eq. (9.134)] readily attack benzene to form the **94** hydroxycyclohexadienyl radical (Scheme 9.19):

$$H_2O_2 + Fe^{2+} \longrightarrow Fe^{3+} + HO^- + HO\cdot \tag{9.134}$$

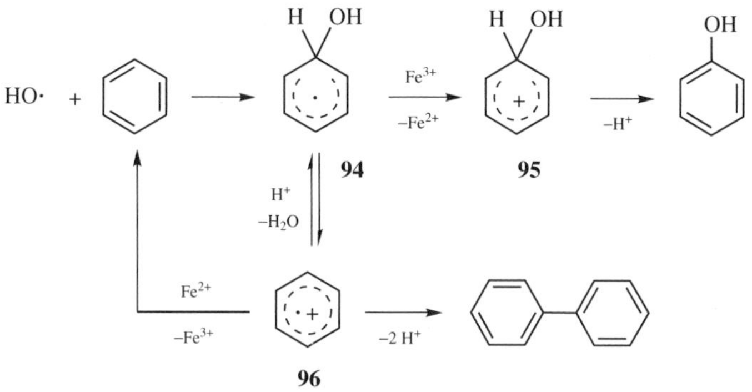

Scheme 9.19

It is then oxidized to afford phenol through cation **95**. The pH of the reaction medium has a strong effect on conversion and product distribution since **94** may participate in an acid-catalyzed competitive dehydration to yield a radical cation (**96**). Compound **96** may be reduced to the starting material. If there is no other oxidant present, dimerization may also occur. This is the case when benzene undergoes radiolysis in aqueous solution.[739,741]

The same radical cation is believed to be the intermediate when oxidation of aromatics is carried out with peroxydisulfate[739,745] or peroxydiphosphate.[746] In the presence of suitable oxidants such as Cu(II), phenolic products are formed with high selectivity. This supports the suggestion that initial attack on the aromatic ring gives rise to **96**, which, in turn, reacts rapidly with water to yield the hydroxycyclohexadienyl radical:

$$(9.135)$$

Depending on the pH and whether $HO^{\bullet}$ or $SO_4^{-\bullet}$ is the oxidant, substantially different product distributions may be observed.

Similar systems studied are the Undenfriend reagent[747] (Fe^{2+} + EDTA + ascorbic acid + O_2 or H_2O_2) and the Hamilton reagent[748] (Fe^{3+} + catechol + H_2O_2). The Undenfriend reagent attracted special interest because it resembles the enzymatic hydroxylation of aromatic compounds. Copper(I)-catalyzed oxidations with dioxygen were studied, too, mainly to model the copper-containing monooxygenase tyrosinase.[744] Hydroxy radicals formed according to Eqs. (9.136) and (9.137) attack benzene as depicted in Scheme 9.19:

$$O_2 \; + \; 2\,Cu^+ \; + \; 2\,H^+ \longrightarrow H_2O_2 \; + \; 2\,Cu^{2+} \qquad (9.136)$$

$$H_2O_2 \; + \; Cu^+ \; H^+ \longrightarrow HO\cdot \; + \; H_2O \; + \; Cu^{2+} \qquad (9.137)$$

In the presence of oxygen, however, intermediate **94** may further react to form a peroxy radical (**97**), which is oxidized to hydroquinone[749,750] (Scheme 9.20). Each intermediate may be transformed to phenol as well. Selective formation of either phenol or hydroquinone was achieved by running the oxidation at proper pH (phenol : hydroquinone = 71 : 29 at pH = 1.3, and 14 : 86 at pH = 3.5).[750] A direct route to hydroquinone was also recognized when the oxidation of benzene was performed in a H_2SO_4–CH_3CN–H_2O solution.[751]

On the basis of differences in the product distributions in oxidations by the Fenton reagent, and by the Udenfriend reagent a different mechanism for the latter was suggested. In aprotic solvents, however, the Fenton reagent gives results similar to those in enzymatic hydroxylations. It was observed that the hydrogen displaced

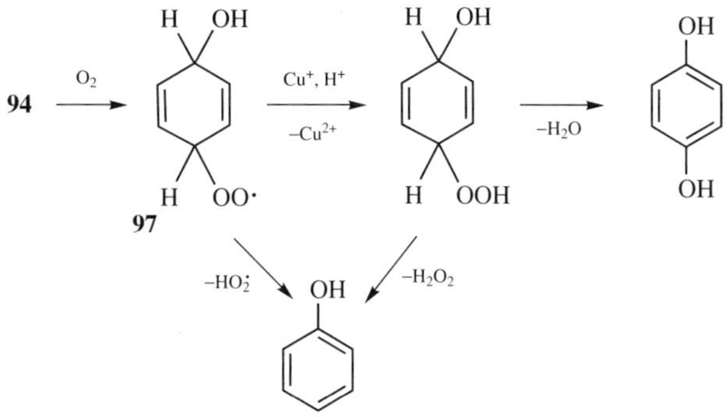

Scheme 9.20

by the hydroxy function in enzymatic aromatic hydroxylation migrates to an adjacent carbon atom in the ring:[752,753]

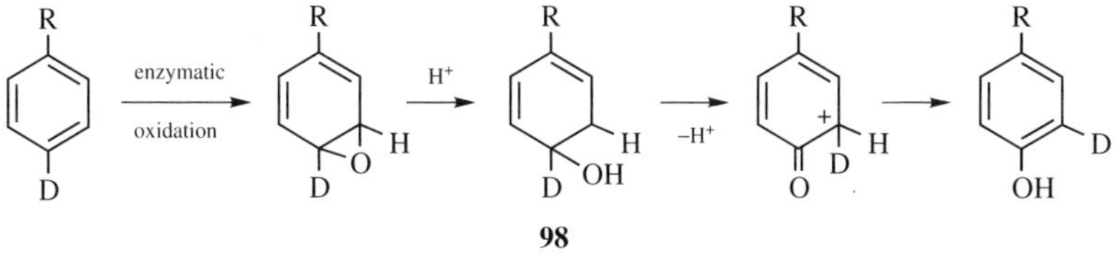

$$(9.138)$$

This hydroxylation-induced intramolecular migration, known as the *NIH shift*, was explained by the involvement of arene oxides formed by the attack of electrophilic oxoiron(V) porphyrin on the aromatic ring.[753] Intermediate **98** was also suggested to be formed in hydroxylation by the Fenton and related reagents in aprotic media after initial oxidation with an oxoiron(V) complex followed by electron transfer.[744,754]

Many variations in the conventional oxidizing systems have been made, and new reagent combinations have been introduced to improve conversion and selectivity. Ascorbic acid, for instance, can be replaced by other hydrogen donors in the Udenfriend system,[744] and Fe^{2+} may be replaced by metallic copper.[755] Platinum[756] and iron[172] complexes, as well as vanadium(V) peroxo complexes,[757,758] were also tested in the oxidation of benzene. Many of these reagents were studied to find a viable one-step oxidation of benzene to phenol to replace the now existing industrial synthesis of phenol by the cumene process. Of the heterogeneous systems studied,[759–763] titanium silicalites (TS), new synthetic zeolites are effective and selective for the oxidation with H_2O_2 of several classes of organic compounds. Although selectivites in the hydroxylation of benzene vary with the formation of hydroquionone, selective monohydroxylation of alkylbenzenes can be achieved under appropriate reaction conditions[761–763] or by carrying out the reaction over a palladium-containing titanium silicalite.[190] More recently developed catalysts, titanium-containing hexagonal mesoporous molecular sieves, have been found to exhibit even higher activity and selectivity in hydroxylation of benzene.[764] Gas-phase oxidation of benzene over Pd–Cu-on-silica catalysts in the presence of water yields phenol with more than 90% selectivity, although at very low (less than 0.1%) conversion.[765]

One-step electrophilic hydroxylation of aromatic compounds using various peroxide reagents in the presence of acid catalysts has been achieved. The systems studied include hydrogen peroxide in the presence of sulfuric acid,[766] hydrogen fluoride,[767] Lewis acids,[768,769] and pyridinium poly(hydrogen fluoride).[770] Lewis acid–promoted electrophilic hydroxylation with peracids,[771,772] di-*tert*-butyl peroxide,[773] and diisopropyl peroxydicarbonate[774,775] were also described. A common feature of these reagents is the formation of monohydroxylated compounds in low yields.

Electrophilic hydroxylation of aromatics with hydrogen peroxide in superacid media has been reported by Olah and coworkers.[776–778] Phenols formed in these systems are protonated by the superacid and are thus deactivated against further electrophilic attack, usually resulting in >50% yields. Benzene and alkylbenzenes are effectively hydroxylated with 98% H_2O_2 in HSO_3F–SO_2ClF or Magic Acid,[776] or with 30% H_2O_2 in HF–BF_3.[777] Similar good yields were achieved in the hydroxylation of benzene and alkylbenzenes with sodium perborate–triflic acid at $-10°C$,[779] whereas sodium percarbonate proved to be a less effective hydroxylating agent.[210] In these reactions $H_3O_2^+$ is assumed to be the electrophilic hydroxylating agent:

$$Ar-H \; + \; H_3O_2^+ \; \xrightarrow[-H_3O^+]{} \; Ar-OH \qquad (9.139)$$

Uniquely regioselective hydroxylation of naphthalene was also achieved in acidic systems.[778] Regioselectivity was found to depend markedly on the acidity of the system and the solvent used:

$$\qquad (9.140)$$

HF, -10 to $0°C$, or		
7:3 HF–pyridine, 0 to 20°C,	98.4	1.6
in HF, SO_2ClF or CH_2Cl_2		
HF–SbF_5, $-78°C$	1.8	98.2

In hydroxylation of alkylbenzenes in Magic Acid and that of naphthalene in hydrogen fluoride or HF–solvent systems the actual electrophilic hydroxylating agent is the hydroperoxonium ion ($H_3O_2^+$). It attacks the aromatic ring and results in product formation by the usual aromatic electrophilic substitution mechanism. 2-Naphthol formation in superacids (HF–BF_3, HF–SbF_5, HF–TaF_5, HSO_3F, HSO_3F–SbF_5), in contrast, results from the attack of hydrogen peroxide to protonated naphthalene[778] according to Eq. (9.141):

$$(9.141)$$

Ring Acyloxylation. Soluble and supported palladium catalysts are suitable reagents used to achieve ring acyloxylation of aromatic compounds.[499,503,693,780] Reaction conditions, however, are critical since coupling and side chain substitution also occur.

Pd(OAc)$_2$ catalyzes the formation of aryl acetates in the presence of strong oxidants[780,781] such as $Cr_2O_7^{2-}$, $S_2O_8^{2-}$, MnO_4^-, and NO_3^-. Added alkali acetates inhibit coupling but bring about increasing side-chain substitution of alkyl-substituted benzenes. Electrophilic substitution to form arylpalladium(II) acetate, which is oxidized to an arylpalladium(IV) intermediate followed by decomposition, was invoked to interpret ring acetoxylation.[782] In contrast, an intermediate formed by electrophilic addition was suggested to account for the characteristic *meta* selectivity in acetoxylation of benzenes substituted with electron-donating substituents.[783,784] Transformation in the presence of $S_2O_8^{2-}$ was explained to occur through a radical cation.[785]

Acetoxylation of naphthalene was achieved with stoichiometric amount of Pd(OAc)$_2$ or in a catalytic reaction using the $Cu^{2+} + O_2$ system for the reoxidation of Pd(0):[786]

$$(9.142)$$

Highly selective formation of phenyl acetate was observed in the oxidation of benzene with palladium promoted by heteropoly acids.[694] Lead tatraacetate, in contrast, usually produces acetoxylated aromatics in low yields due to side reactions.[491,693] Electrochemical acetoxylation of benzene and its derivatives[787,788] and alkoxylation of polycyclic aromatics[789,790] are also possible. Thermal or photochemical decomposition of diacyl peroxides, when carried out in the presence of polycyclic aromatic compounds, results in ring acyloxylation.[688] The less reactive benzene is phenylated when reacts with benzoyl peroxide.[791] A dramatic increase in the yield of acyloxylated products is observed in the presence of oxygen[792] [Eq. (9.143)] or other additives, such as iodine[793] [Eq. (9.143)] and $CuCl_2$:[794]

$$(9.143)$$

hν under N_2, 50–60 h	13%
hν in O_2, 50–60 h	58%
I_2, 90°C, 20 h	62%

Near-quantitative yields of aryl isopropyl carbonates were realized in a similar process when diisopropyl peroxydicarbonate was reacted with alkyl-substituted aromatics.[794] Substitution is effected by a radical entity possessing a considerable ionic character.

Oxidation to Quinones. Direct oxidation of arenes to quinones can be accomplished by a number of reagents.[795,796] Very little is known, however, about the mechanism of these oxidations. Benzene exhibits very low reactivity, and its alkyl-substituted derivatives undergo benzylic oxidation. Electrochemical methods appear to be promising in the production of *p*-benzoquinone.[797] In contrast, polynuclear aromatic compounds are readily converted to the corresponding quinones.

Among the many different oxidizing agents, chromium(VI) compounds under acidic conditions are the most widely used reagents.[641,798] Although these oxidants are also effective in side-chain oxidation, selective formation of quinones can be achieved under appropriate reaction conditions. Ceric ammonium sulfate in H_2SO_4–CH_3CN,[799] $Cu(NO_3)_2$ or $Zn(NO_3)_2$ supported on silica gel,[800] [$RhCl(PPh_3)_3$] with O_2 or *tert*-BuOOH[801] are reagents that afford quinones in excellent yields under mild conditions:

$$(9.144)$$

	R = H	R = Me
[$RhCl(PPh_3)_3$], *tert*-BuOOH	91%	87%
$VO(acac)_2$, *tert*-BuOOH	48%	64%
$Cu(NO_3)_2$ or $Zn(NO_3)_2$	80–82%	68–73%

The Rh-catalyzed reaction is more efficient than the vanadium-catalyzed process or stoichiometric oxidation with CrO_3, but fails for benzene and naphthalene.

A unique way to convert arenes to quinones is to form arylthallium(III) trifluoroacetates followed by oxidation with H_2O_2:[802]

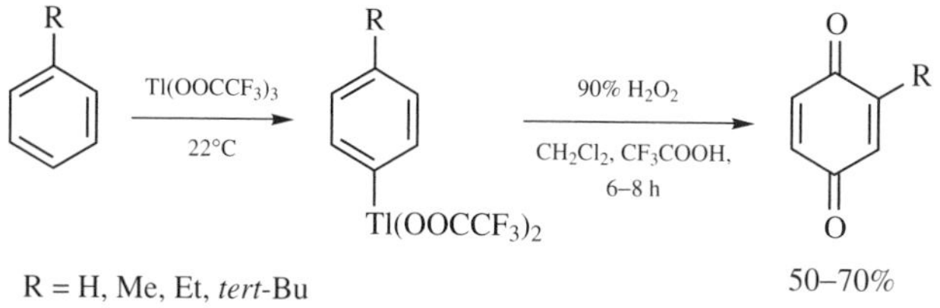

R = H, Me, Et, *tert*-Bu

50–70%

(9.145)

In certain cases (*tert*-butylbenzene, mesitylene) dealkylated products were also isolated.

Oxidation to Arene Oxides and Arene Diols. Arene oxides and arene diols have been widely studied particularly concerning the metabolism of polycyclic aromatic compounds and their implication as intermediates responsible for the carcinogenicity and mutagenicity of polycyclic aromatic hydrocarbons.[803,804]

Some of the reagents used in olefin epoxidation can be applied in the direct oxidation of arenes to arene oxides. Benzene oxide, however, like other arene oxides, exists in equilibrium with oxepin, its valence tautomer, and has not been isolated. Existence of benzene oxides as intermediates can be concluded from observations like the NIH shift discussed above.[752,753]

Direct oxidation of aromatics with *m*-CPBA in a two-phase system affords the most stable so-called K-region arene oxides in moderate yields.[805] Careful control of pH is necessary to avoid acid-catalyzed rearrangement of the acid-sensitive product epoxides. The new powerful oxidants, dimethyldioxiranes,[81,82] have also been used to convert arenes to arene oxides.[806]

Various arene oxides can be prepared in high yield with NaOCl in the presence of a phase-transfer agent, under mild conditions, at controlled pH:[807]

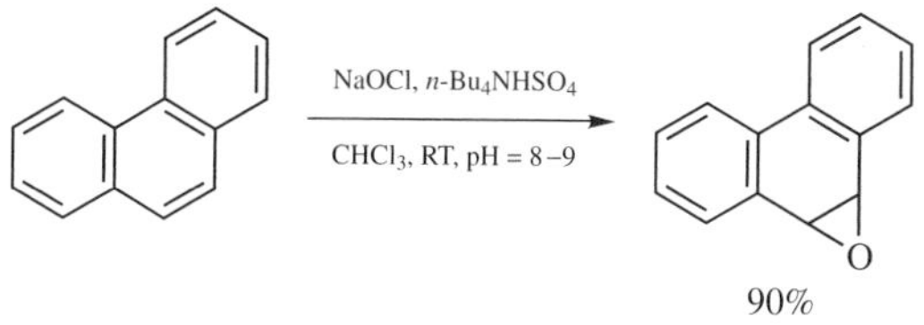

90%

(9.146)

At higher pH, very little epoxidation occurs, and at lower pH other oxidation products are observed.

Polynuclear aromatics that cannot form K-region epoxides such as anthracene may be oxidized to the corresponding quinones[806] by these reagents. Others, such as naphthalene and triphenylene, may be converted to polyepoxides under carefully controlled reaction conditions and workup procedures.[808,809]

OsO_4, which is the reagent of choice to convert alkenes to *cis*-1,2-diols, is also the most convenient reagent to carry out oxidation of aromatics to *cis*-dihydrodiols.[804] A highly specific and selective method is microbial oxidation. Different microorganisms have been used to convert benzene and its substituted derivatives and polycyclic aromatics to the corresponding *cis*-dihydrodiols,[810–812] usually with high enantioselectivity.[813] A mutant strain of *Pseudomonas putida* has found particularly widespread application in the oxidation of substituted benzenes:[449]

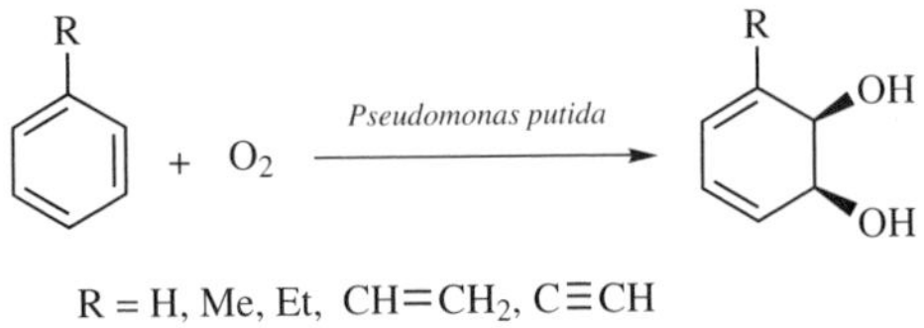

$$R = H, Me, Et, CH{=}CH_2, C{\equiv}CH$$

(9.147)

Both oxygen atoms in the product proved to originate from atmospheric oxygen.[810]

Oxidation with Singlet Oxygen. Polynuclear aromatic hydrocarbons, namely, anthracene and the higher homologs, have long been known to undergo self-sensitized photooxygenation to form endoperoxides through 1,4-cycloaddition:[360,364,372,373,814,815]

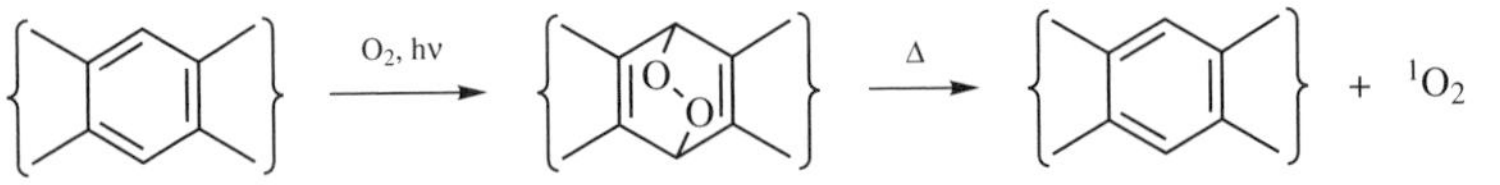

(9.148)

The main interest originated from the easy reversibility of the reaction using it as a singlet oxygen source.

The lower homologs, benzene and naphthalene, as well as phenanthrene, are not reactive toward singlet oxygen. Similar to that of other singlet oxygen oxygenations, however, the reactivity of aromatic hydrocarbons can be enhanced by introducing electron-donating substituents into the ring.[816–820] Pentamethyl- and hexamethylbenzene do react with singlet oxygen.[817] They form unstable endoperoxides detected by NMR spectroscopy that are prone to participate in the ene reaction with a second molecule of oxygen. The benzene ring in strained systems, such as certain cyclophanes, can also react to form endoperoxides.[821,822] Similar two-step singlet oxygenations may also take place in alkenylarenes[425,823,824] [Eq. (9.149)[424]]:

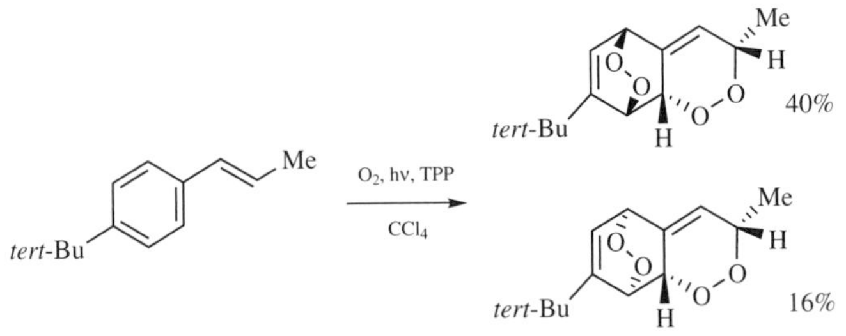

(9.149)

Methyl-substituted naphthalenes react with singlet oxygen,[818–820,825,826] but the reactivity depends on the position of the substituents. Whereas 2-methylnaphthalene is unreactive, other derivatives exhibit the following reactivity pattern: 1,4-dimethyl > 1,2-dimethyl > 1-methyl > 1,3-dimethyl.[820]

Anthracene and its 9,10-disubstituted derivatives add singlet oxygen to form the 9,10-endoperoxide,[825–827] but strong directing effect is displayed by substituents in the 1,4 positions:[827]

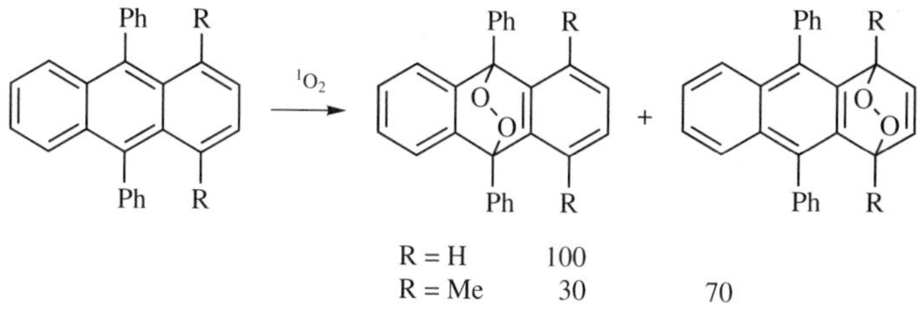

$$(9.150)$$

R = H	100	
R = Me	30	70

9.4.2. Oxidative Ring Cleavage

Sufficiently strong oxidants are able to cleave the carbon–carbon bonds in aromatic rings. Ozone, although less reactive toward arenes than alkenes, is effective in such ring fission transformations.[584,828] Ozonation of o-xylene gives a 3 : 2 : 1 mixture of glyoxal, methylglyoxal, and biacetyl (Scheme 9.21). This product distribution reflects a statistical bond cleavage.[829] Ozonation may afford compounds of different degrees of oxidation depending on workup conditions. In the transformation of phenanthrene,[830] for example, the cleavage of the 9,10 double bond can lead to dialdehyde, aldehyde acid or dicarboxylic acid.

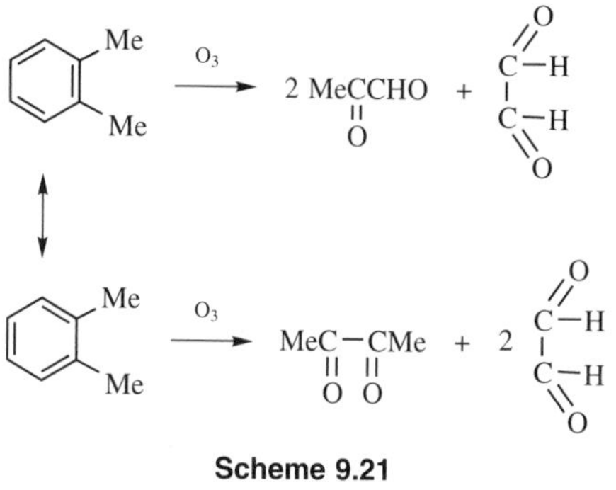

Scheme 9.21

Arenes can also be cleaved with peroxyacetic acid[831] and permanganate.[636,832] RuO$_4$ effectively oxidizes arenes to cleavage products with the use of a cooxidant such as NaIO$_4$.[648,659] Violent reaction with benzene[833] and immediate oxidation of several other aromatics such as triphenylmethane and tetralin[834] take place, but oxidation products could not be isolated. Since alkanes are resistant to RuO$_4$, phenylcyclohexane is oxidized to yield cyclohexanecarboxylic acid.[835] No systematic study is available in this field.

By far the most important aromatic cleavage reactions are the conversion on vanadia of benzene to maleic anhydride and that of naphthalene to phthalic anhydride. These oxdiations are treated in Section 9.5.4.

9.4.3. Benzylic Oxidation

Oxidation of Methyl-Substituted Aromatics. Autoxidation of alkyl-substituted aromatics possessing benzylic hydrogen affords benzylic hydroperoxides (**100**) as the primary product formed with high selectivity according to a free radical mechanism[836] (Scheme 9.22).

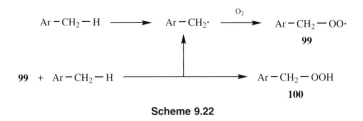

Scheme 9.22

The ready formation of benzylic hydroperoxides is used in industrial oxidations, as in the synthesis of propylene oxide and phenol (see Sections 9.5.2 and 9.5.4, respectively). In contrast with autoxidation of alkenes, where various secondary processes may follow, autoxidation of arenes is less complicated. Chain termination of **99** may lead to an alcohol and aldehyde [Eq. (9.151)], and the rapid autoxidation of the latter may produce the corresponding carboxylic acid [Eq. (9.152)]:

$$2\,Ar-CH_2-OO\cdot \longrightarrow Ar-CH_2-OH + Ar-CHO + O_2 \qquad (9.151)$$
$$\text{\textbf{99}}$$

$$Ar-CHO + O_2 \longrightarrow Ar-COOH \qquad (9.152)$$

In the noncatalyzed autoxidation the further oxidation of the aldehyde is much faster than the hydrogen abstraction to form the benzylic radical. As a result, carboxylic acids become the main product.

Metal ions [Co(III), Mn(III), Cu(II)] have a pronounced effect on the rate and product distribution of autoxidation. In the presence of metal ions, secondary transformations become dominant and selectivity in the formation of benzylic hydroperoxide **100** decreases.

Most studies of the metal-catalyzed oxidation of methyl-substituted arenes focused on the effect of Co(III) acetate[837] and Mn(III) acetate[838] on the selectivity of the transformation of methylbenzenes to aldehydes. While metal ion catalysis usually favors the formation of carboxylic acids, aldehyde oxidation becomes the rate-determining step in the presence of a high concentration of Co(III) ions, resulting in the selective oxidation of methylbenzenes to aldehydes.[836] The main role of the metal is the decomposition of the hydroperoxide [Eqs (9.153) and (9.154)], but it also catalyzes hydrogen abstraction[839,840] [Eq. (9.155)].

$$Co^{2+} \; + \; Ar-CH_2-OOH \longrightarrow Co^{3+} \; + \; Ar-CH_2-O\cdot \qquad (9.153)$$

100

$$Co^{3+} \; + \; \mathbf{100} \longrightarrow Co^{2+} \; + \; Ar-CH_2-OO\cdot \; + \; H^+ \qquad (9.154)$$

99

$$Co^{3+} \; + \; Ar-CH_2-H \longrightarrow Ar-CH_2\cdot \; + \; Co^{2+} \; + \; H^+ \qquad (9.155)$$

The order of reactivity of different alkyl-substituted benzenes is toluene > ethylbenzene > isopropylbenzene, which is opposite to the general reactivity expected from the C—H bond energies. This was explained in terms of an initiating electron transfer equilibrium between $Co(OAc)_3$ and the arene as the rate-determining step.[840]

Studies of the internal selectivities of oxidation of methyl and ethyl groups in the same molecule led to a similar conclusion suggesting a radical cation intermediate.[841] It was observed that selectivities depend on the nature of the metal ion:

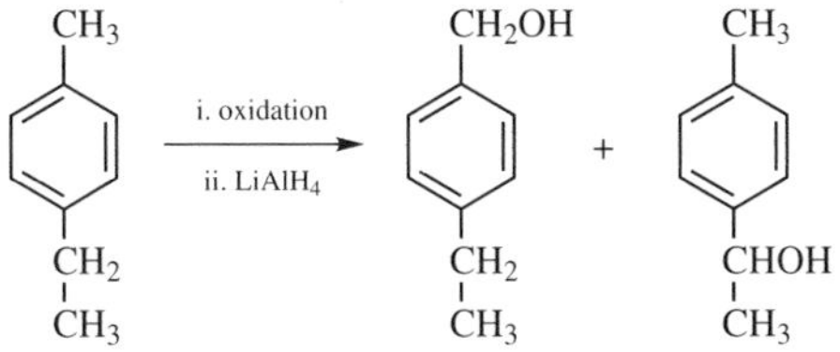

(9.156)

Co(OAc)$_3$, AcOH, 60°C	53.4	46.6
Ce(NH$_3$)$_2$(NO$_3$)$_6$, AcOH, 60°C	22.4	77.6
Mn(OAc)$_3$, H$_2$SO$_4$, AcOH, 20°C	9.2	90.8

Strong acids enhance oxidation, and extensive transformation can be achieved at room temperature [see Eq. (9.160)]. In the absence of oxygen, side-chain acetoxylation and hydroxylation take place.[842]

Stoichiometric oxidations may also lead to aldehydes from methyl-substituted arenes. Chromyl chloride applied in the classical Étard reaction[843] forms a 2 : 1 insoluble adduct with the possible structure **101**. The aldehyde is isolated after treatment with water:

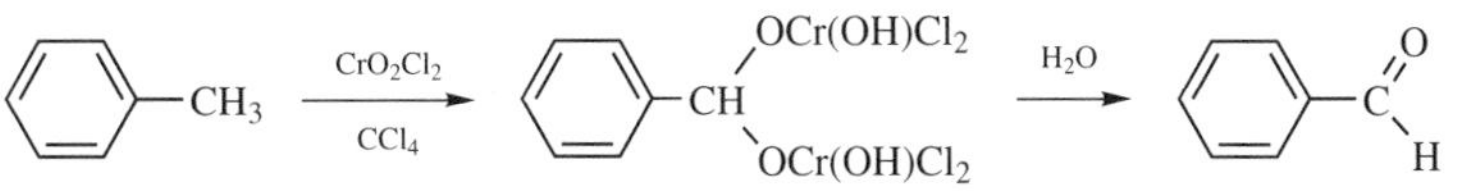

(9.157)

A free-radical process seems to account for the experimental observations.[110,844]

In a reaction somewhat similar to the CrO_2Cl_2 oxidation, CrO_3 in AcOH in the presence of mineral acids produces benzylidene diacetates, which can then be hydrolyzed to the corresponding aldehydes.[110,693,798] Cr(V) and Cr(IV) species were shown to be involved in the reaction.[798] Silver(II) in acidic solution,[845] Ce(NH$_3$)$_2$(NO$_3$)$_6$,[846] and photooxidation in the presence of TiO_2 and Ag_2SO_4[847] are also used to accomplish aldehyde formation.

Methyl-substituted aromatics may also be converted to the corresponding carboxylic acids. As was mentioned above, a shift in the selectivity of the autoxidation of methyl-substituted arenes is brought about by metal ions eventually producing carboxylic acids. Carboxylic acids may also be formed by the base-catalyzed autoxidation of methyl-substituted arenes.[848] Oxidation with *tert*-BuOK in HMPA affords carboxylic acids in moderate yields with very high selectivities.

Oxidation of Other Arenes. Aromatic compounds with longer alkyl side chains can be converted to ketones or carboxylic acids. All the previously discussed reagents except CrO_2Cl_2 usually afford the selective formation of ketones from alkyl-substituted arenes. Oxidation with CrO_2Cl_2 usually gives a mixture of products. These include compounds oxidized in the β position presumably formed via an alkene intermediate or as a result of the rearrangement of an intermediate epoxide.[110,705]

Permanganate ion is a strong oxidant and tends to cleave aromatic side chains. However, when used under phase-transfer conditions,[109,849] it converts benzylic methine groups into alcohols, and benzylic methylene compounds to ketones. Convenient oxidations can be carried out when $KMnO_4$ is applied in Et$_3$N, an organic solvent (CHCl$_3$), and traces of water at room temperature.[850]

The Jones reagent[851] and *tert*-BuOOH in the presence of chromium(VI) complexes[852,853] were found to be particularly useful in the oxidation of tetralins and indans. Oxidation with 2,3-dichloro-5,6-dicyanobenzoquinone (DDQ) occurs with an exceptional mechanism.[854] In contrast with the radical processes observed in other oxidations DDQ generates a carbocation by hydride abstraction that is trapped by water to form an alcohol:

$$(9.158)$$

Further oxidation of the latter furnishes the product ketone.

Selective formation of ketones may be achieved through base-catalyzed oxidations.[848] Transformation of 9,10-dihydroanthracene catalyzed by benzyltrimethylammonium hydroxide (Triton B)[855] starts with proton abstraction:

$$(9.159)$$

Subsequent one-electron transfer and intramolecular hydrogen migration lead to radical **102** followed by reaction with O_2 to yield hydroperoxide radical **103**. Radical **103** is further oxidized to a dihydroperoxide (**104**), which decomposes to anthraquinone. Alternatively, **103** may be transformed to a diradical that eventually gives anthracene as a byproduct. The ratio of the two products strongly depends on the solvent used. The highest yield of anthraquinone (85% at 100% conversion) was achieved in 95% aqueous pyridine.

Aromatic carboxylic acids may be formed with the oxidative cleavage of the side chain. Oxidation of diarylmethanes, however, usually stops at the ketone stage. Such oxidations are usually carried out with permanganate, hexavalent chromium, or nitric acid. Since CrO_3 in acidic medium readily attacks the ring in polycyclic aromatics to yield quinones, only alkyl-substituted benzenes and isolated ring systems are oxidized exclusively at the benzylic position.[110,705,798] Working under neutral conditions, as with aqueous sodium dichromate, can allow one to avoid ring oxidation, and selective benzylic oxidation of polynuclear aromatics may occur. The same can be applied for permanganate oxidations.[832]

Benzylic Acetoxylation. The reaction of $Mn(OAc)_3$ in AcOH with methylbenzenes when carried out in the absence of oxygen yields products of side-chain acetoxylation and hydroxylation:[842]

in O_2	71%	21%	3%	66% conversion
in N_2	25%	62%	12%	61% conversion

$$(9.160)$$

The oxidation occurs in a one-electron process through a radical cation [Eq. (9.161)]. Bromide ions have a catalytic effect with the involvement of radical intermediates[856] [Eq. (9.162)]:

$$\text{ArCH}_3 \xrightleftharpoons[]{Mn(OAc)_3, AcOH} \text{ArCH}_3^+\cdot \xrightarrow{\text{AcOH}} \text{ArCH}_2\cdot \xrightarrow[-Mn(OAc)_2]{Mn(OAc)_3} \text{ArCH}_2\text{OAc} \qquad (9.161)$$

$$\text{ArCH}_3 \xrightarrow[\text{AcOH, KBr}]{Mn(OAc)_3} \text{ArCH}_2\cdot \xrightarrow[-Mn^{2+}]{Mn^{3+},\ Br^-} \text{ArCH}_2\text{Br} \xrightarrow[-Br^-]{AcO^-} \text{ArCH}_2\text{OAc} \qquad (9.162)$$

Several studies have been reported about Pd(II)-catalyzed acetoxylation. Pd(OAc)$_2$, which catalyzes aromatic coupling and nuclear acyloxylation as well (see Section 9.4.1), is an efficient catalyst of benzylic acetoxylation in the presence of alkali acetates. Stoichiometric oxidation in the absence of oxygen[857,858] and a catalytic method[858] are known. A twofold rate increase was observed when a mixture of Pd(OAc)$_2$, Sn(OAc)$_2$, and charcoal was employed in acetoxylation.[859] Xylenes are converted to the α,α'-diacetyloxy compounds in high yields.[860]

Oxidation with peroxydisulfate in AcOH in the presence of catalytic amounts of iron and copper salts gives benzylic acetates in good yields.[785,861] The reaction of lead tetraacetate with alkylarenes in AcOH provides benzylacetates in moderate yields.[693] Most of these oxidations usually involve methyl-substituted benzenes since aromatics with longer chain produce different side products.

9.5. PRACTICAL APPLICATIONS

9.5.1. Oxidation of Alkanes

Acetic Acid. Although at the time of this writing Monsanto's Rh-catalyzed methanol carbonylation (see Section 7.2.4) is the predominant process in the manufacture of acetic acid, providing about 95% of the world's production, some acetic acid is still produced by the air oxidation of *n*-butane or light naphtha.[862–867] *n*-Butane is used mainly in the United States, whereas light naphtha fractions from petroleum refining are the main feedstock in Europe.

Noncatalytic oxidation to produce acetic acid can be carried out in the gas phase (350–400°C, 5–10 atm) or in the liquid phase (150–200°C). Liquid-phase catalytic oxidations are operated under similar mild conditions. Conditions for the oxidation of naphtha are usually more severe than those for *n*-butane, and the process gives more complex product mixtures.[865–869] Cobalt and other transition-metal salts (Mn, Ni, Cr) are used as catalysts, although cobalt acetate is preferred. In the oxidation carried out in acetic acid solution at almost total conversion, carbon oxides, carboxylic acids and esters, and carbonyl compounds are the major byproducts. Acetic acid is produced in moderate yields (40–60%) and the economy of the process depends largely on the sale of the byproducts (propionic acid, 2-butanone).

A radical chain reaction to yield *sec*-butyl hydroperoxide (**105**) as intermediate is operative[17,870] [Eq. (9.163)]. Metal ions play a role in decomposition of the latter by β cleavage [Eq. (9.164)]:

$$n\text{-}C_4H_{10} \longrightarrow CH_3CH_2\overset{\centerdot}{C}HCH_3 \xrightarrow{O_2} \underset{OO\centerdot}{CH_3CH_2CHCH_3} \xrightarrow[-n\text{-}C_4H_9\centerdot]{n\text{-}C_4H_{10}} \underset{\textbf{105} \quad OOH}{CH_3CH_2CHCH_3} \tag{9.163}$$

$$\textbf{105} \xrightarrow[-Mn^{(n+1)+},\ -OH^-]{Mn^{n+}} \underset{O\centerdot}{CH_3CH_2CHCH_3} \longrightarrow CH_3CHO + CH_3\overset{\centerdot}{C}H_2 \tag{9.164}$$

Acetaldehyde thus formed is oxidized in situ to acetic acid. Decomposition may also take place through other pathways. Ethyl alcohol can be an important primary product which cooxidizes rapidly.[870,871]

Oxidation of butane and butenes to maleic anhydride is discussed in connection with the synthesis of maleic anhydride (see Section 9.5.4.).

Oxidation of Cyclohexane. The synthesis of cyclohexanol and cyclohexanone is the first step in the transformation of cyclohexane to adipic acid, an important compound in the manufacture of fibers and plastics. Cyclohexane is oxidized industrially by air in the liquid phase to a mixture of cyclohexanol and cyclohexanone.[866,872–877] Cobalt salts (naphthenate, oleate, stearate) produce mainly cyclohexanone at about 100°C and 10 atm. The conversion is limited to about 10% to avoid further oxidation by controlling the oxygen content of the reaction mixture. Combined yields of cyclohexanol and cyclohexanone are about 60–70%.

The cobalt-catalyzed oxidation of cyclohexane takes place through cyclohexyl hydroperoxide with the cobalt catalyst acting primarily in the decomposition of the hydroperoxide to yield the products:[870,877]

$$\tag{9.165}$$

Oxidation catalyzed with boric acid (160–175°C, 8–10 atm) results in the formation of cyclohexyl esters.[878,879] After hydrolysis a product mixture containing mainly cyclohexanol is formed.[878] Since borate esters are less sensitive to further oxidation, better yields are usually achieved (85% at 12% conversion).[872]

Oxidation of Cyclododecane. 1,12-Cyclododecanedioic acid used in the production of polymers is synthesized in a two-step process[864,866] similar to the manufacture of adipic acid. Cyclododecane is first oxidized to a mixture of cyclododecanol and cyclododecanone. Both the cobalt-catalyzed and the borate processes (Huels) are used. Further oxidation of the product mixture leads to 1,12-cyclododecanedioic acid.

sec-Alcohols. At present, methods to manufacture detergent-range straight-chain (C_{10}–C_{16}) secondary alcohols through the oxidation of *n*-alkanes[880–883] are practiced in Japan and the former Eastern Block countries. Boric acid-modified air oxidation (the Bashkirov method[884]) at low conversion levels (20%) to maintain suitable selectivity is applied. An optimal concentration of boric acid is required, since it inhibits oxidation in high concentration, but increases alcohol yields. Aromatic compounds also act as inhibitors, necessitating the use of high-purity feeds (<0.1% aromatics). It was found, however, that ammonia and amine compound cocatalysts exclude inhibition problems and allow higher conversions with the same selectivity (Nippon Shokubai process).[885,886] The oxidation product is hydrolyzed, and unreacted paraffins and recovered boric acid are recycled. Hydrogenation of the hydrolysis product (70% alcohols, 20% ketones, 10% carboxylic acids) increases yields to better than 90% by converting ketones to alcohols. The alcohol mixture is purified by distillation and reacted with ethylene oxide to achieve better biodegradability. The product *sec*-alcohol ethoxylates are the main ingredients in liquid detergents.

9.5.2. Oxygenation of Alkenes and Dienes

Ethylene Oxide. Essentially all ethylene oxide produced industrially is manufactured by the direct oxidation of ethylene catalyzed by silver:[867,887–889]

$$CH_2{=}CH_2 \ + \ 0.5\,O_2 \ \xrightarrow{\text{Ag}} \ \underset{O}{\triangle} \quad \Delta H = -25.1 \text{ kcal/mol} \qquad (9.166)$$

This oxidation is unique, since only silver is capable of epoxidizing ethylene, and silver is active only in the oxidation of ethylene. A low-surface-area α-alumina is usually applied to support about 10–15% silver. Organic halides (1,2-dichloroethane, ethyl and vinyl chloride) are added as moderators, and additives (Cs, Ba) are also used to increase selectivity. At present selectivity in industrial oxidations is about 80%.

Several processes based on air or oxygen have been developed.[890–895] Oxidation with air (260–280°C) or oxygen (230°C) is carried out at about 15–25 atm at a limited conversion (about 10–15%) to achieve the highest selectivity.[896–898] High-purity, sulfur-free ethylene is required to avoid poisoning of the catalyst. Ethylene concentration is about 20–30 vol% or 5 vol% when oxygen or air, respectively, is used as oxidants. The main byproducts are CO_2 and H_2O, and a very small amount of acetaldehyde is formed via isomerization of ethylene oxide. Selectivity to ethylene oxide is 65–75% (air process) or 70–80% (O_2 process).[867]

Numerous papers and several review articles[889,899–907] deal with adsorption studies and discuss the kinetics and mechanism of the silver-catalyzed epoxidation of ethylene. A simple triangular kinetic scheme of first-order reactions satisfies the experimental observations (Scheme 9.23). On the best industrial catalysts k_1/k_2 is 6, and k_2/k_3 is 2.5.

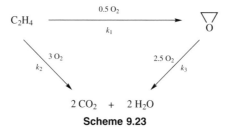

Scheme 9.23

Both associative [Eq. (9.167)] and dissociative [Eq. (9.168)] adsorptions of oxygen were found to occur on silver:

$$O_2 + Ag \longrightarrow O_{2\ ads}^- + Ag^+ \tag{9.167}$$

$$O_2 + 4\,Ag_{adj} \longrightarrow 2\,O_{ads}^{2-} + 4\,Ag_{adj}^+ \tag{9.168}$$

Dissociative adsorption occurring on four adjacent silver atoms (Ag_{adj}) is responsible for nonselective oxidation to yield CO_2. An optimum chlorine coverage of about 25% of surface silver atoms effectively blocks dissociative adsorption of oxygen by physical blocking ensuring increased selectivity.

On the basis of IR studies and site-blocking experiments, Sachtler et al. suggested that diatomic adsorbed oxygen species react selectively with ethylene to yield ethylene oxide[904,908] [Eq. (9.169)]; atomic oxygen formed, in turn, reacts with ethylene to result in complete oxidation [Eq. (9.170)]:

$$CH_2{=}CH_2 + O_{2,\ ads} \longrightarrow \triangledown\!\!\!\!_O + O_{ads} \tag{9.169}$$

$$CH_2{=}CH_2 + O_{ads} \longrightarrow \triangledown\!\!\!\!_O \xrightarrow{5\,O_{ads}} 2\,CO_2 + 2\,H_2O \tag{9.170}$$

Steady-state conditions require the combination of Eqs. (9.168) and (9.170) giving Eq. (9.171). It also indicates the formation of vacant surface sites ($\square$) capable of further oxygen adsorption. If certain further assumptions hold, Eq. (9.171) determines a limiting selectivity of 6/7 or 85.7%:

$$7\,CH_2{=}CH_2 + 6\,O_{2,\ ads} \longrightarrow 6\,\triangledown\!\!\!\!_O + 2\,CO_2 + 2\,H_2O + 6\,\square \tag{9.171}$$

Selectivities exceeding 90% were, however, later observed.[909,910] Bell et al. argued in favor of atomic oxygen as the species that forms ethylene oxide in a reaction with gaseous ethylene.[911] Ethylene adsorbed on silver vacancies at low oxygen coverage undergoes total combustion. The role of chlorine is to suppress surface silver vacancies, preventing ethylene adsorption and C–H activation.

This view supported by surface science studies has found wide acceptance.[889] Under working conditions a silver oxide–chlorine surface phase is formed, which may contain ionic silver (Ag^+, Ag^{3+}). Oxygen and chlorine occupy subsurface positions and act as electron-withdrawing ligands allowing the adsorption of electrophilic oxygen atoms. Electrophilic attack of such oxygen on one of the sp^2-hybridized carbon atoms (**106**) leads to epoxidation.[889] Nonselective oxidation, in contrast, is brought about by attack of strongly bound, bridging oxygen to hydrogen atoms (**107**):

| **106** | **107** |

The unique ability of silver to epoxidize ethylene lies in the fact that it adsorbs oxygen dissociatively, and that atomic oxygen formed at high oxygen coverage is weakly bound. The low activity of silver to activate C—H bonds is the key factor in the selectivity of epoxidation.

Propylene Oxide. Unlike ethylene, propylene cannot be selectively transformed to propylene oxide by silver-catalyzed oxidation. Instead, indirect oxidations (the peracid and the hydroperoxide routes) are employed.[912–915]

In the peracid process (Bayer–Degussa technology[916]) propionic acid is oxidized by hydrogen peroxide in the presence of H_2SO_4 to yield perpropionic acid, which, in turn, is used to oxidize propylene to propylene oxide. The peracetic acid process (Daicel technology[917,918]) employs a mixture of acetaldehyde, ethyl acetate, and compressed air in the presence of a metal ion catalyst to produce the reagent (30–50°C, 25–40 atm). After separation of the unreacted acetaldehyde a 30% solution of peracetic acid in ethyl acetate containing 10–15% acetic acid is used to epoxidize propylene (50–80°C, 9–12 atm). A 90–92% yield of propylene oxide is reached at 97–98% conversion of peracetic acid.

About 50% of the current worldwide propylene oxide capacity is based on a newer process called *hydroperoxide* (*coproduct* or *oxirene*) route (Halcon–Arco technology[919,920]). According to this process, hydroperoxides synthesized by the oxidation of certain hydrocarbons, such as ethylbenzene or isobutane, are used for epoxidation of propylene in the presence of a transition-metal catalyst:

$$R-H \xrightarrow{O_2} R-OOH \xrightarrow{C_3H_6} \underset{O}{\overset{Me}{\triangle}} + R-OH \qquad (9.172)$$

The additional advantage of this process is the possibility of the utilization of the alcohol products. *tert*-BuOH (and MTBE, its methyl ether) is used as a gasoline additive. It can be dehydrated to yield isobutylene, which is used to produce high-octane alkylates (see Section 5.5.1). 1-Phenylethanol is dehydrated to styrene over TiO_2 on alumina.[919]

In the C_4 coproduct route isobutane is oxidized with oxygen at 130–160°C and under pressure to *tert*-BuOOH, which is then used in epoxidation. In the styrene coproduct process ethylbenzene hydroperoxide is produced at 100–130°C and at lower pressure (a few atmospheres) and is then applied in isobutane oxidation. Epoxidations are carried out in high excess of propylene at about 100°C under high pressure (20–70 atm) in the presence of molybdenum naphthenate catalyst. About 95% epoxide selectivity can be achieved at near complete hydroperoxide and 10–15% propylene conversions. Shell developed an alternative, heterogeneous catalytic system (TiO_2 on SiO_2), which is employed in a styrene coproduct process.[913,914]

Acetaldehyde and Acetone. The direct palladium-catalyzed oxidation of ethylene to acetaldehyde[507,867,921–926] is an important industrial processes. The similar oxidation of propylene to acetone,[915,927,928] though practiced, is less important,[915,929] due to other large-scale processes for the manufacture of acetone (dehydrogenation of 2-propanol, the cumene oxidation process; see Section 9.5.4). A single-stage technology (Hoechst process,[922,925] for oxidation with oxygen) and a two-stage process (Wacker process,[922,925] for oxidation with air) have been developed.

The homogeneous oxidation of alkenes in aqueous solution to carbonyl compounds by palladium salts results in the concomitant reduction of Pd(II) to metallic palladium. In the catalytic process a suitable oxidant, usually $CuCl_2$ is used to regenerate Pd(II), and Cu_2Cl_2 thus formed is reoxidized with oxygen (see Section 9.2.4). In the single-stage operation the alkene and oxygen are fed into the aqueous solution of $PdCl_2$ containing a large excess of $CuCl_2$. Oxidation of the alkene to the carbonyl compound and regeneration of the catalyst system occur in the same reactor (120–130°C, 3 atm). This process requires the use of high-purity ethylene (99.7%) and oxygen (99%).

In the two-stage process the alkene is reacted with the catalyst system (100–110°C, 10 atm). The reduced catalyst solution containing Cu_2Cl_2 is reoxidized with air in a second reactor under the same conditions. In both cases about 95% oxo yield is achieved at 95–99% alkene conversion. Because of the explosion hazard of mixing ethylene or propylene with pure oxygen, most commercial operations favor the less hazardous two-stage process.

Vinyl Acetate. Two homogeneous liquid-phase processes for the production of vinyl acetate have been developed.[244,567,930,931] In the ICI process[931,932] a suspension of $PdCl_2$, KCl, $Cu(OAc)_2$ and KOAc in acetic acid is reacted with ethylene at 40–50 atm and 120°C in the presence of a small amount of oxygen. At 2–3% conversion per pass a significant amount of acetaldehyde is obtained as well, which is oxidized to produce the required acetic acid for the process. In the other, two-stage process[567] stoichiometric oxidation with $PdCl_2$ and $CuCl_2$ is carried out, followed by reoxidation with air of the catalyst mixture in a second reactor. Serious drawbacks of both processes such as explosion hazard, corrosion problems, and decreased selectivity in the presence of $CuCl_2$ resulted in the development of heterogeneous gas-phase processes.[244,922,931]

In the process jointly developed by Bayer and Hoechst,[933–935] a palladium–gold-on-silica or alumina catalyst impregnated with KOAc is used. A mixture of ethylene, acetic acid, and oxygen is converted at 150–170°C and about 5–10 atm to produce vinyl acetate with about 91–92% selectivity at about 10% conversion in a highly exothermic reaction. The only major byproduct is CO_2. KOAc requires continuous replenishment. A similar process was independently developed by U.S.I. Chemicals.[932,936]

1,4-Diacetoxy-2-butene. Mitsubishi commercialized a new proces, the acetoxylation of 1,3-butadiene, as an alternative to the Reppe (acetylene–formaldehyde) process for the production of 1,4-diacetoxy-2-butene.[937,938] 1,4-Diacetoxy-2-butene is tranformed to 1,4-butanediol used in polymer manufacture (polyesters, polyurethanes). Additionally, 1,4-butanediol is converted to tetrahydrofuran, which is an important solvent and also used in polymer synthesis.

The homogeneous palladium-catalyzed process for acetoxylation was never commercialized because of low selectivity and the difficulty in separating the catalyst from the reaction mixture. Heterogeneous palladium catalysts applied in the gas phase, in turn, quickly lose activity caused by buildup of polybutadiene. The Mitsubishi process uses a Pd–Te-on-active-carbon catalyst in the liquid phase. Tellurium apparently prevents palladium elution to acetic acid.

The acetoxylation reaction is carried out at 70°C under 70 atm pressure by reacting 1,3-butadiene and acetic acid in the presence of air and a small amount of polymerization inhibitor. A special three-step activation (reduction–oxidation–reduction), also used in regeneration of the used catalyst, ensures high activity and selectivity. Regeneration of the catalyst is necessary after about one year of operation.

Acrolein and Acrylic Acid. Acrolein and acrylic acid are manufactured by the direct catalytic air oxidation of propylene. In a related process called *ammoxidation*, heterogeneous oxidation of propylene by oxygen in the presence of ammonia yields acrylonitrile (see Section 9.5.3). Similar catalysts based mainly on metal oxides of Mo and Sb are used in all three transformations. A wide array of single-phase systems such as bismuth molybdate or uranyl antimonate; and multicomponent catalysts, such as iron oxide–antimony oxide or bismuth oxide-molybdenum oxide with other metal ions (Ce, Co, Ni), may be employed.[939] The first commercial process to produce acrolein through the oxidation of propylene, however, was developed by Shell applying cuprous oxide on Si–C catalyst in the presence of I_2 promoter.[915,940]

When a mixture of propylene, air, and steam (1 : 8–10 : 2–6) is reacted in the vapor phase (300–400°C, 1–3 atm) acrolein is formed in about 85% yield in an exothermic process[898,915] [Eq. (9.173)]; 5–10% of acrylic acid may also be isolated:

$$CH_2{=}CH{-}CH_3 \ + \ O_2 \ \longrightarrow \ CH_2{=}CH{-}CHO \ + \ H_2O \qquad \Delta H = -87.8 \ \text{kcal/mol}$$

$$(9.173)$$

Oxidation in the original Sohio process[941,942] was carried out over a bismuth molybdate catalyst, which was later superseded by bismuth phosphomolybdate with various amounts of additional metal ions (Ce, Co, Ni), and multicomponent metal oxides based on Mo, Fe, and Bi supported on silica.

A single-stage or two-stage technology may be used to produce acrylic acid.[898,915,943–948] In the two-stage process multicomponent metal oxides similar to the catalysts used in the synthesis of acrolein are used to transform the propylene–air–steam mixture into a mixture of acrolein and acrylic acid. This product mixture is further oxidized in a second step under different operating conditions. The Toya Soda technology,[943,946] for instance, carries out the two steps slightly above atmospheric pressure, at 330–370°C and 260–300°C, respectively, and achieves a 67% overall yield of acrylic acid. Because of the higher yields achieved as a result of the use of optimal catalysts and temperatures in each reaction step, the two-stage technology is preferred.

Methacrolein and Methacrylic Acid. A two-stage technology, essentially the same as the propylene oxidation process for the manufacture of acrolein and acrylic acid, was developed to oxidize isobutylene to methacrolein and methacrylic acid.[949–951] Two different molybdenum-based multicomponent catalysts are used. In a typical procedure[949] isobutylene is reacted with excess steam and air (5 : 30 : 65) at about 350°C to produce a mixture of methacrolein and methacrylic acid with 80–85% selectivity at a conversion of 98%. In the second stage this reaction mixture is oxidized at slightly above 300°C to yield methacrylic acid (80% selectivity at >90% conversion).

9.5.3. Ammoxidation

Acrylonitrile. Ammoxidation of propylene to produce acrylonitrile [Eq. (9.174)] is usually carried out in fluidized-bed reactors converting a nearly stoichiometric mixture of propylene, air, and ammonia at about 400–500°C, slightly above atmospheric pressure:[952–954]

$$CH_2{=}CH{-}CH_3 \ + \ 1.5\,O_2 \ + \ NH_3 \ \xrightarrow{\text{catalyst}} \ CH_2{=}CH{-}CN \ + \ 3\,H_2O \quad \Delta H = -123.2 \text{ kcal/mol}$$

$$(9.174)$$

Practically complete conversion of propylene and ammonia is achieved to produce acrylonitrile in 65–70% yield. Acetonitrile and HCN are the main byproducts. The Sohio process originally used oxides of Bi, Co, and Mo, and bismuth and cobalt molybdates.[898,915,941,953] Other catalysts developed later (uranyl antimonate; antimony oxide–iron oxide; oxides of Fe, Ce, and Mo; mixed oxides of Sb and Sn)[898,915,939,953,955,956] produce fewer byproducts and ensure higher yields of acrylonitrile.

Intensive studies on the mechanism of oxidation and ammoxidation have been carried out.[900,939,957–960] All selective oxidation and ammoxidation catalysts are redox systems. They are capable of oxidizing propylene with the concomitant reduction of the surface-active sites. The catalyst subsequently undergoes reoxidation; that is, incorporation of molecular oxygen into the lattice takes place.

It is well established that the first step in the oxidation of propylene is α-hydrogen abstraction in a rate-determining step with the formation of a symmetric π-allyl intermediate (**108**) (Scheme 9.24). The active site consists of a bismuth–molybdenum pair; the surface Bi : Mo ratio is 1. The hydrogen abstraction occurs on propylene adsorbed on a coordinatively unsaturated Mo center by Bi oxygen.[961] A σ-*O*-allyl molybdate (**109**) formed in the next step decomposes to the product acrolein, creating an oxygen vacancy. The latter is replenished by gaseous oxygen re-forming the original active site.

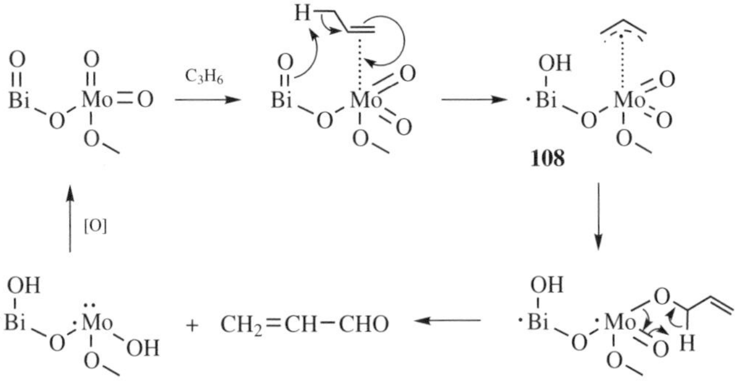

Scheme 9.24

In summary, oxygen associated with Bi is responsible for allylic hydrogen abstraction, whereas oxygen polyhedra around Mo are inserted into the allylic intermediate.[962–964] Nitrogen insertion in ammoxidation takes place in a similar way[965] on an active site containing Mo=NH instead of Mo=O.

Processes based on propane ammoxidation to manufacture acrylonitrile have also been developed,[915,966] and BP has announced commercialization.[966] Dehydrogenation at high reaction temperature (485–520°C), which is about 100°C higher than for propylene ammoxidation, results in the formation of propylene, which subsequently undergoes normal ammoxidation. Despite higher investments and the markedly lower selectivity (30–40%), the process can be economical because of the price difference between propylene and propane.[966] Better selectivites can be achieved at lower (40–60%) conversions.

Other Processes. Other similar technologies also exist to manufacture less important products such as benzonitrile[967] and isomeric dicyanobenzenes from the corresponding methylbenzenes.[968–970] In the Lummus process to produce terephthalonitrile, ammoxidation is carried out without free oxygen. *p*-Xylene and ammonia are reacted over a V_2O_5-on-alumina catalyst at 400–450°C in a fluidized-bed reactor. The catalyst is reduced during the process and reoxidized in a separate step. The product terephthalonitrile is subsequently hydrolized to terephthalic acid.

9.5.4. Oxidation of Arenes

Phenol and Acetone. Cumene oxidation is presently the only commercially significant route for the production of phenol.[927,971–973] In a two-step process[974,975] cumene is first oxidized in an autocatalytic, free-radical reaction to cumene hydroperoxide[976] [**110**, Eq. (9.175)]:

$$(9.175)$$

The oxidation is carried out in the liquid phase, at relatively low temperature (90–120°C), pressure (5–7 atm), and conversion rate (20–40%) to achieve high selectivity. Cumene hydroperoxide is concentrated (70–85%) and then further converted. This second step includes an acid-catalyzed rearrangement and cleavage[977] yielding phenol and acetone:

$$(9.176)$$

Byproducts are α-methylstyrene, acetophenone and cumene.

Phenol can also be prepared by the decomposition of benzoic acid prepared by the oxidation of toluene.[927,978] The process is an oxidative decarboxylation catalyzed by copper(II). An interesting feature of this reaction is that the phenolic hydroxyl group enters into the position *ortho* to the carboxyl group as was proved by ^{14}C labeling.[979] In the Dow process[980] molten benzoic acid is transformed with steam and air in the presence of Cu(II) and Mg(II) salts at 230–240°C. A copper oxide catalyst is used in a vapor-phase oxidation developed by Lummus.[981]

Benzoic Acid. The industrial-scale oxidation of toluene to benzoic acid is carried out with cobalt catalysts.[973,978,982,983] The process, a free-radical autoxidation, is significantly promoted by bromide ions.[984] Under these conditions bromine atoms rather than alkylperoxy radicals serve as a regenerable source of chain-initiating free radicals. Substantial rate increase can be achieved by the addition of manganese(II) ions.[984]

Liquid-phase oxidation of toluene to benzoic acid (Dow process,[980,985,986] Mid-Century process[987]) may be carried out in acetic acid or without solvent. Cobalt(II) naphthenate or cobalt(II) 2-ethylhexanoate promoted by bromine is used as the catalyst at 140–190°C, at about 8–10 atm pressure. The highly exothermic oxidation is run until about 50% conversion. In a process developed by SNIA Viscosa,[988,989] Co(II) heptanoate or Co(OAc)₃ in water is used. Benzaldehyde is always the major byproduct.

Terephthalic Acid. The oxidation of *p*-xylene to terephthalic acid is a more difficult reaction.[866] Once one of the methyl groups is oxidized, the second methyl group in *p*-toluic acid formed is deactivated by the electron-withdrawing effect of the carboxylic group. Bromine-promoted catalytic processes and cooxidation in the presence of suitable additives are usually practiced.[970,978,990]

The Mid-Century–Amoco process[987] employs a mixture of Co(OAc)₂ and Mn(OAc)₂ as the catalyst and bromine compound cocatalysts (NH₄Br and tetrabromoethane) in acetic acid solvent. Since terephthalic acid is not soluble in acetic acid it crystallizes in high purity during oxidation (220°C, about 20 atm). The oxidation is run to almost complete conversion with better than 90% selectivity. Certain organic compounds, when cooxidized together with *p*-xylene, promote the formation of terephthalic acid. These promoters, such as acetaldehyde (Eastman-Kodak process[991]), paraldehyde (Toray process[992]), and 2-butanone (Mobil process[993,994]), supply oxygen-containing radicals which regenerate the Co(III) catalyst.

The Dynamit Nobel–Hercules process[995] produces dimethyl terephthalate in a more complex synthesis (Scheme 9.25). To overcome the difficulty in the oxidation of the second methyl group, *p*-xylene and recycled methyl *p*-toluate formed after

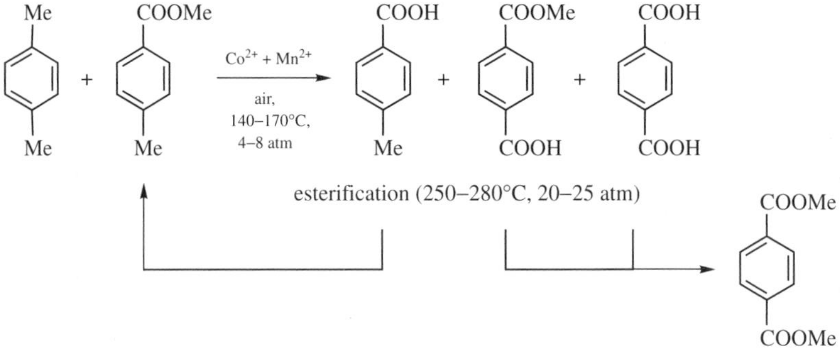

Scheme 9.25

esterification are cooxidized without solvent at 140–170°C and 4–8 atm. The catalyst is a mixture of Co(II) and Mn(II) 2-ethylhexanoate.

The product mixture, consisting mainly of *p*-toluic acid, monomethyl terephthalate, and some terephthalic acid, is esterified and methyl *p*-toluate is recycled after separation from the end-product dimethyl terephthalate. The special advantage of cooxidation is that the hydroperoxy radical (**111**) formed in the autoxidation of *p*-xylene participates in initiation through hydrogen abstraction from methyl *p*-toluate, thus promoting oxidation of the latter:

$$(9.177)$$

In the manufacture of terephthalic acid by the oxidation of *p*-xylene, separation of the xylene from its isomeric mixture is necessary (see Section 2.5.2). An alternative process introduced in Japan uses the oxidation of *p*-tolualdehyde, which is obtained in good regioselectivity by the HF–BF$_3$ catalyzed carbonylation of toluene without the necessity of separation of the isomers.

Maleic Anhydride. Gas-phase catalytic oxidation of benzene or *n*-butane is the principal process for the industrial production of maleic anhydride.[973,996–999] Until the 1970s commercial production was based predominantly on benzene. Because of its more favorable economics, a switch to butane as an alternative feedstock has taken place since then.[966,999–1002] At present almost all new facilities use *n*-butane as the starting material. Smaller quantities of maleic anhydride may be recovered as a byproduct of phthalic anhydride manufacture (about 5–6%).[1003,1004]

The Halcon process, modified for *n*-butane oxidation, was originally developed for the oxidation of benzene.[1005] Preheated air in a large excess to keep the composition below the explosion limit is mixed with benzene (1.7 mol%), and the mixture is then fed to tubular reactors. The catalyst is V$_2$O$_5$ and MoO$_3$ on an inert support containing promoters. A large amount of heat is produced during the oxidation, which is carried out at 350–400°C and slightly above atmospheric pressure:

$$+ \ 4.5 \ O_2 \longrightarrow \quad + \ 2 \ CO_2 \ + \ 2 \ H_2O \quad \Delta H = -448.6 \ \text{kcal/mol} \quad (9.178)$$

More than 70% of benzene is converted to maleic anhydride. Maleic anhydride is partially condensed, and the remaining quantity is absorbed in water to form maleic acid. It is converted to maleic anhydride by azeotropic distillation with o-xylene. Thin-film evaporation under vacuum to avoid isomerization to fumaric acid may be used.

A large number of studies have been published about the mechanism of oxidation of benzene to maleic anhydride, and the structure and role of catalysts and promoters.[999,1006,1007] At present hydroquinone formed through a peroxidic adduct is well established to be the intermediate in selective oxidation[1008] [Eq. (9.179)], whereas p-benzoquinone is the intermediate in the nonselective oxidation pathway [Eq. (9.180)]:

$$(9.179)$$

$$(9.180)$$

Oxidation of benzene takes place through its interaction with adsorbed O_2 molecules.

Fully oxidized vanadia is a poor, nonselective catalyst for benzene oxidation. Reduction to lower oxides brings about a large increase in catalytic activity and some improvement in selectivity. Substantial enhancement of selectivity can be achieved by the addition of molybdena,[1009] reaching a maximum at a MoO_3 content of about 30%. Promoters (Ag, Ti, Ni, Co) lead to further improvements.[1006]

In the process based on n-butane feedstock, vanadium phosphorous oxides (V–P–O) catalysts are mainly used.[1010–1012] Processes for the oxidation of low-cost C_4 fraction from naphtha cracker consisting mainly of butenes have also been developed.[1013,1014] In contrast with benzene oxidation where two carbon atoms are lost in the form of ethylene no carbon is lost in the oxidation of C_4 hydrocarbons:

$$C_4H_{10} \; + \; 3.5\,O_2 \; \longrightarrow \; \text{[maleic anhydride]} \; + \; 2\,H_2O \quad \Delta H = -301.4 \text{ kcal/mol} \qquad (9.181)$$

Oxidation of n-butane and butenes requires higher reaction temperatures (400–480°C). Since more water is produced in this reaction, most of the product is recovered in the form of maleic acid. The currently best process is the Alma (Alusuisse) process.[1015–1018] Its main features are a fluidized-bed reactor and an anhydrous product recovery. Because of the better temperature control. a lower air : hydrocarbon ratio can be employed (4 mol% of n-butane). Instead of absorption in water, maleic anhydride is recovered from the reactor effluent gas by a high-boiling organic solvent. Most commercial processes are operated at 80% conversion, yielding maleic anhydride with about 70% selectivity.

The oxidation of n-butane to maleic anhydride is a very complex reaction. Because of its industrial importance, however, thorough studies have been carried out.[899,900,999,1007,1010,1011,1019] More recent findings indicate that the active phase of the VPO catalyst is vanadyl phosphate $[(VO)_2P_2O_7]$.[900,1020–1022] Both activated (O*) and surface lattice oxygens (O_{sl}) play an important role in oxidation[1023] (Scheme 9.26). The former participates in the oxidation of the C—H bonds in butane and furan. Lattice oxygen, in turn, is responsible for allylic oxidation and ring insertion. Each intermediate may be oxidized to carbon oxides. These inter-mediates, however, are not detectable in the gas phase during a normal oxidation process,[1011] but could be indentified under extreme conditions by spectroscopic methods as surface-bound species. The key factor in selective oxidation to maleic anhydride is the fast replenishment of oxygen consumed at the reaction site. If this process is not fast enough, intermediates may desorb, resulting in decreased selectivity.

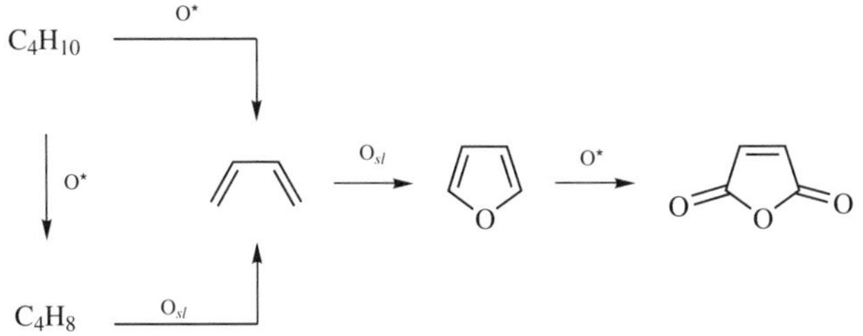

Scheme 9.26

In industrial procedures catalyst preparation in organic solvents is preferred, pro-ducing a more amorphous precursor with high density of active sites.[1022] In addi-tion, phosphorous compounds are usually injected into the feed to replenish phosphorous lost through hydrolysis. In this way an appropriate surface phosphor-ous : vanadium ratio is maintained ensuring prolonged catalytic activity and selectivity.

Phthalic Anhydride. Phthalic anhydride is manufactured by the catalytic vapor-phase oxidation of o-xylene [Eq. (9.182)] or naphthalene [Eq. (9.183)]:

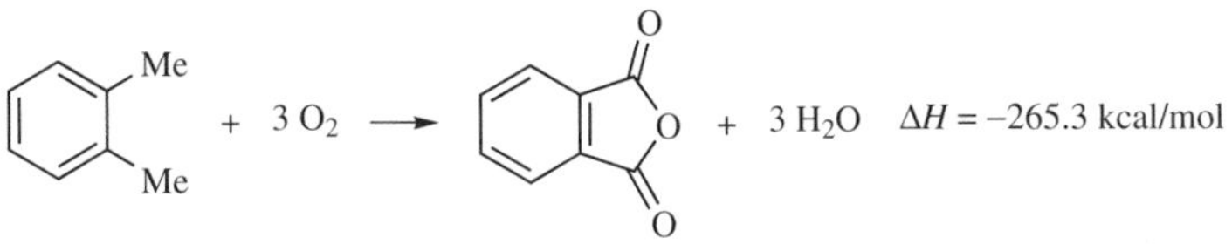

$$\text{(9.182)}$$

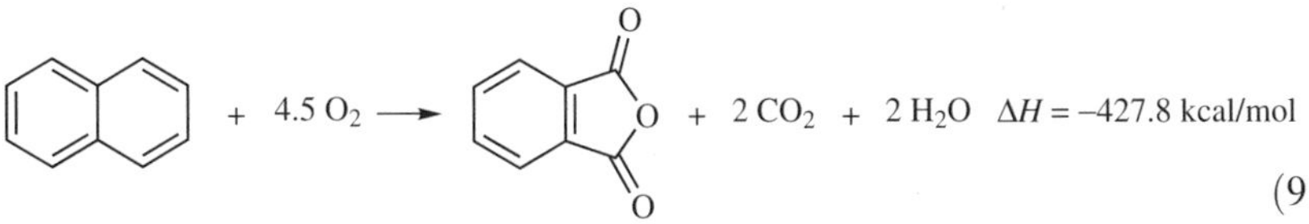

$$\text{(9.183)}$$

Both oxidations are highly exothermic and carried out almost exclusively in tubular reactors cooled by a molten salt.[1024] Supported vanadium oxide with additives to improve activity, selectivity, and stability usually serves as the catalyst.[970,990,1025] Because of its more favorable stoichiometry (no carbon is lost in oxidation), most new plants use o-xylene as the starting material.

Compressed air and o-xylene or naphthalene vapor are mixed in an approximately 20 : 1 ratio and oxidized at ~400°C. The product phthalic anhydride is recovered by condensation and purified by vacuum distillation. Maleic anhydride, a byproduct in both oxidations may be recovered from the waste gases in the form of maleic acid after water scrubbing.

The BASF process[1026] was developed to use o-xylene exclusively. Most processes, however, such as the Wacker–von Heyden process[1027,1028] are capable of operating on both feedstocks or may use a mixture of the two compounds. Fluidized-bed operations were developed in the 1960s (Badger process[1029,1030]) but have been replaced by improved, more economical fixed-bed processes. The Alusuisse low-air-ratio (LAR) process, for instance, allows the use of a 9.5 : 1 air : o-xylene mixture and achieves an increased catalyst productivity.[1031,1032]

Detailed mechanistic studies about both oxidations have been carried out.[901,1006,1033] On the basis of the observed intermediates, a reaction network [Eq. (9.184)] was suggested in the oxidation of o-xylene indicating o-tolualdehyde as the first intermediate:[1034,1035]

$$(9.184)$$

The most stable, active, and selective catalyst in the oxidation of o-xylene is V_2O_5 supported on TiO_2. It is well established that anatase is the suitable support. For example, when the vanadia on anatase catalyst is treated above 500°C, anatase is transformed to rutile and a significant decrease in catalyst performance is observed.[1006] It is speculated that the special nonmicroporous structure of anatase[1036] and the spreading of vanadia on the surface of anatase[1037] are important features responsible for the observed differences. The highest catalyst activity is exhibited by the catalyst that contains 2 mol% of vanadia corresponding to the monolayer coverage.[1006] The active center consists of a reducible V=O bond and an acidic center (oxohydroxy–vanadium group) forming a monolayer on the surface.[1035] The latter catalyzes dehydration of noncyclic precursors to five-membered ring compounds.

A typical reaction network for the oxidation of naphthalene includes naphthoquinones as intermediates[1006,1033] (Scheme 9.27).

Oxidation via 1,4-naphthoquinone is the major pathway, but 1,2-naphthoquinone may be a significant intermediate.[1038,1039] Industrial catalysts for the oxidation

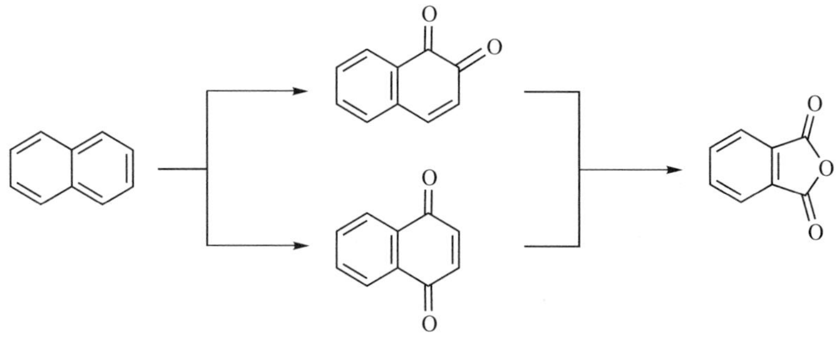

Scheme 9.27

of naphthalene consist of vanadia supported on silica promoted with alkali sulfates.[1033] SO_2 was found to increase activity and selectivity.[1040]

Anthraquinone. Anthraquinone an important intermediate used extensively in the dye industry can be manufactured through the oxidation of anthracene.[1041,1042] Almost all anthraquinone is currently produced by oxidation with CrO_3 in the liquid phase (50–100°C) with selectivity better than 90% at complete conversion.

9.6. RECENT DEVELOPMENTS

The selective oxidation of hydrocarbons, particularly that of alkanes, remains a challenge. It is not surprising, therefore, that the problems of oxidation processes are addressed in several books,[1043–1045] reviews,[1046–1057] and a journal special issue,[1058] as well as in international conferences[1059–1064] devoted to the topic. For the advances in chirally catalyzed oxidation processes, including asymmetric epoxidation and osmylation, Sharpless was one of the recipients of the 2001 Nobel Prize in Chemistry.

9.6.1. Oxidation of Alkanes

Oxygenation of methane and higher hydrocarbons, that is, the production of valuable products through C—H activation, is an area of continuing interest.[1065–1070]

Oxidation of Methane. A variety of new catalyst systems have been disclosed, and new reagents were developed with the aim to perform selective transformation of methane to methanol, methyl esters, and formaldehyde. Much work was carried out in strongly acidic solutions, which enhances the electrofilicity of the metal ion catalyst, and the ester formed is prevented from further oxidation.[1071,1072] An important advance in the selective oxidation of methane to methanol is Periana's 70% one-pass yield with high selectivity in sulfuric acid solution under moderate conditions.[1073] The most effective catalyst is a Pt–bipyrimidine complex. Pt(II) was shown to be the most active oxidation state generating a Pt–methyl intermediate that is oxidized to yield the product methyl ester. A density functional study supports this mechanism.[1074] A highly catalytic bimetallic system consists of $CuCl_2$ and Pd on C, operates in a 3 : 1 mixture of CF_3COOH acid and water in the presence of O_2 and CO, and selectively yields methanol.[1075] The $RhCl_3–NaCl–KI$ system in aqueous CF_3COOH is also effective in the formation of methanol and methyl trifluoroacetate.[1076]

Gas-phase oxidation of methane could be enhanced by the addition of a small amount of NO or NO_2 in the feed gas.[1077] Addition of methanol to the $CH_4–O_2–NO_2$ mixture results in a further increase in methane reactivity.[1078] Photocatalytic conversion of methane to methanol is accomplished in the presence of water and a semiconductor photocatalyst (doped WO_3) at 94°C and atmospheric pressure.[1079] The yield of methanol significantly increased by the addition of H_2O_2 consistent with the postulated mechanism that invokes hydroxyl radical as an intermediate in the reaction.

A systematic study to identify solid oxide catalysts for the oxidation of methane to methanol resulted in the development of a $Ga_2O_3–MoO_3$ mixed metal oxide catalyst showing an increased methanol yield compared with the homogeneous gas-phase reaction.[1080,1081] Fe-ZSM-5 after proper activation (pretreatment under vacuum at 800–900°C and activation with N_2O at 250°C) shows high activity in the formation of methanol at 20°C.[1082] Density functional theory studies were conducted for the reaction pathway of the methane to methanol conversion by first-row transition-metal monoxide cations (MO^+).[1083] These are key to the mechanistic aspects in methane hydroxylation, and CuO^+ was found to be a likely excellent mediator for the reaction. A mixture of vanadate ions and pyrazine-2-carboxylic acid efficiently catalyzes the oxidation of methane with O_2 and H_2O_2 to give methyl hydroperoxide and, as consecutive products, methanol and formaldehyde.[1084,1085]

The activity of a modified W/HZSM-5 catalyst in the partial oxidation of methane to formaldehyde was 3 times higher than that of the conventional impregnated catalysts.[1086,1087] Precipitated silica[1088] and V_2O_5 on silica[1089] allowed for high formaldehyde yields. Of a series of supported $FePO_4$ catalysts the silica-supported sample produced the highest yield of formaldehyde.[1090,1091] Low iron loading with Fe in fivefold coordination sites, isolated by phosphate groups and enrichment of the surface in phosphorus, are characteristic of the selective catalyst.

Effective catalytic hydroxylations of alkanes can be accomplished by metalloporphyrins.[1092–1094] Electron-deficient porphyrin ligands containing, for example, Ru,[1095] Fe,[1096,1097] and Co[1098] ions, were used mainly in applying various oxidants. Immobilized catalysts, such as iron and manganese porphyrins covalently bound to silica,[1096,1099,1100] encapsulated in X zeolite,[1101] and nanocrystalline TiO_2 modified with Fe(III) porphyrin,[1102] also showed favorable characteristics. Mn(III) porphyrins may be used for the formation of ketones and lactones with Oxone ($KHSO_5$) in a solid–liquid system.[1103] New evidence of the high-valence oxo–metal radical cation intermediate and hydrogen radical abstraction mechanism was found in alkane hydroxylation with metalloporphyrins.[1104]

Oxidation of Higher Alkanes. New, selective and active catalyst systems were also developed for the metal-catalyzed oxidation of higher alkanes. A carboxylate-bridged dimanganase(III) complex was shown to be active in the transformation of cyclohexane and adamantane to alcohols and ketones with *tert*-BuOOH and O_2.[1105] $OsCl_3$ is effective in the oxygenation of alkanes with H_2O_2 in the presence of acetonitrile and a N-containing heterocycle.[1106] Ti-containing mesoporous molecular sieves modified by trimethylsilylation exhibit enhanced catalytic activity in the oxidation of hexane to hexanols with H_2O_2.[1107] Similar selective hydroxylation was achieved with a mononuclear iron carboxylate catalyst immobilized on a modified silica surface,[1108] dinuclear Fe(II) macrocyclic complexes by O_2 with H_2S as a two-electron reductant,[1109] and synthetic amavadine complexes.[1110] In contrast, Mn-substituted peroxometalates are effective catalyst for the oxidation of alkanes to ketones with ozone.[1111] Evidence for a Fe(V)=O active species[1112] and a low-spin Fe(III)OOH intermediate[1113] was found for the stereospecific alkane hydroxylation by various nonhem iron complex catalysts.

Alkanes undergo oxidation with H_2O_2 in the presence of methylrhenium trioxide and pyrazine-2-carboxylic acid to give alcohols as main products. The oxygen insertion is stereospecific with retention of configuration.[1056] Stoichiometric oxidation is mediated by a highly oxidizing Ru(VI)–oxo complex using *tert*-BuOOH.[1114] The reaction is highly stereoselective: *cis*-decalin gives exclusively *cis*-decalol in high yield. The Ru on C/peracetic acid system affords the corresponding ketones and alcohols highly efficiently at room temperature in CF_3COOH.[1115] Cyclohexane, for example, gives cyclohexyl trifluoroacetate and cyclohexanone with 90% conversion and 90% selectivity (85 : 15). Urea–H_2O_2 in CF_3COOH is a convenient oxidant to form trifluoroacetates of cyclic alcohols without a transition-metal catalyst.[1116] Adamantane can be oxidized to yield hydroxylated products with in situ–formed peracids or by metal-assisted aerobic oxidation using $M(acac)_2$ (M = Pd, Co, Fe, Ni, Mn, Cu).[1117] Oxyfunctionalization of cyclohexane with *tert*-BuOOH was achieved using Mn(II) complexes included in Y zeolite.[1118] Oxidation of alkanes (cyclohexane, *n*-hexane, adamantane) was performed with $Ir(acac)_3$ supported on a carbon fiber anode.[1119] Strong electrophilic oxygen species generated from H_2O was suggested to be involved in oxygenation. In contrast to an oxenoid insertion, a novel mechanism, a free-radical process involving acylperoxyl radicals, was found to be operative in the oxidation of alkanes by aromatic peracids.[1120]

Efficient oxidation of alkanes with molecular oxygen can be attained using *N*-hydroxyphthalimide combined with $Co(acac)_n$ (n = 2,3).[1121] According to a new concept, photocatalyzed selective oxidation of small hydrocarbons at room temperature is carried out over alkali or alkaline-earth ion-exchanged zeolites.[1122]

Information of the Gif system has been summarized,[1055,1123] and new results, including new oxidants such as bis(trimethylsilyl) peroxide,[1124] the synergistic oxidation of saturated hydrocarbons and H_2S,[1125] studies with the Fe^{3+}–picolinate complex encapsulated within zeolites,[1126] and the use of Udenfriend's system under Gif conditions[1127] were disclosed. Gif-type oxidations were found to be moderately stereoselective.[1128] Iron/zinc-containing species involved in Gif-type chemistry were synthesized, and their reactivity and catalytic behavior were studied.[1129]

The radical or nonradical chemistry of the Fenton-like systems, including the Fenton reaction and the Fe(III)—ROOH (or H_2O_2) system, is still hotly debated.[1123,1130–1134] New information supporting a free-radical mechanism with carbon- and oxygen-centered radicals[1135–1138] and new evidence for a nonradical process[1139,1140] have been published.

A new and very successful approach is the application of molecular sieve catalysts with designed, atomically engineered active centers for the oxidation of alkanes in air.[1141–1144] A few percent of the Al in framework-substituted molecular sieves is replaced by transition-metal ions, thereby generating the catalytically active sites, which, in combination with shape-selective properties, allow regio- and shape-selective oxyfunctionalization. To prevent leaching of the metals the transformations are usually carried out at low conversions. ALPO molecular sieve samples containing isolated, four-coordianted Co(III) or Mn(II) ions substituted into the framework work as regioselective catalyst in the oxidation of linear alkanes to the corresponding monocarboxylic acids.[1145] CoALPO-18 catalyzes the direct conversion of hexane to adipic acid in substantial amounts.[1146] Cyclohexane can be transformed into various products depending on the catalyst applied. FeALPO-5 is quite selective in the aerial oxidation of cyclohexane to cyclohexanol,[1147] whereas adipic acid is the main product over FeALPO-31.[1148] A monohydroxyco-balt complex immobilized into MCM-41 shows high selectivity in the formation of cycylohexanone.[1149] Selective oxidation of cyclohexane, though, is still a challenge.[1150]

Oxidations Relevant to Industrial Processes. The oxidation of propane to acrylic acid has been reviewed, including the three leading catalysts—vanadium pyrophosphate, heteropoly acids and salts, and mixed metal oxides—and discussing the oxidation pathways and structural properties.[1151] New results were also disclosed on the oxidation of butane to maleic anhydride. The application of membrane reactors allows the use of very high butane concentrations.[1152] The addition of CO_2 to the reaction mixture gives rise to a significantly improved yield,[1153] whereas the VPO catalysts can be completely rehydrated in the solid phase, allowing reuse of the spent catalyst.[1154] Evidence for two competing mechanisms, namely, the alkoxide and the olefinic intermediate routes, was found by labeling studies.[1155] Developments of Dupont's new circulating fluidized-bed technology including commercialization have been summarized.[1156]

9.6.2. Oxidation of Alkenes

Epoxidation. Alkene epoxidation is in the focus of vigorous research activity because of the high chemical versatility of epoxides for laboratory and industrial manufacturing of a wide variety of products via highly selective reactions. The use of hydrogen peroxide, a cheap and environmentally friendly oxidizing agent, has been attracted particular attention. Numerous new catalysts were developed that are active and selective in the transformation of various alkenes to the corresponding epoxides. Information accumulated with respect to mechanistic studies and new catalysts and methods, including the heterogenization of homogeneous catalysts, are summarized in review papers.[1055–1057,1157–1169]

Epoxidation by dioxiranes can be carried out by applying ketones and Oxone, thereby generating dioxiranes the active oxidizing agent. The method is highly selective for *trans*-alkenes. More recently, however, epoxidation of *cis*-alkenes was achieved with high ee values by applying the fructose-derived **112** chiral dioxirane.[1170] Epoxidation by Oxone may also be mediated by iminium salts.[1171–1175] Dioxiranes derived from chiral ketones are promising reagents for catalytic asymmetric epoxidation.[1176,1177] The best dioxirane precursors are ketones with an electronegative heteroatom α to the carbonyl group.[1178–1180] Fluorine is a particularly strong activator (see, e.g., **113**).[1180–1183]

112   **113**

Various new observations, including the higher reactivity of *cis*-alkenes, isotope effects, and *syn* stereospecificity, confirm a concerted mechanism with a transition state of spiro geometry. Theoretical studies provided support to this view.[1184–1188] Radical pathways were shown by computational methods and experiments to also be possible.[1189,1190]

Alkenes may also be epoxidized by perborates or percarbonates.[1191] When applying H_2O_2 with sodium or ammonium bicarbonate, the active oxidant is the HCO_4^- ion.[1192] H_2O_2 can be used in combination with alumina,[1193] trifluoroacetone at high pH,[1194] perfluoroacetone,[1195] or a mixture of perfluoroacetone and hexafluoro-2-propanol.[1196] Silica functionalized with octafluoroacetophenone[1197] was also tested. Fluorinated alcohols were shown to be effective solvents for uncatalyzed epoxidations with aqueous H_2O_2.[1198,1199] Effective epoxidation could be performed in supercritical CO_2 using O_2 in the presence of aldehydes as sacrificial oxidants.[1200]

Newer studies provide numerous examples of the use of metal catalysts exhibiting high activity and selectivity in epoxidation with various oxidants mainly with H_2O_2.[1201–1210] Catalysts developed by combinatorial methods[1211,1212] and heterogenized catalysts[1213] were also applied.

Highly enantioselective epoxidation of unfunctionalized alkenes was developed by using chiral metalloporphyrin catalysts.[1214–1218] Remarkable anion axial ligand effects were observed with [Fe(TPFPP)X] complexes (X = triflate, perchlorate, nitrate).[1219] Hexafluoroacetone was found to be an efficient cocatalyst with H_2O_2,[1220] and alkenes could be epoxidized by ozone at ambient temperature.[1221]

Since the porphyrin ring is liable to oxidative self-destruction and catalyst recovery is problematic, studies have focused on the development of porphyrins with electron-withdrawing substituents[1222] and immobilization on organic solids. Silicas,[1100,1223] Nb-doped silicates,[1224] molecular sieves,[1225,1226] zirconium phosphate,[1227] and zeolite X[1101] were used. The problems encountered with these immobilized catalysts are catalyst leaching and lower activity and selectivity. In contrast, polymer-supported Ru porphyrins are of high activity and selectivity and can be recycled.[1228] Reactive metal–oxo (M=O) complexes are involved as reactive intermediates in these oxidations[1222,1229] although radical intermediates were also suggested.[1230,1231]

Methyltrioxorhenium (MTO) is a highly efficient alkene epoxidation catalyst in the presence of H_2O_2.[1056,1162,1232,1233] Addition of a catalytic amount of pyridine to the MTO/H_2O_2 system greatly improves the reactivity; specifically, it enhances the rate, prevents the hydrolysis of the epoxide to diol and promotes the catalyst lifetime.[1234] 3-Cyanopyridine proved to be the best additive.[1235,1236] This system has a high selectivity and wider scope, can be applied for the synthesis of acid-sensitive epoxides, requires less solvent, and involves easier product workup and isolation. Pyrazol is a similarly effective additive.[1237] Further developments are the use of MTO supported on silica tethered with polyethers, which catalyzes the reaction with increased selectivity without the use of organic solvent.[1238] MTO with 85% H_2O_2 is highly efficient in the presence of NaY zeolite.[1239] The oxidative species is located inside the zeolite supercages, thereby preventing the Lewis acid–assisted hydrolysis of the epoxide to the diol. In addition to H_2O_2, the urea–H_2O_2 adduct,[1240] bis(trimethylsilyl)peroxide,[1241] and sodium percarbonate[1242] were successfully applied. ReO_4 supported on alumina is active in the epoxidation of cycloalkenes with anhydrous H_2O_2.[1243] The MTO-catalyzed epoxidation of terminal alkenes is significantly improved by conducting the transformation in trifluoroethanol.[1244]

Mechanistic studies showed that epoxidation catalyzed by MTO/H_2O_2 occurs as direct oxygen transfer via transition states of spiro structure likely involving both mono- and bisperoxo complexes.[1233,1245–1247]

Since discovery of the selective epoxidation catalyst titanium silicalite (TS-1),[184,185] numerous studies appeared in attempts to acquire information for a better understanding of the system[1160] and to improve its properties. New information by X-ray absorption fine-structure (EXAFS) measurements established that the catalyst has single-site active centers [tripodally attached (Si–O)$_3$–TiOH groups] allowing investigators to discount the participation of a three-coordinated titanyl group.[1248] The problem of intracrystallite diffusion limitations owing to the small size of the zeolite micropores was solved by developing a mesoporous titanium-containing zeolite,[1249] which exhibits significantly higher activity in the epoxidation of cyclohexene. Other Ti-containing zeolites, such as Ti-Beta,[1250] Ti-ITQ-7,[1251] and Ti-MWW,[1252] were also successfully prepared and used. The latter catalyst is unique in its ability to selectively epoxidize *trans* isomers from a mixture of *cis–trans* alkenes. The small amount of Al used in the preparation of these catalysts as crystallization-promoting agent gives rise to the formation of acid sites, decreasing the selectivity by promoting the ring opening of the product epoxide. A dramatic improvement of epoxide selectivity was reported by using a

TiAl-Beta catalyst ion-exchanged with quaternary ammonium salts.[1253] This is due to the selective poisoning of the acid sites without suppressing the oxidation activity of Ti sites.

Sol-gel-derived titania–silica mixed oxides are excellent epoxidation catalysts with alkyl hydroperoxides. They exhibit outstanding activity and selectivity in the epoxidation of bulky olefins[1254–1257] when compared to ultra-large-pore Ti-substituted molecular sieves.[1258] Supported Ti catalysts prepared by using amorphous silica or MCM-41 are also useful epoxidation catalysts.[1259–1263] A highly active single-site epoxidation catalyst was prepared by reacting [Cp_2TiCl_2] with the silanol groups of MCM-41.[1248] Modification (replacement of one Si by Ge), thereby changing hydrophobicity, resulted in a further improvement in catalyst performance. The synthesis of Ti-containing polysiloxane,[1264] metallosilicates prepared by using dendrimeric silanes,[1265] and a titanium-silsesquioxane catalyst with good characteristics, optimized by parallel combinatorial chemistry,[1266] was reported.

Gas-phase epoxidation of propylene with O_2/H_2 mixtures was accomplished over Ag[1267] or Au[1268] catalysts dispersed on TS-1 or other Ti-containing supports and Ti-modified high-silica zeolites.[1269] Sodium ions were shown to be beneficial on the selectivity of propylene epoxidation with H_2O_2 over titanium silicalite.[1270] A chromia–silica catalyst is active in the visible light-induced photoepoxidation of propylene by molecular oxygen.[1271]

New metalloporphyrins with electrondeficient porphyrin ligands are efficient either in the homogeneous phase[1096,1097,1099] or in immobilized form.[1096,1099–1102]

Metallosalen complexes have assumed considerable importance as asymmetric epoxidation catalysts.[1159,1272–1274] This catalyst system is widely applicable for unfunctionalized olefins and optimal for *cis* internal alkenes, particularly for cycloalkenes. The most useful monooxygen sources are PhIO, NaOCl, and aromatic peracids. The latter may be generated in situ.[1275] Good results were also reported for *trans*-alkenes using Mn(III) salen[1276] and Ru(II) salen[1277] complexes with naphthalene ligands. Recycling of the complex catalyst may also be possible by using the [bmim][PF_6] ionic liquid.[1278] Fluorous chiral salen manganese complexes were also developed and applied in enantioselective epoxidation under fluorous biphasic conditions, affording good epoxide yields (77–98%) and ee values ranging from 50 to 87%.[1279]

The exact mechanism of the oxygen transfer step is still debated,[1280] but there is general agreement that the transformation involves both Mn(III) and Mn(V) oxo species. The role of the stoichiometric oxidant was proved to form a Mn(V) oxo species.[1281] New experimental results show that the Jacobsen–Katsuki epoxidation occurs by direct attack of the alkene at the oxo ligand of the Mn(salen) complex and is likely to involve radical intermediates.[1282] More recent theoretical calculations provided new insight into the origin of the reactivity and selectivity of Mn(salen) complexes, emphasizing the importance of the axial ligand and ensuring higher rates and increased enantioselectivity.[1283] In a 2001 study,[1284] three specific features were connected to the origin of chirality: (1) the chiral diimine bridge induces folding of the salen ligand to form a chiral pocket; (2) bulky groups at the

3,3′-positions, which causes a preferential approach from the side of the aromatic rings; and (3) π participation of the olefinic bond. Laser flash photolysis[1285] and dual-mode EPR studies[1286] gave valuable information recently.

The presence of both Mn(III) and Mn(V) oxo species in the process permitting the formation of oxo-bridged dimers is harmful for the catalysis. Immobilization offers a solution to this problem, namely, the site isolation of the complex on a surface. Promising results of heterogenization of various salen complexes were obtained with inorganic solids[1167,1287–1289] and polymers.[1165,1167,1290–1294]

The regioselectivity in the ene reaction has been reviewed.[1295] The intrazeolitic photooxygenation of alkenes to produce hydroperoxides has attracted considerable attention primarily due to the significant enhancement of product selectivity.[1296,1297]

Bis-hydroxylation. Molecular oxygen or air was used as the terminal oxidant in the osmium-catalyzed oxidation of alkenes to form *cis*-diols with high conversions at low catalyst amount.[1298] In a triple catalytic system using H_2O_2 as the terminal oxidant, a cinchona alkaloid ligand has a dual function—it provides stereocontrol and acts as reoxidant via its *N*-oxide. The formation of the latter is catalyzed by a biomimetic flavin component.[1299]

The toxicity of osmium, the problematic recovery, and the high cost of osmium and the ligands prevent the industrial use of this efficient method. Soluble and insoluble polymer-supported[1300–1302] and silica-anchored[1303] ligands have been developed. Microencapsulated[1304,1305] and silica-bound[1306] osmium reagents ensure complete osmium recovery. Strong evidence for the [3 + 2] mechanism was provided by kinetic isotope effects[1307] and by high-level density functional theory calculations.[1308]

Vinylic Oxidation. Various alkenes are oxidized to the corresponding ketones using *tert*-BuOOH in the presence of Pd(II) catalysts bearing perfluorinated ligands in a fluorous biphasic system.[1309] The catalyst can be reused, but progressively longer reaction times are required.

It was observed that the substitution of a methyl group on the terminal carbon of styrene or adamantane [Eq. (9.185)] switches the regioselectivity from Markovnikov to anti-Markovnikov in the Wacker oxidation:[1310]

R = H	1	
R = Me	1	2.2

$$(9.185)$$

The availability of an allylic hydrogen through the formation of the **114** agostic intermediate or an enyl (σ + π) complex was suggested to account for the observed selectivity:

114

To overcome the problems associated with the Wacker process, namely, the corrosive nature of the catalyst and the formation of chlorinated byproducts when higher alkene homologs are oxidized, new catalysts and heterogeneous catalyst systems were developed.

In new studies heteropoly acids as cocatalysts were found to be very effective in combination with oxygen in the oxidation of ethylene.[1311] Addition of phosphomolybdic acid to a chloride ion-free Pd(II)–Cu(II) catalyst system results in a great increase in catalytic activity and selectivity.[1312] Aerobic oxidation of terminal alkenes to methyl ketones can be performed with Pd(OAc)$_2$[1313] or soluble palladium complexes.[1314] Modified cyclodextrins accelerates reaction rates and enhance selectivities in two-phase systems under mild conditions.[1315,1316]

Heterogeneous catalysts that exhibit good characteristics are silica-supported mixed Mo–V heteropoly acids and their Pd salts,[1317] Pd on titania,[1318] supported H$_3$PMO$_{12}$O$_{40}$,[1319] and heteropoly acids and salts with Pd(OAc)$_2$[1320] or PdCl$_2$.[1321]

Selective oxidation of ethylene to acetaldehyde was carried out over carbon-supported Pd and Pt membrane catalysts.[1322] The concept of supported liquid-phase catalysis was also successfully applied in the Wacker oxidation.[1323] The Wacker reaction can be performed in alcohol–supercritical CO$_2$.[1324] CO$_2$ as cosolvent accelerates reaction rates and remarkably affects the selectivity towards methyl ketone in the presence of an alcohol.

Oxidative Cleavage. Efficient oxidative cleavage of various alkenes to carboxylic acids, including the transformation of cyclohexene to adipic acid, is accomplished with H$_2$O$_2$ catalyzed by a peroxotungsten complex.[1325] Clean synthesis of adipic acid via the oxidation of cyclohexene with H$_2$O$_2$ can also be accomplished by using Na$_3$WO$_3$ and a phase-transfer catalyst[1326] or over peroxytunstate–organic complex catalysts.[1327] Various oxidation protocols were developed to cleave alkenes to aldehydes with RuCl$_3$ catalyst. When combined with Oxone or NaIO$_4$ in different solvents (Cl$_3$CH–H$_2$O or 1,2-dichloroethane–H$_2$O), it is useful to transform both aliphatic and aromatic alkenes to the corresponding aldehydes in excellent yields.[1328]

Ozonolysis of cyclic olefins in the presence of carbonyl compounds gives the corresponding cross-ozonides.[1329] In the ozonation of 1,2,4,5-tetramethyl-1,4-cyclohexadiene, oxidative dehydrogenation (formation of 1,2,4,5-tetramethylbenzene) was found to compete with oxidative cleavage because of steric hindrance.[1330] Secondary ozonides (the **76** 1,2,4-trioxolanes) are formed in high yields in the gas-phase, low-temperature ozonation of terminal and disubstituted alkenes.[1331]

Kinetic measurements and theoretical calculations provided evidence that carbonyl oxide intermediates (**75**) are an important source of HO radicals in gas-phase ozonolysis.[1332,1333] New studies with a wide range of variously substituted alkenes,[1334] including tetraphenylethylene,[1335] suggest that ozonation proceeds via an electron transfer mechanism.

Allylic Oxidation. The molybdovanadophosphate–hydroquinone–O_2 system is an efficient reoxidation system in the acetoxylation of cyclic alkenes with $Pd(OAc)_2$ to form 3-acetoxy-1-cycloalkenes in good yields.[1336,1337] Low-valence Pd nanoclusters afford the highly selective oxidation of cyclic alkenes into allylic esters.[1338] Oxidation of alkenes were performed to yield α,β-unsaturated ketones as the main product with iridium nitro complexes[1339] or photoexcited Pd and Fe porphyrins caged into a Nafion membrane.[1340] An organic sensitizer within Ti zeolites as photocatalyst is able to catalyze the oxidation of cyclohexene to the corresponding α,β-unsaturated alcohol and ketone.[1341] Similar oxidations were performed with Fe,[1342] Mn,[1343,1344] Co,[1345] and Ru[1346] complexes. With increasing ratio of cyclohexene to the Ru=O complex, 2-cyclohexene-1-ol becomes the dominant product.

Copper-catalyzed allylic oxidation allows the functionalization of unactivated alkenes into chiral allylic carboxylates.[1347] The use of oxazoline-containing ligands give good enantioselectivities, but the reaction is extremely slow.[1348–1351] Chiral bipyridine complexes, in turn, are much more active and give products in good yields and enantioselectivities up to 70% when applied in benzoyloxylation of cycloalkenes with *tert*-butyl perbenzoate.[1352,1353]

The technology of the acrylic acid manufacturing process developed in Japan has been disclosed.[1354]

9.6.3. Oxidation of Alkynes

Novel practical methods using various reagents, such as [Co(OAc)Br],[1355] sulfur trioxide,[1356] or *cis*-dioxoruthenium complexes,[1357] were developed to transform alkynes to 1,2-diketones. Radical-catalyzed aerobic oxidation using *N*-hydroxyphthalimide combined with a transition metal (Co, Cu, or Mn) affords α,β-acetylenic ketones in good yields.[1358] Oxidation by the HOF• acetonitrile complex yields diketones, ketoepoxides, or cleavage products.[1359] Ozonolysis of acetylenes combined with trapping techniques affords to isolate various derivatives.[1360,1361] New information for the ozonolysis of acetylene was acquired by quantum-chemical investigatons.[1362]

9.6.4. Oxidation of Aromatics

Ring Oxygenation. Hydroxylation of benzene and toluene with H_2O_2 can be catalyzed by vanadium(V) polyoxometalates.[1363–1365] H-Ga-FER zeolites can be used in the gas phase to perform hydroxylation by N_2O.[1366] Various Cu-containing catalysts are effective in the liquid phase using molecular oxygen and ascorbic acid as a reductant.[1367,1368] Direct hydroxylation of benzene was performed with oxygen and hydrogen on silica-supported noble metals modified with V_2O_5[1369] or $V(acac)_3$.[1370] High selectivity was found for Cu-substituted molecular sieves using H_2O_2.[1371] Methylrhenium trioxide, in turn, is effective in the oxidation of alkyl-substituted benzenes and naphthalenes to quinones.[1056,1372]

Benzylic Oxidation. Enantioselective hydroxylation of benzylic C—H bonds was performed by chiral dioxoruthenium(VI) porphyrin.[1373] The highest ee (76%) was attained in the catalytic oxidation of 4-ethyltoulene with 2,6-dichloropyridine *N*-oxide as terminal oxidant. A cytochrome-450-type monooxygenase activity was observed with the microorganism *Bacillus megaterium* to mediate highly chemoselective and enantioselective hydroxylation.[1374] Oxidation with *tert*-BuOOH by Ru-exchanged K10[1375] or Ni(acac)$_2$[1376] can give aryl *tert*-butyl peroxides or hydroperoxides, respectively.

Oxidation at the benzylic position to the corresponding ketones is achieved by metal ion catalysts. Aldehydes and ketones are formed in similar amounts in the first-ever application of corrole metal complexes.[1377] Mn(III) porphyrins with Ph$_4$PHSO$_5$ as oxidant afford better ketone yields.[1103,1378] The alcohol : ketone chemoselectivity was found to be determined by the association of the alcohol intermediate to the catalyst prior to the second oxidation step leading to ketone.[1378] Highly selective ketone formation can also be performed by other manganese complexes and *tert*-BuOOH,[1105,1344] and using Mn-substituted polyoxometalates with ozone.[1111] Cr-MCM-41[1379] and [RuCl$_2$(PPh$_3$)$_3$][1115] combined with *tert*-BuOOH are also effective in ketone formation. Tetralin and indan can be transformed into the corresponding ketones in oxidation by the O$_2$/3-methylbutanal system in the presence of the same M(acac)$_2$ catalysts used in the hydroxylation of alkanes.[1117] Ketone formation can be accomplished by the photochemical oxidation with nitropyridinium salts[1380] and with *m*-CPBA and oxygen.[1381] An easy-to-   perform one-pot reaction for the room-temperature synthesis of aromatic ketones and aldehydes is possible by Ce(IV) triflate.[1382] A unique selectivity is observed in the oxidation with supported KMnO$_4$; in contrast to the oxidation in homogeneous solution, C—C bond cleavage does not occur.[1383]

A new reagent, *N*-hydroxyphthalimide combined with Co(acac)$_n$ ($n = 2,3$), transforms alkylbenzenes to ketones, whereas methylbenzenes give the corresponding carboxylic acids.[1121] Phthalimide *N*-oxyl was found to be the key intermediate. Novel oxoperoxo Mo(VI) complexes mediate the cost-effective and environmentally benign oxidation of methylbenzenes to carboxylic acids.[1384] Similar green oxidation of *p*-xylene to terephthalic acid was reported by using a Ru-substituted heteropolyanion.[1385]

9.6.5. Ammoxidation

Ammoxidation processes require catalysts with rather complex composition and physical makeup. Our understanding of their actions is continually increasing,[1054,1386–1388] and this results in the development of even more complex and more efficient catalysts. The two major types of catalyst systems studied most are V–Sb oxides and V–Te–Nb–Mo oxides,[1054] with information published more recently.[1389–1393] New, improved preparation methods and characterization were disclosed for other catalysts, including Re-containig oxides,[1394,1395] Fe antimonate,[1396] and multicomponent BiPO$_4$.[1397] The presence of Brønsted acid sites acting in synergy with redox sites in Ga-modified MFI zeolites can catalyze the ammoxidation of propane.[1398]

REFERENCES

1. G. A. Olah, D. G. Parker, and N. Yoneda, *Angew. Chem., Int. Ed. Engl.* **17**, 909 (1978).

2. R. A. Sheldon and J. K. Kochi, *Adv. Catal.* **25**, 272 (1976).

3. J. A. Howard, in *Free Radicals*, Vol. 2, J. K. Kochi, ed., Wiley, New York, 1973, Chapter 12, p. 3.

4. R. Hiatt, in *Organic Peroxides*, Vol. 2, D. Swern, ed., Wiley-Interscience, New York, 1971, Chapter 1, p. 1.

5. R. A. Sheldon and J. K. Kochi, *Metal-Catalyzed Oxidations of Organic Compounds*, Academic Press, New York, 1981, Chapter 2, p. 17.

6. R. Criegee, *Chem. Ber.* **77**, 22 (1944).

7. G. S. Fisher, J. S. Stinson, and L. A. Goldblatt, *J. Am. Chem. Soc.* **75**, 3675 (1953).

8. W. Pritzkow and K.-H. Gröbe, *Chem. Ber.* **93**, 2156 (1960).

9. D. E. Winkler and G. W. Hearne, *Ind. Eng. Chem.* **53**, 655 (1961).

10. D. L. Allara, T. Mill, D. G. Hendry, and F. R. Mayo, *Adv. Chem. Ser.* **76**, 40 (1968).

11. F. R. Mayo, *Acc. Chem. Res.* **1**, 193 (1968).

12. R. A. Sheldon and J. K. Kochi, *Metal-Catalyzed Oxidations of Organic Compounds*, Academic Press, New York, 1981, Chapter 11, p. 340.

13. E. R. Bell, F. H. Dickey, J. H. Raley, F. F. Rust, and W. E. Vaughan, *Ind. Eng. Chem.* **41**, 2597 (1949).

14. G. Sosnovsky and D. J. Rawlinson, in *Organic Peroxides*, Vol. 2, D. Swern, ed., Wiley-Interscience, New York, 1971, Chapter 2, p. 153.

15. F. F. Rust, *J. Am. Chem. Soc.* **79**, 4000 (1957).

16. R. Criegee and P. Ludwig, *Erdöl Kohle* **15**, 523 (1962).

17. N. M. Emanuel, Z. K. Maizus, and I. P. Skibida, *Angew. Chem., Int. Ed. Engl.* **8**, 97 (1969).

18. N. R. Foster, *Appl. Catal.* **19**, 1 (1985).

19. H. D. Gesser, N. R. Hunter, and C. B. Prakash, *Chem. Rev.* **85**, 235 (1985).

20. H. D. Gesser and N. R. Hunter, in *Methane Conversion by Oxidative Processes: Fundamental and Engineering Aspects*, E. E. Wolf, ed., Van Nostrand Reinhold, New York, 1992., Chapter 12, p. 403.

21. R. Pitchai and K. Klier, *Catal. Rev.-Sci. Eng.* **28**, 13 (1986).

22. M. S. Scurrell, *Appl. Catal.* **32**, 1 (1987).

23. G. J. Hutchings, M. S. Scurrell, and J. R. Woodhouse, *Chem. Soc. Rev.* **18**, 251 (1989).

24. M. Yu. Sinev, V. N. Korshak, and O. V. Krylov, *Russ. Chem. Rev.* (Engl. transl.) **58**, 22 (1989).

25. A. Bielanski and J. Haber, *Oxygen in Catalysis*, Marcel Dekker, New York, 1991, Chapter 9, p. 423.

26. M. J. Brown and N. D. Parkyns, *Catal. Today* **8**, 305 (1991).

27. R. D. Srivastava, P. Zhou, G. J. Stiegel, V. U. S. Rao, and G. Cinquegrane, in *Catalysis, A Specialist Periodical Report*, Vol. 9, J. J. Spivey, senior reporter, Royal Society of Chemistry, Thomas Graham House, Cambridge, 1992, Chapter 4, p. 183.

28. P. S. Yarlagadda, L. A. Morton, N. R. Hunter, and H. D. Gesser, *Ind. Eng. Chem. Res.* **27**, 252 (1988).

29. S. Ahmed and J. B. Moffat, *Catal. Lett.* **1**, 141 (1988).

30. T. R. Baldwin, R. Burch, G. D. Squire, and S. C. Tsang, *Appl. Catal.* **75**, 153 (1991).

31. Z. Sojka, R. G. Herman, and K. Klier, *J. Chem. Soc., Chem. Commun.* 185 (1991).

32. K. Otsuka and M. Hatano, *Chem. Lett.* 2397 (1992).

33. K. Otsuka, T. Komatsu, K. Jinno, Y. Uragami, and A. Morikawa, in *Proc. 9th Int. Congress Catalysis*, Calgary, 1988, M. J. Phillips and M. Ternan, eds., The Chemical Institute of Canada, Ottawa, 1988, Vol. 2, p. 915.

34. N. D. Spencer and C. J. Pereira, *J. Catal.* **116**, 399 (1989).

35. N. D. Spencer, *J. Catal.* **109**, 187 (1988).

36. D. Miceli, F. Arena, A. Parmaliana, M. S. Scurrell, and V. Sokolovskii, *Catal. Lett.* **18**, 283 (1993).

37. A. Parmaliana, F. Frusteri, A. Mezzapica, M. S. Scurrell, and N. Giordano, *J. Chem. Soc., Chem. Commun.* 751 (1993).

38. B. Kartheuser and B. K. Hodnett, *J. Chem. Soc., Chem. Commun.* 1093 (1993).

39. A. Parmaliana, F. Frusteri, D. Miceli, A. Mezzapica, M. S. Scurrell, and N. Giordano, *Appl. Catal.* **78**, L7 (1991).

40. Q. Sun, R. G. Herman, and K. Klier, *Catal. Lett.* **16**, 251 (1992).

41. J. H. Lunsford, in *Methane Conversion*, Studies in Surface Science and Catalysis, Vol. 36, D. M. Bibby, C. D. Chang, R. F. Howe, and S. Yurchak, eds., Elsevier, Amsterdam, 1988, p. 359.

42. V. Amir-Ebrahimi and J. J. Rooney, *J. Mol. Catal.* **50**, L17 (1989).

43. E. MacGiolla Coda and B. K. Hodnett, in *New Developments in Selective Oxidation*, Studies in Surface Science and Catalysis, Vol. 55, G. Centi and F. Trifirò, eds., Elsevier, Amsterdam, 1990, p. 459.

44. T. R. Baldwin, R. Burch, G. D. Squire, and S. C. Tsang, *Appl. Catal.* **74**, 137 (1991).

45. N. R. Hunter, H. D. Gesser, L. A. Morton, P. S. Yarlagadda, and D. P. C. Fung, *Appl. Catal.* **57**, 45 (1990).

46. R.-S. Liu, M. Iwamoto, and J. H. Lunsford, *J. Chem. Soc., Chem. Commun.* 78 (1982).

47. H.-F. Liu, R.-S. Liu, K. Y. Liew, R. E. Johnson, and J. H. Lunsford, *J. Am. Chem. Soc.* **106**, 4117 (1984).

48. M. M. Khan and G. A. Somorjai, *J. Catal.* **91**, 263 (1985).

49. K. J. Zhen, M. M. Khan, C. H. Mak, K. B. Lewis, and G. A. Somorjai, *J. Catal.* **94**, 501 (1985).

50. Y. Barbaux, A. R. Elamrani, E. Payen, L. Gengembre, J. P. Bonnelle, and B. Grzybowska, *Appl. Catal.* **44**, 117 (1988).

51. F. Solymosi, I. Tombácz, and Gy. Kutsán, *J. Chem. Soc., Chem. Commun.* 1455 (1985).

52. S. Ahmed and J. B. Moffat, *Appl. Catal.* **40**, 101 (1988).

53. S. Kasztelan and J. B. Moffat, *J. Catal.* **116**, 82 (1989).

54. H. D. Gesser, N. R. Hunter, and P. A. Das, *Catal. Lett.* **16**, 217 (1992).

55. Q. Sun, J. I. Di Cosimo, R. G. Herman, K. Klier, and M. M. Bhasin, *Catal. Lett.* **15**, 371 (1992).

56. K. Ogura and M. Kataoka, *J. Mol. Catal.* **43**, 371 (1988).

57. K. Ogura, C. T. Migita, and M. Fujita, *Ind. Eng. Chem. Res.* **27**, 1387 (1988).

58. C. T. Migita, S. Chaki, and K. Ogura, *J. Phys. Chem.* **93**, 6368 (1989).

59. T. Suzuki, K. Wada, M. Shima, and Y. Watanabe, *J. Chem. Soc., Chem. Commun.* 1059 (1990).

60. R. Burch, G. D. Squire, and S. C. Tsang, *J. Chem. Soc., Faraday Trans. 1* **85**, 3561 (1989).

61. G. A. Olah, *Acc. Chem. Res.* **20**, 422 (1987).

62. G. A. Olah, N. Yoneda, and D. G. Parker, *J. Am. Chem. Soc.* **99**, 483 (1977).

63. E. Gretz, T. F. Oliver, and A. Sen, *J. Am. Chem. Soc.* **109**, 8109 (1987).

64. L.-C. Kao, A. C. Hutson, and A. Sen, *J. Am. Chem. Soc.* **113**, 700 (1991).

65. M. N. Vargaftik, I. P. Stolarov, and I. I. Moiseev, *J. Chem. Soc., Chem. Commun.* 1049 (1990).

66. R. A. Periana, D. J. Taube, E. R. Evitt, D. G. Löffler, P. R. Wentrcek, G. Voss, and T. Masuda, *Science* **259**, 340 (1993).

67. G. A. Olah, N. Yoneda, and D. G. Parker, *J. Am. Chem. Soc.* **98**, 5261 (1976).

68. D. L. Heywood, B. Phillips, and H. A. Stansbury, Jr., *J. Org. Chem.* **26**, 281 (1961).

69. Y. Mazur, *Pure Appl. Chem.* **41**, 145 (1975).

70. J. Fossey, D. Lefort, M. Massoudi, J.-Y. Nedelec, and J. Sorba, *Can. J. Chem.* **63**, 678 (1985).

71. N. Takaishi, Y. Fujikura, and Y. Inamoto, *Synthesis* 293 (1983).

72. M. Tori, M. Sono, and Y. Asakawa, *Bull. Chem. Soc. Jpn.* **58**, 2669 (1985).

73. W. Müller and H.-J. Schneider, *Angew. Chem., Int. Ed. Engl.* **18**, 407 (1979).

74. H.-J. Schneider and W. Müller, *Angew. Chem., Int. Ed. Engl.* **21**, 146 (1982).

75. H.-J. Schneider and W. Müller, *J. Org. Chem.* **50**, 4609 (1985).

76. U. Frommer and V. Ullrich, *Z. Naturforsch. B* **26**, 322 (1971).

77. N. C. Deno, E. J. Jedziniak, L. A. Messer, M. D. Meyer, S. G. Stroud, and E. S. Tomezsko, *Tetrahedron* **33**, 2503 (1977).

78. D. Lefort, J. Fossey, M. Gruselle, J.-Y. Nedelec, and J. Sorba, *Tetrahedron* **41**, 4237 (1985).

79. P. D. Bartlett, *Rec. Chem. Prog.* **11**, 47 (1950).

80. S. Rozen, M. Brand, and M. Kol, *J. Am. Chem. Soc.* **111**, 8325 (1989).

81. R. W. Murray, *Chem. Rev.* **89**, 1187 (1989).

82. W. Adam, R. Curci, and J. O. Edwards, *Acc. Chem. Res.* **22**, 205 (1989).

83. W. Adam and L. Hadjiarapoglou, *Top. Curr. Chem.* **164**, 45 (1993).

84. R. Mello, M. Fiorentino, O. Sciacovelli, and R. Curci, *J. Org. Chem.* **53**, 3890 (1988).

85. R. Mello, M. Fiorentino, C. Fusco, and R. Curci, *J. Am. Chem. Soc.* **111**, 6749 (1989).

86. R. W. Murray, R. Jeyaraman, and L. Mohan, *J. Am. Chem. Soc.* **108**, 2470 (1986).

87. R. Mello, L. Cassidei, M. Fiorentino, C. Fusco, and R. Curci, *Tetrahedron Lett.* **31**, 3067 (1990).

88. D. D. DesMarteau, A. Donadelli, V. Montanari, V. A. Petrov, and G. Resnati, *J. Am. Chem. Soc.* **115**, 4897 (1993).

89. G. A. Olah, G. K. S. Prakash, and J. Sommer, *Superacids*, Wiley-Interscience, New York, 1985, Chapter 5, p. 311.

90. G. A. Olah and G. K. S. Prakash, in *The Chemistry of Alkanes and Cycloalkanes*, S. Patai and Z. Rappoport, eds., Wiley, Chichester, 1992, Chapter 13, p. 635.

91. N. Yoneda and G. A. Olah, *J. Am. Chem. Soc.* **99**, 3113 (1977).

92. P. S. Bailey, *Ozonation in Organic Chemistry*, Vol. 2: *Nonolefinic Compounds*, Academic Press, New York, 1982, Chapter 9, p. 255.

93. E. Keinan and H. T. Varkony, in *The Chemistry of Peroxides*, S. Patai, ed., Wiley, Chichester, 1983, Chapter 19, p. 649.

94. G. A. Hamilton, B. S. Ribner, and T. M. Hellman, *Adv. Chem. Ser.* **77**, 15 (1968).

95. H. Varkony, S. Pass, and Y. Mazur, *J. Chem. Soc., Chem. Commun.* 437 (1974).

96. T. M. Hellman and G. A. Hamilton, *J. Am. Chem. Soc.* **96**, 1530 (1974).

97. P. S. Nangia and S. W. Benson, *J. Am. Chem. Soc.* **102**, 3105 (1980).

98. M. C. Whiting, A. J. N. Bolt, and J. H. Parish, *Adv. Chem. Ser.* **77**, 4 (1968).

99. L. V. Ruban, S. K. Rakovski, and A. A. Popov, *Bull. Acad. Sci. USSR, Div. Chem. Sci.* (Engl. transl.) **25**, 1834 (1976).

100. D. H. Giamalva, D. F. Church, and W. A. Pryor, *J. Am. Chem. Soc.* **108**, 7678 (1986).

101. D. H. Giamalva, D. F. Church, and W. A. Pryor, *J. Org. Chem.* **53**, 3429 (1988).

102. Z. Cohen, E. Keinan, Y. Mazur, and T. H. Varkony, *J. Org. Chem.* **40**, 2141 (1975).

103. Z. Cohen, H. Varkony, E. Keinan, and Y. Mazur, *Org. Synth., Coll. Vol.* **6**, 43 (1988).

104. E. Proksch and A. de Meijere, *Angew. Chem., Int. Ed. Engl.* **15**, 761 (1976).

105. D. Tal, E. Keinan, and Y. Mazur, *J. Am. Chem. Soc.* **101**, 502 (1979).

106. E. Zadok and Y. Mazur, *Angew. Chem., Int. Ed. Engl.* **21**, 303 (1982).

107. F. Cacace and M. Sperenza, *Science* **265**, 208 (1994).

108. R. Stewart and U. A. Spitzer, *Can. J. Chem.* **56**, 1273 (1978).

109. H.-J. Schmidt and H. J. Schäfer, *Angew. Chem., Int. Ed. Engl.* **18**, 68 (1979).

110. K. B. Wiberg, in *Oxidation in Organic Chemistry*, Part A, K. B. Wiberg, ed., Academic Press, New York, 1965, Chapter 2, p. 69.

111. J. Rocek and F. Mares, *Coll. Czech. Chem. Commun.* **24**, 2741 (1959).

112. R. C. Bingham and P. v. R. Schleyer, *J. Org. Chem.* **36**, 1198 (1971).

113. K. B. Wiberg and G. Foster, *J. Am. Chem. Soc.* **83**, 423 (1961).

114. J. Rocek, *Tetrahedron Lett.* 135 (1962).

115. A. Onopchenko and J. G. D. Schulz, *J. Org. Chem.* **38**, 3729 (1973).

116. A. Onopchenko and J. G. D. Schulz, *J. Org. Chem.* **40**, 3338 (1975).

117. J. Hanotier, Ph. Camerman, M. Hanotier-Bridoux, and P. de Radzitzky, *J. Chem. Soc., Perkin Trans. 2* 2247 (1972).

118. D. B. Clark, M. Fleischmann, and D. Pletcher, *J. Chem. Soc., Perkin Trans. 2*, 1578 (1973).

119. B. Meunier, *Bull. Soc. Chim. Fr.* 578 (1986).

120. T. Mlodnicka, *J. Mol. Catal.* **36**, 205 (1986).

121. T. J. McMurry and J. T. Groves, in *Cytochrome P-450: Structure, Mechanism, and Biochemistry*, P. R. Ortiz de Montellano, ed., Plenum Press, New York, 1986, Chapter 1, p. 1.

122. D. Mansuy, *Pure Appl. Chem.* **59**, 759 (1987); **62**, 741 (1990).

123. L. I. Simándi, *Catalytic Activation of Dioxygen by Metal Complexes*, Kluwer, Dordrecht, 1992, Chapter 2, p. 74.

124. I. Tabushi, *Coord. Chem. Rev.* **86**, 1 (1988).

125. B. Meunier, *New J. Chem.* **16**, 203 (1992).

126. D. Mansuy and P. Battioni, in *Activation and Functionalization of Alkanes*, C. L. Hill, ed., Wiley-Interscience, New York, 1989, Chapter 6, p. 195.

127. B. De Poorter, M. Ricci, O. Bartolini, and B. Meunier, *J. Mol. Catal.* **31**, 221 (1985).

128. J. P. Collman, H. Tanaka, R. T. Hembre, and J. I. Brauman, *J. Am. Chem. Soc.* **112**, 3689 (1990).

129. M. W. Nee and T. C. Bruice, *J. Am. Chem. Soc.* **104**, 6123 (1982).

130. R. B. Brown, Jr., M. M. Williamson, and C. L. Hill, *Inorg. Chem.* **26**, 1602 (1987).

131. L.-C. Yuan and T. C. Bruice, *J. Chem. Soc., Chem. Commun.* 868 (1985).

132. J. T. Groves and T. E. Nemo, *J. Am. Chem. Soc.* **105**, 6243 (1983).

133. J. R. Lindsay Smith and P. R. Sleath, *J. Chem. Soc., Perkin Trans. 2* 1165 (1983).

134. J. T. Groves, W. J. Kruper, Jr., and R. C. Haushalter, *J. Am. Chem. Soc.* **102**, 6375 (1980).

135. J. A. Smegal and C. L. Hill, *J. Am. Chem. Soc.* **105**, 3515 (1983).

136. C. L. Hill, in *Activation and Functionalization of Alkanes*, C. L. Hill, ed., Wiley-Interscience, New York, 1989, Chapter 8, p. 243.

137. J. T. Groves, G. A. McClusky, R. E. White, and M. J. Coon, *Biochem. Biophys. Res. Commun.* **81**, 154 (1978).

138. F. P. Guengerich and T. L. Macdonald, *Acc. Chem. Res.* **17**, 9 (1984).

139. D. Mansuy, J.-F. Bartoli, J.-C. Chottard, and M. Lange, *Angew. Chem., Int. Ed. Engl.* **19**, 909 (1980).

140. D. Mansuy, J.-F. Bartoli, and M. Momenteau, *Tetrahedron Lett.* **23**, 2781 (1982).

141. P. Battioni, J.-P. Renaud, J. F. Bartoli, and D. Mansuy, *J. Chem. Soc., Chem. Commun.* 341 (1986).

142. C. K. Chang and F. Ebina, *J. Chem. Soc., Chem. Commun.* 778 (1981).

143. P. S. Traylor, D. Dolphin, and T. G. Traylor, *J. Chem. Soc., Chem. Commun.* 279 (1984).

144. M. J. Nappa and C. A. Tolman, *Inorg. Chem.* **24**, 4711 (1985).

145. V. S. Belova, A. M. Khenkin, and A. E. Shilov, *Kinet. Catal.* (Engl. transl.) **28**, 897 (1987).

146. J. F. Bartoli, O. Brigaud, P. Battioni, and D. Mansuy, *J. Chem. Soc., Chem. Commun.* 440 (1991).

147. K. Suslick, B. Cook, and M. Fox, *J. Chem. Soc., Chem. Commun.* 580 (1985).

148. B. R. Cook, T. J. Reinert, and K. S. Suslick, *J. Am. Chem. Soc.* **108**, 7281 (1986).

149. K. S. Suslick, in *Activation and Functionalization of Alkanes*, C. L. Hill, ed., Wiley-Interscience, New York, 1989, Chapter 7, p. 219.

150. P. Battioni, J.-P. Lallier, L. Barloy, and D. Mansuy, *J. Chem. Soc., Chem. Commun.* 1149 (1989).

151. L. Barloy, J. P. Lallier, P. Battioni, D. Mansuy, Y. Piffard, M. Tournoux, J. B. Valim, and W. Jones, *New J. Chem.* **16**, 71 (1992).

152. P. Battioni, J. F. Bartoli, P. Leduc, M. Fontecave, and D. Mansuy, *J. Chem. Soc., Chem. Commun.* 791 (1987).

153. E. I. Karasevich, A. M. Khenkin, and A. E. Shilov, *J. Chem. Soc., Chem. Commun.* 731 (1987).

154. M. Fontecave and D. Mansuy, *Tetrahedron* **40**, 4297 (1984).

155. P. E. Ellis, Jr. and J. E. Lyons, *J. Chem. Soc., Chem. Commun.* 1187, 1189, 1315 (1989).

156. J. E. Lyons, P. E. Ellis, Jr., and V. A. Durante, in *Structure-Activity and Selectivity Relationships in Heterogeneous Catalysis*, Studies in Surface Science and Catalysis, Vol. 67, R. K. Grasselli, and A. W. Sleight, eds., Elsevier, Amsterdam, 1991, p. 99.

157. D. H. R. Barton and N. Ozbalik, in *Activation and Functionalization of Alkanes*, C. L. Hill, ed., Wiley-Interscience, New York, 1989, Chapter 9, p. 281.

158. D. H. R. Barton and D. Doller, *Acc. Chem. Res.* **25**, 504 (1992).

159. D. H. R. Barton, M. J. Gastiger, and W. B. Motherwell, *J. Chem. Soc., Chem. Commun.* 41 (1983).

160. D. H. R. Barton, M. J. Gastiger, and W. B. Motherwell, *J. Chem. Soc., Chem. Commun.* 731 (1983).

161. D. H. R. Barton, J. Boivin, M. Gastiger, J. Morzycki, R. S. Hay-Motherwell, W. B. Motherwell, N. Ozbalik, and K. M. Schwartzentruber, *J. Chem. Soc., Perkin Trans. 1* 947 (1986).

162. G. Balavoine, D. H. R. Barton, J. Boivin, A. Gref, N. Ozbalik, and H. Rivière, *J. Chem. Soc., Chem. Commun.* 1727 (1986).

163. S. Murahashi, Y. Oda, and T. Naota, *J. Am. Chem. Soc.* **114**, 7913 (1992).

164. D. H. R. Barton and D. Doller, in *Dioxygen Activation and Homogeneous Catalytic Oxidation*, Studies in Surface Science and Catalysis, Vol. 66, L. I. Simándi, ed., Elsevier, Amsterdam, 1991, p. 1.

165. D. H. R. Barton and D. R. Hill, *Tetrahedron Lett.* **35**, 1431 (1994).

166. D. H. R. Barton, F. Halley, N. Ozbalik, E. Young, G. Balavoine, A. Gref, and J. Boivin, *New J. Chem.* **13**, 177 (1989).

167. D. H. R. Barton, S. D. Bévière, W. Chavasiri, D. Doller, W.-G. Liu, and J. H. Reibenspies, *New J. Chem.* **16**, 1019 (1992).

168. D. H. R. Barton, S. D. Bévière, W. Chavasiri, E. Csuhai, D. Doller, and W.-G. Liu, *J. Am. Chem. Soc.* **114**, 2147 (1992).

169. J. B. Vincent, J. C. Huffman, G. Christou, Q. Li, M. A. Nanny, D. N. Hendrickson, R. H. Fong, and R. H. Fish, *J. Am. Chem. Soc.* **110**, 6898 (1988).

170. C. Sheu, S. A. Richert, P. Cofré, B. Ross, Jr., A. Sobkowiak, D. T. Sawyer, and J. R. Kanofsky, *J. Am. Chem. Soc.* **112**, 1936 (1990).

171. R. H. Fish, M. S. Konings, K. J. Oberhausen, R. H. Fong, W. M. Yu, G. Christou, J. B. Vincent, D. K. Coggin, and R. M. Buchanan, *Inorg. Chem.* **30**, 3002 (1991).

172. N. Kitajima, M. Ito, H. Fukui, and Y. Moro-oka, *J. Chem. Soc., Chem. Commun.* 102 (1991).

173. H.-C. Tung, C. Kang, and D. T. Sawyer, *J. Am. Chem. Soc.* **114**, 3445 (1992).

174. Y. V. Geletii, V. V. Lavrushko, and G. V. Lubimova, *J. Chem. Soc., Chem. Commun.* 936 (1988).

175. T.-C. Lau, C.-M. Che, W.-O. Lee, and C.-K. Poon, *J. Chem. Soc., Chem. Commun.* 1406 (1988).

176. S.-I. Murahashi, Y. Oda, T. Naota, and T. Kuwabara, *Tetrahedron Lett.* **34**, 1299 (1993).

177. M. Bressan, A. Morvillo, and G. Romanello, *J. Mol. Catal.* **77**, 283 (1992).

178. M. M. Taqui Khan, R. S. Shukla, and A. Prakash Rao, *Inorg. Chem.* **28**, 452 (1989).

179. S. Perrier, T. C. Lau, and J. K. Kochi, *Inorg. Chem.* **29**, 4190 (1990).

180. S. Davis and R. S. Drago, *J. Chem. Soc., Chem. Commun.* 250 (1990).

181. K. Nomura and S. Uemura, *J. Chem. Soc., Chem. Commun.* 129 (1994).

182. A. Tenaglia, E. Terranova, and B. Waegell, *J. Chem. Soc., Chem. Commun.* 1344 (1990).

183. R. Neumann and C. Abu-Gnim, *J. Chem. Soc., Chem. Commun.* 1324 (1989).

184. B. Notari, in *Innovations in Zeolite Materials and Science*, Studies in Surface Science and Catalysis, Vol. 37, P. J. Grobet, W. J. Mortier, E. F. Vansant, and G. I. Schulz-Ekloff, eds., Elsevier, Amsterdam, 1988, p. 413.

185. B. Notari, in *Structure-Activity and Selectivity Relationship in Heterogeneous Catalysis*, Studies in Surface Science and Catalysis, Vol. 67, R. K. Grasselli and A. W. Sleight, eds., Elsevier, Amsterdam, 1991, p. 243.

186. T. Tatsumi, M. Nakamura, S. Negishi, and Tominaga, *J. Chem. Soc., Chem. Commun.* 476 (1990).

187. D. R. C. Huybrechts, L. De Bruycker, and P. A. Jacobs, *Nature* (London) **345**, 240 (1990).

188. D. R. C. Huybrechts, P. L. Buskens, and P. A. Jacobs, *J. Mol. Catal.* **71**, 129 (1992).

189. M. G. Clerici, *Appl. Catal.* **68**, 249 (1991).

190. T. Tatsumi, K. Yuasa, and H. Tominaga, *J. Chem. Soc., Chem. Commun.* 1446 (1992).

191. J. S. Reddy, S. Sivasanker, and P. Ratnasamy, *J. Mol. Catal.* **70**, 335 (1991).

192. K. Ogura and H. Arima, *Chem Lett.* 311 (1988).

193. S. H. Sharma, H. R. Sonawane, and S. Dev, *Tetrahedron* **41**, 2483 (1985).

194. E. Baciocchi, T. Del Giacco, and G. V. Sebastiani, *Tetrahedron Lett.* **28**, 1941 (1987).

195. K. Wada, K. Yoshida, Y. Watanabe, and T. Suzuki, *Appl. Catal.* **74**, L1 (1991).

196. J. W. Weller and G. A. Hamilton, *J. Chem. Soc., Chem. Commun.* 1390 (1970).

197. K. Otsuka, I. Yamanaka, and K. Hosokawa, *Nature* (London) **345**, 697 (1990).

198. A. Onopchenko and J. G. D. Schulz, *J. Org. Chem.* **38**, 909 (1973).

199. J. G. D. Schulz and A. Onopchenko, *J. Org. Chem.* **45**, 3716 (1980).

200. K. Tanaka, *Hydrocarbon Process.* **53**(11), 114 (1974); *Chemtech* **4**, 555 (1974).

201. G. R. Steinmetz, N. L. Lafferty, and C. E. Sumner, Jr., *J. Mol. Catal.* **49**, L39 (1988).

202. R. Kuhn and H. Roth, *Chem. Ber.* **66**, 1274 (1933).

203. H. Roth, in *Methoden der Organischen Chemie (Houben-Weyl)*, Vol. 2: *Analitische Methoden*, Thieme, Stuttgart, 1953, p. 273.

204. E. Wiesenberger, *Mikrochemie ver. Mikrochim. Acta* **33**, 51 (1947); *Chem. Abstr.* **41**, 3017d (1947).

205. G. Ingram, in *Comprehensive Analytical Chemistry*, Vol. 1B: *Classical Analysis*, C. L. Wilson and D. W. Wilson, eds., Elsevier, Amsterdam, 1960, p. 654.

206. V. S. Pansare and S. Dev, *Tetrahedron* **24**, 3767 (1968).

207. A. K. Awasthy, R. Belcher, and A. M. G. Macdonald, *J. Chem. Soc. C* 799 (1967).

208. T. Preuss, E. Proksch, and A. de Meijere, *Tetrahedron Lett.* 833 (1978).

209. G. A. Olah, T. D. Ernst, C. B. Rao, and G. K. S. Prakash, *New J. Chem.* **13**, 791 (1989).

210. G. A. Olah, Q. Wang, N. Krass, and G. K. S. Prakash, *Rev. Roum. Chim.* **36**, 567 (1991).

211. N. Prileschajew, *Chem. Ber.* **42**, 4811 (1909).

212. D. Swern, *Org. React.* (NY) **7**, 378 (1953).

213. D. Swern, in *Organic Peroxides*, Vol. 2, D. Swern, ed., Wiley-Interscience, New York, 1971, Chapter 5, p. 355.

214. D. I. Metelitsa, *Russ. Chem. Rev.* (Engl. transl.) **41**, 807 (1972).

215. G. Bouillon, C. Lick, and K. Schank, in *The Chemistry of Peroxides*, S. Patai, ed., Wiley, Chichester, Chapter 10, p. 279.

216. B. Plesnicar, in *Oxidation in Organic Chemistry*, Part C, W. S. Trahanovsky, ed., Academic Press, New York, 1978, Chapter 3, p. 211.

217. B. Plesnicar, in *The Chemistry of Peroxides*, S. Patai, ed., Wiley, Chichester, 1983, Chapter 17, p. 521.

218. M. Bartók and K. L. Láng, in *The Chemistry of Heterocyclic Compounds*, A. Weissberger and E. C. Taylor, eds., Vol. 42: *Small Ring Heterocycles*, Part 3, A. Hassner, ed., Wiley, New York, 1985, Chapter 1, p. 1.

219. A. S. Rao, S. K. Paknikar, and J. G. Kirtane, *Tetrahedron* **39**, 2323 (1983).

220. A. S. Rao, in *Comprehensive Organic Synthesis*, B. M. Trost and I. Fleming, eds., Vol. 7: *Oxidation*, S. V. Ley, ed., Pergamon Press, Oxford, 1991, Chapter 3.1, p. 357.

221. Y. Sawaki, in *The Chemistry of Functional Groups*, Supplement E2: *The Chemistry of Hydroxyl, Ether and Peroxide Groups*, S. Patai, ed., Wiley, Chichester, 1993, Chapter 11, p. 587.

222. H. Heaney, *Top. Curr. Chem.* **164**, 1 (1993).

223. B. M. Lynch and K. H. Pausacker, *J. Chem. Soc.* 1525 (1955).

224. G. Berti, *Top. Stereochem.* **7**, 93 (1973).

225. D. J. Goldsmith, *J. Am. Chem. Soc.* **84**, 3913 (1962).

226. T. Sato and E. Murayama, *Bull. Chem. Soc. Jpn.* **47**, 1207 (1974).

227. H. Kwart and D. M. Hoffman, *J. Org. Chem.* **31**, 419 (1966).

228. H.-J. Schneider, N. Becker, and K. Philippi, *Chem. Ber.* **114**, 1562 (1981).

229. R. D. Bach, C. L. Willis, and J. M. Domaglia, in *Applications of MO Theory in Organic Chemistry*, I. G. Csizmadia, ed., Elsevier, Amsterdam, 1977, p. 221.

230. B. Plesnicar, M. Tasevski, and A. Azman, *J. Am. Chem. Soc.* **100**, 743 (1978).

231. M. S. Stytilin, *Russ. J. Phys. Chem.* (Engl. transl.) **51**, 284 (1977).

232. R. M. Makitra and Ya. N. Pirig, *Russ. J. Phys. Chem.* (Engl. transl.) **52**, 451 (1978).

233. J. Skerjanc, A. Regent, and B. Plesnicar, *J. Chem. Soc., Chem. Commun.* 1007 (1980).

234. V. G. Dryuk, *Tetrahedron* **32**, 2855 (1976).

235. J. G. Traynham, G. R. Franzen, G. A. Knesel, and D. J. Northington, Jr., *J. Org. Chem.* **32**, 3285 (1967).

236. H. C. Brown J. H. Kawakami, and S. Ikegami, *J. Am. Chem. Soc.* **92**, 6914 (1970).

237. G. B. Payne, P. H. Deming, and P. H. Williams, *J. Org. Chem.* **26**, 659 (1961).

238. J. Rebek, Jr., *Heterocycles* **15**, 517 (1981).

239. R. D. Bach and J. W. Knight, *Org. Synth.* **60**, 63 (1981).

240. J. Rebek, Jr., R. McCready, and R. Wolak, *J. Chem. Soc., Chem. Commun.* 705 (1980).

241. J. Rebek, Jr. and R. McCready, *J. Am. Chem. Soc.* **102**, 5602 (1980).

242. H. Mimoun, in *The Chemistry of Peroxides*, S. Patai, ed., Wiley, Chichester, 1983, Chapter 15, p. 463.

243. P. A. Chaloner, *Handbook of Coordination Catalysis in Organic Chemistry*, Butterworths, London, 1986, Chapter 6, p. 451.

244. H. Mimoun, in *Comprehensive Coordination Chemistry*, Vol. 6, G. Wilkinson, R. D. Gillard, and J. A. McCleverty, eds., Pergamon Press, Oxford, 1987, Chapter 61.3, p. 317.

245. K. A. Jørgensen, *Chem. Rev.* **89**, 431 (1989).

246. N. Miyaura and J. K. Kochi, *J. Am. Chem. Soc.* **105**, 2368 (1983).

247. H. Mimoun, I. Seree de Roche, and L. Sajus, *Tetrahedron* **26**, 37 (1970).

248. H. Arakawa, Y. Moro-Oka, and A. Ozaki, *Bull. Chem. Soc. Jpn.* **47**, 2958 (1974).

249. L. I. Simándi, É. Záhonyi-Budó, and J. Bodnár, *Inorg. Chim. Acta* **65**, L181 (1982).

250. F. P. Ballistreri, A. Bazzo, G. A. Tomaselli, and R. M. Toscano, *J. Org. Chem.* **57**, 7074 (1992).

251. A. Arcoria, F. P. Ballistreri, E. Spina, G. A. Tomaselli, and R. M. Toscano, *Gazz. Chim. Ital.* **120**, 309 (1990).

252. K. B. Sharpless, J. M. Townsend, and D. R. Williams, *J. Am. Chem. Soc.* **94**, 295 (1972).

253. G. Amato, A. Arcoria, F. P. Ballistreri, G. A. Tomaselli, O. Bortolini, V. Conte, F. Di Furia, G. Modena, and G. Valle, *J. Mol. Catal.* **37**, 165 (1986).

254. C. Venturello, R. D'Aloisio, J. C. Bart, and M. Ricci, *J. Mol. Catal.* **32**, 107 (1985).

255. K. U. Ingold, *Acc. Chem. Res.* **2**, 1 (1969).

256. T. V. Filippova and E. A. Blyumberg, *Russ. Chem. Rev.* (Engl. transl.) **51**, 582 (1982).

257. S. W. Benson, *J. Am. Chem. Soc.* **87**, 972 (1965).

258. M. Flowers, L. Batt, and S. W. Benson, *J. Chem. Phys.* **37**, 2662 (1962).

259. F. R. Mayo, *J. Am. Chem. Soc.* **80**, 2497 (1958).

260. F. R. Mayo, A. A. Miller, and G. A. Russel, *J. Am. Chem. Soc.* **80**, 2500 (1958).

261. W. Pritzkow, R. Radeglia, and W. Schmidt-Renner, *J. Prakt. Chem.* **321**, 813 (1979).

262. F. Tsuchiya and T. Ikawa, *Can. J. Chem.* **47**, 3191 (1969).

263. G. Montorsi, G. Caprara, G. Pregaglia, and G. Messina, *Int. J. Chem. Kinet.* **5**, 777 (1973).

264. R. R. Diaz, K. Selby, and D. J. Waddington, *J. Chem. Soc., Perkin Trans. 2* 758 (1975).

265. K. Kaneda, S. Haruna, T. Imanaka, M. Hamamoto, Y. Nishiyama, and Y. Ishii, *Tetrahedron Lett.* **33**, 6827 (1992).

266. N. Shimizu and P. D. Bartlett, *J. Am. Chem. Soc.* **98**, 4193 (1976).

267. Y. Sawaki and Y. Ogata, *J. Am. Chem. Soc.* **103**, 2049 (1981).

268. Y. Sawaki and Y. Ogata, *J. Org. Chem.* **49**, 3344 (1984).

269. P. S. Bailey, *Ozonation in Organic Chemistry*, Vol. 1: *Olefinic Compounds*, Academic Press, New York, 1978, Chapter 11, p. 197.

270. P. S. Bailey, J. W. Ward, R. E. Hornish, and F. E. Potts, III, *Adv. Chem. Ser.* **112**, 1 (1972).

271. R. W. Murray, W. Kong, and S. N. Rajdhyaksha, *J. Org. Chem.* **58**, 315 (1993).

272. R. Curci, M. Fiorentino, L. Troisi, J. O. Edwards, and R. H. Pater, *J. Org. Chem.* **45**, 4758 (1980).

273. A. L. Baumstark and P. C. Vasquez, *J. Org. Chem.* **53**, 3437 (1988).

274. R. W. Murray and R. Jeyaraman, *J. Org. Chem.* **50**, 2847 (1985).

275. J. K. Crandall and D. J. Batal, *J. Org. Chem.* **53**, 1338 (1988).

276. R. W. Murray, M. K. Pillay, and R. Jeyaraman, *J. Org. Chem.* **53**, 3007 (1988).

277. S. Rozen and M. Kol, *J. Org. Chem.* **55**, 5155 (1990).

278. R. A. Sheldon and J. K. Kochi, *Metal-Catalyzed Oxidations of Organic Compounds*, Academic Press, New York, 1981, Chapter 3, p. 33.

279. Y. Ogata and Y. Sawaki, in *Organic Syntheses by Oxidation with Metal Compounds*, W. J. Mijs and C. R. H. I. de Jonge, eds., Plenum Press, New York, 1986, Chapter 16, p. 839.

280. C. Venturello, E. Alneri, and M. Ricci, *J. Org. Chem.* **48**, 3831 (1983).

281. J. Prandi, H. B. Kagan, and H. Mimoun, *Tetrahedron Lett.* **27**, 2617 (1986).

282. V. Conte and F. Di Furia, in *Catalytic Oxidations with Hydrogen Peroxide as Oxidant*, G. Strukul, ed., Kluwer, Dordredcht, 1992, Chapter 7, p. 223.

283. C. Venturello and R. D'Aloisio, *J. Org. Chem.* **53**, 1553 (1988).

284. Y. Ishii, K. Yamawaki, T. Ura, H. Yamada, T. Yoshida, and M. Ogawa, *J. Org. Chem.* **53**, 3587 (1988).

285. R. A. Sheldon, *J. Mol. Catal.* **7**, 107 (1980).

286. J. Sobczak and J. J. Ziólkowski, *J. Mol. Catal.* **13**, 11 (1981).

287. K. B. Sharpless and T. R. Verhoeven, *Aldrichim. Acta* **12**, 63 (1979).

288. R. A. Sheldon and J. A. Van Doorn, *J. Catal.* **31**, 427 (1973).

289. Z. Dawoodi and R. L. Kelly, *Polyhedron* **5**, 271 (1986).

290. M. N. Sheng and J. G. Zajacek, *J. Org. Chem.* **35**, 1839 (1970).

291. V. P. Yur'ev, I. A. Gailyunas, L. V. Spirikhin, and G. A. Tolstikov, *J. Gen. Chem. USSR* (Engl. transl.) **45**, 2269 (1975).

292. H. Mimoun, *J. Mol. Catal.* **7**, 1 (1980).

293. H. Mimoun, *Angew. Chem., Int. Ed. Engl.* **21**, 734 (1982).

294. P. Chaumette, H. Mimoun, L. Saussine, J. Fischer, and A. Mitschler, *J. Organomet. Chem.* **250**, 291 (1983).

295. R. A. Sheldon, *Recl. Trav. Chim. Pays-Bas* **92**, 253 (1973).

296. A. O. Chong and K. B. Sharpless, *J. Org. Chem.* **42**, 1587 (1977).

297. K. A. Jørgensen and R. Hoffmann, *Acta Chem. Scand., Ser B* **40**, 411 (1986).

298. T. Yamada, T. Takai, O. Rhode, and T. Mukaiyama, *Bull. Chem. Soc. Jpn.* **64**, 2109 (1991).

299. T. Takai, E. Hata, T. Yamada, and T. Mukaiyama, *Bull. Chem. Soc. Jpn.* **64**, 2513 (1991).

300. T. Punniyamurthy, B. Bhatia, and J. Iqbal, *J. Org. Chem.* **59**, 850 (1994).

301. E. Bouhlel, P. Laszlo, M. Levart, M.-T. Montaufier, and G. P. Singh, *Tetrahedron Lett.* **34**, 1123 (1993).

302. P. Laszlo and M. Levart, *Tetrahedron Lett.* **34**, 1127 (1993).

303. S.-I. Murahashi, Y. Oda, T. Naota, and N. Komiya, *J. Chem. Soc., Chem. Commun.* 139 (1993).

304. M. G. Clerici, G. Bellussi, and U. Romano, *J. Catal.* **129**, 159 (1991).

305. M. G. Clerici and P. Ingallina, *J. Catal.* **140**, 71 (1993).

306. T. Tatsumi, M. Nakamura, K. Yuasa, and H. Tominaga, *Chem. Lett.* 297 (1990).

307. L. I. Simándi, *Catalytic Activation of Dioxygen by Metal Complexes*, Kluwer, Dordrecht, 1992, Chapter 3, p. 109.

308. J. A. Smegal, B. C. Schardt, and C. L. Hill, *J. Am. Chem. Soc.* **105**, 3510 (1983).

309. A. W. van der Made, J. W. H. Smeets, R. J. M. Nolte, and W. Drenth, *J. Chem. Soc., Chem. Commun.* 1204 (1983).

310. B. Meunier, E. Guilmet, M.-E. De Carvalho, and R. Poilblanc, *J. Am. Chem. Soc.* **106**, 6668 (1984).

311. K. S. Suslick and B. R. Cook, *J. Chem. Soc., Chem. Commun.* 200 (1987).

312. J. P. Collman, J. I. Brauman, J. P. Fitzgerald, P. D. Hampton, Y. Naruta, and T. Michida, *Bull. Chem. Soc. Jpn.* **61**, 47 (1988).

313. J. P. Collman, X. Zhang, R. T. Hembre, and J. I. Brauman, *J. Am. Chem. Soc.* **112**, 5356 (1990).

314. J. P. Collman, X. Zhang, V. Lee, R. T. Hembre, and J. I. Brauman, *Adv. Chem. Ser.* **230**, 153 (1992).

315. P. R. Cooke and J. R. L. Smith, *Tetrahedron Lett.* **33**, 2737 (1992).

316. J. T. Groves and T. E. Nemo, *J. Am. Chem. Soc.* **105**, 5786 (1983).

317. J. T. Groves and R. S. Myers, *J. Am. Chem. Soc.* **105**, 5791 (1983).

318. K. Tatsumi and R. Hoffmann, *Inorg. Chem.* **20**, 3771 (1981).

319. B. Chevrier, R. Weiss, M. Lange, J.-C. Chottard, and D. Mansuy, *J. Am. Chem. Soc.* **103**, 2899 (1981).

320. J. P. Collman, J. I. Brauman, B. Meunier, T. Hayashi, T. Kodadek, and S. A. Raybuck, *J. Am. Chem. Soc.* **107**, 2000 (1985).

321. B. Meunier, M.-E. De Carvalho, and A. Robert, *J. Mol. Catal.* **41**, 185 (1987).

322. B. Meunier, *Gazz. Chim. Ital.* **118**, 485 (1988).

323. D. Ostovic and T. C. Bruice, *J. Am. Chem. Soc.* **111**, 6511 (1989).

324. D. Mohajer and S. Tangestaninejad, *J. Chem. Soc., Chem. Commun.* 240 (1993).

325. P. Battioni, J. P. Renaud, J. F. Bartoli, M. Reina-Artiles, M. Fort, and D. Mansuy, *J. Am. Chem. Soc.* **110**, 8462 (1988).

326. P. L. Anelli, S. Banfi, F. Montanari, and S. Quici, *J. Chem. Soc., Chem. Commun.* 779 (1989).

327. B. Meunier, in *Catalytic Oxidations with Hydrogen Peroxide as Oxidant*, G. Strukul, ed., Kluwer, Dordredcht, 1992, Chapter 5, p. 153.

328. I. Tabushi and A. Yazaki, *J. Am. Chem. Soc.* **103**, 7371 (1981).

329. H. J. Ledon, P. Durbut, and F. Varescon, *J. Am. Chem. Soc.* **103**, 3601 (1981).

330. W. Nam, R. Ho, and J. S. Valentine, *J. Am. Chem. Soc.* **113**, 7052 (1991).

331. T. Yamamoto and M. Kimura, *J. Chem. Soc., Chem. Commun.* 948 (1977).

332. T. Katsuki and K. B. Sharpless, *J. Am. Chem. Soc.* **102**, 5974 (1980).

333. A. Pfenninger, *Synthesis* 89 (1986).

334. R. A. Johnson and K. B. Sharpless, in *Comprehensive Organic Synthesis*, B. M. Trost and I. Fleming, eds., Vol. 7: *Oxidation*, S. V. Ley, ed., Pergamon Press, Oxford, 1991, Chapter 3.2, p. 389.

335. K. Tani, M. Hanafusa, and S. Otsuka, *Tetrahedron Lett.* 3017 (1979).

336. C. Bolm, *Angew. Chem., Int. Ed. Engl.* **30**, 403 (1991).

337. E. Höft, *Top. Curr. Chem.* **164**, 63 (1993).

338. V. Schurig, K. Hintzer, U. Leyrer, C. Mark, P. Pitchen, and H. B. Kagan, *J. Organomet. Chem.* **370**, 81 (1989).

339. H. B. Kagan, H. Mimoun, C. Mark, and V. Schurig, *Angew. Chem., Int. Ed. Engl.* **18**, 485 (1979).

340. F. A. Davis and S. Chattopadhyay, *Tetrahedron Lett.* **27**, 5079 (1986).

341. J. T. Groves and P. Viski, *J. Org. Chem.* **55**, 3628 (1990).

342. Y. Naruta, F. Tani, N. Ishihara, and K. Maruyama, *J. Am. Chem. Soc.* **113**, 6865 (1991).

343. Y. Naruta, N. Ishihara, F. Tani, and K. Maruyama, *Bull. Chem. Soc. Jpn.* **66**, 158 (1993).

344. R. Irie, K. Noda, Y. Ito, N. Matsumoto, and T. Katsuki, *Tetrahedron Lett.* **31**, 7345 (1990).

345. W. Zhang, J. L. Loebach, S. R. Wilson, and E. N. Jacobsen, *J. Am. Chem. Soc.* **112**, 2801 (1990).

346. W. Zhang and E. N. Jacobsen, *J. Org. Chem.* **56**, 2296 (1991).

347. T. Yamada, K. Imagawa, T. Nagata, and T. Mukaiyama, *Chem. Lett.* 2231 (1992).

348. S. W. May, M. S. Steltenkamp, R. D. Schwartz, and C. J. McCoy, *J. Am. Chem. Soc.* **98**, 7856 (1976).

349. M.-J. de Smet, B. Witholt, and H. Wynberg, *J. Org. Chem.* **46**, 3128 (1981).

350. V. Schurig and D. Wistuba, *Angew. Chem., Int. Ed. Engl.* **23**, 796 (1984).

351. H. Ohta and H. Tetsukawa, *J. Chem. Soc., Chem. Commun.* 849 (1978).

352. O. Takahashi, J. Umezawa, K. Furuhashi, and M. Takagi, *Tetrahedron Lett.* **30**, 1583 (1989).

353. E. J. Allain, L. P. Hager L. Deng, and E. N. Jacobsen, *J. Am. Chem. Soc.* **115**, 4415 (1993).

354. W. Adam and M. J. Richter, *Acc. Chem. Res.* **27**, 57 (1994).

355. E. S. Huyser, *Free-Radical Chain Reactions*, Wiley-Interscience, New York, 1970, Chapter 11, p. 293.

356. V. V. Voronenkov, A. N. Vinogradov, and V. A. Belyaev, *Russ. Chem. Rev.* (Engl. transl.) **39**, 944 (1970).

357. I. Rosenthal, in *Singlet O$_2$*, Vol. 1: *Physical-Chemical Aspects*, A. A. Frimer, ed., CRC Press, Boca Raton, FL, 1985, Chapter 2, p. 13.

358. R. W. Murray, in *Singlet Oxygen*, H. H. Wasserman and R. W. Murray, eds., Academic Press, New York, 1979, Chapter 3, p. 59.

359. E. A. Ogryzlo, in *Singlet Oxygen*, H. H. Wasserman and R. W. Murray, eds., Academic Press, New York, 1979, Chapter 2, p. 35.

360. D. R. Kearns, *Chem. Rev.* **71**, 395 (1971).

361. G. O. Schenck, *Angew. Chem.* **69**, 579 (1957).

362. K. Gollnick, *Adv. Photochem.* **6**, 1 (1968).

363. R. W. Denny and A. Nickon, *Org. React. (NY)* **20**, 133 (1973).

364. G. Ohloff, *Pure Appl. Chem.* **43**, 481 (1975).

365. K. Gollnick and H. J. Kuhn, in *Singlet Oxygen*, H. H. Wasserman and R. W. Murray, eds., Academic Press, New York, 1979, Chapter 8, p. 287.

366. A. A. Frimer, *Chem. Rev.* **79**, 359 (1979).

367. A. A. Frimer and L. M. Stephenson, in *Singlet O₂*, Vol. 2: *Reaction Modes and Products*, Part 1, A. A. Frimer, ed., CRC Press, Boca Raton, 1985, Chapter 3, p. 67.

368. A. P. Schaap and K. A. Zaklika, in *Singlet Oxygen*, H. H. Wasserman and R. W. Murray, eds., Academic Press, New York, 1979, Chapter 6, p. 173.

369. P. D. Bartlett and M. E. Landis, in *Singlet Oxygen*, H. H. Wasserman and R. W. Murray, eds., Academic Press, New York, 1979, Chapter 7, p. 243.

370. A. L. Baumstark, in *Singlet O₂*, Vol. 2: *Reaction Modes and Products*, Part 1, A. A. Frimer, ed., CRC Press, Boca Raton, FL, 1985, Chapter 1, p. 1.

371. Yu. A. Arbuzov, *Russ. Chem. Rev.* (Engl. transl.) **34**, 558 (1965).

372. I. Sato and S. S. Nittala, in *The Chemistry of Peroxides*, S. Patai, ed., Wiley, Chichester, 1983, Chapter 11, p. 311.

373. A. J. Bloodworth and H. J. Eggelte, in *Singlet O₂*, Vol. 2: *Reaction Modes and Products*, Part 1, A. A. Frimer, ed., CRC Press, Boca Raton, FL, 1985, Chapter 4, p. 93.

374. W. Adam and F. Yany, in *The Chemistry of Heterocyclic Compounds*, A. Weissberger and E. C. Taylor, eds., Vol. 42: *Small Ring Heterocycles*, Part 3, A. Hassner, ed., Wiley, New York, 1985, Chapter 4, p. 351.

375. H. H. Wasserman and J. L. Ives, *Tetrahedron* **37**, 1825 (1981).

376. M. Balci, *Chem. Rev.* **81**, 91 (1981).

377. M. Matsumoto, in *Singlet O₂*, Vol. 2: *Reaction Modes and Products*, Part 1, A. A. Frimer, ed., CRC Press, Boca Raton, FL, 1985, Chapter 5, p. 205.

378. C. S. Foote, S. Wexler, W. Ando, and R. Higgins, *J. Am. Chem. Soc.* **90**, 975 (1968).

379. G. O. Schenck, *Naturwissenschaften* **35**, 28 (1948).

380. E. Koch, *Tetrahedron* **24**, 6295 (1968).

381. S.-K. Chung and A. I. Scott, *J. Org. Chem.* **40**, 1652 (1975).

382. K. H. Schulte-Elte and V. Rautenstrauch, *J. Am. Chem. Soc.* **102**, 1738 (1980).

383. G. Rousseau, P. Le Perchec, and J. M. Conia, *Tetrahedron Lett.* 45 (1977).

384. L. M. Stephenson, M. J. Grdina, and M. Orfanopoulos, *Acc. Chem. Res.* **13**, 419 (1980).

385. M. Orfanopoulos, M. Stratakis, and Y. Elemes, *Tetrahedron Lett.* **30**, 4875 (1989).

386. G. O. Schenck, H. Eggert, and W. Denk, *Justus Liebigs Ann. Chem.* **584**, 177 (1953).

387. C. S. Foote, S. Wexler, and W. Ando, *Tetrahedron Lett.* 4111 (1965).

388. A. Nickon and J. F. Bagli, *J. Am. Chem. Soc.* **83**, 1498 (1961).

389. E. W. H. Asveld and R. M. Kellogg, *J. Org. Chem.* **47**, 1250 (1982).

390. W. Fenical, D. R. Kearns, and P. Radlick, *J. Am. Chem. Soc.* **91**, 7771 (1969).

391. L. B. Harding and W. A. Goddard, III, *J. Am. Chem. Soc.* **102**, 439 (1980).

392. W. Fenical, D. R. Kearns, and P. Radlick, *J. Am. Chem. Soc.* **91**, 3396 (1969).

393. D. R. Kearns, *J. Am. Chem. Soc.* **91**, 6554 (1969).

394. F. A. Litt and A. Nickon, *Adv. Chem. Ser.* **77**, 118 (1968).

395. K. R. Kopeczky and J. H. van de Sande, *Can. J. Chem.* **50**, 4034 (1972).

396. C. W. Jefford, *Chem. Soc. Rev.* **22**, 59 (1993).

397. L. M. Stephenson, *Tetrahedron Lett.* **21**, 1005 (1980).

398. G. O. Schenck, K. Gollnick, G. Buchwald, S., Schroeter, and G. Ohloff, *Justus Liebigs Ann. Chem.* **674**, 93 (1964).

399. C. S. Foote, *Acc. Chem. Res.* **1**, 104 (1968).

400. J. H. Wieringa, J. Strating, H. Wynberg, and W. Adam, *Tetrahedron Lett.* 169 (1972).

401. G. B. Schuster, N. J. Turro, H.-C. Steinmetzer, A. P. Schaap, G. Faler, W. Adam, and J. C. Liu, *J. Am. Chem. Soc.* **97**, 7110 (1975).

402. P. D. Bartlett and M. S. Ho, *J. Am. Chem. Soc.* **96**, 627 (1974).

403. F. McCapra and I. Beheshti, *J. Chem. Soc., Chem. Commun.* 517 (1977).

404. A. Krebs, H. Schmalstieg, O. Jarchow, and K.-H. Klaska, *Tetrahedron Lett.* **21**, 3171 (1980).

405. N. M. Hasty and D. R. Kearns, *J. Am. Chem. Soc.* **95**, 3380 (1973).

406. P. A. Burns and C. S. Foote, *J. Am. Chem. Soc.* **96**, 4339 (1974).

407. H. E. Zimmerman and G. E. Keck, *J. Am. Chem. Soc.* **97**, 3527 (1975).

408. P. D. Bartlett and A. P. Schaap, *J. Am. Chem. Soc.* **92**, 3223 (1970).

409. A. P. Schaap and G. R. Faler, *J. Am. Chem. Soc.* **95**, 3381 (1973).

410. T. Kondo, M. Matsumoto, and M. Tanimoto, *Tetrahedron Lett.* 3819 (1978).

411. J.-P. Hagenbuch and P. Vogel, *Tetrahedron Lett.* 561 (1979).

412. M. Matsumoto and K. Kondo, *J. Org. Chem.* **40**, 2259 (1975).

413. G. O. Schenck and D. E. Dunlap, *Angew. Chem.* **68**, 248 (1956).

414. A. C. Cope, T. A. Liss, and G. W. Wood, *J. Am. Chem. Soc.* **79**, 6287 (1957).

415. Y. Kayama, M. Oda, and Y. Kitahara, *Chem. Lett.* 345 (1974).

416. A. Horinaka, R. Nakashima, M. Yoshikawa, and T. Matsuura, *Bull. Chem. Soc. Jpn.* **48**, 2095 (1975).

417. W. Adam and H. J. Eggelte, *Angew. Chem., Int. Ed. Engl.* **16**, 713 (1977).

418. G. O. Schenck and K. Ziegler, *Naturwissenschaften* **32**, 157 (1944).

419. G. O. Schenck, K. G. Kinkel, and H.-J. Mertens, *Justus Liebigs Ann. Chem.* **584**, 125 (1953).

420. B. M. Monroe, *J. Am. Chem. Soc.* **103**, 7253 (1981).

421. W. Adam and M. Balci, *J. Am. Chem. Soc.* **101**, 7537 (1979).

422. W. Adam and H. Rebollo, *Tetrahedron Lett.* **22**, 3049 (1981).

423. W. Adam and H. Rebollo, *Tetrahedron Lett.* **23**, 4907 (1982).

424. M. Matsumoto, S. Dobashi, and K. Kondo, *Tetrahedron Lett.* 2329 (1977).

425. J. D. Boyd, C. S. Foote, and D. K. Imagawa, *J. Am. Chem. Soc.* **102**, 3641 (1980).

426. M. Matsumoto and K. Kondo, *Tetrahedron Lett.* 3935 (1975).

427. M. Matsumoto, S. Dobashi, and K. Kondo, *Bull. Chem. Soc. Jpn.* **51**, 185 (1978).

428. G. Rio and J. Berthelot, *Bull. Soc. Chim. Fr.* 1664 (1969).

429. J. Rigaudy, P. Capdevielle, S. Combrisson, and M. Maumy, *Tetrahedron Lett.* 2757 (1974).

430. R. D. Ashford and E. A. Ogryzlo, *Can. J. Chem.* **52**, 3544 (1974).

431. G. Rio and J. Berthelot, *Bull. Soc. Chim. Fr.* 2938 (1971).

432. L. A. Paquette, R. V. C. Carr, E. Arnold, and J. Clardy, *J. Org. Chem.* **45**, 4907 (1980).

433. M. J. S. Dewar and W. Thiel, *J. Am. Chem. Soc.* **99**, 2338 (1977).

434. A. J. Fatiadi, *Synthesis* 85 (1987).

435. A. H. Haines, in *Comprehensive Organic Synthesis*, B. M. Trost and I. Fleming, eds., Vol. 7: *Oxidation*, S. V. Ley, ed., Pergamon Press, Oxford, 1991, Chapter 3.3, p. 437.

436. J. E. Taylor, *Can. J. Chem.* **62**, 2641 (1984).

437. T. Ogino, *Tetrahedron Lett.* **21**, 177 (1980).

438. W. P. Weber and J. P. Shepherd, *Tetrahedron Lett.* 4907 (1972).

439. T. A. Foglia, P. A. Barr, and A. J. Malloy, *J. Am. Oil. Chem. Soc.* **54**, 858A (1977).

440. T. Ogino and K. Mochizuki, *Chem. Lett.* 443 (1979).

441. V. Bhushan, R. Rathore, and S. Chandrasekaran, *Synthesis* 431 (1984).

442. F. Freeman and L. Y. Chang, *J. Am. Chem. Soc.* **108**, 4504 (1986).

443. F. Freeman, L. Y. Chang, J. C. Kappos, and L. Sumarta, *J. Org. Chem.* **52**, 1460 (1987).

444. T. Ogino, K. Hasegawa, and E. Hoshino, *J. Org. Chem.* **55**, 2653 (1990).

445. T. Ogino and N. Kikuiri, *J. Am. Chem. Soc.* **111**, 6174 (1989).

446. D. G. Lee and T. Chen, *J. Am. Chem. Soc.* **111**, 7534 (1989).

447. M. Schröder, *Chem. Rev.* **80**, 187 (1980).

448. H. S. Singh, in *Organic Syntheses by Oxidation with Metal Compounds*, W. J. Mijs and C. R. H. I. de Jonge, eds., Plenum Press, New York, 1986, Chapter 12, p. 633.

449. Á. Molnár, in *The Chemistry of Functional Groups*, Supplement E2: *The Chemistry of Hydroxyl, Ether and Peroxide Groups*, S. Patai, ed., Wiley, Chichester, 1993, Chapter 18, p. 937.

450. N. A. Milas and S. Sussman, *J. Am. Chem. Soc.* **58**, 1302 (1936).

451. N. A. Milas, J. H. Trepagnier, J. T. Nolan, Jr., and M. I. Iliopulos, *J. Am. Chem. Soc.* **81**, 4730 (1959).

452. R. A. Sheldon, in *The Chemistry of Peroxides*, S. Patai, ed., Wiley, Chichester, 1983, Chapter 6, p. 161.

453. V. VanRheenen, D. Y. Cha, and W. M. Hartley, *Org. Synth., Coll. Vol.* **6**, 342 (1988).

454. M. Minato, K. Yamamoto, and J. Tsuji, *J. Org. Chem.* **55**, 766 (1990).

455. S. G. Hentges and K. B. Sharpless, *J. Am. Chem. Soc.* **102**, 4263 (1980).

456. B. B. Lohray, T. H. Kalantar, B. M. Kim, C. Y. Park, T. Shibata, J. S. M. Wai, and K. B. Sharpless, *Tetrahedron Lett.* **30**, 2041 (1989).

457. K. Tomioka, *Synthesis* 541 (1990).

458. B. B. Lohray, *Tetrahedron: Asymmetry* **3**, 1317 (1992).

459. K. Tomioka, M. Nakajima, and K. Koga, *J. Am. Chem. Soc.* **109**, 6213 (1987).

460. T. Oishi and M. Hirama, *J. Org. Chem.* **54**, 5834 (1989).

461. E. N. Jacobsen, I. Markó, W. S. Mungall, G. Schröder, and K. B. Sharpless, *J. Am. Chem. Soc.* **110**, 1968 (1988).

462. J. S. M. Wai, I. Markó, J. S. Svendsen, M. G. Finn, E. N. Jacobsen, and K. B. Sharpless, *J. Am. Chem. Soc.* **111**, 1123 (1989).

463. H.-L. Kwong, C. Sorato, Y. Ogino, H. Chen, and K. B. Sharpless, *Tetrahedron Lett.* **31**, 2999 (1990).

464. Y. Ogino, H. Chen, E. Manoury, T. Shibata, M. Beller, D. Lübben, and K. B. Sharpless, *Tetrahedron Lett.* **32**, 5761 (1991).

465. D. Xu, G. A. Crispino, and K. B. Sharpless, *J. Am. Chem. Soc.* **114**, 7570 (1992).

466. K. B. Sharpless, W. Amberg, Y. L. Bennani, G. A. Crispino, J. Hartung, K.-S. Jeong, H.-L. Kwong, K. Morikawa, Z.-M. Wang, D. Xu, and X.-L. Zhang, *J. Org. Chem.* **57**, 2768 (1992).

467. M. P. Arrington, Y. L. Bennani, T. Göbel, P. Walsh, S.-H. Zhao, and K. B. Sharpless, *Tetrahedron Lett.* **34**, 7375 (1993).

468. W. Amberg, Y. L. Bennani, R. K. Chadha, G. A. Crispino, W. D. Davis, J. Hartung, K.-S. Jeong, Y. Ogino, T. Shibata, and K. B. Sharpless, *J. Org. Chem.* **58**, 844 (1993).

469. C. P. Casey, *J. Chem. Soc., Chem. Commun.* 126 (1983).

470. E. J. Corey and G. I. Lotto, *Tetrahedron Lett.* **31**, 2665 (1990).

471. K. A. Jørgensen, *Tetrahedron Lett.* **31**, 6417 (1990).

472. K. Tomioka, M. Nakajima, Y. Itaka, and K. Koga, *Tetrahedron Lett.* **29**, 573 (1988).

473. M. Schröder and E. C. Constable, *J. Chem. Soc., Chem. Commun.* 734 (1982).

474. E. J. Corey, P. D. Jardine, S. Virgil, P.-W. Yuen, and R. D. Connell, *J. Am. Chem. Soc.* **111**, 9243 (1989).

475. M. R. Sivik, J. C. Gallucci, and L. A. Paquette, *J. Org. Chem.* **55**, 391 (1990).

476. K. A. Jørgensen, *Tetrahedron: Asymmetry* **2**, 515 (1991).

477. Y. Ogino, H. Chen, H.-L. Kwong, and K. B. Sharpless, *Tetrahedron Lett.* **32**, 3965 (1991).

478. C. Venturello and M. Gambaro, *Synthesis* 295 (1989).

479. S. Warwel, M. Rüsch gen. Klaas, and M. Sojka, *J. Chem. Soc., Chem. Commun.* 1578 (1991).

480. R. B. Woodward and F. V. Brutcher, Jr., *J. Am. Chem. Soc.* **80**, 209 (1958).

481. L. Mangoni, M. Adinolfi, G. Barone, and M. Parrilli, *Tetrahedron Lett.* 4485 (1973).

482. R. C. Cambie and P. S. Rutledge, *Org. Synth., Coll. Vol.* **6**, 348 (1988).

483. F. Fringuelli, R. Germani, F. Pizzo, and G. Savelli, *Synth. Commun.* **19**, 1939 (1989).

484. G. Xie, L. Xu, J. Hu, S. Ma, W. Hou, and F. Tao, *Tetrahedron Lett.* **29**, 2967 (1988).

485. M. Mugdan and D. P. Young, *J. Chem. Soc.* 2988 (1949).

486. G. B. Payne and C. W. Smith, *J. Org. Chem.* **22**, 1682 (1957).

487. A. Stoll, A. Lindenmann, and E. Jucker, *Helv. Chim. Acta* **36**, 268 (1953).

488. N. Sonoda and S. Tsutsumi, *Bull. Chem. Soc. Jpn.* **38**, 958 (1965).

489. R. T. Taylor and L. A. Flood, *J. Org. Chem.* **48**, 5160 (1983).

490. M. Lj. Mihailovic, Z. Cekovic, and L. Lorenc, in *Organic Syntheses by Oxidation with Metal Compounds*, W. J. Mijs and C. R. H. I. de Jonge, eds., Plenum Press, New York, 1986, Chapter 14, p. 741.

491. R. Criegee, in *Oxidation in Organic Chemistry*, Part A, K. B. Wiberg, ed., Academic Press, New York, 1965, Chapter 5, p. 277.

492. F. V. Brutcher, Jr. and F. J. Vara, *J. Am. Chem. Soc.* **78**, 5695 (1956).

493. C. V. Wilson, *Org. React. (NY)* **9**, 332 (1957).

494. P. M. Maitlis, *The Organic Chemistry of Palladium*, Vol 2: *Catalytic Reactions*, Academic Press, New York, 1971, Chapter 2, p. 77.

495. R. Jira and W. Freiesleben, *Organomet. React.* **3**, 1 (1972).

496. W. Kitching, *Organomet. React.* **3**, 319 (1972).

497. P. M. Henry, *Adv. Organomet. Chem.* **13**, 363 (1975).

498. P. M. Henry, *Palladium Catalyzed Oxidation of Hydrocarbons*, Reidel, Dordrecht, 1980, Chapter 2, p. 41.

499. J. Tsuji, *Organic Synthesis with Palladium Compounds*, Springer, Berlin, 1980, Chapter 3, p. 4.

500. J. Tsuji, in *Comprehensive Organic Synthesis*, B. M. Trost and I. Fleming, eds., Vol. 7: *Oxidation*, S. V. Ley, ed., Pergamon Press, Oxford, 1991, Chapter 3.4, p. 449.

501. R. A. Sheldon and J. K. Kochi, *Metal-Catalyzed Oxidations of Organic Compounds*, Academic Press, New York, 1981, Chapter 7, p. 189.

502. B. M. Trost and T. R. Verhoeven, in *Comprehensive Organometallic Chemistry*, Vol. 8, G. Wilkinson, F. G. A. Stone, and E. W. Abel, eds., Pergamon Press, Oxford, 1982, Chapter 57, p. 799.

503. S. F. Davison and P. M. Maitlis, in *Organic Syntheses by Oxidation with Metal Compounds*, W. J. Mijs and C. R. H. I. de Jonge, eds., Plenum Press, New York, 1986, Chapter 9, p. 469.

504. J. Tsuji, *Synthesis* 369 (1984).

505. J. Tsuji, H. Nagashima, and H. Nemoto, *Org. Synth.* **62**, 9 (1984).

506. F. C. Phillips, *Am. Chem. J.* **16**, 255 (1894).

507. J. Smidt, W. Hafner, R. Jira, J. Sedlmeier, R. Sieber, R. Rüttinger, and H. Kojer, *Angew. Chem.* **71**, 176 (1959).

508. J. Smidt, W. Hafner, R. Jira, R. Sieber J. Sedlmeier, and A. Sabel, *Angew. Chem., Int. Ed. Engl.* **1**, 80 (1962).

509. J. E. Bäckvall, B. Åkermark, and S. O. Ljunggren, *J. Am. Chem. Soc.* **101**, 2411 (1979).

510. J. K. Stille and R. Divakaruni, *J. Organomet. Chem.* **169**, 239 (1979).

511. A. Harada, Y. Hu, and S. Takahashi, *Chem. Lett.* 2083 (1986).

512. T. T. Wenzel, *J. Chem. Soc., Chem. Commun.* 862 (1993).

513. H. H. Horowitz, *J. Appl. Electrochem.* **14**, 779 (1984).

514. J. Tsuji and M. Minato, *Tetrahedron Lett.* **28**, 3683 (1987).

515. M. Kolb, E. Bratz, and K. Dialer, *J. Mol. Catal.* **2**, 399 (1977).

516. D. G. Miller and D. D. M. Wayner, *J. Org. Chem.* **55**, 2924 (1990).

517. J. Tsuji, H. Nagashima, and K. Hori, *Chem. Lett.* 257 (1980).

518. H. Mimoun, R. Charpentier, A. Mitschler, J. Fischer, and R. Weiss, *J. Am. Chem. Soc.* **102**, 1047 (1980).

519. K. I. Matveev, *Kinet. Catal.* (Engl. transl.) **18**, 716 (1977).

520. I. V. Kozhevnikov and K. I. Matveev, *Appl. Catal.* **5**, 135 (1983).

521. H. Ogawa, H. Fujinami, K. Taya, and S. Teratani, *Bull. Chem. Soc. Jpn.* **57**, 1908 (1984).

522. S. F. Davison, B. E. Mann, and P. M. Maitlis, *J. Chem. Soc., Dalton Trans.* 1223 (1984).

523. K. Takehira, I. H. Oh, V. C. Martinez, R. S. Chavira, T. Hayakawa, H. Orita, M. Shimidzu, and T. Ishikawa, *J. Mol. Catal.* **42**, 237 (1987).

524. K. Takehira, H. Orita, I. H. Oh, C. O. Leobardo, G. C. Martinez, M. Shimidzu, T. Hayakawa, and T. Ishikawa, *J. Mol. Catal.* **42**, 247 (1987).

525. W. G. Lloyd and B. J. Luberoff, *J. Org. Chem.* **34**, 3949 (1969).

526. W. H. Clement and C. M. Selwitz, *J. Org. Chem.* **29**, 241 (1964).

527. D. R. Fahey and E. A. Zuech, *J. Org. Chem.* **39**, 3276 (1974).

528. J.-E. Bäckvall and R. B. Hopkins, *Tetrahedron Lett.* **29**, 2885 (1988).

529. J. E. Bäckvall, R. B. Hopkins, H. Grennberg, M. M. Mader, and A. K. Awasthi, *J. Am. Chem. Soc.* **112**, 5160 (1990).

530. B. R. James and M. Kastner, *Can. J. Chem.* **50**, 1698, 1708 (1972).

531. H. Mimoun, M. M. P. Machirant, and I. S. de Roch, *J. Am. Chem. Soc.* **100**, 5437 (1978).

532. R. S. Drago, A. Zuzich, and E. D. Nyberg, *J. Am. Chem. Soc.* **107**, 2898 (1985).

533. C. W. Dudley, G. Read, and P. J. C. Walker, *J. Chem. Soc., Dalton Trans.* 1926 (1974).

534. G. Read and P. J. C. Walker, *J. Chem. Soc., Dalton Trans.* 883 (1977).

535. R. Tang, F. Mares, N. Neary, and D. E. Smith, *J. Chem. Soc., Chem. Commun.* 274 (1979).

536. F. Igersheim and H. Mimoun, *J. Chem. Soc., Chem. Commun.* 559 (1978).

537. R. Sugimoto, H. Suzuki, Y. Moro-oka, and T. Ikawa, *Chem. Lett.* 1863 (1982).

538. M. T. Atlay, L. Carlton, and G. Read, *J. Mol. Catal.* **19**, 57 (1983).

539. O. Bortolini, F. di Furia, G. Modena, and R. Seraglia, *J. Mol. Catal.* **22**, 313 (1984).

540. H. Mimoun, *Pure Appl. Chem.* **53**, 2389 (1981).

541. L. Carlton, G. Read, and M. Urgelles, *J. Chem. Soc., Chem. Commun.* 586 (1983).

542. H. A. Zahalka, K. Januszkiewicz, and H. Alper, *J. Mol. Catal.* **35**, 249 (1986).

543. D. E. Hamilton, R. S. Drago, and A. Zombeck, *J. Am. Chem. Soc.* **109**, 374 (1987).

544. Y. Matsushita, T. Matsui, and K. Sugamoto, *Chem. Lett.* 1381 (1992).

545. B. S. Tovrog, F. Mares, and S. E. Diamond, *J. Am. Chem. Soc.* **102**, 6616 (1980).

546. D. A. Muccigrosso, F. Mares, S. E. Diamond, and J. P. Solar, *Inorg. Chem.* **22**, 960 (1983).

547. M. A. Andrews and K. P. Kelly, *J. Am. Chem. Soc.* **103**, 2894 (1981).

548. K. Fujimoto, H. Takeda, and T. Kunugi, *Ind. Eng. Chem., Prod. Res. Dev.* **13**, 237 (1974).

549. K. Fujimoto, O. Kuchi-ishi, and T. Kunugi, *Ind. Eng. Chem., Prod. Res. Dev.* **15**, 259 (1976).

550. I. S. Shaw, J. S. Dranoff, and J. B. Butt, *Ind. Eng. Chem. Res.* **27**, 935 (1988).

551. P. H. Espeel, M. C. Tielen, and P. A. Jacobs, *J. Chem. Soc., Chem. Commun.* 669 (1991).

552. T. G. Alkhazov, K. Yu. Adzhamov, and A. K. Khanmamedova, *Russ. Chem. Rev.* (Engl. transl.) **51**, 542 (1982).

553. S. Tan, Y. Moro-oka, and A. Ozaki, *J. Catal.* **17**, 132 (1970).

554. Y. Takita, Y. Moro-oka, and A. Ozaki, *J. Catal.* **52**, 95 (1978).

555. J. M. López Nieto, G. Kremenic', and J. L. G. Fierro, *Appl. Catal.* **61**, 235 (1990).

556. A. Lethbridge, R. O. C. Norman, and C. B. Thomas, *J. Chem. Soc., Perkin Trans. 1*, 35 (1973).

557. A. McKillop, J. D. Hunt, F. Kienzle, E. Bigham, and E. C. Taylor, *J. Am. Chem. Soc.* **95**, 3635 (1973).

558. H. Hart and L. R. Lerner, *J. Org. Chem.* **32**, 2669 (1967).

559. H. R. Rogers, J. X. McDermott, and G. M. Whitesides, *J. Org. Chem.* **40**, 3577 (1975).

560. F. Freeman, P. J. Cameron, and R. H. DuBois, *J. Org. Chem.* **33**, 3970 (1968).

561. F. Freeman, R. H. DuBois, and T. G. McLaughlin, *Org. Synth., Coll. Vol.* **6**, 1028 (1988).

562. N. S. Srinivasan and D. G. Lee, *Synthesis* 520 (1979).

563. S. Baskaran, J. Das, and S. Chandrasekaran, *J. Org. Chem.* **54**, 5182 (1989).

564. K. B. Sharpless, R. F. Lauer, O. Repic, A. Y. Teranishi, and D. R. Williams, *J. Am. Chem. Soc.* **93**, 3303 (1971).

565. H. P. Jensen and K. B. Sharpless, *J. Org. Chem.* **39**, 2314 (1974).

566. D. G. Lee and V. S. Chang, *J. Org. Chem.* **43**, 1532 (1978).

567. G. W. Parshall and S. D. Ittel, *Homogeneous Catalysis*, Wiley-Interscience, New York, 1992, Chapter 6, p. 137.

568. I. I. Moiseev, M. N. Vargaftik, and Ya. K. Syrkin, *Dokl. Chem.* (Engl. transl.) **133**, 801 (1960).

569. E. W. Stern and M. L. Spector, *Proc. Chem. Soc.* 370 (1961).

570. R. van Helden, C. F. Kohll, D. Medema, G. Verberg, and T. Jonkhoff, *Recl. Trav. Chim. Pays-Bas* **87**, 961 (1968).

571. W. Kitching, Z. Rappoport, S. Winstein, and W. G. Young, *J. Am. Chem. Soc.* **88**, 2054 (1966).

572. R. G. Brown, R. V. Chaudhari, and J. M. Davidson, *J. Chem. Soc., Dalton Trans.* 183 (1977).

573. S. Winstein, J. McCaskie, H.-B. Lee, and P. M. Henry, *J. Am. Chem. Soc.* **98**, 6913 (1976).

574. P. M. Henry, *J. Org. Chem.* **32**, 2575 (1967).

575. C. B. Anderson and S. Winstein, *J. Org. Chem.* **28**, 605 (1963).

576. J.-E. Bäckvall, *Acc. Chem. Res.* **16**, 335 (1983).

577. J.-E. Bäckvall and R. E. Nordberg, *J. Am. Chem. Soc.* **103**, 4959 (1981).

578. J.-E. Bäckvall, S. E. Byström, and R. E. Nordberg, *J. Org. Chem.* **49**, 4619 (1984).

579. J.-E. Bäckvall and A. Gogoll, *J. Chem. Soc., Chem. Commun.* 1236 (1987).

580. K. Takehira, H. Mimoun, and I. S. De Roch, *J. Catal.* **58**, 155 (1979).

581. H. Shinohara, *Appl. Catal.* **24**, 17 (1986); **30**, 203 (1987).

582. P. S. Bailey, *Chem. Rev.* **58**, 925 (1958).

583. P. S. Bailey, *Ozonation in Organic Chemistry*, Vol. 1: *Olefinic Compounds*, Academic Press, New York, 1978.

584. J. S. Belew, in *Oxidation*, R. L. Augustine, ed., Marcel Dekker, New York, 1969, Chapter 6, p. 259.

585. R. W. Murray, *Acc. Chem. Res.* **1**, 313 (1968).

586. S. D. Razumovskii and G. E. Zaikov, *Russ. Chem. Rev.* (Engl. transl.) **49**, 1163 (1980).

587. V. N. Odinokov and G. A. Tolstikov, *Russ. Chem. Rev.* (Engl. transl.) **50**, 636 (1981).

588. P. M. Henry and G. L. Lange, in *The Chemistry of Functional Groups, Supplement A: The Chemistry of Double-Bonded Functional Groups*, S. Patai., ed., Wiley, London, 1977, Chapter 11, p. 965.

589. R. L. Kuczkowski, *Acc. Chem. Res.* **16**, 42 (1983).

590. R. L. Kuczkowski, *Chem. Soc. Rev.* **21**, 79 (1992).

591. G. Zvilichovsky and B. Zvilichovsky, in *The Chemistry of Functional Groups, Supplement E2: The Chemistry of Hydroxyl, Ether and Peroxide Groups*, S. Patai, ed., Wiley, Chichester, 1993, Chapter 13, p. 687.

592. D. G. Lee and T. Chen, in *Comprehensive Organic Synthesis*, B. M. Trost and I. Fleming, eds., Vol. 7: *Oxidation*, S. V. Ley, ed., Pergamon Press, Oxford, 1991, Chapter 3.8, p. 541.

593. R. Criegee, *Angew. Chem., Int. Ed. Engl.* **14**, 745 (1975).

594. R. Criegee, *Justus Liebigs Ann. Chem.* **583**, 1 (1953).

595. R. Criegee and G. Schröder, *Chem. Ber.* **93**, 689 (1960).

596. F. L. Greenwood, *J. Org. Chem.* **29**, 1321 (1964).

597. R. Criegee and G. Lohaus, *Chem. Ber.* **86**, 1 (1953).

598. L. A. Hull, I. C. Hisatsune, and J. Heicklen, *J. Am. Chem. Soc.* **94**, 4856 (1972).

599. G. W. Fischer and T. Zimmermann, in *Comprehensive Heterocyclic Chemistry*, Vol. 6, A. R. Katritzky and C. W. Rees, eds., Pergmon Press, Oxford, 1984, Chapter 4.33, p. 851.

600. P. S. Bailey, J. A. Thompson, and B. A. Shoulders, *J. Am. Chem. Soc.* **88**, 4098 (1966).

601. L. J. Durham and F. L. Greenwood, *J. Chem. Soc., Chem. Commun.* 843 (1967).

602. J. Z. Gillies, C. W. Gillies, R. D. Suenram, and F. J. Lovas, *J. Am. Chem. Soc.* **110**, 7991 (1988).

603. R. Criegee and H. Korber, *Chem. Ber.* **104**, 1812 (1971).

604. R. Criegee, A. Kerckow, and H. Zinke, *Chem. Ber.* **88**, 1878 (1955).

605. K. Griesbaum, W. Volpp, R. Greinert, H.-J. Greunig, J. Schmid, and H. Henke, *J. Org. Chem.* **54**, 383 (1989).

606. L. D. Loan, R. W. Murray, and P. R. Story, *J. Am. Chem. Soc.* **87**, 737 (1965).

607. O. Lorenz and C. R. Parks, *J. Org. Chem.* **30**, 1976 (1965).

608. G. D. Fong and R. L. Kuczkowski, *J. Am. Chem. Soc.* **102**, 4763 (1980).

609. S. Fliszár and J. Carles, *J. Am. Chem. Soc.* **91**, 2637 (1969).

610. C. W. Gillies and R. L. Kuczkowski, *J. Am. Chem. Soc.* **94**, 7609 (1972).

611. D. P. Higley and R. W. Murray, *J. Am. Chem. Soc.* **98**, 4526 (1976).

612. G. Schröder, *Chem. Ber.* **95**, 733 (1962).

613. F. L. Greenwood, *J. Am. Chem. Soc.* **88**, 3146 (1966).

614. R. W. Murray, R. D. Youssefyeh, and P. R. Story, *J. Am. Chem. Soc.* **88**, 3143 (1966).

615. N. L. Bauld, J. A. Thompson, C. E. Hudson, and P. S. Bailey, *J. Am. Chem. Soc.* **90**, 1822 (1968).

616. P. S. Bailey and T. M. Ferrell, *J. Am. Chem. Soc.* **100**, 899 (1978).

617. R. P. Lattimer, R. L. Kuczkowski, and C. W. Gillies, *J. Am. Chem. Soc.* **96**, 348 (1974).

618. V. Ramachandran and R. W. Murray, *J. Am. Chem. Soc.* **100**, 2197 (1978).

619. R. W. Murray and J.-S. Su, *J. Org. Chem.* **48**, 817 (1983).

620. J. A. Sousa and A. L. Bluhm, *J. Org. Chem.* **25**, 108 (1960).

621. L. A. Flippin, D. W. Gallagher, and K. Jalali-Araghi, *J. Org. Chem.* **54**, 1430 (1989).

622. J. J. Pappas, W. P. Keaveney, E. Gancher, and M. Berger, *Tetrahedron Lett.* 4273 (1966).

623. D. Gupta, R. Soman, and S. Dev, *Tetrahedron* **38**, 3013 (1982).

624. V. N. Odinokov, L. P. Zhemaiduk, and G. A. Tolstikov, *J. Org. Chem. USSR* (Engl. transl.) **14**, 48 (1978).

625. V. N. Odinokov, L. P. Botsman, G. Yu. Ishmuratov, and G. A. Tolstikov, *J. Org. Chem. USSR* (Engl. transl.) **16**, 453 (1980).

626. R. M. Habib, C.-Y. Chiang, and P. S. Bailey, *J. Org. Chem.* **49**, 2780 (1984).

627. J. Neumeister, H. Keul, M. P. Saxena, and K. Griesbaum, *Angew. Chem., Int. Ed. Engl.* **17**, 939 (1978).

628. I. E. Den Besten and T. H. Kinstle, *J. Am. Chem. Soc.* **102**, 5968 (1980).

629. S. L. Schreiber, R. E. Claus, and J. Reagan, *Tetrahedron Lett.* **23**, 3867 (1982).

630. R. E. Claus and S. L. Schreiber, *Org. Synth.* **64**, 150 (1986).

631. K. Griesbaum, I. C. Jung, and H. Mertens, *J. Org. Chem.* **55**, 6024 (1990).

632. Z. Wang and G. Zvlichovsky, *Tetrahedron Lett.* **31**, 5579 (1990).

633. K. Griesbaum, H. Mertens, and I. C. Jung, *Can. J. Chem.* **68**, 1369 (1990).

634. T. Veysoglu, L. A. Mitscher, and J. K. Swayze, *Synthesis* 807 (1980).

635. M. I. Fremery and E. K. Fields, *J. Org. Chem.* **28**, 2537 (1963).

636. R. Stewart, in *Oxidation in Organic Chemistry*, Part A, K. B. Wiberg, ed., Academic Press, New York, 1965, Chapter 1, p. 1.

637. D. G. Lee, *The Oxidation of Organic Compounds by Permanganate Ion and Hexavalent Chromium*, Open Court, La Salle, 1980, Chapter 1, p. 11.

638. P. Viski, Z. Szeverényi, and L. I. Simándi, *J. Org. Chem.* **51**, 3213 (1986).

639. R. Rathore and S. Chandrasekaran, *J. Chem. Res. (S)*, 458 (1986).

640. D. G. Lee, K. C. Brown, and H. Karaman, *Can. J. Chem.* **64**, 1054 (1986).

641. G. Cainelli and G. Cardillo, *Chromium Oxidations in Organic Chemistry*, Springer, Berlin, 1984, Chapter 3, p. 59.

642. H. Schildknecht and W. Föttinger, *Justus Liebigs Ann. Chem.* **659**, 20 (1962).

643. L. M. Baker and W. L. Carrick, *J. Org. Chem.* **33**, 616 (1968).

644. L. M. Baker and W. L. Carrick, *J. Org. Chem.* **35**, 774 (1970).

645. T. K. Chakraborty and S. Chandrasekaran, *Org. Prep. Proced. Int.* **14**, 362 (1982); *Chem. Abstr.* **97**, 215073q (1982).

646. R. Pappo, D. S., Allen, Jr., R. U. Lemieux, and W. S. Johnson, *J. Org. Chem.* **21**, 478 (1956).

647. K. Inomata, Y. Nakayama, and H. Kotake, *Bull. Chem. Soc. Jpn.* **53**, 565 (1980).

648. D. G. Lee and M. van den Engh, in *Oxidation in Organic Chemistry*, Part B, W. S. Trahanovsky, ed., Academic Press, New York, 1973, Chapter 4, p. 177.

649. G. Barak and Y. Sasson, *J. Chem. Soc., Chem. Commun.* 1266 (1987).

650. R. Neumann and C. Abu-Gnim, *J. Am. Chem. Soc.* **112**, 6025 (1990).

651. E. Steckhan and C. Kandzia, *Synlett* 139 (1992).

652. C. M. Starks, *J. Am. Chem. Soc.* **93**, 195 (1971).

653. A. P. Krapcho, J. R. Larson, and J. M. Eldridge, *J. Org. Chem.* **42**, 3749 (1977).

654. D. G. Lee, S. E. Lamb, and V. S. Chang, *Org. Synth.* **60**, 11 (1981).

655. D. J. Sam and H. F. Simmons, *J. Am. Chem. Soc.* **94**, 4024 (1972).

656. J. T. B. Ferreira, W. O. Cruz, P. C. Vieira, and M. Yonashiro, *J. Org. Chem.* **52**, 3698 (1987).

657. D. G. Lee, T. Chen, and Z. Wang, *J. Org. Chem.* **58**, 2918 (1993).

658. R. U. Lemieux and E. von Rudloff, *Can. J. Chem.* **33**, 1701 (1955).

659. J. L. Courtney, in *Organic Syntheses by Oxidation with Metal Compounds*, W. J. Mijs and C. R. H. I. de Jonge, eds., Plenum Press, New York, 1986, Chapter 8, p. 445.

660. S. Wolfe, S. K. Hasan, and J. R. Campbell, *J. Chem. Soc., Chem. Commun.* 1420 (1970).

661. P. H. J. Carlsen, T. Katsuki, V. S. Martin, and K. B. Sharpless, *J. Org. Chem.* **46**, 3936 (1981).

662. D. G. Lee and U. A. Spitzer, *J. Org. Chem.* **41**, 3644 (1976).

663. Y. Ogata, in *Oxidation in Organic Chemistry*, Part C, W. S. Trahanovsky, ed., Academic Press, New York, 1978, Chapter 4, p. 259.

664. F. Freeman, in *Organic Syntheses by Oxidation with Metal Compounds*, W. J. Mijs and C. R. H. I. de Jonge, eds., Plenum Press, New York, 1986, Chapter 1, p. 1.

665. J. E. Franz, J. F. Herber, and W. S. Knowles, *J. Org. Chem.* **30**, 1488 (1965).

666. G. Gut and W. Lindenmann, *Chimia* **22**, 307 (1968).

667. T. Oguchi, T. Ura, Y. Ishii, and M. Ogawa, *Chem. Lett.* 857 (1989).

668. N. Rabjohn, *Org. React.* (NY) **5**, 331 (1949); **24**, 261 (1976).

669. E. N. Trachtenberg, in *Oxidation*, R. L. Augustine, ed., Marcel Dekker, New York, 1969, Chapter 3, p. 119.

670. C. Paulmier, *Selenium Reagents and Intermediates in Organic Synthesis*, Pergamon Press, Oxford, 1986, Chapter 12, p. 353.

671. P. C. B. Page and T. J. McCarthy, in *Comprehensive Organic Synthesis*, B. M. Trost and I. Fleming, eds., Vol. 7: *Oxidation*, S. V. Ley, ed., Pergamon Press, Oxford, 1991, Chapter 2.1, p. 83.

672. A. Guillemonat, *Ann. Chim.* **11**, 143 (1939); *Chem. Abstr.* **33**, 4579^2 (1939).

673. G. Büchi and H. Wüest, *Helv. Chim. Acta* **50**, 2440 (1967).

674. U. T. Bhalerao and H. Rapoport, *J. Am. Chem. Soc.* **93**, 4835 (1971).

675. A. F. Thomas and W. Bucher, *Helv. Chim. Acta* **53**, 770 (1970).

676. H. P. Jensen and K. B. Sharpless, *J. Org. Chem.* **40**, 264 (1975).

677. K. B. Wiberg and S. D. Nielsen, *J. Org. Chem.* **29**, 3353 (1964).

678. J. P. Schaefer, B. Horvath, and H. P. Klein, *J. Org. Chem.* **33**, 2647 (1968).

679. E. N. Trachtenberg, C. H. Nelson, and J. R. Carver, *J. Org. Chem.* **35**, 1653 (1970).

680. K. B. Sharpless and R. F. Lauer, *J. Am. Chem. Soc.* **94**, 7154 (1972).

681. D. Arigoni, A. Vasella, K. B. Sharpless, and H. P. Jensen, *J. Am. Chem. Soc.* **95**, 7917 (1973).

682. L. M. Stephenson and D. R. Speth, *J. Org. Chem.* **44**, 4683 (1979).

683. W.-D. Woggon, F. Ruther, and H. Egli, *J. Chem. Soc., Chem. Commun.* 706 (1980).

684. J. M. Coxon, E. Dansted, and M. P. Hartshorn, *Org. Synth., Coll. Vol.* **6**, 946 (1988).

685. M. A. Umbreit and K. B. Sharpless, *J. Am. Chem. Soc.* **99**, 5526 (1977).

686. M. A. Warpehoski and B. Chabaud, *J. Org. Chem.* **47**, 2897 (1982).

687. G. Sosnovsky and S.-O. Lawesson, *Angew. Chem., Int. Ed. Engl.* **3**, 269 (1964).

688. D. J. Rawlinson and G. Sosnovsky, *Synthesis* 1 (1972).

689. K. Pedersen, P. Jakobsen, and S.-O. Lawesson, *Org. Synth., Coll. Vol.* **5**, 70 (1973).

690. G. Sosnovsky, *Tetrahedron* **21**, 871 (1965).

691. C. Walling and A. A. Zavitsas, *J. Am. Chem. Soc.* **85**, 2084 (1963).

692. J. Muzart, *Bull. Soc. Chim. Fr.* 65 (1986).

693. D. J. Rawlinson and G. Sosnovsky, *Synthesis* 567 (1973).

694. J. E. Lyons, *Catal. Today* **3**, 245 (1988).

695. A. D. Ketley and J. Braatz, *J. Chem. Soc., Chem. Commun.* 169 (1968).

696. D. R. Chrisope and P. Beak, *J. Am. Chem. Soc.* **108**, 334 (1986).

697. B. M. Trost and P. J. Metzner, *J. Am. Chem. Soc.* **102**, 3572 (1980).

698. J. E. McMurry and P. Kocovsky, *Tetrahedron Lett.* **25**, 4187 (1984).

699. S. Uemura, S. Fukuzawa, A. Toshimitsu, and M. Okano, *Tetrahedron Lett.* **23**, 87 (1982).

700. S. Hansson, A. Heumann, T. Rein, and B. Åkermark, *J. Org. Chem.* **55**, 975 (1990).

701. S. E. Byström, E. M. Larsson, and B. Åkermark, *J. Org. Chem.* **55**, 5674 (1990).

702. H. Grennberg, V. Simon, and J.-E. Bäckvall, *J. Chem. Soc., Chem. Commun.* 265 (1994).

703. E. M. Larsson and B. Åkermark, *Tetrahedron Lett.* **34**, 2523 (1993).

704. T. Seiyama, N. Yamazoe, J. Hojo, and M. Hayakawa, *J. Catal.* **24**, 173 (1972).

705. G. Cainelli and G. Cardillo, *Chromium Oxidations in Organic Chemistry*, Springer, Berlin, 1984, Chapter 2, p. 8.

706. W. G. Dauben, M. Lorber, and D. S. Fullerton, *J. Org. Chem.* **34**, 3587 (1969).

707. J. Muzart, *Tetrahedron Lett.* **28**, 4665 (1987).

708. N. Chidambaram and S. Chandrasekaran, *J. Org. Chem.* **52**, 5048 (1987).

709. B. M. Choudary, A. D. Prasad, V. Swapna, V. L. K. Valli, and V. Bhuma, *Tetrahedron* **48**, 953 (1992).

710. L. I. Simándi, in *The Chemistry of Functional Groups, Supplement C: The Chemistry of Triple-Bonded Functional Groups*, S. Patai and Z. Rappoport, eds., Wiley, Chichester, 1983, Chapter 13, p. 513.

711. H. Gopal and A. J. Gordon, *Tetrahedron Lett.* 2941 (1971).

712. P. Müller and A. Godoy, *Helv. Chim. Acta* **64**, 2531 (1981).

713. D. G. Lee and V. S. Chang, *Synthesis* 462 (1978).

714. N. S. Srinivasan and D. G. Lee, *J. Org. Chem.* **44**, 1574 (1979).

715. D. G. Lee and V. S. Chang, *J. Org. Chem.* **44**, 2726 (1979).

716. A. McKillop, O. H. Oldenziel, B. P. Swann, E. C. Taylor, and R. L. Robey, *J. Am. Chem. Soc.* **95**, 1296 (1973).

717. S. Wolfe, W. R. Pilgrim, T. F. Garrard, and P. Chamberlain, *Can. J. Chem.* **49**, 1099 (1971).

718. M. Schröder and W. P. Griffith, *J. Chem. Soc., Dalton Trans.* 1599 (1978).

719. N. Sonoda, Y. Yamamoto, S. Murai, and S. Tsutsumi, *Chem. Lett.* 229 (1972).

720. M. S. Yusybov and V. D. Filimonov, *Synthesis* 131 (1991).

721. E. B. Merkushev, L. E. Karpitskaya, and G. I. Novosel'tseva, *Dokl. Chem.* (Engl. transl.) **245**, 140 (1979).

722. F. P. Ballistreri, S. Failla, and G. A. Tomaselli, *Tetrahedron* **48**, 9999 (1992).

723. F. P. Ballistreri, S. Failla, and G. A. Tomaselli, *J. Org. Chem.* **53**, 830 (1988).

724. F. P. Ballistreri, S. Failla, G. A. Tomaselli, and R. Curci, *Tetrahedron Lett.* **27**, 5139 (1986).

725. B. Chabaud and K. B. Sharpless, *J. Org. Chem.* **44**, 4202 (1979).

726. J. E. Shaw and J. J. Sherry, *Tetrahedron Lett.* 4379 (1971).

727. W. B. Sheats, L. K. Olli, R. Stout, J. T. Lundeen, R. Justus, and W. G. Nigh, *J. Org. Chem.* **44**, 4075 (1979).

728. Y. Ishii and Y. Sakata, *J. Org. Chem.* **55**, 5545 (1990).

729. R. Curci, M. Fiorentino, C. Fusco, R. Mello, F. P. Ballistreri, S. Failla, and G. A. Tomaselli, *Tetrahedron Lett.* **33**, 7929 (1992).

730. R. M. Moriarty, R. Penmasta, A. K. Awasthi, and I. Prakash, *J. Org. Chem.* **53**, 6124 (1988).

731. F. P. Ballistreri, S. Failla, E. Spina, and G. A. Tomaselli, *J. Org. Chem.* **54**, 947 (1989).

732. W. B. DeMore and C.-L. Lin, *J. Org. Chem.* **38**, 985 (1973).

733. N. C. Yang and J. Libman, *J. Org. Chem.* **39**, 1782 (1974).

734. S. Jackson and L. A. Hull, *J. Org. Chem.* **41**, 3340 (1976).

735. R. E. Keay and G. A. Hamilton, *J. Am. Chem. Soc.* **98**, 6578 (1976).

736. R. N. McDonald and P. A. Schwab, *J. Am. Chem. Soc.* **86**, 4866 (1964).

737. J. K. Stille and D. D. Whitehurst, *J. Am. Chem. Soc.* **86**, 4871 (1964).

738. P. W. Concannon and J. Ciabattoni, *J. Am. Chem. Soc.* **95**, 3284 (1973).

739. D. F. Sangster, in *The Chemistry of the Hydroxyl Group*, S. Patai, ed., Wiley, London, 1971, Chapter 3, p. 133.

740. G. Sosnovsky and D. J. Rawlinson, in *Organic Proxides*, Vol. 2, D. Swern, ed., Wiley-Interscience, New York, 1971, Chapter 3, p. 269.

741. D. I. Metelitsa, *Russ. Chem. Rev.* (Engl. transl.) **40**, 563 (1971).

742. C. Walling, *Acc. Chem. Res.* **8**, 125 (1975).

743. R. A. Sheldon and J. K. Kochi, *Metal-Catalyzed Oxidations of Organic Compounds*, Academic Press, New York, 1981, Chapter 8, p. 215.

744. L. I. Simándi, *Catalytic Activation of Dioxygen by Metal Complexes*, Kluwer, Dordrecht, 1992, Chapter 4, p. 181.

745. C. Walling, D. M. Camaioni, and S. S. Kim, *J. Am. Chem. Soc.* **100**, 4814 (1978).

746. K. Tomizawa and Y. Ogata, *J. Org. Chem.* **46**, 2107 (1981).

747. S. Udenfriend, C. T. Clark, J. Axelrod, and B. B. Brodie, *J. Biol. Chem.* **208**, 731 (1954).

748. G. A. Hamilton and J. P. Friedman, *J. Am. Chem. Soc.* **85**, 1008 (1963).

749. S. Ito, T. Yamasaki, H. Okada, S. Okino, and K. Sasaki, *J. Chem. Soc., Perkin Trans. 2*, 285 (1988).

750. S. Ito, A. Kunai, H. Okada, and K. Sasaki, *J. Org. Chem.* **53**, 296 (1988).

751. J. van Gent, A. A. Wismeijer, A. W. P. G. P. Rit, and H. van Bekkum, *Tetrahedron Lett.* **27**, 1059 (1986).

752. G. Guroff, J. W. Daly, D. M. Jerina, J. Renson, B. Witkop, and S. Udenfriend, *Science* **157**, 1524 (1967).

753. D. M. Jerina and J. W. Daly, in *Oxidases and Related Redox Systems*, T. E. King, H. S. Mason, and M. Morrison, eds., University Park Press, Baltimore, 1973, p. 143.

754. T. Kurata, Y. Watanabe, M. Katoh, and Y. Sawaki, *J. Am. Chem. Soc.* **110**, 7472 (1988).

755. H. Orita, T. Hayakawa, M. Shimizu, and K. Takehira, *J. Mol. Catal.* **42**, 99 (1987).

756. K. Sakai and K. Matsumoto, *J. Mol. Catal.* **67**, 7 (1991).

757. H. Mimoun, L. Saussine, E. Daire, M. Postel, J. Fischer, and R. Weiss, *J. Am. Chem. Soc.* **105**, 3101 (1983).

758. M. Bonchio, V. Conte, F. Di Furia, and G. Modena, *J. Org. Chem.* **54**, 4368 (1989).

759. K. Sasaki, S. Ito, and A. Kunai, in *New Developments in Selective Oxidation*, Studies in Surface Science and Catalysis, Vol. 55, G. Centi and F. Trifirò, eds., Elsevier, Amsterdam, 1990, p. 125.

760. T. Tagawa, Y.-J. Seo, and S. Goto, *J. Mol. Catal.* **78**, 201 (1991).

761. U. Romano, A. Esposito, F. Maspero, C. Neri, and M. G. Clerici, in *New Developments in Selective Oxidation*, Studies in Surface Science and Catalysis, Vol. 55, G. Centi and F. Trifirò, eds., Elsevier, Amsterdam, 1990, p. 33.

762. A. Thangaraj, R. Kumar, and P. Ratnasamy, *Appl. Catal.* **57**, L1 (1990).

763. J. S. Reddy and R. Kumar, *J. Catal.* **130**, 440 (1991).

764. P. T. Tanev, M. Chibwe, and T. J. Pinnavaia, *Nature* (London) **368**, 321 (1994).

765. T. Kitano, Y. Kuroda, M. Mori, S. Ito, K. Sasaki, and M. Nitta, *J. Chem. Soc., Perkin Trans. 2* 981 (1993).

766. D. H. Derbyshire and W. A. Waters, *Nature* (London) **165**, 401 (1950).

767. J. A. Vesely and L. Schmerling, *J. Org. Chem.* **35**, 4028 (1970).

768. J. D. McClure and P. H. Williams, *J. Org. Chem.* **27**, 24 (1962).

769. M. E. Kurz and G. J. Johnson, *J. Org. Chem.* **36**, 3184 (1971).

770. G. A. Olah, T. Keumi, and A. P. Fung, *Synthesis* 536 (1979).

771. H. Hart and C. A. Buehler, *J. Org. Chem.* **29**, 2397 (1964).

772. H. Hart, C. A. Buehler, and A. J. Waring, *Adv. Chem. Ser.* **51**, 1 (1965).

773. S. Hashimoto and W. Koike, *Bull. Chem Soc. Jpn.* **43**, 293 (1970).

774. P. Kovacic and S. T. Morneweck, *J. Am. Chem. Soc.* **87**, 1566 (1965).

775. P. Kovacic and M. E. Kurz, *J. Am. Chem. Soc.* **87**, 4811 (1965); *J. Org. Chem.* **31**, 2011 (1966).

776. G. A. Olah and R. Ohnishi, *J. Org. Chem.* **43**, 865 (1978).

777. G. A. Olah, A. P. Fung, and T. Keumi, *J. Org. Chem.* **46**, 4305 (1981).

778. G. A. Olah, T. Keumi, J. C. Lecoq, A. P. Fung, and J. A. Olah, *J. Org. Chem.* **56**, 6148 (1991).

779. G. K. S. Prakash, N. Krass, Q. Wang, and G. A. Olah, *Synlett* 39 (1991).

780. P. M. Henry, *Palladium Catalyzed Oxidation of Hydrocarbons*, Reidel, Dordrecht, 1980, Chapter 6, p. 306.

781. P. M. Henry, *J. Org. Chem.* **36**, 1886 (1971).

782. L. M. Stock, K.-T. Tse, L. J. Vorvick, and S. A. Walstrum, *J. Org. Chem.* **46**, 1757 (1981).

783. L. Eberson and L. Gomez-Gonzales, *J. Chem. Soc., Chem. Commun.* 263 (1971).

784. L. Eberson and L. Jönsson, *J. Chem. Soc., Chem. Commun.* 885 (1974).

785. F. Minisci, A. Citterio, and C. Giordano, *Acc. Chem. Res.* **16**, 27 (1983).

786. G. G. Arzoumanidis and F. C. Rauch, *J. Org. Chem.* **38**, 4443 (1973).

787. L. Eberson and L. Gomez-Gonzalez, *Acta Chem. Scand.* **27**, 1162 (1973).

788. Z. Blum, L. Cedheim, and K. Nyberg, *Acta Chem. Scand., Ser. B* **29**, 715 (1975).

789. V. D. Parker, J. P. Dirlam, and L. Eberson, *Acta Chem. Scand.* **25**, 341 (1971).

790. G. Bockmair and H. P. Fritz, *Electrochim. Acta.* **21**, 1099 (1976).

791. D. R. Augood and G. H. Williams, *Chem. Rev.* **57**, 123 (1957).

792. T. Nakata, K. Tokumaru, and O. Simamura, *Tetrahedron Lett.* 3303 (1967).

793. P. Kovacic, C. G. Reid, and M. J. Brittain, *J. Org. Chem.* **35**, 2152 (1970).

794. P. Kovacic, C. G. Reid, and M. E. Kurz, *J. Org. Chem.* **34**, 3302 (1969).

795. R. H. Thomson, in *The Chemistry of the Quinonoid Compounds*, S. Patai, ed., Wiley, London, 1974, Chapter 3, p. 111.

796. Y. Naruta and K. Maruyama, in *The Chemistry of the Quinonoid Compounds*, Vol. 2, S. Patai and Z. Rappoport, eds., Wiley, Chichester, 1988, Chapter 8, p. 241.

797. D. Degner, *Top. Curr. Chem.* **148**, 1 (1988).

798. F. Freeman, in *Organic Syntheses by Oxidation with Metal Compounds*, W. J. Mijs and C. R. H. I. de Jonge, eds., Plenum Press, New York, 1986, Chapter 2, p. 41.

799. M. Periasamy and M. V. Bhatt, *Synthesis* 330 (1977).

800. M. Anastasia, P. Allevi, C. Bettini, A. Fiecchi, and A. M. Sanvito, *Synthesis* 1083 (1990).

801. P. Müller and C. Bobillier, *Tetrahedron Lett.* **22**, 5157 (1981); **24**, 5499 (1983).

802. G. K. Chip and J. S. Grossert, *J. Chem. Soc., Perkin Trans. 1* 1629 (1972).

803. R. G. Harvey, *Acc. Chem. Res.* **14**, 218 (1981).

804. R. G. Harvey, *Synthesis* 605 (1986).

805. K. Ishikawa, H. C. Charles, and G. W. Griffin, *Tetrahedron Lett.* 427 (1977).

806. R. Jeyaraman and R. W. Murray, *J. Am. Chem. Soc.* **106**, 2462 (1984).

807. S. Krishnan, D. G. Kuhn, and G. A. Hamilton, *J. Am. Chem. Soc.* **99**, 8121 (1977).

808. K. Ishikawa and G. W. Griffin, *Angew. Chem., Int. Ed. Engl.* **16**, 171 (1977).

809. C. Charles, R. J. Baker, L. M. Trefonas, and G. W. Griffin, *J. Chem. Soc., Chem. Commun.* 1075 (1980).

810. D. T. Gibson, G. E. Cardini, F. C. Maseles, and R. E. Kallio, *Biochemistry* **9**, 1631 (1970).

811. D. T. Gibson, V. Mahadevan, D. M. Jerina, H. Yagi, and H. J. C. Yeh, *Science* **189**, 295 (1975).

812. R. A. Johnson, in *Oxidation in Organic Chemistry*, Part C, W. S. Trahanovsky, ed., Academic Press, New York, 1978, Chapter 2, p. 131.

813. D. R. Boyd, M. R. J. Dorrity, M. V. Hand, J. F. Malone, H., Dalton, D. J. Gray, and G. N. Sheldrake, *J. Am. Chem. Soc.* **113**, 666 (1991).

814. J. Rigaudy, *Pure Appl. Chem.* **16**, 169 (1968).

815. I. Saito and T. Matsuura, in *Singlet Oxygen*, H. H. Wasserman and R. W. Murray, eds., Academic Press, New York, 1979, Chapter 10, p. 511.

816. B. Stevens, S. R. Perez, and J. A. Ors, *J. Am. Chem. Soc.* **96**, 6846 (1974).

817. C. J. M. van den Heuvel, A. Hofland, H. Steinberg, and Th. J. de Boer, *Recl. Trav. Chim. Pays-Bas* **99**, 275 (1980).

818. H. H. Wasserman and D. L. Larsen, *J. Chem. Soc., Chem. Commun.* 253 (1972).

819. H. Hart and A. Oku, *J. Chem. Soc., Chem. Commun.* 254 (1972).

820. C. J. M. van den Heuvel, H. Steinberg, and Th. J. de Boer, *Recl. Trav. Chim. Pays-Bas* **99**, 109 (1980).

821. R. Gray and V. Boekelheide, *J. Am. Chem. Soc.* **101**, 2128 (1979).

822. I. Erden, P Gölitz, R. Näder, and A. de Meijere, *Angew. Chem., Int. Ed. Engl.* **20**, 583 (1981).

823. P. A. Burns, C. S. Foote, and S. Mazur, *J. Org. Chem.* **41**, 899 (1976).

824. P. A. Burns and C. S. Foote, *J. Org. Chem.* **41**, 908 (1976).

825. O. Chalvet, R. Daudel, G. H. Schmid, and J. Rigaudy, *Tetrahedron* **26**, 365 (1970).

826. J. Rigaudy, J. Guillaume, and D. Maurette, *Bull. Soc. Chim. Fr.* 144 (1971).

827. J. Santamaria, *Tetrahedron Lett.* **22**, 4511 (1981).

828. P. S. Bailey, *Ozonation in Organic Chemistry*, Vol. 2: *Nonolefinic Compounds*, Academic Press, New York, 1982, Chapters 3–5.

829. J. P. Wibaut and P. W. Haayman, *Nature* (London) **144**, 290 (1939); *Science* **94**, 49 (1941).

830. P. S. Bailey and R. E. Erickson, *Org. Synth., Coll. Vol.* **5**, 489, 493 (1973).

831. S. N. Lewis, in *Oxidation*, R. L. Augustine, ed., Marcel Dekker, New York, 1969, Chapter 5, p. 213.

832. D. G. Lee, in *Oxidation*, R. L. Augustine, ed., Marcel Dekker, New York, 1969, Chapter 1, p. 1.

833. C. Djerassi and R. R. Engle, *J. Am. Chem. Soc.* **75**, 3838 (1953).

834. L. M. Berkowitz and P. N. Rylander, *J. Am. Chem. Soc.* **80**, 6682 (1958).

835. J. A. Caputo and R. Fuchs, *Tetrahedron Lett.* 4729 (1967).

836. R. A. Sheldon and J. K. Kochi, *Metal-Catalyzed Oxidations of Organic Compounds*, Academic Press, New York, 1981, Chapter 10, p. 315.

837. F. Freeman, in *Syntheses by Oxidation with Metal Compounds*, W. J. Mijs and C. R. H. I. de Jonge, eds., Plenum Press, New York, 1986, Chapter 5, p. 315.

838. W. J. de Klein, in *Syntheses by Oxidation with Metal Compounds*, W. J. Mijs and C. R. H. I. de Jonge, eds., Plenum Press, New York, 1986, Chapter 4, p. 261.

839. Y. Kamiya and M. Kashima, *J. Catal.* **25**, 326 (1972).

840. Y. Kamiya and M. Kashima, *Bull. Chem. Soc. Jpn.* **46**, 905 (1973).

841. E. Baciocchi, L. Mandolini, and C. Rol, *J. Org. Chem.* **45**, 3906 (1980).

842. J. Hanotier, M. Hanotier-Bridoux, and P. de Radzitzky, *J. Chem. Soc., Perkin Trans. 2*, 381 (1973).

843. W. H. Hartford and M. Darrin, *Chem. Rev.* **58**, 1 (1958).

844. K. B. Wiberg and R. Eisenthal, *Tetrahedron* **20**, 1151 (1964).

845. L. Syper, *Tetrahedron Lett.* 4193 (1967).

846. W. S. Trahanovsky and L. B. Young, *J. Org. Chem.* **31**, 2033 (1966).

847. E. Baciocchi, G. C. Rosato, C. Rol, and G. V. Sebastiani, *Tetrahedron Lett.* **33**, 5437 (1992).

848. H. Pines and W. M. Stalick, *Base-Catalyzed Reactions of Hydrocarbons and Related Compounds*, Academic Press, New York, 1977, Chapter 13, p. 508.

849. S. M. Gannon and J. G. Krause, *Synthesis* 915 (1987).

850. W.-S. Li and L. K. Liu, *Synthesis* 293 (1989).

851. R. Rangarajan and E. J. Eisenbraun, *J. Org. Chem.* **50**, 2435 (1985).

852. A. J. Pearson and G. R. Han, *J. Org. Chem.* **50**, 2791 (1985).

853. J. Muzart, *Tetrahedron Lett.* **27**, 3139 (1986).

854. H. Lee and R. G. Harvey, *J. Org. Chem.* **48**, 749 (1983).

855. J. O. Hawthorne, K. A. Schowalter, A. W. Simon, M. H. Wilt, and M. S. Morgan, *Adv. Chem. Ser.* **75**, 203 (1968).

856. J. R. Gilmore and J. M. Mellor, *J. Chem. Soc., Chem. Commun.* 507 (1970).

857. J. M. Davidson and C. Triggs, *J. Chem. Soc. A* 1331 (1968).

858. D. R. Bryant, J. E. McKeon, and B. C. Ream, *Tetrahedron Lett.* 3371 (1968).

859. D. R. Bryant, J. E. McKeon, and B. C. Ream, *J. Org. Chem.* **33**, 4123 (1968).

860. D. R. Bryant, J. E. McKeon, and B. C. Ream, *J. Org. Chem.* **34**, 1106 (1969).

861. L. Jönsson and L.-G. Wistrand, *J. Chem. Soc., Perkin Trans. 1* 669 (1979).

862. R. P. Lowry and A. Aguilo, *Hydrocarbon Process.* **53**(11), 103 (1974).

863. M. Sittig, *Hydrocarbon Process Petrol. Refiner* **41**(4), 157 (1962).

864. G. W. Parshall, *J. Mol Catal.* **4**, 243 (1978).

865. K. S. McMahon, in *Encyclopedia of Chemical Processing and Design*, Vol. 1, J. J. McKetta and W. A. Cunningam, eds., Marcel Dekker, New York, 1976, p. 216.

866. G. W. Parshall and S. D. Ittel, *Homogeneous Catalysis*, Wiley-Interscience, New York, 1992, Chapter 10, p. 237.

867. K. Weissermel and H.-J. Arpe, *Industrial Organic Chemistry*, 2nd ed., VCH, Weinheim, 1993, Chapter 7, p. 141.

868. *Chem. Eng. News* **40**(46), 50 (1962).

869. *Hydrocarbon Process.* **54**(11), 102 (1975).

870. H. Pines, *The Chemistry of Catalytic Hydrocarbon Conversions*, Academic Press, New York, 1981, Chapter 5, p. 213.

871. C. C. Hobbs, T. Horlenko, H. R. Gerberich, F. G. Mesich, R. L. Van Duyne, J. A. Bedford, D. L. Storm, and J. H. Weber, *Ind. Eng. Chem., Proc. Des. Dev.* **11**, 59 (1972).

872. V. D. Luedeke, in *Encyclopedia of Chemical Processing and Design*, Vol. 2, J. J. McKetta and W. A. Cunningam, eds., Marcel Dekker, New York, 1977, p. 128.

873. W. B. Fisher and J. F. Van Peppen, in *Kirk-Othmer Encyclopedia of Chemical Technology*, 3rd ed., Vol. 7, M. Grayson and D. Eckroth, eds., Wiley-Interscience, New York, 1979, p. 410.

874. D. D. Davis and D. R. Kemp, in *Kirk-Othmer Encyclopedia of Chemical Technology*, 4th ed., Vol. 1, J. I. Kroschwitz and M. Howe-Grant, eds., Wiley-Interscience, New York, 1991, p. 466.

875. K. Weissermel and H.-J. Arpe, *Industrial Organic Chemistry*, 2nd ed., VCH, Weinheim, 1993, Chapter 10, p. 238.

876. M. Sittig, *Hydrocarbon Process Petrol. Refiner* **41**(6), 175 (1962).

877. H. J. Boonstra and P. Zwietering, *Chem. Ind.* (London) 2039 (1966).

878. J. Alagy, L. Asselineau, C. Busson, B. Cha, and H. Sandler, *Hydrocarbon Process.* **47**(12), 131 (1968).

879. *Hydrocarbon Process.* **48**(11), 170 (1969).

880. N. J. Stevens and J. R. Livingston, Jr., *Chem. Eng. Prog.* **64**(7), 61 (1968).

881. N. Kurata, K. Koshida, H. Yokoyama, and T. Goto, *ACS Symp. Ser.* **159**, 113 (1981).

882. J. D. Wagner, G. R. Lappin, and J. R. Zietz, in *Kirk-Othmer Encyclopedia of Chemical Technology*, 4th ed., Vol. 1, J. I. Kroschwitz and M. Howe-Grant, eds., Wiley-Interscience, New York, 1991, p. 893.

883. K. Weissermel and H.-J. Arpe, *Industrial Organic Chemistry*, 2nd ed., VCH, Weinheim, 1993, Chapter 8, p. 205.

884. A. N. Bashkirov, V. V. Kamzolkin, *Khim. Nauka i Prom.* **4**, 607 (1959); *Chem. Abstr.* **54**, 7187d (1960).

885. N. Kurata and K. Koshida, *Hydrocarbon Process.* **57**(1), 145 (1978).

886. *Hydrocarbon Process., Int. Ed.* **62**(11), 106 (1983).

887. F. Asinger, *Mono-Olefins, Chemistry and Technology*, Pergamon Press, Oxford, 1968, Chapter 6, p. 568.

888. J. M. Berty, in *Applied Industrial Catalysis*, Vol. 1, B. E. Leach, ed., Academic Press, New York, 1983, Chapter 8, p. 207.

889. R. A. Van Santen and H. P. C. E. Kuipers, *Adv. Catal.* **35**, 265 (1987).

890. I. Kiguchi, T. Kumazawa, and T. Nakai, *Hydrocarbon Process., Int. Ed.* **55**(3), 69 (1976).

891. M. Gans and B. J. Ozero, *Hydrocarbon Process.* **55**(3), 73 (1976).

892. B. DeMaglie, *Hydrocarbon Process.* **55**(3), 78 (1976).

893. *Hydrocarbon Process., Int. Ed.* **64**(11), 138, 139 (1985).

894. J. V. Porcelli, *Catal. Rev.-Sci. Eng.* **23**, 151 (1981).

895. J. C. Zomerdijk and M. W. Hall, *Catal. Rev.-Sci. Eng.* **23**, 163 (1981).

896. J. N. Cawse, J. P. Henry, M. W. Swartzlander, and P. H. Waida, in *Kirk-Othmer Encyclopedia of Chemical Technology*, 3rd ed., Vol. 9, M. Grayson and D. Eckroth, eds., Wiley-Interscience, New York, 1980, p. 432.

897. B. J. Ozero and R. Landau, in *Encyclopedia of Chemical Processing and Design*, Vol. 20, J. J. McKetta and W. A. Cunningam, eds., Marcel Dekker, New York, 1984, p. 274.

898. C. N. Satterfield, *Heterogeneous Catalysis in Industrial Practice*, McGraw-Hill, New York, 1991, Chapter 8, p. 267.

899. H. H. Voge and C. R. Adams, *Adv. Catal.* **17**, 151 (1967).

900. A. Bielanski and J. Haber, *Oxidation in Catalysis*, Marcel Dekker, New York, 1991, Chapter 7, p. 277.

901. L. Ya. Margolis, *Adv. Catal.* **14**, 429 (1963).

902. W. M. H. Sachtler, *Catal. Rev.-Sci. Eng.* **4**, 27 (1971).

903. D. J. Hucknall, *Selective Oxidation of Hydrocarbons*, Academic Press, London, 1974, Chapter 2, p. 6.

904. P. A. Kilty and W. M. H. Sachtler, *Catal. Rev.-Sci. Eng.* **10**, 1 (1975).

905. X. E. Verykios, F. P. Stein, and R. W. Coughlin, *Catal. Rev.-Sci. Eng.* **22**, 197 (1980).

906. R. W. Clayton and S. V. Norval, in *Catalysis, A Specialist Periodical Report*, Vol. 3, C. Kemball and D. A. Dowden, senior reporters, The Chemical Society, Burlington House, London, 1980, Chapter 3, p. 70.

907. W. M. H. Sachtler, C. Backx, and R. A. Van Santen, *Catal. Rev.-Sci. Eng.* **23**, 127 (1981).

908. P. A. Kilty, N. C. Rol, and W. M. H. Sachtler, *Proc. 5th Int. Congr. Catalysis*, Miami Beach, 1972, J. W. Hightower, ed., North-Holland, Amsterdam, 1973, Vol. 2, p. 929.

909. D. Bruce-Smith, E. T. Blues, and B. Griffe de Martinez, *Chem. Ind.* (London) 717 (1983).

910. D. W. Park, S. Ghazali, and G. Gau, *Appl. Catal.* **6**, 175 (1983).

911. E. L. Force and A. T. Bell, *J. Catal.* **40**, 356 (1975); **44**, 175 (1976).

912. A. J. Gait, in *Propylene and its Industrial Derivatives*, E. G. Hancock, ed., Wiley, New York, 1973, Chapter 7, p. 273.

913. R. O. Kirk and T. J. Dempsey, in *Kirk-Othmer Encyclopedia of Chemical Technology*, 3rd ed., Vol. 19, M. Grayson and D. Eckroth, eds., Wiley-Interscience, New York, 1982, p. 246.

914. J. R. Valbert, J. G. Zajacek, and D. I. Orenbuch, in *Encyclopedia of Chemical Processing and Design*, Vol. 45, J. J. McKetta, ed., Marcel Dekker, New York, 1993, p. 88.

915. K. Weissermel and H.-J. Arpe, *Industrial Organic Chemistry*, 2nd ed., VCH, Weinheim, 1993, Chapter 11, p. 263.

916. *Hydrocarbon Process., Int. Ed.* **66**(11), 86 (1987).

917. K. Yamagishi, O. Kageyama, H. Haruki, and Y. Numa, *Hydrocarbon Process.* **55**(11), 102 (1976).

918. *Hydrocarbon Process., Int. Ed.* **62**(11), 146 (1983).

919. *Hydrocarbon Process.* **46**(4), 141 (1967).

920. R. Landau, G. A. Sullivan, and D. Brown, *Chemtech* **9**, 602 (1979).

921. F. Asinger, *Mono-Olefins, Chemistry and Technology*, Pergamon Press, Oxford, 1968, Chapter 8, p. 753.

922. G. Szonyi, *Adv. Chem. Ser.* **70**, 53 (1968).

923. A. Aguliό and J. D. Penrod, in *Encyclopedia of Chemical Processing and Design*, Vol. 1, J. J. McKetta and W. A. Cunningam, eds., Marcel Dekker, New York, 1976, p. 114.

924. H. J. Hagemeyer, in *Kirk-Othmer Encyclopedia of Chemical Technology*, 4th ed., Vol. 1, J. I. Kroschwitz and M. Howe-Grant, eds., Wiley-Interscience, New York, 1991, p. 94.

925. R. Jira, W. Blau, and D. Grimm, *Hydrocarbon Process.* **55**(3), 97 (1976).

926. *Hydrocarbon Process., Int. Ed.* **62**(11), 69 (1983).

927. J. C. Fielding, in *Propylene and its Industrial Derivatives*, E. G. Hancock, ed., Wiley, New York, 1973, Chapter 6, p. 234.

928. *Hydrocarbon Process., Int. Ed.* **62**(11), 72 (1983).

929. W. L. Howard, in *Kirk-Othmer Encyclopedia of Chemical Technology*, 4th ed., Vol. 1, J. I. Kroschwitz and M. Howe-Grant, eds., Wiley-Interscience, New York, 1991, p. 176.

930. *Hydrocarbon Process.* **46**(4), 146 (1967).

931. W. Daniels, in *Kirk-Othmer Encyclopedia of Chemical Technology*, 3rd ed., Vol. 23, M. Grayson and D. Eckroth, eds., Wiley-Interscience, New York, 1983, p. 820.

932. R. B. Stobaugh, W. C. Allen, Jr., and Van R. H. Sternberd, *Hydrocarbon Process.* **51**(5), 153 (1972).

933. W. Schwerdtel, *Chem. Ind.* (London) 1559 (1968).

934. W. Schwerdtel, *Hydrocarbon Process.* **47**(11), 187 (1968).

935. *Hydrocarbon Process. Int. Ed.* **68**(11), 114 (1989).

936. *Hydrocarbon Process. Int. Ed.* **62**(11), 154 (1983).

937. Y. Tanabe, *Hydrocarbon Process., Int Ed.* **60**(9), 187 (1981).

938. *Chemtech* **18**, 759 (1988).

939. D. J. Hucknall, *Selective Oxidation of Hydrocarbons*, Academic Press, London, 1974, Chapter 3, p. 23.

940. W. G. Etzkorn, J. J. Kurland, and W. D. Neilsen, in *Kirk-Othmer Encyclopedia of Chemical Technology*, 4th ed., Vol. 1, J. I. Kroschwitz and M. Howe-Grant, eds., Wiley-Interscience, New York, 1991, p. 232.

941. D. J. Hadley and E. M. Evans, in *Propylene and its Industrial Derivatives*, E. G. Hancock, ed., Wiley, New York, 1973, Chapter 10, p. 367.

942. *Hydrocarbon Process.* **54**(11), 105 (1975).

943. H. Nakatani, *Hydrocarbon Process.* **48**(5), 152 (1969).

944. G. E. Schaal, *Hydrocarbon Process.* **52**(9), 218 (1973).

945. W. Bauer, Jr., in *Kirk-Othmer Encyclopedia of Chemical Technology*, 4th ed., Vol. 1, J. I. Kroschwitz and M. Howe-Grant, eds., Wiley-Interscience, New York, 1991, p. 287.

946. *Hydrocarbon Process.* **52**(11), 96 (1973).

947. S. Sakuyama, T. Ohara, N. Shimizu, and K. Kubota, *Chemtech* **3**, 350 (1973).

948. *Hydrocarbon Process.* **58**(11), 123 (1979).

949. J. W. Nemec and L. S. Kirch, in *Kirk-Othmer Encyclopedia of Chemical Technology*, 3rd ed., Vol. 15, M. Grayson and D. Eckroth, eds., Wiley-Interscience, New York, 1981, p. 346.

950. Y. Oda, I. Gotoh, K. Uchida, T. Morimoto, J. Endoh, and T. Ueno, *Hydrocarbon Process.* **54**(10), 115 (1975).

951. *Hydrocarbon Process., Int. Ed.* **62**(11), 116 (1983).

952. F. Veatch, J. L. Callahan, J. D. Idol, and E. C. Milberger, *Chem. Eng. Prog.* **56**(10), 65 (1960); *Hydrocarbon Process Petrol. Refiner* **41**(11), 187 (1962).

953. R. B. Stobaugh, S. G. McH. Clark, G. D. Camirand, *Hydrocarbon Process.* **50**(1), 109 (1971).

954. J. F. Brazdil, in *Kirk-Othmer Encyclopedia of Chemical Technology*, 4th ed., Vol. 1, J. I. Kroschwitz and M. Howe-Grant, eds., Wiley-Interscience, New York, 1991, p. 352.

955. *Hydrocarbon Process.* **48**(11), 147 (1969).

956. P. R. Pujado, B. V. Vora, and A. P. Krueding, *Hydrocarbon Process.* **56**(5), 169 (1977).

957. R. Higgins and P. Hayden, in *Catalysis, A Specialist Periodical Report*, Vol. 1, C. Kemball, senior reporter, The Chemical Society, Burlington House, London, 1977, Chapter 5, p. 168.

958. R. K. Grasselli and J. D. Burrington, *Adv. Catal.* **30**, 133 (1981).

959. T. P. Snyder and C. G. Hill, Jr., *Catal. Rev.-Sci. Eng.* **31**, 43 (1989).

960. C. F. Cullis and D. J. Hucknall, in *Catalysis, A Specialist Periodical Report*, Vol. 5, G. C. Bond, G. Webb, senior reporters, The Royal Society of Chemistry, Burlington House, London, 1982, Chapter 7, p. 273.

961. J. D. Burrington, C. T. Kartisek, and R. K. Grasselli, *J. Catal.* **63**, 235 (1980).

962. B. Grzybowska, J. Haber, and J. Janas, *J. Catal.* **49**, 150 (1977).

963. J. D. Burrington and R. K. Grasselli, *J. Catal.* **59**, 79 (1979).

964. I. Matsuura, R. Schut, and K. Hirakawa, *J. Catal.* **63**, 152 (1980).

965. J. D. Burrington, C. T. Kartisek, and R. K. Grasselli, *J. Catal.* **81**, 489 (1983).

966. G. Centi, *Catal. Lett.* **22**, 53 (1993).

967. *Hydrocarbon Process., Int. Ed.* **66**(11), 66 (1987).

968. A. P. Gelbein, M. C. Sze, R. and T. Whitehead, *Hydrocarbon Process.* **52**(9), 209 (1973).

969. M. C. Sze and A. P. Gelbein, *Hydrocarbon Process.* **55**(2), 103 (1976).

970. K. Weissermel and H.-J. Arpe, *Industrial Organic Chemistry*, 2nd ed., VCH, Weinheim, 1993, Chapter 14, p. 381.

971. C. Thurman, in *Kirk-Othmer Encyclopedia of Chemical Technology*, 3rd ed., Vol. 17, M. Grayson and D. Eckroth, eds., Wiley-Interscience, New York, 1982, p. 373.

972. P. R. Pujado and S. Sifniades, in *Encyclopedia of Chemical Processing and Design*, Vol. 35, J. J. McKetta and W. A. Cunningham, eds., Marcel Dekker, New York, 1990, p. 372.

973. K. Weissermel and H.-J. Arpe, *Industrial Organic Chemistry*, 2nd ed., VCH, Weinheim, 1993, Chapter 13, p. 333.

974. *Hydrocarbon Process., Int. Ed.* **70**(3), 168 (1991); **64**(11), 155 (1985); **60**(11), 196, 198 (1981).

975. P. R. Pujado, J. R. Salazar, and C. V. Berger, *Hydrocarbon Process.* **55**(3), 91 (1976).

976. H. Hock and H. Kropf, *Angew. Chem.* **69**, 313 (1957).

977. F. H. Seubold, Jr. and W. E. Vaughan, *J. Am. Chem. Soc.* **75**, 3790 (1953).

978. D. McNeil, in *Toluene, the Xylenes and Their Industrial Derivatives*, E. G. Hancock, ed., Elsevier, Amsterdam, 1982, Chapter 6, p. 172.

979. W. Schoo, J. U. Veenland, J. A. Bigot, and F. L. J. Sixma, *Rec. Trav. Chim. Pays-Bas* **80**, 134 (1961).

980. W. W. Kaeding, R. O. Lindblom, R. G. Temple, and H. I. Mahon, *Ind. Eng. Chem., Proc. Des. Dev.* **4**, 97 (1965).

981. A. P. Gelbein and A. S. Nislick, *Hydrocarbon Process.* **57**(11), 125 (1978).

982. R. W. Ingwalson and G. D. Kyker, in *Encyclopedia of Chemical Processing and Design*, Vol. 4, J. J. McKetta and W. A. Cunningham, eds., Marcel Dekker, New York, 1977, p. 296.

983. A. E. Williams, in *Kirk-Othmer Encyclopedia of Chemical Technology*, Vol. 3, M. Grayson and D. Eckroth, eds., Wiley-Interscience, New York, 1978, p. 778.

984. R. Landau and A. Saffer, *Chem. Eng. Prog.* **64**(10), 20 (1968).

985. W. W. Kaeding, *Hydrocarbon Process Petrol. Refiner* **43**(11), 173 (1964).

986. *Hydrocarbon Process Petrol. Refiner* **44**(11), 255 (1965).

987. P. H. Towle and R. H. Baldwin, *Hydrocarbon Process Petrol. Refiner* **43**(11), 149 (1964).

988. G. Messina, *Hydrocarbon Process Petrol. Refiner* **43**(11), 191 (1964).

989. *Hydrocarbon Process.* **56**(11), 134 (1977).

990. A. G. Bemis, J. A. Dindorf, B. Horwood, and C. Samans, in *Kirk-Othmer Encyclopedia of Chemical Technology*, 3rd ed., Vol. 17, M. Grayson and D. Eckroth, eds., Wiley-Interscience, New York, 1982, p. 732.

991. *Hydrocarbon Process.* **52**(11), 184 (1973).

992. *Hydrocarbon Process.* **56**(11), 230 (1977).

993. H. S. Bryant, C. A. Duval, L. E. McMakin, and J. I. Savoca, *Chem. Eng. Prog.* **67**(9), 69 (1971).

994. Y. Ichikawa and Y. Takeuchi, *Hydrocarbon Process.* **51**(11), 103 (1972).

995. *Hydrocarbon Process.* **56**(11), 147 (1977).

996. W. D. Robinson and R. A. Mount, in *Kirk-Othmer Encyclopedia of Chemical Technology*, 3rd ed., Vol. 14, M. Grayson and D. Eckroth, eds., Wiley-Interscience, New York, 1981, p. 770.

997. S. D. Cooley and J. D. Powers, in *Encyclopedia of Chemical Processing and Design*, Vol. 29, J. J. McKetta and W. A. Cunningham, eds., Marcel Dekker, New York, 1988, p. 35.

998. K. Lohbeck, H. Haferkorn, W. Fuhrmann, and N. Fedtke, in *Ullmann's Encyclopedia of Industrial Chemistry*, 5th ed., Vol. A16, B. Elvers, S. Hawkins, and G. Schulz, eds., VCH, New York, 1990, p. 53.

999. G. Chinchen, P. Davies, and R. J. Sampson, in *Catalysis: Science and Technology*, Vol. 8, J. R. Anderson and M. Boudart, eds., Springer, Berlin, 1987, Chapter 1, p. 1.

1000. M. Malow, *Hydrocarbon Process., Int. Ed.* **59**(11), 149 (1980).

1001. D. A. DeMaio, *Chem. Eng.* (NY), **87**(10), 104 (1980).

1002. J. C. Burnett, R. A. Keppel, and W. D. Robinson, *Catal. Today* **1**, 537 (1987).

1003. E. Weyens, *Hydrocarbon Process.* **53**(11), 132 (1974).

1004. F. Wirth, *Hydrocarbon Process.* **54**(8), 107 (1975).

1005. *Hydrocarbon Process., Int. Ed.* **60**(11), 179 (1981).

1006. A. Bielanski and J. Haber, *Oxidation in Catalysis*, Marcel Dekker, New York, 1991, Chapter 8, p. 371.

1007. D. J. Hucknall, *Selective Oxidation of Hydrocarbons*, Academic Press, London, 1974, Chapter 4, p. 75.

1008. R. W. Petts and K. C. Waugh, *J. Chem. Soc., Faraday Trans. 1* **78**, 803 (1982).

1009. M. Blanchard, G. Louguet, G. K. Boreskov, V. S. Muzykantov, and G. I. Panov, *Bull. Soc. Chim. Fr.* 814 (1971).

1010. B. K. Hodnett, *Catal. Rev.-Sci. Eng.* **27**, 373 (1985).

1011. G. Centi and F. Trifirò, J. R. Ebner, M. Franchetti, *Chem. Rev.* **88**, 55 (1988).

1012. G. J. Hutchings, *Appl. Catal.* **72**, 1 (1991).

1013. *Hydrocarbon Process.* **48**(11), 199 (1969).

1014. *Hydrocarbon Process., Int. Ed.* **60**(11), 180 (1981).

1015. S. C. Arnold, G. D. Suciu, L. Verde, and A. Neri, *Hydrocarbon Process., Int. Ed.* **64**(9), 123 (1985).

1016. *Hydrocarbon Process., Int. Ed.* **64**(11), 142 (1985).

1017. F. Budi, A. Neri, and G. Stefani, *Hydrocarbon Process., Int. Ed.* **61**(1), 159 (1982).

1018. G. Stefani, F. Budi, C. Fumagalli, and G. D. Suciu, in *New Developments in Selective Oxidation*, Studies in Surface Science and Catalysis, Vol. 55, G. Centi and F. Trifirò, eds., Elsevier, Amsterdam, 1990, p. 537.

1019. R. L. Varma and D. N. Saraf, *Ind. Eng. Chem., Prod. Res. Dev.* **18**, 7 (1979).

1020. E. Bordes and P. Courtine, *J. Chem. Soc., Chem. Commun.* 294 (1985).

1021. G. Busca, F. Cavani, G. Centi, and F. Trifirò, *J. Catal.* **99**, 400 (1986).

1022. F. Cavani and F. Trifirò, *Chemtech* **24**(4), 18 (1994).

1023. J. T. Gleaves, J. R. Ebner, and T. C. Kuechler, *Catal. Rev.-Sci. Eng.* **30**, 49 (1988).

1024. F. K. Towae, W. Enke, R. Jäckh, and N. Bhargawa, in *Ullmann's Encyclopedia of Industrial Chemistry*, 5th ed., Vol. A20, B. Elvers, S. Hawkins, and G. Schulz, eds., VCH, New York, 1992, p. 181.

1025. H. P. Dengler, in *Encyclopedia of Chemical Processing and Design*, Vol. 36, J. J. McKetta and W. A. Cunningham, eds., Marcel Dekker, New York, 1991, p. 33.

1026. *Hydrocarbon Process., Int. Ed.* **62**(11), 124 (1983).

1027. O. Wiedemann and W. Gierer, *Chem. Eng.* (NY) **86**(3), 62 (1979).

1028. *Hydrocarbon Process., Int. Ed.* **68**(11), 107 (1989).

1029. *Hydrocarbon Process.* **46**(11), 215 (1967).

1030. N. P. Chopey, *Chem. Eng.* (NY) **69**(2), 104 (1962).

1031. L. Verde and A. Neri, *Hydrocarbon Process., Int. Ed.* **63**(11), 83 (1984).

1032. *Hydrocarbon Process., Int. Ed.* **64**(11), 156 (1985).

1033. M. S. Wainwright and N. R. Foster, *Catal. Rev.-Sci. Eng.* **19**, 211 (1979).

1034. R. Y. Saleh and I. E. Wachs, *Appl. Catal.* **31**, 87 (1987).

1035. G. C. Bond, *J. Catal.* **116**, 531 (1989).

1036. G. C. Bond, S. Flamerz, and R. Shukri, *Faraday Discuss. Chem. Soc.* **87**, 65 (1989).

1037. M. Gasior, J. Haber, and T. Machej, *Appl. Catal.* **33**, 1 (1987).

1038. K. A. Shelstad, J. Downie, and W. F. Graydon, *Can. J. Chem. Eng.* **38**, 102 (1960).

1039. M. S. Wainwright, H. Ali, A. J. Bridgewater, and R. P. Chaplin, *J. Mol. Catal.* **38**, 383 (1986).

1040. D. W. B. Westerman, N. R. Foster, and M. S. Wainwright, *Appl. Catal.* **3**, 151 (1982).

1041. R. H. Chung, in *Kirk-Othmer Encyclopedia of Chemical Technology*, 3rd ed., Vol. 2, M. Grayson and D. Eckroth, eds., Wiley-Interscience, New York, 1978, p. 700.

1042. K. Weissermel and H.-J. Arpe, *Industrial Organic Chemistry*, 2nd ed., VCH, Weinheim, 1993, Chapter 12, p. 324.

1043. *Metalloporphyrins Catalyzed Oxidations*. F. Montanari and L. Casella, eds., Kluwer, Dordrehct, 1994.

1044. A. E. Shilov, *Metal Complexes in Biomimetic Chemical Reactions*, CRC Press, Boca Raton, 1997, Chapter 2, pp. 101–213.

1045. G. Centi, F. Cavani, and F. Trifirò, *Selective Oxidation by Heterogeneous Catalysis*, Kluwer Academic/Plenum Publishers, New York, 2001.

1046. I. E. Wachs, in *Catalysis, a Specialist Periodical Report*, Vol. 13, J. J. Spivey, senior reporter, The Royal Society of Chemistry, Cambridge, UK, 1997, Chapter 2, p. 37.

1047. M. G. Clerici and P. Ingallina, *Catal. Today* **41**, 351 (1998).

1048. Y. Moro-oka and M. Akita, *Catal. Today* **41**, 327 (1998).

1049. R. A. Sheldon, M. Wallau, I. W. C. E. Arends, and U. Schuchardt, *Acc. Chem. Res.* **31**, 485 (1998).

1050. G. Centi and S. Perathoner, *Curr. Opin. Solid State Mat.* **4**, 74 (1999).

1051. B. Grzybowska-Świerkosz, *Top. Catal.* **11/12**, 23 (2000).

1052. C. Limberg, *Chem. Eur. J.* **6**, 2083 (2000).

1053. J. S. Rafelt and J. H. Clark, *Catal. Today* **57**, 33 (2000).

1054. R. K. Grasselli, *Catal. Today* **49**, 141 (1999); *Top. Catal.* **15**, 93 (2001).

1055. E. I. Karasevich, V. S. Kulikova, A. E. Shilov, and A. A. Shteinman, *Russ. Chem. Rev.* **67**, 335 (1998).

1056. G. S. Owens, J. Arias, and M. M. Abu-Omar, *Catal. Today* **55**, 317 (2000).

1057. W. Adam, C. R. Saha-Möller, and P. A. Ganeshpure, *Chem. Rev.* **101**, 3499 (2001).

1058. *Catalytic Oxidation for the Synthesis of Specialty and Fine Chemicals*, T. Mallat and A. Baiker, eds.; *Catal. Today* **57**, 1–166 (2000).

1059. *New Developments in Selective Oxidation II*, Studies in Surface Science and Catalysis, Vol. 82, V. Cortés Corberán and S. Vic Bellón, eds., Elsevier, Amsterdam, 1994.

1060. *Heterogeneous Hydrocarbon Oxidation*, ACS Symp. Ser. Vol. 638, B. K. Warren and S. T. Oyama, eds., American Chemical Society, Washington, DC, 1996.

1061. *6th Int. Symp. Activation of Dioxygen and Homogeneous Catalytic Oxidation*, Noordwijkerhout, 1996, R. A. Sheldon, ed.; *J. Mol. Catal. A: Chem.* **117**, 1–478 (1997).

1062. *3rd World Congr. Oxidation Catalysis*, Studies in Surface Science and Catalysis, Vol. 110, R. K. Grasselli, S. T. Oyama, A. M. Gaffney, and J. E. Lyons, eds., Elsevier, Amsterdam, 1997.

1063. *ISO'99, Innovation in Selective Oxidation by Solid Catalysts*, Rimini, 1999, G. Centi, V. C. Corberán, S. Perathoner, and P. Ruiz, eds.; *Catal. Today* **61**, 1–382 (2000).

1064. *The Role of Site Isolation and Phase Cooperation in Selective Oxidation Catalysis*, R. K. Grasselli and J. M. Thomas, eds., Kloster Irsee, 2000; *Top. Catal.* **15**, 83–289 (2001).

1065. K. Otsuka and Y. Wang, *Appl. Catal., A* **222**, 145 (2001).

1066. S. S. Stahl, J. A. Labinger, and J. E. Bercaw, *Angew. Chem., Int. Ed. Engl.* **37**, 2181 (1998).

1067. H. D. Gesser and N. R. Hunter, *Catal. Today* **42**, 183 (1998).

1068. J. H. Lunsford, *Catal. Today* **63**, 165 (2000).

1069. *Advances in Natural Gas Conversion, 4th European Congr. Catalysis*, F. Basile, G. Fornasari, J. R. Rostrup-Nielsen, and A. Vaccari, eds.; *Catal. Today* **64**, 1–138 (2001).

1070. *Selective Oxidation Workshop 2000*, W. Ueda, ed., *Catal. Today* **71**, 1–223 (2001).

1071. A. Sen, *Acc. Chem. Res.* **31**, 550 (1998).

1072. D. Wolf, *Angew. Chem., Int. Ed. Engl.* **37**, 3351 (1998).

1073. R. A. Periana, D. J. Taube, S. Gamble, H. Taube, T. Satoh, and H, Fuji, *Science* **280**, 560 (1998).

1074. K. Mylvaganam, G. B. Bacskay, and N. S. Hush, *J. Am. Chem. Soc.* **121**, 4633 (1999).

1075. M. Lin, T. Hogan, and A. Sen, *J. Am. Chem. Soc.* **119**, 6048 (1997).

1076. E. G. Chepaikin, A. P. Bezruchenko, A. A. Leshcheva, G. N. Boyko, I. V. Kuzmenkov, E. H. Grigoryan, and A. E. Shilov, *J. Mol. Catal. A: Chem.* **169**, 89 (2001).

1077. T. Takemoto, K. Tabata, Y. Teng, A. Nakayama, and E. Suzuki, *Appl. Catal., A* **205**, 51 (2001).

1078. Y. Teng, Y. Yamaguchi, T. Takemoto, L. Dai, K. Tabata, and E. Suzuki, *Chem. Commun.* 371 (2000).

1079. C. E. Taylor and R. P. Noceti, *Catal. Today* **55**, 259 (2000).

1080. S. H. Taylor, J. S. J. Hargreaves, G. J. Hutchings, R. W. Joyner, and C. W. Lembacher, *Catal. Today* **42**, 217 (1998).

1081. G. J. Hutchings and S. H. Taylor, *Catal. Today* **49**, 105 (1999).

1082. P. P. Knops-Gerrits and W. A. Goddard, III, *J. Mol. Catal. A: Chem.* **166**, 135 (2001).

1083. Y. Shiota and K. Yoshizawa, *J. Am. Chem. Soc.* **122**, 12317 (2000).

1084. G. Süss-Fink, G. V. Nizova, S. Stanislas, and G. B. Shul'pin, *J. Mol. Catal. A: Chem.* **130**, 163 (1998).

1085. G. Süss-Fink, S. Stanislas, G. B. Shul'pin, and G. V. Nizova, *Appl. Organomet. Chem.* **14**, 623 (2000).

1086. A. de Lucas, J. L. Valverde, P. Cañizares, and L. Rodriguez, *Appl. Catal., A* **184**, 143 (1999).

1087. A. de Lucas, J. L. Valverde, L. Rodriguez, P. Sanchez, and M. T. Garcia, *J. Mol. Catal. A: Chem.* **171**, 195 (2001).

1088. A. Parmaliana, F. Arena, F. Frusteri, and A. Mezzapica, in *Natural Gas Conversion V*, Studies in Surface Science and Catalysis, Vol. 119, A. Parmaliana, D. Sanfilippo, F. Frusteri, A. Vaccari, and F. Arena, eds., Elsevier, Amsterdam, 1998, p. 551.

1089. R. G. Herman, Q. Sun, C. Shi, K. Klier, C.-B. Wang, H. Hu, I. E. Wachs, and M. M. Bhasin, *Catal. Today* **37**, 1 (1997).

1090. R. L. McCormick and G. O. Alptekin, *Catal. Today* **55**, 269 (2000).

1091. R. L. McCormick, G. O. Alptekin, D. L. Williamson, and T. R. Ohno, *Top. Catal.* **10**, 115 (2000).

1092. A. E. Shilov and E. I. Karasevich, in *Metalloporphyrins Catalyzed Oxidations*, F. Montanari and L. Casella, eds., Kluwer, Dordrehct, 1994, pp. 87–120.

1093. J. L. McLain, J. Lee, and J. T. Groves, in *Biomimetic Oxidations Catalyzed by Transition Metal Complexes*, B. Meunier, ed., Imperial College Press, London, 2000, pp. 91–169.

1094. G. B. Maravin, M. V. Avdeev, and E. I. Bagrii, *Petrol. Chem.* **40**, 1 (2000).

1095. J. T. Groves, M. Bonchio, T. Carofiglio, and K. Shalyaev, *J. Am. Chem. Soc.* **118**, 8961 (1996).

1096. M. A. Schiavon, Y. Iamamoto, O. R. Nascimento, and M. das Dores Assis, *J. Mol. Catal. A: Chem.* **174**, 213 (2001).

1097. J.-F. Bartoli, K. Le Barch, M. Palacio, P. Battioni, and D. Mansuy, *Chem. Commun.* 1718 (2001).

1098. W. Nam, I. Kim, Y. Kim, and C. Kim, *Chem. Commun.* 1262 (2001).

1099. F. G. Doro, J. R. L. Smith, A. G. Ferreira, and M. D. Assis, *J. Mol. Catal. A: Chem.* **164**, 97 (2000).

1100. F. S. Vinhado, C. M. C. Prado-Manso, H. C. Sacco, and Y. Iamamoto, *J. Mol. Catal. A: Chem.* **174**, 279 (2001).

1101. I. L. V. Rosa, C. M. C. P. Manso, O. A. Serra, and Y. Iamamoto, *J. Mol. Catal. A: Chem.* **160**, 199 (2000).

1102. A. Molinari, R. Amadelli, L. Antolini, A. Maldotti, P. Battioni, and D. Mansuy, *J. Mol. Catal. A: Chem.* **158**, 521 (2000).

1103. L. Cammarota, S. Campestrini, M. Carrieri, F. Di Furia, and P. Ghiotti, *J. Mol. Catal. A: Chem.* **137**, 155 (1999).

1104. C.-C. Guo, J.-X. Song, X.-B. Chen, and G.-F. Jiang, *J. Mol. Catal. A: Chem.* **157**, 31 (2000).

1105. G. Blay, I. Fernández, T. Giménez, J. R. Pedro, R. Ruiz, E. Pardo, F. Lloret, and M. C. Muñoz, *Chem. Commun.* 2102 (2001).

1106. G. B. Shul'pin, G. Süss-Fink, and L. S. Shul'pina, *Chem. Commun.* 1131 (2000).

1107. T. Tatsumi, K. A. Koyano, and N. Igarashi, *Chem. Commun.* 325 (1998).

1108. K. Miki and T. Furuya, *Chem. Commun.* 97 (1998).

1109. Z. Wang, A. E. Martell, and R. J. Motekaitis, *Chem. Commun.* 1523 (1998).

1110. P. M. Reis, J. A. L. Silva, J. J. R. F. da Silva, and A. J. L. Pombeiro, *Chem. Commun.* 1845 (2000).

1111. R. Neumann and A. M. Khenkin, *Chem. Commun.* 1967 (1998).

1112. K. Chen and L. Que, Jr., *J. Am. Chem. Soc.* **123**, 6327 (2001).

1113. G. Roelfes, M. Lubben, R. Hage, L. Que, Jr., and B. L. Feringa, *Chem. Eur. J.* **6**, 2152 (2000).

1114. C.-M. Che, K.-W. Cheng, M. C. W. Chan, T.-C. Lau, and C.-K. Mak, *J. Org. Chem.* **65**, 7996 (2000).

1115. S. Murahashi, N. Komiya, Y. Oda, T. Kuwabara, and T. Naota, *J. Org. Chem.* **65**, 9186 (2000).

1116. C. J. Moody and J. L. O'Connell, *Chem. Commun.* 1311 (2000).

1117. M. M. Dell'Anna, P. Matsrorilli, and C. F. Nobile, *J. Mol. Catal. A: Chem.* **130**, 65 (1998).

1118. F. Farzaneh, M. Majidian, and M. Ghandi, *J. Mol. Catal. A: Chem.* **148**, 227 (1999).

1119. I. Yamanaka, T. Furukawa, and K. Otsuka, *Chem. Commun.* 2209 (2000).

1120. A. Bravo, H.-R. Bjorsvik, F. Fontana, F. Minisci, and A. Serri, *J. Org. Chem.* **61**, 9409 (1996).

1121. Y. Ishii, *J. Mol. Catal. A: Chem.* **117**, 123 (1997).

1122. F. Blatter, H. Sun, S. Vasenkov, and H. Frei, *Catal. Today* **41**, 297 (1998).

1123. D. H. R. Barton, *Chem. Soc. Rev.*, **25**, 237 (1996); *J. Mol. Catal. A: Chem.* **117**, 3 (1997).

1124. D. H. R. Barton and B. M. Chabot, *Tetrahedron* **53**, 487 (1997).

1125. D. H. R. Barton, T. Li, and J. MacKinnon, *Chem. Commun.* 557 (1997).

1126. M. Álvaro, B. Ferrer, H. García, and A. Sanjuán, *Tetrahedron* **55**, 11895 (1999).

1127. D. H. R. Barton and N. C. Delanghe, *Tetrahedron* **54**, 4471 (1998).

1128. D. H. R. Barton and W. Chavasiri, *Tetrahedron* **53**, 2997 (1997).

1129. B. Singh, J. R. Long, F. F. de Biani, D. Gatteschi, and P. Stavropoulos, *J. Am. Chem. Soc.* **119**, 7030 (1997).

1130. D. T. Sawyer, A. Sobkowiak, and T. Matsushita, *Acc. Chem. Res.* **29**, 409 (1996).

1131. C. Walling, *Acc. Chem. Res.* **31**, 155 (1998).

1132. P. A. MacFaul, D. D. M. Wayner, and K. U. Ingold, *Acc. Chem. Res.* **31**, 159 (1998).

1133. M. J. Perkins, *Chem. Soc. Rev.* **25**, 229 (1996).

1134. F. Gozzo, *J. Mol. Catal. A: Chem.* **171**, 1 (2001).

1135. F. Minisci, F. Fontana, S. Araneo, F. Recupero, and L. Zhao, *Synlett* 119 (1996).

1136. S. Kiani, A. Tapper, R. J. Staples, and P. Stavropoulos, *J. Am. Chem. Soc.* **122**, 7503 (2000).

1137. A. E. Tapper, J. R. Long, R. J. Staples, and P. Stavropoulos, *Angew. Chem., Int. Ed. Engl.* **39**, 2343 (2000).

1138. P. Stavropoulos, R. Çelenligil-Çetin, and A. E. Tapper, *Acc. Chem. Res.* **34**, 745 (2001).

1139. D. H. R. Barton and T. Li, *Chem. Commun.* 821 (1998).

1140. U. Schuchardt, M. J. D. M. Jannini, D. T. Richens, M. C. Guerreiro, and E. V. Spinacé, *Tetrahedron* **57**, 2685 (2001).

1141. J. M. Thomas, *Angew. Chem., Int. Ed. Engl.* **38**, 3589 (1999).

1142. J. M. Thomas, R. Raja, G. Sankar, B. F. G. Johnson, and D. W. Lewis, *Chem. Eur. J.* **7**, 2973 (2001).

1143. J. M. Thomas, R. Raja, G. Sankar, and R. G. Bell, *Acc. Chem. Res.* **34**, 191 (2001).

1144. J. M. Thomas and R. Raja, *Chem. Commun.* 675 (2001).

1145. J. M. Thomas, R. Raja, G. Sankar, and R. G. Bell, *Nature* **398**, 227 (1999).

1146. R. Raja, G. Sankar, and J. M. Thomas, *Angew. Chem., Int. Ed. Engl.* **39**, 2313 (2000).

1147. R. Raja, G. Sankar, and J. M. Thomas, *J. Am. Chem. Soc.* **121**, 11926 (1999).

1148. M. Dugal, G. Sankar, R. Raja, and J. M. Thomas, *Angew. Chem., Int. Ed. Engl.* **39**, 2311 (2000).

1149. T. Maschmeyer, R. D. Oldroyd, G. Sankar, J. M. Thomas, I. J. Shannon, J. A. Klepetko, A. F. Masters, J. K. Beattie, and C. R. A. Catlow, *Angew. Chem., Int. Ed. Engl.* **36**, 1639 (1997).

1150. U. Schuchardt, D. Cardoso, R. Sercheli, R. Pereira, R. S. da Cruz, M. C. Guerreiro, D. Mandelli, E. V. Spinacé, and E. L. Pires, *Appl. Catal., A* **211**, 1 (2001).

1151. M. M. Lin, *Appl. Catal., A* **207**, 1 (2001).

1152. R. Mallada, M. Menéndez, and J. Santamaría, *Catal. Today* **56**, 191 (2000).

1153. E. Xue, J. R. H. Ross, R. Mallada, M. Menendez, J. Santamaria, J. Perregard, and P. E. H. Nielsen, *Appl. Catal., A* **210**, 271 (2001).

1154. G.-U. Wolf, B. Kubias, B. Jacobi, and B. Lücke, *Chem. Commun.* 1517 (2000).

1155. B. Chen and E. J. Munson, *J. Am. Chem. Soc.* **121**, 11024 (1999).

1156. R. M. Contractor, *Chem. Eng. News* **54**, 5627 (1999).

1157. B. Meunier, in *Metalloporphyrins Catalyzed Oxidations*, F. Montanari and L. Casella, eds., Kluwer, Dordrehct, 1994, pp. 4–15.

1158. G. G. A. Balavoine and E. Manouty, *Appl. Organomet. Chem.* **9**, 199 (1995).

1159. E. N. Jacobsen, in *Comprehensive Organometallic Chemistry II*, Vol. 12, G. Wilkinson, F. G. A. Stone, E. W. Abel, and L. S. Hegedus, eds., Pergamon Press, Oxford, 1995, Chapter 11.1.

1160. B. Notari, *Adv. Catal.* **41**, 253 (1996).

1161. R. A. Sheldon, in *Applied Homogeneous Catalysis with Organometallic Complexes*, B. Cornils and W. A. Herrmann, eds., VCH, Weinheim, 1996, Chapter 2.4.3, p. 411.

1162. K. P. Gable, *Adv. Organomet. Chem.* **41**, 127 (1997).

1163. R. A. Sheldon, I. W. C. E. Arends, and H. E. B. Lempers, *Catal. Today* **41**, 387 (1998).

1164. A. Archelas and R. Furstoss, *Top. Curr. Chem.* **200**, 160 (1999).

1165. D. C. Sherrington, *Catal. Today* **57**, 87 (2000).

1166. M. Dusi, T. Mallat, and A. Baiker, *Catal. Rev.-Sci. Eng.* **42**, 213 (2000).

1167. D. E. De Vos, I. F. J. Vankelecom, and P. A. Jacobs, *Chiral Catalyst Immobilization and Recycling*, Wiley-VCH, Weinheim, 2000, Chapter 10, p. 235.

1168. I. W. C. E. Arends and R. A. Sheldon, *Appl. Catal., A* **212**, 175 (2001).

1169. R. A. Sheldon and M. C. A. van Vliet, in *Fine Chemicals through Heterogeneous Catalysis*, R. A. Sheldon and H. van Bekkum, eds., Wiley–VCH, Weinheim, 2001, Chapter 9.1, p. 473.

1170. H. Tian, X. She, L. Shu, H. Yu, and Y. Shi, *J. Am. Chem. Soc.* **122**, 11551 (2000).

1171. A. Armstrong, G. Ahmed, I. Garnett, and K. Goacolou, *Synlett* 1075 (1997).

1172. X. Lusinchi and G. Hanquet, *Tetrahedron* **53**, 13727 (1997).

1173. P. C. B. Page, G. A. Rassias, D. Bethell, and M. B. Schilling, *J. Org. Chem.* **63**, 2774 (1998).

1174. A. Armstrong, G. Ahmed, I. Garnett, K. Goacolou, and J. S. Wailes, *Tetrahedron* **55**, 2341 (1999).

1175. S. Minakata, A. Takemiya, K. Nakamura, I. Ryu, and M. Komatsu, *Synlett* 1810 (2000).

1176. S. E. Denmark and Z. Wu, *Synlett* 847 (1999).

1177. M. Frohn and Y. Shi, *Synthesis* 1979 (2000).

1178. Z. X. Wang, Y. Tu, M. Frohn, J. R. Zhang, and Y. Shi, *J. Am. Chem. Soc.* **119**, 11224 (1997).

1179. D. Yang, Y. C. Yip, J. Chen, and K. K. Cheung, *J. Am. Chem. Soc.* **120**, 7659 (1998).

1180. Y. Tu, Z.-X. Wang, M. Frohn, M. He, H. Yu, Y. Tang, and Y. Shi, *J. Org. Chem.* **63**, 8475 (1998).

1181. S. E. Denmark, Z. Wu, C. M. Crudden, and H. Matsuhashi, *J. Org. Chem.* **62**, 8288 (1997).

1182. A. Armstrong and B. R. Hayter, *Chem. Commun.* 621 (1998).

1183. D. S. Brown, B. A. Marples, P. Smith, and L. Walton, *Tetrahedron* **51**, 3587 (1995).

1184. R. D. Bach, M. N. Glukhovtsev, C. Gonzalez, M. Marquez, C. M. Estévez, A. G. Baboul, and H. B. Schlegel, *J. Phys. Chem. A* **101**, 6092 (1997).

1185. C. Jenson, J. Liu, K. N. Houk, and W. L. Jorgensen, *J. Am. Chem. Soc.* **119**, 12982 (1997).

1186. X. Du and K. N. Houk, *J. Org. Chem.* **63**, 6480 (1998).

1187. D. V. Deubel, G. Frenking, H. M. Senn, and J. Sundermeyer, *Chem. Commun.* 2469 (2000).

1188. D. V. Deubel, *J. Org. Chem.* **66**, 3790 (2001).

1189. J. Liu, K. N. Houk, A. Dinoi, C. Fusco, and R. Curci, *J. Org. Chem.* **63**, 8565 (1998).

1190. A. Bravo, F. Fontana, G. Fronza, F. Minisci, and L. Zhao, *J. Org. Chem.* **63**, 254 (1998).

1191. A. McKillop and W. R. Sanderson, *Tetrahedron* **51**, 6145 (1995).

1192. H. Yao and D. E. Richardson, *J. Am. Chem. Soc.* **122**, 3220 (2000).

1193. D. Mandelli, M. C. A. van Vliet, R. A. Sheldon, and U. Schuchardt, *Appl. Catal., A* **219**, 209 (2001).

1194. L. Shu and Y. Shi, *J. Org. Chem.* **65**, 8807 (2000).

1195. P. A. Ganespure and W. Adam, *Synthesis* 179 (1996).

1196. M. C. A. van Vliet, I. W. C. E. Arends, and R. A. Sheldon, *Synlett* 1305 (2001).

1197. K. Neimann and R. Neumann, *Chem. Commun.* 487 (2001).

1198. K. Neimann and R. Neumann, *Org. Lett.* **2**, 2861 (2000).

1199. M. C. A. van Vliet, I. W. C. E. Arends, and R. A. Sheldon, *Synlett* 248 (2001).

1200. F. Loeker and W. Leitner, *Chem. Eur. J.* **6**, 2011 (2000).

1201. R. Neumann and M. Dahan, *Nature* **388**, 353 (1997).

1202. K. Sato, M. Aoki, M. Ogawa, T. Hashimoto, D. Panyella, and R. Noyori, *Bull. Chem. Soc. Jpn.* **70**, 905 (1997).

1203. X.-G. Zhou, X.-Q. Yu, J.-S. Huang, S.-G. Li, L.-S. Li, and C.-M. Che, *Chem. Commun.* 1789 (1999).

1204. R. I. Kureshy, N. H. Khan, S. H. R. Abdi, S. T. Patel, P. Iyer, E. Suresh, and P. Dastidar, *J. Mol. Catal. A: Chem.* **160**, 217 (2000).

1205. R. Ben-Danial, A. M. Khenkin, and R. Neumann, *Chem. Eur. J.* **6**, 3722 (2000).

1206. R. I. Kureshy, N. H. Khan, S. H. R. Abdi, S. T. Patel, and R. V. Jasra, *Tetrahedron: Asymmetry* **12**, 433 (2001).

1207. J. M. Mitchell and N. S. Finney, *J. Am. Chem. Soc.* **123**, 862 (2001).

1208. B. S. Lane and K. Burgess, *J. Am. Chem. Soc.* **123**, 2933 (2001).

1209. M. C. White, A. G. Doyle, and E. N. Jacobsen, *J. Am. Chem. Soc.* **123**, 7194 (2001).

1210. A. M. Daly, M. F. Renehan, and D. G. Gilheany, *Org. Lett.* **3**, 663 (2001).

1211. M. Havranek, A. Singh, and D. Sames, *J. Am. Chem. Soc.* **121**, 8965 (1999).

1212. M. B. Francis and E. N. Jacobsen, *Angew. Chem., Int. Ed. Engl.* **38**, 937 (1999).

1213. J. Chisem (neé Bovey), I. C. Chisem, J. S. Rafelt, D. J. Macquarrie, and J. H. Clark, *Chem. Commun.* 2203 (1997).

1214. J. P. Collman, Z. Wang, A. Straumanis, and M. Quelquejeu, *J. Am. Chem. Soc.* **121**, 460 (1999).

1215. Z. Gross and S. Ini, *J. Org. Chem.* **62**, 5514 (1997).

1216. A. K. Mandal and J. Iqbal, *Tetrahedron* **53**, 7641 (1997).

1217. R. Zhang, W.-Y. Yu, T.-S. Lai, and C.-M. Che, *Chem. Commun.* 409 (1999).

1218. T.-S. Lai, R. Zhang, K.-K. Cheung, H.-L. Kwong, and C.-M. Che, *Chem. Commun.* 1583 (1998).

1219. W. Nam, M. H. Lim, S.-Y. Oh, J. H. Lee, H. J. Lee, S. K. Woo, C. Kim, and W. Shin, *Angew. Chem., Int. Ed. Engl.* **39**, 3646 (2000).

1220. S. Campestrini and U. Tonellato, *J. Mol. Catal. A: Chem.* **171**, 37 (2001).

1221. F. J. Waller, A. J. Bailey, W. P. Griffith, S. P. Marsden, and E. H. Smith, *J. Mol. Catal. A: Chem.* **154**, 85 (2000).

1222. D. Dolphin, T. G. Traylor, and L. Y. Xie, *Acc. Chem. Res.* **30**, 251 (1997).

1223. M. das Dores Assis and J. R. L. Smith, *J. Chem. Soc., Perkin Trans. 2* 2221 (1998).

1224. L. Zhang, T. Sun, and J. Y. Ying, *Chem. Commun.* 1103 (1999).

1225. C.-J. Liu, S.-G. Li, W.-Q. Pang, and C.-M. Che, *Chem. Commun.* 65 (1997).

1226. J. Poltowicz, E. M. Serwicka, E. Bastardo-Gonzalez, W. Jones, and R. Mokaya, *Appl. Catal., A* **218**, 211 (2001).

1227. M. E. Niño, S. A. Giraldo, and E. A. Páez-Mozo, *J. Mol. Catal. A: Chem.* **175**, 139 (2001).

1228. X.-Q. Yu, J.-S. Huang, W.-Y. Yu, and C.-M. Che, *J. Am. Chem. Soc.* **122**, 5337 (2000).

1229. J. T. Groves, J. Lee, and S. S. Marla, *J. Am. Chem. Soc.* **119**, 6269 (1997).

1230. W.-H. Fung, W.-Y. Yu, and C.-M. Che, *J. Org. Chem.* **63**, 7715 (1998).

1231. C.-J. Liu, W.-Y. Yu, C.-M. Che, and C.-H. Yeung, *J. Org. Chem.* **64**, 7365 (1999).

1232. W. A. Herrmann and F. E. Kühn, *Acc. Chem. Res.* **30**, 169 (1997).

1233. J. H. Espenson, *Chem. Commun.* 479 (1999).

1234. J. Rudolph, K. L. Reddy, and J. P. Chiang, *J. Am. Chem. Soc.* **119**, 6189 (1997).

1235. C. Copéret, H. Adolfsson, and K. B. Sharpless, *Chem. Commun.* 1565 (1997).

1236. H. Adolfsson, C. Copéret, J. P. Chiang, and A. K. Yudin, *J. Org. Chem.* **65**, 8651 (2000).

1237. W. A. Herrmann, R. M. Kratzer, H. Ding, W. R. Thiel, and H. Glas, *J. Organomet. Chem.* **555**, 293 (1998).

1238. R. Neumann and T.-J. Wang, *Chem. Commun.* 1915 (1997).

1239. W. Adam, C. R. Saha-Möller, and O. Weichold, *J. Org. Chem.* **65**, 2897 (2000).

1240. W. Adam and C. M. Mitchell, *Angew. Chem., Int. Ed. Engl.* **35**, 533 (1996).

1241. A. K. Yudin and K. B. Sharpless, *J. Am. Chem. Soc.* **119**, 11536 (1997).

1242. A. R. Vaino, *J. Org. Chem.* **65**, 4210 (2000).

1243. D. Mandelli, M. C. A. van Vliet, U. Arnold, R. A. Sheldon, and U. Schuchardt, *J. Mol. Catal. A: Chem.* **168**, 165 (2001).

1244. M. C. A. van Vliet, I. W. C. E. Arends, and R. A. Sheldon, *Chem. Commun.* 821 (1999).

1245. P. Gisdakis, S. Antonczak, S. Köstlmeier, W. A. Herrmann, and N. Rosch, *Angew. Chem., Int. Ed. Engl.* **37**, 2211 (1998).

1246. A. M. Al-Ajlouni and J. H. Espenson, *J. Org. Chem.* **61**, 3969 (1996).

1247. C. C. Romão, F. E. Kühn, and W. A. Herrmann, *Chem. Rev.* **97**, 3197 (1997).

1248. J. M. Thomas and G. Sankar, *Acc. Chem. Res.* **34**, 571 (2001).

1249. I. Schmidt, A. Krogh, K. Winnberg, A. Carlsson, M. Brorson, and C. J. H. Jacobsen, *Chem. Commun.* 2157 (2000).

1250. N. Jappar, Q. Xia, and T. Tatsumi, *J. Catal.* **180**, 132 (1998).

1251. M. J. Diaz-Cabañas, L. A. Villaescusa, and M. A. Camblor, *Chem. Commun.* 761 (2000).

1252. P. Wu and T. Tatsumi, *Chem. Commun.* 897 (2001).

1253. Y. Goa, P. Wu, and T. Tatsumi, *Chem. Commun.* 1714 (2001).

1254. Z. Liu, G. M. Crumbaugh, and R. J. Davis, *J. Catal.* **159**, 83 (1996).

1255. S. Klein and W. F. Maier, *Angew. Chem., Int. Ed. Engl.* **35**, 2230 (1996).

1256. H. Kochkar and F. Figueras, *J. Catal.* **171**, 420 (1997).

1257. C. A. Müller, M. Schneider, T. Mallat, and A. Baiker, *J. Catal.* **189**, 221 (2000).

1258. R. Hutter, T. Mallat, D. C. M. Dutoit, and A. Baiker, *Top. Catal.* **3**, 421 (1996).

1259. L. Y. Chen, G. K. Chauch, and S. Jaenicke, *Catal. Lett.* **50**, 107 (1998).

1260. M. C. Capel-Sanchez, J. M. Campos-Martin, J. L. G. Fierro, M. P. de Frutos, and A. P. Polo, *Chem. Commun.* 855 (2000).

1261. J. M. Fraile, J. I. García, J. A. Mayoral, and E. Vispe, *J. Catal.* **189**, 40 (2000).

1262. J. M. Fraile, J. I. García, J. A. Mayoral, E. Vispe, D. R. Brown and M. Naderi, *Chem. Commun.* 1510 (2001).

1263. M. P. Attfield, G. Sankar, and J. M. Thomas, *Catal. Lett.* **70**, 155 (2000).

1264. G. Blanco-Brieva, J. M. Campos-Martín, M. P. de Frutos, and J. L. G. Fierro, *Chem. Commun.* 2228 (2001).

1265. D. Juwiler and R. Neumann, *Catal. Lett.* **72**, 241 (2001).

1266. P. P. Pescarmona, J. C. van der Waal, I. E. Maxwell, and T. Maschmeyer, *Angew. Chem., Int. Ed. Engl.* **40**, 740 (2001).

1267. A. L. de Oliveria, A. Wolf, and F. Schüth, *Catal Lett.* **73**, 157 (2001).

1268. T. A. Nijhuis, B. J. Huizinga, M. Makkee, and J. A. Moulijn, *Ind. Eng. Chem. Res.* **38**, 884 (1999).

1269. K. Murata and Y. Kiyozumi, *Chem. Commun.* 1356 (2001).

1270. G. Li, X. S. Wang, H. S. Yan, Y. Y. Chen, and Q. S. Su, *Appl. Catal., A* **218**, 31 (2001).

1271. C. Murata, H. Yoshida, and T. Hattori, *Chem. Commun.* 2412 (2001).

1272. C. T. Dalton, K. M. Ryan, V. M. Wall, C. Bousquet, and D. G. Gilheany, *Top. Catal.* **5**, 75 (1998).

1273. K. Katsuki, *Coord. Chem. Rev.* **140**, 189 (1995).

1274. Y. N. Ito and T. Katsuki, *Bull. Chem. Soc. Jpn.* **72**, 603 (1999).

1275. P. Pietikäinen, *J. Mol. Catal. A: Chem.* **165**, 73 (2001).

1276. H. Nishikori, C. Ohta, and T. Katsuki, *Synlett* 1557 (2000).

1277. T. Takeda, R. Irie, Y. Shinoda, and T. Katsuki, *Synlett* 1157 (1999).

1278. C. E. Song and E. J. Roh, *Chem. Commun.* 837 (2000).

1279. M. Cavazzini, A. Manfredi, F. Montanari, S. Quici, and G. Pozzi, *Chem. Commun.* 2171 (2000).

1280. T. Linker, *Angew. Chem., Int. Ed. Engl.* **36**, 2060 (1997).

1281. D. Feichtinger and D. A. Plattner, *Angew. Chem., Int. Ed. Engl.* **36**, 1718 (1997).

1282. L. Cavallo and H. Jacobsen, *Angew. Chem., Int. Ed. Engl.* **39**, 589 (2000).

1283. J. El-Bahraoui, O. Wiest, D. Feichtinger, and D. A. Plattner. *Angew. Chem., Int. Ed. Engl.* **40**, 2073 (2001).

1284. H. Jacobsen and L. Cavallo, *Chem. Eur. J.* **7**, 800 (2001).

1285. M. J. Sabater, M. Álvaro, H. García, E. Palomares, and J. C. Scaiano, *J. Am. Chem. Soc.* **123**, 7074 (2001).

1286. K. A. Campbell, M. R. Lashley, J. K. Wyatt, M. H. Nantz, and R. D. Britt, *J. Am. Chem. Soc.* **123**, 5710 (2001).

1287. J. M. Fraile, J. I. García, J. Massam, and J. A. Mayoral, *J. Mol. Catal. A: Chem.* **136**, 47 (1998).

1288. P. Piaggio, C. Langham, P. McMorn, D. Bethell, P. C. Bulman-Page, F. E. Hancock, C. Sly, and G. J. Hutchings, *J. Chem. Soc., Perkin Trans. 2* 143 (2000).

1289. B. M. Choudary, M. L. Kantam, B. Bharathi, P. Sreekanth, and F. Figueras, *J. Mol. Catal. A: Chem.* **159**, 417 (2000).

1290. L. Canali, E. Cowan, H. Deleuze, C. L. Gibson, and D. C. Sherrington, *Chem. Commun.* 2561 (1998).

1291. M. D. Angelino and P. E. Laibinis, *J. Polym. Sci. A: Polym. Chem.* **37**, 3888 (1999).

1292. D. A. Annis and E. N. Jacobsen, *J. Am. Chem. Soc.* **121**, 4147 (1999).

1293. T. S. Reger and K. D. Janda, *J. Am. Chem. Soc.* **122**, 6929 (2000).

1294. P. K. Dhal, B. B. De, and S. Sivaram, *J. Mol. Catal. A: Chem.* **177**, 71 (2001).

1295. M. Stratakis and M. Orfanopoulos, *Tetrahedron* **56**, 1595 (2000).

1296. J. Shailaja, J. Sivaguru, R. J. Robbins, V. Ramamurthy, R. B. Sunoj, and J. Chandrasekhar, *Tetrahedron* **56**, 6927 (2000).

1297. E. L. Clennan, *Tetrahedron* **56**, 9151 (2000).

1298. C. Döbler, G. M. Mehltretter, U. Sundermeier, and M. Beller, *J. Am. Chem. Soc.* **122**, 10289 (2000).

1299. S. Y. Jonsson, H. Adolfsson, and J.-E. Bäckvall, *Org. Lett.* **3**, 3463 (2001).

1300. C. Bolm and A. Gerlach, *Eur. J. Org. Chem.* 21 (1998).

1301. C. Bolm and A. Maischak, *Synlett* 93 (2001).

1302. P. Salvadori, D. Pini, and A. Petri, *Synlett* 1181 (1999).

1303. C. Bolm, A. Maischak and A. Gerlach, *Chem. Commun.* 2353 (1997).

1304. S. Nagayama, M. Endo, and S. Kobayashi, *J. Org. Chem.* **63**, 6094 (1998).

1305. S. Kobayashi, M. Endo, and S. Nagayama, *J. Am. Chem. Soc.* **121**, 11229 (1999).

1306. A. Severeyns, D. E. De Vos, L. Fiermans, F. Verpoort, P. J. Grobet, and P. A. Jacobs, *Angew. Chem., Int. Ed. Engl.* **40**, 586 (2001).

1307. K. N. Houk and T. Strassner, *J. Org. Chem.* **64**, 800 (1999).

1308. A. J. DelMonte, J. Haller, K. N. Houk, K. B. Sharpless, D. A. Singleton, Th. Strassner, and A. A. Thomas, *J. Am. Chem. Soc.* **119**, 9907 (1997).

1309. B. Betzemeier, F. Lhermitte, and P. Knochel, *Tetrahedron Lett.* **39**, 6667 (1998).

1310. M. J. Gaunt, J. Yu, and J. B. Spencer, *Chem. Commun.* 1844 (2001).

1311. K. Nowińska and D. Dudko, *Appl. Catal., A* **159**, 75 (1997).

1312. Y. Kim, H. Kim, J. Lee, K. Sim., Y. Han, and H. Paik, *Appl. Catal., A* **155**, 15 (1997).

1313. T. Nishimura, N. Kakiuchi, T. Onoue, K. Ohe, and S. Uemura, *J. Chem. Soc., Perkin Trans. 1* 1915 (2000).

1314. G.-J. ten Brink, I. W. C. E. Arends, G. Papadogianakis, and R. A. Sheldon, *Appl. Catal., A* **194–195**, 435 (2000).

1315. E. Monflier, S. Tilloy, E. Blouet, Y. Barbaux, and A. Mortreux, *J. Mol. Catal. A: Chem.* **109**, 27 (1996).

1316. E. Karakhanov, A. Maximov, and A. Kirillov, *J. Mol. Catal. A: Chem.* **157**, 25 (2000).

1317. A. W. Stobbe-Kreemers, R. B. Dielis, M. Makkee, and J. J. F. Scholten, *J. Catal.* **154**, 175 (1995).

1318. A. W. Stobbe-Kreemers, M. Makkee, and J. J. F. Scholten, *Appl. Catal., A* **156**, 219 (1997).

1319. M. Li, J. Shen, X. Ge, and X. Chen, *Appl. Catal., A* **206**, 161 (2001).

1320. A. Kishi, T. Higashino, S. Sakaguchi, and Y. Ishii, *Tetrahedron Lett.* **41**, 99 (2000).

1321. K. Nowińska, M. Sopa, D. Dudko, and M. Mocna, *Catal. Lett.* **49**, 43 (1997).

1322. F. Frusteri, A. Parmaliana, N. M. Ostrovskii, A. Iannibello, and N. Giordano, *Catal. Lett.* **46**, 57 (1997).

1323. C. R. Reilly and J. J. Lerou, *Catal. Today* **41**, 433 (1998).

1324. H. Jiang, L. Jia, and J. Li, *Green Chem.* **2**, 161 (2000).

1325. E. Antonelli, R. D'Aloisio, M. Gambaro, T. Fiorani, and C. Venturello, *J. Org. Chem.* **63**, 7190 (1998).

1326. K. Sato, M. Aoki, and R. Noyori, *Science* **281**, 1646 (1998).

1327. Y. Deng, Z. Ma, K. Wang, and J. Chen, *Green Chem.* **1**, 275 (1999).

1328. D. Yang and C. Zhang, *J. Org. Chem.* **66**, 4814 (2001).

1329. H. S. Shin, C. W. Lee, J. Y. Lee, and T. S. Huh, *Eur. J. Org. Chem.* 335 (2000).

1330. I. C. Jung, *Eur. J. Org. Chem.* 1899 (2001).

1331. R. Fajgar, J. Vitek, Y. Haas, and J. Pola, *J. Chem. Soc., Perkin Trans. 2* 239 (1999).

1332. R. Gutbrod, E. Kraka, R. N. Schindler, and D. Cremer, *J. Am. Chem. Soc.* **119**, 7330 (1997).

1333. J. M. Anglada, R. Crehuet, and J. M. Bofill, *Chem. Eur. J.* **5**, 1809 (1999).

1334. X.-M. Zhang and Q. Zhu, *J. Org. Chem.* **62**, 5934 (1997).

1335. K. Schank, H. Beck, M. Buschlinger, J. Eder, T. Heisel, S. Pistorius, and C. Wagner, *Helv. Chim. Acta* **83**, 801 (2000).

1336. H. Grennberg, K. Bergstad, and J.-E. Bäckvall, *J. Mol. Catal. A: Chem.* **113**, 355 (1996).

1337. T. Yokota, S. Fujibayashi, Y. Nishiyama, S. Sakaguchi, and Y. Ishii, *J. Mol. Catal. A: Chem.* **114**, 113 (1996).

1338. N. Yu. Kozitsyna, M. N. Vargaftik, and I. I. Moiseev, *J. Organomet. Chem.* **593–594**, 274 (2000).

1339. P. J. Baricelli, V. J. Sánchez, A. J. Pardey, and S. A. Moya, *J. Mol. Catal. A: Chem.* **164**, 77 (2000).

1340. A. Maldotti, L. Andreotti, A. Molinari, S. Borisov, and V. Vasil'ev, *Chem. Eur. J.* **7**, 3564 (2001).

1341. A. Sanjuán, M. Alvaro, A. Corma, and H. García, *Chem. Commun.* 1641 (1999).

1342. Z. Lei, *React. Funct. Polym.* **39**, 239 (1999).

1343. T. Matsushita, D. T. Sawyer, and A. Sobkowiak, *J. Mol. Catal. A: Chem.* **137**, 127 (1999).

1344. J.-F. Pan and K. Chen, *J. Mol. Catal. A: Chem.* **176**, 19 (2001).

1345. T. Punniyamurthy, B. Bhatia, M. M. Reddy, G. C. Maikap, and J. Iqbal, *Tetrahedron* **53**, 7649 (1997).

1346. L. K. Stultz, M. H. V. Huynh, R. A. Binstead, M. Curry, and T. J. Meyer, *J. Am. Chem. Soc.* **122**, 5984 (2000).

1347. J. Eames and M. Watkinson, *Angew. Chem., Int. Ed. Engl.* **40**, 3567 (2001).

1348. K. Kawasaki and T. Katsuki, *Tetrahedron* **53**, 6337 (1997).

1349. G. Sekar, A. DattaGupta, and V. K. Singh, *J. Org. Chem.* **63**, 2961 (1998).

1350. J. S. Clark, K. F. Tolhurst, M. Taylor, and S. Swallow, *J. Chem. Soc., Perkin Trans. 1*, 1167 (1998).

1351. M. B. Andrus and D. Asgari, *Tetrahedron* **56**, 5775 (2000).

1352. W.-S. Lee, H.-L. Kwong, H.-L. Chan, W.-W. Choi, and L.-Y. Ng, *Tetrahedron: Asymmtery* **12**, 1007 (2001).

1353. A. V. Malkov, I. R. Baxendale, M. Bella, V. Langer, J. Fawcett, D. R. Russel, D. J. Mansfield, M. Valko, and P. Kočovsky, *Organometallics* **20**, 673 (2001).

1354. N. Nojiri, Y. Sakai, and Y. Watanabe, *Catal. Rev.-Sci. Eng.* **37**, 145 (1995).

1355. P. Li, F. H. Cheong, L. C. F. Chao, Y. H. Lin, and I. D. Williams, *J. Mol. Catal. A: Chem.* **145**, 111 (1999).

1356. V. O. Rogatchov, V. D. Filimonov, and M. S. Yusubov, *Synthesis* 1001 (2001).

1357. C.-M. Che, W.-Y. Yu, P.-M. Chan, W.-C. Cheng, S.-M. Peng, K.-C. Lau, and W.-K. Lee, *J. Am. Chem. Soc.* **122**, 11380 (2000).

1358. S. Sakaguchi, T. Takase, T. Iwahama, and Y. Ishii, *Chem. Commun.* 2037 (1998).

1359. S. Dayan, I. Ben-David, and S. Rozen, *J. Org. Chem.* **65**, 8816 (2000).

1360. K. Griesbaum, Y. Dong, and K. J. McCullough, *J. Org. Chem.* **62**, 6129 (1997).

1361. K. Griesbaum and Y. Dong, *Liebigs Ann./Receuil* 753 (1997).

1362. D. Cremer, R. Crehuet, and J. Anglada, *J. Am. Chem. Soc.* **123**, 6127 (2001).

1363. N. A. Alekar, V. I. S. B. Halligudi, D. Srinivas, S. Gopinathan, and C. Gopinathan, *J. Mol. Catal. A: Chem.* **164**, 181 (2000).

1364. K. Nomiya, S. Matsuoka, T. Hasegawa, and Y. Nemoto, *J. Mol. Catal. A: Chem.* **156**, 143 (2000).

1365. K. Nomiya, K. Hashino, Y. Nemoto, and M. Watanabe, *J. Mol. Catal. A: Chem.* **176**, 79 (2001).

1366. S. S. Shevade and B. S. Rao, *Catal. Lett.* **66**, 99 (2000).

1367. J. Okamura, S. Nishiyama, S. Tsuruya, and M. Masai, *J. Mol. Catal. A: Chem.* **135**, 133 (1998).

1368. T. Miyahara, H. Kanzaki, R. Hamada, S. Kuroiwa, S. Nishiyama, and S. Tsuruya, *J. Mol. Catal. A: Chem.* **176**, 141 (2001).

1369. T. Miyake, M. Hamada, Y. Sasaki, and M. Oguri, *Appl. Catal., A* **131**, 32 (1995).

1370. T. Miyake, M. Hamada, H. Niwa, M. Nishizuka, and M. Oguri, *J. Mol. Catal. A: Chem.* **178**, 199 (2002).

1371. B. Chou, J.-L. Tsai, and S. Cheng, *Microp. Mesop. Mat.* **48**, 309 (2001).

1372. J. Jacob and J. H. Espenson, *Inorg. Chim. Acta* **270**, 55 (1998).

1373. R. Zhang, W.-Y. Yu, T.-S. Lai, and C.-M. Che, *Chem. Commun.* 1791 (1999).

1374. W. Adam, Z. Lukacs, D. Harmsen, C. R. Saha-Möller, and P. Schreier, *J. Org. Chem.* **65**, 878 (2000).

1375. M. D. Nikalje and A. Sudalai, *Tetrahedron* **55**, 5903 (1999).

1376. R. Alcantara, L. Canoira, P. Guilherme-Joao, and J. P. Perez-Mendo, *Appl. Catal., A* **218**, 269 (2001).

1377. Z. Gross, L. Simkhovich, and N. Galili, *Chem. Commun.* 599 (1999).

1378. A. Cagnina, S. Campestrini, F. Di Furia, and P. Ghiotti, *J. Mol. Catal. A: Chem.* **130**, 221 (1998).

1379. T. K. Das, K. Chaudhari, E. Nandanan, A. J. Chandwadkar, A. Sudalai, T. Ravindranathan, and S. Sivasanker, *Tetrahedron Lett.* **38**, 3631 (1997).

1380. S. Negele, K. Wieser, and T. Severin, *J. Org. Chem.* **63**, 1138 (1998).

1381. D. Ma, C. Xia, and H. Tian, *Tetrahedron Lett.* **40**, 8915 (1999).

1382. K. K. Laali, M. Herbert, B. Cushnyr, A. Bhatt, and D. Terrano, *J. Chem. Soc., Perkin Trans. 1* 578 (2001).

1383. N. A. Noureldin, D. Zhao, and D. G. Lee, *J. Org. Chem.* **62**, 8767 (1997).

1384. R. Bandyopadhyay, S. Biswas, S. Guha, A. K. Mukherjee, and R. Bhattacharyya, *Chem. Commun.* 1627 (1999).

1385. M. Higashijima, *Chem. Lett.* 1093 (1999).

1386. A. Andersson, S. Hansen, and A. Wickman, *Top. Catal.* **15**, 103 (2001).

1387. E. Bordes, *Top. Catal.* **15**, 131 (2001).

1388. S. B. Derouane-Abd Hamid, G. Centi, P. Pal, and E. G. Derouane, *Top. Catal.* **15**, 161 (2001).

1389. S. Albonetti, G. Blanchard, P. Burattin, F. Cavani, S. Masetti, and F. Trifirò, *Catal. Today* **42**, 283 (1998).

1390. H. Watanabe and Y. Koyasu, *Appl. Catal., A* **194–195**, 479 (2000).

1391. H. W. Zanthoff, W. Grünert, S. Buchholz, M. Heber, L. Stievano, F. E. Wagner, and G. U. Wolf, *J. Mol. Catal. A: Chem.* **162**, 443 (2000).

1392. S. Masetti, F. Trifirò, and G. Blanchard, *Appl. Catal., A* **217**, 119 (2001).

1393. G. Centi, F. Cavani, and F. Trifirò, *Selective Oxidation by Heterogeneous Catalysis*, Kluwer Academic/Plenum Publishers, New York, 2001, Chapter 4.3, p. 171.

1394. N. Viswanadham, T. Shido, and Y. Iwasawa, *Appl. Catal., A* **219**, 223 (2001).

1395. H. Liu, E. M. Gaigneaux, H. Imoto, T. Shido, and Y. Iwasawa, *Catal. Lett.* **71**, 75 (2001).

1396. Y. Sasaki, *Appl. Catal., A* **194–195**, 497 (2000).

1397. T.-S. Chang, L. Guijia, C.-H. Shin, Y. K. Lee, and S.-S. Yun, *Catal. Lett.* **68**, 229 (2000).

1398. S. B. Derouane-Abd Hamid, P. Pal, H. He, and E. G. Derouane, *Catal. Today* **64**, 129 (2001).

10

HETEROSUBSTITUTION

Substitutions such as alkylation (Chapter 5) and oxygenation (Chapter 9) are fundamental transformations essential to the chemistry of hydrocarbons. Other heterosubstitutions (i.e., formation of carbon-heteroatom bonds), such as halogenation, nitration, or sulfuration (sulfonation), are also widely used reactions. It is outside the aim of our book to discuss comprehensively the wide variety of substitution reactions (for a scope, see, e.g., March's *Advanced Organic Chemistry*), but it is considered useful to briefly review some of the most typical selected heterosubstitutions of hydrocarbons.

The usual way to achieve heterosubstitution of saturated hydrocarbons is by free-radical reactions. Halogenation, sulfochlorination, and nitration are among the most important transformations. Superacid-catalyzed electrophilic substitutions have also been developed. This clearly indicates that alkanes, once considered to be highly unreactive compounds (paraffins), can be readily functionalized not only in free-radical from but also via electrophilic activation. Electrophilic substitution, in turn, is the major transformation of aromatic hydrocarbons.

When heterosubstitution is carried out with highly reactive electrophilic reagents such as halogens or hydrogen peroxide in the presence of strong Lewis acids, ozone in superacids, and nitronium salts, the usual C—H insertion is accompanied by C—C insertion, resulting in C—C bond cleavage and the formation of heterosubstituted products of lower molecular weight. A few characteristic examples of such reactions (halogenolysis, nitrolysis) are also presented.

10.1. ELECTROPHILIC (ACID-CATALYZED) SUBSTITUTION

10.1.1. Substitution of Alkanes

Halogenation. Fluorination, chlorination, and bromination of alkanes catalyzed by superacids have been reported.[1,2] Reactions may be carried out in the liquid phase, or in the gas phase over solid superacids or supported noble metal catalysts. High selectivity and relatively mild reaction conditions are the main features of these transformations.

Fluorine[3,4] and fluoroxytrifluoromethane[3] were found to fluorinate tertiary carbon atoms in the liquid phase in halogenated solvents (CH_2Cl_2, $CHCl_3$, $CFCl_3$). The reaction with fluorine was shown to occur in a regio- and stereoselective fashion.[5] Experimental observations were interpreted by invoking strongly polarized fluorine species attacked by the electrons of the tertiary C—H bond to give the fluorinated product through intermediate **1**:

$$\left[\begin{array}{c} \diagdown \overset{\displaystyle |}{\underset{\displaystyle H}{\overset{\displaystyle C}{\diagup}}}\overset{\displaystyle +}{} \\ \overset{\diagup}{} \;\; F \quad F^{-\cdots}HCCl_3 \end{array} \right]$$

1

Chloroform plays a crucial role in the reaction. Among others its somewhat electrophilic hydrogen serves as an acceptor for the fluoride ion.

Electrophilic fluorination of methane itself was effected with good selectivity to methyl fluoride (accompanied by minor amounts of methylene fluoride) by reacting it with $NF_4^+SbF_6^-$ or $N_2F^+AsF_6^-$ in liquid HF or pyridinium poly(hydrogen fluoride) solution.[6]

Electrophilic chlorination of alkanes[7] may be carried out in the presence of SbF_5, $AlCl_3$, and $AgSbF_6$. The Cl_2–SbF_5–SO_2ClF system is the strongest reagent producing electrophilic chlorine ("Cl^+"). With this reagent both carbon—hydrogen and carbon–carbon σ-bond attacks can occur resulting in substitution and carbon–carbon bond cleavage (chlorolysis), respectively (Scheme 10.1). Under the conditions used a reversible dialkylchloronium ion (**2**, **3**) formation takes place, too. Butane and higher alkanes give alkylcarbocations as stable ionization products.

The weaker electrophile formed in the presence of $AlCl_3$ attacks exclusively the carbon–hydrogen bond.[7] Chlorine in the presence of $AgSbF_6$ is the weaker, and therefore the more selective chlorinating agent, which reacts only with strained systems (cyclopropane) and with alkanes possessing tertiary carbon–hydrogen bonds. $AgSbF_6$ effects the electrophilic bromination with bromine in methylene chloride of isoparaffins and cycloalkanes.[8]

Electrophilic chlorination and bromination of methane over supported noble metals[9–11] (Pt on Al_2O_3, Pd on $BaSO_4$) and solid superacid catalysts[9–13] (e.g., $TaOF_3$ on alumina, Nafion-H, zeolites, SbF_5–graphite, sulfated zirconia) have been studied (see Section 3.4.2). Monosubstitution with selectivities better than

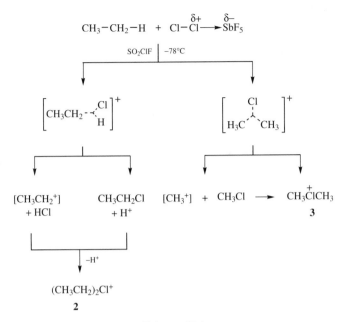

Scheme 10.1

90% are usually achieved. Insertion of electrophilic halogen species into the methane carbon–hydrogen bond may be invoked to interpret these results.[2,9]

Nitration. Electrophilic nitration of alkanes may be brought about by stable nitronium salts[14] generated by reacting nitric acid or nitrates with Lewis acids (PF_5, SbF_5) or with HF and BF_3. $NO_2^+PF_6^-$, $NO_2^+SbF_6^-$, and $NO_2^+BF_4^-$ thus prepared are powerful reagents in methylene chloride–sulfolane solution at room temperature.[15,16] Their reactivity further increases when nitration is carried out in superacid solution (HF, HSO_3F). Both substitution and nitrolysis (carbon–carbon bond cleavage) take place; the latter is usually the dominating process. Tertiary carbon–hydrogen and carbon–carbon bonds are usually the most reactive. An insertion mechanism involving two-electron, three-center-bonded five-coordinate carbocations as shown in the nitration of ethane gives both nitroethane (C—H insertion) [Eq. (10.1)] and nitromethane (C—C insertion) [Eq. (10.2)]:

$$CH_3-CH_3 + NO_2^+\,PF_6^- \longrightarrow \begin{cases} \left[\begin{array}{c} H \\ \lambda \\ H_3CH_2C \quad NO_2 \end{array}\right]^+ \longrightarrow CH_3CH_2NO_2 \quad (10.1) \\[20pt] \left[\begin{array}{c} H_3C \quad CH_3 \\ Y \\ NO_2 \end{array}\right]^+ \longrightarrow CH_3NO_2 + CH_3F \quad (10.2) \end{cases}$$

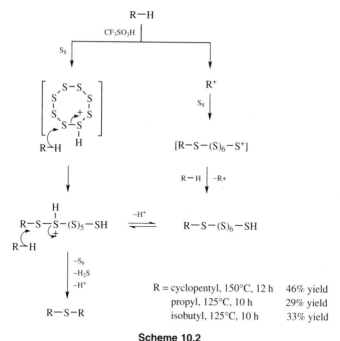

Scheme 10.2

Sulfuration. Selective sulfuration of alkanes and cycloalkanes to produce dialkyl sulfides has been reported.[17] Two mechanistic paths depicted in Scheme 10.2 may account for the product formation.

10.1.2. Substitution of Aromatic Hydrocarbons

Since numerous monographs and review papers cover in detail nearly all aspects of electrophilic aromatic substitutions, only a brief overview of some important reactions resulting in heterosubstitution is given here. Electrophilic alkylation of aromatics, the most important electrophilic substitution with regard to hydrocarbon chemistry, including new C—C bond formation, in turn, is discussed in Section 5.1.4.

Electrophilic substitutions are the most characteristic transformations of aromatic hydrocarbons.[18–26] A two-step mechanism is generally accepted to interpret the formation of heterosubstituted aromatics, although with reactive systems an initial substrate–reagent interaction step may also be involved, allowing independent substrate and positional (regio) selectivities.[25] The attack of an electrophile on the aromatic ring eventually leads to the **4** arenium ion intermediate[27–30] [Eq. (10.3)], called the *Wheland intermediate* or σ *complex*, which undergoes deprotonation to yield the substitution product:

$$
\text{[benzene]} + \text{E}^+ \rightleftharpoons \overset{\text{H}\quad\text{E}}{\text{4}}\left[\text{ring}^+\right] \xrightarrow[-\text{H}^+]{} \overset{\text{E}}{\text{[benzene]}} \tag{10.3}
$$

Depending on the nature of the reagent and, consequently, the electrophilic character (degree of ionization) of the reacting electrophile, a great variation of this basic mechanism may be observed. Carbon–hydrogen bond breaking is not significant in the rate-determining step, and no primary kinetic isotope effects, therefore, are found. This is one of the main arguments against a single-step mechanism once favored conserving aromaticity during the whole reaction. Aromatic electrophilic substitutions usually require the use of a catalyst to aid the necessary polarization of the reagent. Reactive aromatics, such as alkylbenzenes and polynuclear aromatic hydrocarbons, however, may react with certain reagents without a catalyst. This is because not only the electrophilicity of the reagent but also the nucleophilicity of the aromatic reactant affect the reaction.

Halogenation. Chlorination and bromination of aromatic hydrocarbons are usually carried out with chlorine and bromine in the presence of a catalyst.[31] Lewis acids, preferentially aluminum chloride, ferric chloride, or bromide, may be used. The latter compounds may be formed in situ by using metallic iron. Iodine may also be employed as a catalyst. Other effective halogenating agents include HOCl, HOBr, and N-chloro and N-bromo amides usually in the presence of an acid.[31,32] Zeolites have recently been shown to exert beneficial effects on electrophilic halogenation with SO_2Cl_2.[33] The high activity of ZF520 zeolite in such ring chlorinations was attributed to its high Brønsted acidity.[33–35]

Since alkyl groups are activating and *ortho–para* directing substituents, alkylbenzenes exhibit increased reactivity yielding mixtures of *ortho-* and *para*-substituted isomers.[36,37] Product distribution may be affected by the reagent and reaction conditions. Highly *para*-selective monochlorination was achieved under mild conditions with *tert*-BuOCl catalyzed by a proton-exchanged X zeolite.[38] A mixture of thallium acetate and bromine was found to be a mild and efficient reagent for almost exclusive *para* monobromination.[39] An interesting example of halogenation is transfer bromination.[40] It is based on the observation that o- and p-bromotoluene, but not the *meta* isomer, isomerize through intermolecular bromine transfer under Friedel–Crafts isomerization conditions (water-promoted $AlCl_3$).

Alkylbenzenes and other activated aromatics, such as polynuclear compounds, may react without a catalyst. Halogenation, in this case, occurs with the involvement of molecular halogens.[32,37,41] Polarization of chlorine and bromine molecules is brought about by their interaction with the aromatic ring. For example, bromination in the presence of acetic acid is suggested to take place through the transition state **5**, leading to the formation of the σ complex:

$$R-\langle\cdots\overset{\delta+}{\cdots}\rangle\cdots\overset{H}{\underset{\delta-}{Hlg}}\quad\underset{\overset{\delta-}{Hlg}}{}\quad 5$$

The kinetic order in bromine, which is often higher than unity, indicates the involvement of a second molecule of bromine in the rate-determining step assisting ionization.

Iodine, the least reactive halogen, requires the use of strong Lewis acids, such as silver ion, often in the presence of sulfuric acid.[42] Iodination is most often performed by iodine in the presence of various oxidizing agents that convert iodine to a better electrophile. Nitric acid–sulfuric acid, iodic acid, and periodic acid are used most frequently.[31,42] In contrast, with chlorination and bromination, the rate-determining step in iodination usually lies late on the reaction coordinate and, consequently, often involves carbon–hydrogen bond breaking. Competitive iodination of polymethylbenzenes with both oxidizing and nonoxidizing reagents led to the conclusion that a common intermediate, most likely the I^+ ion, is involved in these iodinations.[43]

Among the fluorinating agents of practical significance,[44] only fluoroxytrifluoromethane[45] and *N*-fluoroperfluoroalkylsulfonimides[46] were shown to fluorinate aromatic hydrocarbons by an electrophilic mechanism.

Nitration. Nitration of aromatic hydrocarbons is a well-studied area covered in monographs and review papers.[47–56] Nitration may be accomplished by a number of reagents, most commonly by a mixture of nitric acid and sulfuric acid (mixed acid). Additional important reagents are dinitrogen tetroxide in the presence of acidic catalysts, dinitrogen pentoxide under anhydrous conditions,[57] alkyl and acyl nitrates, and nitronium salts. Metallic nitrates impregnated on K10 montmorillonite were shown to nitrate alkylbenzenes to yield *p*-nitroalkylbenzenes with high selectivity.[34,35,58] The highest *para* selectivity was, however, achieved by 9-nitroanthracene. The 9-nitroanthracenium ion formed under superacidic catalysis (Nafion-H) effects transfer nitration of toluene to give *p*-nitrotoluene with >95% selectivity albeit in low yields (<5%).[59] As the use of mixed acid results in spent (diluted) acid, causing serious environmental disposal problems in industrial applications, solid acid catalysts, such as clays[34,35] or high-acidity resins (Nafion-H,[60–63] etc.), are gaining importance.

Our basic knowledge of electrophilic nitration of aromatic hydrocarbons results from the work of Ingold,[24] who established that the nitronium cation is the true reactive electrophile in these reactions.[64–67] In the traditional nitration process with the mixed acid the nitronium cation is generated according to Eqs. (10.4) and (10.5):

$$HNO_3 + H_2SO_4 \rightleftharpoons H_2NO_3^+ + HSO_4^- \tag{10.4}$$

$$H_2NO_3^+ + H_2SO_4 \rightleftharpoons NO_2^+ + H_3O^+ + HSO_4^- \tag{10.5}$$

In mineral acids, in general, these two reactions are fast, rendering the subsequent formation of the σ complex [Eq. (10.3)] to be rate-determining. The electrophilic nature of nitration is supported by kinetic studies, which give a rate expression first order in both nitric acid and the reacting aromatic compound.[24,64]

Under certain conditions, such as in organic media, the formation of the NO_2^+ ion [Eqs. (10.6) and (10.7)] becomes rate-determining, resulting in a rate zero order in the aromatic:

$$2\,HNO_3 \rightleftharpoons H_2NO_3^+ + NO_3^- \qquad (10.6)$$

$$H_2NO_3^+ \rightleftharpoons NO_2^+ + H_2O \qquad (10.7)$$

Substituent effects which indicate increased reactivity of aromatics with electron-donating substituents directing the incoming nitro group to the *ortho* and *para* positions are additional strong evidence for an electrophilic mechanism. The NO_2^+ ion was also identified and characterized by cryoscopic and spectroscopic studies[68,69] and, subsequently, stable nitronium salts were prepared[70] and employed[56] in electrophilic nitration.

Studies with preformed nitronium salts led eventually to the establishment of a modified mechanism necessitating two intermediates. Olah and coworkers studied the competitive nitration of benzene and alkylbenzenes with nitronium tetrafluoroborate[71,72] and with other nitronium salts.[71,73] They observed high positional selectivities (high *ortho/para* substitution with minor amounts of the *meta* isomer) similar to those in nitration with nitric acid–sulfuric acid. Substrate selectivities, namely, relative rates of nitration of alkylbenzenes versus benzene, were found to be close to unity, which is in sharp contrast to the results in the mixed acid. These observations—specifically, the loss of substrate selectivity and the maintained high positional selectivity—were interpreted to indicate that positional and substrate selectivities are determined in two separate steps in nitration of these reactive aromatics.[25,56,73] Olah proposed that substrate selectivity is determined in a step leading to the formation of a Dewar-type π complex.[74] Positional selectivity, in turn, is determined in the next step, when the π complex is converted into σ complexes corresponding to the individual regioisomers. The assumption of a step preceding the formation of the σ complex was also found to be necessary by Schofield and coworkers.[51,75–77] On the basis of kinetic studies, the formation of an encounter pair consisting of the aromatic substrate and the NO_2^+ ion held together by a solvent shell was suggested. It was reasoned, however, that no π-bonded interaction between the aromatic compound and the ion exists in this intermediate.

A further significant mechanistic pathway for aromatic nitration can involve a single electron-transfer reaction to an initial radical ion intermediate:

$$ArH + NO_2^+ \longrightarrow [ArH^+ \cdot NO_2 \cdot] \longrightarrow Ar\overset{+}{\underset{H}{\big\langle}}{}^{NO_2} \overset{}{\underset{-H^+}{\longrightarrow}} ArNO_2 \qquad (10.8)$$

Perrin,[78] and subsequently Eberson[79] and Kochi[80,81] and others[82–85] carried out extensive studies to probe this mechanism and its scope. Whereas its importance is now well recognized, only polycyclic or reactive systems with lower first ionization potential seem to react by this mechanism, but it is not a general case in most nitrations such as in the reaction of benzene, toluene, ethylbenzene, and similar compounds. Nitrous acid-catalyzed nitrations, in turn, normally occur through the radical cation pathway.[83]

Sulfonation. The basic sulfonating agent to convert aromatic hydrocarbons to arenesulfonic acids is sulfur trioxide and reagents derived from it.[86–91] The reaction of sulfur trioxide with Brønsted acids such as H_2O, HF, HCl, and organic sulfonic acids generates acidic sulfonating agents such as sulfuric acid, fluorosulfuric acid, chlorosulfuric acid, and pyrosulfonic acids. Since sulfonation is a reversible reaction, removal of water or a large excess of the acid may facilitate complete conversion when sulfuric acid is used. Good results may be achieved by chlorosulfonic acid in a slight excess, at 0–30°C in a solvent.

Fuming sulfuric acid and sulfur trioxide are more reactive than sulfuric acid, frequently leading to polysulfonation. Formation of addition compounds of sulfur trioxide with suitable Lewis bases (nitromethane, pyridine, 1,4-dioxane), however, moderates its reactivity allowing clean monosulfonation.

Detailed studies, mainly by Cerfontain and coworkers, have been carried out on the mechanism of sulfonation.[91,92] They found that isomer distributions and relative rates of sulfonation of alkylbenzenes versus benzene are strongly dependent on the concentration of sulfuric acid. The electrophile involved in sulfonation[93,94] at concentrations below 80–85% was shown to be $H_3SO_4^+$. With increasing sulfuric acid concentration there is a gradual change in the sulfonating entity[93,95] from $H_3SO_4^+$ to $H_2S_2O_7$. The concentration-dependent changeover from one electrophile to the other depends on the reacting aromatic substrate. Both species on attacking the aromatic ring form the σ-complex intermediate **6**:

$$\text{(10.9)}$$

6

Product formation in both cases results from deprotonative transformation of the σ complex [Eq. (10.10)]. At very high sulfuric acid concentrations (>95%) the product sulfonic acid may be formed by the direct removal of the aromatic ring proton[93] [Eq. (10.11)]:

$$\text{(10.10)}$$

$$\text{(10.11)}$$

In fuming sulfuric acid the electrophile may be $H_3S_2O_7^+$ or, at concentrations higher than 104%, $H_2S_4O_{13}$ ($H_2SO_4 + 3$ SO_3).[91] Sulfur trioxide itself may also act as the electrophile in aprotic solvents.[90] On the basis of a second-order rate dependence on SO_3 the primary sulfonation product was suggested to be an arenepyrosulfonic acid intermediate (**7**) that is converted in secondary sulfonation steps into the product:

$$Ar-SO_2O-SO_3H$$

7

10.1.3. Practical Applications

Chlorobenzene. Chlorobenzene is an important solvent and intermediate in the production of chemicals and dyes. Its use in phenol manufacture, however, was superseded by the introduction of the cumene process.

Chlorobenzene is produced by the electrophilic chlorination of benzene in the liquid phase[96,97] catalyzed by $FeCl_3$ at 25–50°C. As a result of corrosion problems and deactivation of the catalyst, dry reactants must be used. A few percent of dichlorobenzene is formed as byproduct.

Nitration of Benzene and Toluene. Nitration of benzene[97,98] can be carried out batchwise or continuously using mixed acid of the composition 35% HNO_3, 55% H_2SO_4, and 10% H_2O. Extensive stirring ensures thorough mixing of the heterogeneous reaction mixture and proper heat transfer. At a reaction temperature of 50–55°C, high selectivity is achieved with the formation of 2–3% of m-dinitrobenzene. In continuous operation two to four well-agitated tank reactors (nitrators) are used each operating at a temperature higher than the previous one. A highly economical adiabatic process uses the reaction heat generated to reconcentrate the spent acid.[99]

Toluene, which is more reactive than benzene, may be nitrated with a more dilute mixed acid containing about 23% of water.[97,98,100] Typical isomer compositions are 57–60% *ortho*, 37–40% *para*, and 3–4% *meta* isomers. Although 97–98% overall yields may be achieved with less than 0.5% of dinitrotoluenes, oxidative byproducts, mainly dinitrocresols, are formed.[101]

Nitrobenzene and nitrotoluenes that are manufactured are converted to aniline, isocyanates, and other products used in the production of dyes and pharmaceuticals.

Sulfonation of Benzene and Alkylbenzenes. Since the main utilization of benzenesulfonic acid was its transformation to phenol, the importance of the sulfonation of benzene has diminished. The process, however, is still occasionally utilized since it is a simple and economical procedure even on a small scale. Excess sulfuric acid or oleum is used at 110–150°C to produce benzenesulfonic acid.[97,102] Sulfonation of toluene under similar conditions yields a mixture of isomeric toluenesulfonic acids rich in the *para* isomer. This mixture is transformed directly to cresols by alkali fusion.

More important is the transformation of benzene to 1,3-benzenedisulfonic acid utilized in the manufacture of resorcinol.[97] According to an older method benzene is sulfonated in two steps: first by 100% sulfuric acid at 100°C, and then by 65% oleum at 80–85°C. The Hoechst new continuous one-step process applies sulfur trioxide at 140–160°C in molten 1,3-benzenedisulfonic acid. *m*-Diisopropylbenzene, via its dihydroperoxide is, however, displacing this process.

Sulfonation of detergent alkylates (long-chain linear *sec*-alkylbenzenes) clearly is the most important industrial sulfonation process.[103,104] The product alkylbenzenesulfonic acids produced by using 100% H_2SO_4 or oleum are used in biodegradable detergents. H_2SO_4 is used in a large excess to counterbalance its dilution with water formed. A smaller excess is satisfactory in sulfonation with oleum. The yield of sulfonates is about 85–90%. Batch sulfonation was later replaced by continuous oleum sulfonation and subsequently by SO_3 sulfonation. Since water is not produced in the latter reaction, only a slight excess of SO_3 is required to produce high-purity (97–99%) alkylarenesulfonic acids. A continuous falling SO_3 film sulfonation is practiced the most.[105]

10.2. FREE-RADICAL SUBSTITUTION

10.2.1. Halogenation of Alkanes and Arylalkanes

Free-radical halogenation of hydrocarbons induced thermally or photochemically can be performed with all four halogens, each exhibiting certain specificities. Because of the thermodynamics of the process, however, only chlorination (and bromination) are of practical importance.[31,106–108] Fluorination with elemental fluorine is also possible. This reaction, as discussed above (see Section 10.1.1), follows an electrophilic mechanism in the solution phase.[109,110] Under specific conditions, however, free-radical fluorination can be performed.

A general free-radical chain reaction with the thermal or photoinduced dissociation of halogen as the initiation step is operative in radical halogenation [Eqs. (10.12)–(10.14)].

$$Hlg_2 \xrightleftharpoons{\text{heat or h}\nu} 2\,Hlg\cdot \qquad (10.12)$$

$$R{-}H \;+\; Hlg\cdot \;\rightleftharpoons\; R\cdot \;+\; HHlg \qquad (10.13)$$

$$R\cdot \;+\; Hlg_2 \;\longrightarrow\; R{-}Hlg \;+\; Hlg\cdot \qquad (10.14)$$

Both propagation steps, hydrogen atom abstraction [Eq. (10.13)] and atom transfer [Eq. (10.14)], may significantly affect the outcome of halogenation. The chain may be terminated by the coupling of any two radicals. The most important termination step in gas-phase halogenation is halogen atom coupling.

Free-radical side-chain chlorination of arylalkanes in the benzylic position can be effected by chlorine or SO_2Cl_2. Phosphorus pentachloride is also capable of effectively catalyzing side-chain chlorination.

Chlorination of Alkanes. Free-radical chlorination is the most commonly used method for the chlorination of a saturated hydrocarbon.[31,106–108,111,112] Both thermal and photochemical processes may be carried out in the liquid or vapor phase. The liquid-phase photochemical procedure is preferred for polychlorination; gas-phase photochemical reactions can yield either mono- or polychlorinated product.

Chlorine atoms are highly energetic and reactive species, bringing about extremely long kinetic chain length in free-radical chlorination. Both propagation steps are fairly facile, and reversal of hydrogen abstraction is negligible. Hydrogen abstraction by a chlorine atom from any alkane is exothermic. Since the transition state in exothermic chlorination resembles the reactants more than the products, resonance stabilization does not play a significant role in the reaction.

On the basis of their extensive studies, Hass, McBee, and coworkers formulated some simple general rules of free-radical chlorination.[113] For example, no carbon skeleton rearrangements occur during chlorination if catalysts and excessively high temperatures are avoided. The relative reactivity of substitution of hydrogens is tertiary > secondary > primary. This is due to the strongly electrophilic character of the chlorine atom reacting more readily with hydrogen bonded to a tertiary carbon of higher electron density. Activation energies, however, are very similar, and reactivity differences further diminish with increasing temperature. The relative reactivity ratio of 4.4 : 3.3 : 1 at 300°C, for example, approaches 1 : 1 : 1 with increasing temperature.[113] Because of the high dissociation energy of the C—H bond, methane is quite unreactive toward radical attacks compared with other alkanes.[114]

As a result, free-radical chlorination of alkanes is a nonselective process. Except when only one type of replaceable hydrogen is present (methane, ethane, neopentane, unsubstituted cycloalkanes), all possible monochlorinated isomers are usually formed. Although alkyl chlorides are somewhat less reactive than alkanes, di- and polychlorinations always occur. The presence of a chlorine atom on a carbon atom tends to hinder further substitution at that carbon. The one exception is ethane that yields more 1,1-dichloroethane than 1,2-dichloroethane. The reason for this is that chlorination of an alkyl chloride occurs extremely slowly on the carbon atom adjacent to the one bearing the chlorine atom (vicinal effect).[115]

During slow thermal chlorination, elimination of HCl from the monochloride with the resultant formation of an alkene followed by chlorine addition may be the dominant route to yield dichloroalkanes. This mechanism, however, is negligible in rapid thermal or photochemical reactions.

Substantially increased selectivity of monochlorination were observed in certain solvents.[114,116–118] This is explained as a result of the interaction of the electrophilic (i.e., electron-deficient) chlorine atom with solvent molecules of Lewis base character. Such complex-forming interaction may decrease the reactivity of chlorine to an extent that allows more selective substitution to take place. Relative reactivities of the tertiary hydrogen with respect to the primary hydrogen of 2,3-dimethylbutane, for instance, may vary between 9 (in chlorobenzene at 25°C) and 225 (in high concentration of CS_2) compared with 4.2 in neat chlorination.[117] Substituents of benzene affecting electron density of the aromatic ring may bring about marked changes in selectivities.[119,120]

Surprising observations concerning multiple substitution were made in noncomplexing solvents. In CCl_4 at low substrate concentration, di- and trichloro compounds are formed in unexpectedly high yields, even at very low conversions.[121-123] These results are attributed to cage effects. The geminate pair $Cl^{\bullet}/$ RCl is generated during chlorination, which allows a second, in-cage hydrogen abstraction competing with the diffusion of $Cl^{\bullet}$ out of the cage. The new geminate pair $Cl^{\bullet}/R_{(-H)}Cl_2$ may produce a trichloride in a similar way. Participation of CCl_4 and $Cl_3C^{\bullet}$ in the overall process and residual O_2 and peroxyl radicals may also contribute to unusual selectivities at very low chlorine concentration.[124]

A uniquely high degree of terminal (methyl) selectivities in photochlorination of *n*-alkanes adsorbed on pentasil zeolites was reported.[125] Monochlorination at the terminal methyl group was found to be a function of conversion, water content, Si/Al ratio of the zeolite, and loading of the alkane.

In addition to molecular halogens, numerous other reagents may be used to carry out halogenation of alkanes in radical processes. Most of these were discovered in the search for more effective reagents. Chlorinating agents operating with less energetic hydrogen-atom-abstracting radicals can be expected to furnish chloroalkanes with higher selectivities.

Sulfuryl chloride is a convenient reagent[106,107] requiring initiation by heat, light, organic peroxides [Eq. (10.15)], or metal chlorides. The transformation involves the chlorosulfonyl radical as the hydrogen abstracting species [Eqs. (10.16)–(10.18)] ensuring higher selectivity than in chlorination with elemental chlorine:[126]

$$(PhCOO)_2 \longrightarrow 2\,Ph\cdot \;+\; 2\,CO_2 \tag{10.15}$$

$$Ph\cdot \;+\; SO_2Cl_2 \longrightarrow Ph-Cl \;+\; \cdot SO_2Cl \tag{10.16}$$

$$R-H \;+\; \cdot SO_2Cl \longrightarrow R\cdot \;+\; SO_2 \;+\; HCl \tag{10.17}$$

$$R\cdot \;+\; SO_2Cl_2 \longrightarrow R-Cl \;+\; \cdot SO_2Cl \tag{10.18}$$

The 2-chloro/1-chloro ratio in chlorination of 2,3-dimethylbutane, for instance, is 12.0 with sulfuryl chloride versus 4.2 in photochlorination. The chlorosulfonyl radical, however, easily decomposes under reaction conditions to yield a chlorine atom [Eq. (10.19)], which also participates in chlorination:

$$\cdot SO_2Cl \rightleftharpoons Cl\cdot \;+\; SO_2 \tag{10.19}$$

Hence, two hydrogen abstraction steps are operative, which severely limits the overall selectivity of the reaction. This is supported by the observation that there is not much difference in the selectivities of the two chlorination processes in aromatic solvents.[127] The corresponding selectivity values in benzene, for example, are 53 and 49. Presumably, chlorine atoms are complexed with the solvent in both cases and act as more selective chlorinating agents. The chlorosulfonyl radical generated from SO_2 and Cl_2 under proper conditions is also involved in the chlorosulfonation of alkanes (see Section 10.2.3).

Trichloromethanesulfonyl chloride is a more selective reagent.[107,108] The reaction is initiated by light or peroxides and involves hydrogen abstraction by the trichloromethanesulfonyl radical:[128]

$$R-H \ + \ Cl_3CSO_2\cdot \ \longrightarrow \ R\cdot \ + \ SO_2 \ + \ HCCl_3 \qquad (10.20)$$

$$R\cdot \ + \ Cl_3CSO_2Cl \ \longrightarrow \ R-Cl \ + \ Cl_3CSO_2\cdot \qquad (10.21)$$

The *tert*-BuO· radical is the hydrogen atom abstracting species in halogenations with *tert*-butyl hypohalites.[129,130] *tert*-BuOCl is more selective and less reactive than the chlorine atom.[107,108] *N*-Chloroamines are fairly selective in chlorination of linear alkanes to form the corresponding 2-chloroalkanes.[107,131] This so-called $\omega - 1$ selectivity is attributed to steric effects brought about by the bulky amino cation (aminium) radical. This is clearly indicated by changes in selectivities when chloroamines of different steric demands are used. *N*-Chlorodimethylamine and *N*-chlorodi-*tert*-butylamine yield 2-chloroheptane with 56% and 64% selectivities, respectively.[132] 2-Chlorohexane, in turn, is produced with 83% selectivity when hexane is chlorinated with *N*-chloro-2,2,6,6-tetramethylpiperidine.[133] Even higher selectivities are observed with tertiary amine radicals.[134] Additionally, di- and polychlorinations are usually negligible with these reagents, and both bicyclo[2.2.2]octane and adamantane are chlorinated at bridgehead positions.[135]

Fluorination of Alkanes. Fluorination of alkanes is extremely difficult to control. The reaction usually results in substantial C—C bond rupture and can readily lead to explosion.[136] However, several methods for controlled direct radical fluorination of hydrocarbons have been developed. The key and obvious observation was that the only reaction sufficiently exothermic to cause fragmentation is the termination step between a carbon radical and a fluorine atom. Consequently, if the atomic fluorine population and the mobility of hydrocarbon radicals are minimized, controlled fluorination becomes feasible.

The first successful method was developed by Margrave and Lagow (the LaMar technique) to carry out fluorination with a highly diluted fluorine stream and keeping the organic molecules in the solid state.[136] Reaction conditions were controlled by initial high (infinite) fluorine dilution followed by a continuous increase of fluorine concentration and long reaction time. A further development was a cryogenic apparatus featuring a reactor filled with high-surface-area fluorinated copper turnings (low-temperature-gradient method).[137] Reactants were volatilized into the cooled reaction zone, resulting in higher yields. The aerosol technique is the latest development.[138–140] Reactant molecules, in this case, are adsorbed onto airborne NaF particles, necessitating much shorter reaction times (1–5 min compared with several days). A photochemical reaction stage added after the aerosol fluorination reduces the need for excess fluorine and allows for high fluorine efficiency.[138,140] By these methods cyclic compounds (cyclohexane,[138] cyclooctane,[137] norbornane[137]), highly branched hydrocarbons (neopentane,[137,138] 2,2,3,3-tetramethylbutane[137]), and polymers (polyethylene, polystyrene)[141] could be perfluorinated.

Side-Chain Chlorination of Arylalkanes. Alkyl-substituted aromatic compounds can easily be halogenated at the benzylic position with chlorine or bromine.[107,108,142,143] α-Monohaloalkylaromatics are usually the main products with both reagents.

Toluene is readily transformed to benzyl chloride on illumination. In a thermal process below 125°C, benzyl chloride was prepared with 93.6% selectivity.[144] Peroxides and phosphorous chlorides were found to have advantageous effects on the reaction. Multiple chlorination takes place in consecutive irreversible steps, and selectivity depends only on the total amount of chlorine consumed.[145] In liquid-phase chlorination in the presence of PCl_3, 70% yields of benzyl chloride or benzal chloride were achieved, respectively, with 1.1 or 2.1 equiv of chlorine reacted. Considering the fact that resonance stabilization does not play a significant role in free-radical chlorination with molecular chlorine, unexpectedly high selectivities are observed in the chlorination of other alkylaromatics. The α-versus- β selectivities for ethylbenzene and cumene, for instance, are 14.5 and 42.2, respectively.[146] These are, however, apparent selectivities, since the aromatic ring in the reacting molecules acts as a complexing agent. The complexed chlorine atom, in turn, displays greater selectivity in hydrogen atom abstraction, as was discussed in the chlorination of alkanes. This was proved by diluting the reaction mixture with a noncomplexing solvent (nitrobenzene), resulting in substantial drop of selectivities. The corresponding values for ethylbenzene and cumene extrapolated for zero substrate concentration are 2.70 and 3.50, respectively. Under certain conditions (low temperature, high chlorine concentration) ring chlorination (radical addition) may be competitive with side-chain substitution.[135]

Side-chain chlorination of alkylaromatics by SO_2Cl_2 pioneered by Kharasch and Brown[126] may be initiated by organic peroxides and follows the free-radical chain mechanism depicted in Eqs. (10.15)–(10.18). Light,[147] AIBN [azobis(isobutylnitrile)],[148] and transition-metal complexes[149,150] were reported to promote the reaction. NaX zeolite has been shown to effect highly selective chlorination on illumination or heating.[33–35]

Olah[151] and later Messina[152] found that PCl_5 is a highly effective catalyst for the free-radical side-chain chlorination of arylalkanes. Chlorination in hydrocarbon or chlorinated solvents or without solvent at room temperature yields α monochlorinated products with high selectivity. When the reaction was carried out in polar solvents such as nitromethane and nitrobenzene, only ring chlorination occurred.

The ability of PCl_5 to act as catalyst for free-radical chlorination results from its interaction with chlorine and subsequently the homolysis of the weakened chlorine–chlorine bond:

$$PCl_5 + Cl_2 \rightleftharpoons \underset{\overset{|}{Cl}}{\overset{\overset{Cl}{|}}{\underset{Cl}{\overset{Cl}{P}}}}\begin{smallmatrix}Cl\\Cl\end{smallmatrix}Cl \rightleftharpoons PCl_4\cdot + Cl_2 + Cl\cdot \quad (10.22)$$

The formed $PCl_4\cdot$ and $Cl\cdot$ radicals subsequently can initiate free-radical chlorination reactions. In contrast, a predominant heterolytic cleavage and consequently ionic chlorination takes place in polar solvents.

10.2.2. Allylic Chlorination

When chlorination or bromination of alkenes is carried out in the gas phase at high temperature, addition to the double bond becomes less significant and substitution at the allylic position becomes the dominant reaction.[153–155] In chlorination studied more thoroughly a small amount of oxygen and a liquid film enhance substitution, which is a radical process in the transformation of linear alkenes. Branched alkenes such as isobutylene behave exceptionally, since they yield allyl-substituted product even at low temperature. This reaction, however, is an ionic reaction.[156] Despite the possibility of significant resonance stabilization of the allylic radical, the reactivity of different hydrogens in alkenes in allylic chlorination is very similar to that of alkanes. This is in accordance with the reactivity of benzylic hydrogens in chlorination.

The selectivity greatly depends on the reaction temperature. Propylene,[153] for example, gives allyl chloride in 25% yield at 210°C, whereas the yield is 96% at 400°C. This is explained by the reversibility of radical addition of chlorine to the double bond at high temperature, in contrast to radical substitution, which is not an equilibrium process.[156]

10.2.3. Sulfochlorination

Alkanesulfonyl chlorides are prepared in good yields by the photoreaction of sulfur dioxide and chlorine with saturated hydrocarbons.[89,157,158] This sulfochlorination, known as the *Reed reaction*,[159] presumably proceeds by a free-radical mechanism:[158,160]

$$Cl\cdot \ + \ R{-}H \ \longrightarrow \ R\cdot \ + \ HCl \tag{10.23}$$

$$R\cdot \ + \ SO_2 \ \longrightarrow \ R{-}SO_2\cdot \tag{10.24}$$

$$R{-}SO_2\cdot \ + \ Cl_2 \ \longrightarrow \ RSO_2Cl \ + \ Cl\cdot \tag{10.25}$$

Organic peroxides and azo compounds may be used as alternative methods for initiation.

The free-radical-induced reaction of alkanes with sulfuryl chloride characteristically results in the chlorination of hydrocarbons. However, when pyridine is added to the irradiated reactants, sulfochlorination occurs in quite satisfactory yield. For example, the irradiated reaction of cyclohexane and sulfuryl chloride in the presence of pyridine resulted in a 54.8% yield of cyclohexanesulfonyl chloride and only 9.4% of chlorocyclohexane.[161]

10.2.4. Nitration

The nitration of alkanes and that of alkylaromatics at the side chain can be accomplished with nitric acid or nitrogen oxides in either the liquid or the vapor phase;[162-165] the latter process is industrially more important. The two processes yield somewhat different products. Radical nitration, however, is not a selective transformation. At present only nitration of propane is used as an industrial process.

In the 1890s Konovalov observed[166] that alkanes undergo nitration in the liquid phase by dilute nitric acid (specific gravity 1.075) in a sealed tube (120–130°C). Other Russian chemists made additional important basic observations of the reaction.[162,167,168]

Hydrogen attached to a tertiary hydrogen is most readily substituted; secondary hydrogens are only very slowly replaced, while primary hydrogens are quite unreactive. Alkylaromatics react preferentially at the benzylic position. The reaction is accompanied by oxidation to yield carbonyl compounds, acids, and carbon oxides. Other main byproducts are nitrites, alcohols, and large quantities of poly-nitro compounds.

The mechanism of these nitrations involves free radicals formed by dissociation of nitric acid[169] [Eqs. (10.26) and (10.27)]. In contrast with the formation of alkyl radicals, the actual nitration step is not a chain reaction but a radical coupling [Eq. (10.28)]:

$$HO-NO_2 \longrightarrow HO\cdot + \cdot NO_2 \tag{10.26}$$

$$R-H \quad \begin{array}{|c|} \hline HO\cdot \\ -H_2O \\ \hline \cdot NO_2 \\ -HNO_2 \\ \hline \end{array} \quad R\cdot \tag{10.27}$$

$$R\cdot + \cdot NO_2 \longrightarrow R-NO_2 \tag{10.28}$$

The nitration of chiral alkanes producing racemic nitro compounds, and that of *cis*- and *trans*-decalin, both leading to the formation of *trans*-9-nitrodecalin,[170] support the involvement of alkyl radicals.

Mononitration generally occurs when a mixture of vapors of an alkane and nitric acid is heated at 400–500°C under atmospheric or superatmospheric pressure.[162,164,165] To overcome the danger of explosion, a molar excess of the hydrocarbon over the acid is used. Oxidation accompanies nitration, which occurs at an increased rate when the reaction temperature is increased.[171] The principal nitro compounds produced are those in which the nitro group is on a primary or a secondary carbon:[172]

$$CH_3CH_2CH_2CH_3 \xrightarrow[420°C]{HNO_3} CH_3NO_2 + CH_3CH_2NO_2 + CH_3CH_2CH_2NO_2 +$$

$$\phantom{CH_3CH_2CH_2CH_3 \xrightarrow[420°C]{HNO_3}} 10.5\% \qquad\quad 15.8\% \qquad\qquad 5.3\%$$

$$CH_3CH_2CH_2CH_2NO_2 + CH_3CH_2\underset{\underset{NO_2}{|}}{C}HCH_3 \tag{10.29}$$

$$ 24.2\% \qquad\qquad 44.2\%$$

Extensive carbon–carbon bond breaking takes place, resulting in the formation of nitro compounds with fewer carbon atoms than were present in the starting alkane, but no skeletal rearrangement occurs[172] [Eq. (10.29)]. Highly branched hydrocarbons are less likely to undergo fission during nitration than are less branched compounds.

Much experimental evidence established that the reaction occurs by a free-radical mechanism[164,173] similar to that suggested above [Eqs. (10.26)–(10.28)] for liquid-phase nitration. The nitrous acid produced during the transformation is unstable under the reaction conditions and decomposes to yield nitric oxide, which also participates in nitration, although less effectively. It was found that nitric acid and nitrogen dioxide yield identical products but that the former gives better yields and higher rates.[172]

All the nitroalkanes that can be formed by replacement of any hydrogen atom or alkyl group in the reacting alkane are formed. Although each bond in the alkane is subject to cleavage during nitration, only one per molecule actually does. Addition of oxygen to the reaction mixture increases the yield of low-molecular-weight nitro compounds. This indicates that the mechanism of carbon–carbon bond breaking is similar to that of oxidation of hydrocarbons.[171,174] Indeed, bond rupture was found to occur through alkyloxy radicals formed in the decomposition of alkyl nitrite intermediates[175] [Eq. (10.30)]. Alkyloxy radicals may also be transformed to carbonyl compounds, which are the principal byproducts in vapor-phase nitration[173] [Eq. (10.31)]:

$$R-CH_2-\dot{C}H_2 \ + \ NO_2 \ \longrightarrow \ R-CH_2-CH_2-O-N=O \ \xrightarrow{-\,^{\bullet}NO} \ R-CH_2-CH_2-O\cdot$$
$$(10.30)$$

$$R-CH_2-CH_2-O\cdot \ \overbrace{\xrightarrow{\ \ \ \ \ \ \ \ \ \ }}^{RCH_2CH_2ONO}_{-RCH_2CH_2OH}
\begin{cases}
R-\dot{C}H_2 \ + \ CH_2O \\
RCH_2\dot{C}HO-NO \ \xrightarrow{-\,^{\bullet}NO} \ RCH_2CHO
\end{cases}$$
$$(10.31)$$

10.2.5. Practical Applications

Chlorination of Alkanes. The most direct and economical method for the manufacture of chloromethanes is the thermal free-radical chlorination of methane.[176,177] Whereas in the 1940s and 1950s photochlorination was practiced in some plants, thermal chlorination is the principal industrial process today. The product chloromethanes are important solvents and intermediates. Commercial operations perform thermal chlorination at about 400–450°C. Vapor-phase photochemical chlorination of methane may be accomplished at 50–60°C. Fast and effective removal of heat associated with thermally induced free-radical substitution is a crucial point. Inadequate heat control may lead to explosion attributed to the uncontrollable pyrolysis liberating free carbon and much heat:

$$CH_4 \ + \ 2\,Cl_2 \ \longrightarrow \ C \ + \ 4\,HCl \qquad \Delta H = -70.5 \ \text{kcal/mol} \qquad (10.32)$$

Thermal chlorination requires high-purity methane and chlorine. The presence of higher hydrocarbons forming chlorinated products raises the difficult problem of product purification. Oxygen and sulfur compounds terminating the radical chain must also be removed. Removal of water vapor is critical to prevent corrosion problems, as well as hydrolysis and decomposition of the product chloromethanes. Inert gases such as nitrogen are not tolerated, either, since their concentration builds up with recycling.

In the manufacture of methyl chloride a mixture of methane and chlorine is mixed with recycled gas and fed into a reactor. The $CH_4 : Cl_2$ ratio is 4–5. The reactor temperature is maintained by regulating the gas feed rate. In methane chlorination plants, 95–96% chlorine yields and 90–92% methane yields are typical.[176] All four chloromethanes are formed with a typical composition of 35 wt% methyl chloride, 45 wt% methylene dichloride, 20 wt% chloroform, and a small amount of carbon tetrachloride.[178] The reaction, however, may be regulated so that either mono- or tetrachlorination predominates.[177,179,180]

Methylene dichloride and chloroform may be produced by modified methods using a mixture of chlorine, methane, and methyl chloride as feed. Chlorination is run at 350–400°C reactor temperature at slightly above atmospheric pressure. A 2.6 : 1 chlorine : methane ratio results in an optimal yield of chloroform. Alternatively, excess methane is reacted with chlorine at 485–510°C to produce methylene dichloride as the main product.[181] The predominant method, however, still is the chlorination of methyl chloride manufactured by the reaction of methyl alcohol and hydrogen chloride.[181]

In the production of carbon tetrachloride, chlorination is carried out in excess chlorine. The lower-boiling, partially chlorinated products then enter into a series of reactors where they react with added chlorine to achieve almost full chlorination of methane. In another process called *chlorinolysis*, higher aliphatic hydrocarbons undergo exhaustive chlorination at pyrolytic temperature ($>600°C$).[177,182,183] Under such conditions carbon–carbon bond fission and simultaneous chlorination occur. Aliphatic hydrocarbon wastes are the preferred feedstock, as they react with about 20% excess chlorine.

The direct chlorination of ethane[184–186] is used for the manufacture of two ethane derivatives, ethyl chloride and 1,1,1-trichloroethane. Both compounds are important solvent and 1,1,1-trichloroethane is also used as a degreasing agent.

Since free-radical chlorination is a nonselective process, overchlorination may be a problem in the manufacture of ethyl chloride. Temperature-induced pyrolysis to yield ethylene and hydrogen chloride may occur, too. A fluidized-bed thermal chlorination reactor may be used to overcome these problems. The best selectivity achieved in the temperature range of 400–450°C is 95.5% with a chlorine to ethane ratio of 1 : 5.

The so-called integrated ethyl chloride process combines the abovementioned synthesis with an addition reaction. Hydrogen chloride formed in the thermal chlorination process is used in a separate step to add to ethylene, making the manufacture of ethyl chloride more economical. 1,1,1-Trichloroethane is an exceptional product in free-radical chlorination of higher hydrocarbons since the same carbon atom is polychlorinated. It is manufactured[187] by feeding chlorine, ethane, and recycled mono- and dichloroethane into a reactor at 370–400°C. Since product distribution is basically determined by the reagent ratio, a higher than twofold excess of chlorine is used. Under such conditions 1,1,1-trichloroethane is formed as the main product.

Chlorinated alkanes were used in the production of chlorofluoroalkanes (CFCs).[188] Because of their atmospheric ozone-depleting ability, the use of CFCs is being phased out. Hydrochlorofluorocarbons, containing at least one hydrogen atom (HCFCs), are introduced as safer alternatives.[189,190] In the long run, however, hydrofluorocarbons (HFCs) will be the ideal substitutes to CFCs.[190]

Side-Chain Chlorination of Toluene. Benzyl chloride, used mainly in the manufacture of plasticizers, may be prepared by the thermal or photochemical chlorination of toluene.[191,192] In the thermal process chlorine is passed through toluene at 65–100°C. To minimize the formation of benzal chloride and benzotrichloride, the conversion is limited to about 50%. Since the density of the reaction mixture increases linearly with the formation of benzyl chloride,[145] measurement of density is used to monitor the progress of the reaction. The overall yield based on toluene is about 90%, and the maximum conversion to benzyl chloride is above 70%. Higher yields in photochemical chlorination may be achieved.

A higher than twofold chlorine excess in this reaction results in the formation of benzal chloride in about 70% yield. When the chlorination is carried out at higher temperature (100–140°C) with illumination, the synthesis of benzotrichloride in 95% yield may be accomplished.

Allyl Chloride. The manufacture of allyl chloride, commercialized in 1945, was the first of several modern high-temperature chlorination technologies applied for hydrocarbons. Preheated dry propylene mixed with dry chlorine in a ratio of 4 : 1 is reacted at 500–510°C to produce allyl chloride.[193–195] Chlorine reacts quantitatively in a few seconds. The main byproducts are isomeric monochloropropenes and dichloropropenes. Allyl chloride is used mainly in the manufacture of allyl alcohol and glycerol via epichlorohydrin.

Sulfochlorination of Alkanes. Sulfochlorination of straight-chain alkane mixtures with SO_2 and Cl_2 under light illumination slightly above ambient temperature yields *sec*-alkyl sulfochlorides.[103,196] Polysulfochlorination is prevented by running the reaction at low conversions (about 30%). The product is hydrolyzed to yield sodium alkylsulfonates used in detergents.

Nitroalkanes. A process was developed and operated for a time for the manufacture of ε-caprolactam based on the nitration of cyclohexane.[197] Nitrocyclohexane thus prepared was transformed to ε-caprolactam via cyclohexanone oxime. At present the only industrial process to produce nitroalkanes by direct nitration is the manufacture of nitromethane, nitroethane, 1-nitropropane, and 2-nitropropane.

All four compounds are produced by the vapor-phase nitration of propane.[197–199] Nitroalkanes are important as solvents and intermediates.[199]

The nitration of propane is operated at 370–450°C and at about 10 atm. The use of about a fourfold excess of propane, 60–70% nitric acid, and short residence time ensures good temperature control.[199] The conversion of nitric acid to nitroalkanes is less than 50% (the balance are nitrogen oxides). Excess propane is usually recycled to achieve about 60–80% yield of nitroalkanes. Under these conditions nitropropanes are the main products.

10.3. AMINATION

Considering the ample possibilities to generate new carbon–hereoatom bonds, particularly the wide range of oxygenation reactions, only limited attempts have been made to carry out direct amination of hydrocarbons. A few examples of amination of alkanes, alkenes, and aromatics are discussed briefly here.

Amination of alkanes (cyclohexane, heptane, adamantane) was achieved with iron and manganese porphyrin catalysts by tosylimidoiodobenzene to yield tosylamino derivatives.[200] Selective 1-substitution of adamantane (56% yield) and 2-substitution of heptane (66% selectivity) were reported.

Transition-metal-catalyzed allylic amination with phenylhydroxylamine can be carried out with high selectivity.[201,202] Iron complexes, particularly iron phthalocyanine, exhibit the best properties, and alkenes possessing an aromatic ring in conjugation with the double bond give the best yields (up to 60%).[202] The reagent is believed to be coordinated to the metal in the catalyst, and the C—N bond forming step involves an ene reaction between the alkene and the activated reagent.

The direct one-step amination of aromatics by hydroxylamine derivatives (salts, esters), hydrazoic acid and azides, is known to give aromatic amines in moderate yields.[203] Improvements can be achieved when aryl azides[204,205] or N-arylhydroxylamines[206] are used in the presence of a proton source (phenol, CF$_3$COOH). Phenyl azide combined with triflic acid has been shown to be highly efficient for high-yield electrophilic phenylamination.[207] Amination of aromatics to yield arylamines is also possible with the trimethylsilyl azide/triflic acid reagent [Eq. (10.33)], which involves the in situ–formed aminodiazonium ion (H$_2$NN$_2^+$) as the aminating agent:[208]

$$\text{R = H, Me} \qquad \qquad 93\text{–}96\% \tag{10.33}$$

10.4. HETEROSUBSTITUTION THROUGH ORGANOMETALLICS

10.4.1. Transition Metals

Much effort has been devoted to activate hydrocarbons, particularly saturated compounds, through the formation of organometallic compounds and transformation of the latter to substituted derivatives. A number of transition-metal complexes have been found to insert into carbon–hydrogen bonds leading to stable alkyl metal hydrides:

$$M + R-H \longrightarrow R-M-H \tag{10.34}$$

Such oxidative additions can yield stable organometallic intermediates if the sum of the M—H and M—C bond dissociation energies are comparable (or larger) than the dissociation energy of the C—H bond of the reacting hydrocarbon (about 100 kcal/mol). Since the typical M—H and M—C bond strengths are near 60 and 30 kcal/mol, respectively, metal complexes capable of reacting with hydrocarbons are the exception rather than the rule.[209] Predominantly third-row transition metals form stable compounds, but ligands also affect the stability of the formed complexes.

Stoichiometric reactions and catalytic transformations have been reported, including H/D (hydrogen/deuterium) exchange, and dehydrogenation and carbonylation of alkanes. Review papers,[210–213] books,[214–216] and a collection of papers[217] give detailed information of these reactions. There are also numerous metal-catalyzed oxidations of hydrocarbons with the possible involvement of organometallic intermediates (see Chapter 9). The transformation of organometallic derivatives to other heterosubstituted products, however, is less successful, and only a handful of examples are known. Shilov was the first to observe that C_1–C_6 alkanes react in an aqueous solution of H_2PtCl_6–$PtCl_4^{2-}$ at 110–$120°C$ to form monochlorinated derivatives and alcohols.[218–221] Mechanistic studies with methane revealed an autocatalytic reaction with Pt(IV) as the stoichiometric oxidant and Pt(II) the catalyst and final oxidant:[222]

$$Pt(II) + CH_4 \longrightarrow Pt(II)-CH_3 + H^+ \tag{10.35}$$

$$Pt(II)-CH_3 + Pt(IV) \longrightarrow Pt(IV)-CH_3 + Pt(II) \tag{10.36}$$

$$Pt(IV)-CH_3 \longrightarrow CH_3Cl + CH_3OH + Pt(II) \tag{10.37}$$

The complex $[Cl_5PtCH_3]^{2-}$ was isolated and found to be intermediate. The products (methyl chloride and methanol) are formed in parallel reactions most likely through the reductive elimination of methyl and neighboring Cl or OH group in the coordination sphere.

Methyl chloride and methanol are similarly formed in oxidation of methane with air with the Pt(IV)–Pt(II) pair in the presence of a heteropoly acid ($120°C$, 60–100 atm).[223] The reaction is catalytic with respect to both platinum and the heteropoly acid. A system composed of K_2PtCl_4 and H_2PtCl_6 with $HgSO_4$ as the chlorine source was described, which is very active in the formation of methyl chloride either in the aqueous phase or when supported on silica.[224]

Supported diallylrhodium complexes bound to oxide (silica) surface can be transformed to complex **8**, which reacts with methane in a stoichiometric reaction to form methyl chloride:[221,225,226]

$$(10.38)$$

Chlorination can also be catalytic: **8, 9**, or oxide-bound allylrhodium hydride catalyzes the reaction between methane and chlorine to yield chlorinated methanes predominantly methyl chloride.

The dihydrido complex $[RhH_2(\eta^5\text{-}C_5Me_5)(PMe_3)]$ forms C—H insertion products when irradiated in the presence of alkanes (ethane, propane).[227,228] Reaction with $CHBr_3$ leads to bromoalkylrhodium complexes, which on treatment with bromine give ethyl bromide or 1-bromopropane in 70–85% yield. The less stable iridium complex formed with neopentane could not be converted directly to neopentyl bromide.[229] It gave, however, a mercury derivative that yielded the bromide after treatment with bromine.

Silylation of arenes was reported to take place with rhodium and iridium complexes as the catalyst. The Vaska complex effects the reaction of benzene with pentamethyldisiloxane,[230] whereas $[RhCl(CO)(PMe_3)_2]$ catalyzes the formation of C—Si bonds under irradiation[231] [Eq. (10.39)]:[232]

$$(10.39)$$

$Me_3SiSiMe_3$ reacts similarly with benzene and toluene to yield the corresponding trimethylsilyl derivatives. The reductive elimination of arylsilane from the arylsilylrhodium intermediate was postulated.

Gif systems were shown to be effective in the direct conversion of unactivated C—H bonds (adamantane, cyclohexane, cyclododecane) into phenylseleno, phenylthio, and halogen derivatives.[233] For example, using the Gif[III] system (iron powder, acetic acid, pyridine, H_2S, oxygen) and $(PhSe)_2$ as the reagent, adamantane is transformed to a roughly 1:1 mixture of 1- and 2-phenylselenoadamantane in 35% yield. Halogenation of cyclohexane with polyhalomethanes ($CHCl_3$, $CHBr_3$, CBr_4) gives monohalocyclohexanes as the sole product in low yield. Product formation is interpreted to take place in the reaction of an alkyliron intermediate with the reagent.

10.4.2. Alkali Metals and Magnesium

Organoalkali compounds, particularly lithium and sodium derivatives[234,235] and Grignard reagents,[236] are capable of forming the corresponding organometallic intermediates by removing protons from activated (benzylic, allylic, propargylic, acetylenic) or nonactivated carbons of hydrocarbons. Mixtures of lithium alkyls with alkali alkoxides, particularly the BuLi–*tert*-BuOK combination (Lochmann–Schlosser reagent or LICKOR superbase),[237–239] are the most frequently used reagents. Subsequent transformations with suitable electrophiles can lead to hetero-substituted products.

Allylmetal intermediates are readily formed and then transformed to boron and silicon derivatives by reacting with FB(OMe)$_2$ and Me$_3$SiCl, respectively.[239,240] The structure of the product depends on the organometallics. Organolithium and organomagnesium derivatives of 2-alkenes usually give the corresponding 3-hetero-substituted 1-alkenes formed with double-bond shift[240] [Eq. (10.40)]. Allylpotas-sium derivatives, in contrast, are transformed preferentially to 1-heterosubstituted 2-alkenes[240] [Eq. (10.41)]:

$$RCH=CHCH_2MgBr \xrightarrow{FB(OMe)_2} \underset{90\% \text{ selectivity}}{RCHCH=CH_2} \quad (10.40)$$

$$RHC \overset{K}{\underset{\underset{H}{C}}{\diamond}} CH_2 \xrightarrow{FB(OMe)_2} \underset{>95\% \text{ selectivity}}{RCH=CHCH_2B(OMe)_2} \quad (10.41)$$

Allylpotassium derivatives exist in two stereoisomeric (*endo* or *exo*) forms.[239,240] Although the *endo* isomer is more stable, the equilibration is slow enough to allow the synthesis of stereoisomeric heterosubstituted derivatives [Eqs. (10.42) and (10.43)]. On the other hand, even the *E*-alkenes after appropriate equilibration can be transformed to *Z*-boron compounds **10**:

$$(10.42)$$

$$(10.43)$$

Bromoalkynes, which are important intermediates in coupling reactions,[241] can be synthesized through the corresponding metalated alkynes:[241,242]

$$R-\!\!\!\equiv\!\!\!-H \xrightarrow{R'M} R-\!\!\!\equiv\!\!\!-M \xrightarrow{\text{"Br"}} R-\!\!\!\equiv\!\!\!-Br \qquad (10.44)$$

$$M = \text{Li, MgHlg}$$

Heterosubstitution at nonactivated positions is also possible by using highly reactive metalating agents such as pentylsodium. When used in the presence of *tert*-BuOK, it readily and selectively forms vinylsodium derivatives that can subsequently be converted to organosilicium compounds.[243] Norbornadiene and bicyclo-[3.2.0]hepta-2,6-diene are also readily silylated via the corresponding sodium compounds formed in the reaction with BuLi–*tert*-BuONa.[244] Similar transformations of strained polycyclic cyclopropanes were also achieved[243] [Eq. (10.45)[245]]:

$$\text{(structure)} \xrightarrow[25°C, 24 \text{ h}]{n\text{-}C_5H_{11}Na,\ tert\text{-BuOK}} \text{(structure)}_{Na} \xrightarrow{Me_3SiCl} \text{(structure)}_{SiMe_3} \qquad (10.45)$$

$$\text{60\% yield}$$

Excellent selectivity was observed in the silylation of 1,3,5-cycloheptatriene giving 2-(trimethylsilyl)-1,3,5-cycloheptatriene as the sole product:[244]

$$\text{(structure)} \xrightarrow{n\text{-BuLi,}\ tert\text{-BuONa}} \text{(structure)}{-}Na \xrightarrow{Me_3SiCl} \text{(structure)}{-}SiMe_3 \qquad (10.46)$$

With the extremely efficient metalating agent BuLi–*tert*-BuOK–TMEDA even direct metallation of ethylene and the transformation of the intermediate to heterosubstituted product were accomplished:[246]

$$CH_2{=}CH_2 \xrightarrow[\substack{ii.\ LiBr,\ THF,\ -30°C}]{\substack{i.\ n\text{-BuLi,}\ tert\text{-BuOK}\\ TMEDA,\ -25°C}} CH_2{=}CHLi \xrightarrow{PhSSPh} CH_2{=}CHSPh \qquad (10.47)$$

$$\text{75\%}$$

Selective heterosubstitution of phenylacetylene at the acetylenic carbon or at *ortho* position was achieved through different organometallic intermediates[247,248] (Scheme 10.3).

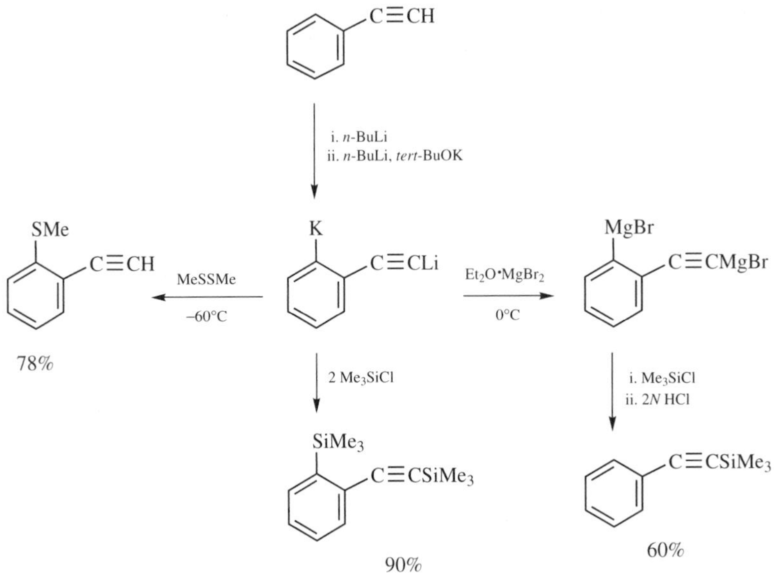

Scheme 10.3

10.5. RECENT DEVELOPMENTS

10.5.1. Electrophilic (Acid-Catalyzed) Substitution

Alkanes. Earlier observations had demonstrated the possibility of fluorination at tertiary C—H sites (see Section 10.1.1). New studies have shown that electrophilic fluorination at both secondary and tertiary carbon atoms may be carried out by elemental fluorine or Selectfluor {1-(chloromethyl)-4-fluoro-1,4-diazabicyclo[2.2.2] octane bis(tetrafluoroborate)} in acetonitrile:[249]

$$\text{(10.48)}$$

10% F_2 in N_2, MeCN, 0 °C   63% yield, 53% conversion
Selectfluor, MeCN, reflux   22% yield, 100% conversion

Aromatics. The application of solid acid catalysts provides excellent possibilities to carry out aromatic electrophilic substitutions in an environmentally friendly way. Various zeolites were found by Smith and coworkers to exhibit high activities and selectivities.[250] Acetyl nitrate generated in situ from acetic anhydride and HNO_3 transforms alkylbenzenes to the corresponding *para*-nitro derivatives in high yield (92–99%) and with excellent selectivity (79–92%) when applied in the presence of large-pore H-Beta zeolites.[251] Lattice flexibility and the coordination of acetyl nitrate to lattice aluminum is suggested to account for the high regioselectivity.[252] Sulfuric acid supported on silica is less effective, although a threefold increase in the amount of the solid acid is sufficient to change highly selective mononitration of toluene into dinitration.[253] Double nitration of toluene with a mixture of HNO_3 and trifluoroacetic anhydride in the presence of zeolite H Beta also occurs with high efficiency (92% yield, 2,4-dinitro- to 2,6-dinitrotoluene ratio $= 25 : 1$).[254]

Zeolite Beta is also highly effective in the bromination of toluene.[255] Reactions of monoalkylbenzenes with bromine in the presence of stoichiometric amounts of zeolite NaY proceed in high yield and with high selectivity to give the corresponding *para*-bromo products:[256]

$$\text{(10.49)}$$

R = Me, Et, isoPr, *tert*-Bu   99–100% selectivity

The zeolite can easily be regenerated by heating. Similar high *para* selectivity can be achieved in the case of toluene by use of *tert*-butyl hypobromite as reagent with zeolite HX in a solvent mixture (CCl_4 and ether). $ZnBr_2$ supported on mesoporous silica or acid-activated montmorillonite is a fast, effective, reusable catalyst for the *para*-bromination of alkylbenzenes.[257]

Partially proton-exchanged Na faujasite X, in turn, is the best catalyst for selective monochlorination with *tert*-butyl hypochlorite.[258] NaX, NaY, and NaKL zeolites used in the chlorination of toluene with sulfuryl chloride undergo rapid deactivation because of the accumulation of polychlorinated toluenes in the pores of the catalysts and dealumination.[259, 260] Direct electrophilic fluorination of aromatics can be effected by using Selectfluor in the presence of triflic acid.[261] Electrophilic fluorination may also be carried out by R_2NF and $R_3N^+FA^-$ reagents.[262] Elemental fluorine may also act as a powerful electrophile in acidic media (sulfuric acid, trifluoroacetic acid, or formic acid), but monosubstituted aromatics give isomeric mixtures.[263–265]

Two mechanisms involving either single- [Eq. (10.50)] or two-electron transfer [Eq. (10.51)] from the aromatic substrate to fluorine may be involved in the reaction with elemental fluorine:[264]

$$\text{(10.50)}$$

$$\text{(10.51)}$$

Fluorination of toluene, for example, gives a mixture of *ortho*- and *para*-fluorotoluene, as expected for an electrophilic process, but the very high ratio indicates that the radical process is also involved. In addition, fluorination of the methyl group also occurs in increasing yield with increasing reaction temperature.

The use of various nitrogen oxides provides a new possibility to achieve aromatic nitration. The so-called *Kyodai* nitration (named for Japanese name of Kyoto University) uses nitric oxide, dinitrogen trioxide, and nitrogen dioxide activated by ozone to react smoothly with arenes, giving the corresponding nitro derivatives in moderate to good yield.[266,267] The reaction is usually carried out at 0°C or below in dichloromethane. The addition of protic acid as catalyst enhances the process, giving good yields of higher nitro compounds. The reaction is generally clean and proceeds rapidly without any accompanying side reaction:

$$R = Me, Et$$
$$isoPr, \textit{tert}\text{-Bu} \quad 13\text{--}57 \qquad\qquad 2\text{--}6 \qquad\qquad 41\text{--}81 \quad 99\% \text{ yield}$$

Exclusive ring-nitration occurs with alkylbenzenes. The nitration of toluene in the presence of H-ZSM-5 and molecular oxygen shows a remarkable enhancement of para selectivity (*ortho : para* ratio = 0.08).[268] A review is available for the nitration of aromatics by nitrogen oxides on zeolite catalysts.[269]

A mechanism involving NO_3 as the initial electrophile initiating single-electron oxidation of the aromatic ring followed by coupling between the nitrate anion and the cation radical to form the σ complex has been suggested to be operative:[270]

This and other novel pathways in aromatic nitration are discussed in a review paper.[271]

In addition, the utility of ionic liquids for electrophilic nitration of simple aromatics (benzene, toluene, 1,3,5-trimethylbenzene, and *tert*-butylbenzene) has been found to be quite promising.[272] Preferred systems are [emim][OTf], [emim][CF$_3$OO], and [HNEtisoPr$_2$][CF$_3$OO] with NH$_4$NO$_3$/CF$_3$OOH, and isoamyl nitrate/TfOH or isoamyl nitrate/BF$_3$–OEt$_2$ as the nitrating agents.

Triflate salts, mainly scandium triflate[273] and lanthanide(III) triflates,[274,275] have found increasing use in various aromatic substitutions. They have some unusual properties; specifically, they may be used in catalytic quantity and active even in the presence of water. They are also easily recyclable via aqueous workup.

Lanthanide(III) triflates catalyze the nitration of a range of simple aromatic compounds (benzene, toluene, biphenyl, *m*- and *p*-xylene, naphthalene) in good to excellent yield using stoichiometric quantities of 69% nitric acid.[276] Bismuth(III) triflate was found to catalyze sulfonylation of aromatics with aromatic sulfonyl chlorides with similar high efficiency.[277]

Similar good results were obtained with Fe^{3+} montmorillonite and zeolite Beta with sulfonic acids, chlorides, or anhydrides as reagents.[278] Iron(III) chloride is also very effective in this reaction under microwave irradiation.[279] Sulfonylation by sulfonic acids catalyzed by Nafion-H allows the synthesis of both symmetric and unsymmetric sulfones.[280]

Protonated polymethylbenzenes[281] and the chlorohexamethylbenzenium cation,[282] intermediates in aromatic electrophilic substitutions known as *Wheland intermediates*, have been isolated as crystalline salts, allowing investigators to obtain their X-ray crystal structure. Nitrosoarenium σ complexes of various arenes were directly observed by transient absorption spectroscopy.[283] Kochi presented a method combining appropriate instrumental techniques (X-ray crystallography, NMR, time-resolved UV–vis spectroscopy) for the observation, identification, and structural characterization of reactive intermediates (σ and π complexes) in electrophilic aromatic substitution.[284]

10.5.2. Free-Radical Substitution

Alkanes. The chlorination of ethane known to produce more 1,1-dichloroethane than 1,2-dichloroethane is explained by the so-called vicinal effect.[115] One study revealed[285] that this observation may be explained by the precursor 1,2-dichloroethane radical (the **11** 2-chloroethyl radical) thermally dissociating into ethylene and a chlorine atom [Eq. (10.54)]. Indeed, this radical is the major source of ethylene under the conditions studied. At temperatures above 300°C, the dissociation dominates over the chlorination reaction [Eq. (10.55)], resulting in a high rate of ethylene formation with little 1,2-dichloroethane:

$$ClCH_2-CH_2\cdot \longrightarrow CH_2=CH_2 + Cl\cdot \qquad (10.54)$$

$$\mathbf{11}$$

$$ClCH_2-CH_2\cdot + Cl_2 \longrightarrow ClCH_2-CH_2Cl + Cl\cdot \qquad (10.55)$$

An unusual radical halogenation of hydrocarbons by phase-transfer catalysis may be performed by reacting alkanes with tetrahalomethanes. The reaction is initiated by single-electron oxidation of OH^- by $CHlg_4$. The tetrahalomethane radical anion formed decomposes to the **12** trihalomethyl radical [Eq. (10.56)], which is then involved in C–H activation and propagation steps:[286]

$$CHlg_4 + OH^- \rightleftharpoons CHlg_4\cdot^- + OH\cdot^-$$

$$\mathbf{12} \quad \Big\downarrow \qquad\qquad\qquad (10.56)$$

$$Hlg_3C\cdot + Cl^-$$

Reactions are rather slow because the slow single-electron transfer initiation equilibrium lies far to the left.

The process is characterized by unusually high regioselectivities; specifically, only secondary and tertiary positions are involved. Furthermore, in the bromination of adamantane and methyl-substituted adamantanes, substitution occurs almost exclusively at the tertiary carbon atom.[286] The same reaction with iodoform may be used to carry out an efficient direct iodination of aliphatic hydrocarbons.[287] The process is quite general; it may be applied to straight-chain, cyclic, cyclic branched, and cage alkanes. In addition, in marked contrast to free-radical halogenation with halogens, this process allows to perform halogenation of cubane without rearrangement.[288]

A more recent, direct, simple iodination was also disclosed using perfluoroalkyl iodides.[289] When cyclohexane, for example, is refluxed with C_4F_9I in acetic acid in the presence of catalytic amounts of *tert*-BuOOH and $Fe(OAc)_3$, iodocyclohexane is formed in 70% yield. Mechanistic consideration of this Gif–Barton catalysis is given in Section 9.6.1.

As discussed in Section 10.2.1, selectivity in the free-radical chlorination significantly increases when the reaction is conducted in an aromatic solvent due to complex formation between the chlorine atom and the solvent. There has been considerable discussion regarding the precise nature of the chlorine atom–benzene complex. Complexes of chlorine or bromine atoms and CS_2 have been reported.[290,291] Enhanced selectivities were also observed in halogenated solvents.[292] Selectivities decrease with increasing chlorine content of the solvent and are greater in bromoalkane solvents compared to chloroalkanes. This suggests that donor-acceptor-type complexes are involved. Radical chlorination in supercritical CO_2, in turn, did not provide any indication of an enhanced cage effect.[293] Radical bromination of alkylaromatics in water showed that the reaction takes place in the organic phase and water exerts no significant effect on the bromination itself.[294]

Nanocrystalline MgO and CaO with high surface area are able to absorb large amounts of chlorine, which undergo dissociative chemisorption. These can serve as rather selective, catalytic alkane chlorination reagents, which suggests that trapped Cl atoms are involved in the reaction.[295] Liquid-phase low-temperature chlorination of alkanes is also possible in the presence of various alkenes as inductors and AIBN [azobis(isobutyronitrile)].[296]

A synthetic octacoordianted vanadium(IV)-containing complex called *avamandine* catalyzes the selective peroxidative monobromination of cyclohexane and benzene.[297]

Efficient catalytic alkane nitrations can be performed with the assistance of *N*-hydroxyphthalimide (NHPI) with NO_2 in air [Eq. (10.57)][298] or with HNO_3.[299] NHPI may also be used to form alkanesulfonic acids by reacting alkanes with SO_2:[300]

$$(10.57)$$

70%

Adamantane can be nitrated with NO_2 in the presence of ozone (Kyodai nitration) to afford 1-nitroadamantane with high regioselectivity.[301] Product formation is interpreted by hydrogen abstraction by NO_3 followed by rapid trapping of the resulting admantyl radical by NO_2.

The first catalytic sulfoxidation of saturated hydrocarbons with SO_2/O_2 by vanadium catalysts to give alkanesulfonic acids has been reported.[302] The best results were achieved with $VO(acac)_2$:

$$R-H \xrightarrow[\text{acetic acid, 40°C, 2 h}]{SO_2/O_2, \text{ VO(acac)}_2} R-SO_3H \qquad (10.58)$$

R—H = octane, cyclohexane,
cyclooctane, adamantane,
1,3-dimethyladamantane

Commercialization of a new vinyl chloride process has been announced. Instead of the traditional three-step production (see Section 6.3.4), it is based on ethane oxychlorination using HCl, O_2, and Cl_2 carried out over a CuCl-based catalysts.[285] Overchlorinated products are dehydrochlorinated and hydrogenated (together with overchlorinated alkenes) in separate reactors; these product streams are then led back to the oxychlorination reactor.

Alkenes. The reaction of NO with alkenes having a double bond in terminal position or conjugated with an aromatic ring is a simple and convenient method for the regioselective synthesis of nitroalkenes.[303] An addition–elimination mechanism with the involvement of intermediate nitroso nitro compound **13** was suggested. The latter undergoes a fast decomposition to give nitro alkenes in good to high yields:

$$R \diagup \hspace{-0.2em}= \xrightarrow[\text{CH}_2\text{Cl}_2, \text{ RT}]{\text{1 atm NO}} \underset{\mathbf{13}}{R \diagup \hspace{-0.4em}\overset{\text{NO}}{\diagdown} \hspace{-0.4em} NO_2} \xrightarrow[\text{−NO, −HNO}_2]{} R \diagup\hspace{-0.4em}\diagdown NO_2 \qquad (10.59)$$

R = n-C_6H_{17} 92%
Ph   95%
2-naphthyl   90%
$PhCH_2CH_2$ 90%

Novel results were reported for allylic bromination. In radical bromination of cyclohexene in CCl_4 under light the selectivity of substitution over addition was shown to be controlled by bromine concentration.[304] Substitution via the corresponding allyl radical, while relatively slow, is irreversible and fast enough to maintain the concentration of bromine at a sufficiently low level to prevent significant addition. The reaction of two strained alkenes, (Z)-1,2-dimethyl-1,2-di-*tert*-butylethylene and the E-isomer (**14**), leads to the corresponding bromosubstituted product, instead of addition:[305]

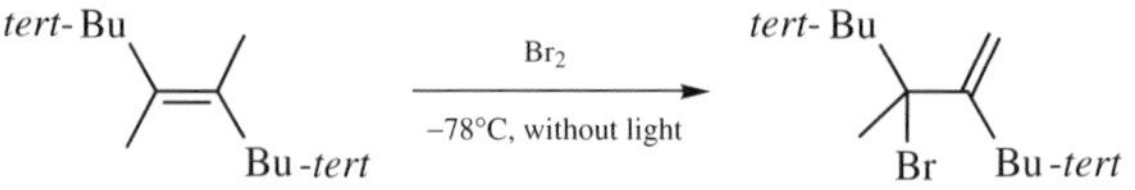

$$
\text{14} \qquad\qquad \xrightarrow[\text{$-78°C$, without light}]{Br_2} \qquad\qquad\qquad (10.60)
$$

14

10.5.3. Amination

Amination of hydrocarbons in allylic, benzylic, or tertiary position can be mediated by various transition metals in high yields.[306] Metal (Co, Ru, Mn) porphyrins[307–311] or metal salts and other metal complexes[312–317] may be used as catalysts. The reagent is mostly *N*-tosylimidoiodobenzene (TsN=IPh); therefore, the reaction may also be called *amidation*. Other reagents are phenylazides,[307,308] aromatic nitro compounds,[312,315] and arylhydroxyalmines.[316] Asymmetric amination is also possible.[309] When alkenes are reacted aziridation may also take place that can even become the main transformation direction. The diimido selenium reagent TsN=Se=NTs is capable of the allylic amination of alkenes without metal assistance.[317]

An example of the oxidative amination of vinylarenes applying various Rh complexes that allows the formation of the corresponding enamines was reported:[318–320]

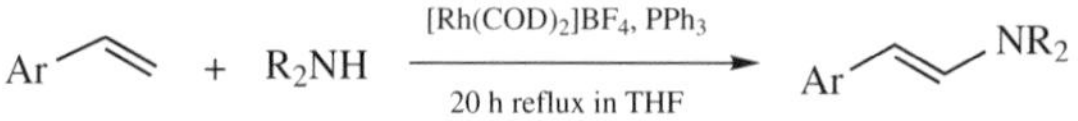

$$
Ar \diagup\!\!\!\diagdown\!\!\!\diagup + R_2NH \xrightarrow[\text{20 h reflux in THF}]{[Rh(COD)_2]BF_4,\ PPh_3} Ar \diagup\!\!\!\diagdown\!\!\!\diagup NR_2 \qquad (10.61)
$$

$$
R_2NH = Et_2NH,\ Bu_2NH,
$$
piperidine

Ar = Ph,
3- and 4-MeC$_6$H$_4$,
1-naphthyl

40–99%

The reaction is completely selective, yielding exclusively the anti-Markovnikov regioisomer.

10.5.4. Heterosubstitution through Organometallics

Activation of the C–H bond of alkanes to form substituted products selectively remains a major challenge. Numerous reagents and processes, both catalytic and stoichiometric, applying metal complexes, have emerged, and the results were reviewed.[321–326] Most of the work was, however, related to oxidation processes, and only a few new transformations are known when organometallics are involved in heterosubstitution.

The Shilov method, discussed in Section 10.4.1, has been accomplished as a catalytic process.[327] Aqueous platinum chlorides can be used as catalysts for the homogeneous catalytic chlorination of methane. The reaction is carried out at 100–125°C with chlorine to yield methyl chloride that is partially hydrolyzed to methanol. $Na_2[PtCl_4]$ reacts with chlorine at room temperature to give $Na_2[PtCl_6]$ which is the major species even at 125°C.

Selective o,o'-dimetallation of diphenylacetylene was successfully performed with the LICKOR superbase.[328] The dimetallo product thus formed reacts smoothly with various electrophiles to give o,o'-disubstituted derivatives in good yields:

RX = MeI, Me₃SiCl, MeSSMe,   47–68% yield
Ph₂PCl, n-Bu₃SnCl

$$(10.62)$$

REFERENCES

1. G. A. Olah, G. K. S. Prakash, and J. Sommer, *Superacids*, Wiley-Interscience, New York, 1985, Chapter 5, p. 243.

2. G. A. Olah and G. K. S. Prakash, in *The Chemistry of Alkanes and Cycloalkanes*, S. Patai and Z. Rappoport, eds., Wiley, Chichester, 1992, Chapter 13, p. 624.

3. D. Alker, D. H. R. Barton, R. H. Hesse, J. Lister-James, R. E. Markwell, M. M. Pechet, S. Rozen, T. Takeshita, and H. T. Toh, *New J. Chem.* **4**, 239 (1980).

4. C. Gal and S. Rozen, *Tetrahedron Lett.* **25**, 449 (1984).

5. S. Rozen and C. Gal, *J. Org. Chem.* **52**, 2769 (1987).

6. G. A. Olah, N. Hartz, G. Rasul, Q. Wang, G. K. S. Prakash, J. Casanova, and K. D. Christie, *J. Am. Chem. Soc.* **116**, 5671 (1994).

7. G. A. Olah, R. Renner, P. Schilling, and Y. K. Mo, *J. Am. Chem. Soc.* **95**, 7686 (1973).

8. G. A. Olah and P. Schilling, *J. Am. Chem. Soc.* **95**, 7680 (1973).

9. G. A. Olah, *Acc. Chem. Res.* **20**, 422 (1987).

10. G. A. Olah, B. Gupta, M. Farina, J. D. Felberg, W. M. Ip, A. Husain, R. Karpeles, K. Lammertsma, A. K. Melhotra, and N. J. Trivedi, *J. Am. Chem. Soc.* **107**, 7097 (1985).

11. G. A. Olah, U. S. Patent 4,523,040 (1985).

12. I. Bucsi and G. A. Olah, *Catal. Lett.* **16**, 27 (1992).

13. P. Batamack, I. Bucsi, Á. Molnár, and G. A. Olah, *Catal. Lett.* **25**, 11 (1994).

14. G. A. Olah, R. Malhotra, and S. C. Narang, *Nitration: Methods and Mechanisms*, VCH, New York, 1989, Chapter 2, p. 57.

15. G. A. Olah and H. C. Lin, *J. Am. Chem. Soc.* **93**, 1259 (1971).

16. G. A. Olah, R. Malhotra, and S. C. Narang, *Nitration: Methods and Mechanisms*, VCH, New York, 1989, Chapter 4, p. 233.

17. G. A. Olah, Q. Wang, and G. K. S. Prakash, *J. Am. Chem. Soc.* **112**, 3697 (1990).

18. C. C. Price, *Chem. Rev.* **29**, 37 (1941).

19. E. Berliner, *Prog. Phys. Org. Chem.* **2**, 253 (1964).

20. *Friedel-Crafts and Related Reactions*, Vol. 3, G. A. Olah, ed., Wiley-Interscience, New York, 1964.

21. R. O. C. Norman and R. Taylor, *Electrophlic Substitution in Benzenoid Compounds*, Elsevier, Amsterdam, 1965, Chapter 2, p. 25.

22. L. M. Stock, *Aromatic Substitution Reactions*. Prentice-Hall, Englewood Cliffs, NJ, 1968, Chapter 2, p. 21.

23. R. Breslow, *Organic Reaction Mechanisms*, Benjamin, New York, 1969, Chapter 5, p. 147.

24. C. K. Ingold, *Structure and Mechanism in Organic Chemistry*, Cornell Univ. Press, Ithaca, NY, 1969, Chapter 6, p. 264.

25. G. A. Olah, *Acc. Chem. Res.* **4**, 240 (1971).

26. R. Taylor, *Chem. Kinet.* **13**, 1 (1972).

27. P. Pfeiffer and R. Wizinger, *Justus Liebigs Ann. Chem.* **461**, 132 (1928).

28. G. W. Wheland, *J. Am. Chem. Soc.* **64**, 900 (1942).

29. H. C. Brown and J. D. Brady, *J. Am. Chem. Soc.* **74**, 3570 (1952).

30. W. v. E. Doering, M. Saunders, H. G. Boyton, H. W. Earhart, E. F. Wadley, W. R. Edwards, and G. Laber, *Tetrahedron* **4**, 178 (1958).

31. M. Hudlicky and T. Hudlicky, in *The Chemistry of Functional Groups*, Supplement D: *The Chemistry of Halides, Pseudo-Halides and Azides*, S. Patai and Z. Rappoport, eds., Wiley, Chichester, 1983, Chapter 22, p. 1021.

32. H. P. Braendlin and E. T. McBee, in *Friedel-Crafts and Related Reactions*, Vol. 3, G. A. Olah, ed., Wiley-Interscience, New York, 1964, Chapter 46, p. 1517.

33. L. Delaude and P. Laszlo, *J. Org. Chem.* **55**, 5260 (1990).

34. M. Balogh and P. Laszlo, *Organic Chemistry Using Clays*, Springer, Berlin, 1993, Chapter 1, p. 1.

35. L. Delaude, P. Laszlo, and K. Smith, *Acc. Chem. Res.* **26,** 607 (1993).

36. L. M. Stock and H. C. Brown, *Adv. Phys. Org. Chem.* **1**, 35 (1963).

37. E. Baciocchi and G. Illuminati, *Prog. Phys. Org. Chem.* **5**, 7 (1967).

38. K. Smith, M. Butters, and B. Nay, *Synthesis* 1157 (1985).

39. A. McKillop, D. Bromley, and E. C. Taylor, *J. Org. Chem.* **37**, 88 (1972).

40. G. A. Olah and M. W. Meyer, *J. Org. Chem.* **27**, 3464 (1962).

41. P. B. D. de la Mare and B. E. Swedlund, in *The Chemistry of the Carbon-Halogen Bond*, S. Patai, ed., Wiley, London, 1973, Chapter 7, p. 490.

42. E. B. Merkushev, *Russ. Chem. Rev.* (Engl. transl.) **53**, 343 (1984); *Synthesis* 923 (1988).

43. C. Galli, *J. Org. Chem.* **56**, 3238 (1991).

44. J. A. Wilkinson, *Chem. Rev.* **92**, 505 (1992).

45. M. J. Fifolt, R. T. Olczak, and R. F. Mundhenke, *J. Org. Chem.* **50**, 4576 (1985).

46. S. Singh, D. D. DesMarteau, S. S. Zuberi, M. Witz, and H.-N. Huang, *J. Am. Chem. Soc.* **109**, 7194 (1987).

47. A. V. Topchiev, *Nitration of Hydrocarbons and Other Organic Compounds*, Pergamon Press, New York, 1959, Chapters 1 and 2.

48. G. A. Olah and S. J. Kuhn, in *Friedel-Crafts and Related Reactions*, Vol. 3, G. A. Olah, ed., Wiley-Interscience, New York, 1964, Chapter 43, p. 1393.

49. W. M. Weaver, in *The Chemistry of the Nitro and Nitroso Groups*, Part 2, H. Feuer, ed., Interscience, New York, 1970, Chapter 1, p. 1.

50. J. H. Ridd, *Acc. Chem. Res.* **4**, 248 (1971); *Adv. Phys. Org. Chem.* **16**, 1 (1978).

51. J. G. Hoggett, R. B. Moodie, J. R. Penton, and K. Schofield, *Nitration and Aromatic Reactivity*, Cambridge Univ. Press, Cambridge, UK, 1971.

52. K. Schofield, *Aromatic Nitration*, Cambridge Univ. Press, Cambridge, UK, 1980.

53. L. M. Stock, *ACS Symp. Ser.* **22**, 48 (1976).

54. G. A. Olah, *ACS Symp. Ser.* **22**, 1 (1976).

55. G. A. Olah, S. C. Narang, J. A. Olah, and K. Lammertsma, *Proc. Natl. Acad. Sci. USA* **79**, 4487 (1982).

56. G. A. Olah, R. Malhotra, and S. C. Narang, *Nitration: Methods and Mechanisms*, VCH, New York, 1989, Chapter 3, p. 117.

57. J. W. Fischer, in *Nitro Compounds, Recent Advances in Synthesis and Chemistry*, H. Feuer and A. T. Nielsen, eds., VCH, New York, 1990, Chapter 3, p. 279.

58. A. Cornélis, L. Delaude, A. Gerstmans, and P. Laszlo, *Terahedron Lett.* **29**, 5657 (1988).

59. G. A. Olah, S. C. Narang, R. Malhotra, and J. A. Olah, *J. Am. Chem. Soc.* **101**, 1805 (1979).

60. G. A. Olah, P. S. Iyer, and G. K. S. Prakash, *Synthesis* 513 (1986).

61. F. J. Waller, *ACS Symp. Ser.* **308**, 42 (1986).

62. S. J. Sondheimer, N. J. Bunce, and C. A. Fyfe, *J. Macromol. Sci., Rev. Macromol. Chem. Phys.* **C26**, 353 (1986).

63. G. K. S. Prakash and G. A. Olah, in *Acid-Base Catalysis*, K. Tanabe, H. Hattori, T. Yamaguchi, and T. Tanaka, eds., Kodansha, Tokyo, 1989, p. 59.

64. E. D. Hughes, C. K. Ingold, and R. I. Reed, *Nature* (London) **158**, 448 (1946); *J. Chem. Soc.* 2400 (1950).

65. E. S. Halberstadt, E. D. Hughes, and C. K. Ingold, *J. Chem. Soc.* 2441 (1950).

66. V. Gold, E. D. Hughes, C. K. Ingold, and G. H. Williams, *J. Chem. Soc.* 2452 (1950).

67. V. Gold, E. D. Hughes, and C. K. Ingold, *J. Chem. Soc.* 2467 (1950).

68. R. J. Gillespie, J. Graham, E. D. Hughes, C. K. Ingold, and E. R. A. Peeling, *Nature* (London), **158**, 480 (1946); *J. Chem. Soc.* 2504 (1950).

69. C. K. Ingold, D. J. Millen, and H. G. Poole, *Nature* (London) **158**, 480 (1946); *J. Chem. Soc.* 2576 (1950).

70. D. R. Goddard, E. D. Hughes, and C. K. Ingold, *Nature* (London) **158**, 480 (1946); *J. Chem. Soc.* 2559 (1950).

71. G. Oláh, S. Kuhn, and A. Mlinkó, *J. Chem. Soc.* 4257 (1956).

72. G. A. Olah, S. J. Kuhn, and S. H. Flood, *J. Am. Chem. Soc.* **83**, 4571 (1961).

73. S. J. Kuhn and G. A. Olah, *J. Am. Chem. Soc.* **83**, 4564 (1961).

74. M. J. S. Dewar, *Nature* (London) **156**, 784 (1946); *J. Chem. Soc.* 777 (1946).

75. R. G. Coombes, R. B. Moodie, and K. Schofield, *J. Chem. Soc. B* 800 (1968).

76. J. G. Hoggett, R. B. Moodie, and K. Schofield, *J. Chem. Soc. B* 1 (1969).

77. R. B. Moodie, K. Schofield, and J. B. Weston, *J. Chem. Soc., Chem Commun.* 382 (1974).

78. C. L. Perrin, *J. Am. Chem. Soc.* **99**, 5516 (1977).

79. L. Eberson and F. Radner, *Acta Chim. Scand., Ser. B* **38**, 861 (1984); *Acc. Chem. Res.* **20**, 53 (1987).

80. J. K. Kochi, *Acta Chim. Scand.* **44**, 409 (1990).

81. E. K. Kim, T. M. Bockman, and J. K. Kochi, *J. Chem. Soc., Perkin Trans. 2* 1879 (1992).

82. J. F. Johnston, J. H. Ridd, and J. P. B. Sandall, *J. Chem. Soc., Chem Commun.* 244 (1989).

83. J. H. Ridd, *Chem. Soc. Rev.* **20**, 149 (1991).

84. J. P. B. Sandall, *J. Chem. Soc., Perkin Trans. 2* 1689 (1992).

85. A. Boughriet and M. Wartel, *J. Chem. Soc., Chem Commun.* 809 (1989).

86. K. L. Nelson, in *Friedel-Crafts and Related Reactions*, Vol. 3, G. A. Olah, ed., Wiley-Interscience, New York, 1964, Chapter 42, p. 1355.

87. E. E. Gilbert, *Sulfonation and Related Reactions*, Interscience, New York, 1965.

88. E. E. Gilbert, *Synthesis* 3 (1969).

89. J. Hoyle, in *The Chemistry of Sulphonic Acids, Esters and their Derivatives*, S. Patai and Z. Rappoport, eds., Wiley, Chichester, 1991, Chapter 10, p. 351.

90. H. Cerfontain, *Recl. Trav. Chim. Pays-Bas* **104**, 153 (1985).

91. H. Cerfontain and C. W. F. Kort, *Int. J. Sulfur Chem., Part C* **6**, 123 (1971).

92. H. Cerfontain, *Mechanistic Aspects in Aromatic Sulfonation and Desulfonation*, Interscience, New York, 1968, Chapters 1 and 2.

93. A. W. Kaandorp, H. Cerfontain, and F. L. J. Sixma, *Recl. Trav. Chim. Pays-Bas* **81**, 969 (1962).

94. A. W. Kaandorp and H. Cerfontain, *Recl. Trav. Chim. Pays-Bas* **88**, 725 (1969).

95. C. W. F. Kort and H. Cerfontain, *Recl. Trav. Chim. Pays-Bas* **87**, 24 (1968).

96. B. E. Kurt and E. W. Smalley, in *Encyclopedia of Chemical Processing and Design*, Vol. 8, J. J. McKetta and W. A. Cunningham, eds., Marcel Dekker, New York, 1979, p. 117.

97. K. Weissermel and H.-J. Arpe, *Industrial Organic Chemistry*, 2nd ed., VCH, Weinheim, 1993, Chapter 13, p. 333.

98. A. A. Guenkel, in *Encyclopedia of Chemical Processing and Design*, Vol. 31, J. J. McKetta and W. A. Cunningham, eds., Marcel Dekker, New York, 1990, p. 165.

99. A. A. Guenkel, H. C. Prime, and J. M. Rae, *Chem. Eng.* (NY) **88**(16), 50 (1981).

100. T. K. Wright and R. Hurd, in *Toluene, the Xylenes and Their Industrial Derivatives*, E. G. Hancock, ed., Elsevier, Amsterdam, 1982, Chapter 9, p. 234.

101. C. Hanson, T. Kaghazchi, and M. W. T. Pratt, *ACS Symp. Ser.* **22**, 132 (1976).

102. D. McNeil, in *Toluene, the Xylenes and their Industrial Derivatives*, E. G. Hancock, ed., Elsevier, Amsterdam, 1982, Chapter 8, p. 209.

103. E. A. Knaggs, M. L. Nussbaum, and A. Shultz, in *Kirk-Othmer Encyclopedia of Chemical Technology*, 3rd ed., Vol. 22, M. Grayson and D. Eckroth, eds., Wiley-Interscience, New York, 1983, p. 1.

104. *Hydrocarbon Process., Int. Ed.* **70**(3), 188 (1991).

105. E. A. Knaggs, *Chemtech* **22**, 436 (1992).

106. E. S. Huyser, *Free-Radical Chain Reactions*, Wiley-Interscience, New York, 1970, Chapter 5, p. 89.

107. E. S. Huyser, in *The Chemistry of the Carbon-Halogen Bond*, S. Patai, ed., Wiley, London, 1973, Chapter 8, p. 549.

108. M. L. Poutsma, in *Free Radicals*, Vol. 2, J. K. Kochi, ed., Wiley-Interscience, New York, 1973, Chapter 15, p. 159.

109. S. Rozen, *Acc. Chem. Res.* **21**, 307 (1988).

110. S. Rozen, in *Synthetic Fluorine Chemistry*, G. A. Olah, R. D. Chambers, and G. K. S. Prakash, eds., Wiley, New York, 1992, Chapter 7, p. 143.

111. A. S. Bratolyubov, *Uspekhi Khim.* **30**, 1391 (1961).

112. G. Chiltz, P. Goldfinger, G. Huybrechts, G. Martens, and G. Verbeke, *Chem. Rev.* **63**, 355 (1963).

113. H. B. Hass, E. T. McBee, and P. Weber, *Ind. Eng. Chem.* **27**, 1190 (1935); **28**, 333 (1936).

114. G. A. Russel, in *The Chemistry of Alkanes and Cycloalkanes*, S. Patai and Z. Rappoport, eds., Wiley, Chichester, 1992, Chapter 21, p. 963.

115. F. F. Rust and W. E. Vaughan, *J. Org. Chem.* **6**, 479 (1941).

116. G. A. Russel, *J. Am. Chem. Soc.* **79**, 2977 (1957); **80**, 4997 (1958).

117. G. A. Russel, *J. Am. Chem. Soc.* **80**, 4987 (1958).

118. C. Walling and M. F. Mayahi, *J. Am. Chem. Soc.* **81**, 1485 (1959).

119. J. M. Tanko and F. E. Anderson, III, *J. Am. Chem. Soc.* **110**, 3525 (1988).

120. K. D. Raner, J. Lusztyk, and K. U. Ingold, *J. Am. Chem. Soc.* **111**, 3652 (1989).

121. P. S. Skell and H. N. Baxter, III, *J. Am. Chem. Soc.* **107**, 2823 (1985).

122. K. D. Raner, J. Lusztyk, and K. U. Ingold, *J. Am. Chem. Soc.* **110**, 3519 (1988).

123. K. U. Ingold, J. Lusztyk, and K. D. Raner, *Acc. Chem. Res.* **23**, 219 (1990).

124. K. D. Raner, J. Lusztyk, and K. U. Ingold, *J. Org. Chem.* **53**, 5220 (1988).

125. N. J. Turro, J. R. Fehlner, D. P. Hessler, K. M. Welsh, W. Ruderman, D. Firnberg, and A. M. Braun, *J. Org. Chem.* **53**, 3731 (1988).

126. M. S. Kharasch and H. C. Brown, *J. Am. Chem. Soc.* **61**, 2142 (1939).

127. G. A. Russel, *J. Am. Chem. Soc.* **80**, 5002 (1958).

128. E. S. Huyser and B. Giddings, *J. Org. Chem.* **27**, 3391 (1962).

129. C. Walling and B. B. Jacknow, *J. Am. Chem. Soc.* **82**, 6108 (1960).

130. C. Walling and A. Padwa, *J. Org. Chem.* **27**, 2976 (1962).

131. F. Minisci, *Synthesis* 1 (1973).

132. R. Bernardi, R. Galli, and F. Minisci, *J. Chem. Soc. B* 324 (1968).

133. N. C. Deno, E. J. Gladfelter, and D. G. Pohl, *J. Org. Chem.* **44**, 3728 (1979).

134. S. E. Fuller, J. R. L. Smith, R. O. C. Norman, and R. Higgins, *J. Chem. Soc., Perkin Trans. 2* 545 (1981).

135. C. V. Smith and W. E. Billups, *J. Am. Chem. Soc.* **96**, 4307 (1974).

136. R. J. Lagow and J. L. Margrave, *Prog. Inorg. Chem.* **26**, 161 (1979).

137. N. J. Maraschin, B. D. Catsikis, L. H. Davis, G. Jarvinen, and R. J. Lagow, *J. Am. Chem. Soc.* **97**, 513 (1975).

138. J. L. Adcock, K. Horita, and E. B. Renk, *J. Am. Chem. Soc.* **103**, 6937 (1981).

139. J. L. Adcock and M. L. Cherry, *Ind. Eng. Chem. Res.* **26**, 208 (1987).

140. J. L. Adcock, in *Synthetic Fluorine Chemistry*, G. A. Olah, R. D. Chambers, and G. K. S. Prakash, eds., Wiley, New York, 1992, Chapter 6, p. 127.

141. J. L. Margrave and R. J. Lagow, *J. Polym. Sci., Polym. Lett. Ed.* **12**, 177 (1974).

142. M. S. Kharasch and M. G. Berkman, *J. Org. Chem.* **6**, 810 (1941).

143. M. S. Kharasch, P. C. White, and F. R. Mayo, *J. Org. Chem.* **3**, 33 (1939).

144. A. Scipioni, *Ann. Chim.* (Rome) **41**, 491 (1951).

145. E. Borello and D. Pepori, *Ann. Chim.* (Rome) **45**, 449 (1955).

146. G. A. Russel, A. Ito, and D. G. Hendry, *J. Am. Chem. Soc.* **85**, 2976 (1963).

147. M. G. Voronkov, É. P. Popova, É.É. Liepin', V. A. Pestunovich, L. F. Bulenkova, M. M. Kalnin', and G. G. Konstante, *J. Org. Chem. USSR* (Engl. transl.) **8**, 1917 (1972).

148. M. C. Ford and W. A. Waters, *J. Chem. Soc.* 1851 (1951).

149. H. Matsumoto, T. Nakano, M. Kato, and Y. Nagai, *Chem. Lett.* 223 (1978).

150. R. Davis, J. L. A. Durrant, and C. C. Rowland, *J. Organomet. Chem.* **315**, 119 (1986).

151. G. A. Olah, P. Schilling, R. Renner, and I. Kerekes, *J. Org. Chem.* **39**, 3472 (1974).

152. G. Messina, M. D. Moretti, P. Ficcadenti, and G. Cancellieri, *J. Org. Chem.* **44**, 2270 (1979).

153. H. P. A. Groll and G. Hearne, *Ind. Eng. Chem.* **31**, 1530 (1939).

154. J. Burgin, W. Engs, H. P. A. Groll, and G. Hearne, *Ind. Eng. Chem.* **31**, 1413 (1939).

155. F. F. Rust and W. E. Vaughan, *J. Org. Chem.* **5**, 472 (1940).

156. M. L. Poutsma, *J. Am. Chem. Soc.* **87**, 2172 (1965).

157. F. Asinger, W. Schmidt, and F. Ebeneder, *Chem. Ber.* **75B**, 34 (1942).

158. W. H. Lockwood, *Chem. Ind.* **62**, 760 (1948).

159. H. Eckoldt, in *Methoden der Organischen Chemie (Houben-Weyl)*, Vol. 9: *Schwefel-, Selen-, Tellur-Verbindungen*, Thieme, Stuttgart, 1955, p. 407.

160. E. E. Gilbert, *Sulfonation and Related Reactions*, Interscience, New York, 1965, Chapter 3, p. 125.

161. M. S. Kharasch and A. T. Read, *J. Am. Chem. Soc.* **61**, 3089 (1939).

162. A. V. Topchiev, *Nitration of Hydrocarbons and Other Organic Compounds*, Pergamon Press, New York, 1959, Chapter 3, p. 144.

163. O. v. Schickh, H. G. Padeken, and A. Segnitz, in *Methoden der Organischen Chemie (Houben-Weyl)*, Vol. 10/1: *Stickstoff-Verbindungen*, Thieme, Stuttgart, 1971, p. 12.

164. A. P. Ballod and V. Ya. Shtern, *Russ. Chem. Rev.* (Engl. transl.) **45**, 721 (1976).

165. G. A. Olah, R. Malhotra, and S. C. Narang, *Nitration: Methods and Mechanisms*, VCH, New York, 1989, Chapter 4, p. 220.

166. M. Konowalow, *Zhur. Russ. Khim. Obshch.* **19**, 157 (1887); **25**, 389, 472, 509 (1893); **27**, 421 (1895); *Chem. Ber.* **28**, 1850, 1852 (1895).

167. W. Markownikoff, *Justus Liebigs Ann. Chem.* **302**, 15 (1898); *Chem. Ber.* **33**, 1905, 1908 (1900).

168. S. Nametkin, *Chem. Ber.* **42**, 1372 (1909).

169. A. I. Titov, *Uspekhi Khim.* **21**, 881 (1952); *Tetrahedron* **19**, 557 (1963).

170. H. Shechter and D. K. Brain, *J. Am. Chem. Soc.* **85**, 1806 (1963).

171. Y. Ogata, in *Oxidation in Organic Chemistry*, Part C, W. S. Trahanovsky, ed., Academic Press, New York, 1978, Chapter 4, p. 298.

172. H. B. Hass and H. Shechter, *Ind. Eng. Chem.* **39**, 817 (1947).

173. G. B. Bachman, L. M. Addison, J. V. Hawett, L. Kohn, and A. Millikan, *J. Org. Chem.* **17**, 906 (1952).

174. G. B. Bachman and M. Pollack, *Ind. Eng. Chem.* **46**, 713 (1954).

175. C. Matasa and H. B. Hass, *Can. J. Chem.* **49**, 1284 (1971).

176. E. M. DeForest, in *Encyclopedia of Chemical Processing and Design*, Vol. 8, J. J. McKetta and W. A. Cunningham, eds., Marcel Dekker, New York, 1979, p. 214.

177. K. Weissermel and H.-J. Arpe, *Industrial Organic Chemistry*, 2nd ed., VCH, Weinheim, 1993, Chapter 2, p. 50.

178. R. C. Ahlstrom, Jr. and J. M. Steele, in *Kirk-Othmer Encyclopedia of Chemical Technology*, 3rd. ed., Vol. 5, M. Grayson and D. Eckroth, eds., Wiley-Interscience, New York, 1979, p. 677.

179. E. T. McBee, H. B. Hass, C. M. Neher, and H. Strickland, *Ind. Eng. Chem.* **34**, 296 (1942).

180. *Hydrocarbon Process., Int. Ed.* **58**(11), 147 (1979); **60**(11), 144 (1981).

181. T. Anthony, in *Kirk-Othmer Encyclopedia of Chemical Technology*, 3rd ed., Vol. 5, M. Grayson and D. Eckroth, eds., Wiley-Interscience, New York, 1979, p. 686.

182. D. Rebhan, in *Encyclopedia of Chemical Processing and Design*, Vol. 8, J. J. McKetta and W. A. Cunningham, eds., Marcel Dekker, New York, 1979, p. 206.

183. *Hydrocarbon Process.* **56**(11), 140 (1977).

184. R. G. Stirling, in *Encyclopedia of Chemical Processing and Design*, Vol. 20, J. J. McKetta and W. A. Cunningham, eds., Marcel Dekker, New York, 1984, p. 68.

185. K. Weissermel and H.-J. Arpe, *Industrial Organic Chemistry*, 2nd ed., VCH, Weinheim, 1993, Chapter 8, p. 193.

186. *Hydrocarbon Process., Int. Ed.* **70**(3), 150 (1991).

187. J. I. Jordan, Jr., in *Encyclopedia of Chemical Processing and Design*, Vol. 8, J. J. McKetta and W. A. Cunningham, eds., Marcel Dekker, New York, 1979, p. 134.

188. G. W. Parshall and S. D. Ittel, *Homogeneous Catalysis*, Wiley-Interscience, New York, 1992, Chapter 12, p. 297.

189. L. E. Manzer, *Science* **249**, 31 (1990).

190. L. E. Manzer and V. N. M. Rao, *Adv. Catal.* **39**, 329 (1993).

191. S. Gelfand, in *Kirk-Othmer Encyclopedia of Chemical Technology*, 3rd ed., Vol. 5, M. Grayson and D. Eckroth, eds., Wiley-Interscience, New York, 1979, p. 828.

192. H. G. Haring, in *Toluene, the Xylenes and their Industrial Derivatives*, E. G. Hancock, ed., Elsevier, Amsterdam, 1982, Chapter 7, p. 197.

193. H. H. Beacham, in *Kirk-Othmer Encyclopedia of Chemical Technology*, 3rd ed., Vol. 2, M. Grayson and D. Eckroth, eds., Wiley-Interscience, New York, 1978, p. 97.

194. A. DeBenedictis, in *Kirk-Othmer Encyclopedia of Chemical Technology*, 3rd ed., Vol. 5, M. Grayson and D. Eckroth, eds., Wiley-Interscience, New York, 1979, p. 763.

195. K. Weissermel and H.-J. Arpe, *Industrial Organic Chemistry*, 2nd ed., VCH, Weinheim, 1993, Chapter 11, p. 291.

196. K. Weissermel and H.-J. Arpe, *Industrial Organic Chemistry*, 2nd ed., VCH, Weinheim, 1993, Chapter 3, p. 81.

197. P. J. Baker, Jr. and A. F. Bollmeier, Jr., in *Kirk-Othmer Encyclopedia of Chemical Technology*, 3rd ed., Vol. 15, M. Grayson and D. Eckroth, eds., Wiley-Interscience, New York, 1981, p. 969.

198. R. F. Purcell, in *Encyclopedia of Chemical Processing and Design*, Vol. 31, J. J. McKetta and W. A. Cunningham, eds., Marcel Dekker, New York, 1990, p. 267.

199. S. B. Markofsky and W. G. Grace, in *Ullmann's Encyclopedia of Industrial Chemistry*, 5th ed., Vol. A17, B. Elvers, S. Hawkins, and G. Schulz, eds., VCH, Weinheim, 1991, p. 401.

200. J.-P. Mahy, G. Bedi, P. Battioni, and D. Mansuy, *New J. Chem.* **13**, 651 (1989).

201. A. Srivastava, Y. Ma, R. Pankayatselvan, W. Dinges, and K. M. Nicholas, *J. Chem. Soc., Chem. Commun.* 853 (1992).

202. M. Johannsen and K. A. Jørgensen, *J. Org. Chem.* **59**, 214 (1994).

203. P. Kovacic, in *Friedel-Crafts and Related Reactions*, Vol. 3, G. A. Olah, ed., Wiley-Interscience, New York, 1964, Chapter 44, p. 1493.

204. K. Nakamura, A. Ohno, and S. Oka, *Synthesis* 882 (1974).

205. H. Takeuchi, K. Takano, and K. Koyama, *J. Chem. Soc., Chem. Commun.* 1254 (1982).

206. K. Shudo, T. Ohta, and T. Okamoto, *J. Am. Chem. Soc.* **103**, 645 (1981).

207. G. A. Olah, P. Ramaiah, Q. Wang, and G. K. S. Prakash, *J. Org. Chem.* **58**, 6900 (1993).

208. G. A. Olah and T. D. Ernst, *J. Org. Chem.* **54**, 1203 (1989).

209. J. F. Liebman and C. D. Hoff, in *Selective Hydrocarbon Activation*, J. A. Davies, P. L. Watson, A. Greenberg, and J. F. Liebman, eds., VCH, New York, 1990, Chapter 6, p. 149.

210. E. L. Muetterties, *Chem. Soc. Rev.* **11**, 283 (1982).

211. R. H. Crabtree, *Chem. Rev.* **85**, 245 (1985).

212. M. L. H. Green and D. O'Hare, *Pure Appl. Chem.* **57**, 1897 (1985).

213. J. Halpern, *Inorg. Chim. Acta* **100**, 41 (1985).

214. A. E. Shilov, *Activation of Saturated Hydrocarbons by Transition Metal Complexes*, Reidel, Dordrecht, 1984.

215. *Activation and Functionalization of Alkanes*, C. L. Hill, ed., Wiley-Interscience, New York, 1989.

216. *Selective Hydrocarbon Activation*, J. A. Davies, P. L. Watson, A. Greenberg, and J. F. Liebman, eds., VCH, New York, 1990.

217. *Alkane Activation and Functionalization*, C. L. Hill, ed., Wiley-Interscience, New York, 1989; *New J. Chem.* **13**(10–11), 645–793 (1989).

218. N. F. Gol'dshleger, V. V. Es'kova, A. E. Shilov, and A. A. Shteinman, *Russ. J. Phys. Chem.* (Engl. transl.) **46**, 785 (1972).

219. A. E. Shilov, *Activation of Saturated Hydrocarbons by Transition Metal Complexes*, Reidel, Dordrecht, 1984, Chapter 5, p. 163.

220. A. E. Shilov, in *Activation and Functionalization of Alkanes*, C. L. Hill, ed., Wiley-Interscience, New York, 1989, Chapter 1, p. 1.

221. W. D. Jones, in *Selective Hydrocarbon Activation*, J. A. Davies, P. L. Watson, J. F. Liebman, and A. Greenberg, eds., VCH, New York, 1990, Chapter 5, p. 113.

222. L. A. Kushch, V. V. Lavrushko, Yu. S. Misharin, A. P. Moravsky, and A. E. Shilov, *New J. Chem.* **7**, 729 (1983).

223. Yu. V. Geletii and A. E. Shilov, *Kinet. Catal.* (Engl. transl.), **24**, 413 (1983).

224. V. P. Tretyakov, G. P. Zimtseva, E. S. Rudakov, and A. N. Osetskii, *React. Kinet. Catal. Lett.* **12**, 543 (1979).

225. N. Kitajima and J. Schwartz, *J. Am. Chem. Soc.* **106**, 2220 (1984).

226. J. Schwartz, *Acc. Chem. Res.* **18**, 302 (1985).

227. R. A. Periana and R. G. Bergman, *Organometallics* **3**, 508 (1984).

228. R. G. Bergman, *Adv. Chem. Ser.* **230**, 211 (1992).

229. A. H. Janowicz and R. G. Bergman, *J. Am. Chem. Soc.* **105**, 3929 (1983).

230. W. A. Gustavson, P. S. Epstein, and M. D. Curtis, *Organometallics* **1**, 884 (1982).

231. M. Tanaka and T. Sakakura, *Adv. Chem. Ser.* **230**, 181 (1992).

232. T. Sakakura, Y. Tokunaga, T. Sodeyama, and M. Tanaka, *Chem. Lett.* 2375 (1987).

233. G. Balavoine, D. H. R. Barton, J. Boivin, P. Lecoupanec, and P. Lelandais, *New J. Chem.* **13**, 691 (1989).

234. J. L. Wardell, in *Comprehensive Organometallic Chemistry*, Vol. 1, G. Wilkinson, F. G. A. Stone, and E. W. Abel, eds., Pergamon Press, Oxford, 1982, Chapter 2, p. 54.

235. J. L. Wardell, in *The Chemistry of the Metal-Carbon Bond*, Vol. 4, F. R. Hartley, ed., Wiley, Chichester, 1987, Chapter 1, p. 27.

236. W. E. Lindsell, in *Comprehensive Organometallic Chemistry*, Vol. 1, G. Wilkinson, F. G. A. Stone, and E. W. Abel, eds., Pergamon Press, Oxford, 1982, Chapter 4, p. 163.

237. L. Lochmann, J. Pospíšil, and D. Lím, *Tetrahedron Lett.* 257 (1966).

238. M. Schlosser, *J. Organomet. Chem.* **8**, 9 (1967).

239. A. Mordini, *Adv. Carbanion Chem.* **1**, 1 (1992).

240. M. Schlosser, *Pure Appl. Chem.* **60**, 1627 (1988).

241. L. Brandsma, *Preparative Acetylenic Chemistry*, 2nd ed., Elsevier, Amsterdam, 1988, Chapter 10, p. 209.

242. L. Brandsma and H. D. Verkruijsse, *Synthesis* 984 (1990).

243. M. Schlosser, J. Hartmann, M. Stähle, J. Kramar, A. Walde, and A. Mordini, *Chimia* **40**, 306 (1986).

244. M. Stähle, R. Lehmann, J. Kramar, and M. Schlosser, *Chimia* **39**, 229 (1985).

245. J. Hartmann and M. Schlosser, *Helv. Chim. Acta* **59**, 453 (1976).

246. L. Brandsma, H. D. Verkruijsse, C. Schade, and P. v. R. Schleyer, *J. Chem. Soc., Chem. Commun.* 260 (1986).

247. H. Hommes, H. D. Verkruijsse, and L. Brandsma, *J. Chem. Soc., Chem. Commun.* 366 (1981).

248. L. Brandsma, H. Hommes, H. D. Verkruijsse, and R. L. P. de Jong, *Recl. Trav. Chim. Pays-Bas* **104**, 226 (1985).

249. R. D. Chambers, M. Parsons, G. Sandford, and R. Bowden, *Chem. Commun.* 959 (2000).

250. K. Smith, in *Supported Reagents and Catalysts in Chemistry*, B. K. Hodnett, A. P. Kybett, J. H. Clark, and K. Smith, eds., The Royal Society of Chemistry, Cambridge, UK, 1998, p. 31.

251. K. Smith, A. Musson, and G. A. DeBoos, *J. Org. Chem.* **63**, 8448 (1998).

252. M. Haouas, A. Kogelbauer, and R. Prins, *Catal. Lett.* **70**, 61 (2000).

253. A. Kogelbauer, D. Vassena, R. Prins, and J. N. Armor, *Catal. Today* **55**, 151 (2000).

254. K. Smith, T. Gibbins, R. W. Millar, and R. P. Clardige, *J. Chem. Soc., Perkin Trans. 1*, 2753 (2000).

255. A. P. Singh, S. P. Mirajkar, and S. Sharma, *J. Mol. Catal. A: Chem.* **150**, 241 (1999).

256. K. Smith, G. A. El-Hiti, M. E. W. Hammond, D. Bahzad, Z. Li, and C. Siquet, *J. Chem. Soc., Perkin Trans. 1* 2745 (2000).

257. J. H. Clark, J. C. Ross, D. J. Macquarrie, S. J. Barlow, and T. W. Bastock, *Chem. Commun.* 1203 (1997).

258. K. Smith, M. Butters, W. E. Paget, D. Goubet, E. Fromentin, and B. Nay, *Green Chem.* **1**, 83 (1999).

259. M. C. Hausladen and C. R. F. Lund, *Appl. Catal., A* **190**, 269 (2000).

260. M. C. Hausladen, R. C. Cynagovich, H. Y. Huang, and C. R. F. Lund, *Appl. Catal., A* **219**, 1 (2001).

261. T. Shamma, H. Buchholz, G. K. S. Prakash, and G. A. Olah, *Isr. J. Chem.* **39**, 207 (1999).

262. G. S. Lal, G. P. Pez, and R. G. Syvret, *Chem. Rev.* **96**, 1737 (1996).

263. L. Conte, G. P. Gambaretto, M. Napoli, C. Fraccaro, and L. E. Legnaro, *J. Flourine Chem.* **70**, 175 (1995).

264. J. Hutchinson and G. Sandford, *Top. Curr. Chem.* **193**, 1 (1997).

265. R. D. Chambers, J. Hutchinson, and G. Sandford, *J. Fluorine Chem.* **100**, 63 (1999).

266. T. Mori and H. Suzuki, *Synlett* 383 (1995).

267. T. Suzuki and R. Noyori, *Chemtracts–Org. Chem.* **10**, 813 (1997).

268. X. Peng, H. Suzuki, and C. Lu, *Tetrahedron Lett.* **42**, 4357 (2001).

269. L. V. Malysheva, E. A. Paukshtis, and K. G. Ione, *Catal. Rev.-Sci. Eng.* **37**, 179 (1995).

270. H. Suzuki and T. Mori, *J. Chem. Soc., Perkin Trans. 2* 41 (1995).

271. J. H. Ridd, *Acta Chim. Scand.* **52**, 11 (1998).

272. K. K. Laali and V. J. Gettwert, *J. Org. Chem.* **66**, 35 (2001).

273. S. Kobayashi, *Eur. J. Org. Chem.* 15 (1999).

274. S. Kobayashi, *Synlett* 689 (1994).

275. R. W. Marshman, *Aldrichimica Acta* **28**, 77 (1995).

276. F. J. Waller, A. G. M. Barrett, D. C. Braddock, R. M. McKinnell, and D. Ramprasad, *J. Chem. Soc., Perkin Trans. 1* 867 (1999).

277. S. Répichet, C. Le Roux, P. Hernandez, and J. Dubac, *J. Org. Chem.* **64**, 6479 (1999).

278. B. M. Choudary, N. S. Chowdari, and M. L. Kantam, *J. Chem. Soc., Perkin Trans. 1*, 2689 (2000).

279. J. Marquié, A. Laporterie, J. Dubac, N. Roques, and J.-R. Desmurs, *J. Org. Chem.* **66**, 421 (2001).

280. G. A. Olah, T. Mathew, and G. K. S. Prakash, *Chem. Commun.* 1696 (2001).

281. C. A. Reed, N. L. P. Fackler, K.-C. Kim, D. Stasko, D. R. Evans, P. D. W. Boyd, and C. E. F. Rickard, *J. Am. Chem. Soc.* **121**, 6314 (1999).

282. R. Rathore, J. Hecht, and J. K. Kochi, *J. Am. Chem. Soc.* **120**, 13278 (1998).

283. S. M. Hubig and J. K. Kochi, *J. Am. Chem. Soc.* **122**, 8279 (2000).

284. S. M. Hubig and J. K. Kochi, *J. Org. Chem.* **65**, 6807 (2000).

285. I. M. Dahl, E. M. Myhrvold, U. Olsbye, F. Rohr, O. A. Rokstad, and O. Swang, *Ind. Eng. Chem. Res.* **40**, 2226 (2001).

286. P. R. Schreiner, O. Lauenstein, I. V. Kolomitsyn, S. Nadi, and A. A. Fokin, *Angew. Chem., Int. Ed. Engl.* **37**, 1895 (1998).

287. P. R. Schreiner, O. Lauenstein, E. D. Butova, and A. A. Fokin, *Angew. Chem., Int. Ed. Engl.* **38**, 2786 (1999).

288. A. A. Fokin, O. Lauenstein, P. A. Gunchenko, and P. R. Schreiner, *J. Am. Chem. Soc.* **123**, 1842 (2001).

289. L. Liguori, H.-R. Bjørsvik, A. Bravo, F. Fontana, and F. Minisci, *Chem. Commun.* 1501 (1997).

290. J. E. Chateauneuf, *J. Am. Chem. Soc.* **115**, 1915 (1993).

291. M. Sadeghipour, K. Brewer, and J. M. Tanko, *J. Org. Chem.* **62**, 4185 (1997).

292. A. S. Dneprovskii, D. V. Kuznetsov, E. V. Eliseenkov, B. Fletcher, and J. M. Tanko, *J. Org. Chem.* **63**, 8860 (1998).

293. B. Fletcher, N. K. Suleman, and J. M. Tanko, *J. Am. Chem. Soc.* **120**, 11839 (1998).

294. H. Shaw, H. D. Pearlmutter, C. Gu, S. D. Arco, and T. O. Quibuyen, *J. Org. Chem.* **62**, 236 (1997).

295. N. Sun and K. J. Klabunde, *J. Am. Chem. Soc.* **121**, 5587 (1999).

296. Yu. A. Strizhakova, S. V. Levanova, A. B. Sokolov, M. G. Pechatnikov, and A. T. Saed, *Russ. J. Org. Chem.* **36**, 670 (2000).

297. P. M. Reis, J. A. L. Silva, J. J. R. Fraústo da Silva, and A. J. L. Pombeiro, *Chem. Commun.* 1845 (2000).

298. S. Sakaguchi, Y. Nishiwaki, T. Kitamura, and Y. Ishii, *Angew. Chem., Int. Ed. Engl.* **40**, 222 (2001).

299. S. Isozaki, Y. Nishiwaki, S. Sakaguchi, and Y. Ishii, *Chem. Commun.* 1352 (2001).

300. Y. Ishii, *J. Synth. Org. Chem. Jpn.* **59**, 2 (2001).

301. H. Suzuki and N. Nonoyama, *J. Chem. Soc., Perkin Trans. 1* 2965 (1997).

302. Y. Ishii, K. Matsunaka, and S. Sakaguchi, *J. Am. Chem. Soc.* **122**, 7390 (2000).

303. E. Hata, T. Yamada, and T. Mukaiyama, *Bull. Chem. Soc. Jpn.* **68**, 3629 (1995).

304. D. W. McMillen and J. B. Grutzner, *J. Org. Chem.* **59**, 4516 (1994).

305. R. Bianchini, C. Chiape, and B. Meyer, *Gazz. Chim. Ital.* **125**, 453 (1995).

306. M. Johannsen and K. A. Jørgensen, *Chem. Rev.* **98**, 1689 (1998).

307. S. Cenini, S. Tollari, A. Penoni, and C. Cereda, *J. Mol. Catal. A: Chem.* **137**, 135 (1999).

308. S. Cenini, E. Gallo, A. Penoni, F. Ragaini, and S. Tollati, *Chem. Commun.* 2265 (2000).

309. X.-G. Zhou, X.-Q. Yu, J.-S. Huang, and C.-M. Che, *Chem. Commun.* 2377 (1999).

310. S.-M. Au, J.-S. Huang, W.-Y. Yu, W.-H. Fung, and C.-M. Che, *J. Am. Chem. Soc.* **121**, 9120 (1999).

311. X.-Q. Yu, J.-S. Huang, X.-G. Zhou, and C.-M. Che, *Org. Lett.* **2**, 2233 (2000).

312. F. Ragaini, S. Cenini, S. Tollari, G. Tummolillo, and R. Beltrami, *Organometallics* **18**, 928 (1999).

313. I. Nägeli, C. Baud, G. Bernardinelli, Y. Jacquier, M. Moran, and P. Müller, *Helv. Chim. Acta* **80**, 1087 (1997).

314. S.-M. Au, S.-B. Zhang, W.-H. Fung, W.-Y. Yu, C.-M. Che, and K.-K. Cheung, *Chem. Commun.* 2677 (1998).

315. R. S. Srivastava and K. M. Nicolas, *Chem. Commun.* 2705 (1998).

316. R. S. Srivastava and K. M. Nicolas, *J. Am. Chem. Soc.* **119**, 3302 (1997).

317. M. Bruncko, T.-A. V. Khuong, and K. B. Sharpless, *Angew. Chem., Int. Ed. Engl.* **35**, 454 (1996).

318. M. Beller, M. Eichberger, and H. Trauthwein, *Angew. Chem., Int. Ed. Engl.* **36**, 2225 (1997).

319. M. Beller, H. Trauthwein, M. Eichberger, C. Breindl, T. E. Müller, and A. Zapf, *J. Organomet. Chem.* **566**, 277 (1998).

320. M. Beller, H. Trauthwein, M. Eichberger, C. Breindl, J. Herwig, T. E. Müller, and O. R. Thiel, *Chem. Eur. J.* **5**, 1306 (1999).

321. B. A. Arndtsen, R. G. Bergman, T. A. Mobley, and T. H. Peterson, *Acc. Chem. Res.* **28**, 154 (1995).

322. R. H. Crabtree, *Chem. Rev.* **95**, 987 (1995).

323. Y. Fujiwara, K. Takaki, and Y. Taniguchi, *Synlett* 591 (1996).

324. A. Sen, in *Applied Homogeneous Catalysis with Organometallic Complexes*, B. Cornils and W. A. Herrmann, eds., VCH, Weinheim, 1996, Chapter 3.3.6, p. 1081.

325. A. Sen, *Acc. Chem. Res.* **31**, 550 (1998).

326. A. E. Shilov and G. B. Shul'pin, *Activation and Catalytic Reactions of Saturated Hydrocarbons in the Presence of Metal Complexes*, Kluwer, Dordrecht, 2000.

327. I. T. Horváth, R. A. Cook, J. M. Millar, and G. Kiss, *Organometallics* **12**, 8 (1993).

328. J. Kowalik and L. M. Tolbert, *J. Org. Chem.* **66**, 3229 (2001).

11

REDUCTION–
HYDROGENATION

The addition of hydrogen across multiple bonds is one of the most widely studied of catalytic reactions. Alkenes and alkynes, as well as di- and polyunsaturated systems can all be hydrogenated, provided the suitable experimental conditions are used. Studies on the ways in which these compounds react with hydrogen have revealed very complex reaction patterns. Because of their resonance stabilization, carbocyclic aromatic hydrocarbons are more difficult to hydrogenate than are other unsaturated compounds.

Heterogeneous catalytic hydrogenation was discovered in 1900,[1] while a breakthrough in the development of homogeneous or soluble-metal-complex-catalyzed hydrogenation came only in the 1960s.[2–4] Catalytic hydrogenation of multiple bonds is a versatile and most useful technique in organic synthesis. It frequently gives high yields and is particularly important if chemo-, regio- or stereoselective transformations are desirable. Because of their great practical and theoretical importance, a large number of papers, books, and review articles on both heterogeneous[5–20] and homogeneous[16,21–33] catalytic hydrogenations are available. Much less information is available on chemical reductions of hydrocarbons.[34–37] Most significant of these methods is the Birch reduction of aromatics. Ionic hydrogenation, in turn, is a new and vigorously developing field.[38–40]

Besides the addition to multiple bonds, hydrogen can add across carbon–carbon single bonds to bring about bond cleavage. Such hydrogenolysis reactions can occur during hydrocracking (see Section 2.1). Because of its practical importance and theoretical significance, metal-catalyzed hydrogenolysis is studied extensively and will be treated briefly.

11.1. HETEROGENEOUS CATALYTIC HYDROGENATION

11.1.1. Catalysts

In general, the saturation of simple olefinic and acetylenic multiple bonds occurs readily in the presence of any typical hydrogenation catalyst. Platinum metals, mainly platinum, palladium, and rhodium, are active at low pressures and temperatures. Nickel, a cheap and frequently used substitute, usually requires elevated pressures and temperatures. Concerning the hydrogenation of aromatics, rhodium, ruthenium, and platinum can be used most effectively, while nickel requires vigorous conditions. Palladium exhibits rather low activity under mild conditions. These catalysts, when used in hydrogenations, are reduced to the metal form; thus, it is the metal that is the actual catalyst. Other catalysts, such as copper chromite[41] (a specially prepared copper oxide–chromium oxide), zinc oxide,[14] and metal sulfides, which are also active in hydrogenations, participate in the catalytic reaction in the oxide or sulfide form. Although most metals are readily poisoned by sulfur compounds,[42–44] controlled partial sulfur treatment of metals can result in sulfided catalysts that exhibit specific activity and selectivity in hydrogenative transformation of hydrocarbons.[44–46]

The choice of catalyst applied in the hydrogenation of a certain unsaturated hydrocarbon depends on several factors, such as the reactivity of the substrate and the experimental conditions (pressure, temperature, solvent, liquid- or gas-phase reaction). Multiply unsaturated compounds may require the use of a selective catalyst attaining the reduction of only one multiple bond. The use of suitable selective catalysts and reaction conditions is also necessary to achieve stereospecific hydrogenations.

Since hydrogenation over heterogeneous catalysts is a surface reaction, all these catalysts have a large surface area per unit weight. Metal catalysts are prepared from their salts, oxides, or hydroxides (metal "black" catalysts, e.g., Adams platinum[47]), or the metal is deposited on a high-surface-area inert material (support) such as carbon, alumina, or silica. Another common technique to prepare a finely divided large-surface-area metal catalyst is the leaching of one component of a bimetallic alloy. Of the so-called skeletal catalysts thus prepared, Raney nickel is one of the oldest and most common hydrogenation catalysts.[48]

11.1.2. Hydrogenation of Alkenes

Alkenes usually are reduced very easily. The ease of hydrogenation of double bonds depends on the number and nature of substituents attached to the sp^2 carbon atoms. In general, increasing substitution results in decreasing rate of hydrogenation known, as the *Lebedev rule*.[49] Terminal olefins exhibit the highest reactivity, and the rate of hydrogenation decreases in the order $RCH{=}CH_2 > R_2C{=}CH_2 > RCH{=}CHR > R_2C{=}CHR > R_2C{=}CR_2$. In these hydrogenations, *cis* isomers are hydrogenated in preference to the corresponding *trans* compounds. Since the rate of hydrogenation is sensitive to operating conditions, relative rates based on the measurements of competitive hydrogenation of binary mixtures are most often determined instead of measuring individual rates.

Mechanism. The generally accepted mechanism for the hydrogenation of double bonds over heterogeneous catalysts was first proposed by Horiuti and Polanyi,[50,51] and was later supported by results of deuteration experiments. It assumes that both hydrogen and alkene are bound to the catalyst surface. The hydrogen molecule undergoes dissociative adsorption [Eq. (11.1)], while the alkene adsorbs associatively [Eq. (11.2)]. Addition of hydrogen to the double bond occurs in a stepwise manner [Eqs (11.3) and (11.4)]:

$$H_2 + 2M \rightleftharpoons H-M \tag{11.1}$$

$$\underset{/}{\overset{\backslash}{C}}=\underset{\backslash}{\overset{/}{C}} + 2M \rightleftharpoons -\underset{\underset{M}{|}}{\overset{|}{C}}-\underset{\underset{M}{|}}{\overset{|}{C}}- \tag{11.2}$$

$$-\underset{\underset{M}{|}}{\overset{|}{C}}-\underset{\underset{M}{|}}{\overset{|}{C}}- + H-M \underset{}{\overset{-2M}{\rightleftharpoons}} -\underset{\underset{M}{|}}{\overset{|}{C}}-\underset{\underset{H}{|}}{\overset{|}{C}}- \tag{11.3}$$

$$-\underset{\underset{M}{|}}{\overset{|}{C}}-\underset{\underset{H}{|}}{\overset{|}{C}}- + H-M \underset{}{\overset{-2M}{\rightleftharpoons}} -\underset{\underset{H}{|}}{\overset{|}{C}}-\underset{\underset{H}{|}}{\overset{|}{C}}- \tag{11.4}$$

While the last step [Eq. (11.4)] is virtually irreversible under hydrogenation conditions, both the adsorption of alkene [Eq. (11.2)] and the formation of alkyl intermediate (half-hydrogenated state) [Eq. (11.3)] are reversible. The reversibility of these steps accounts for the isomerization of alkenes accompanying hydrogenation (see Section 4.3.2). Isomerizations, either double-bond migration or *cis–trans* isomerization, may not be observable, unless the isomer is less reactive, or the isomerization results in other structural changes in the molecule, such as racemization.

Further studies led to the suggestion of other types of surface species, such as the π-adsorbed intermediate (**1**) and dissociatively adsorbed alkenes [σ-vinyl (**2**), σ-allyl (**3**) and π-allyl (**4**)]:

$$
\begin{array}{cccc}
H_2C=CH_2 & HC=CH_2 & H_2C\overset{\overset{\displaystyle H}{\diagup C\diagdown}}{}CH_2 & H_2C\overset{\overset{\displaystyle H}{\diagup C\diagdown}}{}CH_2 \\
\downarrow & \downarrow & | & | \\
M & M & M & M \\
\mathbf{1} & \mathbf{2} & \mathbf{3} & \mathbf{4}
\end{array}
$$

When deuterium instead of hydrogen is used, reversal of the half-hydrogenated intermediate accounts for deuterium exchange in alkenes and the formation of saturated hydrocarbons with more than two deuterium atoms in the molecule[18,52] (Scheme 11.1).

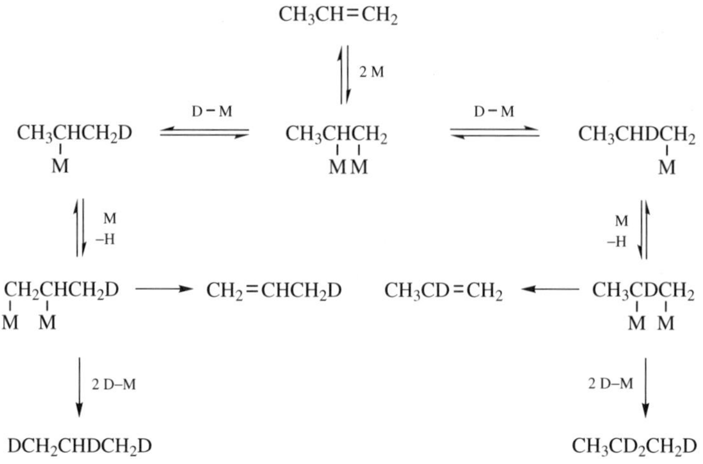

$$CH_3CH=CH_2$$

$$\Big\Uparrow \ 2\,M$$

$$CH_3CHCH_2D \underset{D-M}{\rightleftharpoons} CH_3CHCH_2 \underset{D-M}{\rightleftharpoons} CH_3CHDCH_2$$

(with M, MM, M labels beneath respectively)

$$\Big\Uparrow \begin{smallmatrix}M\\-H\end{smallmatrix} \qquad\qquad\qquad\qquad \begin{smallmatrix}M\\-H\end{smallmatrix}\Big\Downarrow$$

$$CH_2CHCH_2D \longrightarrow CH_2=CHCH_2D \qquad CH_3CD=CH_2 \longleftarrow CH_3CDCH_2$$

(M M ... M M labels)

$$\Big\downarrow \ 2\,D\text{–}M \qquad\qquad\qquad\qquad\qquad\qquad 2\,D\text{–}M \ \Big\downarrow$$

$$DCH_2CHDCH_2D \qquad\qquad\qquad\qquad CH_3CD_2CH_2D$$

Scheme 11.1

Metals differ in their ability to catalyze isomerization. Both the relative rate of transformation of the individual isomers and the initial isomer distribution vary with the metal. The order of decreasing activity of platinum metals in catalyzing the isomerization of alkenes was found to be Pd ≫ Rh, Ru, Pt > Os > Ir.[53] Platinum is generally the preferable catalyst if isomerization is to be avoided. The most active Raney nickel preparations rival palladium in their activity of isomerization.

According to the Horiuti–Polanyi mechanism, isomerization requires the participation of hydrogen. The first addition step, formation of the half-hydrogenated state [Eq. (11.3)], cannot take place without hydrogen. Numerous investigations have supported the role of hydrogen in these so-called hydroisomerizations.

Another possible mechanism interprets *cis–trans* isomerization in combination with double-bond migration assuming the participation of π-allyl intermediates[54,55] (Scheme 11.2). Such dissociatively adsorbed species were originally suggested by Farkas and Farkas.[56]

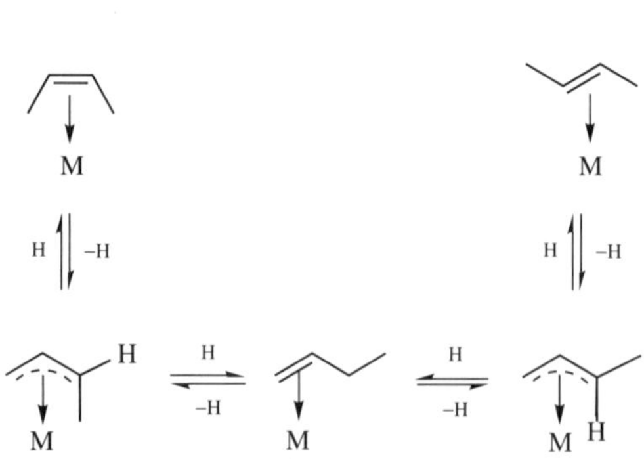

Scheme 11.2

Dissociatively adsorbed surface intermediates were also suggested to account for the selfhydrogenation of ethylene.[52] It occurs on most transition metals and yields ethane without added hydrogen. Highly dehydrogenated surface species may also be formed under normal hydrogenation conditions. The observation by Thomson and Webb showed that many different metal surfaces give nearly equal rate of alkene hydrogenation.[57] They reasoned that a monolayer of adsorbed species represented as $M-C_xH_y$ was involved as hydrogen transfer agent. More recent extensive surface science studies of ethylene adsorption on single crystals revealed the formation of ethylidyne moiety (**5**) on platinum:[58]

$$
\begin{array}{c}
CH_3 \\
| \\
C \\
/\,|\,\backslash \\
M\ M\ M
\end{array}
$$

5

These results suggest that alkene hydrogenation reactions occur on metal surfaces covered with carbonaceous overlayer. Such hydrogen-deficient carbonaceous deposits seem to play an important role in hydrogenation reactions, serving as hydrogen source and providing desorption sites for intermediates and product molecules. Adsorption of reacting molecules, in turn, can lead to surface restructuring during catalytic reactions.[59]

Stereochemistry. It was already recognized in the early 1930s that addition of hydrogen to an alkene is characterized by *syn* (*cis*) stereochemistry, that is, that both hydrogen atoms add from the same face of the double bond. The highest selectivity was observed in the hydrogenation of the isomeric 2,3-diphenyl-2-butenes[60] [Eqs (11.5) and (11.6)]; the *cis* isomer gave the *meso* compound in 98% yield, whereas the racemic compound (the product of *trans* isomer) was isolated in 99% yield:

$$
\begin{array}{c}
Ph \quad\ \ Ph \\
\diagdown\ \diagup \\
C\!=\!C \qquad\xrightarrow[\text{AcOH}]{\text{Pd, H}_2} \quad meso\text{-2,3-diphenylbutane} \\
\diagup\ \diagdown \\
Me \quad\ \ Me
\end{array}
\qquad (11.5)
$$

$$
\begin{array}{c}
Ph \quad\ \ Me \\
\diagdown\ \diagup \\
C\!=\!C \qquad\xrightarrow[\text{AcOH}]{\text{Pd, H}_2} \quad racem.\text{-2,3-diphenylbutane} \\
\diagup\ \diagdown \\
Me \quad\ \ Ph
\end{array}
\qquad (11.6)
$$

Useful data concerning the mechanism of hydrogenation were provided by studying the transformation of cycloalkenes.[61,62] Although *cis*-1,2-dimethylcyclohexane resulting from the expected *syn* addition of hydrogen to 1,2-dimethylcyclohexene is the dominant product, varying amounts of the *trans* isomer were always formed over Pt and Pd. It was found, however, that instead of *anti* (*trans*) addition, the *trans* isomer is formed through 1,6-dimethylcyclohexene (**6**) (Scheme 11.3). The latter

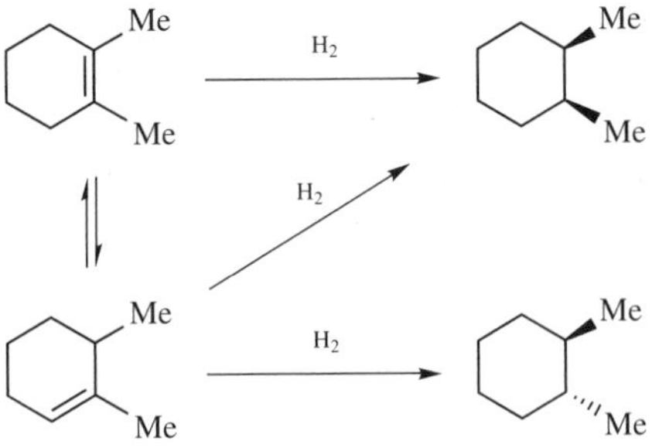

Scheme 11.3

results from isomerization and can yield both *cis* and *trans* saturated products via *syn* addition.

In accordance with this observation, the fraction of the *cis* isomer increases with increasing hydrogen pressure. Since an increase in the hydrogen partial pressure affects step 3 [Eq. (11.3)] in the Horiuti–Polanyi mechanism by shifting the equilibrium to the formation of the half-hydrogenated state, isomerization is suppressed. Palladium, in turn, which exhibits the highest tendency to isomerization among platinum metals, may yield the *trans* isomer as the major product under certain conditions.

Steric effects affecting the mode of adsorption as well as the type of intermediates formed on different metals are decisive factors in determining product distributions. 1,3-Dimethylcyclohexene and 1,5-dimethylcyclohexene yield an identical mixture of *cis*- and *trans*-1,3-dimethylcyclohexane on palladium:[63]

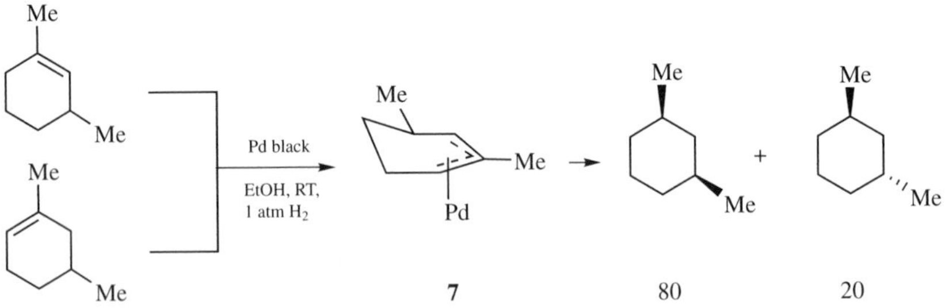

$$(11.7)$$

A common π-allyl intermediate (**7**) adsorbed primarily in the less hindered *cis* mode is invoked to interpret this result.[64]

On platinum, in contrast, the transformation of the isomers takes place through different π-adsorbed species. The favored adsorption modes are indicated in the following equation:

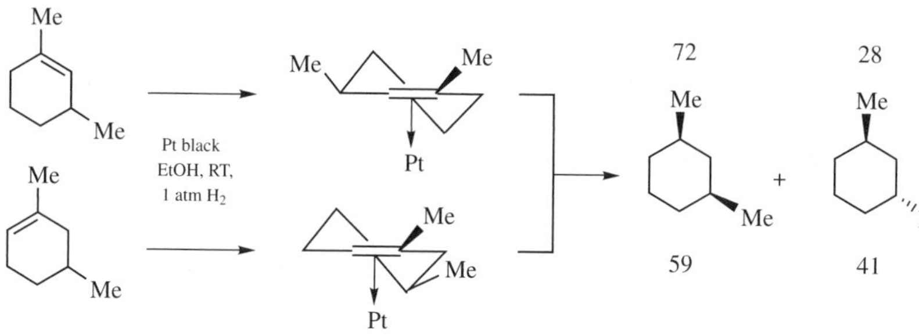

(11.8)

The nearly equal amount of *cis* and *trans* products formed from 1,5-dimethylcyclohexene is explained by the almost equal degree of hindrance of the homoallylic methyl group with the catalyst surface in the alternate adsorption modes.[63,64]

Another interesting example is the platinum-catalyzed hydrogenation of isomeric octalins.[65–67] If *syn* addition to the double bond is assumed, in principle, both *cis-* and *trans*-decalin are expected to result from 1(9)-octalin, but only the *cis* isomer from 9(10)-octalin. In contrast with these expectations, the isomers are produced in nearly the same ratio from both compounds. Transformation in the presence of deuterium revealed that most of the products contained three deuterium atoms. This was interpreted to prove that the very slow rate of hydrogenation of 9(10)-octalin [Eq. (11.9)] permits its isomerization to the 1(9) isomer. As a result, most of the products are formed through 1(9)-octalin [Eq. (11.10)]:

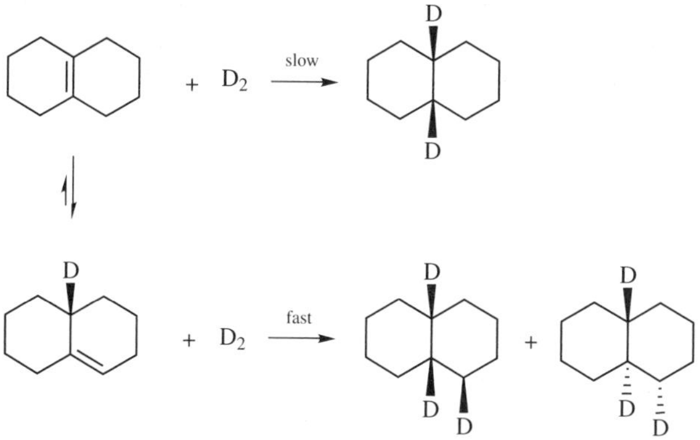

(11.9)

(11.10)

11.1.3. Hydrogenation of Dienes

The hydrogenation of dienes to monoenes introduces several problems of selectivity. Regioselective saturation of one of the double bonds is governed basically by the same effects that determine the relative reactivities of monoalkenes in a binary mixture; that is, a terminal double bond is reduced preferentially to other, more substituted double bonds. During the reduction of a diene, a new competition also emerges since the newly formed monoene and the unreacted diene compete for the same active sites. The simplest situation arises if, for some (e.g., steric) reason, adsorption can occur on only one of the double bonds. In general, however, an equilibrium is established between the species adsorbed on the two olefinic bonds. Judicious choice of catalysts and experimental conditions can allow highly selective monohydrogenation of dienes.

Nonconjugated dienes, namely, allenes and isolated dienes, react preferentially on the terminal double bond.[10] Hydrogenation of 1,2-butadiene over palladium yields 1-butene and *cis*-2-butene as the main products with moderate discrimination of the two double bonds.[68] Deuteration experiments indicated that the dominant *syn* addition to either the 1,2- or the 2,3-olefinic bond occurs. Different vinyl and π-allyl intermediates were invoked to interpret the results.[69,70]

The monohydrogenation of conjugated dienes can occur by either 1,2 or 1,4 addition. 1-Butene (53%) and *trans*-2-butene (42%) are the main products in the hydrogenation of 1,3-butadiene on palladium with only a small amount of *cis*-2-butene.[68] Deuterium distribution reveals that the *trans* isomer is produced by 1,4 addition.

On other Group VIII metals,[71,72] and on copper[71] and gold,[73] the *trans* : *cis* ratio is close to 1. This is very similar to the ratio that arises in the isomerization of 1-butene under the same reaction conditions. It follows from this that common intermediates may be involved in the two processes. According to the accepted mechanism,[71,72] product distributions are determined by the possibility of interconversion between the different adsorbed intermediates (Scheme 11.4). With the exception of palladium, other metals ensure a high degree of interconversion, thus leading to a similar quantity of the 2-butene isomers. On palladium, the large *trans* : *cis* ratio reflects the conformational characteristics of 1,3-butadiene in the gas phase; the ratio of the transoid to cisoid conformation is about 10–20—thus, this ratio is essentially unchanged in the adsorbed phase.

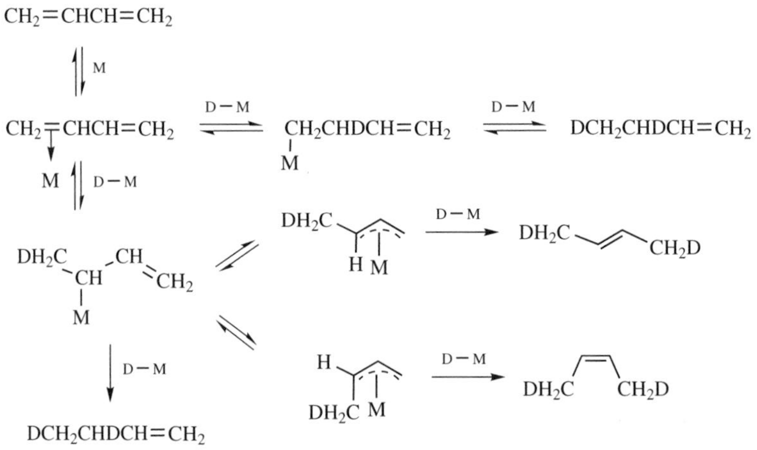

Scheme 11.4

Cyclopentene is produced in excellent yield by the selective hydrogenation of cyclopentadiene. Both Raney nickel in cold ethanol solution[74] and Cu/Al$_2$O$_3$ aerogel in the gas phase[75] are selective catalysts. The cyclopentene is converted to cyclopentane at a rate much lower than that of diene transformation to cyclopentene. Other catalytic processes for the hydrogenation of cyclopentadiene yield cyclopentane as the principal product.

Selective hydrogenation of diolefins and alkenylaromatics in steam-cracked gasoline is of industrial importance. Specific refining by selective hydrogenation of these polymerizable hydrocarbons without hydrogenating other unsaturated compounds (alkenes, aromatics) is required to increase the stability of gasoline (see Section 11.6.1).

11.1.4. Transfer Hydrogenation

An interesting, convenient, and useful alternative to conventional catalytic hydrogenation is transfer hydrogenation. Named as such by Braude and Linstead,[76] the process involves hydrogen transfer from a hydrogen donor, in most cases an organic molecule, to an unsaturated compound. Usually carried out at reflux temperature, it does not require an external hydrogen source or a special apparatus. A special case is disproportionation, specifically, hydrogen transfer between identical donor and acceptor molecules.

Palladium is the most active and most frequently used catalyst in transfer hydrogenations.[77,78] Cyclohexene, a cheap, readily available, highly reactive molecule, is the preferred donor compound. Alternatively, tetralin and monoterpenes and, in general, any hydroaromatic compound, may be used. Mainly alcohols are employed as the donor with Raney Ni.

According to a possible mechanism, transfer hydrogenation requires a catalyst-mediated formation of a donor–acceptor complex, followed by a direct hydrogen transfer. An alternative possibility is a simple consecutive dehydrogenation–hydrogenation process. While the former mechanism on palladium is supported by numerous experimental evidence,[78] direct hydrogen transfer on nickel was disproved.[79]

A unique example,[80] the reduction via transfer hydrogenation of aryl-substituted alkenes, is facilitated by AlCl$_3$:

$$
\begin{array}{c}
R^1 \quad\quad R^2 \\
\diagdown\,\diagup \\
C = C \\
\diagup\,\diagdown \\
Ph \quad\quad R^3
\end{array}
\quad
\xrightarrow[\text{cyclohexene, reflux, 46 h}]{\text{10\% Pd on C, AlCl}_3}
\quad
\begin{array}{c}
R^1 \quad\quad R^2 \\
\diagdown\,\diagup \\
CH - CH \\
\diagup\,\diagdown \\
Ph \quad\quad R^3
\end{array}
\qquad (11.11)
$$

$R^1, R^2, R^3, = H, alkyl, Ph$ \qquad 77–93%

The role of Lewis acid is to activate the substrate alkenes and initiate an acid-catalyzed hydride transfer from an allylic position.

11.1.5. Hydrogenation of Alkynes

Similarly to olefinic double bonds, the carbon–carbon triple bond can be hydrogenated with ease over a wide variety of catalysts. Bulky groups, however, may hinder hydrogenation.

The complete hydrogenation of alkynes to alkanes occurs quite readily under the usual alkene hydrogenation conditions. This may or may not involve an intermediate alkene:[81]

$$RC \equiv CR \xrightarrow{\text{H}_2} RCH = CHR \xrightarrow{\text{H}_2} RCH_2CH_2R \qquad (11.12)$$

$$\underset{2\,\text{H}_2}{\longleftarrow}$$

A more challenging task is the selective partial hydrogenation (semihydrogenation) of alkynes to yield alkenes. This is a selectivity problem similar to the hydrogenation of dienes in that that the alkyne is hydrogenated preferentially in the presence of an alkene. The possibility of the formation of geometric isomers from nonterminal acetylenes raises the problem of stereoselective semihydrogenation.

It was shown in 1899 by Sabatier and Senderens that acetylene can be selectively hydrogenated to ethylene at atmospheric pressure in the presence of palladium and nickel, but not with platinum.[82] The selectivity for formation of *cis* alkenes was also established quite early.[83] The preferential formation of *cis* compounds provides further evidence for the *syn* addition of hydrogen during catalytic hydrogenation.[81] Since a small amount of *trans* alkene may be formed even before the total consumption of alkyne, it is reasonable to assume that the *trans* alkene may also be an initial product.[81] The formation of the *trans* isomer is generally interpreted as a result of isomerization of the *cis* compound. Multiple-bound surface species such as surface ethylidyne[84] and hydrogen-deficient surface layer (C_2H_x, $x < 2$)[85] are suggested to account for nonselective direct hydrogenation of acetylene to ethane.

The acetylene hydrogenation reaction acquired commercial importance because it afforded a means of converting acetylene in cracker gases to ethylene (using the hydrogen already a constituent of the gas) and a means for oil-deficient countries to obtain ethylene starting with calcium carbide produced from available coke and lime. During World War II, ethylene was prepared from acetylene in 90% yield on a large scale over a palladium-on-silica gel catalyst;[86,87] the reaction temperature is raised from 200 to 300°C during 8 months before it became necessary to regenerate the catalyst by burning off carbon with air at 600°C. Since the synthesis of acetylene for chemical use diminished and most acetylene plants were shut down, this method for ethylene production has been abandoned.

The high selectivity of alkene formation is not the result of a large rate difference in the hydrogenation of the triple and double bonds. Rather, it is ensured by the strong adsorption of alkynes compared with that of alkenes. The alkyne displaces the alkene from the surface or blocks its readsorption.

For the selective hydrogenation of alkynes to alkenes and, if possible, to *cis* olefins, palladium is by far the most selective metal.[6,81,88,89] Some nickel catalysts also appear to exhibit reasonable selectivity.[81,90] In practice, palladium is always used in conjugation with different modifiers to increase selectivity. A highly selective palladium catalyst supported on $CaCO_3$ poisoned with lead acetate is known as the *Lindlar catalyst*.[91,92] By adding a nitrogen base to the reaction mixture, a sudden break in the rate of hydrogen consumption generally occurs after the adsorption of 1 mol of hydrogen. Low temperature and pressure are also recommended to achieve optimal selectivity. Palladium modified with Pb or Hg and used in transfer hydrogenation,[93] the P-2 nickel catalyst prepared by the reduction of nickel acetate with $NaBH_4$,[90] and a Pd–Hg-on-SiO_2 catalyst prepared by a novel method (treatment of silica gel–immobilized silyl hydride with $PdCl_2$ then Hg_2SO_4)[94] also ensure similar selectivities.

A simple alternative to the Lindlar reduction process has been recognized.[95] The palladium(II) acetate-catalyzed polymerization of triethoxysilane in water produces finely divided palladium metal dispersed on a polysiloxane matrix with concomitant hydrogen evolution. Addition of $(EtO)_3SiH$ to the mixture of an alkyne and all other necessary constituents allows facile and selective reduction of the alkyne without an external hydrogen source:

$$n\text{-Bu}\!-\!\!\equiv\!\!-\!\text{Bu-}n \quad \xrightarrow[\text{THF, H}_2\text{O, (5 : 1), RT}]{\text{Pd(OAc)}_2,\ (\text{EtO})_3\text{SiH}} \quad \underset{n\text{-Bu}\qquad\text{Bu-}n}{\diagup\!\!=\!\!\diagdown} \qquad (11.13)$$

>96% selectivity
at 100% conversion

Concerning chemical selectivities in multiple unsaturated acetylenic compounds, conjugated enynes are reduced with modest selectivity. The most difficult case involves the selective reduction of an enyne with a terminal double bond and an internal triple bond, because the difference in the rates of their hydrogenation is minimal. The hydrogenation of nonterminal, unconjugated diynes to (Z,Z)-dienes can be achieved in good yield.

One important application of selective hydrogenation of alkynes is their removal from the industrial steam cracker products. These can contain several percents of alkynes as byproducts. They are particularly unwelcome in that they poison the catalysts used for the downstream polymerization of the olefins. Selective hydrogenation of these steam cracker cuts has two advantages. It removes acetylenes and converts them to desired alkenes, thereby increasing the overall yields (see Section 11.6.1).

11.1.6. Hydrogenation of Aromatics

The hydrogenation of benzene and its alkyl-substituted derivatives takes place stepwise (Scheme 11.5). On the basis of instrumental and isotope exchange studies, the involvement of π-adsorbed (**8, 9**) and σ-adsorbed (**10**) species was suggested:[96–98]

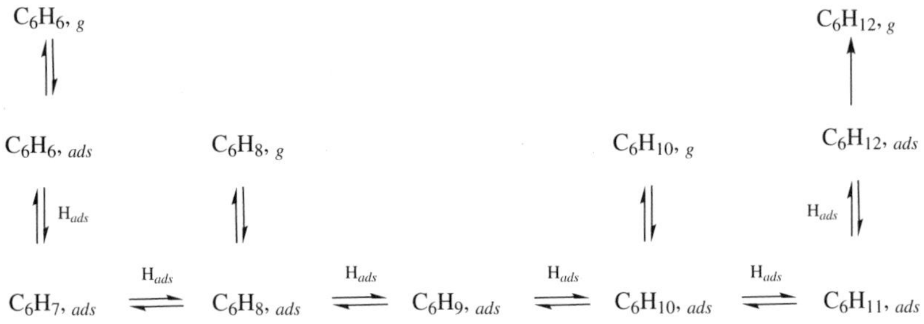

$C_6H_{6,\,g}$ $\updownarrow$

$C_6H_{12,\,g}$ $\uparrow$

$C_6H_{6,\,ads}$ $C_6H_{8,\,g}$ $C_6H_{10,\,g}$ $C_6H_{12,\,ads}$

$\updownarrow H_{ads}$ $\updownarrow$ $\updownarrow$ $H_{ads}\updownarrow$

$C_6H_{7,\,ads} \;\underset{H_{ads}}{\rightleftharpoons}\; C_6H_{8,\,ads} \;\underset{H_{ads}}{\rightleftharpoons}\; C_6H_{9,\,ads} \;\underset{H_{ads}}{\rightleftharpoons}\; C_6H_{10,\,ads} \;\underset{H_{ads}}{\rightleftharpoons}\; C_6H_{11,\,ads}$

Scheme 11.5

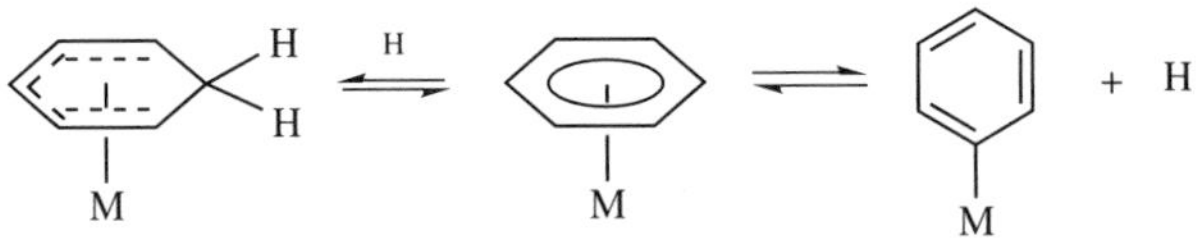

$$\text{(11.14)}$$

8   **9**   **10**

Under the usual hydrogenation conditions the intermediate cyclohexadienes have not been isolated since they are hydrogenated at much higher rate than benzene itself. In contrast, cyclohexenes can be detected in varying amounts during the hydrogenation of benzene and benzene derivatives. The amount of intermediate olefins depends on the structure of the starting compound, the catalyst and reaction conditions. Highly strained systems, such as *o*-di-*tert*-butylbenzene, yield characteristically high amount of olefins:[99]

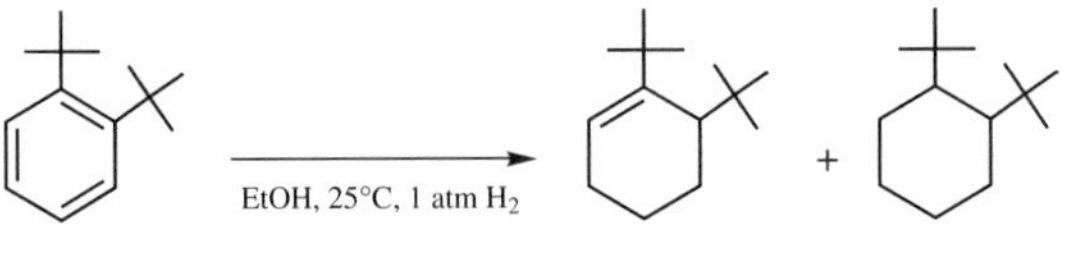

$$\text{(11.15)}$$

		cis : trans
5% Rh on C	30	70 71 : 29
5% Pt on C	6	94 92 : 8

Ruthenium is the most selective metal to produce intermediate olefins. Certain catalyst additives and water[100–102] increase the yield of cyclohexene from benzene up to 48% at 60% conversion.[103] Alkylbenzenes are hydrogenated at somewhat lower rates then benzene itself. As a general rule the rate of hydrogenation decreases as the number of substituents increases, and the more symmetrically substituted compounds react faster than those with substituents arranged with less symmetry.[9,10] Highly substituted strained aromatics tend to undergo ready saturation, even over the less active palladium.

   The intervention of olefin intermediates has significant effects on the stereochemistry of the hydrogenation of aromatics. Similarly to the reduction of alkenes, *cis* isomers are the main products during the hydrogenation of substituted benzenes,

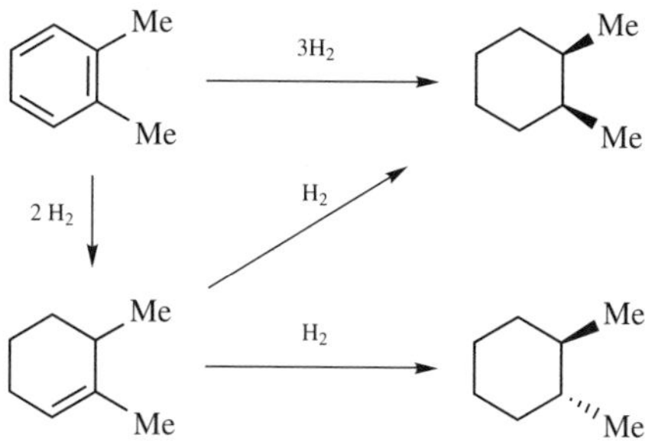

Scheme 11.6

but *trans* isomers are also formed (see also Scheme 11.3). *o*-Xylene, for instance, gives 98.6% and 96.8% *cis*-1,2-dimethylcyclohexane on iridium and osmium, respectively,[104] but the yield of the *cis* compound is only 57.7% on palladium.[10] Experimental observations support the view that *trans* compounds are formed through desorbed cyclohexene intermediates. The ratio of the isomers is determined by a competition between two processes: the addition of six hydrogen atoms during a single period of adsorption on the catalyst surface yields the *cis* compound, while the addition of two hydrogen atoms to the readsorbed cyclohexene intermediate can lead to both the *cis* and *trans* products (Scheme 11.6).

Aromatic hydrocarbons with two or more rings are hydrogenated in a stepwise manner permitting regioselective saturation of one of the rings. Individual metals have a marked influence on selectivity. Biphenyl, for example, is transformed to cyclohexylbenzene with 97% selectivity on palladium, which is the least active to catalyze saturation of benzene under mild conditions.[11,12] Even better selectivities are achieved in transfer hydrogenation.[105]

Hydrogenation of 4-methylbiphenyl with platinum or palladium takes place predominantly in the unsubstituted phenyl ring:[106]

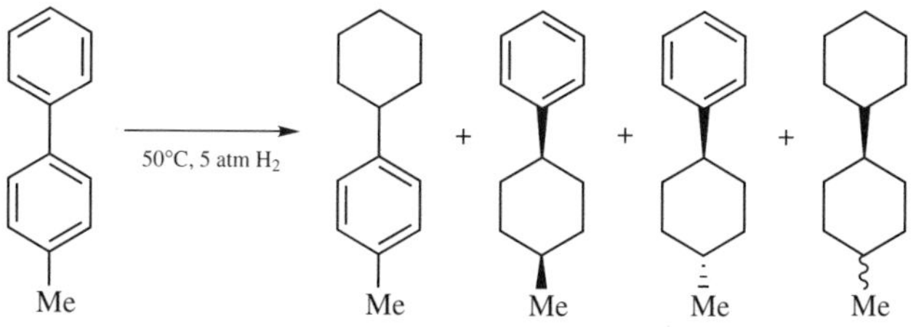

$$(11.16)$$

5% Pd on C, EtOAc	76	17	7	<1
Pt black, EtOAc	77	10	2	11
Raney Ni, EtOH	41	36	23	<1

On the other hand, hydrogenation with Raney nickel causes reduction mainly in the substituted aromatic ring [Eq. (11.16)]. Differences in product composition brought about by the different metals are explained in terms of steric hindrance of the substituted ring (Pt, Pd) versus the anchor effect of the methyl substituent (Ni).[106]

Detailed studies have been conducted on the selectivity of hydrogenation of naphthalene.[67] The reaction proceeds via tetrahydronaphthalene as an intermediate and eventually yields a mixture of the isomeric decalins:

$$(11.17)$$

Traces of 1(9)- and 9(10)-octalin are also detected. Under suitable conditions, the reaction may stop at tetralin on nickel and palladium.[67]

Good selectivities for *cis*-decalin are exhibited by all platinum metals except palladium that is unique in the high yield (52%) of *trans*-decalin.[66,67] It was proved that isomeric octalins are involved in determining product selectivities[66,67] (see Section 11.1.2).

Monosubstituted naphthalene derivatives are usually hydrogenated in the unsubstituted ring:[67,107]

$$(11.18)$$

Rh on Al$_2$O$_3$, 100°C, 59 atm H$_2$	1-Me, 2-Me	20	80
10% Pd on C, 80°C, 1 atm H$_2$	1-Me	34	66
	1-*tert*-Bu	97	3

1-Substituted derivatives with bulky groups, however, are reduced at higher rates and with opposite selectivity on palladium[107] [Eq. (11.18)]. This was explained by the partial release of *peri* strain in the transition state during hydrogenation.

Anthracene and phenanthrene are converted to their 9,10-dihydro derivatives in high yield. Other partially hydrogenated intermediates formed by further stepwise hydrogen addition can also be isolated under suitable experimental conditions.[10,18]

Platinum and palladium exhibit different selectivities in the partial hydrogenation of polycyclic aromatics under mild conditions:[108]

$$(11.19)$$

$$(11.20)$$

11.2. HOMOGENEOUS CATALYTIC HYDROGENATION

11.2.1. Catalysts

A wide range of soluble metal complexes has proved to exert catalytic activity in hydrogenation of unsaturated molecules. Most attention, however, has focused on Group VIII elements since they give rise to the most active catalytic systems.

Homogeneous catalysts have certain advantages compared with heterogeneous catalysts. They contain only one type of active site and as a result are more specific. Since all these active sites are available, homogeneous complexes are potentially more efficient than heterogeneous catalysts. The selectivity of homogeneous catalysts is much more easily modified by selective ligand exchange. In contrast, thermal stability and suitable solvents are often limiting factors in homogeneous hydrogenations.

Possibly the best known, most thoroughly studied, and most widely used catalyst for the reduction of a range of unsaturated compounds is chlorotris(triphenylphosphine)rhodium(I) [RhCl(PPh$_3$)$_3$], known as the *Wilkinson catalyst*.[2–4,109,110] Reduction is generally achieved under mild conditions (room temperature, atmospheric or subatmospheric hydrogen pressure). Since aromatics are inert under these conditions, benzene is often used as solvent. The Wilkinson catalyst, just as many other soluble complexes, selectively catalyzes the reduction of unsaturated carbon–carbon double bonds in the presence of other reducible groups. In general, side reactions (hydrogenolysis, disproportionation) do not occur. Although carbonyl ligands usually tend to deactivate rhodium hydrogenation catalysts, [RhH(CO)(PPh$_3$)$_3$], better known as a *hydroformylation catalyst*, also hydrogenates alkenes.[111,112]

An early and popular complex, [IrCl(CO)(PPh$_3$)$_2$], called the *Vaska complex*,[113] and its derivatives can be used for the hydrogenation of unsaturated hydrocarbons.[26,31]

Very high rates can be achieved for hydrogenations with dichlorotris(triphenylphosphine)ruthenium(II) [RuCl$_2$(PPh$_3$)$_3$], which is complementary to the two important rhodium complexes mentioned before.[114,115] In catalytic hydrogenations it is transformed to [RuHCl(PPh$_3$)$_3$], which is often the true catalyst.

Cyanide-containing cobalt catalysts, particularly potassium pentacyanocobaltate(II) K$_3$[Co(CN)$_5$], are used in the reduction of activated alkenes (conjugated dienes).[26,31] [Co(CO)$_4$]$_2$ is best known as a hydroformylation catalyst, but hydrogenation is also possible under specific conditions. Phosphine-substituted analogs are more successful.

Platinum complexes have been shown to hydrogenate terminal alkenes, dienes, and polyenes. Most generally [PtCl$_2$(PPh$_3$)$_3$] with SnCl$_2$ is used. They form the complex [PtCl(SnCl$_3$)(PPh$_3$)$_2$], which then gives the hydride [PtH(SnCl$_3$)(PPh$_3$)$_2$] with hydrogen.[26,31]

A number of Ziegler-type catalysts based on early transition metals and trialkylaluminum or similar organometallic compounds are soluble in hydrocarbon solvents and function as homogeneous hydrogenation catalysts.[24,26] Many of the soluble transition-metal catalysts can be employed in catalytic transfer hydrogenations.[77,78]

Attempts to utilize commercially some of the extremely effective homogeneous catalysts discovered since the 1960s led to the development of supported transition-metal complex catalysts also known as *heterogenized homogeneous complexes*.[116–120] One important disadvantage of homogeneous catalysts is the difficulty of their separation and recovery after the catalytic reaction is completed. Binding homogeneous catalysts to solid supports such as inorganic oxides, organic polymers, or more recently zeolites via ionic or covalent bonds solves this problem. Extremely active supported organometallic complexes have been reported.[121–123]

11.2.2. Hydrogenation of Alkenes and Dienes

The hydrogenation of many different alkenes, dienes, polyenes, and alkynes may be catalyzed by homogeneous complex catalysts. Many of the soluble complexes have the ability to reduce one particular unsaturated group in the presence of other reducible groups. The selectivity rather than their universality makes these catalysts particularly useful in synthetic organic chemistry. With careful choice of catalyst and reaction conditions, remarkable selectivities are attainable. Most of the practically useful catalysts work under ambient conditions.

Mechanism. Hydrogen must first be activated before it can be added to an unsaturated bond, and generally both hydrogen and the organic substrate must be brought into the coordination sphere of the metal catalyst. The most important property of a homogeneous catalyst is a vacant coordination site, which may be occupied by a readily displaceable ligand.

A common, quite general route for activation of hydrogen is the oxidative addition by homolytic cleavage to form a dihydride complex[24,124,125] [Eq. (11.21)]. It involves an increase of 2 in the formal oxidation state of the metal. In turn, monohydrido complexes can be formed through both homolytic [Eq. (11.22)] and heterolytic [Eq. (11.23)] cleavages of hydrogen:[109]

$$M(I) + H_2 \longrightarrow M(III)H_2 \tag{11.21}$$

$$2\,M + H_2 \longrightarrow 2\,MH \tag{11.22}$$

$$MX + H_2 \longrightarrow MH + H^+ + X^- \tag{11.23}$$

Catalytic cycles operating via the initial formation of a hydridometal complex followed by coordination of the unsaturated compound (S) are termed the *hydride route*[23] (Scheme 11.7). In contrast, the *unsaturate route*[23] involves prior binding of the organic compound (Scheme 11.7). In many cases the hydride route was proved to be more efficient.

Activation of hydrogen through oxidative addition is best exemplified by the Wilkinson catalyst. The hydrogenation mechanism characteristic of dihydride complexes was originally suggested by Wilkinson[2,109] and was later further

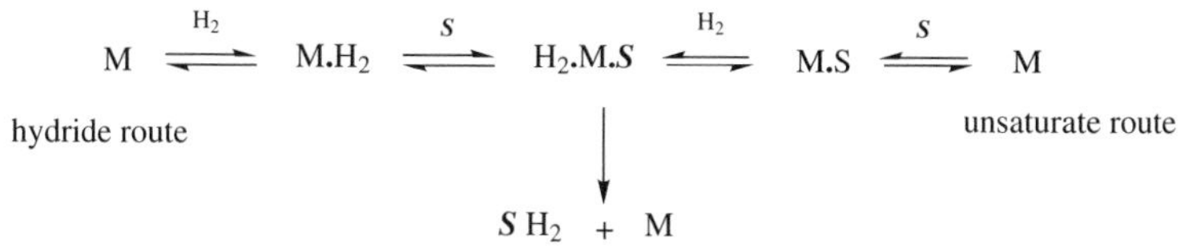

$$M \underset{H_2}{\rightleftharpoons} M.H_2 \underset{S}{\rightleftharpoons} H_2.M.S \underset{H_2}{\rightleftharpoons} M.S \underset{S}{\rightleftharpoons} M$$

hybride route   unsaturate route

$$S\,H_2 \;+\; M$$

Scheme 11.7

elucidated.[126–130] The oxidative addition of hydrogen brings about the transformation of the coordinatively unsaturated square planar rhodium(I) species (**11**) into a coordinatively saturated octahedral rhodium(III) *cis* dihydride complex (**12**). This then participates in a series of simple steps eventually resulting in the formation of the saturated end product. Of the several mechanistic pathways demonstrated for the Wilkinson catalyst, the kinetically dominant is shown in Scheme 11.8. The solvent, not shown, may also play an important role as a coordinating ligand. Dinuclear complexes also observed under reaction conditions were not found to be very effective.

In the octahedral complex (**12**) (Scheme 11.8) the hydride ligand exerts a large *trans* effect labilizing the *trans* ligand. This ligand readily dissociates in solution, creating a vacant site (**13**) that is able to form a π complex (**14**) with the alkene. This substrate association brings the alkene into a suitable environment for addition. The next stage in the catalytic cycle is the migratory insertion of the alkene into the rhodium–hydrogen bond. The σ-alkylrhodium hydride complex **15** thus formed undergoes a rapid reductive elimination. This leads to the saturated product and regenerates the catalytically active rhodium(I) species to complete the cycle.

There are many examples where monohydridometal complexes are formed via homolytic or heterolytic hydrogen splitting.[24,31] The basic difference between the

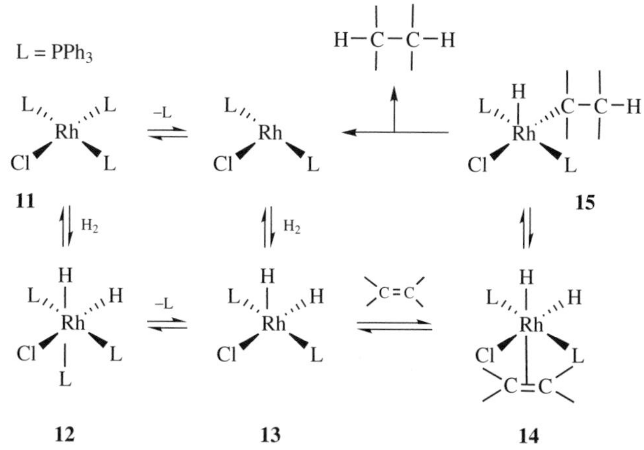

Scheme 11.8

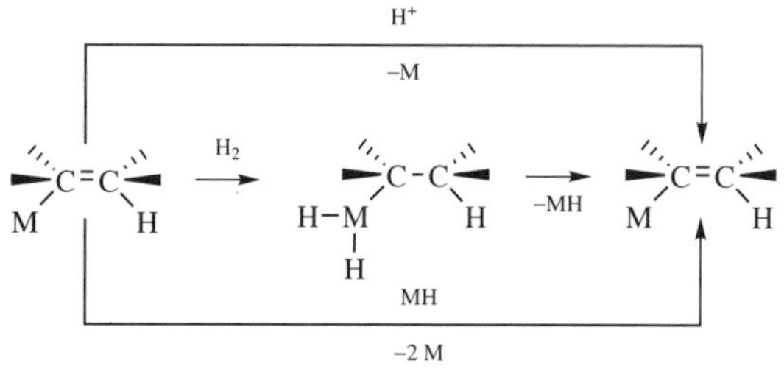

$$H^+$$

Scheme 11.9

mechanism demonstrated for the Wilkinson catalyst and the overall hydrogenation mechanism in the presence of a monohydride complex lies in the final product forming step. The alkylmetal intermediate, in this case, can yield the saturated hydrocarbon via hydrogenolysis using H_2, or as a result of protolysis, or by reacting with a second hydrido complex[24] (Scheme 11.9).

The formation of a metal–substrate bond of intermediate stability is of key importance in a successful hydrogenation. If this bond is too strong, the subsequent migratory insertion may be hindered and hydrogenation cannot take place. Strongly chelating dienes, such as 1,5-cyclooctadiene (COD) or cyclopentadiene, form stable complexes that, in turn, are used as catalysts in hydrogenations.

Strikingly high activities of supported homogeneous complexes have been observed. Whereas organoactinide complexes such as $[(\eta^5\text{-Me}_5\text{C}_5)\text{ThMe}_2]$ and $[(\eta^5\text{-Me}_5\text{C}_5)\text{ZrMe}_2]$ exhibit only marginal activities in alkene hydrogenation, adsorption on dehydroxylated γ-alumina results in dramatic enhancement in catalytic activity.[121,122] Evolution of CH_4 was observed in adsorption studies indicating alkyl anion abstraction as the major adsorptive pathway.[131] The activity of the resulting catalyst possessing cation-like species with lower coordinative saturation and greater electrophilic character is comparable to that of the traditional supported transition-metal catalysts.[121,122]

Selectivity and Stereochemistry. An important property of transition-metal complexes is that they coordinate groups in a specific manner permitting high regio- and stereoselectivity in the catalytic reaction. The migratory insertion step is a highly stereospecific transformation. The four-center transition state **16** illustrated for the Wilkinson catalyst requires a coplanar arrangement of metal, hydride, and alkene π bond:

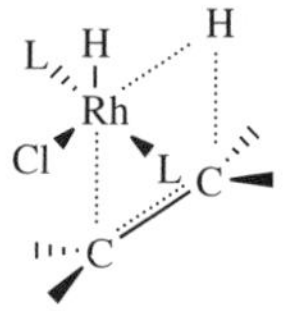

16

The coplanar migratory insertion and the subsequent reductive elimination occurring with retention of configuration at the M—C bond ensure an overall *syn* addition of hydrogen. This was demonstrated in nearly all cases studied. Isotope labeling studies were mainly carried out with functionalized alkenes.[109,132]

Since the π and σ complexes (**14** and **15**, Scheme 11.8) participating in the hydrogenation mechanism are rather crowded species, the hydrogenation reaction is expected to be very susceptible to steric parameters. Steric interaction of the phosphine groups may hinder substrate bonding or prevent hydrogen transfer to the coordinated alkenes, especially with bulky substituents. Hydride alkene complexes of nonreducible olefins were isolated in several cases. Because of these interactions, the relative rate of reduction of simple alkenes depends almost entirely on steric factors. Many of the homogeneous complex catalysts are very selective in the hydrogenation of terminal alkenes. The general reactivity is 1-alkenes > *cis* alkenes > *trans* alkenes > cycloalkenes. The Wilkinson catalyst is not particularly selective, but [RhH(CO)(PPh$_3$)$_3$] and [RuHCl(PPh$_3$)$_3$] exhibit excellent selectivities in reducing terminal alkenes preferably.

The complexation of an alkene to a metal ion strongly affects its reactivity; namely, the double bond is activated toward nucleophilic attack. Ligands decreasing the electron density on the double bond accelerate the hydrogenation reaction. Electron-withdrawing substituents similarly facilitate hydrogenation of alkenes (PhCH=CH$_2$ > RCH=CH$_2$).[29]

The hydrogenation mechanisms discussed above indicate the reversible formation of intermediates. The degree of reversibility depends on the nature of the catalyst and the alkene, and on reaction conditions. Some nonreducible internal alkenes undergo slow hydrogen exchange and isomerization, indicating that reversible steps can occur in these cases. The reversible formation of alkylmetal intermediates provides a ready explanation for the isomerization of alkenes in the presence of certain metal complexes (see Section 4.3.2).

The regioselectivity exhibited in the hydrogenation of monoenes also holds for the hydrogenation of dienes; specifically, the terminal double bond in nonconjugated dienes is usually reduced selectively:[111]

$$(11.24)$$

When isotetralin or 1,4-dihydrotetralin is reduced in the presence of the Wilkinson complex, an identical mixture of 9(10)-octalin and 1(9)-octalin is formed [Eq. (11.25)], requiring the involvement of double-bond migration:[133]

$$(11.25)$$

Since isomerization, in general, is not characteristic of the Wilkinson catalyst, it is believed to be caused by a small amount of oxygen impurity.[29]

Certain monohydride catalysts exhibit high selectivity toward hydrogenation of conjugated dienes and polyenes to give monoenes. Potassium pentacyanocobaltate prepared from $CoCl_2$ and KCN is one of the most active catalysts, but reduces only conjugated double bonds. The reduction of dienes with nonequivalent double bonds follows the usual substitution trend noted for monoenes. 1,3-Pentadiene, for example, is converted selectively to *trans*-2-pentene via 1,2 addition:[134]

$$
\underset{\substack{\text{KOH, }\beta\text{-cyclodextrin,} \\ C_6H_6,\ H_2O,\ 1\ atm\ H_2 \\ RT,\ 24\ h}}{\overset{[CoH(CN)_5]^{3-}}{\xrightarrow{\hspace{3cm}}}}
\quad \underset{96\%}{} \quad + \quad \underset{2\%}{} \tag{11.26}
$$

Lanthanide ions (La^{3+}, Ce^{3+}) were found to promote the cobalt-catalyzed reaction in a two-phase system with β-cyclodextrin or PEG-400 (polyethylene glycol 400) as phase-transfer catalysts.[134]

The hydrogenation of 1,3-butadiene, in contrast, yields a mixture of the isomeric butenes with product distributions highly depending on reaction conditions (nature of solvent, cyanide : cobalt ratio):[134–136]

$$
\underset{}{} \xrightarrow{\hspace{3cm}} \quad \underset{}{} \quad + \quad \underset{}{} \quad + \quad \underset{}{} \tag{11.27}
$$

		CN : Co			
$[CoH(CN)_5]^{3-}$	H_2O	6.0	95%	4%	1%
20°C, 1 atm H_2	glycerol, MeOH	5.0	9%	87%	4%
$[Pd_2(Ph_2PCH_2PPh_2)_3]$	toluene		63%	28%	6%
RT, 6 atm H_2	AcOH		3%	68%	23%

This was interpreted by invoking 1,2 and 1,4 addition occurring on different intermediates participating in a complex equilibrium. Similar changes observed with a palladium complex were explained by 1,2 addition followed by isomerization with the participation of palladium π-allyl species.[137]

In contrast, a heterogenized molybdenum complex showed high activity for the selective hydrogenation of 1,3-butadiene to *cis*-2-butene:[138]

$$
\underset{150°C}{\overset{Mo(CO)_6-LiY}{\xrightarrow{\hspace{2.5cm}}}} \quad \underset{\substack{96\%\ selectivity \\ at\ 97.2\%\ conversion}}{} \tag{11.28}
$$

The $Mo(CO)_3$ subcarbonyl formed under reaction conditions encaged in the zeolite was found to be the active catalytic species.

Certain chromium complexes also exhibit similar selectivities in the hydrogenation of conjugated dienes.[139,140] Coordination of the diene in the cisoid conformation and 1,4 addition can account for the selectivity.

A most unusual and unexpected, completely regioselective 1,2 addition is catalyzed by a montmorillonite-supported bipyridinepalladium(II) acetate complex:

$$\text{Ph}\overset{\text{heterogenized Pd(II)}}{\underset{\text{THF, 1 atm } H_2, \text{ RT}}{\longrightarrow}} \text{PhCH}_2\text{CH}_2 \qquad (11.29)$$

95% yield

The selective η^2 coordination of the diene to the palladium center is suggested.[141]

The site-selective reduction of α,ω-dienes, in which the two double bonds are differentiated only by the presence of an allylic substituent, is a challenging task. A pentamethylyttrocene complex that is extremely selective in hydrogenating terminal alkenes exhibits good selectivity in this reaction:[142]

$$\qquad (11.30)$$

1 h	64%	23%
1.5 h		92%

Of interest, 1,5,9-cyclododecatriene prepared by trimerization of 1,3-butadiene undergoes only hydrogenation to cyclododecene. This reaction has potential since cyclododecene is a precursor to 1,12-dodecanedioic acid, a commercial polyamide intermediate. $[\text{RuCl}_2(\text{CO})_2(\text{PPh}_3)_2]$ gives 98–99% yield of cyclododecene under mild conditions (125–160°C, 6–12 atm, in benzene with added PPh_3 or in N,N-dimethylformamide).[143]

$[\text{PtH}(\text{SnCl}_3)(\text{PPh}_3)_2]$ is one of the most selective catalysts to reduce conjugated and nonconjugated dienes and trienes to monoenes at elevated pressure. This behavior is best explained by supposing that the isomerization of nonconjugated bonds into conjugated position. The high selectivity for hydrogenation of dienes in the presence of monoenes arises from the exceptional stability of the π-allyl complex of this catalyst.

Asymmetric Hydrogenation. Asymmetric hydrogenation with good enantioselectivity of unfunctionalized prochiral alkenes is difficult to achieve.[144,145] Chiral rhodium complexes, which are excellent catalysts in the hydrogenation of activated multiple bonds (first, in the synthesis of α-amino acids by the reduction of α-*N*-acylamino-α-acrylic acids), give products only with low optical yields.[144,146–149] The best results (∼60% ee) were achieved in the reduction of α-ethylstyrene by a rhodium catalyst with a diphosphinite ligand.[150] Metallocene complexes of titanium,[151–155] zirconium,[155–157] and lanthanides[158] were used in recent studies to reduce the disubstituted C—C double bond with medium enantioselectivity.

Trisubstituted alkenes (1,2-diphenylpropene, 1-methyl-3,4-dihydronaphthalene) have been hydrogenated with excellent optical yields (83–99% ee) in the presence of a chiral bis(tetrahydroindenyl)titanocene catalyst.[159]

11.2.3. Hydrogenation of Alkynes

Mechanistic aspects of alkene hydrogenation are by now known in considerable detail, but far less information is available on the mechanistic course of the alkyne hydrogenation reaction.[160] In general, data on the hydrogenation of alkynes are rather limited.[24]

Most homogeneous catalysts that are active in the hydrogenation of alkenes can be used for the reduction of alkynes. In many cases alkynes are hydrogenated less rapidly than alkenes. In binary mixtures, however, alkynes may be reduced preferentially since the binding of alkynes, in general, is favored over alkenes. Strong complexation may prevent hydrogenation, as observed for 1-hexyne in the presence of $[RhH(CO)(PPh_3)_3]$.[111]

In the presence of the Wilkinson catalyst, 1-alkynes are reduced slowly and non-selectively to give saturated hydrocarbons.[30] In contrast, internal alkynes give *cis* alkenes that are resistant to further reduction. Other catalysts can reduce 1-alkynes selectively[161,162] [Eqs (11.31) and (11.32)] and be more active in the hydrogenation of internal alkynes:

$$Ph-\!\!\!\equiv\!\!\!-H \quad \xrightarrow[\text{toluene, 1 atm } H_2\text{, RT}]{[OsHBr(CO)(PPh_3)_3]} \quad PhCH\!=\!CH_2 \qquad (11.31)$$
$$100\%$$

$$\xrightarrow[\substack{\text{benzene, < 1 atm } D_2\text{, 25°C,}\\ \text{PhCOOH, Et}_3\text{N}}]{[RhOOCPh(COD)PPh_3]} \quad \underset{Ph \quad\quad H}{\overset{D \quad\quad D}{>\!\!=\!\!<}} \qquad (11.32)$$
$$>95\%$$

Chromium complexes,[163] cationic homogeneous complexes of rhodium[164] and ruthenium,[165,166] and supported palladium complexes[141,167] have been found to be extremely selective in reduction of alkynes to cis alkenes:

$$Ph-\!\!\!\equiv\!\!\!-R \quad \xrightarrow[\text{THF, 1 atm } H_2\text{, RT}]{\text{heterogenized Pd(II)}} \quad \underset{Ph \quad\quad R}{\overset{H \quad\quad H}{>\!\!=\!\!<}} \qquad (11.33)$$
$$R = Me, Pr \qquad\qquad\qquad 97\text{–}98\%$$

Quite remarkably partial hydrogenation of alkynes to *trans* alkenes is also possible with homogeneous rhodium complexes:[168,169]

$$Ph-\!\!\!\equiv\!\!\!-Ph \quad \xrightarrow[\text{toluene, 1 atm } H_2\text{, 20°C}]{[RhH_2(OOCOH(PisoPr_3)_2]} \quad Ph\diagup\!\!\diagdown\!\!\diagup Ph \qquad (11.34)$$
$$91\% \text{ selectivity}$$

When palladium ions are bound to poly(ethyleneimine), a soluble polychelatogen, selective conversion of 2-pentyne to *cis*-2-pentene is achieved under mild conditions (20°C, 1 atm H_2, 98% stereoselectivity).[170] A dramatic reversal of stereoselectivity (selective formation of *trans*-2-pentene) is observed, however, if a spacious ligand such as benzonitrile is bound to palladium.

11.2.4. Hydrogenation of Aromatics

Hydrogenation of aromatics with homogeneous catalysts is not easy to achieve, and only relatively few complexes were found to be effective.

The partial hydrogenation of benzene to cyclohexene is difficult to carry out with heterogeneous metal catalysts, but it has been achieved by using pentammineosmium(II).[171] The reaction requires the presence of Pd(0), which itself is ineffective as a catalyst for the hydrogenation of benzene:

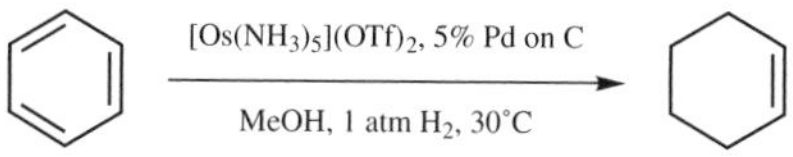

$$\text{[Os(NH}_3)_5\text{](OTf)}_2, \text{5\% Pd on C} \quad \text{MeOH, 1 atm H}_2, \text{30°C} \tag{11.35}$$

The process, however, is not catalytic. Cyclohexene forms the stable $[Os(NH_3)_5(\eta^2\text{-}C_6H_{10})](OTf)_2$ complex from which it is liberated through oxidation.

A number of transition-metal compounds, preferably nickel and cobalt derivatives combined with triethylaluminum give active Ziegler-type catalysts.[172,173] They are efficient in the hydrogenation of different aromatics, but they are not stereoselective [Eq. (11.36)]) and catalyze only partial hydrogenation [Eq. (11.37)]:

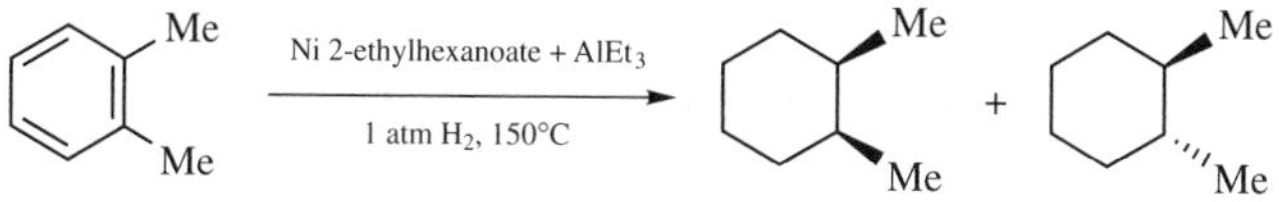

$$\text{Ni 2-ethylhexanoate + AlEt}_3 \quad \text{1 atm H}_2, \text{150°C} \tag{11.36}$$

64.5   35.5

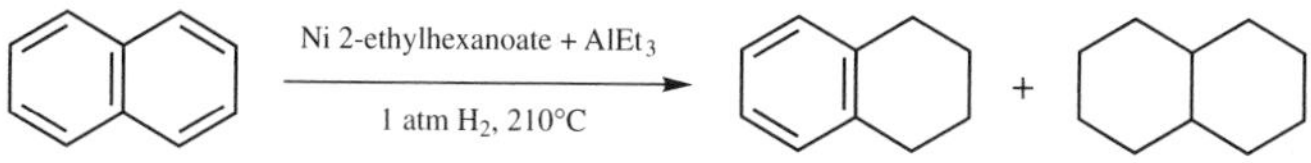

$$\text{Ni 2-ethylhexanoate + AlEt}_3 \quad \text{1 atm H}_2, \text{210°C} \tag{11.37}$$

84   13

Partial hydrogenation of polycyclic aromatics can also be accomplished with $[Co(CO)_4]_2$ under hydroformylation conditions.[24,26] Isolated benzene rings are generally resistant under these conditions. Polycyclic aromatics, consequently, are only partially hydrogenated[174] [Eq. (11.38)], and highly condensed systems usually form phenanthrene derivatives:

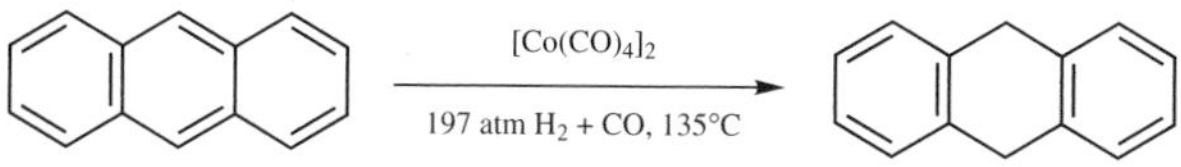

$$\text{[Co(CO)}_4]_2 \quad \text{197 atm H}_2 + \text{CO, 135°C} \tag{11.38}$$

99% selectivity
at 100% conversion

The reaction is thought to involve free-radical intermediates rather than organo-cobalt complexes.[175]

An allylcobalt(I) complex $[Co(\eta^3\text{-}C_3H_5)\{P(OMe)_3\}_3]$ catalyzes the hydrogenation of aromatics at ambient temperature.[160,176] Distinctive features of this catalyst are the high stereoselectivity [only *cis* products are formed from xylenes and naphthalene; see Eq. (11.39)] and complete hydrogenation of aromatic systems[177] [Eqs (11.39) and (11.40)]:

$$\tag{11.39}$$

$$\tag{11.40}$$

76%

Surprisingly, arenes and alkenes are reduced at very similar rates, whereas most hydrogenation catalysts are more active in the hydrogenation of alkenes. No unsaturated compounds were detected in the hydrogenation of aromatics, and no competing isotope exchange was observed. Additionally, the all-*cis* $C_6H_6D_6$ isomer was formed in deuteration of benzene with high selectivity (>95%); thus, *syn* addition of all six deuterium atoms occurs.[176] All these observations indicate that intermediates are not displaced from the metal center until complete hydrogenation takes place. The disadvantages of such catalysts are the limited lifetime and low activity. A mechanism in which the C_6 moiety remains attached at the metal ion until the formation of a cyclohexyl ion accounts for the results. Species such as $[CoH_2(OMe)(\eta^1\text{-}C_3H_5)(\eta^4\text{-}C_6H_6)]$ (**17**) and the π-allyl complex **18** are suggested as intermediates.[177]

The $(\eta^6\text{-}C_6H_6)$ and $(\eta^2\text{-}C_6H_6)$ arene ligands in some metal–arene and allylmetal–arene complexes are very difficult to hydrogenate. This low reactivity suggests that the formation of intermediates with the noncoplanar, highly perturbed aromatic η^4-arene binding as in **17** is necessary for catalytic hydrogenation of aromatic substrates. Stable arene–metal complexes such as $[Ru(\eta^6\text{-}C_6Me_6)(\eta^4\text{-}C_6Me_6)]$ can be used to hydrogenate other aromatic compounds. In contrast with the cobalt system, however, this complex produces significant amounts of cyclohexenes from methyl-substituted benzenes, catalyzes extensive isotope exchange, and exhibits only limited stereoselecitvity.[178]

Some rhodium complexes are more promising in the hydrogenation of arenes. The pentamethylcyclopentadienyl rhodium complex [{RhCl$_2$(η^5-C$_5$Me$_5$)}$_2$], for instance, is stereoselective in the reduction of C$_6$D$_6$, yielding the all-*cis* product almost exclusively, without the formation of cyclohexene.[179]

The binuclear chloro(1,5-hexadiene)rhodium is an extremely mild and selective catalyst in the presence of phase transfer catalysts:[180]

$$
\begin{array}{c}
\xrightarrow[\substack{\text{cetyltrimetylammonium bromide} \\ \text{1 atm H}_2,\ \text{RT, pH} = 7.6}]{[1,5\text{-HDRhCl}]_2}
\end{array}
\tag{11.41}
$$

in benzene	73%	trace
in hexane	20	80

Unique activities and selectivities are exhibited by a simple catalytic system prepared from RuCl$_3$ and Aliquat 336. The [(C$_8$H$_{17}$)$_3$NMe]$^+$[RhCl$_4$]$^-$ solvated ion pair formed is a perfectly stable arene hydrogenation catalyst. It gives decalin-free tetralin under very mild conditions[181] [Eq. (11.42)] and catalyzes the highly selective partial hydrogenation of polycyclic aromatic compounds. In contrast with other heterogeneous and homogeneous catalysts, the terminal ring is hydrogenated in linear polynuclear aromatics [Eq. (11.43)], while phenanthrene is converted to the expected 9,10-dihydro derivative:[182]

$$
\xrightarrow[\substack{\text{Aliquat 336, H}_2\text{O, CH}_2\text{Cl}_2, \\ \text{1 atm H}_2,\ 30°\text{C}}]{\text{RhCl}_3 \cdot 3\text{H}_2\text{O}}
\qquad
\begin{array}{c}
\text{100\% selectivity} \\
\text{at 81\% conversion}
\end{array}
\tag{11.42}
$$

$$
\xrightarrow[\substack{\text{Aliquat 336, H}_2\text{O, CH}_2\text{Cl}_2, \\ \text{1 atm H}_2,\ 30°\text{C}}]{\text{RhCl}_3 \cdot 3\text{H}_2\text{O}}
\qquad
\begin{array}{c}
\text{100\% selectivity} \\
\text{at 20\% conversion}
\end{array}
\tag{11.43}
$$

Organothorium complexes such as [Th(η^3-allyl)$_4$] supported on dehydroxylated γ-alumina have been shown to exhibit activities rivaling those of the most active platinum metal catalysts.[123] Thorium maintains its original +4 oxidation states at all times; that is, the mechanism does not follow the usual oxidative addition–reductive elimination pathway. Partially hydrogenated products cannot be detected at partial conversions. Deuterium is not delivered to a single benzene face, but a $1:3$ mixture of isotopomers is formed:

$$(11.44)$$

$$1:3$$

Accordingly, hydrogenation of *p*-xylene yields a $1:3$ mixture of *cis-* and *trans-*1,4-dimethylcyclohexane. Both observations suggest that dissociation and readsorption of the partially reduced arene take place.

11.3. CHEMICAL AND ELECTROCHEMICAL REDUCTION

11.3.1. Reduction of Alkenes

The carbon–carbon double bond can be reduced by diimide prepared in solution in a number of ways.[34,183,184] Oxidation of hydrazine with oxygen (air) or H_2O_2 in the presence of a catalytic amount of Cu(II) ion was the first method to generate and use diimide in hydrogenation.[183–185] Acid-catalyzed decomposition of alkali azido-dicarboxylates,[185,186] as well as thermal or base-catalyzed decomposition of aromatic sulfonyl hydrazides,[183,184] are also useful methods for preparing the diimide reducing agent.

The relative reactivity of alkenes toward reduction by diimide depends on the degree of substitution. Increasing alkyl substitution results in decreasing reactivity, and strained alkenes exhibit higher reactivity than nonstrained compounds:[187]

| relative rates | 20.2 | 2.65 | 0.50 | 15.5 | 1 | 12.1 | 450 |

The steric course of the addition of the two hydrogen atoms is generally *syn.* Accordingly, the reduction of *cis-* and *trans-*stilbenes with dideuterodiimide affords *meso-* and *racem.*-1,2-dideutero-1,2-diphenylethanes, respectively, with 97% selectivity:[188]

$$(11.45)$$

$$(11.46)$$

It was found that most synthetic processes that are employed to prepare the diimide reagent generate *trans*-diimide, but *cis*-diimide undergoes faster hydrogen atom transfer to a double bond than does the *trans* isomer. It follows that a fast *trans–cis* isomerization precedes reduction. The transfer of hydrogen atoms takes place in a synchronous process[188] via the transition state **19**:

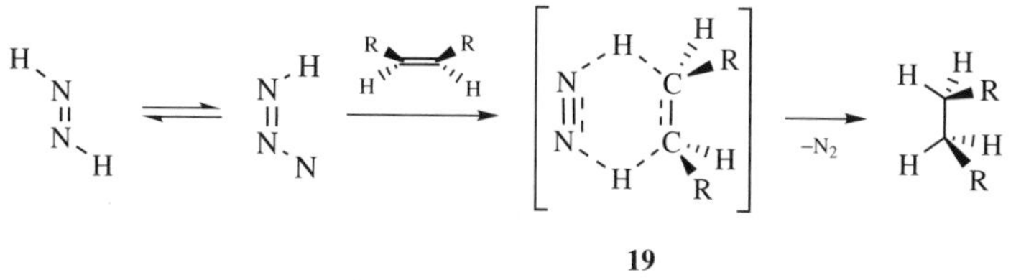

$$(11.47)$$

19

Many studies demonstrated that in the approach of diimide to the double bond, steric hindrance is the decisive factor in determining stereochemistry. Highly selective additions of hydrogen (deuterium) from the sterically less hindered side[189,190] are illustrated in Eqs. (11.48) and (11.49):

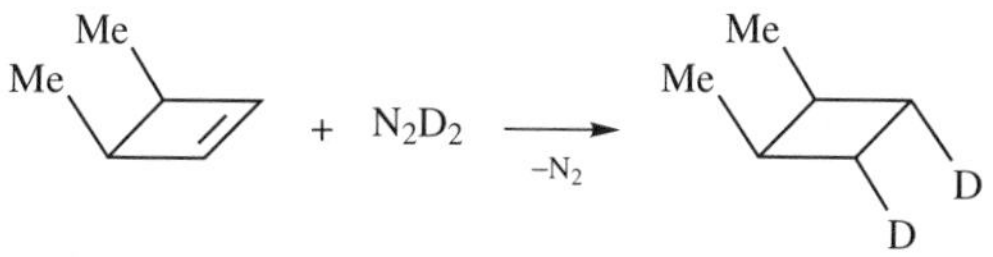

$$(11.48)$$

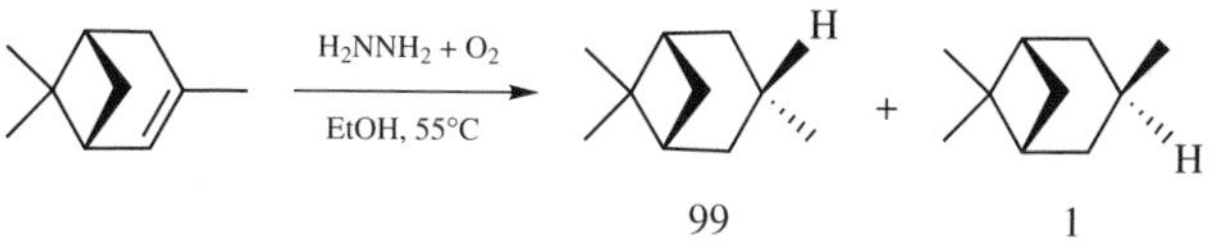

$$(11.49)$$

Less substituted alkenes undergo facile reduction by the reagent LiAlH$_4$ in admixture with transition-metal salts[191] or by NaBH$_4$ with CoCl$_2$.[192] The yields are very high, with CoCl$_2$, NiCl$_2$, and TiCl$_3$ also effective when used in catalytic amounts. 1-Methylcyclohexene, although a trisubstituted alkene, is reduced in quantitative yield[191] by LiAlH$_4$–CoCl$_2$ and LiAlH$_4$–NiCl$_2$.

Fairly selective reduction of certain alkenes can be achieved using the sodium–hexamethylphosphoramide (HMPA)–*tert*-butyl alcohol system,[193] despite the general trend, that nonconjugated alkenes are usually quite resistant to the dissolving metal reduction method (see Section 11.3.2). In the case of 9(10)-octalin, the transformation leads to a nearly equilibrium product distribution:

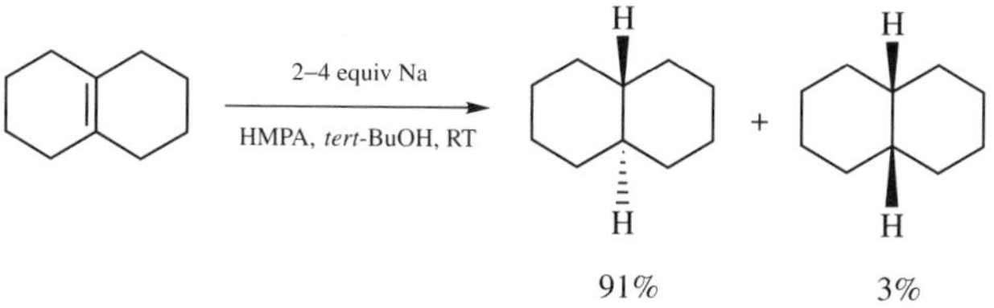

$$(11.50)$$

91%   3%

Magnesium in methanol, another dissolving-metal system, is known to be only active in the reduction of activated multiple bonds. The reactivity, however, can be dramatically enhanced by the addition of catalytic amount of palladium, thus allowing rapid and facile reduction of nonactivated acyclic and cyclic alkenes.[194]

11.3.2. Reduction of Alkynes

As was discussed in Sections 11.1.5 and 11.2.3, the stereoselective partial hydrogenation of alkynes to either *cis* or *trans* alkenes is of key importance. Chemical reductions can also be applied to achieve both selective transformations.

It was observed in 1941 that with sodium in liquid ammonia, called *dissolving-metal reduction*, different dialkylacetylenes were converted to the corresponding *trans* alkenes in good yields and with high selectivity:[195]

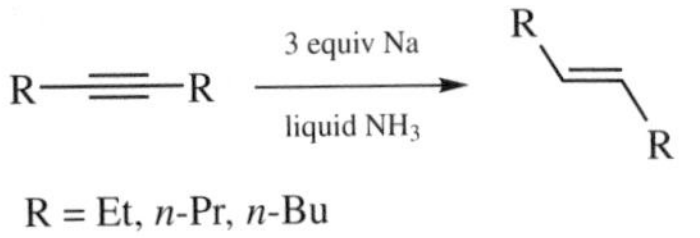

$$R\!\!-\!\!\!\equiv\!\!\!-\!\!R \xrightarrow[\text{liquid NH}_3]{\text{3 equiv Na}} \text{structure} \tag{11.51}$$

R = Et, *n*-Pr, *n*-Bu

The observation was a significant finding since at the time, when the only synthetic method to reduce alkynes selectively was their conversion by heterogeneous catalytic hydrogenation (Raney nickel) to *cis* alkenes. The dissolving-metal reduction provided easy access to high-purity *trans* alkenes since the latter do not readily react further under the conditions used. The efficient reduction of 1-alkynes in this system requires the presence of ammonium ion.[196]

Since that time several other metal–solvent systems have been found useful, effective, and selective in the partial hydrogenation of alkynes. The mechanism of the transformation depends on the nature of the solvent and the substrate. Detailed studies were carried out with the Na–HMPA–*tert*-BuOH system,[197] resulting in the suggestion of the mechanism depicted in Scheme 11.10.

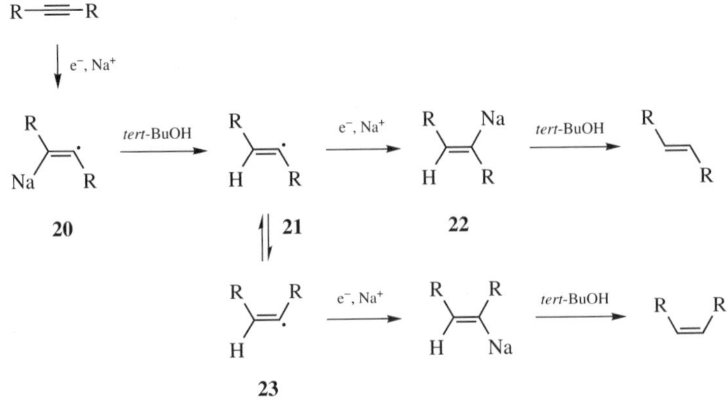

Scheme 11.10

The electron transfer to the acetylenic bond forms the *trans*-sodiovinyl radical **20** that, after protonation, produces *trans* radical **21**. At low temperature ($-33°C$) in the presence of excess sodium, the conversion of the *trans* radical to sodiovinyl intermediate **22** is slightly more rapid than the conversion of the *trans* radical to the cis radical **23** ($21 \rightarrow 22 > 22 \rightarrow 23$). As a result, protonation yields predominantly the *trans* alkene. However, low sodium concentration and increased temperature lead to increasing proportion of the *cis* alkene. Although other dissolving-metal reductions are less thoroughly studied, a similar mechanism is believed to be operative.[34] Another synthetically useful method for conversion of alkynes to *trans* alkenes in excellent yields is the reduction with $CrSO_4$ in aqueous dimethylformamide.[198]

Besides heterogeneous and homogeneous catalytic hydrogenations, chemical reductions can also transform alkynes to *cis* alkenes. Interestingly, activated zinc in the presence of a proton donor (alcohol), although a dissolving-metal reagent, reduces disubstituted alkynes to *cis* alkenes:[199]

$$Ph\text{---}\!\equiv\!\text{---}Ph \xrightarrow[\text{MeOH, reflux, 5-6 h}]{\text{Zn-Cu}} \underset{\underset{\text{100\% selectivity}}{>95\% \text{ conversion}}}{Ph\diagdown\diagup Ph} \qquad (11.52)$$

The metal hydride–transition-metal halide combinations predominantly produce *cis* alkenes; however, significant amounts of *trans* alkenes are also formed.[34] As a notable exception, *cis* alkenes are the sole products in the reduction with MgH_2 in the presence of Cu(I) ions:[200]

$$R^1\text{---}\!\equiv\!\text{---}R^2 \xrightarrow[\text{THF, }-78°C\text{ to RT}]{MgH_2,\ CuI} \underset{80\text{--}90\% \text{ yield}}{R^1\diagdown\diagup R^2} \qquad (11.53)$$

$$R^1, R^2 = H, Me, \textit{n}\text{-Pr}, \textit{n}\text{-Bu}, Ph$$

11.3.3. Reduction of Aromatics

Alkali metals in liquid ammonia represent the most important class of the so-called dissolving-metal reductions of aromatics. First described in 1937, it is a highly efficient and convenient process to convert aromatic hydrocarbons to partially reduced derivatives.[201] The recognition and extensive development of this electron-transfer reduction came from A. J. Birch,[202,203] and the reaction bears his name.

Sodium and lithium are the most common metals, and diethyl ether, tetrahydrofuran, and HMPA are often used as cosolvents to overcome solubility problems. The Benkeser method,[37,204] employing lithium in low-molecular-weight amines (methylamine, ethylamine, ethylenediamine), is a more powerful reducing agent

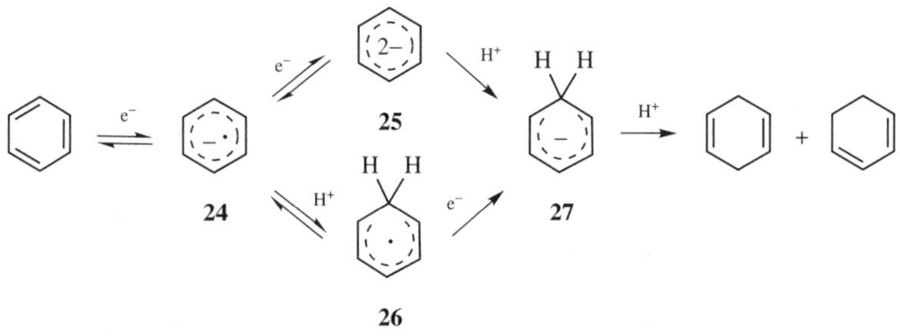

Scheme 11.11

capable of achieving more extensive reductions. Alcohols, usually ethanol or *tert*-butyl alcohol, can serve as proton sources. This topic has been well covered in books and reviews.[35–37,202–207] Its main application in the field of hydrocarbons is the partial reduction of bi- and polycyclic aromatics.

Mechanism. The deep blue solution of alkali metals in liquid ammonia containing solvated electrons is a powerful reducing agent. Electron transfer to an aromatic molecule first forms radical anion **24**. If the substrate has high electron affinity (biphenyl, polycyclic aromatics), it reacts with a second electron to form dianion **25**[35–37,207] (Scheme 11.11). Benzene and its nonactivated derivatives (alkylsubstituted benzenes) require a proton source (alcohol) to shift the equilibrium to the right by forming radical **26**. Both **25** and **26** are transformed to final monoanion **27** by protonation or additional electron transfer, respectively. Ammonia itself may protonate highly reactive substrates in a slow process. Product distribution is kinetically controlled, with the nonconjugated 1,4-diene always favored.[35]

Selectivity. Benzene itself is transformed into 1,4-cyclohexadiene by employing sodium and ethanol in liquid ammonia[207] [Eq. (11.54)]; the Benkeser method affords further reduction to cyclohexene and cyclohexane[208] [Eq. (11.55)]:

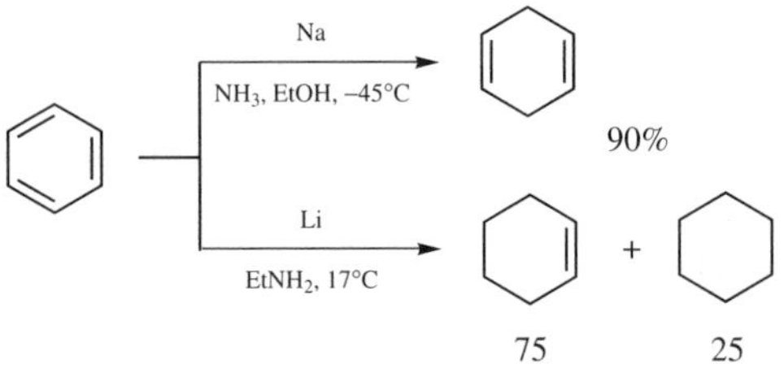

$$(11.54)$$

$$(11.55)$$

Substituent effects are in accord with the proposed mechanism. The Birch rule[202] states that electron-releasing substituents (alkyl groups) deactivate the ring and direct reduction so that the major product has the maximum number of groups attached to the residual double bonds, that is, to vinyl positions. In other words, monoalkylbenzenes are protonated at the 2,5 positions [Eq. (11.56)]. Electron-withdrawing groups enhance rate and occupy the saturated allylic positions in the reduced product [Eq. (11.57)]:

$$(11.56)$$

$$(11.57)$$

By proper choice of reaction conditions (metal, solvent, the order of addition of reagent and reactant), fused polycyclic aromatics can be converted to different partially reduced derivatives with high selectivity. When the red complex of naphthalene, for example, formed in the Na–NH$_3$ solution, is quenched with aqueous ammonium chloride solution, 1,4-dihydronaphthalene is formed[209] [Eq. (11.58)]. Reaction in the presence of an alcohol yields 1,4,5,8-tetrahydronaphthalene[210] [Eq. (11.59)]. The Benkeser reduction affords further saturation to the isomeric octalins[211] [Eq. (11.60)] or even to decalin[212] [Eq. (11.61)]:

$$(11.58)$$

$$(11.59)$$

$$(11.60)$$

$$(11.61)$$

Similar partial hydrogenations were described to take place by electrochemical reductions.[213,214]

Reduction of other polynuclear aromatics can similarly be controlled to yield different products. Although the site of reduction is difficult to predict, the first step in reduction of such simple compounds as phenanthrene and anthracene occurs at the 9 and 10 positions. Formation of product mixtures of different degrees of saturation, however, is rather common.

Cathodic reduction of anthracene using different electrodes and varying experimental conditions (electrode potential, current density) permits the synthesis of the partially saturated di-, tetra- and hexahydroderivatives from good to excellent selectivities:[215]

	vitreous carbon	Al	Pb	
	8	45	100	(11.62)
	17	24		(11.63)
	75	31		(11.64)

11.4. IONIC HYDROGENATION

In contrast to catalytic hydrogenations where dihydrogen is activated by heterogeneous or homogeneous catalysts, mostly by transition metals, hydrogenation through activation of the unsaturated hydrocarbon substrate is also possible. These reactions, called *ionic hydrogenations*, require a hydrogenating system consisting of a protic acid and a hydride ion donor.[38–40] In ionic hydrogenations a sufficiently acidic proton source initially generates a carbocation by protonation of the unsaturated hydrocarbon:

$$(11.65)$$

The carbocation subsequently is quenched by a hydride donor or hydrogen to yield the saturated product. A sufficient stability of the intermediate carbocation and proper choice of the hydrogenating reagent pair are prerequisites to a successful ionic hydrogenation.

Catalytic and ionic hydrogenations may be combined to carry out catalytic ionic hydrogenation.[216] In these cases the catalyst, either a heterogeneous or a homogeneous transition metal, serves as a hydride ion transfer agent.

Ionic hydrogenations have substantial interest as, in contrast to metal-catalyzed hydrogenation, they are not affected by many impurities poisoning catalysts, which is of significance in fuels processing, particularly in coal liquefaction.[40]

11.4.1. Hydrogenation of Alkenes and Dienes

CF_3COOH in combination with silanes (mainly with Et_3SiH) is most frequently used in ionic hydrogenation of alkenes.[39,40] This system offers high selectivity at moderate temperature and ensures high yields:[217]

$$
\begin{array}{ccc}
\text{(Me-cyclohexene)} & \xrightarrow[20°C, 120\ h]{2\ equiv\ CF_3COOH,\ Et_3SiH} & \text{(Me-cyclohexane)} \\
& & 73\%
\end{array}
\tag{11.66}
$$

The reagent pair itself, containing the relatively weak trifluoroacetic acid, does not react significantly at the used temperatures. Subsequently, however, much stronger acids, including CF_3SO_3H and other superacids, were introduced to activate more-difficult-to-reduce substrates. In these cases hydrogen is used, or the hydrogen donor is continuously added.

An important unique feature of ionic hydrogenation is that the reactivity of alkenes is opposite that observed in other hydrogenation methods. Tetrasubstituted alkenes that are unreactive or require forcing conditions in catalytic hydrogenations readily undergo saturation in ionic hydrogenation because of the facile formation of the tertiary carbocation.[39,40] Compounds forming a benzylic cation such as 1,2-dihydronaphthalene (**28**) may be saturated with similar ease[218] [Eq. (11.67)]. In contrast, the isomeric 1,4-dihydronaphthalene (**29**) is not hydrogenated at all under these conditions, but it may serve as hydrogen donor[218] [Eq. (11.68)]:

$$
\begin{array}{ccc}
\textbf{28} & \xrightarrow[50–60°C,\ 1\ h]{CF_3COOH,\ Et_3SiH} & 90\%
\end{array}
\tag{11.67}
$$

$$
\begin{array}{ccc}
\text{(Me-cyclohexene)} + \textbf{29} & \xrightarrow[50–60°C,\ 60–70\ h]{CF_3COOH} & 80\% + 90\%
\end{array}
\tag{11.68}
$$

The selectivity of ionic hydrogenation strongly depends on the stability of the carbocation intermediate formed. It is possible to carry out selective monodeuteration by deuterated silanes of trialkylsusbtituted alkenes at the tertiary carbon atom.[219] The transformation of the optically active $(-)$-**30** (63% optical purity) yielded the cyclized saturated product with the same optical purity as the starting diene:[220]

$$(11.69)$$

(−)-**30**   **31**   73% yield

This indicates that only carbocation **31** of high configurational stability is present in the reaction mixture.

Other alkenes, however, may not react selectively. Product distribution in the transformation of **32** [Eq. (11.70)] is consistent with the equilibrium of the tertiary and secondary cations **33** and **34**, or the presence of hydrogen-bridged ion **35**:[221]

$$(11.70)$$

32   25   75

33   **34**   **35**

The difference in reactivity of different double bonds allows selective hydrogenation of dienes. Conjugated dienes with at least one trisubstituted double bond may be completely reduced by using excess reagents[222] [Eq. (11.71)]. Regioselective hydrogenation of the more substituted double bond occurs in 1,5-dienes. The hydrogenation of 1,4-dienes, however, is not selective [Eq. (11.72)]:

$$(11.71)$$

diene : Et_3SiH : CF_3COOH = 1 : 2 : 3   70%

$$(11.72)$$

diene : Et_3SiH : CF_3COOH = 1 : 2 : 3

The formation of the fully saturated byproduct was considered to take place through hydrogenation of the corresponding 1,3-diene formed by isomerization.[222]

Besides the widely used trifluoroacetic acid, polyphosphoric acid,[223] H_2SO_4[220,224], $HClO_4$,[225] cation exchange resin in formic acid,[226] and Et_2O–

HBF_4[227] were also used in ionic hydrogenation. Et_2O-BF_3, when added to the $CF_3COOH + Et_3SiH$ system, was found to greatly enhance the reaction rate without affecting selectivity.[228] An improved and modified procedure for ionic hydrogenation by adding NH_4F to the $CF_3COOH + Et_3SiH$ reagent has been described.[229] The formed volatile Et_3SiF makes product separation and workup more convenient.

Hydrocarbons with easily abstractable hydride ion such as 1,4-cyclohexadiene,[227] cycloheptatriene,[223] or compounds possessing tertiary or benzylic hydrogen[217,218,223,230,231] are suitable substitutes for silanes. In fact, this was the case in the first ionic hydrogenation[232,233] observed by Ipatieff and coworkers. In an attempt to carry out alkylation of p-cymene with 3-methylcyclohexene, hydrogenation of the latter occurred, where p-cymene was the hydride ion source:

$$(11.73)$$

As mentioned earlier (see Section 11.1.4), cycloalkenes (primarily cyclohexene) can be used to carry out transfer hydrogenations. Similar processes during which the alkene itself serves as hydride ion donor may be observed under ionic hydrogenation conditions:[223]

$$(11.74)$$

The transformation can also be considered as disproportionation.

Catalytic ionic hydrogenations[216] were carried out by means of homogeneous iridium,[234,235] platinum,[235,236] and ruthenium,[235,236] as well as heterogeneous platinum and palladium catalysts.[237] A clear indication of the involvement of ionic hydrogenation is the large rate enhancement of the reduction of tri- and tetrasubstituted alkenes in the presence of the stronger acid CF_3COOH as compared to acetic acid:[234]

$$(11.75)$$

Stereochemical differences were observed in the hydrogenation of 9(10)-octalin when different ionic hydrogenation methods were applied:[223,235,238]

$$(11.76)$$

CF₃COOH + *n*-BuSiH₃	22	78
CF₃COOH + *tert*-Bu₃SiH	93	7
[IrH₃(PPh₃)₃] + CF₃COOH + H₂	83	17
polyphosphoric acid, disproportionation	86	14
CF₃COOH, disproportionation	95	5

The isomer ratio can be controlled by the steric size of the hydride donor silanes, and the highest selectivity is achieved when 9(10)-octalin itself is the hydride donor. None of the catalytic hydrogenations can produce such a high amount of the less stable *cis* isomer.

Selective reduction of alkynes to *cis* alkenes using hydrosilane functions immobilized on silica gel in acetic acid in the presence of a Pd(0) catalyst system has been reported.[239]

11.4.2. Hydrogenation of Aromatics

It was found that the Et_3SiH and boron trifluoride monohydrate (a strong acid compatible with Et_3SiH) ionic hydrogenation system produces partially hydrogenated anthracene [Eq. (11.77)] and naphthacene in excellent yield under mild conditions:[240]

$$(11.77)$$

89%

Since phenanthrene is not hydrogenated, a carbocation stabilized by two phenyl rings is required for reduction. Benzene, naphthalene, and their alkylsubstituted derivatives are not reduced by this reagent, and the $CF_3COOH + Et_3SiH$ system is not active, either.[230] These compounds, however, readily exchange hydrogen with D_2O–BF_3,[241] demonstrating that the hydride abstraction step must be rate-determining.

In contrast, a reaction system consisting of H_2O–BF_3, the soluble transition-metal reagent $[PtCl_2(MeCN)_2]$, and hydrogen gas is active even in the hydrogenation of benzene and naphthalene:[242]

$$ (11.78) $$

It converts anthracene, however, into a mixture of partially and completely reduced products.

It is possible to hydrogenate aromatics in the superacids HF–TaF$_5$, HF–SbF$_5$, or HBr–AlBr$_3$ in the presence of hydrogen. The reduction of benzene was shown to give an equilibrium mixture of cyclohexane and methylcyclopentane.[243,244] Reduction was postulated to proceed via initial protonation of benzene followed by hydride transfer:

$$ (11.79) $$

It was found[244] that a tertiary hydride ion source (isoalkanes formed under reaction conditions through isomerization) is necessary to yield the products. However, when isopentane and hydrogen pressures greater than 14.6 atm are used, the reduction becomes catalytic. The function of hydrogen under such conditions is to reduce the formed isopentyl cation, thereby regenerating isopentane and yielding a proton.

An extreme case of the effectiveness of ionic hydrogenation is the HF–BF$_3$ superacid-promoted hydrogenation of coals at 170°C with 36-atm hydrogen giving efficiently liquid products under these relatively mild conditions.[245,246]

11.5. HYDROGENOLYSIS OF SATURATED HYDROCARBONS

11.5.1. Metal-Catalyzed Hydrogenolysis

When saturated hydrocarbons are treated with metal catalysts in the presence of hydrogen, carbon–carbon σ bonds are cleaved, with the concomitant formation of new carbon–hydrogen bonds called *hydrogenolysis*. Hydrogenolysis is not identical with hydrocracking. The latter is the very complex refinery operation that is used to convert high-molecular-weight hydrocarbons to lower-boiling products (see Section 2.1). In the hydrocracking process bifunctional catalysts consisting of a hydrogenation-dehydrogenation component (a metal) and an acidic component are employed and most cracking takes place on the acidic sites. In the present discussion only the main features of hydrogenolytic transformations (i.e., C—C bond cleavage over monofunctional metal catalysts) are summarized briefly.

Most transition metals are capable of cleaving C—C bonds. The most important transformations are the cracking of alkanes, the ring opening of cycloalkanes, and the hydrogenative dealkylation of aromatics.[16,247–254] In contrast to hydrocracking, little or no isomerization occurs under the usual hydrogenolysis conditions. Different metals have different activity in hydrogenolysis with activities differing in several (6–8) orders of magnitude. The activity also strongly depends on metal dispersion, the support used and reaction conditions (temperature, hydrogen pressure). In most early studies unsupported metal powders (metal blacks) and metal films and foils were used to avoid unwanted support effects (i.e., bifunctional catalysis). The use of proper supports and metal precursors, and careful catalyst pretreatment, however, can provide reliable data on supported catalysts as well. Useful information were also obtained by studying hydrogenolysis over alloy catalysts and recently on single crystals.

Metals can give strikingly different product distributions called the *cracking pattern*, since different carbon–carbon bonds in hydrocarbons (primary–primary, primary–secondary, primary–tertiary, etc.) tend to undergo cleavage at different rates on different metals. The cracking pattern can be defined and quantitatively expressed, for example, as the reactivity factor[255] (the actual rate of cleavage divided by expected rate based on random fission). A further characteristic is the depth of hydrogenolysis[256] (the number of fragment molecules per reacting hydrocarbon molecule).

In the simplest possible reaction, the hydrogenolysis of ethane to methane, Sinfelt found a correlation between the specific activity of second- and third-row Group VIII metals and their percent d character (where Ru and Os are the most active).[249] He also pointed out, however, that percent d character is not the only factor determining activity in hydrogenolysis since a similar correlation does not exist for first-row elements. Later, hydrogenolysis of ethane was well correlated with its adsorption properties. It was shown that metals that catalyze multiple exchange best (Ir, Rh, Ru) are the most active in hydrogenolysis.[257] At the same time, these are the metals that exhibit medium to strong interaction during adsorption with ethane. Adsorption on Pt and Pd is almost reversible, whereas Ni and Co promote excessively dissociative adsorption leading to highly hydrogen-deficient surface adsorbed species.[258,259] The mechanism suggested for ethane hydrogenolysis (Scheme 11.12) involves multiple adsorbed surface species.

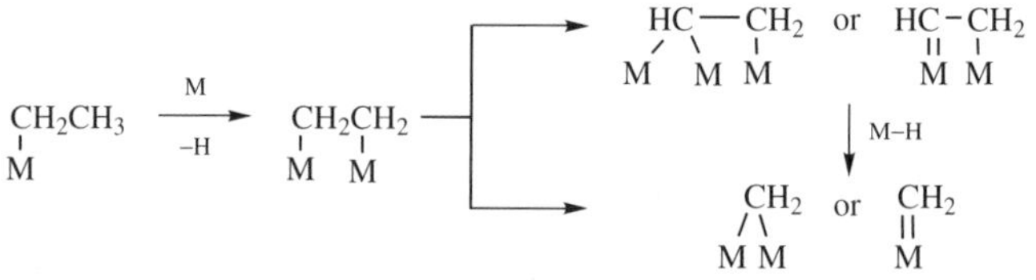

Scheme 11.12

The approach of finding correlation between adsorption properties and hydrogenolysis led to the interpretation of cracking patterns. Later, the realization of relationships between hydrogenolysis and other metal-catalyzed reactions (isomerization) resulted in a much better understanding of the characteristics of hydrogenolysis reactions.

Higher hydrocarbon molecules allow study of the unique cracking pattern of metals. These studies are usually carried out at low conversion to observe only primary hydrogenolysis. Nickel exhibits high selectivity to cleave terminal C—C bonds leading to demethylation; that is, it cleaves only bonds that involve at least one primary carbon atom. For example, in the transformation of *n*-hexane, only methane and *n*-pentane are formed (180°C, Ni-on-silica catalyst, 0.3% conversion), whereas 2-methylpentane and 3-methylpentane yield methane, *n*-pentane, and isopentane.[260] In the transformation of 2-methylpentane, the *n*-pentane : isopentane ratio is close to 2, which corresponds to the statistical value. Under more forcing conditions, successive demethylations lead eventually to methane as the only product.

Similar selectivities were observed in the hydrogenolysis of *n*-heptane over Pd and Rh[261] (Table 11.1). In contrast, on Ru, and particularly on Pt and Ir, hydrogenolysis is not selective, leading to products in roughly equal amounts, indicating that both types of bonds (primary–secondary and secondary–secondary) are cleaved at comparable rates. Data in Table 11.1 also show that Pt and Pd are much less active in hydrogenolysis than are the other three metals studied and that Ru exhibits the highest activity. In a more comprehensive study,[256] in the hydrogenolysis of 3-methylpentane, the following activity order was observed: Rh > Ir > Ru = Os > Ni > Co = Re > Pt > Pd > Cu > Ag. Similar activity orders (with Os, Ru, Rh, and Ir as the most active, and Pt, Pd, and Group IB metals as the least active) were found in the hydrogenolysis of ethane,[249] *n*-pentane,[262] and cyclopentane.[263] Os, Co, and Ru are usually the most active to catalyze deep hydrogenolysis.

As mentioned before, nickel usually catalyzes demethylation. For example, besides methane, mainly neopentane (as well as much less ethane and isobutane) is formed when neohexane undergoes hydrogenolysis on nickel.[251,264] In contrast,

Table 11.1. Product distribution of *n*-heptane[a] hydrogenolysis[261]

Metal	Temperature (°C)	Conversion (%)	Product Distribution (mol%)					
			C_1	C_2	C_3	n-C_4	n-C_5	n-C_6
Pd	300	6.4	46	4			4	46
Rh	113	2.9	42	5	4	3	5	41
Ru	88	4.0	28	12	13	12	10	25
Pt	275	2.3	31	13	17	16	9	14
Ir	125	1.5	21	21	15	14	14	15

[a] The hydrogen : *n*-heptane molar ratio is 5.

certain metals tend to cleave only one particular bond in a reacting molecule. For example, platinum strongly favors cleavage of the secondary–quaternary bond in neohexane to yield ethane and isobutane.[260,264]

Forge and Anderson studied so-called archetypal molecules to establish the cleavage mode of individual metals.[251,265–267] Ethane is an archetype of C_2-unit hydrogenolysis (i.e., the cleavage of primary–secondary and secondary–secondary bonds). Neopentane, in turn, is an archetypal molecule for iso-unit hydrogenolysis (i.e., the hydrogenolysis of bonds with at least one tertiary or quaternary carbon atom).

On iridium[266] ethane, butane and neohexane react via C_2-unit mode (demethylation, deethylation) with an activation energy of about 42 kcal/mol. Isobutane, neopentane, and 2,3-dimethylbutane, in turn, react in an iso-unit mode with a much higher activation energy (~56 kcal/mol). These and other results strongly indicate that the cracking patterns of different molecules depend on their ability to undergo certain types of adsorption. Butane, for example, can easily form $1,2(\alpha,\beta)$-diadsorbed species resulting in cleavage of the corresponding C–C bond. In contrast, only $1,3(\alpha,\gamma)$-diadsorbed species are formed from neopentane. Neohexane may adsorb in either the C_2-unit or iso-unit mode. On iridium, the former is favored (α,β adsorption), whereas on platinum α,γ adsorption is responsible for the characteristic iso-unit cleavage. In accordance with this cleavage pattern, the activation energy for ethane adsorption on platinum was found to be much higher than for neopentane.[251] Further studies with platinum indicated that on this metal β-position bonds to a tertiary carbon atom are exclusively broken. In contrast, hydrogenolysis of bonds α to a quaternary carbon atom are more rapid.[255]

Since in C_2-unit hydrogenolysis both carbon atoms of the C–C bond to be broken must be primary or secondary (isobutane cannot cleave in C_2-unit mode through adsorption of its tertiary carbon atom), Anderson formulated the cleavage of neohexane according to Eq. (11.80) involving carbon–metal double bonds (1,2-dicarbene mechanism):[267]

$$CH_3{-}\underset{\underset{CH_3}{|}}{\overset{\overset{CH_3}{|}}{C}}{-}CH_2{-}CH_3 \longrightarrow Me_3C{-}\underset{M}{\overset{||}{C}}{-}\underset{M}{\overset{||}{CH}} \longrightarrow Me_3C{-}\underset{M}{\overset{|||}{C}} + \underset{M}{\overset{|||}{CH}} \xrightarrow{[H]} Me_3CMe + CH_4$$

$$(11.80)$$

It was also shown that bond breaking does not necessarily require direct bonding of the carbon atoms of the bond to be broken to surface metal atoms. For example, neither 1,2- nor 1,3-diadsorbed species can explain the favored cleavage of 2,2,3,3-tetramethylbutane to yield isobutane. A 1,4-diadsorbed intermediate, however, could account for the hydrogenolysis of the quaternary–quaternary bond.[255]

Adsorption modes and hydrogenolysis were also correlated with other metal-catalyzed reactions. Gault noticed striking similarities in product distributions of isomerization and ring opening of cycloalkanes. Kinetic and tracer studies provided useful data[252,268] to arrive at the conclusion that a common surface intermediate is involved in bond-shift isomerization and iso-unit hydrogenolysis.[252,269] This intermediate is a quasimetallacyclobutane (suggested earlier as the intermediate in isomerization),[268,270] that is, a 1,3-diadsorbed species bound to a single surface metal atom (**36**) that decomposes to give a surface carbene and adsorbed alkene:

$$
\begin{array}{ccccc}
\underset{\underset{\displaystyle M}{|}}{\underset{\displaystyle H_2C}{\overset{\displaystyle Me}{\diagup}}\overset{\displaystyle C}{\underset{\displaystyle \diagdown}{\diagup}}\overset{\displaystyle Me}{\overset{\displaystyle \diagdown}{CH_2}}} & \longrightarrow & H_2C\overset{Me\diagdown\diagup Me}{\underset{\displaystyle M}{\overset{\displaystyle C}{\smile}}}CH_2 & \xrightarrow{\text{[H]}} & Me_2CHMe \; + \; CH_4
\end{array} \tag{11.81}
$$

36

This mechanism is similar to the olefin metathesis reaction. When the molecule structure permits formation of 1,3-diadsorbed species, this reaction can occur via π-allyl adsorbed complexes.[271]

The favored cleavage of bonds β to a tertiary carbon on platinum (e.g., in 2-methylpentane to yield isobutane and ethane) is accounted for by α,α,γ-triadsorbed species **37**:

$$
\underset{\underset{\displaystyle M \quad M}{}}{\overset{\displaystyle Me}{\underset{\displaystyle Me}{\diagdown}}}\!\!\!\!\bigwedge\!\!\!\!\overset{Me}{\parallel}
$$

37

On palladium, which effectively catalyzes demethylation of both 2- and 3-methylpentane [and therefore, the 1,2-dicarbene mechanism as in Eq. (11.80) cannot be operative], an 1,3-diadsorbed surface species attached to two metal atoms was suggested.[252]

The high selectivity of nickel for demethylation is explained by its high affinity for the adsorption of primary carbon atoms and its high ability to form 1,2-diadsorbed species.[272] On cobalt, where excessive methane formation is characteristic, consecutive C—C bond ruptures are believed to occur via species with multiple attachment to the metal.[252]

The adsorbed species discussed above are simplified representations of the actual surface complexes since the number of participating surface metal atoms and the number of bonds between the molecule and the surface are not known. Hydrogenolysis reactions, however, are believed to take place on multiple surface sites, specifically, on a certain ensemble of atoms. For example, on the basis of kinetic data, Martin concluded[273] that at least 12 neighboring nickel atoms free from adsorbed hydrogen are required in the complete cleavage of ethane:

$$
CH_3CH_3 \; + \; 12\,Ni \; \longrightarrow \; 2\underset{\underset{\displaystyle Ni\,Ni\,Ni}{\diagup|\diagdown}}{C} \; + \; 6\underset{\underset{\displaystyle Ni}{|}}{H} \tag{11.82}
$$

As a result, alloying an active metal with an inactive metal is expected to effect hydrogenolysis. This phenomenon (i.e., dilution of active atoms on the surface) was observed, among others, on Ni–Cu, the most studied bimetallic cata-lyst.[274-276] When increasing amounts of copper are added to nickel, the probability of the formation of surface complexes requiring several neighboring nickel atoms decreases. Because of this dilution effect, the specific activity (and selectivity) of the alloy catalyst also decreases. Since α,β-diadsorbed surface species are affected most by alloying, they appear to require the largest ensemble (the most nickel atoms). Similar results on platinum-containing alloys are also well documented.[277] One peculiar observation is, however, the unexpected increase in activity with increasing copper content of bimetallic copper-containing catalysts (Pt, Ir, Ni, Pd) in the hydrogenolysis of n-pentane.[277,278] The explanation given is the so-called mixed ensemble, suggesting the bonding of the hydrocarbon to both metals.[278] Dehydrogenation on the transition metal and hydrogenolysis of the formed alkene on copper can be a viable alternative.

Metal dispersion (i.e., the number of available surface metal atoms compared to the total number of atoms) can also affect hydrogenolysis. The most widely studied reaction is the hydrogenolysis of ethane,[279] which was shown to be structure sen-sitive (i.e., the specific activity was found to change with changing dispersion). On Pt and Ir, increasing dispersion brings about marked decrease in hydrogenolysis activity.[280] This was attributed to some unique geometric configuration of metal atoms, which is less probable for small crystallites. On nickel[281,282] and rho-dium,[283] in contrast, the rate of hydrogenolysis increases with increasing disper-sion. Structure sensitivity in hydrogenolysis of higher hydrocarbons is much less pronounced. On rhodium, however, large variations in rate in the hydrogenolysis of 2,2,3,3-tetramethylbutane was observed.[271]

Since the ratio of different crystallographic faces changes with dispersion, struc-ture sensitivity reflects the difference in the ability of different crystal faces to catalyze hydrogenolysis. This phenomenon, however, can be best studied on single crystals. For example, in the hydrogenolysis of ethane, Ni(111) was found to exhi-bit limited activity compared to the Ni(100) surface.[284] The differences in spacing between bonding sites on the two surfaces could account for the differences in activity. On Ni(111) the 1.4-Å spacing between high coordination (threefold hol-low) bonding sites is ideally suited to maintaining the C–C bond intact. The spa-cing between fourfold hollow sites on Ni(100) is approximately 2.5 Å, which is much higher than the bond length of the C–C σ bond bringing about bond scission. Different platinum crystal faces were also shown to exhibit different reactivities in the hydrogenolysis of isobutane and n-heptane.[285,286] Changes in selectivity (term-inal vs. internal C–C bond cleavage) were observed in the hydrogenolysis of n-butane[287] and n-hexane[288] on iridium and platinum single crystals, respectively.

Hydrogenative ring opening of cycloalkanes is also a well-studied area.[16,252,253,289-292] Mainly cyclopropanes and cyclopentanes were studied, since three- and five-membered adsorbed carbocyclic species are believed to be intermediates in metal-catalyzed isomerization of alkanes (see Section 4.3.1). Ring-opening reactivity of different ring systems decreases in the order cyclopro-pane > cyclobutane > cyclopentane > cyclohexane.[251] Cyclopropane and its subs-tituted derivatives usually react below 100°C.

Pt and Pd, which exhibit low activity in the hydrogenolysis of alkanes, show high selectivity in the ring opening of cyclopropanes; in contrast to most other metals bringing about hydrocracking and deep hydrogenolysis (mainly Fe, Co, and Ni), Pt and Pd catalyze exclusive ring opening. Alkyl-substituted cyclopropanes usually open through the rupture of the C—C bond opposite to the substituent,[293,294] whereas electron-withdrawing groups (phenyl) bring about ring cleavage at the bond adjacent to the substituent. Adsorbed cyclopropane (**38**) formed through dissociative adsorption (characteristically on Pt) and α,γ-diadsorbed cyclopropane (**39**, the result of associative adsorption) are the main surface species:[292,293]

Studies of the hydrogen pressure dependence of the ring opening of substituted derivatives, particularly those of stereoisomers, allowed for conclusions on the nature of the adsorbed species. Depending on the metal, the hydrogen partial pressure, and the steric structure of the reacting molecules, the involvement of surface species formed via flat or edgewise adsorption determining ring-opening selectivities was suggested.[295,296]

Cyclobutanes are less reactive compounds, undergoing ring opening according to the 1,2-dicarbene mechanism through α,α,β, β-tetraadsorbed intermediates (**40**) when CH_2-CH_2 bonds are present.[297,298] One particular exception is *cis*-1,2-dimethylcyclobutane, which always yields 2,3-dimethylbutane as the main product. This was interpreted to occur via a π-adsorbed alkene that is readily formed from the *cis* compound [Eq. (11.83)] due to the presence of vicinal *cis* hydrogens, but not from the *trans* isomer:

$$(11.83)$$

More recent observations revealed strong evidence showing that propylcyclobutane is adsorbed on platinum simultaneously on both the ring and the side chain resulting in unusually high selectivity in the ring opening of the secondary–tertiary bond.[299]

Two competing ring-opening mechanisms can be observed in the hydrogenolysis of cyclopentanes.[252] In the transformation of methylcyclopentane over Pt-on-alumina catalysts of low dispersion, selective ring opening through the cleavage of secondary–secondary bonds takes place to yield 2- and 3-methylpentanes.[297] On highly dispersed Pt catalysts (and on Pd), the product distribution is near-statistical. The 1,2-dicarbene mechanism can account for selective ring opening. The contribution of this mechanism increases in the order Pt < Ni < Rh, which is also the order for multiple exchange of methane. The nonselective mechanism involves π-adsorbed olefinic species similar to those participating in the ring opening of cyclobutanes.[252,300] A third mechanism with the involvement of a metallacyclobutane intermediate was also proposed to interpret enhanced cleavage of bonds involving tertiary carbon atoms on 10% Pt-on-alumina catalysts.[300] The formation of increased amounts of *n*-heptane from 1,2-dimethylcyclopentane can be envisioned with the participation of surface species **41**.

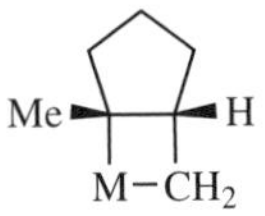

Me—[pentagon]—H
M–CH₂

41

The product distributions in the ring opening of substituted cyclopentanes show striking similarities to those in isomerization of the corresponding alkanes (*n*-hexane, 2- and 3-methylpentane),[301] which led to the formulation of the cyclic mechanism involving adsorbed cyclopentane intermediates to interpret alkane isomerization[15,251,252] (see Section 4.3.1).

Since the hydrogenolysis of cyclohexane and cyclohexane derivatives is less probable than the thermodynamically favored dehydrogenation to form aromatic compounds, most studies address hydrogenolysis only in connection with aromatization as an unwanted side reaction. An interesting observation by Somorjai showed, however, that hydrogenolysis of cyclohexane to form *n*-hexane becomes competitive with aromatization on Pt single crystals with increasing kink density.[302] On a Pt surface where approximately 30% of the atoms in the steps are in kink positions, benzene and *n*-hexane are formed in 1 : 1 ratio.

11.5.2. Ionic Hydrogenolysis

Treating alkanes with superacids such as $FSO_3H–SbF_5$, Olah and coworkers observed both C—H and C—C bond protolysis:[303]

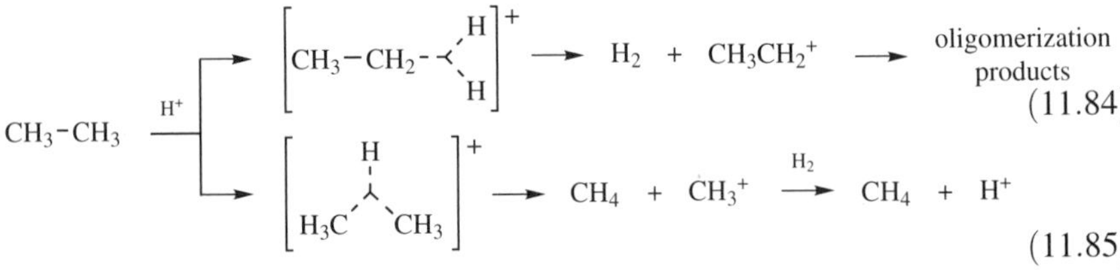

$$CH_3–CH_3 \xrightarrow{H^+} \left[CH_3–CH_2–\overset{H}{\underset{H}{\diamond}} \right]^+ \longrightarrow H_2 + CH_3CH_2^+ \longrightarrow \begin{matrix} \text{oligomerization} \\ \text{products} \end{matrix} \quad (11.84)$$

$$\left[\overset{H}{\underset{H_3C \diamond CH_3}{\diamond}} \right]^+ \longrightarrow CH_4 + CH_3^+ \xrightarrow{H_2} CH_4 + H^+ \quad (11.85)$$

As shown in the case of ethane, the formation of methane is about 15 times higher than that of hydrogen, the byproduct of C—H bond protolysis. The reason can be that H_2 readily quenches carbenium ions to their hydrocarbons. Thus ethane preferentially is hydrogenolyzed to methane according to the following overall reaction:

$$CH_3CH_3 \;+\; H_2 \xrightarrow{\text{``H}^+\text{''}} 2\,CH_4 \tag{11.86}$$

Similarly, superacid-catalyzed hydrogenolysis [Eq. (11.87)] was observed for a series of homologous alkanes. Indeed, as reported by Hogeveen and his colleagues,[304–306] the reaction of long-lived alkylcarbenium ions with hydrogen gives saturated hydrocarbons (i.e., the reverse reaction of the superacidic ionization of hydrocarbons):

$$R-R \xrightarrow{\text{H}^+,\,\text{H}_2} 2\,R-H \tag{11.87}$$

This was shown for *tert*-alkyl, norbornyl, and methylcyclohexyl cations.

Siskin studied the hydrogenolysis of cycloalkanes (and alkanes) in HF–TaF$_5$ system with 200-psi (pounds per square inch; lb/in.^{2}) hydrogen.[307] He was able to show acid-catalyzed hydrogenolysis (ring or chain cleavage) and hydrogenation leading to open chain alkanes:

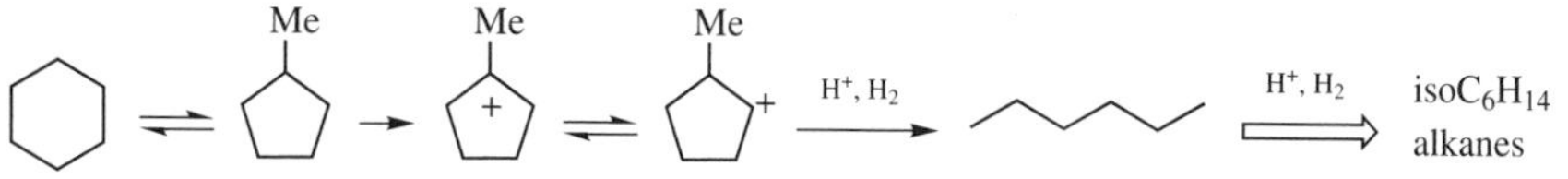

$$\tag{11.88}$$

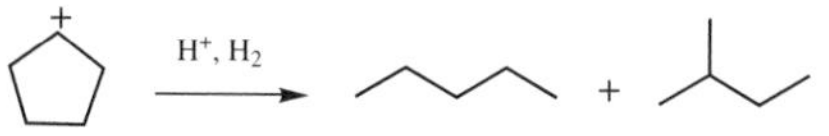

$$\tag{11.89}$$

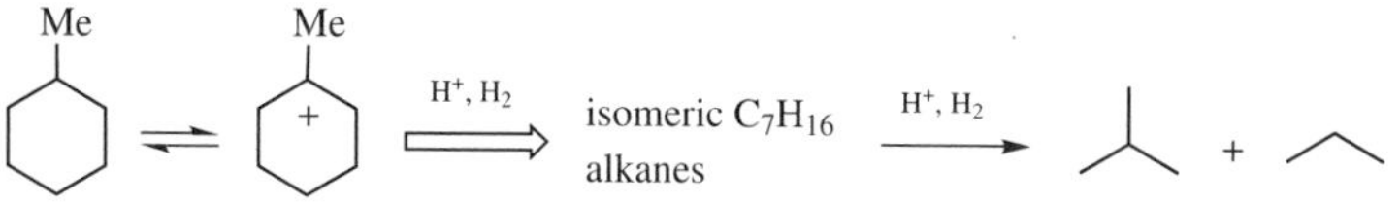

$$\tag{11.90}$$

Clearly these examples also are in line with what is happening in hydrocracking. In acid-catalyzed hydrocracking cleavage of the larger saturated hydrocarbon chains can take place via β cleavage induced by carbenium cations formed via C—H protolysis, and also via direct C—C bond protolysis (Scheme 11.13).

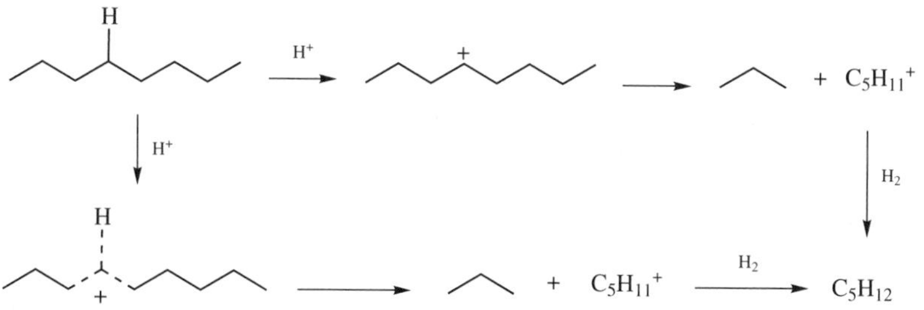

Scheme 11.13

11.6. PRACTICAL APPLICATIONS

11.6.1. Selective Hydrogenation of Petroleum Refining Streams

Cracking processes (catalytic and steam cracking) are used extensively to produce basic raw materials for the petrochemical industry (see Section 2.1). The cracker streams thus produced contain hydrocarbons with multiple unsaturation. These impurities need to be removed to increase productivity of the downstream processes and avoid operating problems (catalyst poisoning and fouling, gum formation) as well as contamination of the end product. Selective hydrogenation processes called *catalytic hydrorefining* are applied.[308] This term is also used to include all industrial hydrotreatment (hydroprocessing) operations, such as hydrodeoxygenation, hydro-denitrogenation, hydrodesulfurization, and hydrodemetalation.[309]

The most selective and widely used catalyst is palladium, usually on an alumina support. A bimetallic palladium catalyst has also been developed.[310] Palladium is more selective and less sensitive to sulfur poisoning than are nickel-based catalysts. Additionally, sulfides can also be employed.

C₂ Hydrorefining. The selective hydrogenation of acetylene in the presence of large amounts of ethylene is a process of considerable importance for the manufacture of polymer-grade ethylene[311,312] and vinyl chloride.[313] Alumina-supported palladium catalysts with low (0.04%) metal content are used widely with carbon monoxide continuously added in trace amounts to poison the catalyst in order to avoid overhydrogenation.[308] Carbon monoxide also inhibits the formation of oligomeric material referred to as *green oil* or *foulant*, which brings about catalyst deactivation with time on stream.

C₃ Hydrorefining. The aim of C_3 hydrorefining is to hydrogenate methylacetylene and propadiene present in the cut. Efficient liquid-phase processes were developed by Bayer[314–316] (cold hydrogenation process carried out at 10–20°C) and IFP,[317] but hydrogenation in the gas phase is also practiced.

C₄ Hydrorefining. The main components of typical C_4 raw cuts of steam crackers are butanes (4–6%), butenes (40–65%), and 1,3-butadiene (30–50%). Additionally, they contain vinylacetylene and 1-butyne (up to 5%) and also some methylacetylene and propadiene. Selective hydrogenations are applied to transform vinylacetylene to 1,3-butadiene in the C_4 raw cut or the acetylenic cut (which is a fraction recovered by solvent extraction containing 20–40% vinylacetylene), and to hydrogenate residual 1,3-butadiene in butene cuts. Hydrogenating vinylacetylene in these cracked products increases 1,3-butadiene recovery ratio and improves purity necessary for polymerization.[308]

Supported palladium and copper catalysts are usually used. A serious problem of this reaction is that palladium forms a complex with vinylacetylene below 100°C. This complex is soluble in the hydrocarbon mixture undergoing hydrorefining and, consequently, palladium is eluted from the catalyst. Operating at temperatures above 100°C or the use of bimetallic palladium catalysts[310] solves this problem.

The cuts after the extraction of 1,3-butadiene (butene cut or raffinate I) and the subsequent removal of isobutylene (raffinate II) contain residual 1,3-butadiene, which must be removed to achieve maximum butene yields. Since the demand for polymer-grade 1-butene has increased, hydrogenation of 1,3-butadiene must be carried out under conditions to avoid isomerization of 1-butene to 2-butenes. Several technologies are available.[308,315,316,318] Hydrorefining of C_4 cuts is also important to improve operating conditions and product quality of alkylation.[319–323] In this case the removal of selective hydrogenation of residual 1,3-butadiene decreases the acid consumption of the alkylation step. Additionally, hydroisomerization of 1-butene to 2-butenes upgrades the alkylation feed, resulting in better alkylates. A similar purification (removal of 1,3-butadiene) is suggested in etherification (oxygenate production) of isoamylenes[324] and higher alkenes[325] with the additional advantage of hydroisomerization with the formation of branched (trisubstituted) alkenes.

Gasoline Hydrorefining. Steam-cracked gasoline contains relatively high amounts of dienes (isoprene, cyclopentadiene) and alkenylarenes (styrene, indene), which readily polimerize, resulting in gum formation. Partial hydrogenation of dienes to monoalkenes and the saturation of the olefinic double bond in alkenylarenes without the hydrogenation of the aromatic ring are necessary. The catalysts are palladium, nickel, or Ni–W sulfides operated at 40–100°C under pressure of 25–30 atm. The most information is available on processes developed by Bayer,[314,315] IFP,[317] and Engelhard.[326]

11.6.2. Cyclohexane

Besides the hydrogenation processes of the petroleum industry discussed above, the hydrogenation of benzene to cyclohexane is the most important large-scale industrial hydrogenation reaction.[327–329] Most of the cyclohexane thus produced is used in the manufacture of polyamides.

Nickel as well as platinum and palladium are the most important catalysts used in the liquid or gas phase usually in the temperature range of 170–230°C and pressure of 20–40 atm. Since the reaction is highly exothermic [Eq. (11.91)], efficient heat removal is required to ensure favorable equilibrium conditions and prevent isomerization of cyclohexane to methylcyclopentane:

$$\text{benzene} \;+\; 3\,H_2 \;\xrightleftharpoons{\text{catalyst}}\; \text{cyclohexane} \qquad \Delta H = -49 \text{ kcal/mol} \qquad (11.91)$$

One of the most important processes is IFP's liquid-phase hydrogenation technology[330–332] applying a Raney nickel catalyst under mild conditions (200–240°C, 15–30 atm). The liquid phase with the suspended catalyst is circulated with an external pump to improve heat removal, attain good dispersion of hydrogen in the liquid phase, and ensure a good suspension of the catalyst. The product containing some unreacted benzene (<5%) is charged into a coupled fixed-bed reactor where gas-phase hydrogenation over a Ni-on-Al$_2$O$_3$ catalyst ensures virtually complete conversion. If high-purity benzene is used, the purity of cyclohexane exceeds 99%.

Numerous other technologies, mainly gas-phase hydrogenations, were developed over the years (Sinclair-Engelhard,[333,334] Hytoray,[335,336] Arco,[337] DSM,[338] UOP[339]). The Arco and DSM processes operate at 400°C and under 25–30 atm. Despite the high temperature, isomerization is negligible because of the very short contact time.

The soluble nickel catalyst developed by IFP for the oligomerization of alkenes applied in different Dimersol processes (see Section 13.1.3) can also be used in benzene hydrogenation to replace Raney nickel (IFP cyclohexane process).[340]

11.6.3. Various Hydrogenations

A substantial amount of α-methylstyrene is produced during the cumene oxidation step in the production of phenol and acetone. Slurry processes applying Raney nickel and a fixed-bed operation with palladium developed by Engelhard[326,341] are used to hydrogenate and recycle α-methylstyrene to produce more phenol and acetone.

Sumimoto introduced a new sebacic acid process including several catalytic hydrogenation reactions.[342] The synthesis starts with naphthalene, which is first partially hydrogenated to tetralin over cobalt oxide or molybdenum oxide, then to decalin over ruthenium or iridium on carbon. The selectivity to *cis*-decalin is better than 90%. In a later phase of the synthesis 5-cyclododecen-1-one is hydrogenated over Raney nickel to obtain a mixture of cyclododecanone and cyclododecanol in a combined yield of 90%. The selectivity of this step is not crucial since subsequent oxidation of either compound leads to the endproduct sebacic acid.

A low-cost route to 1,4-butanediol and tetrahydrofuran based on maleic anhydride has been disclosed (Davy process).[343,344] Here dimethyl or diethyl maleate is hydrogenated over a copper catalyst. Rapid saturation of the C—C double bond forms diethyl succinate, which subsequently undergoes further slower transformations (ester hydrogenolysis and reduction as well as dehydration) to yield a mixture of γ-butyrolactone, 1,4-butanediol, tetrahydrofuran, and ethanol. After separation both ethanol and γ-butyrolactone are recycled.

A somewhat similar hydrogenation problem arose in a different approach to 1,4-butanediol and tetrahydrofuran.[345] In the process developed by Mitsubishi, 1,3-butadiene first undergoes Pd-catalyzed diacetoxylation to yield 1,4-diacetoxy-2-butene. To avoid the further transformation of the diol as in the abovementioned process, 1,4-diacetoxy-2-butene is directly hydrogenated in the liquid phase (60°C, 50 atm) on traditional hydrogenation catalysts to produce 1,4-diacetoxybutane in 98% yield, which is then hydrolyzed to 1,4-butanediol.

11.7. RECENT DEVELOPMENTS

A plethora of more efficient, highly active and selective catalysts has been developed since the early 1990s, and important findings in the understanding of the mechanism of hydrogenation have been reported. Significant new observations have been made in the asymmetric hydrogenation of prochiral compounds even though it still remains a great challenge to achieve high enantioselectivites in the hydrogenation of nonfunctionalized unsaturated hydrocarbons. Two of the recipients of the 2001 Nobel Prize in Chemistry, namely, Knowles and Noyori, made pioneering work in the development of chiral soluble transition-metal catalysts, which led to their application in a number of industrial syntheses of pharmaceutical products.

11.7.1. Heterogeneous Catalytic Hydrogenation

Alkenes. Three review articles summarize the latest results in alkene hydrogenation.[346–348] In addition, comprehensive treatments of hydrogenation of alkenes as a synthetic tool are found in two monographs.[349,350]

The mechanism proposed by Horiuti and Polanyi for the hydrogenation of alkenes (see Section 11.1.2) requiring the surface reaction of the adsorbed alkene and an adsorbed hydrogen atom corresponds to the Langmuir–Hinshelwood mechanism. An alternative but rare possibility is the reaction of an adsorbate with a gas-phase species, the so-called the Eley–Rideal pathway. In fact, this was only observed in the hydrogenation of ethylene over Cu(111)[351] and cyclohexene over Cu(100).[352] The latter study demonstrated that hydrogen may add from the gas phase to the adsorbed alkene. This topside addition results in the same *syn* stereochemistry generally observed in alkene hydrogenation. When a prochiral alkene is hydrogenated, the alkanes formed by bottom-face and top-face additions are mirror images of each other (Scheme 11.14).

Study of the addition of gas-phase D atoms to 1-alkene monolayers adsorbed on Cu(100) suggested that addition to the terminal carbon predominates to generate the corresponding secondary alkyl group (half-hydrogenated intermediate).[353]

In contrast to the participation of surface-bond and gas-phase hydrogen, bulk hydrogen was shown to be active in the hydrogenation of ethylene and cyclohexene on Ni catalysts.[354,355] The use of new, highly sophisticated experimental techniques permitted Somorjai to learn further details of the mechanism of alkene hydrogenation.[356] Infrared–visible *sum frequency generation* (SFG) is a surface-specific vibrational spectroscopy that can operate in a pressure range from ultrahigh vacuum to atmospheric pressure.[356,357] Studies with Pt(111) and Pt(100) single crystals allowed investigators to propose a new model to interpret ethylene and propylene hydrogenation. Surface ethylidyne covering the metal surface was proved to be compressed, allowing the adsorption of ethylene directly to the metal surface to form π-bonded ethylene, which is stepwise hydrogenated through the ethyl intermediate.[358–360] In the case of propylene, both 1-propyl and 2-propyl moieties are formed but hydrogenation proceeds faster through the 2-propyl intermediate.[361] A

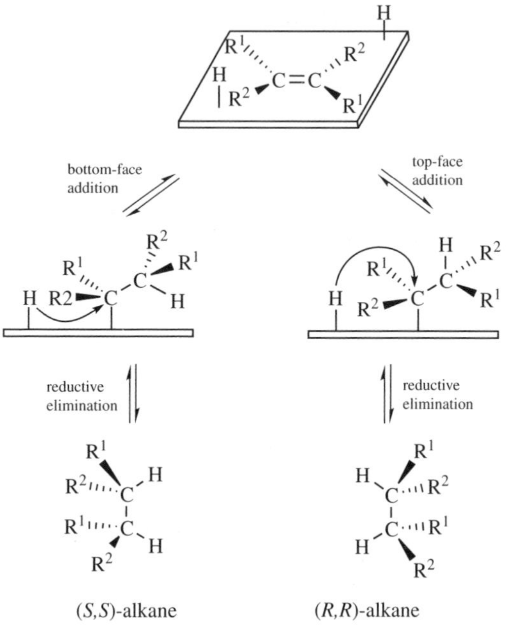

Scheme 11.14

surprising feature of cyclohexene hydrogenation is that it proceeds through a 1,3-cyclohexadiene intermediate.[360]

Numerous examples were reported describing new ways for the preparation of catalysts with improved characteristics. Graphimets, a special class of graphite intercalation compound, contain metals between the layers of graphite either in atomic dispersion or as small clusters. A considerable amount of metal, however, may also be on the graphite surface. Detailed studies suggested that interlayer metal (Pt, Pd, Rh) particles do not participate in alkene hydrogenation, since these are inaccessible for molecules such as 1-butene[362] and cyclohexene.[363,364] According to a recent observation, however, quasi-two-dimensional Pd nanoparticles encapsulated in graphite showed high activity in the hydrogenation of 1-butene.[365]

Highly dispersed palladium nanoparticles immobilized on alumina were prepared via the sonochemical reduction by alcohols of Pd(II).[366] The activity of these catalysts was 3–7 times higher in the hydrogenation of alkenes than that of conventional catalysts, and they showed a high preference to hydrogenate 1-hexene over *trans*-3-hexene.

Supported metal clusters, a class of catalysts usually prepared by the decomposition of metal carbonyl clusters using porous supports, have been studied in hydrogenations.[367,368] They are highly stable materials; characterization data show the lack of significant changes in the cluster nucleations as a results of catalysis. Ag–Ru,[369] Cu–Ru,[370] and Pd–Ru[371] bimetallic clusters anchored inside the mesoporous channels of MCM-41 silica were shown to be highly active in alkene hydrogenation.

Novel platinum-containing heteropolyoxometallic compounds have been shown to exhibit shape-selective hydrogenation activity.[372] The activity of the ultramicroporous $Pt/Cs_{2.1}H_{0.9}[PW_{12}O_{40}]$ and microporous $Pt/Cs_{2.5}H_{0.5}[PW_{12}O_{40}]$ was comparable in the hydrogenation of ethylene, whereas activity of the latter was about one order of magnitude higher in the hydrogenation of cyclohexene.

Interesting observations have been made about metal oxides. Hydrogen without dissociative adsorption is believed to participate in hydrogenation on ZrO_2. In addition, the ethyl species is not involved in the process. As a result, propane-d_2 is selectively formed in deuteration, with no H/D exchange in propylene.[373] New features appear, however, when zirconia is dispersed on other oxides; the rate of hydrogenation is increased considerably, and hydrogen exchange in propylene proceeds simultaneously with addition via the common intermediate 1-propyl and 2-propyl species.[373,374]

The hydrogenolysis of Et_3SiH over silica-supported Pd and Pt catalysts resulted in significant poisoning, specifically, the loss of activity in the hydrogenation of cyclohexene.[375] Oxidation, however, fully restored the activity of catalysts with small metal particles (>50% dispersion) as a result of surface reconstruction.

Dienes and Alkynes. The last decade has seen a burgeoning of literature data providing further insight into the fine details of the chemo-, regio- and stereoselective partial hydrogenation of alkynes and dienes catalyzed by palladium.[376] According to recent observations the surface of the working catalyst is covered by an array of carbon deposits also referred to as *hydrocarbon* or *carbonaceous overlayer* formed under reaction conditions. These surface deposits exerts a significant influence on activity and selectivities.

The high selectivity in the hydrogenation of phenylacetylene over a Pd-on-pumice catalyst was interpreted by the formation polymeric or other species of low reactivity.[377] These occupy active sites and thereby inhibit the adsorption of the intermediate alkene hindering complete hydrogenation. In contrast, it was also shown that over Pd-on-Al_2O_3 catalysts carbon deposits substantially decrease effective diffusivity by blocking the catalyst pores.[378,379] Mass transfer limitations severely hinder intraparticle diffusion of acetylene and, consequently, an increased rate of ethylene hydrogenation in the interior of catalyst particles will result. The overall effect is decreasing alkene selectivity.

Peculiar observations were made in studying the hydrogenation of 1,3-butadiene over a Pd–Ni-on-Nb_2O_5 catalyst prepared from an amorphous $Pd_2Ni_{50}Nb_{48}$ ribbon that showed high semihydrogenation selectivity.[380,381] Over the self-poisoned surface, hydrogenation and isomerization of butenes were remarkably inhibited; nevertheless the catalyst remained active in the hydrogenation of 1,3-butadiene. It was proposed that the unusually high alkene selectivity after depletion of butadiene was ensured by firmly held adspecies formed from the diene. Similar features were observed on Pd–Ni samples prepared by vapor phase deposition of $Pd(acac)_2$ and $Ni(CO)_4$ and then submitted to calcination, reduction and partial deactivation by hydrogenation of 1,3-butadiene.[382]

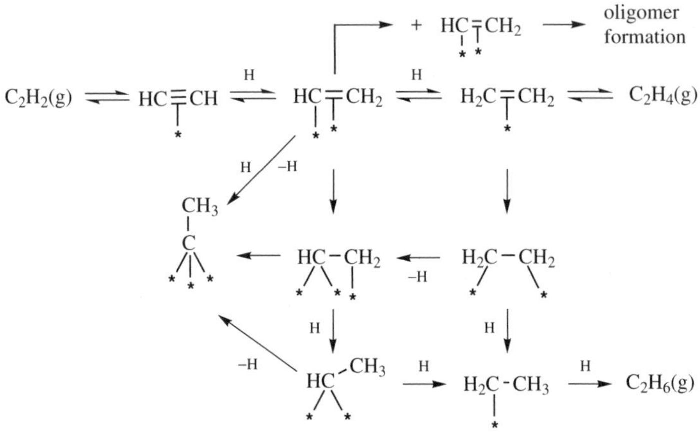

$$C_2H_2(g) \rightleftharpoons HC{\equiv}CH \rightleftharpoons HC{=}CH_2 \rightleftharpoons H_2C{=}CH_2 \rightleftharpoons C_2H_4(g)$$

$$\longrightarrow \; + \; HC{=}CH_2 \longrightarrow \text{oligomer formation}$$

$$CH_3{-}C \quad HC{-}CH_2 \quad H_2C{-}CH_2$$

$$HC{-}CH_3 \xrightarrow{H} H_2C{-}CH_3 \xrightarrow{H} C_2H_6(g)$$

Scheme 11.15

Because of its significance in the petrochemical industry, a large body of work has accumulated about the hydrogenation of the parent compound acetylene. This allows us to summarize the reaction paths leading to all possible products involving π- and σ-bonded species (Scheme 11.15).

The activity of palladium catalysts in the hydrogenation of alkynes and conjugated dienes has been shown to exhibit a strong antipathetic behavior as a function of dispersion.[383–387] This phenomenon—that is, the decrease in activity with increasing dispersion—is believed to be brought about by electronic and geometric effects.[388] Specifically, strong complexation of the highly unsaturated alkyne and diene to electron-deficient atoms of low coordination number on small metallic particles is usually invoked to explain the diminishing activity of highly dispersed palladium catalysts.[389]

It has been known that the intrinsic ability of palladium to yield alkene selectively can be further improved by promoters (modifiers) and additives. Adding a second metal to palladium usually results in a decrease in catalytic activity and an increase in semihydrogenation selectivity. The improvement of selectivity is brought about by altering electronic or geometric properties of palladium.[388,389] Modifiers may change the electron density of palladium affecting the relative adsorption strength of the reactant, intermediates, and hydrogen, or they may block part of the surface, thereby affecting the geometry of the active site. The beneficial effect of Cu,[390,391] Ag,[392] and Pb[393] has been studied in detail. The performance of silica-supported Pd–Cu bimetallic catalysts is also outstanding giving high yields of the alkene (99%) and showing good selectivity towards *cis*-olefins.[394]

Significant improvement of the activity of selectivity of Pd on SiO_2 could be achieved in the hydrogenation of acetylene by adding Ti, Nd, or Ce oxides to the catalyst.[395] The metal oxides modify both geometrically and electronically the Pd surface. They retard the sintering of the dispersed Pd particles, suppress the formation of multiply bound ethylene, and facilitate the desorption of ethylene. The beneficial effect of lead in the hydrogenation of 1,3-butadiene over a Pd–Pb-on-α-Al$_2$O$_3$ catalyst was attributed to a decrease in hydrogen coverage brought about by an alloying effect.[396]

Detailed studies have been carried out with various Ni–B catalysts. Ni$_2$B prepared on borohydride exchange resins hydrogenates acetylenes quantitatively to the corresponding *cis* alkenes in methanol.[397] Amorphous NiP- and NiB-on-SiO$_2$ catalysts show high stability and selectivity in the hydrogenation of cyclopentadiene to cyclopentene.[398,399] The NiB-on-SiO$_2$ catalyst exhibits high resistance to sulfur poisoning as a result of electron donation from B to Ni, causing Ni to adsorb sulfur less readily.[400] Selective, solvent-free hydrogenation of dienes and trienes was described by using catalysts containing Ru–Sn nanoparticles anchored along the interior surface of MCM-41.[401]

Low-loaded, organophilic Pd–montmorillonites were shown to exhibit high *cis*-selectivity in the hydrogenation of 1-phenyl-1-alkynes working with high (=5000) substrate : catalyst ratio.[402] Studies with respect to the use palladium membrane catalysts[403–405] and polymeric hollow fiber reactors[406–408] were reported.

Aromatics. High aromatic content in diesel fuel lowers fuel quality, resulting in low cetane numbers and is associated, for instance, with high particulate emission. A possible solution of this problem, the removal of aromatic hydrocarbons by hydrogenation, therefore, has received considerable attention.[409,410] The main focus is on developing more active noble metal catalysts with higher thermal stability and increased sulfur tolerance. Possible solutions are alloying or selecting supports with appropriate acid–base properties.[409,410] Sulfur tolerance of Pt or Pd catalysts can be greatly increased by using acidic supports such as zeolites attributed to the formation of electron-deficient metal species.[411,412] Pt–Pd bimetallic catalysts show the best characteristics at a Pt : Pd molar ratio of 1 : 4. The high sulfur tolerance was correlated with high catalyst dispersion[411] or electronic effects.[413] The Pt-on-aluminum borate catalyst was found to show improved characteristics with this respect.[414] Hydrogen storage alloys were also tested.[415]

Unsupported Ru showed increased selectivity when treated with NaOH, which was attributed to enhanced hydrophilicity and an electronic promotion effect.[416] The activity of a Ni–B amorphous alloy could be increased with the addition of some Co.[417] Other bicomponent amorphous alloys also showed interesting characteristics. Rare-earth metals, especially samarium, are effective to promote the catalytic properties of Ni–B on alumina.[418] M–Zr (M = Pt, Pd, Rh, Ni) alloys undergo activation during hydrogenation of benzene to form active catalytic materials composed of highly dispersed metal particles on zirconia.[419] The activity of Pt or Pd deposited on an acidic HFAU zeolite is higher in the hydrogenation of toluene than that supported on alumina.[420] Bifunctional catalysis is suggested to account for the observations that hydrogenation of the aromatic molecule adsorbed on acidic sites occurs with the participation of hydrogen spilled over from metallic sites. The activity of Pd–Pt nanoclusters deposited on alumina did not show any synergism but rather a perfect additivity of their individual catalytic properties.[421]

A rather unique approach, the preparation and application of combined homogeneous and heterogeneous catalysts, was described by Angelici and coworkers.[422–424]

These were prepared by tethering Rh and Pt complexes to silica-supported metal catalysts (metal = Pd, Ni, Ru, Au). The catalysts are very active in the hydrogenation of benzene derivatives to the corresponding substituted cyclohexanes under mild conditions. The activities are higher than those of the separate homogeneous complexes, complexes just tethered to silica, or the silica-supported heterogeneous catalysts. When the sol-gel-entrapped $[Rh_2Co_2(CO)_{12}]$ complex was heat-treated at 100°C, immobilized metallic nanoparticles were formed.[425] The catalyst thus prepared efficiently catalyzed substituted benzene derivatives.

The partial hydrogenation of benzene to cyclohexene is of great industrial importance. A possible application of the product cyclohexene is its transformation to cyclohexanone. It was shown that ruthenium may give rise to increased selectivity toward cyclohexene.[426] In addition, studies with various Ru–B catalysts demonstrated that boron effectively promotes the activity and selectivity of ruthenium for cyclohexene.[427,428]

11.7.2. Homogeneous Catalytic Hydrogenation

Two reviews give a broad coverage of hydrogenation induced by metal complexes.[347,429] Therefore only a brief overview seems warranted here.

Alkenes and Dienes. As discussed in Section 11.2.1, Rh complexes are the most widely studied catalysts for homogeneous hydrogenations, and they are still in the focus of new research. In addition, complexes of many other metals, mainly those of Ir, Ru, and Os, have been shown to exhibit high potential as homogeneous catalysts.[347,430]

Metallocene complexes of early transition metals $[Cp_2MR^1R^2]$ (R^1, R^2 = H, alkyl, M = Zr, Ti, Hf) are active and selective catalysts in semihydrogenation of dienes.[431] Two ruthenium–carbene complexes display high activity in alkene hydrogenation, which may be further enhanced by the addition of HBF_4–OEt_2.[432] A turnover number of 12,000 h^{-1} was obtained.

C_1-symmetric organolanthanide complexes exhibit moderate to good enantioselectivities in the hydrogenation and deuteration of styrene and 2-phenyl-1-butene.[433] Cationic iridium–phosphanodihydrooxazole complexes are more efficient catalysts for the asymmetric hydrogenation of unfunctionalized aryl-substituted alkenes. The best catalyst (**42**) gives high yield (>99%) and excellent enantioselectivity (97% ee) in the hydrogenation of (E)-1,2-diphenyl-1-propene:[434]

42

A simple system, $NiBr_2$ in alkaline 2-propanol, has been reported to exhibit high activity in the transfer hydrogenation of 1-octene.[435]

An important finding is the observation that substrate polarity strongly influences mechanism of homogeneous hydrogenation.[436] The results with the catalyst bis[1,2-bis(diphenylphosphino)ethane]palladium(0) indicated that the palladium–hydrogen bond is polarized as $Pd^{\delta-}–H^{\delta+}$. The results with the Wilkinson catalyst, in turn, revealed that, in this case, the addition occurs predominantly as $Rh^{\delta+}–H^{\delta-}$. These observation suggest that catalytic hydrogenation with homogeneous metal complexes is most likely a two-electron process, which, for Pd, is strongly influenced by Coulombic properties of the substrate.

The application of water-soluble transition-metal catalysts in hydrogenation offers a solution to one of the major problems of homogeneous catalysis, namely, the separation of the catalyst from the product. Complexes with water-soluble ligands may be used in aqueous biphasic systems allowing catalyst separation simply by decantation. The application of this concept in homogeneous hydrogenations has been reviewed.[437–439] New examples include the application in alkene hydrogenation of Rh complexes with novel carboxylated phosphines[440] and Ru clusters with the widely used TPPTN [tris(3-sulfonatophenyl) phosphine] ligand.[441] Biphasic catalytic hydrogenation of alkenes with Rh(II) and Co(II) complexes dissolved in $[bmi]^+[BF_4]^-$ (bmim = 1-n-butyl-3-methylimidazolium) ionic liquid has also been described.[442] The Co complex hydrogenates 1,3-butadiene to 1-butene with complete selectivity.

The fluorous biphase concept has been successfully applied in the hydrogenation of alkenes. 1-Dodecene or cyclododecene was hydrogenated in a biphase system composed of toluene and $CF_3C_6F_{11}$ in the presence of $RhCl[P(CH_2CH_2(CF_2)_5CF_3)_3]_3$, the fluorous analog of the Wilkinson catalyst.[443] The catalyst is, however, not comparable to the best homogeneous catalysts when activity is considered. However, from a recycling standpoint it offers fundamentally new extraction possibilities. In a similar study $RhCl[P\{C_6H_4-p-SiMe_2(CH_2)_2C_nF_{n+1}\}_3]_3$ complexes showed high activity in hydrogenation of 1-alkenes under single-phase fluorous conditions and could be recycled using fluorous biphasic separation.[444] The Wilkinson catalyst has also been bound to a fluoroacrylate polymer.[445] The resulting catalyst has excellent activity in the hydrogenation of 1-octene, cyclohexene, and bicyclo[2.2.1]hept-2-ene and can be reused following fluorous biphasic liquid/liquid separation and extraction.

The use of supported (i.e., heterogenized) homogeneous catalysts offers another possibility for easy catalyst separation. New examples include polymer-anchored Schiff-base complexes of Pd(II),[446] $PdCl_2(PhCN)_2$ supported on heterocyclic polyamides,[447] various Pd complexes supported on crosslinked polymers,[448] sol-gel-encapsulated Rh–quaternary ammonium ion-pair catalysts,[449] and zwitterionic Rh(I) catalysts immobilized on silica with hydrogen bonding.[450]

A Ni(salen) complex [salen = bis(salicylidene)ethylenediamine] encapsulated in zeolite is highly efficient in the hydrogenation of simple alkenes (cyclohexene, cyclooctene, 1-hexene).[451] This method, which is the encapsulation of metal complexes into the cavities of zeolites, offers the unique possibility of shape-selective hydrogenation.[452] The hydrogenation of 1-decene, for example, was found to be 18 times faster than that of *trans*-5-decene over Pt-on-zeolite BEA, whereas the ratio was about 2 over Pt on nonmicroporous supports.[453] This was explained by the steric constraints imposed by the microporous structure of the zeolite.

There has been increasing interest in the catalytic application of nanoscale colloidal metals and alloys and nanoclusters because they offer the potential to serve as soluble analogs of heterogeneous catalysts. When properly stabilized by organic ligands or polymers to prevent aggregation, these may be used as liquid-phase catalysts in solution.[454–457] Cluster size and the stabilizing ligand can have a significant effect on the catalytic activities.[458] The longest catalytic lifetimes have been reported for polyoxoanion- and tetrabutylammonium-stabilized nanoclusters; turnovers in alkene hydrogenation are 18 000 for $Ir(0)_n$[459] and $\geq$ 193 000 for $Rh(0)_n$.[460] Alternatively, nanoclusters may be deposited on solid supports and used as heterogeneous catalysts (see Section 11.7.1).

Alkynes. Normally, homogeneous transition-metal complexes hydrogenate alkynes mainly to *cis*-alkenes. The first example of a Pd(0) catalyst bearing a bidentate N ligand shows high activity to homogeneously hydrogenate a wide variety of alkynes to the corresponding *cis*-alkenes.[461] Selectivities are often close to 100%. Additionally, enynes are selectively converted into dienes. The cationic ruthenium complex $[Cp^*Ru(\eta^4\text{-}CH_3CH{=}CHCH{=}COOH)][CF_3SO_3]$ represents one of the few examples that induces the direct formation of *trans*-alkenes from internal alkynes.[462] The parahydrogen-induced polarization (PHIP) NMR technique a powerful tool for mechanistic investigations allowed to show the involvement of a binuclear Rh complex with a bridged alkyne–dihydrogen complex. In the Pt(II)/Sn(II) dichloride–mediated hydrogenation of alkynes to yield *cis*-alkenes, in turn, synchronous dihydrogen addition was proved.[463]

Cluster complexes have also found application in alkyne hydrogenation.[464,465] A layer-separated Pt–Ru cluster shows unusually high activity in the hydrogenation of diphenylacetylene to *cis*-stilbene.[464] The catalytic activity is superior to those of the individual metal complexes. The bifunctional character, specifically, Pt atoms activating the hydrogen and Ru atoms activating the alkyne, is suggested as a possible explanation. Metal segregation and its possible role in synergism has also been discussed.[466]

Some of the supported (heterogenized) metal complex catalysts discussed above show good performance in the selective semihydrogenation of alkynes.[446–449]

Aromatics. A new generation of homogeneous arene hydrogenation catalysts was developed by Rothwell and coworkers. These are hydride derivatives of niobium and tantalum with bulky ancillary aryloxide ligands (see, i.e., **43**) exhibiting high regio- and stereoselectivities:[467]

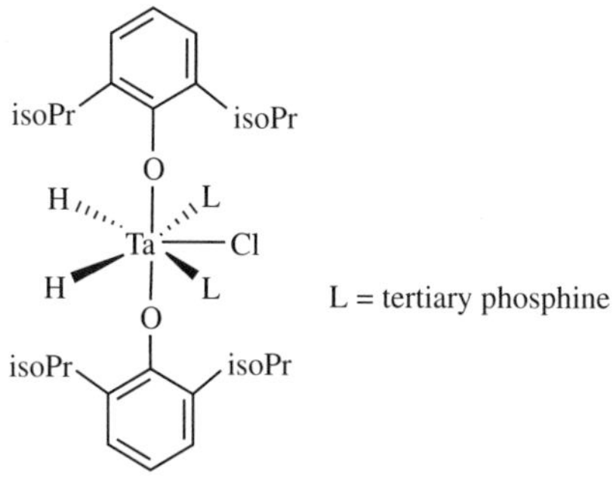

L = tertiary phosphine

43

Almost exclusive formation of the corresponding tetrahydro derivatives in the partial hydrogenation of naphthalene and anthracene are observed with all-*cis* addition of the hydrogens.

Attempts have also been made to develop biphasic methodologies for the hydrogenation of aromatics. The hydrogenation of benzene derivatives was studied using various Ru complexes.[468,469] A trinuclear cluster cationic species was isolated as the tetrafluoroborate salt and showed increased activity in hydrogenation.[470] The air/moisture-stable [bmim][BF$_4$] ionic liquid and water with [Rh$_4$(η^6-C$_6$H$_6$)$_4$][BF$_4$] as the catalyst precursor is an effective system under usual conditions (90°C, 60 atm).[471] A system composed of stabilized Rh(0) nanoparticles proved to be an efficient catalyst in the hydrogenation of alkylbenzenes.[472]

New information has been obtained with respect to the real nature of the [(C$_8$H$_{17}$)NCH$_3$]$^+$[RhCl$_4$]$^-$ ion-pair catalyst (see Section 11.2.4). Unequivocal evidence was presented to show that in benzene hydrogenation—but not in naphthalene and anthracene hydrogenation—the true catalyst is soluble Rh(0) nanoclusters formed under the reaction conditions.[473]

The hydrogenation of aromatics (benzene, toluene, α-methylstyrene) can be carried out under very low (1 : 12,000) catalyst : substrate ratio, and mild conditions on Rh and Ni organometallic complexes anchored to USY zeolites.[474] A Rh complex anchored to functionalized MCM-41 exhibits excellent performance in the hydrogenation of arenes (benzene, toluene, *p*-xylene, mesitylene) under mild conditions (45°C and 1 atm).[475] A uniquely selective hydrogenation of acenaphthene and acenaphthylene was performed by using a triruthenium carbonyl cluster:[476]

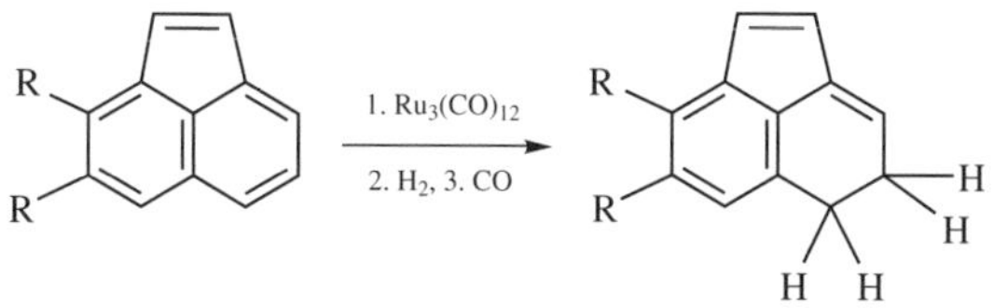

(11.92)

R, R = H
R–R = −CH=CH−CH=CH−

The observation was interpreted as resulting from the effective blocking of the most reactive C–C double bond at the *peri* position through π-cyclopentadienyl coordination to the Rh atom.

11.7.3. Chemical and Electrochemical Hydrogenation

Electrodes prepared by depositing metal particles onto polymer-modified electrodes has been successfully applied in hydrogenations. Palladium microparticles deposited in a viologen-containing polymer layer on the surface of a graphite felt electrode were effective in the electrocatalytic hydrogenation of various alkenes[477] and alkynes[478] to the corresponding saturated hydrocarbons.

Water electrolysis and electrochemical hydrogenation of benzene on a Rh–Pt electrode formed on a polymer electrolyte (Nafion) were combined to produce cyclohexane as a chemical hydrogen carrier.[479]

Hydrogenation of aromatics in the ionic liquid system [emim]Cl–AlCl₃ ([emim]$^+$ = 1-ethyl-3-methylimidazolium cation) with electropositive metals (Li, Zn, Al) and a proton source can be performed with excellent yields and stereoselectivity:[480]

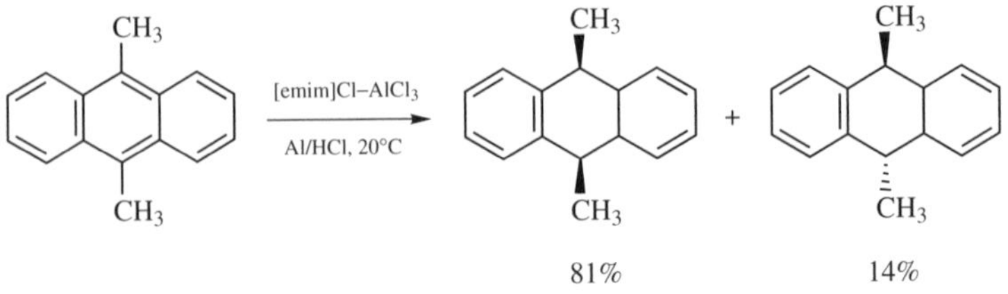

$$(11.93)$$

The reaction differs from other dissolving-metal reductions, such as the Birch reduction (see Section 11.3.3), in that products containing unconjugated double bonds are not observed.

New information about the mechanism of reduction of diphenylacetylene by metallic lithium has been disclosed.[481] Reduction by sodium was shown to result in the formation of the **44a** *trans*-disodium adduct in THF detected by UV spectroscopy yielding *trans*-stilbene on protonation. In contrast, *cis*-stilbene is produced when lithium is used. The **44a** disodium adduct is suggested to be more stable compared to the corresponding *cis* adduct because of conjugation. The **44b** dilithium adduct forms a double-bridged structure allowing each lithium atom to interact with both phenyl groups, which provides even greater stabilization than conjugation:

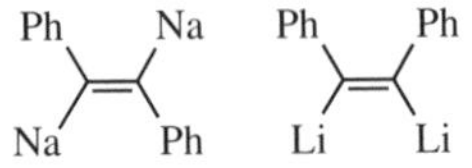

44a   **44b**

In fact, reduction by lithium of dialkylacetylenes without the possibility of such stabilization effect yields exclusively the *trans* alkenes.[482]

11.7.4. Ionic Hydrogenation

Sterically hindered alkenes can be hydrogenated at $-50°C$ using triflic acid and a hydride donor.[483] In addition to Et_3SiH, transition-metal carbonyl hydrides such as $HM(CO)_3Cp$ (M = W, Mo, Os) and $HMn(CO)_5$ (M = Mn, Re) are suitable hydride donors. Alkenes that form tertiary carbocation on protonation are hydrogenated in high yields (90–100%), whereas yields for styrenes are lower (50–60%). Alkynes are converted to the corresponding saturated hydrocarbon by using $HW(CO)_3Cp$ in combination with triflic acid.[484]

11.7.5. Hydrogenolysis of Saturated Hydrocarbons

The hydrogenolysis of alkanes over supported metal catalysts continues to attract attention because of the industrial interest in cracking, hydroforming, and reforming as well as its basic chemical importance. The ultimate aim is to develop catalysts for commercial reforming of hydrocarbons with higher product yields and selectivities. Many factors are known to exert strong influence on hydrogenolysis activity and selectivity, among others, metal–support interactions, modifiers present or added to the catalyst, metal dispersion, and pretreatment and reaction conditions. New details of this complex situation were discussed in more recent publications. Methylcyclopentane as a model compound has been especially widely studied in this context. The relative significance of the two major transformations, cracking (deep hydrogenolysis), that is, the formation of products with less than six carbons and ring opening, may reveal important information on catalyst structure. Additional information may be acquired by analyzing the ratio of 2-methylpentane and 3-methylpentane (selective ring opening) versus *n*-hexane (nonselective ring opening).

The surface carbon formed and accumulated during the transformation of any hydrocarbon on the catalyst surface was shown to affect the transformation of methylcyclopentane. The significant difference between Pt and Rh in methylcyclopentane hydrogenolysis was attributed to the involvement of surface carbonaceous residues. Pt easily builds up a significant amount of surface carbon, thus allowing selective ring opening.[485] Rh, in turn, loses carbon impurities rather easily and this carbon-free state is favorable for cracking. The buildup of carbonaceous matter on Ni on SiO_2 is suggested to result in high activity in the hydrogenolysis of C_2–C_4 alkanes, whereas the low activity of Ni on MoO_3 is attributed to strong metal–support interaction.[486] The presence of oxidized carbonaceous entities are thought to be responsible for highly selective ring opening on Pt black and EUROPT-1.[487] Electron deficient Pt atoms existing on support boundary (adlineation) sites on acidic supports[488] or generated by chlorine impurities[489] significantly influence catalyst activity and bring about nonselective ring opening. High 2-methylpentane :

3-methylpentane ratios over Pt on sulfated zirconia was also attributed to electron deficient Pt sites.[490] The characteristics of supported Pt,[491] Rh,[492] and Ru[493] catalysts in the hydrogenolysis of methylcyclopentane may be modified by appropriate pretreatments or additives.

Nickel catalysts prepared by depositing metal vapor onto supports show high activity, stability, and selectivity in hydrogenolysis, yielding high-liquid products and low-gas products from higher alkanes.[494] Because of high dispersion C–C bond cleavage occurs at relatively low temperature (190–220°C). Ni on ZSM-5 shows the best catalyst performance; it converts n-heptane to isobutene and propane only.

Detailed investigations with substituted cyclobutanes showed that the anchoring effect of the propyl substituent found earlier for Pt[299] also exists on Pd.[495] Hydrogen-pressure-dependence and temperature-dependence studies allowed to gain information on the role of carbonaceous deposits and various surface-adsorbed species in ring-opening processes. In the transformation of methylcyclobutane close to statistical product distribution, approximately 1 : 1 ratios of 2-methylbutane : n-pentane were observed. This indicates that dissociatively adsorbed, flat-lying intermediates are involved in ring opening.[496] Important contributions with respect to the elementary C–H and C–C bond activation reactions over transition-metal surfaces come from the so-called trapping-mediated activation studies. A few studies with C_3–C_8 cycloalkanes over Ru(001)[497–499] and Ir(111)[499,500] surfaces have revealed that activation of these cyclic molecules occurs via initial C–C bond cleavage with surprisingly low activation barriers compared to the corresponding linear alkanes.

A unique information with respect to the use Pd on HZSM-5 in a selective hydrogenolysis has been disclosed.[501] The transformation of methylcyclohexane to n-alkanes with two or more carbon atoms is a useful transformation since these products are desirable components in synthetic steam-cracker feedstock. It was shown that these compounds are not obtained on catalysts with high (0.2 or 1%) Pd loading or without Pd. But on Pd on H-ZSM-5, with Pd content in the range of 10–100 ppm, the desired products are formed with high (~78%) selectivity.

Catalytic hydrogenolysis of light alkanes (propane, butanes, pentanes) with the exception of ethane has been accomplished under very mild conditions over silica-supported hydride complexes.[502] The hydrogenolysis proceeds over $(\equiv SiO)_3$ ZrH,[503] $(\equiv SiO)_3HfH$,[504] and $(\equiv SiO)_3TiH$[505] by stepwise cleavage of carbon–carbon bonds by β-alkyl elimination from surface metal–alkyl intermediates.

Superacid-induced hydrocracking of heavy oils and bitumes, including cleavage and ionic hydrogenation, was developed using preferentially the HF–BF_3 system.[506] The superacid system is not sensitive to many of the impurities present in these heavy oils. Similarly the upgrading of Alberta heavy oils derived from tar sands was effectively carried out using similar HF–BF_3 superacid systems.[507] These superacidic hydrogenolytic transformations involving ionic cleavage–hydrogenation are similar to the process developed by Olah and coworkers,[508] the HF–BF_3-catalyzed depolymerization–ionic hydroliquefaction of coals under mild conditions, and Olah's superacidic upgrading of natural-gas liquids to gasoline.[509]

REFERENCES

1. P. Sabatier and J. B. Senderens, *Compt. Rend.* **124**, 616, 1358 (1897); **131**, 267 (1900).

2. J. F. Young, J. A. Osborn, F. H. Jardine, and G. Wilkinson, *J. Chem. Soc., Chem. Commun.* 131 (1965).

3. M. A. Bennett and P. A. Longstaff, *Chem. Ind.* (London) 846 (1965).

4. R. S. Coffey, Brit. Patent 1,121,642 (1965).

5. R. L. Burwell, Jr., *Chem. Rev.* **57**, 895 (1957).

6. G. C. Bond and P. B. Wells, *Adv. Catal.* **15**, 91 (1964).

7. R. L. Augustine, *Catalytic Hydrogenation: Techniques and Applications in Organic Syntheses*, Marcel Dekker, New York, 1965.

8. R. L. Augustine, *Catal. Rev.-Sci. Eng.* **13**, 285 (1976).

9. S. Siegel, *Adv. Catal.* **16**, 123 (1966).

10. S. Siegel, in *Comprehensive Organic Synthesis*, B. M. Trost and I. Fleming, eds., Vol. 8: *Reduction*, I. Fleming, ed., Pergamon Press, Oxford, 1991, Chapter 3.1, p. 417.

11. P. N. Rylander, *Catalytic Hydrogenation over Platinum Metals*, Academic Press, New York, 1967.

12. P. N. Rylander, *Catalytic Hydrogenation in Organic Syntheses*, Academic Press, New York, 1979.

13. P. Rylander, *Hydrogenation Methods*, Academic Press, London, 1985.

14. R. J. Kokes and A. L. Dent, *Adv. Catal.* **22**, 1 (1972).

15. J. K. A. Clarke and J. J. Rooney, *Adv. Catal.* **25**, 125 (1976).

16. A. P. G. Kieboom and F. van Rantwijk, *Hydrogenation and Hydrogenolysis in Synthetic Organic Chemistry*, Delft Univ. Press, Delft, 1977.

17. M. Freifelder, *Catalytic Hydrogenation in Organic Synthesis. Procedures and Commentary*, Wiley, New York, 1978.

18. H. Pines, *The Chemistry of Catalytic Hydrocarbon Conversions*, Academic Press, New York, 1981, Chapter 3, p. 156.

19. M. Bartók, J. Czombos, K. Felföldi, L. Gera, Gy. Göndös, Á. Molnár, F. Notheisz, I. Pálinkó, G. Wittmann, and Á. G. Zsigmond, *Stereochemistry of Heterogeneous Metal Catalysis*, Wiley, Chichester, 1985.

20. K. Tanaka, *Adv. Catal.* **33**, 99 (1985).

21. R. E. Harmon and S. K. Gupta, D. J. Brown, *Chem. Rev.* **73**, 21 (1973).

22. B. R. James, *Adv. Organomet. Chem.* **17**, 319 (1979).

23. B. R. James, *Homogeneous Hydrogenation*, Wiley, New York, 1973.

24. B. R. James, in *Comprehensive Organometallic Chemistry*, G. Wilkinson, F. G. A. Stone, and E. W. Abel, eds., Pergamon Press, Oxford, 1982, Vol. 8, Chapter 51, p. 285.

25. F. J. McQuillin, *Homogeneous Hydrogenation in Organic Chemistry*, Reidel, Dordrecht, 1976.

26. A. J. Birch and D. H. Williamson, *Org. React.* (NY) **24**, 1 (1976).

27. A. Nakamura and M. Tsutsumi, *Principles and Applications of Homogeneous Catalysis*, Wiley, New York, 1980, p. 125.

28. C. Masters, *Homogeneous Transition-Metal Catalysis: a Gentle Art*, Chapman & Hall, London, 1981, p. 40.

29. R. S. Dickson, *Homogeneous Catalysis with Compounds of Rhodium and Iridium*, Reidel, Dordrecht, 1985, Chapter 3, p. 40.

30. P. A. Chaloner, *Handbook of Coordination Catalysis in Organic Chemistry*, Butterworths, London, 1986, Chapter 2, p. 9.

31. A. Spencer, in *Comprehensive Coordination Chemistry*, Vol. 6: *Applications*, G. Wilkinson, R. G. Gillard, and J. A. McCleverty, eds., Pergamon Press, Oxford, 1987, Chapter 61.2, p. 229.

32. H. Takaya and R. Noyori, in *Comprehensive Organic Synthesis*, B. M. Trost and I. Fleming, eds., Vol. 8: *Reduction*, I. Fleming, ed., Pergamon Press, Oxford, 1991, Chapter 3.2, p. 443.

33. G. W. Parshall and S. D. Ittel, *Homogeneous Catalysis*, Wiley-Interscience, New York, 1992, Chapter 3, p. 25.

34. D. J. Pasto, in *Comprehensive Organic Synthesis*, B. M. Trost and I. Fleming, eds., Vol. 8: *Reduction*, I. Fleming, ed., Pergamon Press, 1991, Chapter 3.3, p. 471.

35. P. W. Rabideau, *Tetrahedron* **45**, 1579 (1989).

36. P. W. Rabideau and Z. Marcinow, *Org. React.* (NY) **42**, 1 (1992).

37. L. N. Mander, in *Comprehensive Organic Synthesis*, B. M. Trost and I. Fleming, eds., Vol. 8: *Reduction*, I. Fleming, ed., Pergamon Press, 1991, Chapter 3.4, p. 489.

38. D. N. Kursanov and Z. N. Parnes, *Russ. Chem. Rev.* (Engl. transl.) **38**, 812 (1969).

39. D. N. Kursanov and Z. N. Parnes, N. M. Loim, *Synthesis* 633 (1974).

40. D. N. Kursanov, Z. N. Parnes, M. I. Kalinkin, and N. M. Loim, *Ionic Hydrogenation and Related Reactions*, Harwood, Chur, 1985.

41. W. A. Lazier and H. R. Arnold, *Org. Synth., Coll. Vol.* **2**, 142 (1943).

42. J. Oudar, *Catal. Rev.-Sci. Eng.* **22**, 171 (1980).

43. C. H. Bartholomew, P. K. Agrawal, and J. R. Katzer, *Adv. Catal.* **31**, 135 (1982).

44. J. Barbier, E. Lamy-Pitara, P. Marecot, J. P. Boitiaux, J. Cosyns, and F. Verna, *Adv. Catal.* **37**, 279 (1990).

45. W. J. Kirkpatrick, *Adv. Catal.* **3**, 329 (1951).

46. P. C. H. Mitchell, in *Catalysis, A Specialist Periodical Report*, Vol. 1, C. Kemball, senior reporter, The Chemical Society, Burlington House, London, 1977, Chapter 6, p. 204; Vol. 4, C. Kemball and D. A. Dowden, senior reporters, The Royal Society of Chemistry, Burlington House, London, 1981, Chapter 7, p. 175.

47. R. Adams, V. Voorhees, and R. L. Shriner, *Org. Synth., Coll. Vol.* **1**, 463 (1941).

48. H. R. Billica and H. Adkins, *Org. Synth., Coll. Vol.* **3**, 176 (1955).

49. S. V. Lebedev, G. G. Kobliansky, and A. O. Yakubchik, *J. Chem. Soc.* 417 (1925).

50. J. Horiuti, G. Ogden, and M. Polanyi, *Trans. Faraday Soc.* **30**, 663 (1934).

51. J. Horiuti and M. Polanyi, *Trans. Faraday Soc.* **30**, 1164 (1934).

52. M. Bartók and Á. Molnár, in *Stereochemistry of Heterogeneous Metal Catalysis*, Wiley, Chichester, 1985, Chapter 3, p. 53.

53. S. Nishimura, H. Sakamoto, and T. Ozawa, *Chem. Lett.* 855 (1973).

54. J. J. Rooney and G. Webb, *J. Catal.* **3**, 488 (1964).

55. P. B. Wells and G. R. Wilson, *Disc. Faraday Soc.* **41**, 237 (1966).

56. A. Farkas, L. Farkas, and E. K. Rideal, *Proc. Roy. Soc. London, Ser. A* **146**, 630 (1934).

57. S. J. Thomson and G. Webb, *J. Chem. Soc., Chem. Commun.* 526 (1976).

58. F. Zaera, A. J. Gellman, and G. A. Somorjai, *Acc. Chem. Res.* **19**, 24 (1986).

59. G. A. Somorjai, *Catal. Lett.* **12**, 17 (1992).

60. F. von Wessely and H. Welleba, *Chem. Ber.* **74**, 777 (1941).

61. S. Siegel and G. V. Smith, *J. Am. Chem. Soc.* **82**, 6082, 6087 (1960).

62. S. Siegel and B. Dmuchovsky, *J. Am. Chem. Soc.* **86**, 2192 (1964).

63. S. Mitsui, S. Imaizumi, A. Nanbu, and Y. Senda, *J. Catal.* **36**, 333 (1975).

64. R. L. Augustine and F. Yaghmaie, *J. Org. Chem.* **52**, 1862 (1987).

65. G. V. Smith and R. L. Burwell, Jr., *J. Am. Chem. Soc.* **84**, 925 (1962).

66. A. W. Weitkamp, *J. Catal.* **6**, 431 (1966).

67. A. W. Weitkamp, *Adv. Catal.* **18**, 1 (1968).

68. E. F. Meyer and R. L. Burwell, Jr., *J. Am. Chem. Soc.* **85**, 2881 (1963).

69. J. Grant, R. B. Moyes, R. G. Oliver, and P. B. Wells, *J. Catal.* **42**, 213 (1976).

70. L. Crombie, P. A. Jenkins, and D. A. Mitchard, *J. Chem. Soc., Perkin Trans. 1*, 1081 (1975).

71. J. J. Phillipson, P. B. Wells, and G. R. Wilson, *J. Chem. Soc. A* 1351 (1969).

72. A. J. Bates, Z. K. Leszczynski, J. J. Phillipson, P. B. Wells, and G. R. Wilson, *J. Chem. Soc. A* 2435 (1970).

73. D. A. Buchanan and G. Webb, *J. Chem. Soc., Faraday Trans. 1* **71**, 134 (1975).

74. S. David, G. Dupont, and C. Paquot, *Bull. Soc. Chim. Fr.* **11**, 561 (1944).

75. J. Chaouki, C. Chavarie, D. Klvana, and G. M. Pajonk, *Appl. Catal.* **21**, 187 (1986).

76. E. A. Braude and R. P. Linstead, *J. Chem. Soc.* 3544 (1954).

77. G. Brieger and T. J. Nestrick, *Chem. Rev.* **74**, 567 (1974).

78. R. A. W. Johnstone, A. H. Wilby, and I. D. Entwistle, *Chem. Rev.* **85**, 129 (1985).

79. H. Hintze and A. Heesing, *Chem. Ber.* **121**, 1133 (1988).

80. G. A. Olah and G. K. S. Prakash, *Synthesis* 397 (1978).

81. M. Bartók and J. Czombos, in *Stereochemistry of Heterogeneous Metal Catalysis*, Wiley, Chichester, 1985, Chapter 4, p. 211.

82. P. Sabatier and J. B. Senderens, *Compt. Rend.* **128**, 1173 (1899); **131**, 40 (1900).

83. C. Kelber and A. Schwarz, *Chem. Ber.* **45**, 1946 (1912).

84. J. M. Moses, A. H. Weiss, K. Matusek, and L. Guczi, *J. Catal.* **86**, 417 (1984).

85. A. S. Al-Ammar and G. Webb, *J. Chem. Soc., Faraday Trans. 1* **74**, 195, 657 (1978).

86. C. J. O'Boyle, *Ind. Eng. Chem.* **42**, 1705 (1950).

87. F. Asinger, *Mono-Olefins, Chemistry and Technology*, Pergamon Press, Oxford, 1968, Chapter 2, p. 175.

88. G. C. Bond and P. B. Wells, *J. Catal.* **5**, 65 (1966).

89. E. N. Marvell and T. Li, *Synthesis* 457 (1973).

90. C. A. Brown and V. K. Ahuja, *J. Org. Chem.* **38**, 2226 (1973).

91. H. Lindlar, *Helv. Chim. Acta* **35**, 446 (1952).

92. H. Lindlar and R. Dubuis, *Org. Synth., Coll. Vol.* **5**, 880 (1973).

93. R. A. W. Johnstone and A. H. Wilby, *Tetrahedron* **37**, 3667 (1981).

94. D. V. Nadkarni and J. L. Fry, *J. Chem. Soc., Chem. Commun.* 997 (1993).

95. J. M. Tour, J. P. Cooper, and S. L. Pendalwar, *J. Org. Chem.* **55**, 3452 (1990).

96. J. L. Garnett, *Catal. Rev.-Sci. Eng.* **5**, 229 (1972).

97. R. B. Moyes and P. B. Wells, *Adv. Catal.* **23**, 121 (1973).

98. M. Bartók and J. Czombos, in *Stereochemistry of Heterogeneous Metal Catalysis*, Wiley, Chichester, 1985, Chapter 5, p. 251.

99. B. van de Graaf, H. van Bekkum, and B. M. Wepster, *Recl. Trav. Chim. Pays-Bas* **87**, 777 (1968).

100. J. A. Don and J. J. F. Scholten, *Faraday Discuss. Chem. Soc.* **72**, 145 (1982).

101. S. Niwa, F. Mizukami, M. Kuno, K. Takeshita, H. Nakamura, T. Tsuchiya, K. Shimizu, and J. Imamura, *J. Mol. Catal.* **34**, 247 (1986).

102. P. J. van der Steen and J. J. F. Scholten, *Appl. Catal.* **58**, 281 (1990).

103. H. Nagahara and M. Konishi, Jpn. Patent JP 6281,332 [8781,332] (1987).

104. S. Nishimura, F. Mochizuki, and S. Kobayakawa, *Bull. Chem. Soc. Jpn.* **43**, 1919 (1970).

105. M. J. Andrews and C. N. Pillai, *Indian J. Chem., Sect. B* **16**, 465 (1978).

106. M. Minabe, K. Watanabe, Y. Ayabe, M. Yoshida, and T. Toda, *J. Org. Chem.* **52**, 1745 (1987).

107. T. J. Nieuwstad, P. Klapwijk, and H. van Bekkum, *J. Catal.* **29**, 404 (1973).

108. P. P. Fu, H. M. Lee, and R. G. Harvey, *J. Org. Chem.* **45**, 2797 (1980).

109. J. A. Osborn, F. H. Jardine, J. F. Young, and G. Wilkinson, *J. Chem. Soc. A* 1711 (1966).

110. F. H. Jardine, *Prog. Inorg. Chem.* **28**, 63 (1981).

111. C. O'Connor and G. Wilkinson, *J. Chem. Soc. A* 2665 (1968).

112. F. H. Jardine, *Polyhedron* **1**, 569 (1982).

113. L. Vaska, *Acc. Chem. Res.* **1**, 335 (1968).

114. D. Evans, J. A. Osborn, F. H. Jardine, and G. Wilkinson, *Nature* (London) **208**, 1203 (1965).

115. F. H. Jardine, *Prog. Inorg. Chem.* **31**, 265 (1984).

116. J. C. Bailar, Jr., *Catal. Rev.-Sci. Eng.* **10**, 17 (1974/75).

117. Yu.I. Yermakov, *Catal. Rev.-Sci. Eng.* **13**, 77 (1976).

118. F. R. Hartley and P. N. Vezey, *Adv. Organomet. Chem.* **15**, 189 (1977).

119. C. U. Pittman, Jr., in *Comprehensive Organometallic Chemistry*, Vol. 8, G. Wilkinson, F. G. A. Stone, and E. W. Abel, eds., Pergamon Press, Oxford, 1982, Chapter 55, p. 553.

120. F. R. Hartley, *Supported Metal Complexes: A New Generation of Catalysts*. Dordrecht, Reidel, 1985, Chapter 6, p. 149.

121. T. J. Marks, *Acc. Chem. Res.* **25**, 57 (1992).

122. R. D. Gillespie, R. L. Burwell, Jr., and T. J. Marks, *Langmuir* **6**, 1465 (1990).

123. M. S. Eisen and T. J. Marks, *J. Am. Chem. Soc.* **114**, 10358 (1992).

124. J. Halpern, *Acc. Chem. Res.* **3**, 386 (1970).

125. J. Halpern, *J. Organomet. Chem.* **200**, 133 (1980).

126. C. A. Tolman, P. Z. Meakin, D. L. Lindner, and J. P. Jesson, *J. Am. Chem. Soc.* **96**, 2762 (1974).

127. Y. Ohtani, M. Fujimoto, and A. Yamagishi, *Bull. Chem. Soc. Jpn.* **50**, 1453 (1977).

128. J. Halpern, T. Okamoto, and A. Zakhariev, *J. Mol. Catal.* **2**, 65 (1977).

129. J. Halpern, *Inorg. Chim. Acta* **50**, 11 (1981).

130. C. Daniel, N. Koga, J. Han, X. Y. Fu, and K. Morokuma, *J. Am. Chem. Soc.* **110**, 3773 (1988).

131. K.-H. Dahmen, D. Hedden, R. L. Burwell, Jr., and T. J. Marks, *Langmuir* **4**, 1212 (1990).

132. Y. Senda, S. Mitsui, and H. Sugiyama, S. Seto, *Bull. Chem. Soc. Jpn.* **45**, 3498 (1972).

133. J. J. Sims, V. K. Honwad, and L. H. Selman, *Tetrahedron Lett.* 87 (1969).

134. J.-T. Lee and H. Alper, *J. Org. Chem.* **55**, 1854 (1990).

135. T. Funabiki and K. Tarama, *Bull. Chem. Soc. Jpn.* **44**, 945 (1971).

136. T. Funabiki, M. Mohri, and K. Tarama, *J. Chem. Soc., Dalton Trans.* 1813 (1973).

137. E. W. Stern and P. K. Maples, *J. Catal.* **27**, 120 (1972).

138. Y. Okamoto, A. Maezawa, H. Kane, and T. Imanaka, *J. Chem. Soc., Chem. Commun.* 380 (1988).

139. M. A. Schroeder and M. S. Wrighton, *J. Organomet. Chem.* **74**, C29 (1974).

140. E. N. Frankel and R. O. Butterfield, *J. Org. Chem.* **34**, 3930 (1969).

141. B. M. Choudary, G. V. M. Sharma, and P. Bharathi, *Angew. Chem., Int. Ed. Engl.* **28**, 465 (1989).

142. G. A. Molander and J. O. Hoberg, *J. Org. Chem.* **57**, 3266 (1992).

143. D. R. Fahey, *J. Org. Chem.* **38**, 80 (1973).

144. H. B. Kagan, in *Comprehensive Organometallic Chemistry*, Vol. 8, G. Wilkinson, F. G. A. Stone, and E. W. Abel, eds., Pergamon Press, Oxford, 1982, Chapter 53, p. 470.

145. H. Takaya and R. Noyori, in *Comprehensive Organic Synthesis*, B. M. Trost and I. Fleming, eds., Vol. 8: *Reduction*, I. Fleming, ed., Pergamon Press, Oxford, 1991, Chapter 3.2, p. 463.

146. W. Dumont, J.-C. Poulin, T.-P. Dang, and H. B. Kagan, *J. Am. Chem. Soc.* **95**, 8295 (1973).

147. M. Tanaka and I. Ogata, *J. Chem. Soc., Chem. Commun.* 735 (1975).

148. T. Hayashi, M. Tanaka, and I. Ogata, *Tetrahedron Lett.* 295 (1977).

149. O. Samuel, R. Couffignal, M. Lauer, S. Y. Zhang, and H. B. Kagan, *New J. Chem.* **5**, 15 (1981).

150. Y. Kawabata, M. Tanaka, and I. Ogata, *Chem. Lett.* 1213 (1976).

151. E. Cesarotti, R. Ugo, and H. B. Kagan, *Angew. Chem., Int. Ed. Engl.* **18**, 779 (1979).

152. L. A. Paquette, J. A. McKinney, M. L. McLaughlin, and A. L. Rheingold, *Tetrahedron Lett.* **27**, 5599 (1986).

153. R. L. Halterman, K. P. C. Vollhardt, M. E. Welker, D. Bläser, and R. Boese, *J. Am. Chem. Soc.* **109**, 8105 (1987).

154. R. L. Halterman and K. P. C. Vollhardt, *Organometallics* **7**, 883 (1988).

155. J. C. Gallucci, B. Gautheron, M. Gugelchuk, P. Meunier, and L. A. Paquette, *Organometallics* **6**, 15 (1987).

156. R. Waymouth and P. Pino, *J. Am. Chem. Soc.* **112**, 4911 (1990).

157. R. B. Grossman, R. A. Doyle, and S. L. Buchwald, *Organometallics* **10**, 1501 (1991).

158. V. P. Conticello, L. Brard, M. A. Giardello, Y. Tsuji, M. Sabat, C. L. Stern, and T. J. Marks, *J. Am. Chem. Soc.* **114**, 2761 (1992).

159. R. D. Broene and S. L. Buchwald, *J. Am. Chem. Soc.* **115**, 12569 (1993).

160. E. L. Muetterties, *Inorg. Chim. Acta* **50**, 1 (1981).

161. R. A. Sánchez-Delgado, A. Andriollo, E. González, N. Valencia, V. León, and J. Espidel, *J. Chem. Soc., Dalton Trans.* 1859 (1985).

162. R. H. Crabtree, *J. Chem. Soc., Chem. Commun.* 647 (1975).

163. M. Sodeoka and M. Shibasaki, *J. Org. Chem.* **50**, 1147, 3246 (1985).

164. R. R. Schrock and J. A. Osborn, *J. Am. Chem. Soc.* **98**, 2143 (1976).

165. M. O. Albers, E. Singleton, and M. M. Viney, *J. Mol. Catal.* **30**, 213 (1985).

166. B. S. Nkosi and N. J. Coville, *J. Mol. Catal.* **39**, 313 (1987).

167. G. V. M. Sharma, B. M. Choudary, M. R. Sarma, and K. K. Rao, *J. Org. Chem.* **54**, 2997 (1989).

168. R. R. Burch, E. L. Muetterties, R. G. Teller, and J. M. Williams, *J. Am. Chem. Soc.* **104**, 4257 (1982).

169. T. Yoshida, W. J. Youngs, T. Sakaeda, T. Ueda, S. Otsuka, and J. A. Ibers, *J. Am. Chem. Soc.* **105**, 6273 (1983).

170. E. Bayer and W. Schumann, *J. Chem. Soc., Chem. Commun.* 949 (1986).

171. W. D. Harman and H. Taube, *J. Am. Chem. Soc.* **110**, 7906 (1988).

172. S. J. Lapporte and W. R. Schuett, *J. Org. Chem.* **28**, 1947 (1963).

173. V. G. Lipovich, F. K. Shmidt, and I. V. Kalechits, *Kinet. Catal.* (Engl. transl.) **8**, 812 (1967).

174. S. Friedman, S. Metlin, A. Svedi, and I. Wender, *J. Org. Chem.* **24**, 1287 (1959).

175. H. M. Feder and J. Halpern, *J. Am. Chem. Soc.* **97**, 7186 (1975).

176. L. S. Stuhl, M. Rakowski DuBois, F. J. Hirsekorn, J. R. Bleeke, A. E. Stevens, and E. L. Muetterties, *J. Am. Chem. Soc.* **100**, 2405 (1978).

177. E. L. Muetterties and J. R. Bleeke, *Acc. Chem. Res.* **12**, 324 (1979).

178. J. W. Johnson and E. L. Muetterties, *J. Am. Chem. Soc.* **99**, 7395 (1977).

179. M. J. Russell, C. White, and P. M. Maitlis, *J. Chem. Soc., Chem. Commun.* 427 (1977).

180. K. R. Januszkiewicz and H. Alper, *Organometallics* **2**, 1055 (1983).

181. I. Amer, H. Amer, and J. Blum, *J. Mol. Catal.* **34**, 221 (1986).

182. I. Amer, H. Amer, R. Ascher, and J. Blum, *J. Mol. Catal.* **39**, 185 (1987).

183. F. Aylward and M. Sawistowska, *Chem. Ind.* (London) 484 (1962).

184. S. Hünig, H. R. Müller, and W. Thier, *Angew. Chem., Int. Ed. Engl.* **4**, 271 (1965).

185. E. J. Corey, W. L. Mock, and D. J. Pasto, *Tetrahedron Lett.* 347 (1961).

186. E. E. van Tamelen, R. S. Dewey, and R. J. Timmons, *J. Am. Chem. Soc.* **83**, 3725 (1961).

187. E. W. Garbisch, Jr., S. M. Schildcrout, D. B. Patterson, and C. M. Sprecher, *J. Am. Chem. Soc.* **87**, 2932 (1965).

188. E. J. Corey, D. J. Pasto, and W. L. Mock, *J. Am. Chem. Soc.* **83**, 2957 (1961).

189. R. Srinivasan and J. N. C. Hsu, *J. Chem. Soc., Chem. Commun.* 1213 (1972).

190. E. E. van Tamelen and R. J. Timmons, *J. Am. Chem. Soc.* **84**, 1067 (1962).

191. E. C. Ashby and J. J. Lin, *J. Org. Chem.* **43**, 2567 (1978).

192. S.-K. Chung, *J. Org. Chem.* **44**, 1014 (1979).

193. G. M. Whitesides and W. J. Ehmann, *J. Org. Chem.* **35**, 3565 (1970).

194. G. A. Olah, G. K. S. Prakash, M. Arvanaghi, and M. R. Bruce, *Angew. Chem., Int. Ed. Engl.* **20**, 92 (1981).

195. K. N. Campbell and L. T. Eby, *J. Am. Chem. Soc.* **63**, 216 (1941).

196. A. L. Henne and K. W. Greenlee, *J. Am. Chem. Soc.* **65**, 2020 (1943).

197. H. O. House and E. F. Kinloch, *J. Org. Chem.* **39**, 747 (1974).

198. C. E. Castro and R. D. Stephens, *J. Am. Chem. Soc.* **86**, 4358 (1964).

199. B. L. Sondengam, G. Charles, and T. M. Akam, *Tetrahedron Lett.* **21**, 1069 (1980).

200. E. C. Ashby, J. J. Lin, and A. B. Goel, *J. Org. Chem.* **43**, 757 (1978).

201. C. B. Wooster, and K. L. Godfrey, *J. Am. Chem. Soc.* **59**, 596 (1937).

202. A. J. Birch, *Quart. Rev.* (London) **4**, 69 (1950).

203. A. J. Birch and H. Smith, *Quart. Rev.* (London) **12**, 17 (1958).

204. E. M. Kaiser, *Synthesis* 391 (1972).

205. H. Smith, *Chemistry in Anhydrous Liquid Ammonia*, Vol. 1, Part 2: *Organic Reactions in Liquid Ammonia*, Vieweg, Braunschweig, Wiley-Interscience, New York, 1963.

206. M. Smith, in *Reduction; Techniques and Applications in Organic Synthesis*, R. L. Augustine, ed., Marcel Dekker, New York, 1968, Chapter 2, p. 95.

207. R. G. Harvey, *Synthesis* 161 (1970).

208. R. A. Benkeser, R. E. Robinson, D. M. Sauve, and O. H. Thomas, *J. Am. Chem. Soc.* **77**, 3230 (1955).

209. P. W. Rabideau and E. G. Burkholder, *J. Org. Chem.* **43**, 4283 (1978).

210. A. J. Birch, A. R. Murray, and H. Smith, *J. Chem. Soc.* 1945 (1951).

211. R. A. Benkeser and E. M. Kaiser, *J. Org. Chem.* **29**, 955 (1964).

212. J. D. Brooks, R. A. Durie, and H. Silberman, *Aust. J. Chem.* **17**, 55 (1964).

213. D. Degner, *Top. Curr. Chem.* **148**, 1 (1988).

214. E. Kariv-Miller, R. I. Pacut, and G. K. Lehman, *Top. Curr. Chem.* **148**, 97 (1988).

215. L. A. Avaca and A. Bewick, *J. Chem. Soc., Perkin Trans. 2* 1709 (1972).

216. M. I. Kalinkin, G. D. Kolomnikova, Z. N. Parnes, and D. N. Kursanov, *Russ. Chem. Rev.* (Engl. transl.) **48**, 332 (1979).

217. D. N. Kursanov, Z. N. Parnes, G. I. Bassova, N. M. Loim, and V. I. Zdanovich, *Tetrahedron* **23**, 2235 (1967).

218. Z. N. Parnes, E. Yu. Beilinson, and D. N. Kursanov, *J. Org. Chem. USSR* (Engl. transl.) **6**, 2579 (1970).

219. Z. N. Parnes, G. A. Khotimskaya, Yu.I. Lyakhovetskii, and P. V. Petrovskii, *Bull. Acad. Sci. USSR, Div. Chem. Sci.* (Engl. transl.) **20**, 1463 (1971).

220. P. A. Krasutskii, N. V. Ambrosienko, V. N. Rodionov, A. G. Yurchenko, Z. N. Parnes, and G. I. Bolestova, *J. Org. Chem. USSR* (Engl. transl.) **21**, 1333 (1985).

221. V. A. Tsyryapkin, N. M. Loim, Z. N. Parnes, and D. N. Kursanov, *J. Org.Chem. USSR* (Engl. transl.) **8**, 2390 (1972).

222. D. N. Kursanov, Z. N. Parnes, and G. I. Bolestova, *Dokl. Chem.* (Engl. transl.) **181**, 726 (1968).

223. R. M. Carlson and R. K. Hill, *J. Org. Chem.* **34**, 4178 (1969).

224. G. D. Kolomnikova, M. I. Kalinkin, Z. N. Parnes, and D. N. Kursanov, *Dokl. Chem.* (Engl. transl.) **265**, 216 (1982).

225. D. N. Kursanov, G. D. Kolomnikova, S. A. Goloshchapova, M. I. Kalinkin, and Z. N. Parnes, *Bull. Acad. Sci. USSR, Div. Chem. Sci.* (Engl. transl.) **33**, 446 (1984).

226. G. D. Kolomnikova, M. I. Kalinkin, Z. N. Parnes, and D. N. Kursanov, *Bull. Acad. Sci. USSR, Div. Chem. Sci.* (Engl. transl.) **30**, 921 (1981).

227. S. F. Nelsen and M. F. Teasley, *J. Org. Chem.* **51**, 3474 (1986).

228. Z. N. Parnes, G. I. Bolestova, and D. N. Kursanov, *J. Org. Chem. USSR* (Engl. transl.) **13**, 434 (1977).

229. G. A. Olah, Q. Wang, and G. K. S. Prakash, *Synlett* 647 (1992).

230. D. N. Kursanov, Z. N. Parnes, and G. I. Bolestova, *Bull. Acad. Sci. USSR, Div. Chem. Sci.* (Engl. transl.) **17**, 1107 (1968).

231. G. I. Bolestova, F. M. Latypova, Z. N. Parnes, and D. N. Kursanov, *Bull. Acad. Sci. USSR, Div. Chem. Sci.* (Engl. transl.) **31**, 1179 (1982).

232. V. N. Ipatieff, H. Pines, and R. C. Olberg, *J. Am. Chem. Soc.* **70**, 2123 (1948).

233. H. Pines, D. R. Strehlau, and V. N. Ipatieff, *J. Am. Chem. Soc.* **72**, 1563, 5521 (1950).

234. S. M. Markosyan, M. I. Kalinkin, Z. N. Parnes, and D. N. Kursanov, *Dokl. Chem.* (Engl. transl.) **255**, 533 (1980).

235. M. I. Kalinkin, Z. N. Parnes, S. M. Markosyan, L. I. Ovsyannikova, and D. N. Kursanov, *Dokl. Chem.* (Engl. transl.) **254**, 439 (1980).

236. M. I. Kalinkin, Z. N. Parnes, D.Kh. Shaapini, N. P. Shevlyakova, and D. N. Kursanov, *Dokl. Chem.* (Engl. transl.) **229**, 492 (1976).

237. G. D. Kolomnikova, M. I. Kalinkin, Z. N. Parnes, and D. N. Kursanov, *Bull. Acad. Sci. USSR, Div. Chem. Sci.* (Engl. transl.) **28**, 738 (1979).

238. M. P. Doyle and C. C. McOsker, *J. Org. Chem.* **43**, 693 (1978).

239. A. D. Kini, D. V. Nadkarni, and J. L. Fry, *Tetrahedron Lett.* **35**, 1507 (1994).

240. J. W. Larsen and L. W. Chang, *J. Org. Chem.* **44**, 1168 (1979).

241. J. W. Larsen and L. W. Chang, *J. Org. Chem.* **43**, 3602 (1978).

242. J. C. Cheng, J. Maioriello, and J. W. Larsen, *Energy Fuels* **3**, 321 (1989).

243. M. Siskin, *J. Am. Chem. Soc.* **96**, 3641 (1974).

244. J. Wristers, *J. Am. Chem. Soc.* **97**, 4312 (1975).

245. G. A. Olah, M. R. Bruce, E. H. Edelson, and A. Husain, *Fuel* **63**, 1130 (1984).

246. G. A. Olah and A. Husain, *Fuel* **63**, 1427 (1984).

247. J. E. Germain, *Catalytic Conversion of Hydrocarbons*, Academic Press, London, 1969, Chapter 3, p. 118.

248. J. H. Sinfelt, *Catal. Rev.-Sci. Eng.* **3**, 175 (1969); **9**, 147 (1974).

249. J. H. Sinfelt, *Adv. Catal.* **23**, 91 (1973).

250. J. H. Sinfelt, in *Catalysis, Science and Technology*, Vol. 1, J. R. Anderson, M. Boudart, eds., Springer, Berlin, 1982, Chapter 5, p. 257.

251. J. R. Anderson, *Adv. Catal.* **23**, 1 (1973).

252. F. G. Gault, *Adv. Catal.* **30**, 1 (1981).

253. Z. Paál and P. Tétényi, in *Catalysis, A Specialist Periodical Report*, Vol. 5, G. C. Bond and G. Webb, senior reporters, The Royal Society of Chemistry, Burlington House, London, 1982, Chapter 3, p. 106.

254. M. Bartók and I. Pálinkó, in *Stereochemistry of Heterogeneous Metal Catalysis*, Wiley, Chichester, 1985, Chapter 1, p. 1.

255. G. Leclercq, L. Leclercq, and R. Maurel, *J. Catal.* **50**, 87 (1977).

256. Z. Paál and P. Tétényi, *Nature* (London) **267**, 234 (1977).

257. P. Tétényi, L. Guczi, and A. Sárkány, *Acta Chim. Acad. Sci. Hung.* **97**, 221 (1978).

258. L. Babernics, L. Guczi, K. Matusek, A. Sárkány, and P. Tétényi, *Proc. 6th Int. Congr. Catalysis*, London, 1976, G. C. Bond, P. B. Wells, and F. C. Tompkins, eds., The Chemical Society, Burlington House, London, 1977, p. 456.

259. A. Sárkány and P. Tétényi, *React. Kinet. Catal. Lett.* **9**, 315 (1978).

260. H. Matsumoto, Y. Saito, and Y. Yoneda, *J. Catal.* **19**, 101 (1970).

261. J. L. Carter, J. A. Cusumano, and J. H. Sinfelt, *J. Catal.* **20**, 223 (1971).

262. R. Montarnal and G. Martino, *Rev. Inst. Fr. Petrol.* **32**, 367 (1977).

263. R. Maurel and G. Leclercq, *Bull Soc. Chim. Fr.* 1234 (1971).

264. J. R. Anderson and B. G. Baker, *Proc. Roy. Soc. London, Ser. A* **271**, 402 (1963).

265. K. Foger and J. R. Anderson, *J. Catal.* **54**, 318 (1978).

266. K. Foger and J. R. Anderson, *J. Catal.* **59**, 325 (1979).

267. K. Foger and J. R. Anderson, *J. Catal.* **64**, 448 (1980).

268. F. Garin and F. G. Gault, *J. Am. Chem. Soc.* **97**, 4466 (1975).

269. F. G. Gault, *Gazz. Chim. Ital.* **109**, 255 (1979).

270. J. J. Rooney, *J. Catal.* **58**, 334 (1979).

271. B. Coq and F. Figueras, *J. Mol. Catal.* **40**, 93 (1987).

272. G. Leclercq, S. Pietrzyk, M. Peyrovi, and M. Karroua, *J. Catal.* **99**, 1 (1986).

273. G. A. Martin, *J. Catal.* **60**, 345 (1979).

274. J. A. Dalmon and G. A. Martin, *J. Catal.* **66**, 214 (1980).

275. V. Ponec, *Adv. Catal.* **32**, 149 (1983).

276. W. M. H. Sachtler and R. A. van Santen, *Adv. Catal.* **26**, 69 (1967).

277. F. Garin, L. Hilaire, and G. Maire, in *Catalytic Hydrogenation*, Studies in Surface Science and Catalysis, Vol. 27, L. Cerveny, ed., Elsevier, Amsterdam, 1986, Chapter 5, p. 145.

278. H. C. de Jongste and V. Ponec, *J. Catal.* **63**, 389 (1980).

279. M. Che and C. O. Bennett, *Adv. Catal.* **36**, 55 (1989).

280. J. Barbier and P. Marecot, *New J. Chem.* **5**, 393 (1981).

281. J. L. Carter, J. A. Cusumano, and J. H. Sinfelt, *J. Phys. Chem.* **70**, 2257 (1966).

282. G. A. Martin, *J. Catal.* **60**, 452 (1979).

283. D. J. C. Yates and J. H. Sinfelt, *J. Catal.* **8**, 348 (1967).

284. D. W. Goodman, *Surf. Sci.* **123**, L679 (1982).

285. G. A. Somorjai, in *Proc. 8th Int. Congr. Catalysis*, West-Berlin, 1984. DECHEMA, Frankfurt am Main, 1984, Vol. 1, p. 113.

286. W. D. Gillespie, R. K. Herz, E. E. Petersen, and G. A. Somorjai, *J. Catal.* **70**, 147 (1981).

287. J. R. Engstrom, D. W. Goodman, and W. H. Weinberg, *J. Am. Chem. Soc.* **108**, 4653 (1986).

288. S. M. Davis, F. Zaera, and G. A. Somorjai, *J. Catal.* **85**, 206 (1984).

289. J. Newham, *Chem. Rev.* **63**, 123 (1963).

290. O. V. Bragin and S. A. Krasavin, *Russ. Chem. Rev.* (Engl. transl.) **52**, 625 (1983).

291. E. Reimann, in *Methoden der Organischen Chemie* (*Houben-Weyl*), Vol. 4/1c: *Reduktion*, Thieme, Stuttgart, 1980, p. 404.

292. M. Bartók and Á. G. Zsigmond, in *Stereochemistry of Heterogeneous Metal Catalysis*, Wiley, Chichester, 1985, Chapter 2, p. 17.

293. T. Chevreau and F. G. Gault, *J. Catal.* **50**, 124 (1977).

294. J. C. Schlatter and M. Boudart, *J. Catal.* **25**, 93 (1972).

295. I. Pálinkó, F. Notheisz, and M. Bartók, *Catal. Lett.* **1**, 127 (1988); *J. Mol. Catal.* **63**, 43 (1990); **68**, 237 (1991).

296. F. Notheisz, I. Pálinkó, and M. Bartók, *Catal. Lett.* **5**, 229 (1990).

297. G. Maire, G. Plouidy, J. C. Prudhomme, and F. G. Gault, *J. Catal.* **4**, 556 (1965).

298. G. Maire and F. G. Gault, *Bull. Soc. Chim. Fr.* 894 (1967).

299. B. Török, I. Pálinkó, Á. Molnár, and M. Bartók, *J. Catal.* **143**, 111 (1993).

300. V. Amir-Ebrahimi, A. Choplin, P. Parayre, and F. G. Gault, *New J. Chem.* **4**, 431 (1980).

301. Y. Barron, G. Maire, J. M. Muller, and F. G. Gault, *J. Catal.* **5**, 428 (1966).

302. G. A. Somorjai and D. W. Blakely, *Nature* (London) **258**, 580 (1975).

303. G. A. Olah, Y. Halpern, J. Shen, and Y. K. Mo, *J. Am. Chem. Soc.* **95**, 4960 (1973).

304. A. F. Bickel, C. J. Gaasbeek, H. Hogeveen, J. M. Oelderik, and C. J. Platteeuw, *J. Chem. Soc., Chem. Commun.* 634 (1967).

305. H. Hogeveen and A. F. Bickel, *J. Chem. Soc., Chem. Commun.* 635 (1967).

306. H. Hogeveen, C. J. Gaasbeek, and A. F. Bickel, *Recl. Trav. Chim. Pays-Bas* **88**, 703 (1969).

307. M. Siskin, *J. Am. Chem. Soc.* **100**, 1838 (1978).

308. M. L. Derrien, in *Catalytic Hydrogenation* Studies in Surface Science and Catalysis, Vol. 27, L. Cerveny, ed., Elsevier, Amsterdam, 1986, Chapter 18, p. 613.

309. D. C. McCulloh, in *Applied Industrial Catalysis*, Vol. 1, B. E. Leach, ed., Academic Press, New York, 1983, Chapter 4, p. 69.

310. J.-P. Boitiaux, J. Cosyns, M. Derrien, and G. Léger, *Hydrocarbon Process., Int. Ed.* **64**(3), 51 (1985).

311. F. Asinger, *Mono-Olefins, Chemistry and Technology*, Pergamon Press, Oxford, 1968, Chapter 3, p. 225.

312. R. E. Reitmeier and H. W. Fleming, *Chem. Eng. Prog.* **54**(12), 48 (1958).

313. H. Mueller, K. M. Deller, G. Vollheim, and W. Kuehn, *Chem.-Ing.-Techn.* **59**, 645 (1987).

314. F. Asinger, *Mono-Olefins, Chemistry and Technology*, Pergamon Press, Oxford, 1968, Chapter 3, p. 228.

315. *Petrol. Refiner* **40**(11), 261 (1961).

316. W. Krönig, *Hydrocarbon Process.* **49**(3), 121 (1970).

317. M. L. Derrien, J. W. Andrews, P. Bonnifay, and J. Leonard, *Chem. Eng. Prog.* **70**(1), 74 (1974).

318. M. Derrien, C. Bonner, J. Cosyns, and G. Leger *Hydrocarbon Process.* **58**(5), 175 (1979).

319. C. L. Rogers, *Oil Gas J.* **69**(45), 60 (1971).

320. A. E. Eleazar, R. M. Heck, and M. P. Witt, *Hydrocarbon Proc., Int. Ed.* **58**(5), 112 (1979).

321. R. M. Heck, G. R. Patel, W. S. Breyer, and D. S. Merril, *Oil Gas J.* **81**(3), 103 (1983).

322. S. Novalany and R. G. McClung, *Hydrocarbon Process., Int. Ed.* **68**(9), 66 (1989).

323. G. Chaput, J. Laurent, J. P. Boitiaux, J. Cosyns, and P. Sarrazin, *Hydrocarbon Process., Int. Ed.* **71**(9), 51 (1992).

324. K. Rock, *Hydrocarbon Process., Int. Ed.* **71**(5), 86 (1992).

325. E. Pescarollo, R. Trotta, and P. R. Sarathy, *Hydrocarbon Process., Int. Ed.* **72**(2), 53 (1993).

326. *Hydrocarbon Process.* **55**(9), 144 (1976).

327. H. J. Boonstra and P. Zwietering, *Chem. Ind.* (London) 2039 (1966).

328. R. L. Marcell and H. M. Sachs, in *Encyclopedia of Chemical Processing and Design*, Vol. 14, J. J. McKetta and W. A. Cunningham, eds., Marcel Dekker, New York, 1982, p. 61.

329. K. Weissermel and H.-J. Arpe, *Industrial Organic Chemistry*, 2nd ed., VCH, Weinheim, 1993, Chapter 13, p. 343.

330. F. A. Dufau, F. Eschard, A. C. Haddad, and C. H. Thonon, *Chem. Eng. Prog.* **60**(9), 43 (1964).

331. B. Cha, J. Cosyns, M. Derrien, and A. Forge, *Oil Gas J.* **72**(23), 64 (1974).

332. *Hydrocarbon Process., Int. Ed.* **68**(11), 99 (1989).

333. S. Field and M. H. Dalson, *Hydrocarbon Process.* **46**(5), 169 (1967).

334. *Hydrocarbon Process.* **52**(11), 116 (1973).

335. *Oil Gas J.* **67**(33), 82 (1969).

336. *Hydrocarbon Process., Int. Ed.* **62**(11), 89 (1983).

337. *Hydrocarbon Process.* **56**(11), 143 (1977).

338. *Hydrocarbon Process.* **58**(11), 150 (1979).

339. *Hydrocarbon Process., Int. Ed.* **60**(11), 148 (1981).

340. Y. Chauvin, J. Gaillard, J. Leonard, P. Bonnifay, and J. W. Andrews, *Hydrocarbon Process., Int. Ed.* **61**(5), 110 (1982).

341. J. C. Bonacci, R. M. Heck, R. K. Mahendroo, G. R. Patel, and E. D. Allan, *Hydrocarbon Process., Int. Ed.* **59**(11), 179 (1980).

342. K. Ito, I. Dogane, and K. Tanaka, *Hydrocarbon Process., Int. Ed.* **64**(10), 83 (1985).

343. N. Harris and M. W. Tuck, *Hydrocarbon Process., Int. Ed.* **69**(5), 79 (1990).

344. *Hydrocarbon Process., Int. Ed.* **72**(3), 170 (1993).

345. Y. Tanabe, *Hydrocarbon Process., Int. Ed.* **60**(9), 187 (1981).

346. H. Arnold, F. Döbert, and J. Gaube, in *Handbook of Heterogeneous Catalysis*, G. Ertl, H. Knözinger, and J. Weitkamp, eds., Wiley-VCH, Weinheim, 1997, Chapter 4.4.1, p. 2165.

347. M. Wills, in *The Chemistry of Functional Groups, Supplement A3: The Chemistry of Double-Bonded Functional Groups*, S. Patai and Z. Rappoport, eds., Wiley, Chichester, 1997, Chapter 15, p. 781.

348. M. Bartók and Á. Molnár, in *The Chemistry of Functional Groups, Supplement A3: The Chemistry of Double-Bonded Functional Groups*, S. Patai and Z. Rappoport, eds., Wiley, Chichester, 1997, Chapter 16, p. 843.

349. R. L. Augustine, *Heterogeneous Catalysis for the Synthetic Chemist*, Marcel Dekker, New York, 1996, Chapter 15.

350. G. V. Smith and F. Notheisz, *Heterogeneous Catalysis in Organic Chemistry*, Academic Press, San Diego, 1999, Chapter 2.

351. M. Xi and B. E. Bent, *J. Vac. Sci. Technol. B* **10**, 2440 (1992).

352. A. V. Teplyakov and B. E. Bent, *J. Chem. Soc., Faraday Trans.* **91**, 3645 (1995).

353. M. X. Yang, A. V. Teplyakov, and B. E. Bent, *J. Phys. Chem. B* **102**, 2985 (1998).

354. S. P. Daley, A. L. Utz, T. R. Trautman, and S. T. Ceyer, *J. Am. Chem. Soc.* **116**, 6001 (1994).

355. K.-A. Son, M. Mavrikakis, and J. L. Gland, *J. Phys. Chem.* **99**, 6270 (1995).

356. G. A. Somorjai, *J. Phys. Chem. B* **104**, 2969 (2000).

357. G. A. Somorjai and K. R. McCrea, *Adv. Catal.* **45**, 385 (2000).

358. P. S. Cremer and G. A. Somorjai, *J. Chem. Soc., Faraday Trans.* **91**, 3671 (1995).

359. P. S. Cremer, X. Su, Y. R. Shen, and G. A. Somorjai, *J. Am. Chem. Soc.* **118**, 2942 (1996).

360. K. R. McCrea and G. A. Somorjai, *J. Mol. Catal. A: Chem.* **163**, 43 (2000).

361. P. S. Cremer, X. Su, Y. R. Shen, and G. A. Somorjai, *J. Phys. Chem.* **100**, 16302 (1996).

362. Á. Mastalir, F. Notheisz, and M. Bartók, *Catal Lett.* **35**, 119 (1995).

363. Á. Mastalir, F. Notheisz, M. Bartók, Z. Király, and I. Dékány, *J. Mol. Catal. A: Chem.* **99**, 115 (1995).

364. Á. Mastalir, Z. Király, I. Dékány, and M. Bartók, *Coll. Surfaces, A* **141**, 397 (1998).

365. Á. Mastalir, J. Walter, F. Notheisz, and M. Bartók, *Langmuir* **17**, 3776 (2001).

366. K. Okitsu, A. Yue, S. Tanabe, and H. Matsumoto, *Chem. Mater.* **12**, 3006 (2000).

367. M. Ichikawa, *Adv. Catal.* **38**, 283 (1992).

368. B. C. Gates, *Chem. Rev.* **95**, 511 (1995).

369. D. S. Shephard, T. Maschmeyer, B. F. G. Johnson, J. M. Thomas, G. Sankar, D. Ozkaya, W. Z. Zhou, R. D. Oldroyd, and R. G. Bell, *Angew. Chem., Int. Ed. Engl.* **36**, 2242 (1997).

370. D. S. Shephard, T. Maschmeyer, G. Sankar, J. M. Thomas, D. Ozkaya, B. F. G. Johnson, R. Raja, R. D. Oldroyd, and R. G. Bell, *Chem. Eur. J.* **4**, 1214 (1998).

371. R. Raja, G. Sankar, S. Hermans, D. S. Shephard, S. Bromely, J. M. Thomas, and B. F. G. Johnson, *Chem. Commun.* 1571 (1999).

372. Y. Yoshinaga, K. Seki, T. Nakato, and T. Okuhara, *Angew. Chem., Int. Ed. Engl.* **36**, 2833 (1997).

373. S. Naito, M. Tanimoto, M. Soma, and Y. Udagawa, in *New Frontiers in Catalysis*, Studies in Surface Science and Catalysis, Vol. 75, L. Guczi, F. Solymosi, and P. Tétényi, eds., Elsevier, Amsterdam, 1993, p. 2043.

374. S. Naito and M. Tanimoto, *J. Catal.* **154**, 306 (1995).

375. G. V. Smith, S. Tjandra, M. Musoiu, T. Wiltowski, F. Notheisz, M. Bartók, I. Hannus, D. Ostgard, and V. Malhotra, *J. Catal.* **161**, 441 (1996).

376. Á. Molnár, A. Sárkány, and M. Varga, *J. Mol. Catal. A: Chem.* **173**, 185 (2001).

377. D. Duca, L. F. Liotta, and G. Deganello, *J. Catal.* **154**, 69 (1995); *Catal. Today* **24**, 15 (1995).

378. S. Asplund, *J. Catal.* **158**, 267 (1996).

379. S. Asplund, C. Fornell, A. Holmgren, and S. Irandoust, *Catal. Today* **24**, 181 (1995).

380. A. Sárkány, Z. Schay, Gy. Stefler, L. Borkó, J. W. Hightower, and L. Guczi, *Appl. Catal. A* **124**, L181 (1995).

381. A. Sarkany, *Appl. Catal. A* **165**, 87 (1997).

382. A. Sarkany, in *12th Int. Congr. Catalysis*, Studies in Surface Science and Catalysis, Vol. 130, A. Corma, F. V. Melo, S. Mendioroz, and J. L. G. Fierro, eds., Elsevier, Amsterdam, 2000, p. 2081.

383. Ph. Maetz and R. Touroude, *Appl. Catal. A* **149**, 189 (1997).

384. S. D. Jackson and N. J. Casey, *J. Chem. Soc., Faraday Trans.* **91**, 3269 (1995).

385. J. Goetz, M. A. Volpe, and R. Touroude, *J. Catal.* **164**, 369 (1996).

386. G. C. Bond and A. F. Rawle, *J. Mol. Catal. A: Chem.* **109**, 261 (1996).

387. L. F. Liotta, A. M. Venezia, A. Martorana, and G. Deganello, *J. Catal.* **171**, 177 (1997).

388. L. Guczi and A. Sárkány, in *Catalysis*, J. J. Spivey and S. K. Agarwal, senior reporters, Vol. 11, The Royal Society of Chemistry, Thomas Graham House, Cambridge, UK, 1994, Chapter 8, p. 318.

389. V. Ponec and G. C. Bond, *Catalysis by Metals and Alloys*, Studies in Surface Science and Catalysis, Vol. 95, Elsevier, Amsterdam, 1995, Chapter 11.

390. L. Guczi, Z. Schay, Gy. Stefler, L. F. Liotta, G. Deganello, and A. M. Venezia, *J. Catal.* **182**, 456 (1999).

391. B. K. Furlong, J. W. Hightower, T. Y.-L. Chan, A. Sarkany, and L. Guczi, *Appl. Catal. A* **117**, 41 (1994).

392. Q. Zhang, J. Li, X. Liu, and Q. Zhu, *Appl. Catal. A* **197**, 221 (2000).

393. M. A. Volpe, P. Rodriguez, and C. E. Gigola, *Catal. Lett.* **61**, 27 (1999).

394. M. P. R. Spee, J. Boersma, M. D. Meijer, M. Q. Slagt, G. van Koten, and J. W. Geus, *J. Org. Chem.* **66**, 1647 (2001).

395. J. H. Kang, E. W. Shin, W. J. Kim, J. D. Park, and S. H. Moon, *Catal. Today* **63**, 183 (2000).

396. J. Goetz, M. A. Volpe, C. E. Gigola, and R. Touroude, *J. Catal.* **199**, 338 (2001).

397. J. Choi and N. M. Yoon, *Tetrahedron Lett.* **37**, 1057 (1996).

398. W.-J. Wang, M.-H. Quiao, H. Li, and J.-F. Deng, *Appl. Catal., A* **166**, L243 (1998).

399. W.-J. Wang, M.-H. Quiao, J. Yang, S.-H. Xie, and J.-F. Deng, *Appl. Catal., A* **163**, 101 (1997).

400. W.-J. Wang, H.-X. Li, and J.-F. Deng, *Appl. Catal., A* **203**, 293 (2000).

401. S. Hermans, R. Raja, J. M. Thomas, B. F. G. Johnson, G. Sankar, and D. Gleeson, *Angew. Chem., Int. Ed. Engl.* **40**, 1211 (2001).

402. Á. Mastalir, Z. Király, Gy. Szöllösi, and M. Bartók, *J. Catal.* **194**, 146 (2000); *Appl. Catal., A* **213**, 133 (2001).

403. S. Ziegler, J. Theis, and D. Fritsch, *J. Membrane Sci.* **187**, 71 (2001).

404. V. Gryaznov, *Catal. Today* **51**, 391 (1999).

405. C. K. Lambert and R. D. Gonzalez, *Catal. Lett.* **57**, 1 (1999).

406. C. Liu, Y. Xu, S. Liao, and D. Yu, *Appl. Catal., A* **172**, 23 (1998).

407. C. Liu, Y. Xu, S. Liao, D. Yu, Y. Zhao, and Y. Fan, *J. Membrane Sci.* **137**, 139 (1997).

408. N. Itoh, W.-C. Xu, and A. M. Sathe, *Ind. Eng. Chem. Res.* **32**, 2614 (1993).

409. A. Stanislaus and B. H. Cooper, *Catal. Rev.-Sci. Eng.* **36**, 75 (1994).

410. B. H. Cooper and B. B. L. Donnis, *Appl. Catal., A* **137**, 203 (1996).

411. H. Yasuda and Y. Yoshimura, *Catal. Lett.* **46**, 43 (1997).

412. A. Corma, A. Martínez, and V. Martínez–Soria, *J. Catal.* **169**, 480 (1997).

413. R. M. Navarro, B. Pawelec, J. M. Trejo, R. Mariscal, and J. L. G. Fierro, *J. Catal.* **189**, 184 (2000).

414. T. C. Huang and B.-C. Kang, *J. Mol. Catal. A: Chem.* **103**, 163 (1995).

415. S. Nakagawa, T. Ono, S. Murata, M. Nomura, and T. Sakai, in *Catalysts in Petroleum Refining and Petrochemical Industries 1995*, Studies in Surface Science and Catalysis, Vol. 100, M. Absi-Halabi, J. Beshara, H. Qabazard, and A. Stanislaus, eds., Elsevier, Amsterdam, 1996, p. 499.

416. L. Ronchin and L. Toniolo, *Appl. Catal., A* **208**, 77 (2001).

417. Z.-B. Yu, M.-H. Qiao, H.-X. Li, and J.-F. Deng, *Appl. Catal., A* **163**, 1 (1997).

418. R. Zhang, F. Li, Q. Shi, and L. Luo, *Appl. Catal., A* **205**, 279 (2001).

419. T. Takahashi and T. Kai, *Mat. Sci. Eng. A* **267**, 207 (1999).

420. J. Chupin, N. S. Gnep, S. Lacombe, and M. Guisnet, *Appl. Catal., A* **206**, 43 (2001).

421. J. L. Rousset, L. Stievano, F. J. C. S. Aires, C. Geantet, A. J. Renouprez, and M. Pellarin, *J. Catal.* **197**, 335 (2001).

422. H. Yang, H. Gao, and R. J. Angelici, *Organometallics* **18**, 989 (1999).

423. H. Gao and R. J. Angelici, *J. Mol. Catal. A: Chem.* **149**, 63 (1999).

424. H. Yang, H. Gao, and R. J. Angelici, *Organometallics* **19**, 622 (2000).

425. F. Gelman, D. Avnir, H. Schumann, and J. Blum, *J. Mol. Catal. A: Chem.* **171**, 191 (2001).

426. P. Kluson and L. Cerveny, *Catal Lett.* **23**, 299 (1994); *Appl. Catal., A* **128**, 13 (1995).

427. S. Xie, M. Qiao, H. Li, W. Wang, and J.-F. Deng, *Appl. Catal., A* **176**, 129 (1999).

428. Z. Liu, W.-L. Dai, B. Liu, and J.-F. Deng, *J. Catal.* **187**, 253 (1999).

429. H. Brunner, in *Applied Homogeneous Catalysis with Organometallic Complexes*, B. Cornils and W. A. Herrmann, eds., VCH, Weinheim, 1996, Chapter 2.2, p. 201.

430. R. A. Sánchez-Delgado, M. Rosales, M. A. Esteruelas, and L. A. Oro, *J. Mol. Catal. A: Chem.* **96**, 231 (1995).

431. J. P. Maye and E. Negishi, *J. Chem. Soc. Chem. Commun.* 1830 (1993).

432. H. M. Lee, D. C. Smith, Jr., Z.He, E. D. Stevens, C. S. Yi, and S. P. Nolan, *Organometallics* **20**, 794 (2001).

433. M. A. Giardello, V. P. Conticello, L. Brard, M. R. Gagné, and T. J. Marks, *J. Am. Chem. Soc.* **116**, 10241 (1994).

434. A. Lightfoot, P. Schnider, and A. Pfaltz, *Angew. Chem., Int. Ed. Engl.* **37**, 2897 (1998).

435. M. D. Le Page and B. R. James, *Chem. Commun.* 1647 (2000).

436. J. Yu and J. B. Spencer, *J. Am. Chem. Soc.* **119**, 5257 (1997).

437. F. Joó and Á. Kathó, *J. Mol. Catal. A: Chem.* **116**, 3 (1997).

438. F. Joó and Á. Kathó, in *Aqueous-Phase Organometallic Catalysis*, B. Cornils and W. A. Herrmann, eds., Wiley-VCH, Weinheim, 1998, Chapter 6.2, p. 340.

439. K. Nomura, *J. Mol. Catal. A: Chem.* **130**, 1 (1998).

440. D. C. Mudalige and G. L. Rempel, *J. Mol. Catal. A: Chem.* **116**, 309 (1997).

441. D. J. Ellis, P. J. Dyson, D. G. Parker, and T. Welton, *J. Mol. Catal. A: Chem.* **150**, 71 (1999).

442. P. A. Z. Suarez, J. E. L. Dullius, S. Einloft, R. F. de Souza, and J. Dupont, *Inorg. Chim. Acta* **255**, 207 (1997).

443. D. Rutherford, J. J. J. Juliette, C. Rocaboy, I. T. Horváth, and J. A. Gladysz, *Catal. Today* **42**, 381 (1998).

444. B. Richter, A. L. Spek, G. van Koten, and B.-J. Deelman, *J. Am. Chem. Soc.* **122**, 3945 (2000).

445. D. E. Bergbreiter, J. G. Franchina, and B. L. Case, *Org. Lett.* **2**, 393 (2000).

446. S. M. Islam, A. Bose, B. K. Palit, and C. R. Saha, *J. Catal.* **173**, 268 (1998).

447. Z. M. Michalska, B.Ostaszewski, J. Zientarska, and J. W. Sobczak, *J. Mol. Catal. A: Chem.* **129**, 207 (1998).

448. M. M. Dell'Anna, M. Gagliardi, P. Mastrorilli, G. P. Suranna, and C. F. Nobile, *J. Mol. Catal. A: Chem.* **158**, 515 (2000).

449. J. Blum, A. Rosenfeld, N. Polak, O. Israelson, H. Schumann, and D. Anvir, *J. Mol. Catal. A: Chem.* **107**, 217 (1996).

450. C. Bianchini, D. G. Burnaby, J. Evans, P. Frediani, A. Meli, W. Oberhauser, R. Psaro, L. Sordelli, and F. Vizza, *J. Am. Chem. Soc.* **121**, 5961 (1999).

451. D. Chatterjee, H. C. Bajaj, A. Das, and K. Bhatt, *J. Mol. Catal.* **92**, L235 (1994).

452. E. J. Creyghton and R. S. Downing, *J. Mol. Catal. A: Chem.* **134**, 47 (1998).

453. E. J. Creyghton, R. A. W. Grotenbreg, R. S. Downing, and H. van Bekkum, *J. Chem. Soc., Faraday Trans.* **92**, 871 (1996).

454. G. Schmid, *Chem. Rev.* **92**, 1709 (1992).

455. H. Bönnemann and W. Brijoux, in *Advanced Catalysts and Nanostructured Materials*, W. Moser, ed., Academic Press, New York, 1996, Chapter 7, p. 165.

456. H. Bönnemann, G. Braun, W. Brijoux, R. Brinkmann, A. Schulze Tilling, K. Seevogel, and K. Siepen, *J. Organomet. Chem.* **520**, 143 (1996).

457. J. D. Aiken, III and R. G. Finke, *J. Mol. Catal. A: Chem.* **145**, 1 (1999).

458. N. Tishima, Y. Shiraishi, T. Teranishi, M. Miyake, T. Tominaga, H. Watanabe, W. Brijoux, H. Bönnemann, and G. Schmid, *Appl. Organomet. Chem.* **15**, 178 (2001).

459. J. D. Aiken, III, Y. Lin, and R. G. Finke, *J. Mol. Catal. A: Chem.* **114**, 29 (1996).

460. J. D. Aiken, III and R. G. Finke, *J. Am. Chem. Soc.* **121**, 8803 (1999).

461. M.W. van Laren and C. J. Elevier, *Angew. Chem., Int. Ed. Engl.* **38**, 3715 (1999).

462. D. Schleyer, H. G. Niessen, and J. Bargon, *New J. Chem.* **25**, 423 (2001).

463. C. Deibele, A. B. Permin, V. S. Petrosyan, and J. Bargon, *Eur. J. Inorg. Chem.* 1915 (1998).

464. R. D. Adams, T. S. Barnard, Z. Li, W. Wu, and J. H. Yamamoto, *J. Am. Chem. Soc.* **116**, 9103 (1994).

465. J. A. Cabeza, J. M. Fernández-Colinas, and A. Llamazares, *Synlett* 579 (1995).

466. R. D. Adams, *J. Organomet. Chem.* **600**, 1 (2001).

467. I. P. Rothwell, *Chem. Commun.* 1331 (1997).

468. E. G. Fidalgo, L. Plasseraud, and G. Süss-Fink, *J. Mol. Catal. A: Chem.* **132**, 5 (1998).

469. L. Plasseraud and G. Süss-Fink, *J. Organomet. Chem.* **539**, 163 (1997).

470. M. Faure, A. T. Vallina, H. Stoeckli-Evans, and G. Süss-Fink, *J. Organomet. Chem.* **621**, 103 (2001).

471. P. J. Dyson, D. J. Ellis, D. G. Parker, and T. Welton, *Chem. Commun.* 25 (1999).

472. J. Schulz, A. Roucoux, and H. Patin, *Chem. Eur. J.* **6**, 618 (2000).

473. K. S. Weddle, J. D. Aiken, III, and R. G. Finke, *J. Am. Chem. Soc.* **120**, 5653 (1998).

474. A. Corma, M. Iglesias, and F. Sánchez, *Catal. Lett.* **32**, 313 (1995).

475. A. M. Liu, K. Hidajat, and S.Kawi, *J. Mol. Catal. A: Chem.* **168**, 303 (2001).

476. H. Nagashima, A. Suzuki, M. Nobata, and K. Itoh, *J. Am. Chem. Soc.* **118**, 687 (1996).

477. Y. Kashiwagi, N. Shibayama, J. Anzai, and T. Osa, *Electrochemistry* **68**, 42 (2000).

478. N. Takano, A. Nakade, and N. Takeno, *Bull. Chem. Soc. Jpn.* **70**, 837 (1997).

479. N. Itoh, W. C. Xu, S. Hara, and K. Sakaki, *Catal. Today* **56**, 307 (2000).

480. C. J. Adams, M. J. Earle, and K. R. Seddon, *Chem. Commun.* 1043 (1999).

481. A. Maercker, M. Kemmer, H. C. Wang, D.-H. Dong, and M. Szwarc, *Angew. Chem., Int. Ed. Engl.* **37**, 2136 (1998).

482. A. Maercker and U. Girreser, *Tetrahedron* **50**, 8019 (1994).

483. R. M. Bullock and J.-S. Song *J. Am. Chem. Soc.* **116**, 8602 (1994).

484. L. Luan, J.-S. Song, and R. M. Bullock, *J. Org. Chem.* **60**, 7170 (1995).

485. U. Wild, D. Teschner, R. Schlögl, and Z. Paál, *Catal. Lett.* **67**, 93 (2000).

486. S. D. Jackson, G. J. Kelly, and G. Webb, *Phys. Chem. Chem. Phys.* **1**, 2581 (1999).

487. Z. Paál, *J. Mol. Catal.* **94**, 225 (1994).

488. K. Hayek, R. Kramer, and Z. Paál, *Appl. Catal., A* **162**, 1 (1997).

489. Y. Zhuang and A. Frennet, *Appl. Catal., A* **134**, 37 (1996).

490. M. R. Smith, J. K. A. Clarke, G. Fitzsimons, and J. J. Rooney, *Appl. Catal., A* **165**, 357 (1997).

491. W. E. Alvarez and D. E. Resasco, *J. Catal.* **164**, 467 (1996).

492. D. Teschner, K. Matusek, and Z. Paál, *J. Catal.* **192**, 335 (2000).

493. G. C. Bond and A. D. Hooper, *Appl. Catal., A* **191**, 69 (2000).

494. V. M. Akhmedov, S. H. Al-Khowaiter, E. Akhmedov, and A. Sadikhov, *Appl. Catal., A* **181**, 51 (1999).

495. B. Török, I. Pálinkó, and M. Bartók, *Catal. Lett.* **31**, 421 (1995).

496. B. Török, J. T. Kiss, and M. Bartók, *Catal. Lett.* **46**, 169 (1997).

497. T. A. Jachimowski and W. H. Weinberg, *Surf. Sci.* **370**, 71 (1997).

498. C. J. Hagedorn, M. J. Weiss, C.-H. Chung, P. J. Mikesell, R. D. Little, and W. H. Weinberg, *J. Chem. Phys.* **110**, 1745 (1999).

499. C. J. Hagedorn, M. J. Weiss, T. W. Kim, and W. H. Weinberg, *J. Am. Chem. Soc.* **123**, 929 (2001).

500. D. Kelly and W. H. Weinberg, *J. Chem. Phys.* **105**, 7171 (1996).

501. A. Raichle, Y. Traa, F. Fuder, M. Rupp, and J.Weitkamp, *Angew. Chem., Int. Ed. Engl.* **40**, 1243 (2001).

502. F. Lefebvre, J. Thivolle-Cazat, V. Dufaud, G. P. Niccolai, and J.-M. Basset, *Appl. Catal., A* **182**, 1 (1999).

503. J. Corker, F. Lefebvre, C. Lécuyer, V. Dufaud, F. Quignard, E. Choplin, J. Evans, and J.-M. Basset, *Science* **271**, 966 (1996).

504. L. d'Ornelas, S. Reyes, F. Quignard, E. Choplin, and J.-M. Basset, *Chem. Lett.* 1931 (1993).

505. C. Rosier, G. P. Niccolai, and J.-M. Basset, *J. Am. Chem. Soc.* **119**, 12408 (1997).

506. O. P. Strausz, T. W. Mojelsky, W. Thomas, J. D. Payzant, G. A. Olah, and G. K. S. Prakash, U. S. Patent 5,290,428 (1994).

507. O. P. Strausz, T. W. Mojelsky, J. D. Payzant, G. A. Olah, and G. K. S. Prakash, *Energy Fuels* **13**, 558 (1999).

508. G. A. Olah, U. S. Patent 4,472,268 (1984).

509. G. A. Olah, U.S. Patent 4,508,618 (1985).

12

METATHESIS

An unanticipated catalytic reaction of olefinic hydrocarbons was described in 1964 by Banks and Bailey.[1,2] They discovered that C_3–C_8 alkenes disproportionate to homologs of higher and lower molecular weight in the presence of alumina-supported molybdenum oxide [Eq. (12.1)], cobalt oxide–molybdenum oxide, molybdenum hexacarbonyl, or tungsten hexacarbonyl at 100–200°C, under about 30 atm pressure:

$$2\ CH_3CH{=}CH_2 \xrightarrow[\text{163°C, 32 atm}]{\text{MoO}_3 \text{ on Al}_2\text{O}_3} CH_3CH{=}CHCH_3 \ + \ CH_2{=}CH_2 \qquad (12.1)$$

$$\text{94\% efficiency}$$
$$\text{at 43\% conversion}$$

Ethylene gave cyclopropane and methylcyclopropane in very low yields. Isomerization was also observed. The term *disproportionation* was used for the reaction because of the analogy with disproportionation of metal ions into different oxidation states. The reaction has a more general nature, however, and consequently the terms *dismutation*[3] and *olefin metathesis*[4] were introduced.

In the metathesis reaction, alkenes are converted to mixtures of lower and higher alkenes through the interchange of groups on the double bonds in an equilibrium process. The number of moles of product lighter than the feed is roughly the same as the number of moles of heavier alkenes. Olefin metathesis is a thermoneutral reaction and produces an equilibrium mixture of products. In simple cases this mixture is close to the statistical distribution of substituent groups.

The reaction, catalyzed by a variety of soluble complexes and supported catalysts, has been found to be general for a large number of unsaturated compounds. The reaction of two identical alkenes, called *homometathesis*, can yield new alkenes as in Eq. (12.1) (productive metathesis). If the reaction does not produce new compounds, it is called *degenerative* or *nonproductive*. This may also be the case in the cross-metathesis of any two terminal alkenes:

$$
\begin{array}{c}
R^1CH=CH_2 \\
+ \\
CH_2=CHR^2
\end{array}
\longrightarrow
\begin{array}{c}
R^1CH \\
\| \\
CH_2
\end{array}
+
\begin{array}{c}
CH_2 \\
\| \\
CHR^2
\end{array}
\qquad (12.2)
$$

Cross-metathesis, however, is usually a nonselective reaction. Transformation of two terminal alkenes in the presence of a metathesis catalyst, for instance, can give six possible products (three pairs of *cis/trans* isomers) since self-metathesis of each alkene and cross-metathesis occur in parallel. It has been observed, however, that terminal olefins when cross-metathesized with styrene yield *trans*-β-alkylstyrenes with high selectivity.[5] A useful synthetic application of cross-metathesis is the cleavage of internal alkenes with ethylene called *ethenolysis* to yield terminal olefins:

$$
\begin{array}{c}
R^1CH \\
\| \\
R^2CH
\end{array}
+
\begin{array}{c}
CH_2 \\
\| \\
CH_2
\end{array}
\longrightarrow
\begin{array}{c}
R^1CH=CH_2 \\
+ \\
R^2CH=CH_2
\end{array}
\qquad (12.3)
$$

Acyclic dienes are the products in cross-metathesis of cycloalkenes and acyclic alkenes. With ethylene, α,ω-dienes are formed:

$$
\text{(cyclopentene)} \; + \; CH_2=CH_2 \longrightarrow
\begin{array}{c}
CH=CH_2 \\
CH=CH_2
\end{array}
\qquad (12.4)
$$

A closely related special case of metathesis is the conversion of cycloalkenes under metathesis conditions, known as *ring-opening polymerization* or metathesis polymerization (see Section 12.2). Macrocyclic polyenes and polymers, called polyalkenamers, are formed with high stereoselectivity:

$$
2n \; \text{(cyclopentene, CH=CH)} \longrightarrow
\left[=CH \quad CH=CH \quad CH= \right]_n
\qquad (12.5)
$$

The only exception is cyclohexene which, for thermodynamic reasons, does not participate in ring-opening metathesis polymerization.

Nonconjugated dienes may undergo polymerization under metathesis conditions. Metathesis, in turn, can be used for structure determination of polyalkenylenes via metathetical degradation with a low-molecular-weight olefin.[6]

Metathesis of alkynes is a much less general reaction, and only a few catalysts were found to be active. Terminal alkynes characteristically undergo cyclotrimerization to yield benzene derivatives in the presence of most metathesis catalysts[7,8] (see Section 13.1.3). Heterogeneous metathesis catalysts[9] [Eq. (12.6)] and molybdenum carbonyls,[10–12] in turn, promote metathesis of internal alkynes. Molybdenum hexacarbonyl is active with added phenolic compounds[12] [Eq. (12.7)]:[11]

$$2 \ Et \text{——}\!\!\equiv\!\!\text{——} Me \xrightarrow[350°C]{\text{6.8\% WO}_3 \text{ on silica}} Me \text{——}\!\!\equiv\!\!\text{——} Me \ + \ Et \text{——}\!\!\equiv\!\!\text{——} Et \qquad (12.6)$$

53% selectivity
44% conversion

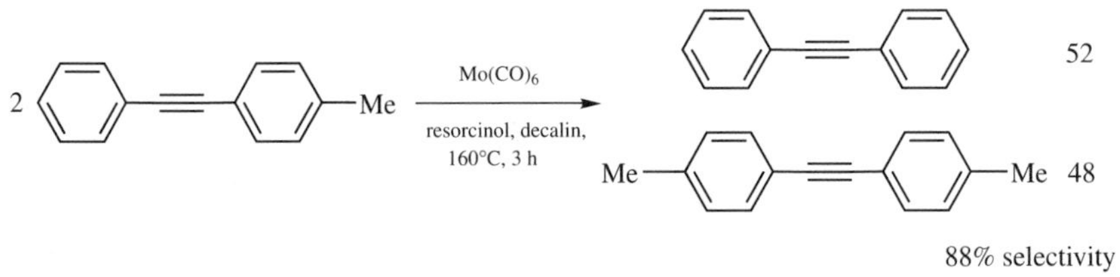

$$\text{(12.7)}$$

88% selectivity

The few known examples indicate that more severe reaction conditions are usually required than in the metathesis of alkenes.

Although the actual mechanism of olefin metathesis is different, formally it can be considered a [2 + 2]-cycloaddition. In a broader sense, several other processes may also be viewed as metathetical transformations. For example, metathesis may be extended to include transformations of metal–carbon multiple-bonded compounds with molecules containing unsaturated sulfur–oxygen, phosphorus–carbon, and carbon–oxygen linkages.[13,14] This is a newly developing field mostly with examples for the reactions of carbonyl compounds with metallocarbenes,[14] a transformation analogous to the Wittig reaction.

Similarly, oxidation of alkenes resulting in the formation of carbonyl compounds through the cleavage of the carbon–carbon double bond [Eq. (12.8)] can be considered as a metathetic transformation:

$$\begin{matrix} CH_2 \\ \| \\ CH_2 \end{matrix} \ + \ \begin{matrix} O \\ \| \\ O \end{matrix} \ \longrightarrow \ 2\,CH_2\!=\!O \qquad (12.8)$$

In fact, the oxidation with singlet oxygen of alkenes not possessing allylic hydrogens is known to involve a 1,2-dioxetane intermediate[15,16] (see Section 9.2.2).

Olefin metathesis has been extensively reviewed,[6,7,14,17–32] and regular symposia have been devoted to the subject.[33–37]

12.1. ACYCLIC ALKENES

Both homogeneous and heterogeneous catalysts can be used in olefin metathesis.[6,26,28,29,31] In general, tungsten or molybdenum halides or carbonyl complexes and a non-transition-metal cocatalyst (main-group I–IV organometallics or hydrides) in an organic solvent result in catalyst systems that are the most effective. These catalysts are extremely active at ambient temperature. The only other metal that shows general reactivity is rhenium. The active component in these catalyst systems is soluble in most cases; however, sometimes it may separate as an insoluble solid. Protic solvents such as ethanol, phenol, water, and acetic acid when added to WCl_6 improve the activity of the resulting catalyst. Typical systems are $WCl_6 + EtOH + EtAlCl_2$, $WCl_6 + BuLi$, $WCl_6 + LiAlH_4$, $ReCl_5 + Bu_4Sn$, and $MoCl_2(NO)_2(PPh_3)_2 + EtAlCl_2$. Certain catalysts prepared from tungsten, molybdenum, and rhenium complexes, such as $W(CO)_5PPh_3 + EtAlCl_2$ or $ReCl_5 + Et_3Al$, require oxygen as an activator.[26,28,30]

The heterogeneous catalysts are mainly oxides, carbonyls, or sulfides of Mo, W, and Re deposited on high-surface-area supports (oxides and phosphates, mainly silica and alumina). Heterogeneous catalysts such as the most widely employed WO_3 on silica and MoO_3 on alumina or silica are usually less active than their homogeneous counterparts and require the use of elevated temperature and pressure. Re_2O_7 on alumina, however, is an outstanding catalyst that is active under milder conditions (room temperature, atmospheric pressure). It is highly selective and resistant to catalyst poisoning. The first vanadium-based metathesis catalyst (Me_4Sn–V_2O_5 on alumina) was reported in 1993.[38] The addition of alkyltin derivatives to some supported heterogeneous catalysts was shown to greatly increase catalytic activity.[39,40]

Most heterogeneous metathesis catalysts and some soluble homogeneous complexes, such as $PhWCl_2 + AlCl_3$ or $WCl_6 + AlCl_3$, do not require the use of a cocatalyst. $AlCl_3$ and other Lewis acids, whenever present, play an important role in catalyst systems by further increasing the activity of homogeneous catalysts.

Immobilized complexes anchored to the surface of oxide supports may serve as active metathesis catalysts. In fact, $Mo(CO)_6$ on alumina was reported in the first paper on metathesis by Banks and Bailey.[1] Molybdenum and tungsten hexacarbonyls and alkylmolybdenum and alkyltungsten complexes have been studied most.[6,24,26] Anchored organomolybdenum complexes are particularly effective catalysts, even at temperatures below 0°C; however, they quickly lose activity.[41–43]

Some catalysts, most notably $W(CO)_6$ in a CCl_4 solution[44,45] and systems based on molybdena,[46] when irradiated, are effective in photoinduced metathesis. In fact, a photoreduced silica-supported molybdena proved to be the most active olefin metathesis catalyst.[47]

A special group of catalysts exerting great impacts on metathesis research, particularly on mechanistic studies, is carbene (alkylidene) complexes. Only in 1972 was a stable rhodium metallocarbene intermediate complex isolated for the first time during a metathesis reaction.[48] The complex also acted as a catalyst. Since then many other preformed carbene complexes have been used as catalysts in olefin metathesis.[7,14,30,49] In addition to carbenes of molybdenum,[50] tungsten,[51–54] and rhenium,[55] carbene complexes of niobium and tantalum[56] also proved to be active catalysts. Alkoxy group ligands have a marked effect on the activity of carbene complexes.[49,54]

With some catalyst systems, selectivity to primary metathesis products is near 100%, but side reactions (double-bond migration, dimerization, cyclopropanation, polymerization) often reduce selectivity. Such side reactions, such as oligomerization and double-bond shift over oxide catalysts, may be eliminated by treatment with alkali and alkaline-earth metal ions.[26]

Numerous studies have dealt with the mechanism of metathesis.[6,7,14,20–31] The first important question to answer in an understanding of the mechanism is whether the carbon–carbon single bonds or the carbon–carbon double bonds are cleaved during the reaction. The corresponding transformations taking place are transalkylation [Eq. (12.9)] or transalkylidenation [Eq. (12.10)], respectively:

$$R^1CH=CH{\dotplus}R^2 \\ R^1{\dotplus}CH=CHR^2 \quad \rightleftharpoons \quad R^1CH=CH{\atop R^1} + R^2{\atop CH=CHR^2} \qquad (12.9)$$

$$R^1CH{\doteqdot}CHR^2 \\ R^1CH{\doteqdot}CHR^2 \quad \rightleftharpoons \quad \begin{matrix}R^1\\CH\\||\\CH\\|\\R^1\end{matrix} + \begin{matrix}R^2\\CH\\||\\CH\\|\\R^2\end{matrix} \qquad (12.10)$$

Early studies with deuterium-labeled[57,58] and ^{14}C-labeled compounds[59,60] established that cleavage of the double bonds occurs. For example, the degenerate metathesis of 2-butene and perdeutero-2-butene[57] results in the formation of $[1,1,1,2{-}^2H_4]$-2-butene as the only new labeled compound [Eq. (12.11)] which is consistent with alkylidene exchange [Eq. (12.10)]:

$$\begin{matrix}CH_3CH=CHCH_3\\+\\CD_3CD=CDCD_3\end{matrix} \quad \xrightarrow[\text{benzene}]{WCl_6+EtAlCl_2+EtOH} \quad CD_3CD=CHCH_3 \qquad (12.11)$$

Similar results were obtained with a mixture of C_2H_4 and C_2D_4 on a heterogeneous Re_2O_7 on Al_2O_3 catalyst.[58]

A concerted pairwise exchange of alkylidene moieties via a "quasicyclobutane" transition state, also known as the *diolefin mechanism*, was first suggested to interpret the transformation:[3]

$$\begin{matrix}R^1CH{=}CHR^2\\\uparrow\\M\\\downarrow\\R^1CH{=}CHR^2\end{matrix} \quad \rightleftharpoons \quad \begin{matrix}R^1 \qquad R^2\\ CH{-}CH\\ |\ M\ |\\ CH{-}CH\\ R^1 \qquad R^2\end{matrix} \quad \rightleftharpoons \quad \begin{matrix}R^1 \qquad R^2\\ CH \qquad CH\\ ||{\leftarrow}M{\rightarrow}||\\ CH \qquad CH\\ |\qquad\ |\\ R^1 \qquad R^2\end{matrix} \qquad (12.12)$$

Although the noncatalytic thermal disproportionation of alkenes is symmetry-forbidden, the formation of metal complex **1** with the four-centered cyclic intermediate makes the reaction symmetry-allowed.[61]

Further suggestions included a tetracarbene intermediate[7,26,62] depicted as **2** or **3**, and a nonconcerted pairwise exchange of alkylidenes via metallacyclopentane intermediates:[63]

$$2 \qquad\qquad 3$$

$$(12.13)$$

A major objection to these mechanisms was raised by Hérisson and Chauvin,[64] who found that cross-metathesis between a cycloalkene and an unsymmetric alkene resulted in a statistical distribution of cross-products even at very low conversion, whereas a simple pairwise mechanism would lead to a single product. It is also important to point out that cyclobutanes are not isolated as intermediates and are unreactive under metathesis conditions.[30]

Conclusive evidence for nonpairwise exchange of alkylidene units was presented by labeling studies.[65–67] The reaction of a mixture of 1,7-octadiene and $[1,1,8,8-^2H_4]$-1,7-octadiene[66] [Eq. (12.14)] was shown to give at very low conversion (2%) d_0-, d_2-, and d_4-ethylenes in a ratio of $1 : 2 : 1$:

$$(12.14)$$

This corresponds to the statistical distribution that is consistent with the carbene mechanism discussed below. A pairwise exchange was calculated to result in a $1 : 1.6 : 1$ ratio of the three labeled ethylenes. The original $1 : 0 : 1$ ratio of the three isotopomer 1,7-octadienes was found to be unchanged in the reaction mixture recovered at the end of the reaction. No isotope scrambling occurred, and isotope effects were found to be negligible. An analogous result was obtained with 2,2′-divinylbiphenyl and the terminally labeled tetradeutero-2,2′-divinylbiphenyl using homogeneous molybdenum and tungsten catalysts.[67]

At present the accepted mechanism is a nonconcerted, nonpairwise process involving metallocarbenes that reversibly react with alkenes to yield metallacyclobutane intermediates:[48,64,68]

$$
\begin{array}{c}
R^1CH \\
\| \\
M \longleftarrow \| \\
CHR^2
\end{array}
\begin{array}{c}
CHR^2
\end{array}
\rightleftharpoons
\begin{array}{c}
R^1CH-CHR^2 \\
| \quad\quad | \\
M-CHR^2
\end{array}
\rightleftharpoons
\begin{array}{c}
R^1CH=CHR^2 \\
\downarrow \\
M=CHR^2
\end{array}
\qquad (12.15)
$$

According to this mechanism, olefin metathesis is a chain reaction with the carbene as chain carrier. It predicts complete randomization of alkylidene fragments from the first turnover. An important additional feature is that both metathesis and ring-opening polymerization of cycloalkenes can be explained by this mechanism, which provides ready explanation of polymer molecular weight.

Important evidence in support of the chain-reaction theory was the synthesis of carbene complexes, such as diphenylcarbenepentacarbonyl–tungsten(0) (**4**), and their reaction with olefins. The results in Eq. (12.16) demonstrate[69] the scission of the carbon–carbon double bond of the alkene and the combination of the ethylidene fragment with the diphenylcarbene group of **4**:

$$
(CO)_5W=C\!\!\begin{array}{c}Ph\\Ph\end{array} + \;\diagdown\!\!\diagup\!\!\diagdown \xrightarrow[\;50°C,\,4\,h\;]{} \begin{array}{c}Ph\\Ph\end{array}\!\!C=CHMe \; + \; W(CO)_6 \; + \; \begin{array}{c}Me\\ \triangle\\Ph\;\;\;Me\\Ph\end{array}
$$

| **4** | | 54% | 41% | trace |

$$
\qquad (12.16)
$$

Further support came from the isolation and use of metallocarbene complexes[7,14,30,48,49] and metallacyclobutanes[54,70] as catalysts in olefin metathesis.

The two key steps in this above mechanism are the formation of the carbene species (initiation) and the propagation through carbene–olefin interchange. Carbene generation in cocatalyst systems can be easily explained. Here the organometallic component reacts with the transition-metal halide to give a σ-alkyl complex, which, in turn, generates the active species via α-hydrogen abstraction:[71]

$$
M-X \; + \; M'-CH_3 \longrightarrow M-CH_3 \rightleftharpoons \overset{\overset{\displaystyle H}{|}}{M=CH_2} \qquad (12.17)
$$

In nonalkylation systems, such as the classic heterogeneous catalysts [alumina-supported MoO_3, Re_2O_7, and $W(CO)_6$] or combinations such as $WCl_6 + AlCl_3$, the carbene must come from the alkene, which is a more difficult problem to explain.

Of the numerous proposals, the metal hydride route with α-hydride abstraction (β elimination) [Eq. (12.18)] has gained substantial experimental support:[7]

$$\underset{\overset{|}{X}}{M}-H + RCH=CHR' \rightleftharpoons \underset{\overset{|}{X}\ \overset{|}{R}}{M}-CHCH_2R' \xrightarrow[-HX]{} M=C\overset{\displaystyle R}{\underset{\displaystyle CH_2R'}{\big<}} \qquad (12.18)$$

The formation of metallacyclobutane through a π-allyl complex without involvement of carbene species [Eq. (12.19)] was also suggested as the initiation step for systems where carbene formation from the alkene is difficult to occur for structural reasons:[72–74]

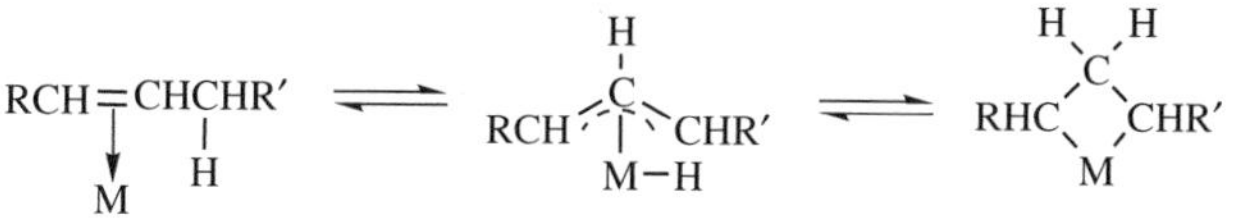

$$\underset{\overset{\downarrow}{M}\ \overset{|}{H}}{RCH=CHCHR'} \rightleftharpoons RCH\overset{\overset{\textstyle H}{\overset{|}{C}}}{\underset{M-H}{\big<}}CHR' \rightleftharpoons RHC\overset{\overset{\textstyle H\ \ H}{C}}{\underset{M}{\big<}}CHR' \qquad (12.19)$$

More recent results support the view that Lewis acids, on the interaction with metal carbonyls, may facilitate dissociation of a ligand to leave a vacant site for alkene coordination.[75] Photochemical generation of carbene species [Eq. (12.20)], the crucial step in photoinduced metathesis, is a complicated, multistep process:[74,76,77]

$$W(CO)_6 + CCl_4 \xrightarrow{h\nu} Cl_2C=W(CO)_3Cl_2 \qquad (12.20)$$

Concerning the propagation step, a series of olefin–carbene interchanges must take place according to Scheme 12.1. This means, for instance, that the productive metathesis of propylene to yield ethylene and 2-butenes [Eq. (12.1)] is brought about by alternative reactions of propylene with methylidene and ethylidene intermediates. Interestingly, for the degenerative metathesis of propylene, the ethylidene route was found to be the predominant path.[78,79] A large body of experimental evidence demonstrates the feasibility of these reactions.[30]

It is generally accepted that ring-opening polymerization of cycloalkenes proceeds via metallocarbenes. Although less extensively studied, metathesis of alkynes

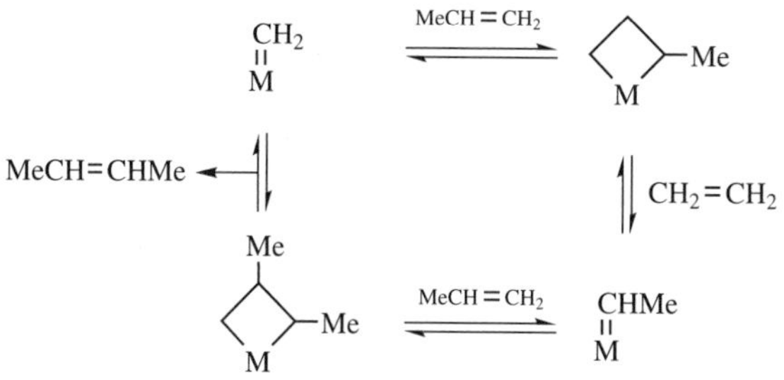

Scheme 12.1

is also believed to occur by an analogous mechanism with the involvement of carbyne species. Labeling studies proved the splitting of the triple bond,[80,81] and carbyne complexes were shown to catalyze metathesis of acetylenes.[81,82] The copolymerization by Group VI transition-metal catalysts of norbornene with acetylenes was described for the first time in 1991.[83] The successful reaction was taken as evidence verifying that this process also proceeds with the metal carbene mechanism.

The relative rate of metathesis, in general, decreases with increasing substitution of the double bond. It is surprising, however, that in the presence of tungsten catalysts internal alkenes produce products faster than do terminal olefins.[84] Moreover, metathesis of a mixture of an internal and a terminal alkene initially yields cross-metathesis products.[84] This indicates that an internal alkene can compete with a terminal alkene in the cross-reaction but does not undergo metathesis with itself. The solution to this paradox is that the terminal alkene undergoes a fast degenerate metathesis that was proved by labeling experiments.[84–86] It was shown that the non-productive metathesis between 1-hexene and $[1,1-^2H_2]$-1-heptene [Eq. (12.21)], that is, that the interchange of the CH_2 and CD_2 groups, was 10^3 faster than productive metathesis into the internal alkene and ethylene:[86]

$$n\text{-}C_4H_9CH{=}CH_2$$
$$+ \qquad \xrightarrow[\text{24 h}]{WCl_6+2\,n\text{-BuLi}} \qquad \begin{matrix} n\text{-}C_4H_9CH \\ \| \\ CD_2 \end{matrix} + \begin{matrix} CH_2 \\ \| \\ CHC_5H_{11}\text{-}n \end{matrix} \qquad (12.21)$$
$$CD_2{=}CHC_5H_1\text{-}n$$

Supposedly, the association constants for these olefin–tungsten complexes are larger for terminal than for internal olefins.

The trend of structural selectivity can be summarized as degenerate metathesis of terminal alkenes (exchange of methylene groups) > cross-metathesis of terminal and internal alkenes > metathesis of internal alkenes > productive metathesis of terminal alkenes (formation of internal alkene and ethylene).[87] Since different catalyst systems exhibit different selectivities, a simple general picture accounting for all stereochemical phenomena of metathesis is not feasible.

In the steric course of olefin metathesis, acyclic and cyclic alkenes exhibit opposite behavior. The stereoselectivity of the transformation of most acyclic olefins is low and usually goes to equilibrium. In some systems there is a strong initial preference for retention of stereochemistry, that is for the transformation of *cis* to *cis* and that of *trans* to *trans* isomers.[88] This means that geometric isomers of an olefin give different isomeric mixtures of the same product:

	cis : *trans* ratio	4.4	28.8	2.3
Mo(NO)$_2$Py$_2$Cl$_2$ +EtAlCl$_2$, chlorobenzene, 0°C, 2 min		2-butene + 2-pentene + 3-hexene		(12.22)
	cis : *trans* ratio	0.1	0.04	0

However, rapid geometric isomerization consistent with the carbene mechanism obscures selectivity.

In one interpretation a perpendicular olefin to carbene approach and a highly puckered metallacyclobutane intermediate were assumed.[27,89,90] The favored pathway leads to conformations with the fewest 1,2 and 1,3 metal/ligand–substituent interactions:

$$
\begin{array}{c}
\text{(structure)} \rightleftharpoons \text{(structure)}
\end{array}
\qquad (12.23)
$$

In another model, the stereoselectivity is predetermined by steric interactions between the metal and/or its ligands and alkene substituents arising during the olefin coordination to the metallocarbene.[91] These steric factors govern the orientation of alkene approaching the metallocarbene in a way to minimize alkyl–ligand repulsion. Compounds **5** and **6** depict favored *cis–cis* and *trans–trans* orientations, respectively.

cis-favored coordination   trans-favored coordination
of a *cis* alkene   of a *trans* alkene

5   **6**

Not readily reconcilable with these models is the finding by Calderon and co-workers, who observed that *cis*-4-methyl-2-pentene, a hindered alkene, underwent a faster isomerization than metathesis[21,92] (formation of 40% *trans* isomer at 4% conversion). The most striking feature of the transformation is the highly stereoselective formation of *trans*-2,5-dimethyl-3-hexene throughout the course of reaction regardless of the structure of the starting alkene:

$$(12.24)$$

It seems that substituent interactions on adjacent carbons in the metallacyclobutane dominate the stereochemistry of the transformation in this case.

12.2. RING-OPENING METATHESIS POLYMERIZATION

Ring-opening metathesis polymerization of cycloolefins,[93–99] a reaction of significant practical importance (see Section 12.3), is catalyzed by a number of well-
defined transition-metal complexes.[100–102] Alkylidene and metallacyclobutane
complexes of molybdenum,[103,104] tungsten,[52,104–109] titanium,[110] tantalum,[111,112]
and rhenium[113] are active catalysts. The first example of ring-opening polymerization of norbornene with titanium alkylidenes generated by thermolysis of dimethyltitanocene was reported in 1993.[114] Ring-opening metathesis polymerization of
strained cycloalkenes, in turn, can also be induced with simple transition-metal
halogenides without cocatalysts.[29,98,99] Particular attention has been paid to the
synthesis of polyacetylene a novel conducting polymer.

   Whereas only limited stereoselectivity is characteristic of the metathesis of
acyclic olefins, ring-opening metathesis polymerization of cycloalkenes may be
highly stereoselective provided the proper catalysts and reaction conditions are
selected. Cyclopentene, for instance, is transformed to either all-*cis* [Eq. (12.25)]
or all-*trans* polypentenamers [Eq. (12.26)] in the presence of tungsten catalysts:[21,92]

$$
n \; \square \quad
\begin{array}{c}
\xrightarrow[\text{0°C}]{\text{WF}_6+\text{EtAlCl}_2}
\end{array}
\left[
\begin{array}{c}
\text{H} \quad \text{H} \\
\diagdown \; / \\
\text{C}=\text{C} \\
/ \qquad \diagdown \\
\text{CH}_2 \qquad \text{CH}_2\text{CH}_2
\end{array}
\right]_n
\qquad (12.25)
$$

97.5% at low conversion
95.9% at high conversion

$$
\xrightarrow[\text{0°C}]{\text{WCl}_6+\text{NC(CH}_2)_2\text{OH}+\text{EtAlCl}_2}
\left[
\begin{array}{c}
\text{CH}_2 \qquad \text{H} \\
\diagdown \quad / \\
\text{C}=\text{C} \\
/ \qquad \diagdown \\
\text{H} \qquad \text{CH}_2\text{CH}_2
\end{array}
\right]_n
\qquad (12.26)
$$

95% at low conversion
88% at high conversion

Metallocarbenes such as **4** were found to exhibit remarkable stereoselectivities producing *cis* polyalkenamers from cycloalkenes of different ring sizes.[115] Catalysts
promoting the formation of *cis* polypentenamer do not catalyze isomerization of
the product even after long periods, and they are very ineffective in metathesis of
acyclic olefins, in sharp contrast to the behavior of *trans*-selective catalysts. Two
different chelating models were proposed; one with a highly chelated tridentate
complexation ensures *cis*-selective ring-opening metathesis, while a less demanding two-ligand sequence results in a less selective transformation to the *trans*
polymer.[21,92]

   Because of its favorable kinetic and thermodynamic properties, norbornene can
be polymerized with simple transition-metal halogenides.[99,116] Even RuCl$_3$, which
is not active in other metathesis reactions, is effective in the ring-opening polymerization of norbornene.[117,118] Developments in the early 1990s include the use of

methyltrioxorhenium[119] and a single crystal of $(Bu_4N)_2[Mo_6O_{19}]$, which was shown to be an efficient, completely recoverable heterogeneous catalyst.[120]

Ring opening of norbornene by Mo, W, Re, Os, Ru, and Ir compounds taking place at the double bond results in polymers with a complete range of *cis/trans* character, but the cyclopentane rings are always *cis*-1,3 enchained:

$$(12.27)$$

Successive rings, however, may have two possible configurations resulting in different tacticities.[98,100] ReCl$_5$, for example, gives an all-*cis* syndiotactic polymer, while RuCl$_3$ yields an all-*trans* nearly atactic product.[117] This means that the syndiotactic polymer contains alternating configurations of the two chiral carbon centers of the cyclopentane ring along the chain [(R,S)-(S,R)-(R,S)-···], while the atactic structure lacks any definite order. In an isotactic polymer identical configurational arrangements are repeated [(R,S)-(R,S)-(R,S)-···].

The polymerization of substituted norbornenes may result in new possibilities for the arrangement of the repeat units in the chain (head-to-head, head-to-tail, tail-to-tail), thus creating very complex stereochemistry. On the basis of the microstructure of the ring-opened polymer of different methylnorbornenes, the widely noted stereoselective *cis* giving *cis* and *trans* giving *trans* in olefin metathesis was attributed to different forms of metallocarbene propagating species rather than to different configurations of metallacyclobutanes.[118]

Much less is known about the ring-opening polymerization of norbornadiene. Surprisingly, OsCl$_3$, a nonselective catalyst in the polymerization of norbornene, was found to yield an all-*cis* polymer.[121]

Exclusive ring opening of the *anti* compound of a mixture of *syn*- and *anti*-7-methylbicyclo[2.2.1]heptene-2 in the presence of Ru-, Os-, Ir-, W-, and Re-based catalysts was observed[116] proving that the *exo* face of norbornene is highly reactive while the *endo* face is inert toward metathesis.

Even though it is a trisubstituted alkene, owing to its strained structure, 1-methylcyclobutene reacts under mild metathesis conditions:[122]

$$(12.28)$$

The polymer formed has a structure resembling that of polyisoprene.

An exciting development in polymer chemistry has been the synthesis of highly conjugated polymers, including polyacetylene. This is due to their unique optical and electrical properties and, consequently, the importance of their possible practical applications.[123] In addition to the obvious way of polymerizing acetylene (see Section 13.2.5), polyacetylene may be synthesized indirectly by ring-opening metathesis polymerization of 1,3,5,7-cyclooctatetraene.[101,102,124] Although the first successful reaction to produce polyacetylene in this way yielded an insoluble nonconducting powder,[125] another method (neat polymerization) allows for production of a conducting polyacetylene film:[126]

$$n \ \bigotimes \ \xrightarrow[\text{RT}]{\text{alkylidene W complexes}} \ \text{\{CHCH=CHCH=CHCH=CHCH\}}_n \qquad (12.29)$$

Substituted polyacetylenes may be produced through the ring-opening metathesis polymerization of substituted cyclooctatetraenes.[127]

12.3. PRACTICAL APPLICATIONS

Olefin metathesis is the subject of strong commercial interest, but its industrial application depends on feedstock availability and economic considerations.[128] Although it is applied in the manufacture of smaller-volume, high-value specialty chemicals, it is utilized in only two large-scale industrial syntheses.[129–132] Heterogeneous catalysts are used in the latter processes, whereas homogeneous catalysts have gained increasing importance in other technologies. Metathesis of functionalized alkenes appears to be a promising field of further industrial applications.[32,133,134] This, however, requires the development of new catalysts since only a few traditional metathesis catalysts tolerate heteroatom functions in functionalized alkenes.

The first commercial application of olefin metathesis was the so-called triolefin process developed by Phillips[135,136] to transform propylene on a heterogeneous metathesis catalyst (WO_3 on SiO_2, 370–450°C) [Eq. (12.1)]. The propylene content of a steam cracker's output was converted in this rather inexpensive way to high-purity ethylene for polymerization. 2-Butenes formed were used in the production of high-octane fuel through alkylation[137] (see Section 5.5.1) or dehydrogenated to 1,3-butadiene (see Section 2.3.3). The economic feasibility of this process depended on cheap propylene, and the process was abandoned in the 1970s because of other, more economical use of propylene.

Technologies for the reverse process to produce propylene from ethylene and 2-butenes were also developed.[138,139] The Arco process[139] dimerizes ethylene to 2-butenes, which, in turn, are metathesized with ethylene to yield propylene. The process is not practiced at present, but it is a potential technology in case of a propylene shortage.

By far the largest application of metathesis is a principal step in the Shell higher olefin process (SHOP),[24,140,141] with the aim of producing detergent-range olefins as feed to the oxo reaction and in the manufacture of linear alkylbenzenes. First, terminal alkenes with a broad distribution range (C_4–C_{30}) are synthesized by the oligomerization of ethylene with a homogeneous nickel complex (see Section 13.1.3). The C_{10}–C_{18} fraction after separation is used directly in benzene alkylation to produce linear alkylbenzenes (see Section 5.5.6), or in the production of fatty alcohols via hydroformylation (see Section 7.1.3). The lower (C_4–C_8) and higher (C_{20+}) fractions, in turn, are isomerized into internal olefins. Metathesis disproportionation in the latter mixture of higher olefins with lower olefins yields a broad spectrum of internal alkenes. The detergent-range C_{11}–C_{14} fraction, again, is transformed by hydroformylation and the rest is recycled through isomerization and metathesis up to total consumption. Since hydroformylation is carried out in the presence of hydrogen and a cobalt catalyst, which causes double-bond migration and has a strong hydrogenating activity, primary alcohols are eventually formed (see Section 7.1.3). In the isomerization and metathesis stage a linear internal alkene yield better than 96% is achieved.

FEAST (further exploitation of advanced shell technology), another Shell process[142–144] commercialized in 1986, utilizes a highly active promoted rhenium-on-alumina catalyst (100°C, 2 atm) to synthesize 1,5-hexadiene from 1,5-cyclooctadiene [Eq. (12.30)] and 1,9-decadiene from cyclooctene [Eq. (12.31)]:

$$(12.30)$$

$$(12.31)$$

Since not only the starting compounds but also the product dienes undergo metathesis, a complex product mixture is formed in each reaction. The byproducts, after the fast removal of the desired dienes, however, are recycled, allowing virtually quantitative yields of the required products.

A number of new processes exploiting metathesis have been developed by Phillips. A novel way to manufacture lubricating oils has been demonstrated.[145] The basic reaction is self-metathesis of 1-octene or 1-decene to produce C_{14}–C_{28} internal alkenes. The branched hydrocarbons formed after dimerization and hydrogenation may be utilized as lubricating oils. Metathetical cleavage of isobutylene with propylene or 2-butenes to isoamylenes has a potential in isoprene manufacture.[136,146] High isoamylene yields can be achieved by further metathesis of C_{6+} byproducts with ethylene and propylene. Dehydrogenation to isoprene is already practiced in the transformation of isoamylenes of FCC C_5 olefin cuts.

In the Phillips neohexene process[147] 2,4,4-trimethyl-2-pentene (**8**) is converted by cleavage with ethylene to neohexene (**9**) used in the production of a perfume musk. The starting material is commercial diisobutylene. Since it is a mixture of positional isomers (2,4,4-trimethyl-2-pentene and 2,4,4-trimethyl-1-pentene) and the latter (**7**) participates in degenerative metathesis, effective utilization of the process requires the isomerization of **7** into **8**. A bifunctional catalyst system consisting of an isomerization catalyst (MgO) and a heterogeneous metathesis catalyst is employed:[131]

$$(12.32)$$

The reaction is run with excess ethylene to drive the equilibrium to product formation. The byproduct isobutylene is recycled to produce diisobutylene via dimerization (see Section 13.1.1).

The so-called toluene-to-styrene process [Eq. (12.33)], first reported in a patent by Monsanto,[148] applies bismuth or lead oxides in dehydrodimerization of toluene to stilbene, which, in turn, is cleaved with ethylene to styrene:[149]

$$\text{Ph-Me} \xrightarrow{[O]} \text{PhCH=CHPh} + \underset{\text{(excess)}}{CH_2=CH_2} \xrightarrow[\text{482°C, 5 atm}]{WO_3 \text{ on } SiO_2} 2\text{ Ph-CH=CH}_2 \quad (12.33)$$

The necessity of the use of stoichiometric amount of oxides in the first step and the difficulty to prevent polymerization of the product styrene in the metathesis step are the major drawbacks preventing commercialization.[130]

The highly stereoselective ring-opening metathesis polymerization of cycloalkenes discussed above is another promising field of industrial application.[150] The living nature of this polymerization process ensures the formation of high-molecular-weight polymers (polyalkenamers), especially if accurately tuned alkylidene complexes are used as catalysts.[100] Polypentenamer has a potential commercial value as a general-purpose rubber comparable with polybutadiene and polyisoprene.[99,151–154] The source of the starting material is C_5 steam-cracker cuts. Cyclopentene separated or produced in selective hydrogenation of cyclopentadiene undergoes ring-opening metathesis polymerization in the presence of tungsten or rhodium catalysts.[130,154]

Tungsten-based catalysts with a Lewis base promoter are used in the polymerization of dicyclopentadiene also isolated from the C_5 stream-cracker fraction.[150,155] The product poly(dicyclopentadiene) or Metton resin has unusual physical and chemical properties. It is processed together with elastomers to increase viscosity of the end product by using reaction injection molding (RIM) technology.[155]

The synthesis of polyoctenamer has been commercialized by Huels.[150] In contrast with the transformation of cyclooctene to 1,9-decadiene [Eq. (12.31)], homogeneous catalyst compositions, such as $WCl_6 + EtAlCl_2$, are used to promote ring-opening metathesis polymerization of cyclooctene. A polymer of narrow molecular-weight distribution with high *trans* content (55–85%) called *Vestenamer* is produced and used as blend component in different rubbers and thermoplastics.

Polynorbornene the oldest polyalkenamer is manufactured by a method developed by CDF Chimie[156,157] since 1976. Tungsten or rhodium catalysts are used to yield a highly *trans* polymer. This is capable of rapidly absorbing hydrocarbon oils and can be crosslinked to produce a very soft rubber.

12.4. RECENT DEVELOPMENTS

Three comprehensive accounts of the subject, two reviews,[158,159] and the monograph by Ivin and Mol[160] were published in 1997. Other reviews summarize current developments such as metal carbenes as intermediates in ring-closing metathesis[161,162] and synthesis of metathesis catalysts.[163–165] The main emphasis has been on the synthesis and application of new catalysts and further enhancement of our understanding of the mechanism of metathesis transformations by studying metal carbene complexes. These latter catalysts are also used increasingly in synthetic applications. By now olefin metathesis can also be considered a powerful tool in synthetic organic chemistry to prepare complex natural products and novel materials[166–174] including asymmetric metathesis.[175] Relevant information can also be found in the proceedings of symposia devoted to metathesis.[176–178]

12.4.1. New Catalysts and Mechanistic Studies

Novel noncarbene metathesis catalysts, $WOCl_3(OAr)$, $WOCl_2(OAr)_2$, or $WOCl(OAr)_3$ in conjunction with R_3SnH, are very effective in the metathesis of 2-pentene when supported on silica.[179] A new tungsten complex, $[WCl_2(CO)_4]_2$, prepared by the photochemical oxidation of $W(CO)_6$ with CCl_4, is an active and stable metathesis catalyst not requiring the use of a Lewis acid.[180] Cobalt acetylacetonate and neodecenoate in combination with R_3Al are the first cobalt catalysts active in metathesis.[181]

Because of their high potential for practical applications, heterogeneous metathesis catalysts, such as MoO_3-on-Al_2O_3,[182–184] MoO_3 supported on hexagonal mesoporous silica,[185] methyltrioxorhenium on silica–alumina,[186] and Re_2O_7 on Al_2O_3 have been studied extensively. The latter catalyst exhibits high activity and high selectivity, and modifiers, such as alkyltin promoters, are able to induce even better performances.[187] Two additional catalyst systems are unusual in the sense that they do not contain a transition metal. Al_2O_3/Me_4Sn has been reported to bring about metathesis of propene, 1-hexene, and 2-pentene.[188] SiO_2, in turn, shows activity in the metathesis of ethene and propene under irradiation when pretreated at high temperature under vacuum.[189] Heterogenization by attaching the Grubbs catalyst to a phosphinated mesoporous support (MCM-41) was also successful.[190]

Electron-deficient (having less than 18 electrons) and coordinatively unsaturated (containing less than 6, usually only 4 or 5 ligands) metal carbene complexes are highly effective in metathesis.[191–194]

Particular attention has been paid to the tetracoordinated alkylidene catalysts $M(=CHCMe_2Ph)(=NAr)(OR)_2$ (M = Mo, W) with bulky substituents Ar and R on the imido and alkoxide ligands developed by Schrock and coworkers.[195] Similarly, ruthenium complexes $Ru(=CHR)(X)_2(PR'_3)_2$ (X = Cl, Br, OOCCF$_3$) developed by Grubbs and coworkers[194,196,197] have attracted much attention. Among them, complex **10** turned out to have high activity, but it is quite sensitive toward oxygen and moisture. Grubbs' catalysts, including the most active **11**, however, generally tolerate various functional groups, are stable to air and water, and are easy to prepare and capable of inducing the metathesis of all kinds of alkenes. Some even catalyze the metathesis of 2-pentene and cycloalkenes of low strain:[192,193]

10   **11**   **12**

They are highly efficient initiators since they are all converted to propagating species at the very early stage of the reaction. Theoretical treatments,[198] and the use of advanced analytical techniques, for example, electrospray ionization tandem mass spectrometry[199] have provided useful information about the nature and properties of intermediates involved in metathesis.

New catalysts of unique structural characteristics or ensuring high efficiency continue to appear in the literature.[200–202] Catalytic performance could be improved by replacing the phosphane ligand with chelating Schiff bases[203] or by the addition of Lewis acids.[204,205] Dramatically improved metathesis activity can be achieved by replacing one phosphine ligand with an N-heterocyclic carbene ligand (e.g., compound **12**).[206,207] In early studies the high activity of the N-heterocyclic carbene complex catalysts was attributed to their ability to promote dissociation of the phosphine ligand. Grubbs, however, has found that the high activity is due to their improved selectivity for binding π-acidic olefinic substrates.[208]

Ruthenium carbenes with phosphane ligands with rigid *cis* stereochemistry[209] and cationic Ru allenylidene complexes also show high catalytic activity.[210] An efficient, but remarkably simple procedure for the one-pot synthesis of Ru catalysts using $RuCl_3$, PCy_3, 1-alkynes, Mg, hydrogen, and water has been described.[211] The synthesis and synthetic utility of binaphthyl-based chiral Mo catalysts has been disclosed.[212] It may be prepared from commercially available starting materials and can be used in situ to effect enantioselective olefin metathesis not requiring the use of specialty glassware.

A surprising new observation made by Rooney and coworkers shows that complex **12** generates radical anions when treated with π-acceptors even dienes and simple alkenes.[213]

12.4.2. Metathesis of Alkynes

Metathesis of alkynes is a much less general reaction and has not yet found the attention that alkene metathesis has received.[158–160] Nonetheless, a few observations have shown that it has the potential of useful applications emerging as a new synthetic tool.[214–216]

Mechanistic studies revealed that alkyne metathesis and ring-opening metathesis polymerization of cycloalkynes proceed via metal carbyne complexes,[217,218] which is also supported by theoretical studies.[219] The polymerization of PhC≡CMe with NbCl$_5$ or TaCl$_5$ yields a polymer that degrades to oligomers as a result of secondary metathesis reaction. A stable polymer, however, may be synthesized with TaCl$_5$ and Ph$_4$Sn as a cocatalyst.[220]

A particularly interesting example is the synthesis of alkyne-bridged oligomers and polymers, which are attractive materials for optical and electronic applications. Bunz and coworkers were able to modify the Mortreux catalyst system [Mo(CO)$_6$ and a suitable phenol][11] and reaction conditions to perform acyclic alkyne metathesis of 1,4-dipropynylated benzenes to produce high-molecular-weight poly(p-phenyleneethynylene)s in excellent to quantitative yields and high purity:[221,222]

$$(12.34)$$

A tungsten carbyne complex was similarly effcetive.[223]

12.4.3. Ring-Closing Metathesis

Several examples of ring-closing metathesis indicate the highly selective nature of this reaction.[161,162,224–226] The application of various Mo and Ru carbene catalysts allow formation of five-, six-, and seven-membered mono- and bicyclic rings.[167,168,171,227] Formation of substituted cyclopentenes, for example, was shown to occur in quantitative yield and with complete stereoselectivity. Cycloisomerization of 1,6- and 1,7-enynes may be used to synthesize cyclopentyl ring derivatives.[228,229]

Cycloalkynes may be formed by applying ring-closing metathesis to diynes.[214] Partial hydrogenation of the product provides a stereoselective route to *cis* cycloalkenes that cannot be directly prepared in pure form by ring-closing metathesis:

$$\text{(12.35)}$$

12.4.4. Ring-Opening Metathesis

The interesting problems of ring-chain equilibria, *cis/trans* blockiness, head-to-tail bias, and the various methods used to determine tacticity are treated in detail in three reviews.[158–160] Consequently, only some more interesting observations are discussed here.

It is quite surprising that $MgCl_2$ is capable of metathesis polymerization of strained cycloalkenes such as norbornene.[230] Norbornene can be transformed to a high-*trans* polymer by the hydrate of $Ru(OTs)_2$ in protic solvents,[231] whereas a high-*cis* polymer is formed in supercritical carbon dioxide.[232] The same catalyst may induce copolymerization of cyclopentene and norbornene, just as the Grubbs catalyst with PCy_3 ligand.[233] In contrast, only a homopolymer was isolated when the ligand was PPh_3. Ruthenium–arene complexes bearing *N*-heterocyclic carbene ligands with aryl substituent are active catalyst precursors for the ring-opening metathesis polymerization of cyclooctene.[234] The process occurs at room temperature and is induced by visible light.

A unique example is the production of a polymer that contains C_{60}. The fullerene monomer was prepared by cycloaddition of quadricyclane followed by polymerization in norbornene at room temperature.[235] The product has *cis* : *trans* ratios between 3 : 1 and 6 : 1 depending on the concentration of the monomer. A processable film containing 1% C_{60} exhibits electronic and electrochemical properties that are typical of the carbon cluster.

12.4.5. Practical Applications

As discussed in Section 12.3, the triolefin process to transform propylene to ethylene and 2-butene developed by Phillips[135,136] is not practiced at present because of the increased demand for propylene. The reverse process, that is, cross-metathesis of ethylene and 2-butene, however, can contribute to satisfy the global demand for propylene. Lyondell Petrochemical operates a 136,000-t/y (ton/year) plant for the production of propylene.[236] In a joint project by BASF and FINA, Phillips metathesis technology will be used to enhance propylene production.[237] A similar project was also announced by DEA.[238] In a continuous process jointly developed by IFP and Chines Petroleum Corporation, cross-metathesis of ethylene and 2-butene is carried out in the liquid phase over Re_2O_7-on-Al_2O_3 catalyst (35°C, 60 bar).[239,240]

REFERENCES

1. R. L. Banks and G. C. Bailey, *Ind. Eng. Chem., Prod. Res. Dev.* **3**, 170 (1964).

2. R. L. Banks, *ACS Symp. Ser.* **222**, 403 (1983).

3. C. P. C. Bradshaw, E. J. Howman, and L. Turner, *J. Catal.* **7**, 269 (1967).

4. N. Calderon, H. Y. Chen, and K. W. Scott, *Tetrahedron Lett.* 3327 (1967).

5. W. E. Crowe and Z. J. Zhang, *J. Am. Chem. Soc.* **115**, 10998 (1993).

6. R. L. Banks, in *Catalysis, A Specialist Periodical Report*, Vol. 4, C. Kemball and D. A. Dowden, senior reporters, The Royal Society of Chemistry, Burlington House, London, 1981, Chapter 4, p. 100.

7. R. H. Grubbs, in *Comprehensive Organometallic Chemistry*, Vol. 8, G. Wilkinson, F. G. A. Stone, and E. W. Abel, eds., Pergamon Press, Oxford, 1982, Chapter 54, p. 499.

8. J. A. Moulijn, H. J. Reitsma, and C. Boelhouwer, *J. Catal.* **25**, 434 (1972).

9. F. Pennella, R. L. Banks, and G. C. Bailey, *J. Chem. Soc., Chem. Commun.* 1548 (1968).

10. S. Devarajan and D. R. M. Walton, *J. Organomet. Chem.* **181**, 99 (1979).

11. A. Mortreux and M. Blanchard, *J. Chem. Soc., Chem. Commun.* 786 (1974).

12. J. A. K. Du Plessis and H. C. M. Vosloo, *J. Mol. Catal.* **65**, 51 (1991).

13. A. F. Hill, *J. Mol. Catal.* **65**, 85 (1991).

14. R. H. Grubbs and S. H. Pine, in *Comprehensive Organic Synthesis*, B. M. Trost and I. Fleming, eds., Vol. 5: *Combining C–C π-Bonds*, L. A. Paquette, ed., Pergamon Press, 1991, Chapter 9.3, p. 1115.

15. P. D. Bartlett and M. E. Landis, in *Singlet Oxygen*, H. H. Wasserman and R. W. Murray, eds., Academic Press, New York, 1979, Chapter 7, p. 243.

16. A. L. Baumstark, in *Singlet O_2*, Vol. 2: *Reaction Modes and Products*, Part 1, A. A. Frimer, ed., CRC Press, Boca Raton, FL, 1985, Chapter 1, p. 1.

17. G. C. Bailey, *Catal. Rev.-Sci. Eng.* **3**, 37 (1970).

18. M. L. Khidekel', A. D. Shebaldova, and I. V. Kalechits, *Russ. Chem. Rev.* (Engl transl.) **40**, 669 (1971).

19. N. Calderon, *Acc. Chem. Res.* **5**, 127 (1972).

20. N. Calderon, in *The Chemistry of Functional Groups*, Supplement A: *The Chemistry of Double-Bonded Functional Groups*, S. Patai, ed., Wiley, London, 1977, Chapter 10, p. 913.

21. N. Calderon, J. P. Lawrence, and E. A. Ofstead, *Adv. Orgnomet. Chem.* **17**, 449 (1979).

22. R. L. Banks, *Top. Curr. Chem.* **25**, 39 (1972).

23. J. C. Mol and J. A. Moulijn, *Adv. Catal.* **24**, 131 (1975).

24. J. C. Mol and J. A. Moulijn, in *Catalysis, Science and Technology*, Vol. 8, J. R. Anderson and M. Boudart, eds., Springer, Berlin, 1987, Chapter 2, p. 69.

25. R. J. Haines and G. J. Leigh, *Chem. Soc. Rev.* **4**, 155 (1975).

26. J. J. Rooney and A. Stewart, in *Catalysis, a Specialist Periodical Report*, Vol. 1, C. Kemball, senior reporter, The Chemical Society, Burlington House, London, 1977, Chapter 8, p. 277.

27. T. J. Katz, *Adv. Orgnometallic Chem.* **16**, 283 (1977).

28. R. H. Grubbs, *Prog. Inorg. Chem.* **24**, 1 (1978).

29. J. K. Ivin, *Olefin Metathesis*, Academic Press, London, 1983.

30. P. A. Chaloner, *Handbook of Coordination Catalysis in Organic Chemistry*, Butterworths, London, 1986, Chapter 9, p. 921.

31. V. Dragutan, A. T. Balaban, and M. Dimonie, *Olefin Metathesis and Ring-Opening Polymerization of Cyclo-Olefins*, Editura Academiei, Bucuresti, and Wiley, Chichester, 1985.

32. W. J. Feast and V. C. Gibson, in *The Chemistry of the Metal-Carbon Bond*, Vol. 5, F. R. Hartley, ed., Wiley, Chichester, 1989, Chapter 6, p. 199.

33. *Proc. 5th Int. Symp. Olefin Metathesis*, Graz, 1983, K. Hummel, F. Stelzer, and A. Moisi, eds.; *J. Mol. Catal.* **28**, 1–402 (1985).

34. *Proc. 6th Int. Symp. Olefin Metathesis*, Hamburg, 1985, E. Thorn-Csányi, ed.; *J. Mol. Catal.* **36**, 1–199 (1986).

35. *Proc. 7th Int. Symp. Olefin Metathesis*, Kingston upon Hall, 1987, A. Ellison, ed.; *J. Mol. Catal.* **46**, 1–444 (1988).

36. *Proc. 8th Int. Symp. Olefin Metathesis*, Bayreuth, 1989, H. L. Krauss, ed.; *J. Mol. Catal.* **65**, 1–285 (1991).

37. *Proc. 9th Int. Symp. Olefin Metathesis*, Collegeville, 1991, A. E. Martin, ed.; *J. Mol. Catal.* **76**, 1–385 (1992).

38. H.-G. Ahn, K. Yamamoto, R. Nakamura, and H. Niiyama, *Chem. Lett.* 141 (1993).

39. K. Tanaka and K. Tanaka, *J. Chem. Soc., Chem. Commun.* 748 (1984).

40. A. Andreini, X. Xiaoding, and J. C. Mol, *Appl. Catal.* **27**, 31 (1986).

41. B. N. Kuznetsov, A. N. Startsev, and Yu.I. Yermakov, *J. Mol. Catal.* **8**, 135 (1980).

42. Y. Iwasawa, H. Kubo, and H. Hamamura, *J. Mol. Catal.* **28**, 191 (1985).

43. A. N. Startsev, O. V. Klimov, and E. A. Khomyakova, *J. Catal.* **139**, 134 (1993).

44. A. Agapiou and E. McNelis, *J. Chem. Soc., Chem. Commun.* 187 (1975).

45. P. Krausz, F. Garnier, and J. E. Dubois, *J. Am. Chem. Soc.* **97**, 437 (1975).

46. M. Anpo, I. Tanahashi, and Y. Kubokawa, *J. Chem. Soc., Faraday Trans. 1* **78**, 2121 (1982).

47. K. A. Vikulov, I. V. Elev, B. N. Shelimov, and V. B. Kazansky, *J. Mol. Catal.* **55**, 126 (1989).

48. D. J. Cardin, M. J. Doyle, and M. F. Lappert, *J. Chem. Soc., Chem. Commun.* 927 (1972).

49. G. W. Parshall and S. D. Ittel, *Homogeneous Catalysis*, Wiley-Interscience, New York, 1992, Chapter 9, p. 217.

50. J. S. Murdzek and R. R. Schrock, *Organometallics* **6**, 1373 (1987).

51. J. Kress and J. A. Osborn, *J. Am. Chem. Soc.* **105**, 6346 (1983).

52. F. Quignard, M. Leconte, and J.-M. Basset, *J. Chem. Soc., Chem. Commun.* 1816 (1985).

53. J. Kress, A. Aguero, and J. A. Osborn, *J. Mol. Catal.* **36**, 1 (1986).

54. R. R. Schrock, R. T. DePue, J. Feldman, C. J. Schaverien, J. C. Dewan, and A. H. Liu, *J. Am. Chem. Soc.* **110**, 1423 (1988).

55. R. Toreki and R. R. Schrock, *J. Am. Chem. Soc.* **112**, 2448 (1990).

56. R. R. Schrock, *Acc. Chem. Res.* **12**, 98 (1979).

57. N. Calderon, E. A. Ofstead, J. P. Ward, and W. A. Judy, K. W. Scott, *J. Am. Chem. Soc.* **90**, 4133 (1968).

58. A. A. Olsthoorn and C. Boelhouwer, *J. Catal.* **44**, 207 (1976).

59. J. C. Mol, J. A. Moulijn, and C. Boelhouwer, *J. Chem. Soc., Chem. Commun.* 633 (1968).

60. A. Clark and C. Cook, *J. Catal.* **15**, 420 (1969).

61. F. D. Mango, *Adv. Catal.* **20**, 291 (1969).

62. G. S. Lewandos and R. Pettit, *J. Am. Chem. Soc.* **93**, 7087 (1971).

63. R. H. Grubbs and T. K. Brunck, *J. Am. Chem. Soc.* **94**, 2538 (1972).

64. J.-L. Hérisson and Y. Chauvin, *Makromol. Chem.* **141**, 161 (1971).

65. R. H. Grubbs, P. L. Burk, and D. D. Carr, *J. Am. Chem. Soc.* **97**, 3265 (1975).

66. R. H. Grubbs, D. D. Carr, C. Hoppin, and P. L. Burk, *J. Am. Chem. Soc.* **98**, 3478 (1976).

67. T. J. Katz and R. Rothchild, *J. Am. Chem. Soc.* **98**, 2519 (1976).

68. DJ. Cardin, B. Cetinkaya, M. J. Doyle, and M. F. Lappert, *Chem. Soc. Rev.* **2**, 99 (1973).

69. C. P. Casey and T. J. Burkhardt, *J. Am. Chem. Soc.* **96**, 7808 (1974).

70. K. C. Wallace, J. C. Dewan, and R. R. Schrock, *Organometallics* **5**, 2162 (1986).

71. E. L. Muetterties, *Inorg. Chem.* **14**, 951 (1975).

72. M. Ephritikhine, B. R. Francis, M. L. H. Green, R. E. Mackenzie, and M. J. Smith, *J. Chem. Soc., Dalton Trans.* 1131 (1977).

73. R. H. Grubbs and S. J. Swetnick, *J. Mol. Catal.* **8**, 25 (1980).

74. T. Szymanska-Buzar, *J. Mol. Catal.* **48**, 43 (1988).

75. T. Szymanska-Buzar, *J. Mol. Catal.* **68**, 177 (1991).

76. F. Garnier, P. Krausz, and H. Rudler, *J. Organomet. Chem.* **186**, 77 (1980).

77. D. Borowczak, T. Szymanska-Buzar, and J. J. Ziólkowski, *J. Mol. Catal.* **27**, 355 (1984).

78. M. Sasaki, K. Tanaka, K. Tanaka, and I. Toyoshima, *J. Chem. Soc., Chem. Commun.* 764 (1986).

79. K. Tanaka, K. Tanaka, H. Takeo, and C. Matsumura, *J. Am. Chem. Soc.* **109**, 2422 (1987).

80. A. Mortreux and M. Blanchard, *Bull. Soc. Chim. Fr.* 1641 (1972).

81. G. J. Leigh, M. T. Rahman, and D. R. M. Walton, *J. Chem. Soc., Chem. Commun.* 541 (1982).

82. J. H. Wengrovius, J. Sancho, and R. R. Schrock, *J. Am. Chem. Soc.* **103**, 3932 (1981).

83. T. Masuda, T. Yoshida, H. Makio, M. Z. Ab. Rahman, and T. Higashimura, *J. Chem. Soc., Chem. Commun.* 503 (1991).

84. W. J. Kelly and N. Calderon, *J. Macromol. Sci., Chem.* **A9**, 911 (1975).

85. J. McGinnis, T. J. Katz, and S. Hurwitz, *J. Am. Chem. Soc.* **98**, 605 (1976).

86. M. T. Mocella, M. A. Busch, and E. L. Muetterties, *J. Am. Chem. Soc.* **98**, 1283 (1976).

87. C. P. Casey, H. E. Tuinstra, and M. C. Saeman, *J. Am. Chem. Soc.* **98**, 608 (1976).

88. W. B. Hughes, *J. Chem. Soc., Chem. Commun.* 431 (1969).

89. T. J. Katz and W. H. Hersh, *Tetrahedron Lett.* 585 (1977).

90. C. P. Casey, L. D. Albin, and T. J. Burkhardt, *J. Am. Chem. Soc.* **99**, 2533 (1977).

91. M. Leconte and J. M. Basset, *J. Am. Chem. Soc.* **101**, 7296 (1979).

92. E. A. Ofstead, J. P. Lawrence, M. L. Senyek, and N. Calderon, *J. Mol. Catal.* **8**, 227 (1980).

93. G. Dall'Asta, *Makromol. Chem.* **154**, 1 (1972).

94. N. C. Billingham, in *Developments in Polymerisation—1*, R. N. Haward, ed., Applied Science Publishers, London, 1979, Chapter 4, p. 147.

95. V. Dragutan, A. T. Balaban, and M. Dimonie, *Olefin Metathesis and Ring-Opening Polymerization of Cyclo-Olefins*, Editura Academiei, Bucuresti, and Wiley, Chichester, 1985, Chapter 4, p. 155.

96. R. Grubbs and B. M. Novak, in *Encyclopedia of Polymer Science and Engineering*, 2nd ed., Supplement Volume, J. I. Kroschwitz, ed., Wiley-Interscience, New York, 1989, p. 420.

97. R. H. Grubbs and W. Tumas, *Science* **243**, 907 (1989).

98. A. J. Amass, in *Comprehensive Polymer Science*, G. Allen, J. C. Bevington, eds., Vol. 4: *Chain Polymerization II*, G. C. Eastmond, A. Ledwith, S. Russo, and P. Sigwalt, eds., Pergamon Press, Oxford, 1989, Chapter 6, p. 109.

99. H. R. Kricheldorf, in *Handbook of Polymer Synthesis*, H. R. Kricheldorf, ed., Marcel Dekker, New York, 1992, Chapter 7, p. 433.

100. R. R. Schrock, *Acc. Chem. Res.* **23**, 158 (1990).

101. J. Feldman and R. R. Schrock, *Prog. Inorg. Chem.* **39**, 1 (1991).

102. B. M. Novak, W. Risse, and R. H. Grubbs, *Adv. Polym. Sci.* **102**, 47 (1992).

103. G. C. Bazan, R. R. Schrock, H.-N. Cho, and V. C. Gibson, *Macromolecules* **24**, 4495 (1991).

104. R. R. Schrock, S. A. Krouse, K. Knoll, J. Feldman, J. S. Murdzek, and D. C. Yang, *J. Mol. Catal.* **46**, 243 (1988).

105. J. Kress, J. A. Osborn, R. M. E. Greene, K. J. Ivin, and J. J. Rooney, *J. Chem. Soc., Chem. Commun.* 874 (1985).

106. R. R. Schrock, J. Feldman, L. F. Cannizzo, and R. H. Grubbs, *Macromolecules* **20**, 1169 (1987).

107. L. L. Blosch, K. Abboud, and J. M. Boncella, *J. Am. Chem. Soc.* **113**, 7066 (1991).

108. J.-L. Couturier, C. Paillet, M. Leconte, J.-M. Basset, and K. Weiss, *Angew. Chem., Int. Ed. Engl.* **31**, 628 (1992).

109. P. A. van der Schaaf, W. J. J. Smeets, A. L. Spek, and G. van Koten, *J. Chem. Soc., Chem. Commun.* 717 (1992).

110. L. R. Gilliom and R. H. Grubbs, *J. Am. Chem. Soc.* **108**, 733 (1986).

111. K. C. Wallace and R. R. Schrock, *Macromolecules* **20**, 448 (1987).

112. K. C. Wallace, A. H. Liu, J. C. Dewan, and R. R. Schrock, *J. Am. Chem. Soc.* **110**, 4964 (1988).

113. R. Toreki, R. R. Schrock, and M. G. Vale, *J. Am. Chem. Soc.* **113**, 3610 (1991).

114. N. A. Petasis and D.-K. Fu, *J. Am. Chem. Soc.* **115**, 7208 (1993).

115. T. J. Katz, S. J. Lee, and N. Acton, *Tetrahedron Lett.* 4247 (1976).

116. J. G. Hamilton, K. J. Ivin, and J. J. Rooney, *J. Mol. Catal.* **28**, 255 (1985).

117. H. H. Thoi, K. J. Ivin, and J. J. Rooney, *J. Mol. Catal.* **15**, 245 (1982).

118. J. G. Hamilton, K. J. Ivin, G. M. McCann, and J. J. Rooney, *J. Chem. Soc., Chem. Commun.* 1379 (1984).

119. W. A. Herrmann, W. Wagner, U. N. Flessner, U. Volkhardt, and H. Komber, *Angew. Chem., Int. Ed. Engl.* **30**, 1636 (1991).

120. M. McCann and D. McDonnell, *J. Chem. Soc., Chem. Commun.* 1718 (1993).

121. B. Bell, J. G. Hamilton, O. N. D. Mackey, and J. J. Rooney, *J. Mol. Catal.* **77**, 61 (1992).

122. T. J. Katz, J. McGinnis, and C. Altus, *J. Am. Chem. Soc.* **98**, 606 (1976).

123. C. B. Gorman and R. H. Grubbs, in *Conjugated Polymers: The Novel Science and Technology of Conducting and Nonlinear Optically Active Materials*, J. L. Brédas and R. Silbey, eds., Kluwer, Dordrecht, 1991, p. 1.

124. E. J. Ginsburg, C. B. Gorman, R. H. Grubbs, F. L. Klavetter, N. S. Lewis, S. R. Marder, J. W. Perry, and M. J. Sailor, in *Conjugated Polymeric Materials: Opportunities in Electronics, Optoelectronics, and Molecular Electronics*, J. L. Brédas and R. R. Chance, eds., Kluwer, Dordrecht, 1990, p. 65.

125. Yu. V. Korshak, V. V. Korshak, G. Kanischka, and H. Höcker, *Makromol. Chem., Rapid Commun.* **6**, 685 (1985).

126. F. L. Klavetter and R. H. Grubbs, *J. Am. Chem. Soc.* **110**, 7807 (1988).

127. C. B. Gorman, E. J. Ginsburg, and R. H. Grubbs, *J. Am. Chem. Soc.* **115**, 1397 (1993).

128. R. Streck, *J. Mol. Catal.* **76**, 359 (1992).

129. R. L. Banks, in *Applied Industrial Catalysis*, Vol. 3, B. E. Leach, ed., Academic Press, New York, 1984, Chapter 6, p. 215.

130. R. Streck, *J. Mol. Catal.* **46**, 305 (1988).

131. R. Streck, *Chemtech* **19**, 498 (1989).

132. K. Weissermel and H.-J. Arpe, *Industrial Organic Chemistry*, 2nd ed., VCH, Weinheim, 1993, Chapter 3, p. 85.

133. R. Streck, *J. Mol. Catal.* **15**, 3 (1982).

134. J. C. Mol, *Chemtech* **13**, 250 (1983).

135. *Hydrocarbon Process.* **46**(11), 232 (1967).

136. R. L. Banks, *Chemtech* **9**, 494 (1979); **16**, 112 (1986).

137. R. S. Logan and R. L. Banks, *Hydrocarbon Process.* **47**(6), 135 (1968); *Oil Gas J.* **66**(21), 131 (1968).

138. P. Amigues, Y. Chauvin, D. Commereuc, C. C. Lai, Y. H. Liu, and J. M. Pan, *Hydrocarbon Process., Int. Ed.* **69**(10), 79 (1990).

139. H. A. Wittcoff, *Chemtech* **20**, 48 (1990).

140. E. R. Freitas and C. R. Gum, *Chem. Eng. Prog.* **75**(1), 73 (1979).

141. G. R. Lappin, in *Alpha Olefins Applications Handbook*, G. R. Lappin and J. D. Sauer, eds., Marcel Dekker, New York, 1989, Chapter 3, p. 35.

142. H. Short and R. Remírez, *Chem. Eng.* (NY) **94**(11), 22 (1987).

143. G. W. Parshall and W. A. Nugent, *Chemtech* **18**, 314 (1988).

144. P. Chaumont and C. S. John, *J. Mol. Catal.* **46**, 317 (1988).

145. W. T. Nelson and L. F. Heckelsberg, *Ind. Eng. Chem., Prod. Res. Dev.* **22**, 178 (1983).

146. R. L. Banks and R. B. Regier, *Ind. Eng. Chem., Prod. Res. Dev.* **10**, 46 (1971).

147. R. L. Banks, D. S. Banasiak, P. S. Hudson, and J. R. Norell, *J. Mol. Catal.* **15**, 21 (1982).

148. Ph.D. Montgomery, R. N. Moore, and W. R. Knox, U. S. Patent 3,965,206 (1976).

149. R. A. Innes and H. E. Swift, *Chemtech* **11**, 244 (1981).

150. E. A. Ofstead, in *Encyclopedia of Polymer Science and Engineering*, 2nd ed., Vol. 11, J. I. Kroschwitz, ed., Wiley-Interscience, New York, 1988, p. 287.

151. W. Graulich, W. Swodenk, and D. Theisen, *Hydrocarbon Process.* **51**(12), 71 (1972).

152. N. Calderon and R. L. Hinrichs, *Chemtech* **4**, 627 (1974).

153. W. J. Feast, in *Comprehensive Polymer Science*, G. Allen and J. C. Bevington, eds., Vol. 4: *Chain Polymerization II*, G. C. Eastmond, A. Ledwith, S. Russo, and P. Sigwalt, eds., Pergamon Press, Oxford, 1989, Chapter 7, p. 135.

154. E. A. Ofstead, in *Kirk-Othmer Encyclopedia of Chemical Technology*, 3rd ed., Vol. 8, M. Grayson and D. Eckroth, eds., Wiley-Interscience, New York, 1979, p. 592.

155. D. S. Breslow, *Chemtech* **20**, 540 (1990).

156. R. F. Ohm, *Chemtech* **10**, 183 (1980).

157. R. Ohm and C. Stein, in *Kirk-Othmer Encyclopedia of Chemical Technology*, 3rd ed., Vol. 18, M. Grayson, and D. Eckroth, eds., Wiley-Interscience, New York, 1982, p. 436.

158. K. J. Ivin, in *The Chemistry of Functional Groups, Supplement A3: The Chemistry of Double-Bonded Functional Groups*, S. Patai and Z. Rappoport, eds., Wiley, Chichester, 1997, Chapter 24, p. 1497.

159. J. C. Mol, in *Handbook of Heterogeneous Catalysis*, G. Ertl, H. Knözinger, and J. Weitkamp, eds., Wiley-VCH, Weinheim, 1997, Chapter 4.12.2, p. 2387.

160. K. J. Ivin and J. C. Mol, *Olefin Metathesis and Metathesis Polymerization*, Academic Press, London, 1997.

161. H.-G. Schmalz, *Angew. Chem., Int. Ed. Engl.* **34**, 1833 (1995).

162. R. H. Grubbs, S. J. Miller, and G. C. Fu, *Acc. Chem. Res.* **28**, 446 (1995).

163. C. Pariya, K. N. Jayaprakash, and A. Sarkar, *Coord. Chem. Rev.* **168**, 1 (1998).

164. J. W. Herndon, *Coord. Chem. Rev.* **181**, 177 (1999).

165. M. R. Buchmeiser, *Chem. Rev.* **100**, 1565 (2000).

166. A. Fürstner, *Top. Catal.* **4**, 285 (1997); *Angew. Chem., Int. Ed. Engl.* **39**, 3012 (2000).

167. M. Schuster and S. Blechert, *Angew. Chem., Int. Ed. Engl.* **36**, 2037 (1997).

168. S. K. Armstrong, *J. Chem. Soc., Perkin Trans. 1* 371 (1998).

169. K. J. Ivin, *J. Mol. Catal. A: Chem.* **133**, 1 (1998).

170. M. L. Randall and M. L. Snapper, *J. Mol. Catal. A: Chem.* **133**, 29 (1998).

171. R. H. Grubbs and S. Chang, *Tetrahedron* **54**, 4413 (1998).

172. *Alkene Metathesis in Organic Synthesis*, A. Fürstner, ed., Springer, Berlin, 1998.

173. D. L. Wright, *Curr. Org. Chem.* **3**, 211 (1999).

174. *Metathesis*, E. M. Carreira, ed.; *Synthesis* 857–903 (2000).

175. A. H. Hoveyda and R. R. Schrock, *Chem. Eur. J.* **7**, 945 (2001).

176. *Proc. 10th Int. Symp. Olefin Metathesis and Polymerization*, Tihany, 1993, L. Bencze, ed.; *J. Mol. Catal. A: Chem.* **90**, 1–224 (1994).

177. *Proc. 11th Int. Symp. Olefin Metathesis and Related Chemistry*, Durham, 1995, W. J. Feast and E. Khosravi, eds.; *J. Mol. Catal. A: Chem.* **115**, 1–228 (1997).

178. *Proc. 13th Int. Symp. Metathesis and Related Chemistry*, Kerkrade, 1999, J. C. Mol, ed.; *J. Mol. Catal. A: Chem.* **160**, 1–197 (2000).

179. F. Verpoort, A. R. Bossuyt, L. Verdonck, and B. Coussens, *J. Mol. Catal., A: Chem.* **115**, 207 (1997).

180. T. Szymańska-Buzar, *J. Mol. Catal. A: Chem.* **123**, 113 (1997).

181. B. L. Goodall, L. H. McIntosh, and L. F. Rhodes, *Macromol. Symp.* **89**, 421 (1995).

182. K. A. Vikulov, B. N. Shelimov, V. B. Kazansky, and J. C. Mol, *J. Mol. Catal.* **90**, 61 (1994).

183. I. Rodríguez-Ramos, A Guerrero-Ruiz, N. Homs, P. R. de la Piscina, and J. L. G. Fierro, *J. Mol. Catal. A: Chem.* **95**, 147 (1995).

184. J. Handzlik, J. Stoch, J. Ogonowski, and M. Mikoajczyk, *J. Mol. Catal. A: Chem.* **157**, 237 (2000).

185. T. Ookoshi and M. Onaka, *Chem. Commun.* 2399 (1998).

186. T. M. Mathew, J. A. K. du Plessis, and J. J. Prinsloo, *J. Mol. Catal. A: Chem.* **148**, 157 (1999).

187. J. C. Mol, *Catal. Today* **51**, 289 (1999).

188. H. G. Ahn, K. Yamamoto, R. Nalamura, and H. Niiyama, *Chem. Lett.* 503 (1992).

189. H. Yoshida, T. Tanaka, S. Matsuo, T. Funabiki, and S. Yoshida, *J. Chem. Soc., Chem. Commun.* 761 (1995).

190. K. Melis, D. De Vos, P. Jacobs, and F. Verpoort, *J. Mol. Catal. A: Chem.* **169**, 47 (2001).

191. P. A. van der Schaaf, R. A. T. M. Abbenhuis, W. P. A. van der Noort, R. de Graaf, D. M. Grove, W. J. J. Smeets, A. L. Spek, and G. van Koten, *Organometallics* **13**, 1433 (1994).

192. A. M. LaPointe and R. R. Schrock, *Organometallics* **14**, 1875 (1995).

193. R. H. Grubbs, M. Hillmyer, A. Benedicto and Z. Wu, *Am. Chem. Soc., Polym. Prep.* **35**(1), 688 (1994).

194. P. Schwab, R. H. Grubbs, and J. W. Ziller, *J. Am. Chem. Soc.* **118**, 100 (1996).

195. J. H. Oskam, H. H. Fox, K. B. Yap, D. H. McConville, R. O'Dell, B. J. Lichtenstein, and R. R. Schrock, *J. Organomet. Chem.* **459**, 185 (1993).

196. P. Schwab, M. B. France, J. W. Ziller, and R. H. Grubbs, *Angew. Chem., Int. Ed. Engl.* **34**, 2039 (1995).

197. T. N. Trnka and R. H. Grubbs, *Acc. Chem. Res.* **34**, 18 (2001).

198. L. Bencze and R. Szilágyi, *J. Organomet. Chem.* **465**, 211 (1994); **475**, 183 (1994); **505**, 81 (1995).

199. C. Adlhart, C. Hinderling, H. Baumann, and P. Chen, *J. Am. Chem. Soc.* **122**, 8204 (2000).

200. T. Weskamp, W. C. Schattenmann, M. Spiegler, and W. A. Herrmann, *Angew. Chem., Int. Ed. Engl.* **37**, 2490 (1998).

201. M. S. Sanford, L. M. Henling, M. W. Day, and R. H. Grubbs, *Angew. Chem., Int. Ed. Engl.* **39**, 3451 (2000).

202. J. Louie and R. H. Grubbs, *Angew. Chem., Int. Ed. Engl.* **40**, 247 (2001).

203. S. Chang, L. Jones, II, C. Wang, L. M. Henling, and R. H. Grubbs, *Organometallics* **17**, 3460 (1998).

204. A. Fürstner and K. Langemann, *J. Am. Chem. Soc.* **119**, 9130 (1997).

205. E. L. Dias and R. H. Grubbs, *Organometallics* **17**, 2758 (1998).

206. T. Weskamp, F. J. Kohl, W. Hieringer, D. Gleich, and W. A. Herrmann, *Angew. Chem., Int. Ed. Engl.* **38**, 2416 (1999).

207. C. W. Bielawski and R. H. Grubbs, *Angew. Chem., Int. Ed. Engl.* **39**, 2903 (2000).

208. M. S. Sanford, M. Ulman, and R. H. Grubbs, *J. Am. Chem. Soc.* **123**, 749 (2001).

209. S. M. Hansen, M. A. O. Volland, F. Rominger, F. Eisenträger, and P. Hofmann, *Angew. Chem., Int. Ed. Engl.* **38**, 1273 (1999).

210. A. Fürstner, M. Liebl, C. W. Lehmann, M. Picquet, R. Kunz, C. Bruneau, D. Touchard, and P. H. Dixneuf, *Chem. Eur. J.* **6**, 1847 (2000).

211. J. Wolf, W. Stüer, C. Grünwald, H. Werner, P. Schwab, and M. Schulz, *Angew. Chem., Int. Ed. Engl.* **37**, 1124 (1998).

212. S. L. Aeilts, D. R. Cefalo, P. J. Bonitatebus, Jr., J. H. Houser, A. H. Hoveyda, and R. R. Schrock, *Angew. Chem., Int. Ed. Engl.* **40**, 1452 (2001).

213. V. Amir-Ebrahimi, J. G. Hamilton, J. Nelson, J. J. Rooney, J. M. Thompson, A. J. Beaumont, A. D. Rooney, and C. J. Harding, *Chem. Commun.* 1621 (1999).

214. A. Fürstner and G. Seidel, *Angew. Chem., Int. Ed. Engl.* **37**, 1734 (1998).

215. U. H. F. Bunz and L. Kloppenburg, *Angew. Chem., Int. Ed. Engl.* **38**, 478 (1999).

216. U. H. F. Bunz, *Chem. Rev.* **100**, 1605 (2000).

217. X.-P. Zhang and G. C. Bazan, *Macromolecules* **27**, 4627 (1994).

218. A. Mortreux, F. Petit, M. Petit, and T. Szymanska-Buzar, *J. Mol. Catal. A: Chem.* **96**, 95 (1995).

219. Z. Lin and M. B. Hall, *Organometallics* **13**, 2878 (1994).

220. H. Kouzai, T. Masuda, and T. Higashimura, *J. Polym. Sci., A. Polym. Chem.* **32**, 2523 (1994); *Polymer* **35**, 4920 (1994).

221. L. Kloppenburg, D. Jones, and U. H. F. Bunz, *Macromolecules* **32** 4194 (1999).

222. U. H. F. Bunz, *Acc. Chem. Res.* **34**, 998 (2001).

223. K. Weiss, A. Michel, E.-M. Auth, U. H. F. Bunz, T. Mangel, and K. Müllen, *Angew. Chem., Int. Ed. Engl.* **36**, 506 (1997).

224. L. R. Sita, *Macromolecules* **28**, 656 (1995).

225. G. W. Coates and R. H. Grubbs, *J. Am. Chem. Soc.* **118**, 229 (1996).

226. W. A. Nugent, J. Feldman and J. C. Calabrese, *J. Am. Chem. Soc.* **117**, 8992 (1995).

227. A. J. Phillips and A. D. Abell, *Aldrichimica Acta* **32**, 75 (1999).

228. B. M. Trost and M. J. Krische, *Synlett*, 1 (1998).

229. B. M. Trost and F. D. Toste, *J. Am. Chem. Soc.* **121**, 9728 (1999).

230. P. Buchacher, W. Fischer, K. D. Aichholzer, and F. Stelzenr, *J. Mol. Catal. A: Chem.* **115**, 163 (1997).

231. A. Mühlebach, P. Bernhard, N. Bühler, T. Karlen, and A. Ludi, *J. Mol. Catal. A: Chem.* **90**, 143 (1994).

232. C. D. Mistele, H. H. Thorp, and J. M. DeSimone, *Am. Chem. Soc., Polym. Prep.* **36**(1), 507 (1995).

233. M. A. Hillmyer, A. D. Benedicto, S. B. Nguyen, Z. Wu, and R. H. Grubbs, *Macromol. Symp.* **89**, 411 (1995).

234. L. Delaude, A. Demonceau, and A. F. Noels, *Chem. Commun.* 986 (2001).

235. N. Zhang, S. R. Schricker, F. Wudl, M. Prato, M. Magginin, and G. Scorrano, *Chem. Mat.* **7**, 441 (1995).

236. *Chem. Weeks* (Nov. 20, 1985).

237. *European Chem. News* p. 57 (Sept. 29, 1997).

238. *European Chem. News* p. 33 (Oct. 6, 1997).

239. P. Amigues, Y. Chauvin, D. Commereuc, C. C. Lai, Y. H. Liu, and J. M. Pan, *Hydrocarbon Process.* **69**(10), 79 (1990).

240. J. Cosyns, J. Chodorge, D. Commereuc, and B. Torck, *Hydrocarbon Process., Int. Ed.* **77**(3), 61 (1998).

13

OLIGOMERIZATION AND POLYMERIZATION

13.1. OLIGOMERIZATION

According to the IUPAC nomenclature, oligomers[1] are molecules containing a few constitutional units repetitively linked to each other. This does not define any specific molecular weight or degree of polymerization. It states, however, that the physical properties of oligomers vary with the addition or removal of one or several units.

Oligomerization has great practical significance. Dimerization of simple alkenes to produce gasoline blending components, production of terminal alkenes (α-olefins) for use as comonomers in polymerization and as intermediates in detergent manufacture, and cyclodimerization and cyclotrimerization of 1,3-dienes are the most important examples. Oligomerization may be achieved thermally or catalytically; the latter having greater practical importance.

13.1.1. Acid-Catalyzed Oligomerization

Lewis acid halides and Brønsted acids, that is, common Friedel–Crafts catalysts, may effect oligomerization of alkenes.[2,3] Two types of transformation may occur: true oligomerization and conjunct polymerization. In true oligomerization the products are alkenes having molecular weights that are integral multiples of the molecular weight of the monomer alkene. Conjunct polymerization, also called *hydropolymerization*, in contrast, yields a product that is a complex mixture of saturated (alkanes and cycloalkanes) and unsaturated (alkenes, alkapolyenes, cycloalkenes, and cycloalkapolyenes) hydrocarbons, and occasionally even aromatic compounds; moreover, the products may include compounds containing less carbon atoms than the multiples in the monomer used. The term *conjunct polymerization* was coined by Ipatieff in 1936 to describe the joint (simultaneous) formation of saturated and polyunsaturated hydrocarbons during polymerization.

The type of oligomerization that occurs depends largely on the catalyst and reaction conditions. With aluminum chloride promoted by hydrogen chloride or with boron trifluoride promoted by hydrogen fluoride, the oligomerization of ethylene occurs at room temperature under pressure to yield conjunct polymers consisting of two layers: (1) a clear liquid upper layer consisting chiefly of alkanes and cycloalkanes having a wide molecular weight range and (2) the lower layer, a viscous red or red-brown liquid, consisting of polyunsaturated aliphatic and cyclic hydrocarbons complexed with the catalyst. Similarly, most alkenes, when dissolved in concentrated sulfuric acid at room temperature, slowly become turbid and finally separate into two layers. The polyunsaturated hydrocarbons dissolve in the acid, yielding a less active catalyst known as "sludge" and separate when diluted with water.

When 91% sulfuric acid is used as catalyst, ethylene undergoes no oligomerization because stable ethyl hydrogen sulfate and diethyl sulfate are formed. When the acid is used for the oligomerization of propylene, conjunct polymer is only very slowly obtained, due to the formation of fairly stable isopropyl hydrogen sulfate. Isobutylene, in contrast, readily undergoes dimerization with dilute (65–70%) sulfuric acid as catalyst. This reaction is used commercially to obtain diisobutylene (isomeric trimethylpentenes) for use as motor gasoline additive. Use of sulfuric acid at greater concentration than about 90% brings about conjunct polymerization of isobutylene at low temperatures.[4]

With solid phosphoric acid catalyst, ethylene at about 300°C and 37 atm yields a mixture of gaseous and liquid conjunct polymers, including butenes, isobutane, and liquid compounds consisting of alkenes, alkanes, cycloalkanes, and aromatic hydrocarbons.[4] Propene, on the other hand, undergoes little conjunct polymerization when passed over solid phosphoric acid at 150–250°C and 10 atm pressure. Only true oligomer is obtained. It consists of a small amount of dimer, much trimer, and a small amount of higher-boiling product, mainly propene tetramers and pentamers. Conjunct polymerization, however, does occur at higher temperatures. True oligomerization occurs with n-butenes at 170°C or isobutylene at 95–120°C.

It seems certain that acid-catalyzed true oligomerization occurs via a carbocationic mechanism as first suggested by Whitmore.[5] Thus, the formation of diisobutylenes by dimerization of isobutylene in the presence of acidic catalysts occurs according to the following equation:

$$
\underset{\overset{||}{CH_2}}{\underset{|}{Me-C}}\text{-Me} + H^+ \;\rightleftharpoons\; \underset{\underset{Me}{|}}{Me-\overset{+}{C}}\text{-Me} \;\overset{isoC_4H_8}{\rightleftharpoons}\; \underset{Me\;\;Me}{Me-\underset{|}{\overset{|}{C}}CH_2\overset{+}{\underset{|}{C}}Me}
$$

1

$$
\overset{-H+}{\rightleftharpoons}\; \underset{Me\;\;Me}{Me-\underset{|}{\overset{|}{C}}CH_2\underset{|}{C}=CH_2} \;+\; \underset{Me\;\;Me}{Me-\underset{|}{\overset{|}{C}}CH=\underset{|}{C}Me} \qquad (13.1)
$$

82   18

2

Friedel–Crafts catalysts and hydrogen halides, protic acids, silica–alumina-type catalysts, or other protic catalysts are effective.

Carbocation **1** can undergo hydride and methyl shifts to yield 2,3,4-trimethyl-pentenes:

$$
\mathbf{1} \;\rightleftharpoons\; \underset{\underset{\text{Me \quad Me}}{|\quad\;\;|}}{\overset{\overset{\text{Me}}{+\;|}}{\text{Me}\overset{}{\text{C}}\text{CHCHMe}}} \;\xrightarrow{-H^+}\; \underset{\underset{\text{Me \quad Me}}{|\quad\;\;|}}{\overset{\overset{\text{Me}}{|}}{\text{MeC}=\text{CCHMe}}} \tag{13.2}
$$

3

Trimers may be formed through the addition of a *tert*-butyl carbocation to **2** or via addition of carbocation **1** to isobutylene. The trimer fraction mainly consists of 4-methylene-2,2,6,6-tetramethylheptane and 2,2,4,6,6-pentamethyl-3-heptene.

Since isobutylene is a very reactive olefin, its oligomerization can be promoted by almost any electrophilic catalyst. More recently fluorinated alumina,[6] cation-exchange resins,[7] benzylsulfonic acid siloxane,[8] pentasil zeolites,[9] and perfluori-nated resinosulfonic acids[10] were studied. Some of these catalysts may bring about improved oligomerization.

This reaction mechanism also explains why oligomerization of propylene, as mentioned, tends to yield much more trimer than dimer. The dimeric carbocation (**4**) formed readily undergoes a hydride shift to yield tertiary cation **5**:

$$
\underset{\mathbf{4}}{\overset{\overset{\text{Me}}{|}}{\text{MeCHCH}_2\overset{+}{\text{C}}\text{HMe}}} \;\rightleftharpoons\; \underset{\mathbf{5}}{\overset{\overset{\text{Me}}{|}}{\underset{+}{\text{Me}\text{C}\text{CH}_2\text{CH}_2\text{Me}}}} \;\xrightarrow{\text{MeCH}=\text{CH}_2}\; \underset{\overset{|}{\text{Me}}}{\overset{\overset{\text{Me}}{|}}{\text{MeCH}_2\text{CH}_2\overset{+}{\text{C}}\text{CH}_2\overset{+}{\text{C}}\text{HMe}}} \tag{13.3}
$$

The latter, however, is more reactive then secondary cation **4** and reacts with propylene more rapidly than it eliminates a proton.

Conjunct polymerization involves hydride abstraction from an olefin or oligomer by a carbocation. For example, during the oligomerization of isobutylene by dissolving it in 96% sulfuric acid, octene products may react with carbocation **1** [Eq. (13.4)] before they eliminate a proton to form trimethylpentenes:[11]

$$
\mathbf{1} + \mathbf{3} \longrightarrow \underset{\underset{\text{Me \quad Me}}{|\quad\;\;|}}{\overset{\overset{\text{Me}}{|}}{\text{MeCCH}_2\text{CHMe}}} + \underset{\underset{\text{Me \quad Me}}{|\quad\;\;|}}{\overset{\overset{\text{Me}}{|\;+}}{\text{MeC}=\text{CCMe}}} \;\xrightarrow{-H+}\; \underset{\underset{\text{Me \quad Me}}{|\quad\;\;|}}{\overset{\overset{\text{Me}}{|}}{\text{MeC}=\text{CC}=\text{CH}_2}} \tag{13.4}
$$

Much of the polyunsaturated material includes polymethylated cyclopentadienes regardless of the olefin undergoing oligomerization. Formation of such products results from further hydride abstraction from an alkadiene:[11]

$$
\mathbf{6} \;\underset{\xleftarrow{R^+,\,-RH}}{\rightleftharpoons}\; \underset{\underset{\text{Me \quad Me}}{|\quad\;\;|}}{\overset{\overset{\text{Me}}{+\;|}}{\text{CH}_2\text{C}=\text{CC}=\text{CH}_2}} \longrightarrow \; \text{image} \;\xrightarrow{-H^+}\; \text{image} \tag{13.5}
$$

7

Carbocation **7**, which is an allylic ion, is quite reactive and undergoes intramolecular self-condensation.

Acidic solids can also be used as oligomerization catalysts.[12] Zeolites, the most widely studied catalysts, usually exhibit poor selectivity in the oligomerization of ethylene to linear α-olefins. High-silica zeolites with strong Brønsted acidic sites, however, were shown to give linear oligomers.[13] H-ZSM-5 is used in an industrial process developed by Mobil to oligomerize higher alkenes.

Practical Applications. Acid-catalyzed oligomerizations are practiced industrially.[14–17] Isobutylene dimerization is a very important process. It yields diisobutylene, mainly a 4 : 1 mixture of 2,4,4-trimethyl-1-pentene and 2,4,4-trimethyl-2-pentene,[18,19] which may be catalytically hydrogenated to 2,2,4-trimethylpentane (isooctane). The reaction can be carried out with sulfuric acid. The so-called hotacid process (68% H_2SO_4, 80°C)[20] has the advantage over the cold-acid method that in addition to dimerization of isobutylene, codimerization of isobutylene and *n*-butylenes also occurs. In this way higher dimer yields (87–90%) are achieved. An acidic ion-exchange resin at 100°C and about 20 atm (Bayer process)[17] yields dimers and trimers formed in a ratio of 3 : 1 at almost complete conversion. Oligomer distribution can be affected by temperature; increasing temperature brings about a decrease in molecular weight.

Solid or liquid phosphoric acid is used in industry to produce gasoline polymers from C_3–C_4 alkenes.[21–23] The most prevalent is UOP's solid phosphoric acid catalyst. Also called *silicophosphoric acid*, it is a pelletized calcined mixture of H_3PO_4 and kieselguhr.[21] This catalyst was widely used in the oligomerization of propylene, a once important large-scale industrial process to produce propylene trimer and tetramer.[15,16] Oligomerization can be accomplished at 170–220°C and 14–42 atm. The once-through tetramer yield is relatively low ($\sim$ 20%), but the dimer and trimer are recycled to attain high overall yields. Both the trimer and tetramer were used to alkylate benzene in the production of detergents. Their use, however, has diminished because of the poor biodegradability of the product branched alkylbenzenesulfonate.

A solid acid catalyst introduced in 1985 is used in UOP's Hexall process to selectively dimerize propylene at high conversion.[24] The C_6 olefin content of the product can reach higher than 90% with an octane number of 95–96.

Oligomerization promoted by H-ZSM-5 is a crucial step in the Mobil olefin-to-distillate-and-gasoline process.[25] When operated at relatively low temperature and high pressure (200–300°C, 20–105 atm), called the *distillate process*, the products are higher-molecular-weight isoolefins.[26,27] Under these conditions light olefins are first converted to higher oligomers. Subsequent isomerization and cracking give intermediate C_4–C_7 olefins. Finally the latter participate in copolymerization to yield the product. The high selectivity of the process (selective formation of higher isoolefins) is a result of the shape-selective action of ZSM-5 preventing hydrogen transfer reactions.[28] The retardation of conjunct polymerization (formation of polyunsaturated and aromatic compounds) ensures long catalyst life. Changing the operating conditions (higher temperature, lower pressure) leads to the formation of products of lower molecular weight with higher aromatic content (gasoline process).

Oligomerization of 1-decene in the presence of a silica–BF_3–H_2O catalyst was shown to yield a mixture of di-, tri- and tetramers that gives a lubricant after hydrogenation.[29]

13.1.2. Base-Catalyzed Oligomerization

All unsaturated hydrocarbons that possess removable allylic hydrogens or activated double bonds readily undergo oligomerization in the presence of basic reagents.[30,31] Alkali metals, organoalkali compounds, and alkali alcoholates are most frequently applied.

One important early study described the oligomerization of ethylene, propylene, isobutylene, and cyclohexene in the presence of sodium.[32] Ethylene itself is reluctant to form oligomers but readily undergoes cooligomerization with other alkenes containing an allylic hydrogen.[33] A mechanism with the involvement of a resonance-stabilized allylic carbanion (**8**) accounts for the observations [Eq. (13.6)]. Proton transfer between the anion and the original alkene yields the product codimer and generates a new anion [Eq. (13.7)]:

$$MeCH{=}CH_2 \xrightarrow[-RH]{R^-Na^+} \underset{\mathbf{8}}{[CH_2 \text{---} CH \text{---} CH_2]^- \, M^+} \xrightleftharpoons{CH_2{=}CH_2} CH_2{=}CHCH_2CH_2\ddot{C}H_2 \, M^+ \quad (13.6)$$

$$CH_2{=}CHCH_2CH_2\ddot{C}H_2 \, M^+ \;+\; MeCH{=}CH_2 \;\rightleftharpoons\; CH_2{=}CHCH_2CH_2CH_3 \;+\; \mathbf{8} \quad (13.7)$$

Dimerization of propylene leads to the formation of isomeric methylpentenes in the presence of alkali metals.[34] The product distribution strongly depends on the metal used. 4-Methyl-1-pentene is formed with high selectivity in the presence of potassium and cesium. Because of extensive isomerization, an equilibrium mixture of the isomers with 4-methyl-2-pentene and 2-methyl-2-pentene as the main products was isolated in a reaction catalyzed with sodium.

Styrenes dimerize and cyclodimerize depending on the catalyst, the reaction conditions, and the substrate. α-Methylstyrene gives cyclic dimer **9** in 93% yield (*tert*-BuOK, 160°C), whereas linear dimers (isomeric 2,5-dipehylhexenes) are formed in 83% yield at higher temperature (225°C).[35] In contrast, a very low yield (11%) of cyclic dimer **10** was attained in the reaction of styrene:

9	**10**

Of the conjugated dienes, 1,3-butadiene undergoes base-initiated polymerization but does not yield oligomers since there is no proton transfer reaction. Isoprene and 1,3-pentadiene, in contrast, give linear and cyclic dimers.

13.1.3. Metal-Catalyzed Oligomerization

Metal-catalyzed oligomerizations are important reactions concerning both their chemistry and practical applications.[12,36,37] Since the catalysts [(Ti(IV), Zr(IV), Ni(I), and Ni(II) complexes] used in coordination polymerization can also be active in oligomerization, catalyst systems and mechanisms will be discussed in more detail in Section 13.2.4. A short discussion of oligomerization with emphasis on the most important processes of practical significance, however, is appropriate here.

The first metal-catalyzed ethylene dimerization reaction was accidentally discovered by Ziegler and coworkers while studying ethylene oligomerization to long-chain α-olefins in the presence of aluminum alkyls.[38] They observed that nickel impurities from the reactor caused dimerization instead of oligomerization leading to the formation of 1-butene. In fact, the growth reaction (i.e., oligomerization of ethylene) is followed by a displacement reaction by ethylene catalyzed by nickel resulting in the formation of an oligomer and regeneration of the aluminum alkyl. The aluminum-catalyzed chain growth step takes place through insertion into the C–Al bond via a nonsymmetric four-center transition state.[39]

Nickel,[40,41] like almost all metal catalysts (e.g., Ti and Zr) used for alkene dimerization, effects the reaction by a three-step mechanism.[12] Initiation yields an organometallic intermediate via insertion of the alkene into the metal–hydrogen bond followed by propagation via insertion into the metal–carbon bond [Eq. (13.8)]. Intermediate **11** either reacts further by repeated insertion [Eq. (13.9)] or undergoes chain transfer to yield the product and regenerate the metal hydride catalyst through β-hydrogen transfer [Eq. (13.10)]:

$$\backslash\!\!\!\underset{/}{M}\!-\!H \; + \; CH_2\!=\!CH_2 \; \longrightarrow \; \backslash\!\!\!\underset{/}{M}\!-\!CH_2CH_3 \; \xrightarrow{\;CH_2=CH_2\;} \; \backslash\!\!\!\underset{/}{M}CH_2CH_2CH_2CH_3 \qquad (13.8)$$

11

$$\mathbf{11} \; \begin{cases} \xrightarrow{(n-1)\,CH_2=CH_2} \; \backslash\!\!\!\underset{/}{M}CH_2(CH_2)_{2n}CH_3 & (13.9) \\[2em] \xrightarrow{} \; C_4H_8 \; + \; \backslash\!\!\!\underset{/}{M}\!-\!H & (13.10) \end{cases}$$

When the transfer reaction competes successfully with further insertion, as in the case of nickel, dimerization becomes the dominant transformation. When metal hydride elimination, in turn, is slow relative to insertion, polymeric macromolecules are formed. Ligand modification, the oxidation state of the metal, and reaction conditions affect the probability of the two reactions. Since nickel hydride, like other metal hydrides, catalyzes double-bond migration, isomeric alkenes are usually isolated.

The simplest possible alkene oligomerization reaction, the dimerization of ethylene to butenes, is a well-studied reaction, and an industrial process was also developed for the selective formation of 1-butene[42] (IFP Alphabutol process).

A soluble titanium-based modified Ziegler–Natta catalyst [Ti(OR)$_4$–Et$_3$Al, R $= n$-Bu, isoPr] is employed in the reaction.[42] Since similar catalysts may be used for the oligomerization and polymerization of ethylene, the nature and oxidation state of the metal and reaction conditions determine selectivity. Ti^{4+} was found to be responsible for high dimerization selectivity, whereas polymerization was shown to be catalyzed by Ti^{3+}. According to a proposed mechanism,[42,43] this catalyst effects the concerted coupling of two molecules of ethylene to form a metallacyclopentane intermediate that decomposes via an intramolecular β-hydrogen transfer:

$$
\begin{array}{ccc}
\underset{CH_2}{\overset{CH_2}{\underset{|}{\overset{||}{Ti\overset{CH_2}{\underset{CH_2}{}}}}}} & \longrightarrow \quad Ti\!\!\bigcirc \quad \longrightarrow & Ti \;+\; \underset{CH_3}{\overset{CH_2}{\underset{|}{\overset{||}{\underset{CH_2}{CH}}}}}
\end{array}
\qquad (13.11)
$$

Another simple oligomerization is the dimerization of propylene. Because of the formation of a relatively less stable branched alkylaluminum intermediate, displacement reaction is more efficient than in the case of ethylene, resulting in almost exclusive formation of dimers. All possible C$_6$ alkene isomers are formed with 2-methyl-1-pentene as the main product and only minor amounts of hexenes. Dimerization at lower temperature can be achieved with a number of transition-metal complexes, although selectivity to 2-methyl-1-pentene is lower. Nickel complexes, for example, when applied with aluminum alkyls and a Lewis acid (usually EtAlCl$_2$), form catalysts that are active at slightly above room temperature. Selectivity can be affected by catalyst composition; addition of phosphine ligands brings about an increase in the yield of 2,3-dimethylbutenes (mainly 2,3-dimethyl-1-butene).

A new, selective catalyst was reported for the dimerization of propylene to 2,3-dimethylbutenes that are valuable intermediates in the manufacture of specialty chemicals.[44] The catalyst is composed of nickel naphthenate, Et$_3$Al, a phosphine, a diene, and chlorinated phenol. Either 2,3-dimethyl-1-butene or 2,3-dimethyl-2-butene can be selectively produced by controlling the catalyst composition.

Cyclooligomerization. Ever since the first cyclodimerization reaction of 1,3-butadiene to yield 1,5-cyclooctadiene catalyzed by a so-called Reppe catalyst was reported[45] [Eq. (13.12)], cyclooligomerization of conjugated dienes has been intensively studied:

$$
2 \diagup\!\!\!\diagdown\!\!\!\diagup \xrightarrow[\text{benzene, 120–130°C, 4 h}]{[(Ph_3P)_2Ni(CO)_2]} \quad \bigcirc\!\!\!\bigcirc \;+\; \text{(4-vinylcyclohexene)} \qquad (13.12)
$$

30–40%   10%

An industrial process for the synthesis of styrene through the transformation of 4-vinylcyclohexene the other possible cyclodimer has been developed.

Homogeneous nickel complexes proved to be versatile catalysts in dimerization and trimerization of dienes to yield different oligomeric products.[46–55] Depending on the actual catalyst structure, nickel catalyzes the dimerization of 1,3-butadiene to yield isomeric octatrienes, and the cyclodimerization and cyclotrimerization to give 1,5-cyclooctadiene and all-*trans*-1,5,9-cyclododecatriene, respectively[46,56] [Eq. (13.13)]. Ziegler-type complexes may be used to form *cis,trans,trans*-1,5,9-cyclododecatriene[37,57,58] [Eq. (13.14)], which is an industrial intermediate:

$$(13.13)$$

$$(13.14)$$

Formally, although these cyclooligomerizations can be considered as cycloaddition reactions, they are known not to occur through a direct cycloaddition process. In the best understood nickel-catalyzed reaction the dimer **12** was shown to be the key intermediate in cyclization.[46,59,60] Acetylene readily undergoes cyclooligomerization in the presence of metal catalysts to form benzene and cyclooctatetraene [Eq. (13. 15)] as well as higher homologs:

$$(13.15)$$

Both compounds were first synthesized in this way by Reppe: Ni(II) salts were shown to be active in the formation of benzene,[61] whereas cyclooctatetraene was isolated as the main product with zerovalent phosphinonickel complexes.[62] No styrene was, however, obtained in the tetramerization, although it was the sought-after product by Reppe.

The formation of benzene (or substituted benzene derivatives) is a common transformation catalyzed by numerous homogeneous and heterogeneous metal catalysts, mainly Co, Rh, Pd, and Ni.[63–69] Even highly crowded molecules, such as hexaisopropylbenzene, could be synthesized in the presence of metal carbonyls such as $[Co(CO)_4]_2$.[70] A very simple catalyst system, Me_3SiCl and Pd on carbon in refluxing tetrahydrofuran, has been shown to transform symmetrical alkynes into hexaalkylbenzenes in excellent yield.[71]

It appears difficult to propose a unified mechanism to explain all experimental observations of the cyclotrimerization of acetylene. The most common pathway, studied mainly with cobalt complexes,[72,73] involves a metallacyclopentadiene intermediate:

$$(13.16)$$

This reacts further with a third molecule of acetylene via a Diels–Alder reaction (through 7-metallanorbornadiene **13**) or insertion. Experimental evidence[74,75] favors the former pathway.

This mechanism is supported by the transformation of preformed metallacyclopentadienes with alkynes,[73,76–78] and labeling experiments[79] that excluded the involvement of cyclobutadiene intermediates. It also accounts for the observation that terminal alkynes yield 1,2,4- (and 1,3,5-) trisubstituted benzene derivatives as the main product but not 1,2,3 derivatives. In contrast with this picture in cyclotrimerization with $PdCl_2$-based catalysts, stepwise linear insertion of alkynes takes place without the involvement of palladacyclopentadiene.[80]

Cyclotetramerization to form cyclooctatetraene occurs only with nickel.[46,63,68] The best catalysts are octahedral Ni(II) complexes, such as bis(cyclooctatetraene) dinickel.[46] Internal alkynes do not form cyclooctatetraene derivatives but participate in cooligomerization with acetylene. Of the possible mechanistic pathways, results with $[1-^{13}C]$-acetylene[81] favor a stepwise insertion process or a concerted reaction, and exclude any symmetric intermediate (cyclobutadiene, benzene). The involvement of dinuclear species are in agreement with most observations.[46,82–84]

Practical Applications. IFP's Alphabutol process is used to dimerize ethylene selectively to 1-butene.[43,85] The significance of this technology is the use of 1-butene as a comonomer in the polymerization of ethylene to produce linear low-density polyethylene (see Section 13.2.6). Under the reaction conditions applied in industry (50–60°C, 22–27 atm), the selectivity of 1-butene formation is higher than 90% at the conversion of 80–85%. Since no metal hydride is involved in this system, isomerization does not take place and only a small amount of higher-molecular-weight terminal alkenes is formed.

Dimerization of propylene and butenes is also accomplished commercially to produce valuable motor gasoline components. IFP's Dimersol G process[86–88] dimerizes propylene to isohexenes in the presence of a nickel-based Ziegler–Natta catalyst. Codimerization of propylene with butenes is also possible. Dimersol E is designed to cooligomerize ethylene with residual propylene found in FCC unit off-gas.[89] Industrial dimerization of n-butenes to octenes is carried out with a soluble nickel catalyst (Dimersol X process)[90,91] or in the presence of a solid catalyst (Huels Octol process).[92,93] In both cases n-octenes and methylheptenes are the predominant products. The solid catalyst developed jointly by Huels and UOP is highly selective to produce normal and monobranched C_8–C_{12} olefins. These are transformed to C_9–C_{13} alcohols through hydroformylation and hydrogenation and used as plasticizers.[93] Huels, however, developed two catalysts, and one of these can produce more highly branched C_8 alkenes for gasoline blending.[94] The feedstocks in the process are raffinate II or raffinate III. The former is a steam cracker C_4 cut rich in n-butylenes (70–80%) after the removal of 1,3-butadiene (by hydrogenation) and isobutylene (by hydration to yield MTBE). Raffinate III, in turn, is what remains after the removal of 1-butene from raffinate II. The Huels process is run below 100°C under sufficient pressure to ensure liquid-phase operation.

Until the mid-1970s metal-catalyzed propylene dimerization had practical significance in isoprene manufacture. Goodyear developed a process to dimerize propylene in the presence of tri-n-propylaluminum to yield 2-methyl-1-pentene.[16,95,96] This was then isomerized to 2-methyl-2-pentene followed by cracking into isoprene and methane. This and other synthetic pocesses, however, are no longer practiced since they are not competitive with isoprene manufactured by cracking of naphtha or gas oil.

Codimerization of ethylene and 1,3-butadiene is also of commercial significance since the product 1,4-hexadiene is used as a comonomer in the manufacture of ethylene–propylene–diene elastomers. Rhodium and nickel catalysts are used since they are the most active and selective, bringing about the formation of the required *trans* isomer.[97]

Oligomerization of ethylene to α-olefins is undeniably the most important oligomerization reaction.[17,98–101] C_{10}–C_{18} α-olefins thus produced are used mainly as intermediates to manufacture detergents. Three technologies have been commercialized. Two processes (Ethyl and Chevron/Gulf) use triethylaluminum to effect Ziegler chain growth chemistry.[102] In the conventional ethyl process a stoichiometric chain growth reaction leads to the formation of an aluminum alkyl mixture. The chain length of the alkyl groups corresponds to a statistical distribution with a maximum around C_6–C_{10} alkenes. The product is usually recycled to get a better yield of the most desirable C_6–C_{14} alkenes. The ethyl technology is a two-step process since the aluminum alkyl mixture is subsequently reacted in a displacement reaction. During this treatment added excess ethylene displaces the alkyl groups to form terminal alkenes and triethylaluminum. The first step is carried out at relatively low temperature (90–120°C, 100–200 atm) where chain growth predominates, whereas displacement requires more severe reaction conditions (260–310°C, 7–20 atm).

Instead of the displacement reaction, aluminum alkyls may be oxidized with dry air to produce aluminum alkoxides that, after hydrolysis with dilute sulfuric acid or water, yield linear primary alcohols with an even number of carbon atoms (fatty alcohols).[103,104] The technology originally developed by Conoco is known as the *Alfol process.*[105–107]

The Chevron/Gulf process also uses triethylaluminum but in a catalytic reaction at higher temperature. Reaction conditions exert a strong effect on product distribution. Under the proper conditions (200–250°C, 140–270 atm) the rates of insertion and chain transfer (displacement) are comparable, ensuring frequent β-hydrogen elimination. A broader product distribution compared with that of the two-step ethyl process is obtained.

The Shell higher olefin process (SHOP)[98,108] is also a catalytic transformation. It employs a zerovalent nickel catalyst with suitable phosphine ligands prepared by reducing a nickel salt with $NaBH_4$ in 1,4-butanediol solvent.[109] Oligomerization is carried out at relatively low temperature (80–120°C). High ethylene pressure (70–140 atm), however, is necessary to control linearity since a sufficiently high ethylene concentration prevents α-olefins from reinserting into the chain. α-Olefins produced by the Shell process have a very broad chain length distribution. Shell, however, developed processes to utilize lower and higher alkenes through isomerization and metathesis (see Section 12.3). Product distributions[110] of the three industrial oligomerization processes are given in Table 13.1. A characteristic composition of thermal cracking is also included for comparison.

Other ethylene oligomerization processes applying zirconium catalyst systems have been patented.[12]

The synthesis of both 1,5-cyclooctadiene and 1,5,9-cyclododecatriene through the cyclodimerization and cyclotrimerization of 1,3-butadiene is practiced commercially since both compounds have multiple applications. The corresponding tetrahalo- and hexahaloderivatives are used as fire retardants. 1,5,9-Cyclododecatriene can be transformed to 1,12-dodecanedioic acid, which is employed in the synthesis of polyesters and polyamides. 1,13-Tetradecadiene is a crosslinking agent, and lauryl lactame is an intermediate in the production of nylon-12.[111] $TiCl_4$-$EtAlCl_2$-Et_2AlCl is the catalyst in the Huels process, which yields the *cis*, *trans,trans* isomer with 90% selectivity.[112] The all-*trans* isomer is produced in

Table 13.1. Product composition of ethylene oligomerizations and thermal cracking[110]

Compounds	Ethylene Oligomerization			Thermal Cracking
	Ethyl	Gulf	Shell	
α-Olefins	63–97.5	91–97	96–97.5	83–89
Branched olefins	1.9–29.1	1.6–7.8	1.0–3.0	3–12
Internal olefins	0.6–8.2		1.0–2.4	
Paraffins	0.1–0.8	1.4	<0.1	1.5–2
Dienes	—	—	<0.1	3.2–6
Aromatics	—	—	<0.1	0–0.4
Total monoolefins	< 99[a]	98.6	99.9	92–95[a]

[a]Estimated values.

the presence of bis(acrylonitrile)nickel (Toyo Soda process).[56] Other nickel catalysts—[Ni(CH$_2$= CHCN)$_2$(PPh$_3$)] and {Ni(CNPh)$_4$[P(OPh)$_3$]}—are more selective in the formation of 1,5-cyclooctadiene.[56]

The Cu$^+$/zeolite-catalyzed cyclodimerization of 1,3-butadiene at 100°C and 7 atm was found to give 4-vinylcyclohexene [Eq. (13.12)] with high (>99%) selectivity. Subsequent oxidative dehydrogenation over an oxide catalyst in the presence of steam gives styrene. The overall process developed by Dow Chemical[113] offers an alternative to usual styrene processes based on ethylation of benzene (see Section 5.5.2).

13.2. POLYMERIZATION

Polymers are high (or relatively high)-molecular-weight substances composed of repetitive linking together of small units (monomers) in sufficient numbers, but addition or removal of some units, in contrast to oligomers, do not change their properties. The two major types of polymerization are step growth or condensation polymerization and chain growth or addition polymerization. The latter is characteristic of unsaturated hydrocarbons. In the context of our discussion only hydrocarbon polymers will be discussed, excluding their functionalized derivatives.

It follows from the chain-reaction character that a successful polymerization requires initiation, namely, the formation of a reactive species capable of interacting with an unsaturated molecule. Further stages of polymerization are propagation (chain growth) through a chain reaction and termination. An important general feature of addition polymerization reactions is the continuous rapid addition of monomer molecules to the growing macromolecular chain throughout the whole process, but reaction between oligomers and macromolecules do not take place.

On the basis of the nature of the initiation step, polymerization reactions of unsaturated hydrocarbons can be classified as cationic, anionic, and free-radical polymerization. Ziegler–Natta or coordination polymerization, though, which may be considered as an anionic polymerization, usually is treated separately. The further steps of the polymerization process (propagation, chain transfer, termination) similarly are characteristic of each type of polymerization. Since most unsaturated hydrocarbons capable of polymerization are of the structure of CH$_2$=CHR, *vinyl polymerization* as a general term is often used.

Because of its chemical and practical significance, polymerization is exhaustively reviewed and excellent monographs,[114–129] a multivolume encyclopedia,[130] as well as the series Advances in Polymer Science,[131] cover all aspects of the field. Consequently, our aim is not a comprehensive treatment, but to discuss the most important general aspects of the major types of polymerization of unsaturated hydrocarbons and to give selected examples of practical significance.

13.2.1. Cationic Polymerization

Historically acid-catalyzed (cationic) polymerization of alkenes is the oldest polymerization. At the end of the eighteenth century Bishop Watson described resin formation when turpentine was treated with sulfuric acid.[132] Numerous observations were made in the next century, including the polymerization of styrene[133] induced by $SnCl_4$. The sulfuric acid–catalyzed polymerization of styrene and isobutylene was reported by Berthelot[134] and Butlerov,[135] respectively. Industrial processes were developed later. Electrophilic catalysts usually bring about conjunct polymerization of ethylene, which yields, however, low-molecular-weight oils (oligomers) in the presence of $AlCl_3$ under appropriate conditions suitable for lubricating oils.[136] The process was commercialized in the 1930s. At present, polyisobutylene is the only commercial polymer that is produced by cationic polymerization (see Section 13.2.6).

Cationic polymerization of alkenes involves the formation of a reactive carbocationic species capable of inducing chain growth (propagation). The idea of the involvement of carbocations as intermediates in cationic polymerization was developed by Whitmore.[5] Mechanistically, acid-catalyzed polymerization of alkenes can be considered in the context of electrophilic addition to the carbon–carbon double bond. Sufficient nucleophilicity and polarity of the alkene is necessary in its interaction with the initiating cationic species. The reactivity of alkenes in acid-catalyzed polymerization corresponds to the relative stability of the intermediate carbocations (tertiary > secondary > primary). Ethylene and propylene, consequently, are difficult to polymerize under acidic conditions.

Polymerization of isobutylene, in contrast, is the most characteristic example of all acid-catalyzed hydrocarbon polymerizations. Despite its hindered double bond, isobutylene is extremely reactive under any acidic conditions, which makes it an ideal monomer for cationic polymerization. While other alkenes usually can polymerize by several different propagation mechanisms (cationic, anionic, free radical, coordination), polyisobutylene can be prepared only via cationic polymerization. Acid-catalyzed polymerization of isobutylene is, therefore, the most thoroughly studied case. Other suitable monomers undergoing cationic polymerization are substituted styrene derivatives and conjugated dienes. Superacid-catalyzed alkane self-condensation (see Section 5.1.2) and polymerization of strained cycloalkanes are also possible.[118]

In general, any electrophilic reagent may add to a suitably nucleophilic carbon–carbon double bond to form a carbocation. Strong organic and inorganic Brønsted acids (CF_3COOH, CF_3SO_3H, H_2SO_4, HSO_3F, HHlg, $HClO_4$) react according to Eq. (13.17) to form an ion pair:[3,119,133,137–142]

$$HA \ + \ CH_2{=}C\overset{\displaystyle Me}{\underset{\displaystyle Me}{\big|}} \ \longrightarrow \ Me{-}\overset{\displaystyle Me}{\underset{\displaystyle Me}{\overset{+}{C}}} \quad A^- \qquad (13.17)$$

Inorganic Lewis acids, specifically, Friedel–Crafts catalysts, such as $SnCl_4$, $TiCl_4$, $AlCl_3$, $AlBr_3$, and BF_3, usually require a cocatalyst (e.g., H_2O always considered to be present), and in this case, initiation is also induced by the proton formed [Eq. (13.18)]. Pioneering work on the role of proton acids as catalysts and cocatalysts in acid-catalyzed polymerization was carried out by Polanyi and coworkers.[143,144] Lewis acids can also react in combination with other coinitiators, such as alkyl halides serving as carbocation source [Eq. (13.19)] or halogens:[119,138–140]

$$MHlg_n \; + \; H_2O \; \rightleftharpoons \; H^+ \, MHlg_{n+1}OH^- \; \xrightarrow{Me_2CH=CH_2} \; Me-\overset{Me}{\underset{Me}{\overset{+}{C}}} \qquad (13.18)$$

$$RHlg \; + \; MHlg_n \; \rightleftharpoons \; RHlg{\rightarrow}MHlg_n \; \rightleftharpoons \; R^+ \, MHlg_{n+1}{}^- \; \xrightarrow{Me_2C=CH_2} \; RCH_2-\overset{Me}{\underset{Me}{\overset{+}{C}}}$$

$$(13.19)$$

The former, however, is not a readily occurring reaction, except when highly stable carbocations are involved or very strong Lewis acids are applied under suitable reaction conditions (e.g., polymerization induced by alkyl fluoride–BF_3 initiation[145]).

Direct initiation without the involvement of a coinitiator (cocatalyst), a long debated possibility, may result in a polar complex or a zwitterion:

$$MHlg_n \; + \; CH_2{=}\overset{Me}{\underset{Me}{C}} \; \longrightarrow \; Hlg_n\overset{-}{M}-CH_2-\overset{Me}{\underset{Me}{\overset{+}{C}}} \qquad (13.20)$$

The self-ionization equilibrium of dimeric aluminum halides ($AlCl_3$, $AlBr_3$) (Korshak–Plesch–Marek theory[119]) elaborated in detail by Plesch[146,147] also generates a cationic species necessary for initiation [Eq. (13.21)]. A more reasonable proposal postulated by Olah[148] is the less energetic opening of only one of the chlorine bridges in dimeric $AlCl_3$ [Eq. (13.22)]:

$$(AlHlg_3)_2 \; \rightleftharpoons \; AlHlg_2{}^+ \, AlHlg_4{}^- \qquad (13.21)$$

$$(13.22)$$

Alternatively, "neat" $AlBr_3$ can initiate polymerization according to a halometallation mechanism (Olah, Sigwalt)[149,150] (Scheme 13.1). It involves formation of a vicinal bromoaluminated intermediate (**14**), which, with excess $AlBr_3$, ionizes to the corresponding cation (route *a*). Since **14** readily loses HBr (route *b*), the process rapidly becomes a typical conjugate acid–catalyzed reaction.

Organometallic systems[151–153] are also very effective initiators. Trialkylaluminum and dialkylaluminum halides, mostly Et_3Al and Et_2AlCl, were studied extensively.[152,153] Both compounds are inactive themselves without coinitiators. However, in the presence of protic acids (e.g., HCl) or in methyl chloride solvent, they effect the polymerization of isobutylene. Et_3Al and Et_2AlCl are considered by Kennedy[153] as the true catalysts, whereas Gandini and Cheradame[140] are in favor of in situ–formed $AlCl_3$. In fact, it was observed long before these systematic studies

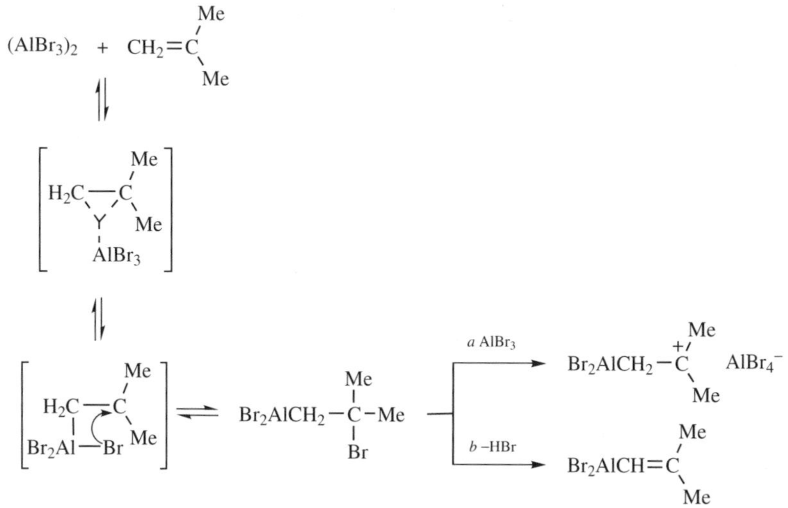

$$(AlBr_3)_2 \;+\; CH_2\!=\!\underset{\displaystyle Me}{\overset{\displaystyle Me}{C}}$$

$$\left[\; \underset{\displaystyle AlBr_3}{\overset{\displaystyle Me}{H_2C\!-\!C}}\!\!\underset{\displaystyle Me}{\;}\;\right]$$

$$\left[\; \underset{\displaystyle Br_2Al\!-\!Br}{\overset{\displaystyle Me}{H_2C\!-\!C}}\!\!\underset{\displaystyle Me}{\;}\;\right] \rightleftharpoons \; Br_2AlCH_2\!-\!\underset{\displaystyle Br}{\overset{\displaystyle Me}{C}\!-\!Me} \;$$

a AlBr$_3$ → $Br_2AlCH_2\!-\!\underset{\displaystyle Me}{\overset{\displaystyle +/Me}{C}} \quad AlBr_4^-$

b –HBr → $Br_2AlCH\!=\!\underset{\displaystyle Me}{\overset{\displaystyle Me}{C}}$

14

Scheme 13.1

by Hall and Nash[154] that polymers formed from ethylene when treated with a mixture of $AlCl_3$ and aluminum powder in the presence of hydrogen under pressure (155°C, 60 atm), had much higher viscosity compared to those prepared with $AlCl_3$ alone. It was established that aluminum sesquichloride ($Et_2AlCl\cdot EtAlCl_2$) was formed in the system.

Preformed or in situ–prepared carbocation salts (tropylium, trityl, etc.) are also active in transforming alkenes to carbocations.[119,138,140] Preformed carbocation salts are the simplest initiators in cationic polymerization and ideal if the cation is identical to the one derived from the momomer (e.g., *tert*-butyl cation in the polymerization of isobutylene).

An alternative way to generate carbocations is the one-electron oxidation of alkenes. Strong oxidants and charge-transfer complexes[119,139] can induce such single-electron transfers from reactive alkenes [Eq. (13.23)]. The resulting cation radical can react with a monomer to form a cation [Eq. (13.24)] or may be complexed to give a bicationic dimer. High-energy ionizing radiation (γ rays),[119,139,155,156] photoinitiation (UV radiation),[119,139,140] and electrochemical oxidation[140,157] are additional possibilities:

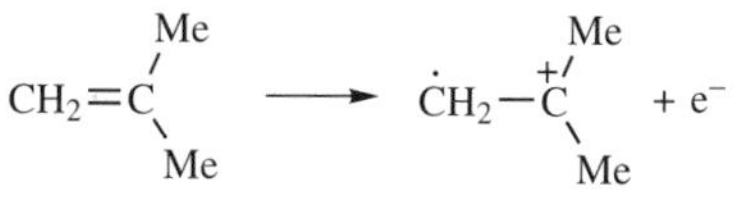

$$CH_2\!=\!\underset{\displaystyle Me}{\overset{\displaystyle Me}{C}} \;\longrightarrow\; \dot{C}H_2\!-\!\underset{\displaystyle Me}{\overset{\displaystyle +/Me}{C}} \;+\; e^- \qquad (13.23)$$

$$\dot{C}H_2\!-\!\underset{\displaystyle Me}{\overset{\displaystyle +/Me}{C}} \;+\; CH_2\!=\!\underset{\displaystyle Me}{\overset{\displaystyle Me}{C}} \;\longrightarrow\; Me\!-\!\underset{\displaystyle Me}{\overset{\displaystyle +/Me}{C}} \;+\; CH_2\!=\!\underset{\displaystyle Me}{\overset{\displaystyle \dot{C}H_2}{C}} \qquad (13.24)$$

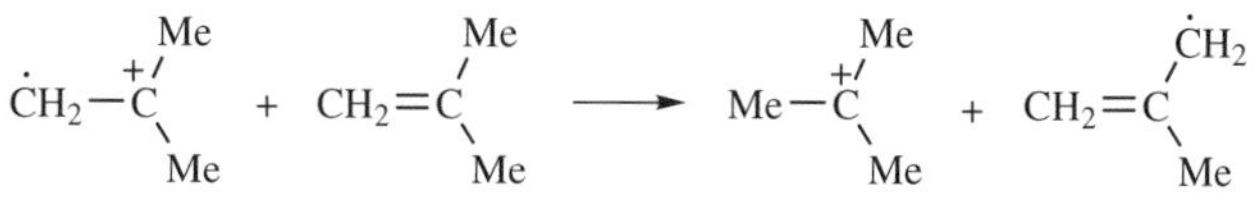

Carbocations formed in the mentioned reactions serve as chain carriers in propagation [Eq. (13.25), where E stands for an electrophilic initiator):

$$ECH_2-\overset{+}{\underset{Me}{\overset{Me}{C}}} \ + \ n+1\ CH_2=\overset{Me}{\underset{Me}{C}} \ \longrightarrow \ ECH_2-\overset{Me}{\underset{Me}{C}}\left[CH_2-\overset{Me}{\underset{Me}{C}}\right]_n CH_2-\overset{+}{\underset{Me}{\overset{Me}{C}}} \qquad (13.25)$$

15

The two basic prerequisites to obtain high-molecular-weight polymers via cationic polymerization are the high nucleophilicity of the monomer and the relative stability of the carbocation to sustain propagation. The difficulty of ethylene and propylene to yield high-molecular-weight polymers in acid-catalyzed polymerization exemplifies this statement; both have relatively low nucleophilicity and the derived ethyl and isopropyl cations have relatively low stability.

Possibilities for chain transfer in cationic polymerization are abundant.[119,138] Proton transfer to the monomer is a rather general transfer reaction leading to two isomeric unsaturated end structures:

$$\mathbf{15} \ + \ CH_2=\overset{Me}{\underset{Me}{C}}$$

$$\longrightarrow ECH_2-\overset{Me}{\underset{Me}{C}}\left[CH_2-\overset{Me}{\underset{Me}{C}}\right]_n CH_2-\overset{CH_2}{\underset{Me}{C}} \ + \ Me-\overset{+}{\underset{Me}{C}} \qquad \mathbf{16}$$

$$\longrightarrow ECH_2-\overset{Me}{\underset{Me}{C}}\left[CH_2-\overset{Me}{\underset{Me}{C}}\right]_n CH=\overset{Me}{\underset{Me}{C}} \ + \ Me-\overset{+}{\underset{Me}{C}} \qquad \mathbf{17}$$

$$(13.26)$$

Hydride abstraction from the monomer, in contrast, is not a well-documented transformation:

$$\mathbf{15} \ + \ CH_2=\overset{Me}{\underset{Me}{C}} \ \longrightarrow \ ECH_2-\overset{Me}{\underset{Me}{C}}\left[CH_2-\overset{Me}{\underset{Me}{C}}\right]_n CH_2-\overset{Me}{\underset{}{CH}} \ + \ CH_2=\overset{\overset{+}{CH_2}}{\underset{Me}{C}} \qquad (13.27)$$

Reactions with the counterion may involve spontaneous acid expulsion [Eq. (13.28)] or ion combination, which is a chain termination reaction [Eq. (13.29)]:

$$\mathbf{15} \ + \ A^-$$

$$\longrightarrow \mathbf{16} \ \text{or} \ \mathbf{17} \ + \ HA \qquad (13.28)$$

$$\longrightarrow ECH_2-\overset{Me}{\underset{Me}{C}}\left[CH_2-\overset{Me}{\underset{Me}{C}}\right]_n CH_2-\overset{Me}{\underset{Me}{C}}-A \qquad (13.29)$$

Halogenation and alkylation at the growing chain end and attack by a solvent molecule or by impurities are also feasible.

Since carbocations are involved in cationic polymerization, a possible side reaction is their isomerization through hydride (alkyde) migration to more stable (less reactive) carbocations. This can lead to a polymer of broad molecular weight distribution or, if the isomerization is irreversible, to termination.

Addition of BF_3 to isobutylene precooled to $-80°C$ results in an immediate, almost explosive reaction and produces polymer having a molecular weight of ∼200,000. On the other hand, if the reaction is carried out at $-10°C$, polymerization is slower and the molecular weight of the polymer is only 10,000. The activity of different Lewis acids decreases over a wide range at $-78°C$ in the series BF_3, $AlBr_3$, $TiCl_4$, $TiBr_4$, $SnCl_4$, BCl_3, and BBr_3, based on the time required to polymerize all the isobutylene. Isobutylene is converted to the polymer within a few seconds after treatment with BF_3. Reaction times of hours are required for complete polymerization of the olefin with $TiCl_4$ and days with $SnCl_4$. In a classic experiment Evans and Polanyi[158] showed that highly purified isobutylene and BF_3 in a carefully dried system, do not polymerize. An instantaneous reaction occurs, however, when trace amounts of water or *tert*-butyl alcohol are introduced in the system. This experiment was key to prove the role of water as cocatalyst in cationic polymerization.[158]

Various Brønsted and Lewis acids can also be used in cationic polymerization of styrene.[118,138,159] The molecular weight, however, is difficult to control. Of the usual transfer processes, transfer to the monomer is the most significant reaction. An additional difficulty, the occurrence of Friedel–Crafts reactions, arises if polymerization is carried out in aromatic solvents. As a result, cationic polymerization of styrene usually leads to ill-defined products and is mainly of academic interest.[159]

One interesting aspect of the polymerization of styrene is its possible pseudocationic propagation. It was observed, for example, that in the presence of $HClO_4$ styrene forms only oligomers and no real cationic species could be found.[160,161] This led to the suggestion that the initiating species is a perchlorate ester stabilized by monomer molecules:

$$5\ PhCH{=}CH_2\ +\ HClO_4\ \underset{\overrightarrow{CH_2Cl_2}}{\rightleftharpoons}\ \underset{\underset{Ph}{|}}{(CH_3CHOClO_3)}\ 4\ PhCH{=}CH_2 \qquad (13.30)$$

Propagation occurs through a six-membered transition state by a pseudocationic mechanism[160] [Eq. (13.31), where P stands for a polymer chain]. Similar observations were made in triflic acid catalyzed polymerization:[162]

$$(13.31)$$

Conjugated dienes (1,3-butadiene, isoprene) have suitable nucleophilicity to undergo cationic polymerization. There is, however, not much practical interest in these processes since the polymers formed are inferior to those produced by other (free-radical, coordination) polymerizations. A significant characteristic of these polymers is the considerably less than theoretical unsaturation due to cyclization processes.[132] A fully cyclized product of isoprene has been synthesized[163] by constant potential electrolysis in CH_2Cl_2.

In cationic polymerization, unlike in anionic polymerization (see Section 13.2.2), it is seldom possible to achieve "living" nature of the polymer by simply using a rate of initiation much faster than that of propagation. Sawamoto and Higashimura[164,165] as well as Kennedy,[166,167] however, more recently found that by using initiator systems such as HI combined with iodine[164] or Lewis acids (e.g., $ZnCl_2$),[165] and complexes of BCl_3 with organic acetates or tertiary alkyl (aryl) ethers,[166,167] living cationic polymerizations can be achieved, allowing polymers to be reinitiated, for example, for copolymerization with other suitable monomers. Isobutylene and styrene were polymerized with BCl_3 complexes.[166,167]

13.2.2. Anionic Polymerization

The first results of anionic polymerization (the polymerization of 1,3-butadiene and isoprene induced by sodium and potassium) appeared in the literature in the early twentieth century.[168,169] It was not until the pioneering work of Ziegler[170] and Szwarc,[171] however, that the real nature of the reaction was understood. Styrene derivatives and conjugated dienes are the most suitable unsaturated hydrocarbons for anionic polymerization. They are sufficiently electrophilic toward carbanionic centers and able to form stable carbanions on initiation. Simple alkenes (ethylene, propylene) do not undergo anionic polymerization and form only oligomers. Initiation is achieved by nucleophilic addition of organometallic compounds or via electron transfer reactions. Hydrocarbons (cylohexane, benzene) and ethers (diethyl ether, THF) are usually applied as the solvent in anionic polymerizations.

A unique characteristic of anionic polymerization is the "living" nature of the process recognized and pioneered by Szwarc.[120,128,129,172] Living polymerization can be achieved when the rate of initiation is very fast and much faster than the rate of propagation. In living polymers the active centers are highly stable and highly reactive, and, in the absence of impurities, no spontaneous transfer or termination occurs. Under such conditions all macromolecules grow proportionally and propagation continues until an equilibrium with the monomer is established. Suitable reagents may be used to annihilate the reactive end known as "killing." Addition of water or any proton donor, for example, converts the active chain ends, resulting in a dead polymer. Living anionic polymerization allows the synthesis of homodisperse polymers (i.e., polymers with very narrow molecular weight distribution) and block copolymers unaccessible by other techniques (see Section 13.2.6). Following the interest and extensive studies of living polymers spurred by the work of Szwarc, anionic polymerization probably became the most studied and best understood of all polymerization methods.

Organolithum compounds (lithium alkyls) are the most valuable initiators in anionic polymerization.[120,168,169,172–175] Since living anionic polymerization requires the fastest possible initiation, *sec*- or *tert*-butyllithium is usually used. Lithium alkyls add readily to the double bond of styrene [Eq. (13.32)] or conjugated dienes and form free ions or an ion pair depending on the solvent:

$$(13.32)$$

Since lithium alkyls are more stable in hydrocarbons than in ethers, hydrocarbon solvents are usually applied.

Electron transfer from polycyclic aromatic radical anions in polar solvents can also initiate propagation.[120,168,169,173] One of the early and best understood systems is naphthalene–sodium, a green solution of stable, solvated naphthalene radical anion.[176,177] The electron transfer from the radical anion to the monomer yields a new radical anion [Eq. (13.33)]. The dominant reaction of the latter is its head-to-head dimerization to the stabile dimeric dicarbanion [Eq. (13.34)], which is the driving force for the electron transfer even when electron affinity of the monomer is less than that of the polycyclic molecule. Propagation proceeds at both ends of the chain:

$$(13.33)$$

$$(13.34)$$

A similar but direct electron transfer from the metal to the monomer is operative when alkali and alkaline-earth metals (e.g., sodium) are used as the initiator.[169,173,177] In this case, however, initiation is slow relative to propagation because of the low metal surface area available, and this method is used only for special purposes. 1,1-Dipehylethylene, for example, forms a dianion that, for steric reasons, is not capable for further head-to-tail addition of the monomer, but it can be used to initiate the polymerization of other monomers.[178]

The propagating active center formed on initiation is an ion pair that can exist as aggregates, contact- or solvent-separated ion pairs, or free ions depending on the solvent, the temperature, and the counterion.[120,128,173,174] Each can add to the monomer but exhibits different reactivity. Polar solvents bring about the disaggregation and stabilization of free ions. Styryllithium, for example, was found to be associated in the dimeric form in hydrocarbon solvents.[179,180] THF breaks up these dimers presumably by means of a solvation equilibrium and accelerates the rate of polymerization.

Living polymers usually require special reagents to achieve termination. Any electrophilic reagent that reacts with the carbanion active center and also allows preparation of polymers with desired terminal functionalities can be used for this purpose.[168,174,181] Hydrogen-terminated polymers can be produced by proton donors, whereas carbon dioxide results in a carboxylate end group. Terminal alcohol functionalities can be formed through reaction with ethylene oxide and carbonyl compounds.

Although the living nature of anionic polymerization of styrene testifies to the relative stability of the carbanion at the chain end, observations indicate that it may undergo spontaneous transformations.[182] Polystyryl sodium in THF was shown to undergo termination involving hydride elimination followed by proton transfer:

$$P-CH_2CH-\overset{\overset{\displaystyle H}{|}}{\underset{\underset{\displaystyle Ph}{|}}{C}}\overset{-}{H}CH\ Na^+ \xrightarrow{\ -NaH\ } P-CH_2CH-CH=CH \tag{13.35}$$

with Ph substituents as shown.

$$P-CH_2\overset{\overset{\displaystyle H}{|}}{\underset{\underset{\displaystyle Ph}{|}}{C}}CH=CH + P-CH_2CHCH_2\overset{-}{C}H \longrightarrow P-CH_2\overset{-}{C}CH=CH + P-CH_2CHCH_2CH_2$$

$$\mathbf{18}$$

$$\tag{13.36}$$

The highly stabilized carbanion **18** thus formed does not participate in propagation.

Lithium and alkyllithiums in aliphatic hydrocarbon solvents are also used to initiate anionic polymerization of 1,3-butadiene and isoprene.[120,183–187] As 1,3-butadiene has conjugated double bonds, homopolymerization of this compound can lead to several polymer structures. 1,4 Addition can produce cis-1,4- or trans-1,4-polybutadiene (**19**, **20**). 1,2 Addition results in a polymer backbone with vinyl groups attached to chiral carbon atoms (**21**). All three spatial arrangements (isotactic, syndiotactic, atactic) discussed for polypropylene (see Section 13.2.4) are possible when polymerization to 1,2-polybutadiene takes place. Besides producing these structures, isoprene can react via 3,4 addition (**22**) to yield polymers with the three possible tacticites:

19	20	21	22

Further variations may arise because of the head-to-tail, head-to-head, and tail-to-tail monomer enchainments. Anionic polymerization is a versatile tool used to convert conjugated dienes into polymers with different microstructure.[183-185] Alkali metals bring about mainly 1,2 addition in the case of polymerization of 1,3-butadiene.[188] Lithium, however, is an exception since it gives polymers with high 1,4 content.[183-185] Dominant 1,2 addition occurs if polymerization is catalyzed by lithium alkyls in ethers.[189] In hydrocarbon solvents, in contrast, 1,4 addition (>90%) prevails with approximately equal amounts of *cis* and *trans* units.[190] Sodium-based catalysts, the so-called Alfin systems[191] (e.g., a mixture of sodium allyl, sodium isopropoxide and NaCl), can be used to synthesize very-high-molecular-weight polybutadiene with dominating *trans* content.[192]

Lithium or lithium alkyls can be used to convert isoprene to *cis*-1,4-polyisoprene with *cis* content up to 96% (a structure similar to that of natural rubber), with the remaining part having 3,4 structure.[193-195] Similar to the changes observed in the polymerizaton of 1,3-butadiene, the dominant *cis*-1,4 selectivity changes to 1,2 and 3,4 addition in ethers.[195,196] Change of solvation of Li^+ in polar solvents affecting the ionic character of the C—Li bond is considered to be an important factor in bringing about these selectivity changes.

The predominant 1,4-addition selectivity with lithium in nonpolar solvents is explained by the localization of Li^+ to the terminal carbon. Morton et al. found[197,198] that lithium is essentially σ-bonded to the terminal carbon (**23**). In polar solvents the negative charge is delocalized (**24**), generating a π-allyl structure that enhances the reactivity of the γ-carbon, affording 1,2 addition:

$$P\text{-}CH_2\text{-}CH=CH\text{-}CH_2Li \qquad \overset{\delta-\quad Li^+\ \ \delta-}{P\text{-}CH_2\text{-}CH\text{·}CH\text{·}CH_2}$$

23   **24**

The high *cis* orientation in isoprene polymerization can be accounted for by initial coordination of cisoid isoprene to the reagent (*n*-BuLi) and the rearrangement of this complex through a six-membered cyclic activated complex (**25**) to a *cis* end unit[195,199,200] (Scheme 13.2). Since the *trans* active center was found to be more stable in nonpolar solvents, isomerization of the newly formed *cis* anionic center to the *trans* center was invoked to account for the formation of the *trans* isomer.[194,200] The high *cis* selectivity in hydrocarbon solutions, therefore, is a kinetic phenomenon. When high monomer : initiator ratios are applied, insertion of the monomer into the propagating *cis* center is fast and occurs before isomerization, affording high *cis* selectivity.[194,201] Decreasing monomer : initiator ratio results in increasing *trans* content.

13.2.3. Free-Radical Polymerization

When ethylene is heated under high pressure in the presence of oxygen or organic peroxides, it is converted in an exothermic process ($\Delta H = -22.5\,kcal/mol$) to a long-chain, white waxy solid polymer containing a relatively small number of short branches. First prepared in 1934 by ICI researchers,[202,203] it soon gained practical importance, first as an electrical insulator and subsequently in many other areas.

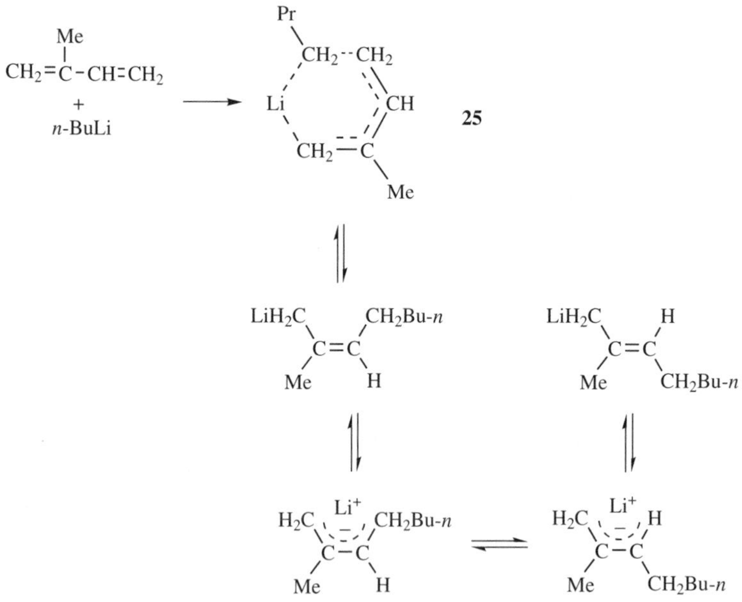

$$CH_2=\overset{\underset{\displaystyle |}{Me}}{C}-CH=CH_2 \; + \; n\text{-BuLi} \longrightarrow$$

25

Scheme 13.2

Because of its commercial importance, the polymerization of ethylene at high pressure has been extensively studied.[204–209] Free-radical polymerization is characteristic of ethylene and vinyl compounds. Simple alkenes, such as 1-butene, however, do not give high-molecular-weight polymers, but they, as well as internal alkenes, can copolymerize with polymerizable monomers.

The radical polymerization of ethylene, in practice, is initiated by free-radical initiators, although radiation-induced,[155,210] photoinduced,[211–213] and thermal[213,214] initiations are also possible. The temperature of high-pressure polymerization should not exceed 350°C since above this temperature a rapid exothermic ($\Delta H = -30.4\,\text{kcal/mol}$) thermal decomposition of ethylene can take place leading to a runaway reaction:

$$CH_2=CH_2 \longrightarrow 2\,C \;+\; 2\,H_2 \quad \text{or} \quad C \;+\; CH_4 \tag{13.37}$$

In early work the initiator was oxygen, either present as impurity or intentionally added. The radicals are formed by decomposition of hydroperoxide or peroxide formed in the reaction of oxygen with ethylene:

$$CH_2=CH_2 \xrightarrow{O_2} CH_2=CH\text{-OOH} \xrightarrow{CH_2=CH_2} CH_2=CHOOCH_2CH_3 \longrightarrow \begin{array}{c} CH_2=CH\text{-}O\cdot \\ + \quad \textbf{26} \\ CH_3CH_2-O\cdot \\ \textbf{27} \end{array}$$

$$\tag{13.38}$$

Commercially available hydroperoxides and peroxides were used subsequently[215–217] with the advantages of lower initiation temperature and better process control. A number of radical initiators (primary radicals) are formed when, for example, a methanol solution of benzoyl peroxide is used:

$$(C_6H_5COO)_2 \longrightarrow C_6H_5COO\cdot \longrightarrow C_6H_5\cdot + CO_2 \quad (13.39)$$
$$\qquad\qquad\qquad \mathbf{28} \qquad\qquad \mathbf{29}$$

$$C_6H_5COO\cdot + CH_3OH \longrightarrow C_6H_5COOH + \dot{C}H_2OH \qquad (13.40)$$
$$\qquad\qquad\qquad\qquad\qquad\qquad\qquad\qquad \mathbf{30}$$

$$C_6H_5\cdot + CH_3OH \longrightarrow C_6H_6 + \mathbf{30} \qquad (13.41)$$

Any of the free radicals (**26–30**, designated as R˙) may initiate the chain reaction [Eq. (13.42)]. The exact nature of R˙ is unimportant since it merely initiates the polymerization chain and reacts only in the first cycle; subsequent cycles are initiated by radicals called *propagating radicals* formed in the reaction with ethylene [Eq. (13.43)]:

$$R\cdot + CH_2{=}CH_2 \longrightarrow RCH_2CH_2\cdot \qquad (13.42)$$

$$RCH_2CH_2\cdot + (n+1)CH_2{=}CH_2 \longrightarrow RCH_2CH_2(CH_2CH_2)_nCH_2CH_2\cdot \quad (13.43)$$

Termination of the propagating chain can occur by a number of reactions. The chain can be terminated by coupling (recombination) [Eq. (13.44)] or disproportionation of two growing radicals [Eq. (13.45)], which may or may not be the same:

$$2\,R(CH_2CH_2)_nCH_2CH_2\cdot \left\{ \begin{array}{l} R(CH_2CH_2)_nCH_2CH_2CH_2CH_2(CH_2CH_2)_nR \qquad (13.44) \\[2mm] R(CH_2CH_2)_nCH{=}CH_2 + R(CH_2CH_2)_nCH_2CH_3 \quad (13.45) \end{array} \right.$$

The growing polymer chain may also be terminated by abstraction of a hydrogen atom from any suitable donor molecule called *telogen* (*S*-H). In these instances, hydrogen transfer from the telogen to the growing radical produces a "dead" polymer macromolecule and a new free radical [Eq. (13.46)]. The latter can add to ethylene and start a new chain [chain termination with kinetic chain transfer; Eq. (13.47)]:

$$R(CH_2CH_2)_nCH_2CH_2\cdot + S{-}H \longrightarrow R(CH_2CH_2)_nCH_2CH_3 + S\cdot \quad (13.46)$$

$$S\cdot + CH_2{=}CH_2 \longrightarrow S{-}CH_2CH_2\cdot \qquad (13.47)$$

In principle, all organic molecules containing hydrogen can serve as hydrogen source in chain-transfer reactions. However, the rate of hydrogen transfer greatly depends on the nature of the carbon–hydrogen bond. Hydrogens attached to tertiary carbon atoms are particularly reactive. Such chain-transfer agents may be added as solvents. High-pressure polymerization of ethylene is generally carried out in the absence of telogens, but added hydrocarbon solvents may serve as telogens. Equations (13.48)–(13.51) illustrate the polymerization of ethylene initiated by di-*tert*-butyl peroxide in cyclohexane:

$$tert\text{-BuOOBu-}tert \longrightarrow tert\text{-BuOO}^{\bullet} \tag{13.48}$$

$$tert\text{-BuOO}^{\bullet} \; + \; \bigcirc \; \longrightarrow \; tert\text{-BuOH} \; + \; \bigcirc^{\bullet} \tag{13.49}$$

$$\bigcirc^{\bullet} \; + \; (n{+}1)\,CH_2{=}CH_2 \; \longrightarrow \; \bigcirc{-}(CH_2CH_2)_n CH_2CH_2{}^{\bullet} \tag{13.50}$$

$$\bigcirc{-}(CH_2)_{2n+1}CH_2{}^{\bullet} \; + \; \bigcirc \; \longrightarrow \; \bigcirc{-}(CH_2)_{2n+1}CH_3 \; + \; \bigcirc^{\bullet} \tag{13.51}$$

The last step [Eq. (13.51)] is chain transfer to the solvent. The product consists of a polymethylene chain with even number of carbon atoms. One end of the chain is attached to a cyclohexyl group and the other end, to a hydrogen atom originally attached to cyclohexane. The process may be called *telomerization*, although the term is generally used to designate processes yielding oligomers of low molecular weight.[218]

Chain transfer to ethylene can take place yielding a polymer with an unsaturated end group and a chain-initiating ethyl radical:

$$R(CH_2CH_2)_n CH_2CH_2{}^{\bullet} \; + \; CH_2{=}CH_2 \; \longrightarrow \; R(CH_2CH_2)_n CH{=}CH_2 \; + \; CH_3CH_2{}^{\bullet} \tag{13.52}$$

Growing polymer chains can also participate in intramolecular or intermolecular hydrogen transfer. Such chain transfers become significant at high polymer concentration. Intramolecular hydrogen transfer, called "backbiting," results in a secondary radical that, after adding to ethylene, forms a side chain on the polyethylene backbone.[219] Since steric factors favor a six-membered transition state of hydrogen transfer (1,5 hydrogen shift), the net result is a four-carbon side chain:

$$P{-}CH_2CH_2CH_2CH_2CH_2{}^{\bullet}$$

$$\left[\begin{array}{c} P \\ \mathrm{H}{\cdots}\mathrm{C} \\ \mathrm{H_2} \end{array} \right]^{\bullet} \; \longrightarrow \; P{-}\overset{\bullet}{C}HCH_2CH_2CH_2CH_3 \; \xrightarrow{CH_2{=}CH_2} \; P{-}\underset{C_4H_9}{CHCH_2CH_2}{}^{\bullet} \tag{13.53}$$

The reaction is called *short-chain branching*.

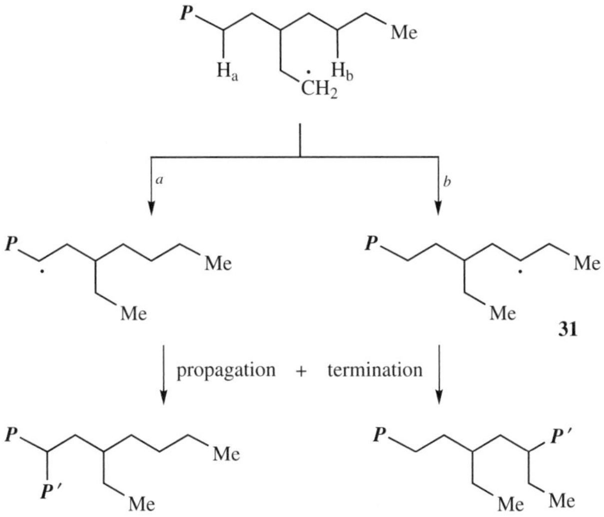

Scheme 13.3

Polyethylene also contains ethyl side chains. These are formed according to the reactions in Scheme 13.3 leading to 2-ethylhexyl and 1,3-paired diethyl branches.[220–222] Formation of such new branching has high probability since the newly formed radical is in close proximity to the chain backbone.

Intermolecular chain transfer between a growing polymer chain (**32**) and a previously formed polymer molecule (**33**) results in long-chain branching:

$$P^1CH_2CH_2\cdot \ + \ P^2CH_2\overset{\overset{H}{|}}{C}HP^3 \ \longrightarrow \ P^1CH_2CH_3 \ + \ P^2CH_2\dot{C}HP^3 \qquad (13.54)$$

32   **33**

$$P^2CH_2\dot{C}HP^3 \ + \ (n+1)\,CH_2{=}CH_2 \ \longrightarrow \ \overset{P^2H_2C}{\underset{P^3}{\diagdown}}CH(CH_2CH_2)_nCH_2CH_2\cdot \qquad (13.55)$$

Besides the chain transfer in Eq. (13.52), another important process called "β scission" can also produce unsaturations in the polymer chain. Secondary radicals (**31**) as well as tertiary radicals such as **34** formed according to Scheme 13.4 may produce low-molecular-weight vinyl or vinylidene compounds, respectively, and new radicals.

Styrene and its derivatives can be polymerized by all possible propagation mechanisms. Free-radical polymerization, however, is the primary process for industrial production of polystyrene.[159,223] Free-radical polymerization of styrene can be carried out without chemical initiators simply by heating the monomer.[223,224] On heating, isomeric 1-phenyltetralins are formed in the Diels–Alder

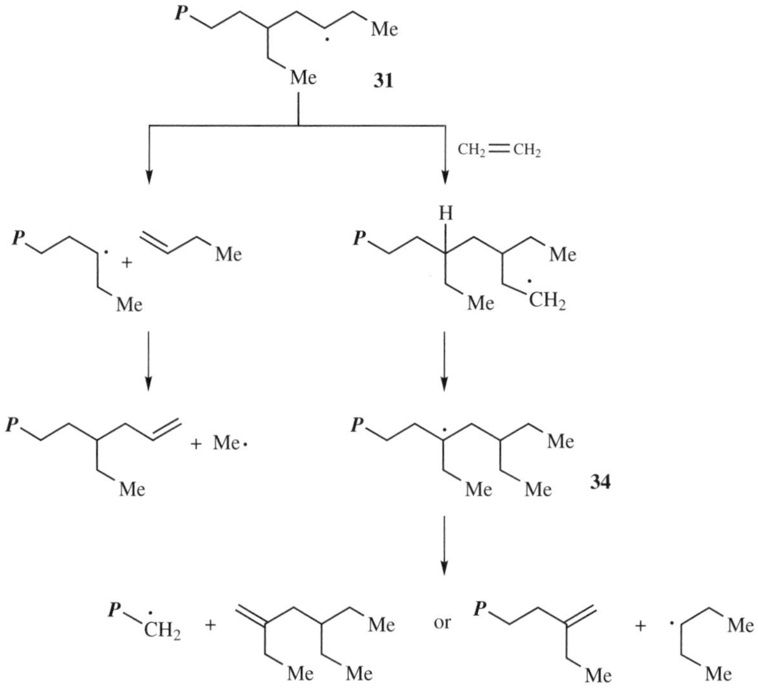

Scheme 13.4

reaction[225] (Scheme 13.5). The less stable isomer with the phenyl group in axial position reacts further with styrene to form two radicals that can start polymerization. Radical formation is facilitated by the reactive tertiary double allylic hydrogen in suitable configuration in the axial isomer.

A number of other dimers are also formed in thermal polymerization of styrene. Two of these (**35**, **36**) can be found as impurities of the highest concentration (up to 1%) in commercial polystyrene.[226,227]

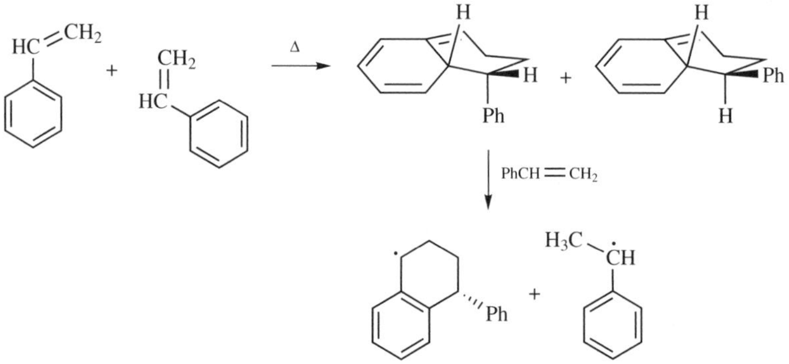

Scheme 13.5

$$
\underset{\textbf{35}}{\overset{\overset{\displaystyle CH_2}{\overset{\|}{PhCH_2CH_2-CPh}}}{}}
\qquad
\underset{\textbf{36}}{\overset{Ph \quad Ph}{\square}}
$$

In industrial styrene polymerization radical initiators are usually used to achieve complete polymerization at reduced temperature in shorter time. Unique initiators with the general formula **37** decomposing into mono and double radicals have been described:[228]

$$
\underset{\textbf{37}}{R'-OO-\overset{\overset{\displaystyle O}{\|}}{C}-X-\overset{\overset{\displaystyle O}{\|}}{C}-OO-R''} \longrightarrow \underset{\textbf{38}}{\cdot O-\overset{\overset{\displaystyle O}{\|}}{C}-X-\overset{\overset{\displaystyle O}{\|}}{C}-O\cdot} \; + \; R'O\cdot \; + \; R''O\cdot \qquad (13.56)
$$

The latter (**38**) can initiate propagation at both radical centers doubling the length of the product polymer molecules.

Propagation in free-radical styrene polymerization proceeds through stable benzylic radicals by head-to-tail addition of the monomer:

$$(13.57)$$

The most important chain-transfer reaction is transfer to the Diels–Alder dimer. Added chemical transfer agents, mainly mercaptans, allow the regulation of molecular weight and molecular weight distribution of commercial polymers. The combination of two growing chains is an almost exclusive way of termination:[229,230]

$$
2\, P\text{-}CH_2\text{-}\overset{\cdot}{C}H \longrightarrow P\text{-}CH_2\text{-}CH\text{-}CH\text{-}CH_2\text{-}P \qquad (13.58)
$$
$$
\quad\;\; \underset{Ph}{|} \qquad\qquad\qquad \underset{Ph\;\;Ph}{|\;\;\;\;|}
$$

13.2.4. Coordination Polymerization

Research efforts in the 1950s led to the discovery of processes capable of polymerizing alkenes at low and medium pressures. Ziegler found that combinations of transition-metal compounds (such as titanium chlorides) and certain metal alkyls, preferentially alkylaluminums, catalyze low-pressure polymerization of ethylene yielding high-density polyethylene (HDPE).[38,231,232] Using similar two-component catalysts Natta could attain the stereoregular polymerization of propylene.[233,234] Other important findings, including the use of oxide-supported transition-metal catalysts, came from industrial laboratories. Phillips disclosed a process using CrO_3 supported on silica or silica–alumina,[235,236] while Mo_2O_3 on alumina was developed by Standard Oil of Indiana.[237,238] The HDPE and stereoregular polypropylene produced by transition-metal based catalysts have important structural characteristics and, consequently, special mechanical and thermal properties, and applications (see Section 13.2.6).

Catalysts. An extremely large number of catalysts and different catalyst combinations are known to catalyze the polymerization of alkenes. In our discussion we will classify them as Ziegler–Natta catalysts and one-component transition-metal catalysts.

The first successful catalyst used by Ziegler and coworkers[38,231,232] for the polymerization of ethylene was $TiCl_4$ and Et_3Al. Other alkylaluminum compounds (Et_2AlCl, $Et_3Al_2Cl_3$, isoBu$_3$Al), also termed *activators* or *cocatalysts*, were employed later. Prepared in situ, the two components form an unstable precipitate in an inert solvent, which is generally a hydrocarbon. In a similar approach Natta and coworkers first used Ziegler's original catalyst combination,[239] then later preformed purple $TiCl_3$ for the polymerization of propylene.[234,240] They found that a mixture of crystalline and amorphous polymers was formed.[233] This observation led to the development of stereoregular polymerization catalysts and processes.

It is of historical interest to note that Max Fischer, working during World War II in the German industry, while investigating how solid waxes were formed in $AlCl_3$-catalyzed oligomerization of ethylene producing synthetic lubricant oils, disclosed in a patent application in 1943 (published only in 1953 because of war aftermath difficulties)[241] that ethylene is polymerized at moderate pressures and temperatures by a catalyst composed of $AlCl_3$, a reducing metal such as Al, and $TiCl_4$ to solid polyethylene. No characterization of the polymer or any suggestions as to the nature of the polymerization was given. A reinvestigation in the 1980s showed that the polyethylene formed in the Max Fischer polymerization, when the temperature is kept at about 150°C, is identical to that of linear Ziegler polyethylene with limited methyl branching.[242] Ethylene, when reacted under similar conditions with $AlCl_3$ and Al, is known to give ethylaluminum sesquichloride.[154] It was subsequently shown[242] that in this system $(Cl_2AlCH_2)_2$ and $(Cl_2Al)_2CHCH_3$ are also formed. Any ethylaluminum compound on addition of $TiCl_4$ forms an active Ziegler catalyst. Fischer thus, not knowing the nature of his polymerization or the linearity of the product, in the early 1940s already produced HDPE and was a forerunner of the Ziegler–Natta polymerization.

Linear polyethylene can further also be considered as polymethylene. Althoguh first prepared by the thermal decomposition of diazomethane,[243,244] Meerwein should be credited to have prepared it by catalytic polymerization of diazomethane effected by boron compounds (esters, halides, alkyls)[245–247] taking place with concomitant dediazotation:

$$n \; CH_2N_2 \xrightarrow{\text{catalyst}} -[CH_2]_n- \; + \; n \; N_2 \qquad (13.59)$$

Meerwein, however, had not explored the nature of the polymer or pursued alternative preparation methods. Since then other catalysts such as diborane, aluminum alkyls, and copper salts were also used, and other diazoalkanes were also poly-merized.[248–250] Polymerization of a C_1 unit considered as incipient CH_2 does not lead to branching, whereas carbocationic or free-radical ethylene polymerization clearly gives a high degree of branching.

Subsequent studies following the original findings of Ziegler and Natta revealed that, in general, combinations of Group IV–VIII metal halides and hydrides, and Group I–III organometallic compounds and halides are all active polymerization catalysts. This catalyst group, called *Ziegler–Natta catalysts*, constitutes a large and important family of olefin polymerization catalysts.[117,125,208,251–256] Ti, Zr, V, Nb, and Cr compounds were used in most studies. The catalysts, however, that gained commercial importance are combinations of $TiCl_4$ or $TiCl_3$ with R_3Al or R_2AlCl.

The role of metal alkyl in forming the active catalyst is illustrated in Eqs. (13.60)–(13.62). After complexation with $TiCl_4$ [Eq. (13.60)] (or, in general, with the transition-metal compound), the metal alkyl brings about the formation of a transition-metal alkyl through halogen–alkyl exchange [Eq. (13.61)]. This is then reduced in a subsequent very complex reduction step including dealkylation to yield low-valence titanium[255] [Eq. (13.62)]:

$$TiCl_4 \ + \ AlEt_3 \ \rightleftharpoons \ TiCl_4.AlEt_3 \qquad (13.60)$$

$$TiCl_4.AlEt_3 \ \rightleftharpoons \ EtTiCl_3.AlEt_2Cl \qquad (13.61)$$

$$2\ EtTiCl_3.AlEt_2Cl \ \longrightarrow \ 2\ TiCl_3.AlEt_2Cl \ + \ C_2H_4 \ + \ C_2H_6 \qquad (13.62)$$

Further reduction can also take place under severe conditions. These reduced tita-nium species are believed to constitute the active catalytic species.

In contrast to Ziegler's catalysts prepared in situ, Natta used preformed solid transition-metal halides in the lower oxidation state. Natta and coworkers discov-ered that of the four crystalline modifications of $TiCl_3$ the three violet forms (α, γ, and δ) exhibit high catalytic activity and stereoselectivity in the polymerization of propylene.[234,240] The exceptional properties of these three crystalline modifications are associated with their layered structure. It was found that the reduction of $TiCl_4$ with hydrogen, aluminum, or aluminum alkyls produces α- or γ-$TiCl_3.0.33AlCl_3$. These solid solutions can be transformed to the more active δ form by ball milling.[257] It is believed, however, that only titanium ions present on the side sur-faces of $TiCl_3$ crystallites constituting only several percentages of the total titanium content participate in polymerization.[258]

The abovementioned so-called first-generation heterogeneous Ziegler–Natta catalysts exhibited rather low activity (low yield of polymer per unit weight of catalyst), and large amounts of the catalyst had to be used. As a result, a tedious and expensive removal of the catalyst was necessary to avoid contamination of the polymer causing color and stability problems. Additionally, these catalysts are very sensitive to water and oxygen, which transform alkylaluminums to alcoho-lates. A number of other chemicals such as CO, CO_2, alkynes, and 1,2- and 1,3-dienes form strong complexes with the transition metals, thus causing catalyst poisoning. Extremely pure solvents, monomers, and stringent reaction conditions were necessary to achieve reliable and reproducible results.

A search for catalysts with higher activity to eliminate catalyst removal in the manufacturing process led to the discovery of the high-activity, high-yield (second-generation) Ziegler–Natta catalysts.[208,252–254,259] Further requirements were to have better control of polymer molecular weight and molecular weight distribution.

It was observed during the early studies with first-generation Ziegler–Natta catalysts that Lewis bases (O-, N-, and P-containing organic compounds) have advantageous effects on propylene polymerization.[253] The best examples are ether-treated $TiCl_3$-based catalysts exhibiting increased catalytic activity and excellent stereoselectivity.[260,261]

An even more important observation was the discovery of supported (third-generation) Ziegler–Natta catalysts. Such catalysts can be prepared by reacting $TiCl_4$ with hydroxylated magnesium derivatives [hydrated MgO, $Mg(OH)_2$, Mg(OH)Cl, hydroxylated $MgCO_3$], magnesium alkoxides, or magnesium organometallics (Grignard reagents, magnesium dialkyls) followed by activation with aluminum alkyls.[262,263] Ball milling[263,264] of $TiCl_4$ or titanium halogen alkoholates with $MgCl_2$ results in catalysts with a disordered structure similar to that of δ-$TiCl_3$. The very-high-surface-area catalysts formed contain isolated and clustered titanium species of various oxidation states.[264] When ball milling is carried out in the presence of Lewis bases (e.g., alcohols and esters), the resulting catalysts are also active in propylene polymerization.[264–267] Ultra-high-activity $MgCl_2$-supported catalysts have been developed by Shell.[266]

Subsequent studies also resulted in the development of homogeneous Ziegler–Natta catalysts. The first soluble homogeneous catalyst system[268,269] was [Cp_2TiCl_2] with Et_3Al or Et_2AlCl. Homogeneous vanadium catalysts[270–273] [VCl_4, $VOCl_3$, $V(acac)_3$] with alkylaluminum cocatalysts are applied in low-temperature syndiotactic polymerization of propylene and in copolymerization of ethylene and propylene.

When trialkylaluminums are treated with water under controlled conditions, linear or cyclic aluminoxanes (**39, 40**) are formed:

$$R_2Al\left[\begin{array}{c}R\\|\\Al-O\end{array}\right]_n R \qquad \left[\begin{array}{c}R\\|\\Al-O\end{array}\right]_n$$

39   **40**

These are excellent cocatalysts to Group IV metallocenes (Ti, Zr, Hf) ensuring prolonged catalytic activities.[274–281] Metallocenes activated by aluminoxanes or noncoordinating anions are extremely powerful polymerization catalysts with activities surpassing those of heterogeneous catalysts. An additional unique property of these systems is that by simply varying the π-bonded ligands they are capable of producing polymers with many different microstructures (isotactic, atactic, syndiotactic, stereoblock, isoblock, discussed later).[208] In addition to bis(cyclopentadienyl) derivatives, bridged bis(cyclopentadienyl), monocyclopentadienyl, indenyl, fluorenyl, and aryloxide complexes have been described. In general, change of the ligand at the transition-metal center have a profound effect on polymerization.[282] Because of the very high productivity and the ability to control molecular weight and molecular weight distribution, these catalysts are possible candidates for future industrial application.[283]

The other major group of alkene polymerization catalysts includes one-component, that is, metal-alkyl-free, transition-metal catalyst systems. The most important of these are the oxide-supported catalysts such as the original Phillips chromium oxide on silica or silica–alumina.[235,236,262,263,284–286] In contrast with the Standard Oil of Indiana catalyst (Mo_2O_3 on alumina), this is a successful commercial catalyst still used in industry. An important characteristic of these catalysts is their specificity. The Phillips catalyst, for instance, although highly efficient for ethylene polymerization, is ineffective for the polymerization of propylene. It is prepared by the impregnation of silica or silica–alumina with CrO_3 dissolved in water and activated in air at 400–800°C. The catalyst contains typically 1% of chromium. Better product characteristics were later achieved by promoting a CrO_3-on-SiO_2 catalyst with titania.[284]

It was later found that stable organometallic compounds of transition metals exhibiting very low polymerization activity could be transformed into high-activity catalysts when deposited on silica, alumina, or silica–alumina.[287–289] Interaction of surface hydroxyl groups with the organometallic compounds such as chromocenes, benzyl, and π-allyl complexes results in the formation of surface-bound organometallic complexes (**41–43**):[289–291]

In addition to these supported transition-metal catalysts, some soluble transition-metal compounds exhibit considerable activity in polymerization without added aluminum alkyls.[292] The most active compounds are σ-organometallics of Ti and Zr with methyl, benzyl, and halogen ligands. π-Allyl compounds of Ti, Zr, and Cr are also useful catalysts.

Active Centers and Mechanisms. Any discussion of the mechanism of coordination polymerization must necessarily recognize the complexity of the reaction. That is why, despite the enormous efforts and the extremely large number of papers published about this topic, the mechanism of coordination polymerization is not known in great detail. Two key steps, however, seem to be well established. First the complexation between the alkene and the active center takes place, which is followed by the insertion of the activated monomer to the growing polymer chain.[117,125,254,255,293-296] The mode of activation of heterogeneous and homogeneous, and metal-alkyl-free and metal-alkyl-promoted catalysts, however, is different. Although some early mechanisms proposed a bound radical or a bound anion as the growth center, propagation at the metal–alkyl bond is favored by most experimental evidence and now commonly accepted. A popular early concept considering propagation at the metal–alkyl bond in Ziegler–Natta catalysis was first proposed by Natta[297,298] suggesting the involvement of the activator–alkyl bond as the growth center. The active catalytic species is formed in a reaction between surface Ti^{3+} ions of $TiCl_3$ crystallites and the cocatalyst:

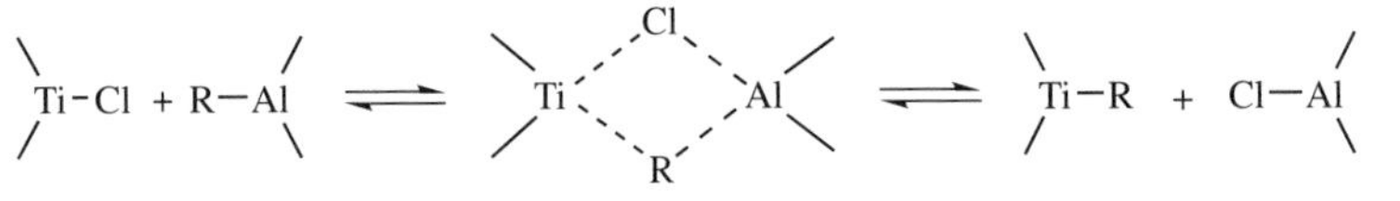

$$\text{\\Ti-Cl + R-Al/} \rightleftharpoons \text{Ti}\overset{Cl}{\underset{R}{\diamond}}\text{Al} \rightleftharpoons \text{\\Ti-R + Cl-Al/}$$ (13.63)

44

This generates a titanium–carbon σ bond through ligand exchange. The active center was believed to be the bimetallic species **44** existing in equilibrium with different monometallic species.

During propagation the alkene molecule forms a π-complex with the transition metal of the active center:

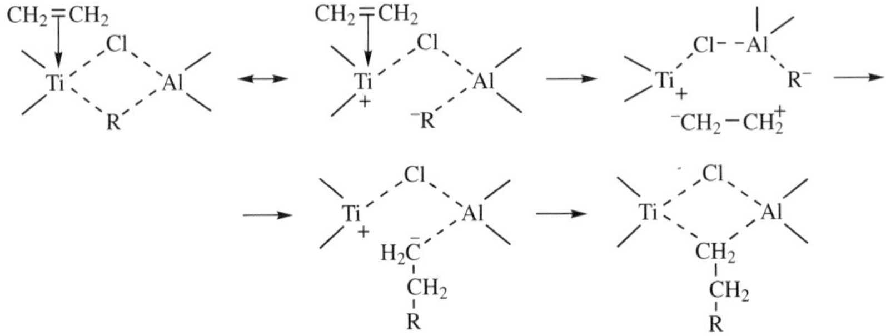

(13.64)

This results in strong polarization of the π bond and dissociation of the Ti–C bond, thus promoting insertion into the activator aluminum–alkyl bond. Repetitive insertions of alkene molecules result in lengthening of the polymer chain. This mechanism is also termed *bimetallic* after the growth center complex species **44**.

Variations of this mechanism included the suggestion of a partially bonded alkene molecule,[299] the participation of a titanium–aluminum ion pair,[300] and a concerted alkene insertion.[301] The development of the activator–alkyl mechanism was probably strongly influenced by the "Aufbau" reaction, studied originally by Ziegle.[102] He observed that Group I–III alkyl compounds such as Et_3Al catalyzed the oligomerization of ethylene to terminal alkenes. Additional evidence of such mechanism comes from the fact that alkylaluminum compounds exist in dimeric forms and from spectroscopic characterization of bimetallic complexes.[297] Other observations, such as that metal alkyls have a pronounced effect on polymer structure observed by Natta, also supported this concept.[302] This effect was, however, later found to be related to the different ability of alkylaluminums to alkylate and reduce the transition metal.

At present, however, most experimental evidence supports the mechanism in which propagation takes place at the transition-metal–alkyl bond. Of the different interpretations, the one proposed by Cossee and Arlman[303–306] is the most widely accepted.[125,254,294]

According to Arlman,[303] the active centers are titanium atoms with chlorine vacancies. Such sites are found along the growth spirals, on lateral faces, and on surface defects.[307] These are the regions where surface titanium atoms are incompletely coordinated. Each such exposed titanium possesses one chlorine vacancy and surrounded by five chlorine atoms (**45**).

45

Four of the chlorines are in the crystal lattice and bridge-bonded with neighboring titaniums, while the fifth single-bonded chlorine (*Cl*) protrudes from the crystal surface. The active center, which is a hexacoordinated transition metal with one vacant octahedral site, is created by replacing the single-bonded chlorine by alkylation with the alkylaluminum:

$$(13.65)$$

Considerable experimental evidence supports such alkylations of soluble titanium complexes[308] and of accessible titanium[309,310] on the surface of $TiCl_3$.

In the propagation step the alkene monomer is complexed on the vacant site with the double bond parallel to an octahedral axis, and insertion takes place:[304]

$$(13.66)$$

The stereoregularity of propylene polymerization is strongly related to the geometry of the catalytic centers and the orientation of complexation.[305,306]

Strong support for the monometallic mechanism was furnished by crystal structure studies[303] and labeling experiments.[311–315] It was observed, for instance, that ethylated titanium reacting with ^{13}C-labeled ethylene yielded polymers with unlabeled chain end.[314,315] In another experiment TiCl$_3$, when methylated with ^{13}C-enriched methyl cocatalyst, reacted to yield polymers with labeled end group.[311]

Although bimetallic active centers with activator molecules complexed to the transition metal were suggested and observed,[307,309,316] experimental evidence indicated the insertion of the monomer into the titanium–carbon bond even in these cases.

The strongest evidence in favor of propagation at the transition metal–alkyl bond is the existence of one-component, that is, metal-alkyl-free polymerization catalysts. Of these systems the Phillips catalyst was studied most thoroughly because of its commercial importance. Originally it was believed that Cr(VI) ions stabilized in the form of surface chromate and perhaps dichromate resulting from the interaction of CrO$_3$ with surface hydroxyl groups above 400°C are the active species in polymerization:[286,294]

$$(13.67)$$

Oxidation states from Cr(V) to Cr(II) were also suggested.

The observation that a long initiation period exists in polymerization with CrO$_3$-on-SiO$_2$ catalyst was thought to be due to the slow reduction of Cr(VI) by ethylene to Cr(II). Catalyst prereduction (CO, 300°C), or addition of reducing agents such as Zn or Al alkyls to form Cr(II) species, eliminates the induction time.[286] XPS data published in 1982 and 1983 furnished direct evidence in support of divalent chromium with high coordinative unsaturation as the active species.[317,318] Subsequent studies on the valence state of the active site found a correlation between the catalytic activity and the intensity of the ESR (electron spin resonance) signal suggesting Cr(III) as the active species.[319] Other observations indicated that both Cr(II) and Cr(III) may be active in polymerization with the maximum activities strongly depending on reaction conditions.[320,321] It seems highly probable that several valence states of chromium may participate in ethylene polymerization.[286] This phenomenon, the existence of different types of active centers, is quite general in heterogeneous catalysis and may be valid for all heterogeneous polymerization catalysts.

Supported organochromium complexes exhibit catalytic behavior very similar to that of the traditional Phillips catalyst, and both systems are very sensitive to minute experimental variables.[322] Modification of the catalyst allows for control of molecular-weight distribution and the structure of polyethtylene. These modifications include pretreatment of the support at high temperature, the use of different supports, and mixed catalysts by depositing an organochromium compound on the calcined chromium oxide catalyst.[322] The latter catalyst yields branched polymers.

A possible initiation on chromium catalysts can take place through the formation of a π-allyl and a hydride species:[286,289]

$$(13.68)$$

Both species can lead to propagation. Many other plausible theories on propagation have been sauggested.[125,286]

A number of processes are available for chain transfer in coordination polymerization.[125,254,255,323] Chain transfer brings about termination of the growing polymer chain, specifically, its separation from the transition metal, which makes the active site available for initiation of a new chain. Since chain transfer with modern, high-activity Ziegler–Natta catalysts occurs very rarely compared to chain growth, polymers with very long chains and of high molecular weight are produced. Since these are unsuitable for processing, effective molecular weight control is necessary. Two transfer reactions are characteristic of certain heterogeneous Ziegler–Natta catalysts. Chain transfer to the monomer [Eq. (13.69)] involves a β-hydrogen transfer that yields a polymer with a methyl group at one end and a vinyl group at the other end. The observation that the monomer concentration has a significant effect on molecular weight is in accord with this mechanism. Chain transfer with hydrogen [Eq. (13.70)] leaves only methyl groups in the polymer chain ends:

$$M-CH_2CH_2-P \; + \; CH_2{=}CH_2 \longrightarrow M-CH_2CH_3 \; + \; P-CH{=}CH_2 \qquad (13.69)$$

$$M-CH_2CH_2-P \; + \; H_2 \longrightarrow M-H \; + \; P-CH_2-CH_3 \qquad (13.70)$$

Addition of hydrogen to terminate the chains is the most important and widely used industrial process to control molecular weight. Chain transfer with organometallic compounds (cocatalysts), which is basically an alkyl exchange, can take place with titanium-based and chromium-based systems:

$$M{-}CH_2CH_2{-}P \; + \; Et_3Al \longrightarrow M{-}CH_2CH_3 \; + \; P{-}CH_2{-}CH_2{-}AlEt_2 \qquad (13.71)$$

It yields a polymer with an organometallic chain end that, being unstable, is hydrolized by moisture.

The basic chain transfer with chromium-based catalysts is a spontaneous β-hydrogen transfer to the transition metal leading to a vinyl chain end:

$$M{-}CH_2CH_2{-}P \longrightarrow M{-}H \; + \; P{-}CH{=}CH_2 \qquad (13.72)$$

Since the significance of this transformation increases with increasing temperature, it provides an easy way of controlling molecular weight in polymerization with the Phillips catalyst.

Stereoregular Polymerization of Propylene. In his pioneering work on polymerization of propylene catalyzed by two-component transition-metal and metal–alkyl catalyst combinations, Natta observed the formation of different types of polymer.[239,324] On the basis of X-ray analysis he proved that the crystalline poly-propylene contains successive head-to-tail monomers with identical configuration. This structure is called *isotactic polymer.* In the amorphous atactic polymer that could be extracted with boiling heptane, monomers with different configurations are randomly distributed along the chain. A third possible structure is the syndio-tactic polymer with regular alternating configuration at adjacent asymmetric centers along the polymer chain.

The structures of the two stereoregular polymers assuming an imaginary planar, fully extended backbone are given in Scheme 13.6. In this representation all methyl groups in the isotactic polymer lie on the same side of the backbone plane, while in the syndiotactic polymer they alternate.

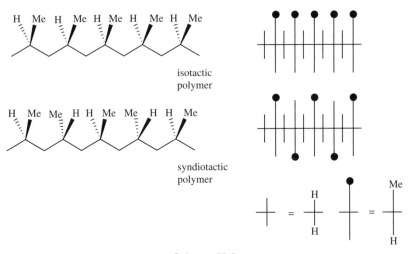

Scheme 13.6

In reality isotactic polypropylene has a helical structure.[325] The helices easily aggregate to form crystalline structures that ensure relatively high melting point and outstanding mechanical properties. Atactic polypropylene is incapable of crys-tallizing because of the statistically irregular sequence of sterical orientation of methyl groups connected to the chiral carbon atoms. On the basis of solubility dif-ferences in boiling heptane, Natta introduced the *isotacticity index,* which refers to the percentage of insoluble, that is, isotactic polymer in the product.[324]

Natta first used the in-situ prepared Ziegler catalyst ($TiCl_4 + Et_3Al$), which produced polymers with low stereoselectivity[324] (isotacticity index = 30–40%).

By assuming that stereoselectivity is connected to the regular heterogeneous catalyst surface, Natta introduced violet $TiCl_3$ and used preformed heterogeneous catalysts in further studies.[234,240] These complexes may act as stereoselective catalysts and polymerize propylene to crystalline stereoregular polymers.

Isospecific polymerization. Important features of stereoregular isospecific polymerization of propylene, and in general, terminal alkenes, are as follows:[125,254,326–328]

1. Isotactic polymerization occurs with practically complete regioselectivity, that is, with a continuous head-to-tail addition, which is a prerequisite to stereoregularity.

2. It was demonstrated by spectroscopic studies (IR, NMR) that in isotactic polymerization of propylene a so-called primary or type 1–2 insertion prevails, resulting in a CH_2 unit as the end group of the growing chain:[329,330]

$$M-P + CH_2=CHCH_3 \longrightarrow M-CH_2CH-P \xrightarrow{\ \ CH_2=CHCH_3\ \ } M-CH_2CH-CH_2CH-P$$
$$\underset{CH_3}{|} \qquad\qquad \underset{CH_3}{|}\ \ \underset{CH_3}{|}$$

$$(13.73)$$

In certain catalyst systems it is suggested to be due to the high activation energy for further olefin insertion following an occasional 2–1 insertion.[331]

3. It was proved that in most cases all propagation and chain-transfer (chain-termination) steps take place without isomerization of the chiral center.[326]

4. Given the high regioselectivity, the type of addition in the insertion step must always be the same. It was found that in the polymerization of *cis*- and *trans*-[1-^{2}H$_1$]-propene a *syn*-type double-bond opening takes place during insertion, eventually resulting in *erythro*-diisotactic and *threo*-diisotactic polymers, respectively:[332–334]

$$(13.74)$$

$$(13.75)$$

The formation of an isotactic polymer requires that insertion always occur at the same prochiral face of the propylene molecule. Theoretically, both a chiral catalytic site (enantiomorphic site control) and the newly formed asymmetric center of the last monomeric unit in the growing polymer chain (chain end control) may determine stereoregulation in stereoselective polymerization. It is generally acknowledged that steric control is governed by the steric environment around the metal center. In other words, isotactic stereoregularity is due to the chiral catalytic site; in other words, enantiomorphic site control prevails. This conclusion is supported by numerous experimental observations.

1. By means of the ^{13}C NMR technique Zambelli et al. proved[311] that in polymerization of propylene catalyzed by $TiCl_3 + (^{13}CH_3)_2AlI$ the stereochemistry of the very first insertion is the same as that of subsequent units despite the lack of asymmetry in the original alkyl group of the activator.

2. Occasional defects in the polymer chain remain isolated. When propylene reacts from the opposite prochiral face bringing about a syndiotactic error (Scheme 13.7, *a*), the original chirality center preceding the error continues to be formed. The same is true for a secondary defect or type 2–1 insertion (Scheme 13.7, *b*) which is a head-to-head addition and creates a CH_2–CH_2 sequence. If the chirality were controlled by the growing chain end, an error would be perpetuated, giving rise to a polymer block with altered chirality. Instead, isotacticity is maintained in both cases.

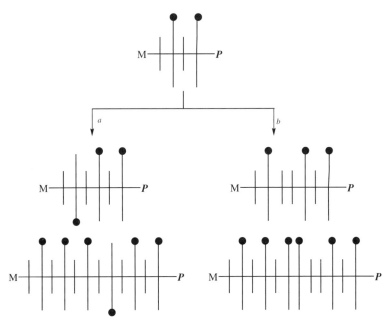

Scheme 13.7

3. The polymerization is stereoselective even when the last unit in the growing chain is achiral. This is the case in isotactic copolymerization of ethylene and propylene when steric order is not affected by ethylene units in the chain.[335–337] It means that after the incorporation of an ethylene monomer, the addition of propylene is still isospecific.

4. Studies on stereoselective polymerization of racemic olefins also support this view.[338] Polymerization of 3,7-dimethyl-1-octene (the chiral center is in α position to the double bond) took place with 90% stereoselectivity yielding an equimolar mixture of homopolymers of the two enantiomers. No stereoselectivity was observed in the polymerization of 5-methyl-1-heptene (the chiral center is in γ position to the double bond). The conclusion is that the ability of a catalytic center to distinguish between the two enantiomers of a monomer required for stereoselective polymerization must arise from its intrinsic asymmetry. The first-ever chiral polypropylene synthesized using a chiral zirconium complex with aluminoxane cocatalyst is the latest evidence to testify the role of the catalyst center in isotactic polymerization.[339]

These observations clearly indicate that heterogeneous Ziegler–Natta catalysts must contain chiral active centers. It is conceivable to assume that the asymmetric ligand environment is responsible for the regio- and stereoselectivity of these systems. Typical isospecific catalysts are the $TiCl_3$-based systems, both the first-generation heterogeneous Ziegler–Natta catalysts and the high-activity $MgCl_2$-supported catalysts.[340] It also follows that the key factor in stereoregulation is the favored discriminative complexation at one of the prochiral faces of the olefin followed by stereospecific insertion. Various models of catalytic centers and mechanisms have been proposed to interpret the origin of stereoregular polymerization.[208,254,295,328,340]

The Cossee–Arlman mechanism discussed before for ethylene polymerization[303,304] offers a conceivable explanation for stereoregulation.[305,306] According to this model, alkylated hexacoordinated titanium atoms with one vacant octahedral site characteristic of α-, γ-, and δ-$TiCl_3$ serve as asymmetric catalytic sites (**45**) [see displayed structure preceding Eq. (13.65)]. These atoms are in fixed positions in the crystal lattice, and they maintain their original chirality during propagation. Because of the nonequivalence of the crystallographic sites of coordination vacancy and the alkyl ligand, isotacticity requires the "back-jump" of the growing chain to its original position to restore the initial active center after each insertion:[305]

$$
\begin{array}{ccc}
& R & \\
& | & \\
\square \quad CHCH_3 & & \\
Cl\!-\!Ti\!-\!CH_2 & \longrightarrow & \\
\end{array}
\qquad (13.76)
$$

Sequences of sterically identical steps lead to isotacticity of the product. Active sites with two vacancies or with one vacancy and a loosely bound chlorine characteristic of β-$TiCl_3$ with a chainlike structure are nonstereospecific and result in polymers with increased atacticity.[306] Molecular mechanics studies on the basis of this model lent additional support for the asymmetric active center being responsible for stereoregulation.[341] It was concluded that the main factor determining stereoselectivity is the steric repulsion arising between the catalyst surface and the first carbon–carbon bond in the growing chain resulting in a fixed chiral orientation of this bond. In other words, stereoselectivity involves chiral orientation of the growing chain.[342]

The weak point of the Cossee–Arlman model is the back-jump of the polymer chain after each insertion step. Modifications to avoid this step have been suggested. Allegra assumes that titanium atoms at the side surfaces of $TiCl_3$ crystals are slightly protrude above the two neighboring layers.[343] Since no unacceptable interatomic contacts with the neighboring layers exist in this case, the two vacancies (one originally existing, the other formed after alkene insertion) are equivalent and alkene coordination can occur alternately. On the basis of steric repulsion arising in the complexation according to the Cossee–Arlman model with the double bond parallel to the titanium–polymer bond (**46**), Allegra considers an outward *trans*-like orientation of the complexed alkene with the double bond perpendicular to the titanium–polymer bond (**47**):

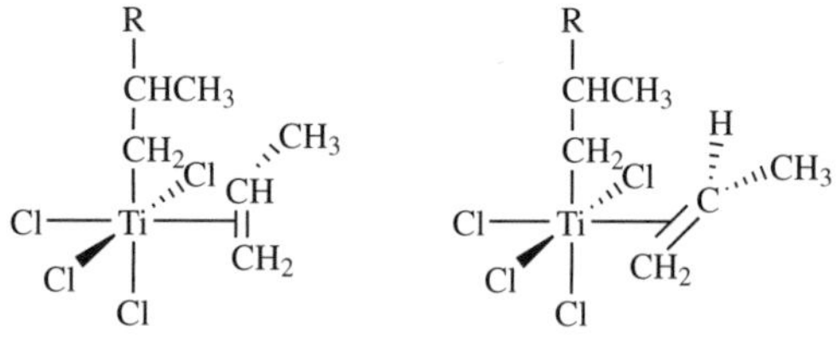

46　　　　　　**47**

During insertion the propylene molecule rotates in order to bring the CH group closer to the CH_2 group bonded to titanium.

A further modification of the active-center model was based on the consideration that the vacancy in the active center is strongly shielded by the polymer chain, mainly by the CH_2 and CH_3 groups of the second monomer unit. As a result, the vacant site is blocked and inaccessible for olefin coordination. Kissin et al. suggest a polymerization center with two vacancies: one shielded and the other open for complexation.[344,345] After each insertion step the end of the growing polymer chain flips from side to side and the two vacant sites are alternately available for alkene coordination.

In a more general picture put forward by Zambelli,[327,346] isotactic propagation arises when bulky substituents of the transition metal (X and Y) in the four-center complex (**48**) hinder secondary insertion:

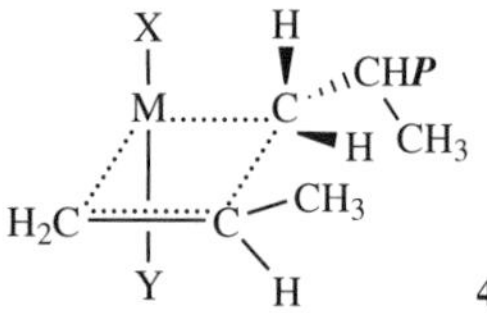

48

If the bulk of the substituents is different (Y bulkier than X), it ensures a constant approach of the monomer from the same side of the active center in constant steric arrangement. Monomer insertion instead of ligand migration is supposed to occur.

Pino and Mülhaupt considered four diastereomeric π complexes formed as a result of the four possible modes of complexation of propylene.[328] They suggested that diastereomeric and rotameric equilibria between these π complexes and/or activation energy for the insertion reaction control the large regioselectivity and enantioface discrimination necessary for stereoregular polymerization.

An interesting suggestion regarding the possibility of a carbene mechanism was also raised[347] but was disproved by NMR studies.[310] In the polymerization of propylene with $TiCl_3 + (^{13}CH_3)_2AlI$, only labeled methyls but no methylenes were observed. The formation of the latter could be expected if the carbene mechanism were operative.

Stimulated by their great industrial importance, high-activity $MgCl_2$-supported titanium catalysts have been extensively studied.[263–267] $MgCl_2 + TiCl_4$ treated with alkylaluminum is a highly active catalyst system with poor stereoselectivity (isotacticity index 50–70%). When treated with suitable electron-donor molecules (preferably with aromatic esters) called *selectivity control agents*, highly selective catalysts are formed. On the basis of structural similarities between $MgCl_2$ and δ-$TiCl_3$, and the similar nonselectivity exhibited by β-$TiCl_3$ and the nontreated $MgCl_2 + TiCl_4$ systems, two types of active centers were suggested:[348] one with two chlorine vacancies (nonstereoselective site), and the other with one vacancy and one loosely bound chlorine (low isospecificity site). Donor molecules (:*D*) transform the nonstereoselective sites into highly isospecific centers [Eq. (13.77)], while the low isospecificity site is poisoned:

$$
\begin{array}{ccc}
\underset{\begin{subarray}{c}\diagup | \\ Mg \\ \underset{Cl}{\overset{\cdots}{\underset{|}{Cl}}} \\ Cl-\underset{\overset{|}{CH_2}}{\overset{|}{Ti}}\!-\!\square \\ \overset{|}{P}\end{subarray}}{} & :D \longrightarrow & \underset{\begin{subarray}{c}\diagup | \\ Mg \\ \underset{Cl}{\overset{\cdots}{\underset{|}{Cl}}} \\ Cl-\underset{\overset{|}{CH_2}:D}{\overset{|}{Ti}}\!-\!\square \\ \overset{|}{P}\end{subarray}}{}
\end{array}
\tag{13.77}
$$

Soluble Ziegler–Natta catalysts can exhibit unique stereochemical properties. Group IV metallocenes in combination with methylaluminoxanes produce isotactic polypropylene with two different isotactic microstructures. The usual enantiomorphic site control is characteristic of enantiomeric racemic titano- and zirconocene complexes (e.g., ethylene-bridged indenyl derivatives[279,349]). In contrast, achiral titanocenes (e.g., [Cp_2TiPh_2]) yield isotactic polypropylene with microstructure **49**, which is consistent with a chain end control mechanism:[279,349–351]

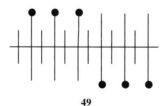

49

It has been found that isotactic polypropylene produced by unbridged biscyclopentadienyl zirconium catalysts is formed through a mixed chain end control and enantiomorphic control.[352] Molecular mechanics calculations led to the development of a model that takes into account nonbonded interactions to rationalize observed selectivities and predict properties of new catalyst systems.[353]

Syndiospecific Polymerization. Syndiotactic polymers have no practical importance, and syndiotactic polymerization is of interest only from the academic point of view. Syndiotactic polypropylene was first isolated by Natta and coworkers using titanium-based catalysts.[354] Since then vanadium-based homogeneous systems proved to be excellent catalysts for syndiotactic polymerization.[271,355–360] A high Al : V ratio (2–10) or a Lewis base (e.g., anisole) and low temperature ($-78°C$) are necessary to achieve high syndioselectivity.[355–357] In the active species trivalent vanadium complexed with the olefin is suggested to be pentacoordinated with one coordinative vacancy and becomes tetracoordianted after insertion.[360] The highly active and versatile metallocene–aluminoxane catalyst systems developed by Ewen and Kaminsky are also capable of producing syndiotactic polypropylene.[280,281,361]

Characteristic features of syndiotactic polymerization are as follows:[125,208,254,327,328,340]

1. The addition of the monomer[362] is *syn*, but in contrast to isotactic polymerization, secondary or type 2–1 insertion is characteristic of syndiotactic polymerization:[358,359]

$$\text{M}-\textbf{\textit{P}} \ + \ \text{CH}_2\text{=CHCH}_3 \ \longrightarrow \ \underset{\underset{\text{CH}_3}{|}}{\text{M}-\text{CHCH}_2}-\textbf{\textit{P}} \ \xrightarrow{\text{CH}_2\text{=CHCH}_3} \ \underset{\underset{\text{CH}_3}{|}}{\text{M}-\text{CHCH}_2}-\underset{\underset{\text{CH}_3}{|}}{\text{CHCH}_2}-\textbf{\textit{P}}$$

$$(13.78)$$

   The different mode of insertion is attributed to the different polarities of the Ti—C and V—C bonds. The regioselectivity of the syndiotactic catalysts, however, is much lower than that of isospecific catalysts; inversion, namely, head-to-head and tail-to-tail additions (type 1–2 addition), are more frequent.[359]

2. The structure of ethylene–propylene copolymers shows that in the case of syndiospecific polymerization the steric control is due to the chirality of the asymmetric carbon of the last unit of the growing chain end.[337]

Considering Zambelli's model,[327,346] syndiotactic propagation is ensured by the higher stability of the intermediate four-center complex **50**, in which the methyl group of the chain end and the methyl group of the complexed monomer are in *trans* position, compared to **51** with *cis* methyl groups:

$$
\begin{array}{cc}
\textbf{50} & \textbf{51}
\end{array}
$$

Soluble catalysts again may exhibit unique selectivities. Hafnium and zirconium fluorenyl metallocenes with methylaluminoxanes give syndiotactic polymers in high yields. The microstructure of the products indicates site stereochemical control with chain migratory insertion resulting in site isomerization with each monomer addition.[280,281]

Stereoregular Polymerization of Dienes. Coordination polymerization exhibiting high degree of chemo- and stereoselectivity is the method of choice to synthesize *cis*-1,4-polybutadiene the commercially most important product[184,187,363] (see Section 13.2.6). The classic Ziegler catalyst (TiCl$_4$ + Et$_3$Al) gives polymers with high proportion of *trans* content.[184,187,363] Depending on the Ti : Al ratio and the polymerization temperature, polymers with *trans* content up to 98% can be prepared.[364] Pure *trans*-1,4-polybutadiene has little application because of its crystallinity and high melting point.

The presence of iodine in Ti–Al catalyst combinations is essential to get *cis*-1,4-polybutadiene. TiI$_4$ + isoBu$_3$Al, for instance, converts 1,3-butadiene to polymers with 92–94% *cis* content.[365–367] Ternary systems, such as the commercially used TiCl$_4$ + I$_2$ + isoBu$_3$Al combination, have similar characteristics. They contain an insoluble component and a soluble component, both essential to activity and selectivity.[368] Both catalytic activity and *cis* selectivity increase with increasing Al/Ti ratio, reaching maximum values at about 3–5 depending on the particular alkylaluminum.[363,367] Both *cis* selectivity and molecular weight are affected by catalyst concentration.

Cobalt compounds,[369–373] particularly salts of organic acids, such as cobalt(II) octanoate[369–371] with aluminum alkyls, are also suitable catalysts in hydrocarbon solvents. An Al : Co ratio higher than 1 and water[370–372] are essential to obtain high catalytic activity and high *cis* selectivity.[184,363] Polybutadiene with a high *cis* content can also be synthesized by nickel-based catalysts.[184,187,363] Of the many different catalysts, fluorine-containing three-component systems[184,374–377] such as nickel naphthenate + Et$_3$Al + Et$_2$O—BF$_3$ are the most widely used commercially, while π-allyl–nickel complexes[47,55] are versatile catalysts to give polybutadienes of different steric structures. The same catalysts employed in the polymerization of 1,3-butadiene can be used to obtain polyisoprenes, except they exhibit lower activity.[186,187,363] Neglecting the structures arising from enchainment of the three different monomers, eight possible polyisoprenes can exist. Of these polymers only the highly stereoregular crystalline *cis*-1,4-polyisoprene, corresponding to natural rubber and *trans*-1,4-polyisoprene, as well as the amorphous atactic 3,4-polyisoprene were synthesized.

The mechanism of coordination polymerization of 1,3-butadiene and, in general, that of conjugated dienes follows the same pathway discussed for alkene polymerization; that is, monomer insertion into the transition metal–carbon bond of the growing polymer chain occurs. One important difference, however, was recognized very early.[47,378,379] In the polymerization of dienes the growing chain end is π-allyl complexed to the transition metal:

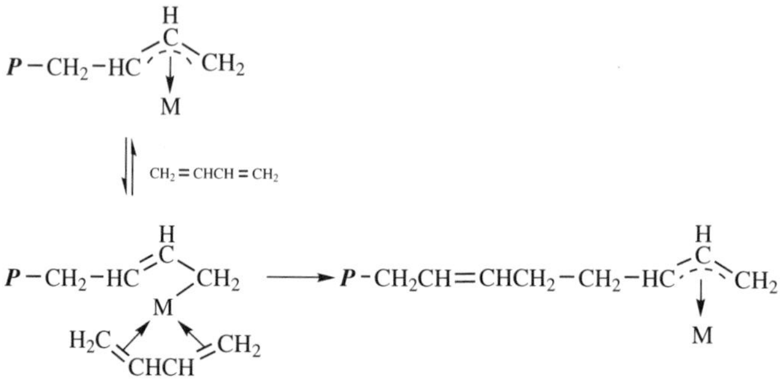

$$P-CH_2-HC \overset{\overset{\displaystyle H}{\underset{\displaystyle |}{C}}}{\underset{\displaystyle M}{\diagup \diagdown}} CH_2$$

$$\Updownarrow \quad CH_2=CHCH=CH_2$$

$$P-CH_2-HC \overset{\overset{\displaystyle H}{C}}{\diagup \diagdown} CH_2 \qquad \longrightarrow \qquad P-CH_2CH=CHCH_2-CH_2-HC \overset{\overset{\displaystyle H}{C}}{\diagup \diagdown} CH_2 \qquad (13.79)$$
$$\underset{\underset{\displaystyle H_2C \diagdown \diagup CH_2}{\underset{\displaystyle CHCH}{}}}{M} \qquad \qquad M$$

Insertion involves a transitory σ-allyl-bonded chain and π-coordinated monomer eventually leading to the re-formation of the π-allyl chain complex.[380–382]

Stereoselectivity, or the formation of a *cis*-1,4 or *trans*-1,4 unit, is connected to the structure of the chain end π-allyl complex (Scheme 13.8). The *syn*-allyl (**52**) and the *anti*-allyl (**53**) form, which are in equilibrium,[383] give rise to the formation of the *trans*-1,4 or *cis*-1,4 groups, respectively.[364,380]

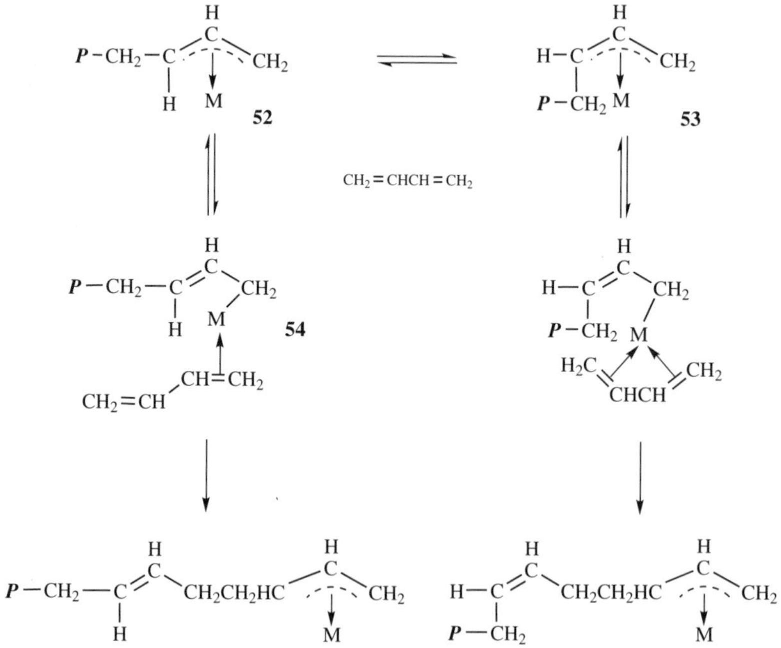

Scheme 13.8

Although not all the governing factors are well understood, it is generally agreed that the mode of coordination of the monomer is the decisive factor in determining the structure of the butenyl group. The coordination is affected by solvents, cocatalysts as ligands, and reaction conditions. A strongly coordinating ligand, for instance, can result in the formation of the monocoordinated complex **54** leading to a *trans*-1,4 unit. The hypothesis of backbiting coordination was advanced by Furukawa to interpret *cis* selective polymerization.[384] The chemoselectivity, specifically, 1,2 versus 1,4 insertion, depends on the attachment of the new monomer to the C(1) or C(3) reaction centers during insertion[385] (Scheme 13.9).

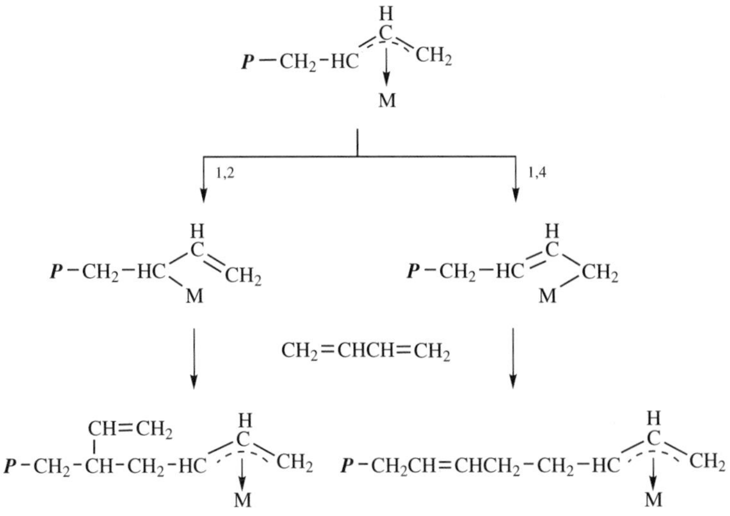

Scheme 13.9

Both catalytic activity and polymer structure strongly depend on the metal involved and the ligands. Ti, Co, and Ni derivatives catalyze the formation of 1,4-polybutadienes, while 1,2-polymer is usually obtained by other metals (V, Cr, Mo) in halide-free systems.[184,363] Nickel complexes of the type (allyl)NiX are excellent examples demonstrating the role of ligands. High-*trans* polymer is formed with the iodide derivative,[386,387] while the chloride[386,387] and trifluoroacetate[387] give high-*cis* product. Addition of electron donors [e.g., P(OPh)$_3$] to the latter catalyst shifts the selectivity to the formation of *trans*-1,4-polybutadiene.[388] Equibinary polymers consisting of *cis* and *trans* units in approximately 1:1 ratio are obtained in benzene or with added CF$_3$COOH.[388]

13.2.5. Conducting Polymers

Polyacetylene and poly(*p*-phenylene) (PPP) are two polymeric hydrocarbons with a unique property; both are conducting polymers on doping. Polyacetylene, a crystalline polymer, has a conjugated π-system with four different structures (**55**–**58**):[389–391]

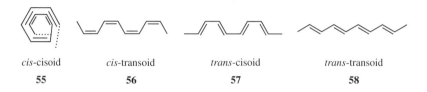

| *cis*-cisoid | *cis*-transoid | *trans*-cisoid | *trans*-transoid |
| **55** | **56** | **57** | **58** |

cis-Polyacetylene is an electrical insulator while *trans*-polyacetylene is a semiconductor. When *cis*-polyacetylene, however, is treated with strong oxidizing or reducing agents called *dopants*, an increase over 10 orders of magnitude in conductivity takes place.[391–394] Radical ions and double-charged species are formed and serve as charge carriers.[391,395]

Numerous catalysts are active in the polymerization of acetylene and susbtituted acetylenes.[389–391,396–399] Natta et al. were the first to prepare polyacetylene as a black powder using the Ti(O-*n*-Bu)$_4$ + Et$_3$Al catalyst system.[400] Later Shirakawa et al. found[401] that high-quality free-standing films could be produced at high catalyst concentrations (Al : Ti ratios of about 3–4) at the gas–liquid interface. Under these conditions polymerization and simultaneous formation of a characteristic crystalline, fibrillar structure take place. Reaction temperature strongly affects the structure of the polymer. The *cis* isomer is formed at low temperature ($-78°$C, 98.1% *cis* content), while the *trans* isomer is the product of high-temperature polymerization (150°C, 100% *trans* content).[401] Later Naarmann[402,403] and others[404,405] found that aging of the catalyst at high temperature (up to 150°C) results in the formation of a highly strechable, highly conducting polymer. A polymer with roughly equal amounts of the *cis* and *trans* forms synthesized in this way at room temperature was reported to exhibit conductivity similar to that of copper.[406] Because of the low ionization potential and low activation energy, heating or high-level doping of the *cis* polymer can readily lead to isomerization to the *trans* polymer. Polyacetylene is sensitive to moisture and oxidation, and its synthesis requires extremely high-purity criteria.[390,391] Besides the Ti(O-*n*-Bu)$_4$ + Et$_3$Al catalyst, which is exclusively used for the polymerization of acetylene, many other catalyst systems were employed in polyacetylene synthesis.[390] In most cases side reactions, first cyclotrimerization to benzene, interfere in polymerization. Nickel halides with metal hydrides (NaBH$_4$, KBH$_4$)[407] or with tertiary phosphines[408] are the best of these catalysts. Ring-opening metathesis polymerization of cyclooctatetraene and substituted derivatives[409–411] is an effective indirect way to produce polyacetylenes (see Section 12.2).

Experimental observations with the Ti(O-*n*-Bu)$_4$ + Et$_3$Al catalyst support the insertion mechanism versus metathesis polymerization.[412] Propagation occurs via the *cis* opening of the triple bond of the coordinated monomer, leading to addition to the Ti—C bond between the growing polymer chain and the catalyst center:

$$(13.80)$$

The process results in a *cis*-transoid structure. The formation of *trans*-polyacetylene is suggested to take place through isomerization of the new segment formed by *cis* insertion before it can crystallize.[412]

Many mono- and a few disubstituted acetylenes were also polymerized.[389,390,411–415] Ziegler–Natta catalysts usually transform monosubstituted acetylenes into benzene derivatives via trimerization or to low-molecular-weight oligomers. Certain catalyst systems may yield polymers if the alkyl substituent is small and noncrowded.[415] In contrast, terminal alkynes with bulky alkyl or aryl substituents can be polymerized with Mo- and W-based systems,[414] such as carbene[416] and carbyne derivatives.[417] $MoCl_5$ and WCl_6 alone[418,419] or with organometallic cocatalysts[420] as well as iron complexes[421,422] are also effective.

Ziegler–Natta catalysts are not active at all in polymerization of disubstituted acetylenes.[415] Mo- and W-based systems (for alkynes with small substituents) and Nb- and Ta-based catalysts (for alkynes with bulky groups), in turn, are very effective catalysts used to convert disubstituted acetylenes into polymers with very high molecular weight.[414,415] A polymerization mechanism similar to that of metathesis polymerization of cycloalkenes are supported by most experimental observations.[414,423,424]

Anionic polymerization of phenylacetylene to a *trans*-cisoid polymer in the presence of crown ether phase-transfer catalysts initiated by sodium amide has been reported.[425] In contrast, the zwitterionic rhodium complex $Rh^+(COD)BPh_4^-$ yields a *cis*-transoid product in the presence of Et_3SiH.[426]

Unlike polyacetylene, substituted polyacetylenes are amorphous, electrically insulator soluble polymers.[413] They are highly stable and not sensitive to oxidation. Since the substituents exert a strong steric effect, the polyene backbone is not coplanar, and as a result, only limited conjugation is possible.

Direct polymerization of benzene through oxidative coupling yields poly(p-phenylene) (PPP) an insoluble polymer of low molecular weight.[427–429] Kovacic's original synthesis[430] using a Lewis acid–oxidant combination [Eq. (13.81)] is the most widely employed and still the most effective procedure for the synthesis of PPP:

$$(13.81)$$

PPP   60% yield

Two principal pathways have been suggested to interprete formation of PPP: the radical cation route and the σ-complex route with intermediates **59** and **60**, respectively.[428,431]

According to a recent view, association of benzene molecules generates **61** resembling a stepladder structure if viewed from the side[432–434] [Eq. (13.82)]. With increasing chain length the structure becomes too delocalized, leading to diminution of reactivity of the chain end groups. This results in covalent bond formation between the associated benzene molecules. Aromatization of the two end rings via a second one-electron oxidation and loss of a proton and a hydrogen atom gives **62**:

$$(13.82)$$

A final oxidative rearomatization of the cyclohexadiene units leads to PPP.

Polymerization of toluene under analogous conditions yields polymer with a poly(*o*-phenylene) backbone,[435] while the polymer with *meta* linkages is synthesized from *m*-terphenyl.[436] Besides the numerous other reagents employed electrochemical methods may also be used in the synthesis of PPP.[429] Free-standing flexible films with structure similar to PPP produced chemically can be prepared in electrochemical oxidation of benzene[437] in the presence of $Et_2O–BF_3$, or in nitrobenzene[438] containing the composite electrolyte $CuCl_2$ and $LiAsF_6$. Polymerization of naphthalene[439] with the latter system was also successful to produce a conductive polymer on doping with I_2.

13.2.6. Practical Applications

Ethylene Polymers. Depending on the polymerization conditions, three major types of polyethylene are manufactured: low-density polyethylene (LDPE) by free-radical polymerization, linear low-density polyethylene (LLDPE) by copolymerization of ethylene with terminal olefins, and high-density polyethylene (HDPE) by coordination polymerization. The processes yield polymers with different characteristics (molecular weight, molecular weight distribution, melt index, strength, crystallinity, density, processability).

Polyethylene produced by oxygen- or peroxide-initiated polymerization at high pressure has a density of about 0.91–0.935 g/cm^3 and is called low-density polyethylene (LDPE).[204,206] The reason for this low density is that chain branching affects the crystallization of polyethylene: increasing short-chain branching (caused by intramolecular chain transfer) brings about lower melting point and crystallinity, and decreasing density. LPDE has a crystallinity about 40–50%, typically contains 15–30 C_2–C_8 alkyl side chains (mostly butyl groups) per 1000 carbon atoms. It also contains long branches formed through (less frequent) intermolecular chain transfer, but these long side chains are more flexible. Typical low-density polyethylene has a molecular weight of about 100,000 to 150,000. The stucture of LDPE is analogous to that of linear low-density polyethylene (LLPDE), which is a copolymer of ethylene with α-olefins (see discussion below).

A number of processes have been developed to obtain products of different physical properties. The nature of the product is affected by the addition of diluents or other additives before carrying out the polymerization. Autoclaves or stirred-tank reactors, and tubular reactors, or their combinations have been developed for the industrial production of high-pressure polyethylene.[206,440] Pressures up to 3500 atm and temperatures near 300°C are typically applied.

Linear low-density polyethylene (LLDPE)[440–442] is a copolymer of ethylene and a terminal alkene with improved physical properties as compared to LDPE. The practically most important copolymer is made with propylene, but 1-butene, 4-methyl-1-pentene, 1-hexene, and 1-octene are also employed.[440] LLDPE is characterized by linear chains without long-chain branches. Short-chain branches result from the terminal alkene comonomer. Copolymer content and distribution as well as branch length introduced permit to control the properties of the copolymer formed. Improvement of certain physical properties (toughness, tensile strength, melt index, elongation characteristics) directly connected to the type of terminal alkene used can be achieved with copolymerization.[442]

In principle, all Ziegler–Natta catalysts and Phillips supported CrO_3 catalyst discussed before are suitable for copolymerization. However, since the reactivity of terminal alkenes is much less than that of ethylene, suitable catalysts are required to produce copolymers with optimum elastomeric properties (i.e., without crystallinity and the presence of homopolymers). The best catalysts are vanadium compounds [$VOCl_3$, VCl_2, $V(acac)_3$] soluble in hydrocarbons with R_2AlCl cocatalysts.[443] A low-pressure, gas-phase fluidized bed process (e.g., UNIPOL[444] and BP Chimie processes[445]) or solution polymerization is employed industrially.[251,446] The latest version of the catalyst applied in Union Carbide's UNIPOL process is CrO_3 on silica containing titanium and a fluorine compound.[447] The role of fluorine is to improve comonomer incorporation and achieve narrower molecular weight distribution by reducing the low-molecular-weight polymer. A slurry process with Montedison's "particle-forming catalyst" produces highly uniform spherical particles (average size 1200 μm) that require no pelletizing.[448]

Incorporation of a small amount of propylene in ethylene polymerization lowers crystallinity, density, and melting point of the product. The copolymer thus formed has a narrower molecular weight distribution and exhibits better optical properties.

With larger amount of propylene a random copolymer known as *ethylene–propylene–monomer* (EPM) copolymer is formed, which is a useful elastomer with easy processability and improved optical properties.[208,449] Copolymerization of ethylene and propylene with a nonconjugated diene [EPDM or *ethylene–propylene–diene–monomer* copolymer] introduces unsaturation into the polymer structure, allowing the further improvement of physical properties by crosslinking (sulfur vulcanization).[443,450] Only three dienes are employed commercially in EPDM manufacture: dicyclopentadiene, 1,4-hexadiene, and the most extensively used 5-ethylidene-2-norbornene.

Compared to LDPE prepared by radical polymerization, HDPE (with densities higher than 0.940 g/cm^3) contains mainly methyl branches, and the branching is much less frequent (0.5–3 per 1000 carbon atoms). As a result, HDPE is characterized by higher crystallinity, higher density, and higher melting point. High-density polyethylene can also be produced by free-radical polymerization by changing process conditions.[203,451] Lower polymerization temperatures and very high pressures (up to 7500 atm) decrease the frequency of short-chain branching resulting in almost linear polyethylene with higher density (0.955 g/cm^3) and with less than 0.8 alkyl group per 1000 carbon atoms.[452] The process, however, commercially is not viable because of the extremely high pressure necessary.

Most of the worldwide production of polyethylene (and polypropylene) is based on low- and medium-pressure catalytic processes. Polymerization is carried out in the liquid or the gas phase.[208,251,323,447,453–455] The catalysts used are unsupported Ziegler catalysts (usually TiCl$_4$ with organomagnesium or oganoaluminum compounds and additives), supported Ziegler catalysts (on MgCl$_2$ support, e.g., Solvay's supported catalyst[456]), supported chromium oxide (Phillips) catalysts, and supported organotransition-metal catalysts.

The slurry process or the solution process is applied in the liquid phase. In slurry processes (e.g., the Hoechst process,[457] which was the first to produce HDPE, and the Montedison process[458]) polymerization is carried out below the melting point of the polymer in a hydrocarbon solvent (isobutane, isopentane, *n*-hexane) in which the product polyethylene is insoluble and obtained as fine free particles. Temperatures of 70–90°C and 7–30 atm pressure are usually applied, and yields of 95–98% are achieved. Slurry processes tend to give polyethylene with very high molecular weight unless chain-transfer agents are used. With Ziegler catalysts hydrogen is commonly used. Solvent operation (e.g., the Stamicarbon process[459,460]), in turn, carried out usually above the melting point of the polymer in cyclohexane, allows better molecular weight control and higher rate (shorter residence time) because of the homogeneous nature and the higher temperature (140–150°C). Compared to slurry processes, however, solvent removal (distillation) is more complicated.

Gas-phase processes commercialized in the late 1960s offer much simpler operation. Since they eliminate solvent, solvent separation, recovery, and purification are unnecessary. Polymerization is carried out in stirred-vessel reactors or in fluidized beds. A very successful fluidized-bed process is Union Carbide's UNIPOL technology[444] designed originally for producing HDPE and extended later to LLDPE. Ethylene is polymerized by injecting the fine catalyst powder (organotitanium or organochromium compounds) into a fluidized bed of polyethylene maintained by circulating ethylene, which also serves to remove the heat generated (75–100°C, 20–25 atm).

A new generation of catalysts, metallocene complexes[282] (with aluminoxane or noncoordinating anion activators), are gaining industrial importance. Because of their single-site character, polyethylene (and polypropylene) with narrow molecular weight and controlled molecular-weight distribution can be produced. Supported catalysts with controlled morphology (narrower particle size distribution) are the latest development.[461]

Polypropylene. Only coordination catalysts are capable of producing crystalline, isotactic polypropylene.[208,256,462–464] The majority of the catalysts used in industrial practice are still based on $TiCl_3$ with alkylaluminum or magnesium compounds. A high-activity unsupported catalyst known as the *Solvay catalyst* is prepared by reducing $TiCl_3$ with Et_2AlCl to produce β-$TiCl_3$ followed by dissolving the complex $3TiCl_4 \cdot AlCl_3$ with diisoamylether.[261] Further treatment with a solution of $TiCl_4$ in hexane results in $TiCl_4$ deposition and formation of δ-$TiCl_3$. Supported Ziegler–Natta catalysts are produced by ball milling $TiCl_4$ and $MgCl_2$ in the presence of Lewis bases (e.g., the Montedison catalyst) or reacting $TiCl_4$, a Grignard reagent and an electron-donor compound.

Slurry, solvent, gas-phase, and bulk processes are used.[251,462,465] The slurry process used most widely is carried out in a hydrocarbon solvent below 90°C under pressure to maintain propylene in the liquid phase. In the bulk process liquefied propylene is the reaction medium, allowing the formation of a slurry since polypropylene is insoluble in liquid propylene. Bulk processes developed by Sumimoto[466] and jointly by Mitsui and Montedison[467,468] employing the high-activity $MgCl_2$-supported catalyst systems with internal modifiers (aromatic esters)[469] are highly efficient technologies. Removal of catalyst residues and extraction of atactic polypropylene are seldom necessary, which simplifies the operation and reduces operating costs significantly.[465] With these catalysts and simplified technologies polypropylene with isotacticity index of 94–99% can be manufactured.[461] The final product contains less than 10 ppm Ti and less than 30 ppm chlorine impurities.

Polybutylenes. Polyisobutylene is the only hydrocarbon polymer of commercial importance that is manufactured by cationic polymerization. Commercial polyisobutylenes with broad molecular weights[208,453,470–472] can be produced with the initiators $AlCl_3$ and BF_3. Molecular weights can be readily regulated by controlling the temperature and the monomer and catalyst concentrations, and by using suitable transfer agents. The rate of transfer reactions decreases at lowering the temperature and as a result, high-molecular-weight polymers are formed. Side reactions (hydride transfer, isomerization) are virtually absent during polymerization. The overall structure of the polymer chain is uniformly linear with head-to-tail enchainment of monomers.

Low- to medium-molecular-weight polyisobutylenes are produced at −10 to −40°C. In the Exxon process 30% isobutylene in hexane is polymerized in the presence of powdered AlCl$_3$ in hexane. Since the polymerization process is exothermic, effective refrigeration is achived by vaporizing ethane in external heat exchangers. In the BASF process the catalyst is BF$_3$ in methanol and the diluent is isobutane, which controls the temperature through evaporation ($-10°C$). The products are lubricating oils with outstanding viscosity–temperature characterisitcs.

In contrast, high-molecular-weight (up to 500,000) elastomeric materials are formed by lowering the polymerization temperature. Two important continuous slurry processes are used: the Exxon (Vistanex) process with AlCl$_3$ catalyst and the BASF process (Oppanol), which polymerizes isobutylene dissolved in ethylene (in a ratio about 1 : 2) in the presence of BF$_3$. Ethylene serves as the internal cooler.

Solution polymerization of refinery C$_4$ streams in the presence of AlCl$_3$ or BF$_3$ catalyst yields liquid polymers called *polybutenes*. Because of the large difference in stability of tertiary and secondary carbocations involved, isobutylene cannot be effectively copolymerized with butenes. As a result, the majority of the product formed is polyisobutylene.

Butyl rubber is one product formed when isobutylene is copolymerized with a few percents of isoprene. In the Exxon process an isobutylene–methyl chloride mixture containing a small amount of isoprene is mixed at $-100°C$ with a solution of AlCl$_3$ in methyl chloride. An almost instantaneous reaction yields the product, which is insoluble in methyl chloride and forms a fine slurry. Molecular weight can be controlled by adding diisobutylene as a chain-transfer agent. Increased catalyst concentration and temperature also result in lowering molecular weight. The product can be vulcanized and is superior to natural rubber. A solution process carried out in C$_5$–C$_7$ hydrocarbons was developed in the former Soviet Union.[471,472]

Huels and Mobil developed technologies[473] to manufacture isotactic poly (1-butene), a less important and more expensive polymer by Ziegler–Natta catalysts. The Mobil process[474] is carried out in excess 1-butene and produces highly isotactic polymer. The Huels technology[475] is a slurry operation and requires removal of the atactic isomer.

Styrene Polymers. Styrene polymers of practical significance include polystyrene produced by free-radical polymerization and numerous copolymers.[223,224] The most important hydrocarbon copolymers are styrene–butadiene rubbers (SBR) produced by free-radical emulsion or anionic polymerization. Anionic polymerization allows the manufacture of styrene–butadiene and styrene–isoprene three-block copolymers.

At present all commercial polystyrene (with average molecular weights between 100,000 and 400,000) is manufactured by radical polymerization, which yields atactic polymers.[476] Peroxides and azo compounds are commonly used initiators. The suspension process (usually as a batch process in water at $80–140°C$) produces a product with relatively high residual monomer content.[223] More important is the continuous solution process (usually in ethylbenzene solvent at $90–180°C$), which yields high-purity product. Styrene can be copolymerized with numerous other monomers.[477] One of these copolymers, the styrene–divinylbenzene copolymer produced by free-radical polymerization, has a crosslinked stucture and is used in ion-exchange resins. Styrene–butadiene rubbers (SBR) produced by free-radical-initiated emulsion polymerization[478] played an important role in World War II to supply synthetic rubber for tire-tread production. Known as Buna-S in Germany and GR-S in the United States it was manufactured in a water–monomer two-phase system that also contained an initiator and emulsifier at 50°C (hot polymerization).[184,223] A molecular weight modifier was also used to avoid the formation of unprocessably high-molecular-weight polymer. Because of the relatively high temperature, crosslinking resulted in significant gel formation. Later the use of more active initiators operating at lower temperatures allowed to perform polymerization at about 5°C, thus significantly reducing crosslinking. Continuous operation was an additional important development.[479]

The nonterminating nature of living anionic polymerization allows the synthesis of block copolymers,[480,481] which are useful thermoplastic elastomers. They have many properties of rubber (softness, flexibility, resilience) but in contrast to rubber can be processed as thermoplastics.[482,483] Block copolymers can be manufactured by polymerizing a mixture of two monomers or by using sequential polymerization.

When a mixture of styrene and 1,3-butadiene (or isoprene) undergoes lithium-initiated anionic polymerization in hydrocarbon solution, the diene polymerizes first. It is unexpected, since styrene when polymerized alone, is more reactive than, for example, 1,3-butadiene. The explanation is based on the differences of the rates of the four possible propagation reactions: the rate of the reaction of the styryl chain end with butadiene (crossover rate) is much faster than the those of the other three reactions[484,485] (styryl with styrene, butadienyl with butadiene or styrene). This means that the styryl chain end reacts preferentially with butadiene.

In hydrocarbon solution butadiene polymerizes first, incorporating only a small amount of styrene; then the residual styrene homopolymerizes into the polybutadienyl-lithium. The resulting copolymer (termed *solution styrene–butadiene rubber*) contains long blocks of polybutadiene and polystyrene. The first such polymer containing 75% butadiene and 25% styrene was marketed by Phillips as Solprene 1205. With increasing relative concentration of styrene, or by the addition of polar substances (ethers, tertiary amines), more styrene can be incorporated from the beginning of polymerization changing the block nature into a more random copolymer.

Styrene–1,3-butadiene–styrene (SBS) or styrene–isoprene–styrene (SIS) triblock copolymers are manufactured by a three-stage sequential polymerization. One possible way of the synthesis is to start with the polymerization of styrene. Since all polystyrene chains have an active anionic chain end, adding butadiene to this reaction mixture resumes polymerization, leading to the formation of a polybutadiene block. The third block is formed after the addition of styrene again. The polymer thus produced contains glassy (or crystalline) polystyrene domains dispersed in a matrix of rubbery polybutadiene.[120,481,486]

Polydienes. The most important diene homopolymers are polybutadiene and polyisoprene produced by anionic or coordination polymerization.[184,186,187,487–489] Highly purified starting materials free from acetylenes, oxygen, and sulfur compounds are required.

The commerical polybutadiene (a highly 1,4 polymer with about equal amounts of *cis* and *trans* content) produced by anionic polymerization of 1,3-butadiene (lithium or organolithium initiation in a hydrocarbon solvent) offers some advantages compared to those manufactured by other polymerization methods (e.g., it is free from metal impurities). In addition, molecular weight distributions and microstructure can easily be modifed by applying appropriate experimental conditions. In contrast with polyisoprene, where high *cis* content is necessary for suitable mechanical properties, these nonstereoselective but dominantly 1,4-polybutadienes are suitable for practical applications.[184,482]

Predominantly *cis*-1,4-polybutadiene is produced by coordination polymerization with mixed catalysts.[187,487,488] Three catalyst systems based on titanium, cobalt, or nickel are used in industrial practice. Iodine is an inevitable component in titanium–alkylaluminum sytems to get high *cis* content. Numerous different technologies are used.[490,491] A unique process was developed by Snamprogetti employing a (π-allyl)uranium halide catalyst with a Lewis acid cocatalyst.[492–494] This catalyst system produces polybutadiene with 1,4-*cis* content up to 99%.

cis-1,4-Polyisoprene corresponds to natural rubber called *synthetic natural rubber*. Goodrich–Gulf Company was the first to produce it in the United States.[495] It can be manufactured by lithium alkyls in hydrocarbon solvents[186,187,496] or by Ziegler–Natta catalysts. Three catalyst systems are employed in the commercial polymerization of isoprene. $TiCl_4 + R_3Al$ and $TiCl_4 + AlH_3$ (alane) are coordination catalysts used in hydrocarbon solvents. Preformed $TiCl_4 + isoBu_3Al$ catalysts are preferred ensuring better product characteristics.[497] The alane catalyst system that does not contain direct metal–carbon bond is often combined with trialkylamines.[498] Polyiiminoalanes ($-[AlHNR]_n-$) with $TiCl_4$ are the best catalyst combinations.[499,500] Polymerization is carried out near room temperature at high monomer conversion to produce a polymer with higher than 95% of 1,4-*cis* content.[501,502]

13.3. RECENT DEVELOPMENTS

13.3.1. Oligomerization

Because of the importance of the product oligomers for the production of detergents, plasticizers, comonomers, and specialty chemicals, oligomerization of simple olefins still attracts considerable attention.[503] The most active catalysts in ethylene oligomerization are Ni-based systems with monoanionic bidentate [P,O][504,505] and [N,O][506] ligands, which work at high temperature and pressure (80–120°C and 70–140 atm). It was also found that neutral [P,O] ligands affording cationic Ni complexes oligomerize ethylene under much milder reaction conditions[507] and similar cationic Pd complexes codimerize ethylene with styrene.[508] More recently, various other neutral chelating [N,N][509–511] and [P,N][512] ligands afforded the synthesis of highly active Ni- and Pd-based oligomerization catalysts. Highly active bis(pyridyl)iron[513,514] and cobalt[514] ethylene oligomerization catalysts have been reported.

Heterogeneous Ni catalysts prepared by ion exchange or impregnation of silica–aluminas are effective, stable, and selective ethylene oligomerization catalysts to produce even-number oligomers in the C_4–C_{20} range.[515] A cationic Ni complex in ionic liquids gives predominantly 1-alkenes with better reactivity and higher selectivity than in conventional solvents.[516]

Dimerization of other simple alkenes such as propylene and butenes is of practical significance because of the problems associated with the use of MTBE the presently widely used high-octane oxygen blending component. The transformation of low-value streams, such as butenes, into valuable gasoline blending components is particularly attractive.

The selective dimerization of propylene into dimethylbutenes may be accomplished by Ni^0 and Ni^{2+} homogeneous complexes[517] or using a heterogenized polymer-supported variant.[518] The multicomponent Ni-naphthenate-based catalyst developed earlier[44] was improved by adding sulfonic acids[519] or using hexafluoro-2-propanol as a new, efficient activator.[520] The regioselective trimeric cyclization of propylene induced by $[Cp_2ZrCl_2]$ and Et_3Al or ethylaluminoxane at high temperature has been reported.[521] 1,1,3-Trimethylcyclopentane and 1,1,3,4-tetramethylcyclopentane were formed with a combined selectivity of 61% under optimized conditions.

Traditionally, solid acidic catalysts are applied in industry for the oligomerization of butenes and are still studied. MTS-type aluminosilicates,[522] a NiCsNaY zeolite,[523] and a silica–alumina containing 13% alumina[524] proved to be active and selective catalysts. Moreover, deactivation rates of these catalysts are also favorable. Sulfated zirconia promoted with Fe and Mn was active and selective to yield primarily dimethylbutene isomers under supercritical conditions.[525] A small amount of water improved productivity and decreased deactivation. A study showed that the blending octane number of C_8 hydrocarbons is directly linked to the number of allylic hydrogens in the molecules.[526]

An unusual selectivity was reported for a zirconocene–MAO (methylaluminoxame) catalyst: it selectively dimerizes 1-butene to 2-ethyl-1-hexene in high yield (85.5% selectivity at 95% conversion).[527] 1,3-Butadiene, when reacted in ionic liquids, can give linear dimers (octatrienes) in the presence of Pd compounds $[PdCl_2, Pd(OAc)_2, Pd (acac)_2,]$[528] or undergo cyclodimerization to form 4-vinylcyclohexene induced by Ti[529] or Fe complexes.[530] Over nanoporous VSB-1 Ni phosphate the product undergoes in situ dehydrogenation to yield ethylbenzene.[531]

Oligomerization of alkynes can lead to linear enyne dimers or afford cyclic oligomers. Dimerization of terminal acetylenes induced by Ru complexes yields stereoisomeric conjugated head-to-head enynes depending on the substituents:[532]

$$R\!-\!C\!\equiv\!C\!-\!H \xrightarrow[\text{toluene, reflux, 20 h}]{\text{Ru catalyst}} \quad\quad + \quad\quad \tag{13.83}$$

R = Ph	91	6	98% conversion
tert-Bu		98	10% conversion

Substituents have a similar strong effect on the selectivity of dimerization induced by organiactinades (Th, U).[533–535] In this case head-to-head or head-to-tail dimers may be formed. Methylaluminoxane (MAO), in turn, induces the selective formation of head-to-tail dimers in very high yields (>97%).[536]

Cyclotrimerization of acetylenes to form benzene derivatives is studied intensively mainly because of the challenge of chemo- and regioselectivities. A regioselective and highly chemoselective method for preparing substituted benzenes was described via $PdCl_2$-catalyzed cyclotrimerization in the presence of $CuCl_2$:[537]

$$R^1-C\equiv C-R^2 \xrightarrow[\text{n-BuOH, benzene, 40°C, 12 h}]{PdCl_2,\ CuCl_2} \qquad (13.84)$$

R^1	R^2		
n-Pent	H	78%	
tert-Bu	H	100%	
Ph	Me		80%

A unique example is a thermally induced cyclotrimerization catalyzed by Si_2Cl_6.[538] In this case, terminal alkynes are transformed to isomeric mixtures.

A survey of the studies of metal-catalyzed acetylene cyclotrimerization on single-crystal, supported, and bimetallic catalysts is available.[539] A new study with size-selected Pd_n clusters found that clusters as small as Pd_7 are able to induce benzene formation at 157°C.[540]

13.3.2. Polymerization

Alkene polymerization is of vast economic importance with an annual production of more than 70 million tons of polyethylene and polypropylene. The majority of these polyolefins are still produced with the Ziegler–Natta catalysts or Phillips chromia-based catalyst but the free-radical process has still maintained its significance.[541] New catalysts and processes, however, are emerging because of the unprecedented development in olefin polymerization catalyzed by organometallic complexes. As a result of the progress in the chemistry of well-defined organometallic catalysts, a wide choice of metal–ligand combinations to control the activity, copolymerization aptitude, and stereoselectivity of the catalytic site became available. Mechanistic and theoretical studies provide insight into the interactions between the substrate, metal, and ligands and how subtle steric and electronic factors determine catalyst performance. Developments in the field of polymerization are summarized in book chapters,[542,543] books,[544–546] and symposia proceedings.[547–553] Specifically, more recent results about organometallic catalysts, mainly metallocenes, are summarized in numerous reviews,[554–561] books,[562–564] and symposia proceedings,[565] and special journal issues are also available.[566–568]

Group IV Metallocene Catalysts. The conventional Ziegler–Natta and CrO_3-based catalysts used in industrial production of polyolefins are *heterogeneous systems*. This term refers to the insolubility of the catalyst in the polymerizing medium, and also to its multisite nature. Heterogeneous catalysts typically have multiple active sites, each with its own rate constant for monomer enchainment, stereoselectivity, comonomer incorporation, and chain transfer. Significant empirical optimization of these catalysts, therefore, is necessary to produce polymers of relatively uniform molecular weight, composition, and stereochemistry.

None of these drawbacks are characteristic of the so-called single-site homogeneous coordination olefin polymerization catalysts, such as metallocenes. The original impetus to study these systems derived from the desire to model the reaction mechanism of heterogeneous polymerization catalyst, which was difficult to investigate in the case of Ti-based heterogeneous Ziegler–Natta catalysts. In homogeneous systems all metal centers could, in principle, equally participate in the reaction, that is, they are suitable as model catalysts. Later research resulted in an extraordinary development in metallocene olefin polymerization, often called the "metallocene revolution," and organometallic complexes are now reaching the early stages of commercialization. In fact, metallocene-catalyzed polypropylene processes by Exxon (Achieve process) and Targor were introduced in 1995 and 1997, respectively.

Group IV metallocenes are the most widely studied and applied polymerization catalyst systems.[569] Developed in the 1980s by Ewen[279–281] and Kaminsky,[274–276] they are extremely powerful and selective catalysts in the polymerization of ethylene and α-olefins. A paper by Kaminsky[278] in 1985 may be considered as the starting point in the upsurge of high interest in metallocene catalysis, more specifically, in search for molecularly defined polymerization catalysts and for new polymeric materials. They are usually dichloride complexes of Ti or Zr containing two 5-member aromatic ring systems (cyclopentadienyl, indenyl or fluorenyl) unbridged (**63**) or tethered (**64**) by a bridge (*ansa*-metallocenes):

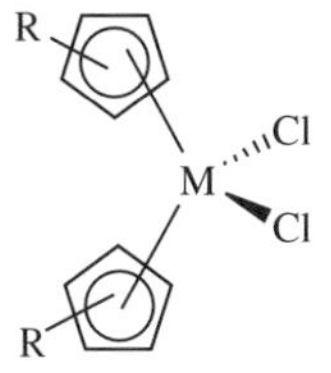

63 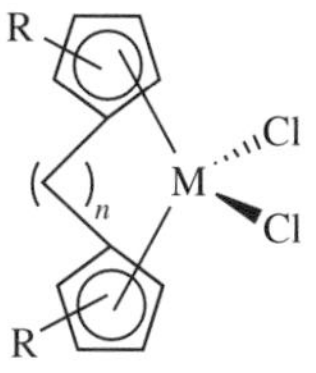   **64**

The presence of substituents on the rings and the bridge can modify the steric and electronic properties and the symmetry of the molecule.

Metallocene complexes require activation to be transformed into active catalysts. This is done by organoaluminoxanes, usually by methylaluminoxane (MAO), which provide maximum activity.[570] During activation first the metal is methylated followed by a carbanion abstraction to form a metallocene monomethyl cation with a free coordination site (**65**), which is the actual active catalytic species:

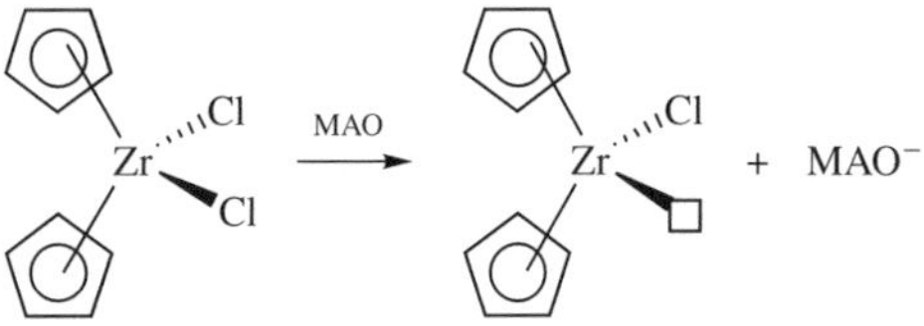

$$(13.85)$$

65

The bulky methylaluminoxane anion stabilizes the coordinatively unsaturated metal cation. Stabilization by noncoordinating anions such as carbosilane dendrimers is also viable.[571] Aluminoxanes, however, are required to be used in large excess to be effective. Alternatively, the active catalyst can also be prepared by reacting a metal dialkyl with fluorinated boranes, borate salts or aluminate salts.

Key steps of the polymerization mechanism are the coordination of the monomer to the coordinatively unsaturated cationic metal center followed by alkyl migration (insertion) into the metal–carbon bond to form the polymer chain and recreating the vacant coordination site, which allows the subsequent coordination of the next monomer:

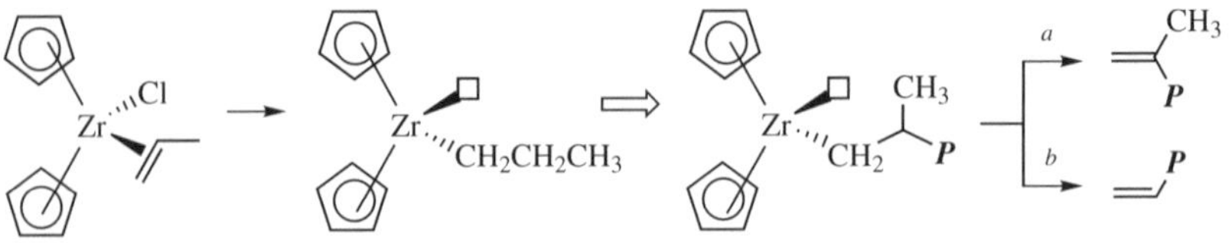

$$(13.86)$$

Termination of the polymer chain occurs predominantly through chain-transfer mechanism involving β-hydrogen (route *a*) or β-methyl (route *b*) elimination. Moreover, adding hydrogen to the reaction mixture can bring about an increase in catalytic activity. Furthermore, since molecular hydrogen can compete with the olefin for the coordination site, a decrease in the molecular weight of the polymer can be achieved.

One of the most active metallocene catalysts is the **66** *ansa*-bis(fluorenylidene) dichloride complex: 1 g of catalyst produces 28 500 kg of polyethylene in 1 h at 60°C.[569]

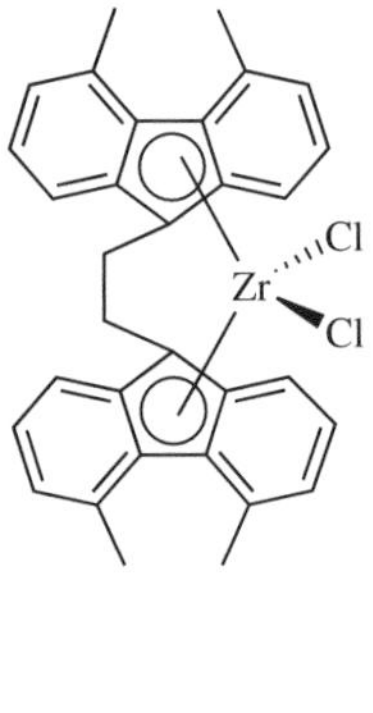

66

The most important characteristics of these catalyst systems can be summarized as follows:

1. The homogeneous single-site nature provides active sites for every molecule in solution ensuring extremely high activities. It also enables to produce uniform homo- and copolymers with narrow molecular weight distribution. Metallocenes can produce an unprecedented variety of polyolefins ranging from stereo- and regioregularities to polydispersities of copolymers and cyclopolymers.

2. They are able to polymerize a large variety of vinyl monomers. The polymer microstructure can be controlled by the symmetry of the catalyst precursor. Prochiral alkenes such as propylene can be polymerized to give stereospecific polymers,[554,572–574] allowing production of polyolefin elastomers. They can give polyolefins with regularly distributed short- and long-chain branches which are new materials for new applications.

3. The termination mechanism involved provides one vinylic chain end group per polymer chain, which can be functionalized.

Some of the drawbacks of the metallocene catalysts are their limited temperature stability and the production of lower-molecular-weight materials under commercial application conditions. It follows that they have a limited possibility for comonomer incorporation due to termination and chain-transfer reactions prohibiting the synthesis of block copolymers by sequential addition of monomers. This led to the development of half-sandwich or constrained geometry complexes, such as *ansa*-monocyclopentadienylamido Group IV complexes (**67**):[575,576]

R = alkyl, aryl
M = Ti, Zr
X = Cl, Me
T = C, Si

A key feature of these catalysts is the open nature of the active sites, which allows them to incorporate other olefins into polyethylene including styrene; that is, these are efficient catalyst for olefin copolymerization.

Commercial application of metallocene catalysts, however, requires the heterogenization of these soluble complexes, which enables to use them in gas phase or slurry polymerization reactors. The particle size of the polymer produced by a homogeneously catalyzed process is very small and, as a result, the polymer is difficult and expensive to process. Other limiting factor is the cocatalyst MAO, which is expensive and has to be applied in large excess (1000–10,000).[577] This excess is needed to prevent the formation of inactive binuclear metallocene species. The major objective, naturally, is to preserve the advantages of homogeneous metallocene catalysts. Heterogenization enables the control of the morphology of the polymer particles and gives a free-flowing polymer resin with minimal reactor fouling. Furthermore, as a result of immobilization, the metallocene to MAO ratio can be decreased by two orders of magnitude; the Al : metal ratio is lowered to approximately 100. This is because immobilization suppresses any interaction between polymerizing sites. The downside, however, is that heterogenized catalysts are less active. This originates form the fact that the metal center is no longer accessible at the surface and the monomer diffusion into the interior pores of the supported catalyst is diminished. Since the molecular weight is seldom affected, the lower number of active centers on the surface is the most probable reason for the decreased activity. The immobilized solid catalysts have active sites different from those in solution, which affects catalyst activity and the properties of the polyolefins formed.

A range of solid carriers including Al_2O_3, SiO_2, MgO, $MgCl_2$, zeolites, and organic support materials such as polystyrene[560,578] have been applied for the heterogenization of metallocene complexes.[577,579] Amorphous silica appears to be the most useful because its high surface area and porosity, suitable mechanical properties, stability, and inertness during reaction and processing. Several methods may be applied. A chemical bond between MAO and the support can be built with the surface OH groups of silica by reacting MAO with the support followed by the adsorption of the complex.[580] Another possibility is supporting the complex first then MAO. Intercalation of the metallocene complex into phyllosilicates by ion-exchange[581] and immobilization through the reaction of an organometallic precursor with functionalized silsesquioxanes were also performed.[582] Furthermore, the self-immobilization of metallocenes can be accomplished.[560] The idea is to prepare a metallocene complex with an alkene or alkyne function that can be used as a comonomer in the polymerization process. Such catalysts then incorporate into the growing polymer chain. Heterogenization on borate-functionalized silica has the advantage of not requiring MAO activation.[575]

Late-Transition-Metal Catalysts. The high oxophilicity of early transition metal complexes (Ti, Zr, Cr) renders them susceptible to poisoning by most functional olefins, particularly the commercially used polar comonomers. Because of their lower oxophilicity, late-transition-metal complexes are possible candidates for the development of new organometallic complex catalysts. Relatively few information, however, are available for such systems.[583–586] The main reason is that late-transition-metal catalysts, in general, exhibit low activity for olefin insertion relative to early transition metals. For these systems, β-hydride elimination typically competes with chain growth resulting in the formation of dimers or oligomers. In fact, this is the basis for the Shell higher olefin process (SHOP), which applies a zerovalent nickel catalyst (see Section 13.1.3). More recently, however, new active and selective catalyst systems based on late transition metals were developed.[585,586]

The use of late transition metals as olefin polymerization catalyst requires the suppression of chain transfer while at the same time a high chain growth rate should be maintained. These new catalysts have an electron-deficient, in most cases, 14-electron and cationic metal center with a vacant coordination site. The most effective systems are based on neutral Ni^{2+} complexes or on cationic Fe, Ni, or Pd complexes with neutral multidentate ligands.[586] The latter catalysts having bulky substituted diimine ligands (**68**) were discovered in 1995 and show a unique property to polymerize ethylene into highly branched, high-molecular-weight homopolymers at high rates:

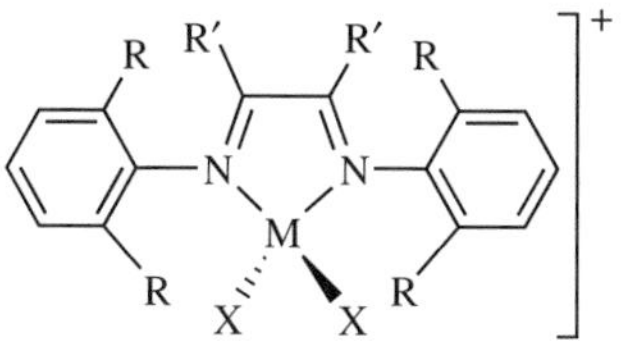

R = Me, isoPr
R′ = H, Me
M = Ni, Pd
L = OEt$_2$, RCN

The key feature of having high-molecular-weight polymers is the retardation of chain transfer by the steric bulk of the *o*-aryl substituent.

Strong interest in late transition metal olefin polymerization catalysts resulted in the development of new five-coordinate Fe and Co systems (**69**) that afford highly linear, crystalline, high-density polyethylene.[587–589] A new class of single-component, neutral Ni catalysts based on salicylaldimine ligands (**70**) was reported to be active in the polymerization of ethylene:[590,591]

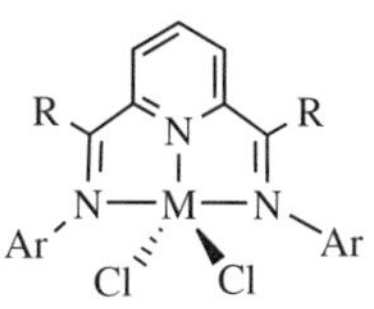

M = Fe, Co
R = Me, isoPr
Ar = substituted Ph

69

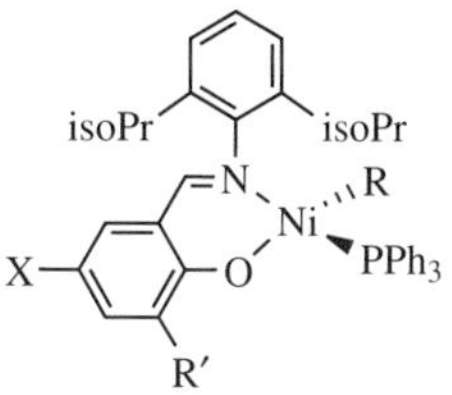

R = Me, Ph
R′= H, *tert*-Bu, Ph
9-anthracenyl
9-phenanthracenyl
trityl
X = OMe, NO$_2$

70

Using some of these catalysts, coordination polymerization of ethylene at high rate were performed in water. A highly branched polymer was obtained with cationic Pd diimio complex,[592,593] whereas a neutral Ni complex with a sulfonate group afforded a linear polymer.[592]

Other Developments. Despite the long history of the Phillips supported CrO_3 catalyst used in the polymerization of olefins, many aspects with respect to the physicochemical state of the surface Cr species still remain controversial. Two phenomena are known to occur during the complex cycles of calcination: calcination-induced reduction of surface-stabilized hexavalent chromate species and the formation of aggregated Cr_2O_3 species. Liu and Terano applied XPS and electron microprobe analysis to acquire more information for a 1% CrO_3 on SiO_2 catalyst.[594] This loading is much lower than the saturation coverage but approximately the same as used in industrial practice. The study showed that, after calcination in dry air at 800°C, three basic kinds of surface species exist: (1) surface-stabilized chromate (70%), (2) Cr_2O_3 mostly chemically bonded to the surface, and (3) Cr_2O_3 existing as crystalline aggregates. The latter are thought to be inactive in polymerization.

Similar ambiguities still exist with respect to the other important polymerization catalyst the Ziegler–Natta $MgCl_2$-supported system. Comilling of $MgCl_2$, $TiCl_4$ and an internal donor followed by treatment with an aluminum trialkyl cocatalyst results in a catalyst with extremely high activity. Although data are scarce, Ti^{4+} is believed to undergo extensive reduction during the activation process. Another study provided quantitative data for the chemical state of titanium by means of redox titration; extensive reduction of titanium was observed with the following distribution: $Ti^{4+} : Ti^{3+} : Ti^{2+} = 9 : 31 : 60$.[595] Furthermore, both Ti^{3+} and Ti^{2+} were found to be active in ethylene polymerization. In contrast, species in higher oxidation state, that is Ti^{3+}, and even possibly Ti^{4+}, are catalytically active in the polymerization of propylene. A patent review of silica-based Ziegler–Natta catalysts has been published [596] and a new method for measuring the number of active centers in heterogeneous Ziegler–Natta catalysts has been disclosed.[597]

One of the recipients of the 2000 Nobel Prize in Chemistry (together with Heeger and MacDiarmid) was Shirakawa who discovered and extensively studied conducting polyacetylene.[401] Polymerization of acetylene with chiral Ti catalysts was reported in 1997.[598] In addition, helical polyacetylene was synthesized under an asymmetric reaction field consisting of chiral nematic liquid crystals.[599] The polyacetylene film was shown by scanning electron microscopy to consist of clockwise or counterclockwise helical structure of fibrils. It has a high *trans* content (90%) and becomes highly electroconducting (1.5–1.8 S cm^{-1}) on iodine doping. These values are comparable to those of metals. Novel magnetic and optical properties of this material are expected.

Well-controlled polymerization of substituted acetylenes was also reported. A tetracoordinate organorhodium complex induces the stereospecific living polymerization of phenylacetylene.[600] The polymerization proceeds via a 2–1 -insertion mechanism to provide stereoregular poly(phenylacetylene) with *cis*–transoidal backbone structure. Rh complexes were also used in the same process in supercritical CO_2[601] and in the polymerization of terminal alkyl- and arylacetylenes.[602] Single-component transition-metal catalysts based on Ni acetylides[603] and Pd acetylides[604] were used in the polymerization of *p*-diethynylbenzene.

REFERENCES

1. IUPAC Commission on Macromolecular Nomenclature, *Pure Appl. Chem.* **40**, 479 (1974).
2. L. Schmerling and V. N. Ipatieff, *Adv. Catal.* **2**, 21 (1950).
3. D. C. Pepper, in *Friedel-Crafts and Related Reactions*, Vol. 2, G. A. Olah, ed., Wiley-Interscience, New York, 1964, Chapter 30, p. 1293.
4. R. E. Schaad, in *The Chemistry of Petroleum Hydrocarbons*, Vol. 2, B. J. Brooks, C. E. Boord, S. S. Kurtz, Jr., and L. Schmerling, eds., Reinhold, New York, 1955, Chapter 48, p. 221.

5. F. C. Whitmore, *Ind. Eng. Chem.* **26**, 94 (1934); *Chem. Eng. News* **26**, 668 (1948).

6. F. P. J. M. Kerkhof, J. C. Oudejans, J. A. Moulijn, and E. R. A. Matulewicz, *J. Colloid Interface Sci.* **77**, 120 (1980).

7. C. T. O'Connor, M. Kojima, and W. K. Schumann, *Appl. Catal.* **16**, 193 (1985).

8. A. Saus and E. Schmidl, *J. Catal.* **94**, 187 (1985).

9. Kh. M. Minachev and D. A. Kondrat'ev, *Russ. Chem. Rev.* (Engl. transl.) **52**, 1113 (1983).

10. I. Bucsi and G. A. Olah, *J. Catal.* **137**, 12 (1992).

11. H. Pines, *The Chemistry of Catalytic Hydrocarbon Conversions*, Academic Press, New York, 1981, 1981, Chapter 1, p. 32.

12. J. Skupinska, *Chem. Rev.* **91**, 613 (1991).

13. L. M. Kustov, V. Yu. Borovkov, and V. B. Kazansky, in *Structure and Reactivity of Modified Zeolites*, Studies in Surface Science and Catalysis, Vol. 18, P. A. Jacobs, N. I. Jaeger, P. Jíru, V. B. Kazansky, and G. Schulz-Ekloff, eds., Elsevier, Amsterdam, 1984, p. 241.

14. E. K. Jones, *Adv. Catal.* **8**, 219 (1956).

15. F. Asinger, *Mono-Olefins, Chemistry and Technology*, Pergamon Press, Oxford, 1968, Chapter 2, p. 185.

16. J. Habeshaw, in *Propylene and its Industrial Derivatives*, E. G. Hancock, ed., Wiley, New York, 1973, Chapter 4, p. 115.

17. K. Weissermel and H.-J. Arpe, *Industrial Organic Chemistry*, 2nd ed., VCH, Weinheim, 1993, Chapter 3, p. 63.

18. R. J. McCubbin and H. Adkins, *J. Am. Chem. Soc.* **52**, 2547 (1930).

19. F. C. Whitmore and J. M. Church, *J. Am. Chem. Soc.* **54**, 3710 (1932).

20. F. Asinger, *Mono-Olefins, Chemistry and Technology*, Pergamon Press, Oxford, 1968, Chapter 5, p. 438.

21. *Petrol. Refiner* **37**(9), 270 (1958).

22. E. D. Kane and G. E. Langlois, *Petrol. Refiner* **37**(5), 173 (1958).

23. *Petrol. Refiner* **37**(9), 267, 269 (1958).

24. D. J. Ward, R. H. Friedlander, R. Frame, and T. Imai, *Hydrocarbon Process., Int. Ed.* **64**(5), 81 (1985).

25. N. Y. Chen, W. E. Garwood, and F. G. Dwyer, *Shape Selective Catalysis in Industrial Applications*, Marcel Dekker, New York, 1989, Chapter 5, p. 195.

26. E. W. Garwood, *ACS Symp. Ser.* **218**, 383 (1983).

27. R. J. Quann, L. A. Green, S. A. Tabak, and P. J. Krambeck, *Ind. Eng. Chem. Res.* **27**, 565 (1988).

28. N. Y. Chen and W. O. Haag, in *Hydrogen Effects in Catalysis*, Z. Paál and P. G. Menon, eds., Marcel Dekker, New York, 1988, Chapter 27, p. 695.

29. A. M. Madgavkar and H. E. Swift, *Ind. Eng. Chem. Res.* **22**, 675 (1983).

30. H. Pines and W. M. Stalick, *Base-Catalyzed Reactions of Hydrocarbons and Related Compounds*, Academic Press, New York, 1977, Chapter 4, p. 206.

31. H. Pines, *The Chemistry of Catalytic Hydrocarbon Conversions*, Academic Press, New York, 1981, Chapter 3, p. 135.

32. V. Mark and H. Pines, *J. Am. Chem. Soc.* **78**, 5946 (1956).

33. W. V. Bush, G. Holzman, and A. W. Shaw, *J. Org. Chem.* **30**, 3290 (1965).

34. A. W. Shaw, C. W. Bittner, W. V. Bush, and G. Holzman, *J. Org. Chem.* **30**, 3286 (1965).

35. A. Zwierzak and H. Pines, *J. Org. Chem.* **28**, 3392 (1963).

36. H. Pines, *The Chemistry of Catalytic Hydrocarbon Conversions*, Academic Press, New York, 1981, Chapter 6, p. 248.

37. G. W. Parshall and S. D. Ittel, *Homogeneous Catalysis*, Wiley-Interscience, New York, 1992, Chapter 4, p. 51.

38. K. Ziegler, E. Holzkamp, H. Breil, and H. Martin, *Angew. Chem.* **67**, 541 (1955).

39. J. J. Eisch, in *Comprehensive Organometallic Chemistry*, Vol. 1, G. Wilkinson, F. G. A. Stone, and E. W. Abel, eds., Pergamon Press, Oxford, 1982, Chapter 6, p. 568.

40. P. W. Jolly and G. Wilke, *The Organic Chemistry of Nickel*, Vol. 2, Academic Press, New York, 1975, Chapter 1, p. 1.

41. B. Bogdanovic, *Adv. Organomet. Chem.* **17**, 105 (1979).

42. A. W. Al-Sa'doun, *Appl. Catal., A* **105**, 1 (1993).

43. D. Commereuc, Y. Chauvin, J. Gaillard, J. Léonard, and J. Andrews, *Hydrocarbon Process., Int. Ed.* **63**(11), 118 (1984).

44. H. Sato, T. Noguchi, and S. Yasui, *Bull. Chem. Soc. Jpn.* **66**, 3069 (1993).

45. H. W. B. Reed, *J. Chem. Soc.* 1931 (1954).

46. G. Wilke, *Angew. Chem., Int. Ed. Engl.* **2**, 105 (1963); **27**, 185 (1988).

47. G. Wilke, B. Bogdanovic, P. Hardt, P. Heimbach, W. Keim, M. Kröner, W. Oberkirch, K. Tanaka, E. Steinrücke, D. Walter, and H. Zimmermann, *Angew. Chem., Int. Ed. Engl.* **5**, 151 (1966).

48. P. Heimbach, P. W. Jolly, and G. Wilke, *Adv. Organomet. Chem.* **8**, 29 (1970).

49. B. Bogdanovic, B. Henc, H.-G. Karmann, H.-G. Nüssel, D. Walter, and G. Wilke, *Ind. Eng. Chem.* **62**(12), 34 (1970).

50. P. W. Jolly and G. Wilke, *The Organic Chemistry of Nickel*, Vol. 2, Academic Press, New York, 1975, Chapter 3, p. 133.

51. C. W. Bird, *Transition Metal Intermediates in Organic Synthesis*, Logos Press, London, 1967, Chapter 2, p. 30.

52. M. F. Schemmelhack, *Org. React. (NY)* **19**, 115 (1972).

53. R. Baker, *Chem. Rev.* **73**, 487 (1973).

54. P. Heimbach and H. Schenkluhn, *Top. Curr. Chem.* **92**, 45 (1980).

55. P. W. Jolly, in *Comprehensive Organometallic Chemistry*, Vol. 8: G. Wilkinson, F. G. A. Stone, and E. W. Abel, eds., Pergamon Press, Oxford, 1982, Chapter 56.4, p. 671.

56. I. Ono and K. Kihara, *Hydrocarbon Process.* **46**(8), 147 (1967).

57. G. W. Parshall, *J. Mol. Catal.* **4**, 243 (1978).

58. G. W. Parshall and W. A. Nugent, *Chemtech* **18**, 314 (1988).

59. P. Heimbach, *Angew. Chem., Int. Ed. Engl.* **12**, 975 (1973).

60. C. R. Graham and L. M. Stephenson, *J. Am. Chem. Soc.* **99**, 7098 (1977).

61. W. Reppe and W. J. Schweckendiek, *Justus Liebigs Ann. Chem.* **560**, 104 (1948).

62. W. Reppe, O. Schlichting, K. Klager, and T. Toepel, *Justus Liebigs Ann. Chem.* **560**, 1 (1948).

63. C. W. Bird, *Transition Metal Intermediates in Organic Synthesis*, Logos Press, London, 1967, Chapter 1, p. 1.

64. C. Hoogzand and W. Hübel, in *Organic Syntheses via Metal Carbonyls*, Vol. 1, I. Wender and P. Pino, eds., Interscience, New York, 1968, p. 343.

65. V. O. Reikhsfel'd and K. L. Makovetskii, *Russ. Chem. Rev.* (Engl. transl.) **35**, 510 (1966).

66. L. P. Yur'eva, *Russ. Chem. Rev.* (Engl. transl.) **43**, 48 (1974).

67. P. W. Jolly and G. Wilke, *The Organic Chemistry of Nickel*, Vol. 2, Academic Press, New York, 1975, Chapter 2, p. 94.

68. M. J. Winter, in *The Chemistry of the Metal–Carbon Bond*, Vol. 3, F. R. Hartley and S. Patai, eds., Wiley, Chichester, 1985, Chapter 5, p. 259.

69. N. E. Schore, in *Comprehensive Organic Synthesis*, B. M. Trost and I. Fleming, eds., Vol. 5: *Combining C–C π-Bonds*, L. A. Paquette, ed., Pergamon Press, Oxford, 1991, Chapter 9.4, p. 1129.

70. E. M. Arnett and J. M. Bollinger, *J. Am. Chem. Soc.* **86**, 4729 (1964).

71. A. K. Jhingan and W. F. Maier, *J. Org. Chem.* **52**, 1161 (1987).

72. K. P. C. Vollhardt, *Acc. Chem. Res.* **10**, 1 (1977); *Angew. Chem., Int. Ed. Engl.* **23**, 539 (1984).

73. Y. Wakatsuki, O. Nomura, K. Kitaura, K. Morokuma, and H. Yamazaki *J. Am. Chem. Soc.* **105**, 1907 (1983).

74. M. A. Bruck, A. S. Copenhaver, and D. E. Wigley, *J. Am. Chem. Soc.* **109**, 6525 (1987).

75. C. Bianchini, K. G. Caulton, C. Chardon, O. Eisenstein, K. Folting, T. J. Johnson, A. Meli, M. Peruzzini, D. J. Rauscher, W. E. Streib, and F. Vizza, *J. Am. Chem. Soc.* **113**, 5127 (1991).

76. E. Müller, *Synthesis* 761 (1974).

77. J. P. Collman, J. W. Kang, W. F. Little, and M. F. Sullivan, *Inorg. Chem.* **7**, 1298 (1968).

78. J. R. Stickler, P. A. Wexler, and D. E. Wigley, *Organometallics* **7**, 2067 (1988).

79. G. M. Whitesides and W. J. Ehmann, *J. Am. Chem. Soc.* **91**, 3800 (1969).

80. P. Maitlis, *Acc. Chem. Res.* **9**, 93 (1976); *J. Organomet. Chem.* **200**, 161 (1980).

81. R. E. Colborn and K. P. C. Vollhardt, *J. Am. Chem. Soc.* **108**, 5470 (1986).

82. G. Wilke, *Pure Appl. Chem.* **50**, 677 (1978).

83. R. Diercks, L. Stamp, J. Kopf, and H. tom Dieck, *Angew. Chem., Int. Ed. Engl.* **23**, 893 (1984).

84. C. J. Lawrie, K. P. Gable, and B. K. Carpenter, *Organometallics* **8**, 2274 (1989).

85. A. Hennico, J. Léonard, A. Forestiere, and Y. Glaize, *Hydrocarbon Process., Int. Ed.* **69**(3), 73 (1990).

86. J. W. Andrews, P. Bonnifay, B. J. Cha, J. C. Barbier, D. Douillet, and J. Raimbault, *Hydrocarbon Process.* **55**(4), 105 (1976).

87. J. Andrews and P. Bonnifay, *ACS Symp. Ser.* **55**, 328 (1977).

88. P. M. Kohn, *Chem. Eng. (NY)*, **84**(11), 114 (1977).

89. Y. Cahuvin, J. F. Gaillard, J. Leonard, P. Bonnifay, and J. W. Andrews, *Hydrocarbon Process., Int. Ed.* **61**(5), 110 (1982).

90. J. Leonard and J. F. Gaillard, *Hydrocarbon Process., Int. Ed.* **60**(3), 99 (1981).

91. *Hydrocarbon Process., Int. Ed.* **64**(11), 152 (1985).

92. *Hydrocarbon Process., Int. Ed.* **64**(11), 151 (1985).

93. R. H. Friedlander, D. J. Ward, F. Obenaus, F. Nierlich, and J. Neumeister, *Hydrocarbon Process., Int. Ed.* **65**(2), 31 (1986).

94. F. Nierlich, *Hydrocarbon Process., Int. Ed.* **71**(2), 45 (1992).

95. N. J. Anhorn, *Petrol. Refiner* **39**(11), 227 (1960).

96. M. L. Senyek, in *Encyclopedia of Polymer Science and Engineering*, 2nd ed., Vol. 8, J. I. Kroschwitz, ed., Wiley-Interscience, New York, 1987, p. 497.

97. A. C. L. Su, *Adv. Organomet. Chem.* **17**, 269 (1979).

98. G. R. Lappin, in *Alpha Olefins Applications Handbook*, G. R. Lappin and J. D. Sauer, eds., Marcel Dekker, New York, 1989, Chapter 3, p. 35.

99. K. L. Lindsay, in *Encyclopedia of Chemical Processing and Design*, Vol. 2, J. J. McKetta and W. A. Cunningham, eds., Marcel Dekker, New York, 1977, p. 482.

100. J. D. Wagner, G. R. Lappin, and J. R. Zietz, in *Kirk-Othmer Encyclopedia of Chemical Technology*, 4th ed., Vol. 1, J. I. Kroschwitz and M. Howe-Grant, eds., Wiley-Interscience, New York, 1991, p. 893.

101. D. G. Demianiw, in *Kirk-Othmer Encyclopedia of Chemical Technology*, 3rd ed., Vol. 16, M. Grayson and D. Eckroth, eds., Wiley-Interscience, New York, 1981, p. 480.

102. K. Ziegler, *Angew. Chem.* **64**, 323 (1952).

103. A. Lundeen and R. Poe, in *Encyclopedia of Chemical Processing and Design*, Vol. 2, J. J. McKetta and W. A. Cunningham, eds., Wiley-Interscience, New York, 1977, p. 465.

104. K. Weissermel and H.-J. Arpe, *Industrial Organic Chemistry*, 2nd ed., VCH, Weinheim, 1993, Chapter 8, p. 206.

105. P. A. Lobo, D. C. Coldiron, L. N. Vernon, and A. T. Ashton, *Chem. Eng. Prog.* **58**(5), 85 (1962).

106. P. H. Washecheck, *ACS Symp. Ser.* **159**, 87 (1981).

107. *Hydrocarbon Process., Int. Ed.* **62**(11), 77 (1983).

108. E. R. Freitas and C. R. Gum, *Chem. Eng. Prog.* **75**(1), 73 (1979).

109. R. C. Wade, in *Catalysis of Organic Reactions*, W. R. Moser, ed., Marcel Dekker, New York, 1981, p. 165.

110. A. H. Turner, *J. Am. Oil Chem. Soc.* **60**, 623 (1983).

111. W. Griehl and D. Ruestem, *Ind. Eng. Chem.* **62**(3), 16 (1970).

112. K. Weissermel and H.-J. Arpe, *Industrial Organic Chemistry*, 2nd ed., VCH, Weinheim, 1993, Chapter 10, p. 241.

113. J. Haggin, *Chem. Eng. News* **72**(25), 31 (1994).

114. *Reactivity, Mechanism and Structure in Polymer Chemistry*, A. D. Jenkins and A. Ledwith, eds., Wiley-Interscience, London, 1974.

115. *The Stereo Rubbers*, W. M. Saltman, ed., Wiley-Interscience, New York, 1977.

116. *Developments in Polymerisation*, R. N. Haward, ed., Applied Science Publishers, London, 1979–1982.

117. J. Boor, Jr., *Ziegler-Natta Catalysis and Polymerization*, Academic Press, New York, 1979.

118. J. P. Kennedy, *Cationic Polymerization of Olefins: A Critical Inventory*, Wiley, New York, 1975.

119. J. P. Kennedy and E. Maréchal, *Carbocationic Polymerization*, Wiley-Interscience, New York, 1982.

120. M. Morton, *Anionic Polymerization: Principles and Practice*, Academic Press, New York, 1983.

121. Y. V. Kissin, *Isospecific Polymerization of Olefins with Heterogeneous Ziegler-Natta Catalysts*, Springer, New York, 1985.

122. *Methoden der Organischen Chemie (Houben-Weyl)*, 4th ed., Vol. E20: *Makromolekulare Stoffe*, H. Bartl and J. Falbe, eds., Thieme, Stuttgart, 1987.

123. *Handbook of Polymer Science and Technology*, N. P. Cheremisinoff, ed., Marcel Dekker, New York, 1989.

124. *Comprehensive Polymer Science*, G. Allen and J. C. Bevington, eds., Pergamon Press, Oxford, 1989.

125. S. van der Ven, *Polypropylene and other Polyolefins. Polymerization and Characterization*, Elsevier, Amsterdam, 1990.

126. *Handbook of Polymer Synthesis*, H. R. Kricheldorf, ed., Marcel Dekker, New York, 1992.

127. S. R. Sandler and W. Karo, *Polymer Syntheses*, 2nd ed., Academic Press, New York, 1992, 1994.

128. M. Szwarc, *Carbanions Living Polymers and Electron Transfer Processes*, Interscience, New York, 1968.

129. M. Szwarc and M. Van Beylen, *Ionic Polymerization and Living Polymers*, Chapman & Hall, New York, 1993.

130. *Encyclopedia of Polymer Science and Engineering*, 2nd ed., J. I. Kroschwitz, ed., Wiley-Interscience, New York, 1985–1990.

131. *Advances in Polymer Science*, Springer, Berlin, since 1958.

132. O. Nuyken and St. D. Pask, in *Comprehensive Polymer Science*, G. Allen and J. C. Bevington, eds., Vol. 3: *Chain Polymerization I*, G. C. Eastmond, A. Ledwith, S. Russo, and P. Sigwalt, eds., Pergamon Press, Oxford, 1989, Chapter 40, p. 619.

133. D. C. Pepper, *Quart. Rev.* (London) **8**, 88 (1954).

134. M. Berthelot, *Bull. Soc. Chim. Fr.* **6**, 294 (1866).

135. W. Goriainov and A. Butlerov, *Justus Liebigs Ann. Chem.* **169**, 146 (1873).

136. H. Zorn, *Angew. Chem.* **60**, 185 (1948).

137. A. G. Oblad, G. A. Mills, and H. Heinemann, in *Catalysis*, Vol. 6, P. H. Emmett, ed., Reinhold, New York, 1958, Chapter 4, p. 341.

138. D. J. Dunn, in *Developments in Polymerisation—1*, R. N. Haward, ed., Applied Science Publishers, London, 1979, Chapter 2, p. 45.

139. G. Sauvet and P. Sigwalt, in *Comprehensive Polymer Science*, G. Allen and J. C. Bevington, eds., Vol. 3: *Chain Polymerization I*, G. C. Eastmond, A. Ledwith, S. Russo, and P. Sigwalt, eds., Pergamon Press, Oxford, 1989, Chapter 39, p. 579.

140. A. Gandini and H. Cheradame, *Adv. Polym. Sci.* **34/35**, 1 (1980).

141. A. Gandini and H. Cheradame, in *Encyclopedia of Polymer Science and Engineering*, 2nd ed., Vol. 2, J. I. Kroschwitz, ed., Wiley-Interscience, New York, 1985, p. 729.

142. A. Ledwith and D. C. Sherrington, in *Reactivity, Mechanism and Structure in Polymer Chemistry*, A. D. Jenkins and A. Ledwith, eds., Wiley-Interscience, London, 1974, Chapter 9, p. 244.

143. A. G. Evans, D. Holden, P. D. Plesch, M. Polanyi, H. A. Skinner, and M. A. Weinberger, *Nature* (London) **157**, 102 (1946).

144. A. G. Evans, G. W. Meadows, and M. Polanyi, *Nature* (London) **158**, 94 (1946).

145. G. A. Olah, H. W. Quinn, and S. J. Kuhn, *J. Am. Chem. Soc.* **82**, 426 (1960).

146. P. H. Plesch, *Makromol. Chem.* **175**, 1065 (1974).

147. D. W. Grattan and P. H. Plesch, *J. Chem. Soc., Dalton Trans.* 1734 (1977).

148. G. A. Olah, *Angew. Chem., Int. Ed. Engl.* **32**, 767 (1993).

149. G. A. Olah, *Makromol. Chem.* **175**, 1039 (1974).

150. P. Sigwalt, *Makromol. Chem.* **175**, 1017 (1974).

151. D. C. Sherrington, in *Catalysis. A Specialist Periodical Report*, Vol. 3, C. Kemball and D. A. Dowden, senior reporters, The Chemical Society, Burlington House, London, 1980, Chapter 9, p. 228.

152. J. P. Kennedy and J. K. Gillham, *Adv. Polym. Sci.* **10**, 1 (1972).

153. J. P. Kennedy and P. D. Trivedi, *Adv. Polym. Sci.* **28**, 83, 113 (1978).

154. F. C. Hall and A. W. Nash, *J. Inst. Petroleum Tech.* **23**, 679 (1937); **24**, 471 (1938).

155. V. T. Stannett, J. Silverman, and J. L. Garnett, in *Comprehensive Polymer Science*, G. Allen and J. C. Bevington, eds., Vol. 4: *Chain Polymerization II*, G. C. Eastmond, A. Ledwith, S. Russo, and P. Sigwalt, eds., Pergamon Press, Oxford, 1989, Chapter 19, p. 317.

156. K. Tanigaki and K. Tateishi, in *Handbook of Polymer Science and Technology*, Vol. 1: *Synthesis and Properties*, N. P. Cheremisinoff, ed., Marcel Dekker, New York, 1989, Chapter 9, p. 307.

157. L. Toppare, in *Handbook of Polymer Science and Technology*, Vol. 1: *Synthesis and Properties*, N. P. Cheremisinoff, ed., Marcel Dekker, New York, 1989, Chapter 8, p. 271.

158. A. G. Evans and M. Polanyi, *J. Chem. Soc.* 252 (1947).

159. K. Matyjaszewski, in *Comprehensive Polymer Science*, G. Allen and J. C. Bevington, eds., Vol. 3: *Chain Polymerization I*, G. C. Eastmond, A. Ledwith, S. Russo, and P. Sigwalt, eds., Pergamon Press, Oxford, 1989. Chapter 41, p. 639.

160. A. Gandini and P. H. Plesch, *J. Chem. Soc.* 4826 (1965).

161. A. Gandini and P. H. Plesch, *J. Polym. Sci. Polym. Lett. Ed.* **3**, 127 (1965).

162. M. Chmelir, N. Cardona, and G. V. Schulz, *Makromol. Chem.* **178**, 169 (1977).

163. U. Akbulut, L. Toppare, and B. Yurttas, *J. Polym. Sci., Polym. Lett. Ed.* **24**, 185 (1986).

164. M. Miyamoto, M. Sawamoto, and T. Higashimura, *Macromolecules* **17**, 265 (1984).

165. K. Kojima, M. Sawamoto, and T. Higashimura, *Macromolecules* **22**, 1552 (1989).

166. R. Faust and J. P. Kennedy, *Polym. Bull.* (Berlin) **15**, 317 (1986); **19**, 21 (1988); *J. Polym. Sci., Polym. Chem. Ed.* **25**, 1847 (1987).

167. M. K. Mishra and J. P. Kennedy, *J. Macromol. Sci., Chem.* **24**, 933 (1987).

168. D. H. Richards, in *Developments in Polymerisation–1*, R. N. Haward, ed., Applied Science Publishers, London, 1979, Chapter 1, p. 1.

169. M. Fontanille, in *Comprehensive Polymer Science*, G. Allen and J. C. Bevington, eds., Vol. 3: *Chain Polymerization I*, G. C. Eastmond, A. Ledwith, S. Russo, and P. Sigwalt, eds., Pergamon Press, Oxford, 1989, Chapter 25, p. 365.

170. K. Ziegler, *Angew. Chem.* **49**, 499 (1936).

171. M. Szwarc, *Nature* (London) **178**, 1168 (1956).

172. M. Szwarc, *Adv. Polym. Sci.* **49**, 1 (1983).

173. S. Bywater, in *Encyclopedia of Polymer Science and Engineering*, 2nd ed., Vol. 2, J. I. Kroschwitz, ed., Wiley-Interscience, New York, 1985, p. 1.

174. R. N. Young, R. P. Quirk, and L. J. Fetters, *Adv. Polym. Sci.* **56**, 1 (1984).

175. A. Parry, in *Reactivity, Mechanism and Structure in Polymer Chemistry*, A. D. Jenkins and A. Ledwith, eds., Wiley-Interscience, London, 1974, Chapter 11, p. 350.

176. M. Szwarc, M. Levy, and R. Milkovich, *J. Am. Chem. Soc.* **78**, 2656 (1956).

177. M. Szwarc, *Makromol. Chem.* **35**, 132 (1960).

178. G. Spach, H. Monteiro, M. Levy, and M. Szwarc, *Trans. Farad. Soc.* **58**, 1809 (1962).

179. M. Morton, L. J. Fetters, R. A. Pett, and J. F. Meier, *Macromolecules* **3**, 327 (1970).

180. D. J. Worsfold and S. Bywater, *Macromolecules* **5**, 393 (1972).

181. M. Fontanille, in *Comprehensive Polymer Science*, G. Allen and J. C. Bevington, eds., Vol. 3: *Chain Polymerization I*, G. C. Eastmond, A. Ledwith, S. Russo, and P. Sigwalt, eds., Pergamon Press, Oxford, 1989, Chapter 27, p. 425.

182. G. M. Spach, M. Levy, and M. Szwarc, *J. Chem. Soc.* 355 (1962).

183. W. Cooper, in *Developments in Polymerisation–1*, R. N. Haward, ed., Applied Science Publishers, London, 1979, Chapter 3, p. 103.

184. D. P. Tate and T. W. Bethea, in *Encyclopedia of Polymer Science and Engineering*, 2nd ed., Vol. 2, J. I. Kroschwitz, ed., Wiley-Interscience, New York, 1985, p. 537.

185. S. Bywater, in *Comprehensive Polymer Science*, G. Allen and J. C. Bevington, eds., Vol. 3: *Chain Polymerization I*, G. C. Eastmond, A. Ledwith, S. Russo, and P. Sigwalt, eds., Pergamon Press, Oxford, 1989, Chapter 28, p. 433.

186. M. L. Senyek, in *Encyclopedia of Polymer Science and Engineering*, Vol. 8, J. I. Kroschwitz, ed., Wiley-Interscience, New York, 1987, p. 487.

187. W. Kaminsky, in *Handbook of Polymer Synthesis*, H. R. Kricheldorf, ed., Marcel Dekker, New York, 1992, Chapter 6, p. 385.

188. A. W. Meyer, R. R. Hampton, and J. A. Davison, *J. Am. Chem. Soc.* **74**, 2294 (1952).

189. S. Bywater, Y. Firat, and P. E. Black, *J. Polymer. Sci., Polym. Chem. Ed.* **22**, 669 (1984).

190. I. Kuntz and A. Gerber, *J. Polymer. Sci.*, **42**, 299 (1960).

191. A. A. Morton, *Adv. Catal.* **9**, 743 (1957).

192. J. D. D'Ianni, P. J. Naples, and J. E. Field, *Ind. Eng. Chem.* **42**, 95 (1950).

193. H. Hsieh, D. J. Kelly, and A. V. Tobolsky, *J. Polym. Sci.* **26**, 240 (1957).

194. D. J. Worsfold and S. Bywater, *Macromolecules* **11**, 582 (1978).

195. S. Bywater, *Adv. Polym. Sci.* **4**, 66 (1965).

196. S. Bywater and D. J. Worsfold, *Can. J. Chem.* **45**, 1821 (1967).

197. M. Morton, R. D. Sanderson, and R. Sakata, *Macromolecules* **6**, 181 (1973).

198. M. Morton, R. D. Sanderson, R. Sakata, and L. A. Falvo, *Macromolecules* **6**, 186 (1973).

199. R. S. Stearns and L. E. Forman, *J. Polym. Sci.* **41**, 381 (1959).

200. W. Gebert, J. Hinz, and H. Sinn, *Makromol. Chem.* **144**, 97 (1971).

201. S. Bywater, *ACS Symp. Ser.* **166**, 71 (1981).

202. E. W. Fawcett, R. O. Gibson, M. W. Perrin, J. G. Paton, and E. G. Williams, Br. Patent 471,590 (1937).

203. H. R. Sailors and J. P. Hogan, *J. Macromol. Sci., Chem.* **15**, 1377 (1981).

204. J. P. Marano, Jr. and J. M. Jemkins, III, in *High Pressure Technology*, Vol. 2, I. L. Spain and J. Paauwe, eds., Marcel Dekker, New York, 1977, Chapter 3, p. 61.

205. Y. Ogo, *J. Macromol. Sci., Rev. Macromol. Chem. Phys.* **C24**, 1 (1984).

206. K. W. Doak, in *Encyclopedia of Polymer Science and Engineering*, Vol. 6, J. I. Kroschwitz, ed., Wiley-Interscience, New York, 1986, p. 386.

207. J. K. Beasley, in *Comprehensive Polymer Science*, G. Allen and J. C. Bevington, eds., Vol. 3: *Chain Polymerization I*, G. C. Eastmond, A. Ledwith, S. Russo, and P. Sigwalt, eds., Pergamon Press, Oxford, 1989, Chapter 21, p. 273.

208. W. Kaminsky, in *Handbook of Polymer Synthesis*, H. R. Kricheldorf, ed., Marcel Dekker, New York, 1992, Chapter 1, p. 1.

209. C. H. Bamford, in *Encyclopedia of Polymer Science and Engineering*, 2nd ed., Vol. 13, J. I. Kroschwitz, ed., Wiley-Interscience, New York, 1988, p. 708.

210. L. E. Kukacka, P. Colombo, and M. Steinberg, *J. Polym. Sci. Polym. Symp.* **16**, 579 (1967).

211. H. Brackemann, M. Buback, and H.-P. Vögele, *Makromol. Chem.* **187**, 1977 (1986).

212. S. P. Pappas, in *Comprehensive Polymer Science*, G. Allen and J. C. Bevington, eds., Vol. 4: *Chain Polymerization II*, G. C. Eastmond, A. Ledwith, S. Russo, and P. Sigwalt, eds., Pergamon Press, Oxford, 1989, Chapter 20, p. 337.

213. M. Buback, C.-R. Choe, and E.-U. Franck, *Makromol. Chem.* **185**, 1685 (1984).

214. F. W. Nees and M. Buback, *Ber. Bunsenges. Phys. Chem.* **80**, 1017 (1976).

215. H. Seidl and G. Luft, *J. Macromol. Sci., Chem.,* **15**, 1 (1981).

216. G. Luft and H. Seidl, *Angew. Macromol. Chem.* **129**, 61 (1985).

217. G. Moad and D. H. Solomon, in *Comprehensive Polymer Science*, G. Allen and J. C. Bevington, eds., Vol. 3: *Chain Polymerization I*, G. C. Eastmond, A. Ledwith, S. Russo, and P. Sigwalt, eds., Pergamon Press, Oxford, 1989, Chapter 8, p. 97.

218. B. Boutevin and Y. Pietrasanta, in *Comprehensive Polymer Science*, G. Allen and J. C. Bevington, eds., Vol. 3: *Chain Polymerization I*, G. C. Eastmond, A. Ledwith, S. Russo, and P. Sigwalt, eds., Pergamon Press, Oxford, 1989, Chapter 14, p. 185.

219. M. J. Roedel, *J. Am. Chem. Soc.* **75**, 6110 (1953).

220. A. H. Willbourn, *J. Polym. Sci.* **34**, 569 (1959).

221. W. L. Mattice, *Macromolecules* **16**, 487 (1983).

222. J. Vile, P. J. Hendra, H. A. Willis, M. E. A. Cudby, and A. Bunn, *Polymer* **25**, 1173 (1984).

223. B. J. Meister and M. T. Malanga, in *Encyclopedia of Polymer Science and Engineering*, 2nd ed., Vol. 16, J. I. Kroschwitz, ed., Wiley-Interscience, New York, 1989, p. 21.

224. O. Nuyken, in *Handbook of Polymer Synthesis*, H. R. Kricheldorf, ed., Marcel Dekker, New York, 1992, Chapter 2, p. 77.

225. O. F. Olaj, H. F. Kauffmann, and J. W. Breitenbach, *Makromol. Chem.* **178**, 2707 (1977).

226. D. J. Stein and H. Mosthaf, *Angew. Macromol. Chem.* **2**, 39 (1968).

227. F. R. Mayo, *J. Am. Chem. Soc.* **90**, 1289 (1968).

228. V. R. Kamath, *Mod. Plast.* **58**(9), 106 (1981).

229. G. Ayrey, F. G. Levitt, and R. J. Mazza, *Polymer* **6**, 157 (1965).

230. J. C. Bevington, H. W. Melville, and R. P. Taylor, *J. Polym. Sci.* **14**, 463 (1954).

231. K. Ziegler, H. Breil, H. Martin, and E. Holzkamp, Germ. Patent 973,626 (1953).

232. K. Ziegler, Belg. Patent 533,362 (1954).

233. G. Natta, P. Pino, P. Corradini, F. Danusso, E. Mantica, G. Mazzanti, and G. Moraglio, *J. Am. Chem. Soc.* **77**, 1708 (1955).

234. G. Natta, *J. Polym. Sci.* **34**, 21 (1959).

235. J. P. Hogan and R. L. Banks, U.S. Patent 2,825,721 (1958).

236. A. Clark, J. P. Hogan, R. L. Banks, and W. C. Lanning, *Ind. Eng. Chem.* **48**, 1152 (1956).

237. A. Zletz, U.S. Patent 2,692,258 (1954).

238. E. F. Peters, A. Zletz, and B. L. Evering, *Ind. Eng. Chem.* **49**, 1879 (1957).

239. G. Natta, P. Pino, and G. Mazzanti, *Gazz. Chim. Ital.* **87**, 528 (1957).

240. G. Natta, P. Corradini, and G. Allegra, *J. Polym. Sci.* **51**, 399 (1961).

241. M. Fischer, Germ. Patent 874,215 (1953).

242. G. A. Olah, M. Bruce, F. Clouet, S. M. F. Farnia, O. Farooq, G. K. S. Prakash, J. Welch, and J. L. Koenig, *Macromolecules* **20**, 2972 (1987).

243. H. v. Pechman, *Chem. Ber.* **31**, 2640 (1898).

244. E. Bamberger and F. Tschirner, *Chem. Ber.* **33**, 955 (1900).

245. H. Meerwein and W. Burneleit, *Chem. Ber.* **61**, 1840 (1928).

246. H. Meerwein and G. Hinz, *Justus Liebigs Ann. Chem.* **484**, 1 (1930).

247. H. Meerwein, *Angew. Chem.* **60**, 78 (1948).

248. R. Wegler, in *Methoden der Organischen Chemie (Houben-Weyl)*, Vol. 14/2: *Makromolekulare Stoffe*, Thieme, Stuttgart, 1963, p. 629.

249. P. Pino, U. Giannini, and L. Porri, in *Encyclopedia of Polymer Science and Engineering*, 2nd ed., Vol. 8, J. I. Kroschwitz, ed., Wiley-Interscience, New York, 1987, p. 152.

250. C. E. H. Bawn, A. Ledwith, and P. Matthies, *J. Polym. Sci.* **34**, 93 (1959).

251. A. D. Caunt, in *Catalysis, a Specialist Periodical Report*, Vol. 1, C. Kemball, senior reporter, The Chemical Society, Burlington House, London, 1977, Chapter 7, p. 234.

252. H. Sinn and W. Kaminsky, *Adv. Organomet. Chem.* **18**, 99 (1980).

253. P. J. T. Tait, in *Comprehensive Polymer Science*, G. Allen and J. C. Bevington, eds., Vol. 4: *Chain Polymerization II*, G. C. Eastmond, A. Ledwith, S. Russo, and P. Sigwalt, eds., Pergamon Press, Oxford, 1989, Chapter 1, p. 1.

254. I. Pasquon and U. Giannini, in *Catalysis, Science and Technology*, Vol. 6, J. R. Anderson and M. Boudart, eds., Springer, Berlin, 1984, Chapter 2, p. 65.

255. Y. V. Kissin, in *Handbook of Polymer Science and Technology*, Vol. 1: *Synthesis and Properties*, N. P. Cheremisinoff, ed., Marcel Dekker, New York, 1989, Chapter 3, p. 103.

256. R. B. Lieberman and P. C. Barbe, in *Encyclopedia of Polymer Science and Engineering*, 2nd ed., Vol. 13, J. I. Kroschwitz, ed., Wiley-Interscience, New York, 1988, p. 464.

257. Z. W. Wilchinsky, R. W. Looney, and E. G. M. Tornqvist, *J. Catal.* **28**, 351 (1973).

258. Yu. I. Yermakov and V. A. Zakharov, in *Coordination Polymerization: A Memorial to Karl Ziegler*, J. C. W. Chien, ed., Academic Press, New York, 1975, p. 91.

259. S. Sivaram, *Ind. Eng. Chem., Prod. Res. Dev.* **16**, 121 (1977).

260. J. P. Hermans and P. Henrioulle, Germ. Patent 2,213,086 (1972); U.S. Patent 4,210,738 (1980).

261. R. P. Nielsen, in *Transtition Metal Catalyzed Polymerizations. Alkenes and Dienes*, R. P. Quirk, ed., Harwood, Chur, 1983, p. 47.

262. H. L. Hsieh, *Catal. Rev.-Sci. Eng.* **26**, 631 (1984).

263. F. J. Karol, *Catal. Rev.-Sci. Eng.* **26**, 557 (1984).

264. J. C. W. Chien, *Catal. Rev.-Sci. Eng.* **26**, 613 (1984).

265. N. Kashiwa, *Polymer J.* **12**, 603 (1980).

266. B. L. Goodall, in *Transtition Metal Catalyzed Polymerizations. Alkenes and Dienes*, R. P. Quirk, ed., Harwood, Chur, 1983, p. 355.

267. P. C. Barbé, G. Cecchin, and L. Noristi, *Adv. Polym. Sci.* **81**, 1 (1986).

268. G. Natta, P. Pino, G. Mazzanti, and U. Giannini, *J. Am. Chem. Soc.* **79**, 2975 (1957).

269. D. S. Breslow and N. R. Newburg, *J. Am. Chem. Soc.* **79**, 5072 (1957).

270. G. Henrici-Olivé and S. Olivé, *Angew. Chem., Int. Ed. Engl.* **10**, 776 (1971).

271. Y. Doi, S. Ueki, and T. Keii, *Macromolecules* **12**, 814 (1979).

272. Y. Doi, S. Suzuki, F. Nozawa, and K. Soga, in *Catalytic Polymerization of Olefins*, Studies in Surface Science and Catalysis, Vol. 25, T. Keii and K. Soga, eds., Kodansha, Tokyo, Elsevier, Amsterdam, 1986, p. 257.

273. G. G. Evens and E. M. J. Pijpers, in *Transtition Metal Catalyzed Polymerizations. Alkenes and Dienes*, R. P. Quirk, ed., Harwood, Chur, 1983, p. 245.

274. A. Andresen, H.-G. Cordes, J. Herwig, W. Kaminsky, A. Merck, R. Mottweiler, J. Pein, H. Sinn, and H.-J. Vollmer, *Angew. Chem., Int. Ed. Engl.* **15**, 630 (1976).

275. H. Sinn, W. Kaminsky, H.-J. Vollmer, and R. Woldt, *Angew. Chem., Int. Ed. Engl.* **19**, 390 (1980).

276. W. Kaminsky, in *Transtition Metal Catalyzed Polymerizations*, R. P. Quirk, ed., Cambridge Univ. Press, Cambridge, UK, 1988, p. 37.

277. K. Reichert, in *Transtition Metal Catalyzed Polymerizations. Alkenes and Dienes*, R. P. Quirk, ed., Harwood, Chur, 1983, p. 465.

278. W. Kaminsky, K. Külper, H. H. Brintzinger, and F. R. W. P. Wild, *Angew. Chem., Int. Ed. Engl.* **24**, 507 (1985).

279. J. A. Ewen, *J. Am. Chem. Soc.* **106**, 6355 (1984).

280. J. A. Ewen, R. L. Jones, A. Razavi, and J. D. Ferrara, *J. Am. Chem. Soc.* **110**, 6255 (1988).

281. J. A. Ewen, M. J. Elder, R. L. Jones, S. Curtis, and H. N. Cheng, in *Catalytic Olefin Polymerization*, Studies in Surface Science and Catalysis, Vol. 56, T. Keii and K. Soga, eds., Kodansha, Tokyo, Elsevier, Amsterdam, 1990, p. 439.

282. F. J. Karol and S.-C. Kao, *New J. Chem.* **18**, 97 (1994).

283. A. A. Montagna and J. C. Floyd, *Hydrocarbon Proc., Int. Ed.* **73**(4), 57 (1994).

284. A. Clark, *Catal. Rev.-Sci. Eng.* **3**, 145 (1970).

285. D. R. Witt, in *Reactivity, Mechanism and Structure in Polymer Chemistry*, A. D. Jenkins and A. Ledwith, eds., Wiley-Interscience, London, 1974, Chapter 13, p. 431.

286. M. P. McDaniel, *Adv. Catal.* **33**, 47 (1985).

287. V. A. Zakharov and Yu. I. Yermakov, *Catal. Rev.-Sci. Eng.* **19**, 67 (1979).

288. Yu. I. Yermakov, B. N. Kuznetsov, and V. A. Zakharov, *Catalysis by Supported Complexes*, Studies in Surface Science and Catalysis, Vol. 8, Elsevier, Amsterdam, 1981.

289. Yu. I. Yermakov and V. Zakharov, *Adv. Catal.* **24**, 173 (1975).

290. F. J. Karol, G. L. Karapinka, C. Wu, A. W. Dow, R. N. Johnson, and W. L. Carrick, *J. Polym. Sci., Polym. Chem. Ed.* **10**, 2621 (1972).

291. Yu. I. Ermakov, B. N. Kuznetsov, L. G. Karakchiev, and S. S. Derbeneva, *Kinet. Catal.* (Engl. trans.) **14**, 611 (1973).

292. D. G. H. Ballard, *Adv. Catal.* **23**, 263 (1973).

293. M. N. Berger, G. Boocock, and R. N. Haward, *Adv. Catal.* **19**, 211 (1969).

294. P. J. T. Tait and N. D. Watkins, in *Comprehensive Polymer Science*, G. Allen and J. C. Bevington, eds., Vol. 4: *Chain Polymerization II*, G. C. Eastmond, A. Ledwith, S. Russo, and P. Sigwalt, eds., Pergamon Press, Oxford, 1989, Chapter 2, p. 533.

295. A. Ledwith and D. C. Sherrington, in *Reactivity, Mechanism and Structure in Polymer Chemistry*, A. D. Jenkins and A. Ledwith, eds., Wiley-Interscience, London, 1974, Chapter 12, p. 384.

296. Ph. Teyssié and F. Dawans, in *The Stereo Rubbers*, W. M. Saltman, ed., Wiley-Interscience, New York, 1977, Chapter 3, p. 79.

297. G. Natta and G. Mazzanti, *Tetrahedron* **8**, 86 (1960).

298. G. Natta and I. Pasquon, *Adv. Catal.* **11**, 1 (1959).

299. P. F. Patat and Hj. Sinn, *Angew. Chem.* **70**, 496 (1958).

300. H. Uelzmann, *J. Org. Chem.* **25**, 671 (1960).

301. J. Boor, Jr., *J. Polym. Sci., Polym. Symp.* **1**, 257 (1963).

302. G. Natta, I. Pasquon, A. Zambelli, and G. Gatti, *J. Polym. Sci.*, **51**, 387 (1961).

303. E. J. Arlman, *J. Catal.* **3**, 89 (1964).

304. P. Cossee, *J. Catal.* **3**, 80 (1964).

305. E. J. Arlman and P. Cossee, *J. Catal.* **3**, 99 (1964).

306. E. J. Arlman, *J. Catal.* **5**, 178 (1966).

307. L. A. M. Rodriguez and H. M. van Looy, *J. Polym. Sci., Polym. Chem. Ed.* **4**, 1971 (1966).

308. W. P. Long and D. S. Breslow, *J. Am. Chem. Soc.* **82**, 1953 (1960).

309. L. A. M. Rodriguez, H. M. van Looy, and J. A. Gabant, *J. Polym. Sci., Polym. Chem. Ed.* **4**, 1905, 1917 (1966).

310. L. A. M. Rodriguez and H. M. van Looy, *J. Polym. Sci., Polym. Chem. Ed.* **4**, 1951 (1966).

311. A. Zambelli, P. Locatelli, M. C. Sacchi, and E. Rigamonti, *Macromolecules* **13**, 798 (1980).

312. A. Zambelli, M. C. Sacchi, P. Locatelli, and G. Zannoni, *Macromolecules* **15**, 211 (1982).

313. A. Zambelli, M. C. Sacchi, and P. Locatelli, in *Transtition Metal Catalyzed Polymerizations. Alkenes and Dienes*, R. P. Quirk, ed., Harwood, Chur, 1983, p. 83.

314. G. Fink and R. Rottler, *Angew. Macromol. Chem.* **94**, 25 (1981).

315. G. Fink, in *Transtition Metal Catalyzed Polymerizations. Alkenes and Dienes*, R. P. Quirk, ed., Harwood, Chur, 1983, p. 495.

316. G. Henrici-Olivé and S. Olivé, *J. Polym. Sci., Polym. Symp.* **22**, 965 (1969).

317. R. Merryfield, M. McDaniel, and G. Parks, *J. Catal.* **77**, 348 (1982).

318. M. P. McDaniel, in *Transtition Metal Catalyzed Polymerizations. Alkenes and Dienes*, R. P. Quirk, ed., Harwood, Chur, 1983, p. 713.

319. D. D. Beck and J. H. Lunsford, *J. Catal.* **68**, 121 (1984).

320. B. Rebenstorf and R. Larsson, *J. Catal.* **84**, 240 (1983).

321. D. L. Myers and J. H. Lunsford, *J. Catal.* **99**, 140 (1986).

322. M. P. McDaniel, *Ind. Eng. Chem. Res.* **27**, 1559 (1988).

323. D. L. Beach and Y. V. Kissin, in *Encyclopedia of Polymer Science and Engineering*, 2nd ed., Vol. 6, J. I. Kroschwitz, ed., Wiley-Interscience, New York, 1986, p. 454.

324. G. Natta, P. Pino, G. Mazzanti, and P. Longi, *Gazz. Chim. Ital.* **87**, 549 (1957).

325. P. L. Luisi and F. Ciardelli, *Reactivity, Mechanism and Structure in Polymer Chemistry*, A. D. Jenkins and A. Ledwith, eds., Wiley-Interscience, London, 1974, Chapter 15, p. 471.

326. Y. V. Kissin, *Isospecific Polymerization of Olefins with Heterogeneous Ziegler-Natta Catalysts*. Springer, New York, 1985, Chapter 3, p. 221.

327. A. Zambelli, in *Coordination Polymerization: A Memorial to Karl Ziegler*, J. C. W. Chien, ed., Academic Press, New York, 1975, p. 15.

328. P. Pino and R. Mülhaupt, *Angew. Chem., Int. Ed. Engl.* **19**, 857 (1980).

329. A. Zambelli, P. Locatelli, and E. Rigamonti, *Macromolecules* **12**, 156 (1979).

330. G. Bucci and T. Simonazzi, *J. Polym. Sci., Polym. Symp.* **7**, 203 (1964).

331. P. Pino, P. Cioni, J. Wei, B. Rotzinger, and S. Arizzi, in *Transtition Metal Catalyzed Polymerizations*, R. P. Quirk, ed., Cambridge Univ. Press, Cambridge, UK, 1988, p. 1.

332. G. Natta, M. Farina, and M. Peraldo, *Chim. Ind.* (Milan) **42**, 255 (1960).

333. T. Miyazawa and Y. Ideguci, *J. Polym. Sci., Polym. Lett. Ed.* **1**, 389 (1963).

334. A. Zambelli, in *Natural and Synthetic High Polymers*, NMR Basic Principles and Progress, Vol. 4, P. Diehl, E. Fluck, and R. Kosfeld, eds., Springer, Berlin, 1971, p. 101.

335. A. Zambelli, G. Bajo, and E. Rigamonti, *Makromol. Chem.* **179**, 1249 (1978).

336. W. O. Crain, Jr., A. Zambelli, and J. D. Roberts, *Macromolecules* **4**, 330 (1971).

337. A. Zambelli, G. Gatti, C. Sacchi, W. O. Crain, Jr., and J. D. Roberts, *Macromolecules* **4**, 475 (1971).

338. P. Pino, A. Oschwald, F. Ciardelli, C. Carlini, and E. Chiellini, in *Coordination Polymerization: A Memorial to Karl Ziegler*, J. C. W. Chien, ed., Academic Press, New York, 1975, p. 25.

339. W. Kaminsky, in *Catalytic Polymerization of Olefins*, Studies in Surface Science and Catalysis, Vol. 25, T. Keii and K. Soga, eds., Kodansha, Tokyo, Elsevier, Amsterdam, 1986, p. 293.

340. P. Corradini, V. Busico, and G. Guerra, in *Comprehensive Polymer Science*, G. Allen and J. C. Bevington, eds., Vol. 4: *Chain Polymerization II*, G. C. Eastmond, A. Ledwith, S. Russo, and P. Sigwalt, eds., Pergamon Press, Oxford, 1989, Chapter 3, p. 29.

341. P. Corradini, V. Barone, R. Fusco, and G. Guerra, *Eur. Polym. J.* **15**, 1133 (1979).

342. P. Corradini, G. Guerra, R. Fusco, and V. Barone, *Eur. Polym. J.* **16**, 835 (1980).

343. G. Allegra, *Makromol. Chem.* **145**, 235 (1971).

344. Y. V. Kissin and N. M. Chirkov, *Eur. Polym. J.* **6**, 525 (1970).

345. Y. V. Kissin, *Isospecific Polymerization of Olefins with Heterogeneous Ziegler-Natta Catalysts*, Springer, New York, 1985, Chapter 5, p. 372.

346. A. Zambelli and C. Tosi, *Adv. Polymer. Sci.* **15**, 31 (1974).

347. K. J. Ivin, J. J. Rooney, C. D. Stewart, M. L. H. Green, and R. Mahtab, *J. Chem. Soc., Chem. Commun.* 604 (1978).

348. M. Kakugo, T. Miyatake, Y. Naito, and K. Mizunuma, *Macromolecules* **21**, 314 (1988).

349. J. A. Ewen, in *Catalytic Polymerization of Olefins*, Studies in Surface Science and Catalysis, Vol. 25, T. Keii, K. Soga, eds., Kodansha, Tokyo, Elsevier, Amsterdam, 1986, p. 271.

350. A. Zambelli, P. Ammendola, A. Grassi, P. Longo, and A. Proto, *Macromolecules* **19**, 2703 (1986).

351. P. Longo, A. Grassi, C. Pallecchia, and A. Zambelli, *Macromolecules* **20**, 1015 (1987).

352. G. Erker, R. Nolte, R. Aul, S. Wilker, C. Krüger, and R. Noe, *J. Am. Chem. Soc.* **113**, 7594 (1991).

353. G. Guerra, L. Cavallo, G. Moscardi, M. Vacatello, and P. Corradini, *J. Am. Chem. Soc.* **116**, 2988 (1994).

354. G. Natta, I. Pasquon, P. Corradini, M. Peraldo, M. Pegoraro, and A. Zambelli, *Atti. accad. nazl. Lincei, Rend., Classe sci. fis., mat. e nat.* **28**, 539 (1961); *Chem. Abstr.* **55**, 8932i (1961).

355. G. Natta, I. Pasquon, and A. Zambelli, *J. Am. Chem. Soc.* **84**, 1488 (1962).

356. J. Boor, Jr. and E. A. Youngman, *J. Polym. Sci., Polym. Chem. Ed.* **4**, 1861 (1966).

357. A. Zambelli, I. Pasquon, R. Signorini, and G. Natta, *Makromol. Chem.* **112**, 160 (1968).

358. T. Suzuki and Y. Takegami, *Bull. Chem. Soc. Jpn.* **43**, 1484 (1970).

359. Y. Doi, T. Koyama, K. Soga, and T. Asakura, *Makromol. Chem.* **185**, 1827 (1984).

360. A. Zambelli and G. Allegra, *Macromolecules* **13**, 42 (1980).

361. A. Galambos, M. Wolkowicz, and R. Zeigler, *ACS Symp. Ser.* **496**, 121 (1992).

362. A. Zambelli, M. G. Giongo, and G. Natta, *Makromol. Chem.* **112**, 183 (1968).

363. L. Porri, A. Giarrusso, in *Comprehensive Polymer Science*, G. Allen and J. C. Bevington, eds., Vol. 4: *Chain Polymerization II*, G. C. Eastmond, A. Ledwith, S. Russo, and P. Sigwalt, eds., Pergamon Press, Oxford, 1989, Chapter 5, p. 53.

364. N. G. Gaylord, T.-K. Kwei, and H. F. Mark, *J. Polymer. Sci.* **42**, 417 (1960).

365. Br. Patent 848,065 to Phillips Petroleum Co. (1960).

366. F. J. Henderson, *J. Polymer. Sci., Polym. Symp.* **4**, 233 (1963).

367. D. H. Richards, *Chem. Soc. Rev.* **6**, 235 (1977).

368. W. M. Saltman and T. H. Link, *Ind. Eng. Chem., Prod. Res. Dev.* **3**, 199 (1964).

369. J. E. Howell, W. G. Curtis, and J. A. Solomon, in *Transtition Metal Catalyzed Polymerizations. Alkenes and Dienes*, R. P. Quirk, ed., Harwood, Chur, 1983, p. 509.

370. W. M. Saltman and L. J. Kuzma, *Rubber Chem. Technol.* **46**, 1055 (1973).

371. S. E. Horne, Jr., in *Transtition Metal Catalyzed Polymerizations. Alkenes and Dienes*, R. P. Quirk, ed., Harwood, Chur, 1983, p. 527.

372. M. Gippin, *Ind. Eng. Chem., Prod. Res. Dev.* **1**, 32 (1962); *Rubber Chem. Technol.* **35**, 1066 (1962).

373. C. Longiave and R. Castelli, *J. Polymer. Sci., Polym. Symp.* **4**, 387 (1963).

374. T. Yoshimoto, K. Komatsu, R. Sakata, K. Yamamoto, Y. Takeuchi, A. Onishi, and K. Ueda, *Makromol. Chem.* **139**, 61 (1970).

375. K. Ueda, A. Onishi, T. Yoshimoto, J. Hosono, and K. Maeda, U.S. Patent 3,170,905 (1965).

376. M. C. Throckmorton and F. S. Farson, *Rubber Chem. Technol.* **45**, 268 (1972).

377. U. Gebauer, J. Ludwig, and K. Gehrke, *Acta Polym.* **39**, 368 (1988).

378. G. Natta, L. Porri, and A. Carbonaro, *Makromol. Chem.* **77**, 126 (1964).

379. B. D. Babitskii, B. A. Dolgoplosk, V. A. Kormer, M. I. Lobach, E. I. Tinyakova, and V. A. Yakovlev, *Bull. Acad. Sci. USSR, Div. Chem. Sci.* (Engl. transl.) 1478 (1965).

380. F. Dawans and Ph. Teyssié, *Ind. Eng. Chem., Prod. Res. Dev.* **10**, 261 (1971).

381. Ph. Teyssié, M. Julémont, J. M. Thomassin, E. Walckiers, and R. Warin, in *Coordination Polymerization: A Memorial to Karl Ziegler*, J. C. W. Chien, ed., Academic Press, New York, 1975, p. 327.

382. M. C. Gallazzi, A. Giarrusso, and L. Porri, *Makromol. Chem., Rapid Commun.* **2**, 59 (1981).

383. C. A. Tolman, *J. Am. Chem. Soc.* **92**, 6777 (1970).

384. J. Furukawa, *Acc. Chem. Res.* **13**, 1 (1980).

385. R. P. Hughes and J. Powell, *J. Am. Chem. Soc.* **94**, 7723 (1972).

386. L. Porri, G. Natta, and M. C. Gallazzi, *J. Polymer. Sci., Polym. Symp.* **16**, 2525 (1967).

387. J. P. Durand, F. Dawans, and Ph. Teyssie, *J. Polymer. Sci., Polym. Lett. Ed.* **5**, 785 (1967).

388. J. C. Marechal, F. Dawans, and Ph. Teyssie, *J. Polymer. Sci., Polym. Chem. Ed.* **8**, 1993 (1970).

389. H. W. Gibson and J. M. Pochan, in *Encyclopedia of Polymer Science and Engineering*, 2nd ed., Vol. 1, J. I. Kroschwitz, ed., Wiley-Interscience, New York, 1985, p. 87.

390. H. Naarmann and P. Strohriegel, in *Handbook of Polymer Synthesis*, H. R. Kricheldorf, ed., Marcel Dekker, New York, 1992, Chapter 21, p. 1353.

391. A. Bolognesi, M. Catellani, and S. Destri, in *Comprehensive Polymer Science*, G. Allen and J. C. Bevington, eds., Vol. 4: *Chain Polymerization II*, G. C. Eastmond, A. Ledwith, S. Russo, and P. Sigwalt, eds., Pergamon Press, Oxford, 1989. Chapter 8, p. 143.

392. R. H. Baughman, J. L. Brédas, R. R. Chance, R. L. Elsenbaumer, and L. W. Shacklette, *Chem. Rev.* **82**, 209 (1982).

393. J. E. Frommer and R. R. Chance, in *Encyclopedia of Polymer Science and Engineering*, 2nd ed., Vol. 5, J. I. Kroschwitz, ed., Wiley-Interscience, New York, 1986, p. 462.

394. S. Pekker and A. Jánossy, in *Handbook of Conducting Polymers*, Vol. 1, T. A. Skotheim, ed., Marcel Dekker, New York, 1986, Chapter 2, p. 45.

395. L. M. Tolbert, *Acc. Chem. Res.* **25**, 561 (1992).

396. M. G. Chauser, Yu. M. Rodionov, V. M. Misin, and M. I. Cherkashin, *Russ. Chem. Rev.* (Engl. transl.) **45**, 348 (1976).

397. W. J. Feast, in *Handbook of Conducting Polymers*, Vol. 1, T. A. Skotheim, ed., Marcel Dekker, New York, 1986, Chapter 1, p. 1.

398. C. Kröhnke and G. Wegner, in *Methoden der Organischen Chemie (Houben-Weyl)*, 4th ed., Vol. E20/2: *Makromolekulare Stoffe*, H. Bartl and J. Falbe, eds., Thieme, Stuttgart, 1987, p. 1319.

399. C. B. Gorman and R. H. Grubbs, in *Conjugated Polymers: the Novel Science and Technology of Highly Conducting and Nonlinear Optically Active Materials*, J. L. Brédas and R. Silbey, eds., Kluwer, Dordrecht, 1991, p. 1.

400. G. Natta, G. Mazzanti, and P. Corradini, *Atti. accad. nazl. Lincei, Rend., Classe sci. fis., mat. e nat.* **25**, 3 (1958); *Chem. Abstr.* **53**, 13985i (1959).

401. T. Ito, H. Shirakawa, and S. Ikeda, *J. Polym. Sci., Polym. Chem. Ed.* **12**, 11 (1974).

402. H. Naarmann, *Synth. Met.* **17**, 223 (1987).

403. H. Naarmann and N. Theophilou, *Synth. Met.* **22**, 1 (1987).

404. G. Luigi, U. Pederetti, and G. Perego, *J. Polym. Sci., Polym. Lett. Ed.* **23**, 129 (1985).

405. K. Akagi, M. Suezaki, H. Shirakawa, H. Kyotani, M. Shimomura, and Y. Tanabe, *Synth. Met.* **28**, D1 (1989).

406. N. Basescu, Z.-X. Liu, D. Moses, A. J. Heeger, H. Naarmann, and N. Theophilou, *Nature* (London) **327**, 403 (1987).

407. L. B. Luttinger, *J. Org. Chem.* **27**, 1591 (1962).

408. W. E. Daniels, *J. Org. Chem.* **29**, 2936 (1964).

409. F. L. Klavetter and R. H. Grubbs, *J. Am. Chem. Soc.* **110**, 7807 (1988).

410. C. B. Gorman, E. J. Ginsburg, and R. H. Grubbs, *J. Am. Chem. Soc.* **115**, 1397 (1993).

411. H. R. Kricheldorf, in *Handbook of Polymer Synthesis*, H. R. Kricheldorf, ed., Marcel Dekker, New York, 1992, Chapter 7, p. 433.

412. M. A. Schen, F. E. Karasz, and J. W. C. Chien, *J. Polym. Sci., Polym. Chem. Ed.* **21**, 2787 (1983).

413. H. W. Gibson, in *Handbook of Conducting Polymers*, Vol. 1, T. A. Skotheim, ed., Marcel Dekker, New York, 1986, Chapter 11, p. 405.

414. T. Masuda and T. Higashimura, *Acc. Chem. Res.* **17**, 51 (1984); *Adv. Polym. Sci.* **81**, 121 (1987).

415. G. Costa, in *Comprehensive Polymer Science*, G. Allen and J. C. Bevington, eds., Vol. 4: *Chain Polymerization II*, G. C. Eastmond, A. Ledwith, S. Russo, and P. Sigwalt, eds., Pergamon Press, Oxford, 1989, Chapter 9, p. 155.

416. T. J. Katz and S. J. Lee, *J. Am. Chem. Soc.* **102**, 422 (1980).

417. T. J. Katz, T. H. Ho, N.-Y. Shih, Y.-C. Ying, and V. I. W. Stuart, *J. Am. Chem. Soc.* **106**, 2659 (1984).

418. T. Masuda, K. Hasegawa, and T. Higashimura, *Macromolecules* **7**, 728 (1974).

419. T. Masuda, Y. Okano, Y. Kuwane, and T. Higashimura, *Polym. J.* **12**, 907 (1980).

420. T. Masuda, K.-Q. Thieu, N. Sasaki, and T. Higashimura, *Macromolecules* **9**, 661 (1976).

421. W. J. Trepka and R. J. Sonnenfeld, *J. Polym. Sci., Polym. Chem. Ed.* **8**, 2721 (1970).

422. F. Ciardelli, S. Lanzillo, and O. Pieroni, *Macromolecules* **7**, 174 (1974).

423. T. Masuda, N. Sasaki, and T. Higashimura, *Macromolecules* **8**, 717 (1975).

424. T. J. Katz, S. M. Hacker, R. D. Kendrick, and C. S. Yannoni, *J. Am. Chem. Soc.* **107**, 2182 (1985).

425. Y.-S. Jane and J.-S. Shih, *J. Mol. Catal.* **89**, 29 (1994).

426. Y. Goldberg and H. Alper, *J. Chem. Soc., Chem. Commun.* 1209 (1994).

427. M. B. Jones and P. Kovacic, in *Encyclopedia of Polymer Science and Engineering*, 2nd ed., Vol. 10, J. I. Kroschwitz, ed., Wiley-Interscience, New York, 1987, p. 670.

428. P. Kovacic and M. B. Jones, *Chem. Rev.* **87**, 357 (1987).

429. M. B. Jones and P. Kovacic, in *Comprehensive Polymer Science*, G. Allen and J. C. Bevington, eds., Vol. 5: *Step Polymerization*, G. C. Eastmond, A. Ledwith, S. Russo, and P. Sigwalt, eds., Pergamon Pres, Oxford, 1989, Chapter 27, p. 465.

430. P. Kovacic and A. Kyriakis, *J. Am. Chem. Soc.* **85**, 454 (1963).

431. H. R. Kricheldorf and G. Schwarz, in *Handbook of Polymer Synthesis*, H. R. Kricheldorf, ed., Marcel Dekker, New York, 1992, Chapter 27, p. 1629.

432. P. Kovacic and W. B. England, *J. Polym. Sci., Polym. Lett. Ed.* **19**, 359 (1981).

433. C.-F. Hsing, I. Khoury, M. D. Bezoari, and P. Kovacic, *J. Polym. Sci., Polym. Chem. Ed.* **20**, 3313 (1982).

434. S. A. Milosevich, K. Saichek, L. Hinchey, W. B. England, and P. Kovacic, *J. Am. Chem. Soc.* **105**, 1088 (1983).

435. P. Kovacic and J. S. Ramsey, *J. Polym. Sci., Polym. Chem. Ed.* **7**, 111 (1969).

436. M. N. Bilow and L. J. Miller, *J. Macromol. Sci., Chem.* **1**, 183 (1967).

437. T. Ohsawa, T. Inoue, S. Takeda, K. Kaneto, and K. Yoshino, *Polym. Commun.* **27**, 246 (1986).

438. M. Satoh, K. Kaneto, and K. Yoshino, *J. Chem. Soc., Chem. Commun.* 1629 (1985).

439. M. Satoh, F. Uesugi, M. Tabata, K. Kaneto, and K. Yoshino, *J. Chem. Soc., Chem. Commun.* 550 (1986).

440. G. L. Dighton, in *Alpha Olefins Application Handbook*, G. R. Lappin and J. D. Sauer, eds., Marcel Dekker, New York, 1989, Chapter 4, p. 63.

441. F. J. Karol, *Chemtech* **13**, 222 (1983).

442. D. E. James, in *Encyclopedia of Polymer Science and Engineering*, 2nd ed., Vol. 6, J. I. Kroschwitz, ed., Wiley-Interscience, New York, 1986, p. 429.

443. G. Crespi, A. Valvassori, and U. Flisi, in *The Stereo Rubbers*, W. M. Saltman, ed., Wiley-Interscience, New York, 1977, Chapter 7, p. 365.

444. I. D. Burdett, *Chemtech* **22**, 616 (1992).

445. C. R. Raufast, *Hydrocarbon Process., Int. Ed.* **63**(5), 105 (1984).

446. J. C. Davis, *Chem. Eng.* (NY) **85**(1), 25 (1978).

447. K.-Y. Choi and W. H. Ray, *J. Macromol. Sci., Rev. Macromol. Chem. Phys.* **C25**, 1 (1985).

448. C. Cipriani and C. A. Trischman, Jr., *Chem. Eng.* (NY) **89**(10), 66 (1982).

449. P. J. T. Tait and I. G. Berry, in *Comprehensive Polymer Science*, G. Allen and J. C. Bevington, eds., Vol. 4: *Chain Polymerization II*, G. C. Eastmond, A. Ledwith, S. Russo, and P. Sigwalt, eds., Pergamon Press, Oxford, 1989, Chapter 4, p. 575.

450. G. Ver Strate, in *Encyclopedia of Polymer Science and Engineering*, 2nd ed., Vol. 6, J. I. Kroschwitz, ed., Wiley-Interscience, New York, 1986, p. 522.

451. A. W. Larchar and D. C. Pease, U.S. Patent 2,816,883 (1957).

452. R. A. Hines, W. M. D. Bryant, A. W. Larchar, and D. C. Pease, *Ind. Eng. Chem.* **49**, 1071 (1957).

453. F. Asinger, *Mono-Olefins, Chemistry and Technology*, Pergamon Press, Oxford, 1968, Chapter 10, p. 866.

454. J. P. Forsman, *Hydrocarbon Process.* **51**(11), 130 (1972).

455. J. N. Short, in *Transtition Metal Catalyzed Polymerizations. Alkenes and Dienes*, R. P. Quirk, ed., Harwood, Chur, 1983, p. 651.

456. J. Stevens, *Hydrocarbon Process.* **49**(11), 179 (1970).

457. H. Kreuter and B. Diedrich, *Chem. Eng.* (NY) **81**(16), 62 (1974).

458. A. Heath, *Chem. Eng.* (NY) **79**(7), 66 (1972).

459. S. D. de Bree, *Chem. Eng.* (NY) **79**(28), 72 (1972).

460. *Hydrocarbon Process.* **53**(2), 115 (1974).

461. G. G. Arzoumanidis and N. M. Karayannis, *Chemtech* **23**(7), 43 (1993).

462. K.-Y. Choi and W. H. Ray, *J. Macromol. Sci., Rev. Macromol. Chem. Phys.* **C25**, 57 (1985).

463. A. Valvassori, P. Longi, and P. Parrini, in *Propylene and Its Industrial Derivatives*, E. G. Hancock, ed., Wiley, New York, 1973, Chapter 5, p. 155.

464. W. Dittmann, in *Methoden der Organischen Chemie (Houben-Weyl)*, 4th ed., Vol. E20/2: *Makromolekulare Stoffe*, H. Bartl and J. Falbe, eds., Thieme, Stuttgart, 1987, p. 722.

465. N. F. Brockmeier, in *Transtition Metal Catalyzed Polymerizations. Alkenes and Dienes*, R. P. Quirk, ed., Harwood, Chur, 1983, p. 671.

466. K. Matsuyama, A. Shiga, M. Kakugo, and H. Hashimoto, *Hydrocarbon Process., Int. Ed* **59**(11), 131 (1980).

467. C. Cipriani and C. A. Trischman, Jr., *Chem. Eng.* (NY) **88**(8), 80 (1981).

468. G. Di Drusco and R. Rinaldini, *Hydrocarbon Process., Int. Ed* **63**(11), 113 (1984).

469. P. Galli, L. Luciani, and G. Cecchin, *Angew. Makromol. Chem.* **94**, 63 (1981).

470. J. P. Kennedy and I. Kirshenbaum, in *High Polymers*, Vol. 24: *Vinyl and Diene Monomers*, Part 2, E. C. Leonard, ed., Wiley-Interscience, New York, 1971, Chapter 3, p. 691.

471. F. P. Baldwin and R. H. Schatz, in *Kirk-Othmer Encyclopedia of Chemical Technology*, 3rd ed., Vol. 8, M. Grayson and D. Eckroth, eds., Wiley-Interscience, New York, 1979, p. 470.

472. E. N. Kresge, R. H. Schatz, and H.-C. Wang, in *Encyclopedia of Polymer Science and Engineering*, 2nd ed., Vol. 8, J. I. Kroschwitz, ed., Wiley-Interscience, New York, 1987, p. 423.

473. A. M. Chatterjee, in *Encyclopedia of Polymer Science and Engineering*, 2nd ed., Vol. 2, J. I. Kroschwitz, ed., Wiley-Interscience, New York, 1985, p. 590.

474. J. C. Davis, *Chem. Eng.* (NY) **80**(3), 32 (1973).

475. M. D. Rosenzweig, *Chem. Eng.* (NY) **80**(3), 56 (1973).

476. J. L. Hahnfeld and B. D. Dalke, in *Encyclopedia of Polymer Science and Engineering*, 2nd ed., Vol. 16, J. I. Kroschwitz, ed., Wiley-Interscience, New York, 1989, p. 62.

477. T. D. Traugott, in *Encyclopedia of Polymer Science and Engineering*, 2nd ed., Vol. 16, J. I. Kroschwitz, ed., Wiley-Interscience, New York, 1989, p. 72.

478. G. W. Poehlein, in *Encyclopedia of Polymer Science and Engineering*, 2nd ed., Vol. 6, J. I. Kroschwitz, ed., Wiley-Interscience, New York, 1986, p. 1.

479. G. W. Poehlein and D. J. Daugherty, *Rubber Chem. Technol.* **50**, 601 (1977).

480. L. J. Fetters, *J. Polym. Sci., Polym. Symp.* **26**, 1 (1969).

481. A. Noshay and J. E. McGrath, *Block Copolymers*. Academic Press, Orlando, FL, 1977, Chapter 5, p. 83, Chapter 6, p. 186.

482. H. L. Hsieh, R. C. Farrar, and K. Udipi, *ACS Symp. Ser.* **166**, 389 (1981).

483. G. Holden and N. R. Legge, in *Thermoplastic Elastomers*, N. R. Legge, G. Holden, and H. E. Schroeder, eds., Hanser Publishers, Munich, 1987, Chapter 3, p. 48.

484. K. F. O'Driscoll and I. Kuntz, *J. Polym. Sci.* **61**, 19 (1962).

485. A. F. Johnson, and D. J. Worsfold, *Makromol. Chem.* **85**, 273 (1965).

486. M. Morton, in *Thermoplastic Elastomers*, N. R. Legge, G. Holden, and H. E. Schroeder, eds., Hanser Publishers, Munich, 1987, Chapter 4, p. 68.

487. G. Sylvester and R. Müller, in *Methoden der Organischen Chemie (Houben-Weyl)*, 4th ed., Vol. E20/2: *Makromolekulare Stoffe*, H. Bartl and J. Falbe, eds., Thieme, Stuttgart, 1987, p. 798.

488. J. Witte, in *Methoden der Organischen Chemie (Houben-Weyl)*, 4th ed., Vol. E20/2: *Makromolekulare Stoffe*, H. Bartl and J. Falbe, eds., Thieme, Stuttgart, 1987, p. 822.

489. E. W. Duck and J. M. Locke, in *The Stereo Rubbers*, W. M. Saltman, ed., Wiley-Interscience, New York, 1977, Chapter 4, p. 139.

490. *Hydrocarbon Process.* **46**(11), 216, 217, 218 (1967); **50**(11), 192 (1971); **54**(11), 176 (1975).

491. *Hydrocarbon Process., Int. Ed.* **60**(11), 203, 204 (1981).

492. M. Bruzzone, A. Mazzei, and G. Giuliani, *Rubber Chem. Technol.* **47**, 1175 (1974).

493. G. Lugli, A. Mazzei, and S. Poggio, *Makromol. Chem.* **175**, 2021 (1974).

494. *Hydrocarbon Process.* **56**(11), 201 (1977).

495. F. C. Price, *Chem. Eng.* (NY) **70**(2), 84 (1963).

496. W. Cooper, in *The Stereo Rubbers*, W. M. Saltman, ed., Wiley-Interscience, New York, 1977, Chapter 2, p. 21.

497. E. Schoenberg, D. L. Chalfant, and R. H. Mayor, *Rubber Chem. Technol.* **37**, 103 (1964).

498. W. Marconi, A. Mazzei, S. Cucinella, and M. De Maldé, *Makromol. Chem.* **71**, 118, 134 (1963).

499. A. Mazzei, S. Cucinella, and W. Marconi, *Makromol. Chem.* **122**, 168 (1969).

500. A. Balducci, M. Bruzzone, S. Cucinella, and A. Mazzei, *Rubber Chem. Technol.* **48**, 736 (1975).

501. M. Bruzzone, W. Marconi, and S. Noe, *Hydrocarbon Process.* **47**(11), 179 (1968).

502. *Hydrocarbon Process.* **50**(11), 200 (1971).

503. C. T. O'Connor, in *Handbook of Heterogeneous Catalysis* G. Ertl, H. Knözinger, and J. Weitkamp, eds., Wiley-VCH, Weinheim, 1997, Chapter 4.12.1, p. 2380.

504. W. Keim and R. P. Schulz, *J. Mol. Catal.* **92**, 21 (1994).

505. P. Braunstein, Y. Chauvin, S. Mercier, L. Saussine, A. De Cian, and J. Fischer, *J. Chem. Soc., Chem. Commun.* 2203 (1994).

506. S. Y. Desjardins, K. J. Cavell, H. Jin, B. W. Skelton, and A. H. White, *J. Organomet. Chem.* **515**, 233 (1996).

507. M. C. Bonnet, F. Dahan, A. Ecke, W. Keim, R. P. Schulz, and I. Tkatchenko, *J. Chem. Soc., Chem. Commun.* 615 (1994).

508. G. J. P. Britovsek, W. Keim, S. Mecking, D. Sainz, and T. Wagner, *J. Chem. Soc., Chem. Commun.* 1632 (1993).

509. S. A. Svejda and M. Brookhart, *Organometallics* **18**, 65 (1999).

510. S. P. Meneghetti, P. J. Lutz, and J. Kress, *Organometallics* **18**, 2734 (1999).

511. T. V. Laine, M. Klinga, A. Maaninen, E. Aitola, and M. Leskelä, *Acta Chem. Scand.* **53**, 968 (1999).

512. E. K. van den Beuken, W. J. J. Smeets, A. L. Spek, and B. L. Feringa, *Chem. Commun.* 223 (1998).

513. B. L. Small and M. Brookhart, *J. Am. Chem. Soc.* **120**, 7143 (1998).

514. G. J. P. Britovsek, S. Mastroianni, G. A. Solan, S. P. D. Baugh, C. Redshaw, V. C. Gibson, A. J. P. White, D. J. Williams, and M. R. J. Elsegood, *Chem. Eur. J.* **6**, 2221 (2000).

515. J. Heveling, C. P. Nicolaides, and M. S. Scurrell, *Appl. Catal., A* **173**, 1 (1998).

516. P. Wasserscheid, C. M. Gordon, C. Hilgers, M. J. Muldoon, and I. R. Dunkin, *Chem. Commun.* 1186 (2001).

517. C. Carlini, M. Marchionna, A. M. R. Galletti, and G. Sbrana, *J. Mol. Catal. A: Chem.* **169**, 19 (2001); **169**, 79 (2001); *Appl. Catal., A* **210**, 173 (2001).

518. F. Benvenuti, C. Carlini, M. Marchionna, R. Patrini, A. M. R. Galletti, and G. Sbrana, *Appl. Catal., A* **204**, 7 (2000).

519. M. Itagaki, G. Suzukamo, and K. Nomura, *Bull. Chem. Soc. Jpn.* **71**, 79 (1998).

520. H. Sato, H. Tojima, and K. Ikimi, *J. Mol. Catal. A: Chem.* **144**, 285 (1999).

521. M. Wang, H. Zhu, M. Qian, C. Jia, and R. He, *Appl. Catal., A* **216**, 131 (2001).

522. B. Chiche, E. Sauvage, F. Di Renzo, I. I. Ivanova, and F. Fajula, *J. Mol. Catal. A: Chem.* **134**, 145 (1998).

523. B. Nkosi, F. T. T. Ng, and G. L. Rempel, *Appl. Catal., A* **158**, 225 (1997).

524. M. Golombok and J. de Bruijn, *Appl. Catal., A* **208**, 47 (2001).

525. A. S. Chellappa, R. C. Miller, and W. J. Thomson, *Appl. Catal., A* **209**, 359 (2001).

526. M. Golombok and J. de Bruijn, *Ind. Eng. Chem. Res.* **39**, 267 (2000).

527. R. S. Karinen, A. O. I. Krause, E. Y. O. Tikkanen, and T. T. Pakkanen, *J. Mol. Catal. A: Chem.* **152**, 253 (2000).

528. S. M. Silva, P. A. Z. Suarez, R. F. de Souza, and J. Dupont, *Polymer Bull.* **40**, 401 (1998).

529. M. D. Spencer, S. R. Wilson, and G. S. Girolami, *Organometallics* **16**, 3055 (1997).

530. R. A. Ligaube, J. Dupont, and R. F. de Souza, *J. Mol. Catal. A: Chem.* **169**, 11 (2001).

531. J.-S. Chang, S.-E. Park, Q. Gao, G. Férey, and A. K. Cheetham, *Chem. Commun.* 859 (2001).

532. C. Slugovc, K. Mereiter, E. Zobetz, R. Schmid, and K. Kirchner, *Organometallics* **15**, 5275 (1996).

533. J. Q. Wang, A. K. Dash, J. C. Berthet, M. Ephritikhine, and M. S. Eisen, *Organometallics* **18**, 2407 (1999).

534. A. Haskel, T. Straub, A. K. Dash, and M. S. Eisen, *J. Am. Chem. Soc.* **121**, 3014 (1999).

535. A. Haskel, J. Q. Wang, T. Straub, T. G. Neyroud, and M. S. Eisen, *J. Am. Chem. Soc.* **121**, 3025 (1999).

536. A. K. Dash and M. S. Eisen, *Org. Lett.* **2**, 737 (2000).

537. J. Li, H. Jiang and M. Chen, *J. Org. Chem.* **66**, 3627 (2001).

538. J. Yang and J. G. Verkade, *Organometallics* **19**, 893 (2000).

539. I. M. Abdelrehim, K. Pelhos, T. E. Madey, J. Eng., Jr., and J. G. Chen, *J. Mol. Catal. A: Chem.* **131**, 107 (1998).

540. S. Abbet, A. Sanchez, U. Heiz, W.-D. Schneider, A. M. Ferrari, G. Pacchioni, and N. Rösch, *J. Am. Chem. Soc.* **122**, 3453 (2000).

541. K. S. Whiteley, T. G. Heggs, H. Koch, R. L. Mawer, and W. Immel, in *Industrial Polymers Handbook: Products, Processes, Applications*, E. S. Wilks, ed., Wiley-VCH, Weinheim, 2001, Vol. 2, Chapter 5.

542. M. P. McDaniel, in *Handbook of Heterogeneous Catalysis*, G. Ertl, H. Knözinger, and J. Weitkamp, eds., Wiley-VCH, Weinheim, 1997, Chapter 4.13.1, p. 2400.

543. W. Kaminsky and M. Arndt, in *Handbook of Heterogeneous Catalysis*, G. Ertl, H. Knözinger, and J. Weitkamp, eds., Wiley-VCH, Weinheim, 1997, Chapter 4.13.2, p. 2405.

544. V. A. Babkin, G. E. Zaikov, and K. S. Minsker, *Quantum Chemical Aspects of Cationic Polymerization of Olefins*, Nova Scinece, Commack, 1997.

545. V. Dragutan and R. Streck, *Catalytic Polymerization of Cycloolefins: Ionic, Ziegler–Natta and Ring-Opening Metathesis Polymerization*, Elsevier, Amsterdam, 2000.

546. *Industrial Polymers Handbook: Products, Processes, Applications*, E. S. Wilks, ed., Wiley-VCH, Weinheim, 2001.

547. *Catalyst Design for Tailor-Made Polyolefins*, Studies in Surface Science and Catalysis, Vol. 89, K. Soga and M. Terano, eds., Kodansha, Tokyo and Elsevier, Amsterdam, 1994.

548. *The Chemistry of Free Radical Polymerization*, G. Moad and D. H. Solomon, eds., Pergamon, Oxford, 1995.

549. *Cationic Polymerization: Fundamentals and Applications*, ACS Symp. Ser. Vol. 665, R. Faust and T. D. Shaffer, eds., American Chemical Society, Washington, DC, 1997.

550. *Application of Anionic Polymerization Research*, ACS Symp. Ser. Vol. 696, R. P. Quirk, ed., American Chemical Society, Washington, DC, 1998.

551. *Controlled Radical Polymerization*, ACS Symp. Ser. Vol. 685, K. Matyjaszewski, ed., American Chemical Society, Washington, DC, 1998.

552. *Olefin Polymerization*, ACS Symp. Ser. Vol. 749, P. Arjunan, J. E. McGrath, and T. L. Hanlon, eds., American Chemical Society, Washington, DC, 2000.

553. *Controlled/Living Radical Polymerization: Progress in ATRP, NMP, and RAFT*, ACS Symp. Ser. Vol. 768, K. Matyjaszewski, ed., American Chemical Society, Washington, DC, 2000.

554. H. H. Brintzinger, D. Fischer, R. Mülhaupt, B. Rieger, and R. M. Waymouth, *Angew. Chem., Int. Ed. Engl.* **34**, 1143 (1995).

555. M. Bochmann, *J. Chem. Soc., Dalton Trans.* 255 (1996).

556. W. Kaminsky and M. Arndt, *Adv. Polym. Sci.* **127**, 143 (1997).

557. M. Hackmann and B. Rieger, *CatTech* **1**, 79 (1997).

558. W. Kaminsky, *J. Chem. Soc., Dalton Trans.* 1413 (1998); *Catal. Today*, **62**, 23 (2000).

559. H. G. Alt and E. Samuel, *Chem. Soc. Rev.* **27**, 323 (1998).

560. H. G. Alt, *J. Chem. Soc., Dalton Trans.* 1703 (1999).

561. W. Kaminsky and A. Laban, *Appl. Catal., A* **222**, 47 (2001).

562. *Ziegler Catalysts: Recent Scientific Innovations and Technological Improvements,* G. Fink, R. Mülhaupt, and H. H. Brintzinger, eds., Springer, Berlin, 1995.

563. *Metallocenes*, A. Togni and R. L. Halterman eds., Wiley, VCH, Weinheim, 1998.

564. *Metallocene-Based Polyolefins: Preparation, Properties, and Technology*, J. Scheirs and W. Kaminsky, eds., Wiley, Chichester, 2000.

565. *Organometallic Catalysts and Olefin Polymerization*, R. Blom, A. Follestad, E. Rytter, M. Tilset, and M. Ystenes, eds., Springer, Berlin, 2001.

566. *Metallocenes and Single Site Olefin Catalysts*, R.-F. Jordan, ed.; *J. Mol. Catal. A: Chem.* **128**, 1–337 (1998).

567. *Advances in Polymerization Catalysis, Catalysts and Processes*, T. J. Marks and J. C. Stevens, eds.; *Top. Catal.* **7**, 1–208 (1999).

568. *Frontiers in Metal-Catalyzed Polymerization*, J. A. Gladysz, ed.; *Chem. Rev.* **100**, 1167–1681 (2000).

569. H. G. Alt and A. Köppl, *Chem. Rev.* **100**, 1205 (2000).

570. E. Y.-X. Chen and T. J. Marks, *Chem. Rev.* **100**, 1391 (2000).

571. M. Mager, S. Becke, H. Windisch, and U. Denninger, *Angew. Chem., Int. Ed. Engl.* **40**, 1898 (2001).

572. J. A. Ewen, *J. Mol. Catal. A: Chem.* **128**, 103 (1998).

573. G. W. Coates, *Chem. Rev.* **100**, 1223 (2000).

574. L. Resconi, L. Cavallo, A. Fait, and F. Piemontesi, *Chem. Rev.* **100**, 1253 (2000).

575. M. Bochmann, G. J. Pindado, and S. J. Lancaster, *J. Mol. Catal. A: Chem.* **146**, 179 (1999).

576. A. L. McKnight and R. M. Waymouth, *Chem. Rev.* **98**, 2587 (1998).

577. G. G. Hlatky, *Chem. Rev.* **100**, 1347 (2000).

578. S. B. Roscoe, J. M. J. Fréchet, J. F. Walzer, and A. J. Dias, *Science* **280**, 270 (1998).

579. G. Fink, B. Steinmetz, J. Zechlin, C. Przybyla, and B. Tesche, *Chem. Rev.* **100**, 1377 (2000).

580. D. Harrison, I. M. Coulter, S. Wang, S. Nistala, B. A. Kuntz, M. Pigeon, J. Tian, and S. Collins, *J. Mol. Catal. A: Chem.* **128**, 65 (1998).

581. J. Tudor, L. Willington, D. O'Hare, and B. Royan, *Chem. Commun.* 2031 (1996).

582. F. T. Edelmann, S. Gießmann, and A. Fischer, *Chem. Commun.* 2153 (2000).

583. A. S. Abu-Surrah and B. Rieger, *Angew. Chem., Int. Ed. Engl.* **35**, 2475 (1996).

584. G. J. P. Britovsek, V. C. Gibson, and D. F. Wass, *Angew. Chem., Int. Ed. Engl.* **38**, 429 (1999).

585. S. D. Ittel, L. K. Johnson, and M. Brookhart, *Chem. Rev.* **100**, 1169 (2000).

586. S. Mecking, *Coord. Chem. Rev.* **203**, 325 (2000); *Angew. Chem., Int. Ed. Engl.* **40**, 534 (2001).

587. G. J. P. Britovsek, V. C. Gibson, B. S. Kimberley, P. J. Maddox, S. J. McTavish, G. A. Solan, A. J. P. White, and D. J. Williams, *Chem. Commun.* 849 (1998).

588. B. L. Small, M. Brookhart, and A. M. A. Bennett, *J. Am. Chem. Soc.* **120**, 4049 (1998).

589. G. J. P. Britovsek, M. Bruce, V. C. Gibson, B. S. Kimberley, P. J. Maddox, S. Mastroianni, S. J. McTavish, C. Redshaw, G. A. Solan, S. Strömberg, A. J. P. White, and D. J. Williams, *J. Am. Chem. Soc.* **121**, 8728 (1999).

590. C. Wang, S. Friedrich, T. R. Younkin, R. T. Li, R. H. Grubbs, D. A. Bansleben, and M. W. Day, *Organometallics* **17**, 3149 (1998).

591. T. R. Younkin, E. F. Connor, J. I. Henderson, S. K. Friedrich, R. H. Grubbs, and D. A. Bansleben, *Science* **287**, 460 (2000).

592. A. Held, F. M. Bauers, and S. Mecking, *Chem. Commun.* 301 (2000).

593. A. Held and S. Mecking, *Chem. Eur. J.* **6**, 4623 (2000).

594. B. Liu and M. Terano, *J. Mol. Catal. A: Chem.* **172**, 227 (2001).

595. D. Fregonese, S. Mortara, and S. Bresadola, *J. Mol. Catal. A: Chem.* **172**, 89 (2001).

596. T. J. Pullukat and R. E. Hoff, *Catal. Rev.-Sci. Eng.* **41**, 389 (1999).

597. Y. V. Kissin, *J. Catal.* **200**, 232 (2001).

598. Y. Aoki, S. Fujita, S. Haramizu, K. Akagi, and H. Shirakawa, *Synth. Metals* **84**, 307 (1997).

599. K. Akagi, G. Piao, S. Kaneko, K. Sakamaki, H. Shirakawa, and M. Kyotani, *Science* **282**, 1683 (1998).

600. Y. Kishimoto, P. Eckerle, T. Miyatake, M. Kainosho, A. Ono, T. Ikariya, and R. Noyori, *J. Am. Chem. Soc.* **121**, 12035 (1999).

601. H. Hori, C. Six, and W. Leitner, *Appl. Organomet. Chem.* **15**, 145 (2001).

602. M. Tabata, T. Sone, and Y. Sadahiro, *Macromol. Chem. Phys.* **200**, 265 (1999).

603. X. Zhan and M. Yang, *J. Mol. Catal. A: Chem.* **169**, 27 (2001).

604. X. Zhan, M. Yang, and H. Sun, *J. Mol. Catal. A: Chem.* **169**, 63 (2001).

EMERGING AREAS
AND TRENDS

Some of the major challenges humanity faces in the new century are diminishing natural resources, environmental pollution, and assumed global climate change, which could seriously affect our everyday lives. Some problems have been created by chemistry itself by creating toxic chemicals and hazardous wastes, but fortunately, chemists can also contribute to find viable solutions. The future goal in the broadest sense is to reduce simultaneously the depletion of raw materials and the generation of waste. An ideal chemical reaction should be selective (chemo-, regio-, stereoselective) and, at the same time, satisfy efficiency in terms of atom economy. *Atom economy* refers to the efficiency of a reaction based on the number of atoms of reactants that are transformed to products.[1] A simple addition requiring any other chemicals only in catalytic amounts is an ideal example. Global warming, considered to be caused in part by excessive burning of fossil fuels, is another issue of substantial importance. Recycling of excess carbon dioxide, that is, the transformation of CO_2 to hydrocarbons and derived useful products, can contribute to replenish diminishing natural resources and, at the same time, mitigate global warming.

Hydrocarbon chemistry is heavily involved in all areas we selected as examples. In this chapter we discuss emerging trends and some of the most important approaches promising new solutions to this field with emphasis on and specific examples relevant to hydrocarbon chemistry.

14.1. GREEN CHEMISTRY

Many of the processes operated in industry generate extensive waste released to the environment as liquid effluents or volatile organic molecules. For humanity to survive in the twenty-first century it is essential to change the present-day industrial technologies used in the refining–petrochemical industry in mass production of various chemicals and products. This, however, will also require societal changes to convert our mass production and mass consumption based society into an environmentally friendly, recycling-based, sustainable society living in harmony with the environment. New technologies using starting materials, reactions, and processes of low environmental loading—that is, with reduced waste amount and energy consumption—must be developed. This has now become known as *green chemistry* and defined by Anastas and Warner in their book[2] as "The utilization of a set of principles that reduces or eliminates the use or generation of hazardous substances in the design, manufacture and application of chemical products." Green chemistry, also called "clean" or "sustainable chemistry," includes green starting materials, reactions, reagents, solvents, reaction conditions, and products. If pursued properly, green chemistry may offer environmental benefits as well as significant savings and economic advantages.

Green chemistry has developed rapidly since the early 1990s and finds increasing attention and application in industry. Its principles and examples of application are summarized in books.[2,3] Green chemistry was the topic of journal special issues[4,5] and various symposia,[6–10] and The Royal Society of Chemistry launched a new periodical with the same title. Issues and actions with respect to the oil industry were discussed.[11]

The field is rapidly developing, and the most important issues also relevant to hydrocarbon chemistry include the development of solventless reactions, use of alternative solvents, and, most important of all, replacement of old technologies with clean catalytic processes. Combinatorial chemistry, a method to produce a large number of chemical compounds rapidly on a small scale, can be regarded as a particularly useful tool from the point of view of environmental protection.

14.1.1. Chemistry in Nontraditional Reaction Media

Chemistry, in a large part besides gas-phase and solid-state chemistry, is solution chemistry. Organic chemistry, including homogeneous catalytic processes and organometallic chemistry, uses organic liquids as the reaction medium. Many of the commonly used solvents, however, are volatile organic compounds and may have environmentally harmful and hazardous properties. Contamination of aqueous effluents and substantial releases into the air are of serious concern. Environmentally benign processes, consequently, should use alternative solvents (nontraditional reaction media) or be run under solventless conditions.

Water. Water long used mainly for aqueous, generally ionic chemistry, more recently became attractive to replace organic solvents particularly under supercritical conditions (see discussion below). It is abundant, easy to purify, not flammable or toxic, and easily available in large quantities, and, therefore, its use is most economical. Product isolation in many cases can be a simple phase separation, and because of its large heat capacity, water offers an easy temperature control. The hydrophobic effect, that is, an increase in reaction rate due to the low solubility of certain organic compounds in water, can be an additional advantage. Developments and trends for the use of water for catalytic processes are found in books and reviews[6,12–17] and thematic issues[18–20] including an issue on water technologies.[20]

A particular example is the Diels–Alder reaction,[21,22] which attracted interest after a publication[23] reporting large rate enhancement and a change in stereochemistry in water for the reaction between cyclopentadiene and methyl vinyl ketone. Hydrophobic interactions between the reacting molecules and hydrogen bonding of water to the carbonyl moiety were found to play the major role in the observed phenomena.[14,24,25]

Fluorous Solvents. The *fluorous phase* is an alternative to the aqueous phase.[26] Fluorous (perfluorinated) solvents, such as perfluoroalkanes, perfluoroalkyl ethers, and perfluorinated tertiary amines, have been recognized to be extremely stable and nontoxic and have high density, low solvent strength, and extremely low solubility in water and organic materials.

Fluorous solvents proved to be highly effective in epoxidation of alkenes. H_2O_2 can be used in combination with trifluoroacetone,[27] perfluoroacetone,[28] or a mixture of perfluoroacetone and hexafluoro-2-propanol.[29] In fluorinated alcohols as solvents uncatalyzed epoxidations with aqueous H_2O_2 are performed.[30,31]

High activity in the hydrogenation of alkenes was found using fluorous RhCl[P$\{C_6H_4$-*p*-SiMe$_2(CH_2)_2C_nF_{n+1}\}_3]_3$ complexes.[32] Perluorinated liquids were found to be a good alternative medium for Lewis acid–catalyzed Friedel–Crafts reactions (acetylation of benzene and alkylbenzenes).[33]

Gladysz introduced perfluorinated "tails" into organometallic catalyst systems, which without a fluorous phase can serve as the equivalent of fluorous catalysts.[34]

Ionic Liquids. Since the late 1990s there has been a growing interest in ionic liquids, as solvent and catalyst for certain organic reactions.[35–38] Ionic liquids are salts of low melting points. They have high chemical and thermal stability and negligible vapor pressure. They are polar but weakly coordinating, and good solvents for varied systems, including transition-metal complexes. The strong polarity and high electrostatic fields of these materials usually bring about enhanced activity. Easy recycling is an additional important benefit.

Pyridinium polyhydrogen fluoride and related polyhydrogen fluorides were developed as a modified liquid complex of anhydrous hydrogen fluoride for fluorination reactions.[39] These complexes are ionic liquids. It was subsequently found[40] that such ionic liquids with higher complexed HF content are advantageous reactions media for alkylation and other reactions.

Significant developments were achieved with the discovery in the 1970s and 1980s of varied room-temperature ionic liquids.[41,42] These were organoaluminate ionic liquids, typically a mixture of quaternary ammonium salts with aluminum chloride. A major breakthrough came in 1992 by the discovery of air- and moisture-stable ionic liquids.[43] 1,3-Dialkylimidazolium cations (**1**), specifically, [bemim]$^+$ (1-butyl-3-methylimidazolium) and [emim]$^+$ (1-ethyl-3-methylimidazolium) ions, with BF_4^- and PF_6^- anions are the most widely explored and used:

$$Me - N \overset{(+)}{\underset{\underset{1}{\smile}}{N}} - R \quad X^-$$

Their utility has been demonstrated in various Friedel–Crafts reactions,[44–47] halogen addition,[48,49] electrophilic nitration of aromatics,[50] and various hydrogenation processes[51–53] including the Ru-catalyzed hydrogenation of CO_2 to N,N-dipropyl formamide in supercritical CO_2 under biphasic conditions.[54] The use of in situ IR spectroscopy allowed Horváth and coworkers to demonstrate that the same substrate–catalyst and product–catalyst complexes are produced in Friedel–Crafts acetylation of benzene in ionic liquids as in CH_2Cl_2.[55]

Promising results were also observed in hydroformylation in ionic liquids.[56,57] Reaction rate, selectivity, and the retention of Rh could be optimized by fitting the nature of the anions and cations of the ionic liquid and the modified ligands.[58] A practical recycling method for Jacobsen's chiral Mn(III) salen epoxidation catalyst involving the use of the [bmim][PF$_6$] was developed.[59] Oligomerization of 1-alkenes with a cationic Ni complex[60] and that of 1,3-butadiene-catalyzed Pd compounds[61] showed specific features.

Supercritical Solvents. Chemistry performed in supercritical fluids represents a green approach, which combined with other environmentally friendly processes can provide great benefits.[62,63] Potential advantages that a supercritical medium may offer are higher diffusivities, better heat transfer, high concentration of reactant gases, elimination of transport limitations existing in multiphase systems, and easier product separation. A particularly important property is the tunability of such systems from gaslike to liquidlike. Disadvantages are materials compatibility and the elevated pressure required. Earlier reviews[64,65] and a thematic issue[66] treat all aspects of the field and reviews focus on both homogeneous[67–69] and heterogeneous[69–71] catalytic applications. The most widely studied media are supercritical CO_2 and water.

A significant observation was made in 1994 to show a marked increase in the rate of hydrogenation of CO_2 to formic acid derivatives in supercritical CO_2,[72] which could be further enhanced by adding appropriate solvents.[73–75] Photoreduction[76] and hydrogenation in supercritical CO_2 combined with ionic liquids[54] were also performed.

Promising results were observed in Friedel–Crafts alkylation[77] and epoxidation.[78] Higher rates or better selectivities were found for hydroformylations in supercritical CO_2.[79–84] Simple trialkyl phosphines, for examples, were shown to provide highly active Rh catalysts.[81] Hydroboration showed enhanced regioselectivity.[85] The Wacker reaction performed in alcohol–supercritical CO_2 exhibits high reaction rates and markedly increased selectivity toward methyl ketone.[86]

Results with respect to polymerization are summarized in a review.[87] Norbornene was shown to give high-*cis* polymer in supercritical carbon dioxide[88] in contrast to the high-*trans* polymer formed in protic solvents. Supercritical conditions enhanced selectivities in oligomerization[89] and polymerization processes.[90]

The yield of the catalytic dehydrogenation of C_{10}–C_{14} *n*-alkanes over Pt–Sn/Al_2O_3 was 2–3 times higher in the supercritical phase of the reactant themselves than in the gas phase. This indicates a strong shift of the thermodynamic equilibrium under supercritical conditions.[91] High catalytic activity was demonstrated for the formation of linear alkylbenzenes (alkylation of *p*-xylene with 1-dodecene) using $CF_3(CF_2)_7SO_3H$ in supercritical fluid reaction media; alkylation activity was highly enhanced in CHF_3 compared to CO_2 (90% conversion vs. 1%).[92]

Three reviews discuss the reactivity of organic compounds in superheated water both sub- and supercritical.[93–95] Oxidation in supercritical water of organic compounds is a safe and effective technology, which has been of increasing interest. The aim is to treat organic wastes and organic-containing wastewaters to achieve total oxidation [supercritical water oxidation (SCWO)]. Under sub- to supercritical conditions, organic materials have greatly enhanced solubility in water, which is due to the loss of the hydrogen bonding structure of water. Chemical synthesis in supercritical water, however, has not been very attractive, which is due mainly to the high investment cost of working at the extreme conditions (the critical point of water is 374°C at 218 atm).[61,96] Examples relevant to hydrocarbon chemistry include heterogeneous catalytic partial oxidation of alkylaromatics,[97] methane,[98] propane,[99] cyclohexane,[100] and hexadecane and polyethylene,[101] and the hydration of phenylacetylene to acetophenone.[102] Very effective catalytic reduction of alkenes and alkynes in subcritical water over Pd/C by formate was demonstrated.[103] The Diels–Alder reaction in supercritical water leads to high yields of clean product without added catalysts.[104]

14.1.2. New Catalyst Immobilization or Recovery Strategies

The main disadvantage of the use of homogeneous catalysts as compared to heterogeneous ones is catalyst recovery. Whereas heterogeneous (solid) catalysts are easily removed from the reaction mixture by simple filtration, homogeneous (soluble) catalysts are difficult to separate from the substrate and reaction product(s). This is particularly important when expensive catalysts are involved, since economical operation necessitates complete catalyst recovery and recycling. A successful solution to this problem is the use of liquid multiphase catalysis. Specifically, the catalyst is dissolved in one of two (or more) immiscible liquid phases, whereas the substrate(s) and product(s) are found in the other phase(es). The idea was first articulated by Manassen[105] and has become an important and attractive research field.[106] The simplest case is a biphasic system such as that illustrated in Figure 14.1.

An easy and simple possibility is the application of water-soluble catalysts in water and an organic phase. The substrate(s) are used as neat liquids or are dissolved in the organic solvent immiscible with water. This protocol is called

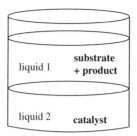

Figure 14.1. General illustration of a liquid–liquid biphasic catalysis system.

aqueous biphasic catalysis.[12,13] After completion of the reaction, the aqueous phase is simply separated (decanted) and recycled. In contrast to the traditional homogeneously catalyzed reactions where distillation is usually necessary to recover the catalyst, aqueous biphasic catalysis offers considerable energy savings. On the other hand, interaction of the catalyst and substrate occurs at the liquid–liquid interface. Efficient stirring or some other method, therefore, is necessary to achieve satisfactory reaction rates. The work by Joó and coworkers[107,108] and the successful industrial application in the hydroformylation of propylene to butanal[109–111] (the Ruhrchemie–Rhône Poulenc process discussed in Section 7.1.3) mark the beginning of this rapidly developing area.[13,14] Historically, however, the Shell higher olefin process (SHOP)[112–114] (see Section 13.1.3) was the first commercial process applying the multiphase catalysis concept. The product α-olefins formed in the oligomerization of ethylene induced by a Ni catalyst with phosphine ligands is not soluble in 1,4-butanediol applied as solvent but separates and is decanted. Appropriately modified water-soluble ligands, specifically, sulfonated triphenylphosphines, are applied in the hydroformylation process. A new type of ligands, namely, water-soluble Ir and Rh complexes with tris(hydroxymethyl)phosphine[115] and a cationic sugar-substituted Rh complex,[116] have been described for hydroformylation.

Aqueous biphasic catalysis is also used in homogeneous hydrogenations.[117–119] In new examples Ru clusters with the widely used TPPTN [tris(3-sulfonatophenyl) phosphine] ligand[120] and Rh complexes with novel carboxylated phosphines[121] were applied in alkene hydrogenation, whereas Ru catalysts were used in the hydrogenation of aromatics.[122,123] Aerobic oxidation of terminal alkenes to methyl ketones was carried out in a biphasic liquid–liquid system by stable, recyclable, water-soluble Pd(II) complexes with sulfonated bidentate diamine ligands.[124]

Fluorous biphasic catalysis is a rapidly developing alternative. Since the initial archival publication by Horváth and Rábai[125] indicating the tremendous potential of this approach, many homogeneous catalytic reactions were carried out in fluorous biphasic system.[126–132] In this case, the homogeneous catalyst is modified with long-chain fluoroponytails to render them soluble in the fluorous phase. To overcome the high electron-withdrawing character of the perfluoroalkyl group attached to the metal center, which may alter electronic properties and reactivities of the

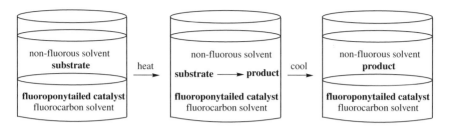

Figure 14.2. Phase changes in a fluorous biphasic catalysis system.

catalyst, the use of a C_2 or C_3 spacer is necessary. Typical fluoroponytails with spacer groups are $CF_3(CF_2)_nCH_2CH_2-$ and $CF_3(CF_2)_nCH_2CH_2CH_2-$. In many cases the fluorous biphasic reaction mixture becomes homogeneous at higher temperature; that is, the catalytic transformation can be accomplished as a truly homogeneous reaction followed by liquid–liquid separation at ambient temperature (Fig. 14.2).

The fluorous biphasic catalysis concept was successfully demonstrated first by hydroformylation of 1-decene carried out in perfluoromethylcyclohexane and toluene, which forms a homogeneous liquid phase at 100°C in the presence of catalyst **2** prepared in situ according to Eq. (14.1):[125,133]

$$\begin{array}{c} PH_3 \\ + \\ CH_2{=}CH(CF_2)_5CF_3 \end{array} \xrightarrow[80-85°C,\ 24\ h]{AIBN} \underset{53\%}{P[CH_2CH_2(CF_2)_5CF_3]_3} \xrightarrow[\substack{2.74\ MPa\ CO/H_2\ (1:1),\\ C_6F_{11}CF_3,\ 80°C,\ 2\ h}]{Rh(CO)_2(acac)}$$

$$\xrightarrow{} \underset{\textbf{2}}{HRh(CO)\{P[CH_2CH_2(CF_2)_5CF_3]_3\}_3} \qquad (14.1)$$

Similar useful characteristics were found for catalysts prepared by fluorous polymer ligands.[134] High reaction rates were found for the Rh-catalyzed hydroformylation of higher alkenes with fluoroponytail-modified arylphosphines in supercritical CO_2.[84]

Hydroboration of a variety of alkenes and terminal alkynes with catecholborane in the fluorous solvent perfluoromethylcyclohexane was performed using fluorous analogs of the Wilkinson catalyst.[135,136] Recycling of a rhodium-based alkene hydrosilylation catalyst was also successful.[137] Activated aromatics and naphthalene showed satisfactory reactivity in Friedel–Crafts acylation with acid anhydrides in the presence of Yb tris(perfluoroalkanesulfonyl)methide catalysts.[138]

More examples are found for varied oxidation processes mainly for various epoxidations carried out by metal catalysts bearing F-modified ligands, such as porphyrins,[139] Ru perfluoroacetylacetonate salt,[140] or salen complexes,[141,142] or using the **3** selenium compound as catalyst.[143] The potential for enantioselective transformations offering an easy way to recover precious chiral reagents and catalysts was demonstrated in enantioselective epoxidation using fluorous chiral salen manganese complexes (**4**).[142] Wacker oxidation of various alkenes to the corresponding ketones was developed using *tert*-BuOOH in the presence of catalyst **5** bearing perfluorinated ligands.[144]

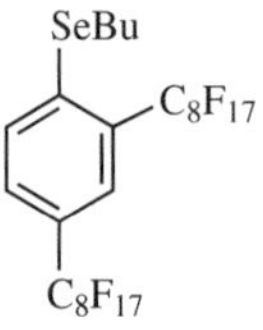

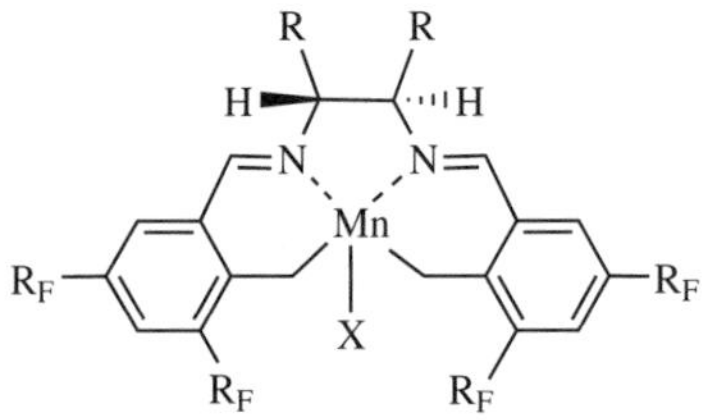

RF = C8F17, 3,5-bis(C8F17)phenyl,

2,3,4-tris[C8F17(CH2)3O]phenyl

X = C7F15COO⁻

4

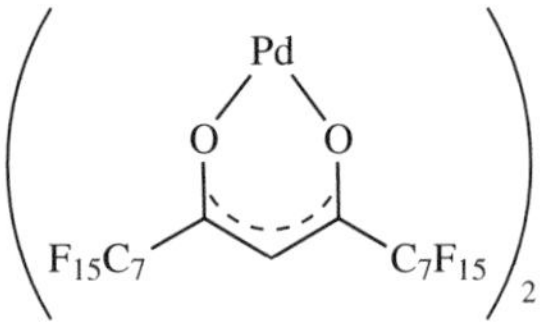

tert-BuOOH and transition-metal salts combined with macrocyclic ligands **6** and **7** are active in allylic oxidations.[145,146]

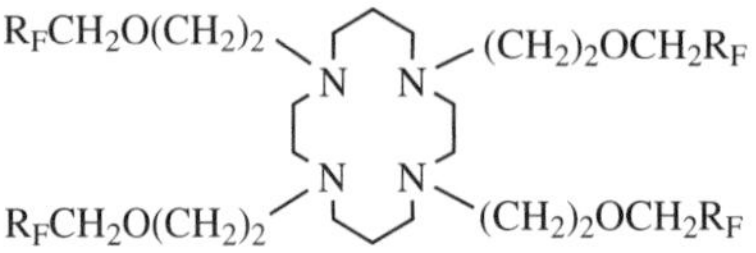

RF = CF2(OCFCF2)q(OCF2)pOCF3
            |
           CF3
        p = 0.11; q = 3.38

6

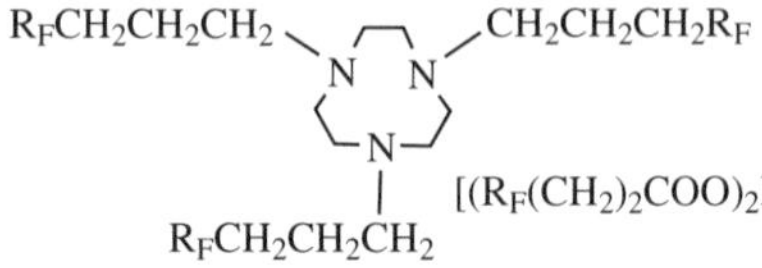

RF = C8F17

7

The fluorous biphase concept has been successfully applied in the hydrogenation of alkenes by various fluorous analogs of the Wilkinson catalyst, although catalyst activities are seldom comparable to the best homogeneous catalysts.[147–149] The usefulness of the concept was also proved in the oligomerization of ethylene in the presence of a Ni catalyst with appropriate fluorinated ligands.[150] In contrast, oligomerization of propylene with fluorinated Ni precursors was less successful because of the migration of the catalyst toward the hydrocarbon phase, generated by the formation of the oligomeric products.[151]

14.1.3. New Catalysts and Catalytic Processes

An efficient way in green chemistry to achieve waste minimization is to replace processes requiring stoichiometric reagents by catalytic methods. This is a particularly important issue in the fine chemical and pharmaceutical industry where the E factor—the amount of waste generated per kilogram of product—is particularly high. Stoichiometric reductions, oxidations, heterosubstitution reactions, and other reactions produce high amounts of inorganic salts. Catalytic methods are environmentally attractive by reducing waste and energy consumption, and offering simpler technologies. Although about 90% of the chemical processes applied in industry are catalytic transformations, there is an ever-increasing demand to develop more active and more selective catalysts.

Environmental catalysis was the topic of a special symposium;[152] a journal is devoted to the subject,[153] and reviews are also available.[154–159]

Examples of catalytic reactions and processes relevant to hydrocarbon chemistry are numerous. The technologies of the oil refinery with extremely low (<0.1) E factors are excellent examples demonstrating the possibilities that can be achieved by the development of selective catalytic methods, particularly by the use of various solid acids (see detailed discussions in Chapter 2). Further examples of commercially highly successful processes are the oxidation catalyst TS-1 developed by Enichem researchers[160,161] (see Sections 9.1.1, 9.2.1, and 9.4.1), the homogeneous aqueous-phase Rh-catalyzed hydroformylation (see Sections 7.1.3 and 7.4.1), and single-site metallocene polymerization catalysts, which allow the preparation of tailored polymers with new properties (see Sections 13.3.2).[162–164]

In addition to large-scale industrial applications, solid acids, such as amorphous silica–alumina, zeolites, heteropoly acids, and sulfated zirconia, are also versatile catalysts in various hydrocarbon transformations. Zeolites are useful catalysts in fine-chemical production (Friedel–Crafts reactions, heterosubstitution).[165–168] Heteropoly compounds have already found industrial application in Japan, for example, in the manufacture of butanols through the hydration of butenes.[169] These are water tolerant, versatile solid-phase catalysts and may be used in both acidic and oxidation processes, and operate as bifunctional catalysts in combination with noble metals.[158,170–174] Sulfated zirconia and its modified versions are promising candidates for industrial processes if the problem of deactivation/reactivation is solved.[175–178]

A monograph by Sheldon and van Bekkum gives a detailed treatment of heterogeneous catalytic processes applied in the fine-chemical industry.[179] Clay materials proved to be extremely useful catalysts in organic chemistry.[180,181] A particular example is clayzic,[161,182] a highly active catalyst in various Friedel–Crafts reactions.

It is appropriate and very instructive to briefly discuss a relatively new and very successful approach, namely, the development of catalysts with designed and atomically engineered active centers. Thomas and coworkers used micro- and mesoporous solids and carried out delicate structural and compositional variations to prepare specific catalysts capable of promoting regioselective, shape-selective, and enantioselective conversions.[183–185] This strategy resulted in the development of framework-substituted CoALPO-18 and MnALPO-18 molecular sieves for the selective aerobic oxidation of linear alkanes to the corresponding monocarboxylic acids,[186] and that of hexane to adipic acid.[187] Framework-substituted MALPO-36

(M = Co or Mn) exhibit good conversions and selectivities in the aerobic epoxidation of alkenes.[188] A uniquely powerful alkene epoxidation catalyst was also developed by grafting Ti(IV) ions on to MCM-41 mesoporous silica.[189,190] Selective oxidation of cyclohexane to cyclohexanol or adipic acid can be accomplished by various FeALPO catalysts.[191,192] A cobalt complex immobilized into MCM-41, in turn, shows high selectivity in the formation of cycylohexanone.[193] Thomas' group also designed durable, high-performance bimetallic catalysts such as Pt–Ru and Ru–Sn for the hydrogenation of carbon–carbon multiple bonds.[194,195]

The increasing demand to develop new, more effective, and more selective catalysts, however, require the use of new, more efficient methods. The traditional trial-and-error approach and individual testing of a large number of catalyst formulations one by one are time-consuming and inefficient. Combinatorial and high-speed screening techniques, which have revolutionized drug discoveries, are increasingly applied in the development of new catalytic materials. Called *combinatorial catalysis*, this is the systematic preparation, processing, and testing of catalyst libraries in highly effective high-throughput fashion allowing characterization and performance analysis of a large number of samples. Reproducibility of the fast parallel or automated sequential synthesis is a key element of high-throughput screening. Furthermore, it requires high-speed screening methods. Three specific techniques have been proposed and found to be appropriate: (1) optical screening using IR thermography or fluorescence indicators, (2) microprobe sampling and scanning mass spectrometry, and (3) resonance enhanced multiphoton ionization. Reviews covering the field of both homogeneous and heterogeneous catalysis[196–207] and thematic issues[208,209] are available.

Array microreactors and mass spectrometry were used to find the best composition of Pt–Pd–Ir for the dehydrogenation of cyclohexane to benzene.[210,211] A catalyst library of 37 binary alloys composed of transition metals, Cu, Zn, and Ti or Si, was screened by IR thermography in the hydrogenation of 1-hexyne, and the oxidation of isooctane and toluene.[212]

Combinatorial methods were also applied in the discovery of new catalysts for the low-temperature oxidation[213] and oxidative dehydrogenation of propane.[214] A 144-member catalyst library was screened by photothermal deflection spectroscopy and mass spectrometry to find the most active compositions of V–Al–Nb and Cr–Al–Nb oxides for the oxidative dehydrogenation of ethane.[215] The ternary combination V(45)–Sn(45)–Mo(10)–O selected by laser-induced fluorescence imaging gave much higher yield than did V_2O_5 in the selective oxidation of naphthalene to naphthoquinone.[216]

A high-throughput colorimetric assay was applied to identify catalysts by combining metals (Pd, Rh, Ru, Ir) and various phosphines for the hydroamination of dienes.[217] Combinatorial catalysis was successfully used to find active catalysts in the Ru-catalyzed ring-closing metathesis reaction[218] and the olefin polymerization by Ni and Pd.[219]

New catalysts were also discovered by combinatorial methods for epoxidation, specifically, diamine Mn complexes,[220] highly active Fe[221] and salen complexes,[222] and titanium–silsesquioxane.[223]

14.2. CARBON DIOXIDE RECYCLING TO HYDROCARBONS

In nature cycles of ice ages were always followed by subsequent warming periods, long before humans appeared. Overpopulation of the Earth (having surpassed 6 billion) certainly has an effect, but human activity is not the only, or even major, cause for climate change. There is a close relationship between CO_2 content of the atmosphere and temperature (going back to Arrhenius' paper of 1898), but this does not necessarily prove that it is mainly CO_2 that drives up the temperature. When we are in a warming cycle (as we fortunately are), there are recognized natural reasons why the CO_2 content increases. At the same time human activity may increase the natural effects. Looking at the composition of air, carbon dioxide, although increased by some 15% in the twentieth century, is only about 350 ppm (or 0.035%). At the same time, argon is 0.9%, water vapor can reach a few percentages. It is not necessarily CO_2 (or methane), but water vapor that causes (due to its high concentration) the dominant greenhouse effect of our atmosphere. We should however, mitigate man-made overproduction of CO_2.

Whereas our fossil fuel (hydrocarbon) resources are going to diminish and will become not only scarcer but also much more expensive, it must be kept in mind that synthetic (human-made) effects due to increasing CO_2 emissions due to burning fossil fuels will also diminish. Atomic energy and some of the alternative energy sources (solar, hydro, wind, etc.) are not producing CO_2; thus the overall effect of climate change by synthetic CO_2 will start to diminish (except by causes of population growth).

The chemical transformation of carbon dioxide including catalytic reductions has attracted increasing attention.[224–234] Fixation of CO_2 by using it as a C_1 building block in chemical synthesis is of great practical significance. Interest in CO_2 chemistry and recycling is stimulated by environmental concerns and the abundant availability of CO_2. Carbon dioxide recycling not only produces useful chemicals but also helps alleviate the ecological problems and environmental impact of excessive burning of fossil fuels.

The hydrogenation of CO_2 to methane is a potentially important reaction relevant to carbon dioxide recycling to hydrocarbons. As discussed in detail in Section 3.2.1, this process, called *methanation*, can be accomplished by applying heterogeneous metal catalysts. Homogeneous hydrogenation by transition-metal complexes, in turn, goes characteristically to formic acid and derivatives. The viability of catalytic reductions, however, is hindered by the high cost of hydrogen. At present, therefore, there is no practical reason for producing methane from carbon dioxide and hydrogen.

The best known heterogeneous catalysts are oxide-supported Ru, Rh, and Ni, and Ru exhibits the highest selectivity.[235–237] Marked support effects are observed and TiO_2 is usually found to be the best support material.[238,239] Pd on zirconia and Ni on zirconia are particularly effective catalysts when prepared using amorphous Pd–Zr, Ni–Zr, and Ni-containing multicomponent alloys by controlled oxidation–reduction treatment[240–242] or generated under reaction conditions.[243–245] Stabilized metal nanoparticles of uniform dispersion embedded into the oxide matrix are the particular characteristics of these highly active and selective catalysts. Hybrid or composite catalyst systems composed of Fischer–Tropsch catalysts and metal oxides or zeolites are also effective in the hydrogenation of CO_2 to higher hydrocarbons. Gasoline-range hydrocarbons can be produced by coupling methanol synthesis and the methanol-to-gasoline reaction.[246–249] The ionic hydrogenation of CO_2 by sodium borohydride–triflic acid yields exclusively methane at ambient temperature and modest pressure (50 atm).[250]

A new promising possibility explored is the electrocatalytic reduction of CO_2 in water by passing through electric current in a reverse fuel cell to produce oxygenated methane derivatives such as formates, formaldehyde, and methanol.[251] Reduction to formates is efficient, but because of problems of overpotential, methane is preferred to methanol. Regardless, new approaches of carbon dioxide reduction, the energy for which is to be provided by atomic energy or alternative energy sources, are promising. Methanol can be considered as a reversible means of energy storage, more convenient to handle than hydrogen itself. Methanol then can also be converted to dimethyl ether, dimethoxymethane, trimethoxymethane, trioxymethylene, dimethyl carbonate, and the like.[252] Methanol or dimethyl ether, in turn, can subsequently also be used to make, via catalytic conversion, ethylene as well as propylene. These allow the preparation of gasoline range or aromatic hydrocarbons.[253]

REFERENCES

1. B. M. Trost, *Science* **254**, 1471 (1991); *Angew. Chem., Int. Ed. Engl.* **34**, 259 (1995).

2. P. T. Anastas and J. C. Warner, *Green Chemistry: Theory and Practice*, Oxford Univ. Press, Oxford, 1998.

3. *Green Chemistry: Frontiers in Benign Chemical Syntheses and Processes*, P. T. Anastas and T. C. Williamson, eds., Oxford Univ. Press, Oxford, 1998.

4. *Environmentally Friendly Reactions and Technologies*, I. Kiricsi and L. Guczi, eds.; *Appl. Catal., A* **189**, 151–276 (1999).

5. *Green Chemistry*, H. Dettwiler, ed.; *Chimia* **54**, 492–530 (2000).

6. *Green Chemistry*. ACS Symp. Ser. Vol. 626, P. T. Anastas and T. C. Williamson, eds., American Chemical Society, Washington, DC, 1996.

7. *Designing Safer Chemicals*, ACS Symp. Ser. Vol. 640, S. C. DeVito and R. L. Garrett, eds., American Chemical Society, Washington, DC, 1996.

8. *2nd World Congress of Environmental Catalysis*, Miami, 1998; papers presented were published in issues of *Catal. Today* [**54**, 379–574 (1999); **55**, 1–204 (2000)] edited by R. B. Subramanian, B. W. L. Jang, and J. J. Sipey.

9. *Intt. Symp. Green Chemistry*, New Delhi, 2001, J. H. Clark, ed.; *Pure Appl. Chem.* **73**, 1–203 (2001).

10. *IUPAC CHEMRAW XIV Confe. Green Chemistry*, Boulder, Co, 2001, D. L. Hjeresen and P. T. Anastas, eds.; *Pure Appl. Chem.* **73**, 1231–1386 (2001).

11. N. Y. Chen, *Chem. Innov.* **31**(4), 10 (2001).

12. *Aqueous Organometallic Chemistry and Catalysis*, I. T. Horváth and F. Joó, eds., Kluwer, Dordrecht, 1995.

13. *Aqueous-Phase Organometallic Catalysis*, B. Cornils and W. A. Herrmann, eds., Wiley-VCH, Weinheim, 1998.

14. C.-J Li, *Chem. Rev.* **93**, 2023 (1993).

15. G. Papadogianakis and R. A. Sheldon, in *Catalysis, a Specialist Periodical Report*, Vol. 13, J. J. Spivey, senior reporter, The Royal Society of Chemistry, Cambridge, UK, 1997, Chapter 5, p. 114.

16. J. B. F. N. Engberts and M. J. Blandamer, *Chem. Commun.* 1701 (2001).

17. F. Joó, *Aqueous Organometallic Catalysis*, Kluwer, Dordrecht, 2001.

18. *Catalysis in Water*, I. T. Horváth, ed.; *J. Mol. Catal. A: Chem.* **116**, 1–316 (1997).

19. *Phase Separable Homogeneous Catalysis*, B. E. Hanson and J. R. Zoeller, eds.; *Catal. Today* **42**, 371–478 (1998).

20. M. Campanati, G. Fornasari, and A. Vaccari, *Catal. Today* **53**, 1–158 (1999).

21. F. Fringuelli and A. Taticchi, *Dienes in the Diels-Alder Reaction*, Wiley-Interscience, New York, 1990.

22. W. Oppolzer, in *Comprehensive Organic Synthesis*, B. M. Trost and I. Fleming, eds., Vol. 5: *Combining C–C π-Bonds*, L. A. Paquette, ed., Pergamon Press, Oxford, 1991, Chapter 4.1.4.2.2, p. 344.

23. D. C. Rideout and R. Breslow, *J. Am. Chem. Soc.* **102**, 7816 (1980).

24. S. Otto and J. B. F. N. Engberts, *Pure Appl. Chem.* **72**, 1365 (2000).

25. U. Pindur, G. Lutz, and C. Otto, *Chem. Rev.* **93**, 741 (1993).

26. D. Curran and Z. Lee, *Green Chem.* **3**, G3 (2001).

27. L. Shu and L. Shi, *J. Org. Chem.* **65**, 8807 (2000).

28. P. A. Ganeshpure and W. Adam, *Synthesis* 179 (1996).

29. M. C. A. van Vliet, I. W. C. E. Arends, and R. A. Sheldon, *Synlett* 1305 (2001).

30. K. Neimann and R. Neumann, *Org. Lett.* **2**, 2861 (2000).

31. M. C. A. van Vliet, I. W. C. E. Arends, and R. A. Sheldon, *Synlett* 248 (2001).

32. B. Richter, A. L. Spek, G. van Koten, and B.-J. Deelman, *J. Am. Chem. Soc.* **122**, 3945 (2000).

33. H. Nakano and T. Kitazume, *Green Chem.* **1**, 179 (1999).

34. M. Wende, R. Meier, and J. A. Gladysz, *J. Am. Chem. Soc.* **123**, 11490 (2001).

35. T. Welton, *Chem. Rev.* **99**, 2071 (1999).

36. P. Wasserscheid and W. Keim, *Angew. Chem., Int. Ed. Engl.* **39**, 3772 (2000).

37. R. Sheldon, *Chem. Commun.* 2399 (2001).

38. C. M. Gordon, *Appl. Catal., A* **222**, 101 (2001).

39. G. A. Olah and X. Y. Li, in *Synthetic Fluorine Chemistry*, G. A. Olah, R. D. Chambers, and G. K. S. Prakash, eds., Wiley, New York, 1992, Chapter 8, and references cited therein.

40. G. A. Olah, U. S. Patent 3,708,553 (1973).

41. H. L. Chum, V. R. Koch, L. L. Miller, and R. A. Osteyrung, *J. Am. Chem. Soc.* **97**, 3264 (1975).

42. J. S. Wilkes, J. A. Levisky, R. A. Wilson, and C. L. Hussey, *Inorg. Chem.* **21**, 1263 (1982).

43. J. S. Wilkes and M. J. Zaworotko, *J. Chem. Soc., Chem. Commun.* 965 (1992).

44. C. E. Song, W. H. Shim, E. J. Roh, and J. H. Choi, *Chem. Commun.* 1695 (2000).

45. K. Qiao and Y. Deng, *J. Mol. Catal. A: Chem.* **171**, 81 (2001).

46. M. H. Valkenberg, C. deCastro, and W. F. Hölderich, *Appl. Catal., A* **215**, 185 (2001).

47. S. J. Nara, J. R. Harjani, and M. M. Salunkhe, *J. Org. Chem.* **66**, 8616 (2001).

48. C. Chiappe, D. Capraro, V. Conte, and D. Pieraccini, *Org. Lett.* **3**, 1061 (2001).

49. T. Schlama, K. Gabriel, V. Gouverneur, and C. Mioskowski, *Angew. Chem., Int. Ed. Engl.* **36**, 2342 (1997).

50. K. K. Laali and V. J. Gettwert, *J. Org. Chem.* **66**, 35 (2001).

51. P. A. Z. Suarez, J. E. L. Dullius, S. Einloft, R. F. de Souza, and J. Dupont, *Inorg. Chim. Acta* **255**, 207 (1997).

52. C. J. Adams, M. J. Earle, and K. R. Seddon, *Chem. Commun.* 1043 (1999).

53. J. Schulz, A. Roucoux, and H. Patin, *Chem. Eur. J.* **6**, 618 (2000).

54. F. Liu, M. B. Abrams, R. T. Baker, and W. Tumas, *Chem. Commun.* 433 (2001).

55. Sz. Csihony, H. Mehdi, and I. T. Horváth, *Green Chem.* **3**, 307 (2001).

56. N. Karodia, S. Guise, C. Newlands, and J.-A. Andersen, *Chem. Commun.* 2341 (1998).

57. P. Wasserscheid and H. Waffenschmidt, *J. Mol. Catal. A: Chem.* **164**, 61 (2000).

58. F. Favre, H. Olivier-Bourbigou, D. Commereuc, and L. Saussine, *Chem. Commun.* 1360 (2001).

59. C. E. Song and E. J. Roh, *Chem. Commun.* 837 (2000).

60. P. Wasserscheid, C. M. Gordon, C. Hilgers, M. J. Muldoon, and I. R. Dunkin, *Chem. Commun.* 1186 (2001).

61. S. M. Silva, P. A. Z. Suarez, R.F. de Souza, and J. Dupont, *Polymer Bull.* **40**, 401 (1998).

62. *Chemical Synthesis Using Supercritical Fluids*, W. Leitner and P. G. Jessop, eds., Wiley-VCH, Weinheim, 1999.

63. P. G. Jessop, T. Ikariya, and R. Noyori, *Science* **269**, 1065 (1995).

64. B. Subramaniam and M. A. McHugh, *Ind. Eng. Chem., Process Des. Dev.* **25**, 1 (1986).

65. P. E. Savage, S. Gopalan, T. I. Mizan, C. J. Martino, and E. E. Brock, *AIChE J.* **41**, 1723 (1995).

66. *Supercritical Fluids*. R. Noyori, ed.; *Chem. Rev.* **99**, 353–633 (1999).

67. P. G. Jessop, T. Ikariya, and R. Noyori, *Chem. Rev.* **99**, 475 (1999).

68. W. Leitner, *C. R. Acad. Sci. II, C* **3**, 595 (2000).

69. J. R. Hyde, P. Licence, D. Carter, and M. Poliakoff, *Appl. Catal., A* **222**, 119 (2001).

70. A. Baiker, *Chem. Rev.* **99**, 453 (1999).

71. P. E. Savage, in *Handbook of Heterogeneous Catalysis*, G. Ertl, H. Knözinger, and J. Weitkamp, eds., Wiley-VCH, Weinheim, 1997, Chapter 8.4, p. 1339.; *Catal. Today* **62**, 167 (2000).

72. P. G. Jessop, T. Ikariya, and R. Noyori, *Nature* **368**, 231 (1994).

73. P. G. Jessop, Y. Hsiao, T. Ikariya, and R. Noyori, *J. Am. Chem. Soc.* **118**, 344 (1996).

74. P. G. Jessop, Y. Hsiao, T. Ikariya, and R. Noyori, *J. Am. Chem. Soc.* **116**, 8851 (1994).

75. P. G. Jessop, Y. Hsiao, T. Ikariya, and R. Noyori, *J. Chem. Soc., Chem. Commun.* 707 (1995).

76. S. Kaneco, H. Kurimoto, Y. Shimizu, K. Ohta, and T. Mizuno, *Energy* **24**, 21 (1999).

77. M. G. Hitzler, F. R. Smail, S. K. Ross, and M. Poliakoff, *Chem. Commun.* 359 (1998).

78. F. Loeker and W. Leitner, *Chem. Eur. J.* **6**, 2011 (2000).

79. J. W. Rathke, R. J. Klingler, and T. R. Krause, *Organometallics* **10**, 1350 (1991).

80. S. Kainz, D. Koch, W. Baumann, and W. Leitner, *Angew. Chem., Int. Ed. Engl.* **36**, 1628 (1997).

81. I. Bach and D. J. Cole-Hamilton, *Chem. Commun.* 1463 (1998).

82. D. R. Palo and C. Erkey, *Ind. Eng. Chem. Res.* **37**, 4203 (1998); *Organometallics* **19**, 81 (2000).

83. N. J. Meehan, A. J. Sandee, J. N. H. Reek, P. C. J. Kamer, P. W. N. M. van Leeuwen, and M. Poliakoff, *Chem. Commun.* 1497 (2000).

84. A. M. B. Osuna, W. Chen, E. G. Hope, R. D. W. Kemmitt, D. R. Paige, A. M. Stuart, J. Xiao, and L. Xu, *J. Chem. Soc., Dalton Trans.* 4052 (2000).

85. C. A. G. Carter, R. T. Baker, S. P. Nolan, and W. Tumas, *Chem. Commun.* 347 (2000).

86. H. Jiang, L. Jia, and J. Li, *Green Chem.* **2**, 161 (2000).

87. J. L. Kendall, D. A. Canelas, J. L. Young, and J. M. DeSimone, *Chem. Rev.* **99**, 543 (1999).

88. C. D. Mistele, H. H. Thorp, and J. M. DeSimone, *Am. Chem. Soc., Polym. Prepr.* **36**(1), 507 (1995).

89. A. S. Chellappa, R. C. Miller, and W. J. Thomson, *Appl. Catal., A* **209**, 359 (2001).

90. H. Hori, C. Six, and W. Leitner, *Appl. Organomet. Chem.* **15**, 145 (2001).

91. W. Wei, Y. Sun, and B. Zhong, *Chem. Commun.* 2499 (1999).

92. M. A. Harmer and K. W. Hutchenson, *Chem. Commun.* 18 (2002).

93. D. Bröll, C. Kaul, A. Krämer, P. Krammer, T. Richter, M. Jung, H. Vogel, and P. Zehner, *Angew. Chem., Int. Ed. Engl.* **38**, 2999 (1999).

94. M. Siskin and A. R. Katritzky, *Chem. Rev.* **101**, 825 (2001).

95. A. R. Katritzky, D. A. Nichols, M. Siskin, R. Murugan, and M. Balasubramanian, *Chem. Rev.* **101**, 837 (2001).

96. P. E. Savage, *Chem. Rev.* **99**, 603 (1999).

97. R. L. Holliday, B. Y. M. Jong, and J. W. Kolis, *J. Supercrit. Fluids* **12**, 255 (1998).

98. J. H. Lee and N. R. Foster, *J. Supercrit. Fluids* **9**, 99 (1996).

99. U. Armbruster, A. Martin, and A. Krepel, *J. Supercrit. Fluids* **21**, 233 (2001).

100. T. Richter and H. Vogel, *Chem.-Ing.-Tech.* **73**, 1165 (2001).

101. M. Watanabe, M. Mochiduki, S. Sawamoto, T. Adschiri, and K. Arai, *J. Supercrit. Fluids* **20**, 257 (2001).

102. J. An, L. Bagnell, T. Cablewski, C. R. Strauss, and R. W. Trainor, *J. Org. Chem.* **62**, 2505 (1997).

103. J. M. Jennings, T. A. Bryson, and J. M. Gibson, *Green Chem.* **2**, 87 (2000).

104. M. B. Korzenski and J. W. Kolis, *Tetrahedron Lett.* **38**, 5611 (1997).

105. Y. Dror and J. Manassen, *J. Mol. Catal. A: Chem.* **2**, 219 (1977).

106. B. Driessen-Hölscher, *Adv. Catal.* **42**, 473 (1998).

107. Z. Tóth, F. Joó, and M. T. Beck, *Inorg. Chim. Acta* **42**, 153 (1980).

108. F. Joó, É. Papp, and Á. Kathó, *Top. Catal.* **5**, 113 (1998).

109. E. G. Kuntz, *Chemtech* **17**, 570 (1987).

110. C. D. Frohning and Ch.W. Kohlpaintner, in *Applied Homogeneous Catalysis with Organometallic Complexes*, B. Cornils and W. A. Herrmann, eds., VCH, Weinheim, 1996, Chapter 2.1.1, p. 29.

111. B. Breit and W. Seiche, *Synthesis* 1 (2001).

112. E. R. Freitas and C. R. Gum, *Chem. Eng. Prog.* **75**(1), 73 (1979).

113. R. C. Wade, in *Catalysis of Organic Reactions*, W. R. Moser, ed., Marcel Dekker, New York, 1981, p. 165.

114. G. R. Lappin, in *Alpha Olefins Applications Handbook*, G. R. Lappin and J. D. Sauer, eds., Marcel Dekker, New York, 1989, Chapter 3, p. 35.

115. A. Fukuoka, W. Kosugi, F. Morishita, M. Hirano, L. McCaffrey, W. Henderson, and S. Komiya, *Chem. Commun.* 489 (1999).

116. S. U. Son, J. W. Han, and Y. K. Chung, *J. Mol. Catal. A: Chem.* **135**, 35 (1998).

117. F. Joó and Á. Kathó, *J. Mol. Catal. A: Chem.* **116**, 3 (1997).

118. F. Joó and Á. Kathó, in *Aqueous-Phase Organometallic Catalysis*, B. Cornils and W. A. Herrmann, eds., Wiley-VCH, Weinheim, 1998, Chapter 6.2, p. 340.

119. K. Nomura, *J. Mol. Catal. A: Chem.* **130**, 1 (1998).

120. D. J. Ellis, P. J. Dyson, D. G. Parker, and T. Welton, *J. Mol. Catal. A: Chem.* **150**, 71 (1999).

121. D.Ch. Mudalige and G. L. Rempel, *J. Mol. Catal. A: Chem.* **116**, 309 (1997).

122. E. G. Fidalgo, L. Plasseraud, and G. Süss-Fink, *J. Mol. Catal. A: Chem.* **132**, 5 (1998).

123. L. Plasseraud and G. Süss-Fink, *J. Organomet. Chem.* **539**, 163 (1997).

124. G.-J. ten Brink, I. W. C. E. Arends, G. Papadogianakis, and R. A. Sheldon, *Chem. Commun.* 2359 (1998).

125. I. T. Horváth and J. Rábai, *Science* **266**, 72 (1994).

126. B. Cornils, *Angew. Chem., Int. Ed. Engl.* **36**, 2057 (1997).

127. I. T. Horváth, *Acc. Chem. Res.* **31**, 641 (1998).

128. R. H. Fish, *Chem. Eur. J.* **5**, 1677 (1999).

129. L. P. Barthel-Rosa and J. A. Gladysz, *Coord. Chem. Rev.* **190–192**, 587 (1999).

130. E. de Wolf, G. van Koten, and B. J. Deelman, *Chem. Soc. Rev.* **28**, 37 (1999).

131. M. Cavazzini, F. Montanari, G. Pozzi, and S. Quici, *J. Fluorine Chem.* **94**, 183 (1999).

132. E. G. Hope and A. M. Stuart, *J. Fluorine Chem.* **100**, 75 (1999).

133. I. T. Horváth, G. Kiss, R. A. Cook, J. E. Bond, P. A. Stevens, J. Rábai, and E. J. Mozeleski, *J. Am. Chem. Soc.* **120**, 3133 (1998).

134. W. Chen, L. Xu, and J. Xiao, *Chem. Commun.* 839 (2000).

135. J. J. J. Juliette, I. T. Horváth, and J. A. Gladysz, *Angew. Chem., Int. Ed. Engl.* **36**, 1610 (1997).

136. J. J. J. Juliette, D. Rutherford, I. T. Horváth, and J. A. Gladysz, *J. Am. Chem. Soc.* **121**, 2696 (1999).

137. E. de Wolf, E. A. Speets, B. J. Deelman, and G. van Koten, *Organometallics* **20**, 3689 (2001).

138. A. G. M. Barrett, D. C. Braddock, D. Catterick, D. Chadwick, J. P. Henschke, and R. M. McKinnell, *Synlett* 847 (2000).

139. G. Pozzi, S. Banfi, A. Manfredi, F. Montarini, and S. Quici, *Tetrahedron* **36**, 11879 (1996).

140. I. Klement, H. Lütjens, and P. Knochel, *Angew. Chem., Int. Ed. Engl.* **36**, 1454 (1997).

141. G. Pozzi, M. Cavazzini, F. Cinato, F. Montanari, and S. Quici, *Eur. J. Org. Chem.* 1947 (1999).

142. X.-Q. Yu, J.-S. Huang, W.-Y. Yu, and C.-M. Che, *J. Am. Chem. Soc.* **122**, 5337 (2000).

143. B. Betzemeier, F. Lhermitte, and P. Knochel, *Synlett* 489 (1999).

144. B. Betzemeier, F. Lhermitte, and P. Knochel, *Tetrahedron Lett.* **39**, 6667 (1998).

145. G. Pozzi, M. Cavazzini, S. Quici, and S. Fontana, *Tetrahedron Lett.* **38**, 7605 (1997).

146. J. M. Vincent, A. Rabion, V. K. Yachandra, and R. H. Fish, *Can. J. Chem.* **79**, 888 (2001).

147. D. Rutherford, J. J. J. Juliette, C. Rocaboy, I. T. Horváth, and J. A. Gladysz, *Catal. Today* **42**, 381 (1998).

148. E. G. Hope, R. D. W. Kemmitt, D. R. Paige, and A. M. Stuart, *J. Fluorine Chem.* **99**, 197 (1999).

149. D. E. Bergbreiter, J. G. Franchina, and B. L. Case, *Org. Lett.* **2**, 393 (2000).

150. W. Keim, M. Vogt, P. Wasserscheid, and B. Drießen-Hölscher, *J. Mol. Catal. A: Chem.* **139**, 171 (1999).

151. F. Benvenuti, C. Carlini, M. Marchionna, A. M. R. Galletti, and G. Sbrana, *J. Mol. Catal. A: Chem.* **178**, 9 (2002).

152. *Environmental Catalysis*, ACS Symp. Ser., Vol. 552, J. Armor, ed., American Chemical Society, Washington, DC, 1994.

153. *Applied Catalysis, B: Environmental* was launched in 1992.

154. J. N. Armor, *Appl. Catal., A* **189**, 153 (1999).

155. R. A. Sheldon and R. S. Downing, *Appl. Catal., A* **189**, 163 (1999).

156. H.-U. Blaser and M. Studer, *Appl. Catal., A* **189**, 191 (1999).

157. P. T. Anastas, L. B. Bartlett, M. M. Kirchhoff, and T. C. Williamson, *Catal. Today* **55**, 11 (2000).

158. M. Misono, *C. R. Acad. Sci. II, C* **3**, 471 (2000).

159. H. Arakawa, M. Aresta, J. N. Armor, M. A. Barteau, E. J. Beckman, A. T. Bell, J. E. Bercaw, C. Creutz, E. Dinjus, D. A. Dixon, K. Domen, D. L. DuBois, J. Eckert, E. Fujita, D. H. Gibson, W. A. Goddard, D. W. Goodman, J. Keller, G. J. Kubas, H. H. Kung, J. E. Lyons, L. E. Manzer, T. J. Marks, K. Morokuma, K. M. Nicholas, R. Periana, L. Que, J. Rostrup-Nielson, W. M. H. Sachtler, L. D. Schmidt, A. Sen, G. A. Somorjai, P. C. Stair, B. R. Stults, and W. Tumas, *Chem. Rev.* **101**, 953 (2001).

160. B. Notari, *Adv. Catal.* **41**, 253 (1996).

161. J. H. Clark and D. J. Macquarrie, *Org. Process Res. Dev.* **1**, 149 (1997).

162. *Metallocene-Based Polyolefins: Preparation, Properties, and Technology*, J. Scheirs and W. Kaminsky, eds., Wiley, Chichester, 2000.

163. S. D. Ittel, L. K. Johnson, and M. Brookhart, *Chem. Rev.* **100**, 1169 (2000).

164. W. Kaminsky and A. Laban, *Appl. Catal., A* **222**, 47 (2001).

165. A. Corma, *Chem. Rev.* **95**, 559 (1995).

166. A. Corma and H. García, *Catal. Today* **38**, 257 (1997).

167. M. G. Clerici, *Top. Catal.* **13**, 373 (2000).

168. *Fine Chemicals through Heterogeneous Catalysis*, R. A. Sheldon and H. van Bekkum, eds., Wiley–VCH, Weinheim, 2001, Chapter 4.

169. N. Mizuno and M. Misono, *Appl. Catal., A* **64**, 1 (1990).

170. T. Okuhara, N. Mizuno, and M. Misono, *Appl. Catal., A* **222**, 63 (2001).

171. M. Misono, *Chem. Commun.* 1141 (2001).

172. N. Mizuno and M. Misono, *Chem. Rev.* **98**, 199 (1998).

173. I. V. Kozhevnikov, *Chem. Rev.* **98**, 171 (1998).

174. T. Okuhara, N. Mizuno, and M. Misono, *Adv. Catal.* **41**, 113 (1996).

175. B. H. Davis, R. A. Keogh, and R. Srinivasan, *Catal. Today* **20**, 219 (1994).

176. X. Song and A. Sayari, *Catal. Rev.-Sci. Eng.* **38**, 329 (1996).

177. K. Tanabe and H. Hattori, in *Handbook of Heterogeneous Catalysis*, G. Ertl, H. Knözinger, and J. Weitkamp, eds., Wiley-VCH, Weinheim, 1997, Chapter 2.4, p. 404.

178. G. D. Yadav and J. J. Nair, *Micropor. Mesopor. Mat.* **33**, 1 (1999).

179. R. A. Sheldon and H. van Bekkum, eds., *Fine Chemicals through Heterogeneous Catalysis*, Wiley–VCH, Weinheim, 2001.

180. M. Balogh and P. Laszlo, *Organic Chemistry Using Clays*. Springer, Berlin, 1993.

181. P. Laszlo, *J. Phys. Org. Chem.* **11**, 356 (1998).

182. J. H. Clark, A. P. Kybett, D. J. Macquarrie, S. J. Barlow, and P. Landon, *J. Chem. Soc., Chem. Commun.* 1353 (1989).

183. J. M. Thomas, *Angew. Chem., Int. Ed. Engl.* **38**, 3589 (1999).

184. J. M. Thomas and R. Raja, *Chem. Commun.* 675 (2001).

185. J. M. Thomas, R. Raja, G. Sankar, B. F. G. Johnson, and D. W. Lewis, *Chem. Eur. J.* **7**, 2973 (2001).

186. J. M. Thomas, R. Raja, G. Sankar, R. G. Bell, *Nature* **398**, 227 (1999).

187. R. Raja, G. Sankar, and J. M. Thomas, *Angew. Chem., Int. Ed. Engl.* **39**, 2313 (2000).

188. J. M. Thomas, R. Raja, G. Sankar, and R. G. Bell, in *12th Int. Congr. Catalysis*, Studies in Surface Science and Catalysis, Vol. 130, A. Corma, F. V. Melo, S. Mendioroz, and J. L. G. Fiero, eds., Elsevier, Amsterdam, 2000, p. 887.

189. M. P. Attfield, G. Sankar, J. M. Thomas, *Catal. Lett.* **70**, 155 (2000).

190. J. M. Thomas and G. Sankar, *Acc. Chem. Res.* **34**, 571 (2001).

191. R. Raja, G. Sankar, and J. M. Thomas, *J. Am. Chem. Soc.* **121**, 11926 (1999).

192. M. Dugal, G. Sankar, R. Raja, and J. M. Thomas, *Angew. Chem., Int. Ed. Engl.* **39**, 2311 (2000).

193. T. Maschmeyer, R. D. Oldroyd, G. Sankar, J. M. Thomas, I. J. Shannon, J. A. Klepetko, A. F. Masters, J. K. Beattie, and C. R. A. Catlow, *Angew. Chem., Int. Ed. Engl.* **36**, 1639 (1997).

194. R. Raja, G. Sankar, S. Hermans, D. S. Shephard, S. Bromley, J. M. Thomas, and B. F. G. Johnson, *Chem. Commun.* 1571 (1999).

195. S. Hermans, R. Raja, J. M. Thomas, B. F. G. Johnson, G. Sankar, and D. Gleeson, *Angew. Chem., Int. Ed. Engl.* **40**, 1211 (2001).

196. B. Jandeleit, D. J. Schaefer, T. S. Powers, H. W. Turner, and W. H. Weinberg, *Angew. Chem., Int. Ed. Engl.* **38**, 2495 (1999).

197. J. Scheidtmann, P. A. Weiss, and W. F. Maier, *Appl. Catal., A* **222**, 79 (2001).

198. R. H. Crabtree, *Chemtech* **21**(4), 21 (1999).

199. S. Senkan, *Angew. Chem., Int. Ed. Engl.* **40**, 313 (2001).

200. M. T. Reetz, *Angew. Chem., Int. Ed. Engl.* **40**, 285 (2001).

201. C. Gennari, H. P. Nestler, U. Piarulli, and B. Salom, *Liebigs Annalen/Receuil* 637 (1997).

202. S. Dahmen and S. Bräse, *Synthesis* 1431 (2001).

203. A. Holzwarth, P. Denton, H. Zanthoff, and C. Mirodatos, *Catal. Today* **67**, 309 (2001).

204. P. P. Pescarmona, J. C. van der Waal, I. E. Maxwell, and T. Maschmeyer, *Catal. Lett.* **63**, 1 (1999).

205. T. Bein, *Angew. Chem., Int. Ed. Engl.* **38**, 323 (1999).

206. F. Gennari, P. Seneci, and S. Miertus, *Catal. Rev.–Sci. Eng.* **42**, 385 (2000).

207. J. A. Loch and R. H. Crabtree, *Pure Appl. Chem.* **73**, 119 (2001).

208. *Combinatorial Chemistry*, J. W. Szostak, ed.; *Chem. Rev.* **97**, 347–509 (1997).

209. V. V. Guliants, ed. *Catal. Today* **67**, 307–409 (2001).

210. S. M. Senkan and S. Ozturk, *Angew. Chem., Int. Ed. Engl.* **38**, 791 (1999).

211. S. Senkan, K. Krantz, S. Ozturk, V. Zengin, and I. Onal, *Angew. Chem., Int. Ed. Engl.* **38**, 2794 (1999).

212. A. Holzwarth, H.-W. Schmidt, and W. F. Maier, *Angew. Chem., Int. Ed. Engl.* **37**, 2644 (1998).

213. U. Rodemerck, D. Wolf, O. V. Buyevskaya, P. Claus, S. Senkan, and M. Baerns, *Chem. Eng. J.* **82**, 3 (2001).

214. D. Wolf, O. V. Buyevskaya, and M. Baerns, *Appl. Catal., A* **200**, 63 (2000).

215. Y. Liu, P. Cong, R. D. Doolen, H. W. Turner, and W. H. Weinberg, *Catal. Today* **61**, 87 (2000).

216. H. Su, Y. J. Hou, R. S. Houk, G. L. Schrader, and E. S. Yeung, *Anal. Chem.* **73**, 4434 (2001).

217. O. Löber, M. Kawatsura, and J. F. Hartwig, *J. Am. Chem. Soc.* **123**, 4366 (2001).

218. M. T. Reetz, M. H. Becker, M. Liebl, and A. Fürstner, *Angew. Chem., Int. Ed. Engl.* **39**, 1236 (2000).

219. T. R. Boussie, C. Coutard, H. Turner, V. Murphy, and T. S. Powers, *Angew. Chem., Int. Ed. Engl.* **37**, 3272 (1998).

220. M. Havranek, A. Singh, and D. Sames, *J. Am. Chem. Soc.* **121**, 8965 (1999).

221. M. B. Francis and E. N. Jacobsen, *Angew. Chem., Int. Ed. Engl.* **38**, 937 (1999).

222. A. Star, I. Goldberg, and B. Fuchs, *J. Organomet. Chem.* **630**, 67 (2001).

223. P. P. Pescarmona, J. C. van der Waal, I. E. Maxwell, and T. Maschmeyer, *Angew. Chem., Int. Ed. Engl.* **40**, 740 (2001).

224. B. Denise and R. P. A. Sneeden, *Chemtech* **12**, 108 (1982).

225. Y. Borodko and G. A. Somorjai, *Appl. Catal., A* **186**, 355 (1999).

226. A. Baiker, *Appl. Organomet. Chem.* **14**, 751 (2000).

227. A. Behr, *Carbon Dioxide Activation by Metal Complexes*, VCH, Weinheim, 1988.

228. *Catalytic Activation of Carbon Dioxide*, ACS Symp. Ser. Vol. 363, W. M. Ayers, ed., American Chemical Society, Washington, DC, 1988.

229. M. M. Halmann, *Chemical Fixation of Carbon Dioxide: Methods for Recycling CO_2 into Useful Products*, CRC Press, Boca Raton, FL, 1993.

230. M. M. Halmann, *Greenhouse Gas Carbon Dioxide Mitigation: Science and Technology*, Lewis Publishers, Boca Raton, FL, 1999.

231. *Carbon Dioxide Chemistry: Environmental Issues*, J. Paul and C.-M. Pradier, eds., Royal Society of Chemistry, Cambridge, UK, 1994.

232. *Advances in Chemical Conversions for Mitigating Carbon Dioxide*, Studies in Surface Science and Catalysis, Vol. 114, T. Inui, M. Anpo, K. Izui, S. Yanagida, and T. Yamaguchi, eds., Elsevier, Amsterdam, 1998.

233. *Greenhouse Gas Control Technologies*, B. Eliasson, P. Riemer, and A. Wokaun, eds., Elsevier, Amsterdam, 1999.

234. *5th Int. Conf. Carbon Dioxide Utilization*, Karlsruhe, 1999, E. Dinjus and O. Walter, eds.; *Appl. Organomet. Chem.* **14**, 749–873 (2000); **15**, 87–150 (2001).

235. F. Solymosi and A. Erdöhelyi, *J. Mol. Catal.* **8**, 471 (1980).

236. I. A. Fisher and A. T. Bell, *J. Catal.* **162**, 54 (1996).

237. M. A. Henderson and S. D. Worley, *J. Phys. Chem.* **89**, 1417 (1985).

238. F. Solymosi, A. Erdöhelyi, and T. Bánsági, *J. Catal.* **68**, 371 (1981).

239. S. Ichikawa, *J. Mol. Catal.* **53**, 53 (1989).

240. M. Yamasaki, H. Habazaki, T. Yoshida, E. Akiyama, A. Kawashima, K. Asami, K. Hashimoto, M. Komori, and K. Shimamura, *Appl. Catal., A* **163**, 187 (1997).

241. M. Yamasaki, H. Habazaki, T. Yoshida, M. Komori, K. Shimamura, E. Akiyama, A. Kawashima, K. Asami, and K. Hashimoto, in *Advances in Chemical Conversions for Mitigating Carbon Dioxide*, Studies in Surface Science and Catalysis, Vol. 114, T. Inui, M. Anpo, K. Izui, S. Yanagida, and T. Yamaguchi, eds., Elsevier, Amsterdam, 1998, p. 451.

242. K. Hashimoto, H. Habazaki, M. Yamasaki, S. Meguro, T. Sasaki, H. Katagiri, T. Matsui, K. Fujimura, K. Izumiya, N. Kumagai, and E. Akiyama, *Mater. Sci. Eng. A* **304–306**, 88 (2001).

243. J. Wambach, A. Baiker, and A. Wokaun, *Phys. Chem. Chem. Phys.* **1**, 5071 (1999).

244. C. Schild, A. Wokaun, and A. Baiker, *J. Mol. Catal.* **69**, 347 (1991).

245. C. Schild, A. Wokaun, R. A. Koeppel, and A. Baiker, *J. Phys. Chem.* **95**, 6341 (1991).

246. J.-K. Jeon, K.-E. Jong, Y.-K. Park, and S.-K. Ihm, *Appl. Catal., A* **124**, 91 (1995).

247. M. Fujiwara, H. Ando, M. Tanaka, and Y. Souma, *Appl. Catal., A* **130**, 105 (1995).

248. Y.-K. Park, K.-C. Park, and S.-K. Ihm, *Catal. Today* **44**, 165 (1998).

249. N. K. Lunev, Yu.I. Shmyrko, N. V. Pavlenko, and B. Norton, *Appl. Organomet. Chem.* **15**, 99 (2001).

250. F. R. Keene and B. P. Sullivan, in *Electrochemical and Electrocatalytic Reactions of Carbon Dioxide*, B. P. Sullivan, K. Krist, and H. E. Guard, eds., Elsevier, New York, 1993, p. 118.

251. G. A. Olah, in *Chemical Research—2000 and Beyond*, P. Barkan, ed., Symposium of the New York Section of the American Chemical Society and Rockefeller University, The American Chemical Society, Washington, DC and Oxford Univ. Press, New York, Oxford, 1998, pp. 40–54.

252. G. A. Olah and G. K. S. Prakash, U. S. Patent 5,928,806 (1999).

253. G. A. Olah, *Acc. Chem. Res.* **20**, 422 (1987).

INDEX

Gold complexes, in carbonylation, 391
Green chemistry, 807
Grignard reagent
  carbomagnesation with, 331
  metallation with, 250, 598
Gross formylation, 416
Group IA cations, as promoters in methane
  coupling, 110
Group IIA oxides, in methane coupling, 110
Group III halides, as additives in
  hydrosilylation, 326
Groups I-IV organometallics and hydrides, as
  cocatalysts in metathesis, 699
Group IV halides, as additives in
  hydrosilylation, 326
Group IVB halides, as additives
  in carboxylation, 382
  in hydroformylation, 372
Group VIA oxides on alumina, in
  butylenes–isobutylene isomerization, 176
Group VIII metals
  activity of, in hydrogenolysis, 656, 657
  in carboxylation of alkenes, 381
  in catalytic hydrosilylation, 323
  in CO reduction, 102
  in dissociative adsorption of CO, 116
  in hydroformylation, 371
  in hydrogenation of dienes, 627
  in isomerization of n-butenes, 187
  methanation activity of, 108
Grubbs catalyst, 712, 714

Halcon–Arco process, for manufacture of
  propylene oxide, 508
Halcon process
  for manufacture of maleic anhydride, 515
  for manufacture of propionic acid, 385
Half-hydrogenated intermediate, in
  CO hydrogenation, 105
Half-hydrogenated state, 185, 621
N-Halo amides, halogenation with, 580
Haloboranes, addition of, to alkynes, 327
Haloboration, 327
Halogenation
  of alkanes, 577, 585, 592
  of alkenes, 304, 337
    addition complexes in, 337, 338
    free-radical, 307
  of alkynes, 310
  allylic, 590, 594
  of aromatics, 580, 584, 585, 589, 594
  of dienes, 308
Halometallation, 327, 345

Halometallation mechanism, in cationic
  polymerization, 736
Halonium ion, in halogenation of dienes,
  309
β-Halovinyl carbocation, in halogenation of
  alkynes, 310
Hamilton reagent, 492
HBr–AlBr₃
  in alkylation, 218
  in isomerization, 166
HBr–GaBr₃, in transalkylation, 247
H₃B–SMe₂, hydroboration with, 316
H₃B–THF, hydroboration with, 316
HCl–AlCl₃
  in alkylation, 216–218
  in formylation, 415
  in isomerization, 166, 171, 192
  in manufacture of p-ethyltoluene, 259
  in oligomerization of ethylene, 724
Heavy oils, 7
  ionic hydrogenation of, 678
Heck–Breslow mechanism, 372, 374
n-Heptane
  dehydrocyclization of, 44, 53
  hydrocracking of, 42–44
  hydrogenolysis of, 657
  hydroxylation of, 441
  isomerization of, 196
[1-¹⁴C]-Heptane, aromatization of, 53
[1,1-²H₂]-Heptene, metathesis of, 704
Heptenes, dehydrocyclization of, 53
Herington–Rideal mechanism, 53
Heteropoly acids, 815
  in acylation, 412
  in alkylation, 261, 262, 264, 265
  in heterosubstitution, 596
  in hydration, 285, 288
  in isomerization, 194
  in methanol conversion, 118
  in oxidation, 432, 455, 472, 483, 495,
    522, 527
  in Prins reaction, 262
Heteropoly compounds
  in hydrogenation, 669
  in methanol conversion, 118
  in oxidation, 443, 482, 489, 522, 527, 529
Hexaalkylbenzenes, formation of, in
  cyclooligomerization, 731
1,4-Hexadiene
  copolymerization of, with ethylene, 772
  manufacture of, 732
1,5-Hexadiene, manufacture of, by metathesis,
  709